Introduction to Theoretical and Computational Fluid Dynamics

Introduction to Theoretical and Computational Fluid Dynamics

C. POZRIKIDIS

New York   Oxford
OXFORD UNIVERSITY PRESS
1997

OXFORD UNIVERSITY PRESS

Oxford   New York
Athens   Auckland   Bangkok   Bogotá   Bombay   Buenos Aires
Calcutta   Cape Town   Dar es Salaam   Delhi
Florence   Hong Kong   Istanbul   Karachi
Kuala Lumpur   Madras   Madrid   Melbourne
Mexico City   Nairobi   Paris   Singapore
Taipei   Tokyo   Toronto

and associated companies in
Berlin   Ibadan

Published by Oxford University Press, Inc.
198 Madison Avenue, New York, New York 10016

Oxford is a registered trademark of Oxford University Press

Library of Congress Cataloging-in-Publication Data
Pozrikidis, C.
    Introduction to theoretical and computational fluid dynamics /
 C. Pozrikidis.
        p.   cm.
    Includes index.
    ISBN 0-19-509320-8
    1. Fluid dynamics.   I. Title.
 QA911.P65   1996
 532'.05–dc20   96-26420

Printing (last digit): 9 8 7 6 5 4 3 2 1

Printed in the United States of America
on acid-free paper

About the Cover: The picture on the cover shows the instantaneous configuration of a random suspension of liquid drops in a channel confined by two moving parallel-sided walls in Couette flow. The computations were conducted using the method of Interfacial Dynamics by the author and *Hua Zhou;* the computer graphics visualization was done by *Richard Charles.*

Contents

Preface ix
Note to the Instructor xi
Note to the Reader xii

1

Kinematics of a Flow 1

1.1 Fluid velocity and motion of fluid parcels 1
1.2 Lagrangian labels 6
1.3 Properties of parcels, conservation of mass, and the continuity equation 10
1.4 Material vectors, material lines, and material surfaces 13
1.5 Differential geometry of surfaces 20
1.6 Description of a material surface in Eulerian form 23
1.7 Streamlines, stream tubes, path lines, and streak lines 24
1.8 Vorticity, vortex lines, vortex tubes, and circulation around loops 30
1.9 Line vortices and vortex sheets 33

2

Analysis of Kinematics 39

2.1 Irrotational flows and the velocity potential 40
2.2 The reciprocal relation for harmonic functions and Green's functions of Laplace's equation 49
2.3 Integral representation and further properties of potential flow 55
2.4 The vector potential for incompressible flow 62
2.5 Representation of an incompressible flow in terms of the vorticity 64
2.6 Representation of a flow in terms of the rate of expansion and the vorticity 70
2.7 Stream functions for incompressible flow 71
2.8 Flow induced by vorticity 76
2.9 Axisymmetric flow induced by vorticity 83
2.10 Two-dimensional flow induced by vorticity 88

3

Stresses, the Equation of Motion, and the Vorticity Transport Equation 94

3.1 Forces acting in a fluid, traction, the stress tensor, and the equation of motion 94
3.2 Constitutive relations for the stress tensor 105
3.3 Traction, force, torque, energy dissipation, and the reciprocal theorem for incompressible Newtonian fluids 109
3.4 The Navier–Stokes, Euler, and Bernoulli equations 114
3.5 Equations and boundary conditions governing the motion of an incompressible Newtonian fluid 120

3.6   Traction, vorticity, and flow kinematics at rigid boundaries, free surfaces, and fluid interfaces   128
3.7   Scaling of the Navier–Stokes equation and dynamic similitude   135
3.8   Change of circulation around material loops and dynamics of the vorticity field   139
3.9   Computation of exact solutions to the equation of motion in two dimensions on the basis of the vorticity transport equation   149

4
Hydrostatics   155

4.1   Pressure distribution within a fluid in rigid-body motion   155
4.2   The Laplace–Young equation   159
4.3   Two-dimensional interfaces   166
4.4   Axisymmetric interfaces   171
4.5   Three-dimensional interfaces   177

5
Computing Incompressible Flows   179

5.1   Steady unidirectional flows   180
5.2   Unsteady unidirectional flows   192
5.3   Stagnation-point flows   203
5.4   Flow due to a rotating disk   211
5.5   Flow in a corner due to a point source   214
5.6   Flow due to a point force   216

6
Flow at Low Reynolds Numbers   222

6.1   Equations and fundamental properties of Stokes flow   222
6.2   Local solutions in corners   227
6.3   Nearly unidirectional flows   243
6.4   Flow due to a point force   253
6.5   Fundamental solutions of Stokes flow   259
6.6   Stokes flow past or due to the motion of rigid bodies and liquid drops   264
6.7   Computation of singularity representations   273
6.8   The Lorentz reciprocal theorem and its applications   276
6.9   Boundary-integral representation of Stokes flow   280
6.10   Boundary-integral-equation methods   287
6.11   Generalized Faxen relations   292
6.12   Formulation of two-dimensional Stokes flow in complex variables   296
6.13   Effects of inertia and Oseen flow   298
6.14   Unsteady Stokes flow   301
6.15   Computation of unsteady Stokes flow past or due to the motion of particles   309

7
Irrotational Flow   314

7.1   Equations and computation of irrotational flow   315
7.2   Flow past or due to the motion of three-dimensional bodies   318
7.3   Force and torque exerted on a three-dimensional body   324
7.4   Flow past or due to the motion of a sphere   332

7.5   Flow past or due to the motion of nonspherical bodies   337
7.6   Flow past or due to the motion of two-dimensional bodies   344
7.7   Computation of two-dimensional flow past or due to the motion of a body   349
7.8   Formulation of two-dimensional flow in complex variables   357
7.9   Conformal mapping   364
7.10   Applications of conformal mapping to flow past two-dimensional bodies   370
7.11   The Schwarz–Christoffel transformation and its applications   377

8

Boundary Layers   385

8.1   Boundary-layer theory   386
8.2   The boundary layer on a semi-infinite flat plate   394
8.3   Boundary layers in accelerating and decelerating flow   401
8.4   Computation of boundary layers around two-dimensional bodies   405
8.5   Boundary layers in axisymmetric and three-dimensional flows   413
8.6   Unsteady boundary layers   416

9

Hydrodynamic Stability   421

9.1   Evolution equations and formulation of the linear stability problem   422
9.2   Solution of the initial-value problem and normal-mode analysis   426
9.3   Normal-mode analysis of unidirectional flows   430
9.4   General theorems on the temporal stability of inviscid shear flows   439
9.5   Stability of a uniform vortex layer subject to spatially periodic disturbances   441
9.6   Numerical solution of the Orr–Sommerfeld and Rayleigh equations   449
9.7   Stability of certain classes of unidirectional flows   455
9.8   Stability of a planar interface in potential flow   458
9.9   Viscous interfacial flows   465
9.10   Capillary instability of a curved interface   473
9.11   Inertial instability of rotating fluids   477

10

Boundary-Integral Methods for Potential Flow   482

10.1   The boundary-integral equation   482
10.2   Boundary-element methods   491
10.3   Generalized boundary-integral representations   495
10.4   The single-layer potential   496
10.5   The double-layer potential   500
10.6   Investigation of integral equations of the second kind   506
10.7   Regularization of integral equations of the second kind   511
10.8   Completed double-layer representation for exterior flow   516
10.9   Iterative solution of integral equations of the second kind   517

11

Vortex Motion   520

11.1   Invariants of the motion   521
11.2   Point vortices   523
11.3   Vortex blobs   532

11.4   Two-dimensional vortex sheets   535
11.5   Two-dimensional flows with distributed vorticity   545
11.6   Two-dimensional vortex patches   548
11.7   Axisymmetric flow   551
11.8   Three-dimensional flow   553

12

Finite-Difference Methods for the Convection-Diffusion Equation   558

12.1   Definitions and procedures   558
12.2   One-dimensional diffusion   562
12.3   Diffusion in two and three dimensions   571
12.4   One-dimensional convection   577
12.5   Convection in two and three dimensions   587
12.6   Convection–diffusion in one dimension   589
12.7   Convection–diffusion in two and three dimensions   596

13

Finite-Difference Methods for Incompressible Newtonian Flow   599

13.1   Methods based on the vorticity transport equation   600
13.2   Velocity–pressure formulation   613
13.3   Implementation of methods in primitive variables   617
13.4   Operator splitting, projection, and pressure-correction methods   623
13.5   Methods of modified dynamics or false transients   627

Appendix A
Index Notation, Differential Operators, and Theorems of Vector Calculus   632

A.1   Index notation   632
A.2   Vector and matrix products, differential operators in Cartesian coordinates   633
A.3   Orthogonal curvilinear coordinates   635
A.4   Differential operators in cylindrical and plane polar coordinates   637
A.5   Differential operators in spherical polar coordinates   638
A.6   Integral theorems of vector calculus   640

Appendix B
Primer of Numerical Methods   641

B.1   Linear algebraic equations   641
B.2   Computation of eigenvalues of a matrix   646
B.3   Nonlinear algebraic equations   647
B.4   Function interpolation   649
B.5   Computation of derivatives   652
B.6   Function integration   652
B.7   Function approximation   657
B.8   Integration of ordinary differential equations   659
B.9   Computation of special functions   660

Index   663

Preface

My goal in this book is to provide a comprehensive and rigorous introduction to the fundamental concepts and basic equations of fluid dynamics and, furthermore, to illustrate the application of numerical methods for solving fundamental and practical problems that involve the flow of incompressible Newtonian fluids. The material is intended to be useful and interesting to advanced undergraduate students, graduate students, and researchers in the various fields of engineering, including chemical, mechanical, and aerospace engineering, applied mathematics, computational science, and other related disciplines. Prerequisites are a basic knowledge of continuum mechanics, intermediate calculus, elementary numerical methods, and a general familiarity with computer programming. The various chapters can be read either as a whole or in parts, according to the reader's experience and needs, and the presentation is designed to be useful as a ready reference.

What distinguishes this book from other texts on theoretical fluid dynamics is that the discussion is taken into the realm of numerical methods and includes the development of specific algorithms for computing a variety of flows of incompressible fluids under diverse conditions. On the other hand, what distinguishes this book from a standard book on computational fluid dynamics is that a full account of the theory is provided with a minimal amount of external references, and no experience in computational fluid dynamics or knowledge of its terminology is assumed. Furthermore, an effort is made to present state-of-the-art numerical methods and computational strategies, and references for specialized and advanced topics are provided.

The level and sequence of presentation should render this book a suitable text for use with advanced undergraduate or introductory graduate courses in fluid mechanics, applied mathematics, applied numerical methods, and computational science. It is inevitable that the topics were selected according to the author's perception of what constitutes essential knowledge of theoretical fluid dynamics and computational methods in science and engineering. The introductory nature of this volume explains the absence of certain advanced theoretical and computational topics, such as compressible and turbulent flows and finite-element, finite-volume, spectral, and spectral-element methods. Furthermore, asymptotic and perturbation methods are only briefly discussed, and the main focus is on analytical and numerical computation. For example, the discussion makes extensive use of the powerful tools of Green's functions and integral representations whose understanding requires a certain degree of patience, but offers a basis for developing efficient methods of numerical computation. To conform with the fundamental objective of this text, subject to the volume constraints, physical descriptions of the motion are kept to a moderate level, and mathematical description and computation are given preferential treatment.

A number of people and institutions assisted and supported me directly or indirectly during the writing of this book. The National Science Foundation, the Office of Scientific Computing of the Department of Energy, the American Chemical Society, and the SUN Microsystems Corporation provided me with funds for sustaining my research program. Madhu Gopalakrishnan's insightful comments on many chapters were invaluable for improving the presentation. William Carlton, Chad Coulliette, Andrjei Domaradzki, Frank Jacobitz, San Minh Le, Xiaofan Li, Kurt Keller, Sherwin Maslowe, Ali Nadim, Howard Stone, and C.-S. Yih read drafts of chapters and provided useful comments. Aris, Chrisi, and Kyriakos Pozrikidis; Audrey Hill; Katherine Kozombolis; Dick Skalak; Sangtae Kim; Bill Schowalter; and Yiannis Tsamopoulos encouraged me to pursue this project, and my editor Robert Rogers readily agreed to publish this book when it was in the early stage of its inception. I thank them all.

It is my contention that future textbooks in the physical and engineering sciences will be written with an increasing emphasis towards balancing fundamental knowledge and numerical computation; this trend will make the computer, along with the textbook, a necessary tool of the learning process. This book represents an early effort in that direction, and I will appreciate the reader's comments at the electronic mail address: *costas@ames.ucsd.edu.*

La Jolla, Calif   C. P.
July 1996

Note to the Instructor

<hr/>

Each chapter is followed by a set of problems and another set of computer problems. The former can and should be solved by paper and pencil, whereas the latter require the use of a computer of modest size, one step above a calculator. Certain computer problems are coordinated, so that a routine in a problem may be called by the main program of a subsequent problem. All of the computer problems were tested and will not present unpleasant surprises.

The first eight chapters, combined with selected material from subsequent chapters, can be used in an upper-level undergraduate or entry-level introductory graduate course in fluid mechanics. The choice of topics will depend upon the students' orientation and field of study. Selected material from Chapters 9–13 may be used in a general or special-topics course in applied mathematics or computational fluids dynamics.

Note to the Reader

Before reading the book, please familiarize yourself with the material in the two appendixes, which will be useful as a reference in following the mathematical derivations and development of numerical methods. In writing computer programs, it is useful to keep in mind that nondimensionalizing physical quantities with respect to a certain variable is effected by setting the value of that variable equal to unity in the program.

1.1 | FLUID VELOCITY AND MOTION OF FLUID PARCELS

Referring to a frame of reference that is fixed in space, let us observe the flow of a homogeneous fluid that is composed of a single chemical species. Let us consider, in particular, the motion of a fluid parcel that, at the particular observation time t, has a spherical shape with radius ε centered at the point $\mathbf{x}$. We define the velocity of translation of the parcel as the average value of the instantaneous velocity of all molecules that reside within the parcel. It is clear that the value of this average velocity will depend upon the parcel radius ε. Taking the limit as ε tends to zero, we find that the average velocity tends to an asymptotic value, until ε becomes comparable to the typical distance between the molecules, whereupon we observe strong oscillations. These are manifestations of random molecular motions.

   We define the velocity of the fluid $\mathbf{u}$ at the position $\mathbf{x}$ at time t as the apparent or outer limit of the velocity of the parcel as the radius of the parcel ε tends to zero, just before the discrete nature of the fluid becomes apparent. Under normal conditions, the velocity $\mathbf{u}(\mathbf{x}, t)$ is an infinitely differentiable function of $\mathbf{x}$ and t, but spatial discontinuities may arise under extreme conditions in high-speed flows, or else emerge as the result of mathematical idealizations associated with asymptotic limits. If a flow is steady, $\mathbf{u}$ is independent of t, $\partial \mathbf{u}/\partial t = \mathbf{0}$, and the velocity at a certain position in space remains constant in time.

   The velocity of a two-dimensional flow in the xy plane is independent of the z coordinate, $\partial \mathbf{u}/\partial z = \mathbf{0}$, and the component of the velocity along the z axis has a constant value that is usually made to vanish by an appropriate choice of the frame of reference. The velocity of an axially symmetric flow, in short called an axisymmetric flow, in the possible presence of swirling motion, is independent of the azimuthal angle φ measured around the axis of the flow.

Subparcels and Point Particles

   To analyze the motion of a fluid parcel, it is helpful to divide it into a collection of subparcels with smaller dimensions, and note that the rate of rotation and deformation of the parcel will depend upon the relative motion of the subparcels. If, for instance, all subparcels move with the same velocity, then the parcel will translate as a rigid body, which means that its rate of rotation and deformation will vanish. Furthermore, it is conceivable that the velocity of the subparcels may be coordinated so that the parent parcel rotates as a rigid body without deforming, that is, without changing its shape.

   If we continue to subdivide a parcel into subparcels of increasingly smaller size, we shall obtain subparcels with infinitesimal dimensions called *point particles*. Each point particle occupies an infinitesimal volume in space, but an infinite collection of point particles that belongs to a finite parcel occupies a finite volume in space. By definition, a point particle located at the point $\mathbf{x}$ moves with the local and current velocity of the fluid $\mathbf{u}(\mathbf{x}, t)$.

Motion of a Parcel

To describe the motion of a fluid parcel in more quantitative terms, we consider the spatial distribution of the velocity $\mathbf{u}$ in the vicinity of a point $\mathbf{x}_0$ that is located somewhere near a designated center of the parcel. Expanding $\mathbf{u}(\mathbf{x}, t)$ in a Taylor series with respect to $\mathbf{x}$ about the point $\mathbf{x}_0$ and retaining only linear terms, we find

$$u_j(\mathbf{x}, t) = u_j(\mathbf{x}_0, t) + \hat{x}_i L_{ij}(\mathbf{x}_0, t) \tag{1.1.1}$$

where $\hat{\mathbf{x}} = \mathbf{x} - \mathbf{x}_0$, and $\mathbf{L}$ is the matrix of *velocity gradients* or *rate of relative displacement*,

$$L_{ij} \equiv \frac{\partial u_j}{\partial x_i} \tag{1.1.2}$$

In physical terms, $\mathbf{L}$ expresses the spatial variation of the velocity of a collection of subparcels or point particles composing a parcel. Later in this section we shall show that $\mathbf{L}$ satisfies a transformation law that renders it a second-order Cartesian tensor.

We proceed by decomposing $\mathbf{L}$ into an antisymmetric component denoted by $\boldsymbol{\Xi}$, a symmetric component with vanishing trace denoted by $\mathbf{E}$, and an isotropic component with finite trace, writing

$$\mathbf{L} = \boldsymbol{\Xi} + \mathbf{E} + \tfrac{1}{3}\,\mathrm{Tr}(\mathbf{L})\,\mathbf{I} \tag{1.1.3}$$

where Tr represents the trace, and where we have introduced the *vorticity tensor*

$$\Xi_{ij} = \frac{1}{2}\left(\frac{\partial u_j}{\partial x_i} - \frac{\partial u_i}{\partial x_j}\right) \tag{1.1.4}$$

the *rate of deformation* or *rate of strain* tensor

$$E_{ij} = \frac{1}{2}\left(\frac{\partial u_j}{\partial x_i} + \frac{\partial u_i}{\partial x_j}\right) - \tfrac{1}{3}\delta_{ij}\frac{\partial u_k}{\partial x_k} \tag{1.1.5}$$

and the identity matrix $\mathbf{I}$. The trace of $\mathbf{L}$ is equal to the divergence of the velocity,

$$\mathrm{Tr}(\mathbf{L}) = \frac{\partial u_i}{\partial x_i} = \nabla \cdot \mathbf{u} \tag{1.1.6}$$

Later in this section we shall show that $\boldsymbol{\Xi}$ and $\mathbf{E}$ are second-order Cartesian tensors.

Vorticity

Because the vorticity tensor is antisymmetric, it has only three independent components and may therefore be expressed in terms of a vector. To implement this simplification, we introduce the vorticity vector $\boldsymbol{\omega}$ and set

$$\Xi_{ij} = \tfrac{1}{2}\varepsilon_{ijk}\omega_k \tag{1.1.7}$$

Conversely, the vorticity vector derives from the vorticity tensor as

$$\omega_k = \varepsilon_{kij}\Xi_{ij} = \tfrac{1}{2}\varepsilon_{kij}\left(\frac{\partial u_j}{\partial x_i} - \frac{\partial u_i}{\partial x_j}\right) = \varepsilon_{kij}\frac{\partial u_j}{\partial x_i}, \tag{1.1.8}$$

which shows that the vorticity is simply equal to the curl of the velocity,

$$\boldsymbol{\omega} = \nabla \times \mathbf{u} \tag{1.1.9}$$

Substituting Eq. (1.1.7) into Eq. (1.1.3) and the resulting expression into Eq. (1.1.1), we obtain an expression that describes the spatial distribution of the velocity in the vicinity of the point $\mathbf{x}_0$ in terms of the vorticity vector, the rate of deformation tensor, and the divergence of the velocity

$$u_j(\mathbf{x}, t) = u_j(\mathbf{x}_0, t) + \tfrac{1}{2}\varepsilon_{jki}\,\omega_k(\mathbf{x}_0, t)\,\hat{x}_i + \hat{x}_i\,E_{ij}(\mathbf{x}_0, t) + \tfrac{1}{3}\hat{x}_j\,\nabla \cdot \mathbf{u} \tag{1.1.10}$$

We proceed now to use this representation in order to analyze the motion of a parcel, reckoning that the point particles execute sequential motions under the influence of each term on the right-hand side.

Translation

The first term on the right-hand side of Eq. (1.1.10) expresses rigid-body translation. Under the influence of this term, the parcel will translate with a velocity that is equal to the velocity of the fluid at its designated center $\mathbf{x}_0$.

Vorticity and rotation

The second term on the right-hand side of Eq. (1.1.10) may be written as $\boldsymbol{\Omega}(\mathbf{x}_0, t) \times \hat{\mathbf{x}}$, where $\boldsymbol{\Omega} = \frac{1}{2}\boldsymbol{\omega}$, and this shows that, under the action of this term, the point particles rotate about the point $\mathbf{x}_0$ with angular velocity that is equal to half the vorticity of the fluid. Conversely, the vorticity vector is parallel to the angular velocity of the point particles, and its magnitude is equal to twice that of the angular velocity of the point particles.

A flow in which the magnitude of the vorticity vanishes everywhere is called an *irrotational* flow, and the corresponding velocity field is irrotational, whereas a flow in which the magnitude of the vorticity has finite values, at least in some regions, is called a *rotational* flow, and the corresponding velocity field is rotational. For example, since the curl of the gradient of any twice differentiable scalar function vanishes, the velocity field $\mathbf{u} = \nabla\phi$ is irrotational for any twice differentiable function ϕ. Sometimes the qualifiers "rotational" and "irrotational" are attributed to the fluid, and one refers to an irrotational fluid to describe a fluid that executes irrotational motion. Strictly speaking, however, irrotationality is not a physical property of the fluid but a kinematic property of the flow.

A number of flows are composed of adjacent regions of nearly rotational and nearly irrotational flow. One example is high-speed flow past a streamlined body, which is irrotational everywhere except within a thin layer that lines the surface of the body and within a slender wake behind the body. Another example is the flow arising when two parallel streams merge at different velocities; this flow is irrotational everywhere except within a shear layer along the stream interface.

Fluid parcels in an irrotational flow may translate, deform, and expand or shrink, but do not rotate. Thus slender fluid parcels resembling needles and rectangular parcels maintain their initial orientation. It is important to note that the fact that a fluid may execute a *global* circulatory motion does not necessarily imply that the individual parcels undergo rotation. For example, the circulatory flow generated by the steady rotation of an infinite cylinder in an ambient fluid of infinite expanse is irrotational, which means that small fluid particles maintain their orientation as they rotate about the cylinder.

Rate of strain and deformation

Let us return to Eq. (1.1.10) and examine the nature of the motion associated with the third term on the right-hand side. Noting that $\mathbf{E}$ is a real symmetric matrix, and therefore has three real eigenvalues λ_1, λ_2, and λ_3 and three mutually orthogonal eigenvectors, suggests that under the action of this term, three infinitesimal fluid parcels resembling slender needles that are initially aligned with the eigenvectors will elongate or compress in their respective directions while remaining orthogonal to each other (Problem 1.1.2). Furthermore, an ellipsoidal fluid parcel with its three axes aligned with the eigenvectors will deform, increasing or reducing its aspect ratio, while maintaining its original orientation. These observations reveal that the third term on the right-hand side of Eq. (1.1.10) expresses *deformation that preserves a parcel's orientation.*

To examine the change in the volume of a parcel, we consider a parcel that has the shape of a rectangular parallelepiped whose sides are aligned with the eigenvectors of $\mathbf{E}$. Let the initial lengths of the sides be δx_1, δx_2, and δx_3. After a small time interval δt has elapsed, the length of the sides will have become equal to $(1 + \lambda_1 \delta t) \, \delta x_1$, $(1 + \lambda_2 \delta t) \, \delta x_2$, and $(1 + \lambda_3 \delta t) \, \delta x_3$. This means that, to first order in δt, the volume of the parcel will be modified by the factor $1 + (\lambda_1 + \lambda_2 + \lambda_3) \delta t$. But because the sum of the eigenvalues of $\mathbf{E}$ is equal to the trace of $\mathbf{E}$, which is equal to zero, the volume of the parcel will be preserved during the deformation.

A flow in which $\mathbf{E}$ is everywhere equal to zero must necessarily express rigid-body motion, including translation and rotation (Problem 1.1.3).

Expansion

The last term on the right-hand side of Eq. (1.1.10) represents isotropic expansion or contraction. Under the action of this term, a small spherical parcel of radius δR centered at the point $\mathbf{x}_0$ will expand or contract isotropically, undergoing neither translation, rotation, nor deformation. To compute the rate of expansion, we note that after a small time interval δt has elapsed, the radius of the parcel will have become equal to $(1 + \frac{1}{3} \nabla \cdot \mathbf{u} \, \delta t) \, \delta R$, and the volume of the parcel will be changed by the differential amount

$$d \, \delta V = \tfrac{4\pi}{3}(1 + \tfrac{1}{3} \nabla \cdot \mathbf{u} \, \delta t)^3 \, \delta R^3 - \tfrac{4\pi}{3} \delta R^3 \approx \delta V \nabla \cdot \mathbf{u} \, \delta t \qquad (1.1.11)$$

which shows that

$$\frac{1}{\delta V} \frac{d \, \delta V}{dt} = \nabla \cdot \mathbf{u} \qquad (1.1.12)$$

This result justifies calling the divergence of the velocity $\nabla \cdot \mathbf{u}$ the *local rate of expansion* or *rate of dilatation* of the fluid.

Second-Order Tensors

Let us consider two Cartesian coordinate systems $\{x_1, x_2, x_3\}$ and $\{y_1, y_2, y_3\}$ whose axes have a common origin. If the $\mathbf{y}$ axes derive from the $\mathbf{x}$ axes by three sequential rotations about the x_1, x_2, and x_3 axes with respective angles equal to φ_1, φ_2, and φ_3, then the $\mathbf{y}$ coordinates of a point in space may be computed from the $\mathbf{x}$ coordinates by means of the relation

$$\mathbf{y} = \mathbf{R}^{(1)} \cdot \mathbf{R}^{(2)} \cdot \mathbf{R}^{(3)} \cdot \mathbf{x} \qquad (1.1.13)$$

where $\mathbf{R}^{(1)}$, $\mathbf{R}^{(2)}$, $\mathbf{R}^{(3)}$ are three *rotation matrices* defined as

$$\mathbf{R}^{(1)} = \begin{bmatrix} 1 & 0 & 0 \\ 0 & \cos\varphi_1 & \sin\varphi_1 \\ 0 & -\sin\varphi_1 & \cos\varphi_1 \end{bmatrix}, \qquad \mathbf{R}^{(2)} = \begin{bmatrix} \cos\varphi_2 & 0 & -\sin\varphi_2 \\ 0 & 1 & 0 \\ \sin\varphi_2 & 0 & \cos\varphi_2 \end{bmatrix},$$

$$\mathbf{R}^{(3)} = \begin{bmatrix} \cos\varphi_3 & \sin\varphi_3 & 0 \\ -\sin\varphi_3 & \cos\varphi_3 & 0 \\ 0 & 0 & 1 \end{bmatrix} \qquad (1.1.14)$$

Each rotation matrix is orthogonal, which means that its transpose is equal to its inverse; furthermore, its determinant as well as the length of any vector represented by a column or row is equal to one.

Carrying out the matrix multiplications in Eq. (1.1.13), we find that the $\mathbf{y}$ and $\mathbf{x}$ coordinates are related by the relations

$$y_i = A_{ij} x_j \qquad \text{and} \qquad x_i = y_j A_{ji} \qquad (1.1.15)$$

where $\mathbf{A} = \mathbf{R}^{(1)} \cdot \mathbf{R}^{(2)} \cdot \mathbf{R}^{(3)}$ is an orthogonal rotation matrix with $\mathbf{A}^T = \mathbf{A}^{-1}$, and the superscripts T and -1 designate, respectively, the transpose and the inverse. The first row of $\mathbf{A}$ contains the cosines of the angles that are subtended by the three $\mathbf{x}$ axes and the unit vector in the direction of the y_1 axis, called the *direction cosines*. The second and third rows contain the corresponding direction cosines for the unit vectors along the y_2 and y_3 axes. The determinant of $\mathbf{A}$ and the length of a vector represented by any row or column of $\mathbf{A}$ is equal to one.

Let us consider now a 3×3 matrix $\mathbf{T}$ whose elements are certain physical variables whose values depend upon position in space and time. If the values of the elements of $\mathbf{T}$ in the $\mathbf{y}$ system, denoted by $\mathbf{T}(\mathbf{y})$, are related to those in the $\mathbf{x}$ system, denoted by $\mathbf{T}(\mathbf{x})$, by means of the equivalent relations

$$T_{ij}(\mathbf{y}) = A_{ik}A_{jl}T_{kl}(\mathbf{x}) \quad \text{and} \quad T_{ij}(\mathbf{x}) = T_{kl}(\mathbf{y})A_{ki}A_{lj} \quad (1.1.16)$$

then the matrix $\mathbf{T}$ is called a second-order tensor. It is worth noting that the transformations shown in Eq. (1.1.16) are special cases of *similarity transformations* of matrix calculus (Problem 1.1.5). Establishing the tensorial nature of a matrix is important for ensuring its admissibility in equations expressing physical laws.

One important property of second-order tensors is that their characteristic polynomial $\text{Det}(\mathbf{T} - \lambda I)$ is independent of the coordinate system in which $\mathbf{T}$ is evaluated. Thus the coefficients and hence the roots of the characteristic polynomial, which are the eigenvalues of $\mathbf{T}$, are invariant under a change of the coordinate system. To show this, we use the fact that the transformation matrix $\mathbf{A}$ is orthogonal and write

$$T_{ij}(\mathbf{x}) - \lambda \delta_{ij} = T_{kl}(\mathbf{y})A_{ki}A_{lj} - \lambda A_{ki}\delta_{kl}A_{lj} = A_{ki}[T_{kl}(\mathbf{y}) - \lambda \delta_{kl}]A_{lj} \quad (1.1.17)$$

In vector notation we obtain the equivalent statement

$$\mathbf{T}(\mathbf{x}) - \lambda \mathbf{I} = \mathbf{A}^T \cdot [\mathbf{T}(\mathbf{x}) - \lambda \mathbf{I}] \cdot \mathbf{A} \quad (1.1.18)$$

Taking the determinant of both sides of Eq. (1.1.18) and expanding out the determinant of the product, we obtain

$$\text{Det}[\mathbf{T}(\mathbf{x}) - \lambda I] = \text{Det}(\mathbf{A}^T) \cdot \text{Det}[\mathbf{T}(\mathbf{y}) - \lambda I] \cdot \text{Det}(\mathbf{A}) = \text{Det}[\mathbf{T}(\mathbf{y}) - \lambda I] \quad (1.1.19)$$

where we have used the fact that $\text{Det}(\mathbf{A}^T) = \text{Det}(\mathbf{A}) = 1$, and this completes the proof.

Furthermore, we recast the characteristic polynomial in the form

$$\text{Det}(\mathbf{T} - \lambda I) = -\lambda^3 + I_3\lambda^2 - I_2\lambda + I_1 \quad (1.1.20)$$

involving the three invariants

$$\begin{aligned} I_1 &= \text{Det}(\mathbf{T}) = \lambda_1\lambda_2\lambda_3 \\ I_2 &= \lambda_1\lambda_2 + \lambda_2\lambda_3 + \lambda_3\lambda_1 = \tfrac{1}{2}\{[\text{Tr}(\mathbf{T})]^2 - \text{Tr}(\mathbf{T}^2)\} \\ I_3 &= \text{Tr}(\mathbf{T}) = \lambda_1 + \lambda_2 + \lambda_3 \end{aligned} \quad (1.1.21)$$

These are defined in terms of the roots of the characteristic polynomial $\lambda_1, \lambda_2, \lambda_3$, which are the eigenvalues of $\mathbf{T}$. We shall see in Section 3.2 that the invariants in Eqs. (1.1.21) play an important role in developing constitutive equations that provide us with the stresses developing in the fluid as a result of the motion.

Kinematic Tensors

Considering now the matrix of velocity gradients $\mathbf{L}$ introduced in Eq. (1.1.2), we use the chain rule of differentiation and the fact that the velocity transforms like the position vector as shown in Eq. (1.1.15), that is,

$$u_i(\mathbf{y}) = A_{ij}u_j(\mathbf{x}), \qquad u_i(\mathbf{x}) = u_j(\mathbf{y})A_{ji} \quad (1.1.22)$$

and write

$$L_{ij} \equiv \frac{\partial u_j(\mathbf{x})}{\partial x_i} = \frac{\partial y_k}{\partial x_i}\frac{\partial u_j(\mathbf{x})}{\partial y_k} = A_{ki}\frac{\partial u_j(\mathbf{x})}{\partial y_k} = A_{ki}A_{lj}\frac{\partial u_l(\mathbf{y})}{\partial y_k} \quad (1.1.23)$$

which shows that

$$L_{ij}(\mathbf{x}) = L_{kl}(\mathbf{y})A_{ki}A_{lj} \quad (1.1.24)$$

Comparing Eq. (1.1.24) with the second equation in (1.1.16) shows that $\mathbf{L}$ is a second-order Cartesian tensor. The third invariant I_3 defined in Eqs. (1.1.21) is equal to the rate of expansion of the fluid $\nabla \cdot \mathbf{u}$, and this verifies that the rate of dilatation of a parcel is independent of the coordinate system in which the flow is described.

It is clear that the symmetric and antisymmetric parts of $\mathbf{L}$ also obey the transformation rules of Eq. (1.1.16). Thus, both the rate of strain matrix $\mathbf{E}$ and vorticity matrix $\mathbf{\Xi}$ are second-order Cartesian tensors. Furthermore, it is a straightforward exercise to show that the sequence of matrices $\mathbf{L}^2, \mathbf{L}^3, \ldots$ and $\mathbf{E}^2, \mathbf{E}^3, \ldots$ are all second-order tensors. These results will be useful in Section 3.2 where we shall develop physical laws relating the motion to the forces exerted on the surface of fluid parcels.

PROBLEMS

1.1.1   **Relative velocity around a point.** (a) Confirm the validity of the decomposition (1.1.3). (b) Show that Eq. (1.1.8) is consistent with Eq. (1.1.7).

1.1.2   **Eigenvalues of a real and symmetric matrix.** Show that a real and symmetric matrix has real eigenvalues and corresponding orthogonal eigenvectors (see, for instance, Atkinson, 1989).

1.1.3   **Flow with vanishing rate of deformation.** Show that if $\mathbf{E}$ vanishes everywhere in the domain of a flow, the flow must express rigid-body motion, including translation and rotation.

1.1.4   **Momentum tensor.** Show that the matrix $\rho u_i u_j$ is a second-order Cartesian tensor. ρ is the density of the fluid.

1.1.5   **Similarity transformations.** Consider a square matrix $\mathbf{A}$, select a non-singular square matrix $\mathbf{P}$ whose size is equal to that of $\mathbf{A}$, and compute the matrix $\mathbf{B} = \mathbf{P}^{-1} \cdot \mathbf{A} \cdot \mathbf{P}$. This is called a *similarity transformation,* and we say that the matrix $\mathbf{B}$ is similar to $\mathbf{A}$. Show that the eigenvalues of the matrix $\mathbf{B}$ are identical to those of $\mathbf{A}$, thereby concluding that similarity transformations preserve the eigenvalues (see, for instance, Atkinson, 1989). Are the eigenvectors also identical?

1.1.6   **Alternative definition of a tensor.** Show that if the two-dimensional matrix $\mathbf{T}$ is a second-order tensor, then $\mathbf{u}(\mathbf{x}) = \mathbf{T}(\mathbf{x}) \cdot \mathbf{x}$ transforms like a vector according to Eqs. (1.1.15) and (1.1.22).

1.2 | LAGRANGIAN LABELS

In Section 1.1 we defined the velocity of the fluid $\mathbf{u}$ in terms of the average velocity of the molecules that reside within a fluid parcel by taking the outer limit as the size of the parcel tends to zero. This point of view led us to regard $\mathbf{u}$ as a field function of position $\mathbf{x}$ and time t.

Computing now the ratio between the mass and volume of the parcel, and then taking the outer limit as the size of the parcel tends to zero, we obtain the density of the fluid as a function of position $\mathbf{x}$ and time t. We can repeat this procedure for any suitable kinematic or intensive thermodynamic variable, such as the spatial or temporal derivatives of the velocity, the kinetic energy, thermal energy, enthalpy, or entropy per unit mass of the fluid, to obtain that variable as a function of $\mathbf{x}$ and t. This point of view establishes an *Eulerian* framework for describing the kinematic structure of the flow, as well as the physical and thermodynamic properties of the fluid.

On many occasions, it is physically appealing and mathematically convenient to describe the state of the fluid and structure of a flow in terms of the state and motion of the point particles. As a first step, we identify the point particles by assigning to each one of them an identification vector $\mathbf{a}$ composed of three scalar variables, called *Lagrangian labels,* that take values over a certain subset of the set of real numbers. For instance, $\mathbf{a}$ may be identified with the Cartesian or some other curvilinear coordinates of the point particles at a specified instant in time. More generally, we assume that, at a particular instant, the intersections of the surfaces of constant a_1, a_2, and a_3 yield a right-handed system of curvilinear axes (Section A.3, Appendix A).

The value of any kinematic, physical, or intensive thermodynamic variable at a particular location $\mathbf{x}$ at the time instant t may be regarded as a property of the point particle that happens to be at that location at that particular instant. Thus, for any appropriate scalar, vectorial, or matrix function f that may be attributed to a point particle, we write

$$f(\mathbf{x}, t) = f(\mathbf{X}(\mathbf{a}, t), t) = cF(\mathbf{a}, t) \qquad (1.2.1)$$

where $\mathbf{X}(\mathbf{a}, t)$ is the position of the point particle labeled $\mathbf{a}$ at the time instant t, c is a constant, and F is a proper variable that expresses a certain property of the point particles.

Equation (1.2.1) states that one way of obtaining the value of the function f at a point $\mathbf{x}$ at time t is to look up the point particle that resides at $\mathbf{x}$ at time t, identify its label $\mathbf{a}$, read the value of the variable F, and multiply this value by the factor c. Often f and F represent the same physical variable, and then the distinction between them is based solely upon the choice of independent variables used to describe the flow. As an example, applying Eq. (1.2.1) for the vorticity, we obtain

$$\boldsymbol{\omega}(\mathbf{x}, t) = \boldsymbol{\omega}(\mathbf{X}(\mathbf{a}, t), t) = 2\boldsymbol{\Omega}(\mathbf{a}, t) \qquad (1.2.2)$$

Here F represents the angular velocity of the point particle labeled $\mathbf{a}$ at the time instant t, $\boldsymbol{\Omega}(\mathbf{a}, t)$, and the constant factor c is equal to two. As a second example, we apply Eq. (1.2.1) for the velocity of the fluid to obtain

$$\mathbf{u}(\mathbf{x}, t) = \mathbf{u}(\mathbf{X}(\mathbf{a}, t), t) = \mathbf{U}(\mathbf{a}, t) \qquad (1.2.3)$$

In this case F represents the velocity of the point particle labeled $\mathbf{a}$ at the time instant t, denoted by $\mathbf{U}(\mathbf{a}, t)$, and the constant factor c is equal to unity. Equation (1.2.3) suggests that one way to obtain the value of the velocity $\mathbf{u}$ at a point $\mathbf{x}$ at time t is to consider the point particle that resides at $\mathbf{x}$ at time t, identify its label $\mathbf{a}$, and measure its velocity $\mathbf{U}$.

As a further example, we consider the velocity gradient tensor introduced in Eq. (1.1.1) and write

$$L_{ij}(\mathbf{x}, t) = L_{ij}(\mathbf{X}(\mathbf{a}, t), t) = \frac{\partial U_j}{\partial X_i}(\mathbf{a}, t) \qquad (1.2.4)$$

The middle term in Eq. (1.2.4) expresses the velocity gradient at the location of the point particle labeled $\mathbf{a}$, which is computed in terms of the spatial distribution of the velocity of the neighboring point particles.

The Material Derivative

Since the velocity of a point particle is equal to the rate of change of its position $\mathbf{X}$, we may write

$$\mathbf{U}(\mathbf{a}, t) = \left(\frac{\partial \mathbf{X}}{\partial t}\right)_{\mathbf{a}}(\mathbf{a}, t) \qquad (1.2.5)$$

The partial derivative with respect to time keeping $\mathbf{a}$ constant is known under the aliases *convective, substantial, substantive,* or *material derivative,* and is denoted by D/Dt. Thus Eq. (1.2.5) may be written in the compact form

$$\mathbf{U}(\mathbf{a}, t) = \frac{D\mathbf{X}}{Dt}(\mathbf{a}, t) \qquad (1.2.6)$$

To this end, we have at our disposal two sets of independent variables that we can use to describe a flow: the Eulerian set $(\mathbf{x}, t)$ and the Lagrangian set $(\mathbf{a}, t)$. The relationship between the partial derivatives of a function f with respect to these two sets may be established by applying the chain rule, writing

$$\frac{Df}{Dt} = \left(\frac{\partial f(\mathbf{X}(\mathbf{a}, t), t)}{\partial t}\right)_{\mathbf{a}} = \left(\frac{\partial f}{\partial t}\right)_{\mathbf{x}} + \left(\frac{\partial f}{\partial X_i}\right)_t \frac{DX_i}{Dt} \tag{1.2.7}$$

For simplicity, we shall drop the parentheses around the Eulerian partial derivatives on the right-hand side of Eq. (1.2.7), and then use Eqs. (1.2.6) and (1.2.3) to obtain

$$\frac{Df}{Dt} = \frac{\partial f}{\partial t} + u_i \frac{\partial f}{\partial x_i} = \frac{\partial f}{\partial t} + \mathbf{u} \cdot \nabla f \tag{1.2.8}$$

which relates the material derivative to the temporal and spatial derivatives with respect to the Eulerian variables. When all point particles maintain their value of f as they move about the domain of flow, then $Df/Dt = 0$ for all values of $\mathbf{a}$, and we say that the field represented by f is *convected by the flow.*

As an example, we apply Eq. (1.2.8) for the velocity and derive an expression for the acceleration of a point particle in the Eulerian form

$$\dot{U}_j \equiv \frac{DU_j}{Dt} = \frac{\partial u_j}{\partial t} + u_i \frac{\partial u_j}{\partial x_i} \tag{1.2.9}$$

In vector notation, Eq. (1.2.9) assumes the form

$$\dot{\mathbf{U}} \equiv \frac{D\mathbf{U}}{Dt} = \frac{\partial \mathbf{u}}{\partial t} + \mathbf{u} \cdot \nabla \mathbf{u} \tag{1.2.10}$$

If a point particle maintains its velocity, which means that it neither accelerates nor decelerates as it moves about the domain of flow, then $D\mathbf{U}/Dt = \mathbf{0}$.

Returning to Eq. (1.2.8), we identify the term $\mathbf{u} \cdot \nabla f$ with $|\mathbf{u}| \, \partial f/\partial l$, where $\partial f/\partial l$ is the rate of change of f with respect to arc length l measured in the direction of the velocity. If the field f is steady, $\partial f/\partial t = 0$, and the point particles maintain their value of f, $Df/Dt = 0$, then $\partial f/\partial l$ must vanish. This requires that the value of f does not change in the direction of the velocity and is therefore constant along the paths traced by the point particles, which are the streamlines of the flow. The value of f, however, is generally different along different streamlines.

Lagrangian Mapping

A point of view was established in this section according to which the fluid is regarded as though it were composed of an infinite collection of point particles that are identified by the value of the vectorial label $\mathbf{a}$. The instantaneous position of a point particle $\mathbf{X}$ is regarded as a function of $\mathbf{a}$ and time t. This functional dependence may be formalized in terms of a time-dependent mapping of the labeling space of $\mathbf{a}$ to the physical space of $\mathbf{X}$, $\mathbf{a} \rightarrow \mathbf{X}$, that may be written in the symbolic form

$$\mathbf{X}(t) = \boldsymbol{\chi}_t(\mathbf{a}) \tag{1.2.11}$$

where the subscript t emphasizes that the mapping function $\boldsymbol{\chi}_t$ changes in time. It is important to note that $\boldsymbol{\chi}_t$ is time dependent even though the velocity field may be steady, and it is constant only when the point particles and hence the fluid are in a state of rest. If we identify $\mathbf{a}$ with the coordinates of the point particles at $t = 0$, then $\boldsymbol{\chi}_0(\mathbf{a}) = \mathbf{a}$.

Differential vectors in physical space are related to differential vectors in labeling space by the equation

$$dX_i = \frac{\partial \chi_{t_i}}{\partial a_j} \, da_j \tag{1.2.12}$$

where $J_{ij} = \partial \chi_{t_i}/\partial a_j$ is the *Jacobian matrix* of the mapping function $\boldsymbol{\chi}_t$. For instance, the differential vector $d\mathbf{a}^{(1)} = (da_1, 0, 0)$ in labeling space corresponds to the differential vector

$$d\mathbf{X}^{(1)} = \frac{\partial \boldsymbol{\chi}_t}{\partial a_1} \, da_1 \tag{1.2.13}$$

in physical space; similar equations define $d\mathbf{X}^{(2)}$ and $d\mathbf{X}^{(3)}$ corresponding to $d\mathbf{a}^{(2)} = (0, da_2, 0)$ and $d\mathbf{a}^{(3)} = (0, 0, da_3)$.

A differential volume $dV(\mathbf{a})$ in labeling space is mapped to a corresponding differential volume $dV(\mathbf{X})$ in physical space. To derive a relationship between the relative magnitude of these two volumes, we write the counterparts of Eq. (1.2.13) for $d\mathbf{a}^{(2)}$ and $d\mathbf{a}^{(3)}$, and form the triple mixed product

$$(d\mathbf{X}^{(1)} \times d\mathbf{X}^{(2)}) \cdot d\mathbf{X}^{(3)} = \left(\frac{\partial \chi_t}{\partial a_1} \times \frac{\partial \chi_t}{\partial a_2}\right) \cdot \frac{\partial \chi_t}{\partial a_3} \, da_1 \, da_2 \, da_3 \qquad (1.2.14)$$

Assuming that the $d\mathbf{X}^{(i)}$ are arranged according to the right-handed rule, we identify the left-hand side of Eq. (1.2.14) with $dV(\mathbf{X})$ and set $da_1 \, da_2 \, da_3 = dV(\mathbf{a})$ to obtain

$$dV(\mathbf{X}) = J \, dV(\mathbf{a}) \qquad (1.2.15)$$

where we have introduced the *Jacobian* of the mapping function

$$J = \text{Det}(\partial \chi_{t_i}/\partial a_j) \qquad (1.2.16)$$

Equation (1.2.15) identifies J with the ratio between two corresponding infinitesimal volumes in physical and labeling space.

Deformation and Relative Deformation Gradient

In the special case where $\mathbf{a}$ is identified with the coordinates of the point particles in a Cartesian or some other curvilinear coordinate system, the Jacobian matrix introduced in Eq. (1.2.12) is called the *deformation gradient* and is denoted by $\mathbf{F}$. Thus, by definition,

$$dX_i = F_{ij} \, da_j \qquad (1.2.17)$$

The *right Cauchy–Green strain tensor* $\mathbf{C}$ and *left Cauchy–Green strain tensor* $\mathbf{B}$ are defined in terms of $\mathbf{F}$ as $\mathbf{C} = \mathbf{F}^T\mathbf{F}$ and $\mathbf{B} = \mathbf{F}\mathbf{F}^T$. An immediate consequence of these definitions is that $\mathbf{B}$ and $\mathbf{C}$ are symmetric and positive real (Problem 1.2.4). These tensors find important applications in the development of constitutive equations that relate the stresses developing in a fluid to the motion and deformation of fluid parcels (Section 3.2).

To avoid the complications arising from the possibility that $\mathbf{a}$ and χ_t may refer to different coordinate systems, it is convenient to identify $\mathbf{a}$ with the coordinates $\mathbf{X}$ of the point particles at time t in some coordinate system, and denote the corresponding coordinates of the point particles in the same system at time τ by $\boldsymbol{\xi}$, thus setting $\boldsymbol{\xi} = \chi_\tau(\mathbf{X})$. Having made this choice, we recast Eq. (1.2.17) into the form

$$d\boldsymbol{\xi} = \mathbf{F}_{(t)}(\tau) \cdot d\mathbf{X} \qquad (1.2.18)$$

where $\mathbf{F}_{(t)}(\tau)$ is the *relative deformation gradient*. One may show that the relative deformation gradient obeys the transformation rules that render it a second-order tensor (Schowalter, 1978).

PROBLEMS

1.2.1 **Lagrangian labeling.** Discuss whether it is possible to label all point particles within a finite three-dimensional parcel using a single scalar variable, or even two scalar variables.

1.2.2 **Material derivative.** Derive an expression for the material derivative $D\dot{\mathbf{U}}/Dt$ in terms of derivatives of the velocity with respect to Eulerian variables, where $\dot{\mathbf{U}}$ is the acceleration of a point particle.

1.2.3 **Relative deformation gradient.** Explain why $\mathbf{F}_{(t)}(t) = \mathbf{I}$, where $\mathbf{I}$ is the identity matrix, and show that for an incompressible fluid $\text{Det}[\mathbf{F}_{(t)}(\tau)] = 1$. A fluid is incompressible when the volume of all parcels remains constant in time (Section 1.3).

1.2.4   **Cauchy-Green strain tensors.** Show that $\mathbf{B}$ and $\mathbf{C}$ are symmetric and positive real. A matrix $\mathbf{A}$ is positive real if $\mathbf{x} \cdot \mathbf{A} \cdot \mathbf{x} > 0$ for any real vector $\mathbf{x}$.

1.2.5   **Flow due to the motion of a rigid body.** Consider the flow due to the steady motion of the rigid body that translates with velocity $\mathbf{U}$ and rotates about the origin with angular velocity $\mathbf{\Omega}$ in an otherwise quiescent fluid of infinite expanse. Show that the velocity field must satisfy the equation

$$\frac{\partial \mathbf{u}}{\partial t} = -(\mathbf{U} + \mathbf{\Omega} \times \mathbf{x}) \cdot \nabla \mathbf{u}$$

provided that, in a frame of reference in which the body appears to be stationary, the flow is steady.

1.3 | PROPERTIES OF PARCELS, CONSERVATION OF MASS, AND THE CONTINUITY EQUATION

Let us consider a fluid parcel and label its constituent point particles using values of the vector label $\mathbf{a}$ that fall within the subset A of the three-dimensional labeling space. Equation (1.2.15) suggests that the volume of the parcel is given by

$$V_P = \int_{\text{Parcel}} dV(\mathbf{X}) = \int_A J \, dV(\mathbf{a}) \tag{1.3.1}$$

which shows that J plays the role of a *volume metric coefficient* or *weighting function*. We note again that $dV(\mathbf{a})$ signifies the differential volume of the parcel in labeling space.

Rate of Change of Volume of a Fluid Parcel

Differentiating Eq. (1.3.1) with respect to time and noting that the domain of integration in labeling space is independent of time, we find that the rate of change of volume of the parcel is given by

$$\frac{dV_P}{dt} = \frac{d}{dt} \int_A J \, dV(\mathbf{a}) = \int_A \frac{DJ}{Dt} \, dV(\mathbf{a}) \tag{1.3.2}$$

Proceeding from a different viewpoint, we recall that the point particles move with the velocity of the fluid, and use the divergence theorem, to write

$$\frac{dV_P}{dt} = \int_{\text{Parcel}} \mathbf{u} \cdot \mathbf{n} \, dS(\mathbf{X}) = \int_{\text{Parcel}} \nabla \cdot \mathbf{u} \, dV(\mathbf{X}) = \int_A \nabla \cdot \mathbf{u} \, J \, dV(\mathbf{a}) \tag{1.3.3}$$

where $\mathbf{n}$ is the unit normal vector pointing outward from the parcel. Comparing Eq. (1.3.2) with Eq. (1.3.3) and noting that the subset A is arbitrary, thereby eliminating the integral signs, we obtain

$$\frac{DJ}{Dt} = J \nabla \cdot \mathbf{u} \tag{1.3.4}$$

which shows that the rate of change of the Jacobian J following a point particle is proportional to the rate of expansion of the fluid.

Reynolds Transport Theorem

Considering next the rate of change of a general scalar, vectorial, or tensorial variable F integrated over the volume of a parcel, we write

$$\frac{d}{dt} \int_{\text{Parcel}} F \, dV(\mathbf{X}) = \frac{d}{dt} \int_A FJ \, dV(\mathbf{a}) = \int_A \frac{D(FJ)}{Dt} \, dV(\mathbf{a}) \tag{1.3.5}$$

To derive the right-hand side of Eq. (1.3.5) we exploited once again the fact that the volume of integration in labeling space is independent of time. Expanding the derivative inside the last integral and using Eq. (1.3.4), we obtain

$$\frac{d}{dt}\int_{\text{Parcel}} F\, dV(\mathbf{X}) = \int_{\text{Parcel}} \left(\frac{DF}{Dt} + F\nabla\cdot\mathbf{u}\right) dV(\mathbf{X})$$

$$= \int_{\text{Parcel}} \left(\frac{\partial F}{\partial t} + \nabla\cdot(F\mathbf{u})\right) dV(\mathbf{X}) \qquad (1.3.6)$$

Finally, applying the divergence theorem for the second term within the last integral we obtain the mathematical statement of the Reynolds transport theorem,

$$\frac{d}{dt}\int_{\text{Parcel}} F\, dV(\mathbf{X}) = \int_{\text{Parcel}} \frac{\partial F}{\partial t}\, dV(\mathbf{X}) + \int_{\text{Parcel}} F\mathbf{u}\cdot\mathbf{n}\, dS(\mathbf{X}) \qquad (1.3.7)$$

where $\mathbf{n}$ is the unit normal vector pointing outwards from the parcel, and the last integral is computed over the surface of the parcel.

The volume integral on the right-hand side of Eq. (1.3.7) represents the rate of accumulation of the quantity F within a control volume that is fixed in space and coincides with the instantaneous position of the parcel; the surface integral represents the rate of convective transport of F outward from the control volume.

Conservation of Mass and the Continuity Equation

In terms of the density of the fluid, the mass of a fluid parcel is given by

$$m_P = \int_{\text{Parcel}} \rho(\mathbf{X}, t)\, dV(\mathbf{X}) = \int_A \rho(\mathbf{a}, t)\, J\, dV(\mathbf{a}) \qquad (1.3.8)$$

Applying Eqs. (1.3.5) and (1.3.6) with $F = \rho$ and requiring that mass neither disappears nor is produced in the flow, or equivalently, fluid parcels maintain their mass as they move about the domain of flow, $dm_P/dt = 0$, we set the integrands equal to zero and obtain the equivalent statements of the *continuity equation expressing conservation of mass,*

$$\frac{D}{Dt}[\rho\, dV(\mathbf{X})] = 0 \qquad (1.3.9)$$

$$\frac{D(\rho J)}{Dt} = 0 \qquad (1.3.10)$$

$$\frac{D\rho}{Dt} + \rho\nabla\cdot\mathbf{u} = 0 \qquad (1.3.11)$$

$$\frac{\partial\rho}{\partial t} + \nabla\cdot(\rho\mathbf{u}) = 0 \qquad (1.3.12)$$

which *are valid for both compressible and incompressible fluids.* Conservation of mass imposes a mathematical constraint by requiring that the structure of the velocity field be such that the fluid parcels do not tend to occupy the same volume in space leaving behind empty holes.

Furthermore, applying the Reynolds transport theorem (1.3.7) with $F = \rho$, and requiring that the left-hand side vanish, we obtain a statement of the continuity equation in Eulerian form,

$$\int_{V_c} \frac{\partial\rho}{\partial t}\, dV(\mathbf{X}) = -\int_{V_c} \rho\mathbf{u}\cdot\mathbf{n}\, dS(\mathbf{X}) \qquad (1.3.13)$$

where V_c is a control volume that is fixed in space and coincides with the instantaneous location of a fluid parcel. Equation (1.3.13) states that the rate of accumulation of mass within the control volume is equal to the rate at which mass enters the control volume through the boundaries.

Incompressible fluids and solenoidal velocity fields

When the fluid is incompressible, the point particles maintain their initial density, $D\rho/Dt = 0$, and Eq. (1.3.11) assumes the simplified form

$$\nabla \cdot \mathbf{u} = 0 \qquad (1.3.14)$$

A vector field with vanishing divergence that satisfies Eq. (1.3.14) is called *solenoidal*. According to our earlier discussion in Section 1.1, a parcel of an incompressible fluid translates, rotates, and deforms while maintaining its original volume. The terms *incompressible fluid, incompressible flow,* and *incompressible velocity field* are sometimes used interchangeably, but one should keep in mind that, strictly speaking, incompressibility is neither a kinematical property of the flow nor a structural property of the velocity field, but a physical property of the fluid.

Combining Eq. (1.3.14) with Eq. (1.3.4) shows that $DJ/Dt = 0$, which states that the Jacobian is convected by the flow. In this case, the mapping function χ_t introduced in Eq. (1.2.11) is called *isochoric,* from the Greek words $\iota\sigma o\varsigma$, which means *same* or *equal,* and $\chi\omega\rho o\varsigma$, which means *space.*

We proceed next to discuss examples of incompressible flows. First we note that the divergence of the curl of any twice differentiable vector field vanishes, and this guarantees that a velocity field that derives from a vector function $\mathbf{A}$, called the vector potential, as $\mathbf{u} = \nabla \times \mathbf{A}$, is solenoidal for any choice of a twice differentiable vector function $\mathbf{A}$ (Section 2.4). As a second example, we consider a velocity field that derives by taking the cross product of two arbitrary vector fields $\mathbf{A}$ and $\mathbf{B}$ as $\mathbf{u} = \mathbf{A} \times \mathbf{B}$. Straightforward differentiation shows that $\nabla \cdot \mathbf{u} = \mathbf{B} \cdot \nabla \times \mathbf{A} - \mathbf{A} \cdot \nabla \times \mathbf{B}$, which suggests that $\mathbf{u}$ will be solenoidal provided that $\mathbf{A}$ and $\mathbf{B}$ are irrotational. In Section 2.1 we shall see that any irrotational vector field may be expressed as the gradient of a scalar function, and this guarantees that the velocity field $\mathbf{u} = \nabla\phi \times \nabla\psi$ is solenoidal for any choice of the twice differentiable functions ϕ and ψ. As a third example, we consider a velocity field that derives by taking the gradient of a scalar function ϕ, called the potential function, as $\mathbf{u} = \nabla\phi$, and find that $\mathbf{u}$ will be solenoidal provided that ϕ is a twice-differentiable harmonic function, $\nabla^2\phi = 0$. In this case ϕ is called a *harmonic potential* (Section 2.1).

Reciprocity of incompressible flows

Let us consider two generally unrelated solenoidal velocity fields $\mathbf{u}$ and $\mathbf{u}'$. Assuming that neither one of them has any singular points within a certain control volume, and using Eq. (1.3.14), we derive the identity

$$\frac{\partial}{\partial x_k}\left(u_i'\frac{\partial u_k}{\partial x_i} - u_i\frac{\partial u_k'}{\partial x_i}\right) = 0 \qquad (1.3.15)$$

Integrating Eq. (1.3.15) over the control volume, and using the divergence theorem, we find

$$\int_D \left(u_i'\frac{\partial u_k}{\partial x_i} - u_i\frac{\partial u_k'}{\partial x_i}\right)n_k\, dS = 0 \qquad (1.3.16)$$

where $\mathbf{n}$ is the unit vector normal to the boundary D of the control volume pointing either into or outwards from the control volume. Equation (1.3.16) places an integral constraint upon the mutual structure of the velocity field of any two flows of incompressible fluids.

Rate of Change of Properties of Parcels

We turn next to discuss the computation of the rate of change of extensive kinematic, physical, or thermodynamic variables of a generally compressible fluid parcel. By definition, an extensive variable is proportional to the parcel's mass. For each extensive variable there is a corresponding intensive variable, so that when the latter is multiplied by the mass of the parcel, and possibly by a physical constant, it produces the extensive variable. Pairs of extensive–intensive variables are momentum and velocity, thermal energy and temperature, kinetic energy and the square of the magnitude of the velocity. In developing dynamical laws for the behavior of the fluid parcels, it is useful to have expressions for the rate of change of the extensive variables in terms of the rate of change of the corresponding intensive variables. This can be done on the basis of the continuity equation expressed by the equivalent forms given in Eqs. (1.3.9)–(1.3.12).

Two examples of extensive variables are the linear momentum and angular momentum of a parcel, given respectively by

$$\mathbf{M}_P = \int_{\text{Parcel}} \rho \mathbf{U} \, dV(\mathbf{X}) = \int_A \rho \mathbf{U} J \, dV(\mathbf{a}) \qquad (1.3.17)$$

$$\mathbf{A}_P = \int_{\text{Parcel}} \rho(\mathbf{X} \times \mathbf{U}) \, dV(\mathbf{X}) = \int_A \rho(\mathbf{X} \times \mathbf{U}) J \, dV(\mathbf{a}) \qquad (1.3.18)$$

Using Eq. (1.3.10), we express the rate of the change of the linear momentum of a parcel in terms of the acceleration of the point particles as

$$\frac{d\mathbf{M}_P}{dt} = \int_A \left(\frac{D\mathbf{U}}{Dt} \rho J + \mathbf{U} \frac{D(\rho J)}{Dt} \right) dV(\mathbf{a}) = \int_{\text{Parcel}} \rho \frac{D\mathbf{U}}{Dt} \, dV(\mathbf{X}) \qquad (1.3.19)$$

Working in a similar manner with the angular momentum and using Eq. (1.2.6), we obtain

$$\frac{d\mathbf{A}_P}{dt} = \int_{\text{Parcel}} \rho \frac{D(\mathbf{X} \times \mathbf{U})}{Dt} \, dV(\mathbf{X}) = \int_{\text{Parcel}} \rho \mathbf{X} \times \frac{D\mathbf{U}}{Dt} \, dV(\mathbf{X}) \qquad (1.3.20)$$

It is worth emphasizing again that Eqs. (1.3.19) and (1.3.20) are valid for both compressible and incompressible fluids.

For a general scalar, vectorial, or tensorial intensive variable f we find

$$\frac{d}{dt} \int_{\text{Parcel}} f \rho \, dV(\mathbf{X}) = \frac{d}{dt} \int_A f \rho J \, dV(\mathbf{a}) = \int_{\text{Parcel}} \rho \frac{Df}{Dt} \, dV(\mathbf{X}) \qquad (1.3.21)$$

Equation (1.3.19) emerges by setting $f = \mathbf{U}$.

PROBLEMS

1.3.1 **Rate of change of extensive variables.** Derive the right-hand sides of Eqs. (1.3.20) and (1.3.21).

1.3.2 **Vector potential.** Show that the vector potentials $\mathbf{A}$ and $\mathbf{A} + \nabla\phi$ generate the same flow, where $\mathbf{A}$ and ϕ are two arbitrary functions. Is this an incompressible flow?

1.3.3 **Reynolds transport theorem.** Show that Eq. (1.3.20) is consistent with Eq. (1.3.7).

1.4 | MATERIAL VECTORS, MATERIAL LINES, AND MATERIAL SURFACES

Analyzing the motion of lines and surfaces composed of a fixed collection of point particles in a prescribed flow is important in developing dynamical laws that govern the kinematical and physical properties of surfaces of fluid parcels and interfaces between two different fluids. The relation between the structure of the velocity field and the motion of material lines and surfaces will be the subject of the present and following two sections.

Material Vectors

We begin by considering the motion of a material vector $\delta\mathbf{X}$, which is a line of infinitesimal length with a designated beginning and end. Applying Eq. (1.2.6) for the point particles that are located at the end points of the material vector, expressing the velocity at the last point in terms of a Taylor series about the first point, and retaining only the linear terms, we obtain

$$\frac{D\,\delta\mathbf{X}}{Dt} = \mathbf{U}^{\text{last}} - \mathbf{U}^{\text{first}} = \delta X_i \frac{\partial \mathbf{u}}{\partial x_i} = \delta\mathbf{X}\cdot\mathbf{L} \tag{1.4.1}$$

where $\mathbf{L}$ is the velocity gradient tensor, introduced in Eq. (1.1.2), evaluated at the position of the material vector. Using Eq. (1.4.1), we find that the rate of change of the differential length of the material vector, $\delta l \equiv |\delta\mathbf{X}|$, is given by

$$\frac{D\,\delta l}{Dt} = \frac{D}{Dt}(\delta\mathbf{X}\cdot\delta\mathbf{X})^{1/2} = \frac{\delta\mathbf{X}}{\delta l}\cdot\frac{D\,\delta\mathbf{X}}{Dt}$$

$$= \frac{\delta\mathbf{X}}{\delta l}\cdot(\delta\mathbf{X}\cdot\mathbf{L}) = \mathbf{t}\cdot\mathbf{L}\cdot\mathbf{t}\,\delta l \tag{1.4.2}$$

where $\mathbf{t} = \delta\mathbf{X}/\delta l$ is the unit vector in the direction of the material vector. The quantity $\mathbf{t}\cdot\mathbf{L}\cdot\mathbf{t}$ on the right-hand side of Eq. (1.4.2) is the rate of extension of the fluid in the direction of the material vector.

To compute the rate of change of the unit vector $\mathbf{t}$ that is tangential to the material vector, we begin by writing

$$\frac{D\,\delta\mathbf{X}}{Dt} = \frac{D(\mathbf{t}\,\delta l)}{Dt} = \delta l\,\frac{D\mathbf{t}}{Dt} + \mathbf{t}\,\frac{D\,\delta l}{Dt} \tag{1.4.3}$$

and then use Eqs. (1.4.1) and (1.4.2) to obtain

$$\frac{D\mathbf{t}}{Dt} = (\mathbf{t}\cdot\mathbf{L})\cdot(\mathbf{I} - \mathbf{tt}) \tag{1.4.4}$$

where $\mathbf{I}$ is the identity matrix. Note that projecting a vector onto $\mathbf{I} - \mathbf{tt}$ eliminates its component in the direction of $\mathbf{t}$.

Material Lines

Next we consider the motion of a material line that is composed of a fixed collection of point particles forming an open or closed loop. To identify the point particles, it is convenient to introduce the scalar label a that takes values within the subset A of the set of real numbers, and regard the position of the point particles $\mathbf{X}$ as a function of a and time t. The unit vector

$$\mathbf{t} = \frac{1}{h}\frac{\partial\mathbf{X}}{\partial a}, \qquad \text{where} \qquad h = \left|\frac{\partial\mathbf{X}}{\partial a}\right| \tag{1.4.5}$$

is tangential to the material line, and the length of an infinitesimal section of the material line is equal to $\delta l = h\,\delta a$. The total arc length of the material line is given by

$$L = \int_A h\,da \tag{1.4.6}$$

which shows that h is a scalar metric coefficient for the arc length, and can be regarded as the counterpart of the Jacobian J for the volume of fluid parcels defined in Eq. (1.2.16). If we identify a with the instantaneous arc length along the material line l, then $h = 1$.

To compute the rate of change of the total length of a material line, we differentiate Eq. (1.4.6) with respect to time and exploit the fact that the limits of integration are fixed to find

$$\frac{dL}{dt} = \int_A \frac{Dh}{Dt}\,da = \int_{\text{Line}} \frac{1}{h}\frac{Dh}{Dt}\,dl \tag{1.4.7}$$

The integrand on the right-hand side of Eq. (1.4.7) expresses the local rate of extension of the material line. Furthermore, taking the material derivative of the second equation in (1.4.5), working as in Eq. (1.4.2), and using the Frenet relation

$$\frac{\partial \mathbf{t}}{\partial l} = -\kappa \mathbf{n} \tag{1.4.8}$$

where κ is the curvature of the material line and $\mathbf{n}$ is the *principal normal unit vector* of the material line, we obtain

$$\frac{Dh}{Dt} = \mathbf{t} \cdot \frac{\partial \mathbf{U}}{\partial a} = \frac{\partial (\mathbf{t} \cdot \mathbf{U})}{\partial a} + \mathbf{U} \cdot \mathbf{n} \kappa \frac{\partial l}{\partial a} \tag{1.4.9}$$

The two terms on the right-hand side of Eq. (1.4.9) express the rate of change of the length of an infinitesimal section of the material line due to stretching in the plane of the line, and extension due to motion normal to the line. These interpretations become more clear by considering the behavior of a circular material line that either exhibits tangential motion with vanishing normal velocity, or expands in the radial direction while remaining in its plane, as discussed in Problem 1.4.1.

Substituting Eq. (1.4.9) into Eq. (1.4.7) yields

$$\frac{dL}{dt} = (\mathbf{t} \cdot \mathbf{U})_{\text{start}}^{\text{end}} + \int_L \mathbf{U} \cdot \mathbf{n} \kappa \, dl \tag{1.4.10}$$

When the material line forms a closed loop, the first term on the right-hand side of Eq. (1.4.10) vanishes, and this shows that the tangential motion does not contribute to the rate of change of its total length.

Material Surfaces

Next we consider an open or closed infinite or finite material surface composed of a fixed collection of point particles. Physically, a material surface may be identified with the boundary of a fluid parcel or with an interface between two fluids.

To describe the shape and motion of a material surface, it is convenient to identify the point particles that compose it in terms of two scalar labels ξ and η, called *surface curvilinear coordinates,* that form a right-handed coordinate system and take values over a specified range in the (ξ, η) parameter plane, as illustrated in Figure 1.4.1. Using these labels we establish a mapping between the curved material surface in physical space and a planar surface in the parametric (ξ, η) plane.

To establish relations between the geometrical properties of the material surface and the position of the point particles in the material surface $\mathbf{X}(\xi, \eta, t)$, we introduce the tangential unit vectors

$$\mathbf{t}_\xi = \frac{1}{h_\xi} \frac{\partial \mathbf{X}}{\partial \xi} \qquad \mathbf{t}_\eta = \frac{1}{h_\eta} \frac{\partial \mathbf{X}}{\partial \eta} \tag{1.4.11}$$

where

$$h_\xi = \left| \frac{\partial \mathbf{X}}{\partial \xi} \right| \qquad h_\eta = \left| \frac{\partial \mathbf{X}}{\partial \eta} \right| \tag{1.4.12}$$

are the metrics associated with the curvilinear coordinates. The arc length of an infinitesimal section of the ξ or η axis is equal to

$$dl_\xi = h_\xi \, d\xi \qquad \text{or} \qquad dl_\eta = h_\eta \, d\eta \tag{1.4.13}$$

Any linear combination of the tangential vectors in Eqs. (1.4.11) is also a tangential vector.

It is evident from the definitions (1.4.11) that

$$\frac{\partial (h_\xi \mathbf{t}_\xi)}{\partial \eta} = \frac{\partial (h_\eta \mathbf{t}_\eta)}{\partial \xi} \tag{1.4.14}$$

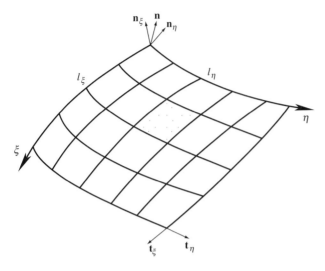

Figure 1.4.1   A system of two curvilinear axes (η, ξ) on a three-dimensional material surface; **n** is the unit vector normal to the material surface, and $\mathbf{n}_\xi$ and $\mathbf{n}_\eta$ are the principal unit vectors normal to the ξ, and to the η axis.

To compute the rate of change of the scaling factor h_ξ, which provides us with the rate of extension of the ξ lines, we take the material derivative of the first equation in (1.4.12), note that by definition $\mathbf{U} = (\partial\mathbf{X}/\partial t)_{\eta,\xi}$, and work as in Eq. (1.4.9) to find

$$\frac{Dh_\xi}{Dt} = \mathbf{t}_\xi \cdot \frac{\partial\mathbf{U}}{\partial\xi} = \frac{\partial(\mathbf{t}_\xi \cdot \mathbf{U})}{\partial\xi} + \mathbf{U}\cdot\mathbf{n}_\xi\kappa_\xi\frac{\partial l_\xi}{\partial\xi} \qquad (1.4.15)$$

where κ_ξ is the curvature of the ξ line, and $\mathbf{n}_\xi$ is the principal unit vector normal to the ξ line shown in Figure 1.4.1, defined by the Frenet formula $\partial\mathbf{t}_\xi/\partial l_\xi = -\kappa_\xi\mathbf{n}_\xi$. The two terms on the right-hand side of Eq. (1.4.15) express, respectively, changes in the length of an infinitesimal section of a ξ line due to stretching in the plane of that ξ line and expansion due to motion in the principal normal direction. The rate of change of h_η is given by Eq. (1.4.15) with ξ replaced throughout by η.

Normal vector
The unit vector normal to the surface is given by

$$\mathbf{n} = \frac{1}{h_S}\frac{\partial\mathbf{X}}{\partial\xi}\times\frac{\partial\mathbf{X}}{\partial\eta}, \qquad \text{where} \qquad h_S = \left|\frac{\partial\mathbf{X}}{\partial\xi}\times\frac{\partial\mathbf{X}}{\partial\eta}\right| \qquad (1.4.16)$$

Combining Eqs. (1.4.11) and (1.4.16), we obtain

$$\mathbf{n} = \frac{h_\xi h_\eta}{h_S}\mathbf{t}_\xi \times \mathbf{t}_\eta \qquad (1.4.17)$$

When $\mathbf{t}_\eta$ is perpendicular to $\mathbf{t}_\xi$, in which case $\mathbf{t}_\xi \cdot \mathbf{t}_\eta = 0$, the system of the surface coordinates (ξ, η) is orthogonal, $\mathbf{n} = \mathbf{t}_\eta \times \mathbf{t}_\xi$ and $h_S = h_\eta h_\xi$.

Since the tangential vectors are perpendicular to the normal vector,

$$\mathbf{n}\cdot\frac{\partial\mathbf{X}}{\partial\xi} = 0 \qquad \text{and} \qquad \mathbf{n}\cdot\frac{\partial\mathbf{X}}{\partial\eta} = 0 \qquad (1.4.18)$$

Differentiating the first equation with respect to η, the second equation with respect to ξ, expanding out the derivatives, and combining the results, we obtain the relation

$$\frac{\partial \mathbf{n}}{\partial l_\xi} \cdot \mathbf{t}_\eta = \frac{\partial \mathbf{n}}{\partial l_\eta} \cdot \mathbf{t}_\xi \tag{1.4.19}$$

which will find use in the imminent computation of surface curvatures.

Surface area

The area of a differential element of the material surface is given by $dS = h_S \, d\xi \, d\eta$, and the total area of the material surface is given by

$$S = \int_H h_S \, d\xi \, d\eta \tag{1.4.20}$$

where H is the domain of definition of ξ and η over the surface. Equation (1.4.20) suggests that h_S serves as a metric coefficient with respect to the surface coordinates and plays a role that is similar to that of the Jacobian J for the Lagrangian labels of a fluid parcel.

To compute the rate of change of the surface area of the material surface, we differentiate Eq. (1.4.20) with respect to time and note that the limits of integration are fixed to obtain

$$\frac{dS}{dt} = \int_H \frac{Dh_S}{Dt} \, d\xi \, d\eta = \int_{\text{Surface}} \frac{1}{h_S} \frac{Dh_S}{Dt} \, dS \tag{1.4.21}$$

The integrand on the right-hand side of Eq. (1.4.21) expresses the local rate of expansion of material patches and is therefore called the *rate of dilatation* of the surface.

Using the definition of the normal vector in Eqs. (1.4.16) and the dynamical law (1.4.1) for the rate of change of a material vector, we compute

$$\frac{D(h_S \mathbf{n})}{Dt} = \frac{D}{Dt}\left(\frac{\partial \mathbf{X}}{\partial \xi}\right) \times \frac{\partial \mathbf{X}}{\partial \eta} + \frac{\partial \mathbf{X}}{\partial \xi} \times \frac{D}{Dt}\left(\frac{\partial \mathbf{X}}{\partial \eta}\right)$$

$$= \left(\frac{\partial \mathbf{X}}{\partial \xi} \cdot \mathbf{L}\right) \times \frac{\partial \mathbf{X}}{\partial \eta} + \frac{\partial \mathbf{X}}{\partial \xi} \times \left(\frac{\partial \mathbf{X}}{\partial \eta} \cdot \mathbf{L}\right) \tag{1.4.22}$$

Switching to index notation and using the rules of repeated multiplication of the alternating matrix (Section A.1, Appendix A), we obtain

$$\frac{D(h_S n_i)}{Dt} = \varepsilon_{ijk} \frac{\partial X_l}{\partial \xi} L_{lj} \frac{\partial X_k}{\partial \eta} + \varepsilon_{ikj} \frac{\partial X_k}{\partial \xi} \frac{\partial X_l}{\partial \eta} L_{lj} = L_{lj}\varepsilon_{ijk}\left(\frac{\partial X_l}{\partial \xi} \frac{\partial X_k}{\partial \eta} - \frac{\partial X_k}{\partial \xi} \frac{\partial X_l}{\partial \eta}\right)$$

$$= L_{lj}\varepsilon_{ijk}\,\varepsilon_{plk}\,\varepsilon_{pmn} \frac{\partial X_m}{\partial \xi} \frac{\partial X_n}{\partial \eta} = L_{lj}(\delta_{ip}\delta_{jl} - \delta_{il}\delta_{jp})\,\varepsilon_{pmn} \frac{\partial X_m}{\partial \xi} \frac{\partial X_n}{\partial \eta}$$

$$= L_{jj}\varepsilon_{imn} \frac{\partial X_m}{\partial \xi} \frac{\partial X_n}{\partial \eta} - L_{ij}\varepsilon_{jmn} \frac{\partial X_m}{\partial \xi} \frac{\partial X_n}{\partial \eta} \tag{1.4.23}$$

Reverting back to vector notation, we express the final result in the form

$$\frac{D(h_S \mathbf{n})}{Dt} = h_S[(\nabla \cdot \mathbf{u})\,\mathbf{I} - \mathbf{L}] \cdot \mathbf{n} \tag{1.4.24}$$

Working in a similar manner with the second of Eqs. (1.4.16) and using Eq. (1.4.24), we find that the rate of dilatation of the material surface is given by

$$\frac{Dh_S}{Dt} = \mathbf{n} \cdot \frac{D(h_S \mathbf{n})}{Dt} = h_S(\nabla \cdot \mathbf{u} - \mathbf{n} \cdot \mathbf{L} \cdot \mathbf{n}) \tag{1.4.25}$$

Since $\mathbf{n} \cdot \mathbf{L} \cdot \mathbf{n}$ represents the normal derivative of the normal component of the velocity, the term within the parentheses on the right-hand side represents the divergence of the velocity in the tangential plane.

Introducing the density of the fluid and combining the preceding two equations with the continuity equation (1.3.11), we obtain the more compact forms

$$\frac{D(\rho h_S \mathbf{n})}{Dt} = -\rho h_S \, \mathbf{L} \cdot \mathbf{n}, \qquad \frac{D(\rho h_S)}{Dt} = -\rho h_S \, \mathbf{n} \cdot \mathbf{L} \cdot \mathbf{n} \qquad (1.4.26a,b)$$

Significance of tangential and normal motion

To illustrate the effect of the tangential and normal motions upon the dilatation of a surface, we take the material derivative of the second equation in (1.4.16) and carry out some straightforward differentiations to find

$$\frac{Dh_S}{Dt} = \mathbf{n} \cdot \left(h_\eta \frac{\partial \mathbf{U}}{\partial \xi} \times \mathbf{t}_\eta - h_\xi \frac{\partial \mathbf{U}}{\partial \eta} \times \mathbf{t}_\xi \right) \qquad (1.4.27)$$

Next we express the velocity over the material surface in terms of the surface coordinates writing

$$\mathbf{U} = U_\xi \mathbf{t}_\xi + U_\eta \mathbf{t}_\eta + U_n \mathbf{n} \qquad (1.4.28)$$

Substituting Eq. (1.4.28) into Eq. (1.4.27) and grouping similar terms we obtain

$$\frac{1}{h_S} \frac{Dh_S}{Dt} = \frac{\partial U_\xi}{\partial l_\xi} + \frac{\partial U_\eta}{\partial l_\eta}$$
$$+ \frac{h_\xi h_\eta}{h_S} \left[U_\xi \left(\frac{\partial \mathbf{t}_\xi}{\partial l_\xi} \times \mathbf{t}_\eta - \frac{\partial \mathbf{t}_\xi}{\partial l_\eta} \times \mathbf{t}_\xi \right) + U_\eta \left(\frac{\partial \mathbf{t}_\eta}{\partial l_\xi} \times \mathbf{t}_\eta - \frac{\partial \mathbf{t}_\eta}{\partial l_\eta} \times \mathbf{t}_\xi \right) \right] \cdot \mathbf{n}$$
$$+ U_n \frac{h_\xi h_\eta}{h_S} \left(\frac{\partial \mathbf{n}}{\partial l_\xi} \times \mathbf{t}_\eta - \frac{\partial \mathbf{n}}{\partial l_\eta} \times \mathbf{t}_\xi \right) \cdot \mathbf{n} \qquad (1.4.29)$$

Rearranging the triple mixed products on the right-hand side and regrouping the terms in the square brackets yields

$$\frac{1}{h_S} \frac{Dh_S}{Dt} = \frac{\partial U_\xi}{\partial l_\xi} + \frac{\partial U_\eta}{\partial l_\eta}$$
$$+ \frac{h_\xi h_\eta}{h_S} \left[\left(U_\xi \frac{\partial \mathbf{t}_\xi}{\partial l_\xi} + U_\eta \frac{\partial \mathbf{t}_\eta}{\partial l_\xi} \right) \cdot (\mathbf{t}_\eta \times \mathbf{n}) - \left(U_\xi \frac{\partial \mathbf{t}_\xi}{\partial l_\eta} + U_\eta \frac{\partial \mathbf{t}_\eta}{\partial l_\eta} \right) \cdot (\mathbf{t}_\xi \times \mathbf{n}) \right]$$
$$+ U_n \frac{h_\xi h_\eta}{h_S} \left(\frac{\partial \mathbf{n}}{\partial l_\xi} \cdot (\mathbf{t}_\eta \times \mathbf{n}) - \frac{\partial \mathbf{n}}{\partial l_\eta} \cdot (\mathbf{t}_\xi \times \mathbf{n}) \right) \qquad (1.4.30)$$

The sum of the terms on the right-hand side of Eq. (1.4.30) involving the tangential velocities is equal to the divergence of the velocity with respect to ξ and η, and thus it expresses dilatation due to expansion in the plane of the surface. The last term expresses dilatation due to normal motion. We shall see in the next section that the term within the last set of large parentheses is equal to twice the mean curvature of the surface κ_m [Eq. (1.5.14)].

These interpretations become more evident by assuming that the surface coordinates are orthogonal. Using the fact that the length of a unit tangential vector is constant, we write $\partial(\mathbf{t}_\xi \cdot \mathbf{t}_\xi)/\partial l_\xi = 2\mathbf{t}_\xi \cdot \partial \mathbf{t}_\xi/\partial l_\xi = 0$ and find that the first and fourth terms within the square brackets on the right-hand side of Eq. (1.4.30) vanish, thereby obtaining the simplified form

$$\frac{1}{h_S} \frac{Dh_S}{Dt} = \frac{\partial U_\xi}{\partial l_\xi} + U_\xi \frac{\partial \mathbf{t}_\xi}{\partial l_\eta} \cdot \mathbf{t}_\eta + \frac{\partial U_\eta}{\partial l_\eta} + U_\eta \frac{\partial \mathbf{t}_\eta}{\partial l_\xi} \cdot \mathbf{t}_\xi + U_n 2\kappa_m \qquad (1.4.31)$$

Next we return to Eq. (1.4.14), expand out the derivatives on both sides, project the resulting expression onto $\mathbf{t}_\eta$ and $\mathbf{t}_\xi$, and simplify to obtain the identities

$$\frac{\partial \mathbf{t}_\xi}{\partial l_\eta} \cdot \mathbf{t}_\eta = \frac{1}{h_\eta} \frac{\partial h_\eta}{\partial l_\xi}, \qquad \frac{\partial \mathbf{t}_\eta}{\partial l_\xi} \cdot \mathbf{t}_\xi = \frac{1}{h_\xi} \frac{\partial h_\xi}{\partial l_\eta} \qquad (1.4.32)$$

Substituting these expressions into Eq. (1.4.31) yields

$$\frac{1}{h_S}\frac{Dh_S}{Dt} = \frac{\partial U_\xi}{\partial l_\xi} + U_\xi \frac{1}{h_\eta}\frac{\partial h_\eta}{\partial l_\xi} + \frac{\partial U_\eta}{\partial l_\eta} + U_\eta \frac{1}{h_\xi}\frac{\partial h_\xi}{\partial l_\eta} + U_n 2\kappa_m$$

$$= \frac{1}{h_\eta h_\xi}\left(\frac{\partial (h_\eta U_\xi)}{\partial \xi} + \frac{\partial (h_\xi U_\eta)}{\partial \eta}\right) + U_n 2\kappa_m \qquad (1.4.33)$$

The first term on the right-hand side of Eq. (1.4.33) involving the tangential velocities is the standard expression for the surface divergence of the velocity in orthogonal curvilinear coordinates (Section A.3, Appendix A).

Rate of change of the flow rate of a vector field through a material surface

Let us consider now the flow rate of a certain vector field $\mathbf{q}$ through a material surface, defined as

$$Q = \int_{\text{Surface}} \mathbf{q} \cdot \mathbf{n}\, dS = \int_H \mathbf{q} \cdot \mathbf{n}\, h_S\, d\xi\, d\eta \qquad (1.4.34)$$

In the various applications of fluid mechanics and transport phenomena, $\mathbf{q}$ may be identified with the velocity, the vorticity, or the gradient of the temperature or concentration of a chemical species. In the first case, Q represents the volumetric flow rate.

Taking the time derivative of Eq. (1.4.34) and using Eq. (1.4.24), we obtain

$$\frac{dQ}{dt} = \int_{\text{Surface}} \left(\frac{D\mathbf{q}}{Dt} + (\nabla \cdot \mathbf{u})\mathbf{q} - \mathbf{q} \cdot \mathbf{L}\right) \cdot \mathbf{n}\, dS \qquad (1.4.35)$$

The second and third terms within the integrand express the effect of dilatation of the surface. Expressing the material derivative $D\mathbf{q}/Dt$ in terms of Eulerian derivatives and using the vector identity (A.2.6) we recast Eq. (1.4.35) into the form

$$\frac{dQ}{dt} = \int_{\text{Surface}} \left(\frac{\partial \mathbf{q}}{\partial t} + (\nabla \cdot \mathbf{q})\mathbf{u} + \nabla \times (\mathbf{q} \times \mathbf{u})\right) \cdot \mathbf{n}\, dS \qquad (1.4.36)$$

If $\mathbf{q}$ happens to be solenoidal, the second term of the integrand makes a vanishing contribution.

PROBLEMS

1.4.1 **Expanding and stretching circle.** Consider a circle with radius R, identify the label a with the polar angle θ, and evaluate the right-hand side of Eq. (1.4.9) for $u_r = U$, $u_\theta = V\cos\theta$, where U, V are two constants.

1.4.2 **Expanding and stretching spherical surface.** Consider a spherical surface of radius a, identify ξ with the meridional polar angle θ, and η with the azimuthal angle φ, and evaluate the right-hand side of Eq. (1.4.31) for $u_r = U$, $u_\theta = V\cos\theta$, $u_\varphi = W\cos\varphi$, where U, V, W are three constants.

1.4.3 **Evolution of the unit vector normal to a material surface.** Combine Eqs. (1.4.24) and (1.4.25) to derive an equation for the rate of change of the unit normal vector following the point particles.

1.4.4 **Change of volume of a parcel resting on a material surface.** Consider a small flattened fluid parcel with surface area δS resting on a material surface. The volume of the parcel is $\delta V = \delta S\, \mathbf{n} \cdot \delta \mathbf{X}$, where $\delta \mathbf{X}$ is the side of the parcel out of the plane of the material surface. Using Eqs. (1.4.1) and (1.4.25), show that

$$\frac{1}{\delta V}\frac{D\delta V}{Dt} = \nabla \cdot \mathbf{u} \qquad (1.4.37)$$

1.5 | DIFFERENTIAL GEOMETRY OF SURFACES

We continue the discussion of the geometry of surfaces with a main objective to develop analytical tools for describing the shape and motion of material surfaces and interfaces between two fluids. Comprehensive treatments of this subject can be found in standard texts of differential geometry, such as the monograph of Struik (1961).

Maintaining the definitions and notation of Section 1.4, we describe a material vector residing in a material surface in the parametric form

$$d\mathbf{l} = \frac{\partial \mathbf{X}}{\partial \xi} d\xi + \frac{\partial \mathbf{X}}{\partial \eta} d\eta = h_\xi d\xi \, \mathbf{t}_\xi + h_\eta d\eta \, \mathbf{t}_\eta \tag{1.5.1}$$

The square of the length of the vector is given by

$$d\mathbf{l} \cdot d\mathbf{l} = \frac{\partial \mathbf{X}}{\partial \xi} \cdot \frac{\partial \mathbf{X}}{\partial \xi} d\xi^2 + 2\frac{\partial \mathbf{X}}{\partial \eta} \cdot \frac{\partial \mathbf{X}}{\partial \xi} d\xi \, d\eta + \frac{\partial \mathbf{X}}{\partial \eta} \cdot \frac{\partial \mathbf{X}}{\partial \eta} d\eta^2 \tag{1.5.2}$$

Introducing the *surface metric tensor* **g**, we recast Eq. (1.5.2) into the compact form

$$d\mathbf{l} \cdot d\mathbf{l} = g_{\xi\xi} d\xi^2 + 2g_{\xi\eta} d\eta \, d\xi + g_{\eta\eta} d\eta^2 \tag{1.5.3}$$

where we have defined

$$g_{\xi\xi} = \frac{\partial \mathbf{X}}{\partial \xi} \cdot \frac{\partial \mathbf{X}}{\partial \xi} = h_\xi^2, \qquad g_{\xi\eta} = g_{\eta\xi} = \frac{\partial \mathbf{X}}{\partial \eta} \cdot \frac{\partial \mathbf{X}}{\partial \xi}, \qquad g_{\eta\eta} = \frac{\partial \mathbf{X}}{\partial \eta} \cdot \frac{\partial \mathbf{X}}{\partial \eta} = h_\eta^2 \tag{1.5.4}$$

When the coordinates ξ and η are orthogonal, the metric tensor is diagonal: $g_{\xi\eta} = 0$. In the nomenclature of differential geometry, Eq. (1.5.3) is called the *first fundamental form* of the surface.

An equivalent form of Eq. (1.5.3) is

$$d\mathbf{l} \cdot d\mathbf{l} = (g_{\xi\xi} + 2\lambda g_{\xi\eta} + g_{\eta\eta}\lambda^2) \, d\xi^2 \tag{1.5.5}$$

where we have defined $\lambda = d\eta/d\xi$. Since the binomial with respect to λ on the right-hand side of Eq. (1.5.5) is positive for any value of λ, the determinant of the metric tensor must also be positive,

$$\text{Det}(\mathbf{g}) = g_{\eta\eta}g_{\xi\xi} - g_{\xi\eta}^2 > 0 \tag{1.5.6}$$

Using the definition (1.4.16), we find $h_S^2 = \text{Det}(\mathbf{g})$, which is consistent with the inequality (1.5.6).

The change of the normal vector across the length of a material vector is given by

$$d\mathbf{n} = \frac{\partial \mathbf{n}}{\partial \xi} d\xi + \frac{\partial \mathbf{n}}{\partial \eta} d\eta \tag{1.5.7}$$

Projecting Eq. (1.5.7) onto Eq. (1.5.1), we obtain the *second fundamental form* of the surface,

$$d\mathbf{n} \cdot d\mathbf{l} = -f_{\xi\xi} d\xi^2 - 2f_{\xi\eta} d\xi \, d\eta - f_{\eta\eta} d\eta^2 \tag{1.5.8}$$

where we have defined

$$f_{\xi\xi} = -\frac{\partial \mathbf{X}}{\partial \xi} \cdot \frac{\partial \mathbf{n}}{\partial \xi} = \frac{\partial^2 \mathbf{X}}{\partial \xi^2} \cdot \mathbf{n}$$

$$f_{\xi\eta} = -\frac{1}{2}\left(\frac{\partial \mathbf{X}}{\partial \xi} \cdot \frac{\partial \mathbf{n}}{\partial \eta} + \frac{\partial \mathbf{X}}{\partial \eta} \cdot \frac{\partial \mathbf{n}}{\partial \xi}\right) = \frac{\partial^2 \mathbf{X}}{\partial \xi \partial \eta} \cdot \mathbf{n} = -\frac{\partial \mathbf{X}}{\partial \xi} \cdot \frac{\partial \mathbf{n}}{\partial \eta} = -\frac{\partial \mathbf{X}}{\partial \eta} \cdot \frac{\partial \mathbf{n}}{\partial \xi} \tag{1.5.9}$$

$$f_{\eta\eta} = -\frac{\partial \mathbf{X}}{\partial \eta} \cdot \frac{\partial \mathbf{n}}{\partial \eta} = \frac{\partial^2 \mathbf{X}}{\partial \eta^2} \cdot \mathbf{n}$$

To derive the last two expressions for $f_{\xi\eta}$ we have made use of Eq. (1.4.19).

The rate of change of the tensors **g** and **f** following a point particle in the surface may be computed from their definitions using the evolution equations discussed in Section 1.4 (Problem 1.5.1).

Curvatures

We proceed next to describe the geometry of a surface in terms of its curvatures. The normal curvature in the direction of the ξ axis, denoted by K_ξ, is equal to the curvature of the trace of the surface in a plane that contains the normal vector **n** and the tangential vector $\mathbf{t}_\xi$. *Meusnier's theorem* states that

$$K_\xi = \kappa_\xi \mathbf{n}_\xi \cdot \mathbf{n} = -\frac{\partial \mathbf{t}_\xi}{\partial l_\xi} \cdot \mathbf{n} = \frac{\partial \mathbf{n}}{\partial l_\xi} \cdot \mathbf{t}_\xi = -\frac{f_{\xi\xi}}{g_{\xi\xi}} \tag{1.5.10}$$

where, by definition, $\partial \mathbf{t}_\xi / \partial l_\xi = -\kappa_\xi \mathbf{n}_\xi$ (Struik, 1961, p. 76). A similar equation may be written for the η axis. The normal curvature in the direction of an arbitrary material vector with $\lambda = d\eta/d\xi$ is given by

$$K(\lambda) = \kappa(\lambda)\,\mathbf{n}(\lambda) \cdot \mathbf{n} = -\frac{\partial \mathbf{t}}{\partial l}(\lambda) \cdot \mathbf{n} = \left(\frac{\partial \mathbf{n}}{\partial l} \cdot \mathbf{t}\right)(\lambda) = -\frac{f_{\xi\xi} + 2f_{\xi\eta}\lambda + f_{\eta\eta}\lambda^2}{g_{\xi\xi} + 2g_{\xi\eta}\lambda + g_{\eta\eta}\lambda^2} \tag{1.5.11}$$

The maximum and minimum values of K over all possible values of λ are called the *principal curvatures* of the surface. On the basis of the last expression in Eq. (1.5.11), we find that if K_{Max} is the maximum principal curvature corresponding to a particular orientation, then K_{Min} will be the minimum principal curvature corresponding to the perpendicular orientation. *Euler's theorem* states that the curvature in an arbitrary direction is related to the principal curvatures by

$$K(\lambda) = K_{\mathrm{Max}} \cos^2 \alpha + K_{\mathrm{Min}} \sin^2 \alpha \tag{1.5.12}$$

where α is the angle subtended by the direction corresponding to λ and the principal direction for the maximum curvature (Struik, 1961, p. 81).

The mean curvature of the surface is equal to mean value of the curvatures in any two perpendicular directions corresponding, for instance, to λ_1 and λ_2, and are given by

$$\kappa_m = \tfrac{1}{2}(K_{\mathrm{Max}} + K_{\mathrm{Min}}) = \tfrac{1}{2}[K(\lambda_1) + K(\lambda_2)] = -\frac{1}{2}\frac{g_{\xi\xi} f_{\eta\eta} - 2g_{\xi\eta} f_{\xi\eta} + g_{\eta\eta} f_{\xi\xi}}{g_{\xi\xi} g_{\eta\eta} - g_{\xi\eta}^2} \tag{1.5.13}$$

An equivalent expression emerging from Eq. (1.4.30) is

$$\kappa_m = \frac{1}{2}\frac{h_\xi h_\eta}{h_S}\left(\frac{\partial \mathbf{n}}{\partial l_\xi} \cdot (\mathbf{t}_\eta \times \mathbf{n}) - \frac{\partial \mathbf{n}}{\partial l_\eta} \cdot (\mathbf{t}_\xi \times \mathbf{n})\right) \tag{1.5.14}$$

One may show by straightforward algebraic manipulation that Eq. (1.5.14) is equivalent to Eq. (1.5.13) (Problem 1.5.2).

Numerical computation of curvatures

To show how the preceding formulae can be put to practice, let us assume that we have available the position of five points $\mathbf{X}_i$, $i = 0, 1, 2, 3, 4$, along the ξ and η axes, as shown in Figure 1.5.1. Using central differences, we find that, at the location of the point $\mathbf{X}_0$,

$$\mathbf{t}_\xi = \frac{\mathbf{X}_2 - \mathbf{X}_1}{|\mathbf{X}_2 - \mathbf{X}_1|}, \qquad \mathbf{t}_\eta = \frac{\mathbf{X}_4 - \mathbf{X}_3}{|\mathbf{X}_4 - \mathbf{X}_3|} \tag{1.5.15}$$

(Section B.5, Appendix B). The accuracy of this approximation is first order with respect to $\Delta\xi$ and $\Delta\eta$, unless the points happen to be evenly spaced with respect to ξ and η, in which case it becomes second order with respect to $\Delta\xi$ and $\Delta\eta$. We then compute

$$\mathbf{n} = \frac{\mathbf{t}_\xi \times \mathbf{t}_\eta}{|\mathbf{t}_\xi \times \mathbf{t}_\eta|} \tag{1.5.16}$$

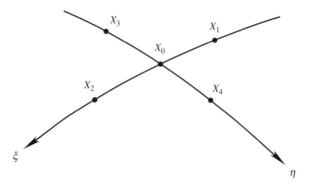

Figure 1.5.1   Five points along two curvilinear axes on a surface. If the
axes are orthogonal, knowledge of the position of five points is sufficient
for computing the mean curvature.

and use a finite-difference approximation to obtain

$$\left(\frac{\partial \mathbf{t}_\xi}{\partial \ell_\xi}\right)_\eta \cong \frac{2}{|\mathbf{X}_2 - \mathbf{X}_1|}\left(\frac{\mathbf{X}_2 - \mathbf{X}_0}{|\mathbf{X}_2 - \mathbf{X}_0|} - \frac{\mathbf{X}_0 - \mathbf{X}_1}{|\mathbf{X}_0 - \mathbf{X}_1|}\right) \tag{1.5.17}$$

The normal vector $\mathbf{n}_\xi$ follows by normalizing the right-hand side of Eq. (1.5.17), and the curvature
κ_ξ is obtained by setting

$$\kappa_\xi = -\left(\frac{\partial \mathbf{t}_\xi}{\partial \ell_\xi}\right)_\eta \cdot \mathbf{n}_\xi \tag{1.5.18}$$

Similar equations are used to compute $\mathbf{n}_\eta$ and κ_η. If the curvilinear axes happen to be orthogonal,
one may substitute the results into Eq. (1.5.10) and then into Eq. (1.5.13) to obtain an approximation of the mean curvature at the point $\mathbf{X}_0$. Thus, in this case, knowledge of the position of five
points in the surface is sufficient for computing the mean curvature.

PROBLEMS

1.5.1   **Rate of change of the surface metric tensor.** Express the material derivative of the surface
metric tensor in terms of the velocity field.

1.5.2   **Mean curvature.** (a) Show that the right-hand side of Eq. (1.5.14) may be reduced to the last
expression in Eq. (1.5.13). Begin with expressing the normal vector in the cross products in
terms of the tangential vectors using Eq. (1.4.17), and then expand the triple cross products. (b)
Derive an expression for the rate of change of the mean curvature following a point particle in
terms of the velocity.

Computer Problem

1.5.3   **Mean curvature of a spheroidal surface.** Consider a spheroidal surface with axes a and b
and introduce its natural orthogonal curvilinear axes ξ and $\eta = \varphi$. The position of a point on
the spheroid is given in the parametric form $X = a\cos\xi$, $Y = b\sin\xi\cos\varphi$, $Z = b\sin\xi\sin\varphi$,
where $0 < \xi < \pi$ and $0 < \varphi < 2\pi$. Use Eqs. (1.5.15)–(1.5.18) with a sufficiently small $\Delta\xi$
and $\Delta\eta$ to compute the mean curvature at a sequence of points that lie in the azimuthal plane
$\varphi = 0$. Verify that when $a = b$, in which case the spheroid reduces to a sphere, the mean
curvature assumes the uniform value $1/a = 1/b$.

1.6 | DESCRIPTION OF A MATERIAL SURFACE IN EULERIAN FORM

It is sometimes convenient, if not necessary, to describe the location of a material surface in an Eulerian parametric form in terms of Cartesian or other global curvilinear coordinates instead of the surface curvilinear coordinates discussed in Sections 1.4 and 1.5. The Eulerian description is particularly useful in studies of two-fluid flow where a material surface is typically identified with a fluid interface or with the free surface between a liquid and a gas.

A function that describes the location of a material surface in Eulerian form satisfies an evolution equation that emerges by requiring that the motion of the point particles in the material surface is consistent with the deformation of the material surface as described in the Eulerian form.

With reference to Figure 1.6.1, let us describe the location of a material surface in Cartesian coordinates in the form $z = f(x, y, t)$. When the surface is evolving, the function f changes in time as indicated by the third of its arguments. To derive an evolution equation for f, we consider the position of a point particle in the material surface at times t and $t + \Delta t$, take into account that the point particle moves with the velocity of the fluid, and use geometrical reasoning to write

$$f(x + u_x \Delta t, y + u_y \Delta t, t + \Delta t) = f(x, y, t) + u_z \Delta t \tag{1.6.1}$$

Expanding the left-hand side of Eq. (1.6.1) in a Taylor series with respect to the first two of its arguments about (x, y, t) yields

$$f(x, y, t + \Delta t) + \frac{\partial f}{\partial x} u_x \Delta t + \frac{\partial f}{\partial y} u_y \Delta t = f(x, y, t) + u_z \Delta t \tag{1.6.2}$$

Taking the limit as Δt tends to zero and rearranging the various terms, we obtain the evolution equation

$$\frac{\partial f}{\partial t} + \frac{\partial f}{\partial x} u_x + \frac{\partial f}{\partial y} u_y - u_z = 0 \tag{1.6.3}$$

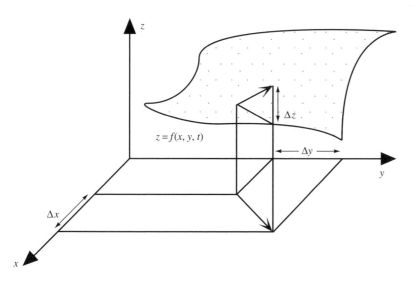

Figure 1.6.1 Describing the location of a material surface in Eulerian form in terms of global Cartesian coordinates.

Describing the material surface by the equivalent form

$$F(x, y, z, t) = f(x, y, t) - z = 0 \tag{1.6.4}$$

allows us to express Eq. (1.6.3) in terms of the material derivative as

$$\frac{DF}{Dt} = \frac{\partial F}{\partial t} + \mathbf{u} \cdot \nabla F = 0 \tag{1.6.5}$$

The implicit function theorem allows us to generalize the preceding results and state that if a material surface is described in a certain parametric form as $F(\mathbf{x}, t) = c$, where c is a constant, then the evolution of F will be governed by Eq. (1.6.5), which is another way of saying that the function F will be convected by the flow.

Furthermore, we note that the gradient ∇F is perpendicular to the material surface, and this allows us to enhance the velocity $\mathbf{u}$ in Eq. (1.6.5) with an arbitrary tangential vector. Maintaining the normal component of $\mathbf{u}$ and enhancing it with a tangential vector $\mathbf{t}$, we obtain

$$\frac{\partial F}{\partial t} + [(\mathbf{u} \cdot \mathbf{n})\mathbf{n} + \mathbf{t}] \cdot \nabla F = 0 \tag{1.6.6}$$

For example, we may set $\mathbf{t} = \beta(\mathbf{I} - \mathbf{nn}) \cdot \mathbf{u} = \beta \mathbf{n} \times \mathbf{u} \times \mathbf{n}$ where β is an arbitrary coefficient whose value may be a time dependent function of position over the material surface. Note that the projection operator $\mathbf{I} - \mathbf{nn}$ extracts the tangential component of a vector that it multiplies. Equation (1.6.6) then suggests that it is kinematically consistent to allow the point particles that lie in the surface to move with the modified velocity

$$\mathbf{v} = (\mathbf{u} \cdot \mathbf{n})\mathbf{n} + \beta(\mathbf{I} - \mathbf{nn}) \cdot \mathbf{u} \tag{1.6.7}$$

When $\beta = 1$, $\mathbf{v}$ reduces to the velocity of the fluid $\mathbf{u}$. When $\beta \neq 1$, the point particles are reduced to interfacial marker points.

PROBLEMS

1.6.1 **Expanding sphere.** Consider a radially expanding spherical surface of radius $a(t)$ described in spherical polar coordinates as $F(r, t) = r - a(t)$, and use Eq. (1.6.5) to compute the radial velocity at the surface.

1.6.2 **Boundary condition on a wavy surface.** Consider a cylindrical material surface whose position is given by $y = a \sin[k(x - ct)]$, where k is the wave number and c is the phase velocity, and use Eq. (1.6.5) to derive a boundary condition for the velocity at the position of the surface.

1.7 | STREAMLINES, STREAM TUBES, PATH LINES, AND STREAK LINES

An instantaneous streamline is a line in the flow whose tangential vector at every point is parallel to the instantaneous velocity vector at that point. A streamline that forms a closed loop is called a *closed streamline,* whereas a streamline that crosses the boundaries of the flow or extends to infinity is called an *open streamline.* A streamline may intersect one or more other streamlines at a *stagnation point.* Stagnation points may occur in the interior of a flow as well as on the boundaries. Since the velocity is a single-valued function of position, the velocity at a stagnation point must necessarily vanish.

One way of describing a streamline is to introduce a variable τ that increases monotonically along the streamline in the direction of the velocity vector. One plausible choice for τ is the time it takes a point particle to move along the streamline from a specified initial position as it is convected by the *frozen* instantaneous velocity field. The streamline is then described by the following autonomous differential equation, with no time dependence on the right-hand side,

$$\frac{d\mathbf{x}}{d\tau} = \mathbf{u}(\mathbf{x}, t = \text{constant}) \tag{1.7.1}$$

Restating Eq. (1.7.1) as

$$\frac{dx}{u_x} = \frac{dy}{u_y} = \frac{dz}{u_z} = d\tau \tag{1.7.2}$$

illustrates that a vector that is tangential to the streamline is parallel to the velocity.

Computation of Streamlines

To compute the streamline that passes through the point $\mathbf{x}_0$ we integrate Eq. (1.7.1) either forward or backward with respect to τ subject to the initial condition $\mathbf{x}(\tau = 0) = \mathbf{x}_0$, using a standard numerical method, such as a Runge–Kutta method (Section B.8, Appendix B). The integration is terminated when the magnitude of the velocity becomes exceedingly small, signaling approach towards a stagnation point. If the integration step $\Delta\tau$ is kept constant during the integration, the travel distance along the streamline over one step is proportional to the magnitude of the local velocity. This has the practical disadvantage that a large number of steps will be required at regions where the flow is slow, but the structure of the streamline pattern is relatively simple.

One way of circumventing this difficulty is to set $\Delta\tau$ inversely proportional to the local magnitude of the velocity, thereby ensuring a nearly constant travel distance during each step. This method, however, has the practical disadvantage that the computed streamline may cross over stagnation points where it must end. To avoid this complication, we may set the travel distance equal to or less than the finest length scale in the flow.

Ideally, $\Delta\tau$ should be adjusted according to both the magnitude of the velocity and the curvature of the streamline, so that sharply turning streamlines are described with sufficient accuracy, and the computation does not stall at regions of slow flow with simple structure. Unless, however, a high degree of accuracy is desired, implementing these criteria makes for computer programming complexity, and the method of constant travel distance works well in most cases.

Stream Surfaces and Stream Tubes

The collection of all streamlines that pass through an open line in a flow forms a *stream surface,* and the collection of all streamlines that pass through a closed loop forms a *stream tube.*

Let us consider two closed loops that wrap around a particular stream tube once, and draw two surfaces D_1 and D_2 that are bounded by each loop, as shown in Figure 1.7.1. The volumetric

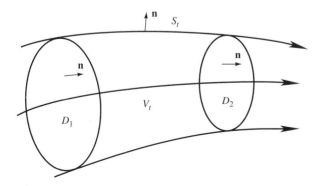

Figure 1.7.1   Illustration of a stream tube in a flow, and two closed loops wrapping around the stream tube.

flow rates across the two surfaces are given by

$$q_i = \int_{D_i} \mathbf{u} \cdot \mathbf{n} \, dS \tag{1.7.3}$$

where $i = 1, 2$ and $\mathbf{n}$ is the unit vector normal to D_1 and D_2 oriented as shown in Figure 1.7.1. Using the divergence theorem, we compute

$$q_2 = q_1 - \int_{S_t} \mathbf{u} \cdot \mathbf{n} \, dS + \int_{V_t} \nabla \cdot \mathbf{u} \, dV \tag{1.7.4}$$

where S_t is the surface of the stream tube subtended between the two loops, and V_t is the volume enclosed by D_1, D_2, and S_t. Since the velocity is tangential to the stream tube and therefore perpendicular to the normal vector on S_t, the surface integral over S_t vanishes, and Eq. (1.7.4) yields

$$q_2 = q_1 + \int_{V_t} \nabla \cdot \mathbf{u} \, dV \tag{1.7.5}$$

Equation (1.7.5) shows that the volumetric flow rate may increase or decrease along a stream tube depending on whether the fluid inside the stream tube is undergoing expansion or contraction. If the fluid is incompressible, the flow rate across any section of a stream tube is constant, $q_1 = q_2$. Consequently, in the absence of singularities, a stream tube that carries a finite amount of incompressible fluid may not collapse down to a nonsingular point where the velocity has finite magnitude. If this occurred, the flow rate at the point of collapse would have to vanish, which contradicts the assumption that the stream tube carries a finite amount of fluid. A similar argument may be used to show that a streamline may not suddenly end in a flow, but must either meet one or more other streamlines at a stagnation point, form a closed loop, extend to infinity, or cross the boundaries.

Another consequence of Eq. (1.7.5) for two-dimensional incompressible flow is that the distance between two adjacent streamlines is inversely proportional to the local magnitude of the velocity of the fluid; the faster the velocity, the closer the streamlines.

Streamline Coordinates

Useful insights into the motion of fluid parcels may be obtained by considering the behavior of point particles with reference to the instantaneous structure of the streamline pattern. The point particles translate tangentially to the streamlines and rotate around the local vorticity vector, which may point in an arbitrary direction with respect to the streamlines. We note parenthetically that a flow in which the vorticity vector is parallel to the velocity vector at every point is called a *Beltrami flow*. Our main goal at present is to establish a relation between the direction and magnitude of the vorticity vector and the structure of the streamline pattern in two-dimensional and three-dimensional flow.

Local curvilinear coordinates

As a preliminary, we introduce a system of orthogonal curvilinear coordinates that are constructed with reference to the streamlines. For this purpose, we consider a family of streamlines in the neighborhood of a particular streamline denoted by L, as shown in Figure 1.7.2(a), label the point particles that lie on L using the arc length l, and note that the unit vector $\mathbf{t} = d\mathbf{X}/dl$ is tangential to L and therefore parallel to the velocity. The *principal unit vector normal to L* is given by

$$\mathbf{n} = -\frac{1}{\kappa} \frac{\partial \mathbf{t}}{\partial l} \tag{1.7.6}$$

where κ is the curvature of the streamline, which is allowed to have a positive or a negative sign.

(a)

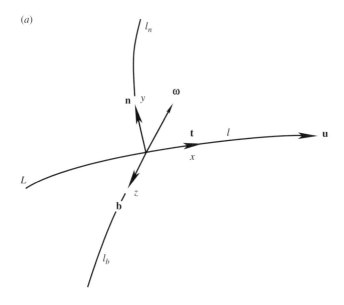

(b)

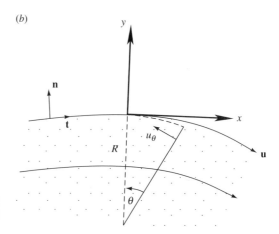

Figure 1.7.2 A streamline and the associated curvilinear axes in (a) a three-dimensional and (b) a two-dimensional flow.

Furthermore, we introduce the *binormal* unit vector defined as

$$\mathbf{b} = \mathbf{t} \times \mathbf{n} \tag{1.7.7}$$

The three unit vectors $\mathbf{t}$, $\mathbf{n}$, and $\mathbf{b}$ define three mutually orthogonal directions that may be used to construct a right-handed, orthogonal, curvilinear system of axes (Section A.3, Appendix A). The three planes containing $\mathbf{t}$ and $\mathbf{n}$, $\mathbf{n}$ and $\mathbf{b}$, $\mathbf{b}$ and $\mathbf{t}$ are called, respectively, the *osculating* plane, the *normal* plane, and the *rectifying* plane.

Next, we differentiate Eq. (1.7.7) with respect to arc length l, expand the derivative on the right-hand side, and use Eq. (1.7.6) to find that the vector $d\mathbf{b}/dl$ is perpendicular to $\mathbf{t}$. Since $\mathbf{b}$ is a unit vector, $d(\mathbf{b} \cdot \mathbf{b})/dl = 2\mathbf{b} \cdot (d\mathbf{b}/dl) = 0$, which shows that $d\mathbf{b}/dl$ is also perpendicular to $\mathbf{b}$. Thus $\partial \mathbf{b}/\partial \ell$ must be parallel to $\mathbf{n}$, and we set

$$\frac{\partial \mathbf{b}}{\partial l} = -\frac{1}{\tau}\mathbf{n} \tag{1.7.8}$$

where τ is the *torsion* of the streamline.

Furthermore, rewriting Eq. (1.7.7) in the form $\mathbf{n} = \mathbf{b} \times \mathbf{t}$, differentiating it with respect to l, and using Eqs. (1.7.6) and (1.7.8), we find

$$\frac{\partial \mathbf{n}}{\partial l} = \frac{1}{\tau}\mathbf{b} + \kappa\mathbf{t} \tag{1.7.9}$$

Equations (1.7.6), (1.7.8), and (1.7.9) are the *Frenet* or *Frenet-Serret* formulae derived by Frenet in 1847 and then by Serret in 1851 (Struik, 1961, p. 19).

Local Cartesian coordinates

Let us now introduce a Cartesian system of coordinates with the origin at a particular point on L, and the x, y, and z axes pointing in the directions of $\mathbf{t}$, $\mathbf{n}$, and $\mathbf{b}$, as shown in Figure 1.7.2(a). At the origin $u_y = 0$, $u_z = 0$, and

$$\frac{\partial u_y}{\partial x} = \frac{\partial \mathbf{u}}{\partial l} \cdot \mathbf{n} = \frac{\partial(\mathbf{u} \cdot \mathbf{n})}{\partial l} - \mathbf{u} \cdot \frac{\partial \mathbf{n}}{\partial l}$$

$$\frac{\partial u_z}{\partial x} = \frac{\partial \mathbf{u}}{\partial l} \cdot \mathbf{b} = \frac{\partial(\mathbf{u} \cdot \mathbf{b})}{\partial l} - \mathbf{u} \cdot \frac{\partial \mathbf{b}}{\partial l} \tag{1.7.10}$$

Because the velocity is tangential to the streamline and thus perpendicular to the normal and binormal vectors, the first terms on the right-hand sides of Eq. (1.7.10) vanish. Using the Frenet formulae (1.7.8) and (1.7.9), we simplify the second terms and obtain

$$\frac{\partial u_y}{\partial x} = -\kappa u_x, \qquad \frac{\partial u_z}{\partial x} = 0 \tag{1.7.11}$$

Vorticity of a two-dimensional flow

Having made the necessary preparations, we turn to relate the vorticity of a two-dimensional flow in the xy plane to the structure of the velocity field around a streamline. Using the first of Eqs. (1.7.11) and Eq. (1.7.6), we find that, at the origin, $\boldsymbol{\omega} = \omega\boldsymbol{\kappa}$, where

$$\omega = \frac{\partial u_y}{\partial x} - \frac{\partial u_x}{\partial y} = -\kappa u_x - \mathbf{n} \cdot \nabla(\mathbf{u} \cdot \mathbf{t}) \tag{1.7.12}$$

Introducing plane polar coordinates with the origin at the center of curvature of a streamline at a point as shown in Figure 1.7.2(b), where $R = 1/\kappa$ is the radius of curvature, we find that, at the origin,

$$\omega = \frac{u_\theta}{R} + \left(\frac{\partial u_\theta}{\partial r}\right)_{r=R} \tag{1.7.13}$$

This expression shows that the point particles spin about the z axis due to the global motion of the fluid associated with the curvature of the streamline as well as due to velocity variations in the normal direction.

Vorticity of a three-dimensional flow

For three-dimensional flow, we write $\mathbf{u} = u\mathbf{t}$, and use the definition of the vorticity, Eq. (1.1.9), to obtain

$$\boldsymbol{\omega} = \nabla \times (u\mathbf{t}) = u\,\nabla \times \mathbf{t} + \nabla u \times \mathbf{t} \tag{1.7.14}$$

Projecting Eq. (1.7.14) onto $\mathbf{t}$ yields the tangential component of the vorticity,

$$\boldsymbol{\omega} \cdot \mathbf{t} = \mathbf{t} \cdot (\nabla \times \mathbf{u}) = (\mathbf{n} \times \mathbf{b}) \cdot (\nabla \times \mathbf{u})$$

$$= \mathbf{n} \cdot (\nabla\mathbf{u}) \cdot \mathbf{b} - \mathbf{b} \cdot (\nabla\mathbf{u}) \cdot \mathbf{n} \tag{1.7.15}$$

Projecting Eq. (1.7.14) onto $\mathbf{n}$, noting that the length of $\mathbf{t}$ is constant, and using Eqs. (1.7.6) and (1.7.7), we obtain

$$\boldsymbol{\omega} \cdot \mathbf{n} = u(\nabla \times \mathbf{t}) \cdot \mathbf{n} + (\nabla u \times \mathbf{t}) \cdot \mathbf{n} = u(\nabla \times \mathbf{t}) \cdot (\mathbf{b} \times \mathbf{t}) + (\mathbf{t} \times \mathbf{n}) \cdot \nabla u$$

$$= u\,\mathbf{b} \cdot (\mathbf{t} \times \nabla \times \mathbf{t}) + \mathbf{b} \cdot \nabla u = u\,\mathbf{b} \cdot [(\nabla \mathbf{t}) \cdot \mathbf{t} - \mathbf{t} \cdot \nabla \mathbf{t}] + \frac{\partial u}{\partial l_b}$$

$$= -u\,\mathbf{b} \cdot \frac{\partial \mathbf{t}}{\partial l} + \frac{\partial u}{\partial l_b} = u\kappa\mathbf{b} \cdot \mathbf{n} + \frac{\partial u}{\partial l_b} = \frac{\partial u}{\partial l_b} \tag{1.7.16}$$

where l_b is the arc length in the direction of the binormal vector. Finally, projecting Eq. (1.7.14) onto $\mathbf{b}$ and working as in Eq. (1.7.16), we obtain

$$\boldsymbol{\omega} \cdot \mathbf{b} = u(\nabla \times \mathbf{t}) \cdot \mathbf{b} + (\nabla u \times \mathbf{t}) \cdot \mathbf{b} = u(\nabla \times \mathbf{t}) \cdot (\mathbf{t} \times \mathbf{n}) + (\mathbf{t} \times \mathbf{b}) \cdot \nabla u$$

$$= -u\mathbf{n} \cdot (\mathbf{t} \times \nabla \times \mathbf{t}) - \mathbf{n} \cdot \nabla u = u\mathbf{n} \cdot \frac{\partial \mathbf{t}}{\partial l} - \frac{\partial u}{\partial l_n} = -\kappa u - \frac{\partial u}{\partial l_n} \tag{1.7.17}$$

where l_n is the arc length in the normal direction. The preceding three equations may be compiled into the unified form

$$\boldsymbol{\omega} = \mathbf{t}\,[\mathbf{n} \cdot (\nabla \mathbf{u}) \cdot \mathbf{b} - \mathbf{b} \cdot (\nabla \mathbf{u}) \cdot \mathbf{n}] + \mathbf{n}\frac{\partial u}{\partial l_b} - \mathbf{b}\left(\kappa u + \frac{\partial u}{\partial l_n}\right) \tag{1.7.18}$$

which reveals that the point particles spin about the tangential vector due to the twisting of the streamline pattern, spin about the normal vector due to velocity variations in the binormal direction, and spin about the binormal vector due to velocity variations in the normal direction, but also due to the global motion of the fluid associated with the curvature of the streamline. In the case of two-dimensinal flow, the vorticity vector points in the binormal direction, and Eq. (1.7.18) reduces to Eq. (1.7.12).

Path Lines

A path line represents the trajectory of a point particle that has been released from a certain position $\mathbf{X}_0$ at some previous time instant t_0. When the flow is steady, a path line coincides with the streamline that passes through $\mathbf{X}_0$. The shape of a path line is described by the generally nonautonomous ordinary differential equation

$$\frac{d\mathbf{X}}{dt} = \mathbf{u}(\mathbf{X}, t) \tag{1.7.19}$$

where $\mathbf{X}$ is the position of the point particle along its path. To compute a path line we select an ejection location and time, and integrate Eq. (1.7.19) forward in time using a standard numerical method such as a Runge–Kutta method. Formal integration yields the position of the point particle at time t as

$$\mathbf{X}(t; t_0) = \mathbf{X}_0(t_0) + \int_{t_0}^{t} \mathbf{u}(\mathbf{X}(t'; t_0), t')\,dt'$$

$$= \mathbf{X}_0(t_0) + \int_{0}^{t-t_0} \mathbf{u}(\mathbf{X}(\tau + t_0; t_0), \tau + t_0)\,d\tau \tag{1.7.20}$$

which may be regarded as a parametric representation of the path line in terms of time t.

Streak Lines

A streak line is the instantaneous trace of a chain of point particles that have been released from the same or different locations at the same or different prior times in a flow. In the laboratory, streak lines may be produced by ejecting a dye from a stationary or moving needle.

Regarding the injection time t_0 as a Lagrangian marker variable, we find that the shape of a streak line at a particular time t is described by Eq. (1.7.20). When the point particles are injected from the same location, the first term on the right-hand side is constant.

PROBLEMS

1.7.1   Beltrami and complex lamellar flows. Explain why a two-dimensional or an axisymmetric flow without swirling motion may not be a Beltrami flow, that is, the vorticity cannot be parallel to the velocity, but is necessarily a *complex lamellar* flow, which means that the velocity is perpendicular to its curl.

1.7.2   Fluid in rigid-body rotation. Use Eq. (1.7.18) to compute the vorticity of a fluid that executes rigid-body rotation, and show that the result is consistent with the definition $\boldsymbol{\omega} = \nabla \times \mathbf{u}$.

1.7.3   Linear flows. Sketch and discuss the streamline pattern of the following linear flows; in all cases k is a constant with dimensions of inverse time: (a) Purely rotational two-dimensional flow: $u_x = -ky$, $u_y = kx$, $u_z = 0$. (b) Two-dimensional extensional flow: $u_x = kx$, $u_y = -ky$, $u_z = 0$. (c) Axisymmetric extensional flow: $u_x = 2kx$, $u_y = -ky$, $u_z = -kz$. What are the components of the velocity in cylindrical polar coordinates?

Computer Problems

1.7.4   Drawing a streamline. Write a routine called *STRLN* that, given the instantaneous velocity field, computes the streamline that passes through a specified point in a flow. The integration should be carried out using the modified Euler method, and the size of the step $\Delta\tau$ should be selected so that the integration at every step proceeds roughly by a preset distance (Section B.8, Appendix B).

1.7.5   Drawing streamlines in a box. Write a routine called *GRIDINT* that returns the velocity at a point within a rectangular domain of flow in the xy plane confined between $a < x < b$, $c < y < d$, where a, b, c, d are four specified constants. The input should include the two components of a two-dimensional velocity field at the nodes of a two-dimensional rectangular grid of size $N + 1$ by $M + 1$, with grid points located at $x_i = a + (i - 1)\Delta x$ and $y_j = b + (j - 1)\Delta y$, where $\Delta x = (b - a)/N$, $\Delta y = (d - c)/M$, $i = 1, \ldots, N + 1$, and $j = 1, \ldots, M + 1$. The velocity between grid points should be computed using bilinear interpolation (Section B.4, Appendix B).

1.7.6   Drawing the streamline pattern in a box. (a) Combine *STRLN* and *GRIDINT* of Problems 1.7.4 and 1.7.5 into a program called *GRIDSTR* that draws streamlines in a rectangular domain. (b) Run *GRIDSTR* to draw the streamline pattern of the flow within the square box $0 < x < 1$, $0 < y < 1$, with $N = 10$, $M = 20$, where the components of the velocity at the grid points are given by

$$u_{ij} = \exp(i\pi\Delta x) - i\pi\Delta x\cos(j\pi\Delta y), \qquad v_{ij} = \sin(j\pi\Delta y) - j\pi\Delta y\exp(i\pi\Delta x)$$

Is this velocity field solenoidal?

1.8 | VORTICITY, VORTEX LINES, VORTEX TUBES, AND CIRCULATION AROUND LOOPS

In Section 1.1 we saw that the vorticity vector at a certain point in a flow is parallel to the instantaneous angular velocity vector of the point particle that happens to be at that location, and its magnitude is equal to twice that of the angular velocity vector of the point particle.

Using the definition (1.1.9) and the fact that the divergence of the curl of any twice differentiable vector field vanishes, we find that the vorticity field is solenoidal,

$$\nabla \cdot \boldsymbol{\omega} = 0 \tag{1.8.1}$$

The vorticity of a two-dimensional flow in the xy plane is oriented along the z axis, and this allows us to write

$$\boldsymbol{\omega}(x, y) = \omega(x, y)\,\mathbf{k} \tag{1.8.2}$$

where ω is the strength of the vorticity and $\mathbf{k}$ is the unit vector along the z axis.

The vorticity of an axisymmetric flow without swirling motion is directed in the azimuthal direction, so that

$$\boldsymbol{\omega}(x, \sigma) = \omega(x, \sigma)\,\mathbf{e}_{\varphi} \tag{1.8.3}$$

where ω is the strength of the vorticity and $\mathbf{e}_{\varphi}$ is the unit vector in the azimuthal direction. In the presence of swirling motion, the vorticity may point in an arbitrary direction.

Vortex Lines

An instantaneous *vortex line* is a line in the flow whose tangential vector at every point is parallel to the vorticity vector and therefore to the point-particle angular velocity vector, evaluated at that particular instant. The collection of all vortex lines that pass through a closed loop generates a surface called a *vortex tube,* as illustrated schematically in Figure 1.8.1. Remembering that the vorticity field is solenoidal and repeating the arguments put forth in Section 1.7 following Eq. (1.7.5), we find that a vortex line may not end in the interior of a flow. It must form a closed loop, meet one or more than one vortex lines at stagnation points of the vorticity field, extend to infinity, or cross the boundaries of the flow.

The vortex tubes of a two-dimensional flow are cylindrical surfaces perpendicular to the plane of the flow. The vortex lines of an axisymmetric flow with no swirling motion are concentric circles, and the vortex tubes form concentric axisymmetric surfaces. The vortex lines of an axisymmetric flow *with* swirling motion are spiral lines.

Circulation

The circulation around a closed loop L that lies in the domain of a flow is defined by the expressions

$$C = \int_{L} \mathbf{u} \cdot d\mathbf{X} = \int_{L} \mathbf{u} \cdot \mathbf{t}\, dl = \int_{A} \mathbf{u} \cdot \mathbf{t}\, h\, da \tag{1.8.4}$$

where l is the arc length along the loop, the parameter a labels the point particles along the loop taking values over the set A, h is the arc length metric coefficient defined in Eq. (1.4.5), and the unit tangential vector $\mathbf{t}$ is oriented in a specified direction. Using Stokes's theorem, we find

$$C = \int_{D} (\nabla \times \mathbf{u}) \cdot \mathbf{n}\, dS = \int_{D} \boldsymbol{\omega} \cdot \mathbf{n}\, dS \tag{1.8.5}$$

where D is an arbitrary surface that is bounded by the material loop as illustrated in Figure 1.8.1, and the direction of the normal vector $\mathbf{n}$ is chosen so that $\mathbf{t}$ and $\mathbf{n}$ form a right-handed system of axes with respect to a designated side of D. Equation (1.8.5) states that the circulation around a loop is equal to the flow rate of the vorticity across any surface that is bounded by the loop.

Consider next two material loops L_1 and L_2 that wrap around the same vortex tube once, as shown in Figure 1.8.1. Using Eq. (1.8.5), we find that the difference in circulation around these loops is given by

$$C_2 - C_1 = \int_{D_2} \boldsymbol{\omega} \cdot \mathbf{n}\, dS - \int_{D_1} \boldsymbol{\omega} \cdot \mathbf{n}\, dS \tag{1.8.6}$$

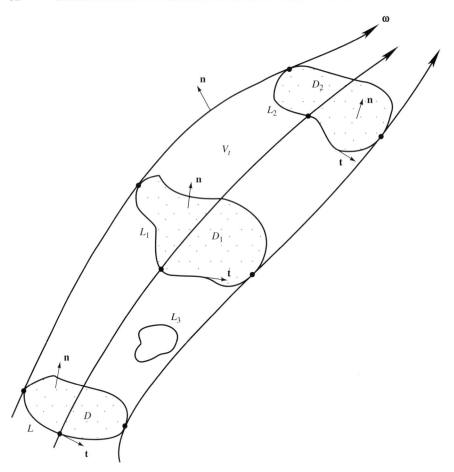

Figure 1.8.1   Schematic illustration of a vortex tube. The circulation around a loop that wraps around the tube once is equal to the flow rate of the vorticity across a surface D that is bounded by the loop.

Since the vorticity is tangential to the vortex tube, it is permissible to enhance the right-hand side of Eq. (1.8.6) with a corresponding integral over the surface of the tube that is subtended between the loops L_1 and L_2. Using the divergence theorem, we then find

$$C_2 - C_1 = \int_{V_t} \nabla \cdot \boldsymbol{\omega} \, dV \tag{1.8.7}$$

where V_t is the volume enclosed by D_1, D_2, and the surface of the vortex tube. Since the vorticity field is solenoidal, the right-hand side of Eq. (1.8.7) vanishes, and this shows that *the circulation around any loop that wraps a vortex tube once is independent of the actual location and shape of the loop around the tube, as long as the loop lies on the tube.*

Using similar arguments, we find that the circulation around a loop that lies on a vortex tube but does not wrap around the tube, such as the loop L_3 shown in Figure 1.8.1, is equal to zero (Problem 1.8.1). Furthermore, we find that the circulation around a closed loop that wraps around a vortex tube m times is equal to $m\kappa$, where κ is the circulation around a closed loop that wraps around the vortex tube once, called the *cyclic constant of the flow around* or *strength* of the vortex tube.

As an application, we consider an infinite flow that is bounded internally by a closed surface D_B, which may be regarded as the surface of a body, and argue that the rate of flow of the vorticity across D_B must be equal to zero. This becomes evident by introducing two surfaces D_1 and D_2 that are bounded by an arbitrary closed loop L and whose union encloses the body, integrating Eq. (1.8.1) over the volume enclosed by D_1, D_2, and D_B, and applying the divergence theorem to convert the volume integral to a surface integral, thereby obtaining

$$\int_{D_1} \boldsymbol{\omega} \cdot \mathbf{n} \, dS - \int_{D_2} \boldsymbol{\omega} \cdot \mathbf{n} \, dS = \int_{D_B} \boldsymbol{\omega} \cdot \mathbf{n} \, dS \qquad (1.8.8)$$

The normal vector $\mathbf{n}$ over D_1 and D_2 is defined such that when these surfaces collapse, $\mathbf{n}$ points in the same direction that depends upon the orientation of $\mathbf{n}$ over D_B. Each integral on the left-hand side of Eq. (1.8.8) is equal to the circulation around L, and this means that the integral on the right-hand side and therefore the rate of flow of vorticity across D_B must necessarily vanish.

Rate of Change of Circulation around a Material Loop

Differentiating the first expression in Eq. (1.8.4) with respect to time and expanding out the derivative, we find

$$\frac{dC}{dt} = \frac{d}{dt} \int_L \mathbf{u} \cdot d\mathbf{X} = \int_L \frac{D\mathbf{u}}{Dt} \cdot d\mathbf{X} + \int_L \mathbf{u} \cdot \frac{d(d\mathbf{X})}{dt} \qquad (1.8.9)$$

Concentrating on the last integral in Eq. (1.8.9), we use Eq. (1.4.1) to write

$$\int_L \mathbf{u} \cdot \frac{d(d\mathbf{X})}{dt} = \int_L (d\mathbf{X} \cdot \mathbf{L}) \cdot \mathbf{u} = \tfrac{1}{2} \int_L d\mathbf{X} \cdot \nabla(\mathbf{u} \cdot \mathbf{u}) = 0 \qquad (1.8.10)$$

where $\mathbf{L}$ is the velocity gradient tensor. The value of zero arises from the fact that L is a closed line. We thus find that Eq. (1.8.9) reduces to the simplified form

$$\frac{dC}{dt} = \int_L \frac{D\mathbf{u}}{Dt} \cdot d\mathbf{X} \qquad (1.8.11)$$

which identifies the rate of change of circulation around a material loop with the circulation of the acceleration field $D\mathbf{u}/Dt$ around the loop. If the acceleration field happens to be irrotational, in which case $D\mathbf{u}/Dt$ may be written as the gradient of a potential function, as will be discussed in Section 2.1, the rate of change of circulation will vanish, and the circulation around the loop will be preserved during the motion.

PROBLEMS

1.8.1   **A loop on a vortex tube.** Show that the circulation around the loop L_3 illustrated in Figure 1.8.1 vanishes.

1.8.2   **Flow within a cylinder due to a rotating lid.** Consider a flow within a cylindrical container that is closed at the bottom, driven by the rotation of the top lid, and sketch the vortex-line pattern.

1.8.3   **Solenoidality of the vorticity.** (a) Show that $\boldsymbol{\omega} = \nabla\psi \times \nabla\chi$, where ψ and χ are two arbitrary functions, is an acceptable vorticity field, in the sense that it is solenoidal. (b) Verify that $\mathbf{u} = \psi\nabla\chi + \nabla\zeta$, where ζ is an arbitrary function, is an acceptable associated velocity field. Show that when ζ is constant, this velocity field is complex lamellar; that is, the velocity is perpendicular to the vorticity at every point in the flow.

1.9 | LINE VORTICES AND VORTEX SHEETS

There is a class of flows whose velocity exhibits sharp variations across thin columns or layers of fluid. Examples are flows containing shear layers developing between two streams that merge at different velocities and around the edges of jets, turbulent flows, and flows due to tornadoes and whirls. The distinguishing feature of these flows is that the support of the vorticity is compact, which means that the magnitude of the vorticity takes significant values only within certain compact regions concisely called *vortices;* the flow outside the vortices is precisely or nearly irrotational.

Line Vortices

Let us consider a flow wherein the vorticity vanishes everywhere except near a vortex tube with a small cross-sectional area ΔS centered around the line L. Taking the limit as ΔS tends to zero while maintaining the circulation C around the vortex tube constant equal to κ, we obtain a vortex tube with infinitesimal cross-sectional area, infinite vorticity, but finite circulation, called a *line vortex.*

Since the vorticity is tangential to the vortex lines, the vorticity field associated with a line vortex is given by the generalized distribution

$$\boldsymbol{\omega}(\mathbf{x}) = \kappa \int_L \mathbf{t}(\mathbf{x}') \delta(\mathbf{x} - \mathbf{x}') dl(\mathbf{x}') \tag{1.9.1}$$

where δ is the three-dimensional delta function, $\mathbf{t}$ is the unit vector tangential to the line vortex, and l is the arc length along the line vortex.

Vortex Sheets

Consider next a flow whose vorticity vanishes everywhere except within a sheet of small thickness h centered around the surface S. Taking the limit as h tends to zero while requiring that the circulation around any loop that pierces the sheet through any two fixed points remains constant, we obtain a *vortex sheet* with infinitesimal cross-sectional area, infinite vorticity, but finite circulation.

The vorticity field associated with a vortex sheet, sometimes called a *sheet vortex,* is given by the generalized distribution

$$\boldsymbol{\omega}(\mathbf{x}) = \int_S \boldsymbol{\zeta}(\mathbf{x}') \delta(\mathbf{x} - \mathbf{x}') dS(\mathbf{x}') \tag{1.9.2}$$

where δ is the three-dimensional delta function. The *strength of the vortex sheet* $\boldsymbol{\zeta}$ is oriented tangentially to S but in an otherwise arbitrary direction. The fact that the vorticity field is solenoidal requires that the divergence of $\boldsymbol{\zeta}$ in the tangential plane vanish, $\nabla' \cdot \boldsymbol{\zeta} = 0$, where ∇' is the tangential gradient $\nabla' = (\mathbf{I} - \mathbf{nn}) \cdot \nabla$, and $\mathbf{n}$ is the unit vector normal to the vortex sheet. As a consequence of this constraint, a vortex sheet must form a closed surface, end at the boundaries, or extend to infinity.

Using Stokes's theorem, we find that the circulation around a loop L that pierces a vortex sheet at the points P and Q, shown in Figure 1.9.1(a), is given by

$$C \equiv \int_L \mathbf{u}(\mathbf{x}) \cdot \mathbf{t}(\mathbf{x}) dl(\mathbf{x}) = \int_D \boldsymbol{\omega}(\mathbf{x}) \cdot \mathbf{n}(\mathbf{x}) dS(\mathbf{x})$$

$$= \int_D \left(\int_S \boldsymbol{\zeta}(\mathbf{x}') \delta(\mathbf{x} - \mathbf{x}') dS(\mathbf{x}') \right) \cdot \mathbf{n}(\mathbf{x}) dS(\mathbf{x}) \tag{1.9.3}$$

where D is an arbitrary surface bounded by L, and $\mathbf{n}$ is the unit vector normal to D pointing in an appropriate direction. Interchanging the order of the integrations on the right-hand side, we obtain

$$C = \int_S \boldsymbol{\zeta}(\mathbf{x}') \cdot \left(\int_D \mathbf{n}(\mathbf{x}) \delta(\mathbf{x} - \mathbf{x}') dS(\mathbf{x}) \right) dS(\mathbf{x}') \tag{1.9.4}$$

Next we write $dS(\mathbf{x}') = dl(\mathbf{x}') dl_\xi(\mathbf{x}')$, where l is the arc length measured along the intersection T of the surface D and the vortex sheet S, l_ξ, is the arc length along the unit vector $\mathbf{e}_\xi$, which lies

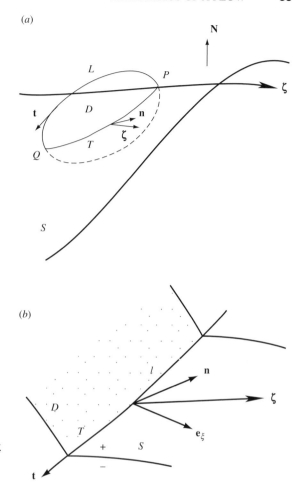

Figure 1.9.1   (a) Schematic illustration of a three-dimensional vortex sheet. (b) Closeup of the intersection between the surface D that is bounded by the loop L and the vortex sheet.

in S and is perpendicular to $\mathbf{t}$, and $\mathbf{t}$ is the unit vector tangential to T, as shown in Figure 1.9.1(b). Using the geometrical relation $dl_n = \mathbf{n} \cdot \mathbf{e}_\xi \, dl_\xi$, where dl_n is the arc length in the direction of $\mathbf{n}$, we rewrite Eq. (1.9.4) as

$$C = \int_S \frac{\boldsymbol{\zeta}(\mathbf{x}')}{\mathbf{n}(\mathbf{x}') \cdot \mathbf{e}_\xi(\mathbf{x}')} \cdot \left(\int_D \mathbf{n}(\mathbf{x}) \, \delta(\mathbf{x} - \mathbf{x}') \, dS(\mathbf{x}) \, dl_n(\mathbf{x}) \right) dl(\mathbf{x}')$$

$$= \int_T \frac{\mathbf{n}(\mathbf{x}') \cdot \boldsymbol{\zeta}(\mathbf{x}')}{\mathbf{n}(\mathbf{x}') \cdot \mathbf{e}_\xi(\mathbf{x}')} \, dl(\mathbf{x}') \tag{1.9.5}$$

Note that we used the distinguishing properties of the delta function to simplify the term within the large parentheses. Now we note that $\boldsymbol{\zeta}$ lies in the plane containing $\mathbf{t}$ and $\mathbf{e}_\xi$, which is perpendicular to the plane containing $\mathbf{e}_\xi$ and $\mathbf{n}$, and this allows us to write $[(\mathbf{n} \times \mathbf{e}_\xi) \times \mathbf{e}_\xi] \cdot \boldsymbol{\zeta} = 0$, which can be shown to be equivalent to $(\mathbf{n} \cdot \mathbf{e}_\xi)(\mathbf{e}_\xi \cdot \boldsymbol{\zeta}) = \mathbf{n} \cdot \boldsymbol{\zeta}$ (Problem 1.9.2). Thus Eq. (1.9.5) simplifies to

$$C = \int_T \mathbf{e}_\xi(\mathbf{x}') \cdot \boldsymbol{\zeta}(\mathbf{x}') \, dl(\mathbf{x}') \tag{1.9.6}$$

Taking the limit as the loop L collapses onto the vortex sheet at both sides while the point Q tends to P, we find

$$(\mathbf{u}^+ - \mathbf{u}^-) \cdot \mathbf{t} = \boldsymbol{\zeta} \cdot \mathbf{e}_\xi \tag{1.9.7}$$

where $\mathbf{t}$ is the unit vector tangential to T. The superscripts plus and minus designate, respectively, the velocity just above and below the vortex sheet.

Equation (1.9.7) shows that the tangential component of the velocity undergoes a discontinuity across a vortex sheet. Conservation of mass requires that the normal component of the velocity remain continuous across the vortex sheet, and this, in conjunction with Eq. (1.9.7), allows us to write

$$\mathbf{u}^+ - \mathbf{u}^- = \boldsymbol{\zeta} \times \mathbf{N} \tag{1.9.8}$$

where $\mathbf{N}$ is the unit vector normal to the vortex sheet as shown in Figure 1.9.1(a). Equation (1.9.8) reveals that $\boldsymbol{\zeta}$, $\mathbf{N}$, and $\mathbf{u}^+ - \mathbf{u}^-$ define three mutually perpendicular directions, and this allows us to write $\boldsymbol{\zeta} = \mathbf{N} \times (\mathbf{u}^+ - \mathbf{u}^-)$.

We thus find that a vortex sheet represents a singular surface across which the tangential component of the velocity changes from one value above to another value below, and the difference between these two values is given by the right-hand side of Eq. (1.9.8) in terms of the strength of the vortex sheet.

The mean value of the velocity just above and below the vortex sheet is called the *principal velocity of the vortex sheet,* or more precisely, the *principal value of the velocity of the vortex sheet,* and will by denoted by $\mathbf{u}^{\mathrm{PV}}$. Thus, by definition,

$$\mathbf{u}^{\mathrm{PV}} = \tfrac{1}{2}(\mathbf{u}^+ + \mathbf{u}^-) \tag{1.9.9}$$

Combining Eqs. (1.9.9) and (1.9.8), we obtain the velocity on either side of the vortex sheet in terms of the strength of the vortex sheet and the principal velocity,

$$\mathbf{u}^{\pm} = \mathbf{u}^{\mathrm{PV}} \pm \tfrac{1}{2}\boldsymbol{\zeta} \times \mathbf{N} \tag{1.9.10}$$

Two-Dimensional Flow

We proceed next to describe the structure of line vortices and vortex sheets in two-dimensional flow in the xy plane, exploiting the fact that the vortex lines are straight lines parallel to the z axis.

Point vortex

A rectilinear line vortex is called a *point vortex.* The location of a point vortex may be identified in terms of its trace in the (x, y) plane, $\mathbf{x}_0 = (x_0, y_0)$. The vorticity in the xy plane is represented by the generalized function

$$\omega(\mathbf{x}) = \kappa\, \delta(\mathbf{x} - \mathbf{x}_0) \tag{1.9.11}$$

where δ is the two-dimensional delta function and κ is the strength of the point vortex.

Vortex sheet

A two-dimensional vortex sheet is a cylindrical vortex sheet whose generators and strength $\boldsymbol{\zeta}$ are oriented along the z axis with corresponding unit vector $\mathbf{k}$. The location of the vortex sheet may be described in terms of its trace T in the xy plane, as illustrated in Figure 1.9.2(a). The strength of the vorticity field in the xy plane is represented by the generalized distribution

$$\omega(\mathbf{x}) = \int_T \gamma(\mathbf{x}')\, \delta(\mathbf{x} - \mathbf{x}')\, dl(\mathbf{x}') \tag{1.9.12}$$

where δ is the two-dimensional delta function and $\boldsymbol{\zeta} = \gamma\mathbf{k}$.

Using Eq. (1.9.8), we find that the discontinuity in the velocity across a two-dimensional vortex sheet is given by

$$\mathbf{u}^+ - \mathbf{u}^- = \gamma\mathbf{t} \tag{1.9.13}$$

where $\mathbf{t}$ is a unit vector tangential to T, as shown in Figure 1.9.2(a). In terms of the principal velocity of the vortex sheet, the velocities on the upper and lower side of the vortex sheet are

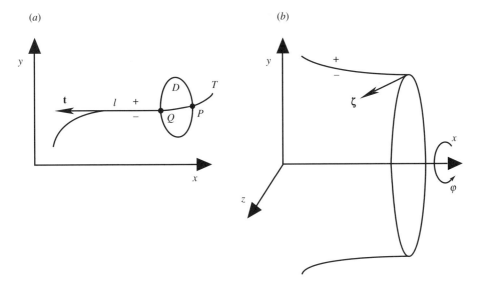

Figure 1.9.2 Schematic illustrations of (a) a two-dimensional vortex sheet, and (b) an axisymmetric vortex sheet.

given by

$$\mathbf{u}^+ = \mathbf{u}^{PV} + \tfrac{1}{2}\gamma\mathbf{t}, \qquad \mathbf{u}^- = \mathbf{u}^{PV} - \tfrac{1}{2}\gamma\mathbf{t} \tag{1.9.14}$$

The circulation around a loop that lies in the xy plane and pierces the vortex sheet at the points P and Q, as illustrated in Figure 1.9.2(a), is given by

$$C = \int_D \omega(\mathbf{x})\,dA(\mathbf{x}) = \int_D \int_{T^{PQ}} \gamma(\mathbf{x}')\,\delta(\mathbf{x} - \mathbf{x}')\,dl(\mathbf{x}')\,dA(\mathbf{x})$$

$$= \int_{T^{PQ}} \gamma(\mathbf{x}')\,dl(\mathbf{x}') \tag{1.9.15}$$

where D is the area in the xy plane enclosed by the loop, and T^{PQ} is the section of T between the points P and Q. Fixing the point P and regarding the circulation C as a function of location of the point Q along T, denoted by Γ, we obtain

$$\frac{d\Gamma}{dl}(\mathbf{x}) = \gamma(\mathbf{x}) \tag{1.9.16}$$

This definition allows us to express the vorticity distribution (1.9.12) in an alternative form in terms of Γ as

$$\omega(\mathbf{x}) = \int_T \delta(\mathbf{x} - \mathbf{x}')\,d\Gamma(\mathbf{x}') \tag{1.9.17}$$

Comparing Eq. (1.9.17) to Eq. (1.9.11) shows that a cylindrical vortex sheet may be regarded as a continuous distribution of point vortices with circulation per unit length equal to γ.

Axisymmetric Flow

The vorticity of an axisymmetric flow without swirling motion is oriented in the azimuthal direction, and the vortex lines are concentric circles. The position of an axisymmetric line vortex, called a *line vortex ring,* may be described in terms of its trace in an azimuthal plane, which is usually taken to be the xy plane corresponding to $\varphi = 0$. The strength of the vorticity is given by Eq. (1.9.11), where δ is the two-dimensional delta function operating in the azimuthal plane.

An axisymmetric vortex sheet may be identified by its trace T in an azimuthal plane. The strength of the vortex sheet ζ is oriented in the azimuthal direction, and the associated vorticity distribution in an azimuthal plane is given by Eq. (1.9.12), where δ is the two-dimensional delta function operating with respect to the planar coordinates that are defined in the azimuthal plane.

PROBLEMS

1.9.1 **A line vortex subtended between two bodies.** Consider an infinite flow that contains two bodies and a single line vortex that begins on the surface of the first body and ends at the surface of the other. Discuss whether this is an acceptable and realizable flow.

1.9.2 **Three-dimensional vortex sheet.** With reference to the discussion of the three-dimensional vortex sheet, show that the equation $[(\mathbf{n} \times \mathbf{e}_\xi) \times \mathbf{e}_\xi] \cdot \boldsymbol{\zeta} = 0$ is equivalent to $(\mathbf{n} \cdot \mathbf{e}_\xi)(\mathbf{e}_\xi \cdot \boldsymbol{\zeta}) = \mathbf{n} \cdot \boldsymbol{\zeta}$.

References

Atkinson, K. E., 1989, *An Introduction to Numerical Analysis.* Wiley.
Schowalter, W. R., 1978, *Mechanics of Non-Newtonian Fluids.* Pergamon.
Struik, D. J., 1961, *Lectures on Classical Differential Geometry.* Dover.

In Chapter 1 we examined the motion of fluid parcels, material vectors, material lines, and material surfaces in a specified field of flow. The velocity field was assumed to be known either in terms of Eulerian variables, including space and time, or Lagrangian variables, including point-particle labels and time. In the present chapter we shall discuss alternative methods of describing the flow in terms of *secondary* scalar or vectorial fields. By definition, the velocity field is related to a secondary field through a differential or integral relationship.

Examples of secondary fields are the vorticity and rate of expansion introduced in Chapter 1. Additional secondary fields, to be introduced in the present chapter, are the velocity potential for irrotational flow, the vector potential for incompressible flow, the stream function for two-dimensional flow, the Stokes stream function for axisymmetric flow, and a pair of stream functions for a general three-dimensional incompressible flow. Certain secondary fields, such as the rate of expansion, the vorticity, and the stream functions, have a clear physical significance, but others are mathematical entities introduced for mere analytical convenience.

Describing a flow in terms of a secondary field is motivated by two reasons. The first one has to do with the fact that *the number of scalar secondary fields that are necessary in order to describe an incompressible flow is less than the dimensionality of the flow by one unit,* and this results in analytical and computational simplifications. For instance, we shall see in this chapter that a two-dimensional flow of an incompressible fluid may be described in terms of a single scalar function, and a three-dimensional flow of an incompressible fluid may be described in terms of two scalar functions called the stream functions. Imposing additional constraints reduces the number of required scalar functions even further. Thus a three-dimensional incompressible *and* irrotational flow may be expressed in terms of a single scalar function called the *harmonic potential.* The reduction in the number of scalar functions with respect to the number of nonvanishing components of the velocity is explained by the fact that the components of the velocity may not be assigned independently, but must be coordinated so that the continuity equation is fulfilled.

Furthermore, expressing the velocity field in terms of certain secondary fields allows us to gain physical insights into the significance of the motion of the fluid parcels on the global structure or evolution of a flow. For instance, representing the velocity field in terms of the vorticity distribution and rate of expansion illustrates the effect of spinning and expansion or contraction of fluid parcels on the overall motion of the fluid.

The secondary fields discussed in the present chapter are introduced with reference to the structure of the velocity field discussed in Chapter 1. There is another class of secondary fields that are defined with respect to the stresses developing in the fluid and with reference to Cauchy's equation that governs the motion of the fluid. Examples of these fields will be discussed in Section 6.12.

2.1 | IRROTATIONAL FLOWS AND THE VELOCITY POTENTIAL

A velocity field whose vorticity vanishes at every point in the flow is called *irrotational*. Small fluid parcels in an irrotational flow translate and deform but do not rotate. Consequently, a small cubical fluid parcel may translate and deform to obtain a rectangular shape, but maintains its initial orientation.

In Chapter 3 we shall examine the mechanisms of production and evolution of vorticity and shall see that, in practice, no flow can be truly irrotational except during an infinitesimal time period when the fluid starts moving from a state of rest. It appears then that the concept of irrotational flow is a mathematical idealization with little physical relevance, and this is often indeed the case.

There are a number of flows, however, that are virtually irrotational or are composed of adjacent regions of nearly irrotational and nearly rotational flow. For example, high-speed unseparated streaming flow past an airfoil is virtually irrotational everywhere except within a thin boundary layer that lines the airfoil and within a slender wake. The flow produced by the propagation of waves on the surface of the ocean is nearly irrotational everywhere except within a thin boundary layer along the surface. In most cases, the conditions under which a flow will be nearly or partially irrotational are not known a priori but must be assessed by carrying out detailed experimental or theoretical studies.

Potential and Irrotational Flows

Since the curl of the gradient of any twice differentiable function vanishes, any *potential flow*, defined as a flow whose velocity derives by

$$\mathbf{u} = \nabla\phi \tag{2.1.1}$$

where ϕ is a twice differentiable scalar function called the *potential function* or *velocity potential*, is irrotational. Different families of irrotational flows may then be derived by making different selections for ϕ. We wish to inquire now whether the inverse is also true, that is, whether an irrotational flow may be expressed in terms of the gradient of a potential function as in Eq. (2.1.1).

To this end, it is necessary to distinguish between singly connected and multiply connected domains of flow. For this purpose, we draw a closed loop within the domain of a flow; if the loop can be shrunk down to a point that lies within the flow without crossing the boundaries, then it is called *reducible;* if, however, the loop must cross one or more boundaries in order to collapse down to a point that lies within the flow, then it is called *irreducible.*

If any loop that can possibly be drawn within a particular domain of flow is reducible, then the domain is called *singly connected;* otherwise it is called *multiply connected.* A loop that wraps around a toroidal or cylindrical boundary of infinite extent with possible wavy corrugations is irreducible, and the corresponding domains are doubly connected. A domain of flow that contains two distinct cylindrical boundaries of infinite extent is triply connected.

Let us consider the reducible loop L illustrated in Figure 2.1.1(a). Applying Stokes's theorem, we find that the circulation around the loop is given by

$$\int_L \mathbf{u} \cdot \mathbf{t}\, dl = \int_D \boldsymbol{\omega} \cdot \mathbf{n}\, dS = 0 \tag{2.1.2}$$

where D is an arbitrary surface bounded by L, $\mathbf{t}$ is the unit vector tangential to L, and $\mathbf{n}$ is the unit vector normal to D oriented according to the right-handed rule with respect to $\mathbf{t}$. Equation (2.1.2) states that the circulation around any reducible loop that lies within an irrotational flow vanishes.

(a)

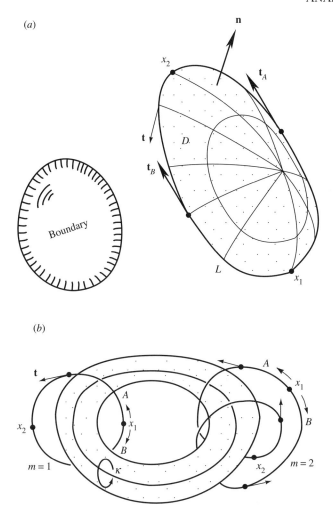

(b)

Figure 2.1.1   (a) A reducible loop within a singly connected domain of flow, showing the decomposition of the circulation integral into two paths. (b) Two irreducible loops in a doubly connected domain of flow with number of turns $m = 1$ and $m = 2$. The flow in a multiply connected domain with finite circulation around the boundaries may be resolved into the flow due to line vortices that lie inside the boundaries and outside the domain of flow, and a complementary irrotational flow described in terms of a single valued potential.

Next we select two points $\mathbf{x}_1$ and $\mathbf{x}_2$ on L and decompose the line integral in Eq. (2.1.2) into two parts, writing

$$\int_{\mathbf{x}_1}^{\mathbf{x}_2} \mathbf{u} \cdot \mathbf{t}_A \, dl = \int_{\mathbf{x}_1}^{\mathbf{x}_2} \mathbf{u} \cdot \mathbf{t}_B \, dl \qquad (2.1.3)$$

The integral on the left-hand side is taken along path A, and the integral on the right-hand side is taken along path B with corresponding tangential vectors $\mathbf{t}_A = \mathbf{t}$ and $\mathbf{t}_B = -\mathbf{t}$, as shown in Figure 2.1.1(a). Equation (2.1.3) states that the circulations around any paths that connect two points $\mathbf{x}_1$ and $\mathbf{x}_2$ on a reducible loop are identical, and this allows us to introduce a single-valued scalar function of position ϕ and write

$$\int_{\mathbf{x}_1}^{\mathbf{x}_2} \mathbf{u} \cdot \mathbf{t}_A \, dl = \phi(\mathbf{x}_2) - \phi(\mathbf{x}_1) \tag{2.1.4}$$

Taking the limit as $\mathbf{x}_2$ tends to $\mathbf{x}_1$ and using the trapezoidal rule to approximate the integral, we find $\mathbf{u} \cdot \mathbf{t}_A = \mathbf{t}_A \cdot \nabla\phi$. Noting that the unit vector $\mathbf{t}_A$ points in an arbitrary direction, we recover Eq. (2.1.1), thereby demonstrating that *for any irrotational flow in a singly connected domain there exists a corresponding single-valued potential function ϕ*. Stated differently, an irrotational flow in a singly connected domain is also a potential flow, and as a result it may be described in terms of *one scalar function*.

A slight complication arises when the domain of flow is multiply connected. The central difficulty is that the circulation around an irreducible loop is not necessarily equal to zero, but may depend on the number of turns m that the loop performs around a boundary. Two loops with $m = 1$ and 2 are illustrated in Figure 2.1.1(b). The circulation around a loop that performs multiple turns is given by

$$\int_{L_m} \mathbf{u} \cdot \mathbf{t} \, dl = m\kappa \tag{2.1.5}$$

where κ is the lowest value of the circulation corresponding to a loop that performs a single turn, called the *cyclic constant* of the flow around the boundary. In this case, the application of Stokes's theorem as in Eq. (2.1.2) is prohibited by the fact that, in general, the surface D must cross the boundaries and thus does not lie entirely within the fluid. Breaking up the circulation integral around the loop into two paths and working as in Eqs. (2.1.3) and (2.1.4), we find that for an irreducible loop

$$\Delta\phi_A - \Delta\phi_B = m\kappa \tag{2.1.6}$$

where Δ denotes the change in the potential function from the beginning to the end of the paths A and B illustrated in Figure 2.1.1(b).

Equation (2.1.6) reveals that the potential function in a multiply connected domain of flow may be a multivalued function of position. To avoid the analytical and computational complications arising from the use of multivalued functions, it is helpful to decompose the velocity field into two components writing $\mathbf{u} = \mathbf{v} + \nabla\phi$, where $\mathbf{v}$ is a known irrotational flow whose cyclic constants around the boundaries are identical to those of the flow $\mathbf{u}$. For instance, $\mathbf{v}$ may be identified with the flow due to a line vortex lying *outside* the domain of flow in the interior of a toroidal boundary, where the strength of the line vortex is equal to cyclic constant of the flow around the boundary, as illustrated in Figure 2.1.1(b). The velocity potential ϕ is a single-valued function of position and may thus be computed using standard analytical and numerical methods without any additional complications. Applications of this decomposition will be discussed in Chapter 7 with reference to flow past two-dimensional airfoils.

Jump in the Potential across a Vortex Sheet

Let us consider a two-dimensional vortex sheet separating two regions of irrotational flow, such as that illustrated in Figure 1.9.2(a), and express the velocity on either side of the vortex sheet in terms of the velocity potentials ϕ^+ and ϕ^-. Substituting Eq. (2.1.1) into Eq. (1.9.13) and using Eq. (1.9.16), we find

$$\nabla\phi^+ - \nabla\phi^- = \gamma\mathbf{t} = \frac{d\Gamma}{dl}\mathbf{t} \tag{2.1.7}$$

where γ is the strength of the vortex sheet, Γ is the circulation along the vortex sheet, and l is the arc length along the vortex sheet measured in the direction of the unit tangential vector $\mathbf{t}$. Projecting Eq. (2.1.7) onto $\mathbf{t}$, and integrating with respect to arc length, we find that the jump in the velocity potential across the vortex sheet is given by

$$\phi^+ - \phi^- = \Gamma \tag{2.1.8}$$

where we have assumed that ϕ^+ and ϕ^- have the same values at the point where we begin measuring the circulation along the vortex sheet. Equation (2.1.8) finds useful applications in the computation of the motion of vortex sheets using numerical methods, to be discussed in Section 11.4.

Considering next two points P and Q on a three-dimensional vortex sheet, we express the velocity on either side of the vortex sheet as the gradient of the corresponding potentials, and integrate Eq. (1.9.8) along a path that connects P and Q and lies in the vortex sheet to obtain

$$(\phi^+ - \phi^-)_Q = (\phi^+ - \phi^-)_P + \int_P^Q (\boldsymbol{\zeta} \times \mathbf{N}) \cdot \mathbf{t} \, dl \tag{2.1.9}$$

where $\mathbf{t}$ is the tangential unit vector along the path. The fact that the divergence of $\boldsymbol{\zeta}$ in the tangential plane vanishes guarantees that the value of the integral is independent of its path (see p. 42). When the integration path coincides with a vortex line, that is, it is tangential to $\boldsymbol{\zeta}$ at every point, the integrand in Eq. (2.1.9) vanishes; this shows that the jump of the velocity potential across the vortex sheet remains constant along the vortex lines.

The Velocity Potential in Terms of the Rate of Expansion

The velocity field corresponding to a potential flow may or may not be solenoidal, and the associated flow may or may not be incompressible. Taking the divergence of Eq. (2.1.1), we find

$$\nabla \cdot \mathbf{u} = \nabla^2 \phi \tag{2.1.10}$$

which may be regarded as a Poisson equation for ϕ forced by the rate of expansion $\nabla \cdot \mathbf{u}$.

Using Poisson's inversion formula, to be discussed in detail in Section 2.2, we obtain an expression for the potential of a three-dimensional flow in terms of the rate of expansion,

$$\phi(\mathbf{x}) = -\frac{1}{4\pi} \int_{\text{Flow}} \frac{1}{r} \nabla' \cdot \mathbf{u}(\mathbf{x}') \, dV(\mathbf{x}') + H(\mathbf{x}) \tag{2.1.11}$$

where $r = |\mathbf{x} - \mathbf{x}'|$, the gradient ∇' involves derivatives with respect to $\mathbf{x}'$, and H is a harmonic function determined by the boundary conditions, $\nabla^2 H = 0$. When the domain of flow extends to infinity, in order to ensure that the volume integral in Eq. (2.1.11) is finite, we require that the rate of expansion decays at a rate that is faster than $1/R^2$, where R is the distance from the origin.

The counterpart of Eq. (2.1.11) for two-dimensional flow in the xy plane is

$$\phi(\mathbf{x}) = \frac{1}{2\pi} \int_{\text{Flow}} \ln r \, \nabla' \cdot \mathbf{u}(\mathbf{x}') \, dA(\mathbf{x}') + H(\mathbf{x}) \tag{2.1.12}$$

where $dA = dx \, dy$ and H is a harmonic function of x and y.

To obtain an expression for the velocity field in terms of the rate of expansion, we take the gradient of both sides of Eqs. (2.1.11) and (2.1.12) and interchange the gradient with the integral, obtaining

$$\mathbf{u}(\mathbf{x}) = \frac{1}{4\pi} \int_{\text{Flow}} \frac{1}{r^3} \hat{\mathbf{x}} \nabla' \cdot \mathbf{u}(\mathbf{x}') \, dV(\mathbf{x}') + \nabla H(\mathbf{x}) \tag{2.1.13}$$

$$\mathbf{u}(\mathbf{x}) = \frac{1}{2\pi} \int_{\text{Flow}} \frac{1}{r^2} \hat{\mathbf{x}} \nabla' \cdot \mathbf{u}(\mathbf{x}') \, dA(\mathbf{x}') + \nabla H(\mathbf{x}) \tag{2.1.14}$$

respectively, for three-dimensional and two-dimensional flow, where $\hat{\mathbf{x}} = \mathbf{x} - \mathbf{x}'$. We shall see later in this section that the integrals on the right-hand sides of these two equations may be interpreted as volume or area distributions of point sources or sinks of mass. The densities of the distributions are equal to the rate of expansion of the fluid.

Incompressible Flows and Harmonic Potentials

When the fluid is incompressible, the rate of expansion on the left-hand side of Eq. (2.1.10) vanishes, and this shows that ϕ satisfies Laplace's equation

$$\nabla^2 \phi = 0 \tag{2.1.15}$$

In this case ϕ is a harmonic function called the *harmonic potential.*

Kinetic energy of the fluid and the significance of normal boundary motion

The kinetic energy K of an incompressible fluid with uniform density executing irrotational motion in a singly connected domain may be expressed in terms of a boundary integral involving the boundary values of the harmonic potential. To derive this expression, we apply the rules of product differentiation and write

$$K \equiv \tfrac{1}{2}\rho \int_{\text{Flow}} \mathbf{u} \cdot \mathbf{u}\, dV = \tfrac{1}{2}\rho \int_{\text{Flow}} \mathbf{u} \cdot \nabla\phi\, dV$$

$$= \tfrac{1}{2}\rho \int_{\text{Flow}} (\nabla \cdot (\phi\mathbf{u}) - \phi\, \nabla \cdot \mathbf{u})\, dV \tag{2.1.16}$$

Because the velocity field is solenoidal, the second term within the integral on the right-hand side vanishes. Applying the divergence theorem, we find

$$K = -\tfrac{1}{2}\rho \int_{\text{Boundaries}} \phi\mathbf{u} \cdot \mathbf{n}\, dS \tag{2.1.17}$$

where $\mathbf{n}$ is the unit vector normal to the boundaries pointing *into* the flow.

Equation (2.1.17) shows that if the normal component of the velocity vanishes over *all* boundaries, then the kinetic energy of the fluid and therefore the magnitude of the velocity will be equal to zero, and the fluid must be in a state of rest. In turn, this requires that the harmonic potential have a constant value throughout the domain of flow. When the velocity potential has the *same* constant value over all boundaries, we invoke conservation of mass to show that the right-hand side of Eq. (2.1.17) must vanish, and thus find that the fluid will be in a state of rest and the potential will be constant, equal to its boundary value, throughout the domain of flow.

We can derive corresponding results for flow in multiply connected domains, in which case the velocity potential may be a multivalued function. Let us consider, for instance, a domain of flow exterior to a toroidal boundary as shown in Figure 2.1.2, and draw an arbitrary surface D that

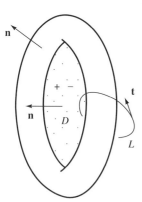

Figure 2.1.2 A doubly connected domain of flow with a toroidal boundary is rendered singly connected by introducing an artificial surface ending at the boundary.

ends at the toroidal boundary. Regarding D as a virtual boundary of the flow renders the domain of flow singly connected. Repeating the above manipulations, we find that the kinetic energy of the fluid is given by

$$K = -\tfrac{1}{2}\rho \int_{\text{Boundaries}} \phi\,\mathbf{u}\cdot\mathbf{n}\,dS - \tfrac{1}{2}\rho \int_D (\phi^+ - \phi^-)\mathbf{u}\cdot\mathbf{n}\,dS \qquad (2.1.18)$$

where $\phi^{\pm}$ are the values of the potential on either side of D, and the normal vector $\mathbf{n}$ over D points into the side corresponding to the plus superscript. Noting that the circulation around the loop L shown in Figure 2.1.2, which is equal to the cyclic constant κ of the flow around the torus, is given by

$$\kappa = \int_L \mathbf{u}\cdot\mathbf{t}\,dl = \int_L \nabla\phi\cdot\mathbf{t}\,dl = \int_L \frac{\partial\phi}{\partial l}\,dl = \phi^+ - \phi^- \qquad (2.1.19)$$

allows us to obtain the expression

$$K = -\tfrac{1}{2}\rho \int_{\text{Boundaries}} \phi\,\mathbf{u}\cdot\mathbf{n}\,dS - \tfrac{1}{2}\rho\kappa Q \qquad (2.1.20)$$

where Q is the volumetric flow rate across D toward the positive side. This expression shows that the kinetic energy of the fluid will vanish provided that (1) either the potential has the same constant value over all boundaries, or the normal component of the velocity vanishes at the boundaries, and (2) either the cyclic constant κ or the flow rate Q are equal to zero.

Uniqueness of solution

One important consequence of the preceding results concerning the kinetic energy of the fluid is that, given boundary conditions for the normal component of the velocity, an incompressible irrotational flow in a singly connected domain is unique, and the corresponding harmonic potential is determined uniquely up to an arbitrary constant. To show this, consider two harmonic potentials representing two distinct flows, and note that their difference is also an acceptable harmonic potential; that is, it describes an incompressible irrotational flow. If the two flows have identical boundary conditions for the normal component of the velocity, the difference flow must vanish, and the corresponding harmonic potential must have a constant value. As a consequence, the two original flows must be identical, and the corresponding harmonic potentials may differ, at most, by a scalar constant. A similar reasoning allows us to conclude that specifying the boundary distribution of the potential determines a flow in a singly connected domain in a unique manner.

For flow in a doubly connected domain, we use Eq. (2.1.20) to find that, given boundary conditions for the normal component of the velocity or specifying the boundary distribution of the potential, and prescribing either the value of the cyclic constant κ or the flow rate Q, determines the flow in a unique manner.

Kelvin's minimum-kinetic-energy theorem

Kelvin (1849) showed that of all solenoidal velocity fields that satisfy a prescribed boundary condition for the normal component of the velocity, the one expressing irrotational motion has the least amount of kinetic energy, provided that the density is uniform throughout the domain of flow.

The proof proceeds by assuming that $\mathbf{u}$ is an irrotational velocity field described by the velocity potential ϕ, and $\mathbf{v}$ is another solenoidal rotational velocity field, subject to the condition that $\mathbf{u}\cdot\mathbf{n} = \mathbf{v}\cdot\mathbf{n}$ at the boundaries. The difference of the kinetic energies of the two flows is

$$K(\mathbf{v}) - K(\mathbf{u}) = \tfrac{1}{2}\rho \int_{\text{Flow}} (\mathbf{v}\cdot\mathbf{v} - \mathbf{u}\cdot\mathbf{u})\,dV$$

$$= \tfrac{1}{2}\rho \int_{\text{Flow}} (\mathbf{v} - \mathbf{u})\cdot(\mathbf{v} - \mathbf{u})\,dV + \rho \int_{\text{Flow}} (\mathbf{v} - \mathbf{u})\cdot\mathbf{u}\,dV \qquad (2.1.21)$$

Manipulating the last integral on the right-hand side and using the divergence theorem, we find

$$\int_{\text{Flow}} (\mathbf{v} - \mathbf{u}) \cdot \nabla \phi \, dV = \int_{\text{Boundaries}} \phi \, (\mathbf{v} - \mathbf{u}) \cdot \mathbf{n} \, dS \tag{2.1.22}$$

which vanishes in view of the prescribed boundary conditions for the normal component of the velocity. We thus conclude that the right-hand side of Eq. (2.1.21) is positive, and this demonstrates that the energy of the rotational flow $\mathbf{v}$ is greater than that of the irrotational flow $\mathbf{u}$.

Singularities of Irrotational–Incompressible Flow

Singular solutions of Laplace's equation for the harmonic potential, representing fundamental irrotational incompressible flows, play an important role in the theory of potential flow, by providing us with building blocks for constructing and analyzing a variety of flows.

Point source

Let us consider the solution of Eq. (2.1.15) in an infinite domain of flow with no interior boundaries, subject to a singular forcing term on the right-hand side,

$$\nabla^2 \phi = m \, \delta(\mathbf{x} - \mathbf{x}_0) \tag{2.1.23}$$

where δ is the three-dimensional or two-dimensional delta function, m is a constant, and $\mathbf{x}_0$ is an arbitrary point in the domain of flow. Using the method of Fourier transforms, or by trial and error, we find that the solution to Eq. (2.1.23) is given by

$$\phi = -\frac{m}{4\pi} \frac{1}{r}, \qquad \phi = \frac{m}{2\pi} \ln r \tag{2.1.24}$$

respectively, for three-dimensional and two-dimensional flow, where $r = |\mathbf{x} - \mathbf{x}_0|$.

Since ϕ is a harmonic function everywhere except at the point $\mathbf{x}_0$, it represents an acceptable incompressible irrotational flow. The corresponding velocity field is given by

$$\mathbf{u} = \frac{m}{4\pi} \frac{\mathbf{x} - \mathbf{x}_0}{r^3}, \qquad \mathbf{u} = \frac{m}{2\pi} \frac{\mathbf{x} - \mathbf{x}_0}{r^2} \tag{2.1.25}$$

respectively, for three-dimensional and two-dimensional flow. The associated streamlines are straight radial lines emanating from the singular point $\mathbf{x}_0$.

It is a straightforward exercise to verify that the flow rate across a spherical surface or circular line that is centered at $\mathbf{x}_0$ is equal to m, and this justifies identifying ϕ with the potential due to a point source of strength m located in an infinite domain of flow. In Section 2.2 we shall see that the potential for $m = -1$ provides us with the *free-space Green's function* of Laplace's equation, and, furthermore, we shall discuss the velocity field due to point sources in bounded domains of flow.

We return now to inspect the integrals on the right-hand sides of Eqs. (2.1.13) and (2.1.14) in conjunction with Eqs. (2.1.25), and interpret them as volume or area distributions of point sources of mass; the densities of the distributions are equal to the rate of expansion.

Point-source dipole

Let us consider two point sources with strengths of equal magnitude and opposite sign located at the two points $\mathbf{x}_0$ and $\mathbf{x}_1$. Due to the linearity of Eq. (2.1.15), the associated potential may be constructed by superposition as

$$\phi = -\frac{m}{4\pi} \frac{1}{|\mathbf{x} - \mathbf{x}_0|} + \frac{m}{4\pi} \frac{1}{|\mathbf{x} - \mathbf{x}_1|}$$

$$\phi = \frac{m}{2\pi} \ln |\mathbf{x} - \mathbf{x}_0| - \frac{m}{2\pi} \ln |\mathbf{x} - \mathbf{x}_1| \tag{2.1.26}$$

(*a*)

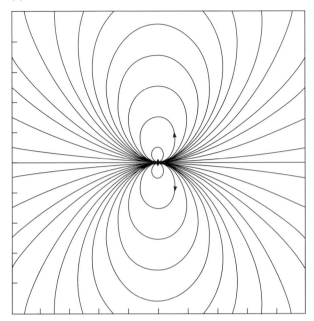

(*b*)

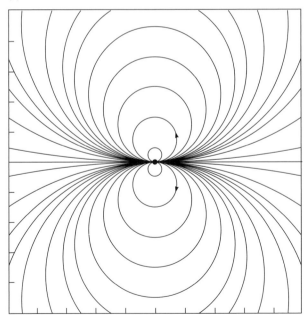

Figure 2.1.3 Streamline pattern due to (a) a three-dimensional, and (b) a two-dimensional potential dipole pointing toward the positive direction of the *x* axis.

respectively, for three-dimensional and two-dimensional flow. Next we place $\mathbf{x}_0$ and $\mathbf{x}_1$ close to one another, expand the potential due to the second point source in a Taylor series with respect to $\mathbf{x}_1$ about $\mathbf{x}_0$, and retain only the linear terms to obtain

$$\phi = \frac{m}{4\pi}(\mathbf{x}_1 - \mathbf{x}_0) \cdot \nabla_0 \frac{1}{|\mathbf{x} - \mathbf{x}_0|}$$

$$\phi = -\frac{m}{2\pi}(\mathbf{x}_1 - \mathbf{x}_0) \cdot \nabla_0 \ln |\mathbf{x} - \mathbf{x}_0|$$

(2.1.27)

where the derivatives of the gradient ∇_0 are taken with respect to $\mathbf{x}_0$. Taking the limit as the distance $|\mathbf{x}_0 - \mathbf{x}_1|$ tends to vanish while $m(\mathbf{x}_0 - \mathbf{x}_1)$ remains constant equal to $\mathbf{d}$, and carrying out the differentiations, we derive the velocity potential due to a three-dimensional or two-dimensional *point-source dipole*,

$$\phi = -\frac{1}{4\pi}\frac{\mathbf{x} - \mathbf{x}_0}{r^3} \cdot \mathbf{d}, \qquad \phi = -\frac{1}{2\pi}\frac{\mathbf{x} - \mathbf{x}_0}{r^2} \cdot \mathbf{d}$$

(2.1.28)

where $r = |\mathbf{x} - \mathbf{x}_0|$. The associated velocity fields are given by

$$\mathbf{u} = \frac{1}{4\pi}\left(-\frac{1}{r^3}\mathbf{I} + 3\frac{\hat{\mathbf{x}}\hat{\mathbf{x}}}{r^5}\right) \cdot \mathbf{d}$$

$$\mathbf{u} = \frac{1}{2\pi}\left(-\frac{1}{r^2}\mathbf{I} + 2\frac{\hat{\mathbf{x}}\hat{\mathbf{x}}}{r^4}\right) \cdot \mathbf{d}$$

(2.1.29)

where $\mathbf{I}$ is the identity matrix and $\hat{\mathbf{x}} = \mathbf{x} - \mathbf{x}_0$. One may readily verify that the flow rate across a spherical surface, and therefore across any other closed surface that encloses a three-dimensional point-source dipole, is equal to zero. Similarly, the flow rate across any closed loop that encloses a two-dimensional point-source dipole is equal to zero.

The streamline pattern in an azimuthal plane due to a three-dimensional dipole, and that in the xy plane due to a two-dimensional dipole, are illustrated in Figure 2.1.3(a,b). In both cases the dipole is oriented toward the positive direction of the x axis.

PROBLEMS

2.1.1   **A harmonic velocity field.** Consider a velocity field $\mathbf{u}$ with the property that each component of the velocity is a harmonic function. Show that, if the rate of expansion is constant or vanishes, the corresponding vorticity field will be irrotational, $\nabla \times \boldsymbol{\omega} = \mathbf{0}$.

2.1.2   **Homogeneous boundary conditions.** Consider an incompressible potential flow in a domain that is bounded by two closed surfaces, as shown in Figure 2.2.1(a), and require that the normal component of the velocity vanishes over the first surface, whereas the tangential component of the velocity vanishes over the second surface. Does this imply that the velocity field must vanish throughout the whole domain of flow?

2.1.3   **Infinite flow.** Consider a three-dimensional incompressible and irrotational flow in an infinite domain where the velocity vanishes at infinity. In Section 2.3 we shall show that the velocity potential at infinity must tend to a constant value [Eq. (2.3.18)]. Based on this observation, show that if the flow has no interior boundaries and no singular points, the velocity field must necessarily vanish throughout the whole domain of flow.

2.1.4   **Irrotational vorticity field.** (a) Show that an irrotational vorticity field, $\nabla \times \boldsymbol{\omega} = \mathbf{0}$, may be expressed as the gradient of a harmonic function. (b) Consider an irrotational vorticity field of an infinite flow with no interior boundaries, where the vorticity vanishes at infinity. Using the results of Problem 2.1.3, show that the vorticity must necessarily vanish throughout the whole domain.

2.1.5   **Point source.** Show that the flow rate across any surface that encloses a three-dimensional or two-dimensional point source is equal to m, but the flow rate across any surface or loop that does not enclose the point source is equal to zero.

2.2 | THE RECIPROCAL RELATION FOR HARMONIC FUNCTIONS AND GREEN'S FUNCTIONS OF LAPLACE'S EQUATION

In the next section we shall develop an integral representation for the velocity potential of an irrotational flow in terms of the rate of expansion of the fluid, the boundary values of the velocity potential, and the boundary distribution of the normal component of the velocity. To prepare the ground for these developments, in the present section we introduce the reciprocal theorem for harmonic functions and discuss the Green's functions of Laplace's equation.

Green's Identities and the Reciprocal Relation

Green's first identity states that any two twice-differentiable functions f and g satisfy the relation

$$f \nabla^2 g = \nabla \cdot (f \nabla g) - \nabla f \cdot \nabla g \tag{2.2.1}$$

whose validity may be demonstrated readily by straightforward differentiation working in index notation. Interchanging the roles of f and g, we obtain

$$g \nabla^2 f = \nabla \cdot (g \nabla f) - \nabla g \cdot \nabla f \tag{2.2.2}$$

Subtracting Eq. (2.2.2) from Eq. (2.2.1), we derive *Green's second identity*

$$f \nabla^2 g - g \nabla^2 \phi = \nabla \cdot (f \nabla g - g \nabla f) \tag{2.2.3}$$

If both functions f and g are harmonic, the left-hand side of Eq. (2.2.3) vanishes, yielding the *reciprocal relation* for harmonic functions

$$\nabla \cdot (f \nabla g - g \nabla f) = 0 \tag{2.2.4}$$

Integrating Eq. (2.2.4) over a certain control volume that is bounded by the singly or multiply connected surface D, and using the divergence theorem to convert the volume integral to a surface integral, we obtain an alternative statement in integral form,

$$\int_D f \nabla g \cdot \mathbf{n} \, dS = \int_D g \nabla f \cdot \mathbf{n} \, dS \tag{2.2.5}$$

where $\mathbf{n}$ is the unit vector normal to D pointing either into or outward from the control volume. Equation (2.2.5) places an integral constraint upon the boundary values and distribution of normal derivatives of any pair of nonsingular harmonic functions.

Green's Functions of the Three-Dimensional Laplace's Equation

Next we turn our attention to a special class of harmonic functions that are singular at a point, called the Green's functions of Laplace's equation. By definition, a three-dimensional Green's function satisfies the singularly forced Laplace's equation

$$\nabla^2 G(\mathbf{x}, \mathbf{x}_0) + \delta(\mathbf{x} - \mathbf{x}_0) = 0 \tag{2.2.6}$$

where δ is the three-dimensional delta function, $\mathbf{x}$ is a *field point,* and $\mathbf{x}_0$ is the location of the Green's function, also called the *pole.* When the domain of flow extends to infinity, the Green's functions are required to decay at least as fast as $1/|\mathbf{x} - \mathbf{x}_0|$.

In addition to satisfying Eq. (2.2.6), a *Green's function of the first kind* is required to vanish over a surface S_B representing the boundary of the flow; that is,

$$G(\mathbf{x}, \mathbf{x}_0) = 0 \qquad (2.2.7)$$

when $\mathbf{x}$ is on S_B. The normal derivative of a *Green's function of the second kind,* also called a *Neumann function,* is required to vanish over the boundary S_B; that is,

$$\nabla G(\mathbf{x}, \mathbf{x}_0) \cdot \mathbf{n}(\mathbf{x}) = 0 \qquad (2.2.8)$$

when $\mathbf{x}$ is on S_B.

Comparing Eq. (2.2.6) with Eq. (2.1.23) shows that, physically, a Green's function may be identified either with the steady temperature field due to a *point source of heat* of unit strength located at the point $\mathbf{x}_0$ in the presence of an isothermal or insulated boundary, or with the harmonic potential due to a *point sink of mass* with strength $m = -1$ located at $\mathbf{x}_0$ in a finite or infinite domain of flow. Specifically, a Green's function of the first kind represents the temperature field due to a point source of heat subject to the condition that the boundaries are kept at a constant temperature, which, for simplicity, is set equal to zero. A Green's function of the second kind represents the harmonic potential due to a point sink of mass subject to the condition that the normal component of the velocity vanishes at the boundaries; that is, the boundaries are impenetrable to the fluid.

Keeping in mind these interpretations, and invoking conservation of heat or mass shows that, when the domain is enclosed completely by a closed surface S_B, a Green's function of the second kind cannot be found: Heat or mass supplied by the point source or withdrawn by the point sink can neither accumulate nor escape through the boundaries.

Free-Space Green's Function

The free-space Green's function corresponds to an infinite domain of flow with no interior boundaries. Solving Eq. (2.2.6) by the method of Fourier transforms, or simply applying the first equation in (2.1.24) with $m = -1$, we obtain

$$G(\mathbf{x}, \mathbf{x}_0) = \frac{1}{4\pi r} \qquad (2.2.9)$$

where $r = |\mathbf{x} - \mathbf{x}_0|$.

Green's Functions in Bounded Domains

As the observation point $\mathbf{x}$ approaches the pole $\mathbf{x}_0$, all Green's functions exhibit a common singular behavior. Specifically, a Green's function is composed of a singular part that is identical to the free-space Green's function given in Eq. (2.2.9), and a complementary part expressed by a harmonic function H that is nonsingular throughout and on the boundaries of the flow, so that

$$G(\mathbf{x}, \mathbf{x}_0) = \frac{1}{4\pi r} + H(\mathbf{x}, \mathbf{x}_0) \qquad (2.2.10)$$

In the case of the free-space Green's function, H vanishes. More generally, the precise form of H depends upon the geometry of the boundary S_B.

There is a limited class of boundary geometries for which the complementary function H may be found by the method of images, that is, by introducing free-space Green's functions and their derivatives at strategically selected locations outside the domain of flow. For instance, for semi-infinite flow bounded by a plane wall located at $x = w$, we obtain

$$G(\mathbf{x}, \mathbf{x}_0) = \frac{1}{4\pi} \left(\frac{1}{r} \pm \frac{1}{R} \right) \tag{2.2.11}$$

where $r = |\mathbf{x} - \mathbf{x}_0|$, $R = |\mathbf{x} - \mathbf{x}_0^{IM}|$, $\mathbf{x}_0^{IM} = (2w - x_0, y_0, z_0)$ is the image of $\mathbf{x}_0$ with respect to the wall, and the minus and plus signs correspond, respectively, to the Green's function of the first or second kind.

The Green's function of the first kind for a flow that is bounded internally or externally by a spherical surface of radius a centered at the point $\mathbf{x}_c$ is given by

$$G(\mathbf{x}, \mathbf{x}_0) = \frac{1}{4\pi} \left(\frac{1}{r} - \frac{a}{|\mathbf{x}_0 - \mathbf{x}_c|} \frac{1}{R} \right) \tag{2.2.12}$$

where $r = |\mathbf{x} - \mathbf{x}_0|$, $R = |\mathbf{x} - \mathbf{x}_0^{IM}|$, and $\mathbf{x}_0^{IM}$ is the inverse point of $\mathbf{x}_0$ with respect to the sphere located at $\mathbf{x}_0^{IM} = \mathbf{x}_c + (\mathbf{x}_0 - \mathbf{x}_c)a^2/|\mathbf{x}_0 - \mathbf{x}_c|^2$. The corresponding Green's function of the second kind, representing the flow due to a point sink of mass in the presence of an interior impermeable spherical boundary, will be derived in Section 7.4. A Green's function of the second kind for the flow in the interior of a spherical boundary does not exist.

Integral Properties of Green's Functions

Let us consider a singly or multiply connected control volume V_c of finite extent that is bounded by one or more closed surfaces collectively denoted by D, as illustrated in Figure 2.2.1(a). The boundary S_B associated with the Green's function may be one of these surfaces. Integrating Eq. (2.2.6) over V_c and using the divergence theorem and the properties of the delta function, we find that G satisfies the integral constraint

$$\int_D \nabla G(\mathbf{x}, \mathbf{x}_0) \cdot \mathbf{n}(\mathbf{x}) \, dS(\mathbf{x}) = \begin{cases} 1, & \text{when } \mathbf{x}_0 \text{ is within } V_c \\ \frac{1}{2}, & \text{when } \mathbf{x}_0 \text{ is on } D \\ 0, & \text{when } \mathbf{x}_0 \text{ is outside } V_c \end{cases} \tag{2.2.13}$$

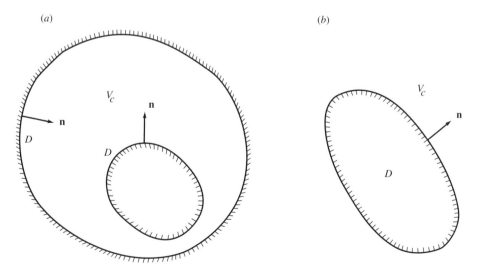

(a)   (b)

Figure 2.2.1   Schematic illustration of (a) a finite control volume in a flow bounded by an exterior and an interior closed surface, (b) an infinite control volume bounded by an interior closed surface.

where the normal vector $\mathbf{n}$ is directed into V_c. When $\mathbf{x}_0$ is located on D, the integral on the left-hand side of Eq. (2.2.13) is an improper but convergent integral called a *principal-value integral.* Using the three relations in (2.2.13), we find

$$\int_{D^\pm} \nabla G(\mathbf{x}, \mathbf{x}_0) \cdot \mathbf{n}(\mathbf{x}) \, dS(\mathbf{x}) = \int_D^{PV} \nabla G(\mathbf{x}, \mathbf{x}_0) \cdot \mathbf{n}(\mathbf{x}) \, dS(\mathbf{x}) \pm \tfrac{1}{2} \tag{2.2.14}$$

where PV denotes the principal-value integral, and the plus or minus sign on the left-hand side apply, respectively, when the point $\mathbf{x}_0$ lies inside or outside the control volume.

Symmetry of Green's Functions and Its Consequences

Let us return to Green's second identity (2.2.3) and identify both f and g with a certain Green's function G with distinct poles placed, respectively, at the points $\mathbf{x}_1$ and $\mathbf{x}_2$. Using the definition (2.2.6), we obtain

$$\begin{aligned} &- G(\mathbf{x}, \mathbf{x}_1) \delta(\mathbf{x} - \mathbf{x}_2) + G(\mathbf{x}, \mathbf{x}_2) \delta(\mathbf{x} - \mathbf{x}_1) \\ &= \nabla \cdot [G(\mathbf{x}, \mathbf{x}_1) \nabla G(\mathbf{x}, \mathbf{x}_2) - G(\mathbf{x}, \mathbf{x}_2) \nabla G(\mathbf{x}, \mathbf{x}_1)] \end{aligned} \tag{2.2.15}$$

Integrating Eq. (2.2.15) over a control volume V_c that is bounded by the surface S_B, and in the case of infinite flow by a surface S_∞ of large dimensions, using the divergence theorem to convert the volume integral on the right-hand side to a surface integral, and making use of the distinguishing properties of the delta function to write, for example,

$$\int_{V_c} G(\mathbf{x}, \mathbf{x}_1) \delta(\mathbf{x} - \mathbf{x}_2) \, dV(\mathbf{x}) = G(\mathbf{x}_2, \mathbf{x}_1) \tag{2.2.16}$$

we find

$$\begin{aligned} G(\mathbf{x}_2, \mathbf{x}_1) - G(\mathbf{x}_1, \mathbf{x}_2) = \int_{S_B, S_\infty} [G(\mathbf{x}, \mathbf{x}_1) \nabla G(\mathbf{x}, \mathbf{x}_2) \cdot \mathbf{n}(\mathbf{x}) \\ - G(\mathbf{x}, \mathbf{x}_2) \nabla G(\mathbf{x}, \mathbf{x}_1) \cdot \mathbf{n}(\mathbf{x})] \, dS(\mathbf{x}) \end{aligned} \tag{2.2.17}$$

Since either the Green's function itself or its normal derivative vanishes over S_B, the integral over S_B on the right-hand side of Eq. (2.2.17) is equal to zero. If the domain of flow is infinite, we let the large surface expand to infinity and find that the integrals over S_∞ make vanishing contributions, for the integrand decays at a rate that is faster than quadratic. We thus find that a Green's function of the first or second kind must satisfy the symmetry property

$$G(\mathbf{x}_2, \mathbf{x}_1) = G(\mathbf{x}_1, \mathbf{x}_2) \tag{2.2.18}$$

which allows us to interchange the observation point with the pole.

In physical terms, Eq. (2.2.18) states that the temperature or velocity potential at the point $\mathbf{x}_2$ due to a point source of heat or mass located at the point $\mathbf{x}_1$ is equal to the temperature or velocity potential at the point $\mathbf{x}_1$ due to a point source located at $\mathbf{x}_2$. In Chapter 10 we shall see that the Green's functions serve as kernels in integral equations for the boundary distribution of the harmonic potential or its normal derivatives, and the symmetry property (2.2.18) has important implications on the properties of the solution.

One noteworthy consequence of Eq. (2.2.18) is that when the pole $\mathbf{x}_0$ of a Green's function of the *first* kind is placed on the boundary S_B where the Green's function is required to vanish, the Green's function must be equal to zero at every point in the domain of flow; that is, $G(\mathbf{x}, \mathbf{x}_0) = 0$ when $\mathbf{x}_0$ is on S_B for any $\mathbf{x}$. This behavior may be understood in physical terms by identifying the Green's function with the temperature field established when a point source of heat is placed on a body of constant temperature, and noting that the heat released by the point source will be instantaneously absorbed by the body without having a chance to generate a field.

Green's Functions with Multiple Poles

Adding N Green's functions with distinct poles, we obtain a Green's function with a multitude of poles,

$$G^{(N)}(\mathbf{x}, \mathbf{x}_0, \mathbf{x}_1, \mathbf{x}_2, \ldots, \mathbf{x}_N) = \sum_{n=1}^{N} G(\mathbf{x}, \mathbf{x}_n) \qquad (2.2.19)$$

Physically, this Green's function may be identified with the temperature field due to a collection of point sources of heat, or with the velocity potential due to a collection of point sinks of mass.

A periodic Green's function represents the temperature field or harmonic potential due to a simply, doubly, or triply periodic array of point sources of heat or point sinks of mass. In certain cases, a periodic Green's function may not be computed simply by adding an infinite number of Green's functions with single poles, for the sum will diverge, but instead, it must be found by solving the defining equation

$$\nabla^2 G^{\text{per}}(\mathbf{x}, \mathbf{x}_0) + \sum_{n=-\infty}^{\infty} \delta(\mathbf{x} - \mathbf{x}_n) = 0 \qquad (2.2.20)$$

where the sum is over the periodic array, using, for instance, expansions in Fourier series in the directions of periodicity.

Multipoles of Green's Functions

Differentiating a Green's function with respect to the pole, we obtain a vectorial singularity called the Green's function dipole, given by

$$\mathbf{G}^D = \nabla_0 G(\mathbf{x}, \mathbf{x}_0) \qquad (2.2.21)$$

The subscript 0 indicates differentiation with respect to $\mathbf{x}_0$. Physically, Eq. (2.2.21) provides us with the temperature field or harmonic potential due to a point-source dipole of heat or mass. Higher derivatives of Eq. (2.2.21) with respect to the pole yield tensorial singularities that are multipoles of the Green's function, the first three of which are the quadruple $\mathbf{G}^Q$, the octuple $\mathbf{G}^O$, and the sextuple $\mathbf{G}^S$.

The free-space Green's function dipole and quadruple are given by

$$\mathbf{G}^D(\mathbf{x}, \mathbf{x}_0) = \frac{1}{4\pi} \frac{\hat{\mathbf{x}}}{|\hat{\mathbf{x}}|^3}$$

$$\mathbf{G}^Q(\mathbf{x}, \mathbf{x}_0) = \nabla_0 \nabla_0 G(\mathbf{x}, \mathbf{x}_0) = \frac{1}{4\pi} \left(-\frac{\mathbf{I}}{|\hat{\mathbf{x}}|^3} + 3 \frac{\hat{\mathbf{x}}\hat{\mathbf{x}}}{|\hat{\mathbf{x}}|^5} \right) \qquad (2.2.22)$$

where $\hat{\mathbf{x}} = \mathbf{x} - \mathbf{x}_0$ and $\mathbf{I}$ is the identity matrix. Comparing the first equation in (2.2.22) with the first equation in (2.1.28) allows us to express the velocity potential due to a point-source dipole of strength $\mathbf{d}$ as

$$\phi^D = -\mathbf{G}^D \cdot \mathbf{d} \qquad (2.2.23)$$

The associated velocity field is given by $\mathbf{u}^D = \mathbf{G}^Q \cdot \mathbf{d}$. Working in a similar manner, we find that the velocity potential due to a point-source quadruple is given by

$$\phi^Q = -G_{ij}^Q q_{ij} \qquad (2.2.24)$$

where $\mathbf{q}$ is a constant matrix expressing the strength and spatial structure of the quadruple. The associated velocity field is given by $\mathbf{u}^D = \mathbf{G}^O \cdot \mathbf{d}$.

Two-Dimensional Green's Functions

The apparatus of harmonic functions may be extended in a straightforward manner to two-dimensional flows. The two-dimensional Green's functions satisfy Eq. (2.2.6), with δ being the two-dimensional delta function. The free-space Green's function is given by

$$G(\mathbf{x}, \mathbf{x}_0) = -\frac{1}{2\pi} \ln r \tag{2.2.25}$$

It is important to note that the free-space Green's function increases at a logarithmic rate, in contrast with its three-dimensional counterpart that decays like $1/r$.

Using the method of images, we find that the Green's function for a semi-infinite domain of flow that is bounded by a plane wall located at $y = w$ is given by

$$G(\mathbf{x}, \mathbf{x}_0) = -\frac{1}{2\pi}(\ln r \pm \ln R) \tag{2.2.26}$$

where $r = |\mathbf{x} - \mathbf{x}_0|$, $R = |\mathbf{x} - \mathbf{x}_0^{\mathrm{IM}}|$, $\mathbf{x}_0^{\mathrm{IM}} = (x_0, 2w - y_0)$ is the image of $\mathbf{x}_0$ with respect to the wall, and the minus or plus signs correspond, respectively, to the Green's function of the first or second kind.

The Green's function of the first kind for a flow that is bounded either internally or externally by a circular boundary of radius a centered at the point $\mathbf{x}_c$ is given by

$$G(\mathbf{x}, \mathbf{x}_0) = -\frac{1}{2\pi} \left[\ln r - \ln \left(\frac{a}{|\mathbf{x}_0 - \mathbf{x}_c|} \frac{1}{R} \right) \right] \tag{2.2.27}$$

where $r = |\mathbf{x} - \mathbf{x}_0|$, $R = |\mathbf{x} - \mathbf{x}_0^{\mathrm{IM}}|$, and $\mathbf{x}_0^{\mathrm{IM}}$ is the inverse point of $\mathbf{x}_0$ with respect to the circle located at $\mathbf{x}_0^{\mathrm{IM}} = \mathbf{x}_c + (\mathbf{x}_0 - \mathbf{x}_c) a^2/|\mathbf{x}_0 - \mathbf{x}_c|^2$. The Green's function of the second kind, representing the flow due to a point sink of mass in the presence of an interior circular boundary, will be derived in Section 7.8. The Green's function of the second kind for flow in the interior of a circular boundary cannot be found.

The reciprocal relation and identities (2.2.13) and (2.2.14) are also valid for two-dimensional flow, provided that the control volume V_c is replaced by the control area A_c, and the boundary D is replaced by the contour C enclosing A. Furthermore, the two-dimensional Green's functions satisfy the symmetry property (2.2.18); the proof is carried out as for three-dimensional flow. Note, however, that for infinite two-dimensional flow, we encounter an apparent complication due to the fact that the Green's function may increase at a logarithmic rate, and, therefore, the integrals of the two terms in Eq. (2.2.17) over the large contour C_∞ that is the counterpart of S_∞ may not vanish in the limit as C_∞ tends to infinity. Expanding, however, the Green's functions within the integrand of Eq. (2.2.17) in a Taylor series with respect to $\mathbf{x}_0$ about the origin, we find that the sum of the two integrals makes a vanishing contribution, and the combined integral over C_∞ may be overlooked.

PROBLEMS

2.2.1   **Free-space Green's function.** (a) Differentiating in index notation, show explicitly that the free-space Green's function satisfies Laplace's equation everywhere except at the pole. (b) Identify D with a spherical surface that is centered at the pole $\mathbf{x}_0$, and show that the free-space Green's function fulfills the first component of the integral constraint (2.2.13).

2.2.2   **Solution of Poisson's equation.** Using the distinguishing properties of the delta function, show that

$$\phi(\mathbf{x}) = -\int_{\mathrm{Flow}} G(\mathbf{x}, \mathbf{x}') \nabla' \cdot \mathbf{u}(\mathbf{x}') \, dV(\mathbf{x}') + H(\mathbf{x}) \tag{2.2.28}$$

provides us with the general solution of Poisson's Equation (2.1.10), where G is a Green's function and H is a nonsingular harmonic function.

2.2.3   Symmetry of Green's functions. (a) Verify that the Green's functions given in Eqs. (2.2.11) and (2.2.12) satisfy the symmetry property (2.2.18). (b) Discuss whether Eq. (2.2.18) implies

$$\nabla G(\mathbf{x}, \mathbf{x}_0) = \nabla G(\mathbf{x}_0, \mathbf{x}) \tag{2.2.29}$$

or

$$\nabla G(\mathbf{x}, \mathbf{x}_0) = \nabla_0 G(\mathbf{x}, \mathbf{x}_0) \tag{2.2.30}$$

where the gradient ∇_0 involves derivatives with respect to $\mathbf{x}_0$.

2.2.4   Green's function octuple. (a) Derive the three-dimensional free-space Green's function octuple and discuss its physical interpretation in terms of point sources and sinks. (b) Repeat for the two-dimensional octuple.

2.3 | INTEGRAL REPRESENTATION AND FURTHER PROPERTIES OF POTENTIAL FLOW

Having introduced the reciprocal theorem and the Green's functions of Laplace's equation, we proceed to develop integral representations for the velocity potential of an irrotational flow of an incompressible or compressible fluid.

Three-Dimensional Incompressible Flow

We begin by considering a three-dimensional incompressible flow in a singly connected domain. Applying the reciprocal relation (2.2.4) with a certain nonsingular single-valued harmonic potential ϕ in place of f and a certain Green's function $G(\mathbf{x}, \mathbf{x}_0)$ in place of g, we obtain

$$\nabla \cdot [\phi(\mathbf{x}) \nabla G(\mathbf{x}, \mathbf{x}_0) - G(\mathbf{x}, \mathbf{x}_0) \nabla \phi(\mathbf{x})] = 0 \tag{2.3.1}$$

Next we select a control volume V_c that is bounded by the collection of surfaces D as illustrated in Figure 2.3.1. When the pole of the Green's function $\mathbf{x}_0$ is placed outside V_c, the left-hand side of Eq. (2.3.1) is nonsingular throughout V_c. Repeating the procedure that led us from Eq. (2.2.15)

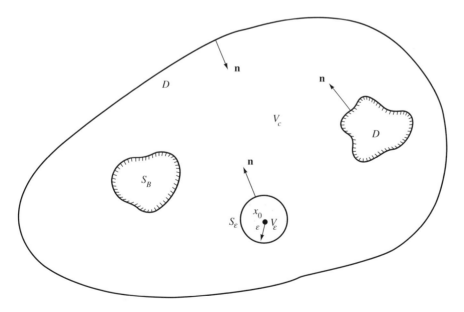

Figure 2.3.1   Control volume in a flow used to derive the boundary integral representation.

to Eq. (2.2.17), we obtain

$$\int_D \phi(\mathbf{x}) \nabla G(\mathbf{x}, \mathbf{x}_0) \cdot \mathbf{n}(\mathbf{x}) \, dS(\mathbf{x}) = \int_D G(\mathbf{x}, \mathbf{x}_0) \nabla \phi(\mathbf{x}) \cdot \mathbf{n}(\mathbf{x}) \, dS(\mathbf{x}) \tag{2.3.2}$$

When the pole $\mathbf{x}_0$ is placed within V_c, the left-hand side of Eq. (2.3.1) is nonsingular at every point within V_c except at $\mathbf{x}_0$. In order to apply the divergence theorem, we exclude from the control volume a small spherical volume of radius ε centered at $\mathbf{x}_0$, as shown in Figure 2.3.1. The result is an equation that is identical to Eq. (2.3.2), except that the boundaries are enhanced with the spherical surface S_ε,

$$\int_{D, S_\varepsilon} \phi(\mathbf{x}) \nabla G(\mathbf{x}, \mathbf{x}_0) \cdot \mathbf{n}(\mathbf{x}) \, dS(\mathbf{x}) = \int_{D, S_\varepsilon} G(\mathbf{x}, \mathbf{x}_0) \nabla \phi(\mathbf{x}) \cdot \mathbf{n}(\mathbf{x}) \, dS(\mathbf{x}) \tag{2.3.3}$$

Considering the integrals over S_ε, we write $dS = \varepsilon^2 \, d\Omega$, where $d\Omega$ is the differential solid angle, that is, the differential area of a sphere of unit radius, and use Eq. (2.2.10) to obtain

$$\nabla G(\mathbf{x}, \mathbf{x}_0) = -\frac{1}{4\pi\varepsilon^2} \mathbf{n} + \nabla H(\mathbf{x}, \mathbf{x}_0) \tag{2.3.4}$$

where the unit normal vector $\mathbf{n} = (\mathbf{x} - \mathbf{x}_0)/\varepsilon$ points into the control volume as shown in Figure 2.3.1. Taking the limit as ε tends to zero, using Eqs. (2.2.10) and (2.3.4), and remembering that the complementary component H is nonsingular, we find

$$\int_{S_\varepsilon} G(\mathbf{x}, \mathbf{x}_0) \nabla \phi(\mathbf{x}) \cdot \mathbf{n}(\mathbf{x}) \, dS(\mathbf{x}) = \int_{S_\varepsilon} \frac{1}{4\pi\varepsilon} \nabla \phi(\mathbf{x}) \cdot \mathbf{n}(\mathbf{x}) \, \varepsilon^2 \, d\Omega \to 0 \tag{2.3.5}$$

$$\int_{S_\varepsilon} \phi(\mathbf{x}) \nabla G(\mathbf{x}, \mathbf{x}_0) \cdot \mathbf{n}(\mathbf{x}) \, dS(\mathbf{x}) \to -\phi(\mathbf{x}_0) \int_{S_\varepsilon} \frac{1}{4\pi\varepsilon^2} \varepsilon^2 \, d\Omega \to -\phi(\mathbf{x}_0) \tag{2.3.6}$$

Substituting Eqs. (2.3.5) and (2.3.6) into Eq. (2.3.3) yields the final result

$$\phi(\mathbf{x}_0) = -\int_D G(\mathbf{x}, \mathbf{x}_0) \nabla \phi(\mathbf{x}) \cdot \mathbf{n}(\mathbf{x}) \, dS(\mathbf{x}) + \int_D \nabla G(\mathbf{x}, \mathbf{x}_0) \cdot \mathbf{n}(\mathbf{x}) \phi(\mathbf{x}) \, dS(\mathbf{x}) \tag{2.3.7}$$

which provides us with a *boundary integral representation* of a harmonic function in terms of its boundary values and the boundary distribution of its normal derivative; the latter is equal to the normal component of the velocity. To compute the value of ϕ at a particular point $\mathbf{x}_0$ within the control volume, we simply evaluate the two boundary integrals on the right-hand side of Eq. (2.3.7).

The symmetry property (2.2.18) allows us to change the order of the arguments of the Green's function in Eq. (2.3.7), obtaining

$$\phi(\mathbf{x}_0) = -\int_D G(\mathbf{x}_0, \mathbf{x}) \nabla \phi(\mathbf{x}) \cdot \mathbf{n}(\mathbf{x}) \, dS(\mathbf{x}) + \int_D \nabla G(\mathbf{x}_0, \mathbf{x}) \cdot \mathbf{n}(\mathbf{x}) \phi(\mathbf{x}) \, dS(\mathbf{x}) \tag{2.3.8}$$

The two integrals on the right-hand side represent boundary distributions of the *Green's function* and of the *Green's function dipole* oriented *perpendicular* to the boundaries of the control volume, amounting to boundary distributions of point sinks and point-source dipoles. By analogy with corresponding results in the theory of electrostatics, concerning distributions of electric charges and charge dipoles, we coin the two integrals in Eq. (2.3.8) the *single-layer* and *double-layer potential*. The densities of these potentials are equal, respectively, to the outward normal derivative and boundary value of the harmonic potential.

It is worth noting that the representation (2.3.7) may be derived in a more direct manner using the properties of the delta function, in a procedure that is similar to the one that led us from

Eq. (2.2.15) to Eq. (2.2.17). The preceding derivation, however, is more comforting in the sense that it bypasses the use of generalized functions.

Green's Third Identity

Applying the boundary integral representation given in Eq. (2.3.8) with the free-space Green's function given in Eq. (2.2.9), we obtain *Green's third identity*

$$\phi(\mathbf{x}_0) = -\frac{1}{4\pi} \int_D \frac{1}{r} \nabla\phi(\mathbf{x}) \cdot \mathbf{n}(\mathbf{x}) \, dS(\mathbf{x}) + \frac{1}{4\pi} \int_D \frac{1}{r^3} (\mathbf{x}_0 - \mathbf{x}) \cdot \mathbf{n}(\mathbf{x}) \, \phi(\mathbf{x}) \, dS(\mathbf{x}) \qquad (2.3.9)$$

where $r = |\mathbf{x} - \mathbf{x}_0|$.

Mean-Value Theorems for Harmonic Functions

An important property of harmonic functions emerges by selecting a spherical control volume of radius a centered at a point $\mathbf{x}_0$ that resides entirely within the domain of a flow. Identifying D with the spherical boundary, applying Eq. (2.3.9), and noting that $r = a$ and $\mathbf{n} = (\mathbf{x}_0 - \mathbf{x})/a$, we obtain

$$\phi(\mathbf{x}_0) = -\frac{1}{4\pi a} \int_{\text{Sphere}} \nabla\phi(\mathbf{x}) \cdot \mathbf{n}(\mathbf{x}) \, dS(\mathbf{x}) + \frac{1}{4\pi a^2} \int_{\text{Sphere}} \phi(\mathbf{x}) \, dS(\mathbf{x}) \qquad (2.3.10)$$

Using the divergence theorem and the fact that ϕ is a harmonic function, we find that the first integral on the right-hand side of Eq. (2.3.10) vanishes, and thereby obtain

$$\phi(\mathbf{x}_0) = \frac{1}{4\pi a^2} \int_{\text{Sphere}} \phi(\mathbf{x}) \, dS(\mathbf{x}) \qquad (2.3.11)$$

which states that the mean value of a harmonic function over the *surface* of a sphere is equal to the value of the function at the center of the sphere.

One interesting consequence of the mean-value theorem is that if the harmonic potential of an infinite flow with no interior boundaries vanishes at infinity, it must vanish throughout the whole space.

Now, using Eq. (2.3.11), we find that the mean value of a harmonic function over the *volume* of a sphere is given by

$$\frac{3}{4\pi a^3} \int_{\text{Sphere}} \phi(\mathbf{x}) \, dV(\mathbf{x}) = \frac{3}{4\pi a^3} \int_0^a \int_{\substack{\text{Sphere} \\ \text{of radius } r}} \phi(\mathbf{x}) \, dS(\mathbf{x}) \, dr$$

$$= \frac{3}{4\pi a^3} \int_0^a \phi(\mathbf{x}_0) \, 4\pi r^2 \, dr = \phi(\mathbf{x}_0) \qquad (2.3.12)$$

Thus, the mean value of a harmonic function over the *volume* of a sphere is also equal to the value of the function at the center of the sphere.

Extrema of Harmonic Functions

The mean-value theorem may be used to show that minima or maxima of nonsingular harmonic functions may occur only at the boundaries of the flow. To see this, assume temporarily that an extremum is located at a certain point $\mathbf{x}_0$ within the domain of a flow. Applying the mean-value theorem we find that there must be at least one point on the surface of a sphere that is centered at the alleged point of extremum where the value of ϕ is higher or lower than $\phi(\mathbf{x}_0)$, so that the mean value of ϕ is equal to $\phi(\mathbf{x}_0)$, but this contradicts the original assumption.

Consider now a domain that is bounded by a closed surface over which ϕ is held at the constant value c. According to this discussion, c must be both the minimum and maximum of ϕ, and this requires that ϕ be equal to c at every point in its domain of definition.

Maximum of the Magnitude of the Velocity

Another consequence of the mean-value theorem is that, in the absence of singularities, the velocity may attain maximum magnitude only at the boundaries of the flow. To see this, let us assume temporarily that the maximum occurs at a point $\mathbf{x}_0$ that is located within the domain of a flow, and introduce the unit vector in the direction of the local velocity $\mathbf{e}_0 = \mathbf{u}(\mathbf{x}_0)/|\mathbf{u}(\mathbf{x}_0)|$. Geometrical arguments require that at any point $\mathbf{x}$ in the domain of flow, $\mathbf{u}(\mathbf{x}) \cdot \mathbf{u}(\mathbf{x}) > |\mathbf{u}(\mathbf{x}) \cdot \mathbf{e}_0|^2$. Noting, on the other hand, that $\mathbf{u}(\mathbf{x}) \cdot \mathbf{e}_0$ is a harmonic function suggests that we can find a point $\mathbf{x}$ near $\mathbf{x}_0$ where $|\mathbf{u}(\mathbf{x}) \cdot \mathbf{e}_0|^2 > |\mathbf{u}(\mathbf{x}_0) \cdot \mathbf{e}_0|^2 = \mathbf{u}(\mathbf{x}_0) \cdot \mathbf{u}(\mathbf{x}_0)$. Combining the two inequalities, we find $\mathbf{u}(\mathbf{x}) \cdot \mathbf{u}(\mathbf{x}) > \mathbf{u}(\mathbf{x}_0) \cdot \mathbf{u}(\mathbf{x}_0)$, which contradicts the original assumption.

One corollary of this result is that if the velocity of an infinite irrotational flow with no interior boundaries vanishes at infinity, it must vanish throughout the whole space.

Unbounded Flow Vanishing at Infinity

Consider now flow in an infinite domain that is enclosed by one or more closed interior boundaries, called a *periphractic* domain from the Greek words $\pi\varepsilon\rho\iota$, which means *about,* and $\varphi\rho\alpha\chi\tau\eta\varsigma$, which means *fence,* subject to the assumption that the velocity vanishes at infinity. An example is shown in Figure 2.2.1(b). We shall show that *the velocity potential at infinity must tend to a constant value.*

We begin by selecting a point $\mathbf{x}_0$ within the domain of flow, and define a control volume that is enclosed by the collection of the interior boundaries B and a spherical surface of large radius R centered at $\mathbf{x}_0$, denoted by S_∞. Applying Green's third identity (2.3.9), we obtain

$$\phi(\mathbf{x}_0) = \frac{1}{4\pi}\frac{Q}{R} + M(R, \mathbf{x}_0) - \frac{1}{4\pi}\int_B \frac{1}{r}\nabla\phi(\mathbf{x}) \cdot \mathbf{n}(\mathbf{x})\, dS(\mathbf{x})$$

$$+ \frac{1}{4\pi}\int_B \frac{(\mathbf{x}_0 - \mathbf{x}) \cdot \mathbf{n}(\mathbf{x})}{r^3}\phi(\mathbf{x})\, dS(\mathbf{x}) \tag{2.3.13}$$

where Q is the flow rate across S_∞ or any other closed surface enclosed by S_∞, defined as

$$Q = -\int_{S_\infty} \nabla\phi(\mathbf{x}) \cdot \mathbf{n}(\mathbf{x})\, dS(\mathbf{x}) = \int_B \nabla\phi(\mathbf{x}) \cdot \mathbf{n}(\mathbf{x})\, dS(\mathbf{x}) \tag{2.3.14}$$

and M is the average value of ϕ over S_∞,

$$M(R, \mathbf{x}_0) = \frac{1}{4\pi R^2}\int_{S_\infty} \phi(\mathbf{x})\, dS(\mathbf{x}) \tag{2.3.15}$$

Considering the flow rate, we write

$$Q = \int_{S_\infty} \frac{\partial\phi(\mathbf{x})}{\partial R}\, dS(\mathbf{x}) = \int_{S_\infty} \frac{\partial\phi(\mathbf{x})}{\partial R}R^2\, d\Omega(\mathbf{x})$$

$$= R^2\frac{\partial}{\partial R}\int_{S_\infty} \phi(\mathbf{x})\, d\Omega(\mathbf{x}) = 4\pi R^2\frac{dM(R)}{dR} \tag{2.3.16}$$

where $d\Omega$ is the differential solid angle, that is, the differential area of a sphere of unit radius centered at $\mathbf{x}_0$. Integrating Eq. (2.3.16) with respect to R and noting that Q is a constant, we obtain $M(R, \mathbf{x}_0) = -Q/(4\pi R) + c(\mathbf{x}_0)$, where c is independent of R. Taking the derivatives of the preceding equation with respect to $\mathbf{x}_0$, keeping R fixed, we obtain

$$\frac{dc}{dx_{0,i}} = \frac{\partial M}{\partial x_{0,i}} = \frac{\partial}{\partial x_{0,i}}\int_{S_\infty} \phi(\mathbf{x})\, d\Omega(\mathbf{x}) = \int_{S_\infty} \frac{\partial\phi(\mathbf{x})}{\partial x_i}\, d\Omega(\mathbf{x}) = 0 \tag{2.3.17}$$

where the value of zero emerges by letting R in the last integral tend to infinity and using the original assumption that the velocity decays at infinity. We have thus shown that c is an absolute constant. Substituting $M(R, \mathbf{x}_0) = -Q/(4\pi R) + c$ back into Eq. (2.3.13) yields a simplified boundary integral representation that neglects the integral over the large surface,

$$\phi(\mathbf{x}_0) = c - \frac{1}{4\pi} \int_B \frac{1}{r} \nabla\phi(\mathbf{x}) \cdot \mathbf{n}(\mathbf{x}) \, dS(\mathbf{x}) + \frac{1}{4\pi} \int_B \frac{(\mathbf{x}_0 - \mathbf{x}) \cdot \mathbf{n}(\mathbf{x})}{r^3} \phi(\mathbf{x}) \, dS(\mathbf{x})$$

(2.3.18)

Behavior of the Potential at Infinity and Kinetic Energy of an Infinite Flow

Letting the point $\mathbf{x}_0$ in Eq. (2.3.18) tend to infinity and noting that r is the distance from the point $\mathbf{x}_0$ to a point $\mathbf{x}$ that is located on the interior boundary B, we find the asymptotic behavior

$$\phi(\mathbf{x}_0) = c - \frac{Q}{4\pi R} + \cdots$$

(2.3.19)

where R is the distance from the origin. We thus find that the potential of an infinite three-dimensional flow that vanishes at infinity must tend to a constant value at large distances from the interior boundaries.

Furthermore, based on Eq. (2.3.19), we find that the expression for the kinetic energy (2.1.17) becomes

$$K = -\tfrac{1}{2}\rho \int_B \phi(\mathbf{x})\, \mathbf{u}(\mathbf{x}) \cdot \mathbf{n}(\mathbf{x})\, dS(\mathbf{x}) + \tfrac{1}{2}\rho c Q$$

(2.3.20)

Following the arguments of Section 2.1, we then deduce that, given the boundary distribution of the normal component of the velocity, a potential flow that vanishes at infinity is unique, and the corresponding harmonic potential is determined uniquely up to an arbitrary constant.

Simplified Boundary Integral Representations

The boundary integral representation (2.3.8) may be simplified by reducing the domain of integration of the hydrodynamic potentials. This may be accomplished by using a Green's function that is designed to observe the symmetry, periodicity, or other topological features of the flow. If, for instance, the Green's function or its normal derivative vanishes over a particular boundary, the corresponding single-layer or double-layer potential will be equal to zero and may be neglected. If the velocity and harmonic potential are periodic in one or more directions, we use a Green's function that follows the periodicity of the flow so that the integrals over the periodic boundaries enclosing one period of the flow cancel each other, and may thus be overlooked.

Compressible Flow

To derive an integral representation for a compressible potential flow, we apply the Green's second identity (2.2.3) with a generally *nonharmonic* potential ϕ in place of f and the Green's function G in place of g, and obtain

$$\nabla \cdot [\phi(\mathbf{x}) \nabla G(\mathbf{x}, \mathbf{x}_0) - G(\mathbf{x}, \mathbf{x}_0) \nabla\phi(\mathbf{x})] = -G(\mathbf{x}, \mathbf{x}_0) \nabla^2\phi(\mathbf{x})$$

(2.3.21)

Following the procedure discussed earlier in this section for incompressible flow, we derive the integral representation

$$\phi(\mathbf{x}_0) = -\int_D G(\mathbf{x}_0, \mathbf{x}) \nabla\phi(\mathbf{x}) \cdot \mathbf{n}(\mathbf{x}) \, dS(\mathbf{x}) + \int_D \nabla G(\mathbf{x}_0, \mathbf{x}) \cdot \mathbf{n}(\mathbf{x}) \phi(\mathbf{x}) \, dS(\mathbf{x})$$

$$- \int_{V_c} G(\mathbf{x}_0, \mathbf{x}) \nabla^2\phi(\mathbf{x}) \, dV(\mathbf{x})$$

(2.3.22)

which is identical to that shown in Eq. (2.3.8) except that the right-hand side is enhanced with a volume integral over the control volume V_c enclosed by the boundary D, called the *Poisson integral* or *volume potential*, representing a volume distribution of point sources.

Using the free-space Green's function, we obtain the counterpart of Green's third identity for compressible flow,

$$
\phi(\mathbf{x}_0) = -\frac{1}{4\pi} \int_D \frac{1}{r} \nabla\phi(\mathbf{x}) \cdot \mathbf{n}(\mathbf{x})\, dS(\mathbf{x}) + \frac{1}{4\pi} \int_D \frac{1}{r^3} (\mathbf{x}_0 - \mathbf{x}) \cdot \mathbf{n}(\mathbf{x})\, \phi(\mathbf{x})\, dS(\mathbf{x})
$$

$$
- \frac{1}{4\pi} \int_{V_c} \frac{1}{r} \nabla^2\phi(\mathbf{x})\, dV(\mathbf{x}) \tag{2.3.23}
$$

The unit normal vector $\mathbf{n}$ points *into* the control volume.

Mean-Value Theorem for Biharmonic Functions

As an application of Eq. (2.3.23), we develop a mean-value theorem for functions Φ that satisfy the biharmonic equation $\nabla^4\Phi = 0$. Working as in the case of harmonic functions, we consider a spherical control volume of radius a that is centered at the point $\mathbf{x}_0$ and resides entirely within the fluid, identify D with the spherical boundary, apply Eq. (2.3.23) with Φ in place of ϕ, and note that $r = a$ and $\mathbf{n} = (\mathbf{x}_0 - \mathbf{x})/a$ to obtain

$$
\Phi(\mathbf{x}_0) = -\frac{1}{4\pi a} \int_{\text{Sphere}} \nabla\Phi(\mathbf{x}) \cdot \mathbf{n}(\mathbf{x})\, dS(\mathbf{x}) + \frac{1}{4\pi a^2} \int_{\text{Sphere}} \Phi(\mathbf{x})\, dS(\mathbf{x})
$$

$$
- \frac{1}{4\pi} \int_{\text{Sphere}} \frac{1}{r} \nabla^2\Phi(\mathbf{x})\, dV(\mathbf{x}) \tag{2.3.24}
$$

Next we use the divergence theorem to set the first surface integral on the right-hand side equal to the negative of the integral of $\nabla^2\Phi$ over the volume of the sphere, note that $\nabla^2\Phi$ is a harmonic function, and use the mean-value theorem (2.3.12) to replace its volume integral over the sphere with the value $(4\pi a^3/3)\,\nabla^2\Phi(\mathbf{x}_0)$. Rearranging the last volume integral in Eq. (2.3.24), we obtain

$$
\Phi(\mathbf{x}_0) = \tfrac{1}{3}a^2\,\nabla^2\Phi(\mathbf{x}_0) + \frac{1}{4\pi a^2} \int_{\text{Sphere}} \Phi(\mathbf{x})\, dS(\mathbf{x})
$$

$$
- \frac{1}{4\pi} \int_0^a \frac{1}{r} \int_{\substack{\text{Sphere} \\ \text{of radius } r}} \nabla^2\phi(\mathbf{x})\, dS(\mathbf{x})\, dr \tag{2.3.25}
$$

Finally, we use the mean-value theorem expressed by Eq. (2.3.11) for the harmonic function $\nabla^2\Phi$ to simplify the last integral in Eq. (2.3.25), and rearrange to obtain the mathematical statement of the mean-value theorem

$$
\frac{1}{4\pi a^2} \int_{\text{Sphere}} \Phi(\mathbf{x})\, dS(\mathbf{x}) = \Phi(\mathbf{x}_0) + \tfrac{1}{6}a^2\,\nabla^2\phi(\mathbf{x}_0) \tag{2.3.26}
$$

which relates the mean value of a biharmonic function over the surface of a sphere to the value of the function and its Laplacian at the center of the sphere.

Working as in Eq. (2.3.12), we find that the mean value of a biharmonic function over the *volume* of a sphere is given by

$$
\frac{3}{4\pi a^3} \int_{\text{Sphere}} \Phi(\mathbf{x})\, dV(\mathbf{x}) = \Phi(\mathbf{x}_0) + \tfrac{1}{10}a^2\,\nabla^2\Phi(\mathbf{x}_0) \tag{2.3.27}
$$

Note that if Φ happens to be a harmonic function, which may be considered to be a special case of a biharmonic function, Eqs. (2.3.26) and (2.3.27) reduce to Eqs. (2.3.11) and (2.3.12).

Two-Dimensional Flow

A boundary integral representation of the single-valued harmonic potential of a two-dimensional flow may be derived working in a manner that is completely analogous to that described previously in this section for three-dimensional flow. Equations (2.3.7) and (2.3.8) remain valid provided the control volume is replaced by a control area, and the boundary D is replaced by the contour C that encloses the control area. *Green's third identity* for a single-valued potential takes the form

$$\phi(\mathbf{x}_0) = \frac{1}{2\pi} \int_C \ln r \, \nabla \phi(\mathbf{x}) \cdot \mathbf{n}(\mathbf{x}) \, dl(\mathbf{x}) + \frac{1}{2\pi} \int_C \frac{1}{r^2} (\mathbf{x}_0 - \mathbf{x}) \cdot \mathbf{n}(\mathbf{x}) \, dl(\mathbf{x}) \quad (2.3.28)$$

The mean-value theorem states that the mean value of a harmonic potential over the arc length of a circle or area of a circular disk are equal to the value of the potential at the center of the circle or disk.

Infinite flow

Considering a flow in a domain of infinite expanse, where the velocity vanishes at infinity, we derive the counterpart of Eq. (2.3.18) with a straightforward change in notation. Letting the point $\mathbf{x}_0$ tend to infinity we find that the velocity potential behaves like

$$\phi(\mathbf{x}_0) = \frac{Q}{2\pi} \ln R + c + \cdots \quad (2.3.29)$$

where R is the distance from the origin, which is assumed to be in the vicinity of the boundary C, and

$$Q = \int_C \nabla \phi(\mathbf{x}) \cdot \mathbf{n}(\mathbf{x}) \, dl(\mathbf{x}) \quad (2.3.30)$$

is the flow rate across the union of the interior boundaries (Problem 2.3.3).

Considering the kinetic energy of the modified flow expressed by the potential $\phi - (Q/2\pi) \ln R$ and repeating the procedure discussed previously for three-dimensional flow, we find that an infinite two-dimensional flow described by a single-valued harmonic potential is determined uniquely by specifying the normal component of the velocity over the interior boundaries.

Compressible flow

The integral representation for compressible flow (2.3.22) remains effective subject to the aforementioned changes in notation.

Working as in the case of three-dimensional flow, we derive mean-value theorems for functions that satisfy the biharmonic equation, expressed by the equations

$$\frac{1}{2\pi a} \int_{\text{Circle}} \Phi(\mathbf{x}) \, dl(\mathbf{x}) = \Phi(\mathbf{x}_0) + \tfrac{1}{4} a^2 \, \nabla^2 \Phi(\mathbf{x}_0) \quad (2.3.31)$$

$$\frac{1}{\pi a^2} \int_{\text{Disc}} \Phi(\mathbf{x}) \, dA(\mathbf{x}) = \Phi(\mathbf{x}_0) + \tfrac{1}{8} a^2 \, \nabla^2 \Phi(\mathbf{x}_0) \quad (2.3.32)$$

where the integration is over the arc length of a circle or area of a disk of radius a centered at the point $\mathbf{x}_0$ (Problem 2.3.5).

PROBLEMS

2.3.1 **The boundary integral equation for a uniform potential.** Apply Eqs. (2.3.2) and (2.3.7) with ϕ equal to a constant to obtain the first and third equations in (2.2.13).

2.3.2 **Boundary integral representation for the velocity.** Derive the following boundary integral representation for the velocity

$$\mathbf{u}(\mathbf{x}_0) = -\frac{1}{4\pi} \int_D \nabla_0 G(\mathbf{x}_0, \mathbf{x}) \, \mathbf{u}(\mathbf{x}) \cdot \mathbf{n}(\mathbf{x}) \, dS(\mathbf{x}) + \int_D [\nabla_0 \nabla G(\mathbf{x}_0, \mathbf{x})] \cdot \mathbf{n}(\mathbf{x}) \, \phi(\mathbf{x}) \, dS(\mathbf{x}) \quad (2.3.33)$$

Note that the integrals on the right-hand side represent boundary distributions of Green's functions and Green's function dipoles.

2.3.3 **Two-dimensional infinite flow.** Derive the asymptotic expression (2.3.29).

2.3.4 **Poisson's integrals.** (a) Using the Green's function for three-dimensional flow bounded by a sphere given in Eq. (2.2.12), show that the harmonic potential at a point $\mathbf{x}_0$ that is located in the interior or exterior of a sphere of radius a centered at the origin is given by *Poisson's integral*

$$\phi(\mathbf{x}_0) = \pm \frac{a^2 - |\mathbf{x}_0|^2}{4\pi a} \int_{\text{Sphere}} \frac{\phi(\mathbf{x})}{|\mathbf{x}_0 - \mathbf{x}|^3} \, dS(\mathbf{x}) \qquad (2.3.34)$$

The plus and minus signs correspond to the interior and exterior problems, respectively (Kellogg, 1953, p. 240). In the latter case, ϕ is assumed to vanish at infinity. (b) Using the Green's function for two-dimensional flow bounded by a circle given in Eq. (2.2.27), show that the harmonic potential at a point $\mathbf{x}_0$ that is located in the interior or exterior of a circle of radius a centered at the origin is given by *Poisson's integral*

$$\phi(\mathbf{x}_0) = \pm \frac{a^2 - |\mathbf{x}_0|^2}{2\pi} \int_{\text{Circle}} \frac{\phi(\mathbf{x})}{|\mathbf{x}_0 - \mathbf{x}|^2} \, d\theta \qquad (2.3.35)$$

The plus and minus signs correspond to the interior and exterior problems, respectively (Dettman, 1965, p. 241). In the latter case, ϕ is assumed to vanish at infinity.

2.3.5 **Mean-value theorems for biharmonic functions in two dimensions.** Prove the mean-value theorems expressed by Eqs. (2.3.31) and (2.3.32).

2.4 | THE VECTOR POTENTIAL FOR INCOMPRESSIBLE FLOW

In Section 1.3 we saw that a velocity field that arises by taking the curl of an arbitrary twice differentiable vector field is solenoidal. We shall demonstrate now that the inverse is also true; that is, given a solenoidal velocity field $\mathbf{u}$, it is always possible to find a vector potential $\mathbf{A}$ so that

$$\mathbf{u} = \nabla \times \mathbf{A} \qquad (2.4.1)$$

We begin by stipulating that the first component of $\mathbf{A}$, denoted by A_1, is a function of x_1 alone, thus setting

$$A_1 = f_1(x_1) \qquad (2.4.2)$$

where f_1 is an arbitrary function. Integrating the second and third scalar components of Eq. (2.4.1) with respect to x_1 from the arbitrary position $x_1 = a$ yields

$$A_2(\mathbf{x}) = \int_a^{x_1} u_3(\mathbf{x}') \, dx_1' + f_2(x_2, x_3)$$

$$A_3(\mathbf{x}) = -\int_a^{x_1} u_2(\mathbf{x}') \, dx_1' + f_3(x_2, x_3) \qquad (2.4.3)$$

where f_2, f_3 are two arbitrary but, as we shall see, somewhat related functions of x_2 and x_3. Substituting Eqs. (2.4.3) into the first component of Eq. (2.4.1) we find

$$u_1(\mathbf{x}) = -\int_a^{x_1} \left(\frac{\partial u_2}{\partial x_2} + \frac{\partial u_3}{\partial x_3} \right)(\mathbf{x}') \, dx_1' + \frac{\partial f_3}{\partial x_2}(x_2, x_3) - \frac{\partial f_2}{\partial x_3}(x_2, x_3) \qquad (2.4.4)$$

Furthermore, using the fact that $\mathbf{u}$ is solenoidal, we simplify the integrand in Eq. (2.4.4), obtaining

$$u_1(\mathbf{x}) = \int_a^{x_1} \frac{\partial u_1}{\partial x_1}(\mathbf{x}')\,dx'_1 + \frac{\partial f_3}{\partial x_2}(x_2, x_3) - \frac{\partial f_2}{\partial x_3}(x_2, x_3) \qquad (2.4.5)$$

which requires that

$$\frac{\partial f_3}{\partial x_2}(x_2, x_3) - \frac{\partial f_2}{\partial x_3}(x_2, x_3) = u_1(a, x_2, x_3) \qquad (2.4.6)$$

Equation (2.4.6) imposes a differential constraint upon the form of the otherwise arbitrary functions f_2 and f_3. We have thus constructed the components of a vector potential $\mathbf{A}$ explicitly in terms of the components of the solenoidal velocity field $\mathbf{u}$, via Eqs. (2.4.2) and (2.4.3), and the proof of the existence of $\mathbf{A}$ is complete.

As a secondary point, we note that the stipulation (2.4.2) and the restriction (2.4.6) may be used to derive the vectorial form

$$\nabla \times \mathbf{f} = (u_1(a, x_2, x_3), 0, 0) \qquad (2.4.7)$$

The vector potential corresponding to a particular flow of an incompressible fluid is not unique. For instance, since the curl of the gradient of any twice differentiable scalar function vanishes, any vector potential may be enhanced with the gradient of an arbitrary such function with no consequences on the velocity field. This observation allows us to assert that *it is always possible to find a solenoidal vector potential* $\mathbf{B}$, with $\nabla \cdot \mathbf{B} = 0$. Indeed, if $\mathbf{A}$ is a certain non-solenoidal potential, then $\mathbf{A} = \mathbf{B} - \nabla F$ will be a solenoidal potential, provided that F is a solution of Poisson's equation $\nabla^2 F = \nabla \cdot \mathbf{B}$.

To this end, the question of what is the minimum number of independent scalar functions that are necessary in order to describe the vector potential of a certain incompressible flow arises in a natural manner. The answer will be given in Section 2.7, where we shall show that the vector potential of a two-dimensional or axisymmetric flow may be described in terms of just one scalar function, called, respectively, the *Helmholtz two-dimensional stream function* and the *Stokes stream function,* whereas the vector potential of a three-dimensional flow may be described in terms of two scalar functions, called the stream functions. The stream function of a two-dimensional or axisymmetric flow is defined uniquely up to an arbitrary scalar constant.

Specifically, in Section 2.7 we shall show that it is always possible to find a vector potential of a three-dimensional flow that takes the form

$$\mathbf{A} = \psi\,\nabla\chi \qquad (2.4.8)$$

where ψ and χ are the stream functions. It is interesting to note that $\nabla \times \mathbf{A} = \nabla\psi \times \nabla\chi$, which suggests that $\mathbf{A} \cdot (\nabla \times \mathbf{A}) = 0$; that is, $\mathbf{A}$ is perpendicular to its curl. A vector field that possesses this property is called a *complex lamellar* field (Aris, 1962, p. 63). In contrast, a vector field that is parallel to its own curl, $\mathbf{A} \times (\nabla \times \mathbf{A}) = 0$, is called a *Beltrami field.*

PROBLEMS

2.4.1 **A vector potential in explicit form.** Show that two acceptable vector potentials for the velocity field $\mathbf{u} = \nabla\psi \times \nabla\chi$ are given by $\mathbf{A} = \psi\,\nabla\chi$ and $\mathbf{A} = -\chi\,\nabla\psi$, where ψ and χ are two arbitrary functions.

2.4.2 **Deriving a vector potential.** Repeat the derivation of the vector potential discussed in the text, but this time assume that $A_1 = f_1(x_2)$ in place of Eq. (2.4.2).

2.4.3 **A property of the vector potential.** Show that the velocity and the vector potential satisfy the symmetry property $\mathbf{u} \cdot \nabla\mathbf{A} = (\nabla\mathbf{A}) \cdot \mathbf{u}$. *Hint:* Begin by writing $\mathbf{u} \times \mathbf{u} = (\nabla \times \mathbf{A}) \times \mathbf{u} = 0$.

2.5 | REPRESENTATION OF AN INCOMPRESSIBLE FLOW IN TERMS OF THE VORTICITY

Continuing the study of the flow of incompressible fluids, we turn to examine the relationship between the vorticity distribution $\boldsymbol{\omega}$ and structure of the velocity field $\mathbf{u}$. Specifically, we seek to derive a representation for the velocity field in terms of the associated vorticity distribution by inverting the equation defining the vorticity, $\boldsymbol{\omega} = \nabla \times \mathbf{u}$. One way of carrying out this inversion is to use Eqs. (2.4.1)–(2.4.3), identifying $\mathbf{u}$ with $\boldsymbol{\omega}$ and $\mathbf{A}$ with $\mathbf{u}$. It will be more appropriate for the present purposes, however, to proceed in an alternative manner by expressing the velocity in terms of a vector potential as shown in Eq. (2.4.1).

Let us thus assume that $\mathbf{A}$ is the most general vector potential that is capable of producing the flow $\mathbf{u}$. Taking the curl of both sides of Eq. (2.4.1) and manipulating the repeated curl on the right-hand side we derive the following differential equation for $\mathbf{A}$

$$\boldsymbol{\omega} = \nabla \times \nabla \times \mathbf{A} = \nabla(\nabla \cdot \mathbf{A}) - \nabla^2 \mathbf{A} \tag{2.5.1}$$

For reasons that will soon become apparent, we decompose $\mathbf{A}$ into the sum

$$\mathbf{A} = \mathbf{B} + \mathbf{C} \tag{2.5.2}$$

where $\mathbf{B}$ is a particular solution of Poisson's equation

$$\nabla^2 \mathbf{B} = -\boldsymbol{\omega} \tag{2.5.3}$$

and the complementary part $\mathbf{C}$ is a solution of the equation

$$\nabla \times \nabla \times \mathbf{C} = -\nabla(\nabla \cdot \mathbf{B}) \tag{2.5.4}$$

To this end, in order to make the derivations more explicit, we confine our attention to three-dimensional flow. If the domain of flow extends to infinity, we require that the velocity decays at least as fast as $1/R$, where R is the distance of a point from the origin, and the corresponding vorticity decays at least as fast as $1/R^2$. If the velocity does not decay at infinity, we subtract off the nondecaying far-field component and consider the vector potential associated with the remaining decaying flow.

Under these conditions, we obtain a particular solution to Eq. (2.5.3) using Poisson's inversion formula with the free-space Green's function,

$$\mathbf{B}(\mathbf{x}) = \frac{1}{4\pi} \int_{\text{Flow}} \frac{1}{r} \boldsymbol{\omega}(\mathbf{x}') \, dV(\mathbf{x}') \tag{2.5.5}$$

where $r = |\mathbf{x} - \mathbf{x}'|$. Substituting Eq. (2.5.5) into the right-hand side of Eq. (2.5.2), taking the curl of the resulting expression, interchanging the curl with the integral operator on the right-hand side, and carrying out the differentiations under the integral sign, we derive an integral representation of the velocity,

$$\mathbf{u}(\mathbf{x}) = -\frac{1}{4\pi} \int_{\text{Flow}} \frac{1}{r^3} \hat{\mathbf{x}} \times \boldsymbol{\omega}(\mathbf{x}') \, dV(\mathbf{x}') + \nabla \times \mathbf{C} \tag{2.5.6}$$

where $\hat{\mathbf{x}} = \mathbf{x} - \mathbf{x}'$. The volume integral on the right-hand side of Eq. (2.5.6) is similar to the *Biot–Savart integral* in electromagnetics expressing the magnetic field due to an electrical current. By analogy, the integral in Eq. (2.5.6) expresses the velocity field induced by the vortex lines, although it should be acknowledged that the relation between the velocity and the vorticity is symbiotic rather than causal. In a more formal sense, the Biot–Savart integral represents a volume distribution of singular fundamental solutions called *rotlets* or *vortons,* where the density of the distribution is equal to the vorticity.

Working in a slightly different manner, we take the curl of Eq. (2.5.5), interchange the curl with the integral sign on the right-hand side, change the variable of differentiation from $\mathbf{x}$ to $\mathbf{x}'$

while simultaneously introducing a minus sign, differentiate by parts under the integral sign, and apply the divergence theorem to find

$$(\nabla \times \mathbf{B})_i = \frac{1}{4\pi} \varepsilon_{ijk} \frac{\partial}{\partial x_j} \int_{\text{Flow}} \frac{1}{r} \omega_k(\mathbf{x}') \, dV(\mathbf{x}')$$

$$= \frac{1}{4\pi} \varepsilon_{ijk} \int_{\text{Flow}} \omega_k(\mathbf{x}') \frac{\partial}{\partial x_j} \left(\frac{1}{r}\right) dV(\mathbf{x}')$$

$$= -\frac{1}{4\pi} \varepsilon_{ijk} \int_{\text{Flow}} \omega_k(\mathbf{x}') \frac{\partial}{\partial x_j'} \left(\frac{1}{r}\right) dV(\mathbf{x}')$$

$$= -\frac{1}{4\pi} \varepsilon_{ijk} \int_{\text{Flow}} \left[\frac{\partial}{\partial x_j'} \left(\frac{1}{r} \omega_k(\mathbf{x}')\right) - \frac{1}{r} \frac{\partial \omega_k(\mathbf{x}')}{\partial x_j'} \right] dV(\mathbf{x}')$$

$$= \frac{1}{4\pi} \int_B \frac{1}{r} \varepsilon_{ijk} \, \omega_k(\mathbf{x}') \, n_j(\mathbf{x}') \, dS(\mathbf{x}')$$

$$+ \frac{1}{4\pi} \int_{\text{Flow}} \frac{1}{r} \varepsilon_{ijk} \frac{\partial \omega_k(\mathbf{x}')}{\partial x_j'} \, dV(\mathbf{x}') \tag{2.5.7}$$

where the unit normal vector $\mathbf{n}$ is directed *into* the flow and B designates the boundaries. Switching back to vector notation and using Eq. (2.5.2), we obtain

$$\mathbf{u}(\mathbf{x}) = \frac{1}{4\pi} \int_B \frac{1}{r} \mathbf{n}(\mathbf{x}') \times \boldsymbol{\omega}(\mathbf{x}') \, dS(\mathbf{x}') + \frac{1}{4\pi} \int_{\text{Flow}} \frac{1}{r} \nabla' \times \boldsymbol{\omega}(\mathbf{x}') \, dV(\mathbf{x}') + \nabla \times \mathbf{C} \tag{2.5.8}$$

The boundary integral on the right-hand side of Eq. (2.5.8) vanishes when the vortex lines intersect the boundaries at a right angle, whereas the volume integral vanishes when the vorticity field is irrotational, which means that all components of the velocity are harmonic functions. The curl of the vorticity of a two-dimensional or axisymmetric flow without swirling motion lies in the plane of the flow or in an azimuthal plane, and the volume integral produces respective motions in these planes.

The Complementary Potential

To solve Eq. (2.5.4) for the complementary vector potential $\mathbf{C}$, we first compute the forcing function expressed by the right-hand side. For this purpose, we take the divergence of Eq. (2.5.5), interchange the order of differentiation and integration on the right-hand side, change the variable of differentiation from $\mathbf{x}$ to $\mathbf{x}'$ while simultaneously introducing a minus sign, differentiate by parts, and use the fact that the vorticity field is solenoidal to write

$$\nabla \cdot \mathbf{B}(\mathbf{x}) = \frac{1}{4\pi} \frac{\partial}{\partial x_i} \int_{\text{Flow}} \frac{1}{r} \omega_i(\mathbf{x}') \, dV(\mathbf{x}') = \frac{1}{4\pi} \int_{\text{Flow}} \omega_i(\mathbf{x}') \frac{\partial}{\partial x_i} \left(\frac{1}{r}\right) dV(\mathbf{x}')$$

$$= -\frac{1}{4\pi} \int_{\text{Flow}} \omega_i(\mathbf{x}') \frac{\partial}{\partial x_i'} \left(\frac{1}{r}\right) dV(\mathbf{x}')$$

$$= -\frac{1}{4\pi} \int_{\text{Flow}} \left[\frac{\partial}{\partial x_i'} \left(\frac{1}{r} \omega_i(\mathbf{x}')\right) - \frac{1}{r} \frac{\partial \omega_i}{\partial x_i'}(\mathbf{x}') \right] dV(\mathbf{x}')$$

$$= \frac{1}{4\pi} \int_B \frac{1}{r} \boldsymbol{\omega}(\mathbf{x}') \cdot \mathbf{n}(\mathbf{x}') \, dS(\mathbf{x}') \tag{2.5.9}$$

where the unit normal vector $\mathbf{n}$ is directed *into* the flow.

The boundary integral on the right-hand side of Eq. (2.5.9) vanishes when the vorticity vector is perpendicular to the normal vector, or, equivalently, the vortex lines are tangential to the boundaries. This feature occurs always in a two-dimensional or axisymmetric flow without swirling motion, but not necessarily in a three-dimensional flow (Problem 2.5.1). When the domain of flow is totally or partially infinite, the boundaries include the whole or part of a spherical surface of large radius R that lies within the fluid. As the radius of the large surface R tends to infinity, the corresponding integral vanishes provided that the vorticity decays faster than $1/R$.

When the boundary integral on the right-hand side of Eq. (2.5.9) vanishes, $\nabla \cdot \mathbf{B} = 0$, and Eq. (2.5.4) shows that the complementary flow $\nabla \times \mathbf{C}$ is irrotational. In that case, $\nabla \times \mathbf{C}$ may be expressed as the gradient of a potential function ϕ, where in order to ensure that $\mathbf{u}$ is solenoidal, we require that ϕ is a harmonic function. More generally, we write

$$\nabla \times \mathbf{C} = \mathbf{v} + \nabla H \tag{2.5.10}$$

where $\mathbf{v}$ is a solenoidal solution of the equation $\nabla \times \mathbf{v} = -\nabla(\nabla \cdot \mathbf{B})$ and H is a harmonic function. Note that if we have a particular solution $\mathbf{v}^P$, we can construct a solenoidal solution $\mathbf{v}$ by setting $\mathbf{v} = \mathbf{v}^P + \nabla F$, where the function F satisfies Poisson's equation $\nabla^2 F = -\nabla \cdot \mathbf{v}^P$. In the case of infinite flow with no interior boundaries, all three vector fields $\mathbf{v}$, ∇H, and $\nabla \times \mathbf{C}$ vanish uniformly throughout the domain of flow.

Interpretation of the Complementary Flow in Terms of the Extended Vorticity

We shall show now that the rotational component of the complementary flow $\mathbf{v}$ may be identified with the flow associated with the extension of the vortex lines outward from the boundaries of the flow, in the sense of the Biot–Savart integral.

Let us consider a flow that is bounded by the closed interior boundary B and introduce a nonsingular solenoidal vector field $\boldsymbol{\zeta}$ within the volume V_B enclosed by B, subject to the constraint that

$$\boldsymbol{\zeta} \cdot \mathbf{n} = \boldsymbol{\omega} \cdot \mathbf{n} \tag{2.5.11}$$

over B, where $\mathbf{n}$ is the unit vector normal to B pointing *into* the flow. A nonsingular solenoidal field $\boldsymbol{\zeta}$ that satisfies Eq. (2.5.11) will exist only if the rate of flow of $\boldsymbol{\zeta}$ across B vanishes, but this is always true: using Eq. (2.5.11) and the fact that $\boldsymbol{\omega}$ is solenoidal, we find

$$\int_B \boldsymbol{\zeta}(\mathbf{x}) \cdot \mathbf{n}(\mathbf{x}) \, dS(\mathbf{x}) = \int_B \boldsymbol{\omega}(\mathbf{x}) \cdot \mathbf{n}(\mathbf{x}) \, dS(\mathbf{x})$$

$$= -\int_{Flow} \nabla \cdot \boldsymbol{\omega}(\mathbf{x}) \, dV(\mathbf{x}) = 0 \tag{2.5.12}$$

Having established the existence of $\boldsymbol{\zeta}$, we recast Eq. (2.5.4) into the form

$$\nabla^2 \mathbf{C} = \nabla(\nabla \cdot \mathbf{B} + \nabla \cdot \mathbf{C}) \tag{2.5.13}$$

A particular solution comprised of three scalar harmonic functions is

$$\mathbf{C}(\mathbf{x}) = \frac{1}{4\pi} \int_{V_B} \frac{1}{r} \boldsymbol{\zeta}(\mathbf{x}') \, dV(\mathbf{x}') \tag{2.5.14}$$

Using Eqs. (2.5.9) and (2.5.11), we find $\nabla \cdot \mathbf{C} = -\nabla \cdot \mathbf{B}$, which verifies that the right-hand side of Eq. (2.5.13) is equal to zero (Problem 2.5.4).

Equation (2.5.14) relates the complementary vector potential $\mathbf{C}$ to the extended field $\boldsymbol{\zeta}$ through a Biot–Savart integral, and this allows us to regard $\boldsymbol{\zeta}$ as the *extension of vorticity field outward from the domain of flow*. This extension may be effected in an arbitrary manner subject to the sole constraint imposed by Eq. (2.5.11).

Taking the curl of Eq. (2.5.14) and repeating the manipulations that led us from Eq. (2.5.6) to Eq. (2.5.8), we derive two equivalent expressions for the complementary flow

$$\mathbf{v}(\mathbf{x}) = -\frac{1}{4\pi} \int_{V_B} \frac{1}{r^3} \hat{\mathbf{x}} \times \boldsymbol{\zeta}(\mathbf{x}') \, dV(\mathbf{x}')$$

$$= -\frac{1}{4\pi} \int_B \frac{1}{r} \mathbf{n}(\mathbf{x}') \times \boldsymbol{\zeta}(\mathbf{x}') \, dS(\mathbf{x}') + \frac{1}{4\pi} \int_{V_B} \frac{1}{r} \nabla' \times \boldsymbol{\zeta}(\mathbf{x}') \, dV(\mathbf{x}') \qquad (2.5.15)$$

where the unit normal vector $\mathbf{n}$ is directed *into* the flow. If the extended vorticity $\boldsymbol{\zeta}$ is irrotational, the volume integral over the body on the second line of Eq. (2.5.15) vanishes, leaving an expression for the complementary flow in terms of a surface integral over B involving the tangential component of $\boldsymbol{\zeta}$ alone. One way to ensure that $\boldsymbol{\zeta}$ is irrotational is to set $\boldsymbol{\zeta} = \nabla H$, where H is a harmonic function, and then compute H by solving Laplace's equation within V_B subject to the Neumann boundary condition (2.5.11).

The Complementary Flow in Terms of the Boundary Velocity

Switching to a different point of view, we consider the vector identity $\nabla^2 \mathbf{u} = -\nabla \times \boldsymbol{\omega}$, which is valid for any incompressible flow, and apply the modified version of Green's third identity (2.3.23) with $\mathbf{u}$ in place of ϕ to obtain

$$\mathbf{u}(\mathbf{x}) = -\frac{1}{4\pi} \int_B \frac{1}{r} \mathbf{n}(\mathbf{x}') \cdot \nabla \mathbf{u}(\mathbf{x}') \, dS(\mathbf{x}') + \frac{1}{4\pi} \int_B \mathbf{u}(\mathbf{x}') \left(\mathbf{n}(\mathbf{x}') \cdot \nabla' \left(\frac{1}{r} \right) \right) dS(\mathbf{x}')$$

$$+ \frac{1}{4\pi} \int_{\text{Flow}} \frac{1}{r} \nabla' \times \boldsymbol{\omega}(\mathbf{x}') \, dV(\mathbf{x}') \qquad (2.5.16)$$

where $r = |\mathbf{x} - \mathbf{x}'|$. Note the change in notation from $\mathbf{x}_0$ to $\mathbf{x}$ and from $\mathbf{x}$ to $\mathbf{x}'$.

Comparing Eq. (2.5.16) with Eq. (2.5.8) suggests a boundary integral representation of the complementary flow in terms of the boundary velocity and the tangential component of the vorticity in the form

$$\nabla \times \mathbf{C}(\mathbf{x}) = -\frac{1}{4\pi} \int_B \left[\frac{1}{r} [\mathbf{n}(\mathbf{x}') \times \boldsymbol{\omega}(\mathbf{x}') + \mathbf{n}(\mathbf{x}') \cdot \nabla' \mathbf{u}(\mathbf{x}')] \right.$$

$$\left. - \mathbf{u}(\mathbf{x}') \left(\mathbf{n}(\mathbf{x}') \cdot \nabla' \left(\frac{1}{r} \right) \right) \right] dS(\mathbf{x}') \qquad (2.5.17)$$

Furthermore, beginning with the definition $\boldsymbol{\omega} = \nabla \times \mathbf{u}$ and working in index notation, we find that the expression within the small square brackets within the integrand is equal to $[\nabla' \mathbf{u}(\mathbf{x}')] \cdot \mathbf{n}(\mathbf{x}')$. This allows us to obtain an expression for the complementary flow in terms of the boundary velocity and velocity gradient tensor,

$$\nabla \times \mathbf{C}(\mathbf{x}) = -\frac{1}{4\pi} \int_B \left(\frac{1}{r} \nabla' \mathbf{u}(\mathbf{x}') - \mathbf{u}(\mathbf{x}') \nabla' \left(\frac{1}{r} \right) \right) \cdot \mathbf{n}(\mathbf{x}') \, dS(\mathbf{x}') \qquad (2.5.18)$$

A more appealing representation emerges by switching to index notation and using the divergence theorem to convert a part of the surface integral to a volume integral, yielding

$$[\nabla \times \mathbf{C}(\mathbf{x})]_i = -\frac{1}{4\pi} \int_B \left[\frac{1}{r} \frac{\partial u_j(\mathbf{x}')}{\partial x_i'} - u_i(\mathbf{x}') \frac{\partial}{\partial x_j'} \left(\frac{1}{r} \right) \right] n_j(\mathbf{x}') \, dS(\mathbf{x}')$$

$$= \frac{1}{4\pi} \int_{\text{Flow}} \frac{\partial}{\partial x_j'} \left(\frac{1}{r} \frac{\partial u_j(\mathbf{x}')}{\partial x_i'} \right) dV(\mathbf{x}')$$

$$+ \frac{1}{4\pi} \int_B u_i(\mathbf{x}') \frac{\partial}{\partial x_j'} \left(\frac{1}{r} \right) n_j(\mathbf{x}') \, dS(\mathbf{x}') \qquad (2.5.19)$$

Next we use the continuity equation to rewrite the penultimate integrand in the form

$$
\frac{\partial}{\partial x_j'}\left(\frac{1}{r}\frac{\partial u_j}{\partial x_i'}\right) = \frac{\partial^2}{\partial x_i'\partial x_j'}\left(\frac{u_j}{r}\right) - \frac{\partial}{\partial x_j'}\left[u_j\frac{\partial}{\partial x_i'}\left(\frac{1}{r}\right)\right]
$$

$$
= \frac{\partial}{\partial x_i'}\left[u_j\frac{\partial}{\partial x_j'}\left(\frac{1}{r}\right)\right] - \frac{\partial}{\partial x_j'}\left[u_j\frac{\partial}{\partial x_i'}\left(\frac{1}{r}\right)\right] \tag{2.5.20}
$$

Substituting this expression into the volume integral of Eq. (2.5.19), and using the divergence theorem to convert it to a surface integral, we obtain

$$
[\nabla \times \mathbf{C}(\mathbf{x})]_i = -\frac{1}{4\pi}\int_B\left[u_j(\mathbf{x}')\frac{\partial}{\partial x_j'}\left(\frac{1}{r}\right)n_i(\mathbf{x}') - u_j(\mathbf{x}')\frac{\partial}{\partial x_i'}\left(\frac{1}{r}\right)n_j(\mathbf{x}')\right.
$$

$$
\left. - u_i(\mathbf{x}')\frac{\partial}{\partial x_j'}\left(\frac{1}{r}\right)n_j(\mathbf{x}')\right]dS(\mathbf{x}') \tag{2.5.21}
$$

The first and third terms of the integrand may be combined to form a double cross product, and the result is an expression for the complementary flow in terms of the tangential and normal components of the boundary velocity,

$$
\nabla \times \mathbf{C}(\mathbf{x}) = \frac{1}{4\pi}\int_B\left[[\mathbf{n}(\mathbf{x}') \times \mathbf{u}(\mathbf{x}')] \times \nabla'\left(\frac{1}{r}\right)\right.
$$

$$
\left. + [\mathbf{u}(\mathbf{x}') \cdot \mathbf{n}(\mathbf{x}')]\nabla'\left(\frac{1}{r}\right)\right]dS(\mathbf{x}') \tag{2.5.22}
$$

The boundary integral of the first term on the right-hand side of Eq. (2.5.22) expresses the velocity field due to a vortex sheet with strength $\mathbf{n} \times \mathbf{u}$ that wraps around the boundaries. The strength of the vortex sheet is equal to the tangential component of the velocity, and this suggests that the vortex sheet acts to make the tangential velocity on the boundary vanish. The integral of the second term in Eq. (2.5.22) expresses a boundary distribution of point sources whose strength is equal to the normal velocity of the fluid.

Two-Dimensional Flow

The counterpart of Eq. (2.5.5) for two-dimensional flow in the xy plane is

$$
\mathbf{B}(\mathbf{x}) = -\frac{1}{2\pi}\int_{\text{Flow}}\boldsymbol{\omega}(\mathbf{x}')\ln r\,dA(\mathbf{x}') \tag{2.5.23}
$$

Note that both $\boldsymbol{\omega}$ and $\mathbf{B}$ are oriented along the z axis. Since the vortex lines do not cross the boundaries, the divergence of $\mathbf{B}$ vanishes, and the velocity field is given by either one of the equivalent representations

$$
\mathbf{u}(\mathbf{x}) = -\frac{1}{2\pi}\int_{\text{Flow}}\frac{1}{r^2}\hat{\mathbf{x}} \times \boldsymbol{\omega}(\mathbf{x}')\,dA(\mathbf{x}') + \nabla H
$$

$$
= -\frac{1}{2\pi}\int_C \mathbf{n}(\mathbf{x}') \times \boldsymbol{\omega}(\mathbf{x}')\ln r\,dl(\mathbf{x}')
$$

$$
-\frac{1}{2\pi}\int_{\text{Flow}}\nabla' \times \boldsymbol{\omega}(\mathbf{x}')\ln r\,dA(\mathbf{x}') + \nabla H \tag{2.5.24}
$$

where C is the boundary of the flow, and H is a harmonic function. The first representation in Eq. (2.5.24) expresses the Biot–Savart integral. In the case of infinite flow with no interior boundaries, both the boundary integral and the gradient ∇H on the right-hand side of Eq. (2.5.24) vanish.

PROBLEMS

2.5.1 **Vortex lines at boundaries.** Show that the vortex lines may not cross a rigid boundary that is either stationary or translates in a viscous fluid, but must necessarily cross a boundary that rotates as a rigid body. The velocity of the fluid at the boundary is assumed to be equal to the velocity of the boundary, conforming with the no-slip and no-penetration condition (Section 3.5).

2.5.2 **Infinite flow.** Show that for infinite flow with no interior boundaries that vanishes at infinity, Eq. (2.5.8) simplifies to

$$\mathbf{u}(\mathbf{x}) = \frac{1}{4\pi} \int_{\text{Flow}} \frac{1}{r} \nabla' \times \boldsymbol{\omega}(\mathbf{x}') \, dV(\mathbf{x}') = -\frac{1}{4\pi} \int_{\text{Flow}} \frac{1}{r} \nabla'^2 \mathbf{u}(\mathbf{x}') \, dV(\mathbf{x}') \qquad (2.5.25)$$

2.5.3 **A vortex line that starts and ends on a body.** Consider an infinite incompressible flow that contains a single line vortex that starts and ends at the surface of a body. There are many ways to extend the line vortex into the body subject to the constraint expressed by Eq. (2.5.11). Show that the flows associated with any two extended line vortices have identical rotational components or, equivalently, the difference between these two flows expresses irrotational motion. *Hint:* Consider the closed loop formed by the two extended vortex lines, and use the Biot–Savart integral to show that the flow induced by this loop is irrotational everywhere except on the loop.

2.5.4 **Complementary flow.** (a) Show that Eq. (2.5.14) gives a particular solution to Eq. (2.5.13). (b) Derive Eq. (2.5.18) from Eq. (2.5.17).

2.5.5 **Reduction of the Biot–Savart integral from three to two dimensions.** Consider an infinite two-dimensional flow in the xy plane that vanishes at infinity. Substituting Eq. (1.8.2) into Eq. (2.5.6), setting $dV = dz \, dA$, where $dA = dx \, dy$, and performing the integral with respect to z, derive the first integral representation in Eq. (2.5.24). *Hint:* Reference to standard tables of definite integrals (Gradshteyn and Ryzhik, 1980, p. 86) shows

$$\int_{-\infty}^{\infty} (x^2 + y^2 + z^2)^{-3/2} \, dz = \frac{2}{x^2 + y^2} \qquad (2.5.26)$$

2.5.6 **Complementary flow for impenetrable boundaries.** Milne-Thomson (1968, p. 570) derives a simplified version of Eq. (2.5.22) for the flow of an incompressible fluid that satisfies the no-penetration condition $\mathbf{u} \cdot \mathbf{n} = 0$ over all boundaries denoted by B. We begin by introducing a solenoidal vector potential $\mathbf{A}$ for the velocity with $\nabla \cdot \mathbf{A} = 0$, as discussed in Section 2.4. Next we set $\mathbf{A} = \nabla \times \mathbf{B}$, where $\mathbf{B}$ is a vector potential of $\mathbf{A}$, and obtain $\mathbf{u} = \nabla \times \nabla \times \mathbf{B} = \nabla(\nabla \cdot \mathbf{B}) - \nabla^2 \mathbf{B}$. Assuming that $\nabla \cdot \mathbf{B} = 0$, we find a particular solution in terms of the Poisson integral

$$\mathbf{B}(\mathbf{x}) = \frac{1}{4\pi} \int_{\text{Flow}} \frac{1}{r} \mathbf{u}(\mathbf{x}') \, dV(\mathbf{x}') \qquad (2.5.27)$$

which is analogous to that given in Eq. (2.5.5). Taking the curl of Eq. (2.5.27) and using Gauss's theorem yields

$$\mathbf{A}(\mathbf{x}) = \nabla \times \mathbf{B}(\mathbf{x}) = \frac{1}{4\pi} \nabla \times \int_{\text{Flow}} \frac{1}{r} \mathbf{u}(\mathbf{x}') \, dV(\mathbf{x}')$$

$$= \frac{1}{4\pi} \int_{\text{Flow}} \mathbf{u}(\mathbf{x}') \times \nabla' \left(\frac{1}{r} \right) dV(\mathbf{x}')$$

$$= \frac{1}{4\pi} \int_{\text{Flow}} \frac{1}{r} \nabla' \times \mathbf{u}(\mathbf{x}') \, dV(\mathbf{x}') - \frac{1}{4\pi} \int_{\text{Flow}} \nabla' \times \left(\mathbf{u}(\mathbf{x}') \frac{1}{r} \right) dV(\mathbf{x}')$$

$$= \frac{1}{4\pi} \int_{\text{Flow}} \frac{1}{r} \boldsymbol{\omega}(\mathbf{x}') \, dV(\mathbf{x}') + \frac{1}{4\pi} \int_{B} \frac{1}{r} \mathbf{n}(\mathbf{x}') \times \mathbf{u}(\mathbf{x}') \, dS(\mathbf{x}') \qquad (2.5.28)$$

(a) Show that

$$\nabla \cdot \mathbf{B}(\mathbf{x}) = \frac{1}{4\pi} \int_B \frac{1}{r} \mathbf{u}(\mathbf{x}') \cdot \mathbf{n}(\mathbf{x}') \, dS(\mathbf{x}') = 0 \tag{2.5.29}$$

where the unit normal vector $\mathbf{n}$ points *into* the flow, so that the condition for the validity of Eq. (2.5.29) is fulfilled. (b) Based on Eq. (2.5.29), obtain the general solution for the velocity field in the form

$$\mathbf{u}(\mathbf{x}) = -\frac{1}{4\pi} \int_{\text{Flow}} \frac{1}{r^3} \hat{\mathbf{x}} \times \boldsymbol{\omega}(\mathbf{x}') \, dV(\mathbf{x}')$$

$$+ \frac{1}{4\pi} \int_B \frac{1}{r^3} \hat{\mathbf{x}} \times [\mathbf{u}(\mathbf{x}') \times \mathbf{n}(\mathbf{x}')] \, dS(\mathbf{x}') \tag{2.5.30}$$

Note that the second integral on the right-hand side is a simplified version of the right-hand side of Eq. (2.5.22).

2.6 | REPRESENTATION OF A FLOW IN TERMS OF THE RATE OF EXPANSION AND THE VORTICITY

In the preceding sections of this chapter, we showed that an irrotational velocity field may be expressed as the gradient of a potential function, whereas a solenoidal velocity field may be expressed as the curl of a vector potential. The velocity field of a flow that is both irrotational *and* incompressible may be expressed either in terms of a harmonic potential function or in terms of a vector potential.

Let us consider now the most general case of a flow that is neither incompressible nor irrotational. If the domain of the flow is infinite, we require that $\mathbf{u}$ decays at a rate that is faster than $1/R$, where R is the distance from the origin, while the rate of expansion and the vorticity decay at a rate that is faster than $1/R^2$. If the velocity does not decay at infinity, we subtract off the nondecaying far-field component and consider the complementary decaying flow.

Under these assumptions, the velocity field is subject to the *fundamental theorem of vector analysis*, also known as the *Hodge* or *Helmholtz decomposition theorem*, stating that $\mathbf{u}$ may be decomposed into two constituents as

$$\mathbf{u} = \nabla F + \mathbf{w} \tag{2.6.1}$$

where ∇F is an irrotational field expressed in terms of the scalar potential F, and $\mathbf{w}$ is a solenoidal field so that

$$\nabla \cdot \mathbf{w} = 0 \tag{2.6.2}$$

The fact that $\mathbf{w}$ is solenoidal allows us to express it in terms of the vector potential $\mathbf{A}$, and therefore recast Eq. (2.6.1) into the form

$$\mathbf{u} = \nabla F + \nabla \times \mathbf{A} \tag{2.6.3}$$

To demonstrate the validity of the Helmholtz decomposition, we take the curl of Eq. (2.6.1) and require that

$$\boldsymbol{\omega} \equiv \nabla \times \mathbf{u} = \nabla \times \mathbf{w} \tag{2.6.4}$$

which necessitates that $\mathbf{u} - \mathbf{w}$ be an irrotational field expressible in terms of a gradient ∇F as discussed in Section 2.1, which is consistent with Eq. (2.6.1).

Now, taking the divergence of Eq. (2.6.1), we find

$$\nabla \cdot \mathbf{u} = \nabla^2 F \tag{2.6.5}$$

which, along with Eq. (2.6.3), shows that the rate of expansion of the flow ∇F and the vorticity of the flow $\mathbf{w}$ are identical to those of the flow $\mathbf{u}$. Combining then Eq. (2.6.1) with Eqs. (2.1.13), (2.5.6), and (2.5.10), we obtain a representation of $\mathbf{u}$ in terms of the rate of expansion, the vorticity, and an unspecified incompressible and solenoidal velocity field described by the harmonic potential H in the form

$$
\mathbf{u}(\mathbf{x}) = \frac{1}{4\pi} \int_{\text{Flow}} \frac{1}{r^3} \hat{\mathbf{x}} \, \nabla' \cdot \mathbf{u}(\mathbf{x}') \, dV(\mathbf{x}')
$$

$$
- \frac{1}{4\pi} \int_{\text{Flow}} \frac{1}{r^3} \hat{\mathbf{x}} \times \boldsymbol{\omega}(\mathbf{x}') \, dV(\mathbf{x}') + \mathbf{v}(\mathbf{x}) + \nabla H \tag{2.6.6}
$$

The complementary rotational velocity $\mathbf{v}$ may be computed by extending the vortex lines outward from the domain of flow across the boundaries as discussed in Section 2.5. For an unbounded flow that vanishes at infinity, the last two terms on the right-hand side of Eq. (2.6.6) are absent.

Specifying the distributions of the rate of expansion $\nabla \cdot \mathbf{u}$ and vorticity $\boldsymbol{\omega}$ throughout the domain of flow renders the first three terms on the right-hand side of Eq. (2.6.6) known; the fourth term is defined uniquely by prescribing one scalar boundary condition, for example, the boundary distribution of the normal velocity component $\mathbf{u} \cdot \mathbf{n}$. An important consequence of this observation is that it is not generally permissible to specify the distributions of both the rate of expansion and vorticity $\boldsymbol{\omega}$ in an arbitrary manner while requiring more than one scalar constraint over the boundaries.

To illustrate the preceding point, let us consider an incompressible flow due to a line vortex in an infinite domain of flow that contains a stationary body. To ensure that the fluid does not penetrate the body, we impose the no-penetration condition requiring that the velocity component normal to the surface of the body be equal to zero; this allows us to compute ∇H in an unambiguous manner. To satisfy a further boundary condition, such as the no-slip condition requiring that the tangential component of the boundary velocity also be equal to zero, we must enhance the vorticity field with a vortex sheet situated over the surface of the body; the velocity induced by the vortex sheet annihilates the tangential velocity induced by the line vortex. *The vorticity distribution associated with the vortex sheet must be taken into account when computing the Biot–Savart integral on the right-hand side of Eq. (2.6.6).*

PROBLEMS

2.6.1   **Integral representations.** Discuss whether it is consistent to introduce a velocity field that satisfies certain required boundary conditions, compute the associated rate of expansion $\nabla \cdot \mathbf{u}$ and vorticity $\boldsymbol{\omega}$, and then use Eq. (2.6.6) to deduce the corresponding $\mathbf{v}$ and H.

2.6.2   **Poincaré decomposition.** Derive an expression for the velocity in terms of the vorticity by applying Eqs. (2.4.1)–(2.4.3) with $\boldsymbol{\omega}$ in place of $\mathbf{u}$, and $\mathbf{u}$ in place of $\mathbf{A}$.

2.7 | STREAM FUNCTIONS FOR INCOMPRESSIBLE FLOW

In the preceding sections we developed differential representations for the velocity field in terms of the potential function and the vector potential, and integral representations in terms of the vorticity, the rate of expansion, and the boundary velocity. The differential representations allowed us to describe a flow in terms of a reduced number of scalar functions; for example, in the case of irrotational flow, instead of considering the three components of the velocity, we describe the flow in terms of the potential function alone. The integral representations allowed us to obtain insights into the effects of the behavior of the fluid parcels on the global structure of a flow.

To this end, we return to address the question of what is the minimum number of scalar functions that are necessary in order to describe a flow. We have already found that an irrotational flow may be described in terms of just one scalar function, the potential function ϕ, which is determined uniquely up to an arbitrary constant. Our present goal is to address the more general class of rotational flows.

Two-Dimensional Flow

Examining the streamline pattern of a two-dimensional flow of an incompressible fluid, schematically illustrated in Figure 2.7.1(a), we find that the volumetric flow rate across a line L that begins at a point P on a particular streamline and ends at another point Q on another streamline is constant, independent of the actual location of the points P and Q along the two streamlines. This suggests that to every streamline we may assign a numerical value of a function ψ so that the difference in the values of ψ corresponding to two different streamlines is equal to the instantaneous flow rate across any line that begins at a point on the first streamline and ends at another point on the second streamline. Accordingly, we write

$$\psi_2 - \psi_1 = \int_P^Q \mathbf{u} \cdot \mathbf{n}\, dl \qquad (2.7.1)$$

where the integral is computed along the line L shown in Figure 2.7.1(a). The right-hand side of Eq. (2.7.1) expresses the flow rate across L. Since the set of all streamlines fill up the entire domain of flow, we may regard ψ a field function of position x and y and time t, called the *Helmholtz stream function* or simply the *stream function*.

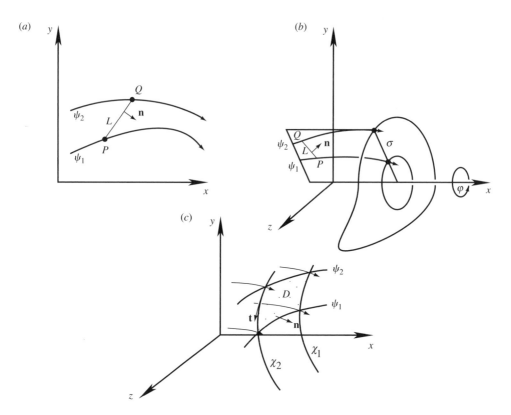

Figure 2.7.1 Schematic illustration of the streamline patterns of (a) a two-dimensional, (b) an axisymmetric, and (c) a three-dimensional flow.

Considering two streamlines that are located very close to one another, we apply the trapezoidal rule to approximate the integral in Eq. (2.7.1) and obtain the differential form

$$d\psi = u_x \, dy - u_y \, dx \tag{2.7.2}$$

Since the right-hand side of Eq. (2.7.2) is a complete differential, we may write

$$u_x = \frac{\partial \psi}{\partial y}, \qquad u_y = -\frac{\partial \psi}{\partial x} \tag{2.7.3}$$

which may be recast into the compact vector form

$$\mathbf{u} = \nabla \psi \times \mathbf{k} = \nabla \times (\psi \mathbf{k}) \tag{2.7.4}$$

where $\mathbf{k}$ is the unit vector along the z axis. It is evident from the definitions in Eq. (2.7.3) that $\nabla \psi \cdot \mathbf{u} = 0$, which shows that $\nabla \psi$ is perpendicular to the streamlines, and therefore ψ is constant along the streamlines.

Equation (2.7.4) shows that an acceptable vector potential of a two-dimensional flow is given by

$$\mathbf{A} = \psi \mathbf{k} \tag{2.7.5}$$

In summary, we have managed to express the velocity field and vector potential of a two-dimensional flow in terms of a single scalar function, the stream function ψ. It is clear from Eqs. (2.7.3) that the stream function of a particular flow is determined uniquely up to an arbitrary scalar constant.

As an example, the stream function associated with a two-dimensional point source introduced in Eq. (2.1.24) is given by

$$\psi = \frac{m}{2\pi} \theta \tag{2.7.6}$$

where θ is the polar angle subtended between the x axis and the vector $\mathbf{x} - \mathbf{x}_0$. This example makes it clear that *the stream function may be a multivalued function of position*; this will be the case when the domain of flow contains point sources or point sinks, or is multiply connected and the flow rate across a surface that encloses a boundary has a finite value. An example of the latter situation is provided by the flow due to the radial expansion of a two-dimensional bubble.

Taking the curl of Eq. (2.7.4), we find that the vorticity is directed along the z axis, $\boldsymbol{\omega} = \omega \mathbf{k}$, and its strength is given by

$$\omega = -\nabla^2 \psi \tag{2.7.7}$$

which shows that, when the flow is irrotational, the stream function is a harmonic function. Using Poisson's formula to invert Eq. (2.7.7), we express the stream function in terms of the vorticity distribution as

$$\psi(\mathbf{x}) = -\frac{1}{2\pi} \int_{\text{Flow}} \ln r \, \omega(x', y') \, dA(\mathbf{x}') + H(\mathbf{x}) \tag{2.7.8}$$

where $\hat{\mathbf{x}} = \mathbf{x} - \mathbf{x}'$, $r = |\hat{x}|$, and H is a harmonic function of x and y, $\nabla^2 H = 0$. For an unbounded flow that vanishes at infinity with no interior boundaries, H is a constant whose value is usually set equal to zero. It is worth observing that Eqs. (2.7.5) and (2.7.8) are consistent with the more general form (2.5.23).

Returning to Eqs. (2.7.4) and (2.7.7), we write the radial and angular polar components of the velocity and magnitude of the vorticity in plane polar coordinates as

$$u_r = \frac{1}{r}\frac{\partial \psi}{\partial \theta}, \qquad u_\theta = -\frac{\partial \psi}{\partial r}, \qquad \omega = -\frac{1}{r}\frac{\partial}{\partial r}\left(r\frac{\partial \psi}{\partial r}\right) - \frac{1}{r^2}\frac{\partial^2 \psi}{\partial \theta^2} \tag{2.7.9}$$

Axisymmetric Flow

Next we consider an axisymmetric flow without swirling motion and introduce cylindrical polar coordinates as illustrated in Figure 2.7.1(b). Repeating the preceding arguments for two-dimensional flow, we find that to every streamline we may assign a numerical value of the axisymmetric scalar function $\Psi(x, \sigma, t)$, called the *Stokes stream function*, so that the difference in the values Ψ between two streamlines is proportional to the instantaneous flow rate across an axisymmetric surface whose trace in an azimuthal plane begins at a point on one streamline and ends at another point on the second streamline. Thus, by definition,

$$\Psi_2 - \Psi_1 = \int_P^Q \mathbf{u} \cdot \mathbf{n} \sigma \, dl \tag{2.7.10}$$

When the integral on the right-hand side of Eq. (2.7.10) is multiplied by 2π, it becomes equal to the volumetric flow rate through an axisymmetric surface whose trace in an azimuthal plane is the line L shown in Figure 2.7.1(b).

Considering two streamlines that lie in the same azimuthal plane and are separated by an infinitesimal distance, we apply the trapezoidal rule and express Eq. (2.7.10) in the differential form

$$d\Psi = u_x \sigma \, d\sigma - u_\sigma \sigma \, dx \tag{2.7.11}$$

which suggests the differential relations

$$u_x = \frac{1}{\sigma} \frac{\partial \Psi}{\partial \sigma}, \qquad u_\sigma = -\frac{1}{\sigma} \frac{\partial \Psi}{\partial x} \tag{2.7.12}$$

Combining these two equations we obtain

$$\mathbf{u} = \nabla \times \left(\frac{\Psi}{\sigma} \mathbf{e}_\varphi \right) \tag{2.7.13}$$

where $\mathbf{e}_\varphi$ is the unit vector in the azimuthal direction. Thus,

$$\mathbf{A} = \frac{\Psi}{\sigma} \mathbf{e}_\varphi \tag{2.7.14}$$

is a vector potential of an axisymmetric flow. Equations (2.7.12) show that $\nabla \Psi \cdot \mathbf{u} = 0$, which reveals that $\nabla \Psi$ is perpendicular to the streamlines, and therefore Ψ is constant along the streamlines.

In summary, we have managed to express the velocity field and vector potential of an axisymmetric flow in terms of a single scalar function, the Stokes stream function Ψ. It is clear from the definitions in Eqs. (2.7.12) that Ψ is determined uniquely to within an arbitrary scalar constant. Note, however, that when the domain of flow is multiply connected, Ψ may be a multivalued function of position. An example pertains to the flow due to the expansion of a toroidal bubble.

As an example, the Stokes stream function associated with a three-dimensional point source introduced in Eq. (2.1.24) is given by

$$\Psi = -\frac{m}{4\pi} \cos\theta = -\frac{m}{4\pi} \frac{x - x_0}{|\mathbf{x} - \mathbf{x}_0|} \tag{2.7.15}$$

where θ is the polar angle subtended between the x axis and the vector $\mathbf{x} - \mathbf{x}_0$.

Taking the curl of Eq. (2.7.13), we find that the vorticity is oriented in the azimuthal direction, $\boldsymbol{\omega} = \omega \mathbf{e}_\varphi$, and its strength is given by

$$\omega = -\frac{1}{\sigma} \left(\frac{\partial^2 \Psi}{\partial x^2} + \frac{\partial^2 \Psi}{\partial \sigma^2} - \frac{1}{\sigma} \frac{\partial \Psi}{\partial \sigma} \right) \tag{2.7.16}$$

The inverse relation expressing the stream function in terms of the vorticity will be given in Section 2.9.

In spherical polar coordinates, the expressions in Eqs. (2.7.14) and (2.7.16) assume the forms

$$u_r = \frac{1}{r^2 \sin\theta} \frac{\partial \Psi}{\partial \theta}, \qquad u_\theta = -\frac{1}{r \sin\theta} \frac{\partial \Psi}{\partial r}$$

$$\omega = -\frac{1}{\sin\theta}\left(\frac{\partial^2 \Psi}{\partial r^2} + \frac{1}{r^2}\frac{\partial^2 \Psi}{\partial \theta^2} - \frac{\cot\theta}{r^2}\frac{\partial \Psi}{\partial \theta}\right) \tag{2.7.17}$$

Three-Dimensional Flow

Finally, we consider the most general case of a genuinely three-dimensional flow. Inspecting the streamline pattern, we identify two distinct families of stream surfaces where each family fills up the entire domain of flow, as illustrated in Figure 2.7.1(c). A stream surface is composed of all streamlines that pass through a specified line. Focusing on the stream tube that is confined between two pairs of stream surfaces, one pair in each family, we note that the flow rate Q across any surface D that is bounded by the stream tube is constant, and assign to the four stream surfaces the four labels $\psi_1, \psi_2, \chi_1, \chi_2$, as shown in Figure 2.7.1(c), so that

$$Q = \int_D \mathbf{u} \cdot \mathbf{n}\, dS = (\psi_2 - \psi_1)(\chi_2 - \chi_1) \tag{2.7.18}$$

In this light, ψ and χ emerge as field functions of space and time, called the *stream functions*. Furthermore, we use Stokes's theorem to write

$$\int_D \nabla \times (\psi\, \nabla\chi) \cdot \mathbf{n}\, dl = \int_C \psi\, \nabla\chi \cdot \mathbf{t}\, dl = (\psi_2 - \psi_1)(\chi_2 - \chi_1) = Q \tag{2.7.19}$$

where C is the boundary of D and $\mathbf{t}$ is the unit vector tangential to C pointing in a counterclockwise direction with respect to $\mathbf{n}$. Comparing Eq. (2.7.19) with Eq. (2.7.18) and remembering that D is an arbitrary surface allows us to write

$$\mathbf{u} = \nabla \times (\psi\, \nabla\chi) = \nabla\psi \times \nabla\chi = -\nabla \times (\chi\, \nabla\psi) \tag{2.7.20}$$

which shows that $\mathbf{A} = \psi\, \nabla\chi$ and $\mathbf{A} = -\chi\, \nabla\psi$ are two acceptable vector potentials of the flow. Since the gradients $\nabla\psi$ and $\nabla\chi$ are normal to the corresponding stream tubes, the functional form of the middle term in Eq. (2.7.20) is consistent with the fact that the intersection of two stream tubes is a streamline.

We have thus managed to express a three-dimensional solenoidal velocity field in terms of two scalar functions, the *stream functions* ψ and χ. The spatial distribution of the stream functions depends upon the choice of the families of stream surfaces used to derive Eq. (2.7.18), but having made this choice, the stream functions are determined uniquely up to an arbitrary scalar constant. The two-dimensional stream function and the Stokes stream function, in particular, derive from Eq. (2.7.20) by setting, respectively, $\chi = z$ and $\chi = \varphi$, where φ is the azimuthal angle (Problem 2.7.1).

Taking the curl of Eq. (2.7.20), we express the vorticity field in terms of the stream functions in the form

$$\boldsymbol{\omega} = \mathbf{L}(\psi) \cdot \nabla\chi - \mathbf{L}(\chi) \cdot \nabla\psi \tag{2.7.21}$$

where the matrix operator $\mathbf{L}$ is defined as

$$\mathbf{L} = -\mathbf{I}\, \nabla^2 + \nabla\,\nabla \tag{2.7.22}$$

and $\mathbf{I}$ is the identity matrix. Each term on the right-hand side of Eq. (2.7.21) represents a solenoidal vector field, and this renders the vorticity field solenoidal for any choice of ψ and χ (Problem 2.7.2).

PROBLEMS

2.7.1 **Stream functions.** Show that the two-dimensional and the Stokes stream functions derive from Eq. (2.7.20) by setting, respectively, $\chi = z$ and $\chi = \varphi$.

2.7.2 **Vorticity and stream functions.** (a) Derive Eq. (2.7.21), and (b) show that $\mathbf{u} = \mathbf{L}(f) \cdot \mathbf{a}$, where f is an arbitrary function and $\mathbf{a}$ is an arbitrary constant, is a solenoidal velocity field. The operator $\mathbf{L}$ is defined in Eq. (2.7.22).

2.7.3 **Point-source dipoles.** Derive the stream functions corresponding to the two-dimensional and the three-dimensional point-source dipole.

2.7.4 **A linear velocity field.** Derive the stream function and sketch the streamlines of a two-dimensional flow with velocity components given by $u = k(x + y)$, $v = k(x - y)$, where k is a constant. Discuss the physical interpretation of this flow.

2.7.5 **Stokes stream function.** Derive the Stokes stream function and sketch the streamlines of an axisymmetric flow whose radial and meridional spherical polar components of the velocity are given by

$$u_r = -U \cos\theta \left(1 - \frac{a^3}{r^3}\right), \qquad u_\theta = \tfrac{1}{2} U \cos\theta \left(2 + \frac{a^3}{r^3}\right) \tag{2.7.23}$$

where U and a are two constants. Verify that the velocity field is solenoidal, compute the vorticity, and discuss the physical interpretation of this flow.

Computer Problem

2.7.6 **Streamlines.** Use the program of Problem 1.7.4 to draw the streamlines of the flows described in Problems (a) 2.7.4 and (b) 2.7.5.

2.8 | FLOW INDUCED BY VORTICITY

We return in this section to discuss the structure and properties of an incompressible flow associated with a certain specified distribution of vorticity with compact support. For simplicity, we shall assume that the flow occurs in an infinite domain with no interior boundaries, and the velocity vanishes far from the region where the magnitude of the vorticity assumes finite values. The presence of interior or exterior boundaries may be taken into account in a straightforward manner by introducing an appropriate complementary flow according to our earlier discussion in Section 2.5.

Our point of departure is the Biot–Savart law expressed by Eqs. (2.5.2), (2.5.5), (2.5.6), and (2.5.8). For infinite flow with no interior boundaries, we obtain simplified expressions for the velocity potential and velocity field, given by

$$\mathbf{A}(\mathbf{x}) = \frac{1}{4\pi} \int_{\text{Flow}} \frac{1}{r} \boldsymbol{\omega}(\mathbf{x}') \, dV(\mathbf{x}') \tag{2.8.1}$$

and

$$\mathbf{u}(\mathbf{x}) = -\frac{1}{4\pi} \int_{\text{Flow}} \frac{1}{r^3} \hat{\mathbf{x}} \times \boldsymbol{\omega}(\mathbf{x}') \, dV(\mathbf{x}') = \frac{1}{4\pi} \int_{\text{Flow}} \frac{1}{r} \nabla' \times \boldsymbol{\omega}(\mathbf{x}') \, dV(\mathbf{x}') \tag{2.8.2}$$

where $\hat{\mathbf{x}} = \mathbf{x} - \mathbf{x}'$, $r = |\mathbf{x} - \mathbf{x}'|$, and the derivatives of the gradient ∇' operate with respect to $\mathbf{x}'$.

Structure of the Far Flow

Let us assume that the vorticity is concentrated within a compact region in the vicinity of the point $\mathbf{x}_0$, concisely called a vortex, and vanishes far from this region. To study the structure of the flow far from the vortex, we select a point $\mathbf{x}$ far from the vortex, expand the first integral in Eq. (2.8.2) in a Taylor series with respect to $\mathbf{x}'$ about the point $\mathbf{x}_0$, and retain only the constant and linear terms, finding

$$\mathbf{u}(\mathbf{x}) = -\frac{1}{4\pi} \frac{\mathbf{x} - \mathbf{x}_0}{|\mathbf{x} - \mathbf{x}_0|^3} \times \int_{\text{Flow}} \boldsymbol{\omega}(\mathbf{x}') \, dV(\mathbf{x}')$$

$$-\frac{1}{4\pi} \int_{\text{Flow}} \left[(\mathbf{x}' - \mathbf{x}_0) \cdot \left(\nabla' \frac{\mathbf{x} - \mathbf{x}'}{r^3} \right) \right]_{\mathbf{x}' = \mathbf{x}_0} \times \boldsymbol{\omega}(\mathbf{x}') \, dV(\mathbf{x}') \qquad (2.8.3)$$

The term before the first integral on the right-hand side of Eq. (2.8.3) represents the flow due to a singularity called the *rotlet* or *vorton* located at the point $\mathbf{x}_0$; the strength of this singularity is equal to the integral of the vorticity over the volume of the flow. Now, using the fact that the vorticity field is solenoidal we write

$$\int_{\text{Flow}} \boldsymbol{\omega} \, dV = \int_{\text{Flow}} \nabla \cdot (\boldsymbol{\omega} \mathbf{x}) \, dV = \int_{S_\infty} \mathbf{x} (\boldsymbol{\omega} \cdot \mathbf{n}) \, dS \qquad (2.8.4)$$

where S_∞ is a surface of large size enclosing the vortex. Assuming that the vorticity decays fast enough so that the last integral in Eq. (2.8.4) vanishes as the size of S_∞ tends to infinity, we find that the leading term on the right-hand side of Eq. (2.8.3) makes a vanishing contribution.

Concentrating on the second term on the right-hand side of Eq. (2.8.3), we change the variable of differentiation of the gradient within the integrand from $\mathbf{x}'$ to $\mathbf{x}$, while simultaneously introducing a minus sign, and obtain

$$\mathbf{u}(\mathbf{x}) = \frac{1}{4\pi} \int_{\text{Flow}} \left[\hat{\mathbf{x}}' \cdot \nabla \left(\frac{\mathbf{x} - \mathbf{x}_0}{|\mathbf{x} - \mathbf{x}_0|^3} \right) \right] \times \boldsymbol{\omega}(\mathbf{x}') \, dV(\mathbf{x}') \qquad (2.8.5)$$

where $\hat{\mathbf{x}}' = \mathbf{x}' - \mathbf{x}_0$. Switching to index notation, we derive the equivalent expression

$$u_k(\mathbf{x}) = -\frac{1}{4\pi} \frac{\partial^2}{\partial x_i \, \partial x_j} \left(\frac{1}{|\mathbf{x} - \mathbf{x}_0|} \right) \varepsilon_{jlk} \int_{\text{Flow}} \hat{x}_i' \, \omega_l(\mathbf{x}') \, dV(\mathbf{x}') \qquad (2.8.6)$$

To simplify the right-hand side of Eq. (2.8.6), we introduce the identity

$$\int_{\text{Flow}} (x_i \omega_l + \omega_i x_l) \, dV = \int_{\text{Flow}} \frac{\partial}{\partial x_k} (\omega_k x_i x_l) \, dV = \int_{S_\infty} x_i x_l \omega_k n_k \, dS \qquad (2.8.7)$$

Assuming that the vorticity over the large surface S_∞ decays sufficiently fast so that the last integral in Eq. (2.8.7) vanishes, we find that the integral on the right-hand side of Eq. (2.8.6) is antisymmetric with respect to the indices i and l, and this allows us to write

$$\int_{\text{Flow}} \hat{x}_i' \, \omega_l(\mathbf{x}') \, dV(\mathbf{x}') = \tfrac{1}{2} \varepsilon_{mil} \varepsilon_{mnk} \int_{\text{Flow}} \hat{x}_n' \, \omega_k(\mathbf{x}') \, dV(\mathbf{x}') \qquad (2.8.8)$$

Substituting the right-hand side of Eq. (2.8.8) for the integral on the right-hand side of Eq. (2.8.6), contracting the repeated multiplications of the alternating matrix, noting that $1/|\mathbf{x} - \mathbf{x}_0|$ is a harmonic function at every point except $\mathbf{x}_0$, and switching to vector notation, we finally obtain

$$\mathbf{u}(\mathbf{x}) = \frac{1}{8\pi} \int_{\text{Flow}} \hat{\mathbf{x}}' \times \boldsymbol{\omega}(\mathbf{x}') \, dV(\mathbf{x}') \cdot \left(\nabla \nabla \frac{1}{|\mathbf{x} - \mathbf{x}_0|} \right) \qquad (2.8.9)$$

which shows that, far from the vortex, the flow is similar to that due to a point-source dipole located at the point $\mathbf{x}_0$ whose strength is proportional to the angular moment of the vorticity, defined as

$$\mathbf{d} = \frac{1}{2} \int_{\text{Flow}} \hat{\mathbf{x}}' \times \boldsymbol{\omega}(\mathbf{x}') \, dV(\mathbf{x}') \tag{2.8.10}$$

Cursory inspection of the volume and surface integrals that appear in the preceding expressions reveals that Eq. (2.8.9) is valid provided that the vorticity decays faster than $1/|\mathbf{x} - \mathbf{x}_0|^4$.

Kinetic Energy of the Fluid

We now develop an expression for the kinetic energy of the fluid in terms of the vorticity distribution, assuming that the density of the fluid is uniform throughout the domain of flow. One way to proceed is to write

$$K = \frac{1}{2} \rho \int_{\text{Flow}} \mathbf{u} \cdot \mathbf{u} \, dV = \frac{1}{2} \rho \int_{\text{Flow}} \mathbf{u} \cdot \nabla \times \mathbf{A} \, dV$$

$$= \frac{1}{2} \rho \int_{\text{Flow}} [\mathbf{A} \cdot \boldsymbol{\omega} - \nabla \cdot (\mathbf{u} \times \mathbf{A})] \, dV \tag{2.8.11}$$

Using the divergence theorem, we convert the volume integral of the second term within the integral on the right-hand side to an integral over the large surface S_∞, and find that the integral vanishes provided that the velocity decays at a sufficiently fast rate. This leaves us with the expression

$$K = \frac{1}{2} \rho \int_{\text{Flow}} \mathbf{A} \cdot \boldsymbol{\omega} \, dV \tag{2.8.12}$$

The vector potential $\mathbf{A}$ can be expressed in terms of the vorticity distribution by means of Eq. (2.8.1). For axisymmetric flow, we express $\mathbf{A}$ in terms of the Stokes stream function and obtain

$$K = \pi \rho \int_{\text{Flow}} \Psi(x, \sigma) \omega(x, \sigma) \, dx \, d\sigma \tag{2.8.13}$$

where $\boldsymbol{\omega} = \omega \mathbf{e}_\varphi$, and $\mathbf{e}_\varphi$ is the unit vector in the azimuthal direction. An analogous expression for two-dimensional flow will be discussed in Section 11.1.

A more useful representation of the kinetic energy in terms of the velocity and the vorticity emerges by using the identity

$$\nabla \cdot [\mathbf{u} (\mathbf{u} \cdot \mathbf{x}) - \tfrac{1}{2}(\mathbf{u} \cdot \mathbf{u}) \, \mathbf{x}] + \tfrac{1}{2} \, \mathbf{u} \cdot \mathbf{u} = \mathbf{u} \cdot (\mathbf{x} \times \boldsymbol{\omega}) \tag{2.8.14}$$

(Batchelor, 1967, p. 520). Solving for the last term on the left-hand side, substituting the result into the first integral in Eq. (2.8.11), and using the divergence theorem to simplify the integral, we find

$$K = \rho \int_{\text{Flow}} \mathbf{u} \cdot (\mathbf{x} \times \boldsymbol{\omega}) \, dV \tag{2.8.15}$$

Expressing the velocity in terms of the vorticity distribution by means of Eq. (2.8.2) provides us with an expression involving the vorticity alone.

Flow Due to a Vortex Sheet

To compute the velocity field due to a three-dimensional vortex sheet, we substitute the vorticity distribution (1.9.2) into the first integral of Eq. (2.8.2) and find

$$\mathbf{u}(\mathbf{x}) = -\frac{1}{4\pi} \int_{\text{Sheet}} \frac{1}{r^3} \hat{\mathbf{x}} \times \boldsymbol{\zeta}(\mathbf{x}') \, dS(\mathbf{x}') \tag{2.8.16}$$

which provides us with an expression for the velocity field in terms of an integral over the surface of the vortex sheet expressing a surface distribution of rotlets or vortons. The coefficient of the dipole, computed from Eq. (2.8.10), is given by

$$\mathbf{d} = \frac{1}{2} \int_{\text{Sheet}} \hat{\mathbf{x}}' \times \boldsymbol{\zeta}(\mathbf{x}') \, dS(\mathbf{x}') \tag{2.8.17}$$

The flow at any point off the vortex sheet is irrotational and may thus be described in terms of a velocity potential ϕ. In Section 10.5 we shall see that ϕ may be represented in terms of a distribution of point-source dipoles oriented normal to the vortex sheet. Combining Eqs. (10.5.1), written for the free-space Green's function, and Eq. (10.5.7), we obtain

$$\phi(\mathbf{x}) = -\frac{1}{4\pi} \int_{\text{Sheet}} \frac{1}{r^3} \hat{\mathbf{x}} \cdot \mathbf{N}(\mathbf{x}')(\phi^+ - \phi^-)(\mathbf{x}') \, dS(\mathbf{x}') \tag{2.8.18}$$

where $\mathbf{N}$ is the unit vector normal to the vortex sheet pointing into the upper side, which corresponds to the plus superscript. Taking the gradient of Eq. (2.8.18), integrating by parts, and using the definition of $\boldsymbol{\zeta}$ recovers Eq. (2.8.16).

Flow Due to a Line Vortex

Flows induced by line vortices are archetypes of flows with concentrated vorticity such as turbulent flows and flows due to tornadoes and whirls. To compute the vector potential and velocity field due to a line vortex L with strength κ illustrated in Figure 2.8.1(a), we substitute

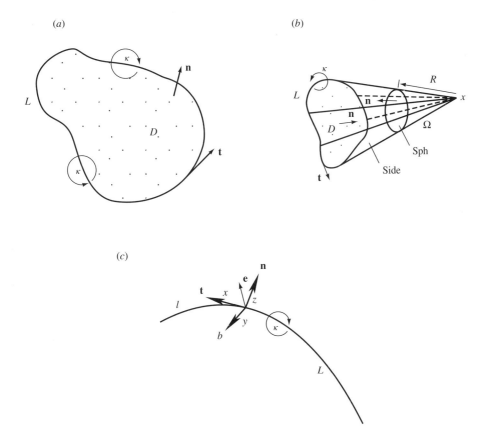

Figure 2.8.1   (a) Schematic illustration of a closed line vortex L and a surface D bounded by the line vortex. (b) The potential function is proportional to the solid angle Ω subtended by the line vortex. (c) Closeup of a line vortex.

the vorticity distribution (1.9.1) into Eq. (2.8.1) and into the first integral of Eq. (2.8.2) finding

$$\mathbf{A}(\mathbf{x}) = \frac{\kappa}{4\pi} \int_{\text{Flow}} \frac{1}{r} \mathbf{t}(\mathbf{x}') \, dl(\mathbf{x}'), \qquad \mathbf{u}(\mathbf{x}) = -\frac{\kappa}{4\pi} \int_{L} \frac{1}{r^3} \hat{\mathbf{x}} \times \mathbf{t}(\mathbf{x}') \, dl(\mathbf{x}') \qquad (2.8.19)$$

where $\mathbf{t}$ is the unit tangential vector along the line vortex.

Furthermore, we use Eq. (2.8.10) and apply Stokes's theorem to find that the associated coefficient of the dipole is given by

$$\mathbf{d} = \tfrac{1}{2}\kappa \int_{\text{Flow}} \hat{\mathbf{x}}' \times \mathbf{t}(\mathbf{x}') \, dl(\mathbf{x}') = \kappa \int_{D} \mathbf{n}(\mathbf{x}') \, dS(\mathbf{x}') \qquad (2.8.20)$$

where D is an arbitrary closed surface bounded by the line vortex as shown in Figure 2.8.1(a). The last integral shows that the flow far from the line vortex is similar to that due to a uniform distribution of point-source dipoles oriented perpendicular to D.

Velocity potential

The flow at any point off a line vortex is irrotational and may thus be expressed in terms of a harmonic potential ϕ. Before attempting to compute ϕ, we acknowledge that, since the circulation around a loop that encloses the line vortex once is finite and equal to κ, ϕ will be a multivalued function of position. Keeping in mind this subtlety, let us consider the closed line vortex depicted in Figure 2.8.1(a), write the second of Eqs. (2.8.19) in index notation, apply Stokes's theorem to convert the line integral along the line vortex to a surface integral over an arbitrary surface D that is bounded by the line vortex, and use the fact that $1/r$ is a harmonic function to obtain

$$u_i(\mathbf{x}) = -\frac{\kappa}{4\pi} \int_{L} \frac{1}{r^3} \varepsilon_{ijk} \hat{x}_j t_k(\mathbf{x}') \, dl(\mathbf{x}')$$

$$= -\frac{\kappa}{4\pi} \int_{D} \varepsilon_{kmn} \frac{\partial}{\partial x'_m} \left(\frac{1}{r^3} \varepsilon_{ijn} \hat{x}_j \right) n_k(\mathbf{x}') \, dS(\mathbf{x}')$$

$$= \frac{\kappa}{4\pi} \int_{D} \varepsilon_{kmn} \varepsilon_{ijn} \frac{\partial^2}{\partial x'_m \, \partial x_j} \left(\frac{1}{r} \right) n_k(\mathbf{x}') \, dS(\mathbf{x}')$$

$$= -\frac{\kappa}{4\pi} \frac{\partial}{\partial x_i} \int_{D} \frac{\partial}{\partial x'_k} \left(\frac{1}{r} \right) n_k(\mathbf{x}') \, dS(\mathbf{x}') \qquad (2.8.21)$$

The right-hand side of Eq. (2.8.21) expresses the velocity as the gradient of the velocity potential

$$\phi(\mathbf{x}) = -\frac{\kappa}{4\pi} \int_{D} \nabla' \left(\frac{1}{r} \right) \cdot \mathbf{n}(\mathbf{x}') \, dS(\mathbf{x}') = -\frac{\kappa}{4\pi} \int_{D} \frac{\mathbf{x} - \mathbf{x}'}{r^3} \cdot \mathbf{n}(\mathbf{x}') \, dS(\mathbf{x}') \qquad (2.8.22)$$

which is identical to that given by the far-field expansion (2.8.9).

A more natural interpretation of ϕ emerges by introducing a conical surface that contains all rays emanating from the point $\mathbf{x}$ and passing through the line vortex, and defining a control volume that is bounded by (1) the surface D, (2) a section of a sphere of radius R centered at the point $\mathbf{x}$ and confined by the conical surface, denoted by Sph, and (3) the section of the conical surface contained between the spherical surface and the line vortex, denoted by Side, as illustrated in Figure 2.8.1(b). Beginning with the right-hand side of Eq. (2.8.22) and using the divergence theorem, we find

$$\phi(\mathbf{x}) = \frac{\kappa}{4\pi} \int_{\text{Sph,Side}} \frac{\mathbf{x} - \mathbf{x}'}{r^3} \cdot \mathbf{n}(\mathbf{x}') \, dS(\mathbf{x}') \qquad (2.8.23)$$

The integral over the conical surface vanishes because the normal vector is perpendicular to the distance $\mathbf{x} - \mathbf{x}'$. The normal vector over the spherical surface is given by $\mathbf{n} = \pm(\mathbf{x}' - \mathbf{x})/R$, where the plus and minus signs apply, respectively, when the sphere is on the right or on the left of the surface D; for the situation depicted in Figure 2.8.1(b), we select the plus sign. Equation (2.8.23) then yields

$$\phi(\mathbf{x}) = \pm \frac{\kappa}{4\pi} \int_{\mathrm{Sph}} \frac{\mathbf{x} - \mathbf{x}'}{R^3} \cdot \frac{\mathbf{x}' - \mathbf{x}}{R} \, dS(\mathbf{x}') = \frac{\kappa}{4\pi} \Omega \tag{2.8.24}$$

where Ω is the solid angle subtended at the point $\mathbf{x}$ by the line vortex; for the situation depicted in Figure 2.8.1(b), Ω has a positive value.

The solid angle Ω, and therefore the potential ϕ, is a multivalued function of position. For example, in the case of flow due to line vortex ring, Ω changes its value from -2π to 2π as the point $\mathbf{x}$ crosses the plane of the ring through its interior. This means that ϕ undergoes a corresponding discontinuity of magnitude equal to κ, in agreement with our earlier discussion in Section 2.1 concerning the behavior of the potential in a multiply connected domain.

Flow near a line vortex and self-induced velocity

If one attempts to use the second of Eqs. (2.8.19) to compute the velocity at a point $\mathbf{x}$ that is located in the vicinity of, or on a line vortex, one encounters a substantial difficulty due to the singular nature of the integrand. To resolve the local structure of the flow, we introduce local Cartesian coordinates with their origin at a point on the line vortex and the x, y, and z axes oriented in the directions of the tangential, binormal, and normal vectors $\mathbf{t}$, $\mathbf{b}$, and $\mathbf{n}$, as shown in Figure 2.8.1(c), and consider the velocity at a point $\mathbf{x}$ that lies in the normal plane containing $\mathbf{n}$ and $\mathbf{b}$, at the position $\mathbf{x} = \sigma\mathbf{e}$, where $\mathbf{e}$ is a unit vector that lies in the normal plane.

As a preliminary towards desingularizing the Biot–Savart integral, we expand the position vector $\mathbf{x}'$ and tangential vector $\mathbf{t}(\mathbf{x}')$ along the line vortex in the vicinity of the origin in Taylor series with respect to arc length l, measured from the origin of the Cartesian axes in the direction of $\mathbf{t}$, and retain two significant terms, obtaining

$$\mathbf{x}' = \frac{\partial \mathbf{x}'}{\partial l}(0)\, l + \frac{1}{2} \frac{\partial^2 \mathbf{x}'}{\partial l^2}(0)\, l^2 + O(l^3) = \mathbf{t}(0)\, l - \tfrac{1}{2} c(0)\, \mathbf{n}(0)\, l^2 + O(l^3) \tag{2.8.25}$$

$$\mathbf{t}(\mathbf{x}') = \mathbf{t}(0) + \frac{\partial \mathbf{t}}{\partial l}(0)\, l + O(l^2) = \mathbf{t}(0) - c(0)\, \mathbf{n}(0)\, l + O(l^2) \tag{2.8.26}$$

where c is the curvature of the line vortex.

Considering the second of Eqs. (2.8.19), we write $\hat{\mathbf{x}} = \sigma\mathbf{e} - \mathbf{x}'$, and subtract off the singularity of the integrand using Eq. (2.8.26), obtaining

$$
\begin{aligned}
\mathbf{u}(\mathbf{x}) = {}& -\frac{\kappa\sigma}{4\pi}\mathbf{e} \times \int_L \frac{\mathbf{t}(\mathbf{x}')}{r^3}\, dl(\mathbf{x}') \\
& + \frac{\kappa}{4\pi} \int_L \frac{1}{r^3}\mathbf{x}' \times [\mathbf{t}(\mathbf{x}') - \mathbf{t}(0) + c(0)\,\mathbf{n}(0)\, l]\, dl(\mathbf{x}') \\
& + \frac{\kappa}{4\pi} \int_L \frac{1}{r^3}\mathbf{x}' \times \mathbf{t}(0)\, dl(\mathbf{x}') - \frac{\kappa}{4\pi} c(0) \int_L \frac{1}{r^3}\mathbf{x}' \times \mathbf{n}(0)\, l\, dl(\mathbf{x}')
\end{aligned}
\tag{2.8.27}
$$

where $r = |\mathbf{x} - \mathbf{x}'|$. The second integral on the right-hand side of Eq. (2.8.27) is nonsingular and makes a finite contribution. To investigate the behavior of the third integral, we use the expansion (2.8.25) and obtain

$$
\begin{aligned}
\int_L \frac{1}{r^3}\mathbf{x}' \times \mathbf{t}(0)\, dl(\mathbf{x}') & \approx \int_L \frac{1}{r^3}[\mathbf{t}(0)\, l - \tfrac{1}{2} c(0)\,\mathbf{n}(0)\, l^2] \times \mathbf{t}(0)\, dl(\mathbf{x}') \\
& \approx -\tfrac{1}{2} c(0)\, \mathbf{b}(0) \int_L \frac{l^2}{r^3}\, dl(\mathbf{x}')
\end{aligned}
\tag{2.8.28}
$$

Working in a similar manner with the fourth integral on the right-hand side of Eq. (2.8.27), we obtain

$$\int_L \frac{1}{r^3} \mathbf{x}' \times \mathbf{n}(0) \, l \, dl(\mathbf{x}') \approx -\mathbf{b}(0) \int_L \frac{l^2}{r^3} \, dl(\mathbf{x}') \tag{2.8.29}$$

Finally, we substitute Eqs. (2.8.28) and (2.8.29) into Eq. (2.8.27) and approximate the tangential vector in the first integral with its value at the origin, and thus find that, to leading order, the velocity field in the vicinity of the origin is given by

$$\mathbf{u}(\mathbf{x}) \approx -\frac{\kappa\sigma}{4\pi} \mathbf{e} \times \mathbf{t}(0) \int_L \frac{dl(\mathbf{x}')}{r^3} + \frac{\kappa}{8\pi} c(0) \, \mathbf{b}(0) \int_L \frac{l^2}{r^3} \, dl(\mathbf{x}') \tag{2.8.30}$$

To assess the local behavior of the flow, we truncate the limits of integration at the values $l = -a$ and $l = a$, set $r^2 = \sigma^2 + l^2$, and thus obtain

$$\mathbf{u}(\mathbf{x}) \approx -\frac{\kappa}{4\pi\sigma} \mathbf{e} \times \mathbf{t}(0) \int_{-a/\sigma}^{a/\sigma} \frac{d\eta}{(1+\eta^2)^{3/2}} + \frac{\kappa}{8\pi} c(0) \, \mathbf{b}(0) \int_{-a/\sigma}^{a/\sigma} \frac{\eta^2 d\eta}{(1+\eta^2)^{1/2}} \tag{2.8.31}$$

where $\eta = l/\sigma$. Evaluating the integrals in the limit as σ/a tends to zero, we obtain the asymptotic form

$$\mathbf{u}(\mathbf{x}) \approx -\frac{\kappa}{2\pi\sigma} \mathbf{e} \times \mathbf{t}(0) + \frac{\kappa}{4\pi} c(0) \, \mathbf{b}(0) \ln \frac{a}{\sigma} \tag{2.8.32}$$

first derived by Da Rios in 1906 (Ricca, 1991).

The first term on the right-hand side of Eq. (2.8.32) expresses the anticipated swirling motion around the line vortex, which is similar to that occurring around a point vortex. In the limit as σ/a tends to zero, the second term becomes infinite, and this reveals that *the self-induced velocity of a curved line vortex assumes an infinite value*. This singular behavior is a reflection of the severe approximations involved in the mathematical fabrication of singular vortex structures. The regularization of Eq. (2.8.32), accounting for the finite size of the core, will be discussed in Chapter 11 in the context of vortex methods.

The local-induction approximation amounts to evaluating Eq. (2.8.32) at the origin, and maintaining only the second term on the right-hand side. We then find that, if the product of the curvature and binormal vector happens to be constant along the line vortex, the line vortex will translate as a rigid body at an infinite speed; this occurs, for example, for a circular vortex ring. Two additional examples of a line vortex that is known to move as a rigid body are provided by (1) a helical vortex that advances along its axis while rotating about it, and (2) a planar nearly rectilinear line vortex with small-amplitude sinusoidal undulations that rotates about its axis as a rigid entity (Problem 2.8.6).

PROBLEMS

2.8.1   **Impulse.** The impulse required to generate the motion of a fluid with uniform density can be expressed in terms of the momentum integral

$$\mathbf{P} = \rho \int_{Flow} \mathbf{u} \, dV \tag{2.8.33}$$

Show that two flows of an incompressible fluid with different vorticity distributions but identical dipole strengths **d** require identical impulses. *Hint:* Consider the counterpart of Eq. (2.8.4) for the velocity.

2.8.2   **Far flow.** Derive the second-order term in the asymptotic expansion (2.8.3) and discuss its physical interpretation. Comment on the asymptotic behavior of the flow when the coefficient of the dipole vanishes.

2.8.3   **Identities.** Prove identities (2.8.7) and (2.8.14).

2.8.4   **A rectilinear line vortex.** Use the second of Eqs. (2.8.19) to compute the velocity field due to a rectilinear line vortex, and compare the results with the velocity field due to a point vortex discussed in Section 2.10.

2.8.5   **A helical line vortex.** A helical line vortex can be described in the parametric form $x = a\cos\varphi$, $y = a\sin\varphi$, $z = k\varphi$, where a is the radius of the circumscribed cylinder, φ is the azimuthal angle, and $2\pi k$ is the helical pitch. Show that the velocity induced by a helical line vortex is given by $\mathbf{u} = (\kappa/4\pi)\nabla\cdot\mathbf{P}$, where $\mathbf{P}$ is an antisymmetric matrix defined as (Hardin, 1982)

$$\mathbf{P} = \begin{bmatrix} 0 & kI_1 & -aI_2 \\ -kI_1 & 0 & -aI_3 \\ aI_2 & aI_3 & 0 \end{bmatrix} \tag{2.8.34}$$

and

$$\begin{bmatrix} I_1 \\ I_2 \\ I_3 \end{bmatrix}(x, y, z) = \int_{-\infty}^{\infty} \begin{bmatrix} 1 \\ \cos\theta \\ \sin\theta \end{bmatrix} \frac{d\theta}{[(x - a\cos\theta)^2 + (y - a\sin\theta)^2 + (y - k\theta)^2]^{1/2}} \tag{2.8.35}$$

2.8.6   **Sinusoidal line vortex.** Show that a planar, nearly rectilinear line vortex with small-amplitude sinusoidal undulations rotates about its axis as a rigid body with infinite angular velocity.

2.9   **AXISYMMETRIC FLOW INDUCED BY VORTICITY**

Considering an axisymmetric flow without swirling motion, we exploit the fact that the vorticity is directed in the azimuthal direction to simplify the Biot–Savart integral by performing the integration in the azimuthal direction using analytical or semianalytical methods.

We begin by introducing cylindrical coordinates (x, σ, φ) with the x axis in the direction of the axis of the flow, substitute Eq. (1.8.3) into the first integral in Eq. (2.8.2), and set $dV = \sigma\, d\varphi\, dA$, where $dA = dx\, d\sigma$, and thus obtain

$$\mathbf{u}(\mathbf{x}) = -\frac{1}{4\pi} \int_{\text{Flow}} \left(\int_0^{2\pi} \frac{\hat{\mathbf{x}} \times \mathbf{e}_\varphi}{r^3} d\varphi' \right) \omega(x', \sigma')\sigma'\, dA(\mathbf{x}') \tag{2.9.1}$$

Next we substitute

$$\hat{\mathbf{x}} = (\hat{x}, \sigma\cos\varphi - \sigma'\cos\varphi', \sigma\sin\varphi - \sigma'\sin\varphi')$$
$$\mathbf{e}_\varphi = (0, -\sin\varphi', \cos\varphi') \tag{2.9.2}$$

into Eq. (2.9.1) and obtain

$$\mathbf{u}(\mathbf{x}) = -\frac{1}{4\pi} \int_{\text{Flow}} \left(\int_0^{2\pi} \frac{1}{r^3} \begin{bmatrix} \sigma\cos\hat{\varphi} - \sigma' \\ -\hat{x}\cos\hat{\varphi} \\ -\hat{x}\sin\hat{\varphi} \end{bmatrix} d\varphi' \right)$$
$$\times \omega(x', \sigma')\sigma'\, dA(\mathbf{x}') \tag{2.9.3}$$

where $\hat{x} = x - x'$, $\hat{\varphi} = \varphi - \varphi'$. Finally, we write

$$r = (\hat{x}^2 + \sigma^2 + \sigma'^2 - 2\sigma\sigma'\cos\hat{\varphi})^{1/2}$$
$$= [\hat{x}^2 + (\sigma + \sigma')^2 - 4\sigma\sigma'\cos^2(\hat{\varphi}/2)]^{1/2} \tag{2.9.4}$$

and $u_\sigma = u_y \cos \varphi' + u_z \sin \varphi'$, and use Eq. (2.9.3) to derive expressions for the axial and radial components of the velocity in the forms

$$
\begin{bmatrix} u_x \\ u_\sigma \end{bmatrix} (\mathbf{x}) = \frac{1}{4\pi} \int_{\text{Flow}} \begin{bmatrix} -\sigma I_{31}(\hat{x}, \sigma, \sigma') + \sigma' I_{30}(\hat{x}, \sigma, \sigma') \\ \hat{x} I_{31}(\hat{x}, \sigma, \sigma') \end{bmatrix}
$$
$$
\times \, \omega(x', \sigma') \sigma' \, dA(\mathbf{x}') \tag{2.9.5}
$$

where we have introduced the integrals

$$
I_{nm}(\hat{x}, \sigma, \sigma') \equiv \int_0^{2\pi} \frac{\cos^m \varphi}{[\hat{x}^2 + (\sigma + \sigma')^2 - 4\sigma \sigma' \cos^2(\varphi/2)]^{n/2}} \, d\varphi \tag{2.9.6}
$$

Working in a similar manner with the expression for the vector potential in terms of the vorticity distribution given in Eq. (2.8.1), and recalling the $\mathbf{A} = (\Psi/\sigma)\mathbf{e}_\varphi$, where Ψ is the Stokes stream function, we derive the corresponding representation

$$
\Psi(x, \sigma) = \frac{\sigma}{4\pi} \int_{\text{Flow}} I_{11}(\hat{x}, \sigma, \sigma') \omega(x', \sigma') \sigma' \, dA(\mathbf{x}') \tag{2.9.7}
$$

Computation of the Integrals I_{nm}

To compute the integrals I_{nm} introduced in Eq. (2.9.6), we write

$$
I_{nm}(\hat{x}, \sigma, \sigma') = \frac{4}{[\hat{x}^2 + (\sigma + \sigma')^2]^{n/2}} J_{nm}(k) \tag{2.9.8}
$$

where

$$
J_{nm}(\hat{x}, \sigma, \sigma') \equiv \int_0^{\pi/2} \frac{(2\cos^2 \eta - 1)^m}{(1 - k^2 \cos^2 \eta)^{n/2}} \, d\eta \tag{2.9.9}
$$

$$
k^2 \equiv \frac{4\sigma \sigma'}{\hat{x}^2 + (\sigma + \sigma')^2} \tag{2.9.10}
$$

The integrals J_{nm} may be expressed in terms of the complete elliptic integrals of the first and second kind, F and E, defined in Eq. (B.9.1), which may be computed using efficient iterative procedures or polynomial approximations as discussed in Section B.9, Appendix B.

For instance, with the help of standard tables of integrals (Gradshteyn and Ryzhik, 1980, p. 590), we find

$$
J_{10} = F(k), \qquad J_{11} = \frac{1}{k^2}\left((2 - k^2)F(k) - 2E(k)\right)
$$
$$
J_{30} = \frac{E(k)}{1 - k^2}, \qquad J_{31} = \frac{1}{k^2}\left(-2F(k) + \frac{2 - k^2}{1 - k^2}E(k)\right) \tag{2.9.11}
$$

Line Vortex Ring

To compute the velocity field due to a circular line vortex ring of radius σ_0, we set $\omega(\mathbf{x}) = \kappa \, \delta(\mathbf{x} - \mathbf{x}_0)$, where δ is the two-dimensional delta function operating in an azimuthal plane and $\mathbf{x}_0$ is the trace of the ring in the azimuthal plane, and use Eq. (2.9.5) to obtain the axial and radial components of the velocity

$$
u_x(x, \sigma) = \frac{\kappa}{4\pi}[-\sigma I_{31}(\hat{x}, \sigma, \sigma_0) + \sigma_0 I_{30}(\hat{x}, \sigma, \sigma_0)]
$$

$$
u_\sigma(x, \sigma) = \frac{\kappa}{4\pi} \hat{x} I_{31}(\hat{x}, \sigma, \sigma_0) \tag{2.9.12}
$$

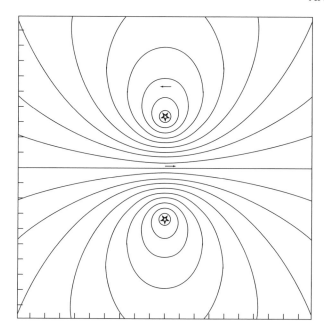

Figure 2.9.1   Streamline pattern in an azimuthal plane due to a line vortex ring.

where $\hat{x} = x - x_0$. The corresponding Stokes stream function is found from Eq. (2.9.7) to be

$$\Psi(x, \sigma) = \frac{\kappa}{4\pi}\sigma\,\sigma_0\,I_{11}(\hat{x}, \sigma, \sigma_0) \tag{2.9.13}$$

The streamline pattern in an azimuthal plane is shown in Figure 2.9.1.

The velocity potential due to a line vortex ring is given in Eqs. (2.8.22) and (2.8.24). Identifying the surface D with a circular disk of radius σ_0 bounded by the vortex ring, we find

$$\phi(\mathbf{x}) = -\frac{1}{4\pi}\Omega_{\text{Ring}}(\mathbf{x}) \tag{2.9.14}$$

where Ω_{Ring} is the angle subtended by the ring at the point $\mathbf{x}$, given by

$$\begin{aligned}
\Omega_{\text{Ring}}(\mathbf{x}) &= \hat{x}\int_0^{2\pi}\int_0^{\sigma_0}\frac{\sigma'\,d\sigma'\,d\varphi}{[\hat{x}^2 + (\sigma - \sigma'\cos\varphi)^2 + \sigma'^2\sin^2\varphi]^{3/2}} \\
&= \hat{x}\int_0^{\sigma_0} I_{30}(\hat{x}, \sigma, \sigma')\,\sigma'\,d\sigma' \\
&= 4\hat{x}\int_0^{\sigma_0}\frac{J_{30}(k)\,\sigma'}{[\hat{x}^2 + (\sigma + \sigma')^2]^{3/2}}\,d\sigma' \\
&= 4\hat{x}\int_0^{\sigma_0}\frac{E(k)\,\sigma'}{(1 - k^2)[\hat{x}^2 + (\sigma + \sigma')^2]^{3/2}}\,d\sigma' \tag{2.9.15}
\end{aligned}$$

The functions I_{30} and J_{30} are defined in Eqs. (2.9.6), (2.9.8) and (2.9.9), k is defined in Eq. (2.9.10), and E is the complete elliptic integral of the second kind defined in Section B.9, Appendix B. An alternative method of computing Ω_{Ring} is discussed by Pozrikidis (1986).

Axisymmetric Vortices with Linear Vorticity Distribution

Consider next a flow that contains an axisymmetric vortex inside of which the strength of the vorticity ω varies in a linear manner with radial distance, $\omega = \alpha\sigma$, where α is a constant. The flow outside the vortex is irrotational. Using Eq. (2.9.7), we find that the associated Stokes stream function is given by

$$\Psi(x, \sigma) = \frac{\alpha}{4\pi}\sigma\int_{A_V} I_{11}(\hat{x}, \sigma, \sigma')\sigma'^2 \, dA(\mathbf{x}') \tag{2.9.16}$$

where A_V is the area occupied by the vortex in an azimuthal plane, and $dA = dx\,d\sigma$.

Using the divergence theorem, we derive a simplified expression for the radial component of the velocity in terms of a line integral along the trace of the vortex in an azimuthal plane C,

$$u_\sigma(x, \sigma) = -\frac{1}{\sigma}\frac{\partial\Psi(x, \sigma)}{\partial x} = \frac{\alpha}{4\pi}\int_{A_V}\frac{\partial}{\partial x'}[I_{11}(\hat{x}, \sigma, \sigma')\sigma'^2]\,dA(\mathbf{x}')$$

$$= \frac{\alpha}{4\pi}\int_C I_{11}(\hat{x}, \sigma, \sigma')\,n_x(\mathbf{x}')\sigma'^2 \, dl(\mathbf{x}') \tag{2.9.17}$$

where $\mathbf{n}$ is the unit normal vector pointing outward from the vortex. To develop a corresponding expression for the axial component of the velocity, we note that the flow outside the vortex is irrotational and introduce the potential function ϕ. By analogy with Eq. (2.9.16), we write

$$\phi(x, \sigma) = -\frac{\alpha}{4\pi}\int_{A_V}\Omega_{\text{Ring}}(\hat{x}, \sigma, \sigma')\sigma' \, dA(\mathbf{x}') \tag{2.9.18}$$

and work as in Eq. (2.9.17) to find

$$u_x(x, \sigma) = \frac{\partial\phi(x, \sigma)}{\partial x} = \frac{\alpha}{4\pi}\int_{A_V}\frac{\partial}{\partial x'}[\Omega_{\text{Ring}}(\hat{x}, \sigma, \sigma')]\sigma' \, dA(\mathbf{x}')$$

$$= \frac{\alpha}{4\pi}\int_C \Omega_{\text{Ring}}(\hat{x}, \sigma, \sigma')\,n_x(\mathbf{x}')\sigma' \, dl(\mathbf{x}') \tag{2.9.19}$$

To compute the right-hand side of Eq. (2.9.19), we must introduce a branch cut in order to render the solid angle a single-valued function of position (Pozrikidis, 1986). One way of bypassing this complication is to start afresh with the right-hand side of Eq. (2.9.1) and express the curl of the vorticity in terms of generalized functions. This approach yields the computationally convenient form

$$u_x(x, \sigma) = -\frac{\alpha}{4\pi}\int_C [\hat{x}\, I_{10}(\hat{x}, \sigma, \sigma')\,n_x(\mathbf{x}') + \sigma\, I_{11}(\hat{x}, \sigma, \sigma')\,n_\sigma(\mathbf{x}')]\sigma' \, dl(\mathbf{x}') \tag{2.9.20}$$

where $\hat{x} = x - x'$ and the integrals I_{nm} are defined in Eq. (2.9.6) (Shariff, Leonard, and Ferziger, 1989).

Hill's vortex

Hill's vortex provides us with a celebrated example of an axisymmetric vortex with linear vorticity distribution. In cylindrical coordinates with the origin at the center of the vortex, and in a frame of reference in which the vortex is stationary, the Stokes stream function inside and outside Hill's vortex of radius a is given by

$$\Psi_{\text{in}} = \alpha\tfrac{1}{10}\sigma^2(a^2 - x^2 - \sigma^2), \quad \Psi_{\text{out}} = -\alpha\tfrac{1}{15}a^2\sigma^2\left(1 - \frac{a^3}{(x^2 + \sigma^2)^{3/2}}\right) \tag{2.9.21}$$

Inspecting the exterior flow shows that the vortex translates steadily along the x axis with velocity

$$U = \tfrac{2}{15}\alpha\, a^2 \tag{2.9.22}$$

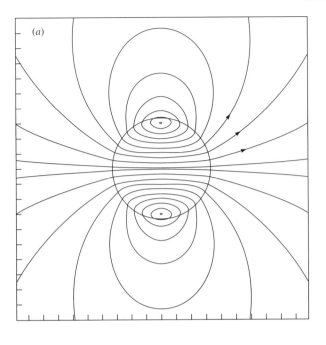

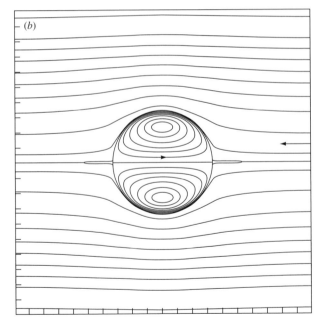

Figure 2.9.2 Streamline pattern in an azimuthal plane associated with Hill's spherical vortex in (a) a stationary frame of reference and, (b) a frame of reference traveling with the vortex.

while maintaining its spherical shape. The streamline pattern in an azimuthal plane, in a stationary frame of reference and in a frame of reference that translates with the vortex, are shown in Figure 2.9.2. Comparing the pattern shown in Figure 2.9.2(a) to that shown in Figure 2.9.1 for a line vortex ring shows that the exterior flows have similar structure, and this suggests that the particular way in which the vorticity is distributed within a vortex plays a secondary role in the structure of the flow far from the vortex.

Vortex rings

Hill's vortex is the limiting member of a family of steadily translating vortex rings parametrized by their cross-sectional area; the other extreme member of the family is a line vortex ring with infinitesimal cross-sectional area. The structure and stability of these rings are discussed by Norbury (1973) and Pozrikidis (1986).

PROBLEM

2.9.1   **Hill's spherical vortex.** On the basis of Eqs. (2.9.21) confirm that (a) the azimuthal component of the vorticity is equal to $\omega = \alpha\sigma$ within Hill's vortex and vanishes in the exterior of the vortex, (b) the velocity is continuous across the boundary of the vortex, (c) the velocity of translation of the vortex is given by Eq. (2.9.22).

Computer Problems

2.9.2   **Complete elliptic integrals.** Write a routine called *ELLINT* that computes the complete elliptic integrals F and E according to the iterative method discussed in Section B.9 of Appendix B.

2.9.3   **A line vortex ring.** (a) Write a program called *VRING* that computes the velocity field induced by a line vortex ring, and reproduce the streamline pattern shown in Figure 2.9.1. The elliptic integrals should be computed using the subroutine *ELLINT* of Problem 2.9.2. (b) Write a program called *PRING* that computes the velocity potential associated with a line vortex ring.

2.9.4   **Streamlines of Hill's spherical vortex.** Use the program *STRLN* of Problem 1.7.4 to reproduce Figure 2.9.2(a). The velocity should be computed by numerical differentiation, setting, for instance,

$$\frac{\partial\Psi}{\partial\sigma} = \frac{\Psi(\sigma + \varepsilon) - \Psi(\sigma - \varepsilon)}{2\varepsilon}$$

where ε is a small number.

2.10 │ TWO-DIMENSIONAL FLOW INDUCED BY VORTICITY

Integral representations of the velocity field of a two-dimensional flow in terms of the vorticity distribution may be derived using the general formulae presented in Section 2.8 for three-dimensional flow, under the additional stipulation that the vortex lines are rectilinear, parallel to the z axis. It is more expedient, however, to begin afresh from Eq. (2.5.24), which provides us with the velocity field in terms of the expressions

$$u_x(\mathbf{x}) = -\frac{1}{2\pi}\int_{\text{Flow}} \frac{\hat{y}}{\hat{x}^2 + \hat{y}^2}\, \omega(\mathbf{x}')\, dA(\mathbf{x}')$$

$$u_y(\mathbf{x}) = \frac{1}{2\pi}\int_{\text{Flow}} \frac{\hat{x}}{\hat{x}^2 + \hat{y}^2}\, \omega(\mathbf{x}')\, dA(\mathbf{x}')$$

$$(2.10.1)$$

and

$$\mathbf{u}(\mathbf{x}) = -\frac{1}{4\pi} \int_{\text{Flow}} \ln(\hat{x}^2 + \hat{y}^2) \nabla' \times [\omega(\mathbf{x}') \, \mathbf{k}] \, dA(\mathbf{x}') \tag{2.10.2}$$

where $\hat{\mathbf{x}} = \mathbf{x} - \mathbf{x}'$, ω is the strength of the vorticity, and $\mathbf{k}$ is the unit vector along the z axis. The associated stream function is found using Eq. (2.5.23) or Eq. (2.7.8) to be

$$\psi(\mathbf{x}) = -\frac{1}{4\pi} \int_{\text{Flow}} \ln(\hat{x}^2 + \hat{y}^2) \, \omega(\mathbf{x}') \, dA(\mathbf{x}') \tag{2.10.3}$$

Point Vortex

To derive the flow due to a point vortex located at the point $\mathbf{x}_0$, we substitute $\omega(\mathbf{x}) = \kappa \, \delta(\mathbf{x} - \mathbf{x}_0)$ into Eqs. (2.10.1) and (2.10.3), where δ is the two-dimensional delta function and κ is the strength of the point vortex, and obtain

$$u_x(\mathbf{x}) = -\frac{\kappa}{2\pi} \frac{y - y_0}{|\mathbf{x} - \mathbf{x}_0|^2}$$

$$u_y(\mathbf{x}) = \frac{\kappa}{2\pi} \frac{x - x_0}{|\mathbf{x} - \mathbf{x}_0|^2} \tag{2.10.4}$$

$$\psi(\mathbf{x}) = -\frac{\kappa}{2\pi} \ln|\mathbf{x} - \mathbf{x}_0|$$

The flow at any point except $\mathbf{x}_0$ is irrotational, and the associated multivalued velocity potential is given by

$$\phi(\mathbf{x}) = \frac{\kappa}{2\pi} \theta \tag{2.10.5}$$

where θ is the polar angle subtended by the x axis and the vector $\mathbf{x} - \mathbf{x}_0$. The magnitude of the velocity decays like $1/|\mathbf{x} - \mathbf{x}_0|$, the streamlines are concentric circles centered at the point $\mathbf{x}_0$, and the cyclic constant of the motion around the point vortex is equal to κ.

Array of Evenly Spaced Point Vortices

To derive the velocity field due to a periodic array of point vortices with identical strengths κ placed along the x axis and separated by a distance a, we introduce the complex variable $Z = X + iY$ and use the identity

$$\sin(\pi Z) = \pi Z \prod_{n=1}^{\infty} \left(1 - \frac{Z^2}{n^2}\right) \tag{2.10.6}$$

where $\prod$ denotes the product (Ahlfors, 1979, p. 197). Applying Eq. (2.10.6) with $Z = (z - z_0)/a$, where z_0 is the position of one arbitrary point vortex in the array, and rearranging, we obtain

$$\sin\frac{\pi(z - z_0)}{a} = \frac{\pi(z - z_0)}{a} \prod_{\substack{n=-\infty \\ n \neq 0}}^{\infty} (-1) \frac{z - z_0 + na}{(na)^2} \tag{2.10.7}$$

Next we note that $|z - z_0 - na| = |\mathbf{x} - \mathbf{x}_n|$ is the distance of the point z from the nth point vortex, and use the expression for the stream function in Eqs. (2.10.4) to find that the stream function due to the array is given by

$$\psi(\mathbf{x}) = -\frac{\kappa}{2\pi} \text{Re} \left\{ \ln\left[\sin\left[\tfrac{1}{2} k(z - z_0)\right]\right]\right\}$$

$$= -\frac{\kappa}{4\pi} \ln\left\{ \cosh[k(y - y_0)] - \cos[k(x - x_0)]\right\} \tag{2.10.8}$$

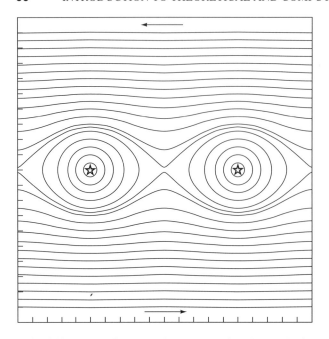

Figure 2.10.1   Streamline pattern due to an array of evenly spaced point vortices with positive strength.

where $k = 2\pi/a$ is the wave number. Differentiating Eq. (2.10.8), we derive the velocity field

$$u_x(\mathbf{x}) = -\frac{\kappa}{2a} \frac{\sinh k\hat{y}}{\cosh k\hat{y} - \cos k\hat{x}}$$

$$u_y(\mathbf{x}) = \frac{\kappa}{2a} \frac{\sin k\hat{x}}{\cosh k\hat{y} - \cos k\hat{x}} \qquad (2.10.9)$$

where $\hat{\mathbf{x}} = \mathbf{x} - \mathbf{x}_0$, and $\mathbf{x}_0$ is the position of an arbitrary point vortex in the array. The associated streamline pattern, shown in Figure 2.10.1, is characterized by a cat's eye pattern. Far above and below the vortex array, the velocity tends to obtain the uniform value $u_x = \pm \kappa/(2a)$, $u_y = 0$; second-order terms decay at an exponential rate. As the wave number k tends to zero, Eqs. (2.10.8) and (2.10.9) reproduce the stream function and velocity field associated with a single point vortex (Problem 2.10.1).

Point Vortex Dipole

Let us consider the flow due to two point vortices with strengths of equal magnitude and opposite sign, located at the two points $\mathbf{x}_0$ and $\mathbf{x}_1$. Taking advantage of the linearity of Eq. (2.10.1) with respect to the vorticity, we construct the associated stream function by superposing the stream functions due to the individual point vortices. Placing, in particular, the point $\mathbf{x}_1$ very close to $\mathbf{x}_0$ and taking the limit as the distance $|\mathbf{x}_0 - \mathbf{x}_1|$ tends to zero while $\kappa(\mathbf{x}_0 - \mathbf{x}_1)$ remains constant equal to $\boldsymbol{\lambda}$, we obtain the stream function due to a point vortex dipole

$$\psi(\mathbf{x}) = -\frac{1}{2\pi} \boldsymbol{\lambda} \cdot \nabla_0 \ln r = \frac{1}{2\pi} \frac{\mathbf{x} - \mathbf{x}_0}{r^2} \cdot \boldsymbol{\lambda} \qquad (2.10.10)$$

where $r = |\mathbf{x} - \mathbf{x}_0|$ and the derivatives in the gradient ∇_0 operate with respect to $\mathbf{x}_0$. The associated velocity field follows readily by differentiation as

$$u_x(\mathbf{x}) = \frac{1}{2\pi r^2} \left(\lambda_y - 2(y - y_0) \frac{\mathbf{x} - \mathbf{x}_0}{r^2} \cdot \boldsymbol{\lambda} \right)$$

$$u_y(\mathbf{x}) = -\frac{1}{2\pi r^2} \left(\lambda_x - 2(x - x_0) \frac{\mathbf{x} - \mathbf{x}_0}{r^2} \cdot \boldsymbol{\lambda} \right)$$

(2.10.11)

The streamline pattern of the flow due to a point vortex dipole pointing toward the positive direction of the y axis is identical to that due to a potential dipole pointing toward the positive direction of the x axis shown in Figure 2.1.3(b).

An alternative method of deriving the flow due to a point vortex dipole involves using the properties of generalized delta functions. We set $\omega(\mathbf{x}) = \boldsymbol{\lambda} \cdot \nabla_0 \delta(\mathbf{x} - \mathbf{x}_0)$ and then use Eqs. (2.10.1) and (2.10.2) to derive Eqs. (2.10.10) and (2.10.11).

Vortex Sheet

To derive the velocity field due to a two-dimensional vortex sheet, we substitute Eq. (1.9.17) into Eq. (2.10.1) and obtain

$$u_x(\mathbf{x}) = -\frac{1}{2\pi} \int_T \frac{\hat{y}}{\hat{x}^2 + \hat{y}^2} \, d\Gamma(\mathbf{x}'), \qquad u_y(\mathbf{x}) = \frac{1}{2\pi} \int_T \frac{\hat{x}}{\hat{x}^2 + \hat{y}^2} \, d\Gamma(\mathbf{x}') \qquad (2.10.12)$$

where $\hat{\mathbf{x}} = \mathbf{x} - \mathbf{x}'$, $d\Gamma = \gamma \, dl$ is the differential circulation along the vortex sheet, and T is the trace of the vortex sheet in the xy plane. Comparing Eqs. (2.10.12) with Eqs. (2.10.4) shows that a vortex sheet may be regarded as a continuous distribution of point vortices.

Periodic Vortex Sheet

Based on identity (2.10.7), we find that the velocity due a vortex sheet that is repeated periodically in the x direction with period equal to a is given by

$$u_x(\mathbf{x}) = -\frac{1}{2a} \int_T \frac{\sinh k\hat{y}}{\cosh k\hat{y} - \cos k\hat{x}} \, d\Gamma(\mathbf{x}')$$

$$u_y(\mathbf{x}) = \frac{1}{2a} \int_T \frac{\sin k\hat{x}}{\cosh k\hat{y} - \cos k\hat{x}} \, d\Gamma(\mathbf{x}')$$

(2.10.13)

where $k = 2\pi/a$, $\hat{\mathbf{x}} = \mathbf{x} - \mathbf{x}'$, and T is the trace of the vortex sheet in the xy plane over one period. As the wave number k tends to zero, Eqs. (2.10.13) reduce to Eqs. (2.10.12).

Vortex Patch

Consider next the flow due to a region of constant vorticity Ω, called a vortex patch, that is embedded in an otherwise stationary fluid. Introducing the two-dimensional vector potential expressed in terms of the stream function as $\mathbf{A} = (0, 0, \psi)$ and using Eq. (2.10.1), we find that the induced velocity field is given by

$$\mathbf{u}(\mathbf{x}) = \frac{\Omega}{4\pi} \int_{A_V} \nabla' \times [\mathbf{k} \ln(\hat{x}^2 + \hat{y}^2)] \, dA(\mathbf{x}') \qquad (2.10.14)$$

where A_V is the area occupied by the patch. Using Stokes's theorem, we convert the area integral to a contour integral around the boundary C of A_V, obtaining

$$\mathbf{u}(\mathbf{x}) = -\frac{\Omega}{4\pi} \int_C \ln(\hat{x}^2 + \hat{y}^2) \, \mathbf{t}(\mathbf{x}') \, dl(\mathbf{x}') \qquad (2.10.15)$$

where $\mathbf{t}$ is the unit tangential vector oriented in the counterclockwise sense around C.

A different way of deriving Eq. (2.10.15) makes use of the identity

$$\nabla \times \boldsymbol{\omega}(\mathbf{x}) = \Omega \int_C \delta(\mathbf{x} - \mathbf{x}') \, \mathbf{t}(\mathbf{x}') \, dl(\mathbf{x}') \qquad (2.10.16)$$

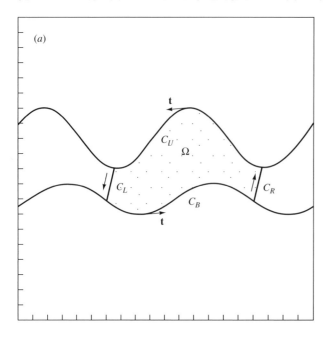

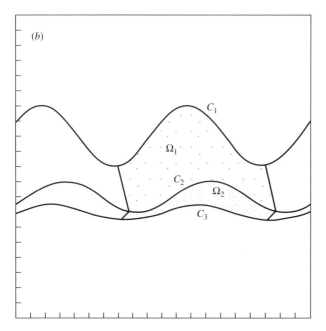

Figure 2.10.2 (a) A periodic vortex layer with constant vorticity. (b) A periodic compound vortex layer composed of two adjacent vortex layers with constant vorticity Ω_1 and Ω_2.

where δ is the two-dimensional delta function. Substituting Eq. (2.10.16) into Eq. (2.10.2) and using the properties of the delta function, we recover Eq. (2.10.15).

If the flow contains a number of disconnected vortex patches with different values of vorticity, the integral in Eq. (2.10.15) is computed over each vortex contour separately, and is then multiplied by the corresponding value of the constant vorticity Ω.

Periodic Vortex Patches

Based on identity (2.10.7), we find that the velocity field due to an infinite array of vortex patches with constant vorticity Ω that is repeated periodically in the x direction with period equal to a is given by

$$\mathbf{u}(\mathbf{x}) = -\frac{\Omega}{4\pi} \int_C \ln(\cosh k\hat{y} - \cos k\hat{x}) \, \mathbf{t}(\mathbf{x}') \, dl(\mathbf{x}') \qquad (2.10.17)$$

where $k = 2\pi/a$, C is the contour of one arbitrary patch in the array, and $\mathbf{t}$ is the unit tangential vector oriented in the counterclockwise sense.

As an application, let us consider the velocity due to a periodic vortex layer illustrated in Figure 2.10.2(a). We select one period of the vortex layer and identify C with the union of the upper and lower contours C_U and C_B, and the left and right contours C_L and C_R. The contributions from the contours C_L and C_R to the integral in Eq. (2.10.17) cancel each other, due to the fact that the tangential vectors point into opposite directions and the logarithmic function within the integral is periodic. In this manner, the contour C reduces to the union of C_U and C_B.

PROBLEMS

2.10.1 **An array of point vortices.** Verify that as the period a tends to infinity, Eqs. (2.10.8) and (2.10.9) yield the velocity and stream function due to a single point vortex.

2.10.2 **Compound vortex layer.** Consider a periodic compound vortex layer composed of two adjacent layers of constant vorticity Ω_1 and Ω_2 illustrated in Figure 2.10.2(b). Derive an expression for the velocity field in terms of contour integrals along C_1, C_2, and C_3.

References

Ahlfors, L. A., 1979, *Complex Analysis.* McGraw–Hill.

Aris, R., 1962, *Vectors, Tensors, and the Basic Equations of Fluid Mechanics.* Dover.

Batchelor, G. K., 1967, *An Introduction to Fluid Dynamics.* Cambridge University Press.

Dettman, J. W., 1965, *Applied Complex Variables.* Dover.

Gradshteyn, I. S., and Ryzhik, I. M., 1980, *Table of Integrals, Series, and Products.* Academic.

Hardin, J. C., 1982, The velocity field induced by a helical vortex filament. *Phys. Fluids* **25**, 1949–52.

Kellogg, O. D., 1953, *Foundations of Potential Theory.* Dover.

Kelvin, Lord, 1849, Notes on hydrodynamics. On the vis-visa of a liquid in motion. *Camb. and Dublin Math. J.* Feb. (Reprinted in *Mathematical and Physical Papers,* vol. I, pp. 107–113, Cambridge, 1882).

Milne-Thomson, L. M., 1968, *Theoretical Hydrodynamics.* MacMillan.

Norbury, J., 1973, A family of steady vortex rings. *J. Fluid Mech.* **57**, 417–31.

Pozrikidis, C., 1986, The nonlinear instability of Hill's spherical vortex. *J. Fluid Mech.* **168**, 337–67.

Ricca, R. L., 1991, Rediscovery of Da Rios equations. *Nature* **352**, 561–62.

Shariff, K., Leonard, A., and Ferziger, J. H., 1989, Dynamics of a class of vortex rings. *NASA TM-102257.*

CHAPTER 3

Stresses, the Equation of Motion, and the Vorticity Transport Equation

In the first two chapters we examined the kinematic structure of a flow and discussed ways of describing the velocity field in terms of secondary variables, such as the velocity potential, the vector potential, and the stream functions, but made no reference to the actual physical process that is responsible for establishing the flow or to the conditions that are necessary in order to sustain the flow. To investigate these issues, in the present chapter we introduce the fundamental concepts and physical variables that are necessary in order to describe and compute the forces developing in a fluid as the result of its motion, and then use them to derive an equation of motion that governs the permanent structure of a steady flow and the evolution of an unsteady flow. The equation of motion will be derived on the basis of *Newton's second law* for the motion of a fluid parcel, stating that the rate of change of momentum of a fluid parcel is equal to the sum of all forces exerted over its volume or on its boundaries.

In developing the dynamical laws, we shall concentrate on a special but common class of incompressible fluids called *Newtonian fluids*. The distinguishing feature of these fluids is that their response to a given type of motion may be described in terms of a simple constitutive equation that relates the forces developing on the surface of the fluid parcels to their deformation, by means of a linear relation. The equation of motion for an incompressible Newtonian fluid takes the form of a second-order partial differential equation in the spatial coordinates for the velocity, called the *Navier–Stokes* equation. Supplementing the Navier–Stokes equation with the continuity equation and requiring appropriate boundary conditions provides us with a complete set of equations that may be used to compute a flow. Analytical, asymptotic, and numerical methods for solving this governing set of equations, under a broad range of conditions, will be discussed throughout the subsequent chapters.

The equation of motion for an unsteady flow may be regarded as a dynamical law for the evolution of the velocity field. To derive a corresponding law for the evolution of the vorticity field, expressing the rate of rotation of fluid parcels, we take the curl of the Navier–Stokes equation and obtain the *vorticity transport equation*. Studying the various forms of the latter equation, in the present and subsequent chapters, will allow us to obtain useful insights into the dynamics of rotational flows, and furthermore, it will provide us with a natural framework for analyzing and computing a class of flows that are dominated by the presence or motion of compact vortex structures, including line vortices and vortex sheets.

3.1 | FORCES ACTING IN A FLUID, TRACTION, THE STRESS TENSOR, AND THE EQUATION OF MOTION

Let us consider a certain control volume that is occupied entirely by a fluid and is fixed in space, as illustrated in Figure 3.1.1(a). As the fluid flows, molecules enter and leave the control volume from all sides carrying momentum and thus imparting to the fluid that occupies the control volume, at a particular instant in time, a normal force. Furthermore, short-range intermolecular forces cause

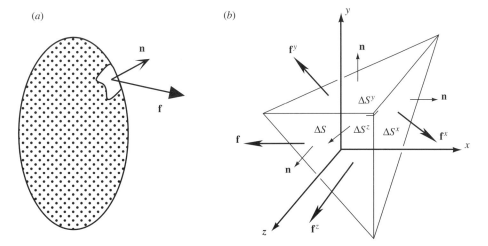

Figure 3.1.1 (a) Schematic illustration of a control volume in a flow. Molecules enter the volume from all sides, imparting to the fluid that occupies the volume a net normal force, as well as a tangential frictional force. (b) Three small triangular surfaces perpendicular to the three Cartesian axes forming a tetrahedral control volume; **n** is the unit vector normal to the control volume, and **f** is the surface traction.

the molecules that are located on either side of the boundary of the control volume to be attracted, thereby generating an effective frictional force.

The force per unit area exerted on an infinitesimal surface element of the boundary of the control volume is called the *surface stress* or *traction,* and will be denoted by **f**. Clearly, the value of **f** will depend upon both the position and orientation of the infinitesimal surface element.

To analyze the behavior of the traction, it is useful to decompose it into a normal component $\mathbf{f}^N$ that points in the direction of the unit normal vector **n**, and a tangential or shearing component $\mathbf{f}^T$, respectively given by

$$\mathbf{f}^N = (\mathbf{f} \cdot \mathbf{n})\,\mathbf{n}, \qquad \mathbf{f}^T = \mathbf{n} \times (\mathbf{f} \times \mathbf{n}) = \mathbf{f} \cdot (\mathbf{I} - \mathbf{nn}) \tag{3.1.1}$$

where **I** is the identity matrix. The projection matrix $\mathbf{I} - \mathbf{nn}$ extracts the tangential component of a vector that it multiplies.

In terms of the traction, the instantaneous hydrodynamic force exerted on the surface of a fluid parcel may be expressed in the form of a surface integral as

$$\mathbf{F} = \int_{\text{Parcel}} \mathbf{f}\, dS \tag{3.1.2}$$

A long-range ambient force field acting upon the molecules, such as the gravitational field or an electromagnetic field, imparts to the fluid that occupies the parcel a *body force* that may be expressed in the form

$$\mathbf{B} = \int_{\text{Parcel}} \kappa \mathbf{b}\, dV \tag{3.1.3}$$

where **b** is the *strength* of the body force field and κ is an appropriate physical constant that may be a function of time as well as position in the domain of flow. In the case of the gravitational field, **b** is the acceleration of gravity **g** and κ is the density of the fluid ρ. For simplicity, in the remainder of this book we shall assume that the body force field is due to gravity alone.

Stresses

Let us introduce a system of Cartesian coordinates and consider a control volume that has the shape of a small tetrahedron with three sides perpendicular to the x, y, and z axes, respectively, as illustrated in Figure 3.1.1(b). We denote the tractions exerted on each of the three planar sides, respectively, by $\mathbf{f}^x$, $\mathbf{f}^y$, and $\mathbf{f}^z$, and stack them above each other to form the matrix of stresses $\sigma_{ij} = f_j^i$,

$$
\boldsymbol{\sigma} \equiv \begin{bmatrix} \leftarrow \mathbf{f}^x \rightarrow \\ \leftarrow \mathbf{f}^y \rightarrow \\ \leftarrow \mathbf{f}^z \rightarrow \end{bmatrix} \tag{3.1.4}
$$

The diagonal elements of $\boldsymbol{\sigma}$ are the normal stresses exerted on the three mutually orthogonal sides of the control volume, and the off-diagonal elements are the tangential stresses.

We shall show that the traction exerted on the slanted side of the infinitesimal tetrahedron illustrated in Figure 3.1.1(b) may be computed from a knowledge of its orientation and the value of the matrix $\boldsymbol{\sigma}$ at the origin. Knowledge of the body force field is not required. For this purpose, we apply Newton's second law for the fluid parcel that is enclosed by the tetrahedron, stating that the rate of change of the linear momentum of the parcel is equal to the sum of the surface and body forces exerted on the parcel, and use Eqs. (3.1.2) and (3.1.3) to obtain

$$
\frac{d}{dt} \int_{\text{Parcel}} \rho \mathbf{u} \, dV = \int_{\text{Parcel}} \mathbf{f} \, dS + \int_{\text{Parcel}} \rho \mathbf{g} \, dV \tag{3.1.5}
$$

In the limit as the size of the tetrahedron becomes infinitesimal, the tractions exerted on each side may be assumed to be constant, and variations in the momentum of the fluid and body force over the volume of the parcel may be neglected. Implementing these simplifications allows us to recast Eq. (3.1.5) into the form

$$
\frac{D(\rho \mathbf{u} \, \Delta V)}{Dt} = \mathbf{f}^x \, \Delta S^x + \mathbf{f}^y \, \Delta S^y + \mathbf{f}^z \, \Delta S^z + \mathbf{f} \, \Delta S + \rho \mathbf{g} \, \Delta V \tag{3.1.6}
$$

where D/Dt is the material derivative, ΔV is the volume of the tetrahedron and the elemental surfaces ΔS^x, ΔS^y, ΔS^z, and ΔS are defined in Figure 3.1.1(b). Dividing Eq. (3.1.6) through by ΔS and rearranging the terms, we obtain

$$
\frac{1}{\Delta S} \left(\frac{D(\rho \mathbf{u} \, \Delta V)}{Dt} - \rho \mathbf{g} \, \Delta V \right) = \mathbf{f}^x \, \frac{\Delta S^x}{\Delta S} + \mathbf{f}^y \, \frac{\Delta S^y}{\Delta S} + \mathbf{f}^z \, \frac{\Delta S^z}{\Delta S} + \mathbf{f} \tag{3.1.7}
$$

In the limit as the size of the parcel tends to zero, the ratio $\Delta V/\Delta S$ vanishes, and the left-hand side of Eq. (3.1.7) becomes infinitesimal. Introducing the unit vector $\mathbf{n}$ that is normal to the transverse side and points out of the tetrahedron, making use of the geometrical relations $n_x = -\Delta S^x/\Delta S$, $n_y = -\Delta S^y/\Delta S$, $n_z = -\Delta S^z/\Delta S$, using the definition of the stress matrix in Eq. (3.1.4), and rearranging Eq. (3.1.7), we obtain

$$
\mathbf{f} = \mathbf{n} \cdot \boldsymbol{\sigma} \tag{3.1.8}
$$

We thus find that the traction is a linear function of the normal vector, with a constant of proportionality that is equal to the stress matrix $\boldsymbol{\sigma}$.

Force on a Fluid Parcel in Terms of Stresses

Combining Eqs. (3.1.2) and (3.1.8) and using the divergence theorem, we find that the hydrodynamic force exerted on the surface of a fluid parcel is given by the equivalent expressions

$$
\mathbf{F} = \int_{\text{Parcel}} \mathbf{n} \cdot \boldsymbol{\sigma} \, dS = \int_{\text{Parcel}} \nabla \cdot \boldsymbol{\sigma} \, dV \tag{3.1.9}
$$

Newton's third law requires that the parcel exerts upon the ambient fluid a force of equal magnitude in the opposite direction.

The total force exerted on the parcel is equal to the sum of the hydrodynamic force given in Eq. (3.1.9) and the body force given in Eq. (3.1.3), so that

$$\mathbf{F}^{\text{Total}} = \int_{\text{Parcel}} (\nabla \cdot \boldsymbol{\sigma} + \rho\mathbf{g}) \, dV \tag{3.1.10}$$

It is worth noting at this point that when the divergence of the stress matrix balances the body force, the total force exerted on any parcel vanishes. We shall see later in this chapter that this occurs for flows in which the effect of inertia on the motion of the fluid is negligibly small.

Hydrodynamic Force Exerted on a Boundary

To compute the hydrodynamic force exerted on a boundary, we consider a fluid parcel that has the shape of a thin sheet lining the boundary. We take the limit as the thickness of the sheet tends to zero and note that, because the mass of the parcel tends to become infinitesimal, the sum of the hydrodynamic forces exerted on both sides of the parcel must add up to zero. Newton's third law requires that the force exerted on the side of the parcel that is adjacent to the boundary be equal and opposite to that exerted by the fluid on the boundary, and this shows that the hydrodynamic force exerted on the boundary is given by

$$\mathbf{F} = \int_{\text{Boundary}} \mathbf{n} \cdot \boldsymbol{\sigma} \, dS \tag{3.1.11}$$

where $\mathbf{n}$ is the unit normal vector pointing into the fluid.

$\boldsymbol{\sigma}$ Is a Tensor

Next we shall show that the matrix of stresses $\boldsymbol{\sigma}$ is a second-order Cartesian tensor. For this purpose, following the standard procedure outlined in Section 1.1, we introduce two Cartesian systems of axes x_i and y_i that are related by the linear transformation $y_i = A_{ij}x_j$ and denote the values of $\boldsymbol{\sigma}$ in the y_i system by $\boldsymbol{\sigma}(\mathbf{y})$ and those in the x_i system by $\boldsymbol{\sigma}(\mathbf{x})$. Denoting the hydrodynamic surface force exerted on a fluid parcel in the y_i system by $\mathbf{F}(\mathbf{y})$ and the corresponding force in the x_i system by $\mathbf{F}(\mathbf{x})$, introducing the vector transformations $F_i(\mathbf{y}) = A_{ij}F_j(\mathbf{x})$ and $n_k(\mathbf{x}) = n_i(\mathbf{y})A_{ik}$, and using Eq. (3.1.8), we obtain

$$\int_{\text{Parcel}} n_i(\mathbf{y})\sigma_{ij}(\mathbf{y}) \, dS = A_{jl} \int_{\text{Parcel}} n_k(\mathbf{x})\sigma_{kl}(\mathbf{x}) \, dS$$
$$= A_{jl}A_{ik} \int_{\text{Parcel}} n_i(\mathbf{y})\sigma_{kl}(\mathbf{x}) \, dS \tag{3.1.12}$$

Comparing the first with the last terms in Eq. (3.1.12) shows that $\boldsymbol{\sigma}$ satisfies the distinguishing property of second-order Cartesian tensors (1.1.16) with $\boldsymbol{\sigma}$ in place of $\mathbf{T}$.

One important consequence of the fact that $\boldsymbol{\sigma}$ is a tensor is that it has three scalar invariants. We note, in particular, that both the trace and the determinant of $\boldsymbol{\sigma}$ are independent of the choice of the Cartesian axes.

Cauchy's Equation of Motion

To derive the equivalent of Newton's second law for the motion of a fluid, we combine Eq. (3.1.5) with Eq. (3.1.10), use Eq. (1.3.19) along with the definition (1.3.17), note that the shape and size of the parcel are arbitrary to discard the integral sign, and thereby obtain *Cauchy's equation of motion*

$$\rho\frac{D\mathbf{u}}{Dt} = \nabla \cdot \boldsymbol{\sigma} + \rho\mathbf{g} \tag{3.1.13}$$

which *is applicable for both compressible and incompressible fluids*. Expressing the material derivative D/Dt in terms of Eulerian derivatives, we obtain

$$\rho\left(\frac{\partial\mathbf{u}}{\partial t} + \mathbf{u} \cdot \nabla\mathbf{u}\right) = \nabla \cdot \boldsymbol{\sigma} + \rho\mathbf{g} \tag{3.1.14}$$

Furthermore, using the continuity equation (1.3.12), we derive the alternative form

$$\frac{\partial(\rho\mathbf{u})}{\partial t} = \nabla\cdot\Sigma + \rho\mathbf{g} \tag{3.1.15}$$

where we have introduced the stress–momentum tensor

$$\Sigma = \sigma - \rho\mathbf{u}\mathbf{u} \tag{3.1.16}$$

When the flow is steady, the left-hand side of Eq. (3.1.15) vanishes, and the divergence of the stress–momentum tensor is balanced by the body force.

Momentum Integral Balance

Integrating Eq. (3.1.15) over a control volume V_c that is fixed in space and lies entirely within the fluid, and using the divergence theorem to convert the volume integral of the divergence of the stress–momentum tensor to a surface integral over the boundary D of V_c, we obtain the momentum integral or macroscopic balance

$$\int_{V_c} \frac{\partial(\rho\mathbf{u})}{\partial t}\, dV = -\int_D \Sigma\cdot\mathbf{n}\, dS + \int_{V_c} \rho\mathbf{g}\, dV \tag{3.1.17}$$

where $\mathbf{n}$ is the unit normal vector pointing *into* the control volume. Equation (3.1.17) states that the rate of change of the momentum of the fluid that occupies a control volume that is fixed in space is balanced by (1) the flow rate of the stress–momentum tensor outward from the boundaries of the control volume and (2) the body force exerted on the fluid that occupies the control volume.

The momentum integral balance finds extensive applications in computing global quantities of a steady or unsteady flow. Typically, it is used to obtain the boundary stresses from the boundary velocity and *vice versa*.

Torque on a Fluid Parcel

The total torque or couple with respect to a point $\mathbf{x}_0$ exerted on a fluid parcel is composed of the torque due to the surface traction, the torque due to the body force, and the torque due to an external torque field with intensity $\mathbf{c}$, and is given by

$$\mathbf{T}^{\text{Total}} = \int_{\text{Parcel}} \hat{\mathbf{x}}\times(\mathbf{n}\cdot\sigma)\, dS + \int_{\text{Parcel}} \hat{\mathbf{x}}\times(\rho\mathbf{g})\, dV + \int_{\text{Parcel}} \lambda\mathbf{c}\, dV \tag{3.1.18}$$

where $\hat{\mathbf{x}} = \mathbf{x} - \mathbf{x}_0$ and λ is an appropriate physical constant associated with $\mathbf{c}$. A torque field may arise in a suspension of bipolar particulates, in the presence of an electric field. The divergence theorem allows us to convert the surface integral on the right-hand side of Eq. (3.1.18) to a volume integral, obtaining

$$T_i^{\text{Total}} = \int_{\text{Parcel}} \left(\varepsilon_{ilk}\sigma_{lk} + \varepsilon_{ijk}\hat{x}_j\frac{\partial\sigma_{lk}}{\partial x_l} + \rho\varepsilon_{ijk}\hat{x}_j g_k + \lambda c_i\right) dV \tag{3.1.19}$$

In the absence of the external torque field $\mathbf{c}$, the torque exerted on the parcel with respect to a point $\mathbf{x}_1$, $\mathbf{T}^{\text{Total}}(\mathbf{x}_1)$, is related to the torque with respect to the point $\mathbf{x}_0$ by $\mathbf{T}^{\text{Total}}(\mathbf{x}_1) = \mathbf{T}^{\text{Total}} + (\mathbf{x}_1 - \mathbf{x}_0)\times\mathbf{F}^{\text{Total}}(\mathbf{x}_0)$. When the total force exerted on the parcel vanishes, which means that the rate of change of the linear momentum of the parcel is negligible, the value of the torque is independent of the location of the point with respect to which it is computed.

Hydrodynamic Torque Exerted on a Boundary

The hydrodynamic torque with respect to a point $\mathbf{x}_0$ exerted on a boundary is given by the simplified version of Eq. (3.1.18),

$$\mathbf{T} = \int_{\text{Boundary}} \hat{\mathbf{x}}\times(\mathbf{n}\cdot\sigma)\, dS \tag{3.1.20}$$

where $\mathbf{n}$ is the unit normal vector pointing *into* the fluid. When the hydrodynamic force exerted on the boundary vanishes, the value of the hydrodynamic torque is independent of the location of the point $\mathbf{x}_0$ (Problem 3.1.3).

Symmetry of the Stress Tensor

The *angular momentum balance* for a fluid parcel requires that the rate of change of the angular momentum with respect to the point $\mathbf{x}_0$ of the fluid that occupies the parcel, be equal to the total torque exerted on the parcel given in Eq. (3.1.18) or (3.1.19),

$$\frac{d}{dt} \int_{\text{Parcel}} \rho \hat{\mathbf{x}} \times \mathbf{u} \, dV = \mathbf{T}^{\text{Total}} \tag{3.1.21}$$

Using Eq. (1.3.21), we simplify the rate of change of angular momentum on the left-hand side of Eq. (3.1.21), switch to index notation, express the total torque in terms of the right-hand side of Eq. (3.1.19), and rearrange to obtain

$$\int_{\text{Parcel}} \left[\varepsilon_{ijk} \hat{x}_j \left(\rho \frac{Du_k}{Dt} - \frac{\partial \sigma_{lk}}{\partial x_l} - \rho g_k \right) - \varepsilon_{ilk} \sigma_{lk} - \lambda c_i \right] dV = 0 \tag{3.1.22}$$

Furthermore, we make use of the fact that the volume of the parcel is arbitrary to discard the integral sign, and use the equation of motion (3.1.13) to simplify the integrand, finding

$$\varepsilon_{ilk} \sigma_{lk} + \lambda c_i = 0 \tag{3.1.23}$$

Multiplying Eq. (3.1.23) by ε_{imn} and manipulating the product of the alternating tensors as discussed in Section A.1, Appendix A, yields

$$\sigma_{mn} - \sigma_{nm} = -\lambda c_i \varepsilon_{imn} \tag{3.1.24}$$

which shows that when an external torque field $\mathbf{c}$ is absent, the stress tensor tensor is symmetric,

$$\sigma_{ij} = \sigma_{ji} \quad \text{or} \quad \boldsymbol{\sigma} = \boldsymbol{\sigma}^T \tag{3.1.25}$$

where the superscript T designates the transpose. In that case, only six out of the nine components of the stress tensor are independent, and the rest of them are equal to their transpose counterparts.

The fact that in the absence of a torque field, $\boldsymbol{\sigma}$ is real and symmetric guarantees that it has three real eigenvalues and corresponding orthogonal eigenvectors. The traction exerted on an infinitesimal planar surface that is perpendicular to an eigenvector is directed in the normal direction; choosing the Cartesian axes in the direction of the eigenvectors renders the stress tensor diagonal. If the fluid is isotropic, that is, it has no favorable directions, the eigenvectors of the stress tensor must coincide with those of the rate-of-deformation tensor, as will be discussed further in Section 3.2.

Energy Dissipation within a Parcel

The total energy of the fluid that resides within a parcel is composed of the *kinetic energy* due to the motion of the fluid, the *potential energy* due to the gravitational body force, and the *internal thermodynamic energy*. The instantaneous kinetic and potential energies are given respectively by

$$K_P = \frac{1}{2} \int_{\text{Parcel}} \rho \mathbf{u} \cdot \mathbf{u} \, dV, \qquad P_P = -\int_{\text{Parcel}} \rho \mathbf{X} \cdot \mathbf{g} \, dV \tag{3.1.26}$$

where $\mathbf{X}$ is the position of the point particles that occupy the parcel.

Taking the material derivative of the first equation in (3.1.26) and using Eq. (1.3.9), we express the rate of change of the kinetic energy of the fluid within the parcel in terms of the acceleration of the point particles as

$$\frac{dK_P}{dt} = \frac{1}{2} \int_{\text{Parcel}} \rho \frac{D(\mathbf{u} \cdot \mathbf{u})}{Dt} \, dV = \int_{\text{Parcel}} \rho \mathbf{u} \cdot \frac{D\mathbf{u}}{Dt} \, dV \tag{3.1.27}$$

Furthermore, using the equation of motion, we express the acceleration in terms of the stress tensor and the body force, obtaining

$$\frac{dK_P}{dt} = \int_{\text{Parcel}} \mathbf{u} \cdot (\nabla \cdot \boldsymbol{\sigma}) \, dV + \int_{\text{Parcel}} \rho \mathbf{u} \cdot \mathbf{g} \, dV \tag{3.1.28}$$

Next we manipulate the term within the first integral on the right-hand side of Eq. (3.1.28) and use the divergence theorem to write

$$\int_{\text{Parcel}} \mathbf{u} \cdot (\nabla \cdot \boldsymbol{\sigma}) \, dV = - \int_{\text{Parcel}} \mathbf{u} \cdot (\boldsymbol{\sigma} \cdot \mathbf{n}) \, dS - \int_{\text{Parcel}} \boldsymbol{\sigma} : \nabla \mathbf{u} \, dV \tag{3.1.29}$$

where the unit normal vector $\mathbf{n}$ points *into* the parcel and the double inner product of two matrices is defined in Section A.1, Appendix A. Working in parallel, we use Eq. (1.3.9) and recast the last integral on the right-hand side of Eq. (3.1.28) into the form

$$\int_{\text{Parcel}} \rho \frac{D\mathbf{X}}{Dt} \cdot \mathbf{g} \, dV = - \frac{dP_P}{dt} \tag{3.1.30}$$

where P_P is the potential energy of the parcel defined in the second of Eqs. (3.1.26). Substituting Eqs. (3.1.29) and (3.1.30) into Eq. (3.1.28) yields the *energy balance*

$$\frac{dK_P}{dt} = - \int_{\text{Parcel}} \mathbf{u} \cdot (\boldsymbol{\sigma} \cdot \mathbf{n}) \, dS - \int_{\text{Parcel}} \boldsymbol{\sigma} : \nabla \mathbf{u} \, dV - \frac{dP_P}{dt} \tag{3.1.31}$$

The first integral on the right-hand side of Eq. (3.1.31) expresses the rate of working of the tractions acting on the surface of the parcel. It then follows from the first law of thermodynamics that the second integral must express the rate of production of internal energy or energy dissipation within the parcel I_P, which is expended towards raising the temperature of the fluid. Thus,

$$\frac{dI_P}{dt} = \int_{\text{Parcel}} \boldsymbol{\sigma} : \nabla \mathbf{u} \, dV \tag{3.1.32}$$

One important message delivered by Eq. (3.1.32) is that the rate of production of internal energy within a fluid depends both upon the structure of the velocity field and distribution of stresses within the fluid. The two, however, are related by a constitutive equation, as will be discussed in Section 3.2.

Energy Integral Balance

Certain global features of a flow may be computed on the basis of an energy integral or macroscopic balance that derives by projecting the equation of motion (3.1.15) onto the velocity, yielding

$$\frac{\partial}{\partial t}(\tfrac{1}{2}\rho |\mathbf{u}|^2) = (\nabla \cdot \boldsymbol{\Sigma}) \cdot \mathbf{u} + \rho \mathbf{g} \cdot \mathbf{u} \tag{3.1.33}$$

Writing out the stress–momentum tensor on the right-hand side in terms of the velocity and stress, we find that for an incompressible fluid,

$$\frac{D}{Dt}(\tfrac{1}{2}\rho |\mathbf{u}|^2) = \frac{\partial}{\partial t}(\tfrac{1}{2}\rho |\mathbf{u}|^2) + \nabla \cdot (\tfrac{1}{2}\rho |\mathbf{u}|^2 \mathbf{u})$$

$$= \nabla \cdot (\boldsymbol{\sigma} \cdot \mathbf{u}) - \boldsymbol{\sigma} : \nabla \mathbf{u} + \rho \mathbf{g} \cdot \mathbf{u} \tag{3.1.34}$$

where D/Dt is the material derivative. Upon integrating Eq. (3.1.34) over a control volume V_c that is fixed in space and is enclosed by the surface D and using the divergence theorem, we obtain the energy integral balance

$$
\int_{V_c} \frac{\partial}{\partial t}(\tfrac{1}{2}\rho\,|\mathbf{u}|^2)\,dV = \int_D (\tfrac{1}{2}\rho\,|\mathbf{u}|^2)\mathbf{u}\cdot\mathbf{n}\,dS - \int_D \mathbf{u}\cdot\mathbf{f}\,dS
$$
$$
- \int_{V_c} \boldsymbol{\sigma}:\nabla\mathbf{u}\,dV + \int_{V_c} \rho\mathbf{g}\cdot\mathbf{u}\,dV
$$

(3.1.35)

where the unit normal vector $\mathbf{n}$ is directed *into* V_c.

The four terms on the right-hand side of Eq. (3.1.35) represent, respectively, the *rate of supply of kinetic energy* into the control volume by convection, the *rate of working of the tractions* at the boundary of the control volume, the rate of *energy dissipation,* and the *rate of working against the body force.*

When the density of the fluid is uniform throughout the domain of flow, the last volume integral in Eq. (3.1.35) may be expressed in terms of a surface integral over D by writing

$$
\int_{V_c} \rho\mathbf{g}\cdot\mathbf{u}\,dV = \rho\int_{V_c} \nabla\cdot[\mathbf{u}(\mathbf{g}\cdot\mathbf{x})]\,dV = -\rho\int_D (\mathbf{g}\cdot\mathbf{x})(\mathbf{u}\cdot\mathbf{n})\,dS \qquad (3.1.36)
$$

If, in addition, the normal component of the velocity over D obeys the no-penetration boundary condition for a translating body, $\mathbf{u}\cdot\mathbf{n} = \mathbf{U}\cdot\mathbf{n}$, where $\mathbf{U}$ is a constant, we may apply the divergence theorem to find that the value of the last integral is equal to $-\rho V_{cv}\mathbf{g}\cdot\mathbf{U}$, where V_{cv} is the volume of the control volume. In this case the last term in Eq. (3.1.35) may be identified with the rate of working that is necessary in order to elevate the fluid within the control volume with velocity $\mathbf{U}$.

The Equation of Motion in a Noninertial Frame

Cauchy's equation of motion was derived under the assumption that the frame of reference was inertial, which requires that the Cartesian axes are either stationary or translate in space at a constant velocity, but exhibit neither acceleration nor rotation. In certain cases, it is convenient to work with a noninertial frame whose origin translates with respect to an inertial frame with a time-dependent velocity $\mathbf{U}(t)$, while its axes rotate about the instantaneous position of the origin with a time-dependent angular velocity $\boldsymbol{\Omega}(t)$. Since Newton's law applies for inertial systems only, some modifications are necessary in order to account for the implicit linear and angular acceleration of the fluid.

Our first task will be to compute the acceleration of a point particle in terms of its coordinates in the noninertial frame. To begin, we introduce the three unit vectors $\mathbf{i}_1, \mathbf{i}_2, \mathbf{i}_3$ associated with the noninertial coordinates y_1, y_2, y_3, and express the position of a point particle in the inertial frame in the form

$$
\mathbf{X} = \mathbf{x}_0 + Y_i\mathbf{i}_i \qquad (3.1.37)
$$

where $\mathbf{x}_0$ is the instantaneous position of the origin of the noninertial system. By definition, we have

$$
\frac{d\mathbf{x}_0}{dt} = \mathbf{U}, \qquad \frac{d\mathbf{i}_i}{dt} = \boldsymbol{\Omega}\times\mathbf{i}_i \qquad (3.1.38)
$$

Taking the material derivative of Eq. (3.1.37) and using Eq. (3.1.38), we find

$$
\frac{D\mathbf{X}}{Dt} = \frac{d\mathbf{x}_0}{dt} + \frac{DY_i}{Dt}\mathbf{i}_i + Y_i\frac{d\mathbf{i}_i}{dt} = \mathbf{U} + \frac{DY_i}{Dt}\mathbf{i}_i + \boldsymbol{\Omega}\times(Y_i\mathbf{i}_i) \qquad (3.1.39)
$$

The second term on the right-hand side is the velocity of a point particle in the noninertial system, $\mathbf{v} = (DY_i/Dt)\mathbf{i}_i$. To compute the acceleration of a point particle, we take the material derivative of Eq. (3.1.39) and obtain

$$\frac{D^2\mathbf{X}}{Dt^2} = \frac{d\mathbf{U}}{dt} + \frac{D^2Y_i}{Dt^2}\mathbf{i}_i + \frac{DY_i}{Dt}\frac{d\mathbf{i}_i}{dt} + \boldsymbol{\Omega} \times \left(\frac{DY_i}{Dt}\mathbf{i}_i\right)$$
$$+ \frac{d\boldsymbol{\Omega}}{dt} \times \mathbf{Y} + \boldsymbol{\Omega} \times \left(Y_i\frac{d\mathbf{i}_i}{dt}\right) \tag{3.1.40}$$

The second term on the right-hand side is the acceleration of the point particles in the noninertial system $D\mathbf{v}/Dt$. Using the second equation in (3.1.38) to simplify the third term on the right-hand side of Eq. (3.1.40), we obtain

$$\frac{D^2\mathbf{X}}{Dt^2} = \frac{d\mathbf{U}}{dt} + \frac{D\mathbf{v}}{Dt} + 2\boldsymbol{\Omega} \times \mathbf{v} + \frac{d\boldsymbol{\Omega}}{dt} \times \mathbf{Y} + \boldsymbol{\Omega} \times (\boldsymbol{\Omega} \times \mathbf{Y}) \tag{3.1.41}$$

Finally, we substitute the right-hand side of Eq. (3.1.41) for the acceleration of the point particles into the equation of motion (3.1.13) and thereby obtain a generalized equation of motion that is applicable for a noninertial Cartesian system

$$\rho\frac{D\mathbf{v}}{Dt} = \nabla \cdot \boldsymbol{\sigma} + \rho\mathbf{g} + \mathbf{F}^{\text{Inertial}} \tag{3.1.42}$$

where we have introduced the *fictitious inertial force*

$$\mathbf{F}^{\text{Inertial}} = \rho\left(-\frac{d\mathbf{U}}{dt} - 2\boldsymbol{\Omega} \times \mathbf{v} - \boldsymbol{\Omega} \times (\boldsymbol{\Omega} \times \mathbf{y}) - \frac{d\boldsymbol{\Omega}}{dt} \times \mathbf{y}\right) \tag{3.1.43}$$

containing four components: (1) the *linear-acceleration force* $-\rho\, d\mathbf{U}/dt$, (2) the *Coriolis force* $-2\rho\,\boldsymbol{\Omega} \times \mathbf{v}$, (3) the *centrifugal force* $-\rho\,\boldsymbol{\Omega} \times (\boldsymbol{\Omega} \times \mathbf{u})$, and (4) the *angular-acceleration force* $-\rho(d\boldsymbol{\Omega}/dt) \times \mathbf{y}$.

Stresses and the Equation of Motion in Curvilinear Coordinates

The stress tensor was introduced in Eq. (3.1.4) in terms of the three tractions $\mathbf{f}^x, \mathbf{f}^y, \mathbf{f}^z$ that are exerted on three mutually perpendicular infinitesimal planar surfaces that are normal to the x, y, and z axes. Working in the same spirit, we introduce the components of the stress tensor with reference to a general curvilinear, orthogonal, or nonorthogonal system of coordinates.

Cylindrical polar coordinates

In cylindrical polar coordinates (x, σ, φ), the components of the stress tensor are defined in terms of the equation

$$\mathbf{f}^\alpha = \sigma_{\alpha\beta}\mathbf{e}_\beta \tag{3.1.44}$$

as shown in Figure 3.1.2(a), where Greek indices stand for x, σ, or φ, with corresponding unit vectors $\mathbf{e}_\beta$, and summation over the index β is implied on the right-hand side. The corresponding components of the equation of motion are shown in Table 3.1.1(a).

The negative of the last term on the right-hand side in the σ equation u_φ^2/σ, and the last term on the right-hand side in the φ equation $u_\sigma u_\varphi/\sigma$, are commonly called the centrifugal force and the Coriolis force per unit mass of the fluid. This terminology is not entirely consistent, for the centrifugal and Coriolis forces are attributed to a noninertial system. To understand the physical motivation for this practice, consider a fluid that executes rigid-body rotation with constant azimuthal velocity. Describing the motion in a stationary frame of reference identifies u_φ^2/σ with the centripetal acceleration of the fluid parcels, which must be balanced by a radially oriented force. Describing the motion in a noninertial frame of reference that rotates with the fluid shows that the Coriolis force vanishes, and the centrifugal force per unit mass of fluid is equal to u_φ^2/σ. The temptation to call the term u_φ^2/σ in the inertial system the centrifugal force per unit mass of the fluid is then apparent.

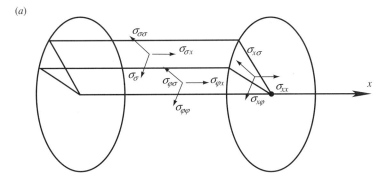

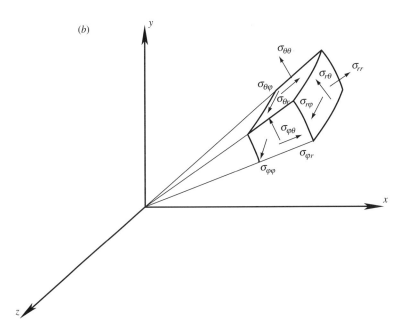

Figure 3.1.2   The components of the stress tensor in (a) cylindrical polar and (b) spherical polar coordinates.

Spherical polar coordinates

In spherical polar coordinates (r, θ, φ), the components of the stress tensor are defined in terms of Eq. (3.1.44), as shown in Figure 3.1.2(b), where Greek indices stand for r, θ, or φ. The corresponding components of the equation of motion are shown in Table 3.1.1(b).

PROBLEMS

3.1.1   **Normal component of the traction.** Verify that, in terms of the stress tensor, the normal component of the traction is given by

$$\mathbf{f}^N = [\boldsymbol{\sigma} : (\mathbf{nn})]\,\mathbf{n} \qquad\qquad (3.1.45)$$

The double dot product is defined in Section A.1, Appendix A.

TABLE 3.1.1

(a) The x, σ, and φ components of the equation of motion in cylindrical polar coordinates; (b) the r, θ, and φ components of the equation of motion in spherical polar coordinates.

(a)

$$\frac{\partial u_x}{\partial t} + u_x \frac{\partial u_x}{\partial x} + u_\sigma \frac{\partial u_x}{\partial \sigma} + u_\varphi \frac{1}{\sigma}\frac{\partial u_x}{\partial \varphi} = \frac{1}{\rho}\left(\frac{\partial \sigma_{xx}}{\partial x} + \frac{1}{\sigma}\frac{\partial(\sigma\sigma_{\sigma x})}{\partial \sigma} + \frac{1}{\sigma}\frac{\partial \sigma_{\varphi x}}{\partial \varphi}\right) + g_x$$

$$\frac{\partial u_\sigma}{\partial t} + u_x \frac{\partial u_\sigma}{\partial x} + u_\sigma \frac{\partial u_\sigma}{\partial \sigma} + u_\varphi \frac{1}{\sigma}\frac{\partial u_\sigma}{\partial \theta} - \frac{u_\varphi^2}{\sigma} = \frac{1}{\rho}\left(\frac{\partial \sigma_{x\sigma}}{\partial x} + \frac{1}{\sigma}\frac{\partial(\sigma\sigma_{\sigma\sigma})}{\partial \sigma} + \frac{1}{\sigma}\frac{\partial \sigma_{\sigma\varphi}}{\partial \varphi} - \frac{\sigma_{\varphi\varphi}}{\sigma}\right) + g_\sigma$$

$$\frac{\partial u_\varphi}{\partial t} + u_x \frac{\partial u_\varphi}{\partial x} + u_\sigma \frac{\partial u_\varphi}{\partial \sigma} + u_\varphi \frac{1}{\sigma}\frac{\partial u_\varphi}{\partial \theta} + \frac{u_\sigma u_\varphi}{\sigma} = \frac{1}{\rho}\left(\frac{\partial \sigma_{x\varphi}}{\partial x} + \frac{1}{\sigma^2}\frac{\partial(\sigma^2\sigma_{\sigma\varphi})}{\partial \sigma} + \frac{1}{\sigma}\frac{\partial \sigma_{\varphi\varphi}}{\partial \varphi}\right) + g_\varphi$$

(b)

$$\frac{\partial u_r}{\partial t} + u_r \frac{\partial u_r}{\partial r} + u_\theta \frac{1}{r}\frac{\partial u_r}{\partial \theta} + u_\varphi \frac{1}{r\sin\theta}\frac{\partial u_r}{\partial \varphi} - \frac{u_\theta^2 + u_\varphi^2}{r} = \frac{1}{\rho}\left(\frac{1}{r^2}\frac{\partial(r^2\sigma_{rr})}{\partial r} + \frac{1}{r\sin\theta}\frac{\partial(\sigma_{\theta r}\sin\theta)}{\partial \theta} + \frac{1}{r\sin\theta}\frac{\partial \sigma_{\varphi r}}{\partial \varphi} - \frac{\sigma_{\theta\theta} + \sigma_{\varphi\varphi}}{r}\right) + g_r$$

$$\frac{\partial u_\theta}{\partial t} + u_r \frac{\partial u_\theta}{\partial r} + u_\theta \frac{1}{r}\frac{\partial u_\theta}{\partial \theta} + u_\varphi \frac{1}{r\sin\theta}\frac{\partial u_\theta}{\partial \varphi} + \frac{u_r u_\theta}{r} - u_\varphi^2 \frac{\cot\theta}{r} = \frac{1}{\rho}\left(\frac{1}{r^2}\frac{\partial(r^2\sigma_{r\theta})}{\partial r} + \frac{1}{r\sin\theta}\frac{\partial(\sigma_{\theta\theta}\sin\theta)}{\partial \theta} + \frac{1}{r\sin\theta}\frac{\partial \sigma_{\varphi\theta}}{\partial \varphi} + \frac{\sigma_{r\theta}}{r} - \frac{1}{r}\sigma_{\varphi\varphi}\cot\theta\right) + g_\theta$$

$$\frac{\partial u_\varphi}{\partial t} + u_r \frac{\partial u_\varphi}{\partial r} + u_\theta \frac{1}{r}\frac{\partial u_\varphi}{\partial \theta} + u_\varphi \frac{1}{r\sin\theta}\frac{\partial u_\varphi}{\partial \varphi} + \frac{u_\varphi}{r}(u_r + u_\theta\cot\theta) = \frac{1}{\rho}\left(\frac{1}{r^2}\frac{\partial(r^2\sigma_{r\varphi})}{\partial r} + \frac{1}{r}\frac{\partial \sigma_{\theta\varphi}}{\partial \theta} + \frac{2\cos\theta}{r}\sigma_{\theta\varphi} + \frac{1}{r\sin\theta}\frac{\partial \sigma_{\varphi\varphi}}{\partial \varphi}\right) + g_\varphi$$

3.1.2 **Hydrodynamic torque on a boundary.** Show that, when the hydrodynamic force exerted on a boundary vanishes, the hydrodynamic torque is independent of the location of the point with respect to which it is computed.

3.1.3 **Stress–momentum tensor.** Show that the stress–momentum matrix is a second-order tensor (see Section 1.1).

3.1.4 **Mean value of the stress tensor over a parcel.** Show that the mean value of the stress tensor over a fluid parcel is given by

$$\int_{\text{Parcel}} \sigma_{ij} \, dV = \int_{\text{Parcel}} \sigma_{ik} n_k x_j \, dS - \int_{\text{Parcel}} \frac{\partial \sigma_{ik}}{\partial x_k} x_j \, dV \qquad (3.1.46)$$

and then use the equation of motion to express the last integral in terms of the point-particle acceleration and body force. The unit vector $\mathbf{n}$ points outward from the parcel.

3.1.5 **Computing the surface force.** Assume that the stress tensor within a fluid is given by

$$\boldsymbol{\sigma} = \begin{bmatrix} x & xy & xyz \\ xy & y & z \\ xyz & z & z \end{bmatrix} \qquad (3.1.47)$$

in units of dyn/cm^2, where lengths are measured in cm. Evaluate the normal and shearing component of the traction (a) over the surface of a sphere centered at the origin and (b) over the surface of a cylinder that is coaxial with the x axis.

3.2 | CONSTITUTIVE RELATIONS FOR THE STRESS TENSOR

If a fluid has been in a macroscopic state of rest for a sufficiently long period of time, one should expect that the molecular motions would have reached a state of dynamical equilibrium in which the stress field assumes the isotropic form

$$\boldsymbol{\sigma} = -p^{\text{Th}} \mathbf{I} \qquad (3.2.1)$$

where $\mathbf{I}$ is the identity matrix and p^{Th} is the thermodynamic pressure that is a function of the density of the fluid and temperature, and depends on the chemical composition of the fluid in a manner that is determined by an appropriate equation of state. For instance, for an ideal gas, $p^{\text{Th}} = \rho R T / M$, where M is the molecular weight and R is the ideal-gas constant.

Physical intuition suggests that the instantaneous structure of the stress field $\boldsymbol{\sigma}$ within a fluid that has been in a state of motion for some time should depend not only upon the current thermodynamic conditions, but also upon the history of motion of all fluid parcels that comprise the fluid, from inception of the motion up to the present time. Specifically, leaving aside physico-chemical interactions that are independent of the motion, one may argue that $\boldsymbol{\sigma}$ must depend upon the structure of the velocity field at all prior times. This argument leads us to introduce a *constitutive equation for the stress tensor* that relates the stress field at a certain point at a particular instant to the structure of the velocity field at all previous times, writing

$$\boldsymbol{\sigma}(t = T) = \mathbf{G}[\mathbf{u}(-\infty < t < T)] \qquad (3.2.2)$$

where $\mathbf{G}$ is an operator that involves derivatives and integrals of the whole velocity field with respect to space and time, and contains a number of coefficients that may be regarded as physical properties of the fluid.

To be admissible, a constitutive equation must satisfy a number of conditions (Schowalter, 1978; Tanner, 1988). The condition of *coordinate invariance* requires the constitutive equation to be valid independently of the coordinate system in which the position vector, velocity, or stress is described. Thus, the functional form (3.2.2) must hold true independently of whether $\boldsymbol{\sigma}$ and $\mathbf{u}$ are expressed in Cartesian, cylindrical polar, spherical polar, or any other type of curvilinear coordinates. The condition of *material objectivity* requires the instantaneous stress field to be independent of the motion of the observer. The condition of *fading memory* requires that the instantaneous structure of the stress field depend upon the recent motion of the fluid more strongly than it does on its ancient history.

Next we argue that the history of rotation, deformation, and expansion of a fluid parcel rather than its velocity is significant as far as determining the stresses on its boundary; it is clear that rigid-body motions will generate no stresses. This suggests replacing the velocity in the arguments of the operator in Eq. (3.2.2) with the deformation gradient $\mathbf{F}$ introduced in Eq. (1.2.17), writing

$$\boldsymbol{\sigma}(t = T) = \mathbf{G}[\mathbf{F}(-\infty < t < T)] \tag{3.2.3}$$

Simple Fluids

A *simple fluid* is a fluid wherein the stress tensor at the position of a point particle labeled $\mathbf{a}$ is a function of the history of the deformation gradient $\mathbf{F}$ evaluated at all prior positions of that particular point particle, over all past times up to the present time (Truesdell, 1974). The constitutive equation for a simple fluid thus takes the form

$$\boldsymbol{\sigma}(\mathbf{a}, t = T) = \mathbf{G}[\mathbf{F}(\mathbf{a}, -\infty < t < T)] \tag{3.2.4}$$

It can be shown that a simple fluid is necessarily isotropic; that is, it has no favorable or unfavorable directions in space (Schowalter, 1978, p. 67).

It is sometimes useful to rewrite Eq. (3.2.4) in the alternative form

$$\boldsymbol{\sigma}(\mathbf{a}, t = T) = -p\mathbf{I} + \boldsymbol{\tau}[\mathbf{F}(\mathbf{a}, -\infty < t < T)] \tag{3.2.5}$$

where p is the *hydrodynamic* or *reaction pressure* defined by $p = -\frac{1}{3}\mathrm{Tr}(\boldsymbol{\sigma})$, and $\boldsymbol{\tau}$ is the *deviatoric* component of the stress tensor. Since $\boldsymbol{\sigma}$ is a tensor, its trace and therefore the pressure are invariant under changes of the axes of the Cartesian coordinates. Physically, $-p$ may be identified with the mean value of the normal component of the traction exerted on a small surface that is located at a certain point, averaged over all possible orientations of the surface, or with the mean value of the normal component of the traction exerted on the surface of a small spherical parcel (Problem 3.2.1).

Purely Viscous Fluids

A *purely viscous fluid* is a simple fluid whose stress at the location of a certain point particle at a certain time instant is a function of the velocity gradient evaluated at the position of the point particle at that particular time instant *only* (Schowalter, 1978, p. 134). Thus, for a purely viscous fluid Eq. (3.2.4) takes the simplified form

$$\boldsymbol{\sigma}(\mathbf{a}, t = T) = \mathbf{G}[\mathbf{L}(\mathbf{a}, t = T)] \tag{3.2.6}$$

Applying the principle of material objectivity, we find that the functional form of the operator $\mathbf{G}$ in Eq. (3.2.6) must be such that the antisymmetric part of $\mathbf{L}$ drops out, which means that the stress is a function of the rate of deformation tensor $\mathbf{E}$, and we write

$$\boldsymbol{\sigma}(\mathbf{a}, t = T) = \mathbf{G}[\mathbf{E}(\mathbf{a}, t = T)] \tag{3.2.7}$$

which was first proposed by Stokes in 1845. Furthermore, using the fact that $\boldsymbol{\sigma}$ as well as $\mathbf{E}, \mathbf{E}^2, \mathbf{E}^3, \ldots$ are all second-order tensors and therefore transform in a similar manner as discussed in Section 1.1, we find that the most general form of Eq. (3.2.7) is

$$\boldsymbol{\sigma}(\mathbf{a}, t = T) = f_0(I_1, I_2, I_3)\mathbf{I} + f_1(I_1, I_2, I_3)\mathbf{E} + f_2(I_1, I_2, I_3)\mathbf{E}^2 + \cdots \tag{3.2.8}$$

where f_i are functions of the three invariants of the rate of deformation tensor introduced in Eqs. (1.1.21). Built into Eq. (3.2.8) is the assumption that the principal directions of the stress tensor are identical to those of the rate of deformation tensor, as required by the stipulation that the fluid is isotropic. By the Cayley–Hamilton theorem (Wilkinson 1965, p. 38), we can express $\mathbf{E}^n$ for $n \geq 3$ as a linear combination of $\mathbf{I}, \mathbf{E},$ and $\mathbf{E}^2$ with coefficients that are functions of the three invariants, and this allows us to retain only the three leading terms on the right-hand side of Eq. (3.2.8). Fluids that obey Eq. (3.2.8) with three terms on the right-hand side are sometimes called *Reiner–Rivlin fluids*.

Generalized Newtonian Fluids

A *generalized Newtonian fluid* is a Reiner–Rivlin fluid for which the function f_2 is equal to zero, while the functions f_0 and f_1 are not necessarily equal to zero. Examples of generalized Newtonian fluids are the power-law and Bingham plastic fluids.

Consider, for instance, a unidirectional flow $\mathbf{u} = (u(y), 0, 0)$ along the x axis in which the shear stress σ_{xy} is given by

$$\sigma_{xy} = \Lambda \left| \frac{du}{dy} \right|^n \frac{du}{dy} \tag{3.2.9}$$

where Λ and n are two physical constants. This scalar constitutive equation describes a generalized Newtonian fluid, called the *power-law fluid*. Setting $n = 0$, we obtain a Newtonian fluid.

Newtonian Fluids

A *Newtonian fluid* is a Reiner–Rivlin fluid whose stress tensor depends in a linear manner upon the rate of deformation tensor. This requires that all functions f_n with $n > 1$ in Eq. (3.2.8) vanish, f_1 be a constant, and f_0 be a linear function of the invariant $I_3 = \nabla \cdot \mathbf{u}$ defined in Eqs. (1.1.21), which is equal to the rate of expansion of the fluid. We thus set

$$f_0 = -p - \tfrac{2}{3}\mu \nabla \cdot \mathbf{u}, \qquad f_1 = 2\mu \tag{3.2.10}$$

where p is the hydrodynamic pressure, allowed to be a linear function of $\nabla \cdot \mathbf{u}$, and μ is a physical constant with dimensions of mass per time and length called the *dynamic viscosity* or simply the *viscosity* of the fluid. The constitutive equation for a Newtonian fluid thus takes the linear form

$$\boldsymbol{\sigma}(\mathbf{a}, t = T) = -p(\mathbf{a}, t = T)\mathbf{I} - \tfrac{2}{3}\mu \nabla \cdot \mathbf{u}(\mathbf{a}, t = T)\mathbf{I} + 2\mu\, \mathbf{E}(\mathbf{a}, t = T) \tag{3.2.11}$$

Note that the trace of the deviatoric component of the stress tensor comprised of the second and third terms on the right-hand side vanishes, as required by the definition of p in Eq. (3.2.5). This becomes more evident by rewriting Eq. (3.2.11) in the form

$$\boldsymbol{\sigma}(\mathbf{a}, t = T) = -p(\mathbf{a}, t = T)\mathbf{I} + 2\mu\, \mathbf{E}^D(\mathbf{a}, t = T) \tag{3.2.12}$$

where $\mathbf{E}^D = \mathbf{E} - \tfrac{1}{3}(\nabla \cdot \mathbf{u})\mathbf{I}$ is the deviatoric component of $\mathbf{E}$.

The dynamic viscosity μ is often measured in terms of the unit *poise*, which is equal to 1 g/(cm sec). In general, the viscosity of gases increases, whereas the viscosity of liquids decreases as the temperature is raised. The first column of Table 3.2.1 summarizes certain values of μ for water and air at different temperatures.

TABLE 3.2.1
Values of the dynamic viscosity μ and kinematic viscosity ν of water and air at several temperatures.

		μ [in centipoise $= 10^{-2}$ g/(cm sec)]	$\nu = \mu/\rho$ (in 10^{-2} cm^2/sec)
Water	20°C	1.002	1.004
	40°C	0.653	0.658
	80°C	0.355	0.365
Air	20°C	0.0181	15.05
	40°C	0.0191	18.86
	80°C	0.0209	20.88

The first term on the right-hand side of Eq. (3.2.11) represents the isotropic component of $\boldsymbol{\sigma}$. If a fluid is left in a macroscopic state of rest for a sufficiently long period of time, one expects that p should become equal to the thermodynamic pressure p^{Th} whose value may be computed in terms of the local density of the fluid and temperature by means of an appropriate equation of state. Under flow conditions, we use the fact that the fluid is isotropic and exploit the required linear dependence of $\boldsymbol{\sigma}$ on $\mathbf{E}$ to write

$$p = p^{\mathrm{Th}} - \kappa \, \nabla \cdot \mathbf{u} \tag{3.2.13}$$

where κ is a physical constant with dimensions of mass per time and length called the *dilatational viscosity, expansion viscosity,* or *second coefficient of viscosity* (Rosenhead, 1954). The Newtonian constitutive equation then obtains the form

$$
\begin{aligned}
\boldsymbol{\sigma}(\mathbf{a}, t = T) &= -p^{\mathrm{Th}}(\mathbf{a}, t = T)\mathbf{I} + (\kappa - \tfrac{2}{3}\mu)\nabla \cdot \mathbf{u}(\mathbf{a}, t = T)\mathbf{I} \\
&\quad + 2\mu\mathbf{E}(\mathbf{a}, t = T) \\
&= -p^{\mathrm{Th}}(\mathbf{a}, t = T)\mathbf{I} + \kappa\,\nabla \cdot \mathbf{u}(\mathbf{a}, t = T)\mathbf{I} + 2\mu\mathbf{E}^{D}(\mathbf{a}, t = T)
\end{aligned}
\tag{3.2.14}
$$

which is known to describe with high accuracy the stress distribution within a broad range of fluids whose molecules have a simple configuration. The coefficient $\mu' = \kappa - \tfrac{2}{3}\mu$ is sometimes called the second coefficient of viscosity, although this term is usually reserved for κ.

Substituting the right-hand side of Eq. (3.2.14) into Eq. (3.1.32), we find that the rate of production of internal energy within a parcel is given by

$$\frac{dI_P}{dt} = -\int_{\mathrm{Parcel}} p^{\mathrm{Th}}\,\nabla \cdot \mathbf{u}\,dV + \int_{\mathrm{Parcel}} \kappa(\nabla \cdot \mathbf{u})^2\,dV + \int_{\mathrm{Parcel}} 2\mu\,\mathbf{E}^{D} : \mathbf{E}^{D}\,dV \tag{3.2.15}$$

The first term on the right-hand side expresses *reversible* production of energy, in the usual sense of thermodynamics. The second term expresses *irreversible* dissipation of energy due to the expansion of the fluid, and this justifies calling κ the expansion viscosity. The third term expresses irreversible dissipation of energy due to motions other than those representing isotropic expansion.

Incompressible Newtonian Fluids

When a Newtonian fluid is incompressible, the rate of expansion vanishes, and Eq. (3.2.11) takes the simplified form

$$\boldsymbol{\sigma}(\mathbf{a}, t = T) = -p(\mathbf{a}, t = T)\mathbf{I} + 2\mu\,\mathbf{E}(\mathbf{a}, t = T) \tag{3.2.16}$$

Note that, because the trace of the rate of deformation tensor $\mathbf{E}$ vanishes, the trace of $\boldsymbol{\sigma}$ is equal to $-3p$.

At this point, it is important to acknowledge that by requiring incompressibility, that is, by assuming that fluid parcels maintain their original volume and density, we essentially abandon the relation between the hydrodynamic and thermodynamic pressure, and we can no longer require Eq. (3.2.13). Alternatively, a Newtonian fluid may be considered to arise in the limit as κ tends to infinity and $\nabla \cdot \mathbf{u}$ tends to vanish, so that their product takes a finite value.

Reverting to Eulerian variables and writing out the rate of deformation tensor in terms of the derivatives of the velocity, we recast Eq. (3.2.16) into the form

$$\boldsymbol{\sigma}(\mathbf{x}, t) = -p(\mathbf{x}, t)\mathbf{I} + \mu\,\nabla\mathbf{u}(\mathbf{x}, t) + \mu[\nabla\mathbf{u}(\mathbf{x}, t)]^{T} \tag{3.2.17}$$

Explicit expressions for the components of the stress tensor in terms of the velocity and pressure in cylindrical and spherical polar coordinates are shown in Table 3.2.2. Note that the stress tensor remains symmetric in these curvilinear coordinates.

Inviscid Fluids

Inviscid fluids, sometimes called *ideal fluids,* are Newtonian fluids with vanishing viscosity. The constitutive equation for an incompressible inviscid fluid derives from Eq. (3.2.17) by setting $\mu = 0$, and this yields

$$\boldsymbol{\sigma}(\mathbf{a}, t = T) = -p(\mathbf{a}, t = T)\mathbf{I} \tag{3.2.18}$$

TABLE 3.2.2
Constitutive relations for the components of the stress tensor of an incompressible Newtonian fluid in (a) cylindrical (x, σ, φ) and (b) spherical (r, θ, φ) polar coordinates. Note that the stress tensor remains symmetric in these curvilinear coordinates.

(a)

$$
\begin{bmatrix} \sigma_{xx} & \sigma_{x\sigma} & \sigma_{x\varphi} \\ \sigma_{\sigma x} & \sigma_{\sigma\sigma} & \sigma_{\sigma\varphi} \\ \sigma_{\varphi x} & \sigma_{\varphi\sigma} & \sigma_{\varphi\varphi} \end{bmatrix} = -p\mathbf{I} + \mu
\begin{bmatrix}
2\dfrac{\partial u_x}{\partial x} & \dfrac{\partial u_x}{\partial \sigma} + \dfrac{\partial u_\sigma}{\partial x} & \dfrac{\partial u_\varphi}{\partial x} + \dfrac{1}{\sigma}\dfrac{\partial u_x}{\partial \varphi} \\[2mm]
\dfrac{\partial u_x}{\partial \sigma} + \dfrac{\partial u_\sigma}{\partial x} & 2\dfrac{\partial u_\sigma}{\partial \sigma} & \sigma\dfrac{\partial}{\partial \sigma}\left(\dfrac{u_\varphi}{\sigma}\right) + \dfrac{2}{\sigma}\dfrac{\partial u_\sigma}{\partial \varphi} \\[2mm]
\dfrac{\partial u_\varphi}{\partial x} + \dfrac{1}{\sigma}\dfrac{\partial u_x}{\partial \varphi} & \sigma\dfrac{\partial}{\partial \sigma}\left(\dfrac{u_\varphi}{\sigma}\right) + \dfrac{2}{\sigma}\dfrac{\partial u_\sigma}{\partial \varphi} & \dfrac{2}{\sigma}\dfrac{\partial u_\varphi}{\partial \varphi} + \dfrac{2}{\sigma}u_\sigma
\end{bmatrix}
$$

(b)

$$
\begin{bmatrix} \sigma_{rr} & \sigma_{r\theta} & \sigma_{r\varphi} \\ \sigma_{\theta r} & \sigma_{\theta\theta} & \sigma_{\theta\varphi} \\ \sigma_{\varphi r} & \sigma_{\varphi\theta} & \sigma_{\varphi\varphi} \end{bmatrix} = -p\mathbf{I}
$$

$$
+\mu
\begin{bmatrix}
2\dfrac{\partial u_r}{\partial r} & r\dfrac{\partial}{\partial r}\left(\dfrac{u_\theta}{r}\right) + \dfrac{1}{r}\dfrac{\partial u_r}{\partial \theta} & \dfrac{1}{r\sin\theta}\dfrac{\partial u_r}{\partial \varphi} + r\dfrac{\partial}{\partial r}\left(\dfrac{u_\varphi}{r}\right) \\[2mm]
r\dfrac{\partial}{\partial r}\left(\dfrac{u_\theta}{r}\right) + \dfrac{1}{r}\dfrac{\partial u_r}{\partial \theta} & \dfrac{2}{r}\dfrac{\partial u_\theta}{\partial \theta} + 2\dfrac{u_r}{r} & \dfrac{\sin\theta}{r}\dfrac{\partial}{\partial \theta}\left(\dfrac{u_\varphi}{\sin\theta}\right) + \dfrac{1}{r\sin\theta}\dfrac{\partial u_\theta}{\partial \varphi} \\[2mm]
\dfrac{1}{r\sin\theta}\dfrac{\partial u_r}{\partial \varphi} + r\dfrac{\partial}{\partial r}\left(\dfrac{u_\varphi}{r}\right) & \dfrac{\sin\theta}{r}\dfrac{\partial}{\partial \theta}\left(\dfrac{u_\varphi}{\sin\theta}\right) + \dfrac{1}{r\sin\theta}\dfrac{\partial u_\theta}{\partial \varphi} & \dfrac{1}{r\sin\theta}\left(2\dfrac{\partial u_\varphi}{\partial \varphi} + u_r\sin\theta + u_\theta\cos\theta\right)
\end{bmatrix}
$$

In real life no fluid can be truly inviscid, and Eq. (3.2.18) must be regarded as a mathematical idealization arising in the limit as the rate of deformation tensor $\mathbf{E}$ tends to become vanishingly small. Note, however, that at very low temperatures, superfluid helium behaves like an inviscid fluid and has been used in the laboratory in order to visualize the structure and dynamics of ideal flows.

Finally, it is instructive to note the similarity in functional form between Eqs. (3.2.18) and (3.2.1), but also emphasize that the pressure in Eq. (3.2.18) is a flow variable, whereas the pressure in Eq. (3.2.1) is a thermodynamic variable determined by an appropriate equation of state.

PROBLEM

3.2.1 **Hydrodynamic pressure.** Show that $-p$ is equal to the average value of the normal component of the traction exerted on the surface of a spherical fluid parcel with infinitesimal dimensions.

3.3 | **TRACTION, FORCE, TORQUE, ENERGY DISSIPATION, AND THE RECIPROCAL THEOREM FOR INCOMPRESSIBLE NEWTONIAN FLUIDS**

In the remainder of this book we shall concentrate on the motion of incompressible Newtonian fluids. In the present section we shall use the constitutive equation (3.2.17) to obtain specific expressions for the traction, force, and torque exerted on a fluid parcel and on a boundary of the flow in terms of the velocity and pressure; assess the rate of change of the internal energy of the fluid due to viscous dissipation; and derive the specific form of the integral energy balance.

Furthermore, we shall derive a reciprocal relation, which is similar to Green's third identity for harmonic functions, that imposes an integral constraint on the mutual structure of the velocity and stress fields of any two unrelated Newtonian flows of the same or different fluids.

Traction

Substituting Eq. (3.2.17) into Eq. (3.1.8), we find that, in terms of the instantaneous pressure and velocity, the traction exerted on the surface of a fluid parcel is given in vector and index notation by the expressions

$$\mathbf{f} = -p\mathbf{n} + 2\mu\mathbf{E} \cdot \mathbf{n} = -p\mathbf{n} + \mu(\nabla\mathbf{u}) \cdot \mathbf{n} + \mu\mathbf{n} \cdot (\nabla\mathbf{u}) \qquad (3.3.1a)$$

$$f_i = -pn_i + \mu \frac{\partial u_j}{\partial x_i} n_j + \mu n_j \frac{\partial u_i}{\partial x_j} \qquad (3.3.1b)$$

The third term on the right-hand sides contains the spatial derivative of the velocity in the direction of the normal vector. When the fluid is inviscid, we obtain a simplified form involving the pressure alone, $\mathbf{f} = -p\mathbf{n}$.

In terms of the vorticity tensor $\boldsymbol{\Xi}$ defined in Eq. (1.1.4), we obtain the equivalent form

$$\mathbf{f} = -p\mathbf{n} + 2\mu\mathbf{n} \cdot (\nabla\mathbf{u}) + 2\mu\boldsymbol{\Xi} \cdot \mathbf{n} = -p\mathbf{n} + 2\mu\mathbf{n} \cdot (\nabla\mathbf{u}) + \mu\mathbf{n} \times \boldsymbol{\omega} \qquad (3.3.2)$$

When the velocity field is irrotational, the last term on the right-hand side vanishes, and the component of the traction corresponding to the deviatoric part of the stress tensor is proportional to the derivative of the velocity in the direction of the normal vector.

Normal and tangential components

It is often useful to decompose the traction into its *normal* component $\mathbf{f}^N$ and *tangential* or *shear* component $\mathbf{f}^T$, so that $\mathbf{f} = \mathbf{f}^N + \mathbf{f}^T$. Projecting the last expression in Eq. (3.3.2) onto the normal vector yields

$$\mathbf{f}^N = [-p + 2\mu\mathbf{n} \cdot (\nabla\mathbf{u}) \cdot \mathbf{n}]\mathbf{n} \qquad (3.3.3)$$

Introducing a local Cartesian coordinate system with two axes perpendicular to $\mathbf{n}$ shows that the second term on the right-hand side of Eq. (3.3.3) expresses the viscous traction associated with the normal derivative of the normal component of the velocity, which is sometimes called the *rate of elongation* or *extension* of the fluid.

Forming the inner product between Eq. (3.3.1a) and the projection matrix $\mathbf{I} - \mathbf{nn}$, we obtain the tangential component of the traction in the form

$$\mathbf{f}^T = 2\mu\mathbf{n} \cdot \mathbf{E} \cdot (\mathbf{I} - \mathbf{nn}) = 2\mu\mathbf{n} \times (\mathbf{n} \cdot \mathbf{E}) \times \mathbf{n} \qquad (3.3.4)$$

In general, both the normal and tangential components of the traction have finite values. Along a *free surface*, which is defined as *a surface that cannot withstand a shear stress*, the tangential component vanishes, and the structure of the flow must be such that the right-hand side of Eq. (3.3.4) is equal to zero.

Force and Torque on a Fluid Parcel

Substituting Eq. (3.3.1a) into Eq. (3.1.9) and adding the body force, we find that the total force exerted on a parcel of an incompressible Newtonian fluid is given by

$$\mathbf{F}^{\text{Total}} = -\int_{\text{Parcel}} p\mathbf{n}\,dS + 2\int_{\text{Parcel}} \mu\mathbf{E} \cdot \mathbf{n}\,dS + \int_{\text{Parcel}} \rho\mathbf{g}\,dV \qquad (3.3.5)$$

where the unit normal vector $\mathbf{n}$ points *outwards* from the parcel.

Working in a similar manner, we substitute one of the expressions given in Eqs. (3.3.1) into Eq. (3.1.18) and find that, in the absence of an external torque field, the total torque with respect to a point $\mathbf{x}_0$ exerted on the parcel is given by

$$
\mathbf{T}^{\text{Total}} = -\int_{\text{Parcel}} p(\mathbf{x} - \mathbf{x}_0) \times \mathbf{n} \, dS + 2\int_{\text{Parcel}} \mu(\mathbf{x} - \mathbf{x}_0) \times (\mathbf{n} \cdot \mathbf{E}) \, dS
$$
$$
+ \int_{\text{Parcel}} \rho(\mathbf{x} - \mathbf{x}_0) \, dV \times \mathbf{g}
\tag{3.3.6}
$$

Modified Pressure and Stress

When the density of the fluid and the acceleration of gravity are uniform, it is beneficial to eliminate the explicit presence of the body force from the expressions (3.3.5) and (3.3.6) by introducing the *modified pressure* and corresponding *modified stress tensor* defined as

$$
P \equiv p - \rho \mathbf{g} \cdot \mathbf{x} + c
\tag{3.3.7}
$$
$$
\boldsymbol{\sigma}^{\text{Mod}} \equiv -P\mathbf{I} + 2\mu\mathbf{E} = \boldsymbol{\sigma} + \rho(\mathbf{g} \cdot \mathbf{x})\mathbf{I} - c\mathbf{I}
\tag{3.3.8}
$$

where c is an arbitrary constant. In terms of modified variables, the total force and torque with respect to the point $\mathbf{x}_0$ exerted on a fluid parcel are given as *surface integrals alone* in the form

$$
\mathbf{F}^{\text{Total}} = \int_{\text{Parcel}} \boldsymbol{\sigma}^{\text{Mod}} \cdot \mathbf{n} \, dS = -\int_{\text{Parcel}} P\mathbf{n} \, dS + 2\int_{\text{Parcel}} \mu\mathbf{E} \cdot \mathbf{n} \, dS
\tag{3.3.9}
$$

$$
\mathbf{T}^{\text{Total}} = -\int_{\text{Parcel}} P(\mathbf{x} - \mathbf{x}_0) \times \mathbf{n} \, dS + 2\int_{\text{Parcel}} \mu(\mathbf{x} - \mathbf{x}_0) \times (\mathbf{n} \cdot \mathbf{E}) \, dS
\tag{3.3.10}
$$

Analogous simplifications apply to the equation of motion to be discussed in Section 3.4.

Force and Torque Exerted on a Boundary

To obtain the hydrodynamic force exerted on a boundary, we substitute Eq. (3.3.1a) into Eq. (3.1.11) and obtain

$$
\mathbf{F} = -\int_{\text{Boundary}} p\mathbf{n} \, dS + 2\int_{\text{Boundary}} \mu\mathbf{E} \cdot \mathbf{n} \, dS
\tag{3.3.11}
$$

where the unit normal vector $\mathbf{n}$ points *into* the fluid. The first term on the right-hand side of Eq. (3.3.11) represents the *form drag,* and the second term represents the *skin friction.*

To obtain the hydrodynamic torque with respect to the point $\mathbf{x}_0$ exerted on a boundary, we substitute Eq. (3.3.1a) into Eq. (3.1.20) and obtain

$$
\mathbf{T} = -\int_{\text{Boundary}} p(\mathbf{x} - \mathbf{x}_0) \times \mathbf{n} \, dS + 2\int_{\text{Boundary}} \mu(\mathbf{x} - \mathbf{x}_0) \times (\mathbf{n} \cdot \mathbf{E}) \, dS
\tag{3.3.12}
$$

When the viscous forces are insignificant, the force and torque may be computed from a knowledge of the pressure distribution over the boundary alone.

Energy Dissipation within a Parcel

Substituting Eq. (3.2.17) into Eq. (3.1.32), we find that the rate of production of internal energy within a fluid parcel is given by

$$
\frac{dI_P}{dt} = \int_{\text{Parcel}} \Phi \, dV
\tag{3.3.13}
$$

where

$$
\Phi \equiv 2\mu\mathbf{E} : \mathbf{E}
\tag{3.3.14}
$$

is a non-negative quantity expressing the rate of *viscous dissipation per unit volume* of the fluid. Viscosity acts to dissipate energy, converting it into thermal energy, thereby raising the temperature of the fluid. When the rate of viscous dissipation is negligible, the sum of the kinetic and potential energy remains constant. It is instructive to note that Eq. (3.3.13) is a special version of Eq. (3.2.15), but also recall that the latter is applicable to generally compressible Newtonian fluids.

Relation between the Rate of Working of Tractions, Vorticity, and Viscous Dissipation

Substituting Eq. (3.2.17) into Eq. (3.1.29), we obtain the identity

$$\int_{\text{Parcel}} \mathbf{u} \cdot (\nabla \cdot \boldsymbol{\sigma}) \, dV = - \int_{\text{Parcel}} \mathbf{u} \cdot \mathbf{f} \, dS - \int_{\text{Parcel}} \Phi \, dV \qquad (3.3.15)$$

where $\mathbf{f} = \boldsymbol{\sigma} \cdot \mathbf{n}$ is the boundary traction and the normal vector $\mathbf{n}$ points *into* the parcel. The two terms on the right-hand side of Eq. (3.3.15) represent, respectively, the rate of working of the surface tractions and rate of viscous dissipation.

Working in a similar manner with the deviatoric part of the stress tensor defined in Eq. (3.2.5), we find

$$\int_{\text{Parcel}} \mathbf{u} \cdot (\nabla \cdot \boldsymbol{\tau}) \, dV = - \int_{\text{Parcel}} \mathbf{u} \cdot \mathbf{t} \, dS - \int_{\text{Parcel}} \Phi \, dV \qquad (3.3.16)$$

where $\mathbf{t} \equiv \boldsymbol{\tau} \cdot \mathbf{n}$. Furthermore, we use the Newtonian constitutive relation (3.2.17) and the continuity equation, and find that when the viscosity of the fluid is uniform, $\nabla \cdot \boldsymbol{\tau} = \mu \nabla^2 \mathbf{u} = -\mu \nabla \times \boldsymbol{\omega}$, which, in conjunction with Eq. (3.3.16), suggests that *when the flow is irrotational or the vorticity is constant, the rate of working of the deviatoric viscous traction is balanced by viscous dissipation,*

$$\int_{\text{Parcel}} \mathbf{u} \cdot \mathbf{t} \, dS = - \int_{\text{Parcel}} \Phi \, dV \qquad (3.3.17)$$

Energy Integral Balance

Substituting Eq. (3.2.17) into Eq. (3.1.35) we derive the explicit form of the energy integral balance over a fixed control volume V_c,

$$\int_{V_c} \frac{\partial}{\partial t} (\tfrac{1}{2} \rho |\mathbf{u}|^2) \, dV = \int_D (\tfrac{1}{2} \rho |\mathbf{u}|^2) \mathbf{u} \cdot \mathbf{n} \, dS - \int_D \mathbf{u} \cdot \mathbf{f} \, dS$$
$$- \int_{V_c} \Phi \, dV + \int_{V_c} \rho \mathbf{g} \cdot \mathbf{u} \, dV \qquad (3.3.18)$$

where the normal vector $\mathbf{n}$ points *into* the control volume. The five integrals in Eq. (3.3.18) represent, respectively, the rate of *accumulation* of kinetic energy within the control volume, the rate of *convection of kinetic energy* into the control volume, the rate of *working of surface forces*, the rate of *viscous dissipation*, and the rate of change of the *potential energy* associated with the body force.

When the flow is irrotational or the vorticity is constant, we use Eq. (3.3.17) to obtain the simplified form

$$\int_{V_c} \frac{\partial}{\partial t} (\tfrac{1}{2} \rho |\mathbf{u}|^2) \, dV = \int_D (\tfrac{1}{2} \rho |\mathbf{u}|^2) \mathbf{u} \cdot \mathbf{n} \, dS + \int_D p \mathbf{u} \cdot \mathbf{n} \, dS + \int_{V_c} \rho \mathbf{g} \cdot \mathbf{u} \, dV \qquad (3.3.19)$$

which is also valid when the vorticity is finite, but the fluid may be considered to be effectively inviscid.

Reciprocity of Newtonian Flows

Let us consider two unrelated incompressible flows of two different Newtonian fluids with densities ρ and ρ', and viscosities μ and μ', and corresponding velocity fields $\mathbf{u}$ and $\mathbf{u}'$ and stress fields $\boldsymbol{\sigma}$ and $\boldsymbol{\sigma}'$, and compute

$$u_i' \frac{\partial \sigma_{ij}}{\partial x_j} = \frac{\partial}{\partial x_j}(u_i' \sigma_{ij}) - \sigma_{ij} \frac{\partial u_i'}{\partial x_j} = \frac{\partial}{\partial x_j}(u_i' \sigma_{ij}) - \left[-p\,\delta_{ij} + \mu \left(\frac{\partial u_i}{\partial x_j} + \frac{\partial u_j}{\partial x_i} \right) \right] \frac{\partial u_i'}{\partial x_j}$$

$$= \frac{\partial}{\partial x_j}(u_i' \sigma_{ij}) - \mu \left(\frac{\partial u_i}{\partial x_j} + \frac{\partial u_j}{\partial x_i} \right) \frac{\partial u_i'}{\partial x_j} \qquad (3.3.20)$$

where we have used the continuity equation to eliminate the term involving the pressure. Note that setting $\mathbf{u}'$ equal to $\mathbf{u}$, integrating Eq. (3.320) over the volume of a parcel, and using the divergence theorem, we recover Eq. (3.3.15). Interchanging the roles of the two flows, we obtain the analogous form

$$u_i \frac{\partial \sigma_{ij}'}{\partial x_j} = \frac{\partial}{\partial x_j}(u_i \sigma_{ij}') - \mu' \left(\frac{\partial u_i'}{\partial x_j} + \frac{\partial u_j'}{\partial x_i} \right) \frac{\partial u_i}{\partial x_j} \qquad (3.3.21)$$

Multiplying Eq. (3.3.20) with μ' and Eq. (3.3.21) with μ, and subtracting corresponding sides of the resulting equations from each other, we obtain the differential statement of the *generalized Lorentz reciprocal identity*

$$\frac{\partial}{\partial x_j}(\mu' u_i' \sigma_{ij} - \mu u_i \sigma_{ij}') = \mu' u_i' \frac{\partial \sigma_{ij}}{\partial x_j} - \mu u_i \frac{\partial \sigma_{ij}'}{\partial x_j} \qquad (3.3.22)$$

which imposes an integral constraint upon the mutual structure of the velocity and stress fields of any two incompressible Newtonian flows. It is a straightforward exercise to show that Eqs. (3.3.20)–(3.3.22) are also applicable for the modified stress tensor defined in Eq. (3.3.8) (Problem 3.3.1).

Expressing the divergence of the stress tensors on the right-hand side of Eq. (3.3.22) in terms of the acceleration of the point particles and body force using the equation of motion yields the new form

$$\frac{\partial}{\partial x_j}(\mu' u_i' \sigma_{ij} - \mu u_i \sigma_{ij}') = \rho \mu' u_i' \frac{Du_i}{Dt} - \rho' \mu u_i \frac{Du_i'}{Dt} - (\rho \mu' u_i' - \rho' \mu u_i) g_i \qquad (3.3.23)$$

In terms of the modified stress tensor, we obtain the simpler form

$$\frac{\partial}{\partial x_j}(\mu' u_i' \sigma_{ij}^{\mathrm{Mod}} - \mu u_i \sigma_{ij}'^{\mathrm{Mod}}) = \rho \mu' u_i' \frac{Du_i}{Dt} - \rho' \mu u_i \frac{Du_i'}{Dt} \qquad (3.3.24)$$

Integrating Eq. (3.3.24) over a certain control volume V_c that is bounded by the surface D and using the divergence theorem to convert the volume integral of the left-hand side into a surface integral over D, we derive the integral form

$$\int_D (\mu' u_i' f_i^{\mathrm{Mod}} - \mu u_i f_i'^{\mathrm{Mod}})\, dS = \int_{V_c} \left(\mu' u_i' \frac{\partial \sigma_{ij}^{\mathrm{Mod}}}{\partial x_j} - \mu u_i \frac{\partial \sigma_{ij}'^{\mathrm{Mod}}}{\partial x_j} \right) dV \qquad (3.3.25)$$

where the normal vector $\mathbf{n}$ points *outward* from the control volume. In Chapter 6 we shall see that the reciprocal identities (3.3.24) and (3.3.25) find extensive applications in the study and computation of Stokes flow, where the inertia of the fluid has a negligible influence on the motion of the fluid parcels.

PROBLEMS

3.3.1   **Reciprocal identity.** (a) Show that Eqs. (3.3.20)–(3.3.22) are also applicable for the modified stress tensor defined in Eq. (3.3.8). (b) Derive Eq. (3.3.25) from Eq. (3.3.23).

3.3.2   **Momentum integral balance.** Derive the explicit form of the momentum integral balance expressed by Eq. (3.1.17) for an incompressible Newtonian fluid, in terms of the velocity and pressure.

3.4 | THE NAVIER–STOKES, EULER, AND BERNOULLI EQUATIONS

Substituting the constitutive equation for an incompressible Newtonian fluid, Eq. (3.2.17), into Cauchy's equation of motion (3.1.13), we obtain the *Navier–Stokes* equation

$$\rho \frac{D\mathbf{u}}{Dt} = -\nabla p + 2 \nabla \cdot (\mu \mathbf{E}) + \rho \mathbf{g} \qquad (3.4.1)$$

The density ρ and viscosity μ are allowed to vary in time and with position in the domain of flow.

When the density of the fluid and the acceleration of gravity are uniform throughout the domain of flow, it is convenient to work with the modified pressure and modified stress tensor introduced in Eqs. (3.3.7) and (3.3.8), obtaining

$$\rho \frac{D\mathbf{u}}{Dt} = \nabla \cdot \boldsymbol{\sigma}^{\text{Mod}} = -\nabla P + 2 \nabla \cdot (\mu \mathbf{E}) \qquad (3.4.2)$$

which is distinguished by the absence of the body force. In solving the Navier–Stokes equation, the distinction between the regular and modified pressure or stress becomes significant only when we require boundary conditions involving the pressure.

Substituting the definition of the rate-of-deformation tensor (1.1.5) into Eq. (3.4.1), expanding out the derivatives, using the continuity equation, and expressing the material derivative of the velocity in terms of Eulerian derivatives, we recast Eq. (3.4.1) into the more explicit form

$$\rho \left(\frac{\partial \mathbf{u}}{\partial t} + \mathbf{u} \cdot \nabla \mathbf{u} \right) = -\nabla p + \mu \nabla^2 \mathbf{u} + 2 \nabla \mu \cdot \mathbf{E} + \rho \mathbf{g} \qquad (3.4.3)$$

The four terms on the right-hand side of Eq. (3.4.3) express, respectively, the *pressure force,* the *viscous force,* a *force due to viscosity variations,* and the *body force.*

Dividing both sides of Eq. (3.4.3) by the density of the fluid, we obtain the new form

$$\frac{\partial \mathbf{u}}{\partial t} + \mathbf{u} \cdot \nabla \mathbf{u} = -\frac{1}{\rho} \nabla p + \nu \nabla^2 \mathbf{u} + \frac{2}{\rho} \nabla \mu \cdot \mathbf{E} + \mathbf{g} \qquad (3.4.4)$$

where $\nu = \mu/\rho$ is a new physical constant with dimensions of length squared over time, called the *kinematic viscosity* of the fluid. Certain values of ν for water and air at different temperatures are given in Table 3.2.1.

TABLE 3.4.1
The x, σ, and φ components of the Navier–Stokes equation for a fluid with uniform viscosity in cylindrical polar coordinates.

$$\frac{\partial u_x}{\partial t} + u_x \frac{\partial u_x}{\partial x} + u_\sigma \frac{\partial u_x}{\partial \sigma} + u_\varphi \frac{1}{\sigma} \frac{\partial u_x}{\partial \varphi}$$

$$= -\frac{1}{\rho} \frac{\partial p}{\partial x} + \nu \left[\frac{\partial^2 u_x}{\partial x^2} + \frac{1}{\sigma} \frac{\partial}{\partial \sigma} \left(\sigma \frac{\partial u_x}{\partial \sigma} \right) + \frac{1}{\sigma^2} \frac{\partial^2 u_x}{\partial \varphi^2} \right] + g_x$$

$$\frac{\partial u_\sigma}{\partial t} + u_x \frac{\partial u_\sigma}{\partial x} + u_\sigma \frac{\partial u_\sigma}{\partial \sigma} + u_\varphi \frac{1}{\sigma} \frac{\partial u_\sigma}{\partial \theta} - \frac{u_\varphi^2}{\sigma}$$

$$= -\frac{1}{\rho} \frac{\partial p}{\partial \sigma} + \nu \left[\frac{\partial^2 u_\sigma}{\partial x^2} + \frac{\partial}{\partial \sigma} \left(\frac{1}{\sigma} \frac{\partial (\sigma u_\sigma)}{\partial \sigma} \right) + \frac{1}{\sigma^2} \frac{\partial^2 u_\sigma}{\partial \varphi^2} - \frac{2}{\sigma^2} \frac{\partial u_\varphi}{\partial \varphi} \right] + g_\sigma$$

$$\frac{\partial u_\varphi}{\partial t} + u_x \frac{\partial u_\varphi}{\partial x} + u_\sigma \frac{\partial u_\varphi}{\partial \sigma} + u_\varphi \frac{1}{\sigma} \frac{\partial u_\varphi}{\partial \varphi} + \frac{u_\sigma u_\varphi}{\sigma}$$

$$= -\frac{1}{\rho} \frac{1}{\sigma} \frac{\partial p}{\partial \varphi} + \nu \left[\frac{\partial^2 u_\varphi}{\partial x^2} + \frac{\partial}{\partial \sigma} \left(\frac{1}{\sigma} \frac{\partial (\sigma u_\varphi)}{\partial \sigma} \right) + \frac{1}{\sigma^2} \frac{\partial^2 u_\varphi}{\partial \varphi^2} + \frac{2}{\sigma^2} \frac{\partial u_\sigma}{\partial \varphi} \right] + g_\varphi$$

TABLE 3.4.2
The r, θ, and φ components of the Navier–Stokes equation for a fluid with uniform viscosity in spherical polar coordinates.

$$\frac{\partial u_r}{\partial t} + u_r \frac{\partial u_r}{\partial r} + u_\theta \frac{1}{r}\frac{\partial u_r}{\partial \theta} + u_\varphi \frac{1}{r\sin\theta}\frac{\partial u_r}{\partial \varphi} - \frac{u_\theta^2 + u_\varphi^2}{r}$$

$$= -\frac{1}{\rho}\frac{\partial p}{\partial r} + \nu\left[\nabla^2 u_r - \frac{2}{r^2}u_r - \frac{2}{r^2}\left(\frac{\partial u_\theta}{\partial \theta} + u_\theta \cot\theta\right) - \frac{2}{r^2\sin\theta}\frac{\partial u_\varphi}{\partial \varphi}\right] + g_r$$

$$\frac{\partial u_\theta}{\partial t} + u_r \frac{\partial u_\theta}{\partial r} + u_\theta \frac{1}{r}\frac{\partial u_\theta}{\partial \theta} + u_\varphi \frac{1}{r\sin\theta}\frac{\partial u_\theta}{\partial \varphi} + \frac{u_r u_\theta}{r} - u_\varphi^2 \frac{\cot\theta}{r}$$

$$= -\frac{1}{\rho}\frac{1}{r}\frac{\partial p}{\partial \theta} + \nu\left(\nabla^2 u_\theta + \frac{2}{r^2}\frac{\partial u_r}{\partial \theta} - \frac{u_\theta}{r^2\sin^2\theta} - \frac{2\cos\theta}{r^2\sin^2\theta}\frac{\partial u_\varphi}{\partial \varphi}\right) + g_\theta$$

$$\frac{\partial u_\varphi}{\partial t} + u_r \frac{\partial u_\varphi}{\partial r} + u_\theta \frac{1}{r}\frac{\partial u_\varphi}{\partial \theta} + u_\varphi \frac{1}{r\sin\theta}\frac{\partial u_\varphi}{\partial \varphi} + u_\varphi \frac{1}{r}(u_r + u_\theta \cot\theta)$$

$$= -\frac{1}{\rho}\frac{1}{r\sin\theta}\frac{\partial p}{\partial \varphi} + \nu\left(\nabla^2 u_\varphi - \frac{u_\varphi}{r^2\sin^2\theta} + \frac{2}{r^2\sin\theta}\frac{\partial u_r}{\partial \varphi} + \frac{2\cos\theta}{r^2\sin^2\theta}\frac{\partial u_\theta}{\partial \varphi}\right) + g_\varphi$$

The components of Eq. (3.4.4) in cylindrical and spherical polar coordinates for a fluid with constant viscosity are shown in Tables 3.4.1 and 3.4.2.

Vorticity and Viscous Forces

The identity $\nabla^2 \mathbf{u} = -\nabla \times \boldsymbol{\omega}$ allows us to express the Laplacian of the velocity on the right-hand side of Eq. (3.4.3) in terms of the vorticity, thereby establishing a relation between the structure of the vorticity field and the direction and magnitude of the viscous force. Assuming, for simplicity, that the viscosity is uniform, we write Eq. (3.4.3) in the form

$$\rho \frac{D\mathbf{u}}{Dt} = -\nabla p - \mu \nabla \times \boldsymbol{\omega} + \rho \mathbf{g} \qquad (3.4.5)$$

which shows that *viscous forces are important only in regions where the curl of the vorticity obtains substantial values.* Thus irrotational flows and flows whose vorticity field is irrotational behave like inviscid flows, and their dynamics is determined by a balance between the rate of change of momentum due to the fluid parcel acceleration, the pressure force, and the body force. For these flows, viscosity is important only insofar as to establish the vorticity distribution: Once this is established, viscosity plays no role in the force balance.

Flows with irrotational vorticity fields include two-dimensional flows with constant vorticity and axisymmetric flows without swirling motion in which the magnitude of the vorticity increases linearly with distance from the axis, $\omega = \alpha\sigma$, where α is a constant.

Euler's Equation

When the viscous force on the right-hand side of Eq. (3.4.5) is negligible, the Navier–Stokes equation reduces to *Euler's equation*

$$\rho \frac{D\mathbf{u}}{Dt} = -\nabla p + \rho \mathbf{g} \qquad (3.4.6)$$

which is strictly applicable for inviscid fluids. An important difference between Euler's equation and the Navier–Stokes equation is that the former is a first-order partial differential equation, whereas the latter is a second-order partial differential equation for the velocity with respect to the spatial coordinates. In Section 3.5 we shall see that this difference has important implications on the number of boundary conditions that are required in order to complete the statement of a fluid flow problem.

Uniform Density and Barotropic Fluids

Next we focus our attention on fluids with uniform density and barotropic fluids in which the pressure is a function of the density alone. The latter stipulation is not entirely consistent with the assumptions underlying the derivation of the Navier–Stokes equation for incompressible fluids, but we allow it for the sake of generality. We may then write $(1/\rho)\nabla p = \nabla F$, where in the case of uniform density fluids $F = p/\rho$, whereas in the case of barotropic fluids F is found by integrating the ordinary differential equation $dF/d\rho = (1/\rho)\,dp/d\rho$. For simplicity, in the ensuing discussion we shall consider uniform-density fluids, bearing in mind that adaptations for barotropic fluids can be made by straightforward modifications.

Bernoulli's function

Writing out the material derivative in Eq. (3.4.5) in terms of Eulerian derivatives and then using the identity

$$\mathbf{u} \cdot \nabla\mathbf{u} = \tfrac{1}{2}\nabla(\mathbf{u} \cdot \mathbf{u}) - \mathbf{u} \times \boldsymbol{\omega} \tag{3.4.7}$$

we obtain the new form of the Navier–Stokes equation,

$$\frac{\partial\mathbf{u}}{\partial t} + \nabla\left(\tfrac{1}{2}|\mathbf{u}|^2 + \frac{p}{\rho} - \mathbf{g} \cdot \mathbf{x}\right) = \mathbf{u} \times \boldsymbol{\omega} - \nu\nabla \times \boldsymbol{\omega} \tag{3.4.8}$$

Nonlinear terms appear within the second term on the left-hand side and the first term on the right-hand side. The term within the parentheses on the left-hand side is called the *Bernoulli function* and is usually denoted by H,

$$H \equiv \tfrac{1}{2}|\mathbf{u}|^2 + \frac{p}{\rho} - \mathbf{g} \cdot \mathbf{x} \tag{3.4.9}$$

Physically, H expresses the total energy per unit mass of the fluid, consisting of the kinetic energy, the internal energy due to the pressure, and the potential energy due to the body force. We shall see later in this section that, under certain conditions, H is constant along the streamlines or even throughout the domain of flow; in these cases H is called *Bernoulli's constant*.

Let us apply Eq. (3.4.8) at a certain point in a *steady flow*, and then project it onto the unit vector $\mathbf{t}$ that is tangential to the streamline that passes through that point, $\mathbf{t} = \mathbf{u}/|\mathbf{u}|$. The projection of the first term on the right-hand side onto $\mathbf{t}$ vanishes, yielding

$$\frac{\partial H}{\partial l} = -\nu\mathbf{t} \cdot (\nabla \times \boldsymbol{\omega}) \tag{3.4.10}$$

where l is the arc length along the streamline measured toward the direction of $\mathbf{t}$. Equation (3.4.10) states that the rate of change of H with respect to arc length is equal to the component of the viscous force that is tangential to the streamline. Thus, when the viscous force opposes the motion of the fluid, H decreases along the streamline.

Vortex force

The first term on the right-hand side of Eq. (3.4.8) is called the *vortex force* per unit mass of the fluid. When the velocity is parallel to the vorticity at every point, in which case we obtain a *Beltrami flow*, the vortex force vanishes, and the nonlinear term $\mathbf{u} \cdot \nabla\mathbf{u}$ is equal to the gradient of the kinetic energy per unit mass of the fluid.

Considering now a two-dimensional or axisymmetric flow, we introduce the stream function ψ or the Stokes stream function Ψ, and express the outer product of the velocity and vorticity as

$$\mathbf{u} \times \boldsymbol{\omega} = -\omega\,\nabla\psi, \qquad \mathbf{u} \times \boldsymbol{\omega} = -\frac{\omega}{\sigma}\nabla\Psi \tag{3.4.11}$$

where σ is the radial distance in cylindrical polar coordinates, and ω is the strength of the vorticity field.

Later in this chapter we shall discuss a class of steady two-dimensional flows in which the strength of the vorticity is and remains constant along the streamlines, and may thus be regarded as a function of the stream function, $\omega = f(\psi)$, as well as a class of axisymmetric flows in which the ratio ω/σ is constant along the streamlines, and may thus be regarded as a function of the Stokes stream function, $\omega/\sigma = f(\Psi)$. In these cases, the right-hand sides of Eqs. (3.4.11) may be written in the form of gradients as

$$\mathbf{u} \times \boldsymbol{\omega} = -\nabla F(\psi), \qquad \mathbf{u} \times \boldsymbol{\omega} = -\nabla F(\Psi) \qquad (3.4.12)$$

where the function F is the indefinite integral of the function f, defined in terms of the differential equations

$$\frac{dF}{d\psi} = f(\psi), \qquad \frac{dF}{d\Psi} = f(\Psi) \qquad (3.4.13)$$

For example, for two-dimensional flow with constant vorticity $f = \Omega$, we find $F = \Omega\psi + c$; for axisymmetric flow with $\omega = \alpha\sigma$, we find $F = \alpha\Psi + c$, where c is constant throughout the domain of flow. Eqs. (3.4.12) show that the cross product $\mathbf{u} \times \boldsymbol{\omega}$ constitutes an irrotational vector field, and may thus be grouped with the gradient on the left-hand side of Eq. (3.4.8).

Flow with negligible viscous forces

When the density of the fluid and the acceleration of gravity are uniform throughout the domain of flow, Euler's equation (3.4.6) yields $\rho\, D\mathbf{u}/Dt = -\nabla P$, where P is the modified pressure, which shows that the point-particle acceleration field $D\mathbf{u}/Dt$ is irrotational. A consequence of this result is that, if the flow is irrotational at the initial instant, it will remain irrotational at all times. The permanence of irrotational motion for a fluid with uniform density and negligible viscous forces will be discussed in detail in Section 3.8 in the context of vorticity dynamics.

Bernoulli's Equations

Bernoulli's equations are integrated forms of simplified versions of the Navier–Stokes equation applicable for certain special classes of flows.

Steady flow with negligible viscous forces

For steady flow with negligible viscous forces, Eq. (3.4.8) assumes the simplified form

$$\nabla H = \mathbf{u} \times \boldsymbol{\omega} \qquad (3.4.14)$$

For a Beltrami flow, distinguished by the fact that the velocity is parallel to the vorticity at every point, the right-hand side of this equation vanishes, and Bernoulli's function H is constant throughout the entire domain of the flow.

Considering the more general case of a flow in which the velocity and vorticity are not aligned, we apply Eq. (3.4.14) at a certain point on a streamline, project it onto the tangential vector, note that the right-hand side is normal to the velocity and therefore also normal to the streamline, and integrate the resulting expression with respect to arc length along the streamline to find that *H is constant along a streamline*. The value of H at a particular streamline is usually computed by applying Eq. (3.4.14) at the entrance or exit of the flow and then requiring appropriate boundary conditions for the velocity or traction.

To investigate the variation of H across the streamlines, we project Eq. (3.4.14) onto the unit vector $\mathbf{n}$ that is normal to a streamline, as well as onto the associated binormal vector $\mathbf{b} = \mathbf{t} \times \mathbf{n}$, where $\mathbf{t}$ is the unit vector tangential to the streamline, $\mathbf{t} = \mathbf{u}/|\mathbf{u}|$; the triplet $(\mathbf{t}, \mathbf{n}, \mathbf{b})$ defines an orthogonal right-handed system of axes. Rearranging the triple scalar product on the right-hand sides of the projected equations yields

$$\frac{1}{|\mathbf{u}|}\frac{\partial H}{\partial l_n} = -\boldsymbol{\omega} \cdot \mathbf{b}, \qquad \frac{1}{|\mathbf{u}|}\frac{\partial H}{\partial l_b} = \boldsymbol{\omega} \cdot \mathbf{n} \qquad (3.4.15)$$

where l_n and l_b are the arc lengths measured in the directions of $\mathbf{n}$ and $\mathbf{b}$. When the vorticity is parallel to the velocity, the right-hand sides of Eqs. (3.4.15) vanish, and this shows that H reaches a local minimum or maximum value along the corresponding streamline.

For two-dimensional or axisymmetric flows that fulfill the prerequisites for Eq. (3.4.12), we obtain the explicit expressions $H = -F(\psi) + c$ and $H = -F(\Psi) + c$, where c is constant throughout the domain of flow. For example, for a two-dimensional flow with constant vorticity Ω, we find $H = -\Omega\psi + c$, and for an axisymmetric flow with $\omega = \alpha\sigma$ we find $H = -\alpha\Psi + c$. Since the curl of the vorticity and thus the magnitude of the viscous forces vanishes, these flows represent *exact* solutions to the Navier–Stokes equation. The structure of the velocity field is computed based upon kinematical considerations alone, and the pressure follows from a knowledge of the stream function using the preceding expressions for H. One example of this kind of flow is the flow within Hill's spherical vortex discussed in Section 2.9.

Unsteady irrotational flow

In this case the whole of the right-hand side of Eq. (3.4.8) vanishes. Introducing the velocity potential ϕ, substituting $\mathbf{u} = \nabla\phi$ into the temporal derivative on the left-hand side, and integrating the resulting equation with respect to the spatial variables yields Bernoulli's equation

$$\frac{\partial\phi}{\partial t} + H = \frac{\partial\phi}{\partial t} + \tfrac{1}{2}|\mathbf{u}|^2 + \frac{p}{\rho} - \mathbf{g}\cdot\mathbf{x} = c(t) \tag{3.4.16}$$

where $c(t)$ is a time-dependent function whose value is uniform throughout the domain of flow. In practice, the value of $c(t)$ is usually determined by applying Eq. (3.4.16) at a certain point on a selected boundary of the flow and then requiring an appropriate boundary condition for the velocity or pressure. It is worth noting that Eq. (3.4.16) may be regarded as an evolution equation for ϕ.

In certain applications involving unsteady flows with free surfaces, it is convenient to work with an alternative form of Eq. (3.4.16) that involves the material derivative of the velocity potential $D\phi/Dt = \partial\phi/\partial t + \mathbf{u}\cdot\nabla\phi$, which is equal to the rate of change of the potential following a point particle. Adding $\mathbf{u}\cdot\nabla\phi$ to the left-hand side of Eq. (3.4.16) and subtracting $\mathbf{u}\cdot\mathbf{u}$ yields

$$\frac{D\phi}{Dt} - \tfrac{1}{2}|\mathbf{u}|^2 + \frac{p}{\rho} - \mathbf{g}\cdot\mathbf{x} = c(t) \tag{3.4.17}$$

Working in a similar manner, we find that the rate of change of the potential $d\phi/dt$ as seen by an observer who travels with an arbitrary velocity $\mathbf{V}$ is given by

$$\frac{d\phi}{dt} + (\tfrac{1}{2}\mathbf{u} - \mathbf{V})\cdot\mathbf{u} + \frac{p}{\rho} - \mathbf{g}\cdot\mathbf{x} = c(t) \tag{3.4.18}$$

The preceding two equations will find applications in Chapter 10 where we shall discuss methods of computing free-surface flows based on boundary-integral representations.

Two-Dimensional Flow with Constant Vorticity

A judicious decomposition of the velocity field of a generally unsteady two-dimensional flow whose vorticity is and remains uniform and constant in time, equal to Ω, allows us to describe the flow in terms of a velocity potential and, furthermore, derive a Bernoulli equation for the pressure. In Section 3.10 we shall see that the physical conditions for the vorticity to be uniform are that (1) it is uniform at the initial instant and (2) the effect of viscosity is sufficiently small.

The key idea is to introduce the stream function and write

$$\mathbf{u}(\mathbf{x}, t) = \nabla \times (\psi\mathbf{k}) = \mathbf{v}(\mathbf{x}) + \nabla\phi(\mathbf{x}, t) \tag{3.4.19}$$

where $\mathbf{v}$ is a certain steady two-dimensional flow with uniform vorticity Ω, such as the simple shear flow $\mathbf{v} = (-\Omega y, 0)$. Substituting Eq. (3.4.19) into Eq. (3.4.8) and noting that the viscous force vanishes, we obtain

$$\nabla\left(\frac{\partial\phi}{\partial t} + H\right) = [\nabla \times (\psi\mathbf{k})] \times (\Omega\mathbf{k}) = -\Omega\,\nabla\psi \tag{3.4.20}$$

Integrating Eq. (3.4.20) with respect to the spatial variables, we derive Bernoulli's equation in terms of the stream function,

$$\frac{\partial \phi}{\partial t} + H + \Omega \dot{\psi} = c(t) \tag{3.4.21}$$

where c is constant throughout the domain of flow.

Bernoulli's Equations in a Noninertial Frame of Reference

The equation of motion in an accelerating and rotating frame of reference contains three types of fictitious forces that must be taken into consideration when integrating the Navier–Stokes equation to derive Bernoulli's equations.

Consider first a noninertial frame of reference that translates with linear velocity $\mathbf{U}(t)$ with respect to an inertial frame, and assume that the velocity $\mathbf{v}$ in the noninertial frame is irrotational and may thus be expressed in terms of the unsteady velocity potential Φ so that $\mathbf{v} = \nabla \Phi$. Taking into account the acceleration–reaction body force, we obtain the modified Bernoulli equation

$$\left(\frac{\partial \Phi}{\partial t} \right)_{\mathbf{y}} + \tfrac{1}{2} |\mathbf{v}|^2 + \frac{p}{\rho} + \left(\frac{d\mathbf{U}}{dt} - \mathbf{g} \right) \cdot \mathbf{y} = c(t) \tag{3.4.22}$$

where $\mathbf{y}$ is the position vector in the noninertial frame.

Consider next a noninertial frame whose axes rotate with constant angular velocity $\boldsymbol{\Omega}$ with respect to the origin of an inertial frame, and a flow that is irrotational in the inertial frame and appears to be steady in the noninertial frame. Taking into account the fictitious inertial forces given in Eq. (3.1.43), we obtain the modified Bernoulli equation

$$\tfrac{1}{2} |\mathbf{v}|^2 + \frac{p}{\rho} - \mathbf{g} \cdot \mathbf{y} - \tfrac{1}{2} |\boldsymbol{\Omega} \times \mathbf{y}|^2 = G(\text{Streamline}) \tag{3.4.23}$$

where the modified Bernoulli function G is constant along the streamlines. Note that Eq. (3.4.23) is an enhanced version of Eq. (3.4.16) for steady flow. In the literature of geophysical fluid dynamics, the last two terms on the left-hand side of Eq. (3.4.23) are termed the *geopotential*.

PROBLEMS

3.4.1 **Flow past a body.** Consider a uniform steady flow of a viscous fluid in the horizontal direction past a stationary body, and discuss whether the Bernoulli function H increases or decreases in the streamwise direction.

3.4.2 **Hydrostatics.** Consider a fluid with uniform density that is either stationary or translates steadily as a rigid body, and show that the general solution of the equation of motion is $p = \rho \mathbf{g} \cdot \mathbf{x} + c(t)$ or $P = c(t)$, where $c(t)$ is an arbitrary function of time, and P is the modified pressure. Discuss the computation of $c(t)$ for a problem of your choice.

3.4.3 **Prandtl–Batchelor theorem.** (a) Consider the curl of the vorticity of a steady flow of a viscous fluid, $\mathbf{q} = \nabla \times \boldsymbol{\omega}$, and show that the circulation of $\mathbf{q}$ around any closed streamline vanishes

$$\int_{\text{Streamline}} \mathbf{q} \cdot \mathbf{t} \, dl = 0 \tag{3.4.24}$$

where l is the arc length along the streamline, and $\mathbf{t}$ is the corresponding unit tangential vector [Hint: use Eq. (3.4.8)]. (b) Show that, for a two-dimensional flow wherein the magnitude of the vorticity ω is constant along the streamlines, Eq. (3.4.24) yields

$$\frac{d\omega}{d\psi} \int_{\text{Streamline}} \mathbf{u} \cdot \mathbf{t} \, dl = 0 \tag{3.4.25}$$

where ψ is the stream function. Explain why Eq. (3.4.25) implies that the vorticity must be constant within regions of steady recirculating flow. In Section 3.8 we shall see that the condition for this result to be valid is that viscous forces be small within these regions. (c) Repeat part (b) for axisymmetric flow where the ratio ω/σ is constant along the streamlines, and make analogous deductions.

3.4.4   **Fluid sloshing in a tank.** Consider fluid sloshing in the xy plane within a two-dimensional tank that executes rotational oscillations around the z axis with angular velocity $\Omega(t)$, and assume that, in a stationary frame of reference (x, y, z), the motion of the fluid is irrotational and may therefore be described in terms of the velocity potential ϕ as $\mathbf{u} = \nabla\phi$. Derive an expression for the rate of change of the potential ϕ in a frame of reference (X, Y, Z), in which the walls of the container appear to be stationary.

3.4.5   **Bernoulli's equations in noninertial frames.** Derive Eqs. (3.4.22) and (3.4.23).

3.5 | EQUATIONS AND BOUNDARY CONDITIONS GOVERNING THE MOTION OF AN INCOMPRESSIBLE NEWTONIAN FLUID

The flow of an incompressible Newtonian fluid is governed by the Navier–Stokes equation (3.4.1), expressing Newton's second law for the motion of a small fluid parcel, and the continuity equation $\nabla \cdot \mathbf{u} = 0$ expressing conservation of mass. Together, these two equations provide us with a system of four scalar partial differential equations, called the *equations of incompressible Newtonian flow,* involving four scalar functions: the three components of the velocity $\mathbf{u}$ and the pressure p.

To complete the mathematical statement of a fluid flow problem, we must supply an appropriate number of boundary conditions for certain flow variables, including the velocity, the pressure, and the traction. When the flow is unsteady, we must also supply a suitable initial condition for the velocity; the associated initial pressure is found by requiring that the velocity is and remains solenoidal at all times, as will be discussed in Sections 9.1 and 13.2. *The initial and boundary conditions arise from considerations that are independent of those that led us to the equation of motion,* and must be stated so that the mathematical problem is defined in a unique and unambiguous fashion.

Since the Navier–Stokes equation is a second-order partial differential equation for the velocity, we must supply three scalar boundary conditions over each boundary of the flow. For instance, we may specify the three components of the velocity, the three components of the traction, or a combination of the velocity and traction. If we specify the normal component of the velocity over all boundaries of a domain of flow whose volume remains constant in time, we must ensure that the total volumetric flow rate into the flow vanishes, so that the continuity equation is fulfilled. In certain computational procedures for solving the equations of incompressible Newtonian flow, the boundary velocity is specified at discrete points and, as a result, the total flow rate is not precisely equal to zero due to numerical error. This occurrence may provide a cause for numerical inaccuracies.

In the particular case of irrotational flow or flow of an inviscid fluid governed by Euler's equation, the order of the Navier–Stokes equation is reduced to one. Consequently, we require only one scalar boundary condition over each boundary of the flow.

The No-Penetration Condition over an Impermeable Boundary

Since the molecules of a fluid cannot penetrate an impermeable surface, the component of the velocity of the fluid normal to an impermeable boundary, $\mathbf{n} \cdot \mathbf{u}$, must be equal to the corresponding normal component of the boundary, $\mathbf{n} \cdot \mathbf{V}$, and this requires the no-penetration condition

$$\mathbf{n} \cdot \mathbf{u} = \mathbf{n} \cdot \mathbf{V} \tag{3.5.1}$$

where $\mathbf{V}$ may be constant or vary over the boundary. If the boundary moves as a rigid body, translating with velocity $\mathbf{U}$ and rotating with angular velocity $\boldsymbol{\Omega}$ about the point $\mathbf{x}_0$, then $\mathbf{V} = \mathbf{U} + \boldsymbol{\Omega} \times (\mathbf{x} - \mathbf{x}_0)$, and the no-penetration condition requires that

$$\mathbf{n} \cdot \mathbf{u} = \mathbf{n} \cdot [\mathbf{U} + \boldsymbol{\Omega} \times (\mathbf{x} - \mathbf{x}_0)] \tag{3.5.2}$$

Note that an impermeable boundary is not necessarily a rigid boundary, as it may represent, for instance, a fluid interface or the surface of a flexible body.

Two-dimensional flow

For two-dimensional flow in the xy plane past an impermeable boundary that executes rigid-body motion, translating and rotating about an axis that is parallel to the z axis and passes through the point $\mathbf{x}_0$, we write $\mathbf{V} = \mathbf{U} + \Omega \mathbf{k} \times (\mathbf{x} - \mathbf{x}_0)$, where $\mathbf{k}$ is the unit vector along the z axis. Expressing the velocity in terms of the stream function and writing $\mathbf{n} = (dy/dl, -dx/dl)$, where dx and dy are differential changes around the boundary corresponding to a differential change in the arc length dl measured in the counterclockwise direction, we obtain

$$
\begin{aligned}
\mathbf{u} \cdot \mathbf{n} &= \frac{\partial \psi}{\partial y} \frac{dy}{dl} + \frac{\partial \psi}{\partial x} \frac{dx}{dl} = \frac{\partial \psi}{\partial l} \\
&= \mathbf{U} \cdot \mathbf{n} + \Omega [\mathbf{k} \times (\mathbf{x} - \mathbf{x}_0)] \cdot \mathbf{n} \\
&= U_x \frac{dy}{dl} - U_y \frac{dx}{dl} - \Omega \left((x - x_0) \frac{dx}{dl} + (y - y_0) \frac{dy}{dl} \right)
\end{aligned}
\tag{3.5.3}
$$

Integrating with respect to l yields the scalar boundary condition

$$\psi = U_x y - U_y x - \tfrac{1}{2} \Omega |\mathbf{x} - \mathbf{x}_0|^2 + c \tag{3.5.4}$$

where c is a constant. Equation (3.5.4) shows that the stream function has a constant value over a stationary impermeable surface. In general, the constant values of the stream function over different disconnected stationary boundaries may not be specified a priori, but must be computed as part of the solution.

Axisymmetric flow

For axisymmetric flow past a body that translates along its axis with velocity $\mathbf{U} = U\mathbf{i}$, where $\mathbf{i}$ is the unit vector along the axis, we introduce the Stokes stream function Ψ and work in a similar manner to derive the scalar boundary condition

$$\Psi = \tfrac{1}{2} U \sigma^2 + c \tag{3.5.5}$$

which shows that the Stokes stream function takes a constant value over a stationary impermeable axisymmetric surface.

No Slip at a Fluid–Fluid or Fluid–Solid Interface

Experimental observations with a broad class of fluids and under a wide range of conditions have shown that the tangential component of the velocity of the fluid remains continuous across a fluid–solid or a fluid–fluid interface, that is, the slip velocity vanishes. It is important to emphasize that the no-slip condition has been shown to be valid independently of the chemical constitution of the boundaries and physical properties of the fluids. One exception arises in the case of rarified gas flow, where the mean free path of the molecules is comparable to the size of the boundaries, and a description of the flow in the context of continuum mechanics is no longer appropriate.

The physical origin of the no-slip condition for gas flow over a solid surface may be traced back to the fact that the molecules are adsorbed onto the surface over a time period that is long enough for thermal equilibrium to be established. There is an analogous explanation for liquids based on the formation of short-lived bonds between the liquid and solid molecules due to weak intermolecular forces. An alternative explanation, elaborated by Richardson (1973), is that the proper boundary condition on a solid surface is that of vanishing tangential traction, and the apparent no-slip condition arises on a macroscopic level due to the inherent boundary irregularities.

The no-slip condition requires that the tangential component of the velocity of a fluid over a stationary solid boundary vanish. As a consequence, the velocity of the fluid $\mathbf{u}$ over an *impermeable* solid boundary must be equal to the velocity of the boundary $\mathbf{V}$. If the boundary executes rigid-body motion translating with velocity $\mathbf{U}$ and rotating with angular velocity $\boldsymbol{\Omega}$ about the point $\mathbf{x}_0$, then $\mathbf{u} = \mathbf{U} + \boldsymbol{\Omega} \times (\mathbf{x} - \mathbf{x}_0)$.

Slip on Fluid–Solid Interfaces

An alternative boundary condition used on occasion in place of the no-slip condition requires that the tangential component of the velocity of the fluid $\mathbf{u}$ relative to the velocity of the boundary $\mathbf{V}$ be proportional to the tangential component of the surface traction,

$$\mathbf{f} \cdot (\mathbf{I} - \mathbf{nn}) = l(\mathbf{u} - \mathbf{V}) \cdot (\mathbf{I} - \mathbf{nn}) \tag{3.5.6}$$

The constant l is called the *slip coefficient,* and its value depends upon the chemical properties and microstructure of the boundary. For gases, one may show on the basis of statistical mechanics that l is equal to the molecular mean free path.

The slip condition has been used with success to describe flow over the surface of a porous medium, in which case it accounts for the presence of pores, and flow in the neighborhood of a moving three-phase contact line, in which case it removes the singular behavior of the traction in the vicinity of the contact line (Dussan, 1979).

Boundary Conditions at a Three-Phase Contact Line

In a number of applications, involving in particular liquid films and droplets, we encounter stationary or moving fluid interfaces that end at stationary or moving solid surfaces; certain examples are illustrated in Figure 3.5.1(a–c). The lines of contact are called *three-phase static* or *dynamic contact lines.*

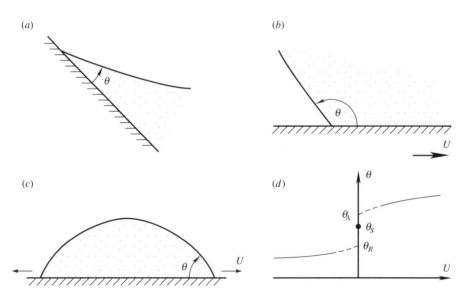

Figure 3.5.1   (a) A stationary interface meets a stationary solid surface at a static contact line. (b) A stationary interface meets a moving solid surface at a dynamic contact line. (c) A contact line moving over a stationary solid surface due to the spreading of a drop. (d) A typical graph of the dynamic contact angle as a function of the velocity of the contact line U. With reference to (c), a positive value of U corresponds to a spreading drop.

The angle that is subtended between (a) the vector that is normal to the contact line and tangential to the interface and (b) the vector that is normal to the contact line and lies on the solid boundary, is called the *static* or *dynamic contact angle.* The contact angle is measured from the side of a specified fluid, as shown in Figure 3.5.1(a); for a liquid–gas interface, the contact angle is measured from the side of the liquid.

The value of the *static contact angle* θ_S, corresponding to a stationary contact line that is pinned on a stationary solid surface, is a physical constant that depends upon the chemical properties of the solid and fluids. The value of the *dynamic contact angle* θ, corresponding to a stationary contact line that lies over a moving solid surface, or to an evolving contact line moving over a stationary solid surface as illustrated in Figure 3.5.1(b, c), depends not only upon the chemistry of the fluids and solid, but also upon the velocity of the surface and the velocity of the contact line (Dussan, 1979).

A typical graph of the dynamic contact angle with respect to the velocity of the contact line U over a stationary surface is shown in Figure 3.5.1(d). Measurements have shown that $\partial\theta/\partial U > 0$ independently of the nature of the fluids. The extrapolated values of θ in the limit as U tends to vanish from positive or negative values are called the *advancing* or *receding contact lines,* and are denoted, respectively, by θ_A and θ_R. The observed dependence of the extrapolated values of θ_A and θ_R on the experimental procedure, and the fact that, when $U = 0$, θ may take any value between θ_A and θ_R, suggests the occurrence of *contact-angle hysteresis.*

A local analysis of the equation of motion in the vicinity of a contact line moving on a solid surface reveals that the tangential component of the surface traction becomes singular (Section 6.2). More importantly, the singularity is nonintegrable, and hence the drag force exerted on the surface assumes an infinite value. This unphysical behavior suggests that either the equation of motion breaks down in the immediate vicinity of the dynamic contact line, or the no-slip boundary condition ceases to be valid. The latter explanation motivates replacing the no-slip boundary condition with the slip condition shown in Eq. (3.5.6). Computational experimentation has shown that this is a reasonable remedy with no significant consequences for the global structure of the flow (Dussan, 1979).

The Jump in Surface Force across a Fluid Interface

In general, the tractions exerted on the two sides of an interface between two fluids labeled 1 and 2 have two different values, with a corresponding discontinuity

$$\Delta\mathbf{f} \equiv \mathbf{f}^{(1)} - \mathbf{f}^{(2)} = (\boldsymbol{\sigma}^{(1)} - \boldsymbol{\sigma}^{(2)}) \cdot \mathbf{n} \tag{3.5.7}$$

where $\mathbf{n}$ is the unit normal vector *pointing into fluid 1,* as shown in Figure 3.5.2, and $\boldsymbol{\sigma}^{(1)}$ and $\boldsymbol{\sigma}^{(2)}$ are the stress tensors within the two fluids evaluated at the interface.

The direction and magnitude of the discontinuity of the interfacial traction $\Delta\mathbf{f}$ depend upon the mechanical properties of the interface, which are determined by the physico-chemical properties of the fluids and molecular structure of the interface, and are therefore affected by the pres-

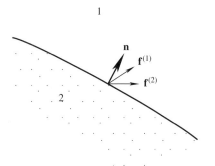

Figure 3.5.2 Schematic illustration of an interface between two fluids.

ence of surface-active substances. An equation that relates $\Delta\mathbf{f}$ to the velocity field, the properties of the fluids, and the shape and thermodynamic properties of the interface, is called a *constitutive equation for the discontinuity in the interfacial traction.*

When it is appropriate to work with the modified pressure and stress defined in Eqs. (3.3.7) and (3.3.8), we introduce the associated jump in modified interfacial traction defined as

$$\Delta\mathbf{f}^{\text{Mod}} \equiv (\boldsymbol{\sigma}^{(1)\,\text{Mod}} - \boldsymbol{\sigma}^{(2)\,\text{Mod}}) \cdot \mathbf{n} = \Delta\mathbf{f} + (\rho_1 - \rho_2)\,(\mathbf{x}\cdot\mathbf{g})\,\mathbf{n} - (c_1 - c_2)\,\mathbf{n} \quad (3.5.8)$$

where c_1 and c_2 are two constants corresponding to the two fluids.

Isotropic tension

The most common type of interfacial behavior pertains to uncontaminated interfaces between two immiscible fluids characterized by isotropic surface tension γ, which may be regarded as a kind of energy per unit surface area or surface pressure. The physical origin of surface tension may be traced to differences in the attraction forces between the molecules of the two liquids. The surface tension of a clean interface between water and air at 20°C is $\gamma = 73$ dyn/cm, whereas that between glycerine and air at the same temperature is $\gamma = 63$ dyn/cm. In general, the surface tension decreases as the temperature is raised, and vanishes when the temperature reaches the boiling point. If an interface is populated by surface-active agents, the surface tension is a function of surfactant concentration c; the derivative $\partial\gamma/\partial c$ is sometimes called the *Gibbs surface elasticity.* Typically, the surface tension decreases as c is increased and reaches a plateau at a saturation concentration where the interface is covered by a monolayer of surfactants.

To derive the constitutive equation for $\Delta\mathbf{f}$, we assume that the interfacial stratum has negligible mass and write a force balance over a small section of the interface D that is bounded by the contour C, as shown in Figure 3.5.3, requiring

$$\int_D \Delta\mathbf{f}\,dS + \int_C \gamma\mathbf{t}\times\mathbf{n}\,dl = \mathbf{0} \quad (3.5.9)$$

where $\mathbf{n}$ is the unit vector normal to D pointing into fluid 1, $\mathbf{t}$ is the unit vector tangential to C, and $\mathbf{t}\times\mathbf{n} \equiv \mathbf{b}$ is the binormal vector as shown in Figure 3.5.3. Furthermore, we extend the domain of definition of the surface tension and normal vector from the interface into the whole space. The first extension can be done in an unrestricted manner, whereas the second extension is done by setting $\mathbf{n} = \nabla F/|\nabla F|$, where the equation $F(x, y, z) = 0$ describes the instantaneous location of the interface. A variation of Stokes's theorem states that for any arbitrary vector function $\mathbf{F}$,

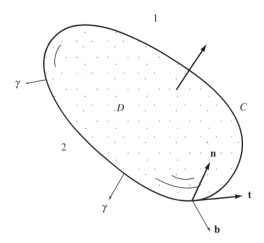

Figure 3.5.3 Surface tension acting along the edges of a test section of an interface.

$$\int_C \mathbf{F} \times \mathbf{t} \, dl = \int_D (\mathbf{n} \nabla \cdot \mathbf{F} - (\nabla \mathbf{F}) \cdot \mathbf{n}) \, dS \qquad (3.5.10)$$

(Section A.6, Appendix A). Applying Eq. (3.5.10) with $\mathbf{F} = \gamma \mathbf{n}$, and combining the resulting expression with Eq. (3.5.9), we obtain

$$\int_D \Delta \mathbf{f} \, dS = \int_D [\mathbf{n} \nabla \cdot (\gamma \mathbf{n}) - (\nabla(\gamma \mathbf{n})) \cdot \mathbf{n}] \, dS \qquad (3.5.11)$$

We now take the limit as D shrinks down to a point, expand out the derivatives within the integrand on the right-hand side, and thus obtain the desired constitutive equation

$$\Delta \mathbf{f} = \gamma \mathbf{n} \nabla \cdot \mathbf{n} - (\mathbf{I} - \mathbf{nn}) \cdot \nabla \gamma = \gamma \mathbf{n} \nabla \cdot \mathbf{n} - (\mathbf{n} \times \nabla \gamma) \times \mathbf{n} \qquad (3.5.12)$$

The first and second terms in the middle and on the right-hand side of Eq. (3.5.12) express, respectively, discontinuities in the normal and tangential directions. When the surface tension is uniform, the tangential component vanishes, and the jump in the interfacial traction points in the normal direction.

The divergence of the normal vector on the right-hand side of Eq. (3.5.12) is equal to twice the mean curvature of the interface, denoted by κ_m,

$$\nabla \cdot \mathbf{n} = 2\kappa_m \qquad (3.5.13)$$

By definition, the mean curvature is positive when the interface has a spherical shape with fluid 2 lying inside and the normal vector pointing outward. The computation of the mean curvature will be discussed in more detail in Section 4.2 in the context of hydrostatics.

Interfaces with involved mechanical properties

The discontinuity in traction across interfaces with complex structure can be described in a phenomenological manner in terms of interfacial shear and dilatational surface viscosities and moduli of elasticity. Appropriate constitutive equations are discussed by Edwards, Brenner, and Wasan (1991) and Pozrikidis (1995).

Jump in Pressure across a Fluid Interface

The normal component of the traction and thus the pressure undergo a discontinuity across an interface. To compute the jump in pressure in terms of the velocity, we resolve $\Delta \mathbf{f}$ into its normal and tangential components, and then identify the normal component with the term within the square brackets on the right-hand side of Eq. (3.3.3), finding

$$\Delta \mathbf{f} = \{-p_1 + p_2 + 2 \mathbf{n} \cdot (\mu_1 \nabla \mathbf{u}^{(1)} - \mu_2 \nabla \mathbf{u}^{(2)}) \cdot \mathbf{n}\} \mathbf{n} + \mathbf{n} \times \Delta \mathbf{f} \times \mathbf{n} \qquad (3.5.14)$$

Projecting Eq. (3.5.14) onto the normal vector $\mathbf{n}$ produces the jump in interfacial pressure in terms of the normal component of $\Delta \mathbf{f}$ and the viscous normal stresses,

$$\Delta p \equiv p_1 - p_2 = -\Delta \mathbf{f} \cdot \mathbf{n} + 2 \mathbf{n} \cdot (\mu_1 \nabla \mathbf{u}^{(1)} - \mu_2 \nabla \mathbf{u}^{(2)}) \cdot \mathbf{n} \qquad (3.5.15)$$

where $\Delta \mathbf{f}$ is given by an appropriate interfacial constitutive equation.

For example, when the fluids are stationary or inviscid, the second and third terms on the right-hand side of Eq. (3.5.15) are absent. Assuming that the interface exhibits uniform isotropic tension, and using Eq. (3.5.12), we find $\Delta p = -2\kappa_m \gamma$.

Boundary Condition for the Velocity at a Fluid Interface

Certain computational methods for solving problems of multifluid flow require restating the dynamic boundary condition for the jump in the interfacial traction $\Delta \mathbf{f}$ in a form that does not involve the interfacial pressures. The tangential component of the dynamic boundary condition involves derivatives of the velocity alone and does not require further manipulations. To eliminate the pressure from the normal component of $\Delta \mathbf{f}$, we apply the equation of motion on either side of the interface, form the difference between the two equations, and then project the difference onto a unit tangential vector $\mathbf{t}$ to obtain

$$
\mathbf{t} \cdot \nabla(p_2 - p_1) = \left(-\rho_2 \frac{D\mathbf{u}^{(2)}}{Dt} + \rho_1 \frac{D\mathbf{u}^{(1)}}{Dt} + \mu_2 \nabla^2 \mathbf{u}^{(2)} \right.
$$
$$
\left. - \mu_1 \nabla^2 \mathbf{u}^{(1)} + (\rho_2 - \rho_1)\,\mathbf{g} \right) \cdot \mathbf{t} \tag{3.5.16}
$$

Substituting the right-hand side of Eq. (3.5.15) into the left-hand side of Eq. (3.5.16) yields the desired boundary condition in terms of the velocity alone.

For example, when the fluids are inviscid, in which case the velocity is allowed to be discontinuous across the interface, we obtain the simplified form

$$
\mathbf{t} \cdot \nabla(\Delta \mathbf{f} \cdot \mathbf{n}) = \left(-\rho_2 \frac{D\mathbf{u}^{(2)}}{Dt} + \rho_1 \frac{D\mathbf{u}^{(1)}}{Dt} + (\rho_2 - \rho_1)\,\mathbf{g} \right) \cdot \mathbf{t} \tag{3.5.17}
$$

which imposes a constraint upon the tangential component of the acceleration of the point particles residing on either side of the interface.

Derivative Boundary Conditions

We have discussed several types of boundary conditions whose origin may be traced back to physical reasoning. *Derivative boundary conditions emerge by combining the physical boundary conditions with the equations governing the motion of the fluid,* including the continuity equation and the equation of motion. One example of such a condition was already presented in Eq. (3.5.16). As a further example, we use the continuity equation and find that the normal derivative of the normal component of the velocity over an impermeable solid surface, where the no-slip condition applies, must vanish, as shown in Eq. (3.6.2). A derivative boundary condition for the pressure over a solid boundary will be discussed in Chapter 13 in connection with finite-difference methods.

Conditions at Infinity

Mathematical idealization of a flow in a domain of large extent often leads to the study of flow in a totally or partially infinite domain. Examples are infinite flow past a stationary object, and infinite shear flow over a plane wall with a cavity or a protrusion. To complete the definition of the mathematical problem, we require a condition for the asymptotic behavior of the flow at infinity. To implement this condition, we resolve the velocity field into the unperturbed and a disturbance component, and require that, as we move to infinity, the ratio between the magnitude of the disturbance velocity and the unperturbed velocity vanish. It should be emphasized that this requirement *does not necessarily imply that the disturbance flow will vanish at infinity.* For example, for uniform flow past a sphere, we require that the disturbance flow due to the sphere vanishes far from the sphere, whereas for parabolic flow past a sphere we allow the disturbance velocity to grow at a rate that is less than quadratic. For simple shear flow over an infinite plane wall with periodic protrusions, we require that the disturbance flow due to the protrusions grows at a rate that is less than linear; the solution reveals that the disturbance flow at infinity tends to a constant value.

Truncated Domains of Flow

In computing external flows or internal flows in infinite domains using numerical methods, it is a standard practice to truncate the *physical domain* of flow at a certain level that allows for an affordable computational cost. For instance, in order to compute the flow past a body in a wind tunnel, we introduce a *computational domain* that is confined by an in-flow plane, an out-flow plane, and the side-flow boundaries. The boundary conditions at the computational boundaries must be derived by means of a far-field asymptotic analysis, which, unfortunately, is available only for a limited number of flows (Gresho, 1991). The assumption of fully developed interior flow is often used in practice. The choice of a far-field boundary condition must be exercised with great care in order to preserve mass and prevent violation of the momentum or energy integral balance. Furthermore, the effect of domain truncation must be assessed carefully before a numerical solution can be claimed to have any degree of physical relevance.

Generalized Equation of Motion for Flow in the Presence of Interfaces

One way of computing the flow of two adjacent fluids with different physical properties is to solve the governing equations within each fluid separately but simultaneously, subject to the interfacial kinematic and dynamic boundary conditions discussed previously in this section.

Another way is to incorporate the interfacial boundary conditions into a generalized equation of motion that is valid within the fluids *as well as at the interface*. The generalized equation of motion emerges by regarding the interface as a *singular surface of concentrated body force*, obtaining

$$\rho \frac{D\mathbf{u}}{Dt} = \nabla \cdot \boldsymbol{\sigma} + \rho \mathbf{g} + \mathbf{F} \tag{3.5.18}$$

where $\mathbf{F}$ is a singular forcing function expressing an interfacial distribution of point forces,

$$\mathbf{F}(\mathbf{x}) = -\int_{\text{Interface}} \Delta \mathbf{f}(\mathbf{x}') \, \delta(\mathbf{x} - \mathbf{x}') \, dS(\mathbf{x}') \tag{3.5.19}$$

δ is the three-dimensional delta function, and $\Delta \mathbf{f}$ is the jump in the interfacial traction. The velocity is assumed to be a continuous function, but the physical properties of the fluids and stress tensor are allowed to undergo step changes across the interface.

To show that Eq. (3.5.18) provides us with a consistent representation of the flow, in the sense that it incorporates the precise form of the dynamic boundary condition (3.5.7), we take its volume integral over a thin sheet of fluid of thickness ε that is centered at the interface, use the divergence theorem to manipulate the integral of the divergence of the stress and the properties of the delta function to manipulate the integral of $\mathbf{F}$, and then take the limit as the sheet collapses to the interface which is denoted by D, to find

$$\int_{\text{Sheet}} \rho \frac{D\mathbf{u}}{Dt} \, dV = \int_{\text{Sheet}} \boldsymbol{\sigma} \cdot \mathbf{n} \, dS + \mathbf{g} \int_{\text{Sheet}} \rho \, dV - \int_{D} \Delta \mathbf{f} \, dS \tag{3.5.20}$$

The volume integrals of the acceleration and gravitational terms are of order ε and may thus be neglected. The rest of the terms may be rearranged to yield

$$0 = \int_{D^+} \boldsymbol{\sigma} \cdot \mathbf{n}^+ \, dS - \int_{D^-} \boldsymbol{\sigma} \cdot \mathbf{n}^+ \, dS - \int_{D} \Delta \mathbf{f} \, dS \tag{3.5.21}$$

where the plus and minus signs signify, respectively, the upper and lower side of D corresponding to fluids 1 and 2, and $\mathbf{n}^+$ is the unit normal vector pointing into fluid 1. It is now evident that Eq. (3.5.21) and therefore Eq. (3.5.18) imply Eq. (3.5.7).

A more formal way of deriving Eq. (3.5.18) involves replacing the step functions that are inherent in the representation of the physical properties of the fluids, and the delta functions that are inherent in the distribution of the interfacial force, with smooth functions that change gradually over a thin interfacial layer of thickness ε. As long as ε is finite, the regular form of the equation of motion is applicable within the bulk of the fluids, as well as within the interfacial layer. Taking the limit as ε tends to zero yields Eq. (3.5.18).

In the case of two-dimensional flow, the interfacial distribution Eq. (3.5.19) takes the form

$$\mathbf{F}(\mathbf{x}) = -\int_{\text{Interface}} \Delta \mathbf{f}(\mathbf{x}') \, \delta(\mathbf{x} - \mathbf{x}') \, dl(\mathbf{x}') \tag{3.5.22}$$

where l is the arc length along the interface and δ is the two-dimensional delta function.

PROBLEMS

3.5.1 **No-penetration condition.** (a) Consider the stream functions of a three-dimensional flow and discuss possible ways of implementing the no-penetration boundary condition. (b) Consider an impenetrable boundary whose position is described by the equation $F(\mathbf{x}, t) = 0$. Show that the no-penetration condition can be stated as $DF/Dt = 0$ evaluated at the boundary, where D/Dt is the material derivative. Furthermore, show that the velocity of the fluid in the material derivative can be amended with the addition of an arbitrary tangential component.

3.5.2 **Interfaces with isotropic tension.** Derive the counterpart of Eq. (3.5.12) for two-dimensional flow and discuss its physical interpretation.

3.5.3 **Generalized equation for interfacial motion in two dimensions.** Write the counterparts of Eqs. (3.5.20) and (3.5.21) for two-dimensional flow.

3.6 | TRACTION, VORTICITY, AND FLOW KINEMATICS AT RIGID BOUNDARIES, FREE SURFACES, AND FLUID INTERFACES

The boundary conditions at an impermeable rigid boundary, free surface, and fluid interface discussed in Section 3.5 allow us to simplify the corresponding expressions for the traction, vorticity, force, and torque, and thereby obtain useful information on the kinematical structure and dynamical behavior of a flow.

Rigid Boundaries

Consider first a flow over an impermeable rigid boundary that is either held stationary or executes rigid-body motion. In the second case, we describe the flow in an appropriate frame of reference in which the boundary appears to be stationary. Concentrating at a particular point on the boundary, we introduce a local Cartesian system of axes with the origin at that point, two axes x and z tangential to the boundary, and the y axis normal to the boundary pointing into the fluid.

Expanding the velocity in a Taylor series with respect to x, y, and z, and requiring the no-slip and no-penetration conditions, we find that all components of the velocity and their first partial derivatives with respect to x and z at the origin must vanish. The continuity equation then requires that the first partial derivative of the normal component of the velocity with respect to z at the origin also vanish. With reference to a general coordinate system, the mathematical expression of these results is

$$\mathbf{u} = \mathbf{0}, \qquad (\mathbf{I} - \mathbf{nn}) \cdot \nabla \mathbf{u} = (\mathbf{n} \times \nabla \mathbf{u}) \times \mathbf{n} = \mathbf{0} \qquad (3.6.1)$$

$$\mathbf{n} \cdot (\nabla \mathbf{u}) \cdot \mathbf{n} = 0 \qquad (3.6.2)$$

all evaluated at the boundary.

Traction, vorticity, force, and torque

Simplifying the expressions given in Eqs. (3.3.3) and (3.3.4) on the basis of Eqs. (3.6.1) and (3.6.2), we find that, at the origin of the local coordinate system, the components of the traction are given by

$$f_x = \mu \frac{\partial u_x}{\partial y}, \qquad f_y = -p, \qquad f_z = \mu \frac{\partial u_z}{\partial y} \qquad (3.6.3)$$

With reference to a general coordinate system, we obtain the compact expression

$$\mathbf{f} = -p\mathbf{n} + \mu \mathbf{n} \cdot (\nabla \mathbf{u}) \cdot (\mathbf{I} - \mathbf{nn}) \qquad (3.6.4)$$

The second term on the right-hand side of Eq. (3.6.4) involves derivatives of the tangential components of the velocity in the direction normal to the boundary.

Using the definition of the vorticity in conjunction with the kinematic constraint expressed by the second of Eqs. (3.6.1), we find that the component of the vorticity vector normal to the boundary vanishes, $\omega_y = 0$. At the origin of the local coordinate system, the tangential components of the vorticity are given by

$$\omega_x = \frac{\partial u_z}{\partial y} = \frac{1}{\mu} f_z, \qquad \omega_z = -\frac{\partial u_x}{\partial y} = -\frac{1}{\mu} f_x \tag{3.6.5}$$

which, along with Eq. (3.6.2), suggest that

$$\mathbf{n} \cdot \nabla \mathbf{u} = \mathbf{n} \cdot (\nabla \mathbf{u}) \cdot (\mathbf{I} - \mathbf{nn}) = \boldsymbol{\omega} \times \mathbf{n} \tag{3.6.6}$$

The last expression in Eq. (3.3.2) may then be simplified to

$$\mathbf{f} = -p\mathbf{n} + \mu\boldsymbol{\omega} \times \mathbf{n} \tag{3.6.7}$$

Substituting Eqs. (3.6.4) and (3.6.7) into Eqs. (3.3.11) and (3.3.12), we obtain simplified expressions for the hydrodynamic force and torque exerted on a rigid boundary in terms of the boundary vorticity or normal derivatives of the tangential components of the velocity.

Skin-friction and surface vortex lines

Equation (3.6.7) shows that the vorticity vector is perpendicular to the *skin-friction vector*, which is defined as the tangential component of $\mathbf{f}$ and is given by the equivalent expressions $\mu\boldsymbol{\omega} \times \mathbf{n}$, $\mathbf{f} \cdot (\mathbf{I} - \mathbf{nn})$, or $(\mathbf{n} \times \mathbf{f}) \times \mathbf{n}$. We thus write

$$\boldsymbol{\omega} \cdot [\mathbf{f} \cdot (\mathbf{I} - \mathbf{nn})] = 0 \qquad \text{or} \qquad \boldsymbol{\omega} \cdot [(\mathbf{n} \times \mathbf{f}) \times \mathbf{n}] = 0 \tag{3.6.8}$$

A line over the boundary whose tangential vector is parallel to the skin friction vector at every point is called a *skin friction line*. A line over the boundary whose tangential vector is parallel to the vorticity at every point is called a *boundary* or *surface vortex line*. Equations (3.6.8) show that *the skin friction lines are orthogonal to the surface vorticity lines.*

Equation (3.6.4) shows that a skin friction line may be described in parametric form by the *autonomous* ordinary differential equation

$$\frac{d\mathbf{x}}{d\tau} = \mathbf{n} \cdot (\nabla \mathbf{u}) \cdot (\mathbf{I} - \mathbf{nn}) = \boldsymbol{\omega} \times \mathbf{n} \tag{3.6.9}$$

where τ is a certain parameter. *Singular points,* around which fluid particles move normal to the boundary at a velocity that is comparable to or higher than that along the boundary, occur when the right-hand side of Eq. (3.6.9) vanishes and therefore the surface vorticity becomes equal to zero.

Singular points are classified into *nodal points of attachment or separation, foci of attachment or separation, and saddle points.* Examples of the corresponding skin friction and vorticity lines are shown, respectively, with dashed and solid lines in Figure 3.6.1(a–c) after Lighthill (1963) and Tobak and Peake (1982). A nodal point belongs to an infinite number of skin friction lines, all of which except for one, the one labeled AA in Figure 3.6.1(a), are tangential to a single line labeled BB. A focal point belongs to an infinite set of skin friction lines that spiral away from or into the focal point. A saddle point is the point of intersection of two skin friction lines. Topological constraints require that the number of nodal points and foci on a boundary exceed the number of saddle points by two. An in-depth discussion of the topography and topology of skin friction lines is presented in the aforementioned two references.

The relation between the skin friction lines and the trajectories of point particles in the vicinity of a boundary becomes evident by introducing a system of *orthogonal* surface curvilinear coordinates (ξ, η) over the boundary, as discussed in Section 1.4. The point particles move with a velocity $\mathbf{U}$ that is nearly tangential to the boundary, except when they are located in the vicinity

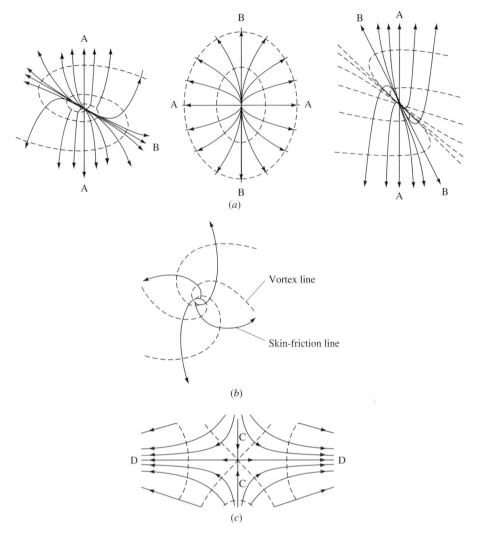

Figure 3.6.1   Illustration of singular points of the skin friction pattern on a solid boundary, after Lighthill (1963). (a) Nodal points, (b) a focus, and (c) a saddle.

of a singular point. Denoting the instantaneous distance of a point particle away from the boundary by ζ and expanding the velocity in a Taylor series in the normal direction, we find that the differential arc lengths traveled by the point particle along the curvilinear axes are given by

$$\frac{dl_\xi}{dt} = \mathbf{U}(\xi, \eta, \zeta) \cdot \mathbf{t}_\xi \cong \zeta (\mathbf{n} \cdot \nabla \mathbf{u})_{\xi, \eta, \zeta = 0} \cdot \mathbf{t}_\xi = \zeta \boldsymbol{\omega} \cdot \mathbf{t}_\eta$$

$$\frac{dl_\eta}{dt} = \mathbf{U}(\xi, \eta, \zeta) \cdot \mathbf{t}_\eta \cong \zeta (\mathbf{n} \cdot \nabla \mathbf{u})_{\xi, \eta, \zeta = 0} \cdot \mathbf{t}_\eta = -\zeta \boldsymbol{\omega} \cdot \mathbf{t}_\xi$$

(3.6.10)

Dividing these equations side by side yields

$$\frac{dl_\xi}{\mathbf{n} \cdot (\nabla \mathbf{u}) \cdot \mathbf{t}_\xi} = \frac{dl_\eta}{\mathbf{n} \cdot (\nabla \mathbf{u}) \cdot \mathbf{t}_\eta}$$

(3.6.11)

which is an alternative expression of Eq. (3.6.9).

Free Surfaces

The tangential component of the velocity over a stationary, moving, or deforming free surface does not necessarily vanish, and Eqs. (3.6.1) and (3.6.2) are not applicable. Since, however, the tangential component of the traction is required to vanish, we obtain the restriction

$$(\boldsymbol{\sigma} \cdot \mathbf{n}) \times \mathbf{n} = \mathbf{0} \qquad \text{or} \qquad (\mathbf{n} \cdot \nabla \mathbf{u}) \times \mathbf{n} = \mathbf{n} \times [(\nabla \mathbf{u}) \cdot \mathbf{n}] \qquad (3.6.12)$$

which may be used to derive a simplified expression for the tangential component of the vorticity at the free-surface given by $\boldsymbol{\omega} \cdot (\mathbf{I} - \mathbf{nn}) = (\mathbf{n} \times \boldsymbol{\omega}) \times \mathbf{n}$. This is done by expressing the vorticity as the curl of the velocity, manipulating the repeated cross product, and using the second of Eqs. (3.6.12) to obtain

$$(\mathbf{n} \times \boldsymbol{\omega}) \times \mathbf{n} = [(\nabla \mathbf{u}) \cdot \mathbf{n} - \mathbf{n} \cdot \nabla \mathbf{u}] \times \mathbf{n} = 2\,[(\nabla \mathbf{u}) \cdot \mathbf{n}] \times \mathbf{n}$$
$$= -2\,\mathbf{n} \times \nabla(\mathbf{u} \cdot \mathbf{n}) + 2\,\mathbf{n} \times [(\nabla \mathbf{n}) \cdot \mathbf{u}] \qquad (3.6.13)$$

The first term on the right-hand side contains derivatives of the normal component of the velocity tangential to the free surface, and, therefore, it vanishes when the free surface is stationary. The derivatives of the normal vector in the second term can be expressed, in part, in terms of the normal curvature of the free surface in the plane that contains the velocity.

Introducing, in particular, a local Cartesian system with its origin at a certain point on a stationary free surface, the x axis in the direction of the velocity vector, and the y axis perpendicular to the free surface, we find that the tangential component of the vorticity is given by $2u\,\partial n_x\,dz\,\mathbf{i} - 2\kappa_{xy}u\,\mathbf{k}$, where κ_{xy} is the principal curvature of the free surface in the xy plane, u is the magnitude of the velocity, and $\mathbf{i}$ and $\mathbf{k}$ are the unit vectors along the x and z axes.

Two-Dimensional Flow

The fact that the streamlines of a two-dimensional flow lie in parallel planes allows us to simplify the expressions for the boundary traction and magnitude of the vorticity, as well as to study the kinematics of the flow at solid boundaries, free surfaces, and fluid interfaces in some more detail.

Rigid boundaries

Considering first the structure of a flow near a stationary rigid boundary, we refer to the local coordinate system illustrated in Figure 3.6.2(a) and use the second of Eqs. (3.6.1) to find that the vorticity and tangential component of the wall shear stress at the origin are given by

$$\omega = -\frac{\partial u_x}{\partial y}, \qquad f_x = \mu\,\frac{\partial u_x}{\partial y} \qquad (3.6.14)$$

Next we concentrate on the structure of the flow in the vicinity of a stagnation point illustrated in Figure 3.6.2(b). Close to the wall, on either side of the stagnation point, the fluid moves into

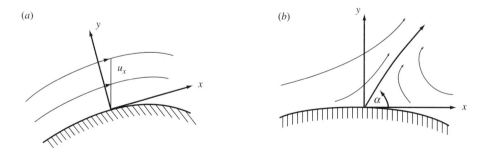

Figure 3.6.2   Illustration of (a) a two-dimensional flow near a solid wall and (b) a stagnation point occurring at a solid wall.

opposite directions, and this requires that the wall shear stress and vorticity have opposite signs. As a consequence, both the wall shear stress and vorticity at the stagnation point must vanish. Conversely, the vanishing of the shear stress provides us with a criterion for the occurrence of a stagnation point.

Another way of showing that the shear stress vanishes at a stagnation point is to consider a point that lies on the dividing streamline close to the stagnation point at (dx, dy), set (dx, dy) parallel to the velocity at that point, and express the velocity in a Taylor series about the stagnation point, finding

$$\tan \alpha \equiv \frac{dy}{dx} = \frac{u_{y,x} \, dx + u_{y,y} \, dy}{u_{x,x} \, dx + u_{x,y} \, dy} = \frac{u_{y,x} + u_{y,y} \tan \alpha}{u_{x,x} + u_{x,y} \tan \alpha} \tag{3.6.15}$$

where the second subscript indicates partial differentiation with respect to the corresponding variable, all partial derivatives are evaluated at the stagnation point, and α is the angle that the dividing streamline forms with the x axis, $dy/dx = \tan \alpha$. We use the continuity equation to write $u_{y,y} = -u_{x,x}$, and then require the no-slip boundary condition to find that all partial derivatives $u_{y,y}, u_{x,x}, u_{y,x}$ must vanish. Since the slope of the dividing streamline is finite, $u_{x,y}$ must also vanish, and we recover the condition of vanishing vorticity and wall shear stress.

Assuming now that the wall is flat, we seek to predict the slope of the dividing streamline in terms of the structure of the velocity. Since the first-order terms in Eq. (3.6.15) vanish, we must retain the second-order contributions, and this yields

$$\tan \alpha \equiv \frac{dy}{dx} = \frac{u_{y,xx} \, dx^2 + 2u_{y,xy} \, dx \, dy + u_{y,yy} \, dy^2}{u_{x,xx} \, dx^2 + 2u_{x,xy} \, dx \, dy + u_{x,yy} \, dy^2}$$

$$= \frac{u_{y,xx} + 2u_{y,xy} \tan \alpha + u_{y,yy} \tan^2 \alpha}{u_{x,xx} + 2u_{x,xy} \tan \alpha + u_{x,yy} \tan^2 \alpha} \tag{3.6.16}$$

Requiring the no-slip condition and using the continuity equation, we find that all three quantities $u_{y,xx}, u_{y,xy} = -u_{x,xx}$, vanish. Writing $u_{y,yy} = -u_{x,yx}$ and rearranging, we obtain

$$\tan \alpha = -\frac{3u_{x,yx}}{u_{x,yy}} = -\frac{3f_{x,x}}{\partial P/\partial x} \tag{3.6.17}$$

where all variables are evaluated at the stagnation point (Oswatitsch, 1958). To derive the expression in the denominator on the right-hand side of Eq. (3.6.17), we applied the Navier–Stokes equation at the stagnation point and used the no-slip condition to set $\mu u_{x,yy} = \partial P/\partial x$, where P is the modifed pressure. Lugt (1987) showed that Eq. (3.6.17) remains valid when the stagnation point is located on a curved wall, provided that the derivatives with respect to x are replaced with derivatives with respect to arc length along the wall l.

Vorticity at an evolving free surface

Consider next the flow below a deforming two-dimensional free surface and introduce Cartesian coordinates with the x axis oriented tangentially to the free surface at a point as shown in Figure 3.6.3(a). The strength of the vorticity and the component of the traction that is tangential to the free surface are given by

$$\omega = \frac{\partial u_y}{\partial x} - \frac{\partial u_x}{\partial y}$$

$$\mathbf{f} \cdot \mathbf{t} = \mu \mathbf{t} \cdot (\nabla \mathbf{u}) \cdot \mathbf{n} + \mu \mathbf{n} \cdot (\nabla \mathbf{u}) \cdot \mathbf{t} = \mu \frac{\partial u_y}{\partial x} + \mu \frac{\partial u_x}{\partial y} \tag{3.6.18}$$

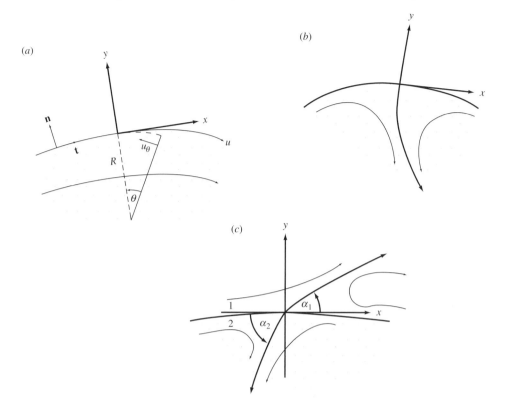

Figure 3.6.3   (a) Illustration of a two-dimensional flow underneath a free surface. (b) A stagnation point occurs at a stationary free surface; the dividing streamline must cross the free surface at a right angle. (c) A stagnation point at a stationary fluid interface.

Setting the shear stress equal to zero and using the Frenet relation (1.7.9), we obtain the simplified version of Eq. (3.6.13),

$$\omega = 2\frac{\partial u_y}{\partial x} = 2\,\mathbf{t}\cdot(\nabla\mathbf{u})\cdot\mathbf{n} = 2\,\mathbf{t}\cdot\nabla(\mathbf{u}\cdot\mathbf{n}) - 2\,\mathbf{t}\cdot(\nabla\mathbf{n})\cdot\mathbf{u} \tag{3.6.19}$$

$$= 2\,\mathbf{t}\cdot\nabla(\mathbf{u}\cdot\mathbf{n}) - 2\kappa u_x$$

where l is the arc length along the free surface and κ is the curvature of the free surface. In terms of the stream function ψ, we obtain

$$\omega = -2\frac{\partial^2\psi}{\partial l\,\partial l_n} - 2\kappa\frac{\partial\psi}{\partial l_n} \tag{3.6.20}$$

where l_n is the arc length measured in the direction of the normal vector $\mathbf{n}$.

Vorticity at a stationary free surface

The fact that the free surface of a steady flow is also a streamline allows us to use the first of Eqs. (1.7.11) to simplify the right-hand side of the expression for the shear stress given in the second of Eqs. (3.6.18) even further, obtaining

$$\mathbf{f}\cdot\mathbf{t} = -\mu\kappa u_x + \mu\frac{\partial u_x}{\partial y} \tag{3.6.21}$$

Introducing plane polar coordinates with their origin at the center of curvature of the free surface at a point as shown in Figure 3.6.3(a), where $R = 1/\kappa$ is the radius of curvature, we obtain

$$\mathbf{f} \cdot \mathbf{t} = \mu \frac{u_\theta}{R} - \mu \left(\frac{\partial u_\theta}{\partial r} \right)_{r=R} = -\mu R \left[\frac{\partial}{\partial r} \left(\frac{u_\theta}{r} \right) \right]_{r=R} \qquad (3.6.22)$$

Setting the shear stress equal to zero and combining Eq. (3.6.21) with Eq. (1.7.12) we obtain a simple expression for the vorticity at a stationary free surface in terms of the magnitude of the velocity u and the curvature of the free surface

$$\omega = -2\kappa u \qquad (3.6.23)$$

which shows that the vorticity vanishes at a stagnation point along a free surface; at an inflection point along a free surface; and over a planar free surface.

Stagnation point at a stationary free surface

A schematic illustration of a stagnation-point flow at a free surface is shown in Figure 3.6.3(b). In terms of the local coordinates, the slope of the dividing streamline at the stagnation point is given in Eq. (3.6.15), where all partial derivatives are evaluated at the stagnation point. Taking into account the first equation in (1.7.11) and Eq. (3.6.21), we find that both $u_{x,y}$ and $u_{y,x}$ at the stagnation point vanish. The continuity equation requires that $u_{x,x} = -u_{y,y}$, and this necessitates that $\tan \alpha$ either vanish, in which case we obtain a degenerate stagnation point, or be equal to $-\infty$, in which case the *dividing streamline meets the free surface at a right angle.*

Stagnation point at a stationary interface

Let us consider now a stagnation point located at the interface between two viscous fluids labeled 1 and 2, as shown in Figure 3.6.3(c). Using Eq. (3.6.21) and requiring the velocity to be continuous across the interface, we find that the jump in the tangential component of the traction across the interface at the stagnation point is given by

$$(\mathbf{f} \cdot \mathbf{t})^{(1)} - (\mathbf{f} \cdot \mathbf{t})^{(2)} = \mu_1 u_{x,y}^{(1)} - \mu_2 u_{x,y}^{(2)} \qquad (3.6.24)$$

Next we apply Eq. (3.6.15) at the dividing streamlines on either fluid, use the first of Eqs. (1.7.11) to obtain $u_{y,x} = 0$ and the continuity equation to write $u_{y,y} = -u_{x,x}$, and thus obtain

$$\tan \alpha_1 = -2 \frac{u_{x,x}^{(1)}}{u_{x,y}^{(1)}} \qquad (3.6.25)$$

Working in a similar manner, we derive a corresponding equation for fluid 2. Dividing the corresponding side of Eq. (3.6.25) and its counterpart for the second fluid, and noting that, because the velocity is continuous across the interface, $u_{x,x}$ has the same value on either side of the interface, we obtain

$$\frac{\tan \alpha_1}{\tan \alpha_2} = \frac{u_{x,y}^{(2)}}{u_{x,y}^{(1)}} \qquad (3.6.26)$$

Finally, requiring that the shear stress be continuous across the interface, we set the left-hand side of Eq. (3.6.24) equal to zero and find the remarkably simple result

$$\frac{\tan \alpha_1}{\tan \alpha_2} = \frac{\mu_1}{\mu_2} \qquad (3.6.27)$$

which may be regarded as a refraction law for the dividing streamlines (Lugt, 1987). It is interesting to note that, when the viscosities of the two fluids are identical, the dividing streamlines join smoothly at the stagnation point.

PROBLEMS

3.6.1   Force on a boundary. Consider a stationary rigid body placed in a flow, and draw a sequence of closed surfaces that enclose the body and tend uniformly to the surface of the body. Denoting the typical separation between a surface and the body by ε, consider the force $\mathbf{F}(\varepsilon)$ that is exerted on the surfaces, and sketch the magnitude of $\mathbf{F}$ as a function of ε. Do you expect that the curve will show a sharp variation as ε becomes very small?

3.6.2   Force on a boundary. Derive Eq. (3.6.22) working in the plane polar coordinates illustrated in Figure 3.6.3(a).

3.6.3   Traction normal to a line in a two-dimensional flow. (a) Show that the component of the traction normal to an arbitrary line in two-dimensional flow is given by

$$\mathbf{f} \cdot \mathbf{n} = -p - 2\mu \frac{\partial(\mathbf{u} \cdot \mathbf{t})}{\partial l} - 2\mu\kappa\mathbf{u} \cdot \mathbf{n} = -p - 2\mu \frac{\partial^2 \psi}{\partial l \, \partial l_n} + 2\mu\kappa \frac{\partial \psi}{\partial l} \qquad (3.6.28)$$

where l is the arc length along the line, $\mathbf{t}$ is the unit tangential vector, κ is the curvature of the line, l_n is the arc length in the direction of the normal vector $\mathbf{n}$, and ψ is the stream function. (b) Use Eq. (3.6.28) to derive a boundary condition for the pressure at a free surface in the presence of surface tension. (c) Use Eq. (3.6.28) to derive a boundary condition for the jump in pressure across a fluid interface in the presence of surface tension.

3.7 | SCALING OF THE NAVIER–STOKES EQUATION AND DYNAMIC SIMILITUDE

Evaluating the various terms of the Navier–Stokes equation at a particular point in a flow, we may find a broad range of magnitudes. Depending upon the structure of the flow and location within the flow, certain terms may make dominant contributions while others may make minor contributions and could thus be neglected without significant consequences. To identify this occurrence and benefit from concomitant simplifications, we inspect the structure of the flow, and typically find that the magnitude of the velocity changes by an amount U over a distance L. This means that the magnitudes of the first and second spatial derivatives of the velocity are, respectively, comparable to the magnitudes of the ratios U/L and U/L^2. Furthermore, we typically find that the magnitude of the velocity at a particular point changes by the amount U over a time period T, which means that the magnitude of the first partial derivative of the velocity with respect to time is comparable to the magnitude of the ratio U/T.

Typically, but not always, the characteristic length L is related to the size of the boundaries, the characteristic velocity U is determined by the particular mechanism that drives the flow, and the characteristic time T is either imposed by external means or simply defined as L/U. For instance, in the case of unidirectional flow through a channel or tube, U may be identified with the maximum velocity within the channel or tube, whereas in the case of uniform flow past a stationary body, U may be identified with the velocity of the incident flow. For a naturally or artificially forced oscillatory flow, T is identified with the period of oscillation.

Next, we reduce all terms in the Navier–Stokes equation using the aforementioned scales and then compare their relative magnitudes. To accomplish the first task, we introduce the dimensionless variables

$$\hat{\mathbf{u}} = \frac{1}{U}\mathbf{u}, \qquad \hat{\mathbf{x}} = \frac{1}{L}\mathbf{x}, \qquad \hat{t} = \frac{t}{T}, \qquad \hat{p} = \frac{L}{\mu U}p \qquad (3.7.1)$$

Expressing the physical variables in terms of the dimensionless variables, and substituting the results back into the Navier–Stokes equation, we obtain the dimensionless form

$$\beta \frac{\partial \hat{\mathbf{u}}}{\partial \hat{t}} + Re\, \hat{\mathbf{u}} \cdot \hat{\nabla}\hat{\mathbf{u}} = -\hat{\nabla}\hat{p} + \hat{\nabla}^2\hat{\mathbf{u}} + \frac{Re}{Fr^2}\frac{\mathbf{g}}{g} \tag{3.7.2}$$

where the gradient $\hat{\nabla}$ involves derivatives with respect to the dimensionless position vector $\hat{\mathbf{x}}$. Since the magnitude of the dimensionless variables and their derivatives in Eq. (3.7.2) is of order unity, and $\mathbf{g}/g$ is a unit vector expressing the direction of the body force, the relative importance of the various terms is determined by the magnitude of their multiplicative factors, which are the *unsteadiness parameter* β, the *Reynolds number Re*, and the *Froude number Fr*, defined, respectively, as

$$\beta = \frac{L^2}{\nu T}, \qquad Re = \frac{UL}{\nu}, \qquad Fr = \frac{U}{(gL)^{1/2}} \tag{3.7.3}$$

The inverse of β is sometimes called the *Stokes number.*

The parameter β expresses the magnitude of the Eulerian inertial–acceleration force relative to the magnitude of the viscous force, or equivalently, the ratio between the characteristic diffusion time L^2/ν and the time scale of the flow T. The Reynolds number expresses the magnitude of the inertial–convective force relative to the magnitude of the viscous force, or equivalently, the ratio between the characteristic diffusion time L^2/ν and convective time L/U. The Froude number expresses the magnitude of the inertial–convective force relative to the magnitude of the body force. Finally, the group $Re/Fr^2 = gL^2/\nu U$ expresses the magnitude of the body forces relative to that of the viscous force. The ratio $\beta/Re = L/UT$ is the *Strouhal number St*. In the absence of external forcing, T may be defined as L/U, in which case β reduces to Re, and the dimensionless Navier–Stokes equation (3.7.2) involves only two independent parameters, Re and Fr.

Steady and Quasisteady Stokes Flow

When $Re \ll 1$ and $\beta \ll 1$, both terms on the left-hand side of Eq. (3.7.2) are small compared to the terms on the right-hand side and may thus be neglected. Reverting to dimensional variables, we find that the rate of change of momentum of the fluid parcels per unit volume $\rho\, D\mathbf{u}/Dt$ is negligible, and the flow is governed by the *Stokes equation*

$$-\nabla p + \mu\, \nabla^2\mathbf{u} + \rho\, \mathbf{g} = 0 \tag{3.7.4}$$

stating that the pressure, viscous, and body forces balance at every instant. A flow that is governed by the Stokes equation is called a *Stokes* or *creeping flow.*

The absence of the temporal derivative in the equation of motion does not necessarily imply that the flow is steady, but merely reflects the fact that the forces exerted on fluid parcels are in a state of dynamic equilibrium. Consequently, the instantaneous structure of the flow depends solely upon the current boundary configuration and boundary conditions, which means that the flow is in a quasisteady state of motion. Stated differently, the history of motion enters the problem only insofar as to determine the current boundary configuration.

Steady and quasisteady Stokes flows are encountered in a variety of natural, biophysical, and engineering applications, including slurry transport, blood flow in the capillaries, flow due to the motion of ciliated microorganisms, atmospheric flow with suspended microscopic aerosol particles, coalescence of drops, and flow in the earth's mantle due to natural convection. In certain cases, the Reynolds number is small due to the fact that the boundary dimensions are small; an example is flow past a red blood cell, which has an average diameter of 5 μm. In other applications, the Reynolds number is small due to the fact that the magnitude of the velocity is small or the kinematic viscosity of the fluid is high. An example is the flow due to the motion of an air bubble in a very viscous liquid such as honey or glycerine.

The linearity of the Stokes equation allows us to conduct extensive theoretical studies of the motion, analyze the properties of the solution, as well as generate desired solutions by linear superposition using a variety of analytical and computational methods. An extensive discussion of the properties and methods of computing Stokes flow will be presented in Chapter 6.

Unsteady Stokes Flow

When $Re \ll 1$ but $\beta \sim 1$, the inertial–convective term on the left-hand side of Eq. (3.7.2) is small compared to the rest of the terms and may be neglected. Reverting to dimensional variables, we find that the nonlinear component of the point particle acceleration $\rho\, \mathbf{u} \cdot \nabla\mathbf{u}$ is negligible, and the rate of change of momentum of a point particle may be approximated with the Eulerian acceleration reaction $\rho\, \partial\mathbf{u}/\partial t$. The motion of the fluid is thus governed by the *unsteady Stokes equation* or *linearized Navier–Stokes equation*

$$\rho \frac{\partial \mathbf{u}}{\partial t} = -\nabla p + \mu\, \nabla^2 \mathbf{u} + \rho\, \mathbf{g} \tag{3.7.5}$$

Because of the presence of the acceleration term, the instantaneous structure of the flow depends not only upon the instantaneous boundary configuration and boundary conditions, but also upon the history of the motion.

Physically, the unsteady Stokes equation governs flows that are characterized by sudden acceleration or deceleration, such as those occurring during hydrodynamic braking, during the impact of a particle on solid surface, or during the initial stages of the flow due to a particle settling from rest in an ambient fluid.

The linearity of the unsteady Stokes equation allows us to compute solutions using a variety of methods, including Fourier and Laplace transforms in time and space, as well as construct desired solutions by linear superposition. The properties and methods of computing unsteady Stokes flows will be discussed in the last two sections of Chapter 6.

Flow at High Reynolds Numbers

It might appear that when $Re \gg 1$ and $\beta \gg 1$ both the pressure and viscous terms on the right-hand side of Eq. (3.7.2) are much smaller than the terms on the left-hand side and may thus be neglected. This may be true in certain cases, but since the pressure was scaled in a rather subjective manner, it is not clear that the magnitude of its dimensionless gradient will remain of order unity in this limit. To allow for this occurrence, we rescale the pressure by introducing the new dimensionless variable

$$\hat{p} = \frac{p}{\rho U^2} = \frac{1}{Re}\, \hat{p} \tag{3.7.6}$$

and work as previously to obtain the new dimensionless form of the Navier–Stokes equation

$$\beta \frac{\partial \hat{\mathbf{u}}}{\partial \hat{t}} + Re\, \hat{\mathbf{u}} \cdot \hat{\nabla}\hat{\mathbf{u}} = -Re\, \hat{\nabla}\hat{p} + \hat{\nabla}^2 \hat{\mathbf{u}} + \frac{Re}{Fr^2}\frac{\mathbf{g}}{g} \tag{3.7.7}$$

Considering now the limit as $Re \gg 1$ and $\beta \gg 1$, we find that the viscous term becomes small compared to the rest of the terms and may thus be neglected, yielding *Euler's equation*

$$\frac{\beta}{Re} \frac{\partial \hat{\mathbf{u}}}{\partial \hat{t}} + \hat{\mathbf{u}} \cdot \hat{\nabla}\hat{\mathbf{u}} = -\hat{\nabla}\hat{p} + \frac{1}{Fr^2}\frac{\mathbf{g}}{g} \tag{3.7.8}$$

which reveals that at high Reynolds numbers a laminar flow, without small-scale turbulent motion, behaves like an inviscid flow.

It should be noted again that by neglecting the viscous term we have reduced the order of the Navier–Stokes equation from two to one. We shall see in subsequent chapters that this occurrence has important implications, not only on the number of required boundary conditions, but also on the structure and properties of the solution.

Dynamic Similitude

Let us consider a steady or unsteady flow in a domain of finite or infinite extent, subject to a certain set of boundary conditions, and introduce the dimensionless variables shown in Eq. (3.7.1) to obtain the dimensionless form of the Navier–Stokes equation (3.7.2). In terms of the dimensionless variables, the continuity equation becomes

$$\hat{\nabla} \cdot \hat{\mathbf{u}} = 0 \qquad (3.7.9)$$

and the boundary conditions may be placed in the symbolic form

$$\hat{\mathbf{F}}(\hat{\mathbf{u}}, \hat{p}, \hat{\boldsymbol{\sigma}}) = \mathbf{0} \qquad (3.7.10)$$

where $\hat{\boldsymbol{\sigma}}$ is the dimensionless stress tensor. The function $\hat{\mathbf{F}}$ may contain a number of dimensionless constants that involve the physical properties of the fluid, the characteristic scales U, L, and T, the body force, and other physical constants that depend upon the nature of the boundaries; examples are the surface tension γ and the slip coefficient l. One dimensionless number pertinent to an interface is the Bond number $Bo = \rho g L^2/\gamma$; another one is the Weber number $We = \rho U^2 L/\gamma$.

In the space of dimensionless variables, the flow is governed by Eq. (3.7.2) and the continuity equation Eq. (3.7.9), and the solution must be found subject to Eq. (3.7.10). It is then evident that the structure of a steady flow or evolution of an unsteady flow will depend upon (1) the values of the dimensionless numbers β, Re, Fr, (2) the functional form of Eq. (3.7.10), and (3) the values of the dimensionless numbers involved in Eq. (3.7.10).

Two flows occurring in two different physical domains will have similar structure provided that the corresponding domains in the space of dimensionless variables coincide, and the associated initial and boundary conditions are identical. Similarity of structure means, for example, that the velocity or pressure field of the first flow may be found from those of the second flow by multiplying with an appropriate factor.

Dynamic similitude may be exploited to study the flow of a particular fluid in a certain domain by studying the flow of another fluid in a similar, larger, or smaller domain. This may be achieved by adjusting the properties of the second fluid so as to match the values of β, Re, Fr, as well as any other dimensionless numbers that appear in the boundary conditions. Miniaturization or amplification of a domain of flow is important in the study of large-scale flows, such as the flow past aircraft, or very-small-scale flows, such as the flow past microorganisms and small biological cells, and the flow over surfaces with small-scale roughness and imperfections; the latter is important in the design of audio magnetic heads and disks.

PROBLEMS

3.7.1   **Scaling for a translating body.** Consider a rigid body that translates steadily with velocity $\mathbf{U}$ within an infinite fluid that moves with uniform velocity $\mathbf{V}$. Discuss the scaling of the various terms in the Navier–Stokes equation and derive the appropriate Reynolds number.

3.7.2   **Oseen flow.** (a) Consider a steady uniform flow with velocity $\mathbf{V}$ past a stationary body. Far from the body, the flow may be decomposed into (1) the incident flow and (2) a disturbance flow with velocity $\mathbf{v}$ due to the body, so that $\mathbf{u} = \mathbf{V} + \mathbf{v}$. Introduce a similar decomposition for the pressure, substitute these expressions into the Navier–Stokes equation, and neglect quadratic terms in $\mathbf{v}$ to derive a linearized form. (b) Repeat part (a) for uniform flow past a semi-infinite plate that is aligned with the flow; that is, derive a linear equation for the disturbance stream function that is applicable far from the plate.

3.7.3   **Dimensionless form of the interfacial dynamic boundary condition.** Express the interfacial boundary condition (3.5.12) in dimensionless form.

3.7.4   **Walking and running.** Compute the Reynolds number of a walking and a running adult person. Repeat for an ant and a stumping elephant.

3.7.5   **Flow past a sphere.** We want to study the structure of uniform flow of water with velocity $U = 40$ km/h past a stationary sphere with diameter $D = 0.5$ cm. For this purpose, we study uniform flow of air past another sphere with larger diameter $D = 10$ cm. What is the appropriate air speed?

3.8 | CHANGE OF CIRCULATION AROUND MATERIAL LOOPS AND DYNAMICS OF THE VORTICITY FIELD

In the preceding sections we discussed the structure and dynamics of incompressible Newtonian flows in terms of the stresses and with reference to the equation of motion. Further insights into the physical mechanisms that govern the motion of the fluid may be obtained by studying the evolution of circulation around material loops and the dynamics of the vorticity field. The study of the latter illustrates the physical processes that contribute to the rate of change of the angular velocity of small fluid parcels. We shall see in this section that the rate of change of vorticity obeys a simplified set of rules that allow for appealing and illuminating physical interpretations.

Change of Circulation around Material Loops

To compute the rate of circulation C around a closed reducible or irreducible material loop L in a Newtonian fluid, we combine Eq. (1.8.11) with the Navier–Stokes equation (3.4.4), and assume that the viscosity of the fluid and acceleration of gravity are uniform throughout the domain of flow to obtain

$$\frac{dC}{dt} = \int_L \left(-\frac{1}{\rho}\nabla p + \nu \nabla^2 \mathbf{u} + \mathbf{g}\right)\cdot d\mathbf{X}$$
$$= -\int_L \frac{dp}{\rho} + \int_L \nu(\nabla^2 \mathbf{u})\cdot d\mathbf{X} + \mathbf{g}\cdot\int_L d\mathbf{X} \qquad (3.8.1)$$

where $\mathbf{X}$ denotes the position of the point particles along the loop. Using the fact that the integral of the derivative of a function is equal to the function itself, we find that the last integral on the right-hand side of Eq. (3.8.1) makes a vanishing contribution. Similar reasoning reveals that, *when the density of the fluid is uniform or the fluid is barotropic,* the integral of the pressure term is also equal to zero. Setting the Laplacian of the velocity equal to the negative of the curl of the vorticity, we finally obtain the simplified form

$$\frac{dC}{dt} = \int_L \nu\,(\nabla^2\mathbf{u})\cdot d\mathbf{X} = -\int_L \nu(\nabla\times\boldsymbol{\omega})\cdot d\mathbf{X} \qquad (3.8.2)$$

It then follows that *when viscous forces are insignificant,*

$$\frac{dC}{dt} = \frac{d}{dt}\int_D \boldsymbol{\omega}\cdot\mathbf{n}\,dS = 0 \qquad (3.8.3)$$

where D is an arbitrary surface bounded by L, and $\mathbf{n}$ is the unit vector normal to D.

Equation Eq. (3.8.3) is the mathematical statement of *Kelvin's circulation theorem for uniform-density and barotropic fluids,* stating that *the circulation around a closed material loop within a flow with negligible viscous forces is preserved during the motion.* As a consequence, the circulation around a loop that wraps around a toroidal boundary, and thus the cyclic constant of the flow around that boundary, remains constant in time.

Considering now a reducible loop, and taking the limit as the loop shrinks down to a point, we find that *when the conditions for Kelvin's circulation theorem to be valid are fulfilled, the point particles maintain their initial vorticity,* which means that they keep spinning at a constant angular velocity as they translate and deform in the domain of flow. The absence of a viscous torque exerted on a fluid parcel guarantees that the angular momentum of the parcel will be preserved. One consequence of this result is *Helmholtz's first theorem,* stating that if the vorticity vanishes at the initial instant, it must vanish at all subsequent times, a behavior that is known as *permanence of irrotational flow.* Another consequence is *Helmholtz's second theorem,* stating that *vortex tubes behave like material surfaces maintaining their initial circulation.* Thus line vortices are composed of the same point particles and may therefore be regarded as material lines.

Dynamics of the Vorticity Field

The equation of motion provides us with an expression for the acceleration of the point particles, as well as with an expression for the rate of change of the velocity at a point in a flow in terms of the instantaneous velocity and pressure. To derive a corresponding expression for the angular velocity of the point particles and rate of change of the vorticity at a point in a flow, we take the curl of the Navier–Stokes equation (3.4.4), assume that the acceleration of gravity is constant, and use the identity

$$\nabla \times (\mathbf{u} \cdot \nabla \mathbf{u}) = \mathbf{u} \cdot \nabla \boldsymbol{\omega} - \boldsymbol{\omega} \cdot \nabla \mathbf{u} \qquad (3.8.4)$$

to obtain the *vorticity transport equation*

$$\frac{D\boldsymbol{\omega}}{Dt} \equiv \frac{\partial \boldsymbol{\omega}}{\partial t} + \mathbf{u} \cdot \nabla \boldsymbol{\omega}$$

$$= \boldsymbol{\omega} \cdot \nabla \mathbf{u} + \frac{1}{\rho^2} \nabla \rho \times \nabla p + \nu \nabla^2 \boldsymbol{\omega} \qquad (3.8.5)$$

$$+ \nabla \nu \times \nabla^2 \mathbf{u} + 2 \nabla \times \left(\frac{1}{\rho} \nabla \mu \cdot \mathbf{E} \right)$$

Projecting both sides of Eq. (3.8.5) onto half the moment of inertia tensor of a small fluid parcel, we obtain an equation expressing the *angular momentum balance*.

Restricting our attention to fluids whose viscosity is uniform throughout the domain of flow, we obtain the simplified equation

$$\frac{D\boldsymbol{\omega}}{Dt} = \frac{\partial \boldsymbol{\omega}}{\partial t} + \mathbf{u} \cdot \nabla \boldsymbol{\omega} = \boldsymbol{\omega} \cdot \nabla \mathbf{u} + \frac{1}{\rho^2} \nabla \rho \times \nabla p + \nu \nabla^2 \boldsymbol{\omega} \qquad (3.8.6)$$

We can solve the Navier–Stokes equation for the pressure gradient and substitute the result into the right-hand side of Eq. (3.8.6) to derive an equivalent expression involving the velocity and vorticity alone, but this will not be necessary for the purposes of the present discussion.

The left-hand side of Eq. (3.8.6) expresses the material derivative of the vorticity, which is equal to the rate of change of the vorticity following a point particle, or twice the rate of change of the angular velocity of the point particle.

The first term on the right-hand side of Eq. (3.8.6) expresses the generation of vorticity due to the interaction between the vorticity field and the velocity gradient tensor $\mathbf{L} = \nabla\mathbf{u}$. Comparing Eq. (3.8.6) with Eq. (1.4.1) shows that, under the action of this term, the vorticity vector evolves as if it were a material vector, rotating with the fluid while being stretched. Thus the generation of a component of the vorticity in a particular direction is due to both the reorientation of the vortex lines under the action of the flow and the compression or stretching of the vorticity vector; the second mechanism is known as *vortex stretching*.

To further illustrate the nature of the nonlinear $\boldsymbol{\omega} \cdot \nabla\mathbf{u}$ term, we begin with the statement $\boldsymbol{\omega} \times \boldsymbol{\omega} = 0$ and use a vector identity to write $\boldsymbol{\omega} \times \boldsymbol{\omega} = (\nabla \times \mathbf{u}) \times \boldsymbol{\omega} = \boldsymbol{\omega} \cdot \nabla\mathbf{u} - \boldsymbol{\omega} \cdot (\nabla\mathbf{u})^T$, where the superscript T denotes the transpose, and thus find $\boldsymbol{\omega} \cdot \nabla\mathbf{u} = \boldsymbol{\omega} \cdot (\nabla\mathbf{u})^T$. This identity suggests that

$$\omega_j \frac{\partial u_i}{\partial x_j} = \omega_j \frac{\partial u_j}{\partial x_i} = \omega_j E_{ji} = \omega_j \left(\beta \frac{\partial u_i}{\partial x_j} + (1 - \beta) \frac{\partial u_j}{\partial x_i} \right) \qquad (3.8.7)$$

where β is an arbitrary constant. Equation (3.8.7) shows that if the vorticity vector happens to be an eigenvector of the rate of deformation tensor $\mathbf{E}$, it will amplify or shrink in its direction, behaving like a material vector.

The second term on the right-hand side of Eq. (3.8.6) expresses the *baroclinic generation of vorticity* due to the interaction between the pressure and density fields. To illustrate the physical process underlying this mechanism, let us consider a vertically stratified fluid that is set in motion by the application of a horizontal pressure gradient. The heavier point particles at the top accelerate more slowly than the lighter fluid particles at the bottom, and this suggests that a fluid vertical column will buckle in the counterclockwise direction, thus generating vorticity of positive sign.

When the density is constant or the pressure is a function of the density alone, the baroclinic production term vanishes. In the second case the gradients of the pressure and density are parallel, and their cross product is equal to zero.

The third term on the right-hand side of Eq. (3.8.6) expresses diffusion of vorticity with a diffusivity that is equal to the kinematic viscosity of the fluid, according to which the components of the vorticity behave like passive scalars.

Eulerian form

An alternative form of the vorticity transport equation (3.8.6) emerges by restating the non-linear term in the equation of motion in terms of the right-hand side of Eq. (3.4.7) before taking its curl. Noting that the curl of the gradient of a twice differentiable function vanishes, we obtain

$$\frac{\partial \boldsymbol{\omega}}{\partial t} + \nabla \times (\boldsymbol{\omega} \times \mathbf{u}) = \frac{1}{\rho^2} \nabla \rho \times \nabla p + \nu \nabla^2 \boldsymbol{\omega} \tag{3.8.8}$$

which is known as the Eulerian form of the vorticity transport equation.

Vorticity transport in uniform-density fluids

When the density of the fluid is uniform or the fluid is barotropic, which means that the pressure is a function of the density alone, the vorticity transport equation (3.8.6) assumes the simplified form

$$\frac{D\boldsymbol{\omega}}{Dt} = \frac{\partial \boldsymbol{\omega}}{\partial t} + \mathbf{u} \cdot \nabla \boldsymbol{\omega} = \boldsymbol{\omega} \cdot \nabla \mathbf{u} + \nu \nabla^2 \boldsymbol{\omega} \tag{3.8.9}$$

and the Eulerian form (3.8.8) becomes

$$\frac{\partial \boldsymbol{\omega}}{\partial t} + \nabla \times (\boldsymbol{\omega} \times \mathbf{u}) = \nu \nabla^2 \boldsymbol{\omega} \tag{3.8.10}$$

Equation (3.8.9) shows that the vorticity at a particular point in the flow evolves due to convection, vortex stretching, and viscous diffusion. The vorticity field, and thus the velocity field, will be at steady state only when the combined effects of these processes balance to zero.

It is instructive to refer back to Eq. (1.4.34) and identify $\mathbf{q}$ with the vorticity, and Q with the circulation around a loop that bounds a material surface. Combining Eq. (1.4.36) with Eq. (3.8.10), and using the fact that the vorticity field is solenoidal, we recover Kelvin's circulation theorem for a flow with negligible viscous forces.

There is a class of flows, called *generalized Beltrami flows*, for which the nonlinear term on the left-hand side of Eq. (3.8.10) vanishes, $\nabla \times (\boldsymbol{\omega} \times \mathbf{u}) = \mathbf{0}$, yielding the vectorial unsteady heat conduction equation. This class of flows includes, as a subset, Beltrami flows where the vorticity is aligned with the velocity at every point, and thus $\boldsymbol{\omega} \times \mathbf{u} = \mathbf{0}$. It is clear that if $\mathbf{u}$ and $\boldsymbol{\omega}$ are the velocity and vorticity of a generalized Beltrami flow, then $-\mathbf{u}$ and $-\boldsymbol{\omega}$ will also satisfy Eq. (3.8.10), which means that the reversed flow will also be an acceptable generalized Beltrami flow (Problem 3.8.8). Since the direction of the velocity along the streamlines is reversed, a generalized Beltrami flow can be called a *two-way flow*.

Using the fact that the vorticity is solenoidal, we write $\nabla^2 \boldsymbol{\omega} = -\nabla \times (\nabla \times \boldsymbol{\omega})$, and rewrite Eq. (3.8.10) as

$$\nabla \times \left(\frac{\partial \mathbf{u}}{\partial t} + \boldsymbol{\omega} \times \mathbf{u} + \nu \nabla \times \boldsymbol{\omega} \right) = \mathbf{0} \tag{3.8.11}$$

which states that the vector within the parentheses remains irrotational at all times.

When viscous forces are insignificant, Eq. (3.8.9) yields $D\boldsymbol{\omega}/Dt = \boldsymbol{\omega} \cdot \nabla \mathbf{u}$, which is the basis for Helmholtz's theorems discussed previously in this section. Reference to Eq. (1.4.1), in particular, shows that the vorticity vector behaves like a material vector: If $d\mathbf{l}$ is a differential material vector oriented along the vorticity vector at some particular instant, then $\boldsymbol{\omega}/|\boldsymbol{\omega}| = d\mathbf{l}/|d\mathbf{l}|$ at all times, where $\boldsymbol{\omega}$ is the vorticity at the location of the material vector.

Burgers' vortex

The flow due to Burgers' vortex provides us with an instructive example of a steady flow where diffusion of vorticity counterbalances convection and stretching (Burgers, 1948, Rott, 1958). The associated flow is constructed by superimposing an axisymmetric uniaxial extensional flow to a swirling flow corresponding to a columnar vortex. Introducing cylindrical polar coordinates with the x axis pointing in the direction of the vortex, we write

$$u_x = kx, \qquad u_\sigma = -\tfrac{1}{2}k\sigma, \qquad u_\varphi = U(\sigma)$$

$$\boldsymbol{\omega} = \omega(\sigma)\,\mathbf{i}, \qquad \omega(\sigma) = \frac{1}{\sigma}\frac{d(\sigma U)}{d\sigma} \tag{3.8.12}$$

where k is the rate of the extensional flow, $U(\sigma)$ is the azimuthal velocity profile, and $\mathbf{i}$ is the unit vector along the x axis. Substituting these expressions into the axial component of Eq. (3.8.10) and setting the temporal derivative equal to zero to ensure steady state, we obtain the homogeneous differential equation

$$k\frac{d(\omega\sigma^2)}{d\sigma} + 2\nu\left(\sigma\frac{d^2\omega}{d\sigma^2} + \frac{d\omega}{d\sigma}\right) = 0 \tag{3.8.13}$$

A nontrivial solution that is finite at the axis is given by

$$\omega = Ak\,\exp\left(-\frac{k}{4\nu}\sigma^2\right) \tag{3.8.14}$$

where A is a dimensionless constant. Equation (3.8.14) shows that the diameter of the vortex is constant along its axis, and its magnitude is comparable to the quantity $(4\nu/k)^{1/2}$. The azimuthal velocity profile is found by integrating the last of Eqs. (3.8.12), and the value of a arises by specifying either the maximum value of the azimuthal velocity $U(\sigma)$ or the circulation around a loop of large radius enclosing the vortex (Problem 3.8.1).

Diffusing vortex sheet under the action of stretching

An example of a flow that continues to evolve until the diffusion of vorticity is balanced by convection is provided by a diffusing vortex sheet separating two streams that merge along the x axis with velocities U_∞ and $-U_\infty$, in the presence of an extensional flow that stretches the vortex lines, as illustrated in Figure 3.8.1. Stipulating that the vortex lines are and remain oriented along the z axis, we express the three components of the velocity and magnitude of the vorticity in the forms

$$u_x = U(y, t) + kx, \qquad u_y = -ky, \qquad u_z = kz, \qquad \omega(y, t) = -\frac{dU}{dy} \tag{3.8.15}$$

where $U(y)$ is the velocity profile across the vortex layer, and k is the rate of the extensional flow. The absence of an intrinsic characteristic length scale, and our hope that the vortex layer will develop in a self-similar manner, suggest setting

$$\omega = \frac{k}{\delta(t)}f(\eta) \tag{3.8.16}$$

where $\delta(t)$ is the effective thickness of the vortex layers that is to be computed as part of the solution, $\eta \equiv y/\delta(t)$ is a similarity variable, and f is a dimensionless function. Substituting the

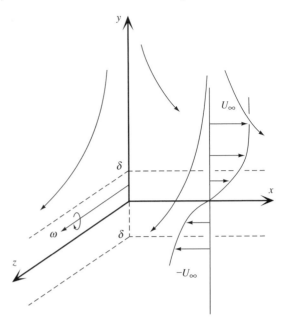

Figure 3.8.1 A diffusing vortex sheet is subjected to a straining flow that stretches the vortex lines and allows for a steady state.

expressions in Eqs. (3.8.15) and (3.8.16) into the z component of the vorticity transport equation and rearranging, we obtain

$$\delta(\delta' + k\delta) = -\nu \frac{f''}{f + \eta f'} \tag{3.8.17}$$

We observe that the left-hand side is a function of t alone, whereas the right-hand side is a function of both t and y, and this requires that each side must be equal to the same constant, which, for future convenience of notation, we set equal to $2\nu B^2$ where B is a dimensionless constant. Solving the resulting ordinary differential equations for δ and f subject to the initial condition $\delta(0) = 0$ and under the stipulation that f is an even function of η, we obtain

$$\delta(t) = B\left(\frac{2\nu}{k}(1 - e^{-2kt})\right)^{1/2}, \qquad f(\eta) = C\exp(-B^2\eta^2) \tag{3.8.18}$$

where C is a new dimensionless constant. Substituting these expressions into Eq. (3.8.16), we obtain the vorticity distribution

$$\omega(y, t) = \frac{kD}{(1 - e^{-2kt})^{1/2}}\exp\left(-\frac{ky^2}{2\nu(1 - e^{-2kt})}\right) \tag{3.8.19}$$

where D is another dimensionless constant. To compute the velocity profile $U(y)$, we substitute Eq. (3.8.19) into the last of Eqs. (3.8.15) and integrate the resulting equation with respect to y to obtain an expression in terms of the error function. The constant D then follows as

$$D = -2U_\infty(\pi\nu k/2)^{-1/2} \tag{3.8.20}$$

(Problem 3.8.2). As time progresses, the thickness of the vortex layer tends to the asymptotic value $(2\nu/k)^{1/2}$, which shows that, at large times, the diffusion of vorticity is balanced by convection. In the absence of the straining flow, the vortex layer diffuses and occupies the whole xy plane (Problem 3.8.3).

Axisymmetric Flow of Uniform-Density Fluids

For axisymmetric flow of a fluid with uniform density without swirling motion, the axial and radial components of Eq. (3.8.9) are satisfied in a trivial manner, and the azimuthal component yields a scalar equation for the strength of the vorticity ω,

$$\frac{D}{Dt}\left(\frac{\omega}{\sigma}\right) = \nu\frac{1}{\sigma}\nabla^2\omega \tag{3.8.21}$$

The vortex stretching term is implicit within the material derivative of ω/σ on the left-hand side. When the strength of the vorticity is proportional to the radial distance at the initial instant, that is, $\omega = \alpha\sigma$, where α is a constant, the Laplacian on the right-hand side vanishes, and this shows that the vorticity remains proportional to the radial distance *at all times.*

Equation (3.8.21) shows that, when viscous effects are insignificant, the point particles move while their vorticity is adjusting so that the ratio ω/σ remains constant, equal to the value at the initial instant. The underlying physical mechanism may be traced back to preservation of circulation around vortex tubes. One consequence of this behavior is that the instantaneous strength of a line vortex ring is proportional to its current radius.

Two-Dimensional Flow of Uniform-Density Fluids

The vorticity transport equation for two-dimensional flow in the xy plane is subject to substantial simplifications due to the absence of the mechanism of vortex stretching. We find that the x and y components of Eq. (3.8.9) are satisfied in a trivial manner, whereas the z component yields a scalar convection–diffusion equation for the strength of the vorticity ω,

$$\frac{D\omega}{Dt} = \frac{\partial\omega}{\partial t} + \mathbf{u}\cdot\nabla\omega = \nu\nabla^2\omega \tag{3.8.22}$$

where ∇^2 is the two-dimensional Laplacian operating in the xy plane. Equation (3.8.22) appears to indicate that ω behaves like a passive scalar. This interpretation, however, is not entirely accurate, for the vorticity has a direct influence on the structure of the velocity field through the Biot–Savart integral.

Let us consider an infinite flow that vanishes at infinity in the absence of interior boundaries, integrate Eq. (3.8.22) over the whole area of the flow, and use the continuity equation to write

$$\frac{d}{dt}\int_{\text{Flow}}\omega\,dA = \int_{\text{Flow}}(-\mathbf{u}\cdot\nabla\omega + \nu\nabla^2\omega)\,dA$$
$$= \int_{\text{Flow}}\nabla\cdot(-\omega\mathbf{u} + \nu\nabla\omega)\,dA \tag{3.8.23}$$

Using the divergence theorem, we convert the last area integral into a line integral over a loop with large dimensions that encloses the flow. Since the velocity is assumed to vanish at infinity, the line integral is equal to zero, and this shows that the total circulation of the flow, defined as the integral of the vorticity over the area of the flow, is conserved. Physically, vorticity neither is produced nor can it escape through boundaries.

When viscous effects are insignificant, the right-hand side of Eq. (3.8.22) vanishes, showing that the point particles move while maintaining their initial vorticity; that is, they spin at a constant angular rate. Two consequences of this behavior are that point vortices maintain their strength, and patches of constant vorticity preserve their initial vorticity. A third consequence is that the vorticity of a steady flow is constant along the streamlines.

Diffusing point vortex

To illustrate the action of viscous diffusion, consider a point vortex with strength κ placed at the origin. The initial vorticity distribution is represented in terms of a two-dimensional delta function in the xy plane as $\omega = \kappa\,\delta(\mathbf{x})$. Assuming that the flow maintains a circular symmetry with respect to the z axis at all times, we introduce plane polar coordinates centered at the origin and find that Eq. (3.8.22) simplifies to the scalar unsteady diffusion equation

$$\frac{\partial \omega}{\partial t} = \nu \frac{1}{r} \frac{\partial}{\partial r}\left(r \frac{\partial \omega}{\partial r}\right) \qquad (3.8.24)$$

which shows that the flow due to a diffusing point vortex is a generalized Beltrami flow. The absence of an intrinsic length scale suggests searching for a solution in the form $\omega = [\kappa/(\nu t)]f(\eta)$, where $\eta \equiv r/(\nu t)^{1/2}$ is a similarity variable and f is a yet unknown function. Substituting this functional form into Eq. (3.8.24), we obtain the second-order linear ordinary differential equation

$$2(\eta f')' + \eta^2 f' + 2f\eta = 0 \qquad (3.8.25)$$

where a prime signifies a derivative with respect to η. Under the stipulation that the derivatives of f at the origin are finite and the total circulation of the flow remains equal to κ, we obtain the solution

$$\omega = \frac{\kappa}{4\pi\nu t} \exp\left(-\frac{r^2}{4\nu t}\right) \qquad (3.8.26)$$

Integrating the definition $\omega = (1/r)\,\partial(ru_\theta)/\partial r$, we find that the azimuthal component of the velocity is given by

$$u_\theta = \frac{\kappa}{2\pi r}\left[1 - \exp\left(-\frac{r^2}{4\nu t}\right)\right] \qquad (3.8.27)$$

which describes the flow due to a *diffusing point vortex* known as the *Oseen vortex*.

Radial profiles of the reduced vorticity $\Omega = \omega a^2/\kappa$ and velocity $U = u_\theta a/\kappa$ with respect to reduced radial distance $R \equiv r/a$ at a sequence of dimensionless times $T \equiv \nu t/a^2$, where a is a certain reference length, are shown in Figure 3.8.2(a, b). It is clear that, as time progresses, the vorticity diffuses away from the point vortex and tends to occupy the whole plane. In this manner, the vorticity distribution is smeared out, and the point vortex reduces to a *vortex blob*.

Source of Vorticity in a Viscous Flow

Equation (3.8.9) reveals that if a flow is irrotational at the initial instant, the initial rate of change of the vorticity will vanish, with the apparent consequence that the flow will remain irrotational at all times. This reasoning, however, ignores the fact that vorticity may enter the flow through the boundaries in a process that is similar to that by which heat enters an initially isothermal domain across the boundaries due to a sudden change in boundary temperature.

The mechanism by which vorticity enters a flow can be illustrated by considering the flow past an impermeable boundary that is introduced suddenly into an incident flow. To satisfy the no-penetration boundary condition, we introduce an appropriate disturbance irrotational flow, but we are still left with a finite slip velocity that amounts to a boundary vortex sheet. Viscosity causes the singular vorticity distribution associated with the vortex sheet to diffuse into the fluid complementing the disturbance irrotational flow, while the vorticity at the boundary is continuously adjusting in response to the developing flow. If the fluid is inviscid, vorticity cannot diffuse into the flow, and the vortex sheet adheres to the boundary at all times.

Generation of Vorticity at an Interface

In Section 3.5 we established a point of view according to which an interface between two fluids represents a surface of distributed point force. Dividing Eq. (3.5.18) by ρ and taking the curl of the resulting equation, we find that the associated rate of production of vorticity is given by

$$\nabla \times \left(\frac{1}{\rho}\mathbf{F}\right) = -\frac{1}{\rho^2}\nabla\rho \times \mathbf{F} + \frac{1}{\rho}\nabla \times \mathbf{F} \qquad (3.8.28)$$

where the singular forcing function $\mathbf{F}$ was defined in Eq. (3.5.22).

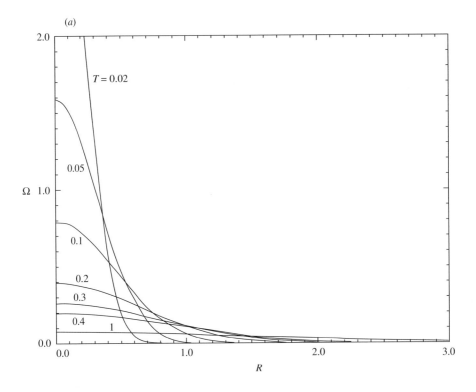

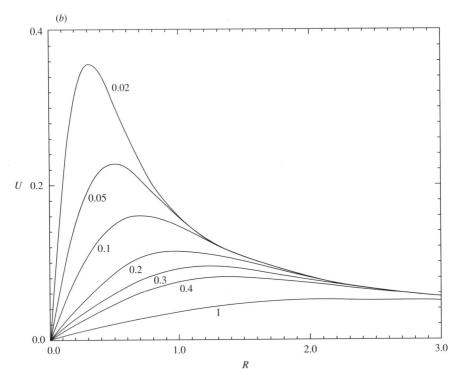

Figure 3.8.2 Profiles of (a) the reduced vorticity $\Omega = \omega a^2/\kappa$ and (b) reduced velocity $U = u_\theta a/\kappa$ of a diffusing point vortex with respect to radial distance $R \equiv r/a$ at a sequence of dimensionless times $T \equiv \nu t/a^2$, where a is a reference length.

The first term of the right-hand side of Eq. (3.8.28) may be combined with the baroclinic production term in the vorticity transport equation. If the discontinuity in the interfacial force is normal to an interface, $\nabla\rho$ and $\mathbf{F}$ are aligned, and the interface makes no contribution to the baroclinic production.

Taking the curl of Eq. (3.5.22), we find

$$(\nabla \times \mathbf{F})_i = \int_{\text{Interface}} \frac{\partial\,\delta(\mathbf{x} - \mathbf{x}')}{\partial x_j}\,\varepsilon_{jki}\,\Delta f_k(\mathbf{x}')\,dS(\mathbf{x}') \tag{3.8.29}$$

which shows that the second term on the right-hand side of Eq. (3.8.28) causes the generation of a sheet of vortex dipoles, with an associated discontinuity in the velocity gradient across the interface.

Two-dimensional flow

To illustrate further the nature of the term on the right-hand side of Eq. (3.8.29), we consider two-dimensional flow in the xy plane, in which case δ is the two-dimensional delta function, and

$$\nabla \times \mathbf{F} = \mathbf{k} \int_{\text{Interface}} \nabla\,\delta(\mathbf{x} - \mathbf{x}') \cdot [\Delta\mathbf{f}(\mathbf{x}') \times \mathbf{k}](\mathbf{x}')\,dl(\mathbf{x}') \tag{3.8.30}$$

where l is the arc length along the interface and $\mathbf{k}$ is the unit vector along the z axis (McCracken and Peskin, 1980). Decomposing the jump in the interfacial traction into its normal and tangential components, respectively designated by the superscripts N and T, we obtain

$$\nabla \times \mathbf{F} = \mathbf{k} \int_{\text{Interface}} \nabla\,\delta(\mathbf{x} - \mathbf{x}') \cdot (\Delta f^T \mathbf{n} - \Delta f^N \mathbf{t})(\mathbf{x}')\,dl(\mathbf{x}') \tag{3.8.31}$$

where we have assumed that the set of unit vectors $\mathbf{n}, \mathbf{t}, \mathbf{k}$ form a right-handed system of axes. The tangential component within the integrand may be manipulated further by writing

$$\int_{\text{Interface}} \Delta f^N(\mathbf{x}')\,\nabla\,\delta(\mathbf{x} - \mathbf{x}') \cdot \mathbf{t}(\mathbf{x}')\,dl(\mathbf{x}')$$

$$= -\int_{\text{Interface}} \nabla'[\Delta f^N(\mathbf{x}')\,\delta(\mathbf{x} - \mathbf{x}')] \cdot \mathbf{t}(\mathbf{x}')\,dl(\mathbf{x}') \tag{3.8.32}$$

$$+ \int_{\text{Interface}} \delta(\mathbf{x} - \mathbf{x}')\,\nabla'[\Delta f^N(\mathbf{x}')] \cdot \mathbf{t}(\mathbf{x}')\,dl(\mathbf{x}')$$

where a prime over the gradient designates differentiation with respect to $\mathbf{x}'$. When the interface is closed or periodic, the first integral on the right-hand side of Eq. (3.8.32) vanishes, and Eq. (3.8.31) obtains the simplified form

$$\nabla \times \mathbf{F} = \mathbf{k}\left(-\int_{\text{Interface}} \nabla'\,\delta(\mathbf{x} - \mathbf{x}') \cdot \mathbf{n}(\mathbf{x}')\,\Delta f^T(\mathbf{x}')\,dl(\mathbf{x}')\right.$$

$$\left.+ \int_{\text{Interface}} \delta(\mathbf{x} - \mathbf{x}')\,d\,\Delta f^N(\mathbf{x}')\right) \tag{3.8.33}$$

The first term on the right-hand side of Eq. (3.8.33) expresses the generation of a sheet of vortex dipoles due to the tangential component of the discontinuity in the interfacial traction. The second term expresses the generation of a vortex sheet due to the normal component of the discontinuity in the interfacial traction.

Vorticity Transport Equation in a Noninertial Frame of Reference

The equation of motion in an accelerating frame of reference that translates with time-dependent velocity $\mathbf{U}(t)$ and rotates about its origin with angular velocity $\mathbf{\Omega}(t)$ includes the fictitious forces shown in Eq. (3.1.43). Adding these forces to the right-hand side of the regular form of the Navier–Stokes equation, taking the curl of the resulting expression under the additional assumption that the density is uniform throughout the domain of flow, and rearranging, we find that the *modified* or *intrinsic vorticity* $\mathbf{W} \equiv \boldsymbol{\omega} + 2\boldsymbol{\Omega}$ satisfies the regular vorticity transport equation; that is,

$$\frac{D\mathbf{W}}{Dt} = \frac{\partial \mathbf{W}}{\partial t} + \mathbf{v} \cdot \nabla \mathbf{W} = \mathbf{W} \cdot \nabla \mathbf{v} + \nu \nabla^2 \mathbf{W} \qquad (3.8.34)$$

where $\mathbf{v}$ is the velocity in the noninertial frame defined after Eq. (3.1.39) (Problem 3.8.4; Speziale, 1987). The simplicity of Eq. (3.8.34) is exploited for the efficient computation of incompressible Newtonian flows in terms of the velocity and intrinsic vorticity (Section 13.1).

PROBLEMS

3.8.1 **Burgers' vortex.** (a) Show that the the azimuthal velocity profile associated with Burgers' vortex is given by

$$U(\sigma) = \frac{2A\nu}{\sigma} \left[1 - \exp\left(-\frac{k}{4\nu}\sigma^2\right) \right] \qquad (3.8.35)$$

(b) Show that $A = C/(4\pi\nu)$, where C is the circulation around a loop of infinite radius enclosing the vortex. (c) Evaluate the constant A in terms of the maximum value of the azimuthal velocity $U(\sigma)$.

3.8.2 **Stretched vortex layer.** Derive the velocity profile associated with the vorticity distribution shown in Eq. (3.8.19), and the value of the constant D shown in Eq. (3.8.20).

3.8.3 **Diffusing vortex layer.** Compute the evolving vorticity distribution associated with a diffusing vortex sheet that separates two streams merging along the x axis with velocities $U(+\infty)$ and $U(-\infty)$ (see Section 5.2).

3.8.4 **Vorticity transport equation in a noninertial frame.** Derive Eq. (3.8.34).

3.8.5 **Vorticity due to the motion of a body.** Discuss the physical process by which vorticity enters the flow due to a body that is suddenly set in motion within a viscous fluid.

3.8.6 **Enstrophy and intensification of the vorticity field.** The enstrophy of a flow is defined as the volume integral of the square of the magnitude of the vorticity. By projecting the vorticity transport equation onto the vorticity vector, show that the enstrophy of a flow with uniform physical properties evolves according to the equation

$$\frac{d}{dt} \int_{\text{Flow}} \boldsymbol{\omega} \cdot \boldsymbol{\omega} \, dV = 2 \int_{\text{Flow}} (\boldsymbol{\omega}\boldsymbol{\omega}) : \mathbf{E} \, dV$$
$$- 2\nu \int_{\text{Flow}} \nabla\boldsymbol{\omega} : \nabla\boldsymbol{\omega} \, dV + \nu \int_{\text{Boundaries}} \nabla(\boldsymbol{\omega}\boldsymbol{\omega}) \cdot \mathbf{n} \, dS \qquad (3.8.36)$$

The terms on the right-hand side represent, respectively, intensification of vorticity due to vortex stretching, the counterpart of viscous dissipation for the vorticity, and surface diffusion across the boundaries. The unit normal vector $\mathbf{n}$ points into the fluid.

3.8.7 **Ertel's theorem.** Consider an arbitrary scalar function of position and time $f(\mathbf{x}, t)$, and show that

$$\frac{D}{Dt}(\boldsymbol{\omega} \cdot \nabla f) = \boldsymbol{\omega} \cdot \nabla \left(\frac{Df}{Dt}\right) \qquad (3.8.37)$$

On the basis of this equation, show that if f is convected by the flow, then $\boldsymbol{\omega} \cdot \nabla f$ will also be convected by the flow.

3.8.8 **Generalized Beltrami flow.** Discuss whether, when we switch the sign of the velocity of a generalized Beltrami flow, we must also switch the sign of the modified pressure gradient in order to satisfy the equation of motion.

3.9 COMPUTATION OF EXACT SOLUTIONS TO THE EQUATION OF MOTION IN TWO DIMENSIONS ON THE BASIS OF THE VORTICITY TRANSPORT EQUATION

The simple form of the vorticity transport equation for two-dimensional flow, shown in Eq. (3.8.22), may be exploited for deriving exact solutions to the equation of motion representing various types of steady and unsteady, viscous or inviscid flows.

Generalized Beltrami Flows

When the gradient of the vorticity is and remains perpendicular to the velocity vector, the nonlinear convective term in Eq. (3.8.22) vanishes, yielding a generalized Beltrami flow whose vorticity evolves according to the unsteady diffusion equation

$$\frac{\partial \omega}{\partial t} = \nu \nabla^2 \omega \tag{3.9.1}$$

Since the gradient of the vorticity is perpendicular to the streamlines, which suggests that the vorticity is constant along the streamlines, the vorticity may be regarded as a function of the stream function alone, and we may write

$$\omega = -\nabla^2 \psi = f(\psi) \tag{3.9.2}$$

Substituting Eq. (3.9.2) into Eq. (3.9.1), we obtain

$$\nabla^2 \left(\frac{\partial \psi}{\partial t} + \nu f(\psi) \right) = 0 \tag{3.9.3}$$

which is fulfilled when the stream function evolves according to the equation

$$\frac{\partial \psi}{\partial t} = -\nu f(\psi) \equiv \nu \nabla^2 \psi \tag{3.9.4}$$

Specifying a certain functional form for f and solving for ψ provides us with various families of two-dimensional generalized Beltrami flows. Setting, in particular, f equal to zero or to a constant value, yields irrotational flows and flows with constant vorticity. Other families of steady and unsteady generalized Beltrami flows are reviewed by Wang (1989, 1991).

Differentiating the first and last terms in Eq. (3.9.4) with respect to x and y, we find that the components of the velocity satisfy the unsteady heat conduction equation,

$$\frac{\partial \mathbf{u}}{\partial t} = \nu \nabla^2 \mathbf{u} \tag{3.9.5}$$

Substituting the first of Eqs. (3.4.12) into the Navier–Stokes equation (3.4.8), replacing $\partial \mathbf{u}/\partial t$ with the right-hand side of Eq. (3.9.5), rewriting the curl of the vorticity on the right-hand side as the negative of the Laplacian of the velocity, and simplifying and integrating the resulting equation with respect to the spatial variables, we derive the associated pressure field

$$p = -\tfrac{1}{2}\rho |\mathbf{u}|^2 + \rho \mathbf{g} \cdot \mathbf{x} - F(\psi) + c \tag{3.9.6}$$

where c is a constant and the function $F(\psi)$ was defined after Eqs. (3.4.12).

Taylor cellular flow

An example of an unsteady generalized Beltrami flow arises by stipulating that the function f defined in Eq. (3.9.2) is a linear function of ψ, setting $f(\psi) = a^2 \psi$, where a is an arbitrary constant. Integrating with respect to time the differential equation that is composed of the first two terms in Eq. (3.9.4) we obtain

$$\psi = g(x, y) \, e^{-a^2 \nu t} \tag{3.9.7}$$

The right-hand side of Eq. (3.9.4) requires that the function g satisfy Helmholtz's equation

$$\nabla^2 g = -a^2 g \tag{3.9.8}$$

The particular solution

$$g = A \cos(bx) \cos(cy) \tag{3.9.9}$$

where $b^2 + c^2 = a^2$ and A is an arbitrary constant, yields a decaying cellular periodic flow with wave numbers in the x and y direction, respectively, equal to b and c (Taylor, 1923). The corresponding pressure distribution is found from Eq. (3.9.6). The streamline pattern for $b = c$ in the plane of the reduced axes $X = bx$ and $Y = by$ is shown in Figure 3.9.1.

Inviscid flows

When the effects of viscosity are insignificant, the right-hand side of Eq. (3.9.4) vanishes, and this shows that any time-independent solution of Eq. (3.9.2) represents an acceptable steady

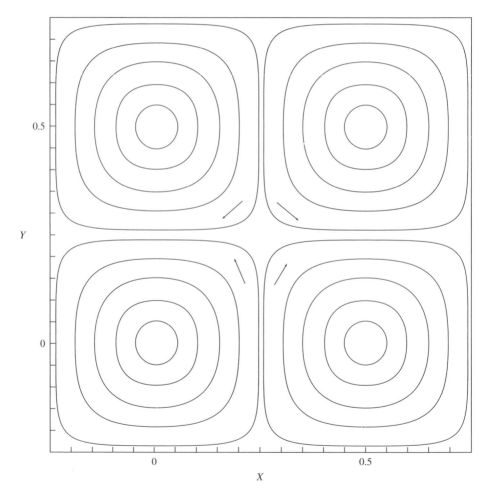

Figure 3.9.1   Streamline pattern of doubly-periodic decaying Taylor cellular flow with square cells.

inviscid flow. Examples are the flow due to a point vortex, the flow due to an infinite array of point vortices, and any unidirectional shear flow with an arbitrary velocity profile. The corresponding pressure distributions are found from Eq. (3.9.6).

Extended Beltrami Flows

Another situation that allows for linearization of the vorticity transport equation arises by stipulating that the vorticity distribution takes the particular form

$$\omega = -\nabla^2\psi = -q(x, y, t) - c\psi \tag{3.9.10}$$

where q is a certain specified function and c is a specified constant with dimensions of inverse squared length. In this case, the vorticity transport equation reduces to a *linear* partial differential equation for ψ,

$$\frac{\partial\nabla^2\psi}{\partial t} + \frac{\partial q}{\partial x}\frac{\partial\psi}{\partial y} - \frac{\partial q}{\partial y}\frac{\partial\psi}{\partial x} - \nu c^2\psi = \nu(\nabla^2 + c)q \tag{3.9.11}$$

which may be solved for a number of cases in closed form (Wang, 1989, 1991).

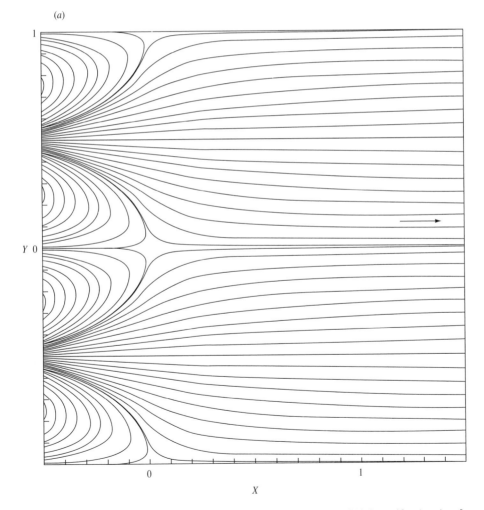

(a)

Figure 3.9.2   Streamline patterns of steady Kovasznay flow for $k\alpha = 1.0$ and (a) $Re = 10$.   (*continued*)

(b)

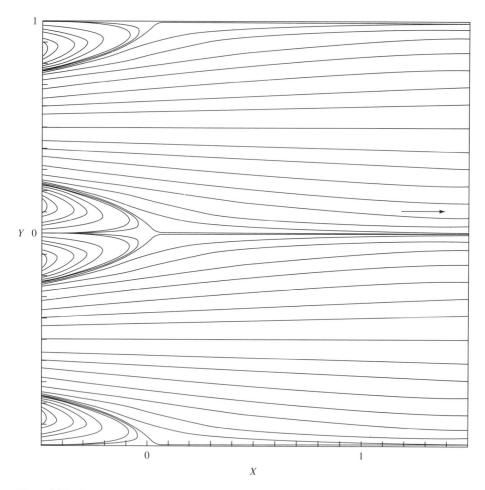

Figure 3.9.2 (*continued*)   Streamline patterns of steady Kovasznay flow for $k\alpha = 1.0$ and (b) $Re = 50$.

Kovasznay flow

As an example, consider a steady flow with $q = -cUy$, where U is a constant, for which Eq. (3.9.10) becomes

$$\omega = -\nabla^2 \psi = cUy - c\psi \tag{3.9.12}$$

(Kovasznay, 1948). The steady version of the vorticity transport equation (3.9.11) obtains the simplified form

$$\frac{U}{\nu c} \frac{\partial \psi}{\partial x} - \psi = -Uy \tag{3.9.13}$$

a solution of which is

$$\psi = U(y - \alpha \sin ky \, e^{Ckx}) \tag{3.9.14}$$

where α is an arbitrary constant, k is an arbitrary wave number in the y direction, and we have introduced the dimensionless number $C = c\nu/(kU)$. It remains to verify that Eq. (3.9.14) is consistent with the assumed functional form of Eq. (3.9.12). Substituting the former into the latter equation, we obtain a quadratic equation for C,

$$C^2 - \frac{Re}{2\pi}C - 1 = 0 \tag{3.9.15}$$

where we have defined the effective Reynolds number $Re = 2\pi U/k\nu$. Retaining the root with the negative value yields

$$\psi = U\left(y - \alpha \sin ky \, \exp\left\{-\frac{1}{2}\left[\left(\frac{Re^2}{4\pi^2} + 4\right)^{1/2} - \frac{Re}{2\pi}\right]kx\right\}\right) \tag{3.9.16}$$

Streamline patterns in the plane of the dimensionless axes $X = x/a$ and $Y = y/a$, where $a = 2\pi/k$ is the separation of two successive cells along the y axis, are illustrated in Figure 3.9.2(a, b) for $k\alpha = 1.0$ and $Re = 10$ and 50. In both cases a stagnation point occurs at the origin. The flow may be identified with that developing at the wake behind an infinite array of cylinders arranged along the y axis, subject to a uniform incident flow in the x direction. It is interesting to observe that, as the Reynolds number is increased, the region of recirculating flow becomes more slender due to the effects of inertia.

PROBLEMS

3.9.1 **Flow over a porous plate with suction.** Consider uniform flow above a porous plate through which fluid is withdrawn at a uniform rate, so that the velocity of the fluid parallel and normal to the surface of the plate have, respectively, the constant values U and V. Show that the velocity field is given by

$$u_x = U\left[1 - \exp\left(-\frac{yV}{\nu}\right)\right], \qquad u_y = V \tag{3.9.17}$$

Interpret the solution in terms of vorticity diffusion toward, and convection away from the plate.

3.9.2 **Extended Beltrami flow.** Show that the flow described by the stream function

$$\psi = -Uy[1 - \exp(-Ux/\nu)] \tag{3.9.18}$$

may be derived as an extended Beltrami flow, and discuss its physical interpretation.

References

Burgers, J. M., 1948, A mathematical model illustrating the theory of turbulence. *Adv. Appl. Mech.* **1,** 171–96. Academic.

Dussan V., E. B., 1979, On the spreading of liquids on solid surfaces: static and dynamic contact lines. *Ann. Rev. Fluid Mech.* **11,** 371–400.

Edwards, D. A., Brenner, H., and Wasan, D. T., 1989, *Interfacial Transport Processes and Rheology.* Butterworth–Heinemann.

Gresho, P. M., 1991, Incompressible fluid dynamics: some fundamental formulation issues. *Annu. Rev. Fluid Mech.* **23,** 413–53.

Kovasznay, L. I. G., 1948, Laminar flow behind a two-dimensional grid. *Proc. Camb. Phil. Soc.* **44,** 58–62.

Lighthill, M. J., 1963, Attachment and separation in three-dimensional flow. In *Laminar Boundary Layers,* Editor: L. Rosenhead, Chapter 2. Oxford University Press.

Lugt, H. J., 1987, Local flow properties at a viscous free surface. *Phys. Fluids* **30,** 3647–52.

McCracken, M. F., and Peskin, C. S., 1980, A vortex method for blood flow through heart valves. *J. Comp. Phys.* **35,** 183–205.

Oswatitsch, K., 1958, Die Ablösungsbedingung von Grenzschichten, in *IUTAM Symp. Boundary Layer Research,* Editor: H. Görtler, Springer–Berlin, 357–67.

Pozrikidis, C., 1995, Stokes flow in the presence of interfaces. *Boundary Element Applications in Fluid Mechanics,* Edited by H. Power. Computational Mechanics Publications, Southampton.

Richardson, S., 1973, On the no-slip boundary condition. *J. Fluid Mech.* **59,** 707–19.

Rosenhead, L., 1954, A discussion on the first and second viscosities of fluids. *Proc. Roy. Soc. London* A **226,** 1–65.

Rott, N., 1958, On the viscous core of a line vortex. *ZAMP* **9,** 543–53.

Schowalter, W. R., 1978, *Mechanics of Non-Newtonian Fluids.* Pergamon.

Speziale, C. G., 1987, On the advantages of the vorticity–velocity formulation of the equations of fluid dynamics. *J. Comp. Phys.* **73,** 476–80.

Tanner, R. I., 1988, *Engineering Rheology.* Oxford University Press.

Taylor, G. I., 1923, On the decay of vortices in a viscous fluid. *Phil. Mag.* **46,** 671–74.

Tobak, M., and Peake, D. J., 1982, Topology of three-dimensional separated flows. *Ann. Rev. Fluid Mech.* **14,** 61–85.

Truesdell, C., 1974, The meaning of viscometry in fluid dynamics. *Annu. Rev. Fluid. Mech.* **6,** 111–46.

Wang, C. Y., 1989, Exact solutions of the unsteady Navier–Stokes equations. *Appl. Mech. Rev.* **42,** S269–82.

Wang, C. Y., 1991, Exact solutions of the steady-state Navier–Stokes equations. *Annu. Rev. Fluid Mech.* **23,** 159–77.

Wilkinson, J. H., 1965, *The Algebraic Eigenvalue Problem.* Oxford University Press.

For a fluid that is stationary or translates as a rigid body the continuity equation is satisfied in a trivial manner and the Navier–Stokes equation reduces to a first-order differential equation expressing a balance between the body force, the gradient of the pressure, and possibly the fictitious force due to the acceleration and rotation of the frame of reference. This simplified force balance may be integrated using elementary methods subject to appropriate boundary conditions to yield the pressure distribution within the fluid. The integration produces a scalar constant whose value is determined by specifying the level of the pressure at some point on a boundary. The implementation of this procedure and its applications will be discussed in Section 4.1 for stationary fluids as well as for fluids executing steady and unsteady rigid-body motion.

The interfacial boundary conditions discussed in Section 3.5 require that the magnitude of the normal component of the traction, which is equal to the pressure, undergo a discontinuity across the interface that is balanced by surface tension. Accordingly, the interface must assume a shape that is compatible with the pressure distribution within the fluids as well as with the boundary conditions for the contact angle or for the location of a three-phase contact line. The consequent mathematical implication is that the pressure distributions within the fluids on either side of the interface may not be computed independently, but must be found simultaneously with the interfacial shape so that all boundary conditions are fulfilled. To this end, it should be noted that a necessary condition for the fluids to be stationary is that the interfacial tension is uniform; variations of interfacial tension over an interface generate shearing tractions that drive a so-called *Marangoni flow*. When the variations of the interfacial tension are due to a temperature field, we obtain a *thermocapillary flow*.

Depending upon the relative location of the contact lines, the magnitude of the contact angle, and the properties of the fluids, a stationary interface may assume a variety of shapes discussed in an illuminating monograph by Isenberg (1992), and its description provides us with a challenging problem (Hartland and Hartley, 1976; Finn, 1985). In this chapter we shall discuss the mathematical formulation and shall develop numerical procedures for computing the shape of interfaces with cylindrical and axisymmetric shapes. Procedures for computing three-dimensional interfaces will be reviewed briefly in the concluding section.

4.1 | PRESSURE DISTRIBUTION WITHIN A FLUID IN RIGID-BODY MOTION

We begin by considering the pressure distribution within a stationary fluid, but since any fluid that moves as a rigid body appears to be stationary in a suitable frame of reference, we also include fluids that execute steady and unsteady translation and rotation.

Stationary and Translating Fluids

The Navier–Stokes equation in an inertial frame of reference for a fluid that is stationary or translates with a uniform velocity $\mathbf{u} = \mathbf{U}(t)$ takes the simplified form

$$\rho \frac{d\mathbf{U}}{dt} = -\nabla p + \rho \mathbf{g} \qquad (4.1.1)$$

Assuming that the density of the fluid is uniform and solving for the pressure, we obtain

$$p = \rho \left(\mathbf{g} - \frac{d\mathbf{U}}{dt} \right) \cdot \mathbf{x} + A(t) \qquad (4.1.2)$$

where $A(t)$ is a time-dependent constant whose value must be found by requiring an appropriate boundary condition.

Force and torque on an immersed body

The force exerted on a body that is immersed in a fluid of uniform density and translates with the fluid is found from Eq. (3.1.11), where the stress tensor assumes the isotropic form $\boldsymbol{\sigma} = -p\mathbf{I}$. Using the pressure distribution given in Eq. (4.1.2) and the divergence theorem we obtain $\mathbf{F} = -\rho V_B(\mathbf{g} - d\mathbf{U}/dt)$, where V_B is the volume of the body, which is simply Archimedes's buoyancy force. Adding to this force the weight of the body, we obtain the total force $\mathbf{F}^{TOT} = (\rho_B - \rho)V_B\mathbf{g} + \rho V_B d\mathbf{U}/dt$, where ρ_B is the density of the body.

The corresponding torque with respect to the point $\mathbf{x}_0$ is found in a similar manner using Eq. (3.1.20). Manipulating the integrand and using the divergence theorem, we obtain $\mathbf{T} = -\rho V_B(\mathbf{X}_c - \mathbf{x}_0) \times (\mathbf{g} - d\mathbf{U}/dt)$, where $\mathbf{X}_c$ is the centroid of the volume of the body defined as

$$\mathbf{X}_c = \frac{1}{V_B} \int_{Body} \mathbf{x} \, dV(\mathbf{x}) = \frac{1}{2V_B} \int_{Body} (x^2 n_x, y^2 n_y, z^2 n_z) \, dS(\mathbf{x}) \qquad (4.1.3)$$

The surface integral representation on the right-hand side facilitates the analytical and numerical computation. Adding to the buoyancy torque the torque due to the weight of the body, we obtain the total torque $\mathbf{T}^{TOT} = (\rho_B - \rho)V_B(\mathbf{X}_c - \mathbf{x}_0) \times \mathbf{g} + \rho V_B(\mathbf{X}_c - \mathbf{x}_0) \times d\mathbf{U}/dt$. It is clear that placing the point $\mathbf{x}_0$ at the centroid $\mathbf{X}_c$ makes the torque vanish.

A floating liquid layer

As a simple application, let us consider the pressure distribution within a stationary liquid layer of a fluid labeled 1 with thickness h resting between air and a pool of a heavier liquid labeled 2, as shown in Figure 4.1.1. We assume that both the free surface and liquid interface are flat, and introduce Cartesian coordinates with the origin at the free surface and the y axis pointing in

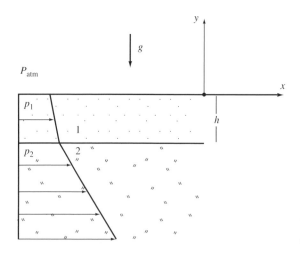

Figure 4.1.1 A liquid layer resting between an underlying heavier liquid and air, and the developing hydrostatic pressure distribution.

the vertical direction upward so that the y component of $\mathbf{g}$ is given by $g_y = -g$, where $g = |\mathbf{g}|$. Applying Eq. (4.1.2) for each fluid individually with $\mathbf{U} = \mathbf{0}$, we obtain the pressure distributions

$$p_1 = -\rho_1 gy + A_1, \qquad p_2 = -\rho_2 gy + A_2 \qquad (4.1.4)$$

where A_1 and A_2 are two constants.

To compute the constant A_1 we require that the pressure p_1 at the free surface of fluid 1 is equal to the atmospheric pressure P_{Atm} and find $A_1 = P_{\text{Atm}}$. To compute the constant A_2 we apply both equations in (4.1.4) at a point on the interface, located at $y = -h$, and subtract the resulting expressions from one another to find $p_2 - p_1 = -\Delta\rho\, gh + A_2 - P_{\text{Atm}}$, where $\Delta\rho = \rho_2 - \rho_1$. Requiring continuity of the normal component of the traction or pressure across the interface yields $A_2 = P_{\text{Atm}} + \Delta\rho\, gh$. Substituting the values of A_1 and A_2 into Eqs. (4.1.4), we obtain the pressure distributions in the explicit forms $p_1 = -\rho_1 gy + P_{\text{Atm}}$ and $p_2 = -\rho_2 gy + P_{\text{Atm}} + \Delta\rho\, gh$.

Pressure within a vibrating container

As a second application, we consider the pressure distribution in a liquid that is placed within a container that vibrates harmonically in the vertical direction with angular frequency Ω. Assuming that the free surface remains flat during the motion, we introduce Cartesian axes with origin at the mean position of the free surface and the y axis pointing against the direction of gravity so that $g_y = -g$. The position of the free surface is described by $y = a\cos(\Omega t)$, where a is the amplitude of the oscillation, and the velocity of the fluid is given by $\mathbf{U}(t) = -a\Omega \sin(\Omega t)\,\mathbf{j}$, where $\mathbf{j}$ is the unit vector along the y axis. Using Eq. (4.1.2), we find that the pressure distribution within the liquid is given by

$$p = -\rho y[g - a\Omega^2 \cos(\Omega t)] + A(t) \qquad (4.1.5)$$

The constant $A(t)$ is evaluated by requiring that the pressure at the free surface be equal to the atmospheric pressure P_{Atm}, and this yields $A(t) = P_{\text{Atm}} + \rho a\cos(\Omega t)[g - a\Omega^2 \cos(\Omega t)]$. Substituting this expression back into Eq. (4.1.5), we obtain

$$p = P_{\text{Atm}} + \rho[y - a\cos(\Omega t)][-g + a\Omega^2 \cos(\Omega t)] \qquad (4.1.6)$$

The term within the first set of square brackets is the vertical component of the position vector in a frame of reference moving with the free surface. The term in the second set of square brackets expresses the difference between the vertical component of the gravitational acceleration and fluid acceleration $-d\mathbf{U}(t)/dt$. It is then evident that Eq. (4.1.6) is consistent with the pressure distribution that would have been computed if we worked in a noninertial frame of reference in which the free surface appears to be stationary, taking into account the fictitious force due to the linear acceleration of the Cartesian axes.

Steadily Rotating Fluids

The velocity distribution within a fluid that executes steady rigid-body rotation around the origin with angular velocity Ω is given by $\mathbf{u} = \Omega \times \mathbf{x}$. Remembering that the Laplacian of the velocity is equal to the negative of the curl of the vorticity, and the vorticity corresponding to rigid-body motion is constant and equal to 2Ω, we find that the viscous force vanishes and the Navier–Stokes equation assumes the simple form

$$\rho\, \Omega \times (\Omega \times \mathbf{x}) = -\nabla p + \rho\, \mathbf{g} \qquad (4.1.7)$$

Next we introduce cylindrical polar coordinates with the x axis pointing in the direction of Ω and rewrite Eq. (4.1.7) as

$$-\rho\, \Omega^2 \sigma \mathbf{e}_\sigma = -\nabla p + \rho\, \mathbf{g} \qquad (4.1.8)$$

where Ω is the magnitude of Ω and $\mathbf{e}_\sigma$ is the unit vector in the radial direction. Expressing the gradient of the pressure in cylindrical polar coordinates and integrating with respect to σ yields

$$p = \rho\, \mathbf{g} \cdot \mathbf{x} + \tfrac{1}{2}\rho\, \Omega^2 \sigma^2 + A(t) = \rho\, \mathbf{g} \cdot \mathbf{x} + \tfrac{1}{2}\rho|\Omega \times \mathbf{x}|^2 + A(t) \qquad (4.1.9)$$

The additional pressure expressed by the second term on the right-hand side is necessary in order to balance the radial centrifugal force due to the rotation. The pressure distribution within a fluid that accelerates with velocity $\mathbf{U}(t)$ while rotating steadily with angular velocity $\mathbf{\Omega}$ is given by the right-hand side of Eq. (4.1.9), where $\mathbf{g}$ is replaced by $\mathbf{g} - d\mathbf{U}/dt$.

Rotating container

As an application, let us consider the pressure distribution within a container that rotates steadily around the horizontal x axis. Setting the y axis in the direction of gravity pointing upward so that $g_y = -g$, we find

$$
\begin{aligned}
p &= -\rho g y + \tfrac{1}{2}\rho\Omega^2(y^2 + z^2) + A(t) \\
&= \tfrac{1}{2}\rho\Omega^2\left[\left(y - \frac{g}{\Omega^2}\right)^2 + z^2\right] - \tfrac{1}{2}\rho\frac{g^2}{\Omega^2} + A(t)
\end{aligned}
\tag{4.1.10}
$$

The expression on the right-hand side shows that the surfaces of constant pressure are concentric horizontal cylinders with axis at the point $z = 0$, $y = g/\Omega^2$. Minimum pressure occurs at the common axis.

Free surface of a rotating fluid

As another application, let us consider the shape of the free surface of a liquid that is placed within a horizontal cylindrical beaker that rotates about its axis with angular velocity Ω. Setting the x axis in the direction of gravity pointing upward, so that $g_x = -g$, we find that the pressure distribution within the fluid is given by

$$
p = -\rho g x + \tfrac{1}{2}\rho\Omega^2\sigma^2 + A
\tag{4.1.11}
$$

Evaluating the pressure at the free surface and neglecting the effects of surface tension, thereby requiring that the pressure at the free surface be equal to the atmospheric pressure, yields an algebraic equation for the shape of the free surface describing a paraboloid.

Rotary Oscillation

Assuming that a fluid rotates about the origin as a rigid body with time-dependent angular velocity $\mathbf{\Omega}(t)$, and substituting the associated velocity field $\mathbf{u} = \mathbf{\Omega}(t) \times \mathbf{x}$ into the Navier–Stokes equation, we obtain the pressure distribution $p = p_1 + p_2$, where p_1 is given by the right-hand side of Eq. (4.1.9) and p_2 satisfies the equation

$$
\rho\frac{d\mathbf{\Omega}}{dt} \times \mathbf{x} = -\nabla p_2
\tag{4.1.12}
$$

Taking the curl of both sides of Eq. (4.1.12) and noting that the curl of the gradient of any twice differentiable function vanishes, we arrive at the solvability condition $d\mathbf{\Omega}/dt = \mathbf{0}$, which contradicts the original assumption of unsteady rotation. Physically this result implies that rotary oscillation requires a velocity field other than that described by rigid-body motion. For instance, subjecting a glass of water to rotary oscillation does not guarantee that the fluid will exhibit rigid-body oscillatory rotation. At high frequencies, in particular, the fluid will remain stationary, and the motion will be confined within a thin boundary layer around the glass surface, as will be discussed in Chapter 5.

Compressible Fluids

The pressure distribution (4.1.2) was derived under the assumption that the density of the fluid is uniform. When the fluid is compressible, p represents the thermodynamic pressure, which is a function of density and temperature T. To compute the pressure distribution, we must introduce an appropriate constitutive equation that relates these three variables.

For instance, for an ideal gas, we have $p = \rho RT/M$, where R is the ideal gas constant and M is the molecular weight. Substituting this constitutive equation into Eq. (4.1.1) and assuming that the fluid is stationary, we obtain the differential equation

$$\frac{1}{p} \nabla p = \nabla \ln p = \frac{M}{RT} \mathbf{g} \tag{4.1.13}$$

When the temperature is constant, the solution is

$$\ln p = \frac{M}{RT} \mathbf{g} \cdot \mathbf{x} + A(t) \tag{4.1.14}$$

which shows the pressure exhibits an exponential dependence on spatial distance, instead of the linear dependence occurring in incompressible fluids.

PROBLEMS

4.1.1 **Two layers resting on a pool.** Consider a liquid layer labeled 2 with thickness h_2 resting upon the surface of a pool of a heavier fluid labeled 3 and underneath a layer of another liquid layer labeled 1 with thickness h_1. The pressure above liquid 1 is atmospheric. Compute the pressure distributions within the two layers and the pool.

4.1.2 **Rotating drop.** Find the shape of an axisymmetric drop pending underneath a horizontal surface and rotating about the vertical axis in the absence of interfacial tension.

4.1.3 **Pressure distribution in the atmosphere.** Derive the pressure distribution in the atmosphere under the assumptions that (a) it behaves like an ideal gas and (b) the temperature distribution has the linear form $T = T_0 - \beta y$, where T_0 is the temperature at sea level, β is a constant called the *lapse rate*, and the y axis points against the direction of gravity.

4.1.4 **Compressible gas in rigid-body motion.** Generalize Eq. (4.1.14) to the case where the ideal gas translates as a rigid body with arbitrary time-dependent velocity and rotates with constant angular velocity.

4.2 THE LAPLACE–YOUNG EQUATION

We proceed in this section to discuss the shape of curved interfaces with uniform tension γ separating two stationary fluids. Using the definition (3.5.7), the pressure distribution given in Eq. (4.1.2) with vanishing acceleration, and the constitutive equation for the jump in the interfacial traction given in Eq. (3.5.12), we write

$$\Delta \mathbf{f} \equiv (\boldsymbol{\sigma}^{(1)} - \boldsymbol{\sigma}^{(2)}) \cdot \mathbf{n} = (p_2 - p_1) \mathbf{n} \tag{4.2.1}$$
$$= [\Delta \rho \, \mathbf{g} \cdot \mathbf{x} + (A_2 - A_1)] \mathbf{n} = \gamma 2 \kappa_m \mathbf{n}$$

where $\Delta \rho = \rho_2 - \rho_1$, A_1 and A_2 are two constants, $\mathbf{n}$ is the unit normal vector pointing into fluid 1, and κ_m is the mean curvature of the interface. Rearranging Eq. (4.2.1), we obtain the Laplace–Young equation

$$2 \kappa_m = \frac{\Delta \rho}{\gamma} \mathbf{g} \cdot \mathbf{x} + B \tag{4.2.2}$$

where $B = (A_2 - A_1)/\gamma$ is a new constant with dimensions of inverse length.

Expressing the mean curvature in a certain parametric form renders Eq. (4.2.2) an ordinary or partial differential equation describing the shape of the interface. The solution is subject to a scalar boundary condition that specifies either the magnitude of the contact angle of the interface at a three-phase contact line, where the interface meets a solid boundary or a third fluid, or the position of the contact line. The first boundary condition is applicable when a solid boundary is perfectly smooth, whereas the second boundary condition is applicable when a solid surface has an appreciable degree of roughness (Dussan, 1985).

Interfaces with Constant Mean Curvature

Under certain conditions, the right-hand side of Eq. (4.2.2) is nearly constant, and the interface assumes a shape with constant mean curvature. Constant-curvature shapes in three dimensions include the sphere, a section of a sphere, and an unduloid. The latter is an axisymmetric surface whose trace in an azimuthal plane coincides with the focus of an ellipse that rolls over the axis. Constant-curvature lines in two dimensions include a circle and sections of a circle.

Meniscus between two vertical plates

Let us consider the shape of an interface between two vertical flat plates that are separated by a distance $2b$, as illustrated in Figure 4.2.1. The height of the meniscus midway between the plates, denoted by h, is called the *capillary rise*. Assuming that the meniscus takes the shape of a circular arc of radius R, we write the geometrical condition $b = R \cos \alpha$, where α is the contact angle. The curvature of the interface is reckoned to be negative when the interface is concave when viewed from fluid 1 as shown in Figure 4.2.1, and positive when the interface is convex.

The pressure distributions within the two fluids are given in Eqs. (4.1.4). Requiring that the pressures within the two fluids at the level of the interface outside the plates at $y = -h$ be equal, we obtain $B = (A_2 - A_1)/\gamma = -\Delta \rho g h/\gamma$, where $\Delta \rho = \rho_2 - \rho_1$. Substituting this value into Eq. (4.2.2), setting $\kappa_m = -1/(2R) = -\cos \alpha/(2b)$, and evaluating the resulting expression at the origin yields

$$h = \frac{\gamma}{\Delta \rho \, g b} \cos \alpha = \frac{l^2}{b} \cos \alpha \qquad (4.2.3)$$

where $l = (\gamma/\Delta \rho \, g)^{1/2}$ is a physical constant with dimensions of length called the *capillary length*. For water and air at 20°C, l is approximately equal to 2.5 mm.

The sign of the capillary rise in Eq. (4.2.3) is determined by the value of the contact angle α. When $\alpha < \pi/2$ the meniscus rises, when $\alpha > \pi/2$ the meniscus submerges, and when $\alpha = \pi/2$ the meniscus remains flat at the level of the free surface outside the plates.

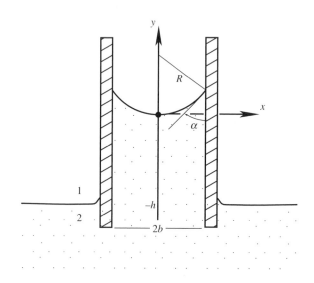

Figure 4.2.1   The meniscus of a fluid interface between two vertical flat plates.

We can now return to Eq. (4.2.2) and establish the conditions under which the assumption that the right-hand side is nearly constant is valid. Requiring that the variation in the elevation of the interface, equal to $R(1 - \sin \alpha)$, be smaller than h, we find

$$\left(\frac{b}{l}\right)^2 \ll 1 + \sin \alpha \tag{4.2.4}$$

Meniscus within a circular tube

As a second application, we consider the shape of an interface within a cylindrical capillary tube of radius a depicted in Figure 4.4.1. Assuming that the meniscus has a constant mean curvature and therefore takes the shape of a section of a sphere with radius R, and working as in the case of the two-dimensional meniscus, we find that the capillary rise is given by

$$h = 2\frac{\gamma}{\Delta \rho \, ga} \cos \alpha = 2\frac{l^2}{a} \cos \alpha \tag{4.2.5}$$

which is identical to that given in Eq. (4.2.3) with $a/2$ in place of b (Problem 4.2.1). The condition for this estimate to be accurate is

$$\left(\frac{a}{l}\right)^2 \ll 2(1 + \sin \alpha) \tag{4.2.6}$$

Computation of the Mean Curvature

In order to derive a differential equation that describes the shape of a stationary interface according to the Laplace–Young equation, we must have available an expression for the mean curvature $\kappa_m = \frac{1}{2}\nabla \cdot \mathbf{n}$ in a convenient parametric form that is amenable to analytical or numerical computation.

One way of computing the divergence of the normal vector is to describe the interface in terms of the equation $F(x, y, z) = 0$, extend the domain of definition of the normal vector from the interface into the whole space by setting $\mathbf{n} = \nabla F / |\nabla F|$, and then compute

$$\begin{aligned}
\kappa_m &= \frac{1}{2}\nabla \cdot \left(\frac{\nabla F}{|\nabla F|}\right) = \frac{1}{2}\frac{\partial}{\partial x_i}\left[\frac{\partial F}{\partial x_i} \Big/ \left(\frac{\partial F}{\partial x_j}\frac{\partial F}{\partial x_j}\right)^{1/2}\right] \\
&= \frac{1}{2}\frac{1}{|\nabla F|^3}\left[|\nabla F|^2 \nabla^2 F - \nabla F \cdot (\nabla \nabla F) \cdot \nabla F\right]
\end{aligned} \tag{4.2.7}$$

The sign of the function F must be chosen so that the normal vector points into fluid 1. For example, when the interface is described in the form $z = f(x, y)$, with fluid 1 residing in the upper half-space, we set $F = z - f(x, y)$. In Table 4.2.1 we summarize expressions for the normal and mean curvatures of two-dimensional, axisymmetric, and three-dimensional interfaces in several parametric forms.

When a global representation of an interface is inconvenient or inappropriate, we use a local representation that is defined in terms of the position of a set of interfacial marker points, and then compute the mean curvature according to Eq. (4.2.7). One way of effecting the local representation is to write $z = f(x, y)$, expand the function f in a Taylor series with respect to its arguments, and then compute the coefficients of the series by requiring that f describes a surface that passes through a collection of marker points; this is an example of a collocation method.

Proceeding with an alternative point of view, we introduce local Cartesian coordinates with the x and z axes tangential to the interface at a point, and the y axis perpendicular to the interface at that point. At the origin, $n_x = n_z = 0$, and $n_y = 1$, and since the length of the normal vector is constant, $\nabla(\mathbf{n} \cdot \mathbf{n}) = \mathbf{0}$, we find $\partial n_y / \partial y = 0$. This leads us to an expression for the mean curvature in terms of tangential derivatives as

TABLE 4.2.1
Principal and normal curvatures of interfaces in several parametric forms.

Two-dimensional interfaces
Fluid 1 is in the upper half-plane.

Description	Curvature in xy plane	Curvature in yz plane	Mean curvature
$y = f(x)$	$\kappa_{xy} = \dfrac{1}{f'}\dfrac{d}{dx}\dfrac{1}{(1+f'^2)^{1/2}}$ $= -\dfrac{f''}{(1+f'^2)^{3/2}}$	$\kappa_{yz} = 0$	$\kappa_m = \tfrac{1}{2}\kappa_{xy}$

Axisymmetric interfaces
Fluid 1 is in outer space.

	Curvature in $x\sigma$ plane	Curvature in conjugate plane	Mean curvature
$x = f(\sigma)$	$\kappa_1 = -\dfrac{f''}{(1+f'^2)^{3/2}}$	$\kappa_2 = -\dfrac{f'}{\sigma(1+f'^2)^{1/2}}$	$\kappa_m = \tfrac{1}{2}(\kappa_1+\kappa_2)$ $= -\dfrac{1}{2}\dfrac{1}{\sigma}\dfrac{d}{d\sigma}\left(\dfrac{\sigma f'}{(1+f'^2)^{1/2}}\right)$
$\sigma = f(x)$	$\kappa_1 = -\dfrac{f''}{(1+f'^2)^{3/2}}$	$\kappa_2 = \dfrac{1}{f(1+f'^2)^{1/2}}$	$\kappa_m = \tfrac{1}{2}(\kappa_1+\kappa_2)$ $= \dfrac{1}{2}\dfrac{1+f'^2-ff''}{f(1+f'^2)^{3/2}}$

Three-dimensional interfaces
Subscripts stand for partial derivatives. Fluid 1 is in the upper half-space.

	Mean curvature	Linearized form
$z = f(x, y)$	$\kappa_m = -\dfrac{1}{2}\dfrac{(1+f_y^2)f_{xx} - 2f_x f_y f_{xy} + (1+f_x^2)f_{yy}}{(1+f_x^2+f_y^2)^{3/2}}$	$\kappa_m = -\tfrac{1}{2}(f_{xx}+f_{yy})$
$r = f(\theta, \varphi)$	$\kappa_m = \tfrac{1}{2}\nabla\cdot\mathbf{n}$ $\mathbf{n} = \dfrac{1}{D}\left(\mathbf{e}_r - \dfrac{f_\theta}{r}\mathbf{e}_\theta - \dfrac{f_\theta}{r\sin\theta}\mathbf{e}_\varphi\right)$ $D = \left(1+\dfrac{f_\theta^2}{r^2} + \dfrac{f_\varphi^2}{r^2\sin^2\theta}\right)^{1/2}$	$\kappa_m = \dfrac{1}{r} - \dfrac{\cot\theta}{r^2}f_\theta - \dfrac{1}{r^2}f_{\theta\theta}$ $- \dfrac{1}{r^2\sin^2\theta}f_{\varphi\varphi}$

$$\kappa_m = \tfrac{1}{2}\left(\frac{\partial n_x}{\partial x} + \frac{\partial n_z}{\partial z}\right) = \tfrac{1}{2}(\kappa_x + \kappa_z) \qquad (4.2.8)$$

where κ_x and κ_z are the normal curvatures of the trace of the interface in the xy and yz planes discussed in Section 1.5.

Equation (4.2.8) suggests computing the mean curvature in terms of the surface curvilinear orthogonal or nonorthogonal coordinates ξ and η discussed in Section 1.5. In practice, this is done by tracing the curvilinear axes with a set of marker points whose position is described as $\mathbf{X}(\xi, \eta)$, and then constructing parametric representations of the ξ and η lines using methods of curve fitting and function interpolation. The required partial derivatives of $\mathbf{X}$ with respect to ξ and η are computed by numerical differentiation as discussed in Section B.5, Appendix B.

For example, for a two-dimensional interface, we parametrize the trace of the interface in the xy plane using a parameter ξ that increases in the direction of the tangential vector $\mathbf{t}$, whose orientation is such that $\mathbf{t}$ and $\mathbf{n}$ form a right-handed system of axes, and regard x and y along the interface as functions of ξ. We write $dy/dx = y_\xi/x_\xi$, where the subscript ξ denotes differentiation with respect to ξ, and use Table 4.2.1 to find that the curvature of the interface in the xy plane is given by

$$\kappa_{xy} = \frac{x_{\xi\xi}\, y_\xi - y_{\xi\xi}\, x_\xi}{(x_\xi^2 + y_\xi^2)^{3/2}}$$ (4.2.9)

The curves $x(\xi)$ and $y(\xi)$ are constructed from the location of marker points along the interface using, for instance, spline interpolation, and their derivatives are computed by numerical differentiation (Problem 4.2.5). If ξ increases in a direction that is opposite to that of $\mathbf{t}$, a minus sign must be introduced in front of the fraction on the right-hand side of Eq. (4.2.9). In practice, ξ may be identified with the arc length of the polygonal line that connects successive marker points.

Computation of the average value of the mean curvature over an interfacial element

An approximate method of computing the surface-average value of the mean curvature $\langle \kappa_m \rangle$ over a small interfacial element E, emerges by applying Eq. (3.5.10) with $\mathbf{F} = \mathbf{n}$, and noting that $(\nabla \mathbf{n}) \cdot \mathbf{n} = 0$, thus obtaining

$$\langle \kappa_m \rangle \langle \mathbf{n} \rangle \equiv \frac{1}{S_E} \int_E \kappa_m \, \mathbf{n} \, dS = \frac{1}{2 S_E} \int_L \mathbf{n} \times \mathbf{t} \, dl$$ (4.2.10)

where $\langle \mathbf{n} \rangle$ is the surface-average value of the unit normal vector over E, S_E is the surface area of E, and the unit vector $\mathbf{t}$ is tangential to the contour L of E; the orientation of $\mathbf{t}$ is such that $\mathbf{t}$ and $\mathbf{n}$ define a right-handed system of axes, as shown in Figure 4.2.2. The contour integral on the right-hand side of Eq. (4.2.10) may be computed by numerical integration. The magnitude of $\langle \kappa_m \rangle$ is equal to that of the vector represented by the right-hand side with an allowance for a plus or minus sign.

To illustrate the practical application of the method, consider the rectangular element shown in Figure 4.2.2, which is defined by the four points A, B, C, D that lie on adjacent ξ and η lines representing surface curvilinear coordinates. Applying the trapezoidal rule to approximate the contour integral in Eq. (4.2.10), we obtain

$$\begin{aligned}
4 \langle \kappa_m \rangle \langle \mathbf{n} \rangle S_E = & \left[\left(\mathbf{n} \times \frac{\partial \mathbf{X}}{\partial \xi} \right)_A + \left(\mathbf{n} \times \frac{\partial \mathbf{X}}{\partial \xi} \right)_B \right] (\xi_B - \xi_A) \\
& + \left[\left(\mathbf{n} \times \frac{\partial \mathbf{X}}{\partial \eta} \right)_B + \left(\mathbf{n} \times \frac{\partial \mathbf{X}}{\partial \eta} \right)_C \right] (\eta_C - \eta_B) \\
& + \left[\left(\mathbf{n} \times \frac{\partial \mathbf{X}}{\partial \xi} \right)_C + \left(\mathbf{n} \times \frac{\partial \mathbf{X}}{\partial \xi} \right)_D \right] (\xi_D - \xi_C) \\
& + \left[\left(\mathbf{n} \times \frac{\partial \mathbf{X}}{\partial \eta} \right)_D + \left(\mathbf{n} \times \frac{\partial \mathbf{X}}{\partial \eta} \right)_A \right] (\eta_A - \eta_D)
\end{aligned}$$ (4.2.11)

The unit normal and tangential vectors on the right-hand side of Eq. (4.2.11) may be computed from a knowledge of the position of interfacial marker points using standard methods of numerical differentiation.

The counterpart of Eq. (4.2.10) for a segment E of a two-dimensional interface with arc length L_E is

$$\langle \kappa_m \rangle \langle \mathbf{n} \rangle \equiv \frac{1}{L_E} \int_E \kappa_m \mathbf{n} \, dl = \frac{1}{2 L_E} \Delta \mathbf{t}$$ (4.2.12)

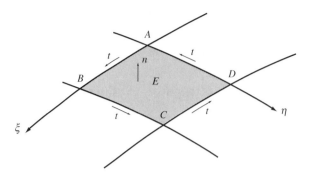

Figure 4.2.2 Approximate computation of the surface-average value of the mean curvature over an interfacial patch in terms of the position of a set of marker points.

where $\Delta \mathbf{t}$ is the difference between the tangential vectors evaluated at the end and at the beginning of the segment.

Small Deformations

When the deformation of an interface from a known equilibrium shape with constant mean curvature is small compared to its overall size, the Laplace–Young equation may be simplified by linearizing the expression for the mean curvature about the known equilibrium position corresponding, for instance, to a flat or spherical shape.

Assume, for example, that the shape of a three-dimensional interface is described in Cartesian coordinates as $y = f(x, z)$, where the magnitude of f is small compared to the global dimensions of the interface, and $f = 0$ yields a flat equilibrium shape. Substituting the linearized expression for the mean curvature with respect to f shown in Table 4.2.1 into Eq. (4.2.2), we obtain a Helmholtz equation for f

$$\frac{\partial^2 f}{\partial x^2} + \frac{\partial^2 f}{\partial z^2} = -\frac{\Delta \rho}{\gamma}(g_x x + g_y f + g_z z) - B \tag{4.2.13}$$

where $\Delta \rho = \rho_2 - \rho_1$, which is to be solved subject to an appropriate boundary condition at the contact line.

Semi-infinite meniscus attached to an inclined plate

As an application, let us consider the shape of a two-dimensional meniscus that is attached to an inclined plate as illustrated in Figure 4.3.1(a). Using the coordinate system depicted in that figure, we describe the interface as $y = f(x)$, assume that the slope of the interface is small, that is, $\alpha + \beta \approx \pi$, note that $\mathbf{g} = (0, -g, 0)$, and thus obtain the simplified version of Eq. (4.2.13)

$$\frac{d^2 f}{dx^2} = \frac{f}{l^2} - B \tag{4.2.14}$$

where $l = (\gamma / \Delta \rho \, g)^{1/2}$ is the capillary length. Requiring that the interface become flat far from the plate, where f tends to vanish, we find $B = 0$. A solution to Eq. (4.2.14) that satisfies the boundary conditions that (1) f vanishes as x tends to infinity and (2) the slope of the interface at the plate is determined by the contact-angle boundary condition $f'(0) = \tan(\alpha + \beta)$ is

$$f = l \tan(\pi - \alpha - \beta) e^{-x/l} \tag{4.2.15}$$

The elevation of the interface at the contact line is then $h \equiv f(0) = l \tan(\pi - \alpha - \beta) \approx l(\pi - \alpha - \beta)$, which represents the asymptotic limit of the exact solution given in Eq. (4.3.6) as the slope of the interface at the plate tends to become small.

PROBLEMS

4.2.1 **Meniscus within a cylindrical capillary.** Derive Eq. (4.2.5) and discuss the physical implications of Eq. (4.2.6).

4.2.2 **Mean curvature of a three-dimensional interface.** Derive the expressions for the mean curvature of a three-dimensional interface in Cartesian, cylindrical, and spherical polar coordinates shown in Table 4.2.1.

4.2.3 **Small deformation of a meniscus between two vertical plates.** Consider the interface of two fluids between the two vertical plates illustrated in Figure 4.2.1. Assuming that the slope of the interface is small, derive an equation that describes the shape of the meniscus in the parametric form $y = f(x)$ and estimate the capillary rise.

Computer Problems

4.2.4 **Catenoid.** (a) Using Table 4.2.1, verify that the solution of the equation $f(x) = c_1[1 + f'(x)^2]^{1/2}$, where c_1 is a constant, provides us with an axisymmetric interface with zero mean curvature, described by the equation $\sigma = f(x)$. This surface is called a *catenoid*, and its trace in an azimuthal plane is called a *catenary*. (b) Show that the catenoid is described by the explicit equation

$$\sigma = c_1 \cosh \frac{x - c_2}{c_1} \qquad (4.2.16)$$

where c_2 is a new constant. (c) Consider a thin liquid film subtended between two coaxial circular rings with identical radii a, separated by a distance h. One of the rings is located at $x = 0$, and the second ring is located at $x = h$. Assuming that gravitational effects are insignificant and requiring that the pressures on either side of the film are identical, we find that the film must assume a shape with vanishing mean curvature. Assuming that the film takes the shape of a catenoid, compute the constants c_1 and c_2 by requiring that the catenoid pass through the rings, and plot the ratios c_1/a and c_2/a against h/a. When $0 < h/a < 1.325$, you will obtain two real solutions for c_1/a; the physically relevant solution is the one with the larger value. For larger values of h/a, the solution is complex, indicating that a catenoid cannot be established. Plot the profiles of the film in an azimuthal plane for $h/a = 0.20, 0.50, 1.00, 1.325$.

4.2.5 **Curvature of a line by parametric representation.** The following set of points, also called *nodes*, represent a smooth loop in the xy plane:

x	y
1.1	1.0
2.2	0.1
3.9	1.1
4.1	3.0
3.0	3.9
2.0	4.1
1.3	3.0

Write a program that uses cubic spline interpolation to compute the Cartesian coordinates of a point and curvature κ_{xy} in the xy plane of the loop as a function of a suitably chosen parameter ξ that increases monotonically from zero to one along the line, from start to finish (Section B.4, Appendix B). One plausible choice for ξ is the polygonal arc length, that is, the length of the polygonal line that connects successive nodes. The program should return the coordinates (x, y)

corresponding to a given value of ξ between zero and unity; $\xi = 0$ and $\xi = 1$ correspond to $x = 1.1$, $y = 1.0$. Construct a table of 21 values of the quadruplets (ξ, x, y, κ_{xy}) at 20 evenly spaced intervals of ξ between zero and unity.

4.3 | TWO-DIMENSIONAL INTERFACES

We proceed now to discuss specific methods of computing the shape of two-dimensional interfaces with uniform surface tension governed by the Laplace–Young equation (4.2.2).

At the outset, we introduce Cartesian coordinates with the y axis pointing against the direction of gravity so that $\mathbf{g} = (0, -g, 0)$, and use the first entry of Table 4.2.1 to find that the shape of an interface that is described in the parametric form $y = f(x)$, with fluid 2 located below fluid 1, is governed by the second-order nonlinear ordinary differential equation

$$
\frac{1}{f'} \frac{d}{dx} \frac{1}{(1 + f'^2)^{1/2}} = -\frac{f''}{(1 + f'^2)^{3/2}} = -\frac{f}{l^2} + B \tag{4.3.1}
$$

where $l = (\gamma/\Delta\rho\, g)^{1/2}$ is the capillary length, $\Delta\rho = \rho_2 - \rho_1$, and B is a constant with dimensions of inverse length. Integrating Eq. (4.3.1) once with respect to x, we obtain

$$
\frac{1}{\sqrt{1 + f'^2}} = -\frac{f^2}{2l^2} + Bf + C \tag{4.3.2}
$$

where C is a dimensionless integration constant.

Expressing the slope of the interface in terms of the angle θ that is subtended between the x axis and the tangent to the interface, so that $f' = \tan\theta$, we recast Eq. (4.3.2) into the form

$$
|\cos\theta| = -\frac{f^2}{2l^2} + Bf + C \tag{4.3.3}
$$

Furthermore, rearranging Eq. (4.3.2), we obtain the standard form of a first-order ordinary differential equation

$$
\frac{df}{dx} = \pm\left[\left(-\frac{f^2}{2l^2} + Bf + C\right)^{-2} - 1\right]^{1/2} \tag{4.3.4}
$$

The plus or minus sign must be selected according to the nature of the expected interfacial shape.

A Semi-Infinite Meniscus Attached to an Inclined Plate

As a first application, we consider the shape of the meniscus of a liquid pinned at an immersed flat plate that is inclined at an angle β with respect to the horizontal, lying below another fluid, as illustrated in Figure 4.3.1(a,b), where $0 < \beta < \pi$. We set the origin of the axes at the level of the undeformed interface right below the contact line, require that as x tends to infinity the interface tends to becomes flat, and therefore both the curvature and slope of the interface tend to vanish, and use Eqs. (4.3.1) and (4.3.3) to obtain $B = 0$ and $C = 1$. Both constants A_1 and A_2 defined in Eqs. (4.1.4) are equal to the pressure at the interface far from the plate. Since both the inclination angle β and the contact angle α range between 0 and π, the angle $\theta_{CL} = \alpha + \beta$ that is subtended between the x axis and the tangent to the interface at the contact line ranges between 0 and 2π.

Monotonic shapes

Assuming that the interface obtains a monotonically varying shape as illustrated in Figure 4.3.1(a), we apply Eq. (4.3.3) at the contact line with $B = 0$ and $C = 1$, and use the boundary

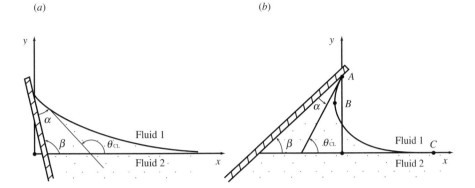

(a)   (b)

(c)

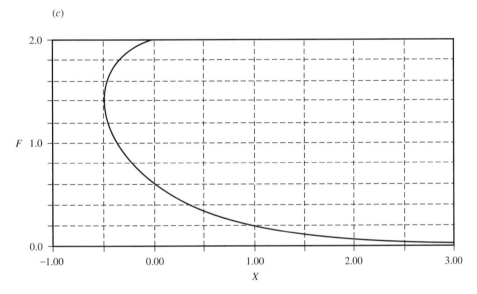

Figure 4.3.1   A two-dimensional meniscus between two fluids pinned at an inclined plate. Schematic illustration of (a) a monotonic and (b) a reentrant shape. The shape of the meniscus for any value of α or β may be deduced from the curve shown in (c). To obtain the shape for a particular value of α and β, we find the intersection between the inclined plate and the curve plotted in (c) subject to the condition that the elevation of the meniscus at the point of contact is given by Eq. (4.3.6). When h is negative, we work with its mirror-image.

condition $\theta_{\mathrm{CL}} = \alpha + \beta$ to find that the capillary rise $h \equiv f(0)$ is given by

$$\frac{h^2}{2l^2} = 1 - |\cos(\alpha + \beta)| \tag{4.3.5}$$

Assuming further that $\pi/2 < \alpha + \beta < 3\pi/2$, and using standard trigonometric identities, we recast Eq. (4.3.5) into the form

$$h = 2l \cos\left(\frac{\alpha + \beta}{2}\right) \tag{4.3.6}$$

which shows that the maximum possible value of $|h|$ is equal to $2^{1/2}l$.

In the limit as θ_{CL} tends to π, $h \to l(\pi - \alpha - \beta)$, in agreement with the asymptotic solution (4.2.15). When $\alpha + \beta = \pi$, we find $h = 0$, which shows that the meniscus is undeformed. This observation provides us with a practical method of measuring the contact angle, according to which the slope of the plate β is varied until the interface appears to be flat; at that point $\alpha = \pi - \beta$.

Having obtained the elevation of the interface at the plate, we proceed to compute its shape. To simplify the notation, we introduce the nondimensional variables $F = f/l$ and $X = x/l$, set $B = 0$ and $C = 1$, and recast Eq. (4.3.4) into the form

$$\frac{dF}{dX} = \pm F \frac{\sqrt{4 - F^2}}{2 - F^2} \tag{4.3.7}$$

where the plus and minus sign must be selected according to the expected interfacial shape. Note that the denominator on the right-hand side of Eq. (4.3.7) vanishes when $F = 2^{1/2}$, at which point the slope of the meniscus becomes infinite, signaling a transition to reentrant shapes. Equation (4.3.7) is accompanied by the boundary condition $F(0) = h/l \equiv H = 2\cos[(\alpha + \beta)/2]$.

Solving Eq. (4.3.7) using a standard numerical method, such as a Runge–Kutta method, provides us with a family of shapes parametrized by $\alpha + \beta$. In practice, it is preferable to work with the inverse of Eq. (4.3.7), integrating X with respect to F from $F = H$ to 0 with the initial condition $X(H) = 0$.

As an alternative to numerical computation, we rearrange Eq. (4.3.7) and obtain the following implicit equation for F

$$X = \pm \int_H^F \frac{2 - \omega^2}{\omega \sqrt{4 - \omega^2}} \, d\omega \tag{4.3.8}$$

Evaluating the integral with the help of tables, we find

$$X = \pm \left[\ln\left(\frac{2 + \sqrt{4 - F^2}}{F} \right) - \ln\left(\frac{2 + \sqrt{4 - H^2}}{H} \right) - \sqrt{4 - F^2} + \sqrt{4 - H^2} \right] \tag{4.3.9}$$

(Gradshteyn and Ryzhik, 1980, pp. 81–85). To obtain the shape of the meniscus, we plot X as a function of F for $0 < F < H$.

Reentrant shapes

The preceding results are applicable in the range $\pi/2 < \alpha + \beta < 3\pi/2$. Outside this range, for $0 < \alpha + \beta < \pi/2$ or $3\pi/2 < \alpha + \beta < 2\pi$, the interface is expected to turn upon itself, as shown in Figure 4.3.1(b). Equation (4.3.1) is still valid between the points B and C, but since fluid 2 lies above fluid 1 between the points A and B, the sign of the mean curvature must be reversed in that section. Repeating the preceding computations, we find that Eq. (4.3.7) is uniformly valid over the whole length of the interface, but Eq. (4.3.5) becomes

$$\frac{h^2}{2l^2} = 1 + |\cos(\alpha + \beta)| \tag{4.3.10}$$

Using standard trigonometric identities, we recover Eq. (4.3.6), which shows that the maximum possible value of $|h|$ is $2l$, corresponding to $\alpha + \beta = 0$ or 2π.

The shape of the free surface in the complete range of $\alpha + \beta$ may be deduced from the curve shown in Figure 4.3.1(c), which was constructed by integrating Eq. (4.3.7) with the plus sign with respect to F from $F = 2$ to 0, with initial condition $X(2) = 0$. To obtain the shape of the meniscus for a particular value of α and β, we locate the intersection of the inclined plate and the depicted curve subject to the condition that the capillary rise is given by Eq. (4.3.6). When h is negative, we work with the mirror-image.

A Meniscus between Two Vertical Plates

Consider next the shape of an interface between two vertical flat plates that are separated by a distance $2b$, as illustrated in Figure 4.2.1. We begin by requiring that the pressure at the level of the undeformed interface outside the plates, located at $y = -h$, is equal to P_0, and use Eqs. (4.1.4) to find $A_1 = P_0 - \rho_1 g h$ and $A_2 = P_0 - \rho_2 g h$. Based on these values, we compute the constant $B \equiv (A_2 - A_1)/\gamma = -\Delta\rho\, gh/\gamma = -h/l^2$, where $l = (\gamma/\Delta\rho\, g)^{1/2}$ is the capillary length, and $\Delta\rho = \rho_2 - \rho_1$.

Furthermore, we require that, because of symmetry, the slope of the interface midway between the plates is equal to zero, and use Eqs. (4.3.1) and (4.3.2) to obtain $B = -f''(0)$ and $C = 1$. The first of these equations yields

$$h = l^2 f''(0) \tag{4.3.11}$$

which provides us with an expression for the capillary rise in terms of the curvature of the meniscus at the centerline.

Substituting the derived values of B and C into Eq. (4.3.3), we obtain a relation between the slope of the interface and the elevation of the meniscus,

$$f(f + 2h) = 2l^2(1 - |\cos\theta|) \tag{4.3.12}$$

Equation (4.3.4) provides us with a first-order differential equation describing the shape of the free surface in terms of the yet unknown capillary rise h

$$\frac{df}{dx} = \pm\left(\frac{l^4}{[l^2 - f(\frac{1}{2}f + h)]^2} - 1\right)^{1/2} \tag{4.3.13}$$

which is accompanied by the two boundary conditions $f(0) = 0$ and $f'(b) = \cot\alpha$.

In practice, it is more convenient to work with the second-order differential equation (4.3.1), which, for $B = -h/l^2$, takes the form

$$f'' = \frac{1}{l^2}(f + h)(1 + f'^2)^{3/2} \tag{4.3.14}$$

This is to be solved subject to the three boundary conditions $f(0) = 0$, $f'(0) = 0$, $f'(b) = \cot\alpha$.

It is convenient to recast Eq. (4.3.14) in dimensionless form in terms of the variables $F = f/b$, $X = x/b$, $H = h/b$ as

$$F'' = Bo\,(F + H)(1 + F'^2)^{3/2} \tag{4.3.15}$$

where we have introduced the dimensionless *Bond number*

$$Bo \equiv \left(\frac{b}{l}\right)^2 = \frac{\Delta\rho\, gb^2}{\gamma} \tag{4.3.16}$$

The boundary conditions are $F(0) = 0$, $F'(0) = 0$, $F'(1) = \cot\alpha$. It is now evident that the shape of the interface depends upon the value of the Bond number and the magnitude of the contact angle α.

A standard method of solving Eq. (4.3.15) proceeds by recasting it into the standard form of a system of two first-order nonlinear ordinary differential equations as

$$\begin{aligned}
\frac{dF}{dX} &= G \\
\frac{dG}{dX} &= Bo\,(F + H)(1 + G^2)^{3/2}
\end{aligned} \tag{4.3.17}$$

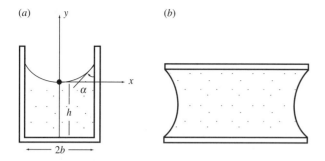

Figure 4.3.2 (a) The meniscus of a liquid within a rectangular container; (b) a two-dimensional liquid bridge subtended between two centered horizontal plates.

with boundary conditions $F(0) = 0$, $G(0) = 0$, $G(1) = \cot\alpha$. Having specified the values of Bo and α, we compute the solution in an iterative manner using a shooting method according to the following steps:

1. Guess the capillary rise H. A suitable guess, provided by Eq. (4.2.3), is $H = (1/Bo)\cos\alpha$.
2. Integrate Eqs. (4.3.17) as though we had an initial-value problem from $X = 0$ to 1 with initial conditions $F(0) = 0$ and $G(0) = 0$.
3. Check to see whether the numerical solution satisfies the third boundary condition $G(1) = \cot\alpha$, and if it does not, repeat the computation with a new and improved value for H. The improvement may be done using, for instance, Newton's method discussed in Section B.3, Appendix B.

Meniscus in a Narrow Container

A related problem concerns the shape of the interface of a fixed volume of a liquid placed within a rectangular container that is closed at the bottom, as shown in Figure 4.3.2(a). Applying Eqs. (4.3.1) and (4.3.2) at the free surface midway between the vertical walls yields $B = -f''(0)$ and $C = 1$, but we can no longer relate the value of the constant B to h.

Working as in the case of the meniscus between two plates, we derive Eqs. (4.3.17) with the dimensionless constant $D = -(l^2/b)B = (l^2/b)f''(0)$ in place of H, subject to the boundary conditions $F(0) = 0, G(0) = 0, G(1) = \cot\alpha$. The solution can be found using the shooting method described previously, where the guess is done with respect to D. Assuming that the meniscus has a circular shape yields the educated guess $D = (1/Bo)\cos\alpha$. Once the shape has been found, H is computed from the requirement that the volume per unit width of the liquid has a specified value v,

$$H = \frac{v}{2b^2} - \int_0^1 F(X)\,dX \qquad (4.3.18)$$

If H turns out to be zero or negative, the meniscus will touch or cross the bottom, and the solution will lose its physical relevance. In that case, the fluid will arrange itself on either corner of the container according to the specified value of the contact angle, and the shape of the meniscus must be found using a different type of parametrization (Problem 4.3.1).

PROBLEMS

4.3.1   **Meniscus within a container.** Assuming that the meniscus crosses the bottom of the container shown in Figure 4.3.2(a), develop an alternative appropriate parametric representation, derive the governing differential equation, and state the accompanying boundary conditions.

4.3.2 **A liquid bridge between two horizontal plates.** Derive the differential equation that describes the free surface of a two-dimensional liquid bridge subtended between the two centered horizontal plates shown in Figure 4.3.2(b) in a suitable parametric form.

Computer Problems

4.3.3 **Meniscus pinned to a wall.** Integrate Eq. (4.3.7) to generate profiles of the meniscus for eight evenly spaced values of $\alpha + \beta$ between 0 and 2π, separated by $\pi/4$.

4.3.4 **Meniscus between two plates.** Write a program called *MEN2D* that computes the shape of a two-dimensional meniscus subtended between two vertical plates for a specified Bond number *Bo* and contact angle α. The integration of the ordinary differential equations should be carried out using the modified Euler method described in Section B.8, Appendix B. Run the program to compute the profile of a meniscus of water between two plates that are separated by a distance $b = 5$ mm, for contact angle $\alpha = \pi/4$, and surface tension $\gamma = 70$ dyn/cm.

4.4 | AXISYMMETRIC INTERFACES

To compute the shape of axisymmetric interfaces, we follow a procedure that is similar to that for two-dimensional interfaces discussed in the preceding section. The problem is reduced to solving a second-order ordinary differential equation involving an unspecified constant that is typically evaluated by requiring an appropriate boundary condition. We shall see, however, that some minor complications arise when the differential equation is applied at the axis of circular symmetry, requiring special attention.

A Meniscus within a Capillary Tube

As a first case study, we consider the shape of an interface within an open vertical cylindrical tube of radius a that is immersed in a quiescent pool as shown in Figure 4.4.1 (Concus, 1968). Describing the interface as $x = f(\sigma)$ and carrying out some preliminary computations that are similar to those discussed in Section 4.3 for the two-dimensional meniscus between two vertical plates, we find $B = -\Delta\rho\, gh/\gamma = -h/l^2$. Using this value and the corresponding

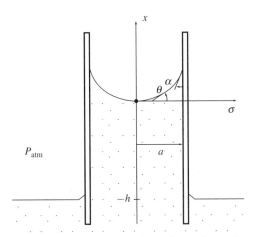

Figure 4.4.1 The meniscus of a liquid within a cylindrical tube immersed in a quiescent pool.

expression for the mean curvature from the second row of Table 4.2.1, we obtain the second-order nonlinear ordinary differential equation

$$f'' = \frac{1}{l^2}(f + h)(1 + f'^2)^{3/2} - \frac{1}{\sigma}f'(1 + f'^2) \tag{4.4.1}$$

which is to be solved subject to the boundary conditions $f(0) = 0$, $f'(0) = 0$, $f'(a) = \cot\alpha$, where α is the contact angle. It is instructive to reflect upon the differences between Eqs. (4.4.1) and (4.3.14), where the latter pertains to the corresponding problem in two dimensions.

An apparent difficulty arises when we attempt to evaluate Eq. (4.4.1) at the center-line of the tube at $\sigma = 0$, due to the fact that the second term on the right-hand side becomes indeterminate. Using, however, the regularity condition $f'(0) = 0$, we find that the mean curvature of the interface at the centerline is equal to $-f''(0)$, which may be substituted into the Laplace–Young equation to give

$$f''(0) = \frac{h}{2l^2} \tag{4.4.2}$$

This expression may be derived in an alternative manner by applying the l'Hôpital rule to evaluate the right-hand side of Eq. (4.4.1).

Parametric description

Following standard procedure, we decompose Eq. (4.4.1) into a system of two first-order differential equations, which we then solve using a shooting method. Unfortunately, the computations exhibit a pronounced sensitivity to the numerical accuracy. To circumvent this difficulty, we introduce a parametric representation of the interface in terms of the slope-angle θ defined by the equation $\tan\theta = f'$, where $0 < \theta < \pi/2 - \alpha$, as shown in Figure 4.4.1. The contact angle boundary condition is automatically satisfied by this particular choice of parametrization. To derive the equations that govern the shape of the interface, we write

$$f'' = \frac{df'}{d\sigma} = \frac{df'}{d\theta}\frac{d\theta}{d\sigma} = \frac{d\tan\theta}{d\theta}\frac{d\theta}{d\sigma} = \frac{1}{\cos^2\theta}\frac{d\theta}{d\sigma} \tag{4.4.3}$$

$$\frac{dx}{d\theta} = \frac{dx}{d\sigma}\frac{d\sigma}{d\theta} = \tan\theta\frac{d\sigma}{d\theta} \tag{4.4.4}$$

Substituting the right-hand side of Eq. (4.4.3) into the left-hand side of Eq. (4.4.1), replacing f' on the right-hand side by $\tan\theta$, and simplifying, we obtain

$$\frac{d\sigma}{d\theta} = \frac{\cos\theta}{Q}, \tag{4.4.5}$$

where

$$Q = \frac{x + h}{l^2} - \frac{\sin\theta}{\sigma}$$

Inserting Eq. (4.4.5) into the right-hand side of Eq. (4.4.4) yields

$$\frac{dx}{d\theta} = \frac{\sin\theta}{Q} \tag{4.4.6}$$

Equations (4.4.5) and (4.4.6) provide us with the desired system of ordinary differential equations describing the shape of the meniscus in a parametric form. To complete the definition of the problem, we must supply three boundary conditions: two because we have a system of two first-order ordinary differential equations, and one more because of the presence of the unspecified constant h. These are

$$\sigma = 0 \quad \text{and} \quad x = 0 \quad \text{at} \quad \theta = 0, \qquad \sigma = a \quad \text{at} \quad \theta = \pi/2 - \alpha \tag{4.4.7}$$

It will be noted that the denominator Q becomes unspecified at the origin, at $\theta = 0$. Combining Eq. (4.4.2) with Eqs. (4.4.3) and (4.4.4), however, we find that at the origin

$$\frac{d\sigma}{d\theta} = \frac{2l^2}{h}, \qquad \frac{dx}{d\theta} = 0 \qquad (4.4.8)$$

which may be used to initialize the computation. The first of Eqs. (4.4.8) may also be derived by applying the l'Hôpital rule to evaluate the right-hand side of Eq. (4.4.5) and then solving for $d\sigma/d\theta$ (Problem 4.4.1).

To this end, we introduce the dimensionless variables $\Sigma = \sigma/a$ and $X = x/a$, and rewrite Eqs. (4.4.5)–(4.4.7) as

$$\frac{d\Sigma}{d\theta} = \frac{\cos\theta}{W}, \qquad \frac{dX}{d\theta} = \frac{\sin\theta}{W}, \qquad (4.4.9)$$

where

$$W = Bo(X + H) - \frac{\sin\theta}{\Sigma}$$

with boundary conditions

$$\Sigma = 0 \quad \text{and} \quad X = 0 \text{ at } \theta = 0, \qquad \Sigma = 1 \text{ at } \theta = \pi/2 - \alpha \qquad (4.4.10)$$

where $H = h/a$ and Bo is the Bond number defined as

$$Bo \equiv \left(\frac{a}{l}\right)^2 = \frac{\Delta\rho\, ga^2}{\gamma} \qquad (4.4.11)$$

At the origin, $\theta = 0$, Eqs. (4.4.8) provide us with

$$\frac{d\Sigma}{d\theta} = \frac{2}{Bo}\frac{1}{H}, \qquad \frac{dX}{d\theta} = 0 \qquad (4.4.12)$$

Having specified the values of Bo and α, we compute the solution using a shooting method that is similar to the one described in Section 4.3 for the analogous problem in two dimensions, according to the following steps:

1. Guess H; an educated choice, provided by Eq. (4.2.5), is $H = (2/Bo)\cos\alpha$.
2. Integrate Eqs. (4.4.9) as though we had an initial-value problem from $\theta = 0$ to $\pi/2 - \alpha$. To initialize the computation, use Eqs. (4.4.12).
3. Check to see whether the numerical solution satisfies the third boundary condition in Eqs. (4.4.10), and if it does not, repeat the computation with a new and improved value for H. The improvement may be done using, for instance, Newton's method described in Section B.3, Appendix B.

A Sessile Drop Resting on a Flat Plate

As a second application, we consider the shape of an axisymmetric drop with a specified volume V resting on a horizontal plane wall, as shown in Figure 4.4.2(a). Because x may not be a single-valued function of σ, the parametric description $x = f(\sigma)$ is no longer appropriate, and we introduce the alternative representation $\sigma = f(x)$. Substituting the appropriate expression for the mean curvature from the third row of Table 4.2.1 into the Laplace–Young equation (4.2.2), we obtain the second-order ordinary differential equation

$$f'' = \left(\frac{x}{l^2} - B\right)(1 + f'^2)^{3/2} + \frac{1 + f'^2}{f} \qquad (4.4.13)$$

Evaluating Eq. (4.2.2) at the origin, we find that B is equal to twice the mean curvature of the interface at the centerline.

To formulate the numerical problem, we introduce the parameter ψ that is defined by the equation $\cot \psi = -f'$, and ranges from zero at the centerline of the drop to the value of the contact angle α at the contact line as shown in Figure 4.4.2(a), and regard x and σ along the interface as functions of ψ. The contact angle boundary condition is satisfied automatically by this particular choice of parametrization. Following the procedure that led us from Eq. (4.4.1) to Eqs. (4.4.5) and (4.4.6), we reduce Eq. (4.4.13) to the following system of two first-order equations

$$\frac{dx}{d\psi} = \frac{\sin \psi}{Q}, \qquad \frac{d\sigma}{d\psi} = -\frac{\cos \psi}{Q}, \qquad (4.4.14)$$

where

$$Q = \frac{\sin \psi}{\sigma} + \frac{x}{l^2} - B$$

which are accompanied by the boundary conditions $\sigma(0) = 0$ and $x(0) = 0$. One more condition emerges by requiring that the volume of the drop have a specified value V, yielding

$$\int_{-d}^{0} \sigma^2 \, dx = \frac{V}{\pi} \qquad (4.4.15)$$

where d is the height of the drop defined in Figure 4.4.2(a).

The denominator Q becomes indeterminate at the centerline, $\psi = 0$. Using the l'Hôpital rule to evaluate the right-hand side of the second of Eqs. (4.1.14), however, we obtain

$$\left(\frac{d\sigma}{d\psi} \right)_0 = -1 \Big/ \left[\left(\frac{d\psi}{d\sigma} \right)_0 - B \right] \qquad (4.4.16)$$

Rearranging, we find that, at the origin,

$$\frac{dx}{d\psi} = 0, \qquad \frac{d\sigma}{d\psi} = \frac{2}{B} \qquad (4.4.17)$$

In terms of the dimensionless variables $X = x/l$ and $\Sigma = \sigma/l$, Eqs. (4.4.14) and (4.4.15) assume the forms

$$\frac{dX}{d\psi} = \frac{\sin \psi}{W}, \qquad \frac{d\Sigma}{d\psi} = -\frac{\cos \psi}{W}, \qquad (4.4.18)$$

where

$$W = \frac{\sin \psi}{\Sigma} + X - C$$

(a)

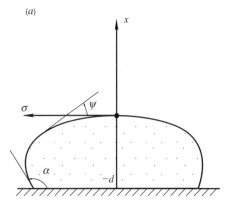

Figure 4.4.2   (a) Schematic illustration of a drop resting on a horizontal plate.   (*continued*)

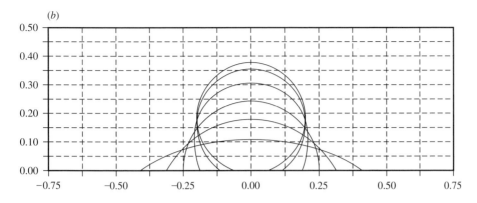

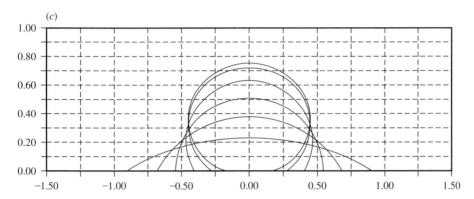

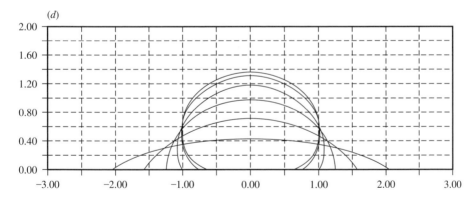

Figure 4.4.2 (*continued*)   Computed drop profiles for reduced volume (b) $E = 0.01$, (c) $E = 0.1$, (d) $E = 1.0$ and several values of the contact angle.   (*continued*)

with boundary conditions $\Sigma(0) = 0$ and $X(0) = 0$, and

$$\int_{-D}^{0} \Sigma^2 \, dX = E \tag{4.4.19}$$

where $C = Bl$, $D = d/l$, and $E = V/(\pi l^3)$. Eqs. (4.4.17) state that, at the origin,

$$\frac{dX}{d\psi} = 0, \qquad \frac{d\Sigma}{d\psi} = \frac{2}{C} \tag{4.4.20}$$

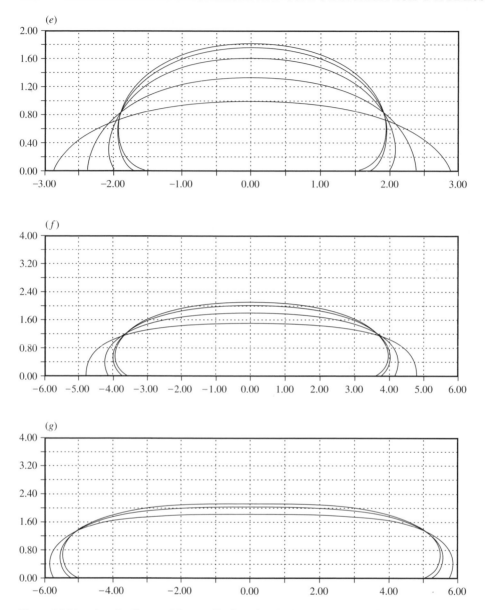

Figure 4.4.2 (*continued*) Computed drop profiles for reduced volume (e) $E = 5.0$, (f) $E = 25.0$, (g) $E = 50.0$, and several values of the contact angle.

It is now clear that the shape of the drop will depend upon the value of the reduced volume E and contact angle α. Given the values of E and α, we compute the drop shape using a shooting method that involves the following steps:

1. Guess the value of the constant C. Under appropriate conditions, a good estimate may be obtained by assuming that the drop has the shape of a section of a sphere of radius a whose volume is equal to V. An expression for a in terms of α and V can be found using elementary geometry. Since B is equal to the mean curvature at the centerline, we set $B = 1/a$ and compute $C = 1/a$.

2. Integrate Eqs. (4.4.18) as if we had an initial-value problem from $\psi = 0$ to α. To initialize the computation use Eqs. (4.4.20).

3. Check to see whether Eq. (4.4.19) is fulfilled, and, if it is not, repeat the computation with a new and improved value of C.

Six families of solutions corresponding to six values of E are shown in Figure 4.4.2(b–g). The members within each family correspond to different contact angles. When the reduced volume E is sufficiently small, the drops take the shape of sections of a sphere. As E is increased, the drops flatten at the top due to the action of gravity.

PROBLEMS

4.4.1   **A sessile drop.** Derive the first of Eqs. (4.4.8) by applying the l'Hôpital rule to evaluate the right-hand side of the first of Eqs. (4.4.5) and then solving for $d\sigma/d\theta$.

4.4.2   **A pendant drop.** Derive, in dimensionless form, the differential equations and boundary conditions that govern the shape of an axisymmetric drop pending underneath a horizontal flat plate.

4.4.3   **A rotating drop.** Derive, in dimensionless form, the differential equations and boundary conditions that govern the shape of an axisymmetric drop resting upon or pending underneath a horizontal flat plate that rotates steadily about the axis of the drop with angular frequency Ω.

Computer Problems

4.4.4   **Meniscus within a capillary tube.** Write a program called *MENT* that computes the axisymmetric meniscus of a liquid within a circular tube for a specified Bond number and contact angle. The integration of the differential equations should be carried out using the modified Euler method. Run the program to compute the profile of a meniscus of water within a capillary tube of radius $a = 2.5$ mm for contact angle $\theta = \pi/4$ and surface tension $\gamma = 70$ dyn/cm. (The solution yields $h = 0.359$ mm.)

4.4.5   **A sessile drop.** Write a program called *DROPSES* that computes the shape of a sessile drop with a specified reduced volume E and contact angle α. The integration of the differential equations should be carried out using the modified Euler method. Run the program to compute the shape of a water drop of volume $V = 2$ ml for contact angle $\alpha = 3\pi/4$ and $\gamma = 70$ dyn/cm.

4.5 │ THREE-DIMENSIONAL INTERFACES

A standard method of computing the shape of a three-dimensional interface proceeds by describing the interface in parametric form in terms of two surface variables or curvilinear coordinates ξ and η, and then computing the position of a collection of marker points that are located at intersections of grid lines. The successful choice of parametrization depends upon the nature of the expected interfacial shape; unwise choices may lead to numerical instabilities. The position of the marker points is computed by solving the Young–Laplace equation, which in this case reduces to a second-order nonlinear partial differential equation for the coordinates of the marker points with respect to ξ and η, using finite-difference, finite-element, variational, or spectral methods. Numerical implementations are discussed by Brown (1979), Milinazzo and Shinbrot (1988), Hornung and Mittelmann (1990), and Li and Pozrikidis (1996).

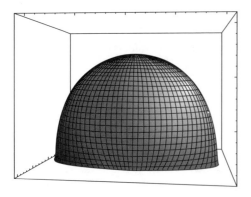

Figure 4.5.1   The shape of a drop adhering to a plane wall with an elliptical contact line in the limit of vanishing Bond number where gravitational effects become insignificant (Li and Pozrikidis, 1996).

As an example, in Figure 4.5.1 we present the shape of a drop attached to a plane wall with an elliptical contact line in the limit of vanishing Bond number where gravitational effects become insignificant, after Li and Pozrikidis (1996).

References

Brown, R. A., 1979, Finite-element methods for the calculation of capillary surfaces. *J. Comp. Phys.* **33,** 217–35.

Concus, P., 1968, Static menisci in a vertical right circular cylinder. *J. Fluid Mech.* **34,** 481–95.

Dussan, V. E. B., 1985, On the ability of drops or bubbles to stick to non-horizontal surfaces of solids. Part 2. Small drops or bubbles having contact angles of arbitrary size. *J. Fluid Mech.* **151,** 1–20.

Finn, R., 1985, *Equilibrium Capillary Surfaces.* Springer–Verlag.

Gradshteyn, I. S., and Ryzhik, I. M., 1980, *Table of Integrals, Series, and Products.* Academic Press.

Hartland, A., and Hartley, R. W., 1976, *Axisymmetric Fluid–Liquid Interfaces.* Elsevier.

Hornung, U., and Mittelmann, H. D., 1990, A finite element method for capillary surfaces with volume constraints. *J. Comp. Phys.* **87,** 126–36.

Isenberg, C., 1992, *The Science of Soap Films and Soap Bubbles.* Dover.

Li, X., and Pozrikidis, C., 1996, Shear flow over a liquid drop adhering to a solid surface. *J. Fluid Mech.* **307,** 167–90.

Milinazzo, F., and Shinbrot, M., 1988, A numerical study of a drop on a vertical wall. *J. Coll. Interf. Sc.* **121,** 254–64.

Computing
Incompressible Flows

■■■■■

Having established the equations that govern the motion of an incompressible Newtonian fluid and the associated boundary conditions, we proceed to discuss their solutions. Not surprisingly, we find that computing analytical solutions in closed or transcendental form is hindered by the presence of the nonlinear term associated with the point-particle acceleration. This term renders the governing system of equations nonlinear with respect to the velocity in a quadratic way. An inevitable consequence is that analytical solutions can be found only for a limited class of flows in the majority of which the nonlinear term either *happens* to vanish or *is assumed* to make an insignificant contribution. Under more general circumstances, the solution must be found using approximate, asymptotic, and numerical methods suitable for solving ordinary and partial differential equations. Fortunately, with the availability of an extensive arsenal of computational methods, to be discussed in the subsequent chapters, we are in a position to tackle a broad range of problems pertinent to a wide range of physical conditions.

In the present chapter we discuss a family of flows whose solution may be found either analytically, in closed or transcendental form, or by numerically solving ordinary and simple one-dimensional partial differential equations. We shall begin by considering unidirectional flows in channels and tubes for which the nonlinear term in the equation of motion vanishes due to the fact that the fluid particles travel along straight paths and the magnitude of the velocity is constant along the streamlines. Swirling flow inside or outside a circular tube and between concentric cylinders will provide us with an example of a flow where the nonlinear term does not vanish but assumes a simple form that allows for analytical treatment. In the subsequent sections we shall discuss a class of flows in totally or partially infinite domains that may be computed by solving ordinary differential equations without any approximations. Reviews and discussions of further exact solutions can be found in the articles of Berker (1963), Whitham (1963), Lagestrom (1964), Rott (1964), and Wang (1989, 1991).

In Chapter 6 we shall discuss flows at low Reynolds number in which the nonlinear acceleration term contributes a negligible amount to the momentum balance, and the flow is governed by a linearized version of the equation of motion. Linearization will allow us to build an extensive theoretical framework, as well as derive exact solutions to a broad range of problems using analytical and efficient numerical methods.

In Chapters 7, 8, and 10 we shall discuss irrotational and nearly irrotational flows in which the vorticity is confined within slender wakes and boundary layers. The velocity field in the main part of the flow is obtained by solving Laplace's equation for the harmonic velocity potential. Once this outer irrotational flow is available, the flow within the boundary layers and wakes is found by solving simplified versions of the equation of motion that result from boundary-layer approximations.

In Chapter 11 we shall discuss methods of computing inviscid or nearly inviscid flows that contain regions of concentrated vorticity including columnar vortices and vortex sheets. Finally, in

Chapters 12 and 13 we shall discuss finite-difference methods for solving the complete system of governing equations without any approximations apart from those involved in the implementation of the numerical method.

5.1 | STEADY UNIDIRECTIONAL FLOWS

The distinguishing feature of unidirectional flows is that, in a certain Cartesian system of coordinates, all components of the velocity except for one vanish, and the nonvanishing component is constant in the direction of the flow. For convenience, we presently extend this family to include swirling flows with circular streamlines. These features allow for substantial simplifications of the equation of motion, leading to the feasibility of analytical solutions in closed or transcendental form.

Despite their simplicity, unidirectional flows are ubiquitous in nature and technology. Their study may be traced back to the seminal works of Rayleigh, Stokes, Couette, Poiseuille, and Nusselt, and the investigation of their properties continues to be a subject of current research.

Rectilinear Flows

One important class of unidirectional flows encompasses steady rectilinear flows of fluids with uniform physical properties through straight channels and tubes, due to the action of an externally imposed pressure gradient, gravity, or longitudinal boundary motion. The streamlines are straight lines, the fluid particles travel along straight paths, and the pressure gradient assumes a uniform value throughout the domain of flow.

Neglecting entrance effects and assuming that the flow occurs along the x axis, we express the streamwise pressure gradient in the convenient form

$$\frac{\partial p}{\partial x} = \rho g_x - G \tag{5.1.1}$$

where G is a constant. When the pressure at the entrance of a channel or tube is equal to that at the exit, $\partial p/\partial x = 0$ and $G = \rho g_x$. When the x axis is oriented in the horizontal direction, $g_x = 0$, which means that $-G$ is the streamwise pressure gradient generated, for example, by a pump.

Setting for simplicity $u_x = u$ and substituting $\partial u/\partial t = 0$, $u_y = u_z = 0$, and $\partial u/\partial x = 0$ into the Navier–Stokes equation, we derive three linear scalar equations corresponding to the x, y, and z directions

$$0 = G + \mu \left(\frac{\partial^2 u}{\partial y^2} + \frac{\partial^2 u}{\partial z^2} \right)$$

$$0 = -\frac{\partial p}{\partial y} + \rho g_y \tag{5.1.2}$$

$$0 = -\frac{\partial p}{\partial z} + \rho g_z$$

The nonlinear inertial terms are absent because the magnitude of the velocity has been assumed to be constant along the streamlines.

The solution for the pressure is readily found from Eq. (5.1.1) and the last two equations (5.1.2) to be

$$p = \rho \mathbf{g} \cdot \mathbf{x} - Gx + P_0 \tag{5.1.3}$$

where P_0 is a reference pressure that is found by requiring an appropriate boundary condition. It is now evident that $-G$ represents the streamwise pressure gradient of the modified pressure, $G = -\partial P/\partial x$. When $G = 0$, the pressure assumes its hydrostatic distribution discussed in Chapter 4.

The first of Eqs. (5.1.2) yields a two-dimensional Poisson equation for the streamwise component of the velocity with a constant forcing term on the right-hand side,

$$\frac{\partial u^2}{\partial y^2} + \frac{\partial u^2}{\partial z^2} = -\frac{G}{\mu} \qquad (5.1.4)$$

which is to be solved subject to the no-slip condition that specifies the distribution of u along the solid boundary of a channel or tube, or the condition of vanishing shear stress $\mathbf{n} \cdot \nabla u$ along a free surface; ∇ is the two-dimensional gradient operating with respect to y and z, and $\mathbf{n}$ is the unit vector normal to the free surface. In the case of two-fluid flow, we require that the shear stress $\mu \mathbf{n} \cdot \nabla u$ remain continuous across the interface, which means that the normal derivative of the velocity $\mathbf{n} \cdot \nabla u$ undergoes a jump whose value is determined by the viscosities of the two fluids.

Pressure-, Gravity-, and Shear-Driven Flows

When a channel or tube is horizontal and the walls are stationary, we obtain *pressure-driven flow* due to an externally imposed pressure gradient $\partial p/\partial x = -G$. When the walls are stationary and the pressure does not change in the direction of the flow, $\partial p/\partial x = 0$, we obtain $G = \rho g_x$, which yields *gravity-driven flow*. Finally, when $G = 0$, which means that the pressure assumes its hydrostatic distribution, we obtain *shear-driven flow* due to the longitudinal translation of the whole or a section of a wall. In the last case, Eq. (5.1.4) reduces to Laplace's equation for u, and the solution is independent of the viscosity μ. Mixed cases of pressure-, gravity-, and shear-driven flow may be constructed by linear superposition.

Flow through a Channel with Parallel-Sided Walls

Consider first two-dimensional flow through a channel bounded by two parallel walls that are separated by a distance equal to h, as shown in Figure 5.1.1(a). The bottom and top walls translate parallel to themselves with respective velocities equal to U_1 and U_2. Requiring that u be a function of y alone, setting the origin of the Cartesian axes at a point on the lower wall, and demanding the boundary conditions $u = U_1$ at $y = 0$ and $u = U_2$ at $y = h$, we find that the velocity profile is given by the parabolic distribution

$$u(y) = U_1 + (U_2 - U_1)\frac{y}{h} + \frac{G}{2\mu}y(h - y) \qquad (5.1.5)$$

The volumetric flow rate per unit width of the channel is given by

$$Q \equiv \int_0^h u\,dy = \tfrac{1}{2}(U_1 + U_2)\,h + \frac{Gh^3}{12\mu} \qquad (5.1.6)$$

When $G = 0$, we obtain *Couette flow* with a linear velocity profile first studied by Couette (1890), also called *simple shear flow,* whereas when $U_1 = 0$ and $U_2 = 0$, we obtain *plane Hagen–Poiseuille flow* (Sutera and Skalak, 1993).

As an application, let us consider a long horizontal channel that is confined between two translating parallel belts and is closed at both ends. Since the flow rate through the channel must vanish, Eq. (5.1.6) predicts the spontaneous onset of a pressure field with $dp/dx = -G = 6\mu(U_1 + U_2)/h^2$.

Figure 5.1.1(a) shows the velocity profiles corresponding to the modular cases of Couette and Poiseuille flow, as well as to a flow with vanishing flow rate, all with $U_1 = 0$.

Flow of a Liquid Film down an Inclined Plane

Gravity-driven flow of a liquid film down an inclined plane is encountered in a variety of engineering applications including the manufacturing of magnetic recording media and photographic films. Under certain conditions, to be discussed in Section 9.10 in the context of

(a)

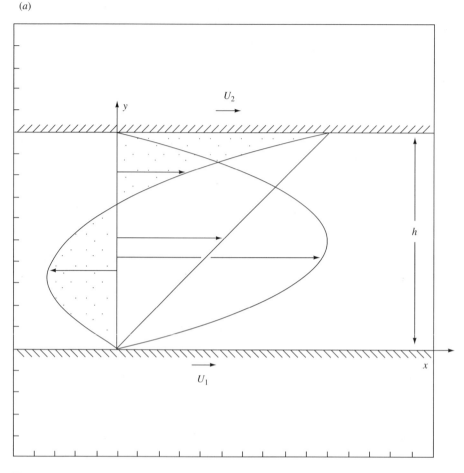

Figure 5.1.1   (a) Velocity profiles of flow through a two-dimensional channel with parallel-sided walls separated by distance h. The linear profile corresponds to Couette flow, the parabolic profile to plane Hagen–
Poiseuille flow, and the intermediate profile to flow with vanishing flow rate, all with $U_1 = 0$.   (*continued*)

hydrodynamic stability, the free surface may be assumed to be flat, yielding a film of uniform
thickness H.

In the inclined coordinate system depicted in Figure 5.1.1(b), the x component of the velocity
is a function of y alone, and the gravity vector is given by $\mathbf{g} = g(\sin\theta_0, -\cos\theta_0, 0)$, where g is
the acceleration of gravity and θ_0 is the inclination of the plane with respect to the horizontal
direction; $\theta_0 = 0$ yields a horizontal plane, and $\theta_0 = \pi/2$ yields a vertical plane. Since the
pressure outside and therefore inside the film is independent of the x coordinate, $G = \rho g_x = \rho g \sin\theta_0$. Integrating Eq. (5.1.4) twice subject to the no-slip boundary condition at the wall, $u = 0$
at $y = 0$, and the condition of vanishing shear stress at the free surface, $\partial u/\partial y = 0$ at $y = H$,
and using the general solution for the pressure given in Eq. (5.1.3), we obtain

$$u = \frac{g\sin\theta_0}{2\nu}y(2H - y), \qquad p = -\rho g\cos\theta_0(y - H) + P_{\text{Atm}} \qquad (5.1.7)$$

which shows that the velocity profile has a semiparabolic shape as deduced by Hopf and Nusselt
in 1910 and 1916 (Fulford, 1964). The velocity at the free surface, volumetric flow rate per unit

(b)

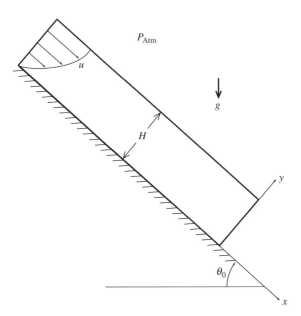

Figure 5.1.1 (*continued*)   (b) Flow of a flat liquid film down an inclined plane wall.

width of the film, and mean velocity of the fluid are given by

$$u_{\text{Max}} = \frac{gH^2 \sin \theta_0}{2v}$$

$$Q \equiv \int_0^H u \, dy = \frac{gH^3 \sin \theta_0}{3v} \qquad (5.1.8)$$

$$u_{\text{Mean}} \equiv \frac{Q}{H} = \tfrac{2}{3} u_{\text{Max}} = \frac{gH^2 \sin \theta_0}{3v}$$

Flow through Tubes

We turn now to discuss the structure and properties of pressure-, gravity-, and shear-driven flow along straight tubes with various cross-sectional shapes.

Pressure- and gravity-driven flow through a circular tube

Pressure-driven flow through a circular tube was first considered by Hagen and Poisseuille, the latter in his treatise of blood flow through vessels (Sutera and Skalak, 1993). Writing Eq. (5.1.4) in cylindrical polar coordinates and assuming that the velocity does not vary in the azimuthal direction φ, which means that the flow is axisymmetric, yields the second-order ordinary differential equation

$$\frac{1}{\sigma} \frac{d}{d\sigma} \left(\sigma \frac{du}{d\sigma} \right) = -\frac{G}{\mu} \qquad (5.1.9)$$

which is to be solved subject to the regularity condition $du/d\sigma = 0$ at the center-line $\sigma = 0$, and the no-slip condition $u = 0$ over the wall at $\sigma = a$ [Fig. 5.1.2(a)]. Two straightforward

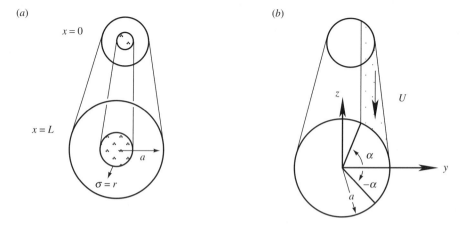

Figure 5.1.2   Illustrations of (a) pressure- and gravity-driven flow and (b) shear-driven along a circular tube.

integrations of Eq. (5.1.9) yield the parabolic profile

$$u = \frac{G}{4\mu}(a^2 - \sigma^2) \qquad (5.1.10)$$

The maximum velocity of the fluid, occurring at the center of the tube at $\sigma = 0$, flow rate through the tube, and mean velocity of the flow are given by

$$u_{\text{Max}} = \frac{Ga^2}{4\mu}$$

$$Q \equiv \int_{\text{Tube}} u \, dy \, dz = 2\pi \int_0^a u\sigma \, d\sigma = \frac{\pi G a^4}{8\mu} \qquad (5.1.11)$$

$$u_{\text{Mean}} \equiv \frac{Q}{\pi a^2} = \tfrac{1}{2} u_{\text{Max}} = \frac{Ga^2}{8\mu}$$

The fact that the flow rate is proportional to the fourth power of the tube radius is sometimes called *Poiseuille's law,* after Poiseuille, who first inferred this relationship on the basis of experimental observation, at a time when neither the validity of the no-slip boundary condition nor the parabolic velocity profile had been established.

It is instructive to rederive the parabolic velocity profile shown in Eq. (5.1.10) by performing a force balance over a cylindrical fluid column of length L and radius $\sigma = r$ shown in Figure 5.1.2(a). Requiring that the sum of surface forces and body forces balance, we obtain

$$\sigma_{xx}(x=L)\pi r^2 - \sigma_{xx}(x=0)\pi r^2 + \sigma_{x\sigma}(\sigma=r)2\pi rL + \rho g_x \pi r^2 L = 0 \qquad (5.1.12)$$

Expressing the stresses in terms of the velocity and pressure yields the equivalent form

$$-p(x=L)(\pi r^2) + p(x=0)(\pi r^2) + \mu \left(\frac{du}{d\sigma}\right)_{\sigma=r} (2\pi rL) + \rho g_x(\pi r^2 L) = 0 \quad (5.1.13)$$

Solving for the first derivative of the velocity, we obtain

$$\mu \left(\frac{du}{d\sigma}\right)_{\sigma=r} = -\tfrac{1}{2}Gr, \qquad \text{where} \quad G \equiv \rho g_x - \frac{p(x=L) - p(x=0)}{L} \qquad (5.1.14)$$

The first of Eqs. (5.1.14) is the first integral of Eq. (5.1.9). Integrating once subject to the boundary condition $u = 0$ at $\sigma = a$ yields the parabolic profile given in Eq. (5.1.10).

The Reynolds number of the flow through a circular tube is usually expressed in terms of either the maximum velocity of the fluid or the flow rate as

$$Re \equiv \frac{2au_{\text{Max}}\rho}{\mu} = \frac{4\rho Q}{\pi\mu a} = \frac{\rho a^3 G}{2\mu^2} \tag{5.1.15}$$

In practice, the rectilinear flow described by Eq. (5.1.10) will be established provided that the Reynolds number is less than a critical value that depends upon the roughness of the wall, and is roughly equal to 2,000, as discussed in Section 9.7. Above this value, the streamlines become wavy, the flow develops random motions, and the motion becomes turbulent at sufficiently high Reynolds numbers.

Shear-driven flow through a circular tube

Consider next flow through a horizontal circular tube driven by the translation of a sector of the tube with aperture equal to 2α, confined between the azimuthal planes $-\alpha < \varphi < \alpha$, in the absence of a pressure drop, as shown in Figure 5.1.2(b). To compute the velocity profile, we solve Laplace's equation (5.1.4) with the right-hand side set equal to zero, and boundary conditions $u = 0$ at $\sigma = a$ except that $u = U$ at $\sigma = a$ for $-\alpha < \varphi < \alpha$. The solution is found using the Poisson integral to be

$$
\begin{aligned}
u(\sigma, \varphi) &= \frac{U}{2\pi}(a^2 - \sigma^2) \int_{-\alpha}^{\alpha} \frac{d\omega}{a^2 + \sigma^2 - 2a\sigma \cos(\varphi - \omega)} \\
&= \frac{U}{\pi}\left[\arctan\left(\frac{a+\sigma}{a-\sigma}\tan\frac{\alpha-\varphi}{2}\right) + \arctan\left(\frac{a+\sigma}{a-\sigma}\tan\frac{\alpha+\varphi}{2}\right)\right]
\end{aligned} \tag{5.1.16}
$$

(Problem 2.3.4(b)). In the limiting case $\alpha = \pi$ the fluid translates as a rigid body in a plug-flow mode with the wall velocity U.

Flow through a tube with elliptical cross-section

The velocity profile for pressure- and gravity-driven flow follows from the observation that the equation of the ellipse is a quadratic function of y and z, and thus it represents a solution of Poisson's equation with a constant term on the right-hand side. Setting the origin at the center of the ellipse, we obtain

$$u(y, z) = \frac{G}{2\mu}\frac{a^2 b^2}{a^2 + b^2}\left(1 - \frac{y^2}{a^2} - \frac{z^2}{b^2}\right) \tag{5.1.17}$$

where a and b are the semiaxes of the elliptical cross section corresponding to the y and z axes, as shown in Figure 5.1.3(a). Maximum velocity occurs at the origin, and the flow rate through the tube is given by

$$Q = \frac{\pi G}{4\mu}\frac{a^3 b^3}{a^2 + b^2} \tag{5.1.18}$$

Setting $a = b$ recovers the earlier results for the circular tube.

Flow through a tube with triangular cross-section

The velocity profile for pressure- and gravity-driven flow through a tube whose cross-section has the shape of an equilateral triangle is given by

$$u(y, z) = \frac{G}{36\mu a}\left(2\sqrt{3}z + a\right)\left(\sqrt{3}z + 3y - a\right)\left(\sqrt{3}z - 3y - a\right) \tag{5.1.19}$$

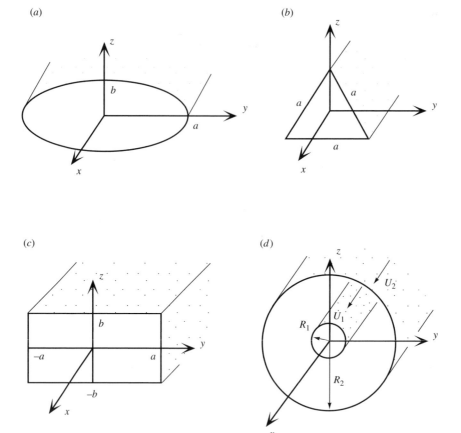

Figure 5.1.3   Illustrations of pressure- and gravity-driven flow along a tube with (a) elliptical, (b) triangular, (c) rectangular, and (d) annular cross-section.

where a is the length of one side of the triangular cross-section, and the origin has been set at the centroid of the tube, as shown in Figure 5.1.3(b) (Sparrow, 1962). Integrating the velocity over the cross-section of the tube, we find the volumetric flow rate

$$Q = \frac{\sqrt{3}}{320} \frac{Ga^4}{\mu} \qquad (5.1.20)$$

Sparrow (1962) derives the velocity profile of flow through a tube whose cross-section is an isosceles triangle in terms of an infinite series.

Flow through a tube with rectangular cross-section
Consider next flow through a tube with rectangular cross-section whose sides along the y and z axes are respectively equal to $2a$ and $2b$ as shown in Figure 5.1.3(c). Setting the origin at the center of the tube, expressing the velocity as the sum of a parabolic component with respect to z that satisfies the Poisson equation (5.1.4) and a homogeneous component that satisfies Laplace's equation, and then using Fourier expansions in the z direction to compute the homogeneous component subject to appropriate boundary conditions, we obtain

$$u(x, y) = \frac{G}{2\mu}\left[b^2 - z^2 + 4b^2\sum_{n=1}^{\infty}\frac{(-1)^n}{\alpha_n^3}\frac{\cosh(\alpha_n y/b)}{\cosh(\alpha_n a/b)}\cos\left(\alpha_n\frac{z}{b}\right)\right] \qquad (5.1.21)$$

where $\alpha_n = (2n - 1)\pi/2$. Integrating over the cross-section, we find that the volumetric flow rate is given by

$$Q = \frac{4Gab^3}{3\mu}F\left(\frac{a}{b}\right), \qquad \text{where} \quad F(x) = 1 - \frac{6}{x}\sum_{n=1}^{\infty}\frac{\tanh(\alpha_n x)}{\alpha_n^5} \qquad (5.1.22)$$

As x tends to infinity, the function $F(x)$ tends to unity, whereas as x tends to zero, $F(x)$ behaves like x^2; both cases correspond to flow through a channel with parallel-sided walls described by Eq. (5.1.5).

Further shapes

Berker (1963) and Shah and London (1978) discuss solutions for pressure- and gravity-driven flow through families of tubes with a variety of cross-sectional shapes including the *half-moon,* the *circular-sector,* the *limacon,* and the *eccentric annulus.* The velocity distribution and flow rate are typically expressed in terms of an infinite series similar to those shown in Eqs. (5.1.21) and (5.1.22).

Solution for arbitrary cross-sections in complex variables

It is possible to derive a general solution for the velocity profile corresponding to pressure-, gravity-, and shear-driven flow through a straight tube with arbitrary cross-section based on the theory of functions of a complex variable.

In the case of pressure- or gravity-driven flow, we decompose the velocity field into a particular component and a homogeneous component, writing $u(y, z) = -(G/\mu) \cdot [f(y, z) + h(y, z)]$, where the function f satisfies the Poisson equation $\nabla^2 f = 1$ and the function h satisfies Laplace's equation $\nabla^2 h = 0$ with boundary condition $h = -f$ around the tube; one simple choice for f is $f = \frac{1}{2}y^2$. The problem is then reduced to computing the harmonic function h.

The theory of functions of a complex variable guarantees that h may be regarded as the real part of an analytic function $G(w)$, $h = \text{Re}[G(w)]$, where $w = y + iz$ is the complex variable in the yz plane. To compute the general solution, let us assume that the function $w = F(\zeta)$ maps a disk of radius a centered at the origin in the ζ complex plane to the cross section of the tube in the physical w plane (see Sections 7.9 and 7.10). Writing $\zeta = \rho e^{i\omega}$ and using Poisson's integral, we obtain

$$h(w) = h[F(\zeta)] = \frac{a^2 - \rho^2}{2\pi}\int_0^{2\pi}\frac{h[F(ae^{it})]}{a^2 + \rho^2 - 2a\rho\cos(\omega - t)}\,dt \qquad (5.1.23)$$

where the brackets enclose the arguments of the preceding variable. The computation of the required mapping function $F(\zeta)$ will be discussed in Section 7.10. In general, the integral in Eq. (5.1.23) must be evaluated using a numerical method.

Flow through a Concentric Annular Tube

Next we consider flow in the annular space confined between two concentric circular cylinders with respective radii equal to R_1 and R_2, where $R_2 > R_1$.

Axial flow

First we consider rectilinear flow generated by the translation of the cylinders along their lengths with respective velocities equal to U_1 and U_2, as shown in Figure 5.1.3(d), possibly in the presence of an axial pressure gradient. Writing Eq. (5.1.4) in cylindrical polar coordinates and integrating twice subject to the boundary conditions $u = U_1$ at $\sigma = R_1$ and $u = U_2$ at $\sigma = R_2$, we derive the velocity profile

$$u(\sigma) = U_2 + (U_1 - U_2)\frac{\ln(R_2/\sigma)}{\ln(R_2/R_1)} + \frac{G}{4\mu}\left(R_2^2 - \sigma^2 - (R_2^2 - R_1^2)\frac{\ln(R_2/\sigma)}{\ln(R_2/R_1)}\right) \quad (5.1.24)$$

The volumetric flow rate through the tube is found by straightforward integration to be

$$Q = \pi(U_2 R_2^2 - U_1 R_1^2) - \frac{\pi}{2}(U_2 - U_1)\frac{R_2^2 - R_1^2}{\ln(R_2/R_1)}$$

$$+ \frac{\pi G}{8\mu}(R_2^2 - R_1^2)\left(R_2^2 + R_1^2 - \frac{R_2^2 - R_1^2}{\ln(R_2/R_1)}\right) \quad (5.1.25)$$

When the channel width $h = R_2 - R_1$ is small compared to R_1, the curvature of the walls becomes insignificant, and Eqs. (5.1.24) and (5.1.25) reduce to Eqs. (5.1.5) and (5.1.6) with $y = \sigma - R_1$ describing flow in a parallel-sided channel (Problem 5.1.5(a)).

Circular Couette flow

Thus far we have considered rectilinear unidirectional flows in which the nonvanishing component of the velocity is directed along the x axis. To this end, we turn our attention to swirling flow occurring when the cylinders rotate about their common axis with angular velocities equal to Ω_1 and Ω_2.

Assuming that the axial and radial components of the velocity vanish and the azimuthal component is independent of the axial and azimuthal coordinates, we set $u_\varphi = u(\sigma)$, $u_x = 0$, $u_\sigma = 0$. The axial component of the equation of motion expresses a balance between the pressure gradient and the gravitational force, whereas the azimuthal and radial components yield the ordinary differential equations

$$\frac{d^2 u}{d\sigma^2} + \frac{1}{\sigma}\frac{du}{d\sigma} - \frac{u}{\sigma^2} = 0, \qquad \rho\frac{u^2}{\sigma} = \frac{dP}{d\sigma} \quad (5.1.26)$$

where P is the modified pressure. The second equation, in particular, states that the centrifugal force is balanced by a developing radial pressure gradient.

Solving the first of Eqs. (5.1.26) subject to the boundary conditions $u = \Omega_1 R_1$ at $\sigma = R_1$ and $u = \Omega_2 R_2$ at $\sigma = R_2$ yields

$$u = \frac{\Omega_2 - \alpha\Omega_1}{1 - \alpha}\sigma - \frac{\Omega_2 - \Omega_1}{1 - \alpha}\frac{R_1^2}{\sigma} \quad (5.1.27)$$

where $\alpha = (R_1/R_2)^2 < 1$. It is interesting to note that when $\Omega_2 = \Omega_1$ the fluid rotates like a rigid body, whereas when $\Omega_2 = \alpha\Omega_1$, the flow resembles that due to a point vortex with circulation equal to $2\pi\Omega_1 R_1^2 = 2\pi\Omega_2 R_2^2$ located at the axis; in the second case the flow is irrotational. In the limit as α approaches unity, the clearance of the channel becomes small compared to the radii of the cylinders, and the flow along the gap resembles plane-Couette flow in a channel of width $h = R_2 - R_1$ (Problem 5.1.5(b)).

The pressure distribution is found by substituting the velocity profile (5.1.27) into the second of Eqs. (5.1.26). In the special cases where the fluid rotates like a rigid body or resembles that due to a point vortex, we obtain

$$P^{\text{rigid body}} = P_0 + \tfrac{1}{2}\rho\Omega_1^2\sigma^2$$

$$P^{\text{point vortex}} = P_0 - \tfrac{1}{2}\rho\Omega_1^2\frac{R_1^4}{\sigma^2} \quad (5.1.28)$$

where P_0 is a constant reference pressure.

The circular Couette flow provides us with a simple device for assessing the viscosity of a fluid, by measuring the torque exerted on either the inner or the outer cylinder, which is given by

$$T = -4\pi\mu \frac{\Omega_2 - \Omega_1}{1 - \alpha} R_1^2 \tag{5.1.29}$$

The stability of the circular Couette flow will be discussed in Section 9.11. The results show that the laminar flow discussed here will be established only when the pair (Ω_1, Ω_2) falls within a particular range.

Approximate Solutions for Nearly Unidirectional Flows

The exact solutions for steady unidirectional flow discussed in the present section may be used to construct approximate solutions to problems involving flows that are dominated by unidirectional motions. The procedure will be illustrated in detail in Section 6.3 in the context of low-Reynolds-number flow, but it is instructive at this point to discuss a characteristic example.

Settling of a slab down a channel with parallel-sided walls

Consider an elongated rectangular solid slab with thickness $2b$ and length L settling under the action of gravity down the centerline of a channel that is confined between two vertical plates separated by distance $2a$ as shown in Figure 5.1.4. The channel is closed at the bottom and open to the atmosphere at the top. Our objective is to estimate the velocity of settling U and the pressure difference between the bottom and top of the slab.

To simplify the problem, we assume that the flow between the slab and the walls can be regarded as unidirectional flow with an associated constant value of G. This means that the velocity profile across the gap is given by Eq. (5.1.5) with $U_1 = U, U_2 = 0$, and $h = a - b$. If p_{Top} is

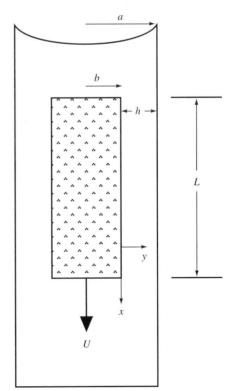

Figure 5.1.4 A rectangular slab settling down the centerline of a two-dimensional channel that is closed at the bottom.

the pressure at the top of the slab, then $p_{Bottom} = p_{Top} + \rho g L - GL$ will be the pressure at the bottom. The constant G and velocity of settling U are two unknowns that must be computed by introducing two requirements.

One requirement arises by stipulating that the rate of displacement of the liquid by the slab is equal to the upward flow rate,

$$Ub = -Q \qquad (5.1.30)$$

where Q is given by Eq. (5.1.6). The second requirement emerges by balancing the forces exerted on the slab in the x direction. Demanding that the pressure forces at the top and bottom of the slab counterbalance its weight and the force due to the shear stress along the sides yields

$$-(\rho g - G) Lb + \rho_s g \, Lb + \mu \left(\frac{\partial u}{\partial y}\right)_{y=0} L = 0 \qquad (5.1.31)$$

where ρ_s is the density of the solid. Substituting Eqs. (5.1.5) and (5.1.6) into Eqs. (5.1.30) and (5.1.31), and solving the resulting equations for U and G, we obtain

$$U = \frac{(\rho_s - \rho)ga^2}{4\mu} \frac{\varepsilon(1 - \varepsilon)^3}{\varepsilon^2 + \varepsilon + 1}$$

$$G = -(\rho_s - \rho)g\frac{3}{2}\frac{\varepsilon(1 + \varepsilon)}{\varepsilon^2 + \varepsilon + 1} \qquad (5.1.32)$$

where $\varepsilon = b/a$. It is worth noting that U vanishes when $\varepsilon = 0$ or 1, in which cases the weight of the slab is infinitesimal or the sides of the slab tend to stick to the side plates, and reaches a maximum at the intermediate value $\varepsilon = 0.204$.

PROBLEMS

5.1.1 **Flow through a triangular tube with partitions.** Consider pressure-driven flow through a channel whose cross-section in the yz plane is parametrized by the index n as follows: $n = 1$ corresponds to a channel whose cross-section is an equilateral triangle with side length equal to a; $n = 2$ corresponds to a partitioned channel that derives from the channel for $n = 1$ by introducing three straight segments connecting the middle points of the sides of the original triangle; each time n is increased by one unit, each triangle is partitioned into four smaller triangles by connecting the middle points of its sides. The number of triangles at the nth level is equal to 4^{n-1}. Show that the flow rate through the nth member of the family is given by

$$Q = \frac{1}{8^{n-1}} \frac{\sqrt{3}}{320} a^4 \frac{G}{\mu} \qquad (5.1.33)$$

Discuss the behavior of the flow rate in the limit as n tends to infinity.

5.1.2 **Taylor's two-dimensional paint brush.** Taylor (1960) developed a simple model for estimating the amount of paint that is deposited onto a plane wall during brushing. The brush is modeled as an infinite sequence of semi-infinite parallel plates that are separated by a distance equal to $2a$, sliding with velocity U at a right angle over a flat painted surface. The space between the brush planes and the painted surface is filled with a liquid. (a) Show that the velocity distribution within a channel of the brush is given by

$$u = U\left[1 - \frac{4}{\pi} \sum_{n=0}^{\infty} \frac{(-1)^n}{2n + 1} \exp\left(-\frac{(2n + 1)\pi z}{2a}\right) \cos\frac{(2n + 1)\pi y}{2a}\right] \qquad (5.1.34)$$

where the origin has been set on the painted surface midway between two plates with the z axis perpendicular to the surface. (b) Show that the amount of paint deposited on the surface per channel is given by

$$Q = \frac{32Ua^2}{\pi^3} \sum_{n=0}^{\infty} \frac{1}{(2n+1)^3} \cong 1.085Ua^2 \qquad (5.1.35)$$

Explain why the thickness of the film left behind the brush is $h = Q/(Ua)$.

5.1.3 **Flow through a tapered tube.** Consider a conical tube with a slowly varying radius $R(x)$. Derive an expression for the pressure drop that is necessary to drive a flow with a given flow rate.

5.1.4 **Flow of two layers.** (a) Consider the unidirectional flow of two adjacent layers of two different fluids in a channel with parallel-sided walls separated by a distance h. Derive the velocity profile across the two fluids in terms of the physical properties and flow rates of the fluids for the mixed case of shear- and pressure-driven flow. (b) Repeat (a) for gravity-driven flow of two layers flowing down an inclined plane.

5.1.5 **Flow between concentric cylinders.** (a) Show that, as the channel width $h = R_2 - R_1$ becomes small compared to the inner diameter R_1, Eqs. (5.1.24) and (5.1.25) reduce to Eqs. (5.1.5) and (5.1.6) with $y = \sigma - R_1$. (b) Show that, in the same limit, the velocity profile across the gap for circular-Couette flow resembles that of plane-Couette flow in a channel of thickness $h = R_2 - R_1$.

5.1.6 **Ekman flow.** Consider a semi-infinite body of a fluid extending from $z = 0$ to $-\infty$, rotating around the z axis with angular velocity Ω. Show that in a noninertial frame of reference that rotates with the fluid, and under the assumptions that (a) the modified pressure that accounts for the effect of gravity and centrifugal force is uniform, and (b) the vertical component u_z vanishes, an exact solution to the equation of motion subject to the boundary condition $u_x = U_x$ and $u_y = U_y$ at $z = 0$ is given by the real or imaginary part of

$$\frac{u_x + iu_y}{U_x + iU_y} = \exp\left(\sqrt{\frac{2i\Omega}{\nu}}\, z\right) \qquad (5.1.36)$$

where i is the imaginary unit. Discuss the structure for this flow (Batchelor 1967, p. 197).

5.1.7 **Free-surface flow in a square duct.** Consider unidirectional gravity-driven flow along a tilted square duct with side length $2a$, inclined at an angle θ_0, whose top is open to the atmosphere. The velocity of the fluid at the bottom, left, and right walls and the shear stress at the top surface are required to vanish. Assuming that the free surface is flat, show that in Cartesian coordinates where the left and right walls are located at $y = \pm a$ and the bottom wall is located at $z = -a$, the velocity field and flow rate are given by

$$u(y, z) = \frac{G}{2\mu}\left[a^2 - y^2 + 4a^2 \sum_{n=1}^{\infty} \frac{(-1)^n}{\alpha_n^3} \frac{\cosh[\alpha_n(1 + z/a)]}{\cosh(2\alpha_n)} \cos\left(\alpha_n \frac{y}{a}\right)\right]$$
$$\qquad (5.1.37)$$

$$Q = 4\frac{Ga^4}{\mu}\left(\frac{1}{3} - \sum_{n=1}^{\infty} \frac{\tanh(2\alpha_n)}{\alpha_n^5}\right)$$

where $G = \rho g \sin\theta_0$, $\alpha_n = (2n-1)\pi/2$.

5.1.8 **Coating a rod.** A cylindrical rod of radius a is being pulled upward with velocity U after it has been coated with a liquid film of thickness h. The coated liquid is draining downward due to the action of gravity. Show that in cylindrical polar coordinates in which the x axis is coaxial with the rod pointing upward, the velocity profile across the film is given by

$$u = U + \frac{g}{4\nu}\left(\sigma^2 - a^2 - 2(a+h)^2 \ln\frac{\sigma}{a}\right) \qquad (5.1.38)$$

What is the proper value of U for the thickness of the film to remain constant?

Computer Problems

5.1.9   **Flow through a flexible hose.** A gardener delivers water through a circular hose that is made of a flexible material. By pinching the end of the hose, she is able to obtain elliptical cross-sectional shapes with variable aspect ratios while the perimeter of the hose remains constant. Compute the delivered flow rate as a function of the aspect ratio of the cross-section for a certain pressure gradient.

5.1.10   **Flow through a rectangular channel.** Compute and plot the function $F(x)$ defined in the second of Eqs. (5.1.22), and comment on its behavior as x tends to zero or to infinity.

5.1.11   **The settling of a cylinder in a tube.** Consider a solid cylinder of radius b settling with velocity U under the action of gravity down the center of a circular tube of radius a that is closed at the bottom and open at the top. Present a plot of the reduced velocity of settling $\mu U/((\rho_y - \rho)ga^2)$ against the radius ratio b/a and discuss its behavior.

5.2 | UNSTEADY UNIDIRECTIONAL FLOWS

Unsteady unidirectional flows retain the distinguishing simplifying features of steady unidirectional flows discussed in Section 5.1; an important new feature is that the pressure gradient and nonvanishing component of the velocity u are allowed to vary in time. The flow may be driven either by an unsteady pressure gradient, expressed by the function $G(t)$, or by unsteady boundary motion.

Rectilinear Flow

In this case, the y and z components of the equation of motion are satisfied by the pressure distribution given in Eq. (5.1.3), and the x component reduces to the unsteady heat conduction equation with a time-dependent source term $G(t)$,

$$\rho \frac{\partial u}{\partial t} = G(t) + \mu \left(\frac{\partial^2 u}{\partial y^2} + \frac{\partial^2 u}{\partial z^2} \right) \tag{5.2.1}$$

Semi-Infinite Flow above a Plane Wall

Flow due to the oscillations of the wall

Consider first the flow generated by a flat plate that is oscillating in its plane along the x axis with angular frequency Ω in an otherwise quiescent semi-infinite fluid, first studied by Stokes in 1845. Assuming that u is a function of distance y from the plate alone, and setting $G(t) = 0$, we obtain the simplified version of Eq. (5.2.1)

$$\frac{\partial u}{\partial t} = \nu \frac{\partial^2 u}{\partial y^2} \tag{5.2.2}$$

which is to be solved subject to the boundary condition $u(y = 0, t) = U \cos(\Omega t)$, where U is the amplitude of the oscillation. For convenience, we write $u(y = 0, t) = U \, \mathrm{Re}[\exp(-i\Omega t)]$, where i is the imaginary unit, and motivated by the linearity of Eq. (5.2.2), we set

$$u(y, t) = U \, \mathrm{Re}[F(y) \exp(-i\Omega t)] \tag{5.2.3}$$

where $F(y)$ is a complex function that satisfies the boundary conditions $F(0) = 1$ and $F(\infty) = 0$. Substituting Eq. (5.2.3) into Eq. (5.2.2), we derive the linear ordinary differential equation

$-i\Omega F = \nu \, d^2 F/dy^2$, whose solution is

$$F(y) = \exp\left[\left(-\frac{i\Omega}{\nu}\right)^{1/2} y\right] \tag{5.2.4}$$

with the understanding that we select the square root with the *negative real part,* that is, $(-i)^{1/2} = e^{3\pi i/4}$. Combining Eq. (5.2.4) with Eq. (5.2.3), we obtain the velocity distribution

$$u(y, t) = U \operatorname{Re}\left\{\exp\left[-i\Omega t + \left(-\frac{i\Omega}{\nu}\right)^{1/2} y\right]\right\}$$

$$= U \exp\left[-\left(\frac{\Omega}{2\nu}\right)^{1/2} y\right] \cos\left[\Omega t - \left(\frac{\Omega}{2\nu}\right)^{1/2} y\right] \tag{5.2.5}$$

We thus find that the velocity profile assumes the form of a damped wave with wavelength equal to $2\pi(2\nu/\Omega)^{1/2}$, propagating in the y direction with phase velocity equal to $(2\nu\Omega)^{1/2}$. The amplitude of the velocity decays exponentially with distance y from the plate and becomes exceedingly small outside a layer with thickness $\delta = (2\nu/\Omega)^{1/2}$, called the *Stokes boundary layer.* Velocity profiles at a sequence of characteristic time instants over one period are shown in Figure 5.2.1(a).

Differentiating Eq. (5.2.5) with respect to y, we find that the shear stress at the plate is given by

$$\sigma_{xy}(y = 0, t) = \mu \frac{\partial u}{\partial y}(y = 0, t) = U(\mu\rho\Omega)^{1/2} \cos\left(\Omega t - \frac{3\pi}{4}\right) \tag{5.2.6}$$

It is interesting to note that the phase shift between the shear stress at the plate and the velocity of the plate is equal to $-3\pi/4$ independently of the angular frequency Ω. This, however, is a unique feature of oscillatory flow driven by the motion by a *flat* surface, in the absence of other boundaries. We shall see later in Section 6.15 that, under more general circumstances, the phase shift is a function of Ω.

Flow due to an oscillatory pressure gradient

The flow due to the oscillations of a flat plate is complementary to the unidirectional flow above a stationary plate due to an oscillatory pressure gradient expressed by the function $G(t) = A \sin(\Omega t)$, where A is the amplitude of the fluctuations. The velocity field of the latter is governed by the simplified version of Eq. (5.2.1)

$$\rho \frac{\partial u}{\partial t} = G(t) + \mu \frac{\partial^2 u}{\partial y^2} \tag{5.2.7}$$

whose solution is

$$u(y, t) = -\frac{A}{\rho\Omega} \cos(\Omega t) + v(y, t) \tag{5.2.8}$$

The complementary velocity v is given by the right-hand side of Eq. (5.2.5) with $U = A/(\rho\Omega)$. The velocity vanishes at the wall, at $y = 0$, and describes uniform oscillatory flow away from the wall. This reveals that the flow is composed of an outer region of plug flow and a Stokes layer of thickness $\delta = (2\nu/\Omega)^{1/2}$ adhering to the wall. Velocity profiles at several characteristic time instants over one period are shown in Figure 5.2.1(b).

(*a*)

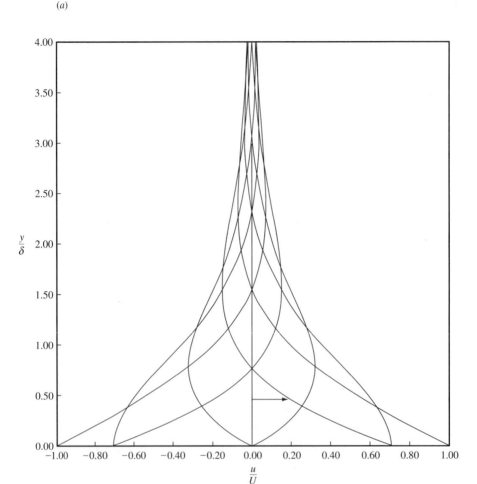

Figure 5.2.1   Velocity profiles for (a) flow due to an oscillating plate.   (*continued*)

Flow due to the sudden translation of a plane wall

Consider next semi-infinite flow above a flat plate that is suddenly set in motion parallel to itself with constant velocity U in an otherwise quiescent fluid. The solution is found by solving Eq. (5.2.2) subject to the initial condition $u(y, t = 0) = 0$ and the boundary conditions $u(y = 0, t > 0) = U$ and $u(y = \infty, t) = 0$.

Because of the linearity of the governing equation and boundary conditions, we expect that the solution will be proportional to U, and thus set $u = UF(y, t, \nu)$, where F is a dimensionless function. The arguments of F must combine in dimensionless groups; the absence of an external length and time scale suggests the combination $\eta = y/(\nu t)^{1/2}$. This means that the velocity as seen by an observer who finds herself at the position $y = (\nu t)^{1/2}$ and is thus traveling with velocity $v = dy/dt = (\nu/4t)^{1/2}$ remains constant. Substituting $u = UF(\eta)$ into Eq. (5.2.2), we obtain

$$\frac{dF}{d\eta}\frac{\partial\eta}{\partial t} = \nu\frac{\partial}{\partial y}\left(\frac{dF}{d\eta}\frac{\partial\eta}{\partial y}\right) \tag{5.2.9}$$

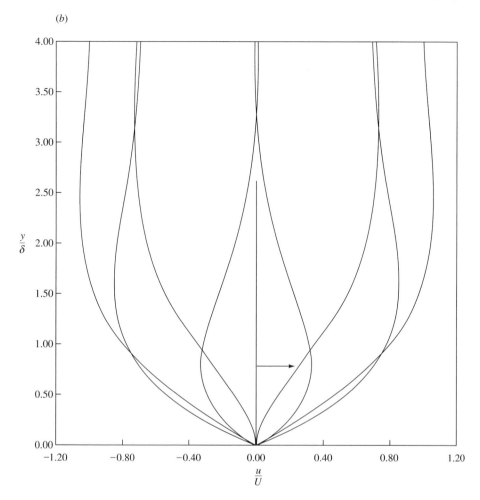

Figure 5.2.1 (*continued*) Velocity profiles for (b) oscillatory flow over a stationary plate at several phases during the oscillation.

Carrying out the differentiations we find the second-order *nonlinear* ordinary differential equation

$$-\tfrac{1}{2}\eta\frac{dF}{d\eta} = \frac{d^2F}{d\eta^2} \tag{5.2.10}$$

which is to be solved subject to the boundary conditions $F(0) = 1$ and $F(\infty) = 0$. Integrating twice, we obtain

$$F(\eta) = \mathrm{erfc}\left(\frac{\eta}{2}\right) = 1 - \mathrm{erf}\left(\frac{\eta}{2}\right) = 1 - \frac{2}{\sqrt{\pi}}\int_0^{\eta/2}\exp(-z^2)\,dz \tag{5.2.11}$$

The complementary error function erfc and error function erf may be computed using polynomial approximations, as discussed in Section B.9, Appendix B. Figure 5.2.2 shows velocity profiles with respect to y/a, where a is an arbitrary length scale, at a sequence of dimensionless times $\nu t/a^2$. At the initial instant, the velocity profile is discontinuous, revealing

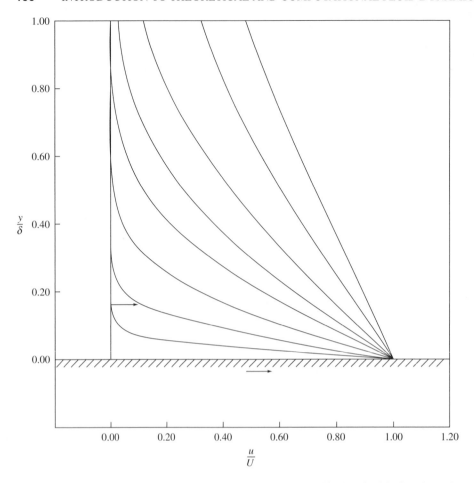

Figure 5.2.2 Velocity profiles with respect to y/a, where a is an arbitrary length scale of the flow, due to the sudden motion of a plate at times $\nu t/a^2 = 0.001, 0.005, 0.02, 0.05, 0.1, 0.2, 0.5, 1.0$.

the presence of a vortex sheet attached to the plate. At later times, the vortex sheet diffuses into the flow transforming into a vortex layer of growing thickness.

The shear stress on the wall is given by

$$\sigma_{xy}(y=0,t) = \mu \frac{\partial u}{\partial y}(y=0,t) = \mu U \frac{df}{d\eta}(\eta=0)\frac{\partial \eta}{\partial y}(y=0) = -\frac{\mu U}{\sqrt{\pi \nu t}} \qquad (5.2.12)$$

We observe a singular behavior at the initial instant, as soon as the plate begins to move. Physically, this implies that one cannot start moving a plate with constant velocity in an impulsive manner, but must gradually increase its velocity from the initial to the final value over a finite time period. Setting the shear stress proportional to $-\mu U/\delta$, where δ is the effective thickness of the vorticity layer, shows that δ increases in time like $t^{1/2}$.

Diffusing vortex sheet

Mere inspection shows that the velocity field $u = U[1 - F(\eta)] = U \operatorname{erf}(\eta/2)$ describes either the flow due to the sudden introduction of an infinite plane wall parallel to a uniform stream flowing along the x axis, or the flow associated with a diffusing vortex sheet that separates two uniform streams translating with velocities U and $-U$ above and below the x axis discussed in Section 3.8

Flow in a Channel with Parallel-Sided Walls

Continuing the study of unsteady unidirectional motion, we concentrate on Couette and Poisseuille flow in a channel with parallel-sided walls separated by distance h, located at $y = 0$ and h. The Couette flow is governed by Eq. (5.2.2) and the Poiseuille flow is governed by Eq. (5.2.7).

Oscillatory Couette flow

First we assume that the upper wall is stationary while the lower wall is oscillating in its plane along the x axis with angular frequency Ω, so that $u(y = h, t) = 0$, and $u(y = 0, t) = U \cos(\Omega t)$, where U is the amplitude of the oscillation. Working as in the case of the single plate, we find that the velocity is given by the real part of

$$u(y, t) = U \exp(-i\Omega t) \frac{\exp[(-i\Omega/\nu)^{1/2} y] - \exp[(-i\Omega/\nu)^{1/2}(2h - y)]}{1 - \exp[2(-i\Omega/\nu)^{1/2} h]} \qquad (5.2.13)$$

where we select the square root with the negative real part, $(-i)^{1/2} = e^{3\pi i/4}$. In the limit as h tends to infinity, we recover the results for the single plate. Comparing the present flow with that due to the motion of a single plate in an infinite fluid, we find that the phase shift between the velocity and shear stress at the lower wall is now a function of Ω, and this demonstrates the subtle hydrodynamic significance of a second boundary.

Transient Couette flow

Next we assume that the upper wall is stationary while the lower wall is suddenly set in motion parallel to itself along the x axis with constant velocity U. Initially, the flow resembles that due to the motion of a flat plate in a semi-infinite body of fluid. At large times, we obtain plane Couette flow with a linear velocity profile. To expedite the solution, it is useful to decompose the flow into the steady plane Couette flow and a transient component that vanishes at long times. Applying the method of separation of variables in y and t, we find that the velocity is given in terms of a Fourier series as

$$u(y, t) = U\left(1 - \frac{y}{h}\right) - \frac{2U}{\pi} \sum_{n=1}^{\infty} \frac{1}{n} \sin\left(\frac{n\pi y}{h}\right) \exp\left(-\frac{n^2 \pi^2 \nu t}{h^2}\right) \qquad (5.2.14)$$

Working in an alternative manner, we apply the Laplace transform and then invert it to obtain the asymptotic series

$$u(y, t) = U\left[\mathrm{erfc}\left(\frac{y}{2\sqrt{\nu t}}\right) - \mathrm{erfc}\left(\frac{2h - y}{2\sqrt{\nu t}}\right) + \mathrm{erfc}\left(\frac{2h + y}{2\sqrt{\nu t}}\right) \right.$$

$$\left. - \mathrm{erfc}\left(\frac{4h - y}{2\sqrt{\nu t}}\right) + \mathrm{erfc}\left(\frac{4h + y}{2\sqrt{\nu t}}\right) - \cdots \right] \qquad (5.2.15)$$

which is more appropriate for computing the velocity at short times.

Oscillatory Poiseuille flow

Consider now the flow due to an oscillatory pressure gradient with $G = A \sin(\Omega t)$, where A is a constant. Straightforward computation shows that the velocity profile is given by the real part of

$$u(y) = -\frac{A}{\rho\Omega} e^{-i\Omega t} \left(1 - \frac{\cosh[(-i\Omega/\nu)^{1/2}(y - h/2)]}{\cosh[(-i\Omega/\nu)^{1/2}h/2]}\right) \qquad (5.2.16)$$

where we select the square root with the negative real part, $(-i)^{1/2} = e^{3\pi i/4}$. It is a common practice to regard the structure of the flow as a function of the *Womersley number* $N_W = h(\Omega/\nu)^{1/2}$. In the limit of low frequencies, we obtain plane Poiseuille flow.

Considering the limit of high frequencies, we replace the hyperbolic cosine in the denominator on the right-hand side of Eq. (5.2.16) with half the exponential of its argument, decompose the hyperbolic cosine in the numerator into its two exponential constituents, and find that the velocity profile is given by the real part of

$$u(y) = -\frac{A}{\rho\Omega} e^{-i\Omega t} \left\{ 1 - \exp\left[\left(-\frac{i\Omega}{\nu}\right)^{1/2} y\right] - \exp\left[\left(-\frac{i\Omega}{\nu}\right)^{1/2}(h - y)\right] \right\} \qquad (5.2.17)$$

Comparing Eq. (5.2.17) with Eq. (5.2.8) shows that the flow is composed of an irrotational core that executes rigid-body motion, and two Stokes boundary layers, one attached to each wall. Furthermore, inspecting the precise form of the velocity profile at high frequencies shows that the amplitude of the velocity may exceed that of the plug flow in the central core by a substantial amount.

Transient Poiseuille flow

As a last application, we consider flow due to the sudden application of a constant pressure gradient. Using the method of separation of variables, as in Eq. (5.2.14), we find

$$u(y, t) = \frac{G}{2\mu} y(h - y) - \frac{4Gh^2}{\mu\pi^3} \sum_{n=1,3,\dots}^{\infty} \frac{1}{n^3} \sin\left(\frac{n\pi y}{h}\right) \exp\left(-\frac{n^2\pi^2\nu t}{h^2}\right) \qquad (5.2.18)$$

which can be shown to yield the steady parabolic profile at long times.

Flow inside a Circular Tube

Next we consider unsteady flow within a circular tube of radius a that is either filled with a viscous fluid or is immersed in an infinite ambient fluid. To facilitate the implementation of the boundary conditions, we express the governing equation (5.2.1) in cylindrical coordinates obtaining

$$\rho\frac{\partial u}{\partial t} = G(t) + \mu\left(\frac{\partial^2 u}{\partial\sigma^2} + \frac{1}{\sigma}\frac{\partial u}{\partial\sigma}\right) \qquad (5.2.19)$$

When the flow is due to the unsteady translation of the cylinder along its length, we obtain the counterpart of Couette flow (Problem 5.2.3). In the remainder of this subsection, we shall concentrate on the case of pressure-driven flow.

Oscillatory Poiseuille flow

Pulsating flow in a tube due to an oscillatory pressure gradient $G = A\sin(\Omega t)$, where A is a constant, has been studied as a model of flow through the large blood vessels (Fung, 1984). Applying the standard method of separation of variables, we find that the solution is given by the real part of the Fourier–Bessel series

$$u(\sigma, t) = -\frac{A}{\rho\Omega} e^{-i\Omega t} \left(1 - \frac{J_0((-i\Omega/\nu)^{1/2}\sigma)}{J_0((-i\Omega/\nu)^{1/2}a)}\right) \qquad (5.2.20)$$

where we select the square root with the negative real part, $(-i)^{1/2} = e^{3\pi i/4}$ (Hildebrand, 1976, p. 226). The Bessel functions having a complex argument may be evaluated by writing

$$J_0(e^{3\pi i/4}x) = \text{ber}_0(x) + i\text{bei}_0(x) \tag{5.2.21}$$

The Kelvin functions of zeroth order ber_0 and bei_0 may be computed using polynomial approximations (Abramowitz and Stegun, 1972, pp. 379, 384; Problem 5.2.8). The angular frequency is often expressed in terms of the *Womersley number* $N_W = \frac{1}{2}a(\Omega/\nu)^{1/2}$. At low values of N_W, we obtain steady Poiseuille flow, whereas at high values we obtain a compound flow that is composed of a plug-flow core and an axisymmetric boundary layer. At high frequencies, the amplitude of the velocity exhibits an overshooting similar to that discussed before for the case of oscillatory plane Poiseuille flow, sometimes called the *annular effect*.

Transient Poiseuille flow

As a further application, we consider the flow generated by the sudden application of a constant pressure gradient, and seek a solution that satisfies the boundary condition $u = 0$ at $\sigma = a$ at all times, and the initial condition $u = 0$ when $t = 0$ for $0 < \sigma < a$. To expedite the computations, we decompose the flow into the steady parabolic Poiseuille flow $u^P = (G/4\mu)(a^2 - \sigma^2)$ that will prevail at long times, and a transient component $v(\sigma, t)$. The transient flow satisfies Eq. (5.2.19) with $G = 0$, observes the boundary condition $v = 0$ at $\sigma = a$ at all times and the initial condition $v = -u^P$ at $t = 0$ for $0 < \sigma < a$, and vanishes at long times. Applying the method of separation of variables, we find

$$u(\sigma, t) = \frac{G}{4\mu}(a^2 - \sigma^2) - \frac{2Ga^2}{\mu}\sum_{n=1}^{\infty}\frac{1}{\alpha_n^3}\frac{J_0(\alpha_n\sigma/a)}{J_1(\alpha_n)}\exp\left(-\frac{\alpha_n^2\nu t}{a^2}\right) \tag{5.2.22}$$

where J_0 and J_1 are the Bessel functions of zeroth and first order, and α_n are the real positive roots of J_0, the first five of which are 2.40482, 5.52007, 8.65372, 11.79153, 14.93091 (Abramowitz and Stegun, 1972, p. 409). The computation of J_0 and J_1 is discussed by Abramowitz and Stegun (1972, p. 369), and computer routines are provided by Press et al. (1986). Velocity profiles at a sequence of dimensionless times $\nu t/a^2$ are shown in Figure 5.2.3. It is instructive to note the presence of a boundary layer near the wall and the occurrence of plug flow at the core at short times. The steady parabolic profile has been virtually established when $t = a^2/\nu$.

Transient Poiseuille flow subject to constant flow rate

Transient flow in a pipe subject to a constant flow rate Q is complementary to starting flow due to the sudden application of a constant pressure gradient. At the initial instant, the velocity profile is flat and the pressure gradient takes an infinite value. At long times, the velocity profile obtains the parabolic shape of Poiseuille flow, and the pressure gradient assumes a corresponding finite value. The solution is found by the method of separation of variables in terms of a Fourier–Bessel series in the form

$$u = \frac{2Q}{\pi a^2}\left[1 - \frac{\sigma^2}{a^2} + 2\sum_{n=1}^{\infty}\frac{1}{\beta_n^2}\left(\frac{J_0(\beta_n\sigma/a)}{J_0(\beta_n)} - 1\right)\exp\left(-\beta_n^2\frac{\nu t}{a^2}\right)\right] \tag{5.2.23}$$

$$G = \frac{8\mu Q}{\pi a^4}\left[1 + \frac{1}{2}\sum_{n=1}^{\infty}\exp\left(-\beta_n^2\frac{\nu t}{a^2}\right)\right] \tag{5.2.24}$$

where β_n, $n = 1, 2, \ldots$, are the real positive roots of the second-order Bessel function J_2.

Solution by Finite-Difference Methods

The analytical solutions derived previously in this section may be approximated with numerical solutions obtained using standard numerical methods. In the remainder of this section,

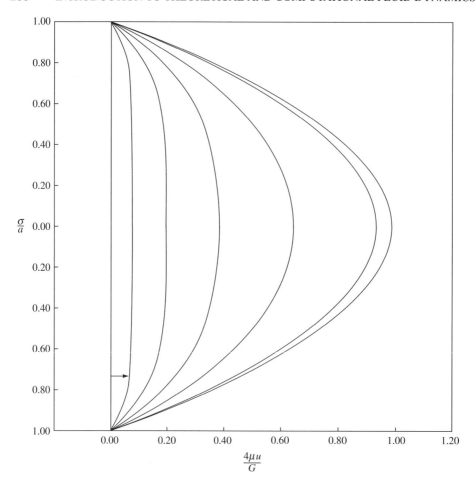

Figure 5.2.3 Velocity profiles of developing flow within a cylindrical tube due to the sudden application of a pressure gradient at times $vt/a^2 = 0.02, 0.05, 0.1, 0.2, 0.5, 1.0, 10.0$.

we shall discuss the application of one particular finite-difference method with reference to transient flow within a circular tube due to a suddenly applied constant pressure gradient. Other finite-difference methods will be discussed in Chapter 12.

We begin with introducing the dimensionless variables $F = u\mu/Ga^2, \xi = tv/a^2$, and $\eta = \sigma/a$, and reduce the governing equation (5.2.19) to the form

$$\frac{\partial F}{\partial \xi} = 1 + \frac{\partial^2 F}{\partial \eta^2} + \frac{1}{\eta}\frac{\partial F}{\partial \eta} \tag{5.2.25}$$

The boundary conditions require that $F = 0$ at $\eta = 1$ at all times, and the initial condition requires that $F = 0$ at $\xi = 0$ for $0 < \eta < 1$. As a first step, we divide the dimensionless radius η into N evenly spaced intervals of uniform size $\Delta\eta$, defined in terms of the grid points $\eta_i = (i-1)/N, i = 1, \ldots, N+1$, and set out to compute the values of the solution at the grid points at a sequence of times with incremental time interval $\Delta\xi$. Applying Eq. (5.2.25) at the ith grid point, where $i = 2, \ldots, N$, and approximating the temporal derivative using backward time differences and the spatial derivatives using central space differences (see Section B.5 of Appendix B, and

Chapter 12), we obtain the *BTCS* finite-difference equation

$$\frac{F_i^{n+1} - F_i^n}{\Delta\xi} = 1 + \frac{F_{i+1}^{n+1} - 2F_i^{n+1} + F_{i-1}^{n+1}}{\Delta\eta^2} + \frac{1}{\eta_i}\frac{F_{i+1}^{n+1} - F_{i-1}^{n+1}}{2\Delta\eta} \tag{5.2.26}$$

where F_i^n designates the value of F at the ith grid point at the time instant $\xi = n\,\Delta\xi$. Rearranging Eq. (5.2.26) yields the linear algebraic equation

$$(\beta_i - \alpha)F_{i-1}^{n+1} + (2\alpha + 1)F_i^{n+1} - (\beta_i + \alpha)F_{i+1}^{n+1} = F_i^n + \Delta\xi \tag{5.2.27}$$

for $i = 2, \ldots, N$, where

$$\alpha \equiv \frac{\Delta\xi}{\Delta\eta^2}, \qquad \beta_i \equiv \frac{\Delta\xi}{2\eta_i\,\Delta\eta} \tag{5.2.28}$$

The no-slip boundary condition at the wall requires that $F_{N+1}^n = 0$ for any value of n. One more equation is required in order to complete the system for the N unknowns F_i^n, $i = 1, \ldots, N$. This equation arises by applying Eq. (5.2.25) at the center of the tube, but we then find that the third term on the right-hand side becomes indeterminate, as β_1 becomes infinite. Using the l'Hôpital rule, however, we find that, at the origin, Eq. (5.2.25) reduces to

$$\frac{\partial F}{\partial\xi} = 1 + 2\frac{\partial^2 F}{\partial\eta^2} \tag{5.2.29}$$

Applying the finite-difference approximation yields the difference equation

$$(4\alpha + 1)F_1^{n+1} - 4\alpha F_2^{n+1} = F_1^n + \Delta\xi \tag{5.2.30}$$

Collecting the unknowns F_i^n, $i = 1, \ldots, N$ into the vector $\mathbf{X}^n$ and appending Eq. (5.2.30) to Eqs. (5.2.27) yields the linear system

$$\mathbf{A} \cdot \mathbf{X}^n = \mathbf{X}^{n-1} + \Delta\xi\,\mathbf{e} \tag{5.2.31}$$

where the vector $\mathbf{e}$ is filled up with ones, $\mathbf{A}$ is a tridiagonal matrix, and $\mathbf{X}^0 = \mathbf{0}$. The computational algorithm proceeds by solving Eq. (5.2.31) for x^n at successive time instants, which can be done using the efficient Thomas algorithm discussed in Section B.1, Appendix B (Problem 5.2.11).

PROBLEMS

5.2.1 **Flow due to the application of a constant shear stress on a planar surface.** Show that the velocity field due to the sudden application of a constant shear stress τ along the planar boundary of a semi-infinite fluid located at $y \geq 0$ is given by

$$u(y, t) = \frac{\tau}{\mu}\sqrt{\nu t}\left[\frac{2}{\sqrt{\pi}}e^{-\eta^2/4} - \eta\,\mathrm{erfc}\left(\frac{\eta}{2}\right)\right] \tag{5.2.32}$$

where $\eta = y/(\nu t)^{1/2}$. Discuss the asymptotic behavior of the flow at long times.

5.2.2 **Flow due to the general motion of a planar surface.** Show that the velocity field due to the translation of a planar boundary of a semi-infinite fluid located at $y \geq 0$ with time-dependent velocity $U(t)$, where $U(t) = 0$ for $t < 0$, is given by (Carslaw and Jaeger, 1959, p. 62)

$$u(y, t) = \frac{y}{2\sqrt{\pi\nu}}\int_0^t \frac{U(\tau)}{(t-\tau)^{3/2}}\exp\left(-\frac{y^2}{4\nu(t-\tau)}\right)d\tau \tag{5.2.33}$$

5.2.3 **Flow due to the axial motion of a circular cylinder.** (a) Derive the velocity profile of the flow inside a circular cylinder that executes axial translational vibrations. (b) Repeat (a) for the flow outside a cylinder that is immersed in an infinite ambient fluid. (c) Consider the flow outside a cylinder that starts moving parallel to its length with constant velocity in an impulsive manner in an otherwise quiescent fluid. Discuss the behavior of the flow and derive an expression for the drag force at short times (Lagestrom, 1964, p. 73; Batchelor, 1954).

5.2.4   **Axial flow within a concentric annular tube.** Derive the velocity field corresponding to axial flow in a channel that is confined between two concentric circular tubes due to sudden application of a constant pressure gradient (Yih, 1979, p. 318).

5.2.5   **Swirling flow outside and inside a circular cylinder.** The azimuthal velocity of an unsteady swirling flow satisfies the linear equation

$$\frac{\partial u_\varphi}{\partial t} = \nu \left(\frac{\partial^2 u_\varphi}{\partial \sigma^2} + \frac{1}{\sigma} \frac{\partial u_\varphi}{\partial \sigma} - \frac{u_\varphi}{\sigma^2} \right) \tag{5.2.34}$$

(a) Consider a solid circular cylinder of radius a immersed in an otherwise quiescent infinite fluid that executes rotational oscillations around its axis with angular frequency Ω, thus generating a swirling flow. Compute the velocity field in terms of Bessel functions (Sherman, 1990, p. 141). (b) A solid circular cylinder of radius a rests immersed in a quiescent infinite fluid. Suddenly, the cylinder starts rotating around its axis with constant angular velocity Ω generating a swirling flow. After a sufficiently long time has elapsed, the flow resembles that due to a rectilinear line vortex with circulation equal to $\kappa = 2\pi\Omega a^2$ located at the center line of the cylinder. Derive an expression for the transient velocity profile in terms of Bessel functions using the method of Laplace transform (Lagestrom, 1964, p. 72; Yih, 1979, p. 316; Sherman, 1990, p. 143). (c) Consider the transient flow inside a hollow circular cylinder of radius a that is filled with a quiescent viscous fluid, as it starts rotating suddenly around its axis with constant angular velocity Ω generating a swirling flow. After a sufficiently long time has elapsed, the fluid executes rigid-body rotation with angular velocity Ω. Show that the transient azimuthal velocity profile is given by

$$u_\varphi(\sigma, t) = \Omega\sigma + 2\Omega a \sum_{n=1}^{\infty} \frac{1}{\alpha_n} \frac{J_1(\alpha_n\sigma/a)}{J_0(\alpha_n)} \exp\left(-\frac{\alpha_n^2 \nu t}{a^2} \right) \tag{5.2.35}$$

where α_n are the positive zeros of the first-order Bessel function J_1 (Batchelor 1967, p. 203). Estimate the time at which rigid-body motion will have virtually been established.

5.2.6   **Swirling flow inside a concentric annular tube.** Consider an annular channel that is confined between two concentric cylinders and contains a viscous fluid. Suddenly, the cylinders start rotating around their axis with different constant angular velocities. Derive an expression for the developing velocity profile in terms of Bessel and related functions (Yih, 1979, p. 316). Discuss the structure of the flow when the cylinders rotate with the same angular velocity.

Computer Problems

5.2.7   **Flow in a two-dimensional channel due to an oscillatory pressure gradient.** Plot the profile of the amplitude of the velocity at a sequence of Womersley numbers, and discuss the occurrence of overshooting at high frequencies.

5.2.8   **Pulsating flow in a circular pipe.** Plot the velocity profiles of pulsating pressure-driven flow in a circular pipe at a sequence of times for several frequencies. Discuss the behavior of your results with reference to the occurrence of boundary layers.

5.2.9   **Transient Couette flow in a parallel-sided channel.** Consider transient Couette flow in a channel with parallel-sided walls discussed in the text. To subtract off the singular behavior at short times, we write $u/U = F(\eta) + Q(y, t)$, where the function F is given in Eq. (5.2.11) and the nonsingular dimensionless function Q satisfies the one-dimensional unsteady diffusion equation (5.2.2) with boundary conditions $Q(0, t) = 0$ and $Q(a, t) = -F[a/(\nu t)^{1/2}]$, and initial condition $Q(y, 0) = 0$. Write a program called *STOKES2P* that advances the function Q in time using a finite-difference method with the *BTCS* discretization discussed in the text. For the computation of the error function, see Section B.9, Appendix B. Plot and discuss the velocity profiles at a sequence of dimensionless times $\xi = t\nu/h^2$.

5.2.10 **Transient swirling flow outside a spinning cylinder.** (a) Consider the transient flow described in Problem 5.2.5(b). To remove the singular behavior at the initial instant, we set

$$u_\varphi = \frac{\Omega a^2}{\sigma} F(\eta, \xi) \tag{5.2.36}$$

where $\eta = (\sigma - a)/(\nu t)^{1/2}$, $\xi = (\nu t)^{1/2}/a$, and F is a dimensionless function (Sherman, 1990, p. 199). Show that F satisfies the convection–diffusion equation

$$F_\xi + \left(\frac{2}{1 + \eta\xi} - \frac{\eta}{\xi} \right) F_\eta = \frac{2}{\xi} F_{\eta\eta} \tag{5.2.37}$$

with boundary conditions $F(0, \xi) = 1$, $F(\infty, \xi) = 0$, and initial condition $F(\eta, 0) = \mathrm{erfc}(\eta/2)$. (b) Write a program called *SWUC* that computes the evolution of F with respect to the dimensionless time-line ξ using a finite-difference method with the *BTCS* discretization discussed in the text. For the computation of the error function, see Section B.9, Appendix B. Plot and discuss the velocity profiles at a sequence of values of ξ.

5.2.11 **Starting flow in a pipe.** Write a program called *PIPETR1* that computes the velocity profiles of starting flow in a circular pipe due to the sudden application of a constant pressure gradient using the finite-difference method discussed in the text. Present profiles of the dimensionless velocity at a sequence of dimensionless times.

5.3 | STAGNATION-POINT FLOWS

There is a class of two-dimensional and axisymmetric flows, the latter in the possible presence of a swirling motion, involving stagnation points, that may be computed by solving systems of ordinary differential equations for certain craftly chosen functions and independent variables. The stagnation points may occur either in the interior of the fluid or at the boundaries, as illustrated in Figure 5.3.1(a,b). The precise location of the stagnation points and the slope of the dividing streamlines or stream surfaces are determined by the global structure of the outer flow, far from the stagnation points, which is specified in the statement of the problem.

The outer flow represents an exact solution to the equation of motion which, however, does not satisfy the required boundary conditions along the dividing streamlines, stream surfaces, or solid boundaries. The problem is reduced to computing a local solution that satisfies the boundary conditions and agrees with the outer solution far from the stagnation points.

Two-Dimensional Oblique Stagnation-Point Flow within a Fluid

Jeffery (1915) derived a class of exact solutions to the equations of two-dimensional incompressible flow. Peregrine (1981) pointed out that one of these solutions represents oblique stagnation-point flow in the interior of a fluid illustrated in Figure 5.3.1(a).

Outer flow

We begin constructing the solution by introducing the outer flow, far from the stagnation point, denoted with the superscript ∞. The stream function, velocity, and vorticity of the outer flow are given by

$$\psi^\infty = k(xy \sin \alpha + \tfrac{1}{2} y^2 \cos \alpha),$$
$$u_x^\infty = k(x \sin \alpha + y \cos \alpha), \qquad u_y^\infty = -ky \sin \alpha, \tag{5.3.1}$$
$$\omega^\infty = -k \cos \alpha$$

(*a*) (*b*)

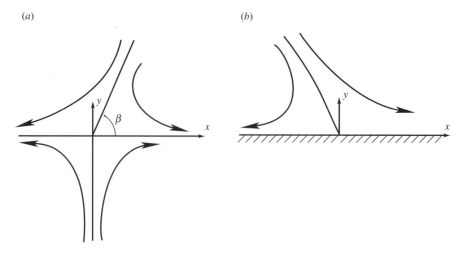

Figure 5.3.1   Schematic illustrations of two-dimensional stagnation-point flow (a) in the interior of a fluid and (b) over a solid wall.

for $y > 0$ and

$$\psi^\infty = kxy \sin \alpha, \qquad u_x^\infty = kx \sin \alpha, \qquad u_y^\infty = -ky \sin \alpha, \qquad \omega^\infty = 0 \qquad (5.3.2)$$

for $y < 0$, where k is a constant; α is a free parameter related to the angle β that is subtended by the x axis and the dividing streamline by $\tan \beta = -2 \tan \alpha$ as shown in Figure 5.3.1(a).

Equations (5.3.1) describe oblique stagnation-point flow with constant vorticity against a planar surface in the upper half-space; $\alpha = 0$ and π correspond, respectively, to unidirectional simple shear flow toward the positive or negative direction of the x axis; $\alpha = \pi/2$ corresponds to irrotational orthogonal stagnation-point flow in the upper half-space with shear rate equal to k. Equations (5.3.2) describe irrotational orthogonal stagnation-point flow with shear rate equal to $k \sin \alpha$ in the lower half-space.

The outer flows in the upper and lower half-spaces constitute exact solutions to the Navier–Stokes equation for a fluid with uniform physical properties, with constant or vanishing vorticity. The velocity is continuous across the dividing streamline at $y = 0$, but the derivatives of the velocity undergo a discontinuity that renders the outer flow admissible only in the context of inviscid fluids. Viscous stresses cause a vortex layer to develop around the dividing streamline along the x axis, so as to render the derivatives of the velocity and therefore the stresses continuous throughout the domain of flow.

Viscous flow

Motivated by the functional form of the stream function of the outer flow given in Eqs. (5.3.1) and (5.3.2), we express the stream function of the viscous flow as

$$\psi = xf(y) + g(y) \qquad (5.3.3)$$

where $f(y)$ and $g(y)$ are two unknown functions that satisfy the far-field boundary conditions $f(\pm\infty) \approx ky \sin \alpha$, $g(+\infty) \approx \frac{1}{2}ky^2 \cos \alpha$, $g(-\infty) = 0$, where $\approx$ denotes the leading-order contributions. Straightforward differentiation yields the components of the velocity and magnitude of the vorticity

$$u_x = xf'(y) + g'(y), \qquad u_y = -f(y), \qquad \omega = -xf''(y) - g''(y) \qquad (5.3.4)$$

Substituting Eqs. (5.3.4) into the steady-state version of the two-dimensional vorticity transport equation (3.8.22),

$$u_x \frac{\partial \omega}{\partial x} + u_y \frac{\partial \omega}{\partial y} = \nu \left(\frac{\partial^2 \omega}{\partial x^2} + \frac{\partial^2 \omega}{\partial y^2} \right) \tag{5.3.5}$$

and factoring out the x variable, λ yields two fourth-order ordinary differential equations for f and g,

$$\nu f'''' + f f''' - f' f'' = 0, \qquad \nu g'''' + f g''' - f'' g' = 0 \tag{5.3.6}$$

The first of Eqs. (5.3.6) involves the function f alone. A solution that satisfies the far-field condition $f(\pm\infty) \approx ky \sin \alpha$ is provided by the far flow itself

$$f(y) = ky \sin \alpha \tag{5.3.7}$$

for which the last of Eqs. (5.3.4) yields $\omega = -g''$. Substituting Eq. (5.3.7) into the second of Eqs. (5.3.6), we obtain an equation for g,

$$g'''' + \frac{k}{\nu} y \sin \alpha \, g''' = 0 \tag{5.3.8}$$

Integrating Eq. (5.3.8) twice subject to the aforementioned far-field conditions and the requirement that the vorticity is continuous across the dividing streamline at $y = 0$, yields the negative of the vorticity $g'' = -\omega$,

$$g'' = \begin{cases} k \cos \alpha + A \operatorname{erfc}\left[y \left(\dfrac{k \sin \alpha}{\nu} \right)^{1/2} \right], & \text{for } y > 0 \\[4mm] (A + k \cos \alpha) \operatorname{erfc}\left[y \left(\dfrac{k \sin \alpha}{\nu} \right)^{1/2} \right], & \text{for } y < 0 \end{cases} \tag{5.3.9}$$

The constant A is found by specifying the rate of decay of vorticity away from the x axis. Further integrations of Eq. (5.3.9) require two additional stipulations regarding the structure of the flow away from the stagnation point.

Two-Dimensional Oblique Stagnation-Point Flow toward a Flat Plate

Consider next an oblique two-dimensional stagnation-point flow against a flat plate illustrated in Figure 5.3.1(b). Far from the plate, the stream function, velocity, and vorticity, assume the far-field distributions given in Eqs. (5.3.1). Since the velocity does not satisfy the no-slip and no-penetration boundary condition $u_x = 0$ and $u_y = 0$ at the plate, it is not an acceptable solution of the equations of viscous flow. A numerical solution that satisfies the full Navier–Stokes equation and boundary conditions was derived by Hiemenz (1911) for the case of orthogonal flow and then by Stuart (1959) for the more general case of oblique flow.

Working as in the case of free stagnation-point flow, we express the stream function as shown in Eq. (5.3.3) and find that the functions f and g satisfy the differential equations (5.3.6). In order to satisfy the no-slip and no-penetration condition on the wall, we require $f'(0) = 0$, $g'(0) = 0$, $f(0) = 0$. Furthermore, we set $\psi(0) = 0$ and obtain $g(0) = 0$. As y tends to infinity, f must behave, to leading order, like $ky \sin \alpha$, and g must behave like $\frac{1}{2} ky^2 \cos \alpha$.

Integrating the first of Eqs. (5.3.6) once, subject to the aforementioned far-field condition, yields

$$\nu f''' + f f'' - f'^2 = -k^2 \sin^2 \alpha \tag{5.3.10}$$

At this point, it is convenient to rewrite Eq. (5.3.10) in terms of the dimensionless variables F and η that are scaled with respect to the forcing function on the right-hand side, defined as

$$F(\eta) = \left(\frac{1}{k\nu \sin \alpha} \right)^{1/2} f(y), \qquad \eta = \left(\frac{k \sin \alpha}{\nu} \right)^{1/2} y \tag{5.3.11}$$

In this manner we obtain the dimensionless equation

$$F''' + FF'' - F'^2 + 1 = 0 \qquad (5.3.12)$$

which is to be solved subject to the boundary conditions $F(0) = 0$ and $F'(0) = 0$, and the far-field condition $F(\infty) \approx \eta$ or $F'(\infty) = 1$.

To solve Eq. (5.3.12), we introduce the auxiliary functions $F' = M$ and $F'' = N$ and rewrite Eq. (5.3.12) in the standard form of a system of three first-order ordinary differential equations as

$$\frac{d}{d\eta} \begin{bmatrix} F \\ M \\ N \end{bmatrix} = \begin{bmatrix} M \\ N \\ M^2 - FN - 1 \end{bmatrix} \qquad (5.3.13)$$

with boundary conditions $F(0) = 0$, $M(0) = 0$, and $M(\infty) = 1$. The solution can be computed using a shooting method that involves the following steps:

1. Guess the value of $N(0)$.
2. Integrate Eqs. (5.3.13) as though we had an initial-value problem from $\eta = 0$ to a, where a is a sufficiently large number; in practice, setting a as low as 3.0 yields satisfactory accuracy.
3. Check to see if $M(a) = 1$ within a preset tolerance; if not, return to step 1 and repeat the computations with a new and improved value for $N(0)$. The new value of $N(0)$ may be found using, for example, Newton's method or the method of false transients applied to the nonlinear function $Q[N(0)] \equiv M(\eta = a) - 1.0 = 0$, as discussed in Section B.3, Appendix B.

Graphs of the functions F, F', and F'' are shown in Figure 5.3.2(a). The numerical solution reveals that $N(0) = 1.2326$. At large values of η, F behaves like

(a)

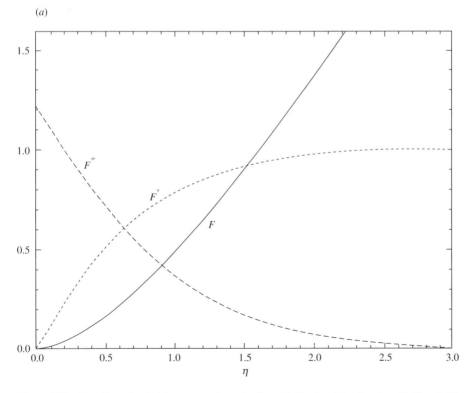

Figure 5.3.2   Two-dimensional oblique stagnation-point flow. (a) Graphs of the functions F, F', and F''. (*continued*)

(b)

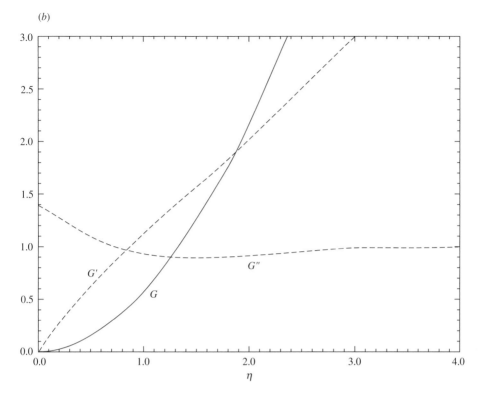

Figure 5.3.2 (*continued*)   Two-dimensional oblique stagnation-point flow. (b) Graphs of the functions G, G', and G''.

$$F(\eta) \cong \eta - b + \cdots \tag{5.3.14}$$

where $b = 0.647900$.

Having computed the function f, we substitute the result back in the second of Eqs. (5.3.6) and obtain a linear homogeneous equation for the function g. One integration yields

$$\nu g''' + f g'' - f' g' = c \tag{5.3.15}$$

where c is a constant. In terms of the dimensionless function F, we obtain

$$\nu g''' + (k\nu \sin\alpha)^{1/2} F(\eta) g'' - k \sin\alpha F'(\eta) g' = c \tag{5.3.16}$$

Letting y or η tend to infinity, and using Eq. (5.3.14) and the far-field condition $g(\infty) = \frac{1}{2} k y^2 \cos\alpha$, we obtain

$$c = -bk \cos\alpha (\nu k \sin\alpha)^{1/2} \tag{5.3.17}$$

In the case of orthogonal stagnation-point flow corresponding to $\alpha = \pi/2$, or simple shear flow corresponding to $\alpha = 0$ or π, we find $c = 0$, which shows that Eq. (5.3.16) has the trivial solution $g = 0$. Considering the more general case where $\alpha \neq 0, \pi/2, \pi$, we express g in terms of the dimensionless function G defined by

$$g(y) = \nu \cot\alpha \, G(\eta) \tag{5.3.18}$$

where η was defined previously in Eqs. (5.3.11). Substituting Eqs. (5.3.18) and (5.3.17) into Eq. (5.3.16), we obtain the dimensionless equation

$$G''' + F G'' - F' G' = -b \tag{5.3.19}$$

which is to be solved subject to the boundary conditions $G(0) = 0$ and $G'(0) = 0$, and the far-field condition that as η tends to infinity G behaves like $\frac{1}{2}\eta^2$. Equation (5.3.12) suggests that $G = bF$ is a particular solution to Eq. (5.3.19), and this allows us to eliminate the forcing function on the right-hand side, thereby obtaining a homogeneous equation for the complementary component. In practice, it is more expedient to compute the solution directly using a numerical method. For this purpose, we introduce the auxiliary functions $G' = K$ and $G'' = L$ and rewrite Eq. (5.3.19) in the standard form of a first-order system of ordinary differential equations as

$$\frac{d}{d\eta}\begin{bmatrix} G \\ K \\ L \end{bmatrix} = \begin{bmatrix} K \\ L \\ -FL + MK - b \end{bmatrix} \tag{5.3.20}$$

which is to be solved subject to the boundary conditions $G(0) = 0$, $K(0) = 0$, and $L(\infty) = 1$. The solution may be found using the shooting method described previously for the function F, where the shooting is now done with respect to $L(0)$; the function F is known at discrete points from the numerical solution of Eq. (5.3.12). Because Eq. (5.3.19) is linear, two shootings followed by linear interpolation are sufficient to complete the solution. The numerical results, plotted in Figure 5.3.2(b), show that $L(0) = 1.406544$ (Dorrepaal, 1986).

In terms of the functions N and L, the magnitude of the vorticity is given by

$$\omega = \omega^\infty + k[1 - L(\eta)]\cos\alpha - kx\sin\alpha\left(\frac{k\sin\alpha}{\nu}\right)^{1/2}N(\eta) \tag{5.3.21}$$

The first term on the right-hand side of Eq. (5.3.21) represents the constant vorticity of the incident flow. The second term represents a vortex layer with uniform thickness attached to the wall. Far from the origin, the third term on the right-hand side dominates the vorticity distribution, and the ratio between the magnitude of the vorticity and the x component of the velocity of the incident flow tends to

$$\left|\frac{\omega}{u_x^\infty}\right| = \left(\frac{k\sin\alpha}{\nu}\right)^{1/2}N(\eta) \tag{5.3.22}$$

which reveals the presence of a vortex layer lining the wall whose thickness is independent of x. In physical terms, diffusion of vorticity from the wall is balanced by convection, and the vorticity gradients are confined within a vortex layer of constant thickness. The thickness of the vortex layer δ may be defined as the point where $N(\eta) = 0.01$. The numerical solution shows that $N(2.4) = 0.01$, which suggests that $\delta = 2.4(\nu/k\sin\alpha)^{1/2}$.

Having computed the velocity field, we substitute the results back into the Navier–Stokes equation and obtain a partial differential equation for the pressure. Straightforward integration yields the modified pressure

$$P = P_0 + c\rho x - \frac{1}{2}\rho k^2 \sin^2\alpha\, x^2 - \mu k\sin\alpha[F'(\eta) + \frac{1}{2}F^2(\eta)] \tag{5.3.23}$$

where the constant c was defined in Eq. (5.3.17), and P_0 is a constant reference pressure.

The stability of the stagnation-point flow has been discussed on a number of occasions, recently by Lasseigne and Jackson (1992). The results show that the behavior of the flow depends upon the structure of the disturbances far from the stagnation-point.

Axisymmetric Orthogonal Stagnation-Point Flow

Consider next an irrotational, axisymmetric, orthogonal stagnation-point flow against a flat plate located at $x = 0$. Far from the plate, as x tends to infinity, the Stokes stream function and velocity components assume the far-field distributions

$$\Psi^\infty = -k\sigma^2 x, \qquad u_x^\infty = -2kx, \qquad u_\sigma^\infty = k\sigma \tag{5.3.24}$$

where k is the rate of extension. Proceeding as in the case of two-dimensional flow discussed previously, we express the Stokes stream function in terms of the dimensionless variable $\eta = x(k/\nu)^{1/2}$, setting

$$\Psi(x, \sigma) = -(k\nu)^{1/2}\sigma^2 F(\eta), \qquad u_x = -2(k\nu)^{1/2}F(\eta)$$
$$u_\sigma = k\sigma F'(\eta), \qquad \omega_\varphi = k\sigma(k/\nu)^{1/2}F''(\eta) \tag{5.3.25}$$

where F is a dimensionless function that is required to satisfy the no-penetration and no-slip boundary conditions $F(0) = 0$, $F'(0) = 0$, and the far-field condition $F'(\infty) = 1$. Substituting these expressions into the vorticity transport equation for axisymmetric flow and integrating once subject to the far-field condition, we obtain the ordinary differential equation

$$F''' + 2FF'' - F'^2 + 1 = 0 \tag{5.3.26}$$

(Homann, 1936). It is worth noting that Eq. (5.3.26) differs from Eq. (5.3.12) only by the value of one coefficient. The solution can be found using a shooting method that is similar to that used to solve Eq. (5.3.13); the results are plotted in Figure 5.3.3. The shootings converge when $F''(0) = 1.3120$.

As in the case of two-dimensional stagnation-point flow, we find that the vorticity is confined within a vortex layer of constant thickness lining the wall. Orthogonal axisymmetric stagnation-point flow provides us with an example of a flow where intensification of the vorticity due to vortex stretching is balanced by diffusion and convection under the influence of the incident flow.

Having computed the velocity field, we substitute the results back into the Navier–Stokes equation and integrate to obtain the modified pressure distribution

$$P = P_0 - \tfrac{1}{2}\rho k^2\sigma^2 - 2\mu k[F'(\eta) + F^2(\eta)] \tag{5.3.27}$$

where P_0 is a constant.

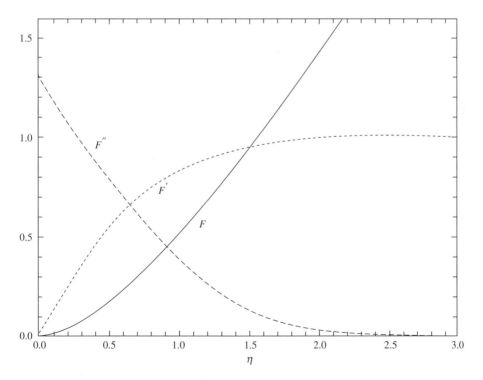

Figure 5.3.3   Graphs of the functions F, F', F'' for axisymmetric orthogonal stagnation-point flow.

Three-Dimensional Orthogonal Stagnation-Point Flow

Generalizing the problem, we consider a three-dimensional orthogonal stagnation-point flow against a flat plate located at $x_3 = 0$. Far from the plate, as x_3 tends to infinity, the velocity components assume the far-field distributions

$$u_1^\infty = k_1 x_1, \qquad u_2^\infty = k_2 x_2, \qquad u_3^\infty = -(k_1 + k_2)x_3 \qquad (5.3.28)$$

where k_1 and k_2 are two independent constant shear rates. When $k_1 = 0$ or $k_2 = 0$, we obtain two-dimensional orthogonal stagnation-point flow in the $x_2 x_3$ or $x_1 x_3$ planes, whereas when $k_1 = k_2$, we obtain axisymmetric orthogonal stagnation-point flow.

Our experience with the two-dimensional and axisymmetric flows discussed previously suggests expressing the velocity in the form

$$u_1 = k_1 x_1 Q'(\eta), \qquad u_2 = k_2 x_2 W'(\eta)$$
$$u_3 = -(k_1 \nu)^{1/2}[Q(\eta) + \alpha W(\eta)] \qquad (5.3.29)$$

where $\eta = x_3(k_1/\nu)^{1/2}$, $\alpha = k_2/k_1$, Q and W are two unknown functions required to satisfy the boundary conditions $Q(0) = 0$, $W(0) = 0$, $Q'(0) = 0$, $W'(0) = 0$, and the far-field conditions $Q'(\infty) = 1$ and $W'(\infty) = 1$; a prime denotes differentiation with respect to η (Howarth, 1951). Note that the functional forms in Eqs. (5.3.29) are consistent with the continuity equation.

Substituting Eqs. (5.3.29) into the x_3 component of the equation of motion shows that $\partial P/\partial x_1$ and $\partial P/\partial x_2$ are independent of the vertical coordinate x_3. This suggests that $\partial P/\partial x_1$ and $\partial P/\partial x_2$ are identical to those of the incident irrotational flow. Using Bernoulli's equation, we find $\partial P/\partial x_1 = -\rho k_1^2 x_1$ and $\partial P/\partial x_2 = -\rho k_2^2 x_2$. Substituting these values along with the functional forms (5.3.29) into the x_1 and x_2 components of the equation of motion we obtain two coupled nonlinear ordinary differential equations for Q and W,

$$Q''' + (Q + \alpha W)Q'' - Q'^2 + 1 = 0$$
$$W''' + (W + \alpha Q)W'' - \alpha W'^2 + \alpha = 0 \qquad (5.3.30)$$

For axisymmetric flow where $\alpha = 1$, both equations reduce to Eq. (5.3.26) with $Q = W = F$. For two-dimensional flow where $\alpha = 0$, the first equation reduces to Eq. (5.3.12) with $Q = F$.

The boundary-value problem expressed by Eqs. (5.3.30) may be solved using a shooting method, where the shootings are done with respect to $Q''(0)$ and $W''(0)$. The numerical solution reveals that for $\alpha = 0^+$, $Q''(0) = 1.233$, $W''(0) = 0.570$; for $\alpha = 0.25$, $Q''(0) = 1.247$, $W''(0) = 0.805$; for $\alpha = 0.5$, $Q''(0) = 1.267$, $W''(0) = 0.998$; for $\alpha = 0.75$, $Q''(0) = 1.288$, $W''(0) = 1.164$; and for $\alpha = 1.0$, $Q''(0) = W''(0) = 1.312$.

Unsteady Flow

The scaling of the velocity shown in Eqs. (5.3.29) is also useful for computing the evolution of an unsteady flow occurring, for instance, during the startup of the stagnation-point flow. Substituting the functional forms

$$u_1 = k_1 x_1 K(x_3, t), \qquad u_2 = k_2 x_2 L(x_3, t), \qquad u_3 = -(k_1 \nu)^{1/2} M(x_3, t)$$
$$P = -\tfrac{1}{2}\rho(k_1^2 x_1^2 + k_2^2 x_2^2) + \mu(k_1 + k_2)N(x_3, t) \qquad (5.3.31)$$

into the continuity equation and equation of motion yields a system of four partial differential equations for the functions K, L, M, and N with respect to x_3 and t, whose solution may be found using a standard numerical method (Sherman, 1990, p. 207). There is a particular protocol of startup for which the flow admits a similarity solution, and the problem is reduced to solving an ordinary differential equation for a carefully crafted similarity variable (Wang, 1989).

PROBLEMS

5.3.1   **Two-dimensional stagnation-point flow against an oscillating plate.** Consider a two-dimensional orthogonal stagnation-point flow against a plate that is oscillating in its plane along the x axis with angular frequency Ω, and describe it in terms of the stream function, given by the real part of

$$(k\nu)^{1/2}x F(\eta) + U\left(\frac{\nu}{k}\right)^{1/2} Q(\eta)\exp(-i\Omega t) \tag{5.3.32}$$

where $\eta = (k/\nu)^{1/2}y$, U is the amplitude of the oscillation, i is the imaginary unit, and F, Q are two unknown complex functions (Rott, 1956; Whitham, 1963, p. 402). The far-field condition requires $F(\infty) = \eta$ or $F'(\infty) = 1$ and $Q'(\infty) = 0$; the no-slip and no-penetration conditions at the plate require that $F(0) = 0$, $F'(0) = 0$, $Q(0) = 0$, and $Q'(0) = 1$. (a) Show that the function F satisfies Eq. (5.3.12) and is therefore identical to that for steady stagnation-point flow, whereas the function Q satisfies the linear equation

$$Q''' + FQ'' - (F' - i\Omega/k)Q' = 0 \tag{5.3.33}$$

(b) Show that the pressure field is not affected by the oscillations. (c) Develop the mathematical formulation for the case where the plate oscillates along the z axis (Whitham, 1963, p. 405). (d) Explain why the present problem is equivalent to that of stagnation-point flow against a stationary flat plate, where the stagnation point oscillates about a mean position.

5.3.2   **Two-dimensional orthogonal stagnation-point flow against a plate that is moving in its plane with a general time-dependent velocity.** Discuss the computation of the flow when the plate executes an arbitrary time-dependent motion in its plane (Watson, 1959).

5.3.3   **Flow due to a stretching sheet issuing from a slit.** Consider the steady two-dimensional flow due to a stretching polymeric sheet issuing along the x axis from a vertical slit into an otherwise quiescent fluid (Crane, 1970). The x component of the velocity at the location of the sheet, at $y = 0$, is $u_x = kx$ where k is the constant rate of extension. Show that the stream function is given by $\psi = (k\nu)^{1/2}x[1 - \exp(-\eta)]$, where $\eta = (k/\nu)^{1/2}y$, compute the x and y components of the velocity, and discuss the structure of the flow.

Computer Problems

5.3.4   **Stagnation-point flows.** (a) Write a program called *SPF1* that uses the shooting method to solve the system of Eqs. (5.3.13) and (5.3.20), and prepare graphs of the functions shown in Figure 5.3.2(a,b). (b) Write a program called *SPF2* that uses the shooting method to solve Eq. (5.3.26) and prepare graphs of the functions shown in Figure 5.3.3. (c) Write a computer program called *SPF3* that uses the shooting method to solve the system of Eqs. (5.3.30) and plot the results for $\alpha = 0.30, 0.60, 0.90$.

5.3.5   **Orthogonal stagnation-point flow against an oscillating plane.** Solve the differential equation (5.3.33) using a shooting method and prepare a graph of the solution.

5.4 | FLOW DUE TO A ROTATING DISK

Consider a horizontal disk of infinite extent immersed in a semi-infinite fluid, rotating in its plane with constant angular velocity Ω, as shown in Figure 5.4.1(a). The rotation of the disk generates a swirling motion that drives a secondary axisymmetric stagnation-point flow against the disk along the axis of rotation. The secondary flow is significant within a boundary layer lining the surface of the disk, but decays far from the disk, as shown in Figure 5.4.1(a). The thickness of the boundary layer is a function of Ω.

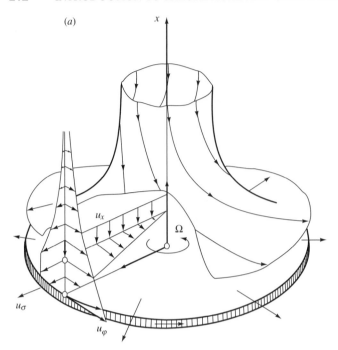

(a)

Figure 5.4.1   (a) Illustration of flow due to a rotating disk.   (*continued*)

Von Kármán (1921) noted that the equation of motion for this flow may be reduced to a system of ordinary differential equations by introducing the transformations

$$u_x = (\nu\Omega)^{1/2}H(\eta), \qquad u_\sigma = \sigma\Omega F(\eta), \qquad u_\varphi = \sigma\Omega G(\eta) \tag{5.4.1}$$

where $\eta = (\Omega/\nu)^{1/2}x$ and H, F, G are three dimensionless functions. The no-slip boundary condition on the disk requires that $H(0) = 0$, $F(0) = 0$, and $G(0) = 1$, and the continuity equation requires that $H'(0) = 0$; the far-field conditions are $F(\infty) = 0$, $G(\infty) = 0$, and $H'(\infty) = 0$. Note that by requiring that $H'(\infty) = 0$ or $H(\infty)$ be a constant, we allow for the occurrence of a uniform axial flow toward the disk.

Substituting Eqs. (5.4.1) into the equation of motion shows that the modified pressure may be consistently assumed to have the functional form

$$P = -\mu\Omega Q(\eta) + P_0 \tag{5.4.2}$$

where Q is a dimensionless function and P_0 is a constant. Note the modified pressure shows no radial dependence due to centripetal acceleration.

Combining the equation of motion with the continuity equation, we obtain a system of four ordinary differential equations for the functions H, F, G, and Q. Combining these equations to eliminate Q and F yields two coupled nonlinear third-order ordinary differential equations for H and G,

$$H''' - H''H + \tfrac{1}{2}H'^2 - 2G^2 = 0 \tag{5.4.3a}$$

$$G'' - G'H + GH' = 0 \tag{5.4.3b}$$

which are to be solved subject to the aforementioned boundary conditions. Once the solution is computed, the functions F and Q arise from the equations

$$F = -\tfrac{1}{2}H', \qquad Q = \tfrac{1}{2}H^2 - H' \qquad\qquad (5.4.4)$$

An approximate solution of Eqs. (5.4.3a,b) was first obtained by Cochran (1934).

To compute a numerical solution, we rewrite Eqs. (5.4.3a,b) as a system of five first-order equations for H, H', H'', G, G' in the usual way, and apply a shooting method in two variables that proceeds according to the following steps:

1. Guess the values of $H''(0)$ and $G'(0)$.

2. Integrate the system from $\eta = 0$ up to a sufficiently large value; in practice $\eta = 10$ yields satisfactory accuracy.

3. Check to see whether the far-field conditions $G(\infty) = 0$ and $H'(\infty) = 0$ are fulfilled. If not, return to step 2 and repeat the procedure with improved guesses for $H''(0)$ and $G'(0)$.

Results computed using this method are plotted in Figure 5.4.1(b). The converged solution shows that $H''(0) = -1.0204$ and $G'(0) = -0.6159$. The thickness of the boundary layer δ over the disk may be defined as the point where the function $F(\eta)$ drops to a small value, for instance, 0.01, and the numerical solution shows that, with this choice, $\delta = 5.4(\nu/\Omega)^{1/2}$. At the edge of the boundary layer, the axial velocity takes the value $-0.89(\nu\Omega)^{1/2}$, and the associated flow serves to feed the boundary layer in order to sustain the radial motion away from the center line.

(b)

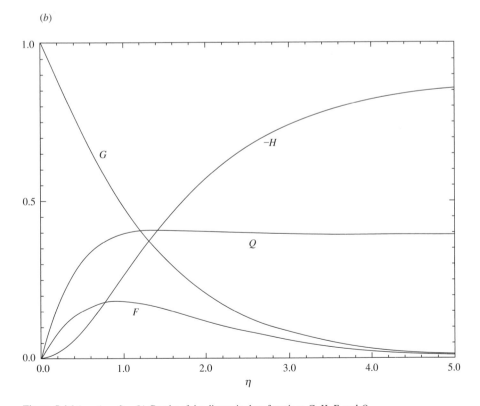

Figure 5.4.1 (*continued*)   (b) Graphs of the dimensionless functions G, H, F, and Q.

Neglecting end effects, we find that the axial component of the torque exerted on a rotating disk of radius a is given by

$$T = \frac{\pi}{2} G'(0) \rho a^4 \sqrt{\nu \Omega^3} \qquad (5.4.5)$$

Observation reveals and stability analysis confirms that the steady flow over the disk is stable as long as the Reynolds number $Re = a(\Omega/\nu)^{1/2}$ is smaller than about 285 (Malik, 1986). Above this value the flow develops nonaxisymmetric waves, yielding spiral vortex patterns.

Unsteady Flow

Von Kármán's scaling of the velocity with respect to radial distance σ shown in Eqs. (5.4.1) is also applicable to unsteady flow due, for instance, to the sudden rotation of a disk with constant or variable angular velocity. Substituting the functional forms

$$
\begin{aligned}
u_x &= (\nu\Omega)^{1/2} H(x, t), & u_\sigma &= \sigma\Omega F(x, t) \\
u_\varphi &= \sigma\Omega G(x, t), & P &= -\mu\Omega Q(x, t) + P_0
\end{aligned}
\qquad (5.4.6)
$$

into the continuity equation and equation of motion yields a system of four partial differential equations for the functions H, F, G, and Q with respect to x and t, whose solution may be found using a standard numerical method (Wang, 1989; Sherman, 1990, p. 204).

PROBLEMS

5.4.1   **Rotary oscillations.** Consider the flow due to a disk that executes rotary oscillations. Derive four differential equations for H, F, G, and Q with respect to x and t, and show that they reduce to Eqs. (5.4.3) and (5.4.4) in the limit of low frequencies.

5.4.2   **Flow due to two rotating disks.** Consider the flow between two parallel disks of infinite extent separated by a distance b, both rotating around their common axis with angular velocities Ω_1 and Ω_2 (Stewartson, 1953). Introduce the functional forms (5.4.1), where Ω is replaced by the angular velocity of the lower disk Ω_1, and show that the functions H, F, G, and Q satisfy Eqs. (5.4.3) and (5.4.4) with boundary conditions $F(0) = 0$, $G(0) = 1$, $H(0) = 0$, $H'(0) = 0$, $F(\eta_1) = 0$, $G(\eta_1) = \Omega_2/\Omega_1$, $H(\eta_1) = 0$, $H'(\eta_1) = 0$, where $\eta_1 = (\Omega_1/\nu)^{1/2} b$.

5.5 │ FLOW IN A CORNER DUE TO A POINT SOURCE

Jeffery (1915) and Hamel (1916) independently considered the two-dimensional flow between two semi-infinite planes intersecting at an angle equal to 2α, driven by a point source or sink of mass located at the apex, as illustrated in Figure 5.5.1. This configuration may be regarded as a model of the local flow in a rapidly converging or diverging channel.

Introducing plane polar coordinates with origin at the apex, assuming that the flow is unidirectional in the radial direction, $u_\theta = 0$, and using the continuity equation, we find that the radial component of the velocity must assume the form

$$u_r = \frac{A}{r} F(\eta) \qquad (5.5.1)$$

where $\eta = \theta/\alpha$, F is a dimensionless function, and A is a dimensional constant that is related to the flow rate Q by means of the equation

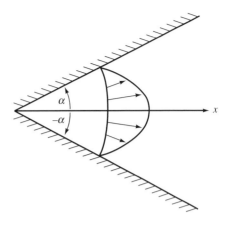

Figure 5.5.1 Illustration of Jeffery–Hamel flow between two intersecting semi-infinite planes driven by a point source of mass located at the apex.

$$Q = A\alpha \int_{-1}^{1} F(\eta)\,d\eta \qquad (5.5.2)$$

We are at freedom to select the relative magnitudes of A and F, and, in order to eliminate this ambiguity, we specify that the maximum value of F is equal to unity, which requires that $dF/d\eta = 0$ when $F = 1$. Furthermore, to satisfy the no-slip boundary condition at the two walls, located at $\theta = \pm\alpha$, we require $F(\pm 1) = 0$.

Since the velocity field is radial, regions of recirculating flow do not occur. It is possible, however, that the flow may reverse its direction within a certain portion of the channel. In that case, the fluid will move against the direction of the primary flow, which is driven by the point source or sink.

Substituting Eq. (5.5.1) into the steady form of the two-dimensional vorticity transport equation, we obtain a third-order nonlinear differential equation for F,

$$F''' + 4\alpha^2 F' + 2\alpha\, Re\, FF' = 0 \qquad (5.5.3)$$

where we have introduced the Reynolds number

$$Re = \frac{\alpha A}{\nu} \qquad (5.5.4)$$

To maintain the notation simple we have allowed Re to be negative for inward flow.

The nonlinear term in Eq. (5.5.3) is due to the fluid acceleration. Neglecting this term yields a linear equation whose solution may be computed readily in closed form. Solving for F and computing A in terms of Q from Eq. (5.5.2) we obtain

$$u_r = \frac{Q}{r} \frac{\cos 2\theta - 2\theta \cos 2\alpha}{\sin 2\alpha - 2\alpha \cos 2\alpha} \qquad (5.5.5)$$

Considering the flow at finite Reynolds numbers, we integrate Eq. (5.5.3) once and obtain the Jeffery–Hamel equation

$$F'' + 4\alpha^2 F + \alpha\, Re\, F^2 = a \qquad (5.5.6)$$

where a is a constant. Substituting Eq. (5.5.1) into the radial component of the two-dimensional Navier–Stokes equation, using Eq. (5.5.6), and integrating once with respect to r, we obtain the distribution of modified pressure

$$P = \frac{2\mu A}{r^2}\left(F(\eta) - \frac{a}{4}\right) \qquad (5.5.7)$$

Proceeding with the computation of F, we multiply Eq. (5.5.6) with F' and rearrange to obtain

$$\tfrac{1}{2}(F'^2)' + 2\alpha^2(F^2)' + \tfrac{1}{3}\alpha\, Re\,(F^3)' = aF' \tag{5.5.8}$$

A further integration subject to the condition $F'(\eta) = 0$ when $F = 1$ yields the first-order equation

$$F' = \pm(1 - F)^{1/2}[\tfrac{2}{3}\alpha\, Re\, F(1 + F) + 4\alpha^2 F + c]^{1/2} \tag{5.5.9}$$

where c is a positive constant. Applying Eq. (5.5.9) at $\eta = \pm1$ reveals that the constant c is related to the magnitude of the shear stress at the walls,

$$|F'(\pm1)| = c \tag{5.5.10}$$

When $A\,F'(1) > 0$ or $A\,F'(-1) < 0$, we obtain a region of reversed flow adjacent to the walls.

Given the values of α and Re, the problem is reduced to solving Eq. (5.5.9) and computing the value of the constant c subject to the boundary condition $F(\pm1) = 0$. Once this is done, the results are substituted into Eq. (5.5.2) to yield the corresponding value of Q. Rosenhead (1940) derived an exact solution in terms of elliptic functions. The results show that the flow exhibits a variety of features, especially at high Reynolds numbers, discussed in detail by a number of authors, including Fraenkel (1962a,b), Whitham (1963), Batchelor (1967), and Hooper, Duffy, and Moffatt (1982).

To compute a solution numerically, we use a shooting method that involves the following steps (see also Hooper et al., 1982):

1. Guess a value for c.

2. Integrate Eq. (5.5.9) with the plus or minus sign and with initial condition $F(-1) = 0$ from $\eta = -1$ to 1.

3. Examine whether $F(1) = 0$ is fulfilled, and if not, repeat the computation with an improved estimate for c.

The stability of the Jeffery–Hamel flow has been addressed on several occasions (Hamadiche et al., 1994). The results show that the critical Reynolds number above which the flow becomes unstable decreases rapidly as the angle α is raised, and the wedge becomes wider.

 ## Computer Problem

5.5.1   **Computing the Jeffery–Hamel flow by a shooting method.** Solve Eq. (5.5.9) for $\alpha = \pi/8$ at a sequence of Reynolds numbers $Re = 1, 10, 20, 50$, and compute the corresponding dimensionless flow rate Q/ν.

5.6 | FLOW DUE TO A POINT FORCE

Consider the flow due to a narrow jet discharging into a large tank that is filled up with the same fluid. As the diameter of the jet becomes increasingly smaller, while the flow rate is kept constant, we obtain a family of flows, the limiting one being the flow due to a point source of momentum or point force located at the point of discharge.

The steady-state velocity and pressure fields due to a point force applied at the stationary point $\mathbf{x}_0$ satisfy the continuity equation and the steady-state version of the Navier–Stokes equation enhanced with a singular forcing term on the right-hand side,

$$\rho\,\mathbf{u}\cdot\nabla\mathbf{u} = -\nabla P + \mu\,\nabla^2\mathbf{u} + \mathbf{b}\,\delta(\mathbf{x} - \mathbf{x}_0) \tag{5.6.1}$$

where the constant vector **b** indicates the magnitude and direction of the point force, and P is the modified pressure. Physically, the solution of Eq. (5.6.1) also describes the flow due to the steady gravitational settling of a particle with infinitesimal dimensions but finite weight in an ambient fluid of infinite expanse, observed in a frame of reference in which the particle appears to be stationary.

Assuming that the flow is axisymmetric, we introduce spherical polar coordinates with the x axis pointing in the direction of the strength of the point force **b**, and describe the flow in terms of the Stokes stream function Ψ. The absence of an intrinsic length scale suggests the functional form

$$\Psi(r, \theta) = \nu r F(\cos \theta) \tag{5.6.2}$$

where F is a dimensionless function, and we have used the argument $\cos \theta$ instead of θ for future convenience. The spherical polar components of the velocity and magnitude of the vorticity are given by

$$u_r = \frac{1}{r^2 \sin \theta} \frac{\partial \Psi}{\partial \theta} = -\frac{\nu}{r} F'$$

$$u_\theta = -\frac{1}{r \sin \theta} \frac{\partial \Psi}{\partial r} = -\frac{\nu}{r \sin \theta} F \tag{5.6.3}$$

$$u_\varphi = 0, \qquad \omega_\varphi = -\frac{\nu}{r} \sin \theta \, F''$$

where a prime designates differentiation with respect to $\cos \theta$. It is clear from these functional forms that the shapes of the streamlines are self-similar, which means that one can be derived from another by a proper adjustment of the radial scale.

One way to proceed is to substitute Eq. (5.6.3) into the vorticity transport equation and derive a fourth-order differential equation for F. It is more expedient, however, to work directly with the generalized equation of motion (5.6.1), thus obtaining a simultaneous solution for the pressure. Motivated again by the absence of an intrinsic length scale, we express the modified pressure in the functional form

$$P(r, \theta) = c + \mu \frac{\nu}{r^2} Q(\cos \theta) \tag{5.6.4}$$

where Q is a dimensionless function and c is a constant. Substituting Eqs. (5.6.3) and (5.6.4) into the radial and angular polar components of Eq. (5.6.1), we obtain two coupled, third-order, nonlinear ordinary differential equations for the functions F and Q,

$$(FF' - \sin^2 \theta \, F'')' + \frac{F^2}{\sin^2 \theta} + 2Q = 0$$

$$F'' + \tfrac{1}{2} \left(\frac{F^2}{\sin^2 \theta} \right)' + Q' = 0 \tag{5.6.5}$$

Integrating the second equation once and combining the result with the first equation to eliminate Q yields an equation for F alone,

$$[(1 - \eta^2)F'' - FF']' + 2F' = a \tag{5.6.6}$$

where $\eta = \cos \theta$ and a is an integration constant. Carrying out one more integration yields the second-order differential equation

$$(1 - \eta^2)F'' - FF' + 2F - a\eta = c \tag{5.6.7}$$

where c is a new integration constant. To evaluate the constants a and c, we note that both the positive and negative parts of the x axis are streamlines, and set $F(\pm 1) = 0$. Furthermore, we require that the radial component of the velocity and vorticity obtain finite values along the x axis, and use Eq. (5.6.5) to find that, as η tends to ± 1, both $F'(\pm 1)$ and the limit of $(1 - \eta^2)^{1/2} F''(\eta)$ must be bounded. Equation (5.6.7) then yields $a = 0$ and $c = 0$. Carrying out one analytical integration of Eq. (5.6.7) yields

$$(1 - \eta^2)F' + 2\eta F - \tfrac{1}{2}F^2 = \alpha \tag{5.6.8}$$

where α is a new integration constant. Applying Eq. (5.6.8) at the x axis, corresponding to $\eta = \pm 1$ and requiring nonsingular behavior, yields $\alpha = 0$. Integrating Eq. (5.6.8), we finally obtain the general solution in closed form

$$F = 2\frac{\sin^2 \theta}{d - \cos \theta} \tag{5.6.9}$$

where d is a new integration constant whose value must be adjusted so that the flow field satisfies Eq. (5.6.1) at the singular point $\mathbf{x}_0$ (Landau, 1944; Squire, 1951). Substituting Eq. (5.6.9) into the first of Eqs. (5.6.5) yields the function Q,

$$Q = 4\frac{d \cos \theta - 1}{(d - \cos \theta)^2} \tag{5.6.10}$$

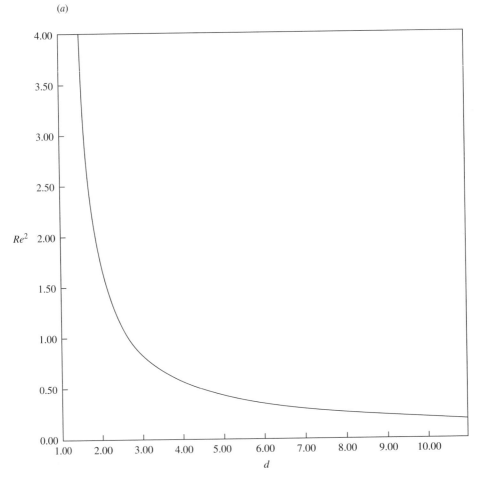

(a)

Figure 5.6.1   (a) The relationship between the effective Reynolds number Re defined in Eq. (5.6.12) and the constant d.   (*continued*)

To evaluate the constant d we apply the momentum integral balance (3.1.17) over a spherical volume of fluid with radius R centered at the point force. Noting that the flow is steady and using the properties of the delta function, we find

$$0 = \int_{\text{Sphere}} (\boldsymbol{\sigma} - \rho\mathbf{uu}) \cdot \mathbf{n}\, dS + \mathbf{b} \tag{5.6.11}$$

where $\boldsymbol{\sigma}$ is the modified stress tensor and $\mathbf{n}$ is the normal vector pointing outward from the spherical surface. Using Eqs. (5.6.9) and (5.6.10) to evaluate the stress tensor and substituting the results into the x component of Eq. (5.6.11), we obtain an implicit equation for d,

$$Re^2 \equiv \frac{|\mathbf{b}|}{8\pi\mu\nu} = \frac{8}{3}\frac{d}{d^2 - 1} + d^2 \ln\frac{d-1}{d+1} + 2d \tag{5.6.12}$$

where Re is an effective Reynolds number of the flow. It is now evident that the structure of the flow will depend upon the value of Re, which determines the intensity of the point force, a fact that could have been predicted a priori on the basis of dimensional analysis. In Figure 5.6.1(a) we present a plot of Re^2 versus the constant d in its range of definition $d > 1$, and in Figure 5.6.1(b) we show the streamline pattern for $d = 1.1$.

(b)

Figure 5.6.1 (*continued*)   (b) Streamline pattern of the flow due to a point source of momentum or point force for $d = 1.1$.

Low-Reynolds-Number Flow

Considering the limit as d becomes large, we expand the right-hand side of Eq. (5.6.12) in a Taylor series with respect to $1/d$ and find $Re^2 = 2/d$, which verifies that Re is small, as shown in Figure 5.6.1(a). Making appropriate approximations, we find that the corresponding stream function and modified pressure are given by

$$\Psi(r, \theta) = \frac{|\mathbf{b}|}{8\pi\mu} r \sin^2 \theta, \qquad P(r, \theta) = \frac{|\mathbf{b}|}{4\pi} \frac{1}{r^2} \cos \theta \qquad (5.6.13)$$

Evaluating the various terms of the Navier–Stokes equation shows that the nonlinear inertial terms are smaller than the pressure and viscous terms by a factor of $1/d$, and the solution (5.6.13) satisfies the equations of Stokes flow. The corresponding streamline pattern, illustrated in Figure 6.4.1(a), is symmetric about the midplane $x = 0$.

High-Reynolds-Number Flow

Considering the opposite limit as d approaches unity, we expand the right-hand side of Eq. (5.6.12) in a Taylor series with respect to $d - 1$, and find $Re^2 = 4/[3(d - 1)]$, which shows that Re becomes large. Making appropriate approximations, we find that as long as $\cos\theta$ is not too close to unity, that is, sufficiently far from the positive part of the x axis, the corresponding stream function and modified pressure are given by

$$\Psi(r, \theta) = 2\nu r(1 + \cos\theta), \qquad P(r, \theta) = -\mu \frac{\nu}{r^2} \frac{1}{1 - \cos\theta} \qquad (5.6.14)$$

independently of the magnitude of Re.

Further Flows Due to a Point Force

The flow due to a point force in an infinite domain has been generalized in several respects to account for the effect of a simultaneous swirling motion and for the presence of a flat or conical boundary. These extensions have been reviewed by Whitham (1963), Goldshtik (1990), and Shtern and Hussain (1993).

PROBLEM

5.6.1 **Edge of the jet.** The edge of the jet due to a point force can be defined as the surface where the associated streamlines are at minimum distance from the x axis. Show that this is a conical surface with $\theta = \theta_0$, where $\cos\theta_0 = 1/d$.

References

Abramowitz, M., and Stegun, I. A., 1972, *Handbook of Mathematical Functions.* Dover.

Batchelor, G. K., 1954, The skin friction on infinite cylinders moving parallel to their length. *Quart. J. Mech. Appl. Math.* **7**, 179–92.

Batchelor, G. K., 1967, *An Introduction to Fluid Dynamics.* Cambridge University Press.

Berker, R., 1963, Intégration des équations du mouvement d'un fluide visqueux incompressible, in *Handbuch der Physik*, **8**(2), 1–384. Springer, Berlin.

Carslaw, H. S., and Jaeger, J. C., 1959, *Conduction of Heat in Solids.* Oxford University Press.

Cochran, W. G., 1934, The flow due to a rotating disk. *Proc. Camb. Phil. Soc.* **30**, 365–75.

Couette, M., 1890, Etudes sur le frottement des liquides. *Ann. Chim. Phys.* **21**, 433–510.

Crane, L. J., 1970, Flow past a stretching plate. *ZAMP* **21**, 645–47.

Dorrepaal, J. M., 1986, An exact solution of the Navier–Stokes equation which describes nonorthogonal stagnation-point flow in two dimensions. *J. Fluid Mech.* **163**, 141–47.

Fraenkel, L. E., 1962a, Laminar flow in symmetrical channels with slightly curved walls. I. On the Jeffery–Hamel solutions for flow between plane walls. *Proc. Roy. Soc. London* A **267**, 119–38.

Fraenkel, L. E., 1962b, Laminar flow in symmetrical channels with slightly curved walls. II. An asymptotic series for the stream function. *Proc. Roy. Soc. London* A **272**, 406–28.

Fulford, G. D., 1964, The flow of liquids in thin films. *Adv. Chem. Eng.* **5,** 151–236.

Fung, Y. C., 1984, *Biodynamics: Circulation.* Springer–Verlag.

Goldshtik, M. A., 1990, Viscous-flow paradoxes. *Annu. Rev. Fluid Mech.* **22,** 441–72.

Hamadiche, M., Scott, J., and Jeandel, D., 1994, Temporal stability of Jeffery–Hamel flow. *J. Fluid Mech.* **268,** 71–88.

Hamel, G., 1916, Spiralförmige Bewegungen zäher Flüssigkeiten. *Jahresb. Deutschen Math. Vereinigung* **25,** 34–60.

Hiemenz, K., 1911, Die Grenzschicht an einem in den gleichförmigen Flüssigkeitsstrom eingetauchten geraden Kreiszylinder. *Dinglers Polyt. J.* **326**(21), 321–410.

Hildebrand, F. B., 1976, *Advanced Calculus for Applications.* Prentice Hall.

Homann, F., 1936, Der Einfluß großer Zähigkeit bei der Strömung um den Zylinder und um die Kugel. *Z. angew. Math. Mech.* **16**(3), 153–64.

Hooper, A., Duffy, B. R., and Moffatt, H. K., 1982, Flow of fluid of non-uniform viscosity in converging and diverging channels. *J. Fluid Mech.* **117,** 283–304.

Howarth, L., 1951, The boundary layer in three-dimensional flow. Part II: The flow near a stagnation point. *Phil. Mag.* **42**(7), 1433–40.

Jeffery, G. B., 1915, The two-dimensional steady motion of a viscous fluid. *Phil. Mag.* **29,** 455–65.

Kármán, T., von, 1921, Z. Über laminare und turbulente Reibung. *Z. angew. Math. Mech.* **1**(3), 153–64.

Lagerstrom, P. A., 1964, Laminar Flow Theory. In *Theory of Laminar Flows,* Edited by F. K. Moore, Princeton University Press.

Landau, L., 1944, An exact solution of Navier–Stokes equations. *Doklady Acad. Nauk SSSR* **43**(7), 299–301.

Lasseigne, D. G., and Jackson, T. L., 1992, Stability of a nonorthogonal stagnation flow to three-dimensional disturbances. *Theoret. Comput. Fluid Dynamics* **3,** 207–18.

Malik, M. R., 1986, The neutral curve for stationary disturbances in rotating-disk flow. *J. Fluid Mech.* **164,** 275–87.

Peregrine, D. H., 1981, The fascination of fluid mechanics. *J. Fluid Mech.* **106,** 59–80.

Press, W. J., Flannery, B. P., Teukolsky, S. A., and Vetterling, W. T., 1986, *Numerical Recipes, The Art of Scientific Computing.* Cambridge.

Rosenhead, L., 1940, The steady two-dimensional radial flow of viscous fluid between two inclined plane walls. *Proc. Roy. Soc. London* A **175,** 436–67.

Rott, N., 1956, Unsteady viscous flow in the vicinity of a stagnation point. *Quart. Appl. Math.* **13,** 444–51.

Rott, N., 1964, *Time-Dependent Solutions of the Navier–Stokes Equations.* In *Theory of Laminar Flows,* Edited by F. K. Moore, Princeton University Press.

Shah, R. K., and London, A. L., 1978, *Laminar Flow Forced Convection in Ducts.* Academic.

Sherman, F. S., 1990, *Viscous Flow.* McGraw Hill.

Shtern, V., and Hussain, F., 1993, Azimuthal instability of divergent flows. *J. Fluid Mech.* **256,** 535–60.

Sparrow, E. M., 1962, Laminar flow in isosceles triangular ducts. *AIChE J.* **8,** 599–604.

Squire, H. B., 1951, The round laminar jet. *Quart. J. Mech. Appl. Math.* **4,** 321–29.

Stewartson, K., 1953, On the flow between two rotating coaxial disks. *Proc. Camb. Phil. Soc.* **49,** 333–41.

Stuart, J. T., 1959, The viscous flow near a stagnation point when the external flow has uniform vorticity. *J. Aero/Space Sc.* **26,** 310–11.

Sutera, S. P., and Skalak, R., 1993, The history of Poiseuille's law. *Annu. Rev. Fluid. Mech.* **25,** 1–19.

Taylor, G. I., 1960, Deposition of a viscous fluid on a plane surface. *J. Fluid Mech.* **9,** 218–24.

Watson, J., 1959, The two-dimensional laminar flow near the stagnation point of a cylinder which has an arbitrary transverse motion. *Quart. J. Mech.* **12,** 175–90.

Whitham, G. B., 1963, The Navier–Stokes equations of motion. *Laminar Boundary Layers,* Edited by L. Rosenhead, Oxford University Press.

Wang, C.-Y., 1989, Exact solutions of the unsteady Navier–Stokes equations. *Appl. Mech. Rev.* **42,** S269–82.

Wang, C.-Y., 1991, Exact solutions of the steady-state Navier–Stokes equations. *Annu. Rev. Fluid Mech.* **23,** 159–77.

Yih, C.-S., 1979, *Fluid Mechanics.* West River Press.

Flow at Low Reynolds Numbers

Flow at low Reynolds numbers is distinguished by the fact that the nonlinear convective term $\rho\mathbf{u} \cdot \nabla\mathbf{u}$ makes a small contribution to the equation of motion, and may thus be neglected. Consequently, the flow is governed by the continuity equation and one of the linear equations of *steady, quasisteady, or unsteady* Stokes flow discussed earlier in Section 3.7. The linear nature of the governing equations allows us to build an extensive theoretical framework regarding the mathematical properties of the solution and physical structure of the flow and, furthermore, derive solutions to a broad range of problems using a variety of analytical and numerical methods, many of them in closed form. Examples include methods based on separation of variables, singularity representations, and boundary integral formulations.

In the present chapter we shall discuss the general properties and illustrate methods of computing steady and unsteady flows at low Reynolds numbers, with reference to specific applications. Classes of flows to be considered include a family of flows near boundary corners, flows of liquid films, lubrication-type flows in confined geometries, flows past or due to the motion of rigid bodies and liquid drops, and oscillatory or transient flows due to particle motions.

The study of uniform flow past a stationary body or flow due to the translation of a body will lead us to examine the structure of the flow far from the body where the assumption of Stokes flow is no longer appropriate. Fortunately, in the far flow, the neglected inertial term $\rho\mathbf{u} \cdot \nabla\mathbf{u}$ may be approximated with the linear form $\rho\mathbf{U} \cdot \nabla\mathbf{u}$, where $\mathbf{U}$ is the uniform velocity of the incident flow, yielding Oseen's equation and corresponding Oseen flow. Computing solutions to Oseen's equation is complicated by the fact that $\mathbf{U}$ and $\mathbf{u}$ are quadratically coupled. Nevertheless, the preserved linearity with respect to $\mathbf{u}$ allows us to carry out some analytical studies and develop efficient methods of numerical computation.

In the final sections of this chapter, we shall discuss unsteady Stokes flow governed by the unsteady Stokes equation (3.7.5). The new feature of time-dependent motion renders the analysis somewhat more involved, but does not prevent us from building a firm theoretical foundation and from deriving solutions using an efficient class of analytical and numerical methods.

6.1 | EQUATIONS AND FUNDAMENTAL PROPERTIES OF STOKES FLOW

The scaling analysis of Section 3.7 revealed that the flow of an incompressible Newtonian fluid with uniform physical properties, at small values of the unsteadiness parameter β and Reynolds number Re, is governed by the continuity equation $\nabla \cdot \mathbf{u} = 0$ and the Stokes equation, written below in the alternative forms

$$\nabla \cdot \boldsymbol{\sigma} + \rho\,\mathbf{g} = \mathbf{0} \qquad \text{or} \qquad -\nabla p + \mu\,\nabla^2\mathbf{u} + \rho\,\mathbf{g} = \mathbf{0} \qquad (6.1.1\text{a,b})$$

In terms of the modified pressure and modified stress tensor, we obtain the equivalent forms

$$\nabla \cdot \boldsymbol{\sigma}^{\text{Mod}} = \mathbf{0} \qquad \text{or} \qquad -\nabla P + \mu\,\nabla^2\mathbf{u} = \mathbf{0} \qquad (6.1.1\text{c,d})$$

A flow that is governed by one of the four equivalent equations (6.1.1a–d) is called a *Stokes* or *creeping flow.*

Combining Eq. (6.1.1a) with (3.1.10), we find that the hydrodynamic force exerted on a fluid parcel of arbitrary size is balanced by that due to the body force, and the total force exerted on the parcel is equal to zero. Thus the assumption of creeping flow guarantees that the rate of change of linear momentum of a fluid parcel is negligibly small. Combining Eq. (6.1.1c) with Eq. (3.1.18) shows that, in the absence of an external torque field, the total torque exerted on a fluid parcel is also equal to zero, and this shows that the rate of change of the parcel's angular momentum is also negligibly small.

Taking the divergence of Eq. (6.1.1b) and using the continuity equation, we find that the pressure is a harmonic function,

$$\nabla^2 p = 0 \qquad (6.1.2)$$

The general properties of harmonic functions discussed in Section 2.1 require that, in the absence of singular points, the pressure attain extreme values at the boundaries of the flow, or else it grows without a bound at infinity. Similar results pertain to the modified pressure.

Consider now a flow in a domain of infinite expanse, possibly in the presence of interior boundaries, under the stipulation that the velocity vanishes and the pressure tends to become constant at infinity. Using the results of Section 2.3, we find that the modified pressure must decay like integral powers of $1/r$, where r is the distance from the origin. But substituting $P = 1/r$ into the Stokes equation, we obtain an equation for the velocity that does not admit a solenoidal solution, and this requires that the pressure decay at least as fast as $1/r^2$. Balancing the orders of decay of the velocity and modified pressure then shows that the velocity must decay at least as fast as $1/r$. Repeating these arguments for two-dimensional flow, we find that the velocity of a two-dimensional Stokes flow must increase at a rate that is less than or equal to $\ln r$ or else decay.

Taking the Laplacian of Eq. (6.1.1b) and using Eq. (6.1.2), we find that the components of the velocity satisfy the biharmonic equation

$$\nabla^4 \mathbf{u} = \mathbf{0} \qquad (6.1.3)$$

and, as a consequence, they are subject to mean-value theorems expressed by the identities

$$\frac{1}{4\pi a^2} \int_{\text{Sphere}} \mathbf{u}\, dS = \mathbf{u}(\mathbf{x}_0) + \tfrac{1}{6} a^2\, \nabla^2 \mathbf{u}(\mathbf{x}_0) \qquad (6.1.4)$$

$$\frac{1}{4\pi a^4} \int_{\text{Sphere}} (\mathbf{x} - \mathbf{x}_0) \times \mathbf{u}\, dS = \tfrac{1}{3} \nabla \times \mathbf{u}(\mathbf{x}_0) \qquad (6.1.5)$$

(Sections 2.3 and 6.9; Problem 6.1.2). The first equation relates the mean value of the velocity over the surface of a sphere of radius a centered at the point $\mathbf{x}_0$ to the values of the velocity and its Laplacian at the center of the sphere, and thus it is the counterpart of the mean-value theorem for the harmonic functions discussed in Section 2.3. The second equation relates the mean value of the angular momentum over the surface of the sphere to the angular velocity of the fluid at the center of the sphere.

Equation (6.1.3) shows that any solenoidal velocity field $\mathbf{u} = \nabla\phi$ satisfies the Stokes equation with a corresponding constant modified pressure. In general, however, irrotational flows cannot be made to satisfy more than one scalar boundary condition and are therefore unable, by themselves, to represent externally or internally bounded flows.

Taking the curl of the Stokes equation and using the fact that the curl of the gradient of any twice-differentiable vector field vanishes, we find that the components of the vorticity are harmonic functions

$$\nabla^2 \boldsymbol{\omega} = \mathbf{0} \qquad (6.1.6)$$

which requires that they attain extreme values at the boundaries of the flow. Another important consequence of Eq. (6.1.6) is that there are no localized concentrations of vorticity in Stokes flow. The onset of regions of recirculating fluid does not imply the presence of compact vortices, as it typically does in the case of high-Reynolds-number flow. One may readily verify that any linear flow with constant vorticity satisfies the equations of Stokes flow with a corresponding constant modified pressure, including the simple shear flow $\mathbf{u} = (kx, 0, 0)$.

Considering next a two-dimensional flow, we use the fact that the vorticity is a harmonic function and find that the stream function ψ satisfies the biharmonic equation

$$\nabla^4 \psi = 0 \tag{6.1.7}$$

Working in a similar manner, we find that the Stokes stream function Ψ of an axisymmetric flow satisfies the linear fourth-order partial differential equation

$$E^2(E^2\Psi) = E^4\Psi = 0 \tag{6.1.8}$$

where the operator E^2 is given in cylindrical and spherical polar coordinates by

$$E^2 \equiv \frac{\partial^2}{\partial x^2} + \frac{\partial^2}{\partial \sigma^2} - \frac{1}{\sigma}\frac{\partial}{\partial \sigma} = \frac{\partial^2}{\partial r^2} + \frac{\sin\theta}{r^2}\frac{\partial}{\partial\theta}\left(\frac{1}{\sin\theta}\frac{\partial}{\partial\theta}\right)$$

$$= \frac{\partial^2}{\partial r^2} + \frac{1}{r^2}\frac{\partial^2}{\partial\theta^2} - \frac{\cot\theta}{r^2}\frac{\partial}{\partial\theta} \tag{6.1.9}$$

The vorticity vector of an axisymmetric flow is given by $\boldsymbol{\omega} = -(1/\sigma)\mathbf{e}_\varphi E^4\Psi$, where $\mathbf{e}_\varphi$ is the unit vector in the azimuthal direction [see Eq. (2.7.16)].

To describe an axisymmetric flow in the presence of swirling motion, such as the flow produced by the axial rotation of a prolate spheroid, it is sometimes convenient to introduce the swirl $\Theta(x, \sigma)$ defined by the equation $u_\varphi = \Theta/\sigma$. The vorticity due to the swirling motion is given by $\boldsymbol{\omega} = \mathbf{i}(1/\sigma)\partial\Theta/\partial\sigma$, where $\mathbf{i}$ is the unit vector along the axis of rotation, and the associated modified pressure is constant. Using the Stokes equation we find that Θ satisfies the second-order partial differential equation

$$E^2\Theta = 0 \tag{6.1.10}$$

where the operator E^2 was defined in Eq. (6.1.9).

Reversibility of Stokes Flow

Let us assume that $\mathbf{u}$ and p are a pair of velocity and pressure fields that satisfy the equations of Stokes flow with a certain body force $\mathbf{g}$. It is evident that $-\mathbf{u}$, $-p$, and $-\mathbf{g}$ will also satisfy the equations of Stokes flow, and therefore *the reversed flow will express a mathematically acceptable and physically viable solution.* It is important to note that the direction of the hydrodynamic force and torque acting on any surface are reversed when the signs of $\mathbf{u}$ and p are switched. The property of reversibility is not generally shared by flow at finite Reynolds numbers, for in that case the sign of the nonlinear term $\rho\mathbf{u}\cdot\nabla\mathbf{u}$ does not necessarily change when the sign of the velocity is reversed.

Reversibility of Stokes flow may be used to deduce the structure of a flow and compute the force and torque exerted on the boundaries, without actually computing the flow. Consider, for instance, a solid sphere moving under the action of a shear flow in the vicinity of a plane wall. In principle, the hydrodynamic force acting on the sphere may have a component perpendicular to the wall and a component parallel to the wall. Let us assume temporarily that the component of the force perpendicular to the wall pushes the sphere away from the wall. Reversing the direction of the shear flow must reverse the direction of the force and thus push the sphere toward the wall. Such an anisotropy, however, is physically unacceptable in view of the fore-and-aft symmetry of the domain of flow. We must conclude that the normal component of the force on the sphere is equal to zero, and thus the sphere must keep moving parallel to the wall.

Using the concept of reversibility, one may infer that the streamline pattern around an axisymmetric and fore-and-aft symmetric object that moves along its axis in an otherwise quiescent fluid must also be axisymmetric and fore-and-aft symmetric. The streamline pattern of simple shear flow over a two-dimensional rectangular cavity must be symmetric with respect to the midplane of the cavity. A neutrally buoyant spherical particle that is convected in a parabolic flow within a cylindrical tube may not move toward the center of the tube or the wall, but must maintain its initial radial position.

Energy Integral Balance

Considering the energy integral balance expressed by Eq. (3.3.18), we note that, due to the absence of inertial forces, the rate of accumulation and flow rate of kinetic energy into a fixed volume V_c that is bounded by the surface D vanishes. Assuming that the physical properties of the fluid are uniform, and using Eq. (3.3.14) we obtain the simplified energy integral balance

$$\int_D \mathbf{u} \cdot \mathbf{f}^{\mathrm{Mod}} \, dS = -2\mu \int_{V_c} \mathbf{E} : \mathbf{E} \, dV \qquad (6.1.11)$$

where the modified traction $\mathbf{f}^{\mathrm{Mod}}$ is defined with respect to the modified stress tensor involving the modified pressure P, and the normal vector $\mathbf{n}$ points into the control volume. Equation (6.1.11) states that the rate of working of the tractions on the boundary is balanced by the rate of viscous dissipation within the control volume.

As an application of Eq. (6.1.11), let us consider the flow produced by the motion of a rigid body that translates with velocity $\mathbf{U}$ and rotates about the origin with angular velocity $\mathbf{\Omega}$ in an ambient fluid of infinite expanse. We select a control volume V_c that is confined by the surface of the body B and a surface of large size S, and apply Eq. (6.1.11) with D representing the union of B and S. Letting the size of S tend to infinity and noting that the velocity decays like $1/r$, we find that the corresponding surface integral on the left-hand side makes a vanishing contribution. Requiring the boundary condition $\mathbf{u} = \mathbf{U} + \mathbf{\Omega} \times \mathbf{x}$ over B yields

$$\mathbf{U} \cdot \mathbf{F}^{\mathrm{Mod}} + \mathbf{\Omega} \cdot \mathbf{T}^{\mathrm{Mod}} = -2\mu \int_{V_c} \mathbf{E} : \mathbf{E} \, dV \qquad (6.1.12)$$

where $\mathbf{F}^{\mathrm{Mod}}$ and $\mathbf{T}^{\mathrm{Mod}}$ are the modified force and torque exerted on the body that exclude the effect of the body force. Equation (6.1.12) expresses a balance between the rate of supply of mechanical energy that is necessary in order to sustain the motion of the body and the rate of viscous dissipation within the fluid. Since the right-hand side and thus the left-hand side are negative or vanish for any combination of $\mathbf{U}$ and $\mathbf{\Omega}$, we conclude that $\mathbf{U} \cdot \mathbf{F}^{\mathrm{Mod}} \leq 0$ and $\mathbf{\Omega} \cdot \mathbf{T}^{\mathrm{Mod}} \leq 0$, independently. Physically, these inequalities show that the drag force exerted on a translating body, and the torque exerted on a rotating body, must resist the motion of the body.

Uniqueness of Solution

Let us assume that the velocity, the modified surface stress, or the projection of the velocity onto the modified surface traction, $\mathbf{u} \cdot \mathbf{f}^{\mathrm{Mod}}$, vanishes over the boundaries of the flow. The left-hand side of Eq. (6.1.11) is equal to zero, and this requires that the rate of deformation tensor $\mathbf{E}$ vanish throughout the flow. As a result, $\mathbf{u}$ must express rigid-body motion, including translation and rotation (Problem 1.1.3). An important consequence of this result is the uniqueness of solution of the equations of Stokes flow, first noted by Helmholtz in 1868.

To prove uniqueness of solution, assume temporarily that there are two solutions corresponding to a given set of boundary conditions on D, either for the velocity or traction. The flow expressed by the difference between these two solutions has homogeneous boundary conditions on D, and must therefore represent rigid-body motion. If the boundary conditions specify the velocity over a portion of D, rigid-body motion is not allowed, the difference flow must vanish, and the solution is unique. If, on the other hand, the boundary conditions specify the traction over all boundaries, any solution may be augmented with the addition of an arbitrary homogeneous solution expressing rigid-body motion.

Minimum Energy Dissipation Principle

A further property of Stokes flow due to Helmholtz is that the *rate of viscous dissipation* in a Stokes flow with velocity **u** is lower than that in any other incompressible flow with velocity **u**′ with identical boundary conditions for the velocity. To show this, we begin with the identity

$$\int_{\text{Flow}} \mathbf{E}' : \mathbf{E}' \, dV = \int_{\text{Flow}} (\mathbf{E}' - \mathbf{E}) : (\mathbf{E}' - \mathbf{E}) \, dV + \int_{\text{Flow}} \mathbf{E} : \mathbf{E} \, dV$$

$$+ 2 \int_{\text{Flow}} (\mathbf{E}' - \mathbf{E}) : \mathbf{E} \, dV \tag{6.1.13}$$

Concentrating on the last integral on the right-hand side, we use the continuity equation and the fact that both velocity fields are solenoidal to write

$$\int_{\text{Flow}} (\mathbf{E}' - \mathbf{E}) : \mathbf{E} \, dV = \int_{\text{Flow}} \frac{\partial(u'_i - u_i)}{\partial x_j} E_{ij} \, dV$$

$$= - \int_{\text{Boundaries}} (u'_i - u_i) \, E_{ij} n_j \, dS - \tfrac{1}{2} \int_{\text{Flow}} (u'_i - u_i) \nabla^2 u_i \, dV \tag{6.1.14}$$

where the unit normal vector **η** points into the fluid. Next we use the Stokes equation and the continuity equation to recast the last integral into the form

$$\frac{1}{\mu} \int_{\text{Flow}} (\mathbf{u}' - \mathbf{u}) \cdot \nabla P \, dV = \frac{1}{\mu} \int_{\text{Flow}} \nabla \cdot [P(\mathbf{u}' - \mathbf{u})] \, dV = -\frac{1}{\mu} \int_{\text{Boundaries}} P(\mathbf{u}' - \mathbf{u}) \cdot \mathbf{n} \, dS \tag{6.1.15}$$

Since **u** and **u**′ have the same values over the boundaries, the right-hand side of Eq. (6.1.14) is equal to zero. Noting that the first two integrals on the right-hand side of Eq. (6.1.14) are non-negative proves the theorem.

One consequence of the minimum energy dissipation principle is that the magnitudes of the drag force and torque exerted on a rigid body that moves steadily in a fluid under conditions of Stokes flow are lower than those for flow at a finite Reynolds number.

PROBLEMS

6.1.1   **Reversibility of Stokes flow.** Show that the drag force exerted on a solid sphere that rotates in the vicinity of a plane wall may not have a component perpendicular to the wall.

6.1.2   **Mean-value theorems.** Using the fact that the components of a Stokes velocity field are bi-harmonic functions, prove Eqs. (6.1.4) and (6.1.5). (*Hint:* expand the velocity field in a Taylor series about the center of the sphere, or see Section 2.3).

6.1.3   **Swirling flow.** An axisymmetric swirling flow may be described in terms of a single scalar function χ defined by the equations (Love, 1944, p. 325)

$$\sigma_{x\varphi} = \mu \frac{1}{\sigma^2} \frac{\partial \chi}{\partial \sigma}, \qquad \sigma_{\sigma\varphi} = -\mu \frac{1}{\sigma^2} \frac{\partial \chi}{\partial x} \tag{6.1.16}$$

(a) Show that the Stokes equation is satisfied for any choice of χ. (b) Show that χ is constant along a line of vanishing traction. (c) Show that the derivative of χ normal to a line C that rotates as a rigid body vanishes. The azimuthal component of the velocity over C is given by $u_\varphi = \Omega \sigma$, where Ω is the angular velocity of rotation. (d) Combine the constitutive equations for a Newtonian fluid given in Table 3.2.2

$$\sigma_{x\varphi} = \mu \frac{\partial u_\varphi}{\partial x}, \qquad \sigma_{\sigma\varphi} = \mu\sigma \frac{\partial}{\partial\sigma}\left(\frac{u_\varphi}{\sigma}\right) \tag{6.1.17}$$

with Eqs. (6.1.16) to show that χ satisfies the second-order linear differential equation

$$\frac{\partial^2\chi}{\partial x^2} + \frac{\partial^2\chi}{\partial\sigma^2} - \frac{3}{\sigma}\frac{\partial\chi}{\partial\sigma} = 0 \tag{6.1.18}$$

(e) Show that $\chi \cos 2\varphi/\sigma^2$ is a three-dimensional harmonic function.

6.1.4   **Shear flow over a wavy wall.** Consider a two-dimensional semi-infinite simple shear flow with shear rate k over a plane wall with sinusoidal corrugations of wavelength λ and amplitude $\varepsilon\lambda$, where ε is a small number compared to unity. The trace of the wall in the xy plane is described by $y = \varepsilon\lambda \cos(2\pi x/\lambda)$. Show that, in the limit of small ε, the stream function is given by the asymptotic expansion

$$\psi = \tfrac{1}{2}ky^2 - \varepsilon ky\lambda \exp(-2\pi y/\lambda)\cos(2\pi x/\lambda) + \cdots \tag{6.1.19}$$

Compute the drag force exerted on the wall over one period accurate to first order in ε.

6.2 | LOCAL SOLUTIONS IN CORNERS

Because the velocity of a viscous fluid over a stationary solid boundary is required to vanish, the Reynolds number of the flow in the vicinity of the boundary is necessarily small, and the structure of the local flow is governed by the equations of Stokes flow. The role of the outer flow is to determine the asymptotic behavior of the flow far from the boundary.

There are other circumstances where the flow is due to the motion of a boundary, but the Reynolds number of the flow in a certain neighborhood of the boundary is small. In these cases, the structure of the Stokes flow in the vicinity of the boundary is determined by the local geometry of the boundary and nature of the required boundary conditions.

In the present section, we examine the structure of a family of two-dimensional flows in wedge-shaped domains that are bounded by intersecting stationary, translating, or rotating walls, as depicted in Figures 6.2.1–6.2.5 and 6.2.7. To bypass the computation of the pressure, and thus reduce the number of unknowns, we describe the flow in terms of the stream function ψ. Noting that the boundary conditions are to be applied at intersecting planes, we introduce plane polar coordinates with the origin at the vertex, and separate the radial from the angular dependence writing

$$\psi = q(r)f(\theta) \tag{6.2.1}$$

Furthermore, we stipulate the power-law functional dependence

$$q(r) = Ar^\lambda \tag{6.2.2}$$

where A and λ are two complex constants; the magnitude of the former is a measure of the intensity of the outer flow; and the latter determines the structure of the inner Stokes flow. The strength of the vorticity is given by

$$\omega(r,\theta) = -\nabla^2\psi = -\frac{1}{r}\frac{\partial}{\partial r}\left(r\frac{\partial\psi}{\partial r}\right) - \frac{1}{r^2}\frac{\partial^2\psi}{\partial\theta^2} = -f\frac{1}{r}\frac{d}{dr}\left(r\frac{dq}{dr}\right) - \frac{q}{r^2}\frac{d^2f}{d\theta^2}$$

$$= -Ar^{\lambda-2}\left(\lambda^2 f + \frac{d^2f}{d\theta^2}\right) \tag{6.2.3}$$

Requiring that the vorticity be a harmonic function, we obtain a fourth-order homogeneous differential equation for f,

$$f'''' + 2(\lambda^2 - 2\lambda + 2)f'' + \lambda^2(\lambda - 2)^2 f = 0 \qquad (6.2.4)$$

where a prime designates a derivative with respect to θ. Assuming an eigensolution of the form

$$f = \exp(\kappa\theta) \qquad (6.2.5)$$

and substituting it into Eq. (6.2.4) yields a quadratic equation for the generally complex constant κ

$$\kappa^4 + 2(\lambda^2 - 2\lambda + 2)\kappa^2 + \lambda^2(\lambda - 2)^2 = 0 \qquad (6.2.6)$$

When κ has an imaginary component, the eigensolutions may be expressed in terms of trigonometric functions. Solving for κ and substituting the result back into Eq. (6.2.5), we obtain the general solution in terms of the real part of the right-hand sides of the equations

$$f = \begin{cases} B\sin(\lambda\theta - \beta) + C\sin[(\lambda - 2)\theta - \gamma], & \text{if } \lambda \neq 0, 1, 2 & (6.2.7a) \\ B\sin(2\theta - \beta) + C\theta + D, & \text{if } \lambda = 0, 2 & (6.2.7b) \\ B\sin(\theta - \beta) + C\theta\sin(\theta - \gamma), & \text{if } \lambda = 1 & (6.2.7c) \end{cases}$$

where B, C are complex and β, γ are real constants. A solution with $\lambda = 0$ was already presented in Eq. (5.5.5) in the context of the Jeffery–Hamel flow.

Two-Dimensional Stagnation-Point Flow on a Plane Wall

As a first application, we consider the flow near a stagnation point on a plane wall illustrated in Figure 6.2.1. The no-slip and no-penetration conditions require that $f = 0$ and $f' = 0$ at $\theta = 0$ and π, and, furthermore, $f = 0$ at the dividing streamline located at $\theta = \alpha$. Making

(a)

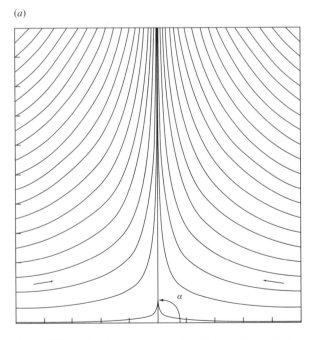

Figure 6.2.1   Stokes flow near a stagnation point on a plane wall when the angle of the dividing streamline is (a) $\alpha = \pi/2$.   (*continued*)

(*b*)

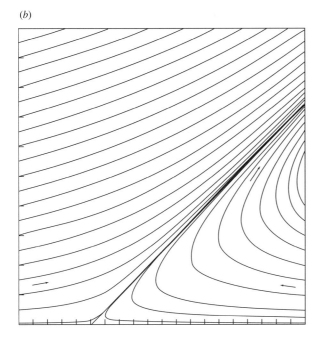

(*c*)

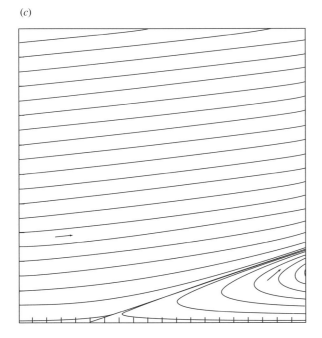

Figure 6.2.1 (*continued*)   Stokes flow near a stagnation point on a plane wall when the angle of the dividing streamline is (b) $\alpha = \pi/4$, (c) $\alpha = \pi/10$.

the judicious choice $\lambda = 3$ and using the general solution given in Eq. (6.2.7a), we obtain

$$f = B \sin(3\theta - \beta) + C \sin(\theta - \gamma) \qquad (6.2.8)$$

Enforcing the boundary conditions, we derive the specific form $f = F \sin^2 \theta \sin(\theta - \alpha)$, which can be substituted into Eq. (6.2.1) to yield the stream function

$$\psi = Gr^3 \sin^2 \theta \sin(\theta - \alpha) = Gy^2(y \cos \alpha - x \sin \alpha) \qquad (6.2.9)$$

where F and G are two real constants whose magnitudes are determined by the intensity of the outer flow that is responsible for the onset of the stagnation point. The Cartesian components of the velocity and modified pressure gradient are given by

$$u_x = Gy(3y \cos \alpha - 2x \sin \alpha)$$
$$u_y = Gy^2 \sin \alpha \qquad (6.2.10)$$
$$\nabla P = 2\mu G(3 \cos \alpha, \sin \alpha)$$

It is worth noting that the modified pressure gradient is constant and points into the same quadrant as the velocity along the dividing streamline. The streamline pattern for the cases $\alpha = \pi/2$, $\pi/4$, and $\pi/10$ are shown in Figure 6.2.1(a–c). As α tends to zero, we obtain unidirectional shear flow with parabolic velocity profile.

Taylor's Scraper

Taylor (1962) studied the flow near the edge of a flat plate that scrapes a plane wall at an angle α, moving along the x axis with velocity V. In a frame of reference moving with the scraper, the scraper appears to be stationary, and the wall appears to move along the x axis with velocity $U = -V$ as shown in Figure 6.2.2. The no-slip boundary condition at the wall requires that $(1/r)\partial\psi/\partial\theta = U$ at $\theta = 0$, and this suggests setting $\lambda = 1$. One interesting consequence

(*a*)

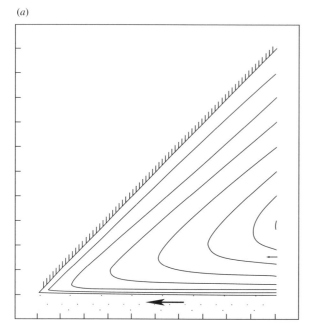

Figure 6.2.2   Streamline pattern of the flow due to a scraper moving over a plane wall in a frame of reference moving with the scraper, for scraping angle (a) $\alpha = \pi/4$.   (*continued*)

(*b*)

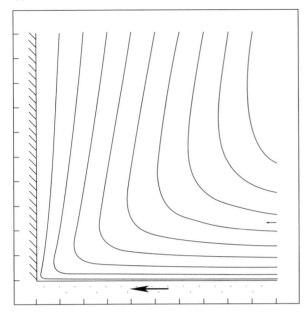

(*c*)

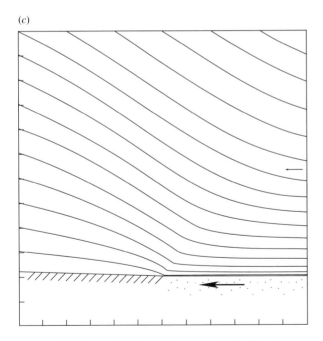

Figure 6.2.2 (*continued*) Streamline pattern of the flow due to a scraper moving over a plane wall in a frame of reference moving with the scraper, for scraping angle (b) $\alpha = \pi/2$, (c) $\alpha = \pi$.

of the choice $\lambda = 1$ is that the radial and azimuthal components of the velocity depend only upon the polar angle θ and are independent of r. Requiring the boundary conditions $f = 0$, $f' = U/A$ at $\theta = 0$ and $f = 0$, $f' = 0$ at $\theta = \alpha$, and using the general solution given in Eq. (6.2.7c), we obtain the stream function

$$\psi = -\frac{U}{\alpha^2 - \sin^2 \alpha} r [\alpha(\theta - \alpha) \sin \theta - \theta \sin \alpha \sin(\theta - \alpha)] \qquad (6.2.11)$$

The streamline patterns for three scraping angles with $V > 0$ and $U < 0$ are shown in Figure 6.2.2(a–c). The distribution of shear stress over the wall is given by

$$\sigma_{r\theta}(\theta = 0) = \frac{\mu}{r} \left(\frac{\partial u_r}{\partial \theta} \right)_{\theta=0} = \frac{\mu}{r^2} \left(\frac{\partial^2 \psi}{\partial \theta^2} \right)_{\theta=0} = \frac{\mu U}{r} \frac{2\alpha - \sin(2\alpha)}{\alpha^2 - \sin^2 \alpha} \qquad (6.2.12)$$

The $1/r$ singularity suggests that an infinite force is required in order to maintain the motion of the scraper. In reality, the corner between the scraper and the wall will have a small but finite curvature, which will modify the flow and hence remove the singular behavior. If the edge of the scraper is perfectly sharp, there will be a small gap between the scraper and the wall that allows for a small amount of leakage.

Flow Due to the Motion of a Plane Wall Parallel to Itself

The case $\alpha = \pi$ corresponds to flow driven by the motion of a semi-infinite belt extending from the origin to infinity, as shown in Figure 6.2.2(c). A slight rearrangement of the general solution (6.2.11) yields the stream function

$$\psi = -U r \frac{\theta - \pi}{\pi} \sin \theta = -U \frac{\theta - \pi}{\pi} y \qquad (6.2.13)$$

Based on this solution, and exploiting the feasibility of linear superposition, we find that the flow due to the motion of a section of the wall subtended between $x = -\varepsilon$ and $+\varepsilon$ is described by

$$\psi = -U y \left(\frac{\theta_- - \pi}{\pi} - \frac{\theta_+ - \pi}{\pi} \right) = U y \frac{\theta_+ - \theta_-}{\pi} \qquad (6.2.14)$$

where $\tan \theta_+ = y/(x - \varepsilon)$ and $\tan \theta_- = y/(x + \varepsilon)$. In the limit as ε/r tends to zero, we obtain the asymptotic solution

$$\psi = -\frac{2\varepsilon U}{\pi} y \frac{\partial \theta}{\partial x} = \frac{2\varepsilon U}{\pi} \frac{y^2}{x^2 + y^2} \qquad (6.2.15)$$

which describes the asymptotic structure of the flow far from the moving section. The streamline pattern close to and far from the moving section are shown in Figure 6.2.3(a,b). The far flow in the first quadrant resembles that due to a point source located at the origin in the presence of two intersecting walls placed along the x and y axes.

Exploiting the feasibility of linear superposition once more, we use Eq. (6.2.15) to derive the stream function of the flow above a plane wall subject to an arbitrary distribution of tangential velocity along the wall $U(x)$,

$$\psi(\mathbf{x}) = \frac{y^2}{\pi} \int_{-\infty}^{+\infty} \frac{U(x')}{(x - x')^2 + y^2} \, dx' \qquad (6.2.16)$$

Flow near a Belt Plunging into a Pool

Akin to the flow due to the motion of a scraper is the flow due to a belt plunging into a liquid pool, as shown in Figure 6.2.4 (Moffatt, 1964). At the surface of the belt we require the no-slip boundary condition $(1/r) \partial \psi / \partial \theta = U$ at $\theta = -\alpha$, and this suggests setting $\lambda = 1$. At the interface we require the no-penetration condition and the condition of vanishing shear

(a)

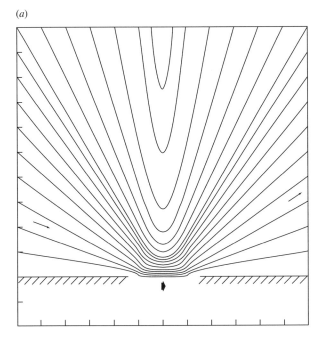

(b)

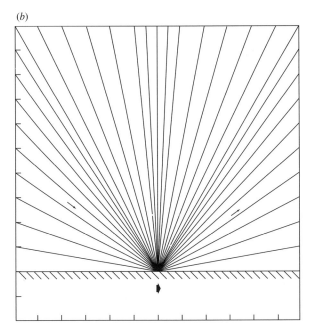

Figure 6.2.3   Streamline pattern of the flow due to the translation of a section of a plane wall (a) close to and (b) far from the moving section.

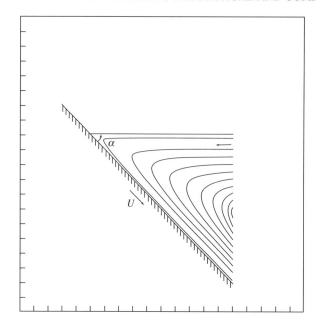

Figure 6.2.4   Flow due to a belt plunging into a pool.

stress. Using the general solution (6.2.7c) and implementing the four boundary conditions $f = 0$, $f' = U/A$ at $\theta = -\alpha$ and $f = 0$, $f'' = 0$ at $\theta = 0$, we obtain the stream function

$$\psi = -\frac{2U}{2\alpha - \sin(2\alpha)} r \left(\theta \cos \theta \sin \alpha - \alpha \sin \theta \cos \alpha \right) \qquad (6.2.17)$$

The streamline pattern for $\alpha = \pi/4$ is shown in Figure 6.2.4. As in the case of the scraper, we find that the radial and azimuthal components of the velocity depend upon the polar angle θ but are independent of r. Furthermore, the shear stress on the belt is equal to $\sigma_{r\theta}(\theta = -\alpha) = -4\mu U \sin^2 \alpha/[r(2\alpha - \sin 2\alpha)]$, and thus it behaves like $1/r$, which suggests that an infinite force is required in order to maintain the motion of the belt. The resolution of this paradoxical behavior was discussed in Section 3.5 in the context of the boundary conditions at a three-phase contact line.

Flow near a Corner

Next we consider the flow within a corner that is confined between two intersecting stationary walls, driven by the motion of the fluid away from the walls, as illustrated in Figures 6.2.5 and 6.2.7 (Dean and Montagnon, 1949; Moffatt, 1964). At the outset, we distinguish between two modular cases according to the structure of the flow far from the corner. The first case corresponds to flow with an antisymmetric streamline pattern, such as that shown in Figure 6.2.5. The second case corresponds to flow with a symmetric streamline pattern, such as that shown in Figure 6.2.7. The physical requirement for the second situation to occur is that the outer flow exhibit a corresponding symmetry, or else the midplane of the flow represents a free surface with vanishing shear stress.

Antisymmetric flow

Concentrating first on the more general case of antisymmetric flow, we require that the velocity vanish at the walls and obtain the boundary conditions $\psi = 0$ and $\partial\psi/\partial\theta = 0$ at $\theta = \pm\alpha$, and $\partial\psi/\partial\theta = 0$ at $\theta = 0$. The fact that all boundary conditions for the stream function are homogeneous forecasts that the solution will be found by solving an algebraic eigenvalue problem,

(a)

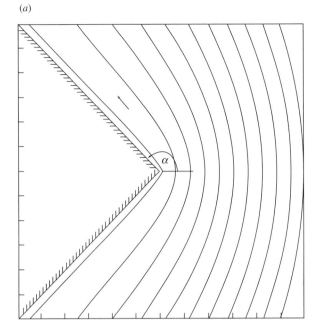

(b)

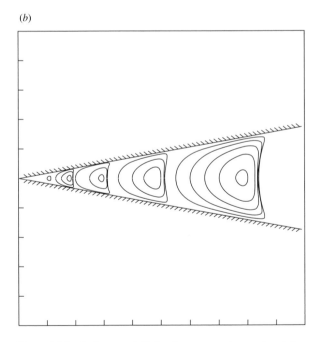

Figure 6.2.5 Antisymmetric Stokes flow near two intersecting station-ary plane walls for corner semiangles (a) $\alpha = 135°$ and (b) $\alpha = 10°$.

which means that the value of λ cannot be assigned *a priori* but must be found as part of the solution. Assuming that $\lambda \neq 0, 1, 2$, and stipulating that the velocity is antisymmetric with respect to the midplane, we use the general solution given in Eq. (6.2.7a) and find

$$f = F\cos(\lambda\theta) + G\cos[(\lambda - 2)\theta] \tag{6.2.18}$$

where F and G are two complex constants. When λ has a complex value, we enhance the right-hand side of Eq. (6.2.18) with its complex conjugate, which amounts to retaining its real part. Implementing the boundary conditions at the wall, we obtain two homogeneous equations comprising the system

$$\begin{bmatrix} \cos(\lambda\alpha) & \cos[(\lambda - 2)\alpha] \\ \lambda\sin(\lambda\alpha) & (\lambda - 2)\sin[(\lambda - 2)\alpha] \end{bmatrix} \begin{bmatrix} F \\ G \end{bmatrix} = 0 \tag{6.2.19}$$

For a nontrivial solution for F and G to exist, the determinant of the coefficient matrix must vanish, and this provides us with the secular equation

$$(\lambda - 2)\cos(\lambda\alpha)\sin[(\lambda - 2)\alpha] - \lambda\cos[(\lambda - 2)\alpha]\sin(\lambda\alpha) = 0 \tag{6.2.20}$$

which may be recast into the more compact form

$$\sin[2\alpha(\lambda - 1)] = (1 - \lambda)\sin(2\alpha) \tag{6.2.21}$$

Equation (6.2.21) provides us with a nonlinear algebraic equation for λ. There is an obvious solution $\lambda = 1$, but for this value, Eq. (6.2.7c) rather than Eq. (6.2.7a) should have been selected, which disqualifies this choice. Dean and Montagnon (1949) noted that, apart from the trivial solution $\lambda = 1$, there are a number of other non-trivial real and positive solutions in the range $73° < \alpha < 180°$ or $0.41\pi < \alpha < \pi$. These may be computed by solving Eq. (6.2.21), using, for instance, Newton's method (Problem 6.2.6). The smallest solution, plotted in Figure 6.2.6(a) with the solid line, corresponds to the strongest velocity field and is therefore expected to prevail in practice. When $\alpha = 90°$, we obtain $\lambda = 2$, corresponding to simple shear flow, whereas when $\alpha = 180°$, we obtain $\lambda = 1.5$, corresponding to flow around a flat plate.

Combining Eqs. (6.2.19), (6.2.18), (6.2.2), and (6.2.1), we find that, in the range $73° < \alpha < 180°$, the stream function is given by

$$\psi = ar^\lambda\{\cos(\lambda\theta)\cos[(\lambda - 2)\alpha] - \cos[(\lambda - 2)\theta]\cos(\lambda\alpha)\} \tag{6.2.22}$$

where a is a real constant. The streamline pattern for $\alpha = 135°$ is shown in Figure 6.2.5(a).

When $\alpha < 73°$, Eq. (6.2.21) has only complex solutions apart from the trivial solution $\lambda = 1$. To study the structure of the flow, we write $\lambda = \lambda_r + i\lambda_r$, and decompose Eq. (6.2.21) into its real and imaginary parts, given by

$$\sin[2\alpha(\lambda_r - 1)]\cosh(2\alpha\lambda_i) = (1 - \lambda_r)\sin(2\alpha)$$
$$\cos[2\alpha(\lambda_r - 1)]\sinh(2\alpha\lambda_i) = -\lambda_i\sin(2\alpha) \tag{6.2.23}$$

Furthermore, to simplify the notation, we introduce the auxiliary variables

$$\xi \equiv 2\alpha(\lambda_r - 1), \qquad \eta \equiv 2\alpha\lambda_i, \qquad k \equiv \frac{\sin(2\alpha)}{2\alpha}, \qquad \text{where } \lambda = 1 + \frac{\xi + i\eta}{2\alpha} \tag{6.2.24}$$

and rewrite Eqs. (6.2.23) in the form

$$\sin\xi\cosh\eta = -k\xi, \qquad \cos\xi\sinh\eta = -k\eta \tag{6.2.25}$$

We note that both the sine and cosine of ξ must be negative, and this suggests that ξ must lie in the range $(2n - 1)\pi < \xi < (2n - \frac{1}{2})\pi$, where n is an integer. The solution may be computed using Newton's method with a suitable initial guess for ξ and η; the solution branch with the smallest real part λ_r is shown in Figure 6.2.6(a,b).

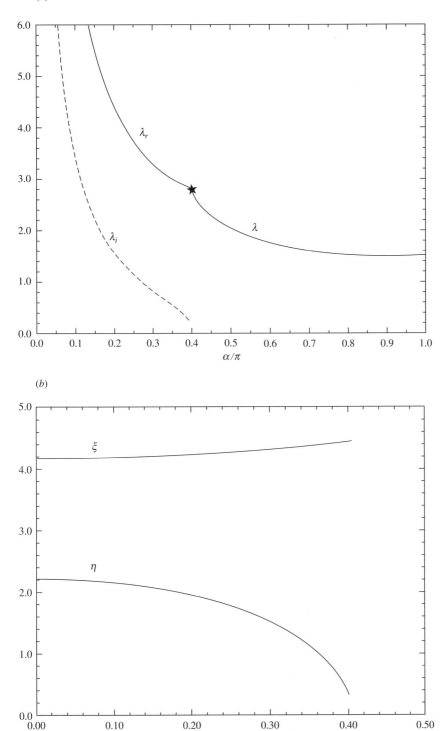

(a)

(b)

Figure 6.2.6 (a) The exponent λ with the smallest real part for antisymmetric flow between two intersecting plane walls as a function of the corner semiangle α. When $0.41\pi < \alpha < \pi$, λ is real. (b) The eigenvalues ξ and η, defined in Eqs. (6.2.24), in the range where eddies form.   (*continued*)

(c)

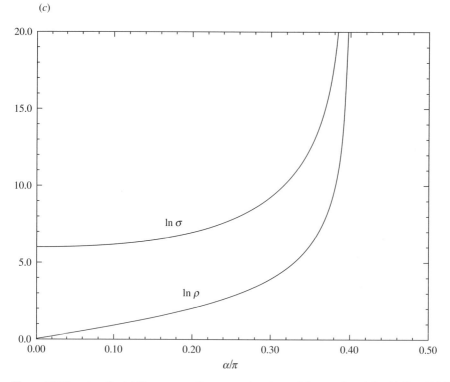

Figure 6.2.6 (*continued*)   (c) The corresponding geometric ratio ρ and decay ratio σ defined in Eqs. (6.2.29) and (6.2.30).

Combining Eqs. (6.2.19), (6.2.18), (6.2.2), and (6.2.1), we find that the stream function for $\alpha < 73°$ is given by

$$
\begin{aligned}
\psi &= ar^{\lambda_r} \operatorname{Re}(e^{i\lambda_i \ln r}Q) \\
&= ar^{\lambda_r}[\operatorname{Re}(Q)\cos(\lambda_i \ln r) - \operatorname{Im}(Q)\sin(\lambda_i \ln r)] \\
&= ar^{\lambda_r}|Q|\cos[\lambda_i \ln r + \operatorname{Arg}(Q)]
\end{aligned}
\tag{6.2.26}
$$

where a is a real constant and

$$
Q \equiv \cos(\lambda\theta)\cos[(\lambda - 2)\alpha] - \cos[(\lambda - 2)\theta]\cos(\lambda\alpha)
\tag{6.2.27}
$$

The streamline pattern for $\alpha = 10°$ is shown in Figure 6.2.5(b).

Moffatt (1964) observed that, as r is increased, the stream function changes sign an infinite number of times, which reveals the occurrence of an infinite sequence of self-similar eddies. Setting $\psi = 0$, we find that the shape of the dividing streamlines is described by the equation

$$
r = \exp\left(-\frac{1}{\lambda_i}[\operatorname{Arg}(Q) + \tfrac{1}{2}(2n + 1)\pi]\right)
\tag{6.2.28}
$$

where n is an integer. Applying Eq. (6.2.28) at a certain value of θ shows that the ratio between the radial positions of two successive dividing streamlines is given by

$$
\rho \equiv \frac{r_n}{r_{n+1}} = \exp\left(\frac{\pi}{\lambda_i}\right)
\tag{6.2.29}
$$

Furthermore, Eq. (6.2.26) shows that the ratio of the magnitude of the corresponding polar component of the velocity is given by

$$\sigma \equiv \frac{(u_\theta)_n}{(u_\theta)_{n+1}} \exp\left(\frac{(\lambda_r - 1)\pi}{\lambda_i}\right) \tag{6.2.30}$$

The geometric ratio ρ and amplification ratio σ are plotted in Figure 6.2.6(c) on a linear-logarithmic scale. As α tends to zero, ρ tends to unity, which means that the eddies tend to obtain a uniform size. Furthermore, ρ takes values that are of order unity for small and moderate corner angles, which means that successive eddies have comparable sizes, but increases rapidly as α approaches the critical value $73°$. On the contrary, σ takes very large values in the whole range of α, which means that the intensity of the flow decays rapidly inside the corner. The minimum value of σ corresponding to the limit of α tending to zero is about 360. When $\alpha = 45°$, σ is equal to about 2000, which means that it takes only one eddy for the magnitude of the velocity to decrease by three orders of magnitude!

In the limit as α tends to zero, the intersecting planes tend to become parallel to each other, yielding a two-dimensional channel with parallel-sided walls. This flow may be identified, for instance, with that within an open rectangular cavity, driven by an overpassing fluid. To derive the appropriate similarity solution, we may examine the limit of the preceding equations as α tends to zero, but it is more expedient to begin afresh working in Cartesian coordinates in which the walls are located at $y = \pm a$. We thus set $\psi = A f(y) \exp(-k|x|)$, where A is a complex constant and k is the complex wave number, require that ψ satisfy the biharmonic equation, and enforce the no-penetration condition $\psi = 0$ at $y = \pm a$ to obtain

$$\psi = B(a \cos ky - y \cot ka \sin ky) \exp(-k|x|) \tag{6.2.31}$$

where B is a new constant. Requiring the no-slip condition $\partial\psi/\partial y = 0$ at $y = \pm a$ yields the algebraic equation $\sin 2ka = -2ka$, which has only nonreal solutions. The solution with the smallest positive real part is given by $2ka = 4.21 + 2.26i$, yielding a periodic eddy pattern with wavelength equal to $4\pi a/2.26 = 5.56a$.

Symmetric flow

Turning next to the case of symmetric flow illustrated in Figure 6.2.7(a,b), we require that the velocity vanish at the walls, located at $\theta = \pm\alpha$, and the shear stress vanish at the midplane $\theta = 0$ and obtain the boundary conditions $\psi = 0$ and $\partial\psi/\partial\theta = 0$ at $\theta = \pm\alpha$, and $\psi = 0$ and $\partial^2\psi/\partial\theta^2 = 0$ at $\theta = 0$. Working as in the case of antisymmetric flow, we set

$$f = F \sin(\lambda\theta) + G \sin[(\lambda - 2)\theta] \tag{6.2.32}$$

where F and G are two complex constants, and enforce the boundary conditions to obtain the secular equation

$$\sin[2\alpha(\lambda - 1)] = (\lambda - 1) \sin(2\alpha) \tag{6.2.33}$$

There are two obvious solutions $\lambda = 1$ and 2, but for these values Eqs. (6.2.7b) and (6.2.7c) rather than Eq. (6.2.7a) should have been selected. Moffatt (1964) noted that in the range $78° < \alpha < 180°$ there are a number of other real and positive solutions; the solution branch with the smaller magnitude is shown in Figure 6.2.8(a) with the solid line. When $\alpha = 90°$, we obtain $\lambda = 3$, which corresponds to orthogonal stagnation-point flow (Eq. (6.2.8)), whereas when $\alpha = 180°$ we obtain $\lambda = 1.5$, which corresponds to flow along a semi-infinite flat plate. When $\alpha < 78° = 0.43\pi$, Eq. (6.2.33) has only complex solutions, which may be found working as in the case of antisymmetric flow in terms of ξ and η; the solution branch with the smallest real part λ_r is shown in Figure 6.2.8(a,b). The corresponding geometric ratio ρ and amplification ratio σ are plotted in Figure 6.2.8(c) on a linear-logarithmic scale. Comparing these results with those shown in Figure 6.2.6(c) reveals that the size of the eddies falls off less rapidly than it does for antisymmetric flow, but the intensity of the eddies decays at an appreciably higher rate.

(a)

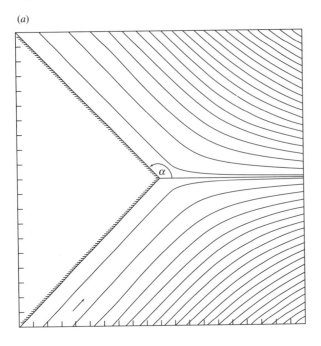

(b)

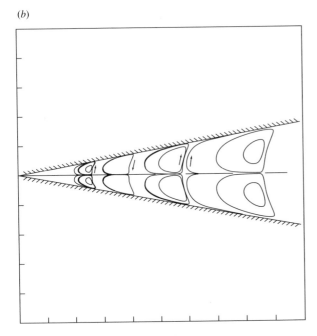

Figure 6.2.7 Symmetric flow near two intersecting stationary plane walls for corner semiangles (a) $\alpha = 135°$ and (b) $\alpha = 10°$.

(a)

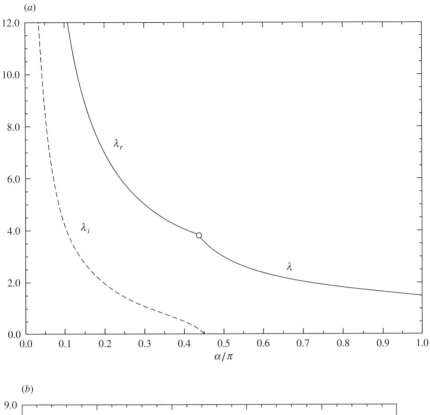

(b)

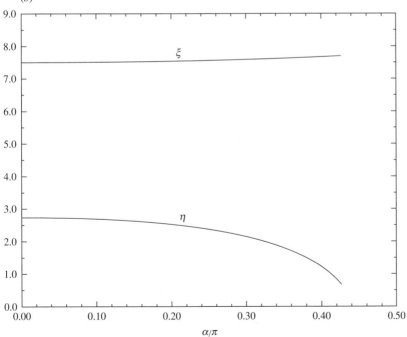

Figure 6.2.8   (a) The exponent λ with the smallest real part for symmetric flow between two intersecting plane walls as a function of the corner semiangle α. When $0.43\pi < \alpha < \pi$, λ is real. (b) The eigenvalues ξ and η defined in Eqs. (6.2.24) in the range where eddies form.   (*continued*)

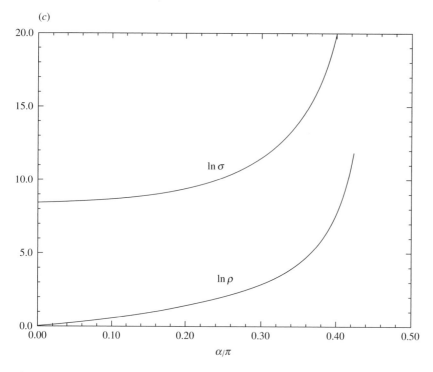

Figure 6.2.8 (*continued*)   (c) The corresponding geometric ratio ρ and decay ratio σ defined in Eqs. (6.2.29) and (6.2.30).

In the limit as α tends to zero, the intersecting planes tend to become parallel to each other, and the asymptotic solution describes Stokes flow within a liquid layer resting on a planar surface. To derive the appropriate similarity solution, it is expedient to introduce Cartesian coordinates in which the lower wall is located at $y = 0$ and the midplane or free surface is located at $y = a$. Working as in the case of antisymmetric flow, we find that the wavelength of the periodic eddy pattern is equal to $2\pi a /2.76 = 2.28a$ (Moffatt, 1964).

Flow due to a disturbance near the corner

In the preceding discussion we considered solutions corresponding to λ with a positive real part, yielding a sequence of eddies that decays rapidly into the corner. To this end, we note that if λ_1 is a solution of Eq. (6.2.21) or (6.2.33) with a positive real part, then $\lambda_2 = 2 - \lambda_1$ is also a solution of these equations with either a positive or a negative real part corresponding to a velocity that behaves like $r^{-\lambda_1+1}$ as r tends to infinity. Thus, the streamline patterns shown in Figures 6.2.5 and 6.2.7 may be reinterpreted as being those due to a disturbance near the corner with the velocity decaying far from the corner.

PROBLEMS

6.2.1   **Flow due to an imposed shear stress near a static contact line.** Consider a flow that is driven by the application of a uniform shear stress τ along a flat interface located at $\theta = 0$, in the presence of a stationary inclined plate located at $\theta = -\alpha$. Show that the corresponding stream function is given by

$$\psi = \frac{\tau}{8\mu} \frac{r^2}{\cos 2\alpha - 1 + \alpha \sin 2\alpha} [(1 - \cos 2\alpha - 2\alpha \sin 2\alpha) \sin 2\theta$$
$$+ (2\alpha \cos 2\alpha - \sin 2\alpha)(\cos 2\theta - 1) + 2\theta(1 - \cos 2\alpha)]$$

(6.2.34)

6.2.2   Tape plunging into a pool. With reference to Figure 6.2.4, derive an expression for the radial component of the velocity at the free surface and compare it to the velocity of the tape.

6.2.3   Flow due to the counter-rotation of two hinged plates. (a) Consider the flow between two hinged plates located at $\theta = \pm\alpha$ that rotate against each other with angular velocities $\pm\Omega$, and show that the stream function is given by (Moffatt, 1964)

$$\psi = \frac{\Omega r^2}{2} \frac{\sin 2\theta - 2\theta \cos 2\alpha}{\sin 2\alpha - 2\alpha \cos 2\alpha}$$

(6.2.35)

Note that the denominator vanishes when $\alpha = 257.45°$, in which case the stream function becomes infinite. Discuss the physical implications of this behavior with reference to the significance of the neglected inertial forces (Moffatt and Duffy, 1980). (b) Generalize the results to arbitrary angular velocities of rotation.

6.2.4   Flow between two free surfaces. Consider the flow between two intersecting free surfaces with vanishing shear stress, driven by the motion of the fluid away from the corner (Moffatt, 1964). This flow occurs, for example, near the cusped interface of a bubble that is subjected to a straining flow. (a) Considering the case of antisymmetric flow described by Eq. (6.2.18), show that λ satisfies the equation

$$(\lambda - 1) \cos(\lambda\alpha) \cos[(\lambda - 2)\alpha] = 0$$

(6.2.36)

which has only real solutions, thereby precluding the occurrence of eddies. Show that the smallest positive solution is given by $\lambda = 2 - \pi/(2\alpha)$ for $\alpha > \pi/2$ corresponding to rotational flow, and $\lambda = \pi/(2\alpha)$ for $\alpha \leq \pi/2$ corresponding to irrotational flow. (b) Repeat (a) for symmetric flow.

6.2.5   Jeffery–Hamel flow. Consider Stokes flow between two intersecting planes due to a point source with strength m located at the apex, and derive the stream function

$$\psi = m \frac{\sin 2\theta - 2\theta \cos 2\alpha}{\sin 2\alpha - 2\alpha \cos 2\alpha}$$

(6.2.37)

Note that the denominator vanishes when $\alpha = 257.45°$, in which case the stream function becomes infinite. Discuss the physical implication of this behavior with reference to the significance of the neglected inertial forces (Moffatt and Duffy, 1980).

Computer Problems

6.2.6   Solution of the eigenvalue problem. (a) Write a program that solves Eqs. (6.2.21) and (6.2.25) and reproduce the plots shown in Figure 6.2.6(a–c). (b) Repeat (a) for the case of symmetric flow.

6.2.7   Streamline patterns. Plot streamline patterns of the flow discussed in Problem 6.2.1 for several values of the inclination angle α.

6.3 NEARLY UNIDIRECTIONAL FLOWS

We turn our attention now to a class of flows in which inertial effects are negligible due to the fact that the curvature of the streamlines is sufficiently small. As a result, the flow may be assumed to be locally unidirectional, which means that the velocity profile may be assumed to depend only upon the *local* pressure gradient, direction of the body force, and geometry of the domain of flow. The union of these assumptions constitutes the premise of *lubrication flow.*

To study the structure of a nearly unidirectional flow in a mathematically rigorous manner, we must develop a formal asymptotic expansion for the flow variables in terms of the magnitude of the curvature of the streamlines, which is assumed to be comparable to that of the boundaries (Langlois, 1964). The results confirm that the assumption of locally unidirectional flow provides us with the correct leading-order approximation to the exact solution.

Flow in the Hele–Shaw Cell

Consider flow in a channel with parallel-sided walls that are separated by a uniform distance h and are confined by side walls, known as the Hele–Shaw cell. The clearance of the channel may be blocked by stationary or moving objects such as disks, or flattened air bubbles and liquid drops, as depicted in Figure 6.3.1. The flow may be driven by an imposed pressure gradient, gravity, or the motion of the objects.

When h is small compared to the global dimensions of the channel and size of the objects, pressure variations across the clearance of the channel may be neglected, and the flow may be approximated with the unidirectional parabolic flow described in Eq. (5.1.5). Setting the z axis perpendicular to the walls, which are located at $z = 0$ and h as shown in Figure 6.3.1, we obtain

$$u_x(x, y, z) = -\frac{1}{2\mu}\frac{\partial P}{\partial x}z(h - z)$$

$$u_y(x, y, z) = -\frac{1}{2\mu}\frac{\partial P}{\partial y}z(h - z) \tag{6.3.1}$$

$$u_z(x, y, z) = 0$$

The modified pressure $P = p - \rho\,\mathbf{g}\cdot\mathbf{x}$ is assumed to be a function of x and y alone. Next we introduce the average velocity of the fluid across the clearance of the channel, $\mathbf{U} = (U_x, U_y)$, and integrate Eqs. (6.3.1) in the z direction across the clearance of the channel to obtain

$$\mathbf{U}(x, y) \equiv \frac{1}{h}\int_0^h \mathbf{u}\,dz = -\frac{h^2}{12\mu}\nabla P \tag{6.3.2}$$

where the gradient ∇ operates with respect to x and y.

It is evident from Eq. (6.3.2) that the modified pressure plays the role of a potential function for the mean velocity $\mathbf{U}$, and this shows that the two-dimensional vector field $\mathbf{U}$ is *irrotational.* Conservation of mass requires that $\mathbf{U}$ be a *solenoidal* vector field with respect to x and y, $\nabla\cdot\mathbf{U} = 0$, and this shows that the modified pressure and therefore the regular pressure are harmonic functions

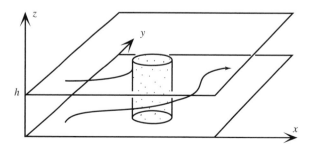

Figure 6.3.1   Schematic illustration of flow in the Hele–Shaw cell past an obstacle.

of x and y,

$$\nabla^2 P = 0, \qquad \nabla^2 p = 0 \qquad (6.3.3)$$

These observations reveal that flow in a Hele–Shaw cell with uniform gap past an obstacle is identical to two-dimensional irrotational flow of an incompressible fluid past a body with corresponding geometry. Interestingly enough, the highly viscous flow within the Hele–Shaw cell provides us with a device for visualizing irrotational flow, whose occurrence requires that the effects of viscosity be negligibly small.

Computing the flow in the Hele–Shaw cell is thus reduced to solving Laplace's equation (6.3.3) subject to (1) the no-penetration condition over the impermeable boundaries of the flow, which provides us with a Neumann boundary condition for the pressure, and (2) a condition that specifies the level of the pressure at the inlet or outlet.

The assumptions that lead us to Eq. (6.3.3) cease to be valid in the vicinity of the boundaries where the flow becomes three-dimensional and the assumption of unidirectional motion is no longer appropriate. Boundary effects in the Hele–Shaw cell, however, are confined within thin layers of fluid whose thickness is of order h and may thus be neglected to a first-order approximation. Neglecting the nonlinear inertial effects due to the curvature of the streamlines introduces an additional error that is small as long as the velocity of the fluid is not exceedingly large.

Flow in a coating die

A coating die is an industrial device that is used to produce thin sheets of a liquid for subsequent coating onto a moving substratum, as illustrated in Figure 6.3.2(a). The die assembly consists of a circular inlet tube that feeds liquid into a Hele–Shaw cell. A schematic illustration of the cross-section of the inlet tube is shown in Figure 6.3.2(b). The die must be designed so as to deliver a uniform flow rate at the outlet of the Hele–Shaw cell, with an objective to manufacture coatings of uniform thickness.

To develop a model for the flow within the die, we assume that the average velocity of the fluid through the channel is related to the modified pressure gradient by Eq. (6.3.2), and the pressure satisfies Laplace's equation (6.3.3). The boundary conditions require that the pressure at the outlet is atmospheric, $p = P_{Atm}$ at $y = L$, and the velocity normal to the side walls vanishes, $\partial P/\partial y = 0$ at $x = 0$ and W. To complete the definition of the problem, we must supply an additional scalar boundary condition for the pressure at the inlet. To derive this condition, we write a mass balance for the flow along the inlet tube, stating

$$\frac{dQ}{dl} = -h\mathbf{U} \cdot \mathbf{n} = \frac{h^3}{12\mu} \nabla P \cdot \mathbf{n} \qquad (6.3.4)$$

where Q is the flow rate along the inlet tube, l is the arc length along the centerline of the inlet tube, $0 < l < D$, and $\mathbf{n}$ is the normal vector in the xy plane pointing into the cell. Assuming further that the flow along the inlet tube is nearly unidirectional Poiseuille flow, we use the second of Eqs. (5.1.11) to write

$$Q = -\frac{\pi a^4}{8\mu} \frac{\partial P}{\partial l} \qquad (6.3.5)$$

Allowing the inlet tube to be tapered, that is, its radius a to be a function of arc length l, we combine Eqs. (6.3.4) and (6.3.5) and derive the inlet boundary condition in the form of the second-order partial differential equation

$$\frac{\partial^2 P}{\partial l^2} + \frac{4}{a} \frac{da}{dl} \frac{\partial P}{\partial l} + \frac{2h^3}{3\pi a^4} \nabla P \cdot \mathbf{n} = 0 \qquad (6.3.6)$$

which is supplemented with the value of the pressure at the inlet $l = 0$, and the condition that the flow rate vanishes at the end of the inlet tube, $\partial P/\partial l = 0$ at $l = D$. This completes the mathematical statement of the problem.

(a)

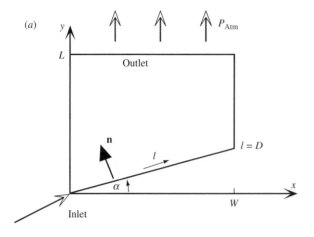

(b)

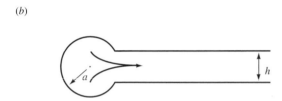

Figure 6.3.2   Flow in a coating die consisting of an inlet tube and a Hele–Shaw cell; (a) top view of the die, (b) cross section of the inlet channel.

Hele–Shaw cell with uneven walls

The analysis of flow in a Hele–Shaw cell with perfectly parallel walls may be extended in a straightforward manner to include circumstances where the clearance of the channel exbibits slow modulations, so that h is a function of x and y. Equations (6.3.1) and (6.3.2) are still valid provided that the origin is set at the lower wall with the upper wall located at $z = h(x, y)$. Conservation of mass, however, requires that Laplace's equation (6.3.3) be replaced by the generalized equation $\nabla \cdot (h\mathbf{U}) = 0$ or $\nabla \cdot (h^3 \nabla P) = 0$. It is worth noting that the last equation describes the steady distribution of temperature in a plate, where the thermal conductivity is proportional to h^3.

Hydrodynamic Lubrication

As a second application, we consider the flow between a horizontal flat plate that moves parallel to itself with velocity V, and a stationary underlying mildly sloped surface representing, for example, the assembly of a rocker bearing, as shown in Figure 6.3.3(a). If the slope of the inclined surface is sufficiently small, the flow at any station across the gap between the planar and the inclined surface may be approximated with plane–Poiseuille–Couette flow between two parallel plates separated by distance h, and the modified pressure may be assumed to be independent of location across the channel. Using Eq. (5.1.6) and assuming, for simplicity, that the effects of gravity are negligible, we find that the flow rate through the channel is given by

$$Q = \tfrac{1}{2}Vh(x) - \frac{dp}{dx}\frac{h^3(x)}{12\mu} \tag{6.3.7}$$

Conservation of mass requires that Q be a constant, independent of x.

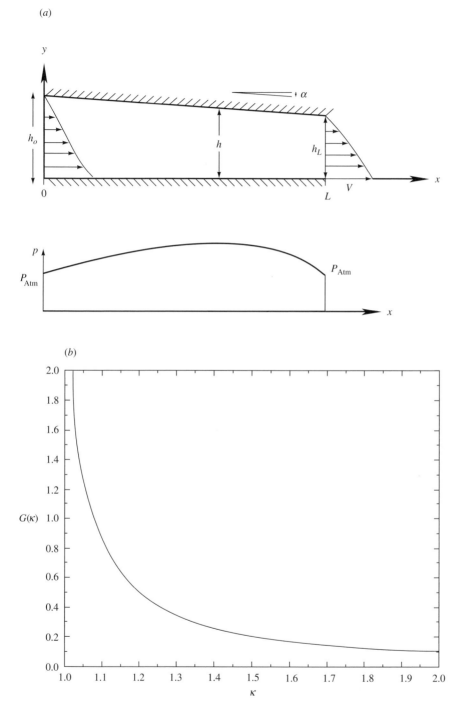

Figure 6.3.3   (a) Lubrication flow between a horizontal and an inclined flat surface; (b) the lift-force function $G(\kappa)$.

To compute the pressure distribution along the gap, we solve Eq. (6.3.7) for dp/dx, obtaining the ordinary differential equation

$$\frac{dp}{dx} = \frac{6\mu V}{h^2(x)} - \frac{12\mu Q}{h^3(x)} \tag{6.3.8}$$

which is to be solved subject to the boundary conditions $p = P_0$ at $x = 0$ and $p = P_L$ at $x = L$, where P_0 and P_L are the pressures away from the lubrication zone, which is confined within $0 < x < L$. One boundary condition is required because Eq. (6.3.8) is a first-order ordinary differential equation, and another boundary condition is required because the flow rate Q is an unknown that must be found as part of the solution.

To make the analysis more tangible, let us consider an inclined wall with a linearly sloped profile as shown in Figure 6.3.3(a). In this case we set $h = h_0 - \alpha x$, where α is the constant slope and h_0 the clearance of the channel at the beginning of the lubrication zone, and assume that the pressure away from the lubrication zone is equal to the atmospheric pressure P_{Atm} setting $P_0 = P_L = P_{\text{Atm}}$. Equation (6.3.8) then becomes

$$\frac{dP}{dx} = \frac{6\mu V}{(h_0 - \alpha x)^2} - \frac{12\mu Q}{(h_0 - \alpha x)^3} \tag{6.3.9}$$

To compute Q, we require that the integral of the right-hand side of Eq. (6.3.9) from $x = 0$ to L, which is equal to the pressure drop across the lubrication zone, vanish. Having obtained Q, we integrate Eq. (6.3.9) with respect to x and thus obtain the pressure distribution. The results are

$$Q = V\frac{h_0 h_L}{h_0 + h_L}$$

$$p = P_{\text{Atm}} + \frac{6\mu V\alpha}{h_0 + h_L}\frac{x(L - x)}{(h_0 - \alpha x)^2} \tag{6.3.10}$$

where $h_L = h_0 - \alpha L$ is the clearance of the channel at the end of the lubrication zone.

It is convenient to introduce the geometric parameter $\kappa \equiv h_0/(\alpha L)$ that takes values within the ranges $(1, \infty)$ and $(-\infty, 0)$. In the first case the inclined wall slopes downward, as shown in Figure 6.3.1(a), whereas in the second case it slopes upward, away from the direction of translation. Eqs. (6.3.10) may now be placed in the forms

$$Q = Vh_0\frac{\kappa - 1}{2\kappa - 1}$$

$$p = P_{\text{Atm}} + \frac{6\mu VL}{h_0^2}\frac{\kappa^2}{2\kappa - 1}\frac{\hat{x}(1 - \hat{x})}{(\kappa - \hat{x})^2} \tag{6.3.11}$$

where $\hat{x} = x/L$. It is a straightforward exercise to show that the maximum pressure

$$p_{\text{Max}} = P_{\text{Atm}} + \frac{6\mu VL}{h_0^2}\frac{\kappa}{(\kappa - 1)(2\kappa - 1)} \tag{6.3.12}$$

occurs at $x = L\kappa/(2\kappa - 1)$. As κ is increased from unity to infinity, the location of maximum pressure is shifted from $x = L$ to $L/2$.

The y component of the hydrodynamic force exerted on the sloped surface can be approximated with the negative of the normal component of the force exerted on the planar surface, which is found by integrating the pressure over the planar surface,

$$F_N = \int_0^L p\, dx = P_{\text{Atm}}L + \frac{6\mu VL^2}{h_0^2}\kappa^2\left(\ln\frac{h_0}{h_L} - 2\frac{h_0 - h_L}{h_0 + h_L}\right)$$

$$= P_{\text{Atm}}L + \frac{6\mu VL^2}{h_0^2}G(\kappa) \tag{6.3.13}$$

where

$$G(\kappa) \equiv \kappa^2 \left(\ln \frac{\kappa}{\kappa - 1} - \frac{2}{2\kappa - 1} \right) \tag{6.3.14}$$

The second term on the right-hand side of Eq. (6.3.13) is the lubrication lifting or *load force* F_L.

In Figure 6.3.3(b) we plot the function $G(\kappa)$ for positive values of κ. The results show that the lubrication force is positive when the planar wall moves in the direction of the minimum gap. In this case, given V, the lift force will be able to balance the weight of an overlying object whose lower surface is represented by the inclined plane, provided that κ is sufficiently close to unity. Alternatively, given κ, the lift force will be able to balance the weight of an overlying object provided that the velocity V is sufficiently large. When $\kappa < 0$, in which case the plane wall moves in the direction of the maximum gap, the lubrication force will pull the object toward to moving plane, thereby closing the gap and chocking the flow.

Industrial applications of hydrodynamic lubrication are discussed in a comprehensive monograph by Hamrock (1994).

Flow of a Film down a Plane Wall

As a further application of the lubrication approximation, we consider the two-dimensional flow of a liquid film down an inclined plane wall; the film thickness $h(x, t)$ is a function of downstream location x and time t, as depicted in Figure 6.3.4. Assuming that the spatial variation of the film thickness is small, $\partial h/\partial x < 1$, and the flow is nearly unidirectional at every position, we use the first of Eqs. (5.1.7) and the second of Eqs. (5.1.8) to write

$$u(x, y, t) = \frac{-p_x + g \sin \theta_0}{2\nu} y(2h - y)$$

$$Q(x, t) = \frac{(-p_x + \rho g \sin \theta_0) h^3}{3\mu} \tag{6.3.15}$$

where the subscript x signifies partial differentiation with respect to x.

Hydrostatic variations aside, the pressure within the film is assumed to be independent of the transverse y position. Taking into consideration the pressure drop across the free surface due to surface tension, we find

$$p(x, y) = \rho g \cos \theta_0 (h - y) + P_{\text{Atm}} + \gamma \kappa$$
$$\cong \rho g \cos \theta_0 (h - y) + P_{\text{Atm}} - \gamma h_{xx} \tag{6.3.16}$$

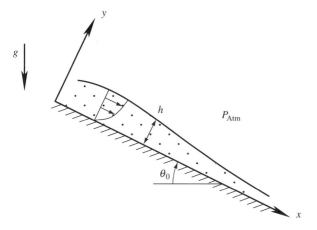

Figure 6.3.4 Evolution of a liquid film flowing down an inclined plane.

where $\kappa \simeq -h_{xx}$ is the curvature of the free surface in the xy plane. Combining the preceding two equations, we find that the flow rate along the x axis is given by

$$Q(x, t) = \frac{gh^3 \sin \theta_0}{3\nu} \left(1 - h_x \cot \theta_0 + \frac{\gamma}{\rho g \sin \theta_0} h_{xxx} \right) \tag{6.3.17}$$

Conservation of mass requires that the film thickness and flow rate satisfy the equation

$$h_t = -Q_x \tag{6.3.18}$$

where the subscripts t and x designate differentiation with respect to t and x. Substituting Eq. (6.3.17) into Eq. (6.3.18), we derive a fourth-order nonlinear partial differential equation describing the evolution of the film thickness

$$h_t + \frac{g \sin \theta_0}{3\nu} \left[h^3 \left(1 - h_x \cot \theta_0 + \frac{\gamma}{\rho g \sin \theta_0} h_{xxx} \right) \right]_x = 0 \tag{6.3.19}$$

In terms of the dimensionless variables $\hat{h} = h/H$, $\hat{x} = x/H$, and $\hat{t} = tgH \sin \theta_0/(2\nu)$, where H is a reference length scale presently identified with the mean film thickness, Eq. (6.3.19) becomes

$$\hat{h}_{\hat{t}} + \tfrac{2}{3}[\hat{h}^3 (1 - \hat{h}_{\hat{x}} \cot \theta_0 + \Gamma \hat{h}_{\hat{x}\hat{x}\hat{x}})]_{\hat{x}} = 0 \tag{6.3.20}$$

where

$$\Gamma \equiv \frac{\gamma}{\rho g H^2 \sin \theta_0} \tag{6.3.21}$$

is the *inverse Bond number* defined with respect to the physical properties of the fluid, mean film thickness, and inclination of the wall.

Film Leveling

In the particular case where the film rests upon a horizontal surface, the evolution equation for the film thickness is subject to substantial simplifications. Let us consider, for example, an uneven coating of a paint resting upon a flat horizontal surface and evolving under the action of gravity and surface tension. Assuming that at every x position the flow is nearly unidirectional and the velocity profile is parabolic, we find that the evolution of the film thickness is governed by the simplified version of Eq. (6.3.19)

$$h_t + \frac{g}{3\nu} \left(-h^3 h_x + \frac{\gamma}{\rho g} h^3 h_{xxx} \right)_x = 0 \tag{6.3.22}$$

In terms of the dimensionless variables $\hat{h} = h/H$, $\hat{x} = x/H$, and $\hat{t} = tgH/2\nu$, we obtain the dimensionless form

$$\hat{h}_{\hat{t}} + \tfrac{2}{3}(-\hat{h}^3 \hat{h}_{\hat{x}} + \Gamma \hat{h}^3 \hat{h}_{\hat{x}\hat{x}\hat{x}})_{\hat{x}} = 0 \tag{6.3.23}$$

where $\Gamma = \gamma/(\rho g H^2)$ is the inverse Bond number.

A numerical method for computing the evolution of a periodic film with dimensionless wavelength $\hat{L} = L/H$ from a certain initial state proceeds by dividing one period of the film located in the range $0 < \hat{x} < \hat{L}$ into N evenly spaced intervals that are separated by the grid points $\hat{x}_i = \hat{L}(i - 1)/N$, where $i = 1, \ldots, N + 1$, and then computing the film thickness at the grid points at a sequence of time instants separated by the time interval $\Delta \hat{t}$. Applying Eq. (6.3.23) at the ith grid point, $i = 1, \ldots, N$, and approximating the temporal derivative using a forward difference, we obtain

$$\hat{h}_i^{n+1} = \hat{h}_i^n - \tfrac{2}{3} \frac{\Delta \hat{t}}{\Delta \hat{x}} (F_{i+1}^n - F_{i-1}^n) \tag{6.3.24}$$

where $\hat{h}_i^n$ designates the value of $\hat{h}$ at the ith grid point at the time instant $\hat{t} = n\,\Delta\hat{t}$, and

$$F = -\hat{h}^3\hat{h}_{\hat{x}} + \Gamma\hat{h}^3\hat{h}_{\hat{x}\hat{x}\hat{x}} \qquad (6.3.25)$$

The function F at the grid points may be computed using the finite-difference approximations compiled in Section B.5, Appendix B. Equation (6.3.24) provides us with an explicit method for updating the film thickness at the grid points. The ratio $\Delta\hat{t}/\Delta\hat{x}$ must be kept sufficiently small in order to prevent the onset of numerical instabilities as discussed in Chapter 12.

Reynolds Lubrication Equation

Let us consider two solid but not necessarily rigid and hence deformable surfaces in close contact, and introduce the Cartesian coordinate system shown in Figure 6.3.5; the gap h between the surfaces is a function of x, y, and time t. The velocities of the upper and lower surfaces, denoted by $\mathbf{V}^{(1)}$ and $\mathbf{V}^{(2)}$, are allowed to vary in space and time reflecting rigid-body motion or deformation. Applying the kinematic boundary condition (1.6.5) at the upper and lower surface, and subtracting corresponding sides of the resulting expressions, we derive an evolution equation for h,

$$\frac{\partial h}{\partial t} = (\mathbf{V}^{(2)} - \mathbf{V}^{(1)}) \cdot \nabla_3[h(x, y, t) - z] \qquad (6.3.26)$$

where ∇_3 is the usual three-dimensional gradient operator. Conservation of mass requires

$$\frac{\partial h}{\partial t} = -\nabla \cdot (h\mathbf{U}) \qquad (6.3.27)$$

where $\mathbf{U}$ is the mean velocity of the fluid along the gap and ∇ is the two-dimensional gradient operator acting in the xy plane. It will be noted that Eq. (6.3.27) is a generalization of Eq. (6.3.18).

To this end, we adopt the approximations of lubrication flow, which amounts to using Eq. (5.1.5) to write

$$\mathbf{U}(x, y, t) = \tfrac{1}{2}(\mathbf{V}^{(1)} + \mathbf{V}^{(2)}) - \frac{h^2}{12\mu}\nabla P \qquad (6.3.28)$$

for the x and y components of $\mathbf{U}$, where P is the modifed pressure. Finally, we combine Eqs. (6.3.26)–(6.3.28), and thereby obtain the *Reynolds lubrication equation* governing the

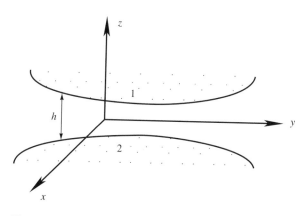

Figure 6.3.5 Lubrication flow between two solid but deformable surfaces in close contact.

distribution of the modified pressure,

$$\nabla \cdot \left(\frac{h^3}{12\mu} \nabla P \right) = \tfrac{1}{2} \nabla \cdot [h(\mathbf{V}^{(1)} + \mathbf{V}^{(2)})] + (\mathbf{V}^{(2)} - \mathbf{V}^{(1)}) \cdot \nabla_3 [h - z] \qquad (6.3.29)$$

Successive terms in Eq. (6.3.29) represent the effect of pressure-driven flow, Couette flow, and normal and translational squeezing motion.

Squeezing flow between two disks

To illustrate the application of the Reynolds lubrication equation, let us consider an axisymmetric flow between two parallel coaxial planar disks representing, for instance, the flattened surfaces of two colliding bodies. The upper disk moves against the stationary lower disk with velocity U. Introducing cylindrical polar coordinates with the origin at the center of the disks located midway between the disks, and the x axis pointing toward the moving disk, we write $\mathbf{V}^{(1)} = -U\mathbf{k}$ and $\mathbf{V}^{(2)} = \mathbf{0}$, where $\mathbf{k}$ is the unit vector along the z axis, and find that Eq. (6.3.29) reduces to

$$\frac{1}{\sigma} \frac{\partial}{\partial \sigma} \left(\sigma \frac{\partial P}{\partial \sigma} \right) = -\frac{12\mu U}{h^3} \qquad (6.3.30)$$

Integrating once with respect to σ and requiring that the pressure gradient be finite at the origin, we obtain

$$\frac{\partial P}{\partial \sigma} = -\frac{6\mu U}{h^3} \sigma \qquad (6.3.31)$$

which shows that the pressure gradient increases linearly with radial distance. The radial velocity profile is found by integrating the lubrication equation $\partial P / \partial \sigma = \mu \, \partial^2 u_\sigma / \partial z^2$, yielding

$$u_\sigma = -\frac{3U\sigma}{h^3} \left(z^2 - \frac{h^2}{4} \right) \qquad (6.3.32)$$

The axial velocity profile arises by integrating the continuity equation subject to Eq. (6.3.32), taking into account the boundary conditions at the upper and lower disks (Problem 6.3.3).

Integrating Eq. (6.3.31) once with respect to σ, we obtain the modified pressure due to the squeezing flow. The modified force exerted on the disk, excluding the buoyancy force, is then found by integrating the corresponding normal stress over the surface of the upper disk from the origin up to $\sigma = a$, yielding

$$\mathbf{F} = \tfrac{3}{2} \frac{\pi \mu U a^4}{h^3} \mathbf{\kappa} \qquad (6.3.33)$$

The sum of this force and the buoyancy force is equal and opposite to the weight of a body whose lower surface is represented by the disk.

The magnitude of the force shown in Eq. (6.3.33) is equal to that of the force that one must apply in order to remove a circular piece of an adhesive tape from a solid surface, pulling it normal to the surface with velocity U. A thin layer of an adhesive viscous liquid is assumed to separate the surface from the tape.

PROBLEMS

6.3.1   **Lubrication flow in a channel.** With reference to Figure 6.3.3(a), compute the maximum value of the lift force on the inclined surface subject to the constraint that the mean clearance of the channel $\tfrac{1}{2}(h_0 + h_L)$ is held constant.

6.3.2   **Lubrication flow between a flat and a curved surface.** Consider the arrangement in Figure 6.3.3(a), but replace the inclined plane with a section of a circular cylinder of radius a with center at $x = L$. Repeat the lubrication analysis discussed in the text and compute the lift force in terms of h_L, L, and a.

6.3.3   **Squeezing flow between two disks.** (a) Compute the axial velocity profile across the gap. (b) Derive Eq. (6.3.33).

Computer Problem

6.3.4   **Film leveling.** Write a program called *FILMLVL* that computes the evolution of a periodic film according to the finite-difference method described in the text for $\hat{L} = 2\pi$, with initial condition $\hat{h} = 1 + a\sin(\hat{x})$. Run the program, compute and plot transient film profiles at a sequence of times $\hat{t}$ for $a = 0.50$ and $\Gamma = 0, 0.5, 1.0$, and discuss the behavior of the solution in each case.

6.4 | FLOW DUE TO A POINT FORCE

The velocity field due to a point force applied at a certain point within a particular domain of a flow plays an important role in the analysis and computation of a variety of Stokes flows, including those due to the motion of particles and the propulsion of microscopic organisms. Physically, the flow due to a point force may be identified with the flow generated by the slow motion of a small particle in an otherwise quiescent fluid, as discussed in Section 5.6 within the more general context of flow at finite Reynolds number.

The velocity and modified pressure fields due to a point force are found by solving the continuity equation $\nabla \cdot \mathbf{u} = 0$ and the singularly forced Stokes equation

$$-\nabla P + \mu \nabla^2 \mathbf{u} + \mathbf{b}\,\delta(\mathbf{x} - \mathbf{x}_0) = 0 \qquad (6.4.1)$$

which is the linearized version of Eq. (5.6.1); $\mathbf{x}_0$ is the location of the point force, the constant $\mathbf{b}$ represents the direction and magnitude of the point force, and δ is the three-dimensional delta function. The solution of Eq. (6.4.1) must be found subject to a boundary condition that requires that the velocity vanish over a designated stationary solid boundary denoted by S_B, that is, $\mathbf{u} = 0$ when $\mathbf{x}$ is on S_B.

To expedite the solution, we introduce the Green's function tensor $\mathbf{G}$ defined by the equation

$$u_i(\mathbf{x}) = \frac{1}{8\pi\mu} G_{ij}(\mathbf{x}, \mathbf{x}_0)\,b_j \qquad (6.4.2)$$

where the factor $1/(8\pi\mu)$ has been introduced for future convenience. To satisfy the condition that the velocity vanishes on S_B, we require that $\mathbf{G}(\mathbf{x}, \mathbf{x}_0) = 0$ when $\mathbf{x}$ is on S_B.

The vorticity, pressure, and stress fields due to the point force may be expressed in terms of the vorticity tensor, pressure vector, and stress tensors $\mathbf{\Omega}$, $\mathbf{\Pi}$, and $\mathbf{T}$ as

$$\omega_i(\mathbf{x}) = \frac{1}{8\pi\mu}\Omega_{ij}(\mathbf{x}, \mathbf{x}_0)\,b_j$$

$$P(\mathbf{x}) = \frac{1}{8\pi}\Pi_j(\mathbf{x}, \mathbf{x}_0)\,b_j \qquad (6.4.3)$$

$$\sigma_{ik}(\mathbf{x}) = \frac{1}{8\pi}T_{ijk}(\mathbf{x}, \mathbf{x}_0)\,b_j$$

The stress tensor $\mathbf{T}$, in particular, is given by

$$T_{ijk}(\mathbf{x}, \mathbf{x}_0) = -\delta_{ik}\Pi_j(\mathbf{x}, \mathbf{x}_0) + \frac{\partial G_{ij}(\mathbf{x}, \mathbf{x}_0)}{\partial x_k} + \frac{\partial G_{kj}(\mathbf{x}, \mathbf{x}_0)}{\partial x_i} \tag{6.4.4}$$

It will be noted that $T_{ijk} = T_{kji}$, as required by the symmetry of the stress tensor $\boldsymbol{\sigma}$. When the domain of flow is unbounded, all $\boldsymbol{\Omega}$, $\boldsymbol{\Pi}$, and $\mathbf{T}$ are required to vanish as $\mathbf{x}$ tends to infinity.

Properties of the Green's Functions

Taking the divergence of Eq. (6.4.2) and using the continuity equation, we obtain the identity

$$\frac{\partial G_{ij}(\mathbf{x}, \mathbf{x}_0)}{\partial x_i} = 0 \tag{6.4.5}$$

Substituting the expressions on the right-hand side of Eqs. (6.4.3) into Eq. (6.4.1), we derive the relations

$$-\frac{\partial \Pi_j(\mathbf{x}, \mathbf{x}_0)}{\partial x_k} + \nabla^2 G_{kj}(\mathbf{x}, \mathbf{x}_0) = -8\pi\,\delta_{kj}\,\delta(\mathbf{x} - \mathbf{x}_0) \tag{6.4.6}$$

$$\frac{\partial T_{ijk}(\mathbf{x}, \mathbf{x}_0)}{\partial x_i} = \frac{\partial T_{kji}(\mathbf{x}, \mathbf{x}_0)}{\partial x_i} = -8\pi\,\delta_{kj}\,\delta(\mathbf{x} - \mathbf{x}_0) \tag{6.4.7}$$

which are equivalent statements of the Stokes equation written for the Green's function.

Point Force in Free Space

One way of computing the flow due to a point force in an infinite domain with no interior boundaries is to express the delta function on the right-hand side of Eq. (6.4.1) as $\delta(\mathbf{x} - \mathbf{x}_0) = -(1/4\pi)\nabla^2(1/r)$, where $r = |\mathbf{x} - \mathbf{x}_0|$. Recalling that the modified pressure is a harmonic function and balancing the dimensions of the pressure term with those of the delta function in Eq. (6.4.1), we set $P = -(1/4\pi)\mathbf{b} \cdot \nabla(1/r)$ and find that Eq. (6.4.1) becomes

$$\mu\,\nabla^2\mathbf{u} = -\frac{1}{4\pi}\mathbf{b} \cdot \left(\nabla\nabla\frac{1}{r} - \mathbf{I}\nabla^2\frac{1}{r}\right) \tag{6.4.8}$$

Next we express the velocity in terms of a scalar function H in the form

$$\mathbf{u} = \mathbf{b} \cdot (\nabla\nabla H - \mathbf{I}\nabla^2 H) \tag{6.4.9}$$

Note that the continuity equation is satisfied for any choice of H. Substituting Eq. (6.4.9) into Eq. (6.4.8) and discarding the arbitrary constant $\mathbf{b}$ yields

$$(\nabla\nabla - \mathbf{I}\nabla^2)\left(\mu\,\nabla^2 H + \frac{1}{4\pi r}\right) = 0 \tag{6.4.10}$$

which is satisfied by any solution of Poisson's equation $\nabla^2 H = -1/(4\pi\mu r)$. Taking the Laplacian of the last equation shows that H is the fundamental solution of the biharmonic equation, $\mu\,\nabla^4 H = \delta(\mathbf{x} - \mathbf{x}_0)$, which is known to be $H = -r/(8\pi\mu)$. Substituting this expression into Eq. (6.4.9), we obtain the flow due to a point force in the form of Eq. (6.4.2) where the free-space Green's function $\mathbf{G}$, also called the *Stokeslet* or the *Oseen–Burgers tensor* and denoted by $\mathbf{S}$, is given by

$$S_{ij}(\hat{\mathbf{x}}) = \frac{\delta_{ij}}{r} + \frac{\hat{x}_i\hat{x}_j}{r^3} \tag{6.4.11}$$

where $r = |\hat{\mathbf{x}}|$, $\hat{\mathbf{x}} = \mathbf{x} - \mathbf{x}_0$. The corresponding vorticity, pressure, and stress fields are given by Eq. (6.4.3), with

$$\Omega_{ij}(\hat{\mathbf{x}}) = 2\varepsilon_{ijl}\frac{\hat{x}_l}{r^3}, \qquad \Pi_i(\hat{\mathbf{x}}) = 2\frac{\hat{x}_i}{r^3}, \qquad T_{ijk}(\hat{\mathbf{x}}) = -6\frac{\hat{x}_i\hat{x}_j\hat{x}_k}{r^5} \qquad (6.4.12)$$

The expression for $\mathbf{T}$ is found by substituting the expressions for $\mathbf{S}$ and $\mathbf{\Pi}$ into Eq. (6.4.4).

The Stokes stream function due to a point force located at the origin and pointing along the x axis with corresponding unit vector $\mathbf{i}$ is given by

$$\Psi = \frac{\mathbf{b}\cdot\mathbf{i}}{8\pi\mu}r\sin^2\theta \qquad (6.4.13)$$

The associated streamline pattern is shown in Figure 6.4.1(a).

As an exercise, let us compute the modfied surface stress exerted on a fluid sphere of radius r centered at a point force. For simplicity, we shall omit the superscript *Mod*. Using Eqs. (6.4.3) and (6.4.12), we find

$$f_i(\mathbf{x}) = \sigma_{ik}(\mathbf{x})\,n_k(\mathbf{x}) = \frac{1}{8\pi}T_{ijk}(\mathbf{x},\mathbf{x}_0)\,n_k(\mathbf{x})\,b_j = -\frac{3}{4\pi}\frac{\hat{x}_i\hat{x}_j}{r^4}b_j \qquad (6.4.14)$$

The corresponding hydrodynamic force exerted on the sphere is given by

$$F_i = \int_{\text{Sphere}} f_i(\mathbf{x})\,dS(\mathbf{x}) = -\frac{3}{4\pi}\frac{1}{r^4}\int_{\text{Sphere}}\hat{x}_i\hat{x}_j\,dS(\mathbf{x})\,b_j \qquad (6.4.15)$$

Using the divergence theorem, we compute

$$\int_{\text{Sphere}}\hat{x}_i\hat{x}_j\,dS(\mathbf{x}) = r\int_{\text{Sphere}}\hat{x}_i n_j\,dS(\mathbf{x}) = r\int_{\text{Sphere}}\frac{\partial\hat{x}_i}{\partial\hat{x}_j}\,dV(\mathbf{x}) = \delta_{ij}\tfrac{4}{3}\pi r^4 \quad (6.4.16)$$

which shows that $\mathbf{F} = -\mathbf{b}$ independently of the radius r. The torque with respect to the location of a point force on any surface that encloses the point force can be shown to vanish (Problem 6.4.3).

Point Force above a Plane Wall

The flow due to a point force located above a plane wall, first computed by Lorentz (1907), arises in the analysis of several types of flows in bounded domains; one example is the flow due to the motion of a cilium that is tethered to a planar surface. Assuming that the wall is located at $x = w$, we require that the corresponding Green's function satisfy the boundary condition $\mathbf{G}(x = w, y, z; \mathbf{x}_0) = \mathbf{0}$. Blake (1971) showed that $\mathbf{G}$ may be constructed from a Stokeslet and a few image singularities including a point force, a potential dipole, and a point-force doublet, as

$$\mathbf{G}(\mathbf{x},\mathbf{x}_0) = \mathbf{S}(\hat{\mathbf{x}}) - \mathbf{S}(\hat{\mathbf{X}}) + 2h_0^2\,\mathbf{D}(\hat{\mathbf{X}}) - 2h_0\,\mathbf{S}^D(\hat{\mathbf{X}}) \qquad (6.4.17)$$

where $\mathbf{S}$ is the Stokeslet, $h_0 = x_0 - w$, $\hat{\mathbf{x}} = \mathbf{x} - \mathbf{x}_0$, $\hat{\mathbf{X}} = \mathbf{x} - \mathbf{x}_0^{\text{Im}}$, and $\mathbf{x}_0^{\text{Im}} = (2w - x_0, y_0, z_0)$ is the image of $\mathbf{x}_0$ with respect to the wall. The tensors $\mathbf{D}$ and $\mathbf{S}^D$ contain, respectively, potential dipoles and Stokeslet doublets (see Section 6.5) and are given by

$$D_{ij}(\mathbf{x}) = \pm\frac{\partial}{\partial x_j}\left(\frac{x_i}{|\mathbf{x}|^3}\right) = \pm\left(\frac{\delta_{ij}}{|\mathbf{x}|^3} - 3\frac{x_ix_j}{|\mathbf{x}|^5}\right)$$

$$S_{ij}^D(\mathbf{x}) = \pm\frac{\partial S_{i1}}{\partial x_j} = x_1 D_{ij}(\mathbf{x}) \pm \frac{\delta_{j1}x_i - \delta_{i1}x_j}{|\mathbf{x}|^3} \qquad (6.4.18)$$

with the minus sign for $j = 1$ corresponding to the x direction, and the plus sign for $j = 2, 3$ corresponding to the y and z directions (Pozrikidis, 1992). By analogy with Eq. (6.4.17), we express

(*a*)

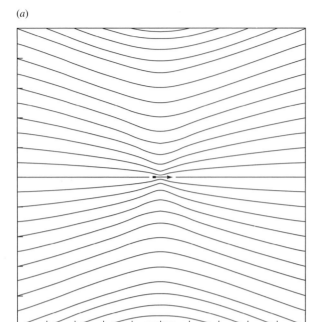

(*b*)

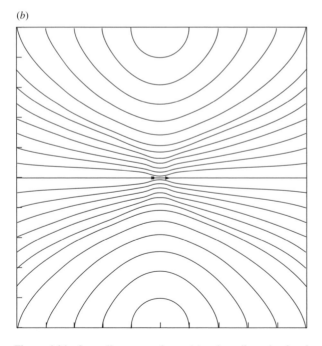

Figure 6.4.1 Streamline pattern due to (a) a three-dimensional and (b) a two-dimensional point force in free space.

the corresponding pressure vector in the form

$$\mathbf{\Pi}(\mathbf{x}, \mathbf{x}_0) = \mathbf{\Pi}^S(\hat{\mathbf{x}}) - \mathbf{\Pi}^S(\hat{\mathbf{X}}) - 2h_0 \mathbf{\Pi}^{SD}(\hat{\mathbf{X}}) \qquad (6.4.19)$$

where $\mathbf{\Pi}^S$ is the pressure vector associated with the Stokeslet given in Eqs. (6.4.12), and

$$\Pi_i^{SD}(\mathbf{x}) = \pm 2 \frac{\partial}{\partial x_i}\left(\frac{x_1}{|\mathbf{x}|^3}\right) = \pm 2\left(\frac{\delta_{i1}}{|\mathbf{x}|^3} - 3\frac{x_1 x_i}{|\mathbf{x}|^5}\right) \qquad (6.4.20)$$

It will be noted that, because the potential dipoles are irrotational singularities, they do not make a contribution to the pressure.

Classification, Computation, and Properties of Green's Functions

The Green's functions may be classified into three categories according to the topology of the domain of flow. First, we have the free-space Green's function for infinite unbounded flow; second, the Green's functions for partially infinite or semi-infinite flow that is bounded by a solid surface; and third, the Green's functions for interior flow in a completely confined domain. As the observation point $\mathbf{x}$ approaches the location of the point force $\mathbf{x}_0$, all Green's functions exhibit a common singular behavior that is identical to that exhibited by the free-space Green's function. The Green's functions for infinite unbounded or bounded flow are required to decay at infinity at a rate that is equal to or lower than that of the free-space Green's function. Explicit expressions of Green's functions for a variety of boundary geometries have been compiled by Pozrikidis (1992).

In Section 6.8 we shall show that the Green's functions satisfy the symmetry property

$$G_{ij}(\mathbf{x}, \mathbf{x}_0) = G_{ji}(\mathbf{x}_0, \mathbf{x}) \qquad (6.4.21)$$

which provides us with a relation between the velocity at the point $\mathbf{x}$ due to a point force placed at the point $\mathbf{x}_0$ and the velocity at the point $\mathbf{x}_0$ due to another point force placed at $\mathbf{x}$. Identity (6.4.21) is useful for checking the correctness of derived Green's functions for domains with involved geometries.

Two-Dimensional Point Force

The flow due to a two-dimensional point force is found by solving Eq. (6.3.1), where δ is now the two-dimensional delta function and ∇^2 is the two-dimensional Laplacian operator. Working as in the case of three-dimensional flow, we express the velocity, vorticity, modifed pressure, and stress in the forms

$$u_i(\mathbf{x}) = \frac{1}{4\pi\mu} G_{ij}(\mathbf{x}, \mathbf{x}_0) b_j \qquad \omega_i(\mathbf{x}) = \frac{1}{4\pi\mu} \Omega_{ij}(\mathbf{x}, \mathbf{x}_0) b_j$$

$$P(\mathbf{x}) = \frac{1}{4\pi} \Pi_j(\mathbf{x}, \mathbf{x}_0) b_j \qquad \sigma_{ik}(\mathbf{x}) = \frac{1}{4\pi} T_{ijk}(\mathbf{x}, \mathbf{x}_0) b_j \qquad (6.4.22)$$

To compute the flow due to a two-dimensional point force in free space we express the two-dimensional delta function on the right-hand side of Eq. (6.4.1) as $\delta(\mathbf{x} - \mathbf{x}_0) = (1/2\pi) \nabla^2 \ln r$. Noting that the pressure is a harmonic function and balancing the dimensions of the modified pressure gradient and the delta function, we set $P = (1/2\pi) \mathbf{b} \cdot \nabla \ln r$. Furthermore, we introduce the auxiliary function H and derive Poisson's equation $\nabla^2 H = (1/2\pi\mu) \ln r$, which shows that H is the Green's function of the biharmonic equation $\mu \nabla^4 H = \delta(\mathbf{x} - \mathbf{x}_0)$; this is known to be $H = (1/8\pi\mu) r^2 (\ln r - 1)$. Substituting this expression into Eq. (6.4.9), we obtain the velocity field in the standard form shown in the first of Eqs. (6.4.22), where $\mathbf{G}$ is the free-space Green's function, or *two-dimensional Stokeslet*, or Oseen–Burgers tensor, denoted by $\mathbf{S}$ and given by

$$S_{ij} = -\delta_{ij} \ln r + \frac{\hat{x}_i \hat{x}_j}{r^2} \qquad (6.4.23)$$

with $r = |\hat{\mathbf{x}}|$ and $\hat{\mathbf{x}} = \mathbf{x} - \mathbf{x}_0$. The associated vorticity, pressure, and stress fields are given by Eqs. (6.4.22) with

$$\Omega_{ij} = 2\varepsilon_{ijk}\frac{\hat{x}_k}{r^2}, \qquad \Pi_i = 2\frac{\hat{x}_i}{r^2}, \qquad T_{ijk} = -4\frac{\hat{x}_i\hat{x}_j\hat{x}_k}{r^4} \tag{6.4.24}$$

The stream functions of the flow due a point force with strength $\mathbf{b}$ pointing in the x or y direction are given, respectively, by

$$\Psi = -\frac{\mathbf{b} \cdot \mathbf{i}}{4\pi\mu}\hat{y}(\ln r - 1), \qquad \Psi = \frac{\mathbf{b} \cdot \mathbf{j}}{4\pi\mu}\hat{x}(\ln r - 1) \tag{6.4.25}$$

where $\mathbf{i}$ and $\mathbf{j}$ are the unit vectors along the x and y axes. The streamline pattern due to a point force oriented along the x axis is depicted in Figure 6.4.1(b).

Using the continuity equation and the Stokes equation, we find that the two-dimensional Green's functions satisfy Eqs. (6.4.5)–(6.4.7), provided that the factor 8π on the right-hand side of the second and third equations is replaced with 4π. The Green's functions $\mathbf{G}$ for partially or totally infinite domains of flow are required to either vanish at infinity or increase, at most, at a logarithmic rate. All Green's functions exhibit a common singular logarithmic behavior at the pole that is identical to that exhibited by the free-space Green's function. A compilation of two-dimensional Green's functions for several boundary geometries has been presented by Pozrikidis (1992).

PROBLEMS

6.4.1   **Three-dimensional Stokeslet via Fourier transforms.** The three-dimensional complex Fourier transform of a function that is defined in the whole three-dimensional space, and its inverse, are related by the equations

$$\hat{f}(\mathbf{k}) = \frac{1}{(2\pi)^{3/2}}\int f(\mathbf{x})\exp(i\,\mathbf{k}\cdot\mathbf{x})\,dV(\mathbf{x}) \tag{6.4.26}$$

$$f(\mathbf{x}) = \frac{1}{(2\pi)^{3/2}}\int \hat{f}(\mathbf{k})\exp(-i\,\mathbf{x}\cdot\mathbf{k})\,dV(\mathbf{k}) \tag{6.4.27}$$

where $\mathbf{k}$ is the wave number. Take the three-dimensional complex Fourier transform of Eq. (6.4.1) and the continuity equation to find

$$\hat{S}_{ij}(\mathbf{k}) = \frac{4}{\sqrt{2\pi}}\frac{1}{|\mathbf{k}|^2}\left(\delta_{ij} - \frac{k_ik_j}{|\mathbf{k}|^2}\right)$$
$$\hat{\Pi}_j(\mathbf{k}) = \frac{4i}{\sqrt{2\pi}}\frac{k_j}{|\mathbf{k}|^2} \tag{6.4.28}$$

Next invert the transformed functions to derive the Stokeslet.

6.4.2   **Two-dimensional Stokeslet via Fourier transforms.** Repeat Problem 6.4.1 for the two-dimensional Stokeslet using the two-dimensional Fourier transform.

6.4.3   **Torque on a surface enclosing a point force.** Using Eq. (6.4.14), show that the torque with respect to the pole of a point force on any surface that encloses the pole of the point force is equal to zero. What is the torque with respect to another point in space?

6.4.4   **Properties of Green's functions.** Prove the identity

$$\frac{\partial}{\partial x_k}[\varepsilon_{ilm}x_lT_{mjk}(\mathbf{x}, \mathbf{x}_0)] = -8\pi\,\varepsilon_{ilj}x_l\,\delta(\mathbf{x} - \mathbf{x}_0) \tag{6.4.29}$$

for three-dimensional flow, and show that it is also valid for two-dimensional flow provided that the factor 8π on the right-hand side is replaced with 4π.

6.4.5   **Properties of Green's functions.** (a) Integrating Eq. (6.4.5) over a volume of fluid that is bounded by the surface D, and using the divergence theorem, derive the identity

$$\int_D G_{ij}(\mathbf{x}, \mathbf{x}_0)\, n_i(\mathbf{x})\, dS(\mathbf{x}) = 0 \tag{6.4.30}$$

which is applicable independently of whether the point $\mathbf{x}_0$ is located inside, on, or outside D. (b) Derive the identities

$$\int_D T_{ijk}(\mathbf{x}, \mathbf{x}_0)\, n_i(\mathbf{x})\, dS(\mathbf{x}) = \int_D T_{kji}(\mathbf{x}, \mathbf{x}_0)\, n_i(\mathbf{x})\, dS(\mathbf{x}) = -\alpha \delta_{jk} \tag{6.4.31}$$

and

$$\int_D \varepsilon_{ilm} x_l T_{mjk}(\mathbf{x}, \mathbf{x}_0)\, n_k(\mathbf{x})\, dS(\mathbf{x}) = -\alpha \varepsilon_{ilj} x_{0,l} \tag{6.4.32}$$

where $\mathbf{n}$ is unit normal vector pointing outward from the control volume, and $x_{0,l}$ on the right-hand side of Eq. (6.4.32) denotes the l component of $\mathbf{x}_0$. The value of the coefficient α on the right-hand sides is equal to 8π, 4π, and 0, depending on whether the point $\mathbf{x}_0$ is located inside, on, or outside the smooth surface D. When $\mathbf{x}_0$ is on D, the integrals are improper but convergent. (c) Show that equations (6.4.30)–(6.4.32) are also valid for the two-dimensional Green's functions provided that the surface integral over D is replaced by a line integral over a closed smooth contour C. The values of the coefficient α on the right-hand sides of Eqs. (6.4.31) and (6.4.32) are equal to 4π, 2π, and 0, depending on whether the point $\mathbf{x}_0$ is located inside, on, or outside C.

Computer Problem

6.4.6   **Flow due to a point force above a plane wall.** Plot the streamline pattern due to a three-dimensional point force located above a plane wall when the point force is oriented (a) perpendicular, and (b) parallel to the wall.

6.5 | FUNDAMENTAL SOLUTIONS OF STOKES FLOW

The linearity of the equations of Stokes flow allows us to construct solutions to particular problems by superposing general fundamental solutions, also called singularities; these satisfy the governing equations but not necessarily the required boundary conditions. The superposition is designed so that flow expressed by a collection or distribution of properly selected fundamental solutions satisfies the boundary conditions in an exact or approximate manner. In the present section we shall develop the fundamental solutions, and in Section 6.6 we shall use them to study several classes of external and internal flows.

Point Source and Derivative Singularities

In Section 6.1 we saw that any irrotational flow satisfies the equations of Stokes flow with an associated constant modified pressure. One example is the flow due to a point source located at the point $\mathbf{x}_0$ described in Eqs. (2.1.25). Differentiating the point source with respect to the singular point $\mathbf{x}_0$, we obtain a sequence of derivative singularities expressing irrotational

TABLE 6.5.1
Velocity and stress fields at the point x due to three-dimensional irrotational singularities of Stokes flow located at the point x_0 in free space; $r = |\hat{x}|$, $\hat{x} = x - x_0$. The associated modified pressure fields are constant.

Point source

$$u_i = \frac{m}{4\pi} S_i, \qquad \sigma_{ik} = \frac{\mu m}{4\pi} T_{ik}^S$$

$$S_i = \frac{\hat{x}_i}{r^3}, \qquad T_{ik}^S = 2\frac{\delta_{ik}}{r^3} - 6\frac{\hat{x}_i \hat{x}_k}{r^5}$$

Potential dipole

$$u_i = \frac{1}{4\pi} D_{ij} d_j, \qquad \sigma_{ik} = \frac{\mu}{4\pi} T_{ijk}^D d_j$$

$$D_{ij} = \frac{\partial S_i}{\partial x_{0,j}} = -\frac{\delta_{ij}}{r^3} + 3\frac{\hat{x}_i \hat{x}_j}{r^5}, \qquad T_{ijk}^D = \frac{\partial T_{ik}^S}{\partial x_{0,j}} = 6\frac{\delta_{ij}\hat{x}_k + \delta_{ik}\hat{x}_j + \delta_{jk}\hat{x}_i}{r^5} - 30\frac{\hat{x}_i\hat{x}_j\hat{x}_k}{r^7}$$

Potential quadruple

$$u_i = \frac{1}{4\pi} Q_{ijl} q_{jl}, \qquad \sigma_{ik} = \frac{\mu}{4\pi} T_{ijlk}^Q q_{jl}$$

$$Q_{ijl} = \frac{\partial D_{ij}}{\partial x_{0,l}} = -3\frac{\delta_{ij}\hat{x}_l + \delta_{il}\hat{x}_j + \delta_{jl}\hat{x}_i}{r^5} + 15\frac{\hat{x}_i\hat{x}_j\hat{x}_l}{r^7}$$

$$T_{ijlk}^Q = \frac{\partial T_{ijk}^D}{\partial x_{0,l}} = -\frac{6}{r^5}(\delta_{ij}\delta_{kl} + \delta_{ik}\delta_{jl} + \delta_{jk}\delta_{il})$$

$$+ \frac{30}{r^7}(\delta_{ij}\hat{x}_k + \delta_{ik}\hat{x}_j + \delta_{jk}\hat{x}_i)\hat{x}_l + \frac{30}{r^7}(\delta_{il}\hat{x}_j\hat{x}_k + \delta_{jl}\hat{x}_i\hat{x}_k + \delta_{kl}\hat{x}_i\hat{x}_j)$$

$$- 210\frac{\hat{x}_i\hat{x}_j\hat{x}_k\hat{x}_l}{r^9}$$

flows, the first three of which are the *potential dipole,* the *potential quadrupole,* and the *potential octuple.*

The velocity and stress fields due to a point source with strength m, a potential dipole with strength $\mathbf{d}$, and a potential quadrupole with strength $\mathbf{q}$, all located at the point x_0, are shown in Table 6.5.1. The associated modified pressure fields are constant and the vorticity vanishes throughout the domain of the flow.

Point Force and Derivative Singularities

Another set of fundamental solutions arises from the point force discussed in Section 6.4. For instance, differentiating the Stokeslet with respect to x_0, we obtain a sequence of derivative singularities representing multipoles of the point force placed in an infinite fluid. The velocity, modified pressure, and stress fields due to a point force with strength equal to $\mathbf{b}$, and a point-force dipole with strength equal to $\mathbf{p}$, both located at the point x_0 in free space, are shown in the first two entries of Table 6.5.2.

Couplet or rotlet

Let us decompose the coefficient of the Stokeslet doublet $\mathbf{p}$ into a symmetric component $\mathbf{s} = \frac{1}{2}(\mathbf{p}+\mathbf{p}^T)$ and an antisymmetric component $\mathbf{r} = \frac{1}{2}(\mathbf{p}-\mathbf{p}^T)$, where the superscript T indicates the transpose, and express the velocity due to the doublet as

$$u_i = \frac{1}{8\pi\mu}(S_{ijl}^{D-S} s_{jl} + S_{ijl}^{D-A} r_{jl}) \qquad (6.5.1)$$

where the superscripts $-S$ and $-A$ denote the symmetric and antisymmetric parts with respect to the indices j and l. Cursory inspection yields

$$S_{ijl}^{D-S} = -\delta_{jl}\frac{\hat{x}_i}{r^3} + 3\frac{\hat{x}_i\hat{x}_j\hat{x}_l}{r^5}, \qquad S_{ijl}^{D-A} = \frac{\delta_{ij}\hat{x}_l - \delta_{il}\hat{x}_j}{r^3} \tag{6.5.2}$$

Exploiting the antisymmetry of **r**, we write

$$r_{jl} = -\tfrac{1}{2}\varepsilon_{jlm}L_m \tag{6.5.3}$$

where the vector **L** is defined as

$$L_m = -\varepsilon_{mjl}\,r_{jl} = -\varepsilon_{mjl}p_{jl} \tag{6.5.4}$$

TABLE 6.5.2
Velocity, modified pressure, and stress fields at the point x due to three-dimensional rotational singularities of Stokes flow located at the point x_0 in free space; $r = |\hat{\mathbf{x}}|$, $\hat{\mathbf{x}} = \mathbf{x} - \mathbf{x}_0$.

Point force

$$u_i = \frac{1}{8\pi\mu}S_{ij}b_j, \qquad P = \frac{1}{8\pi}\Pi_j^S b_j, \qquad \sigma_{ik} = \frac{1}{8\pi}T_{ijk}^S b_j$$

$$S_{ij} = \frac{\delta_{ij}}{r} + \frac{\hat{x}_i\hat{x}_j}{r^3}, \qquad \Pi_j^S = 2\frac{\hat{x}_i}{r^3}, \qquad T_{ijk}^S = -6\frac{\hat{x}_i\hat{x}_j\hat{x}_k}{r^5}$$

Point-force dipole

$$u_i = \frac{1}{8\pi\mu}S_{ijl}^D p_{jl}, \qquad P = \frac{1}{8\pi}\Pi_{jl}^{SD} p_{jl}, \qquad \sigma_{ik} = \frac{1}{8\pi}T_{ijlk}^{SD} p_{jl}$$

$$S_{ijl}^D = \frac{\partial S_{ij}}{\partial x_{0,l}} = \frac{\delta_{ij}\hat{x}_l - \delta_{il}\hat{x}_j - \delta_{jl}\hat{x}_i}{r^3} + 3\frac{\hat{x}_i\hat{x}_j\hat{x}_l}{r^5}$$

$$\Pi_{jl}^{SD} = 2\frac{\partial}{\partial x_{0,l}}\left(\frac{\hat{x}_j}{r^3}\right) = -2\frac{\delta_{jl}}{r^3} + 6\frac{\hat{x}_j\hat{x}_l}{r^5}$$

$$T_{ijlk}^{SD} = -6\frac{\partial}{\partial x_{0,l}}\left(\frac{\hat{x}_i\hat{x}_j\hat{x}_k}{r^5}\right) = 6\frac{\delta_{il}\hat{x}_j\hat{x}_k + \delta_{jl}\hat{x}_i\hat{x}_k + \delta_{kl}\hat{x}_i\hat{x}_j}{r^5} - 30\frac{\hat{x}_i\hat{x}_j\hat{x}_k\hat{x}_l}{r^7}$$

Couplet or rotlet

$$u_i = \frac{1}{8\pi\mu}C_{im}L_m, \qquad P = \text{constant}, \qquad \sigma_{ik} = \frac{1}{8\pi}T_{imk}^C L_m$$

$$C_{im} = \varepsilon_{iml}\frac{\hat{x}_l}{r^3}, \qquad T_{imk}^C = -\tfrac{1}{2}\varepsilon_{jlm}T_{ijlk}^{SD} = 3\frac{\varepsilon_{ijm}\hat{x}_k + \varepsilon_{kjm}\hat{x}_i}{r^5}\hat{x}_j$$

Stresslet

$$u_i = \frac{1}{8\pi\mu}S_{ijl}^{STR} s_{jl}, \qquad P = \frac{1}{8\pi}\Pi_{jl}^{STR} s_{jl}, \qquad \sigma_{ik} = \frac{1}{8\pi}T_{ijlk}^{STR} s_{jl}$$

$$S_{ijl}^{STR} = 3\frac{\hat{x}_i\hat{x}_j\hat{x}_l}{r^5}, \qquad \Pi_{jl}^{STR} = \Pi_{jl}^{SD}$$

$$T_{ijlk}^{STR} \equiv -\delta_{ik}\Pi_{jl}^{STR} + \frac{\partial S_{ijl}^{STR}}{\partial x_k} + \frac{\partial S_{kjl}^{STR}}{\partial x_i} = \tfrac{1}{2}(T_{ijlk}^{SD} + T_{iljk}^{SD}) + \delta_{lj}T_{ik}^S$$

$$= \delta_{ik}\delta_{jl}\frac{2}{r^3} + \frac{3}{r^5}(\delta_{ij}\hat{x}_k\hat{x}_l + \delta_{il}\hat{x}_k\hat{x}_j + \delta_{kj}\hat{x}_i\hat{x}_l + \delta_{kl}\hat{x}_i\hat{x}_j)$$

$$- 30\frac{\hat{x}_i\hat{x}_j\hat{x}_k\hat{x}_l}{r^7}$$

(*CONTINUED*)

T^S is the stress tensor corresponding to the point source

TABLE 6.5.2 (CONTINUED)
Velocity, modified pressure, and stress fields at the point x
due to three-dimensional rotational singularities of Stokes
flow located at the point x_0 in free space; $r = |\hat{x}|$, $\hat{x} = x - x_0$.

Point-force quadruple

$$u_i = \frac{1}{8\pi\mu} S^{SQ}_{ijlm} t_{jlm}, \qquad P = \frac{1}{8\pi} \Pi^{SQ}_{jlm} t_{jlm}, \qquad \sigma_{ik} = \frac{1}{8\pi} T^{SQ}_{ijlmk} t_{jlm}$$

$$S^{SQ}_{ijlm} = \frac{\partial^2 S_{ij}}{\partial x_{0,l}\,\partial x_{0,m}} = \frac{1}{r^3}(\delta_{il}\delta_{jm} + \delta_{im}\delta_{il} - \delta_{ij}\delta_{lm})$$

$$- \frac{3}{r^5}(\delta_{lm}\hat{x}_i\hat{x}_j + \delta_{jm}\hat{x}_i\hat{x}_l + \delta_{jl}\hat{x}_i\hat{x}_m + \delta_{im}\hat{x}_j\hat{x}_l$$

$$+ \delta_{il}\hat{x}_j\hat{x}_m - \delta_{ij}\hat{x}_l\hat{x}_m) + 15\frac{\hat{x}_i\hat{x}_j\hat{x}_l\hat{x}_m}{r^7}$$

$$\Pi^{SQ}_{jlm} = \frac{\partial \Pi^{SD}_{jl}}{\partial x_{0,m}} = -\frac{6}{r^5}(\delta_{jl}\hat{x}_m + \delta_{jm}\hat{x}_l + \delta_{lm}\hat{x}_j) + 30\frac{\hat{x}_j\hat{x}_l\hat{x}_m}{r^7}$$

Stokeson

$$u_i = S^{STN}_{ij} e_j, \qquad P = \mu\Pi^{STN}_j e_j, \qquad \sigma_{ik} = \mu\, T^{STN}_{ijk} e_j$$

$$S^{STN}_{ij} = 2r^2\delta_{ij} - \hat{x}_i\hat{x}_j, \qquad \Pi^{STN}_j = 10\hat{x}_j$$

$$T^{STN}_{ijk} = 3(-4\delta_{ik}\hat{x}_j + \delta_{ij}\hat{x}_k + \delta_{kj}\hat{x}_i)$$

Stokeson dipole

$$u_i = S^{STND}_{ijl} c_{jl}, \qquad P = \mu\Pi^{STND}_{jl} c_{jl}, \qquad \sigma_{ik} = \mu\, T^{STND}_{ijlk} c_{il}$$

$$S^{STND}_{ijl} \equiv \frac{\partial S^{STN}_{ij}}{\partial x_{0,l}} = -4\delta_{ij}\hat{x}_l + \delta_{il}\hat{x}_j + \delta_{jl}\hat{x}_i, \qquad \Pi^{STND}_{jl} = -10\delta_{jl}$$

$$T^{STND}_{ijlk} \equiv \frac{\partial T^{STN}_{ijk}}{\partial x_{0,l}} = 3(-4\delta_{ik}\delta_{jl} + \delta_{ij}\delta_{kl} + \delta_{kj}\delta_{il})$$

---

and recast Eq. (6.5.1) into the form

$$u_i = \frac{1}{8\pi\mu}(S^{D-S}_{ijl} s_{jl} - \tfrac{1}{2}S^{D-A}_{ijl}\varepsilon_{jlm}L_m) \equiv \frac{1}{8\pi\mu}(S^{D-S}_{ijl} s_{jl} + C_{im}L_m) \tag{6.5.5}$$

where we have introduced the new fundamental solution

$$C_{im} \equiv -\tfrac{1}{2}\varepsilon_{jlm}S^{D-A}_{ijl} = -\tfrac{1}{2}\varepsilon_{jlm}S^D_{ijl} = \tfrac{1}{2}\varepsilon_{mlj}\frac{\partial S_{ij}}{\partial x_{0,l}} = \varepsilon_{iml}\frac{\hat{x}_l}{r^3} \tag{6.5.6}$$

termed the *couplet* or *rotlet.* The modified pressure field associated with the rotlet is constant, and the stress field is given in the third entry of Table 6.5.2. Note that the velocity field due to a rotlet of strength **L** may be written as $\mathbf{u} = \mathbf{L} \times \hat{x}/(8\pi\mu r^3)$.

Stresslet

Inspecting the symmetric component of the Stokeslet doublet given in the first of Eqs. (6.5.2), we recognize a point source, and a residual fundamental solution called the *stresslet,* defined as

$$S^{D-S}_{ijl} \equiv \tfrac{1}{2}(S^D_{ijl} + S^D_{ilj}) = -\delta_{jl} S_i + S^{STR}_{ijl} \tag{6.5.7}$$

where **S** is the point source. The velocity, modified pressure, and stress fields due to a stresslet with strength equal to **s** are given in the fourth entry of Table 6.5.2. It worth noting that, apart from a proportionality constant, the stresslet is identical to the stress tensor **T** corresponding to the Stokeslet.

Point-force quadrupole

Differentiating the Stokeslet doublet once with respect to the pole, we obtain the Stokeslet quadrupole shown in the fifth entry of Table 6.5.2. Contracting the last two indices by setting $t_{jlm} = c_j \delta_{lm}$ where **c** is a constant, we obtain the new singularity

$$S_{ijll}^{SQ} = -2\left(-\frac{\delta_{ij}}{r^3} + 3\frac{\hat{x}_i \hat{x}_j}{r^5}\right) \tag{6.5.8}$$

The term in the parentheses on the right-hand side is recognized as the potential dipole **D**, and this shows that

$$D_{ij} = -\tfrac{1}{2}S_{ijll}^{SQ} = -\tfrac{1}{2}\nabla_0^2 S_{ij} \tag{6.5.9}$$

Equation (6.5.9) allows us to express all potential singularities, with the exception of the point source, in terms of derivatives of the Laplacian of the Green's function.

Contribution of Singularities to the Global Properties of a Flow

Inspecting the functional forms of the fundamental solutions discussed previously in this section, we find that the flow rate Q through any closed surface that encloses a singularity vanishes, except for the point source for which $Q = m$ and the stresslet for which $Q = s_{ii}/(2\mu)$.

The modified force **F** exerted on any surface that encloses a singularity vanishes, except for the Stokeslet for which $\mathbf{F} = -\mathbf{b}$. The modified torque **T** with respect to a point $\mathbf{x}_1$ exerted on a spherical surface that encloses a singularity vanishes, except for the point force, the point force dipole, and the couplet, for which

$$\mathbf{T} = (\mathbf{x}_1 - \mathbf{x}_0) \times \mathbf{b}, \qquad T_i = \varepsilon_{ijl}p_{jl}, \qquad \mathbf{T} = -\mathbf{L} \tag{6.5.10}$$

Interior Flow

It is possible to derive families of fundamental solutions for interior flow that have no singular points within the domain of flow but diverge at infinity. To derive these singularities we recall that the pressure is a harmonic function and set

$$P = 10\mu(\mathbf{x} - \mathbf{x}_0) \cdot \mathbf{e} \tag{6.5.11}$$

where **e** is a constant vector. Substituting this expression into the Stokes equation we obtain a Poisson equation for the velocity $\nabla^2 \mathbf{u} = 10\mathbf{e}$, whose solution is $\mathbf{u} = \mathbf{S}^{STN} \cdot \mathbf{e}$ where $\mathbf{S}^{STN}$ is a new singularity called the *Stokeson,* shown in the sixth entry of Table 6.5.2 (Chwang and Wu, 1975).

The derivatives of the Stokeson with respect to $\mathbf{x}_0$ are legitimate fundamental solutions for interior Stokes flow. Differentiating the Stokeson once, we obtain the Stokeson dipole shown in the last entry of Table 6.5.2. The symmetric part of the Stokeson dipole is the *stresson,* and the antisymmetric part is the *roton.* The stresson represents linear, purely straining flow with vanishing vorticity and constant pressure, and the roton represents rigid-body rotation (Problem 6.5.3).

Flow Bounded by Solid Surfaces

The apparatus of fundamental solutions may be extended in a straightforward manner to include flows that are bounded by solid surfaces. The Green's functions discussed in Section 6.4, representing the flow due to a point force, provide us with one such class of fundamental solutions, and other families may be constructed by differentiating them with respect to the pole. It can be shown that the point source for a totally or partially infinite domain of flow is identical to the pressure vector corresponding to the point force **Π**. A further discussion of the properties and explicit forms of fundamental solutions have been compiled by Pozrikidis (1992).

Two-Dimensional Flow

Fundamental solutions for two-dimensional flow may be derived working as for three-dimensional flow. Examples are the Green's functions discussed in Section 6.4, the point source discussed in Section 2.1, and their derivatives. The fact that the velocity due to a point force diverges at infinity places limits on the usefulness of these singularities for constructing solutions to problems of exterior flow (see end of Section 6.6).

PROBLEMS

6.5.1   **Vorticity due to singularities.** Derive the vorticity fields due to the singularities listed in Table 6.5.2.

6.5.2   **Point source above a plane wall.** Using the expression for the point force above a plane wall discussed in Section 6.4 and the fact that the point source for a totally or partially infinite domain of flow is identical to the pressure vector corresponding to the point force, derive the flow due to a point source located above a plane wall. Interpret your results in terms of image singularities for unbounded flow, and assess whether the flow is irrotational (Pozrikidis, 1992).

6.5.3   **Stresson and roton.** Show that the stresson represents purely straining linear flow with vanishing vorticity and constant pressure, and the roton represents rigid-body rotation.

6.5.4   **Two-dimensional flow.** Derive explicit forms for the velocity, modified pressure, and stress fields of the first five singularities shown in Table 6.5.2 for two-dimensional flow.

 ### Computer Problem

6.5.5   **Flow due to singularities.** Draw and compare the streamline patterns due to (a) a potential dipole, (b) a point force, and (c) a Stokeson, all for three-dimensional flow.

6.6 │ STOKES FLOW PAST OR DUE TO THE MOTION OF RIGID BODIES AND LIQUID DROPS

We proceed now to use the fundamental solutions derived in Section 6.5 to compute the velocity field of several types of flows past or due to the motion of spherical and spheroidal rigid bodies and liquid drops. In Section 6.11 we shall see that the representation of these flows in terms of fundamental solutions allows us to develop generalized Faxen relations that yield the force and torque exerted on a body that is immersed in an arbitrary flow in terms of the values of the incident velocity and its derivatives evaluated at the location of the fundamental solutions.

A Translating Solid Sphere and the Stokes Law

Consider first the flow produced by a solid sphere of radius a translating with velocity $\mathbf{U}$ in an ambient fluid of infinite expanse. Inspecting the functional forms of the various singularities listed in Tables 6.5.1 and 6.5.2 suggests representing the flow in terms of a point force and a potential dipole both placed at the center of the sphere $\mathbf{x}_0$, setting

$$u_i(\mathbf{x}) = \frac{1}{8\pi\mu} S_{ij}(\mathbf{x}, \mathbf{x}_0)\, b_j + \frac{1}{4\pi} D_{ij}(\mathbf{x}, \mathbf{x}_0)\, d_j \qquad (6.6.1)$$

Introducing the explicit forms of the singularities, we obtain

$$u_i(\mathbf{x}) = \frac{1}{8\pi\mu}\left(\frac{\delta_{ij}}{r} + \frac{\hat{x}_i\hat{x}_j}{r^3}\right)b_j + \frac{1}{4\pi}\left(-\frac{\delta_{ij}}{r^3} + 3\frac{\hat{x}_i\hat{x}_j}{r^5}\right)d_j \tag{6.6.2}$$

Requiring the boundary condition $\mathbf{u} = \mathbf{U}$ at $r = a$ yields two algebraic equations for the coefficients of the singularities

$$\mathbf{b}a^2 - 2\mu\mathbf{d} = 8\pi\mu\mathbf{U}a^3, \qquad \mathbf{b}a^2 + 6\mu\mathbf{d} = \mathbf{0} \tag{6.6.3}$$

whose solution is

$$\mathbf{b} = 6\pi\mu a\mathbf{U}, \qquad \mathbf{d} = -\pi a^3\mathbf{U} \tag{6.6.4}$$

Substituting Eqs. (6.6.4) into Eq. (6.6.2) and grouping similar terms, we obtain an explicit expression for the velocity field,

$$u_i(\mathbf{x}) = \frac{1}{4}\frac{a}{r}\left(3 + \frac{a^2}{r^2}\right)U_i + \frac{3}{4}\frac{a}{r}\left(1 - \frac{a^2}{r^2}\right)\frac{\hat{x}_i\hat{x}_j}{r^2}U_j \tag{6.6.5}$$

In spherical polar coordinates with the x axis in the direction of translation and corresponding unit vector $\mathbf{i}$, the Stokes stream function is given by

$$\Psi = \mathbf{U}\cdot\mathbf{i}\frac{1}{4}ar\left(3 - \frac{a^2}{r^2}\right)\sin^2\theta \tag{6.6.6}$$

The corresponding streamline patterns in a stationary frame of reference and in a frame of reference moving with the sphere are depicted in Figure 6.6.1(a, b).

(*a*)

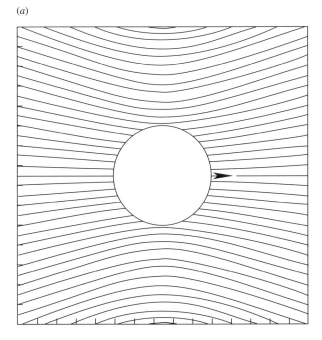

Figure 6.6.1   Streamline pattern of the flow due to a sphere translating under conditions of Stokes flow in (a) a stationary frame of reference. (*continued*)

(*b*)

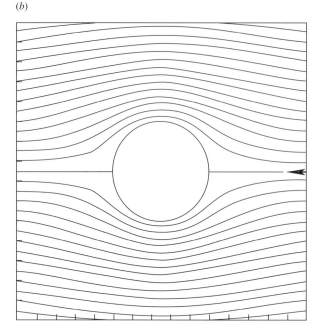

Figure 6.6.1 (*continued*)   Streamline pattern of the flow due to a sphere translating under conditions of Stokes flow in (b) a frame of reference moving with the sphere.

The hydrodynamic traction exerted on the sphere is readily computed from the strength of the singularities using Tables 6.5.1 and 6.5.2, yielding

$$f_i = \frac{1}{8\pi} T^S_{ijk} b_j n_k + \frac{\mu}{4\pi} T^D_{ijk} d_j n_k = -\frac{3}{2} \frac{\mu}{a} U_i \tag{6.6.7}$$

It is interesting to note that $\mathbf{f}$ is a constant vector oriented in the direction of translation; this is a unique and rather fortuitous feature of the spherical geometry.

To compute the hydrodynamic force exerted on the sphere, we can either integrate the traction over the surface of the sphere, or use the properties of the singularities discussed in Section 6.5. Either way we obtain

$$\mathbf{F} = \int_{\text{Sphere}} \mathbf{f} \, dS = -\mathbf{b} = -6\pi\mu a \mathbf{U} \tag{6.6.8}$$

which is known as *Stokes's law*. The torque exerted on the sphere with respect to its center vanishes.

As an application of Stokes's law, we compute the terminal velocity of a sphere that is settling under the action of gravity in an infinite fluid. Requiring that the hydrodynamic drag force exerted on the sphere, given in Eq. (6.6.8), the buoyancy force, and the weight of the sphere balance to zero, we obtain

$$-6\pi\mu a \mathbf{U} + \tfrac{4}{3}\pi a^3 (\rho_s - \rho)\mathbf{g} = \mathbf{0} \tag{6.6.9}$$

Rearranging, we find

$$\mathbf{U} = \frac{2a^2(\rho_s - \rho)}{9\mu} \mathbf{g} \tag{6.6.10}$$

A Sphere in Linear Flow

As a second case study, we consider an infinite linear flow $\mathbf{u}^{\infty} = \mathbf{A} \cdot (\mathbf{x} - \mathbf{x}_0)$ past a stationary solid sphere whose center is located at the point $\mathbf{x}_0$, where $\mathbf{A}$ is a constant velocity gradient tensor with vanishing trace. Inspecting the functional form of the various fundamental solutions discussed in Section 6.5 suggests representing the disturbance flow due to the sphere in terms of a Stokeslet doublet and a potential quadrupole both placed at the center of the sphere, writing

$$u_i(\mathbf{x}) = u_i^{\infty}(\mathbf{x}) + \frac{1}{8\pi\mu} S_{ijl}^{SD}(\mathbf{x}, \mathbf{x}_0)\, p_{jl} + \frac{1}{4\pi} Q_{ijl}(\mathbf{x}, \mathbf{x}_0)\, q_{jl} \qquad (6.6.11)$$

or in explicit form,

$$u_i(\mathbf{x}) = u_i^{\infty}(\mathbf{x}) + \frac{1}{8\pi\mu}\left(\frac{\delta_{ij}\hat{x}_l - \delta_{il}\hat{x}_j - \delta_{jl}\hat{x}_i}{r^3} + 3\frac{\hat{x}_i\hat{x}_j\hat{x}_l}{r^5}\right)p_{jl}$$

$$+ \frac{1}{4\pi}\left(-3\frac{\delta_{ij}\hat{x}_l + \delta_{il}\hat{x}_j + \delta_{jl}\hat{x}_i}{r^5} + 15\frac{\hat{x}_i\hat{x}_j\hat{x}_l}{r^7}\right)q_{jl} \qquad (6.6.12)$$

where $\hat{\mathbf{x}} = \mathbf{x} - \mathbf{x}_0$. Requiring that the velocity at the surface of the sphere vanish yields the equations

$$\mathbf{p} = -\frac{10\mu}{a^2}\mathbf{q}, \qquad \mathbf{A} = \frac{1}{2\pi a^5}[4\mathbf{q} - \mathbf{q}^T - \mathbf{I}\operatorname{Tr}(\mathbf{q})] \qquad (6.6.13)$$

where Tr represents the trace. To solve the second equation for $\mathbf{q}$, we stipulate that $\operatorname{Tr}(\mathbf{q}) = 0$, and split $\mathbf{A}$ and $\mathbf{q}$ into their symmetric and antisymmetric components setting $\mathbf{q} = \mathbf{q}^S + \mathbf{q}^A$, and $\mathbf{A} = \mathbf{E} + \mathbf{\Xi}$ where $\mathbf{E}$ and $\mathbf{\Xi}$ are the rate of deformation and vorticity tensors of the incident flow. Substituting these expressions into the second of Eqs. (6.6.13) yields

$$\mathbf{E} + \mathbf{\Xi} = \frac{1}{2\pi a^5}(3\mathbf{q}^S + 5\mathbf{q}^A) \qquad (6.6.14)$$

which shows that $\mathbf{q}^S = (2\pi/3)a^5\mathbf{E}$, and $\mathbf{q}^A = (2\pi/5)a^5\mathbf{\Xi}$, and therefore,

$$\mathbf{q} = \tfrac{2}{15}\pi a^5(4\mathbf{A} + \mathbf{A}^T) \qquad (6.6.15)$$

Returning to the first of Eqs. (6.6.13), we compute the coefficient of the Stokeslet dipole,

$$\mathbf{p} = -\tfrac{4}{3}\pi\mu a^3(4\mathbf{A} + \mathbf{A}^T) \qquad (6.6.16)$$

The symmetric and antisymmetric components of $\mathbf{p}$ are given by

$$\mathbf{s} = -\tfrac{20}{3}\pi\mu a^3\mathbf{E}, \qquad \mathbf{r} = -4\pi\mu a^3\mathbf{\Xi} \qquad (6.6.17)$$

To compute the coefficient of the couplet that is inherent in the Stokeslet dipole, we use Eq. (6.5.4), obtaining

$$L_m = 4\pi\mu a^3 \varepsilon_{mjl} A_{jl} \qquad (6.6.18)$$

The traction of the disturbance flow exerted on the sphere is given by

$$f_i^{\text{Dist}} = \frac{1}{8\pi} T_{ijlk}^{SD} p_{jl} n_k + \frac{\mu}{4\pi} T_{ijlk}^Q q_{jl} n_k = 3\frac{\mu}{a} A_{ij}\hat{x}_j \qquad (6.6.19)$$

Due to the absence of a point force, the disturbance force exerted on the sphere vanishes. Using the third of Eqs. (6.5.10) and the coefficient of the couplet given in Eq. (6.6.18), we find that the torque with respect to the center of the sphere exerted on the sphere is given by

$$T_m = -L_m = -4\pi\mu a^3 \varepsilon_{mjl} A_{jl} \qquad (6.6.20)$$

Flow Due to a Rotating Sphere

The velocity at the surface of a rigid sphere that rotates about its center with angular velocity Ω is identical to the disturbance velocity due to a sphere that is immersed in a linear flow with

$$A_{ik} = -\varepsilon_{ijk}\Omega_j \qquad (6.6.21)$$

Using the solution for a sphere in linear flow discussed previously and noting that $\mathbf{A}$ is antisymmetric, that is, $\mathbf{A} = -\mathbf{A}^T$, we obtain $\mathbf{q} = (2\pi/5)a^5\mathbf{A}$ and $\mathbf{p} = -4\pi\mu a^3\mathbf{A}$. Furthermore, using the fact that $\mathbf{q}$ is antisymmetric but Q_{ijl} is symmetric with respect to the last two indices, we find that the quadrupole makes a vanishing contribution and may thus be discarded. Finally, using Eq. (6.6.11), we obtain the velocity field

$$u_i = -a^3\tfrac{1}{2}S_{ijl}^{SD}A_{jl} = a^3\tfrac{1}{2}S_{ijl}^{SD}\varepsilon_{jml}\Omega_m = a^3 C_{im}\Omega_m = a^3\varepsilon_{iml}\Omega_m\frac{\hat{x}_l}{r^3} \qquad (6.6.22)$$

which shows that the flow may be represented simply in terms of a couplet with strength $\mathbf{L} = 8\pi\mu a^3\mathbf{\Omega}$ placed at the center of the rotating sphere. The torque exerted on the sphere with respect to its center then follows as

$$\mathbf{T} = -\mathbf{L} = -8\pi\mu a^3\mathbf{\Omega} \qquad (6.6.23)$$

Flow Due to the Rotation or Translation of a Prolate Spheroid

Chwang and Wu (1974) showed that the swirling flow due to a prolate spheroid that rotates with angular velocity Ω about its major axis, which points in the x direction corresponding to the x_1 axis, may be represented in terms of a distribution of couplets over the focal length of the spheroid, as

$$u_i(\mathbf{x}) = \Omega\beta\int_{-c}^{c}(c^2 - x_0^2)\,C_{ix}(\mathbf{x}, \mathbf{x}_0)\,dx_0 \qquad (6.6.24)$$

where

$$\beta = \cfrac{1}{\cfrac{2e}{1-e^2} - \ln\cfrac{1+e}{1-e}} \qquad (6.6.25)$$

$e = c/a$ is the eccentricity of the spheroid, $0 < e < 1$, c is the focal length of the spheroid defined by the equation $c^2 = a^2 - b^2$, and a and b are the major and minor axes of the spheroid. Computing the total strength of the couplets shows that the torque with respect to the origin exerted on the spheroid is equal to $\mathbf{T} = -(32/3)\pi\mu\Omega\beta c^3\mathbf{i}$, where $\mathbf{i}$ is the unit vector in the x direction. In the limit as $e \to 0$, $\beta \to 3/(4e^3)$, yielding the results for the sphere.

Chwang and Wu (1975) showed that the flow generated by the translation of a prolate spheroid may be represented in terms of a distribution of Stokeslets and potential dipoles deployed over the focal length of the spheroid with constant and parabolic densities, respectively, in the form

$$u_i(\mathbf{x}) = U_k\alpha_{kj}\int_{-c}^{c}\left(S_{ij}(\mathbf{x}, \mathbf{x}_0) - \frac{1-e^2}{2e^2}(c^2 - x_0^2)D_{ij}(\mathbf{x}, \mathbf{x}_0)\right)dx_0 \qquad (6.6.26)$$

where $\boldsymbol{\alpha}$ is a diagonal matrix with

$$\alpha_{11} = \cfrac{e^2}{-2e + (1 + e^2)\ln\cfrac{1+e}{1-e}}, \qquad \alpha_{22} = \alpha_{33} = \cfrac{2e^2}{2e - (1 - 3e^2)\ln\cfrac{1+e}{1-e}} \qquad (6.6.27)$$

Note that both the Stokeslets and the potential dipoles in Eq. (6.6.26) point in the direction of translation. Computing the total strength of the point forces shows that the force exerted on the spheroid is given by $\mathbf{F} = -16\pi\mu c\boldsymbol{\alpha} \cdot \mathbf{U}$ in the limit as $e \to 0$, $\alpha_{11}, \alpha_{22}, \alpha_{33} \to 3/(8e)$, recovering the results for the sphere.

A Translating Spherical Liquid Drop

Next we consider the flow due to spherical liquid drop with viscosity $\lambda\mu$ translating with velocity $\mathbf{U}$ in an infinite ambient fluid with viscosity μ. Cursory inspection of the menu of the singularities results in the following selections for the exterior and interior flow

$$u_i^{\text{Ext}}(\mathbf{x}) = \frac{1}{8\pi\mu}S_{ij}(\mathbf{x}, \mathbf{x}_0)\,b_j + \frac{1}{4\pi}D_{ij}(\mathbf{x}, \mathbf{x}_0)\,d_j$$

$$u_i^{\text{Int}}(\mathbf{x}) = c_i + S_{ij}^{\text{STN}}(\mathbf{x}, \mathbf{x}_0)\,e_j \tag{6.6.28}$$

where $\mathbf{S}$, $\mathbf{D}$, and $\mathbf{S}^{\text{STN}}$ are, respectively, the Stokeslet, the potential dipole, and the Stokes on, and $\mathbf{c}$ is a constant. To compute $\mathbf{b}$, $\mathbf{d}$, $\mathbf{c}$, and $\mathbf{e}$ we require four equations.

Requiring that the velocity is continuous across the surface of the drop provides us with the two equations

$$a^2\mathbf{b} - 2\mu a\,\mathbf{d} - 8\pi\mu a^3\mathbf{c} - 16\pi\mu a^5\mathbf{e} = 0$$

$$a^2\mathbf{b} + 6\mu a\,\mathbf{d} + 8\pi\mu a^5\mathbf{e} = 0 \tag{6.6.29}$$

Requiring that, in a frame of reference moving with the drop, the component of the velocity normal to the surface of the drop vanish, that is, $(\mathbf{u}^{\text{Int}} - \mathbf{U}) \cdot \mathbf{n} = 0$, yields the additional equation

$$\mathbf{c} + a^2\mathbf{e} = \mathbf{U} \tag{6.6.30}$$

To obtain the fourth equation, we require that the shear stress is continuous across the interface. Using the formulae in Table 6.5.2, we find that the shear stresses exerted on the exterior and interior side of the interface are given by

$$f_k^{\text{Shear, Ext}} = \frac{1}{4\pi}\left[-3\frac{\hat{x}_i\hat{x}_j}{a^4}b_j + 6\mu\left(\frac{\delta_{ij}}{a^4} - 3\frac{\hat{x}_i\hat{x}_j}{a^6}\right)d_j\right](\delta_{ik} - n_in_k) \tag{6.6.31}$$

$$f_k^{\text{Shear, Int}} = 3\lambda\mu\left(a\delta_{ij} - 3\frac{\hat{x}_i\hat{x}_j}{a}\right)e_j(\delta_{ik} - n_in_k) \tag{6.6.32}$$

Note that the projection operator $\mathbf{I} - \mathbf{nn}$ extracts the tangential component of the surface stress. Setting the right-hand sides of Eqs. (6.6.31) and (6.6.32) equal to one another, we obtain the equation

$$\mathbf{d} = 2\pi\lambda a^5\mathbf{e} \tag{6.6.33}$$

Solving the system of Eqs. (6.6.29), (6.6.30), and (6.6.33) for the coefficients of the fundamental solutions yields

$$\mathbf{b} = 2\pi\mu a\frac{3\lambda + 2}{\lambda + 1}\mathbf{U}, \qquad \mathbf{d} = -\pi a^3\frac{\lambda}{\lambda + 1}\mathbf{U}$$

$$\mathbf{c} = \frac{1}{2}\frac{2\lambda + 3}{\lambda + 1}\mathbf{U}, \qquad \mathbf{e} = -\frac{1}{2}\frac{1}{\lambda + 1}\frac{1}{a^2}\mathbf{U} \tag{6.6.34}$$

In the limit as $\lambda \to \infty$, $\mathbf{b}$ and $\mathbf{d}$ assume the corresponding values for the solid sphere given previously in Eqs. (6.6.4).

Having computed the coefficients of the singularities, we derive the velocity field exterior to and inside the drop,

$$
u_i^{\text{Ext}}(\mathbf{x}) = \frac{1}{4}\frac{1}{\lambda+1}\left[\frac{a}{r}\left(3\lambda+2+\lambda\frac{a^2}{r^2}\right)U_i + \frac{a^3}{r^3}\left(3\lambda+2-3\lambda\frac{a^2}{r^2}\right)\frac{\hat{x}_i\hat{x}_j}{a^2}U_j\right]
$$

$$
u_i^{\text{Int}}(\mathbf{x}) = \frac{1}{2}\frac{1}{\lambda+1}\left[\left(2\lambda+3-2\frac{r^2}{a^2}\right)U_i + \frac{\hat{x}_i\hat{x}_j}{a^2}U_j\right]
$$

(6.6.35)

The streamline patterns for $\lambda = 1$ in a stationary frame of reference and in a frame of reference moving with the drop are depicted in Figure 6.2.2(a, b). Inside the drop, the magnitude of the vorticity increases linearly with distance from the axis, and the flow is identical to that inside Hill's spherical vortex (Figure 2.9.2).

The hydrodynamic drag force exerted on the drop is found from the coefficient of the point force to be

$$
\mathbf{F} = -\mathbf{b} = -2\pi\mu a\frac{3\lambda+2}{\lambda+1}\mathbf{U}
$$

(6.6.36)

This expression was derived independently by Hadamard and Rybczynski in 1911 by solving the appropriate partial-differential equation for the Stokes stream function.

Setting the sum of the hydrodynamic force given in Eq. (6.6.36), the buoyancy force, and the weight of the drop equal to zero, we find that the terminal velocity of a drop that is settling under the action of gravity is given by

$$
\mathbf{U} = \frac{2a^2(\rho_d - \rho)}{3\mu}\frac{\lambda+1}{3\lambda+2}\mathbf{g}
$$

(6.6.37)

As λ tends to infinity we obtain the result for the solid sphere displayed in Eq. (6.6.10).

Drop deformation

The preceding solution was derived without a reference to the distribution of the normal component of the traction at the drop interface. The implicit assumption is that the surface tension is high enough so that even though the difference in the normal component of the traction across the interface may not be uniform, so that it can be balanced by the product of the surface tension and twice the mean curvature $2\gamma/a$, the drop manages to maintain a spherical shape. Using Table 6.5.2 once again, we find that the magnitude of the normal component of the traction on either side of the interface is given by

$$
f^{\text{Normal, Ext}} = -\frac{3}{4\pi a^5}(a^2\mathbf{b} + 4\mu\mathbf{d})\cdot\hat{\mathbf{x}} - \rho\mathbf{g}\cdot\hat{\mathbf{x}} + c_1
$$

$$
f^{\text{Normal, Int}} = -6\lambda\mu\mathbf{e}\cdot\hat{\mathbf{x}} - \rho_d\mathbf{g}\cdot\hat{\mathbf{x}} + c_2
$$

(6.6.38)

where c_1 and c_2 are two constants. Subtracting corresponding sides of these equations, we obtain

$$
f^{\text{Normal, Ext}} - f^{\text{Normal, Int}} = -\frac{3}{4\pi a^5}(a^2\mathbf{b} + 4\mu\mathbf{d} - 8\pi a^5\lambda\mu\mathbf{e})\cdot\hat{\mathbf{x}}
$$
$$
- (\rho - \rho_d)\mathbf{g}\cdot\hat{\mathbf{x}} + c_1 - c_2
$$

(6.6.39)

Substituting Eq. (6.6.37) into Eqs. (6.6.34) and then into Eq. (6.6.39) shows that the jump in the normal component of the traction is constant and equal to $c_1 - c_2 = -2\gamma/a$. Thus the derived solution satisfies *all* required boundary conditions, and *the presence of surface tension is not necessary in order for the drop to maintain the spherical shape.*

(a)

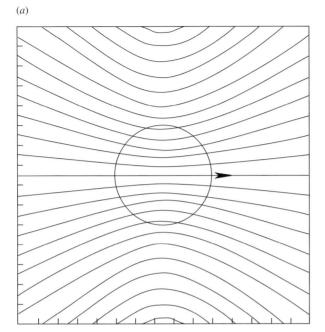

(b)

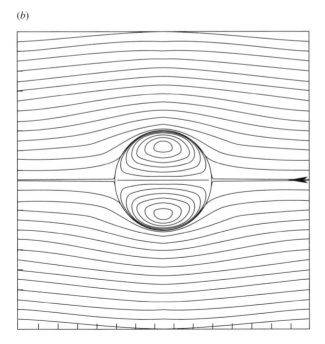

Figure 6.6.2 Streamline pattern due to the translation of a spherical drop in an infinite fluid of same viscosity, $\lambda = 1$, (a) in a stationary frame of reference, and (b) in a frame of reference moving with the drop.

If the drop is perturbed, however, its shape will deform in a manner that depends upon the magnitude of the surface tension and form of the perturbation. When the surface tension is sufficiently large, the drop will restore the spherical shape; otherwise it will either elongate ejecting a tail or transform into a ring (Pozrikidis, 1990).

A Translating Circular Cylinder

The natural choice of singularities for representing the flow due to translating circular cylinder is the two-dimensional Stokeslet **S** and potential dipole **D**, both placed at the center of the cylinder. Since, however, the flow due to a point force increases at a logarithmic rate at infinity, the far-flow condition of vanishing velocity cannot be fulfilled. This is a manifestation of the fact that the Stokes flow approximation ceases to be valid at a certain distance away from the cylinder, and the region of validity of the Stokes flow is confined within a finite region around the cylinder whose size depends upon the magnitude of the Reynolds number.

Deferring the discussion of the singular behavior at infinity to Section 6.13, we note that, since the velocity due to a two-dimensional point force diverges at infinity, it does not do any harm to enhance the singularity representation with a constant term representing a uniform flow, setting

$$u_i(\mathbf{x}) = \frac{1}{4\pi\mu} S_{ij}(\mathbf{x}, \mathbf{x}_0) b_j + \frac{1}{2\pi} D_{ij}(\mathbf{x}, \mathbf{x}_0) d_j + c_j \tag{6.6.40}$$

where $\mathbf{x}_0$ is the instantaneous position of the center of the cylinder, and **c** is a constant. Introducing the explicit forms of the singularities, we obtain

$$u_i(\mathbf{x}) = \frac{1}{4\pi\mu}\left(-\delta_{ij}\ln r + \frac{\hat{x}_i\hat{x}_j}{r^2}\right)b_j + \frac{1}{2\pi}\left(-\frac{\delta_{ij}}{r^2} + 2\frac{\hat{x}_i\hat{x}_j}{r^4}\right)d_j + c_j \tag{6.6.41}$$

Requiring the boundary condition $\mathbf{u} = \mathbf{U}$ at the surface of the cylinder at $r = a$, we obtain an indeterminate system of equations for **b**, **d**, and **c**,

$$-\frac{1}{4\pi\mu}\mathbf{b}\ln a - \frac{1}{2\pi a^2}\mathbf{d} + \mathbf{c} = \mathbf{U}, \qquad a^2\mathbf{b} + 4\mu\mathbf{d} = \mathbf{0} \tag{6.6.42}$$

The solution may be written in the form

$$\mathbf{b} = A\,2\pi\mu\mathbf{U}, \qquad \mathbf{d} = -A\,\tfrac{1}{2}\pi a^2\mathbf{U}, \qquad \mathbf{c} = A\,\tfrac{1}{2}(\ln a - \tfrac{1}{2})\mathbf{U} + \mathbf{U} \tag{6.6.43}$$

where A is an undetermined constant. Substituting these expressions into Eq. (6.6.41), we obtain

$$u_i(\mathbf{x}) = \tfrac{1}{2}A\left(-\ln\frac{r}{a} - \tfrac{1}{2} + \tfrac{1}{2}\frac{a^2}{r^2}\right)U_i + A\,\tfrac{1}{2}\left(1 - \frac{a^2}{r^2}\right)\frac{\hat{x}_i\hat{x}_j}{r^2}U_j + U_i \tag{6.6.44}$$

The force exerted on the cylinder is equal to $\mathbf{F} = -\mathbf{b}$.

In Section 6.11 we shall see that the value of A depends upon the magnitude of the Reynolds number $Re = 2a\rho U/\mu$, and can be determined using the method of matched asymptotic expansions.

PROBLEMS

6.6.1 **Flow due to a translating sphere in terms of the Stokes stream function.** Derive the Stokes stream function given in Eq. (6.6.6) by solving Eq. (6.1.8) subject to appropriate boundary conditions.

6.6.2 **Minimum resistance shape.** (a) Consider a family of spheroids with constant volume and variable aspect ratio b/a executing axial or transverse translation. Compute the aspect ratio of the spheroid with the minimum drag force in each case. (b) Repeat part (a) for the torque in the case of rotation around the major axis.

6.6.3   **Drag coefficients.** (a) Show that, in terms of the drag coefficient $c_D = F/(\pi \rho U^2 a^2)$ and Reynolds number $Re = 2a\rho U/\mu$, Eq. (6.6.36) may be placed in the form

$$c_D = \frac{4}{Re} \frac{3\lambda + 2}{\lambda + 1} \tag{6.6.45}$$

(b) Sketch the functional form of c_D versus Re on a linear-log scale, and investigate and discuss the shape of the function $c_D(Re)$ at finite Reynolds numbers (Clift, Grace, and Weber, 1978, Chapter 5).

6.6.4   **Pressure field in irrotational and Stokes flow.** Discuss the differences in the distribution of pressure around a rigid sphere translating under conditions of (a) potential or (b) Stokes flow.

6.7 | COMPUTATION OF SINGULARITY REPRESENTATIONS

Exact singularity representations are known only for a limited class of flows that are bounded by spherical and spheroidal surfaces. Certain examples were presented in Section 6.6, and further representations are collected and discussed by Chwang and Wu (1974, 1975) and Kim and Karrila (1991). To derive approximate representations for more general flows and arbitrary boundary shapes, we resort to asymptotic and numerical methods. The general strategy is to express the flow in terms of discrete or continuous singularity distributions, and then compute the coefficients of the singularities or densities of the distributions, and possibly their location, so as to satisfy the required boundary conditions in some approximate sense.

Burgers (1938, p. 120) represented the disturbance flow due to a sphere that is held stationary in an incident uniform flow in terms of a point force situated at the center. Computing the strength of the point force by requiring that the mean velocity over the surface of the sphere vanish, he observed that the approximate solution reproduces Stokes's law in its exact form. This perfect agreement is explained by noting that the disturbance flow may be represented exactly in terms of a Stokeslet and a potential dipole, and the average velocity of the dipole over the surface of a sphere is equal to zero. Burgers performed a similar computation for the disturbance flow due to a sphere that is held still in a paraboloidal flow and found that the drag force is also in perfect agreement with the exact solution, which can be found readily using the Faxen relations discussed in Section 6.11.

Numerical Computation of Singularity Representation

To develop a formal procedure for computing singularity representations, we introduce a positive functional that expresses the difference between the required boundary conditions and the corresponding boundary values computed using the singularity representation. For instance, when the problem requires that $\mathbf{u} = \mathbf{U}$ over the boundary D, we introduce a positive functional $F(\mathbf{u} - \mathbf{U})$ that vanishes when $\mathbf{u} = \mathbf{U}$. The general strategy is to represent $\mathbf{u}$ in terms of a certain singularity distribution, and then minimize F with respect to the strength of the singularities, and possibly the location of their poles. A popular choice for F is the *least-squares collocation functional*

$$F(\mathbf{u} - \mathbf{U}) = \sum_{m=1}^{M} [\mathbf{u}(\mathbf{x}_m) - \mathbf{U}(\mathbf{x}_m)] \cdot [\mathbf{u}(\mathbf{x}_m) - \mathbf{U}(\mathbf{x}_m)] \tag{6.7.1}$$

where $\mathbf{x}_m$, $m = 1, \ldots, M$, is a collection of M collocation points over D. Note that minimizing F with respect to the strengths of the singularities yields a linear problem. When the force or torque acting on the boundary D is specified, the strengths of the singularities must satisfy additional linear constraints developed on the basis of Eqs. (6.5.10). The choice of other functionals is discussed by Zhou and Pozrikidis (1995).

The least-squares collocation method has found extensive applications in a variety of numerical studies of Stokes flow involving, in particular, suspended particles. In certain cases, instead of using singularity expansions, it is more expedient to use alternative expansions in terms of spherical harmonic functions (Lamb, 1932, Section 335). The state of the art of these methods is reviewed by Kim and Karrila (1991, Chapter 13). A generalized implementation of the singularity method that allows the singularities to relocate as part of the optimization process is discussed by Zhou and Pozrikidis (1995).

Slender-Body Theory

Flow due to slender bodies finds important applications in the field of biofluid dynamics and, in particular, in studies of flow due to flagella and cilia motion. For instance, the flow due to the motion of a collection of cilia may be represented in terms of singularity distributions over the cilia axes. Requiring the boundary conditions yields a system of Fredholm integral equations of the first kind for the strengths of the singularities.

Axial motion

Let us consider the flow due to motion of a slender axisymmetric cylindrical rod of radius b and length $L = 2a$ subtended between $x = -a$ and a where $a \gg b$. When the rod executes axial translation with velocity U_x, we represent the flow in terms of a distribution of point forces with strength per unit length equal to αU_x, oriented along the x axis. The axial component of the velocity at a point on the surface of the rod, at $\sigma = b$, is given by

$$u_x(x) = \frac{\alpha U_x}{8\pi\mu} \int_{-a}^{a} \left(\frac{1}{[b^2 + (x-x_0)^2]^{1/2}} + \frac{(x-x_0)^2}{[b^2 + (x-x_0)^2]^{3/2}} \right) dx_0 \tag{6.7.2}$$

Introducing the new variable $\omega = (x - x_0)/b$ and rearranging the integrand, we obtain the equivalent form

$$u_x(x) = \frac{\alpha U_x}{8\pi\mu} \int_{-B}^{A} \left(\frac{2}{(1+\omega^2)^{1/2}} - \frac{1}{(1+\omega^2)^{3/2}} \right) d\omega \tag{6.7.3}$$

where $A = (a - x)/b$ and $B = (a + x)/b$. Carrying out the integration, we find

$$u_x(x) = \frac{\alpha U_x}{8\pi\mu} \left(2\ln\frac{\sqrt{1+A^2} + A}{\sqrt{1+B^2} - B} - \frac{A}{\sqrt{1+A^2}} - \frac{B}{\sqrt{1+B^2}} \right) \tag{6.7.4}$$

In the limit as b/a tends to zero, A and B tend to infinity, and Eq. (6.7.4) assumes the asymptotic form

$$u_x(x) \cong \frac{\alpha U_x}{4\pi\mu} \{\ln[(\sqrt{1+A^2} + A)(\sqrt{1+B^2} + B)] - 1\} \cong \frac{\alpha U_x}{4\pi\mu} [\ln(4AB) - 1] \tag{6.7.5}$$

The product AB yields a parabolic distribution with respect to x that vanishes at both ends of the rod. Evaluating Eq. (6.7.5) at the midpoint of the rod, and a point that is located at a quarter of the length of the rod away from the tips, yields

$$u_x(0) \cong \frac{\alpha U_x}{4\pi\mu} \left[2\ln\left(2\frac{a}{b}\right) - 1 \right]$$

$$u_x(\tfrac{1}{2}a) \cong \frac{\alpha U_x}{4\pi\mu} \left[2\ln\left(2\frac{a}{b}\right) + \ln 0.75 - 1 \right] \tag{6.7.6}$$

When the aspect ratio a/b is large, the two velocities in Eqs. (6.7.6) are not too far apart from each other, and the choice

$$\alpha = \frac{4\pi\mu}{2\ln(L/b) - 1} \tag{6.7.7}$$

where $L = 2a$ is the length of the rod, provides us with a reasonable singularity representation. It can be shown that the magnitude of the radial component of the velocity at the surface of the rod corresponding to Eq. (6.7.7) is of order b/a, which is small when the aspect ratio is sufficiently large, corroborating the consistency of the approximate representation. The drag force exerted on the rod is given by $\mathbf{F} = -\alpha U_x L\mathbf{i}$, where $\mathbf{i}$ is the unit vector along the x axis.

Transverse motion

Considering next the flow due to the transverse translation of the rod along the y axis with velocity U_y, we introduce a representation in terms of a distribution of point forces with constant density βU_y and potential dipoles with constant density γU_y, both distributed over the axis of the rod and oriented along the y axis. Repeating the preceding analysis, we find that, to leading order, the velocity at the surface of the rod is given by

$$u_x(x) \cong 0,$$

$$u_y(x) \cong \frac{\beta U_y}{8\pi\mu}\left(2\frac{y^2}{b^2} + \ln(4AB)\right) + \frac{\gamma U_y}{2\pi b^2}\left(-1 + 2\frac{y^2}{b^2}\right) \tag{6.7.8}$$

$$u_z(x) \cong \frac{\beta U_y}{4\pi\mu}\frac{yz}{b^2} + \frac{\gamma U_y}{2\pi}\frac{yz}{b^4}$$

To satisfy the no-slip condition, we set $\gamma = -\beta b^2/(4\mu)$, in which case the second of Eqs. (6.7.8) becomes

$$u_y(x) \cong \frac{\beta U_y}{8\pi\mu}[\ln(4AB) + 1] \tag{6.7.9}$$

Working as in Eq. (6.7.5), we then find that the flow due to a slender rod that translates normal to its axis derives by setting

$$\beta = \frac{8\pi\mu}{2\ln(L/b) + 1} \tag{6.7.10}$$

The drag force exerted on the rod is given by $\mathbf{F} = -\beta U_y L\mathbf{j}$, where $\mathbf{j}$ is the unit vector along the y axis. It is interesting to note that the ratio of the magnitudes of the drag-force coefficient in transverse and axial motion, which is equal to the ratio β/α, is approximately equal to two.

Applications and extensions

As an application, let us consider the flow due to an inclined rod settling under the action of gravity at an angle θ with respect to the vertical direction measured in the counter-clockwise direction; $\theta = 0$ yields a vertical rod, and $\theta = \pi/2$ yields a horizontal rod. The reversibility of Stokes flow requires that the rod maintain its initial inclination. Balancing the drag force with the gravitational force yields $U_P/U_N = 2\cot\theta$, where U_P and U_N are the velocities of the rod parallel and normal to its axis. We then find that the ratio between the horizontal drift velocity U_D and vertical settling velocity U_S is given by $U_D/U_S = \sin\theta\cos\theta/(1 + \cos^2\theta)$.

Hancock (1953) and Gray and Hancock (1955) represented the flow due to a moving flagellum in terms of distributions of point forces and potential dipoles. In the simplest version of their theory, the *resistive-force theory,* they assumed that the strength of the singularities is proportional to the local velocity of translation. A comprehensive account and extensions of their analysis are presented by Lighthill (1975, Chapter 3).

Formal asymptotic analyses for bodies with nonaxisymmetric cross-sectional shapes, flexible bodies, bodies whose volume changes in time, and bodies executing general types of motion have shown that, in general, distributions of point forces, potential dipoles, couplets, and point sources will be necessary to represent an arbitrary flow. Further references are provided by Kim and Karrila (1991) and Pozrikidis (1992).

PROBLEM

6.7.1   **Burgers solution.** (a) Following Burgers (1938), let us represent the flow due to the translation of a sphere as $\mathbf{u}(\mathbf{x}) = \mathbf{S}(\mathbf{x}, \mathbf{x}_0) \cdot \mathbf{b}$, where $\mathbf{S}$ is the Stokeslet and $\mathbf{x}_0$ is the center of the sphere. Compute the coefficient $\mathbf{b}$ by requiring that the average velocity on the surface of the sphere be equal to the velocity of translation. Based on the computed value of $\mathbf{b}$, calculate the drag force exerted on the sphere and compare it with the exact value given by Stokes's law. (b) Perform a similar computation for paraboloidal flow past a stationary sphere, $\mathbf{u}^P = U((y^2 + z^2)/a^2, 0, 0)$ where a is the radius of the sphere and U is a constant, and compare your result for the force exerted on the sphere with the exact value given by the Faxen relation, $\mathbf{F} = 4\pi\mu a(U, 0, 0)$ (Section 6.11).

 Computer Problem ———————————————————

6.7.2   **Exact and approximate representations for prolate spheroids.** Plot the distribution of strength of the point forces given by the exact solution (6.6.26), and compare it with the constant distributions given in Eqs. (6.7.7) and (6.7.10) for a family of prolate spheroids with increasing aspect ratio. Compare the exact with the approximate values of the drag force, and estimate the aspect ratio at which the slender-body solution becomes sufficiently accurate.

6.8 | THE LORENTZ RECIPROCAL THEOREM AND ITS APPLICATIONS

Let us consider two generally unrelated Stokes flows with velocities $\mathbf{u}$ and $\mathbf{u}'$ and associated modified stress tensors $\boldsymbol{\sigma}$ and $\boldsymbol{\sigma}'$. Using Eq. (3.3.22) in conjunction with Eq. (6.1.1c), we obtain the Lorentz reciprocal identity

$$\frac{\partial}{\partial x_j}(\mu' u_i' \sigma_{ij} - \mu u_i \sigma_{ij}') = 0 \qquad (6.8.1)$$

which is the counterpart of Green's second identity for harmonic functions (Lorentz, 1907). For brevity, in the ensuing discussion we shall omit the qualifier "modified" when we refer to the modified pressure and stress.

Assuming that the two flows are free of singular points within a certain control volume, we integrate Eq. (6.8.1) over the control volume and use the divergence theorem to obtain

$$\int_D (\mu' u_i' f_i - \mu u_i f_i') \, dS = 0 \qquad (6.8.2)$$

where D is the boundary of the control volume, and $\mathbf{f} = \boldsymbol{\sigma} \cdot \mathbf{n}$ is the boundary traction; the unit normal vector $\mathbf{n}$ may point either into or out from the control volume.

The major strength of the reciprocal identities (6.8.1) and (6.8.2) is that they allow us to obtain information about a certain flow without having to solve the equations of Stokes flow explicitly, but merely by using information about another flow. This will be demonstrated now by discussing certain characteristic examples.

Flow past a Stationary Particle

Consider the flow around a rigid particle that is held stationary in an incident ambient flow with velocity $\mathbf{u}^\infty$. The particle causes a disturbance flow with velocity $\mathbf{u}^D$, which is added to the ambient flow to give the total flow, $\mathbf{u} = \mathbf{u}^\infty + \mathbf{u}^D$. Turning to Eq. (6.8.2), we identify $\mathbf{u}'$ with the velocity field generated when the particle translates with velocity $\mathbf{U}$ in the same fluid, and exploit the linearity of the Stokes equation to express the traction exerted on the particle in the form

$$\mathbf{f}^T = -\mu \mathbf{R}^T \cdot \mathbf{U} \tag{6.8.3}$$

where $\mathbf{R}^T$ is the *traction resistance matrix* for translation, and the superscript T indicates translation.

Next we select a control volume that is enclosed by the surface of the particle P and a surface S_∞ of large size, and apply Eq. (6.8.2) for the pair of flows $\mathbf{u}^T$ and $\mathbf{u}^D$ with $\mu' = \mu$, obtaining

$$\int_{S_\infty, P} \mathbf{u}^T \cdot \mathbf{f}^D \, dS = \int_{S_\infty, P} \mathbf{u}^D \cdot \mathbf{f}^T \, dS \tag{6.8.4}$$

Letting the size of S_∞ tend to infinity and noting that the disturbance velocity at infinity decays at least as fast as $1/r$ while the traction decays at least as fast as $1/r^2$, where r is the distance from a designated center of the particle, we find that the contributions of the surface integrals over S_∞ disappear. Equation (6.8.4) then yields

$$\mathbf{U} \cdot \mathbf{F}^D = \int_P \mathbf{u}^D \cdot \mathbf{f}^T \, dS \tag{6.8.5}$$

where $\mathbf{F}^D$ is the disturbance force exerted on the particle. Requiring the boundary condition $\mathbf{u} = 0$ or $\mathbf{u}^D = -\mathbf{u}^\infty$ on P, using the results of Section 6.1 to set the disturbance force $\mathbf{F}^D$ equal to the total force $\mathbf{F}$, and introducing the definition in Eq. (6.8.3), we finally obtain

$$\mathbf{F} = \mu \int_P \mathbf{u}^\infty \cdot \mathbf{R}^T \, dS \tag{6.8.6}$$

(Brenner, 1964b). Equation (6.8.6) provides us with an expression for the modified force exerted on the particle in terms of the values of the incident velocity $\mathbf{u}^\infty$ over the surface of the particle and the traction resistance matrix for translation. For a spherical particle, $\mathbf{R}^T$ is isotropic, and the force is simply proportional to the average value of the incident velocity $\mathbf{u}^\infty$ over the particle surface.

Next we consider the torque $\mathbf{T}$ exerted on a stationary rigid particle that is immersed in an ambient flow. Following the preceding steps, we introduce the *traction resistance matrix for rotation* $\mathbf{R}^R$ defined by the equation

$$\mathbf{f}^R = -\mu \mathbf{R}^R \cdot \mathbf{\Omega} \tag{6.8.7}$$

where $\mathbf{f}^R$ is the traction exerted on the particle when it rotates about a certain point $\mathbf{x}_0$ with angular velocity $\mathbf{\Omega}$, and find

$$\mathbf{T} = \mu \int_P \mathbf{u}^\infty \cdot \mathbf{R}^R \, dS \tag{6.8.8}$$

which provides us with an expression for the torque in terms of the values of the incident velocity $\mathbf{u}^\infty$ over the surface of the particle and the traction resistance matrix for rotation.

Equations (6.8.6) and (6.8.8) constitute one version of the *generalized Faxen relations* for the force and torque to be discussed further in Section 6.11. To make these results more tangible, we use Eqs. (6.6.7), (6.6.19), and (6.6.21) and find that the traction resistance matrices for a rigid spherical particle of radius a are given by

$$R_{ij}^T = \frac{3}{2a}\delta_{ij}, \qquad R_{ij}^R = \frac{3}{a}\varepsilon_{ijk}\hat{x}_k \tag{6.8.9}$$

where $\hat{x} = x - x_0$ is the distance from the center of the particle x_0. Substituting these expressions into Eqs. (6.8.6) and (6.8.8) we find

$$\mathbf{F} = \tfrac{3}{2}\frac{\mu}{a}\int_P \mathbf{u}^\infty\, dS, \qquad \mathbf{T} = 3\frac{\mu}{a}\int_P \hat{x}\times\mathbf{u}^\infty\, dS \tag{6.8.10}$$

Force and Torque on a Moving Particle

As a further application of the reciprocal theorem, we consider an incident flow with velocity $\mathbf{u}^\infty$ past a suspended rigid particle that translates with velocity $\mathbf{V}$ while rotating about the point x_0 with angular velocity $\boldsymbol{\omega}$. The presence and motion of the particle cause a disturbance flow with velocity $\mathbf{u}^D$, which is added to the incident flow to yield the total flow, $\mathbf{u} = \mathbf{u}^D + \mathbf{u}^\infty$.

Applying the reciprocal identity for the disturbance flow and the flow produced when the particle translates in an otherwise quiescent fluid, we obtain Eq. (6.8.5). Requiring that on the surface of the particle $\mathbf{u} = \mathbf{V}+\boldsymbol{\omega}\times\hat{x}$, or equivalently, $\mathbf{u}^D = \mathbf{V}+\boldsymbol{\omega}\times\hat{x}-\mathbf{u}^\infty$, where $\hat{x} = x-x_0$, and recalling that the disturbance force is equal to the total force, we obtain

$$\mathbf{V}\cdot\mathbf{F}^T + \boldsymbol{\omega}\cdot\mathbf{T}^T = \int_P \mathbf{u}^\infty\cdot\mathbf{f}^T\, dS + \mathbf{U}\cdot\mathbf{F} \tag{6.8.11}$$

where $\mathbf{F}^T$ and $\mathbf{T}^T$ are the force and torque with respect to x_0 exerted on the particle when it translates with velocity $\mathbf{U}$. Working in a similar manner, we obtain

$$\mathbf{V}\cdot\mathbf{F}^R + \boldsymbol{\omega}\cdot\mathbf{T}^R = \int_P \mathbf{u}^\infty\cdot\mathbf{f}^R\, dS + \boldsymbol{\Omega}\cdot\mathbf{T} \tag{6.8.12}$$

where $\mathbf{F}^R$ and $\mathbf{T}^R$ are the force and torque with respect to x_0 exerted on the particle when it rotates with angular velocity $\boldsymbol{\Omega}$.

Next we take advantage of the linear nature of the equations of Stokes flow with respect to the boundary velocity to write

$$\begin{bmatrix}\mathbf{F}^T\\ \mathbf{T}^T\end{bmatrix} = -\mu\begin{bmatrix}\mathbf{X}\\ \mathbf{P}'\end{bmatrix}\cdot\mathbf{U}, \qquad \begin{bmatrix}\mathbf{F}^R\\ \mathbf{T}^R\end{bmatrix} = -\mu\begin{bmatrix}\mathbf{P}\\ \mathbf{Y}\end{bmatrix}\cdot\boldsymbol{\Omega} \tag{6.8.13}$$

where $\mathbf{X}, \mathbf{P}'$ are resistance matrices for translation and $\mathbf{P}, \mathbf{Y}$ are resistance matrices for rotation. Using Eqs. (6.8.13), (6.8.3), and (6.8.7), we recast Eqs. (6.8.11) and (6.8.12) into the forms

$$\mathbf{V}\cdot\mathbf{X} + \boldsymbol{\omega}\cdot\mathbf{P}' = \int_P \mathbf{u}^\infty\cdot\mathbf{R}^T\, dS - \frac{1}{\mu}\mathbf{F}$$

$$\mathbf{V}\cdot\mathbf{P} + \boldsymbol{\omega}\cdot\mathbf{Y} = \int_P \mathbf{u}^\infty\cdot\mathbf{R}^R\, dS - \frac{1}{\mu}\mathbf{T} \tag{6.8.14}$$

Given the force $\mathbf{F}$ and torque $\mathbf{T}$ exerted on the particle and the resistance matrices for translation and rotation, Eqs. (6.8.14) provide us with a system of linear algebraic equations for the translational and angular velocities $\mathbf{V}$ and $\boldsymbol{\omega}$.

Symmetry of Green's Functions

An important consequence of the reciprocal theorem is the symmetry of the Green's functions stated in Eq. (6.4.21). Analogous symmetry properties exist for the Green's functions of potential flow and linear elastostatics and, more generally, for the Green's function of differential equations involving self-adjoint differential operators [Eq. (2.2.18)].

To prove Eq. (6.4.21), we introduce the velocity fields due to two point forces located at the points $\mathbf{x}_1$ and $\mathbf{x}_2$ with respective strengths equal to $\mathbf{a}$ and $\mathbf{b}$, given by

$$u_i(\mathbf{x}) = \frac{1}{8\pi\mu} G_{ik}(\mathbf{x}, \mathbf{x}_1)\, a_k, \qquad u_i'(\mathbf{x}) = \frac{1}{8\pi\mu} G_{ik}(\mathbf{x}, \mathbf{x}_2)\, b_k \qquad (6.8.15)$$

The corresponding modified stress fields satisfy the equations

$$\frac{\partial \sigma_{ij}}{\partial x_j} = -a_i\, \delta(\mathbf{x} - \mathbf{x}_1), \qquad \frac{\partial \sigma_{ij}'}{\partial x_j} = -b_i\, \delta(\mathbf{x} - \mathbf{x}_2) \qquad (6.8.16)$$

where δ is the three-dimensional delta function. Substituting the preceding equations into the right-hand side of Eq. (3.3.22) with $\mu' = \mu$ yields

$$\frac{\partial}{\partial x_j}(u_i' \sigma_{ij} - u_i \sigma_{ij}')$$

$$= -\frac{1}{8\pi\mu}[G_{ik}(\mathbf{x}, \mathbf{x}_2)\, b_k a_i\, \delta(\mathbf{x} - \mathbf{x}_1) - G_{ik}(\mathbf{x}, \mathbf{x}_1)\, a_k b_i\, \delta(\mathbf{x} - \mathbf{x}_2)] \qquad (6.8.17)$$

Next we integrate Eq. (6.8.17) over a control volume that is confined by the solid boundary S_B, over which the Green's function is required to vanish, two spherical surfaces of infinitesimal radii enclosing $\mathbf{x}_1$ and $\mathbf{x}_2$, and, in the case of infinite flow, a large surface that encloses S_B as well as the points $\mathbf{x}_1$ and $\mathbf{x}_2$. Using the divergence theorem, we convert the volume integral on the left-hand side of the resulting equation to a surface integral over all surfaces enclosing the control volume. The surface integral over S_B vanishes, because the Green's function and hence the velocities $\mathbf{u}$ and $\mathbf{u}'$ vanish over S_B. Letting the radius of the large surface tend to infinity and the radii of the small spheres enclosing $\mathbf{x}_1$ and $\mathbf{x}_2$ tend to zero, we find that the corresponding surface integrals make insignificant contributions. As a result, the whole of the integral of the left-hand side of Eq. (6.8.17) vanishes. Using the properties of the delta function to manipulate the right-hand side, we finally arrive at the equation

$$a_k b_i[G_{ki}(\mathbf{x}_1, \mathbf{x}_2) - G_{ik}(\mathbf{x}_2, \mathbf{x}_1)] = 0 \qquad (6.8.18)$$

The validity of Eq. (6.4.21) becomes evident by noting that $\mathbf{a}$ and $\mathbf{b}$ are arbitrary constant vectors.

Repeating the preceding procedure with minor modifications, one may show that the two-dimensional Green's functions also satisfy the symmetry property Eq. (6.4.21) (Problem 6.8.6).

PROBLEMS

6.8.1 **Alternative statement of the reciprocal theorem.** Show that two Stokes flows of a certain fluid with viscosity μ, with velocities $\mathbf{u}$ and $\mathbf{u}'$ and corresponding modified pressure fields P and P', satisfy the alternative reciprocal relation (Happel and Brenner, 1973)

$$\frac{\partial}{\partial x_k}\left[u_i'\left(-\delta_{ik}P + \mu\frac{\partial u_i}{\partial x_k}\right) - u_i\left(-\delta_{ik}P' + \mu\frac{\partial u_i'}{\partial x_k}\right)\right] = 0 \qquad (6.8.19)$$

6.8.2 **Force on a spherical particle in paraboloidal flow.** Show that the force exerted on a spherical particle of radius a placed at the axis of the paraboloidal flow $U((y/a)^2 + (z/a)^2, 0, 0)$, where U is a constant, is equal to $\mathbf{F} = 4\pi\mu U a\,(1, 0, 0)$.

6.8.3 **Symmetry of the resistance matrices.** Using the reciprocal theorem, show that resistance matrices $\mathbf{X}$ and $\mathbf{Y}$ introduced in Eqs. (6.8.13) are symmetric, and $\mathbf{P}'$ is the transpose of $\mathbf{P}$ (Brenner, 1963, 1964a).

6.8.4   **Force on a particle in a tube.** Use the reciprocal theorem to show that the force and torque exerted on a rigid particle that is held stationary in an incident flow through a tube may be computed from a knowledge of the traction exerted on the particle when it translates or rotates within the tube in an otherwise stationary fluid.

6.8.5   **Force and torque from the mobility problem.** Show that the velocity of translation $\mathbf{U}$ and angular velocity of rotation $\boldsymbol{\omega}$ of a force-free and torque-free rigid particle, with $\mathbf{F} = \mathbf{T} = \mathbf{0}$, suspended in an incident flow with velocity $\mathbf{u}^\infty$, may be computed from the equations

$$\mathbf{U} \cdot \mathbf{Q} = \int_P \mathbf{u}^\infty \cdot \mathbf{f}^F \, dS, \qquad \boldsymbol{\omega} \cdot \mathbf{W} = \int_P \mathbf{u}^\infty \cdot \mathbf{f}^T \, dS \qquad (6.8.20)$$

where $\mathbf{f}^F$ and $\mathbf{f}^T$ are the surface tractions exerted on the particle when it moves under the influence of an external force $\mathbf{Q}$ or an external torque $\mathbf{W}$. An important message delivered by these equations is that the translational and angular velocities of a force-free and torque-free particle that is immersed in an arbitrary ambient flow may be computed from a knowledge of the values of the incident velocity over the surface of the particle and the traction exerted on the surface of the particle when it moves under the influence of an external force or torque.

6.8.6   **Symmetry of the two-dimensional Green's functions.** Show that the two-dimensional Green's functions satisfy the symmetry property (6.4.21). Note that the proof presents an apparent but not essential difficulty due to the fact the Green's functions may diverge at infinity at a logarithmic rate (Pozrikidis, 1992).

6.9 | BOUNDARY INTEGRAL REPRESENTATION OF STOKES FLOW

The solution of linear, elliptic, homogeneous partial differential equations may be represented in terms of boundary integrals involving the unknown function and its derivatives (Stakgold, 1967, 1968). One example is the boundary integral representation of harmonic functions discussed in Section 2.3. Another example is Somigliana's identity for the displacement field in the theory of linear elastostatics (Love, 1944, p. 245; Phan-Thien and Kim, 1994). In the case of Stokes flow, we obtain a boundary integral representation involving the boundary values of the velocity and traction.

A convenient starting point for deriving the boundary integral representation is the Lorentz reciprocal identity (6.8.1) applied for a particular flow of interest with velocity $\mathbf{u}$ and modified stress $\boldsymbol{\sigma}$, and the flow due to a point force with strength $\mathbf{b}$ located at a point $\mathbf{x}_0$, with velocity and modified stress

$$u_i'(\mathbf{x}) = \frac{1}{8\pi\mu} G_{ij}(\mathbf{x}, \mathbf{x}_0)\, b_j, \qquad \sigma_{ik}'(\mathbf{x}) = \frac{1}{8\pi} T_{ijk}(\mathbf{x}, \mathbf{x}_0)\, b_j \qquad (6.9.1)$$

Substituting these expressions into Eq. (6.8.1), setting $\mu' = \mu$ and discarding the arbitrary constant $\mathbf{b}$, we obtain the equation

$$\frac{\partial}{\partial x_k}[G_{ij}(\mathbf{x}, \mathbf{x}_0)\, \sigma_{ik}(\mathbf{x}) - \mu\, u_i(\mathbf{x})\, T_{ijk}(\mathbf{x}, \mathbf{x}_0)] = 0 \qquad (6.9.2)$$

which is valid everywhere except at the singular point $\mathbf{x}_0$.

Let us now select a control volume V_c that is bounded by the closed, singly, or multiply connected surface D, which may be composed of interior fluid surfaces, fluid interfaces, or solid surfaces, as illustrated in Figure 6.9.1, and place the point force outside the control volume. Noting that the function within the square brackets in Eq. (6.9.2) is nonsingular throughout V_c, we integrate both sides over V_c and use the divergence theorem to convert the volume integral over

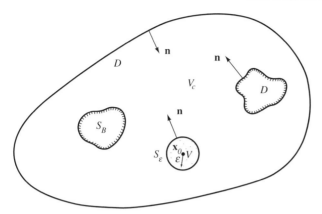

Figure 6.9.1   A control volume used to derive the boundary-integral representation of Stokes flow.

V_c to a surface integral over D, obtaining

$$\int_D [G_{ij}(\mathbf{x}, \mathbf{x}_0)\,\sigma_{ik}(\mathbf{x}) - \mu\,u_i(\mathbf{x})\,T_{ijk}(\mathbf{x}, \mathbf{x}_0)]\,n_k(\mathbf{x})\,dS(\mathbf{x}) = 0 \qquad (6.9.3)$$

In Eq. (6.9.3), as well as in all subsequent equations, the normal vector $\mathbf{n}$ will be assumed to point *into* the control volume V_c.

Next we place the point $\mathbf{x}_0$ inside V_c, and introduce a small spherical volume V_ε with a small radius ε centered at $\mathbf{x}_0$. We note that the function within the square brackets in Eq. (6.9.2) is nonsingular throughout the reduced volume $V_c - V_\varepsilon$, integrate both sides over $V_c - V_\varepsilon$, and use the divergence theorem to convert the volume integral to a surface integral, thereby obtaining

$$\int_{D, S_\varepsilon} [G_{ij}(\mathbf{x}, \mathbf{x}_0)\,\sigma_{ik}(\mathbf{x}) - \mu\,u_i(\mathbf{x})\,T_{ijk}(\mathbf{x}, \mathbf{x}_0)]\,n_k(\mathbf{x})\,dS(\mathbf{x}) = 0 \qquad (6.9.4)$$

where S_ε is the spherical surface enclosing V_ε, as indicated in Figure 6.9.1. Letting the radius ε tend to zero, we find that over S_ε, to leading order in ε, the tensors $\mathbf{G}$ and $\mathbf{T}$ reduce to the Stokeslet and its associated stress tensor

$$G_{ij} \approx \frac{\delta_{ij}}{\varepsilon} + \frac{\hat{x}_i \hat{x}_j}{\varepsilon^3}, \qquad T_{ijk} \approx -6\frac{\hat{x}_i \hat{x}_j \hat{x}_k}{\varepsilon^5} \qquad (6.9.5)$$

where $\hat{\mathbf{x}} = \mathbf{x} - \mathbf{x}_0$. Furthermore, over S_ε, $\mathbf{n} = \hat{\mathbf{x}}/\varepsilon$ and $dS = \varepsilon^2\,d\Omega$, where $d\Omega$ is the differential solid angle. Substituting these expressions along with Eqs. (6.9.5) into Eq. (6.9.4) gives

$$\int_D [G_{ij}(\mathbf{x}, \mathbf{x}_0)\,\sigma_{ik}(\mathbf{x}) - \mu\,u_i(\mathbf{x})\,T_{ijk}(\mathbf{x}, \mathbf{x}_0)]\,n_k(\mathbf{x})\,dS(\mathbf{x})$$

$$= -\int_{S_\varepsilon} \left[\left(\delta_{ij} + \frac{\hat{x}_i \hat{x}_j}{\varepsilon^2} \right)\sigma_{ik}(\mathbf{x}) + 6\mu u_i(\mathbf{x})\frac{\hat{x}_i \hat{x}_j \hat{x}_k}{\varepsilon^4} \right] \hat{x}_k\,d\Omega \qquad (6.9.6)$$

As $\varepsilon \to 0$, $\mathbf{u}$ and $\boldsymbol{\sigma}$ over S_ε tend to their respective values at the center of V_ε, which are equal to $\mathbf{u}(\mathbf{x}_0)$ and $\boldsymbol{\sigma}(\mathbf{x}_0)$. Since $\hat{\mathbf{x}}$ decreases linearly with ε, as $\varepsilon \to 0$ the contribution of the stress term within the integral on the right-hand side of Eq. (6.9.6) decreases linearly with ε, whereas the contribution of the velocity term tends to a constant value. Thus in the limit $\varepsilon \to 0$ Eq. (6.9.6) gives

$$\int_D [G_{ij}(\mathbf{x}, \mathbf{x}_0)\,\sigma_{ik}(\mathbf{x}) - \mu\,u_i(\mathbf{x})\,T_{ijk}(\mathbf{x}, \mathbf{x}_0)]\,n_k(\mathbf{x})\,dS(\mathbf{x}) = -6\mu u_i(\mathbf{x}_0)\frac{1}{\varepsilon^4}\int_{S_\varepsilon} \hat{x}_i \hat{x}_j\,dS(\mathbf{x})$$

$$(6.9.7)$$

Using the divergence theorem, we compute

$$\int_{S_\varepsilon} \hat{x}_i \hat{x}_j \, dS(\mathbf{x}) = \varepsilon \int_{S_\varepsilon} \hat{x}_i n_j \, dS(\mathbf{x}) = \varepsilon \int_{V_\varepsilon} \frac{\partial \hat{x}_i}{\partial \hat{x}_j} \, dV(\mathbf{x}) = \delta_{ij} \tfrac{4}{3} \pi \varepsilon^4 \tag{6.9.8}$$

and substitute the result back into Eq. (6.9.7) to obtain

$$u_j(\mathbf{x}_0) = -\frac{1}{8\pi\mu} \int_D f_i(\mathbf{x}) \, G_{ij}(\mathbf{x}, \mathbf{x}_0) \, dS(\mathbf{x})$$

$$+ \frac{1}{8\pi} \int_D u_i(\mathbf{x}) \, T_{ijk}(\mathbf{x}, \mathbf{x}_0) \, n_k(\mathbf{x}) \, dS(\mathbf{x}) \tag{6.9.9}$$

where $\mathbf{f} = \boldsymbol{\sigma} \cdot \mathbf{n}$ is the modified boundary traction.

Equation (6.9.9) provides us with an integral representation of a Stokes flow in terms of boundary distributions of the Green's function $\mathbf{G}$ and its associated stress tensor $\mathbf{T}$. The densities of the distributions are, respectively, proportional to the boundary values of the traction and velocity. Making an analogy with corresponding results in the theory of electrostatics, we coin the first distribution involving the boundary traction the *single-layer potential,* and the second distribution involving the boundary velocity the *double-layer potential.*

The symmetry property (6.4.21) allows us to switch the order of the indices of the Green's function in the single-layer potential, provided that we also switch the order of the arguments, obtaining

$$u_j(\mathbf{x}_0) = -\frac{1}{8\pi\mu} \int_D G_{ji}(\mathbf{x}_0, \mathbf{x}) \, f_i(\mathbf{x}) \, dS(\mathbf{x})$$

$$+ \frac{1}{8\pi} \int_D u_i(\mathbf{x}) \, T_{ijk}(\mathbf{x}, \mathbf{x}_0) \, n_k(\mathbf{x}) \, dS(\mathbf{x}) \tag{6.9.10}$$

It is now evident that the single-layer potential represents a boundary distribution of point forces with strength per unit surface area equal to $-\mathbf{f}$. The physical interpretation of the double-layer potential will be discussed shortly in this section.

Flow in an Infinite Domain

A number of problems involve flow in a totally or partially infinite domain. Two examples are the flow due to the motion of a small particle in an infinitely dilute suspension, and semi-infinite shear flow over a wall containing a solitary depression or projection. In these cases, in order to develop the boundary-integral representation, we select a control volume that is confined by a solid or fluid boundary B and a large surface S_∞ extending to infinity. If the fluid at infinity is at rest, the velocity must decay at least as fast as $1/r$, and the pressure and stress must decay at least as fast as $1/r^2$, where r is a typical distance from B (Section 6.1). We note that the Green's function decays at least as fast as $1/r$, and its associated stress tensor decays at least as fast as $1/r^2$, and this suggests that as the size of S_∞ tends to infinity, both the single-layer and double-layer potentials over S_∞ make vanishing contributions. As a result, the domain D of the boundary-integral representation is conveniently reduced to B.

Simplification by the Use of Proper Green's Functions

The domain of integration of the single-layer and double-layer potentials consists of all fluid and solid surfaces that enclose an arbitrarily selected volume of flow. If the velocity happens to vanish over a certain portion of the boundary, the corresponding double-layer integral makes a vanishing contribution to the double-layer potential. Similarly, if the traction happens to vanish over another portion of the boundary, the corresponding single-layer integral makes a vanishing contribution to the single-layer potential. A further reduction in the domain of integration may be achieved by using a Green's function that observes the topology, symmetry, or other special features of the flow.

Let us consider, for example, a flow that is bounded by an internal or external solid stationary boundary B. Since the velocity vanishes over B, the corresponding double-layer integral is equal to zero. Using a Green's function that vanishes over B, that is, $\mathbf{G}(\mathbf{x}, \mathbf{x}_0) = \mathbf{0}$ when $\mathbf{x}$ is on B, renders the single-layer integral over B also equal to zero, and reduces the domain of the boundary integral equation from D to D minus B.

Fluid in Rigid-Body Motion

It is useful to apply the boundary integral equation for certain simple flows that are known to be exact solutions of the equations of Stokes flow. For a fluid in rigid-body motion, we set $\mathbf{u} = \mathbf{U} + \mathbf{\Omega} \times \hat{\mathbf{x}}$ where $\hat{\mathbf{x}} = \mathbf{x} - \mathbf{x}_c$ and $\mathbf{x}_c$ is the center of rotation, and $\mathbf{f} = -P\mathbf{n}$, where P is the constant modified pressure, and derive the identities

$$\int_D T_{ijk}(\mathbf{x}, \mathbf{x}_0)\, n_k(\mathbf{x})\, dS(\mathbf{x}) = \alpha\, \delta_{ij}$$

$$\varepsilon_{ilm} \int_D \hat{x}_m T_{ijk}(\mathbf{x}, \mathbf{x}_0)\, n_k(\mathbf{x})\, dS(\mathbf{x}) = \alpha\, \varepsilon_{jlm} \hat{x}_{0,m}$$

(6.9.11)

where D is the boundary of an arbitrary control volume containing fluid, and $\alpha = 8\pi$ or 0 depending on whether the point $\mathbf{x}_0$ is located inside or outside D; $\hat{x}_{0,m}$ denotes the mth component of $\hat{\mathbf{x}}_0$.

Flow past a Translating and Rotating Rigid Body

In the particular case of flow past a stationary or moving rigid body, the boundary integral representation is amenable to an important simplification. This is effected by decomposing the velocity $\mathbf{u}$ into the undisturbed component $\mathbf{u}^\infty$ that would prevail in the absence of the body and the disturbance component $\mathbf{u}^D$ due to the body, applying Eq. (6.9.10) for the disturbance component, and requiring that the velocity of the fluid on the surface of the body is given by $\mathbf{u} = \mathbf{U} + \mathbf{\Omega} \times \hat{\mathbf{x}}$, or, equivalently, the disturbance velocity is given by $\mathbf{u}^D = -\mathbf{u}^\infty + \mathbf{U} + \mathbf{\Omega} \times \hat{\mathbf{x}}$, where $\mathbf{U}$ and $\mathbf{\Omega}$ are the velocity of translation and angular velocity of rotation about the point $\mathbf{x}_c$, and $\hat{\mathbf{x}} = \mathbf{x} - \mathbf{x}_c$. Using the preceding arguments to neglect the integrals at infinity, we find

$$u_j^D(\mathbf{x}_0) = -\frac{1}{8\pi\mu} \int_B f_i^D(\mathbf{x})\, G_{ij}(\mathbf{x}, \mathbf{x}_0)\, dS(\mathbf{x})$$

$$-\frac{1}{8\pi} \int_B u_i^\infty(\mathbf{x})\, T_{ijk}(\mathbf{x}, \mathbf{x}_0)\, n_k(\mathbf{x})\, dS(\mathbf{x})$$

(6.9.12)

where B is the surface of the body. Next we use the fact that the point $\mathbf{x}_0$ is located on the exterior of the volume occupied by the body, and apply the reciprocal identity (6.9.3) for the incident flow $\mathbf{u}^\infty$, obtaining

$$\mu \int_B u_i^\infty(\mathbf{x})\, T_{ijk}(\mathbf{x}, \mathbf{x}_0)\, n_k(\mathbf{x})\, dS(\mathbf{x}) = \int_B f_i^\infty(\mathbf{x})\, G_{ij}(\mathbf{x}, \mathbf{x}_0)\, dS(\mathbf{x})$$

(6.9.13)

Using Eq. (6.9.13) to eliminate the double-layer integral from the right-hand side of Eq. (6.9.12), and adding the incident velocity field $\mathbf{u}^\infty$ to both sides of the resulting equation we derive the single-layer representation

$$u_j(\mathbf{x}_0) = u_j^\infty(\mathbf{x}_0) - \frac{1}{8\pi\mu} \int_B f_i(\mathbf{x})\, G_{ij}(\mathbf{x}, \mathbf{x}_0)\, dS(\mathbf{x})$$

(6.9.14)

Behavior of the flow far from the body

To study the asymptotic behavior of the flow far from the body, we expand the Green's function in a Taylor series with respect to $\mathbf{x}$ about the point $\mathbf{x}_c$ that is located somewhere in the vicinity of or within the body; this provides us with the *multipole expansion* for the disturbance velocity

$$u_j(\mathbf{x}_0) = u_j^\infty(\mathbf{x}_0) - \frac{1}{8\pi\mu}\left(G_{ji}(\mathbf{x}_0, \mathbf{x}_c)\right|_B \int_B f_i(\mathbf{x})\, dS(\mathbf{x})$$

$$+ \frac{\partial G_{ji}}{\partial x_{c,k}}(\mathbf{x}_0, \mathbf{x}_c)\bigg|_B \int_B (x_k - x_{c,k}) f_i(\mathbf{x})\, dS(\mathbf{x}) + \cdots\bigg) \qquad (6.9.15)$$

The first integral on the right-hand side of Eq. (6.9.15) is equal to the force $\mathbf{F}$ exerted on the body. The second integral is equal to the first moment of the surface stress, and subsequent integrals express higher moments of the surface stress.

The second term on the right-hand side of Eq. (6.9.15) represents the flow due to a point force, the third term represents the flow due to a point-force dipole, and subsequent terms represent the flow due to point-force quadrupoles and higher-order multipoles. The coefficient of the point-force dipole $\mathbf{q}$, in particular, may be decomposed into the coefficient of the stresslet $\mathbf{s}$ and an antisymmetric component $\mathbf{r}$, which amounts to a couplet as discussed in Section 6.5.

Equation (6.9.15) shows that, far from the body, the disturbance flow due to the body is similar to that due to a point force with strength equal to $-\mathbf{F}$. When $\mathbf{F} = \mathbf{0}$, the far flow is similar to that produced by a point-force dipole. One consequence of this behavior is that if the domain of flow is infinite, the far disturbance flow decays as $1/r$, unless $\mathbf{F} = \mathbf{0}$, in which case the far flow decays at least as fast as $1/r^2$.

Significance of the Double-Layer Potential

We return now to discuss the physical significance of the double-layer potential. Decomposing the stress tensor $\mathbf{T}$ into its constituents using Eq. (6.4.4), and invoking the symmetry property shown in Eq. (6.4.21), we find

$$\int_D u_i(\mathbf{x})\, T_{ijk}(\mathbf{x}, \mathbf{x}_0)\, n_k(\mathbf{x})\, dS(\mathbf{x}) = -\int_D \Pi_j(\mathbf{x}, \mathbf{x}_0)\, u_i(\mathbf{x})\, n_i(\mathbf{x})\, dS(\mathbf{x})$$

$$+ \int_D \frac{\partial G_{ji}(\mathbf{x}_0, \mathbf{x})}{\partial x_k}(u_i n_k + u_k n_i)(\mathbf{x})\, dS(\mathbf{x}) \qquad (6.9.16)$$

The second integral on the right-hand side of Eq. (6.9.16) represents a distribution of symmetric point-force dipoles amounting to a stresslet. Considering the first integral for a totally or partially infinite domain of flow, we note that the velocity distribution over D may be assigned in an arbitrary manner, and this suggests that the pressure vector $\Pi(\mathbf{x}, \mathbf{x}_0)$ may be identified with the velocity field of a certain Stokes flow. More specifially, inspecting the behavior of the pressure at the location of the point force reveals that $\Pi(\mathbf{x}, \mathbf{x}_0)$ represents the velocity field at the point $\mathbf{x}_0$ due to a point source with strength equal to -8π located at the point $\mathbf{x}$.

It is now evident that the first integral on the right-hand side of Eq. (6.9.16) represents a distribution of point sources whose density is proportional to the normal velocity of the fluid $\mathbf{u} \cdot \mathbf{n}$, and it thus vanishes over a solid boundary where $\mathbf{u} = \mathbf{0}$ or over a stationary fluid interface where $\mathbf{u} \cdot \mathbf{n} = 0$. It is important to note that these results are not valid when the domain of flow is completely enclosed by external boundaries, for then the velocity distribution over D is subject to the constraint that the flow rate across D vanish, and $\Pi(\mathbf{x}, \mathbf{x}_0)$ may no longer be interpreted as the flow due to a point source.

As an example, we consider the flow due to a point force in a semi-infinite domain that is bounded by a plane wall located at $x = w$, and express the pressure vector associated with the

Green's function given in Eq. (6.4.19) in the form

$$\Pi_i(\mathbf{x}, \mathbf{x}_0) = -2\left[\frac{X_i}{|\mathbf{X}|^3} + \frac{Y_i}{|\mathbf{Y}|^3} + 2Y_i\left(-\frac{1}{|\mathbf{Y}|^3} + 3\frac{Y_1^2}{|\mathbf{Y}|^5}\right) - 2h\left(-\frac{\delta_{i1}}{|\mathbf{Y}|^3} + 3\frac{Y_1 Y_i}{|\mathbf{Y}|^5}\right)\right] \quad (6.9.17)$$

where $i = 1$ corresponds to the x axis, $h = x - w$, $\mathbf{X} = \mathbf{x}_0 - \mathbf{x}$, $\mathbf{Y} = \mathbf{x}_0 - \mathbf{x}^{\text{Im}}$, and $\mathbf{x}^{\text{Im}} = (2w - x, y, z)$ is the image of $\mathbf{x}$ with respect to the wall. The first two terms on the right-hand side of Eq. (6.9.17) represent two point sources, each of strength equal to -8π located, respectively, at $\mathbf{x}$ and $\mathbf{x}^{\text{Im}}$. The third term is a Stokeslet doublet, and the fourth term is a potential dipole; each has a pole at $\mathbf{x}^{\text{Im}}$.

Flow Due to a Stresslet

The preceding discussion showed that the stress tensor $\mathbf{T}$ associated with a point force in a totally or partially infinite but not completely enclosed domain of flow represents a fundamental solution of Stokes flow. Specifically,

$$u_j(\mathbf{x}_0) = T_{ijk}(\mathbf{x}, \mathbf{x}_0)\, a_{ik}, \qquad P(\mathbf{x}_0) = \Lambda_{ik}(\mathbf{x}, \mathbf{x}_0)\, a_{ik} \quad (6.9.18)$$

represent the velocity and pressure field at the point $\mathbf{x}_0$ due to a *stresslet* placed at the point $\mathbf{x}$. Here a_{ik} is a constant matrix and Λ is a new pressure tensor. For the free-space Green's function

$$\Lambda_{ik}(\mathbf{x}, \mathbf{x}_0) = 4\left(-\frac{\delta_{ik}}{r^3} + 3\frac{\hat{x}_i \hat{x}_k}{r^5}\right) \quad (6.9.19)$$

Boundary-Integral Representation for the Pressure

The interpretation of the double-layer potential in terms of point sources and point-force dipoles suggests a boundary-integral representation for the pressure in terms of two boundary distributions corresponding to the single-layer and double-layer potential

$$P(\mathbf{x}_0) = -\frac{1}{8\pi}\int_D \Pi_i(\mathbf{x}, \mathbf{x}_0)\, f_i(\mathbf{x})\, dS(\mathbf{x})$$

$$+ \frac{\mu}{8\pi}\int_D u_i(\mathbf{x}) \Lambda_{ik}(\mathbf{x}, \mathbf{x}_0)\, n_k(\mathbf{x})\, dS(\mathbf{x}) \quad (6.9.20)$$

where Π and Λ express the pressure corresponding to the Green's function and its associated stress tensor expressing the stresslet.

Two-Dimensional Flow

To derive the boundary-integral representation of a two-dimensional Stokes flow, we work as before for three-dimensional flow. We thus find that for a point $\mathbf{x}_0$ that is located outside a selected area of flow

$$\int_C f_i(\mathbf{x})\, G_{ij}(\mathbf{x}, \mathbf{x}_0)\, dl(\mathbf{x}) - \mu \int_C u_i(\mathbf{x})\, T_{ijk}(\mathbf{x}, \mathbf{x}_0)\, n_k(\mathbf{x})\, dl(\mathbf{x}) = 0 \quad (6.9.21)$$

where the line C is the boundary of a selected control area, and the normal vector $\mathbf{n}$ points either into or out of the control area. Placing the point $\mathbf{x}_0$ in the interior of the selected control area, we obtain the boundary integral representation

$$u_j(\mathbf{x}_0) = -\frac{1}{4\pi\mu}\int_C f_i(\mathbf{x})\, G_{ij}(\mathbf{x}, \mathbf{x}_0)\, dl(\mathbf{x})$$

$$+ \frac{1}{4\pi}\int_C u_i(\mathbf{x})\, T_{ijk}(\mathbf{x}, \mathbf{x}_0)\, n_k(\mathbf{x})\, dl(\mathbf{x}) \quad (6.9.22)$$

where $\mathbf{n}$ points into the control area.

All properties, interpretations, and simplifications of the boundary integral equation discussed previously in this section for three-dimensional flow also apply to two-dimensional flow. Two exceptions are the results for infinite flow past a body and flow produced by the motion of a body in an infinite fluid. In both cases, if the force acting on the body does not vanish, the disturbance velocity far from the body increases at a logarithmic rate, and the boundary integrals at infinity may not be overlooked.

The pressure vector $\mathbf{\Pi}$ and stress tensor $\mathbf{T}$ associated with a Green's function for a totally or partially infinite domain of flow constitute two fundamental solutions of Stokes flow. Specifically, $\mathbf{\Pi}(\mathbf{x}, \mathbf{x}_0)$ represents the velocity at the point $\mathbf{x}_0$ due to a two-dimensional *point source* with strength equal to -4π located at the point $\mathbf{x}$, whereas the flow given in Eqs. (6.9.18) represents the velocity field due to a two-dimensional *stresslet* placed at $\mathbf{x}$. The corresponding pressure field may be expressed in terms of a pressure tensor $\mathbf{\Lambda}$ as in Eqs. (6.9.18). For flow in infinite space corresponding to the free-space Green's function, we obtain

$$\Lambda_{ik}(\mathbf{x}, \mathbf{x}_0) = 2\left(-\frac{\delta_{ik}}{r^2} + 2\frac{\hat{x}_i \hat{x}_k}{r^4}\right) \tag{6.9.23}$$

The pressure field of a two-dimensional Stokes flow is given in terms of two distributions associated with the single-layer and double-layer potential as

$$P(\mathbf{x}_0) = -\frac{1}{4\pi} \int_C \Pi_i(\mathbf{x}, \mathbf{x}_0)\, f_i(\mathbf{x})\, dl(\mathbf{x})$$
$$+ \frac{\mu}{4\pi} \int_C u_i(\mathbf{x})\, \Lambda_{ik}(\mathbf{x}, \mathbf{x}_0)\, n_k(\mathbf{x})\, dl(\mathbf{x}) \tag{6.9.24}$$

which is analogous to that shown in Eq. (6.9.20).

PROBLEMS

6.9.1   **Reciprocal theorem with a point force.** Derive the boundary-integral representation (6.9.10) on the basis of Eq. (3.3.22) using the properties of the delta function without reference to the small volume V_ε.

6.9.2   **An alternative boundary-integral representation.** Integrate Eq. (6.8.19) over a selected control volume, and use the divergence theorem to derive the alternative boundary-integral representation (Happel and Brenner, 1973, p. 81)

$$u_j(\mathbf{x}_0) = -\frac{1}{8\pi\mu} \int_D G_{ji}(\mathbf{x}_0, \mathbf{x})\left(-Pn_i + \frac{\partial u_i}{\partial x_k}\, n_k\right)(\mathbf{x})\, dS(\mathbf{x})$$
$$+ \frac{1}{8\pi} \int_D u_i(\mathbf{x})\left(-\Pi_j(\mathbf{x}, \mathbf{x}_0)\, n_i(\mathbf{x}) + \frac{\partial G_{ji}(\mathbf{x}_0, \mathbf{x})}{\partial x_k}\, n_k(\mathbf{x})\right) dS(\mathbf{x}) \tag{6.9.25}$$

6.9.3   **Multipole expansion for a fluid particle.** (a) Consider an infinite incident flow with velocity $\mathbf{u}^\infty$ past a fluid particle, and derive the far-field expansion

$$u_j(\mathbf{x}_0) = u_j^\infty(\mathbf{x}_0) - \frac{1}{8\pi\mu}\left(G_{ji}(\mathbf{x}_0, \mathbf{x}_c) \int_{p\mathrm{Ext}} f_i(\mathbf{x})\, dS(\mathbf{x})\right.$$
$$\left. + \frac{\partial G_{ji}(\mathbf{x}_0, \mathbf{x}_c)}{\partial x_{c,k}} \int_{p\mathrm{Ext}} [\hat{x}_k f_i(\mathbf{x}) - \mu(u_k n_i + u_i n_k)(\mathbf{x})]\, dS(\mathbf{x}) + \cdots\right)$$
$$- \frac{1}{8\pi}\Pi_j(\mathbf{x}_c, \mathbf{x}_0) \int_P u_l(\mathbf{x})\, n_l(\mathbf{x})\, dS(\mathbf{x}) + \cdots \tag{6.9.26}$$

where the point $\mathbf{x}_c$ is located somewhere in the vicinity or in the interior of the particle, $\hat{\mathbf{x}} = \mathbf{x} - \mathbf{x}_c$, μ is the viscosity of the ambient fluid, and the superscript Ext indicates that the traction is evaluated on the external side of the particle. The first series on the right-hand side of Eq. (6.9.26) contains multipoles of the point force, and the second series contains multipoles of the pressure. (b) On the basis of the expansion (6.9.26), show that $\mathbf{\Pi}(\mathbf{x}_c, \mathbf{x}_0)$ represents the velocity field at the point $\mathbf{x}_0$ due to a point source of an appropriate strength located at $\mathbf{x}_c$. (c) Show that

$$\frac{\partial \Pi_j(\mathbf{x}_c, \mathbf{x}_0)}{\partial x_{c,k}} = \nabla_c^2 G_{jk}(\mathbf{x}_0, \mathbf{x}_c) \tag{6.9.27}$$

where the subscript c of ∇_c indicates differentiation with respect to $\mathbf{x}_c$. Equation (6.9.27) suggests that all derivatives of the pressure may be expressed in terms of the Laplacian of the Green's function. In turn, this implies that all terms except for the first one in the second series on the right-hand side of Eq. (6.9.26) may be incorporated into the first series, thereby ensuring that the disturbance flow far from the particle may be represented in terms of an expansion of multipoles of the point force enhanced by a point source.

6.9.4   **Flow due to the motion of a thin sheet.** Consider the flow produced by the motion of a piece of paper of infinitesimal thickness in a viscous fluid. Show that the double-layer integral in the boundary integral equation may be eliminated in a natural manner, and the flow may be represented solely in terms of a single-layer integral whose distribution density possesses a simple physical interpretation.

6.9.5   **Flow past a liquid drop.** Consider an infinite incident flow with velocity $\mathbf{u}^\infty$ of a fluid with viscosity μ past a liquid drop with viscosity $\lambda\mu$. (a) Show that the velocity field in the exterior of the drop may be represented as

$$u_j(\mathbf{x}_0) = u_j^\infty(\mathbf{x}_0) - \frac{1}{8\pi\mu} \int_S \Delta f_i(\mathbf{x}) \, G_{ij}(\mathbf{x}, \mathbf{x}_0) \, dS(\mathbf{x})$$

$$+ \frac{1-\lambda}{8\pi} \int_S u_i(\mathbf{x}) \, T_{ijk}(\mathbf{x}, \mathbf{x}_0) \, n_k(\mathbf{x}) \, dS(\mathbf{x}) \tag{6.9.28}$$

where S is the interface, $\Delta f \equiv (\boldsymbol{\sigma}^{\text{Ext}} - \boldsymbol{\sigma}^{\text{Int}}) \cdot \mathbf{n}$ is the discontinuity in modified traction across the interface, and $\mathbf{n}$ is the unit normal vector pointing into the ambient fluid (Pozrikidis, 1992, Chapter 5). (b) Show that the velocity field in the interior of the drop is given by the right-hand side of Eq. (6.9.28) except that all terms, including the velocity of the incident flow, are divided by λ.

6.10 | BOUNDARY-INTEGRAL-EQUATION METHODS

The boundary-integral representation developed in Section 6.9 provides us with a powerful method for computing Stokes flows by solving integral equations for functions that are defined over the boundaries. The important benefit of this approach is that the dimensionality of the computational problem is reduced by one unit: Computing a three-dimensional or a two-dimensional flow reduces to solving an integral equation over a two-dimensional or a one-dimensional domain representing the boundaries.

To derive the boundary-integral equation we examine the behavior of the boundary-integral representation as the field point $\mathbf{x}_0$ approaches the boundary D either from the side of the flow or from the external side. Examining the singularity of the single-layer integral shows that it remains continuous as $\mathbf{x}_0$ crosses D. Considering the double-layer integral, we find that if the boundary D is a Lyapunov surface, which means that it has a continuously varying normal vector, and the velocity over D varies in a continuous manner

$$\lim_{\mathbf{x}_0 \to D} \int_D u_i(\mathbf{x}) \, T_{ijk}(\mathbf{x}, \mathbf{x}_0) \, n_k(\mathbf{x}) \, dS(\mathbf{x})$$

$$= \pm 4\pi u_j(\mathbf{x}_0) + \int_D^{PV} u_i(\mathbf{x}) \, T_{ijk}(\mathbf{x}, \mathbf{x}_0) \, n_k(\mathbf{x}) \, dS(\mathbf{x}) \qquad (6.10.1)$$

where the plus sign applies when the point $\mathbf{x}_0$ approaches D from the side of the flow, which is indicated by the direction of the normal vector, and the minus sign otherwise; PV designates the principal value of the double-layer potential defined as the value of the improper double-layer integral computed when the point $\mathbf{x}_0$ is located on D. One way to derive Eq. (6.10.1) is to use the first integral identity shown in Eq. (6.9.11) as discussed in Problem 6.10.1. Equation (6.10.1) shows that the double-layer potential undergoes a discontinuity of magnitude $8\pi\mathbf{u}$ as the point $\mathbf{x}_0$ crosses D.

Substituting Eq. (6.10.1) with the plus sign into Eq. (6.9.9) or with the minus sign into Eq. (6.9.3), we find that when $\mathbf{x}_0$ is located on D

$$u_j(\mathbf{x}_0) = -\frac{1}{4\pi\mu} \int_D f_i(\mathbf{x}) \, G_{ij}(\mathbf{x}, \mathbf{x}_0) \, dS(\mathbf{x})$$

$$+ \frac{1}{4\pi} \int_D^{PV} u_i(\mathbf{x}) \, T_{ijk}(\mathbf{x}, \mathbf{x}_0) \, n_k(\mathbf{x}) \, dS(\mathbf{x}) \qquad (6.10.2)$$

Equation (6.10.2) imposes an integral constraint on the flow by requiring that the boundary velocity and traction may not be specified independently in an arbitrary manner, but must be such that Eq. (6.10.2) is fulfilled. This is consistent with the fact that, in practice, when we state a particular problem, we specify boundary conditions either for the velocity or for the traction, but not for both as discussed in Section 3.5.

In summary, Eqs. (6.9.3), (6.9.9), and (6.10.2) are valid, respectively, when the point $\mathbf{x}_0$ is located outside, inside, and on the boundary of a selected volume of flow.

Integral Equations

Prescribing the velocity $\mathbf{u}$ over D renders Eq. (6.10.2) a *Fredholm integral equation of the first kind* for the boundary traction $\mathbf{f}$, which may be written in the standard form

$$\int_D f_i(\mathbf{x}) \, G_{ij}(\mathbf{x}, \mathbf{x}_0) \, dS(\mathbf{x}) = -4\pi\mu u_j(\mathbf{x}_0) + \mu I_j^D(\mathbf{x}_0) \qquad (6.10.3)$$

where $\mathbf{I}^D$ represents the known double-layer potential defined as

$$I_j^D(\mathbf{x}_0) \equiv \int_D u_i(\mathbf{x}) \, T_{ijk}(\mathbf{x}, \mathbf{x}_0) \, n_k(\mathbf{x}) \, dS(\mathbf{x}) \qquad (6.10.4)$$

Prescribing instead the traction $\mathbf{f}$ over D renders Eq. (6.10.2) a *Fredholm integral equation of the second kind* for the boundary velocity $\mathbf{u}$, which may be written in the standard form

$$u_j(\mathbf{x}_0) = \frac{1}{4\pi} \int_D^{PV} u_i(\mathbf{x}) \, T_{ijk}(\mathbf{x}, \mathbf{x}_0) \, n_k(\mathbf{x}) \, dS(\mathbf{x}) - \frac{1}{4\pi\mu} I_j^S(\mathbf{x}_0) \qquad (6.10.5)$$

where $\mathbf{I}^S$ represents the known single-layer potential defined as

$$I_j^S(\mathbf{x}_0) \equiv \int_D f_i(\mathbf{x}) \, G_{ij}(\mathbf{x}, \mathbf{x}_0) \, dS(\mathbf{x}) \qquad (6.10.6)$$

Prescribing the velocity over a portion of D and the traction over the remaining part of D, as we do when the flow is bounded by solid boundaries and free surfaces, we obtain a Fredholm integral equation of mixed kind for the unknown boundary distributions. Once these equations are solved, the velocity, pressure, and stress fields may be computed using the boundary integral representations shown in Eqs. (6.9.9) and (6.9.20).

Inspecting the integral equations (6.10.3) and (6.10.5), we observe that the kernels $\mathbf{G}(\mathbf{x}, \mathbf{x}_0)$ and $T_{ijk}(\mathbf{x}, \mathbf{x}_0) n_k(\mathbf{x})$ become singular as the integration point $\mathbf{x}$ approaches the pole $\mathbf{x}_0$. Closer inspection reveals that the singularities of the kernels are not square integrable. This occurrence not only raises computational difficulties, but also prevents us from studying the properties of these equations in the context of the theories of Fredholm and Hilbert-Schmidt, which are applicable for integral equations with square-integrable kernels. Fortunately, when the boundary D is a Lyapunov surface, the kernels are weakly singular, and this ensures that both the single-layer and the double-layer integrals are compact linear operators for which the Fredholm theory remains applicable (Kim and Karrila, 1991; Pozrikidis, 1992).

Two-Dimensional Flow

Repeating the preceding derivations with minor modifications, we arrive at similar equations for two-dimensional flow. More specifically, we find that as the point $\mathbf{x}_0$ approaches the boundary C from either side, the single-layer potential behaves in a continuous manner, but the double-layer potential exhibits a discontinuous behavior. When C is a Lyapunov line, which means that it has a continuously varying normal vector, and the velocity $\mathbf{u}$ varies over C in a continuous manner, we find

$$\lim_{\mathbf{x}_0 \to C} \int_C u_i(\mathbf{x}) \, T_{ijk}(\mathbf{x}, \mathbf{x}_0) \, n_k(\mathbf{x}) \, dl(\mathbf{x})$$

$$= \pm 2\pi u_j(\mathbf{x}_0) + \int_C^{PV} u_i(\mathbf{x}) \, T_{ijk}(\mathbf{x}, \mathbf{x}_0) \, n_k(\mathbf{x}) \, dl(\mathbf{x}) \qquad (6.10.7)$$

where the plus sign applies when $\mathbf{x}_0$ approaches C from the side of the flow, indicated by the normal vector, and the minus sign otherwise (Problem 6.10.1). The principal value of the double-layer integral is defined as the value of the improper double-layer integral computed when the point $\mathbf{x}_0$ is located on C. Equation (6.10.7) shows that the double-layer potential undergoes a discontinuity of magnitude $4\pi\mathbf{u}$ as $\mathbf{x}_0$ crosses C.

Combining Eqs. (6.10.7) and (6.9.22), we find that for a point $\mathbf{x}_0$ that lies on C

$$u_j(\mathbf{x}_0) = -\frac{1}{2\pi\mu} \int_C f_i(\mathbf{x}) \, G_{ij}(\mathbf{x}, \mathbf{x}_0) \, dl(\mathbf{x}) + \frac{1}{2\pi} \int_C^{PV} u_i(\mathbf{x}) \, T_{ijk}(\mathbf{x}, \mathbf{x}_0) \, n_k(\mathbf{x}) \, dl(\mathbf{x})$$

$$(6.10.8)$$

This equation may be used to derive integral equations of the first, second, or mixed kind for the boundary velocity or traction, working as we did before for three-dimensional flow (Pozrikidis, 1992).

Flow in a Rectangular Cavity

As a particular application, let us consider two-dimensional flow in a rectangular cavity driven by a moving lid, illustrated in Figure 6.10.1. We denote the top, left, bottom, and right walls of the cavity respectively by T, L, B, and R, use the representation (6.9.22), and enforce the boundary conditions for the velocity to obtain

$$u_j(\mathbf{x}_0) = -\frac{1}{4\pi\mu} \int_{T,L,B,R} f_i(\mathbf{x}) \, G_{ij}(\mathbf{x}, \mathbf{x}_0) \, dl(\mathbf{x}) + \frac{U}{4\pi} \int_T T_{xjk}(\mathbf{x}, \mathbf{x}_0) \, n_k(\mathbf{x}) \, dl(\mathbf{x})$$

$$(6.10.9)$$

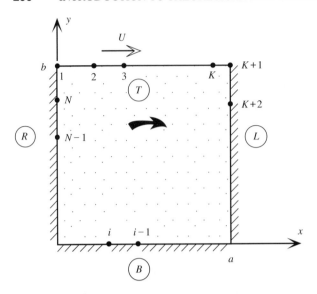

Figure 6.10.1 Implementation of the boundary-integral equation method for two-dimensional flow in a rectangular cavity driven by a moving lid, computed by solving an integral equation of the first kind using a boundary-element method.

which is valid when the point $\mathbf{x}_0$ lies within the domain of flow. The integral equation (6.10.8) takes the specific form

$$\int_{T,L,B,R} f_i(\mathbf{x})\, G_{ij}(\mathbf{x}, \mathbf{x}_0)\, dl(\mathbf{x}) = -2\pi\mu u_j(\mathbf{x}_0) + \mu U \int_T T_{xjk}(\mathbf{x}, \mathbf{x}_0)\, n_k(\mathbf{x})\, dl(\mathbf{x})$$

(6.10.10)

where the point $\mathbf{x}_0$ is located on T, L, B, or R. When $\mathbf{x}_0$ is on T, the last integral is interpreted as a principal value integral. Since the right-hand side of Eq. (6.10.10) is a known function, we have obtained an integral equation of the first kind for the boundary traction $\mathbf{f}$.

Boundary-element method

A simple numerical method for solving Eq. (6.10.10) proceeds by discretizing the boundaries into a collection of straight segments, as shown in Figure 6.10.1, also called *boundary elements,* and assuming that the traction is constant over each element. This approximation allows us to restate Eq. (6.10.10) in the discrete form

$$\sum_{n=1}^{N} f_i^{(n)} A_{ij}^{(n)}(\mathbf{x}_0) = -2\pi\mu u_j(\mathbf{x}_0) + \mu U \sum_{k=1}^{K} b_j^{(k)}(\mathbf{x}_0)$$

(6.10.11)

where $\mathbf{f}^{(n)}$ is the constant value of the traction over the nth segment, N is the total number of segments, and the summation on the right-hand side is over the K segments of the lid. The *influence coefficient* $\mathbf{A}^{(n)}$ and vector $\mathbf{b}^{(n)}$ are defined as

$$A_{ij}^{(n)}(\mathbf{x}_0) = \int_{S^{(n)}} G_{ij}(\mathbf{x}, \mathbf{x}_0)\, dl(\mathbf{x})$$

$$b_j^{(k)}(\mathbf{x}_0) = -\int_{S^{(k)}} T_{xjy}(\mathbf{x}, \mathbf{x}_0)\, dl(\mathbf{x})$$

(6.10.12)

where $S^{(n)}$ signifies the nth segment and we used the fact that $\mathbf{n} = (0, -1)$ over T. Applying Eq. (6.10.11) at the midpoint of the mth segment $\mathbf{X}^{(m)}$, where $m = 1, \ldots, N$, provides us with a system of N linear algebraic equations for $\mathbf{f}^{(n)}$

$$\sum_{n=1}^{N} f_i^{(n)} A_{ij}^{(n)}(\mathbf{X}^{(m)}) = -2\pi\mu u_j(\mathbf{X}^{(m)}) + \mu U \sum_{k=1}^{K} b_j^{(k)}(\mathbf{X}^{(m)}) \qquad (6.10.13)$$

Adopting, for simplicity, the free-space Green's function shown in Eq. (6.4.23), we compute the coefficients in Eqs. (6.10.12) by performing numerical integration over each segment. When $\mathbf{X}^{(m)}$ is located at the lid, the kernel $\mathbf{T}$ vanishes, yielding $\mathbf{b}^{(k)}(\mathbf{X}^{(m)}) = \mathbf{0}$; the associated influence coefficients $\mathbf{A}^{(k)}$ corresponding to the lid segments may be computed exactly by evaluating the integrals

$$A_{xx}^{(k)}(\mathbf{X}^{(m)}) = \int_{S^{(k)}} (-\ln|X^{(m)} - x| + 1)|dx|$$

$$A_{xy}^{(k)}(\mathbf{X}^{(m)}) = A_{yx}^{(k)}(\mathbf{X}^{(m)}) = 0 \qquad (6.10.14)$$

$$A_{yy}^{(k)}(\mathbf{X}^{(m)}) = -\int_{S^{(k)}} \ln|X^{(m)} - x||dx|$$

Similar expressions may be derived for the integrals over the left, bottom, and right wall segments for points $\mathbf{X}^{(m)}$ that lie on the corresponding sides.

Proceeding with the logistics of the computation, we collect the two components of all $\mathbf{f}^{(n)}$ into a large vector $\mathbf{Y}$ of size $2N$ that contains the two components of the unknown traction over the four sides,

$$\mathbf{Y} = (f_x^{(1)}, f_y^{(1)}, f_x^{(2)}, f_y^{(2)}, \ldots, f_x^{(K)}, f_y^{(K)}, \ldots, f_x^{(N)}, f_y^{(N)}) \qquad (6.10.15)$$

and use Eq. (6.10.13) to obtain a system of linear equations

$$\mathbf{C} \cdot \mathbf{Y} = \mathbf{d} \qquad (6.10.16)$$

where $\mathbf{C}$ is an influence coefficient matrix, and $\mathbf{d}$ is a constant vector. The first pair of equations in (6.10.16) correspond to the x and y components of Eq. (6.10.13) for $\mathbf{X}^{(1)}$, the second pair for $\mathbf{X}^{(2)}$, and further pairs correspond to subsequent collocation points. This particular arrangement, however, is not mandatory and can be permutated.

The system (6.10.16) is nearly singular, reflecting the fact that the pressure may be set at an arbitrary level. One way to rectify this deficiency is to specify the value of $f_x^{(N-1)}$ and then solve the first $2N - 1$ equations for the remaining unknowns. A better way is to exploit the anticipated symmetry of the flow, setting, for instance, $f_x^{(1)} = f_x^{(K)}$, $f_y^{(1)} = -f_y^{(K)}$. When K and M are even, implementing the symmetry constraints reduces the size of the final system of equations from $2N$ to N. The simplified linear system contains the equations corresponding to the $N/2$ collocation points $\mathbf{X}^{(m)}$ that are distributed along the left half-part of the boundaries.

Implementations of the boundary-integral-equation method have been reviewed by Pozrikidis (1992).

PROBLEMS

6.10.1 **Discontinuity of the double-layer integral.** (a) Subtract off the singularity of the double-layer integral and use the first of Eqs. (6.9.11) to show the jump condition (6.10.1) (Pozrikidis, 1992, Chapter 2). (b) Repeat part (a) for Eq. (6.10.7).

6.10.2 **Boundary-integral equation.** (a) Following the limiting procedure outlined in Section 6.9, derive the boundary-integral equation for a point $\mathbf{x}_0$ that is located on a smooth boundary D. (*Hint:* consider a point $\mathbf{x}_0$ on D, define a control volume that is enclosed by D but excludes the volume of a half-sphere with radius ε centered at $\mathbf{x}_0$, and let ε tend to zero.) (b) Repeat (a) for a corner. Specifically, show that when a point $\mathbf{x}_0$ is located on a corner, Eq. (6.10.2) remains valid provided that $u_j(\mathbf{x}_0)$ on the left-hand side is replaced by $u_i(\mathbf{x}_0)c_{ij}(\mathbf{x}_0)$, where $c_{ij}(\mathbf{x}_0)$ is a matrix whose value depends on the particular geometry of the corner. Evaluate $c_{ij}(\mathbf{x}_0)$ along the corner of a two-dimensional wedge (Pozrikidis, 1992). (c) Show that the boundary-integral equation for two-dimensional flow (6.10.8) is also valid at a point on a corner, provided that the coefficient $1/2\pi$ in front of the single-layer and double-layer potentials are replaced by $1/(2\alpha)$, where α is the angle subtended by the corner.

Computer Problem

6.10.3   Flow in a cavity. (a) Write a program called *CVT* that computes the boundary tractions of flow in a rectangular cavity driven by a moving lid, using the boundary-element method described in the text. Present plots of the boundary shear stress for cavities with depth ratios $b/a = 0.10, 0.50, 1.0, 2.0$, and discuss your results with reference to flow reversal. (b) Draw streamline patterns for the cases studied in (a) and discuss the kinematic structure of the flow.

6.11 | GENERALIZED FAXEN RELATIONS

In order to compute the velocity of translation and angular velocity of rotation of a body or particle that is suspended in a fluid, we must have available the modified force $\mathbf{F}$, and the torque $\mathbf{T}$ exerted on it when it is held stationary in an incident ambient flow. The generalized Faxen relations, named after Faxen (1924), who first developed relations for the force and torque on a solid spherical particle, provide us with expressions for $\mathbf{F}$ and $\mathbf{T}$ in terms of the values of the incident velocity field and its derivatives evaluated at specific locations in the interior or over the surface of the body.

Force on a Stationary Solid Body

Let us first consider the force $\mathbf{F}$ exerted on a solid body that is held stationary in an ambient flow with velocity $\mathbf{u}^\infty$. Following Kim (1985) and Kim and Lu (1987), we assume that the flow generated when the body translates with velocity $\mathbf{U}$ may be represented in terms of a discrete or continuous distribution of point forces and their multipoles in the form

$$u_i(\mathbf{x}) = U_k M_{kj}^0 \langle G_{ij}(\mathbf{x}, \mathbf{x}_0) \rangle \tag{6.11.1}$$

where $\mathbf{M}^0$ is an integral, differential, or integro-differential operator acting with respect to the point $\mathbf{x}_0$, so that the point forces and their derivatives, expressing multipoles of the point force, are located in the interior of the body. In order to satisfy the boundary condition $\mathbf{u} = \mathbf{U}$ on the surface of the body denoted by B, we require that

$$M_{kj}^0 \langle G_{ij}(\mathbf{x}, \mathbf{x}_0) \rangle = \delta_{ki} \tag{6.11.2}$$

when $\mathbf{x}$ is on B. The modified traction on the surface of the body is given by the corresponding expression

$$f_i^T(\mathbf{x}) = \mu U_k M_{kj}^0 \langle T_{ijl}(\mathbf{x}, \mathbf{x}_0) \rangle n_l(\mathbf{x}) \tag{6.11.3}$$

where the superscript T stands for translation. Combining Eq. (6.11.3) with Eqs. (6.8.3) and (6.8.6), we obtain

$$F_k = -\mu \int_B u_i^\infty(\mathbf{x}) M_{kj}^0 \langle T_{ijl}(\mathbf{x}, \mathbf{x}_0) \rangle n_l(\mathbf{x}) \, dS(\mathbf{x}) \tag{6.11.4}$$

Next we write the boundary integral representation Eq. (6.9.9) for the ambient flow $\mathbf{u}^\infty$ at a point $\mathbf{x}_0$ that is located *within* the volume occupied by the body, and operate on both sides with M^0 to obtain

$$M^0_{kj}\langle u^\infty_j(\mathbf{x}_0)\rangle = \frac{1}{8\pi\mu}\int_B f^\infty_i(\mathbf{x})\, M^0_{kj}\langle G_{ij}(\mathbf{x},\mathbf{x}_0)\rangle\, dS(\mathbf{x})$$

$$- \frac{1}{8\pi}\int_B u^\infty_i(\mathbf{x})\, M^0_{kj}\langle T_{ijl}(\mathbf{x},\mathbf{x}_0)\rangle\, n_l(\mathbf{x})\, dS(\mathbf{x}) \qquad (6.11.5)$$

where the normal vector $\mathbf{n}$ points into the exterior fluid. Incorporating the boundary condition (6.11.2) into the first integral on the right-hand side of Eq. (6.11.5), noting that the modified force due to the ambient flow over the fluid parcel whose instantaneous boundary coincides with that of the body vanishes, and then combining the resulting expression with Eq. (6.11.4), we obtain the *generalized Faxen relation for the modified force on a solid body*

$$F_k = 8\pi\mu\, M^0_{kj}\langle u^\infty_j(\mathbf{x}_0)\rangle \qquad (6.11.6)$$

which provides us with a formula for the force in terms of the velocity of the incident flow.

Torque on a Stationary Solid Body

Considering next the torque exerted on a stationary solid body, we assume that the flow generated when the body rotates with angular velocity $\boldsymbol{\Omega}$ about the origin may be represented in terms of a discrete or continuous distribution of Green's functions as

$$u_i(\mathbf{x}) = \Omega_k N^0_{kj}\langle G_{ij}(\mathbf{x},\mathbf{x}_0)\rangle \qquad (6.11.7)$$

Working as before for the force, we find that the torque with respect to the origin exerted on the body when it is held stationary in an ambient flow with velocity $\mathbf{u}^\infty$ is given by the *generalized Faxen relation for the torque* (Pozrikidis, 1992)

$$T_k = 8\pi\mu\, N^0_{kj}\langle u^\infty_j(\mathbf{x}_0)\rangle \qquad (6.11.8)$$

Higher Moments of the Traction on a Stationary Solid Body

It is possible to develop further generalized Faxen relations pertaining to the higher moments of the traction exerted on the surface of a stationary solid body. The zeroth-order moment is the force, the antisymmetric part of the first-order moment is related to the torque, and the symmetric part is the coefficient of the stresslet $\mathbf{s}$. The latter expresses the symmetric part of the coefficient of the point-force dipole that appears on the right-hand side of the multipole expansion given in Eq. (6.9.15).

Considering the coefficient of the stresslet, we find that if the disturbance flow produced when the body is held stationary in a purely straining flow with velocity $\mathbf{u}^\infty = \mathbf{E} \cdot \mathbf{x}$ is represented as

$$u^D_i(\mathbf{x}) = E_{lk}\, L^0_{lkj}\langle G_{ij}(\mathbf{x},\mathbf{x}_0)\rangle \qquad (6.11.9)$$

where $\mathbf{E}$ is a symmetric matrix with vanishing trace, the coefficient of the stresslet arising when the particle is held still in an arbitrary ambient flow with velocity $\mathbf{u}^\infty$ is given by (Pozrikidis, 1992)

$$s_{lk} = -4\pi\mu[L^0_{lkj}\langle u^\infty_j(\mathbf{x}_0)\rangle + L^0_{klj}\langle u^\infty_j(\mathbf{x}_0)\rangle] \qquad (6.11.10)$$

Faxen's Laws for a Solid Sphere

In Section 6.6 we found that the flow due to the translation of a spherical body in an unbounded fluid may be presented in terms of a point force and a potential dipole, both placed at the center of the body. Using the representation (6.6.1) with the coefficients shown in Eq. (6.6.4), and taking into account Eq. (6.5.9), we write

$$u_i(\mathbf{x}) = U_k\delta_{kj}\left(\tfrac{3}{4}a + \tfrac{1}{8}a^3\nabla^2_0\right)S_{ij}(\mathbf{x},\mathbf{x}_0) \qquad (6.11.11)$$

where $\mathbf{S}$ is the Stokeslet, a is the radius of the body, $\mathbf{x}_0$ is the center of the body, and the Laplacian ∇_0^2 operates with respect to $\mathbf{x}_0$. Faxen's law for the force then follows from Eq. (6.11.6).

$$\mathbf{F} = 6\pi\mu a\, \mathbf{u}^\infty(\mathbf{x}_0) + \pi\mu a^3\, \nabla_0^2 \mathbf{u}^\infty(\mathbf{x}_0) \qquad (6.11.12)$$

Thus the force exerted on a spherical body may be found from a knowledge of the values of the incident velocity and its Laplacian at the center of the body. The latter is proportional to the gradient of the modified pressure.

Furthermore, in Section 6.6 we found that the flow due to the rotation of a spherical body that is immersed in an infinite fluid may be represented in terms of a couplet alone. Combining Eqs. (6.6.22) and (6.5.6) we obtain

$$u_i(\mathbf{x}) = \Omega_k a^3 \tfrac{1}{2} \varepsilon_{klj} \frac{\partial S_{ij}}{\partial x_{0,l}}(\mathbf{x}, \mathbf{x}_0) \qquad (6.11.13)$$

Faxen's law then follows from Eq. (6.11.8) as

$$\mathbf{T} = 4\pi\mu a^3\, \nabla_0 \times \mathbf{u}^\infty(\mathbf{x}_0) \qquad (6.11.14)$$

where the gradient ∇_0 operates with respect to $\mathbf{x}_0$. Thus the torque exerted on a spherical body may be found from a knowledge of the curl of the incident velocity, that is, the vorticity of the incident flow at the center of the body.

Comparing Faxen's relations (6.11.12) and (6.11.14) to those shown in Eqs. (6.8.10) we recover the mean-value theorems expressed by the identities (6.1.4) and (6.1.5).

Combining Eq. (6.6.11) with the values of the coefficients given in Eqs. (6.6.15) and (6.6.16), and the definitions of the singularities discussed in Section 6.5, we find that, for a spherical particle, Eq. (6.11.9) assumes the specific form

$$u_i^D(\mathbf{x}) = -\tfrac{1}{6} E_{jl} \frac{\partial}{\partial x_{0,l}} (5a^3 + \tfrac{1}{2} a^5\, \nabla_0^2) S_{ij}(\mathbf{x}, \mathbf{x}_0) \qquad (6.11.15)$$

Faxen's relation for the coefficient of the stresslet then follows as

$$s_{ij} = \tfrac{10}{3}\pi\mu a^3 \left[\frac{\partial u_j^\infty}{\partial x_i} + \frac{\partial u_i^\infty}{\partial x_j} + \tfrac{1}{10} a^2\, \nabla_0^2 \left(\frac{\partial u_j^\infty}{\partial x_i} + \frac{\partial u_i^\infty}{\partial x_j} \right) \right](\mathbf{x}_0) \qquad (6.11.16)$$

Arbitrary Body Shapes

To derive Faxen's relations for a particular body, we must have available the singularity representations of the flow due to the translation and rotation of the body, as well as the disturbance flow produced when the body is held stationary in a straining ambient flow. A limited number of exact representations were discussed in Section 6.6, and a more comprehensive review has been presented by Kim and Karrila (1991). More generally, the singularity representations must be computed using numerical and approximate methods as discussed in Section 6.7.

Faxen's Relations for a Fluid Particle

Our next objective is to derive Faxen's relations for a fluid particle that is held stationary in an ambient flow with velocity $\mathbf{u}^\infty$. Working as before for the solid particle, we decompose the velocity field as $\mathbf{u} = \mathbf{u}^\infty + \mathbf{u}^D$, where $\mathbf{u}^D$ is the disturbance flow due to the particle. Furthermore, we require that the normal component of the velocity $\mathbf{u}$ over the surface of the particle vanish, and the tangential component of the velocity and interfacial traction remain continuous across the interface. We allow the normal component of the traction to be discontinuous across the interface, assuming that the discontinuity is balanced by surface tension.

Considering first Faxen's relation for the force, we apply the reciprocal theorem for the exterior disturbance flow $\mathbf{u}^D$ and the flow $\mathbf{u}^T$ produced when the particle translates with velocity $\mathbf{U}$, and obtain

$$\int_{pExt} \mathbf{u}^T \cdot \mathbf{f}^D \, dS = \int_{pExt} \mathbf{u}^D \cdot \mathbf{f}^T \, dS \tag{6.11.17}$$

where the superscript T stands for translation and the superscript Ext indicates the exterior side of the interface. Writing $\mathbf{u}^D = \mathbf{u} - \mathbf{u}^\infty$ and making a slight modification on the left-hand side, we obtain

$$\int_{pExt} (\mathbf{u}^T - \mathbf{U}) \cdot \mathbf{f}^D \, dS + \mathbf{U} \cdot \mathbf{F}^D = \int_{pExt} \mathbf{u} \cdot \mathbf{f}^T \, dS - \int_{pExt} \mathbf{u}^\infty \cdot \mathbf{f}^T \, dS \tag{6.11.18}$$

where $\mathbf{F}^D$ is the disturbance force exerted on the exterior side of the interface.

Next we represent the exterior flow $\mathbf{u}^T$ as in Eq. (6.11.1), and project the Eq. (6.11.5) onto $\mathbf{U}$ to obtain

$$U_k \, M^0_{kj} \langle u^\infty_j(\mathbf{x}_0) \rangle = \frac{1}{8\pi\mu} \int_{pExt} f^\infty_i u^T_i \, dS - \frac{1}{8\pi\mu} \int_{pExt} u^\infty_i f^T_i \, dS \tag{6.11.19}$$

where μ is the viscosity of the ambient fluid. Combining the preceding two equations to eliminate the term involving the velocity of the incident flow, and rearranging, we find

$$\int_{pExt} (\mathbf{u}^T - \mathbf{U}) \cdot \mathbf{f} \, dS + \mathbf{U} \cdot \mathbf{F}^D - 8\pi\mu\mathbf{U} \cdot \mathbf{M}^0 \langle \mathbf{u}^\infty(\mathbf{x}_0) \rangle$$

$$= \int_{pExt} \mathbf{u} \cdot \mathbf{f}^T \, dS \tag{6.11.20}$$

Noting that the velocities $\mathbf{u}^T - \mathbf{U}$ and $\mathbf{u}$ are tangential to the surface of the particle, and the tangential components of $\mathbf{f}^T$ and $\mathbf{f}$ are continuous across the interface, allows us to switch the domain of integration of both integrals in Eq. (6.11.20) from the exterior to the interior side of the interface. But then, applying the reciprocal theorem for the flows $\mathbf{u}^T - \mathbf{U}$ and $\mathbf{u}$ over the interior of the particle, we find that the two integrals in Eq. (6.11.20) are identical, and thereby recover the generalized Faxen relation for the force expressed by Eq. (6.11.6).

Following a similar procedure, we find that the generalized Faxen relations for the torque and coefficient of the stresslet for solid particles discussed previously are also valid for liquid drops (Kim and Lu, 1987; Kim and Karrila, 1991; Pozrikidis, 1992).

Spherical drop

As an example, in Section 6.6 we found that the exterior flow produced by the translation of a spherical drop may be written as shown in the first of Eqs. (6.6.28), with the coefficients given in Eqs. (6.6.34), which may be placed in the form

$$u^{Ext}_i(\mathbf{x}) = U_j \, a \frac{1}{8} \left(2 \frac{3\lambda + 2}{\lambda + 1} + a^2 \frac{\lambda}{\lambda + 1} \nabla^2_0 \right) S_{ij}(\mathbf{x}, \mathbf{x}_0) \tag{6.11.21}$$

where λ is the ratio between the viscosities of the drop and ambient fluid. Faxen's relation for the force then follows as

$$\mathbf{F} = \pi\mu a \left(2 \frac{3\lambda + 2}{\lambda + 1} + a^2 \frac{\lambda}{\lambda + 1} \nabla^2_0 \right) \mathbf{u}^\infty(\mathbf{x}_0) \tag{6.11.22}$$

As λ tends to infinity we recover Eq. (6.11.12).

PROBLEMS

6.11.1 **Faxen's relations for a prolate spheroid.** Based on the singularity representations shown in Eqs. (6.6.24) and (6.6.26), show that the Faxen relations for the axial component of the torque and the force on a prolate spheroid are given by

$$T_X(\mathbf{x}) = 4\pi\mu\beta \int_{-c}^{c} (c^2 - x_0^2)(\nabla \times \mathbf{u}^\infty)_x \, dx_0$$

$$F_k(\mathbf{x}) = 8\pi\mu\alpha_{kj} \int_{-c}^{c} \left(u_j^\infty(\mathbf{x}) + \frac{1 - e^2}{4e^2}(c^2 - x_0^2)\nabla^2 u_j^\infty(\mathbf{x}) \right) dx_0$$

6.11.2   Approximate Faxen's relation for a sphere. Based on the approximate singularity representation discussed in Problem 6.7.1(a), derive an approximate form of Faxen's relation for the force on a solid sphere, and compare it with the exact relation given in the text.

6.12 | FORMULATION OF TWO-DIMENSIONAL STOKES FLOW IN COMPLEX VARIABLES

Two-dimensional Stokes flow is amenable to a formulation in complex variables that is analogous to but more complicated than that of two-dimensional potential flow to be discussed in Chapter 7. For potential flow, the complex-variable formulation is based on the observation that the harmonic potential and stream function comprise a pair of conjugate harmonic functions. In the case of Stokes flow, the complex-variable formulation may be developed in two different ways based on either kinematic or dynamic considerations.

Pressure–Vorticity Formulation

Using the identity $\nabla^2\mathbf{u} = -\nabla \times \boldsymbol{\omega}$ along with the Stokes equation $\nabla^2\mathbf{u} = (1/\mu)\nabla P$, where P is the modifed pressure, we obtain

$$\mu \nabla \times \boldsymbol{\omega} = -\nabla P \tag{6.12.1}$$

Writing out the two scalar components of Eq. (6.12.1) shows that the strength of the vorticity ω and the modified pressure P satisfy the Cauchy–Riemann equations

$$\frac{\partial\omega}{\partial x} = \frac{1}{\mu}\frac{\partial P}{\partial y}, \qquad \frac{\partial\omega}{\partial y} = -\frac{1}{\mu}\frac{\partial P}{\partial x} \tag{6.12.2}$$

As a consequence, the complex function

$$f(z) = \omega + \frac{i}{\mu}P \tag{6.12.3}$$

is an analytic function of the complex variable $z = x + iy$, where i is the imaginary unit.

Assigning to the complex function $f(z)$ various forms yields different families of two-dimensional Stokes flows. In order to describe the structure of these flows in terms of the stream function, it is preferable to work with the first integral of $f(z)$, defined by the equation $dF/dz = f$, and observe that

$$\omega = \frac{\partial F_r}{\partial x} = \frac{\partial F_i}{\partial y}, \qquad \frac{P}{\mu} = -\frac{\partial F_r}{\partial y} = \frac{\partial F_i}{\partial x} \tag{6.12.4}$$

To derive an expression for the stream function in terms of F, we note that if H_0, H_1, H_2, H_3 are four harmonic functions and c_0, c_1, c_2, c_3 are four arbitrary constants, then

$$\psi = c_0 H_0 + c_1 x H_1 + c_2 y H_2 + c_3(x^2 + y^2)H_3 \tag{6.12.5}$$

satisfies the biharmonic equation $\nabla^4\psi = 0$, and it is therefore an acceptable stream function. The corresponding vorticity is readily found to be

$$\omega = -\nabla^2\psi = -2c_1\frac{\partial H_1}{\partial x} - 2c_2\frac{\partial H_2}{\partial y} - 4c_3\left(H_3 + x\frac{\partial H_3}{\partial x} + y\frac{\partial H_3}{\partial y}\right) \tag{6.12.6}$$

Comparing Eq. (6.12.6) with the first of Eqs. (6.12.4) suggests setting

$$H_1 = F_r, \qquad H_2 = F_i, \qquad H_3 = 0, \qquad c_1 + c_2 = -\tfrac{1}{2} \qquad (6.12.7)$$

which allows us to express Eq. (6.12.5) as

$$\psi = c_0 H + c_1(x - x_0)F_r - (\tfrac{1}{2} + c_1)(y - y_0)F_i \qquad (6.12.8)$$

where x_0 and y_0 are two arbitrary constants and H is a harmonic function of x and y.

Equations (6.12.4) and (6.12.8) provide us with the vorticity, pressure, and stream function of a two-dimensional Stokes flow in terms of the generating analytic complex function $F(z)$ and the real harmonic function H. Having specified these functions, we obtain a multiplicity of flows with identical vorticity and pressure parametrized by the constants c_0 and c_1. For example, setting F equal to a constant, produces families of irrotational flows with vanishing vorticity and constant pressure.

In the particular case where $c_1 = -\tfrac{1}{4}$, Eq. (6.12.8) assumes the compact form

$$\psi = -\tfrac{1}{4} \operatorname{Re}[(z - z_0)^* F(z) + G(z)] \qquad (6.12.9)$$

where G is an arbitrary analytic function whose real part is equal to $-4c_0 H$, and an asterisk denotes the complex conjugate. Straightforward differentiation yields the associated velocity field in the form

$$u + iv = \tfrac{1}{4} i[F + (z - z_0)F'^* + G'^*] \qquad (6.12.10)$$

where a prime denotes differentiation with respect to z.

To obtain the stream function due to a point force of unit strength located at the point z_0 and oriented along the x axis, we set $c_0 = 1/4\pi$, $H = y - y_0$, $c_1 = 0$, $F(z) = (i/2\pi) \ln|z - z_0|$, and obtain

$$\psi = -\frac{1}{4\pi}(y - y_0)(\ln|z - z_0| - 1) \qquad (6.12.11)$$

The alternative choice $c_0 = -1/4\pi$, $H = x - x_0$, $c_1 = -\tfrac{1}{2}$, $F(z) = -(1/2\pi) \ln|z - z_0|$ yields the stream function due to a point force of unit strength oriented along the y axis,

$$\psi = \frac{1}{4\pi}(x - x_0)(\ln|z - z_0| - 1) \qquad (6.12.12)$$

An extensive discussion and further applications of the pressure–velocity formulation of two-dimensional Stokes flow is given by Langlois (1964, Chapter 7). It is worth noting that this formulation is amenable to a boundary integral representation, which may be used to describe and thus compute a flow by solving boundary integral equations (Pozrikidis, 1992).

Formulation in Terms of the Airy Stress Function

In an alternative formulation, we express a two-dimensional Stokes flow in terms of the Airy stress function Φ defined in terms of the modified stresses by the equations

$$\sigma_{xx} = \frac{\partial^2 \Phi}{\partial y^2}, \qquad \sigma_{xy} = \sigma_{yx} = -\frac{\partial^2 \Phi}{\partial x \, \partial y}, \qquad \sigma_{yy} = \frac{\partial^2 \Phi}{\partial x^2} \qquad (6.12.13)$$

The Stokes equation is satisfied for any choice of Φ. Using the continuity equation and the fact that the pressure is a harmonic function, we find that Φ satisfies the biharmonic equation $\nabla^4 \Phi = 0$. Furthermore, writing out the three independent components of the stress tensor in terms of the pressure and stream function ψ, we obtain

$$\frac{\partial^2 \Phi}{\partial x^2} - \frac{\partial^2 \Phi}{\partial y^2} = -4\mu \frac{\partial^2 \psi}{\partial x \, \partial y}, \qquad \frac{\partial^2 \psi}{\partial x^2} - \frac{\partial^2 \psi}{\partial y^2} = \frac{1}{\mu} \frac{\partial^2 \Phi}{\partial x \, \partial y} \qquad (6.12.14)$$

which can be used to show that the complex function $\chi = \Phi - 2i\mu\psi$ satisfies the differential equation

$$\left(\frac{\partial^2 \chi}{\partial z^{*2}}\right)_z = 0 \tag{6.12.15}$$

Integrating Eq. (6.12.15) twice yields

$$\chi(z) = z^*\chi_1(z) + \chi_2(z) \tag{6.12.16}$$

where χ_1 and χ_2 are two arbitrary analytic functions of z.

Different selections for χ_1 and χ_2 produce various types of two-dimensional Stokes flow. Setting, for instance,

$$\chi = \tfrac{1}{2}\mu[z^2(z^* - \tfrac{1}{3}z) - 4z] \tag{6.12.17}$$

yields two-dimensional Poiseuille flow through a two-dimensional channel of unit half-width.

The formulation in terms of the Airy stress function and the stream function is amenable to a boundary integral representation that provides us with a basis for numerical computation (Pozrikidis, 1992).

PROBLEMS

6.12.1   Pressure–vorticity formulation. With reference to the representation (6.12.8), find the values of c_0 and c_1 and the functions $H(z)$ and $F(z)$ that yield (a) simple unidirectional shear flow, (b) orthogonal stagnation-point flow, (c) oblique stagnation-point flow, and (d) flow due to a point-force dipole.

6.12.2   Airy stress function. Find the functions χ_1 and χ_2 that correspond to the four flows described in Problem 6.12.1.

6.13 | EFFECTS OF INERTIA AND OSEEN FLOW

The solution of the equations of Stokes flow provides us with the leading-order approximation to the structure of a flow at low Reynolds numbers, under the assumption that inertial forces are *uniformly* negligible throughout the domain of flow. For interior flow, this assumption can be made to be valid by making the Reynolds number, defined with respect to the global size of the boundaries, to be sufficiently small. Having obtained the solution of the equations of Stokes flow, we may proceed to study the effects of inertia by expressing the flow variables in terms of regular perturbation series with respect to the Reynolds number as discussed by Van Dyke (1975).

For exterior flow in a partially or totally infinite domain, the assumption that inertial forces are *uniformly* negligible throughout the domain of flow may not be legitimate. Consider, for instance, uniform flow with velocity of magnitude V past a three-dimensional body. The Stokes-flow solution reveals that, when the force exerted on the body is finite, the disturbance flow decays like $1/r$, which means that the magnitude of the left-hand side of the Navier–Stokes equation decays like $\rho V^2/r^2$, whereas the magnitude of the right-hand side decays like $\mu V/r^3$. This means that at a distance $r = \mu/\rho V$ from the body, inertial forces will become important, and the Stokes-flow approximation will cease to be valid.

An important consequence of the failure of the Stokes-flow approximation at large distances is that the Stokes-flow solution will not necessarily satisfy the required far-field boundary conditions. In the case of three-dimensional flow, this difficulty is shielded by the fact that the flow due to a point force decays like $1/r$, but in the case of two-dimensional flow we encounter paradoxical behavior due to the logarithmic divergence of the flow due to a point force discussed in the end of Section 6.6. It is then not surprising that attempting to account for the effects of inertia using regular perturbation expansions leads to expressions that diverge at infinity, a behavior that is known as Whitehead's paradox.

Oseen's Equation

To study steady uniform flow with velocity $\mathbf{V}$ past a stationary body, Oseen (1910) (see also Oseen, 1927) proposed replacing the Stokes equation with the linearized Navier–Stokes equation

$$\rho \mathbf{V} \cdot \nabla \mathbf{u} = -\nabla P + \mu \nabla^2 \mathbf{u} \tag{6.13.1}$$

which is to be solved subject to the conditions that $\mathbf{u}$ vanishes at the surface of the body and tends to $\mathbf{V}$ far from the body. When the Reynolds number, defined with respect to the size of the body, is sufficiently small, the left-hand side of Eq. (6.13.1) is small compared to the right-hand side in the vicinity of the body, and Eq. (6.13.1) describes Stokes flow. Far from the body, the left-hand side captures the dominant contribution of the inertial forces.

Oseen's equation for flow due to a body that translates steadily with velocity $\mathbf{U}$ takes the form of the unsteady linearized Navier–Stokes equation

$$\rho \frac{\partial \mathbf{u}}{\partial t} = -\nabla P + \mu \nabla^2 \mathbf{u} \tag{6.13.2}$$

which is to be solved subject to the conditions that $\mathbf{u}$ is equal to $\mathbf{U}$ at the instantaneous location of the surface of the body, and $\mathbf{u}$ vanishes far from the body. Since in a frame of reference moving with the body the velocity field is steady, we can write

$$\frac{\partial \mathbf{u}}{\partial t} + \mathbf{U} \cdot \nabla \mathbf{u} = \mathbf{0} \qquad \text{or} \qquad \frac{\partial \mathbf{u}}{\partial t} = -\mathbf{U} \cdot \nabla \mathbf{u} \tag{6.13.3}$$

which shows that Eq. (6.13.2) is identical to Eq. (6.13.1) with $\mathbf{V} = -\mathbf{U}$.

Unfortunately, exact solutions to Oseen's equation are not available. An approximate solution that describes the flow due to the translation of a sphere of radius a was developed by Lamb (1911). In spherical polar coordinates with the x axis in the direction of translation, the Stokes stream function is given by

$$\Psi = U a^2 \left\{ -\frac{1}{4} \frac{a}{r} \sin^2 \theta + \frac{3}{Re}(1 - \cos \theta) \left[1 - \exp\left(-Re\, \frac{1}{4} \frac{r}{a}(1 + \cos \theta) \right) \right] \right\} \tag{6.13.4}$$

where $Re = 2a\rho U/\mu$. Close to the sphere, r/a is of order unity, and the argument of the exponential term is small. Expanding the exponential in a Taylor series, we obtain the solution of Stokes flow enhanced by a small correction whose magnitude is proportional to Re. The boundary condition on the surface of the sphere is satisfied with an error of order Re. The streamline pattern corresponding to Eq. (6.13.4), shown in Figure 6.13.1 for $Re = 2$, is no longer symmetric with respect to the midplane of the sphere, as it was for Stokes flow. Far from the sphere, the streamlines are radial lines everywhere except within the wake.

The linearity of Eqs. (6.13.1) and (6.13.2) with respect to $\mathbf{u}$ and P allows us to construct solutions by superposing fundamental solutions represented by Green's functions. The singular fundamental solution of the Oseen equation (6.13.1), called the Oseenlet, satisfies Eq. (6.13.1) with the singular forcing function $\delta(\mathbf{x} - \mathbf{x}_0)\mathbf{b}$ on the right-hand side, where δ is the three-dimensional delta function and $\mathbf{b}$ is a constant vector. The Oseenlet can be derived in a manner that is analogous to that for the Stokeslet. The corresponding velocity field is given in Eq. (6.4.9) with

$$H = -\frac{1}{4\pi \rho V} \int_0^c \frac{1 - e^{-\eta}}{\eta} \, d\eta \tag{6.13.5}$$

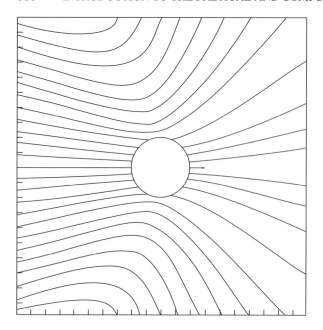

Figure 6.13.1   Streamline pattern of Oseen flow due to the translation of a sphere for $Re = 2$.

where $c = \rho(Vr - \mathbf{V} \cdot \hat{\mathbf{x}})/2\mu$, and $V = |\mathbf{V}|$. At small values of c, after discarding an irrelevant linear term, we find $H \simeq -r/(8\pi\mu)$, yielding Stokes flow. The pressure field of the Oseenlet is identical to that of the Stokeslet. The fundamental solution of Eq. (6.13.2) will be discussed in Section 6.14 in the context of unsteady Stokes flow (Problem 6.14.2).

Inertial effects on the flow due to the motion of rigid particles and liquid drops at small but finite Reynolds numbers have been studied extensively using the method of matched asymptotic expansions and numerical solutions; a wealth of information has been compiled by Clift, Grace, and Weber (1978). For instance, the first inertial correction to the force on a translating solid sphere of radius a yields the drag coefficient

$$c_D \equiv \frac{F}{\pi \rho U^2 a^2} = \frac{12}{Re}(1 + \tfrac{3}{16} Re) \tag{6.13.6}$$

where $Re = 2a\rho U/\mu$. The first-order contribution expresses Stokes's law. Brenner (1961) showed that the first inertial correction to the force on a translating body may be deduced immediately from a knowledge of the drag force in Stokes flow.

Two-Dimensional Flow

A comprehensive discussion of the conceptual, analytical, and computational considerations involved in the analysis and computation of inertial effects in two-dimensional flow in an infinite domain was presented by Van Dyke (1975). For flow due to the translation of a cylinder of radius a with velocity U, Lamb (1911) showed that the Stokes flow solution (6.6.44) is valid provided that the constant A takes the value

$$A = \frac{2}{\ln(7.4/Re)} \tag{6.13.7}$$

where $Re = 2a\rho U/\mu$. The drag force exerted on the cylinder is given by $\mathbf{F} = -\mathbf{b}$, where $\mathbf{b}$ is computed from the first of Eqs. (6.6.43).

PROBLEMS

6.13.1   **Shear flow past a body.** Consider simple shear flow past a stationary body, and explain why the Stokes flow approximation is valid throughout the domain of flow.

6.13.2   **A particle above a plane wall.** Consider a particle translating at a certain distance above an infinite plane wall, and estimate the distance from the particle where inertial effects begin to become important.

6.13.3   **Vorticity in Oseen flow.** Consider Oseen flow due to the translation of a body, and show that the vorticity field satisfies the unsteady heat conduction equation

$$\frac{\partial \boldsymbol{\omega}}{\partial t} = -\mathbf{U} \cdot \nabla \boldsymbol{\omega} = \nu \, \nabla^2 \boldsymbol{\omega} \tag{6.13.8}$$

Explain, in physical terms, why the body acts like an effective source of vorticity.

6.14 | UNSTEADY STOKES FLOW

In the remainder of this chapter, we turn our attention to flows at low Reynolds numbers Re but finite values of the parameter β discussed in Section 3.7 and, more generally, to flows for which Re is much less than β. The motion of the fluid is governed by the continuity equation and the unsteady Stokes equation (3.7.5), repeated here for ready reference

$$\rho \frac{\partial \mathbf{u}}{\partial t} = -\nabla p + \mu \, \nabla^2 \mathbf{u} + \rho \mathbf{g} \tag{6.14.1}$$

Flows that are governed by the unsteady Stokes equation include those generated by the translational or rotational vibrations of solid and liquid particles in a viscous fluid, provided that the amplitude of oscillation is small compared to the size of the particles. Further examples include oscillatory flow past suspended particles due, for instance, to the passage of a sound wave, and the flow due to the rapidly varying transient motion of particles in a quiescent or accelerating ambient fluid.

Properties of Unsteady Stokes Flow

Taking the divergence of Eq. (6.14.1), using the continuity equation, and assuming that the density of the fluid is uniform, we find that, as in the case of steady Stokes flow, the pressure is a harmonic function, $\nabla^2 p = 0$. Furthermore, taking the curl of Eq. (6.14.1), we find that the vorticity evolves according to the vectorial unsteady heat conduction equation

$$\frac{\partial \boldsymbol{\omega}}{\partial t} = \nu \, \nabla^2 \boldsymbol{\omega} \tag{6.14.2}$$

Any irrotational velocity field described in terms of a harmonic potential ϕ as $\mathbf{u} = \nabla \phi$ satisfies the equations of unsteady Stokes flow. The associated pressure is given by the Bernoulli equation

$$p = -\rho \frac{\partial \phi}{\partial t} + \rho \mathbf{g} \cdot \mathbf{x} + c(t) \tag{6.14.3}$$

A potential flow alone, however, is not generally capable of satisfying both the no-penetration and no-slip conditions over the solid boundaries and must be supplemented with a rotational flow.

Oscillatory Flow

The linearity of the equations of unsteady Stokes flow requires that the velocity, vorticity, modified pressure, and modified stress of an oscillatory flow exhibit identical harmonic dependences. We thus write

$$(\mathbf{u}, \boldsymbol{\omega}, P, \boldsymbol{\sigma}^{\text{Mod}})(\mathbf{x}, t) = (\mathbf{v}, \mathbf{w}, q, \boldsymbol{\Sigma})(\mathbf{x})\exp(-i\Omega t) \tag{6.14.4}$$

where Ω is the angular frequency of the oscillations. Substituting these expressions into Eq. (6.14.1) and into the continuity equation, we obtain the equations of oscillatory unsteady Stokes flow

$$-i\Omega\rho\mathbf{v} = -\nabla q + \mu\nabla^2\mathbf{v} \tag{6.14.5}$$

$$\nabla \cdot \mathbf{v} = 0 \tag{6.14.6}$$

Oscillatory Stokes flow shares many of the properties of steady Stokes flow discussed in the preceding sections, including the property of *reversibility*, the *reciprocal identity*, *Faxen's laws*, and the property of *uniqueness subject to specified boundary conditions for the velocity* (Pozrikidis, 1992; Ladyzhenskaya, 1975).

As in the case of Stokes flow, we can construct solutions to particular problems by superposing fundamental solution or developing boundary integral representations. Before we can implement these methods, however, we must have available the Green's functions of the unsteady Stokes equation representing the *flow due to a point force whose strength is a harmonic function of time.*

Oscillatory point force

Consider the flow due to a point force located at the point $\mathbf{x}_0$ whose strength is given by the real or imaginary part of $\mathbf{b} = \mathbf{B}\exp(-i\Omega t)$, where $\mathbf{B}$ is a constant. The corresponding velocity and pressure fields are computed by solving Eq. (6.14.5) where the right-hand side is enhanced with the singular forcing function $\mathbf{B}\,\delta(\mathbf{x} - \mathbf{x}_0)$; δ is the three-dimensional delta function. The solution is found working as in Section 6.4 for Stokes flow, and the final result is Eq. (6.4.9), with $\mathbf{v}$ in place of $\mathbf{u}$ and $\mathbf{B}$ in place of $\mathbf{b}$. The auxiliary function H satisfies the inhomogeneous Helmholtz equation

$$\left(\nabla^2 + i\frac{\Omega}{\nu}\right)H = -\frac{1}{4\pi\mu r} \tag{6.14.7}$$

where $r = |\mathbf{x} - \mathbf{x}_0|$. Taking the Laplacian of Eq. (6.14.7) shows that H is the fundamental solution of the modified biharmonic equation $\mu\nabla^2(\nabla^2 + i\Omega/\nu)H = \delta(\mathbf{x} - \mathbf{x}_0)$, which is known to be

$$H = \frac{1}{4\pi\mu}\left(\frac{i\nu}{\Omega}\right)^{1/2}\frac{1 - e^{-R}}{R} \tag{6.14.8}$$

where $R = (-i\Omega/\nu)^{1/2}r$, and the square root has a positive real part. Note that as $R \to 0$, H assumes the asymptotic form $H \sim c - r/(8\pi\mu)$, where c is a constant, recovering the results for Stokes flow. Substituting Eq. (6.14.8) into the counterpart of Eq. (6.4.9), we derive the velocity field due to an oscillating point force in the standard form

$$v_i(\mathbf{x}, \mathbf{x}_0) = \frac{1}{8\pi\mu}S_{ij}(\hat{\mathbf{x}})\,B_j \tag{6.14.9}$$

where $\hat{\mathbf{x}} = \mathbf{x} - \mathbf{x}_0$ and

$$S_{ij}(\hat{x}) = \frac{\delta_{ij}}{r}A(R) + \frac{\hat{x}_i\hat{x}_j}{r^3}C(R) \tag{6.14.10}$$

is the free-space Green's function of the unsteady Stokes equation, also called the *oscillating Stokeslet*. The functions $A(R)$ and $C(R)$ are given by

$$A(R) = 2e^{-R}\left(1 + \frac{1}{R} + \frac{1}{R^2}\right) - \frac{2}{R^2}$$

$$C(R) = -2e^{-R}\left(1 + \frac{3}{R} + \frac{3}{R^2}\right) + \frac{6}{R^2} \tag{6.14.11}$$

One may readily confirm that $A(0) = C(0) = 1$, which suggests that at small frequencies or close to the point force, the oscillating Stokeslet reduces to the regular Stokeslet for Stokes flow.

The vorticity, pressure, and stress fields associated with the oscillating Stokeslet are given by Eqs. (6.14.4) with

$$w_i = \frac{1}{8\pi\mu}\Omega_{ij} B_j, \qquad q = \frac{1}{8\pi}\Pi_j B_j, \qquad \Sigma_{ik} = \frac{1}{8\pi}T_{ijk} B_j \qquad (6.14.12)$$

where

$$\Omega_{ij} = 2\varepsilon_{ijl}\frac{\hat{x}_l}{r^3}e^{-R}(R+1), \qquad \Pi_j = 2\frac{\hat{x}_j}{r^3}$$

$$T_{ijk} = -\frac{2}{r^3}(\delta_{ij}\hat{x}_k + \delta_{kj}\hat{x}_i)[e^{-R}(R+1) - C] - \frac{2}{r^3}\delta_{ik}\hat{x}_j(1-C) \qquad (6.14.13)$$

$$- 2\frac{\hat{x}_i\hat{x}_j\hat{x}_k}{r^5}[5C - 2e^{-R}(R+1)]$$

and the function C is given in Eqs. (6.14.11).

It is instructive to compute the traction exerted on a spherical surface of radius r that is centered at the oscillating point force. After some algebra we find

$$f_i = \Sigma_{ij}n_j\exp(-i\Omega t) = \frac{1}{8\pi}\left(\frac{\delta_{ij}}{r^2}K(R) + \frac{\hat{x}_i\hat{x}_j}{r^4}L(R)\right)B_j\exp(-i\Omega t) \qquad (6.14.14)$$

where the functions K and L are given by

$$K(R) = 2[C - e^{-R}(R+1)]$$
$$L(R) = 2[e^{-R}(R+1) - 1 - 3C] \qquad (6.14.15)$$

One may show that $K(0) = 0$, $L(0) = -6$, which is consistent with Eq. (6.4.14) for Stokes flow. The force exerted on the spherical surface is found by integrating the traction to be

$$\mathbf{F} = \tfrac{1}{6}(3K + L)\mathbf{B}\exp(-i\Omega t) = -\tfrac{1}{3}[2e^{-R}(R+1) + 1]\mathbf{B}\exp(-i\Omega t) \qquad (6.14.16)$$

We note that the amplitude of the force acting on a small sphere of infinitesimal radius is equal to $-\mathbf{B}\exp(-i\Omega t)$, whereas that on a sphere of large radius is equal to $-\tfrac{1}{3}\mathbf{B}\exp(-i\Omega t)$. The difference between these two values is equal to the rate of change of momentum of the fluid that surrounds the point force.

Asymptotic behavior of the oscillating point force at small and large frequencies

To examine the asymptotic behavior of the flow due to a point force at small frequencies or close to the pole, we expand the unsteady Stokeslet in a Taylor series for small values of R, obtaining

$$\mathbf{S}(\hat{\mathbf{x}}) = \mathbf{S}^{(0)}(\hat{\mathbf{x}}) + \mathbf{S}^{(1)}(\hat{\mathbf{x}}) + R^2\mathbf{S}^{(2)}(\hat{\mathbf{x}}) + R^3\mathbf{S}^{(3)}(\hat{\mathbf{x}}) + \cdots \qquad (6.14.17)$$

where $\mathbf{S}^{(0)}$ is the steady Stokeslet, and

$$S_{ij}^{(1)}(\hat{\mathbf{x}}) = -\tfrac{4}{3}\left(-\frac{i\Omega}{\nu}\right)^{1/2}\delta_{ij}$$

$$S_{ij}^{(2)}(\hat{\mathbf{x}}) = \tfrac{1}{4}\left(3\frac{\delta_{ij}}{r} - \frac{\hat{x}_i\hat{x}_j}{r^3}\right) \qquad (6.14.18)$$

$$S_{ij}^{(3)}(\hat{\mathbf{x}}) = \tfrac{2}{15}\left(2\frac{\delta_{ij}}{r} - \frac{\hat{x}_i\hat{x}_j}{r^3}\right)$$

It is both interesting and significant to note that $\mathbf{S}^{(1)}$ represents uniform streaming flow.

To examine the behavior of the flow at high frequencies or far from the point force, we expand **S** in an asymptotic series for large values of R, obtaining

$$S_{ij} = \frac{2}{R^2} \left(-\frac{\delta_{ij}}{r} + 3 \frac{\hat{x}_i \hat{x}_j}{r^3} \right) + 2 e^{-R} \left(\frac{\delta_{ij}}{r} - \frac{\hat{x}_i \hat{x}_j}{r^3} \right) + \cdots \tag{6.14.19}$$

We observe that the first term on the right-hand side of Eq. (6.14.19) is the steady potential dipole, and this reveals that at high frequencies, or at large distances, the unsteady Stokeslet produces irrotational flow.

Green's functions and the boundary integral equation

The Green's functions of unsteady oscillatory Stokes flow represent the flow due to a point force with oscillating strength in the presence of a stationary solid boundary. When the domain of flow extends to infinity, **G**, **Ω**, **Π**, and **T** are required to decay as the observation point moves far from the point force. Pozrikidis (1989a, 1992) discussed the Green's function for semi-infinite flow bounded by a plane wall.

Using the reciprocal theorem, one may show that the Green's functions for the velocity **G** satisfy the symmetry property (6.4.21). The pressure vector **Π** and stress tensor **T** provide us with two acceptable unsteady Stokes flows representing, respectively, the flow due to a point source and a stresslet with oscillating strengths.

Working as before for steady Stokes flow, we derive a boundary-integral representation that is identical to that discussed in Section 6.9. Numerical methods for solving the associated boundary integral equations have been developed by Pozrikidis (1989b) and Loewenberg (1994).

Flow past a particle

As an application of the boundary integral representation, let us consider an oscillatory Stokes flow past a stationary particle and examine the asymptotic behavior at small frequencies. For this purpose, we expand the free-space Green's function **S** and its associated stress tensor **T** in a Taylor series with respect to R, as shown in Eq. (6.14.17), identify a characteristic length scale a in the flow, form the dimensionless complex frequency $\lambda^2 = -i\Omega a^2/\nu$, and expand the amplitude of the *disturbance* boundary traction **t** defined by $\mathbf{f} = \mathbf{t} \exp(-i\Omega t)$, and amplitude of boundary velocity **v** in Taylor series with respect to λ, writing $\mathbf{t} = \mathbf{t}^{(0)} + \lambda \mathbf{t}^{(1)} + \lambda^2 \mathbf{t}^{(2)} + \cdots$, and $\mathbf{v} = \mathbf{v}^{(0)} + \lambda \mathbf{v}^{(1)} + \lambda^2 \mathbf{v}^{(2)} + \cdots$. Substituting these expansions into the boundary integral equation (6.10.2) and collecting terms of zeroth and first order with respect to λ, we obtain the equations

$$v_j^{(0)}(\mathbf{x}_0) = -\frac{1}{4\pi\mu} \int_P t_i^{(0)}(\mathbf{x}) S_{ij}^{(0)}(\mathbf{x}, \mathbf{x}_0) \, dS(\mathbf{x})$$

$$+ \frac{1}{4\pi} \int_P^{PV} v_i^{(0)}(\mathbf{x}) T_{ijk}^{(0)}(\mathbf{x}, \mathbf{x}_0) n_k(\mathbf{x}) \, dS(\mathbf{x}) \tag{6.14.20}$$

$$v_j^{(1)}(\mathbf{x}_0) - \frac{1}{3\pi\mu a} F_j^{(0)} = -\frac{1}{4\pi\mu} \int_P t_i^{(1)}(\mathbf{x}) S_{ij}^{(0)}(\mathbf{x}, \mathbf{x}_0) \, dS(\mathbf{x})$$

$$+ \frac{1}{4\pi} \int_P^{PV} v_i^{(1)}(\mathbf{x}) T_{ijk}^{(0)}(\mathbf{x}, \mathbf{x}_0) n_k(\mathbf{x}) \, dS(\mathbf{x}) \tag{6.14.21}$$

where $\mathbf{F}^{(0)}$ is the force exerted on the particle in steady Stokes flow, and P represents the surface of the particle. Prescribing the boundary velocity reduces Eqs. (6.14.20) and (6.14.21) to Fredholm integral equations of the first kind for the amplitude of the tractions $\mathbf{t}^{(0)}$ and $\mathbf{t}^{(1)}$. Note that these equations have identical kernels.

Flow due to the motion of a particle

In a related application, we consider the flow due to the translational or rotational vibrations of a rigid particle that is immersed in an infinite fluid. Applying the boundary condition of rigid-body motion at the mean position of the particle, instead of its actual location, introduces an error that is proportional to the Reynolds number, which is negligible as long as the amplitude of the oscillations is small compared to the dimensions of the particle. Working as before for flow past a particle, and requiring that $\mathbf{v}^{(0)} = \mathbf{V} + \mathbf{W} \times \mathbf{x}$ and $\mathbf{v}^{(1)} = \mathbf{v}^{(2)} = \cdots = \mathbf{0}$ on the mean position of the surface of the particle P, where $\mathbf{V}$ and $\mathbf{W}$ are the amplitudes of the translational velocity and angular velocity of oscillation, we obtain two integral equations for $\mathbf{t}^{(0)}$ and $\mathbf{t}^{(1)}$

$$V_j + \varepsilon_{jkl} W_k x_l = -\frac{1}{8\pi\mu} \int_P t_i^{(0)}(\mathbf{x}) S_{ij}^{(0)}(\mathbf{x}, \mathbf{x}_0) \, dS(\mathbf{x}) \tag{6.14.22}$$

$$-\frac{1}{6\pi\mu a} F_j^{(0)} = -\frac{1}{8\pi\mu} \int_P t_i^{(1)}(\mathbf{x}) S_{ij}^{(0)}(\mathbf{x}, \mathbf{x}_0) \, dS(\mathbf{x}) \tag{6.14.23}$$

The solution of the first equation may be expressed in the form

$$\mathbf{t}^{(0)} = -\mu(\mathbf{R}^T \cdot \mathbf{V} + \mathbf{R}^R \cdot \mathbf{W}) \tag{6.14.24}$$

where $\mathbf{R}^T$ and $\mathbf{R}^R$ are, respectively, the traction resistance matrices for steady translation and rotation introduced in Eqs. (6.8.3) and (6.8.7). Furthermore, using the Eqs. (6.8.13), we write the steady force and torque exerted on the particle in the forms

$$\mathbf{F}^{(0)} = -\mu(\mathbf{X} \cdot \mathbf{V} + \mathbf{P} \cdot \mathbf{W})$$
$$\mathbf{T}^{(0)} = -\mu(\mathbf{P}' \cdot \mathbf{V} + \mathbf{Y} \cdot \mathbf{W}) \tag{6.14.25}$$

Comparing the preceding four equations, it becomes evident that the first correction to the traction, force, and torque are given, respectively, by

$$\mathbf{f}^{(1)} = \frac{1}{6\pi a} \mathbf{R}^T \cdot \mathbf{F}^{(0)} \exp(-i\Omega t) = -\frac{\mu}{6\pi a} \mathbf{R}^T \cdot (\mathbf{X} \cdot \mathbf{V} + \mathbf{P} \cdot \mathbf{W}) \exp(-i\Omega t) \tag{6.14.26}$$

$$\mathbf{F}^{(1)} = \frac{1}{6\pi a} \mathbf{X} \cdot \mathbf{F}^{(0)} \exp(-i\Omega t) = -\frac{\mu}{6\pi a} \mathbf{X} \cdot (\mathbf{X} \cdot \mathbf{V} + \mathbf{P} \cdot \mathbf{W}) \exp(-i\Omega t) \tag{6.14.27}$$

$$\mathbf{T}^{(1)} = \frac{1}{6\pi a} \mathbf{P}' \cdot \mathbf{F}^{(0)} \exp(-i\Omega t) = -\frac{\mu}{6\pi a} \mathbf{P}' \cdot (\mathbf{X} \cdot \mathbf{V} + \mathbf{P} \cdot \mathbf{W}) \exp(-i\Omega t) \tag{6.14.28}$$

Remarkably, we find that the first-order corrections may be computed directly from the resistance matrices for steady flow.

Singularities of Oscillatory Flow

The whole apparatus of the singularity method for Stokes flow may be extended in a straightforward manner to unsteady Stokes flow. While the general principles remain the same, the specific expressions for the singularities become quite more involved, and the computation of singularity representations becomes more cumbersome.

In the remainder of the present subsection we discuss a family of singularities describing oscillatory flow in free space, and in the next section we shall illustrate their application to computing exterior flows.

Point source with harmonic rate of discharge

A point source located at the point $\mathbf{x}_0$ whose strength is an oscillatory function of time, $m = M \exp(-i\Omega t)$, where M is a constant, produces irrotational flow with associated velocity and modified pressure fields given by

$$\mathbf{u}(\mathbf{x}) = \frac{M}{4\pi} \frac{\hat{\mathbf{x}}}{r^3} \exp(-i\Omega t)$$

$$P(\mathbf{x}) = -i\rho\Omega \frac{M}{4\pi r} \exp(-i\Omega t) \tag{6.14.29}$$

where $\hat{\mathbf{x}} = \mathbf{x} - \mathbf{x}_0$, $r = |\mathbf{x}|$. As the frequency Ω is decreased, the amplitude of the modified pressure P diminishes, and we recover the results for steady Stokes flow.

Differentiating the point source with respect to its pole $\mathbf{x}_0$, we obtain a sequence of derivative singularities expressing irrotational flow, the first two of which are the potential dipole and the potential quadrupole. The velocity fields are identical to those listed in Table 6.5.1; the pressure fields are derived by differentiating that for the point source.

Multipoles of the oscillating point force

Differentiating the Green's functions with respect to $\mathbf{x}_0$, we obtain a sequence of singularities expressing multipoles of the oscillatory point force. The associated pressure fields are identical to those of the corresponding singularities of steady flow.

Symmetric Stokeslet quadrupole

Taking the Laplacian of the free-space Green's function, in particular, we obtain the symmetric Stokeslet quadrupole given by

$$S_{ij}^{SQ} \equiv -\tfrac{1}{2}\nabla_0^2 S_{ij} = -\frac{\delta_{ij}}{r^3}e^{-R}(1 + R + R^2) + 3\frac{\hat{x}_i\hat{x}_j}{r^5}e^{-R}\left(1 + R + \frac{R^2}{3}\right) \quad (6.14.30)$$

and R was defined after Eq. (6.14.8). One may verify by straightforward algebra that

$$\mathbf{S}^{SQ} = \mathbf{D} + \tfrac{1}{2}\frac{i\Omega}{\nu}\mathbf{S} \quad (6.14.31)$$

where $\mathbf{D}$ is the potential dipole given in Table 6.5.1 and $\mathbf{S}$ is the unsteady Stokeslet given in Eq. (6.14.10). It will be noted that the symmetric Stokeslet quadrupole is identical to the point source doublet *only* in the limit of steady flow.

The velocity field due to a symmetric Stokeslet quadrupole with strength $\mathbf{Q}$ is given by $v_i = (1/4\pi)S_{ij}^{SQ}Q_j$. The associated pressure field vanishes, and the stress field is given by $\Sigma_{ik} = (\mu/4\pi)T_{ijk}^{SQ}Q_j$, where

$$T_{ijk}^{SQ} = e^{-R}\left(\frac{\delta_{ij}\hat{x}_k + \delta_{jk}\hat{x}_i}{r^5}(6 + 6R + 3R^2 + R^3) + 2\frac{\delta_{ik}\hat{x}_j}{r^5}(3 + 3R + R^2)\right.$$
$$\left. -2\frac{\hat{x}_i\hat{x}_j\hat{x}_k}{r^7}(15 + 15R + 6R^2 + R^3)\right) \quad (6.14.32)$$

The traction and force on a spherical surface centered at the symmetric quadrupole are given by

$$f_i = \Sigma_{ij}n_j\exp(-i\Omega t) = \frac{\mu}{4\pi}\left(\frac{\delta_{ij}}{r^4}E(R) + \frac{\hat{x}_i\hat{x}_j}{r^6}H(R)\right)Q_j\exp(-i\Omega t) \quad (6.14.33)$$

$$\mathbf{F} = \frac{\mu}{r^2}[E(R) + \tfrac{1}{3}H(R)]\mathbf{Q}\exp(-i\Omega t) = -\frac{i\Omega}{\rho}\tfrac{2}{3}e^{-R}(1 + R)\mathbf{Q}\exp(-i\Omega t) \quad (6.14.34)$$

where

$$\begin{aligned} E(R) &= e^{-R}(6 + 6R + 3R^2 + R^3) \\ H(R) &= -e^{-R}(18 + 18R + 7R^2 + R^3) \end{aligned} \quad (6.14.35)$$

We observe that the force exerted on a spherical surface centered at the symmetric dipole does not vanish as it does for steady flow. Noting that $E(0) = 6$, $H(0) = -18$, we obtain agreement with the corresponding results for steady Stokes flow.

Couplet

The flow due to an oscillating couplet or rotlet with strength **L** is given in the third entry of Table 6.5.2 with

$$C_{im} = \varepsilon_{iml}\frac{\hat{x}_l}{r^3}e^{-R}(1 + R)\exp(-i\Omega t) \tag{6.14.36}$$

The corresponding pressure is uniform. The torque with respect to the pole on a spherical surface of radius r centered at the couplet is given by

$$\mathbf{T} = -e^{-R}(1 + R + \tfrac{1}{3}R^2)\mathbf{L}\exp(-i\Omega t) \tag{6.14.37}$$

Interior flow

The flow due to an oscillating Stokeson with strength **e** is given in Table 6.5.2 with (Pozrikidis, 1989)

$$S_{ij}^{STN} = [2\delta_{ij}r^2Q(R) - \hat{x}_i\hat{x}_jW(R)]\exp(-i\Omega t) \tag{6.14.38}$$

where

$$Q(R) = -\frac{1}{4R^5}[20R^3 - 30(R^2 + 1)\sinh R + 30R\cosh R] \tag{6.14.39}$$

$$W(R) = \frac{15}{R^5}[(R^3 + 3)\sinh R - 3R\cosh R] \tag{6.14.40}$$

The traction on a spherical surface of radius r centered at the Stokeson is given by

$$f_i = \Sigma_{ij}n_j\exp(-i\Omega t) = \mu\left(\delta_{ij}rG(R) + \frac{\hat{x}_i\hat{x}_j}{r}J(R)\right)e_j\exp(-i\Omega t) \tag{6.14.41}$$

where

$$G(R) = \frac{15}{R^5}[R(R^2 + 6)\cosh R - 3(R^2 + 2)\sinh R]$$

$$J(R) = -10 - \frac{15}{R^5}[R(R^2 + 18)\cosh R - (7R^2 + 18)\sinh R] \tag{6.14.42}$$

In the limit of steady flow we find $G(0) = 3$ and $J(0) = -9$.

Inertial Effects and Steady Streaming

The equations of unsteady Stokes flow describe the structure of a flow in the limit as the parameter β becomes much larger than the Reynolds number Re, whence the nonlinear acceleration term in the equation of motion becomes insignificant. This leading-order solution may be used to assess the effects of inertia on the basis of a regular perturbation expansion.

Let us assume that β is of order one, and introduce a regular asymptotic expansion for the velocity with respect to the Reynolds number Re. In the case of oscillatory flow, the Stokes flow solution is given either by the real or by the imaginary part of the right-hand side of Eq. (6.14.4). Selecting the real part, we write

$$\mathbf{u}(\mathbf{x}, t) = \tfrac{1}{2}(\mathbf{v}e^{-i\Omega t} + \mathbf{v}^*e^{i\Omega t}) + Re\,\mathbf{u}^{(1)}(\mathbf{x}, t) \tag{6.14.43}$$

where $\mathbf{u}^{(1)}$ represents the first inertial correction and an asterisk designates the complex conjugate. The nonlinear convection term on the left-hand side of the equation of motion becomes

$$\mathbf{u}\cdot\nabla\mathbf{u} = \tfrac{1}{4}(e^{-2i\Omega t}\mathbf{v}\cdot\nabla\mathbf{v} + e^{2i\Omega t}\mathbf{v}^*\cdot\nabla\mathbf{v}^*) + \tfrac{1}{4}\nabla\cdot(\mathbf{vv}^* + \mathbf{v}^*\mathbf{v}) + O(Re) \tag{6.14.44}$$

Nondimensionalizing Eqs. (6.14.43) and (6.14.44) as discussed in Section 3.7, substituting the result into the dimensionless form of the Navier–Stokes equation, maintaining terms of first

(a)   (b)

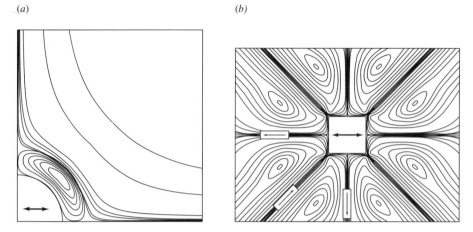

Figure 6.14.1 Streamline pattern of the steady streaming motion due to the vibration of a two-dimensional cylinder with (a) circular and (b) square cross sections (Kim and Troesch, 1989).

order with respect to the Reynolds number, and reverting to the physical variables, we obtain the *inhomogeneous* unsteady Stokes equation

$$\frac{\partial \mathbf{u}^{(1)}}{\partial t} + \frac{1}{\rho} \nabla P^{(1)} - \nu \nabla^2 \mathbf{u}^{(1)}$$

$$= -\tfrac{1}{4}\left(e^{-2i\Omega t}\mathbf{v}\cdot\nabla\mathbf{v} + e^{2i\Omega t}\mathbf{v}^*\cdot\nabla\mathbf{v}^*\right) - \tfrac{1}{4}\nabla\cdot(\mathbf{v}\mathbf{v}^* + \mathbf{v}^*\mathbf{v}) \qquad (6.14.45)$$

It is clear that the nonlinear acceleration term produces second harmonics of the fundamental frequency Ω. The appearance of the last time-independent term on the right-hand side of Eq. (6.14.45) is responsible for the onset of steady streaming motion due to weak inertial forces.

To describe the steady streaming motion, we integrate Eq. (6.14.45) over one period of the oscillation, thus eliminating the harmonically time-dependent functions, and obtain the inhomogeneous Stokes equation

$$\nabla\overline{P^{(1)}} - \mu\nabla^2\overline{\mathbf{u}^{(1)}} = -\tfrac{1}{4}\rho\nabla\cdot(\mathbf{v}\mathbf{v}^* + \mathbf{v}^*\mathbf{v}) \qquad (6.14.46)$$

where an overbar designates the time average over one period. The problem is reduced to solving Eq. (6.14.46) subject to appropriate boundary conditions. At high frequencies, the solution may be found using boundary-layer approximations (Batchelor, 1967, p. 358).

The mathematical analysis and computation of the steady streaming motion due to an oscillatory flow over a curved boundary have been studied extensively with special reference to the flow induced by the transmission of sound waves. Representative analyses, computations, and further references may be found in the articles by Kaneko and Honji (1979), Padmanabhan and Pedley (1987), Kim and Troesch (1989), and Riley (1992). Figure 6.14.1 illustrates computed streamline patterns of the steady streaming motion due to the vibrations of a two-dimensional cylinder with circular and square cross-sectional shapes (Kim and Troesch, 1989).

PROBLEMS

6.14.1 **Oscillating Stokeslet.** Confirm that, as R tends to zero, the vorticity and stress tensors Ω and $\mathbf{T}$ defined in Eqs. (6.14.13) reduce to those for steady flow.

6.14.2   **Impulsive Stokeslet.** (a) The Laplace transform of the velocity field due to the impulsive point force described by $\mathbf{b}\,\delta(\mathbf{x} - \mathbf{x}_0)\,\delta_1(t - t_0)$, where $\mathbf{b}$ is a constant, δ is the three-dimensional delta function, and δ_1 is the one-dimensional delta function, is given by

$$\hat{u}_i(\hat{\mathbf{x}}; s) = \frac{1}{8\pi\mu} S_{ij}(\hat{\mathbf{x}}, s)\, b_j \qquad (6.14.47)$$

where $\hat{\mathbf{x}} = \mathbf{x} - \mathbf{x}_0$ and $\mathbf{S}$ is the unsteady Stokeslet with the Laplace exponent s in place of $-i\Omega$. To compute the long-time behavior of the flow, we use the expansion (6.14.17), finding

$$\hat{u}_i(\hat{\mathbf{x}}; s) = \frac{1}{8\pi\mu}\left[S_{ij}^{(0)}(\hat{\mathbf{x}}) - \tfrac{4}{3}\xi^{1/2}\frac{\delta_{ij}}{r} + \xi\, S_{ij}^{(2)}(\hat{\mathbf{x}}) \right.$$

$$\left. + \xi^{3/2} S_{ij}^{(3)}(\hat{\mathbf{x}}) + \cdots \right] b_j \qquad (6.14.48)$$

where $\xi = sr^2/\nu$. Inverting Eq. (6.14.48), show that the long-time behavior of the flow is described by

$$u_i(\hat{\mathbf{x}}, t) = \frac{\rho^{1/2}}{(4\pi\mu\hat{t})^{3/2}}\left[\tfrac{2}{3}\delta_{ij} + \frac{3r^3}{4\nu\hat{t}} S_{ij}^{(3)}(\hat{\mathbf{x}}) + O\!\left(\frac{r^4}{\nu^2\hat{t}^2}\right) \right] b_j \qquad (6.14.49)$$

where $\hat{t} = t - t_0$. Equation (6.14.49) finds application in the computation of the long-time decay of the angular-velocity autocorrelation function of a rigid Brownian particle (Hocquart and Hinch, 1983). (b) Show that the velocity field due to an impulsive point force is given by Eq. (6.4.9) with

$$H = -\frac{1}{4\pi r}\frac{1}{(\pi\mu\rho\hat{t})^{1/2}}\int_0^r \left[1 - \exp\!\left(-\frac{\eta^2}{4\nu\hat{t}}\right) \right] d\eta \qquad (6.14.50)$$

(Ossen, 1927). On the basis of this solution confirm the validity of the expansion (6.14.49).

6.14.3   **Faxen laws.** Show that the generalized Faxen relations discussed in Section 6.11 are also valid for unsteady oscillatory Stokes flow (Pozrikidis, 1989a).

6.14.4   **Single-layer representation for a translating body.** Show that an unsteady oscillatory Stokes flow with a prescribed constant velocity $\mathbf{v} = \mathbf{U}$ over the surface of a body B may be represented in terms of a single-layer potential as

$$v_j(\mathbf{x}_0) = -\frac{1}{8\pi\mu}\int_B [f_i(\mathbf{x}) + i\Omega\rho\mathbf{U}\cdot\mathbf{x}\,n_i(\mathbf{x})]\, G_{ij}(\mathbf{x}, \mathbf{x}_0)\, dS(\mathbf{x}) \qquad (6.14.51)$$

6.14.5   **Green's functions for two-dimensional flow.** Green's function for two-dimensional unsteady flow may be derived as described in Section 6.4. The results may be expressed in terms of the fundamental solution of the two-dimensional Helmholtz equation (Stakgold, 1968, Vol. II, p. 265). Derive the two-dimensional unsteady Stokeslet, and show that in the limit of small R, it reduces to the steady Stokeslet (Pozrikidis, 1992).

6.15   COMPUTATION OF UNSTEADY STOKES FLOW PAST OR DUE TO THE MOTION OF PARTICLES

Having established the general properties of unsteady Stokes flow, we proceed to derive specific solutions for oscillatory and more general time-dependent motions involving solid and liquid particles.

Singularity Representations for Oscillatory Flow

The analytical and numerical computation of singularity representations for unsteady Stokes flow is analogous to that for the steady Stokes flow discussed in Sections 6.6 and 6.7. When the exterior flow due to the translational oscillations of a rigid or fluid particle is expressed in terms of a collection of singularities, including the oscillating Stokeslet and its derivatives expressing multipoles of the point force, then the force exerted on a particle can be computed directly from the total strength of the Stokeslets $\mathbf{B}$, as

$$\mathbf{F} = -(\mathbf{B} + i\Omega\rho V_p\mathbf{V}) \exp(-i\Omega t) \tag{6.15.1}$$

where V_p is the volume of the particle and $\mathbf{V}$ is the amplitude of the velocity oscillations defined as $\mathbf{U} = \mathbf{V}\exp(-i\Omega t)$, and $\mathbf{U}$ is the particle's velocity of translation (Pozrikidis, 1989a).

Translational oscillations of a sphere

The flow due to the vibrations of a sphere of radius a whose center is oscillating about the mean position $\mathbf{x}_0$ may be represented in terms of an unsteady Stokeslet and a symmetric Stokeslet quadrupole as

$$v_i(\mathbf{x}) = \frac{1}{8\pi\mu}S_{ij}(\mathbf{x}, \mathbf{x}_0)\,B_j + \frac{1}{4\pi}S_{ij}^{SQ}(\mathbf{x}, \mathbf{x}_0)\,Q_j \tag{6.15.2}$$

Requiring the boundary condition $\mathbf{v} = \mathbf{V}$ at $r = a$ yields

$$\mathbf{B} = 6\pi\mu a(1 + \lambda + \tfrac{1}{3}\lambda^2)\mathbf{V}$$

$$\mathbf{Q} = -\pi a^3\frac{6}{\lambda^2}(e^\lambda - 1 - \lambda - \tfrac{1}{3}\lambda^2)\mathbf{V} \tag{6.15.3}$$

where

$$\lambda^2 = -\frac{i\Omega a^2}{\nu} \tag{6.15.4}$$

As $|\lambda|$ tends to vanish, we obtain the expressions in Eqs. (6.6.4) for steady Stokes flow. Using Eqs. (6.14.16) and (6.14.34) or Eq. (6.15.1) along with the coefficients of the singularities in Eqs. (6.15.3), we find that the force exerted on the sphere is given by

$$\mathbf{F} = \left(-\tfrac{1}{3}[2e^{-\lambda}(\lambda + 1) + 1]\mathbf{B} - \frac{i\Omega}{\rho}\tfrac{2}{3}e^{-\lambda}(1 + \lambda)\mathbf{Q}\right)\exp(-i\Omega t)$$

$$= -(\mathbf{B} + i\Omega\rho V_s\mathbf{V})\exp(-i\Omega t)$$

$$= -6\pi\mu a(1 + \lambda + \tfrac{1}{9}\lambda^2)\mathbf{V}\exp(-i\Omega t) \tag{6.15.5}$$

where V_s is the volume of the sphere, deduced independently by Boussinesq (1885) and Basset (1888) (see Vojir and Michaelides, 1994). It is rather remarkable that the expression for the force is a binomial in the complex frequency parameter λ, but this is known to be true only for the spherical shape (Pozrikidis, 1989a). The three terms in the parentheses on the right-hand side of Eq. (6.15.5) represent, respectively, the steady Stokes drag force, the unsteady Boussinesq–Basset force, and the acceleration reaction associated with the virtual mass in potential flow (Chapter 7).

Rotational oscillations of a sphere

The flow due to a sphere of radius a executing rotational oscillations with angular velocity $\mathbf{\Omega} = \mathbf{A}\exp(-i\Omega t)$ may be represented simply in terms of a rotlet with strength $\mathbf{L} = 8\pi\mu a^3 e^\lambda/(1 + \lambda)\mathbf{A}$ placed at the center. The torque exerted on the sphere is given by

$$\mathbf{T} = -8\pi\mu\,a^3\tfrac{1}{3}\left(3 + \frac{\lambda^2}{1 + \lambda}\right)\mathbf{A}\exp(-i\Omega t) \tag{6.15.6}$$

Further singularity representations

Exact singularity representations of unsteady flow are limited to those for flow produced by the translational or rotational oscillations of a solid or liquid sphere, and for the disturbance flow due to a sphere that is held stationary in an oscillating linear ambient flow (Pozrikidis, 1989a; Kim and Karrila, 1991, Chapter 5; Pozrikidis, 1995). It should be noted that since a general oscillatory linear flow is not an exact solution to the equations of unsteady Stokes flow, the last solution must be regarded as the response of the sphere to the linear term of the Taylor series expansion about the center of the sphere of the velocity of a general unsteady incident flow.

Pozrikidis (1989a) computed approximate singularity representations of the flow due to the translational oscillations of a prolate spheroid, in terms of distributions of Green's functions and unsteady Stokeslet quadrupoles dipoles over the focal length of the spheroid. The densities of the distributions range between those for steady flow at low frequencies and those for potential flow at high frequencies.

General Time-Dependent Motion

The solutions of the unsteady Stokes equation for oscillatory motion may be used to describe an arbitrary time-dependent motion using the method of Fourier or Laplace transforms. In the case of the Laplace transform, we assume that the motion has started from rest and the velocity increases at most at an exponential rate in time, and take the Laplace transform of the unsteady Stokes equation to derive Eq. (6.14.5) with the Laplace complex exponent s in place of $-i\Omega$. The solution for the Laplace-transformed variables is identical to that for the corresponding amplitude functions in oscillatory flow subject to the aforementioned substitution. Inverting the Laplace transform allows us to compute general time-dependent motions.

As an example, we begin with Eq. (6.15.5) and find that the force exerted on a solid sphere of radius a and volume V_s, moving with a time-dependent velocity $\mathbf{U}(t)$ is given by

$$\mathbf{F}(t) = -6\pi\mu a \mathbf{U}(t) - 6a^2\sqrt{\pi\rho\mu}\int_{-\infty}^{t}\left(\frac{d\mathbf{U}}{dt}\right)_{t=s}\frac{1}{\sqrt{t-s}}\,ds - \tfrac{1}{2}\rho V_s\frac{d\mathbf{U}(t)}{dt} \qquad (6.15.7)$$

The three terms on the right-hand side of Eq. (6.15.7) represent, respectively, the *Stokes drag force,* the *Boussinesq–Basset viscous memory force,* and the *acceleration reaction.* The inverse square root kernel of the Boussinesq–Basset memory integral reflects the fact that vorticity diffuses away from the particle surface due to the effects of viscosity.

Equation (6.15.7) may be used to derive an integro-differential equation describing the gravitational settling of a sphere that has been released from rest at $t = 0$. Balancing the buoyancy force, the drag force, and the weight of the sphere, with the rate of change of the momentum of the sphere, we obtain

$$(\rho_s + \tfrac{1}{2}\rho)\,V_s\frac{d\mathbf{U}(t)}{dt} = (\rho_s - \rho)\,V_s\,\mathbf{g} - 6\pi\mu a\,\mathbf{U}(t) - 6a^2\sqrt{\pi\rho\mu}\int_{0}^{t}\left(\frac{d\mathbf{U}}{dt}\right)_{t=s}\frac{1}{\sqrt{t-s}}\,ds$$

$$(6.15.8)$$

where ρ_s is the density of the sphere. Equation (6.15.8) can be recast into the equivalent form of a second-order ordinary differential equation in time for $\mathbf{U}$, which may then be integrated using a standard numerical method (see Yih, 1979, p. 376; Clift, Grace, and Weber, 1978, p. 288).

Equations similar to Eqs. (6.15.7) and (6.15.8) have been developed for spherical bubbles and drops, as well as for nonspherical rigid particles, and extensions have been proposed to account for the effects of inertia and for the presence of an unsteady incident flow (Maxey and Riley, 1983). A comprehensive review of these topics is given in the monograph of Clift, Grace, and Weber (1978), and a more recent update of the literature is presented by Vojir and Michaelides (1994) and Pozrikidis (1995).

PROBLEMS

6.15.1 **Force on a sphere in time-dependent motion.** Derive Eq. (6.15.7) from Eq. (6.15.5) using the method of Laplace transform.

6.15.2 **Torque on a sphere in time-dependent rotation.** Starting from Eq. (6.15.6) and using the method of Laplace transform, derive an expression for the torque exerted on a solid sphere that rotates with an arbitrary time-dependent angular velocity $\mathbf{\Omega}(t)$ (Pozrikidis, 1989a).

Computer Problem

6.15.3 **Gravitational settling of a sphere from rest.** Develop and implement a computational procedure for solving the integro-differential equation (6.15.8), and then compute and plot the velocity of the settling sphere as a function of time. To evaluate the singular memory integral, subtract off the singularity and then perform analytical integration.

References

Basset, A. B., 1888, *A Treatise on Hydrodynamics.* Deighton Bell, Cambridge.

Batchelor, G. K., 1967, *An Introduction to Fluid Dynamics.* Cambridge University Press.

Blake, J. R., 1971, A note for the image system for a Stokeslet in a no-slip boundary. *Proc. Camb. Phil. Soc.* **70**, 303–10.

Boussinesq, J., 1885, Sur la résistance qu'oppose un liquide indéfini en repos,.... *Comptes Rendus Academie Science (Paris)* **100**, 935–37.

Brenner, H., 1961, The Oseen resistance of a particle of arbitrary shape. *J. Fluid Mech.* **11**, 604–10.

Brenner, H., 1963, The Stokes resistance of an arbitrary particle. *Chem. Eng. Science* **18**, 1–25.

Brenner, H., 1964a, The Stokes resistance of an arbitrary particle. II. An extension. *Chem. Eng. Science* **19**, 599–629.

Brenner, H., 1964b, The Stokes resistance of an arbitrary particle. IV. Arbitrary fields of flow. *Chem. Eng. Science* **19**, 703–27.

Burgers, J. M., 1938, On the motion of small particles of elongated form, suspended in a viscous liquid. *Second Report on Viscosity and Plasticity.* Kon. Ned. Akad. Wet., Verhand (Eerste sectie), **DI. XVI**, No. 4.

Chwang, A. T., and Wu, T. Y.-T., 1974, Hydromechanics of low-Reynolds-number flow. Part 1. Rotation of axisymmetric prolate bodies. *J. Fluid Mech.* **63**, 607–22.

Chwang, A. T., and Wu, T. Y.-T., 1975, Hydromechanics of low-Reynolds-number flow. Part 2. Singularity methods for Stokes flows. *J. Fluid Mech.* **67**, 787–815.

Clift, R., Grace, J. R., and Weber, M. E., 1978, *Bubbles, Drops, and Particles.* Academic Press.

Dean, W. R., and Montagnon, P. E., 1949, On the steady motion of viscous liquid in a corner. *Proc. Camb. Phil. Soc.* **45**, 389–94.

Faxen, H., 1924, Der Widerstand gegen die Bewegung einer starren Kugel in einer zähen Flüssigkeit, die zwischen zwei parallelen, ebenen Wänden eingeschlossen ist. *Ark. Mat. Astr. Fys.* **18**(29), 3.

Gray, J., and Hancock, G. J., 1955, The propulsion of sea-urchin spermatozoa. *J. Exp. Biol.* **32**, 802–14.

Hamrock, B. J., 1994, *Fundamentals of Fluid Film Lubrication.* McGraw Hill.

Hancock, G. J., 1953, The self-propulsion of microscopic organisms through liquids. *Proc. Roy. Soc. London* A **217**, 96–121.

Happel, J., and Brenner, H., 1973, *Low Reynolds Number Hydrodynamics.* Martinus Nijhoff.

Hocquart, R., and Hinch, E. J., 1983, The long-time tail of the angular-velocity autocorrelation function for a rigid Brownian particle of arbitrary centrally symmetric shape. *J. Fluid Mech.* **137**, 217–20.

Kaneko, A., and Honji, H., 1979, Double structures of steady streaming in the oscillatory viscous flow over a wavy wall. *J. Fluid Mech.* **93**, 727–36.

Kim, S., 1985, A note on Faxen laws for nonspherical particles. *Int. J. Multiphase Flow* **11**, 713–19.

Kim, S., and Karrila, S. J., 1991, *Microhydrodynamics: Principles and Selected Applications.* Butterworth.

Kim, S., and Lu, S.-Y., 1987, The functional similarity between Faxen relations and singularity solutions for fluid–fluid, fluid–solid and solid–solid dispersions. *Int. J. Multiphase Flow* **13**, 837–44.

Kim, S. K., and Troesch, A. W., 1989, Streaming flows generated by high-frequency small-amplitude oscillations of arbitrarily shaped cylinders. *Phys. Fluids* A **1**, 975–85.

Ladyzhenskaya, O. A., 1975, Mathematical analysis of Navier–Stokes equations for incompressible liquids. *Annu. Rev. Fluid Mech.* **7**, 249–72.

Lamb, H., 1911, On the uniform motion of a sphere through a viscous fluid. *Phil. Mag.* **21**, 112–21.

Lamb, H., 1932, *Hydrodynamics.* Dover.

Langlois, W. E., 1964, *Slow Viscous Flow.* MacMillan.

Lighthill, M. J., 1975, *Mathematical Biofluiddynamics.* SIAM.

Loewenberg, M., 1994, Axisymmetric unsteady Stokes flow past an oscillating finite-length cylinder. *J. Fluid Mech.* **265**, 265–88.

Lorentz, H. A., 1907, Ein allgemeiner Satz, die Bewegung einer reibenden Flüssigkeit betreffend, nebst einigen Anwendungen desselben. *Abhand. theor. Phys.* Leipzig, **1**, 23–42. Translated into English in *Lorentz, H. A., Collected Papers, 1937,* Volume IV, 7–14, Martinus Nijhoff.

Love, A. E. H., 1944, *A Treatise on the Mathematical Theory of Elasticity.* Dover.

Maxey, M. R., and Riley, J. J., 1983, Equation of motion for a small rigid sphere in a nonuniform flow. *Phys. Fluids* **26**, 883–89.

Moffatt, H. K., 1964, Viscous and resistive eddies near a sharp corner. *J. Fluid Mech.* **18**, 1–18.

Moffatt, H. K., and Duffy, B. R., 1980, Local similarity solutions and their limitations. *J. Fluid Mech.* **96**, 299-313.

Oseen, C. W., 1910, Über die Stokessche Formel und über eine verwandte Aufgabe in der Hydrodynamik. *Ark. Mat. Astr. Fys.* **6**(29).

Oseen, C. W., 1927, *Neuere Methoden und Ergebnisse in der Hydrodynamik.* Leipzig: Akademische Verlagsgesellschaft.

Padmanabhan, N., and Pedley, T. J., 1987, Three-dimensional steady streaming in a uniform tube with an oscillating elliptical cross-section. *J. Fluid Mech.* **178**, 325–43.

Phan-Thien, N., and Kim, S., 1994, *Microstructures in Elastic Media.* Oxford University Press.

Pozrikidis, C., 1989a, A singularity method for unsteady linearized flow. *Phys. Fluids* A **1**, 1508–20.

Pozrikidis, C., 1989b, A study of linearized oscillatory flow past particles by the boundary integral method. *J. Fluid Mech.* **202**, 17–41.

Pozrikidis, C., 1990, The instability of moving viscous drops. *J. Fluid Mech.* **210**, 1-21.

Pozrikidis, C., 1992, *Boundary Integral and Singularity Methods for Linearized Viscous Flow.* Cambridge University Press.

Pozrikidis, C., 1995, A bibliographical note on the unsteady force on a spherical drop. *Phys. Fluids* **6**, 3209.

Riley, N., 1992, Acoustic streaming about a cylinder in orthogonal beams. *J. Fluid Mech.* **242**, 387–94.

Stakgold, I., 1967, 1968, *Boundary Value Problems of Mathematical Physics.* 2 volumes. Macmillan.

Taylor, G. I., 1962, On scraping viscous fluid from a plane surface. *Miszellaneen Ang. Mech.* Ed. M. Schäfer, Akademie Verlag, Berlin, 313–15.

Van Dyke, M. D., 1975, *Perturbation Methods in Fluid Mechanics.* Parabolic Press.

Vojir, D. R., and Michaelides, E. E., 1994, Effect of the history term on the motion of rigid spheres in a viscous fluid. *Int. J. Multiphase Flow* **20**, 547–56.

Yih, C.-S., 1979, *Fluid Mechanics.* West River Press.

Zhou, H., and Pozrikidis, C., 1995, Adaptive singularity method for Stokes flow past particles. *J. Comp. Phys.* **117**, 79–89.

The vorticity transport equation for barotropic fluids or for fluids whose density is uniform throughout the domain of flow [Eq. (3.8.9)] shows that the rate at which vorticity is produced vanishes at every point within an irrotational flow, and this suggests that vorticity may enter the flow only by means of diffusion across the boundaries. Once vorticity has entered the flow, it is convected by the velocity field while diffusing with a diffusivity that is equal to the kinematic viscosity of the fluid, and intensifies or attenuates due to vortex stretching. An unbounded flow does not have any vorticity entrance ports, and therefore, if it is irrotational at the initial instant, it will remain irrotational at all subsequent times.

Under certain circumstances, the distance across which the vorticity penetrates the fluid from the boundaries is small compared to the overall size of the boundaries, and the bulk of the flow remains nearly irrotational. This occurs, in particular, when the rate of the diffusion of vorticity into the flow is comparable to the rate of convection of vorticity by tangential and normal motions. Balancing the orders of magnitude of the rate of convection and diffusion of vorticity yields the formal requirement that the Reynolds number, defined with respect to the typical size of the boundaries, be sufficiently large. When the necessary conditions are met, the vorticity will be convected along thin layers that wrap around the boundaries, and will then either be channeled into slender wakes or deposited into regions of rotational flow, commonly called *vortices* or *regions of separated flow*. The point where the boundary vortex layer detaches from a boundary and enters the bulk of the flow is the point where a boundary layer is said to separate. A discussion of the prevailing physical mechanisms and mathematical description of these phenomena will be presented in Chapter 8.

Examples of nearly irrotational flows are high-Reynolds-number flows past aircraft wings or streamlined ground vehicles. Under normal operating conditions, these flows are nearly irrotational everywhere except within thin boundary layers that line the surfaces of the wings or vehicles, and within the wakes behind these bodies. Another example is provided by the flow that is generated by the propagation of surface waves in the ocean, which is irrotational everywhere except within a thin boundary layer along the free surface.

Whether the vorticity will remain within boundary layers and slender wakes or generate regions of steady or unsteady recirculating flow is not known a priori, but must be assessed by observation, analysis, or computation. In practice, as the Reynolds number is increased, the onset of hydrodynamic instabilities causes oscillatory and turbulent motions that make the structure of the flow hard to analyze and difficult to describe by analytical or even numerical methods.

Prandtl proposed decoupling the study of the irrotational flow far from the boundary layers from that of the flow within the boundary layers. As a first step we solve the potential flow problem, assuming that its domain extends all the way up to the boundaries, and neglecting the presence of wakes and regions of recirculating flow. As a second step, we use this outer potential flow to study the development and stability of the boundary layers subject to appropriate simplifying assumptions. In the present chapter and in Chapter 10 we shall discuss the properties and computation of the outer potential flow, and in Chapters 8 and 9 we shall discuss the structure and stability of the boundary layers.

7.1 | EQUATIONS AND COMPUTATION OF IRROTATIONAL FLOW

The description and computation of an irrotational flow is simplified substantially by introducing the velocity potential ϕ defined in terms of the equation

$$\mathbf{u} = \nabla\phi \tag{7.1.1}$$

Taking the curl of Eq. (7.1.1), and remembering that the curl of the gradient of a twice differentiable function is equal to zero, shows that the corresponding vorticity vanishes, and this confirms that a potential flow is also an irrotational flow. In Section 2.1 we saw that the converse is also true; that is, an irrotational flow may be described in terms of a singly or multiply valued velocity potential ϕ as shown in Eq. (7.1.1).

The continuity equation requires that, when the fluid is incompressible, $\mathbf{u}$ be a solenoidal vector field, and therefore ϕ be a harmonic function

$$\nabla^2\phi = 0 \tag{7.1.2}$$

The advantages of introducing ϕ become evident by noting that computing the three components of the velocity is reduced to computing a single scalar function that satisfies Laplace's equation (7.1.2), which can be accomplished using a variety of analytical and efficient numerical methods. The velocity vector is perpendicular to the instantaneous surfaces over which ϕ remains constant.

To solve Eq. (7.1.2) we require one appropriate scalar boundary condition over each boundary of the flow. For example, since the normal component of the velocity of the fluid at the edge of a boundary layer along an impermeable stationary surface is small, it is appropriate to require the no-penetration condition $\nabla\phi \cdot \mathbf{n} = 0$. The irrotational flow computed subject to this condition will exhibit a finite tangential velocity, which amounts to a vortex sheet. This is a macroscopic representation of a vortex layer or boundary layer of finite thickness.

We thus find that the velocity field of an irrotational flow may be computed without a reference to the equation of motion, which means that the *kinematics and dynamics of an irrotational flow are decoupled.* The equation of motion provides us with a first-order linear partial differential equation for the pressure, which may be integrated to yield Bernoulli's equation (3.4.16), repeated here for convenient reference

$$\frac{\partial\phi}{\partial t} + \tfrac{1}{2}|\mathbf{u}|^2 + \frac{p}{\rho} - \mathbf{g} \cdot \mathbf{x} = c(t) \tag{7.1.3}$$

where $c(t)$ is a time-dependent constant, and the density has been assumed to be uniform throughout the domain of flow. Once the velocity field is known, the pressure follows from Eq. (7.1.3).

It is worth emphasizing that an irrotational flow and its associated pressure constitute an exact solution to the *full form* of the Navier–Stokes equation, which, however, cannot be made, in general, to satisfy more than one scalar boundary condition. An exception is the two-dimensional flow generated by the steady rotation of a circular cylinder around its center, which is identical to the flow due to a point vortex placed at the center (Section 5.1). Another exception is the flow generated by the radial expansion or contraction of a spherical bubble, which is identical to the flow due to a point source, to be discussed later in this section.

Force and Torque Exerted on a Boundary

Once the pressure field has been found, the force and torque with respect to a certain point $\mathbf{x}_c$ exerted on a boundary can be computed using the simplified versions of Eqs. (3.3.11) and (3.3.12),

$$\mathbf{F} = -\int_{\text{Boundary}} p\,\mathbf{n}\,dS, \qquad \mathbf{T} = -\int_{\text{Boundary}} p\,(\mathbf{x} - \mathbf{x}_c) \times \mathbf{n}\,dS \tag{7.1.4}$$

Strictly speaking, these equations provide us with the force and torque exerted on a fluid surface that encloses the edge of the boundary layer, subject to the assumption that viscous stresses make negligible contributions. These values must be corrected, taking into consideration the viscous stresses along, and the pressure drop across the boundary layers, as will be discussed in Chapter 8.

Viscous Dissipation in Irrotational Flow

The *viscous force* in the equation of motion for irrotational flow vanishes, but the *viscous stresses* and *rate of viscous dissipation* have finite values that depend upon the structure of the flow and viscosity of the fluid (Section 3.3). This observation makes an important distinction between irrotational and inviscid flow. More specifically, beginning with Eq. (3.3.13), exploiting the fact that the flow is irrotational to identify the rate of deformation tensor with the velocity gradient tensor, using the continuity equation, and applying the divergence theorem, we find that the rate of viscous dissipation within an irrotational flow with uniform viscosity is given by

$$\int_{\text{Flow}} \Phi \, dV = 2\mu \int_{\text{Flow}} (\nabla \nabla \phi) : (\nabla \nabla \phi) \, dV$$

$$= -\mu \int_{\text{Boundaries}} \mathbf{n} \cdot \nabla (\mathbf{u} \cdot \mathbf{u}) \, dS \tag{7.1.5}$$

where the unit normal vector $\mathbf{n}$ points *into* the flow. The right-hand side of Eq. (7.1.5) expresses the rate of viscous dissipation in terms of the derivatives of the square of the magnitude of the velocity normal to the boundaries.

Furthermore, working as in Eq. (3.3.16), we find that the rate of viscous dissipation balances the rate of working of the deviatoric viscous traction $\mathbf{t} = \boldsymbol{\tau} \cdot \mathbf{n}$

$$\int_{\text{Flow}} \Phi \, dV = -\int_{\text{Boundaries}} \mathbf{u} \cdot \mathbf{t} \, dS \tag{7.1.6}$$

where $\boldsymbol{\tau}$ is the deviatoric component of the stress tensor and the normal vector $\mathbf{n}$ points into the flow. The relation between the drag force exerted on a boundary and the rate of viscous dissipation is discussed further by Joseph, Liao, and Hu (1993).

A Radially Expanding or Contracting Spherical Bubble

An interesting example of irrotational flow is provided by the radial expansion or contraction of a spherical bubble with time-dependent radius $a(t)$, immersed in a viscous fluid of infinite expanse. The velocity field may be represented in terms of a three-dimensional point source with time-dependent strength $m(t)$ placed at the center of the bubble. In spherical polar coordinates with the origin at the center of the bubble, the velocity potential and radial component of the velocity are given by

$$\phi = -\frac{m(t)}{4\pi} \frac{1}{r}, \qquad u_r = \frac{\partial \phi}{\partial r} = \frac{m(t)}{4\pi} \frac{1}{r^2} \tag{7.1.7}$$

The no-penetration condition at the surface of the bubble requires that $da/dt = u_r(a)$, which may be rearranged to yield the strength of the point source

$$m(t) = 4\pi a^2 \frac{da}{dt} \tag{7.1.8}$$

Far from the bubble, the pressure assumes the hydrostatic distribution $p \approx \rho \mathbf{g} \cdot \mathbf{x} + P_\infty(t)$, where $P_\infty(t)$ is a time-dependent constant. Substituting Eqs. (7.1.7) and (7.1.8) into Eq. (7.1.3), we obtain the distribution of modified pressure $P = p - \rho \mathbf{g} \cdot \mathbf{x} + P_\infty(t)$

$$\frac{P - P_\infty}{\rho} = \frac{1}{3r} \frac{d^2 a^3}{dt^2} - \frac{1}{18r^4} \left(\frac{da^3}{dt} \right)^2 \tag{7.1.9}$$

Next we seek to derive an evolution equation for the radius of the bubble in terms of its internal pressure $p_B(t)$. Neglecting hydrostatic variations across the diameter of the bubble, and requiring the dynamic boundary condition at the interface (3.5.12), we obtain the normal stress balance

$$p_B(t) = p(r = a, t) - 2\mu \left(\frac{\partial u_r}{\partial r}\right)_{r=a} + 2\frac{\gamma}{a} \tag{7.1.10}$$

The surface tension γ is assumed to be uniform over the interface. Rearranging Eq. (7.1.10), we obtain

$$p(r = a, t) = p_B(t) - \frac{4\mu}{a}\frac{da}{dt} - 2\frac{\gamma}{a} \tag{7.1.11}$$

It is worth noting that the normal viscous stress at the interface makes a finite contribution. Evaluating Eq. (7.1.9) at the interface and using Eq. (7.1.11), we derive a second-order nonlinear evolution equation for the radius of the bubble

$$a\frac{d^2a}{dt^2} + \frac{3}{2}\left(\frac{da}{dt}\right)^2 + \frac{4\nu}{a}\frac{da}{dt} = \frac{1}{\rho}\left(P_B(t) - P_\infty(t) - \frac{2\gamma}{a}\right) \tag{7.1.12}$$

which is known as the *generalized Rayleigh equation.*

When the modified pressure within the bubble is equal to that far from the bubble, time does not appear explicitly on the right-hand side of Eq. (7.1.12). In that case, it is possible to reduce the order of the equation by regarding da/dt a function of a, writing $da/dt = f(a)$. Substituting this expression into Eq. (7.1.12) yields the first-order differential equation

$$a\frac{df}{da} + \frac{3}{2}f + \frac{4\nu}{a} = -\frac{2\gamma}{\rho}\frac{1}{af} \tag{7.1.13}$$

When, in addition, viscous stresses and surface tension are negligible, we obtain the exact solution $f = ca^{-3/2}$, where c is a constant. Using the definition of f and integrating once with respect to time, we obtain the exact solution

$$a = a_0\left[1 + \frac{5}{2}\frac{1}{a_0}\left(\frac{da}{dt}\right)_{t=0} t\right]^{2/5} \tag{7.1.14}$$

where a_0 is the radius of the bubble at the initial instant, $a_0 = a(t = 0)$.

In general, the solution of Eq. (7.1.12) subject to a certain initial condition and a specified ambient pressure field or pressure within the bubble must be found using numerical methods (Plesset and Prosperetti, 1977).

The collapse of a bubble near a wall has important consequences on mechanical damage due to cavitation (Blake and Gibson, 1987). Observation and computation have shown that the bubble no longer has a spherical shape, but develops a dimple that may drive a strong jet of fluid toward the wall.

Velocity Variation around a Streamline

The condition of vanishing vorticity imposes a constraint on the structure of the velocity field around a streamline with important implications. Setting the components of the vorticity in Eq. (1.7.18) equal to zero, we find

$$\mathbf{n} \cdot (\nabla\mathbf{u}) \cdot \mathbf{b} = \mathbf{b} \cdot (\nabla\mathbf{u}) \cdot \mathbf{n}, \qquad \frac{\partial u}{\partial l_b} = 0, \qquad \frac{\partial u}{\partial l_n} = -\kappa u \tag{7.1.15}$$

where u is the magnitude of the velocity. The third equation, in particular, shows that when a streamline has a circular shape, and the fluid moves in the clockwise direction, the velocity on the inside of the circle is greater than the velocity on the outside of the circle; the associated counterclockwise rotation of fluid parcels is necessary in order to balance the clockwise rotation due to the global motion of the fluid along the streamline. A consequence of this behavior is that the velocity of an irrotational flow around a sharp corner reaches a maximum value at the salient edge.

Minimum of the Pressure

In Section 2.3 we saw that the velocity potential and magnitude of the velocity must attain extreme values at the boundaries. Here we shall demonstrate that the modified pressure must attain its minimum value at the boundaries. For this purpose, we take the Laplacian of Bernoulli's equation (7.1.3), and find that the modified pressure satisfies Poisson's equation with a negative forcing function,

$$\nabla^2 P = -\tfrac{1}{2}\rho \nabla^2 |\mathbf{u}|^2 = -\rho \nabla \mathbf{u} : \nabla \mathbf{u} \tag{7.1.16}$$

Integrating Eq. (7.1.16) over a volume of fluid V_c that is bounded by a closed surface D and using the divergence theorem, we obtain

$$\int_D \nabla P \cdot \mathbf{n}\, dV = -\rho \int_{V_c} \nabla \mathbf{u} : \nabla \mathbf{u}\, dV < 0 \tag{7.1.17}$$

where the unit normal vector $\mathbf{n}$ points outward from the control volume. Identifying D with a small closed surface enclosing the point where P allegedly attains minimum value, subject to the restriction that P over D is constant, guarantees that $\nabla P \cdot \mathbf{n}$ is positive over the surface of the control volume, but this contradicts the inequality in (7.1.17) (see also Problem 7.1.3).

PROBLEMS

7.1.1   **Expanding or contracting bubble.** (a) Verify that the velocity field given in Eqs. (7.1.7) satisfies Eq. (7.1.6). (b) Present a detailed derivation of Eq. (7.1.14) subject to the underlying assumptions.

7.1.2   **Pressure in a bend.** Consider flow along a straight pipe with a sudden bend. Assuming that the velocity profile along the straight section of the pipe is uniform and the flow around the bend is irrotational, show that maximum velocity and thus minimum pressure occurs at the wall of the pipe on the inside of the bend.

7.1.3   **Minimum of pressure.** Discuss whether the regular pressure p must attain its minimum value at the boundaries.

7.2 | FLOW PAST OR DUE TO THE MOTION OF THREE-DIMENSIONAL BODIES

An important field of study in the area of potential flow concerns the flow past a stationary or moving body, as well as the flow due to the motion and deformation of a body in an effectively inviscid fluid. In this section we shall discuss certain general properties of these flows in singly connected domains, with reference to the behavior of the flow far from the body and the kinetic energy of the fluid. In Section 7.3 we shall discuss the development forces and torques. Interestingly, we shall find that the structure of the far flow, the kinetic energy of the fluid, and the force exerted on the body may be deduced from one another, and this allows for significant analytical and computational simplifications.

Flow past a Stationary or Translating Rigid Body

We begin by considering an incident irrotational flow past a three-dimensional rigid body that translates with velocity $\mathbf{U}$, possibly in the presence of other stationary boundaries, as illustrated in Figure 7.2.1(a). As a first step, we decompose the harmonic potential as $\phi = \phi^\infty + \phi^D$, where ϕ^∞ corresponds to the incident flow that prevails in the absence of the body, and ϕ^D is the disturbance potential due to the body. We then write the boundary integral representation (2.3.8) for the disturbance component ϕ^D at a point $\mathbf{x}_0$ that is located within the flow, where D is composed of the surface of the body B, the surface of a stationary boundary S_B, and a surface of large size extending to infinity S_∞. We note that the integrals over S_B can be made to vanish with the use of an appropriate Green's function of the second kind, and the integrals over the large surface S_∞ are infinitesimal, require the no-penetration condition $\mathbf{u} \cdot \mathbf{n} = 0$ or $\nabla \phi^D \cdot \mathbf{n} = \mathbf{U} \cdot \mathbf{n} - \nabla \phi^\infty \cdot \mathbf{n}$ over B, and thus obtain

$$\phi^D(\mathbf{x}_0) = -\int_B G(\mathbf{x}_0, \mathbf{x}) \, [\mathbf{U} - \nabla \phi^\infty(\mathbf{x})] \cdot \mathbf{n}(\mathbf{x}) \, dS(\mathbf{x})$$

$$+ \int_B \nabla G(\mathbf{x}_0, \mathbf{x}) \cdot \mathbf{n}(\mathbf{x}) \, \phi^D(\mathbf{x}) \, dS(\mathbf{x}) \tag{7.2.1}$$

For flow in an infinite domain with no interior boundaries, G is the free-space Green's function given in Eq. (2.2.9).

Next we apply the integral relation (2.3.2) for the test flow $\mathbf{U} - \nabla \phi^\infty$ corresponding to the potential $\mathbf{U} \cdot \mathbf{x} - \phi^\infty$ and identify the control volume with the volume occupied by the body, and D with the surface of the body B; use of this relation is permitted by the fact that the point $\mathbf{x}_0$ is located in the exterior of the body. We thus obtain

$$\int_B [\mathbf{U} \cdot \mathbf{x} - \phi^\infty(\mathbf{x})] \, \nabla G(\mathbf{x}_0, \mathbf{x}) \cdot \mathbf{n}(\mathbf{x}) \, dS(\mathbf{x})$$

$$= \int_B G(\mathbf{x}_0, \mathbf{x}) \, [\mathbf{U} - \nabla \phi^\infty(\mathbf{x})] \cdot \mathbf{n}(\mathbf{x}) \, dS(\mathbf{x}) \tag{7.2.2}$$

Combining Eqs. (7.2.1) and (7.2.2) to eliminate the single-layer potential we obtain a simplified representation for the disturbance potential in terms of a double-layer potential alone. Adding to both sides of the resulting equation the potential of the incident flow ϕ^∞ yields

$$\phi(\mathbf{x}_0) = \phi^\infty(\mathbf{x}_0) - \int_B \nabla G(\mathbf{x}_0, \mathbf{x}) \cdot \mathbf{n}(\mathbf{x}) \, [\mathbf{U} \cdot \mathbf{x} - \phi(\mathbf{x})] \, dS(\mathbf{x}) \tag{7.2.3}$$

The modular cases of flow past a stationary body and flow due to the translation of a body arise by setting, respectively, $\mathbf{U}$ equal to zero, or ϕ^∞ equal to a constant.

Far-field expansion

The representation (7.2.3) provides us with a convenient starting point for assessing the behavior of the flow far from the body. Assuming that the point $\mathbf{x}_0$ is located far from the body, we select a point $\mathbf{x}_c$ somewhere within or in the vicinity of the body, expand the Green's function dipole in a Taylor series with respect to the point $\mathbf{x}$ about the point $\mathbf{x}_c$, and maintain only the leading term to obtain

$$\phi(\mathbf{x}_0) = \phi^\infty(\mathbf{x}_0) - \nabla_c G(\mathbf{x}_0, \mathbf{x}_c) \cdot \int_B \mathbf{n}(\mathbf{x}) \, [\mathbf{U} \cdot \mathbf{x} - \phi(\mathbf{x})] \, dS(\mathbf{x}) + \cdots \tag{7.2.4}$$

where the subscript c of the gradient signifies differentiation with respect to $\mathbf{x}_c$. Comparing Eq. (7.2.4) with Eq. (2.2.23) shows that, far from the body, the disturbance flow is similar to that

(a)

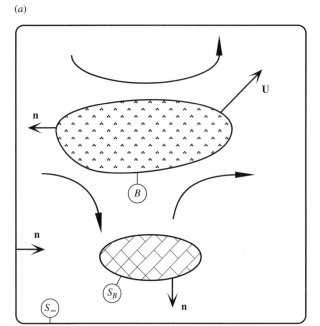

(b)

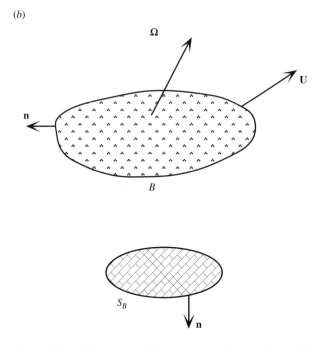

Figure 7.2.1   (a) Illustration of flow past a translating rigid body. (b) Illustration of flow due to the motion or deformation of a body in the presence of a stationary boundary S_B.

due to a point-source dipole located at the point $\mathbf{x}_c$ whose strength is equal to

$$\mathbf{d} = V_B \, \mathbf{U} - \int_B \mathbf{n}(\mathbf{x}) \, \phi(\mathbf{x}) \, dS(\mathbf{x}) \tag{7.2.5}$$

where V_B is the volume of the body. The first term on the right-hand side of Eq. (7.2.5) arises by applying the divergence theorem to manipulate the surface integral in Eq. (7.2.4) involving $\mathbf{U}$. We thus find that the coefficient of the dipole may be computed from a knowledge of the velocity of the body and the distribution of the velocity potential over the surface of the body.

Flow Due to the Motion and Deformation of a Body

Consider next the flow due to the translation, rotation, or deformation of a body in an otherwise quiescent fluid, as illustrated in Figure 7.2.1(b). The body may represent, for instance, a swimming fish or a moving underwater vehicle.

To assess the far-field behavior of the flow, we describe the harmonic potential in terms of the boundary integral representation (2.3.8), assume that the point $\mathbf{x}_0$ is located far from the body, select a point $\mathbf{x}_c$ somewhere in the interior or in the vicinity of the body, and expand the Green's function and its gradient in a Taylor series with respect to $\mathbf{x}$ about the point $\mathbf{x}_c$. In this manner, we obtain an asymptotic expansion for the far flow in terms of a Green's function and its multipoles

$$\phi(\mathbf{x}_0) = -G(\mathbf{x}_0, \mathbf{x}_c) \int_B \nabla\phi(\mathbf{x}) \cdot \mathbf{n}(\mathbf{x}) \, dS(\mathbf{x})$$

$$- \nabla_c G(\mathbf{x}_0, \mathbf{x}_c) \cdot \int_B [-\phi(\mathbf{x}) \, \mathbf{n}(\mathbf{x}) + (\mathbf{x} - \mathbf{x}_c) \nabla\phi(\mathbf{x}) \cdot \mathbf{n}(\mathbf{x})] \, dS(\mathbf{x}) + \cdots \tag{7.2.6}$$

The leading-order term on the right-hand side of Eq. (7.2.6) represents the flow due to a point source whose strength is equal to the flow rate across B. The second integral

$$\mathbf{d} = \int_B [-\phi(\mathbf{x}) \, \mathbf{n}(\mathbf{x}) + (\mathbf{x} - \mathbf{x}_c) \nabla\phi(\mathbf{x}) \cdot \mathbf{n}(\mathbf{x})] \, dS(\mathbf{x}) \tag{7.2.7}$$

is the coefficient of the point-source dipole. Using the divergence theorem, we find that the value of the integral on the right-hand side of Eq. (7.2.7) remains unchanged when the domain of integration B is replaced by any other surface that encloses the body. Furthermore, when the flow across a surface that encloses the body is equal to zero, the value of the integral is independent of the location of $\mathbf{x}_c$.

Motion of a rigid body

As a specific application of Eq. (7.2.7), consider the flow due to a rigid body that translates with velocity $\mathbf{U}$ and rotates with angular velocity $\mathbf{\Omega}$ about the point $\mathbf{x}_c$. The no-penetration condition on the surface of the body requires that

$$\nabla\phi(\mathbf{x}) \cdot \mathbf{n}(\mathbf{x}) = [\mathbf{U} + \mathbf{\Omega} \times (\mathbf{x} - \mathbf{x}_c)] \cdot \mathbf{n}(\mathbf{x}) \tag{7.2.8}$$

Substituting Eq. (7.2.8) into Eq. (7.2.7) and rearranging the triple mixed product, we find that the coefficient of the dipole is given by

$$\mathbf{d} = -\int_B \phi(\mathbf{x}) \, \mathbf{n}(\mathbf{x}) \, dS(\mathbf{x}) + V_B \, \mathbf{U} + \int_B \{\mathbf{\Omega} \cdot [(\mathbf{x} - \mathbf{x}_c) \times \mathbf{n}(\mathbf{x})]\} \, (\mathbf{x} - \mathbf{x}_c) \, dS(\mathbf{x}) \tag{7.2.9}$$

As in the case of flow past a translating body discussed earlier, we find that the strength of the dipole may be computed from a knowledge of the motion of the body and of the distribution of the velocity potential over the surface of the body.

Applying the divergence theorem we find that the second integral on the right-hand side of Eq. (7.2.9) can be made to disappear by selecting the point $\mathbf{x}_c$ to be the center of volume of the body $\mathbf{X}_c$ defined in Eq. (4.1.3).

Decomposition into fundamental modes of rigid-body motion

The linear nature of the equations and boundary conditions governing the flow due to the motion of a rigid body allows us to express the velocity potential as a linear combination of the velocity of translation and angular velocity of rotation about the point $\mathbf{x}_c$ in the form

$$\phi(\mathbf{x}) = U_i(t) \, \Phi_i[\mathbf{x}, \mathbf{x}_c(t), \mathbf{e}(t)] + \Omega_i(t) \, \Phi_{i+3}[\mathbf{x}, \mathbf{x}_c(t), \mathbf{e}(t)] \tag{7.2.10}$$

$\Phi_i, i = 1, \ldots, 6$, are six harmonic potentials corresponding to three fundamental modes of translation and three fundamental modes of rotation, the vector $\mathbf{e}$ is painted on the body, which means that it describes its instantaneous orientation, and the square brackets contain the arguments of Φ_i. The no-penetration boundary condition requires that Φ_i satisfy the boundary conditions

$$\nabla\Phi_i(\mathbf{x}) \cdot \mathbf{n}(\mathbf{x}) = \left\{ \begin{array}{ll} n_i(\mathbf{x}), & \text{for } i = 1, 2, 3 \\[2mm] [(\mathbf{x} - \mathbf{x}_c) \times \mathbf{n}(\mathbf{x})]_{i-3}, & \text{for } i = 4, 5, 6 \end{array} \right. \tag{7.2.11}$$

on the surface of the body, where the subscript of the square brackets denotes the Cartesian coordinate.

Substituting Eq. (7.2.10) into the integrals on the right-hand side of Eq. (7.2.9), we obtain the coefficient of the dipole in the compact form

$$d_j = V_B(U_j + U_i\alpha_{ij} + \Omega_i\beta_{ij} + \Omega_i\gamma_{ij}) \tag{7.2.12}$$

where

$$\alpha_{ij} = -\frac{1}{V_B}\int_B \Phi_i n_j \, dS, \qquad \beta_{ij} = -\frac{1}{V_B}\int_B \Phi_{i+3} n_j \, dS$$

$$\gamma_{ij} = \frac{1}{V_B}\varepsilon_{ilk}\int_B \hat{x}_l\hat{x}_j n_k \, dS = -\frac{1}{V_B}\varepsilon_{ijl}\int_B \hat{x}_l \, dV \tag{7.2.13}$$

$i, j = 1, 2, 3$ and $\hat{\mathbf{x}} = \mathbf{x} - \mathbf{x}_c$. Note that when $\mathbf{x}_c$ is chosen to be the center of volume of the body $\mathbf{X}_c$ defined in Eq. (4.1.3), the coefficients γ_{ij} vanish.

We shall show now that the three-by-three matrix α_{ij} is symmetric. For this purpose, we use the first set of boundary conditions in Eq. (7.2.11) and the fact that Φ_i are harmonic functions and apply the divergence theorem to write

$$\int_B \Phi_i \, n_j \, dS = \int_B \Phi_i \, \nabla\Phi_j \cdot \mathbf{n} \, dS = -\int_{Flow} \nabla\Phi_i \cdot \nabla\Phi_j \, dV \tag{7.2.14}$$

where we have made use of the fact that the Φ_i decay fast enough so that the integral on the left-hand side of Eq. (7.2.14) over a closed surface of large size is infinitesimal. Interchanging the roles of Φ_i and Φ_j, we obtain an expression with an identical right-hand side, and this completes the proof. The symmetry of α_{ij} implies that *the component of the coefficient of the dipole in a particular direction due to translation in another direction is equal to the component of the dipole in the second direction due to translation in the first direction with velocity of equal magnitude.*

For a body that translates but does not rotate, Eq. (7.2.12) obtains the simplified form

$$\mathbf{d} = V_B\mathbf{U} \cdot (\mathbf{I} + \boldsymbol{\alpha}) \tag{7.2.15}$$

where $\mathbf{I}$ is the identity matrix, which shows that the coefficient of the dipole may be computed from knowledge of $\boldsymbol{\alpha}$ and vice versa.

Kinetic energy of the flow due to the motion of a rigid body and the matrix of added mass

Equation (2.1.17) provides us with an expression for the instantaneous kinetic energy of a potential flow in terms of a boundary integral involving the boundary distributions of ϕ and the normal component of the velocity. For flow due to the motion of a rigid body, possibly in the presence of stationary boundaries, we use the boundary condition (7.2.8) to obtain

$$K = -\frac{1}{2}\rho \int_B \phi(\mathbf{x}) \nabla\phi(\mathbf{x}) \cdot \mathbf{n}(\mathbf{x}) \, dS(\mathbf{x}) = -\frac{1}{2}\rho \int_B \phi(\mathbf{x}) \mathbf{u}(\mathbf{x}) \cdot \mathbf{n}(\mathbf{x}) \, dS(\mathbf{x})$$

$$= -\frac{1}{2}\rho \int_B \phi(\mathbf{x}) \mathbf{U} \cdot \mathbf{n}(\mathbf{x}) \, dS(\mathbf{x}) - \frac{1}{2}\rho \int_B \phi(\mathbf{x}) \, [\mathbf{\Omega} \times (\mathbf{x} - \mathbf{x}_c)] \cdot \mathbf{n}(\mathbf{x}) \, dS(\mathbf{x})$$

$$= -\frac{1}{2}\rho\mathbf{U} \cdot \int_B \phi(\mathbf{x}) \, \mathbf{n}(\mathbf{x}) \, dS(\mathbf{x}) - \frac{1}{2}\rho\mathbf{\Omega} \cdot \int_B \phi(\mathbf{x}) \, (\mathbf{x} - \mathbf{x}_c) \times \mathbf{n}(\mathbf{x}) \, dS(\mathbf{x}) \quad (7.2.16)$$

Introducing the six-dimensional vector

$$\mathbf{W} = (U_x, U_y, U_z, \Omega_x, \Omega_y, \Omega_z) \quad (7.2.17)$$

and substituting the decomposition (7.2.10) into the right-hand side of Eq. (7.2.16), we obtain the compact *quadratic form*

$$K = \frac{1}{2}\rho V_B A_{ij} W_i W_j \quad (7.2.18)$$

where $\mathbf{A}$ is the six-by-six *grand added mass matrix*, defined as

$$A_{ij} = -\frac{1}{V_B} \int_B \Phi_i N_j \, dS \quad (7.2.19)$$

$\mathbf{N}$ is a six-dimensional vector whose first and second three-entry blocks contain, respectively, the components of vectors $\mathbf{n}$ and $(\mathbf{x} - \mathbf{x}_c) \times \mathbf{n}$. The matrices $\boldsymbol{\alpha}$ and $\boldsymbol{\beta}$ defined in Eq. (7.2.13) comprise, respectively, the top-diagonal and bottom-left-corner three-by-three blocks of $\mathbf{A}$.

It is evident from the right-hand side of Eq. (7.2.19) that the value of $\mathbf{A}$ depends exclusively upon the instantaneous body shape and orientation and presence of other stationary objects, but is independent of the body's linear or angular velocity or acceleration. Physically, $\mathbf{A}$ expresses the sensitivity of the kinetic energy of the fluid on the velocity of translation and angular velocity of rotation, and may thus be regarded as an influence matrix for the kinetic energy.

Substituting the decomposition (7.2.10) into the first integral in Eq. (7.2.16), we obtain an expression for the grand added mass matrix in the alternative form

$$A_{ij} = -\frac{1}{V_B} \int_B \Phi_i \nabla\Phi_j \cdot \mathbf{n} \, dS \quad (7.2.20)$$

which also follows from Eq. (7.2.19) using the boundary conditions (7.2.11).

An important property of $\mathbf{A}$ is that it is symmetric, which means that the kinetic energy of the fluid produced when the body translates in a particular direction and rotates about another direction is the same as that produced when the body translates in the second direction and rotates around the first direction, with linear and angular velocities of equal magnitude. This property follows immediately by using the divergence theorem to write

$$\int_B \Phi_i \nabla\Phi_j \cdot \mathbf{n} \, dS = -\int_{Flow} \nabla \cdot (\Phi_i \nabla\Phi_j) \, dV$$

$$= -\int_{Flow} \nabla\Phi_i \cdot \nabla\Phi_j \, dV \quad (7.2.21)$$

Interchanging the role of the two potentials on the left-hand side produces the same right-hand side, thereby completing the proof.

Since $\mathbf{A}$ is symmetric, it has only 21 independent components, six of which can be made to vanish by making appropriate choices for the center of rotation $\mathbf{x}_c$ and directions of the Cartesian axes.

Added mass in terms of the coefficient of the dipole

Setting $\mathbf{x}_c$ at the center of volume of the body $\mathbf{X}_c$ defined in Eq. (4.1.3) makes the coefficient $\boldsymbol{\gamma}$ vanish and simplifies Eq. (7.2.12) to

$$d_j = V_B(U_j + U_i\alpha_{ij} + \Omega_i\beta_{ij}) \tag{7.2.22}$$

which provides us with a method of computing eighteen elements of the added mass matrix from a knowledge of the three components of the coefficient of the dipole corresponding to six independent modes of rigid-body motion.

Kinetic energy of the flow due to the translation of a rigid body

Considering a body that translates but does not rotate, we use Eq. (7.2.15) to obtain an expression for the kinetic energy in terms of the coefficient of the dipole

$$K = \tfrac{1}{2}\rho V_B \mathbf{U} \cdot \boldsymbol{\alpha} \cdot \mathbf{U} = \tfrac{1}{2}\rho(\mathbf{d} - V_B\mathbf{U}) \cdot \mathbf{U} \tag{7.2.23}$$

Since α_{ij} is symmetric, we can find three mutually perpendicular principal directions where α_{ij} is diagonal and the quadratic cross terms of the velocity in Eq. (7.2.23) do not appear. If the body is axisymmetric, one principal direction coincides with the axis of the body, and the other two axes lie in the perpendicular plane.

In subsequent sections we shall discuss a method of computing a potential flow in terms of discrete or continuous distributions of point sources and point-source dipoles placed within the volume or over the surface of a body. Equations (7.2.22) and (7.2.23) will provide us then immediately with the added mass matrix $\boldsymbol{\alpha}$ and kinetic energy of fluid in terms of the effective strength of the dipoles, circumventing the need for detailed computations. Furthermore, in Section 7.3 we shall see that the coefficient of the dipole contains information on the force exerted on a rigid body in accelerating motion.

As a final note, we relate the change in kinetic energy of the fluid with respect to the velocity of translation to the coefficient of the dipole. Beginning with Eq. (7.2.23) and using Eq. (7.2.15), we find

$$\frac{\partial K}{\partial U_k} = \tfrac{1}{2}\rho V_B \frac{\partial}{\partial U_k}(U_i\alpha_{ij}U_j) = \rho V_B \alpha_{kj}U_j = \rho(d_k - V_b U_k) \tag{7.2.24}$$

Rearranging gives the relation

$$d_k = \frac{1}{\rho}\frac{\partial K}{\partial U_k} + V_B U_k \tag{7.2.25}$$

PROBLEMS

7.2.1   **Flow past a rotating body.** Discuss whether it is possible to devise a double-layer representation similar to that shown in Eq. (7.2.3) for flow past a rotating rigid body.

7.2.2   **Tensorial nature of the matrix of added masses.** Show that $\mathbf{A}$ is a second-order tensor of the six-dimensional Cartesian space with coordinates $(x_1, x_2, x_3, x_1, x_2, x_3)$ (Yih, 1979, p. 101).

7.2.3   **Elemental potentials.** Derive the boundary conditions (7.2.11).

7.3 | FORCE AND TORQUE EXERTED ON A THREE-DIMENSIONAL BODY

We proceed in this section to compute the force and torque exerted on a three-dimensional moving or stationary body that is immersed in a potential flow, which were given in Eqs. (7.1.4) in terms of the pressure. Our present goal is to develop simplified expressions in terms of the velocity potential and velocity field using Bernoulli's equation (7.1.3). The density will be assumed to be uniform throughout the domain of flow.

Steady Flow past a Stationary Body

We begin by considering steady potential flow past a stationary rigid body, schematically illustrated in Figure 7.3.1. Substituting Bernoulli's equation (7.1.3) into the first of Eqs. (7.1.4) and using the divergence theorem to handle the body-force term we obtain

$$\mathbf{F} = \frac{1}{2}\rho \int_B (\mathbf{u} \cdot \mathbf{u})\,\mathbf{n}\,dS - \rho V_B\,\mathbf{g} \tag{7.3.1}$$

where B is the surface of the body, $\mathbf{n}$ is the unit normal vector pointing *into* the fluid, and V_B is the volume of the body. The second term on the right-hand side of Eq. (7.3.1) expresses the buoyancy force.

To simplify the computation of the integral on the right-hand side of Eq. (7.3.1), we select a control volume V_c that is enclosed by B and one or more closed boundaries D as shown in Figure 7.3.1. We then use the divergence theorem, exploit the fact that $\mathbf{u}$ is irrotational to write $\nabla \mathbf{u} = (\nabla \mathbf{u})^T$, invoke the continuity equation for incompressible fluids, apply the divergence theorem for a second time, and finally use the no-penetration condition $\mathbf{u} \cdot \mathbf{n} = 0$ on B to obtain

$$\int_B (\mathbf{u} \cdot \mathbf{u})\,\mathbf{n}\,dS - \int_D (\mathbf{u} \cdot \mathbf{u})\,\mathbf{n}\,dS = -\int_{V_c} \nabla(\mathbf{u} \cdot \mathbf{u})\,dV$$

$$= -2\int_{V_c} (\nabla \mathbf{u}) \cdot \mathbf{u}\,dV = -2\int_{V_c} \mathbf{u} \cdot \nabla \mathbf{u}\,dV$$

$$= -2\int_{V_c} \nabla \cdot (\mathbf{u}\mathbf{u})\,dV = -2\int_D \mathbf{u}\,(\mathbf{u} \cdot \mathbf{n})\,dS \tag{7.3.2}$$

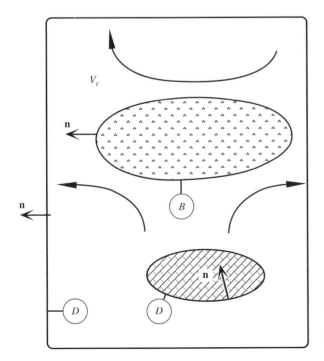

Figure 7.3.1 The force exerted on a stationary rigid body immersed in a potential flow can be computed in terms of surface integrals over the boundaries D that enclose a control volume which is bounded, in part, by the body.

where the unit normal vector $\mathbf{n}$ over D points *outward* from the control volume as shown in Figure 7.3.1. Substituting this result back into Eq. (7.3.1) yields

$$\mathbf{F} = \tfrac{1}{2}\rho \int_D [(\mathbf{u} \cdot \mathbf{u})\mathbf{n} - 2(\mathbf{u} \cdot \mathbf{n})\mathbf{u}]\,dS - \rho V_B\,\mathbf{g} \qquad (7.3.3)$$

The value of the integral on the right-hand side of Eq. (7.3.3) is independent of the particular shape of D. Identifying, in particular, D with B and using the no-penetration condition $\mathbf{u} \cdot \mathbf{n} = 0$ we recover Eq. (7.3.1). The usefulness of the generalized expression shown in Eq. (7.3.3) will become evident when we discuss its practical applications.

Substituting Bernoulli's equation (7.1.3) into the second of Eqs. (7.1.4) and rearranging the various terms, we find that the torque with respect to the point $\mathbf{x}_c$ exerted on the body is given by

$$\mathbf{T} = \tfrac{1}{2}\rho \int_B (\mathbf{x} - \mathbf{x}_c) \times \mathbf{n}\,(\mathbf{u} \cdot \mathbf{u})\,dS - \rho V_B(\mathbf{X}_c - \mathbf{x}_c) \times \mathbf{g} \qquad (7.3.4)$$

where $\mathbf{X}_c$ is the center of the volume of the body defined in Eq. (4.1.3). Note that when the reference point $\mathbf{x}_c$ is placed at $\mathbf{X}_c$, the torque due to the body force vanishes. Working as before for the force, we derive the generalized expression

$$\mathbf{T} = \tfrac{1}{2}\rho \int_D (\mathbf{x} - \mathbf{x}_c) \times [(\mathbf{u} \cdot \mathbf{u})\mathbf{n} - 2(\mathbf{u} \cdot \mathbf{n})\mathbf{u}]\,dS - \rho V_B(\mathbf{X}_c - \mathbf{x}_c) \times \mathbf{g} \qquad (7.3.5)$$

where D is an arbitrary surface, which, along with B, encloses a certain control volume, and the unit normal vector $\mathbf{n}$ over D points outward from the control volume. With the help of the divergence theorem, one may show that the value of the integral on the right-hand side of Eq. (7.3.5) is independent of the choice of D. Identifying, in particular, D with B and using the no-penetration $\mathbf{u} \cdot \mathbf{n} = 0$ we recover Eq. (7.3.4).

Flow due to a point source

As a specific application, let us consider the flow due to a point source with strength m located at the point $\mathbf{x}_S$ in an infinite domain of flow, in the presence of a body but in the absence of any other interior or exterior boundaries. Decomposing the velocity field into a singular component due to the point source and a non-singular disturbance or complementary component due to the presence of the body, denoted by $\mathbf{v}$, we obtain

$$\mathbf{u} = \frac{m}{4\pi}\frac{\hat{\mathbf{x}}}{r^3} + \mathbf{v} \qquad (7.3.6)$$

where $\hat{\mathbf{x}} = \mathbf{x} - \mathbf{x}_S$, and $r = |\hat{\mathbf{x}}|$ is the distance from the point source. Next we identify D with a spherical surface of small radius ε centered at the point source and a surface of large size extending to infinity, and note that the integrals in Eqs. (7.3.3) and (7.3.5) over the large surface make infinitesimal contributions. Over the surface of the small sphere centered at the point source, the unit normal vector is given by $\mathbf{n} = -\hat{\mathbf{x}}/\varepsilon$. Substituting Eq. (7.3.6) into Eqs. (7.3.3) and (7.3.5) and carrying out the integrations in the limit as ε vanishes, we obtain the remarkably simple expressions

$$\mathbf{F} = \rho m\,\mathbf{v}(\mathbf{x}_S) - \rho V_B\,\mathbf{g} \qquad (7.3.7a)$$

$$\mathbf{T} = \rho m\,(\mathbf{x}_S - \mathbf{x}_c) \times \mathbf{v}(\mathbf{x}_S) - \rho V_B\,(\mathbf{X}_c - \mathbf{x}_c) \times \mathbf{g} \qquad (7.3.7b)$$

which provide us with the force and the torque in terms of the complementary velocity at the location of the point source.

Flow due to a point-source dipole

We can derive corresponding results for the flow due to a point-source dipole of strength $\mathbf{d}$ located at the point $\mathbf{x}_D$, by considering the integrals over two small spherical surfaces enclosing a point source and a point sink that are separated by a small distance. The results are (Milne-Thomson, 1968, p. 496)

$$\mathbf{F} = \rho \mathbf{d} \cdot \nabla \mathbf{v}(\mathbf{x}_D) - \rho V_B \, \mathbf{g} \tag{7.3.8a}$$

$$\mathbf{T} = \rho \mathbf{d} \times \mathbf{v}(\mathbf{x}_D) - \rho V_B \, (\mathbf{X}_c - \mathbf{x}_c) \times \mathbf{g} \tag{7.3.8b}$$

where $\mathbf{v}$ is the complementary velocity due to the body (Problem 7.3.1).

Flow due to a collection of singularities

If the flow is due to a collection of singularities including point sources and point-source dipoles, the right-hand sides of Eqs. (7.3.7a,b) and (7.3.8a,b) are summed over all singularities, and $\mathbf{v}$ represents the velocity at the position of one singularity due to all others.

Force and Torque on a Moving Rigid Body

We concentrate now on the force exerted on a rigid body that translates with time-dependent velocity $\mathbf{U}(t)$ and rotates about the point $\mathbf{x}_c(t)$ with time-dependent angular velocity $\mathbf{\Omega}(t)$ in an otherwise quiescent fluid. Substituting Bernoulli's equation (7.1.3) into the first of Eqs. (7.1.4) yields an expression for the force

$$\mathbf{F} = -\int_B p \mathbf{n} \, dS = \rho \int_B \left(\frac{\partial \phi}{\partial t} + \tfrac{1}{2} \mathbf{u} \cdot \mathbf{u} \right) \mathbf{n} \, dS - \rho V_B \, \mathbf{g} \tag{7.3.9}$$

To assess the contribution of the six fundamental modes of rigid-body motion, we express ϕ in terms of the six fundamental potentials introduced in Eq. (7.2.10), and take and expand its time derivative to find

$$\frac{\partial \phi(\mathbf{x})}{\partial t} = \frac{dU_i}{dt} \, \Phi_i + \frac{d\Omega_i}{dt} \, \Phi_{i+3} + U_i \, \frac{\partial \Phi_i(\mathbf{x}, \mathbf{x}_c(t), \mathbf{e}(t))}{\partial t}$$

$$+ \, \Omega_i \, \frac{\partial \Phi_{i+3}(\mathbf{x}, \mathbf{x}_c(t), \mathbf{e}(t))}{\partial t} \tag{7.3.10}$$

where $i = 1, 2, 3$. The material point $\mathbf{x}_c(t)$ marks the instantaneous position of the body, whereas the unit vector $\mathbf{e}(t)$ marks its instantaneous orientation. Substituting Eq. (7.3.10) into Eq. (7.3.9) yields

$$\mathbf{F} = \rho \frac{dU_i}{dt} \int_B \Phi_i \, \mathbf{n} \, dS + \rho \frac{d\Omega_i}{dt} \int_B \Phi_{i+3} \, \mathbf{n} \, dS + \mathbf{F}^s(\mathbf{U}, \mathbf{\Omega}) \tag{7.3.11}$$

For convenience, we have collected all terms that depend upon the instantaneous position, orientation, and linear and angular velocities of the body but are independent of the linear or angular acceleration, into the term $\mathbf{F}^s$ expressing the *force exerted on a body in steady motion.*

The first two terms on the right-hand side of Eq. (7.3.11) express the *acceleration reaction.* In terms of the three-by-three added-mass matrices $\boldsymbol{\alpha}$ and $\boldsymbol{\beta}$ introduced in Eqs. (7.2.13), we obtain the equivalent form

$$F_j = -\rho V_B \left(\frac{dU_i}{dt} \, \alpha_{ij} + \rho \frac{d\Omega_i}{dt} \, \beta_{ij} \right) + F_j^s(\mathbf{U}, \mathbf{\Omega}) \tag{7.3.12}$$

The first term on the right-hand side of Eq. (7.3.12) is a generalized version of Newton's second law, in which the matrix coefficient $\rho V_B \boldsymbol{\alpha}$ plays the role of an effective or virtual mass. Since ρV_B is equal to the mass of the fluid displaced by the body, it is natural to call $\boldsymbol{\alpha}$ the coefficient of *virtual inertia* or *added mass.*

Working in a similar manner, we obtain an expression for the torque with respect to the point $\mathbf{x}_c(t)$ exerted on the moving rigid body in the form

$$\mathbf{T} = \rho \frac{dU_i}{dt} \int_B \Phi_i \, (\mathbf{x} - \mathbf{x}_c) \times \mathbf{n} \, dS + \rho \frac{d\Omega_i}{dt} \int_B \Phi_{i+3} \, (\mathbf{x} - \mathbf{x}_c) \times \mathbf{n} \, dS + \mathbf{T}^s(\mathbf{U}, \mathbf{\Omega}) \tag{7.3.13}$$

The first two terms on the right-hand side express the torque due to unsteady translation and rotation; the steady component expressed by the third term is analogous to the last term on the right-hand side of Eq. (7.3.11).

Equations (7.3.11) and (7.3.13) may be collected in a unified form involving the six-dimensional *force–torque vector* $\mathbf{R} = (\mathbf{F}, \mathbf{T})$ which is composed of the three components of the force and the three components of the torque. Introducing the grand added-mass tensor $\mathbf{A}$ defined in Eq. (7.2.20) and the extended velocity vector $\mathbf{W}$ defined in Eq. (7.2.17), we obtain

$$\mathbf{R} = -\rho V_B \frac{d\mathbf{W}}{dt} \cdot \mathbf{A} + \mathbf{R}^s \tag{7.3.14}$$

where the superscript s denotes the value of $\mathbf{R}$ in steady motion. Since the matrix $\mathbf{A}$ is symmetric, the order of the vector-matrix multiplication in the first term on the right-hand side of Eq. (7.3.14) may be switched. The first term on the right-hand side of Eq. (7.3.14) is a variation of Newton's second law with the matrix coefficient $\rho V_B \mathbf{A}$, playing the role of an effective or virtual mass for the extended velocity vector, and this justifies calling $\mathbf{A}$ the *grand coefficient of virtual inertia* or *grand added mass*.

Force on a Rigid Body Translating in an Infinite Fluid

Next we concentrate on the practically important case of a body that executes *translation without rotation* in an otherwise quiescent fluid of infinite expanse, in the absence of any exterior or interior boundaries. Under these circumstances, Eq. (7.3.12) obtains the simplified form

$$\mathbf{F} = -\rho V_B \frac{d\mathbf{U}}{dt} \cdot \boldsymbol{\alpha} + \rho \int_B \left(\tfrac{1}{2} \mathbf{u} \cdot \mathbf{u} + U_i \frac{\partial \Phi_i[\mathbf{x}, \mathbf{x}_c(t), \mathbf{e}]}{\partial t} \right) \mathbf{n}\, dS - \rho V_B \mathbf{g} \tag{7.3.15}$$

where the material vector $\mathbf{e}$ is independent of time. The three terms on the right-hand side of Eq. (7.3.15) represent, respectively, the *acceleration reaction*, the *steady force*, and the *buoyancy force*.

For a body that translates in a fluid of infinite expanse in the absence of exterior or interior boundaries, the fundamental potentials Φ_i are functions of the distance from the designated center of the body $\mathbf{x}_c$, and this allows us to write

$$\Phi_i[\mathbf{x}, \mathbf{x}_c(t), \mathbf{e}] = \Phi_i[\mathbf{x} - \mathbf{x}_c(t), \mathbf{e}] \tag{7.3.16}$$

where, by definition, $d\mathbf{x}_c/dt = \mathbf{U}$. Concentrating on the computation of the steady force, we write

$$\int_B U_i \frac{\partial \Phi_i[\mathbf{x} - \mathbf{x}_c(t), \mathbf{e}]}{\partial t} \mathbf{n}\, dS = - \int_B U_i \frac{\partial \Phi_i}{\partial x_j} \frac{dx_{c_j}}{dt} \mathbf{n}\, dS = - \int_B (\mathbf{u} \cdot \mathbf{U}) \mathbf{n}\, dS \tag{7.3.17}$$

Substituting this result back into Eq. (7.3.15) yields

$$\mathbf{F} = -\rho V_B \frac{d\mathbf{U}}{dt} \cdot \boldsymbol{\alpha} + \rho \int_B (\tfrac{1}{2} \mathbf{u} \cdot \mathbf{u} - \mathbf{u} \cdot \mathbf{U}) \mathbf{n}\, dS - \rho V_B \mathbf{g} \tag{7.3.18}$$

To simplify the computation of the integral on the right-hand side, we select a control volume V_c that is enclosed by B and a collection of closed boundaries D, use the divergence theorem, note that $\mathbf{u}$ is irrotational to write $\nabla \mathbf{u} = (\nabla \mathbf{u})^T$, invoke the continuity equation, and apply the divergence theorem once more to find

$$\int_B (\mathbf{u} \cdot \mathbf{u}) \mathbf{n}\, dS - \int_D (\mathbf{u} \cdot \mathbf{u}) \mathbf{n}\, dS = - \int_{V_c} \nabla (\mathbf{u} \cdot \mathbf{u})\, dV$$

$$= -2 \int_{V_c} (\nabla \mathbf{u}) \cdot \mathbf{u}\, dV = -2 \int_{V_c} \mathbf{u} \cdot \nabla \mathbf{u}\, dV = -2 \int_{V_c} \nabla \cdot (\mathbf{u}\mathbf{u})\, dV$$

$$= 2 \int_B \mathbf{u}\, (\mathbf{U} \cdot \mathbf{n})\, dS - 2 \int_D \mathbf{u}\, (\mathbf{u} \cdot \mathbf{n})\, dS \tag{7.3.19}$$

where the unit normal vector $\mathbf{n}$ over D points outward from the control volume. Substituting the final result into the right-hand side of Eq. (7.3.18) yields

$$\mathbf{F} = -\rho V_B \frac{d\mathbf{U}}{dt} \cdot \boldsymbol{\alpha} + \rho \int_B [\mathbf{u}(\mathbf{U} \cdot \mathbf{n}) - (\mathbf{U} \cdot \mathbf{u})\mathbf{n}] \, dS$$

$$+ \rho \frac{1}{2} \int_D [(\mathbf{u} \cdot \mathbf{u})\mathbf{n} - 2\mathbf{u}(\mathbf{u} \cdot \mathbf{n})] \, dS - \rho V_B \, \mathbf{g} \qquad (7.3.20)$$

The integrand of the first integral on the right-hand side may be written as $\mathbf{U} \times (\mathbf{u} \times \mathbf{n})$. Using the irrotationality condition, stating that the velocity gradient tensor is symmetric, $\nabla \mathbf{u} = (\nabla \mathbf{u})^T$, one may establish that the value of the corresponding integral over B remains unchanged when the domain of integration is replaced with an arbitrary surface that encloses B. Replacing, in particular, B with D and combining the two integrals on the right-hand side of Eq. (7.3.20), we obtain the more compact form

$$\mathbf{F} = -\rho V_B \frac{d\mathbf{U}}{dt} \cdot \boldsymbol{\alpha} + \rho \int_D \{\mathbf{u}[(\mathbf{U} - \mathbf{u}) \cdot \mathbf{n}] + \tfrac{1}{2}[(\mathbf{u} - 2\mathbf{U}) \cdot \mathbf{u}]\mathbf{n}\} \, dS - \rho V_B \, \mathbf{g} \qquad (7.3.21)$$

Identifying D with B and using the no-penetration condition reproduces Eq. (7.3.18), whereas identifying D with a spherical surface of large radius and noting that far from the body, the velocity behaves like that of the flow due to a point-source dipole and therefore decays like $1/r^3$, shows that the integral on the right-hand side vanishes and thus

$$\mathbf{F} = -\rho V_B \left(\frac{d\mathbf{U}}{dt} \cdot \boldsymbol{\alpha} + \mathbf{g} \right) \qquad (7.3.22)$$

An underlying requirement is that the body have a finite size, which means, for example, that it may not be an infinite cylinder with wavy corrugations.

We have thus arrived at the remarkable conclusion that *the hydrodynamic force exerted on a finite three-dimensional body that translates steadily in a fluid of infinite expanse with no exterior or interior boundaries vanishes.* It is important to emphasize that the presence of boundaries may cause the development of a drag as well as a lift force.

D'Alembert's paradox

The component of the steady force parallel to the direction of translation may be computed in an alternative manner using an energy argument. Considering the energy integral balance (3.3.19), we use the divergence theorem to convert the first boundary integral on the right-hand side to a volume integral and then combine it with the integral on the left-hand side, exploit the fact that density of the fluid is uniform, and express the body-force term as shown in Eq. (3.1.36) to obtain

$$\int_{V_c} \frac{D}{Dt}(\tfrac{1}{2}\rho|\mathbf{u}|^2) \, dV = \int_D \rho \mathbf{u} \cdot \mathbf{n} \, dS - \rho \int_D (\mathbf{g} \cdot \mathbf{x})(\mathbf{u} \cdot \mathbf{n}) \, dS \qquad (7.3.23)$$

where D/Dt is the material derivative. Identifying the control volume with the whole of the domain of flow renders the left-hand side equal to dK/dt where K is the kinetic energy of the fluid; the boundary D is composed of the surface of the body and a surface of large size extending to infinity. Using the fact that, far from the body, the flow behaves like that due to a point-source dipole to eliminate the corresponding integrals over the large surface, enforcing the no-penetration condition over the surface of the body, and then using the divergence theorem to manipulate the surface integral involving the body force, we obtain

$$\frac{dK}{dt} = -\mathbf{U} \cdot (\mathbf{F} + \rho V_B \, \mathbf{g}) \qquad (7.3.24)$$

If the body moves at a constant velocity in the absence of any exterior or interior boundaries, the whole flow pattern is convected with the body, and the kinetic energy of the fluid remains constant, $dK/dt = 0$.

Working independently, we differentiate Eq. (7.2.23) with respect to time, and use the fact that $\boldsymbol{\alpha}$ is symmetric to express the rate of change of the kinetic energy of the fluid in the form

$$\frac{dK}{dt} = \rho V_B \, \alpha_{ij} U_i \, \frac{dU_j}{dt} \tag{7.3.25}$$

Setting the right-hand sides of Eqs. (7.3.24) and (7.3.25) equal to zero and using Eq. (7.2.15), we obtain

$$\mathbf{U} \cdot (\mathbf{F} + \rho V_B \mathbf{g}) = -\rho V_B \, \alpha_{ij} U_i \, \frac{dU_j}{dt} = -\rho(\mathbf{d} - V_B \mathbf{U}) \cdot \frac{d\mathbf{U}}{dt} = 0 \tag{7.3.26}$$

which shows that the component of the hydrodynamic force in the direction of translation is finite only when the body accelerates, and vanishes when the body executes steady motion, a result that is known as *D'Alembert's paradox*. In physical terms, this behavior can be explained by noting that, because the kinetic energy of the fluid is constant and there is no viscous dissipation, a finite rate of working is not necessary in order to maintain the motion.

Force on a body in a uniform unsteady flow

The force exerted on a body that is held stationary in an incident time-dependent uniform flow with velocity $\mathbf{V}(t)$ may be computed from the results for flow due to the translation of the body, working in a frame of reference in which the far flow is and remains stationary at all times, and the body appears to translate with velocity $\mathbf{U}(t) = -\mathbf{V}(t)$ in an otherwise quiescent fluid. Adding to the acceleration reaction given in Eq. (7.3.22), the fictitious inertial force due to the distributed body force field per unit mass $-d\mathbf{V}(t)/dt$, we obtain

$$\mathbf{F} = \rho V_B \left(\frac{d\mathbf{V}}{dt} \cdot \boldsymbol{\alpha} + \frac{d\mathbf{V}}{dt} - \mathbf{g} \right) = \rho V_B \frac{d\mathbf{V}}{dt} \cdot (\mathbf{I} + \boldsymbol{\alpha}) - \rho V_B \, \mathbf{g} \tag{7.3.27}$$

Force on a body translating in a uniform unsteady flow

To compute the force exerted on a body that translates with velocity $\mathbf{U}(t)$ in a uniform unsteady flow with velocity $\mathbf{V}(t)$, we work in a frame of reference in which the far flow is and remains stationary at all times, in which case the body appears to translate with velocity $\mathbf{U}(t) - \mathbf{V}(t)$ in an otherwise quiescent fluid. Repeating the arguments preceding Eq. (7.3.27), we find

$$\mathbf{F} = \rho V_B \left(\frac{d(\mathbf{V} - \mathbf{U})}{dt} \cdot \boldsymbol{\alpha} + \frac{d\mathbf{V}}{dt} - \mathbf{g} \right)$$

$$= -\rho V_B \frac{d\mathbf{U}}{dt} \cdot \boldsymbol{\alpha} + \rho V_B \frac{d\mathbf{V}}{dt} \cdot (\mathbf{I} + \boldsymbol{\alpha}) - \rho V_B \, \mathbf{g} \tag{7.3.28}$$

Acceleration of a body translating in uniform unsteady flow

To compute the acceleration of a body that translates in a uniformly accelerating infinite fluid in the absence of any exterior or interior boundaries, we apply Newton's law for the motion of the body, stating that the change in the linear momentum of the body balances the hydrodynamic and gravitational force, and use Eq. (7.3.28) to obtain

$$\rho_B V_B \frac{d\mathbf{U}}{dt} = -\rho V_B \frac{d\mathbf{U}}{dt} \cdot \boldsymbol{\alpha} + \rho V_B \frac{d\mathbf{V}}{dt} \cdot (\mathbf{I} + \boldsymbol{\alpha}) - \rho V_B \, \mathbf{g} + \rho_B V_B \, \mathbf{g} \tag{7.3.29}$$

where ρ_B is the density of the body. Rearranging, we derive the explicit form

$$\frac{d\mathbf{U}}{dt} = \left[\frac{d\mathbf{V}}{dt} \cdot (\mathbf{I} + \boldsymbol{\alpha}) + \left(\frac{\rho_B}{\rho} - 1 \right) \mathbf{g} \right] \cdot \left(\boldsymbol{\alpha} + \frac{\rho_B}{\rho} \mathbf{I} \right)^{-1} \tag{7.3.30}$$

where the superscript -1 denotes the inverse of the underlying matrix. Applications of Eq. (7.3.30) will be discussed in Section 7.4 with reference to the motion of a sphere.

Torque on a Rigid Body Translating in an Infinite Fluid

Beginning with Eq. (7.3.13) and working as before for the force, we find that the torque with respect to the point $\mathbf{x}_c$ exerted on a body that executes translation without rotation is given by

$$
\mathbf{T} = \rho \frac{dU_i}{dt} \int_B \Phi_i (\mathbf{x} - \mathbf{x}_c) \times \mathbf{n}\, dS + \frac{1}{2} \int_B (\mathbf{u} \cdot \mathbf{u} - 2\mathbf{U} \cdot \mathbf{u})\, (\mathbf{x} - \mathbf{x}_c) \times \mathbf{n}\, dS
$$
$$
- \rho V_B (\mathbf{X}_c - \mathbf{x}_c) \times \mathbf{g} \tag{7.3.31}
$$

which may be recast into the form

$$
\mathbf{T} = \rho \frac{dU_i}{dt} \int_B \Phi_i\, \hat{\mathbf{x}} \times \mathbf{n}\, dS
$$
$$
+ \rho \int_B \phi \mathbf{U} \times \mathbf{n}\, dS + \rho \int_B [\hat{\mathbf{x}} \times \mathbf{u}\, (\mathbf{U} \cdot \mathbf{n}) - \hat{\mathbf{x}} \times \mathbf{n}\, (\mathbf{U} \cdot \mathbf{u}) - \phi \mathbf{U} \times \mathbf{n}]\, dS
$$
$$
+ \rho \frac{1}{2} \int_D \hat{\mathbf{x}} \times [(\mathbf{u} \cdot \mathbf{u})\, \mathbf{n} - 2\mathbf{u}\, (\mathbf{u} \cdot \mathbf{n})]\, dS - \rho V_B (\mathbf{X}_c - \mathbf{x}_c) \times \mathbf{g} \tag{7.3.32}
$$

where $\hat{\mathbf{x}} = \mathbf{x} - \mathbf{x}_c$, D is a surface enclosing B, and $\mathbf{n}$ over D points outward from the control volume that is bounded by D and B. Identifying D with B reproduces Eq. (7.3.31). It can be shown that the value of the penultimate integral on the right-hand side of Eq. (7.3.32) remains unchanged when B is replaced by any closed surface that encloses it (Problem 7.3.4).

Expressing the second integral on the right-hand side of Eq. (7.3.32) in terms of the coefficient of the dipole using Eq. (7.2.9) and restating the integrand of the third integral, yields the new form

$$
\mathbf{T} = \rho \frac{dU_i}{dt} \int_B \Phi_i\, \hat{\mathbf{x}} \times \mathbf{n}\, dS - \rho \mathbf{U} \times \mathbf{d}
$$
$$
+ \rho \int_D \{\hat{\mathbf{x}} \times [\mathbf{U} \times (\mathbf{u} \times \mathbf{n})] - \phi\, \mathbf{U} \times \mathbf{n}\}\, dS
$$
$$
+ \rho \frac{1}{2} \int_D \hat{\mathbf{x}} \times [(\mathbf{u} \cdot \mathbf{u})\, \mathbf{n} - 2\mathbf{u}\, (\mathbf{u} \cdot \mathbf{n})]\, dS - \rho V_B (\mathbf{X}_c - \mathbf{x}_c) \times \mathbf{g} \tag{7.3.33}
$$

Identifying D with a spherical surface of large radius and noting that far from the body the flow is similar to that due to a point-source dipole and the velocity decays like $1/r^3$ shows that the second and third integrals on the right-hand side of Eq. (7.3.33) vanish. We thus obtain the simplified expression

$$
\mathbf{T} = \rho \frac{dU_i}{dt} \int_B \Phi_i\, \hat{\mathbf{x}} \times \mathbf{n}\, dS - \rho\, \mathbf{U} \times \mathbf{d} - \rho V_B\, (\mathbf{X}_c - \mathbf{x}_c) \times \mathbf{g} \tag{7.3.34}
$$

which shows that *the hydrodynamic torque exerted on a steadily translating body vanishes only when the coefficient of the dipole is oriented in the direction of translation.* Expressing $\mathbf{d}$ in terms of $\boldsymbol{\alpha}$ by means of Eq. (7.2.15) and remembering that $\boldsymbol{\alpha}$ is symmetric shows that there exist three mutually perpendicular directions of translation, called the *directions of permanent translation,* for which the coefficient of the dipole is parallel to the direction of the velocity. In these cases, the body will keep translating without showing a tendency for rotation.

PROBLEMS

7.3.1 **Force and torque on a body due to a dipole or a quadrupole.** Derive (a) the expressions in Eqs. (7.3.8) and (b) corresponding expressions for a point-source quadruple.

7.3.2 **Torque on a moving rigid body.** Derive Eq. (7.3.13).

7.3.3   **A body translating in a bounded domain.** (a) Explain why Eq. (7.3.16) remains valid for the two potentials that correspond to translation parallel to a plane wall in a semi-infinite domain of flow. (b) Consider a body that translates parallel to a plane wall in a semi-infinite fluid. Show that Eq. (7.3.21) is valid provided that D is identified with the wall, and simplify the integrand. (c) Consider a body that translates inside or outside a cylindrical boundary, and discuss whether Eq. (7.3.16) is valid for any one of the fundamental potentials.

7.3.4   **Torque on a translating body.** Fill in the steps in the derivation of Eq. (7.3.34) from Eq. (7.3.31).

7.4 | FLOW PAST OR DUE TO THE MOTION OF A SPHERE

Having discussed certain salient features of the flow past or due to the motion of a three-dimensional body, we proceed to consider in more detail the particular case of flow past or due to the motion of an impenetrable sphere.

Flow Due to a Translating Sphere

The flow due to a sphere of radius a translating with velocity $\mathbf{U}$ in an ambient fluid of infinite expanse in the absence of any boundaries may be represented in terms of a source dipole with strength $\mathbf{d} = 2\pi a^3 \mathbf{U}$ placed at the center of the sphere, which is located at $\mathbf{x}_0$. In this case, not only the far flow, but also the whole flow reduces to that due to a potential dipole.

Using Eqs. (2.1.28) and (2.1.29), we find that the harmonic potential and velocity field are given by

$$\phi = -\tfrac{1}{2}\frac{a^3}{r^3}\mathbf{U}\cdot\hat{\mathbf{x}}, \qquad \mathbf{u} = \tfrac{1}{2}a^3\left(-\frac{1}{r^3}\mathbf{U} + \frac{3}{r^5}(\mathbf{U}\cdot\hat{\mathbf{x}})\hat{\mathbf{x}}\right) \qquad (7.4.1)$$

where $\hat{\mathbf{x}} = \mathbf{x} - \mathbf{x}_0$ and $r = |\hat{\mathbf{x}}|$. In spherical polar coordinates with the x axis parallel to $\mathbf{U}$ and corresponding unit vector $\mathbf{i}$, the Stokes stream function is given by

$$\Psi = \tfrac{1}{2}\mathbf{U}\cdot\mathbf{i}\,\frac{a^3}{r}\sin^2\theta \qquad (7.4.2)$$

The streamline pattern is identical to that outside Hill's spherical vortex observed in a stationary frame of reference, illustrated in Figure 2.9.2(a).

Substituting $\mathbf{d} = 2\pi a^3 \mathbf{U}$ into Eqs. (7.2.15), (7.2.23), and (7.3.22), we compute the added mass tensor, kinetic energy of the fluid, and force exerted on the sphere

$$\boldsymbol{\alpha} = \tfrac{1}{2}\mathbf{I}, \qquad K = \tfrac{1}{4}\rho V_B|\mathbf{U}|^2, \qquad \mathbf{F} = -\rho V_B\left(\tfrac{1}{2}\frac{d\mathbf{U}}{dt} + \mathbf{g}\right) \qquad (7.4.3)$$

The last expression shows that the external force that is necessary in order to overcome the hydrodynamic force, and thus accelerate a sphere, is equal to the force necessary to accelerate half the amount of fluid displaced by the sphere.

Setting $\mathbf{x}_c$ at the center of the sphere $\mathbf{x}_0$, using Eq. (7.4.1), and noting that a sphere that rotates about its center does not generate a flow, because the velocity normal to its surface vanishes, shows that the six fundamental potentials defined in Eq. (7.2.10) are given by

$$\Phi_i = -\tfrac{1}{2}\frac{a^3}{r^3}\hat{x}_i, \qquad \text{for } i = 1, 2, 3$$

$$\Phi_i = 0, \qquad \text{for } i = 4, 5, 6 \qquad (7.4.4)$$

If we place $\mathbf{x}_c$ at a point other than the center of the sphere, the first three potentials will be amended, and the last three potentials will express a nontrivial flow.

Uniform Flow past a Sphere

The velocity field corresponding to uniform flow $\mathbf{V}(t)$ past a stationary sphere derives from that for flow due to a translating sphere, by working in a frame of reference in which the fluid far from the sphere appears to be stationary. The harmonic potential and velocity field are given by

$$\phi = \mathbf{V} \cdot \left(\hat{\mathbf{x}} + \frac{1}{2} \frac{a^3}{r^3} \hat{\mathbf{x}} \right)$$

$$\mathbf{u} = \mathbf{V} + \frac{1}{2} \frac{a^3}{r^3} \left(\mathbf{V} - \frac{3}{r^2} (\mathbf{V} \cdot \hat{\mathbf{x}}) \hat{\mathbf{x}} \right)$$

(7.4.5)

In spherical polar coordinates with the x axis parallel to $\mathbf{V}$ and corresponding unit vector $\mathbf{i}$, the Stokes stream function is given by

$$\Psi = \frac{1}{2} \mathbf{V} \cdot \mathbf{i} \left(r^2 - \frac{a^3}{r} \right) \sin^2 \theta$$

(7.4.6)

The streamline pattern is identical to that outside Hills's spherical vortex observed in a frame of reference that translates with the vortex, illustrated in Figure 2.9.2(b).

Unsteady Motion of a Sphere and Its Applications

Consider next a sphere that translates with time-dependent velocity $\mathbf{U}(t)$ in the presence of a uniform unsteady flow $\mathbf{V}(t)$, which may be due, for instance, to the passage of a sound wave. Substituting the expression for the added mass from Eqs. (7.4.3) into Eqs. (7.3.27) and (7.3.30), we find that the force experienced by the sphere is given by

$$\mathbf{F} = -\frac{1}{2} \rho V_B \frac{d\mathbf{U}}{dt} + \frac{3}{2} \rho V_B \frac{d\mathbf{V}}{dt} - \rho V_B \mathbf{g}$$

(7.4.7)

and the acceleration of the sphere is given by

$$\frac{d\mathbf{U}}{dt} = \frac{1}{1 + 2\rho_B/\rho} \left[3 \frac{d\mathbf{V}}{dt} + 2 \left(\frac{\rho_B}{\rho} - 1 \right) \mathbf{g} \right]$$

(7.4.8)

We observe that, in the absence of the gravitational force, the acceleration of a sphere that is made of a material which is heavier or lighter than the fluid is, respectively, lower and higher than that of the fluid.

Gravitational settling or rise of a sphere

As an application, let us consider the gravitational settling or rise of a sphere in a quiescent fluid of infinite expanse. Setting $\mathbf{V} = 0$ reduces Eq. (7.4.8) to

$$\frac{d\mathbf{U}}{dt} = -2 \frac{1 - \rho_B/\rho}{1 + 2\rho_B/\rho} \mathbf{g}$$

(7.4.9)

One interesting result emerging from Eq. (7.4.9) is that *the magnitude of the acceleration of a sphere with negligible mass, such as an air bubble, is equal to twice the acceleration of gravity.*

To this end, we turn to discussing the physical relevance of the potential flow solution and thereby assessing the significance of Eq. (7.4.9). When a sphere starts rising or settling in a viscous fluid, the induced flow is irrotational during an infinitesimal time interval at the beginning of the motion, and Eq. (7.4.9) predicts the exact value of the initial acceleration. At later times, the viscous drag force makes a significant contribution to the force balance, vorticity enters the flow by means of diffusion across boundary layers, and, provided that the motion is stable, the sphere reaches a finite terminal velocity. At steady state, the viscous drag force balances the gravitational force, and the unsteady force in Eq. (7.4.7) makes no contribution.

Terminal velocity of a spherical bubble rising at high Reynolds numbers

When surface tension is sufficiently large, an air bubble rising in an infinite fluid maintains a nearly spherical shape. At high Reynolds numbers, the vorticity is confined within a thin boundary layer that wraps around the interface and within a narrow wake, and the main part of the flow is nearly irrotational (Section 8.1).

To compute the terminal velocity of the bubble, we turn to the energy balance expressed by Eq. (7.3.24), include the viscous dissipation term on the right-hand side, given by Eq. (3.3.13), and note that the kinetic energy of the fluid remains constant in time to obtain

$$\mathbf{U} \cdot (\mathbf{F} + \rho V_B \, \mathbf{g}) = -\int_{\text{Flow}} \Phi \, dV \tag{7.4.10}$$

(Levich, 1962, p. 444; Moore, 1963). In Section 8.1 we shall see that, at sufficiently high Reynolds numbers, the rate of dissipation within the boundary layer around the bubble is negligible compared to that within the bulk of the flow, which means that the total rate of dissipation may be approximated on the basis of Eq. (7.1.5). According to Eq. (7.1.6), the right-hand side of Eq. (7.4.10) is equal to the rate of working of the deviatoric part of the boundary traction computed on the assumption of irrotational flow.

Substituting Eq. (7.4.1) into the last integral in Eq. (7.1.5) and carrying out the integration over the surface of the sphere produces the value $12\pi a U^2$, which shows that the hydrodynamic drag force exerted on the sphere, given by the term in the parentheses on the left-hand side of Eq. (7.4.10), is equal to $\mathbf{D} = -12\pi\mu a \mathbf{U}$. It is interesting to note that the magnitude of $\mathbf{D}$ is equal to twice that given by Stokes's law for a solid sphere moving at low Reynolds numbers given in Eq. (6.6.8). The drag coefficient of a spherical air bubble that moves in a viscous fluid at high Reynolds numbers is then

$$C_D \equiv \frac{D}{\pi\rho U^2 a^2} = \frac{24}{Re} \tag{7.4.11}$$

where $Re = 2aU/\nu$ is the Reynolds number and ν is the kinematic viscosity of the fluid.

Newton's second law requires that $\mathbf{F}$ be equal and opposite to the weight of the bubble, $\mathbf{F} = -\rho_B V_B \mathbf{g}$. Considering the density of the bubble to be negligible, we set $\mathbf{F} = \mathbf{0}$ and use Eq. (7.4.10) with the aforementioned value of the viscous dissipation to find that the magnitude of the terminal velocity is given by

$$U = \frac{1}{9} \frac{g a^2}{\nu} \tag{7.4.12}$$

Weiss's Theorem for Arbitrary Three-Dimensional Flow past a Stationary Sphere

Consider next an arbitrary three-dimensional flow in an infinite domain past a stationary sphere of radius a centered at the point $\mathbf{x}_0$, and assume that the incident flow is described by the harmonic potential $\phi^\infty(\mathbf{x} - \mathbf{x}_0)$ that has no singularities within the volume occupied by the sphere. Weiss's (1945) theorem provides us with the disturbance potential due to the sphere in the form

$$\phi^D(\mathbf{x}) = \frac{a}{r} \phi^\infty\left(\frac{a^2}{r^2}(\mathbf{x} - \mathbf{x}_0)\right) - \frac{2}{ar} \int_0^a \phi^\infty\left(\frac{\eta^2}{r^2}(\mathbf{x} - \mathbf{x}_0)\right)\eta \, d\eta \tag{7.4.13}$$

where $r = |\mathbf{x} - \mathbf{x}_0|$ and the large parentheses enclose the arguments of the potential of the incident flow. Note that the integrand involves the value of ϕ^∞ along a straight segment that connects the center of the sphere to the image of the point $\mathbf{x}$ with respect to the sphere.

Point source exterior to a sphere

As an application, let us consider the flow due to a point source with strength equal to m located at the point $\mathbf{x}_S$ exterior to a stationary sphere, in which case $\phi^\infty(\mathbf{x}) = -m/(4\pi|\mathbf{x} - \mathbf{x}_S|)$.

To apply Weiss's theorem, we must render ϕ^∞ a function of $\mathbf{x} - \mathbf{x}_0$, and this is done by writing $\phi^\infty(\mathbf{x}) = -m/[4\pi|(\mathbf{x} - \mathbf{x}_0) - (\mathbf{x}_S - \mathbf{x}_0)|]$. Straightforward application of Eq. (7.4.13) yields

$$\phi^D(\mathbf{x}) = -\frac{m}{4\pi}\frac{a}{r}\frac{1}{|a^2/r^2(\mathbf{x} - \mathbf{x}_0) - (\mathbf{x}_S - \mathbf{x}_0)|}$$
$$+ \frac{m}{2\pi a r}\int_0^a \frac{\eta\,d\eta}{|\eta^2/r^2(\mathbf{x} - \mathbf{x}_0) - (\mathbf{x}_S - \mathbf{x}_0)|} \tag{7.4.14}$$

Now $\mathbf{x}^{IM}(\eta) = \mathbf{x}_0 + (\eta/r)^2(\mathbf{x} - \mathbf{x}_0)$ is the image of the point $\mathbf{x}$ with respect to a sphere of radius η centered at $\mathbf{x}_0$, and $\mathbf{x}_S^{IM}(\eta) = \mathbf{x}_0 + (\eta/R)^2(\mathbf{x}_S - \mathbf{x}_0)$ is the corresponding image of the point source, where $R = |\mathbf{x}_S - \mathbf{x}_0|$. Using the geometrical reciprocity condition $r|\mathbf{x}^{IM} - \mathbf{x}_S| = R|\mathbf{x}_S^{IM} - \mathbf{x}|$, we recast Eq. (7.4.14) into the form

$$\phi^D(\mathbf{x}) = -\frac{m}{4\pi}\frac{a}{R}\frac{1}{|\mathbf{x} - \mathbf{x}_S^{IM}(a)|} + \frac{m}{2\pi a R}\int_0^a \frac{\eta\,d\eta}{|\mathbf{x} - \mathbf{x}_S^{IM}(\eta)|} \tag{7.4.15}$$

Straightforward manipulation of the integral yields the more useful form

$$\phi^D(\mathbf{x}) = -\frac{m}{4\pi}\frac{a}{R}\frac{1}{|\mathbf{x} - \mathbf{x}_S^{IM}(a)|} + \frac{m}{4\pi a}\int_0^{a^2/R} \frac{d\xi}{|\mathbf{x} - (\mathbf{x}_0 + \xi\,\mathbf{e})|} \tag{7.4.16}$$

where $\mathbf{e}$ is the unit vector directed from the center of the sphere to the point source, $\mathbf{e} = (\mathbf{x}_S - \mathbf{x}_0)/R$. The first term on the right-hand side of Eq. (7.4.16) represents a point source with strength equal to ma/R placed at the image point of the original point source with respect to the sphere. The second term represents a continuous distribution of point sinks with uniform strength extending from the center of the sphere up to the inverse point of the point source with respect to the sphere. Note that the net discharge of the point sources and sinks in the interior of the sphere is equal to zero. The last integral on the right-hand side of Eq. (7.4.16) may be expressed in terms of an incomplete elliptic integral of the first kind or simply computed using numerical integration (Problem 7.4.3).

The force exerted on the sphere may be deduced immediately from the preceding solution using Eq. (7.3.7a). We find that the hydrodynamic component of the force is given by $\rho m^2 a^3/[4\pi R(R^2 - a^2)^2]\mathbf{e}$ and is thus oriented from the center of the sphere toward the point source (Milne-Thomson, 1968, p. 496). The direction of the force may be explained by noting that the magnitude of the velocity at the region of the sphere close to the point source is higher than that at the remote regions; Bernoulli's equation shows that the converse is true for the pressure, and the net result is that the sphere is attracted to the point source.

Butler's Theorem for Axisymmetric Flow past a Stationary Sphere

Consider next an axisymmetric flow past a stationary sphere in a domain of infinite expanse, in the absence of any exterior or interior boundaries, and assume that the incident flow is described in spherical polar coordinates by the Stokes stream function $\Psi^\infty(r, \theta)$ that has no singularities within the volume occupied by the sphere. Furthermore, adjust the level of Ψ^∞ so that it vanishes at the center of the sphere, which is placed at the origin. Since the incident flow is free of singularities inside the sphere, Ψ^∞ will be of order r^2 in the vicinity of the origin. Butler's (1953) sphere theorem states that the disturbance stream function due to the presence of the sphere is given by

$$\Psi^D = -\frac{r}{a}\Psi^\infty\left(\frac{a^2}{r}, \theta\right) \tag{7.4.17}$$

It is reassuring to note that the stream function given in Eq. (7.4.6) for uniform flow past a sphere is in agreement with Eq. (7.4.17).

A second version of Butler's theorem states that, if all singularities of $\Psi^\infty(r, \theta)$ are located inside the sphere and Ψ^∞ decays like $1/r$, then Eq. (7.4.17) provides us with the disturbance flow in the interior of the sphere.

Orthogonal stagnation-point flow past a sphere

As an application of Butler's theorem, consider irrotational orthogonal stagnation-point flow past a sphere, in which case $\Psi^\infty = -kxr^2 \sin^2 \theta$, where k is the constant rate of extension [see Eq. (5.3.24)]. In the presence of the sphere, the stream function becomes

$$\Psi = -kxr^2 \left(1 - \frac{a^5}{r^5}\right) \sin^2 \theta \qquad (7.4.18)$$

Point source exterior to a sphere

Let us return to consider the flow due to a point source in the presence of the sphere discussed previously on the basis of the velocity potential. Assuming that the point source is located on the x axis at the position $\mathbf{x}_S = (b, 0, 0)$, we use Eq. (2.7.15) to find that the incident Stokes stream function to be used with Eq. (7.4.17) is given by

$$\Psi^\infty = -\frac{m}{4\pi} \left(\frac{x - b}{|\mathbf{x} - \mathbf{x}_S|} + \frac{b}{|b|}\right) = -\frac{m}{4\pi} \left(\frac{r\cos\theta - b}{(r^2 + b^2 - 2rb\cos\theta)^{1/2}} + \frac{b}{|b|}\right) \qquad (7.4.19)$$

The disturbance stream function due to the sphere then follows as

$$\Psi^D = \frac{m}{4\pi} \frac{r}{a} \left(\frac{a^2 \cos\theta - rb}{(r^2 b^2 - 2a^2 rb\cos\theta + a^4)^{1/2}} + \frac{b}{|b|}\right) \qquad (7.4.20)$$

Butler (1953) showed that the flow expressed by Eq. (7.4.20) is identical to that corresponding to Eq. (7.4.14) obtained using a different method.

Line vortex ring exterior to a sphere

As a further application, consider the flow due to a line vortex ring of radius c and strength κ placed perpendicular to the x axis at $x = b$, in the presence of a sphere of radius a centered at the origin. The incident Stokes stream function is found from Eq. (2.9.13) to be

$$\Psi^\infty = \frac{\kappa}{4\pi} cr\sin\theta \int_0^{2\pi} \frac{d\varphi}{[(r\cos\theta - b)^2 + r^2 \sin^2\theta + c^2 - 2rc\sin\theta\cos\varphi]^{1/2}} \qquad (7.4.21)$$

Applying Butler's theorem and using the reciprocal theorem for the inverse point with respect to the sphere discussed before Eq. (7.4.15), shows that the disturbance flow may be identified with the flow due to an image line vortex ring located in the interior of the sphere. The strength and position of the image ring are given by

$$\kappa^{Im} = -\kappa \left(\frac{b^2 + c^2}{a^2}\right)^{1/2}, \qquad x^{Im} = \frac{a^2 b}{b^2 + c^2}, \qquad \sigma^{Im} = \frac{a^2 c}{b^2 + c^2} \qquad (7.4.22)$$

PROBLEMS

7.4.1   **Applications of Weiss's theorem.** (a) Use Weiss's theorem to rederive the velocity potential corresponding to uniform flow past a sphere given in Eq. (7.4.1). (b) Consider a rectilinear line vortex with strength κ that is parallel to the z axis and is located at $x = b$, $y = 0$, in the presence of a sphere of radius a centered at the origin. Show that the velocity potential is given by

$$\phi = \frac{\kappa}{2\pi} \tan^{-1} \frac{y}{x - b} + \frac{\kappa a}{2\pi r} \tan^{-1} \frac{y}{x - r^2 b/a^2} - \frac{\kappa}{\pi ar} \int_0^a \tan^{-1} \left(\frac{y}{x - r^2 b/\eta^2}\right) \eta \, d\eta$$

The disturbance potential may be described in terms of an image system that is composed of a line vortex ring of radius $a^2/2b$ with center at $x = a^2/2b$, $z = 0$ lying in the xz plane, and a vortex sheet subtended by the image ring (Weiss, 1945).

7.4.2   **Settling and rising spheres.** Explain in physical terms why the acceleration of a settling rigid sphere is lower than the acceleration of gravity, whereas the acceleration of a rising fluid sphere can be as high as twice the acceleration of gravity.

Computer Problem

7.4.3   **Flow due to a point source in the presence of a sphere.** Draw the streamline patterns of the flow due to a point source located on the exterior of a sphere at a sequence of radial positions $R/a = 1.2, 2.0, 4.0$.

7.5 | FLOW PAST OR DUE TO THE MOTION OF NONSPHERICAL BODIES

Computing the flow past or due to the motion of a spherical body was expedited with the use of certain powerful theorems that, unfortunately, are applicable only to the spherical shape. A standard method of computing the flow past or due to the motion of a nonspherical body is to solve Laplace's equation for the velocity potential in an appropriate curvilinear system of coordinates that conform with the geometry of the body, subject, for example, to the no-penetration condition. The procedure is illustrated in classical texts of fluid mechanics (Lamb, 1932, Chapter 5; Milne-Thomson, 1968), and certain examples will be discussed in the present section.

An alternative method of computing a potential flow involves representing the harmonic potential in terms of a discrete collection or a continuous distribution of singular fundamental solutions of Laplace's equation, including the point source and the point-source dipole, and then computing the strengths of these singularities and possibly their location in order to satisfy the required boundary conditions in an exact or approximate sense, using analytical or numerical methods. One significant advantage of the singularity representation is that it allows us to compute the kinetic energy of the fluid, the added mass tensor, and the acceleration reaction directly from the strength of the singularities without actually computing the velocity field and pressure distributions over the body or within the flow.

Another powerful class of numerical methods for computing potential flow originates from the boundary integral representation discussed in Section 2.3, as well as from similar generalized single-layer or double-layer representations to be discussed in Chapter 10. The numerical procedure involves applying the boundary integral representation at the boundaries of the flow, thereby obtaining Fredholm integral equations of the first or second kind for the boundary distributions of certain unknown functions, such as the harmonic potential and its normal derivative. The integral equations may be solved using well-established boundary-element or panel methods.

With these considerations in mind, we proceed to derive and discuss solutions of a family of problems using singularity representations or working in an appropriate curvilinear system of coordinates. The practical importance and popularity of the boundary-integral methods justifies separate discussion in Chapter 10.

Flow Due to the Translation of a Prolate Spheroid

Chwang and Wu (1974) showed that the flow due to the translation of a prolate spheroid whose major and minor axes are, respectively, equal to a and b, may be represented in terms of a distribution of point-source dipoles over the focal length of the spheroid, which is confined between the foci of the generating ellipse, $-c < x < c$, where $c = ea$ and $e = [1 - (b/a)^2]^{1/2}$ is the eccentricity; the origin has been set at the center of the spheroid. The dipoles are oriented in the direction of translation, and their strengths have parabolic density distributions. The velocity potential for translation with velocity $\mathbf{U}$ is given by

$$\phi(\mathbf{x}) = -\frac{1}{2} \int_{-c}^{c} \left(\frac{U_x}{\delta_x}(x - x_0) + \frac{U_y}{\delta_y}y + \frac{U_z}{\delta_z}z \right) \frac{c^2 - x_0^2}{|\mathbf{x} - \mathbf{x}_0|^3} \, dx_0 \tag{7.5.1}$$

where the point $\mathbf{x}_0 = (x_0, 0, 0)$ lies on the x axis, and δ_x, δ_y, δ_z are dimensionless coefficients whose values depend upon the eccentricity, and are given by

$$\delta_x = \frac{2e}{1 - e^2} - \ln\frac{1 + e}{1 - e}, \qquad \delta_y = \delta_z = e\frac{2e^2 - 1}{1 - e^2} + \frac{1}{2}\ln\frac{1 + e}{1 - e} \tag{7.5.2}$$

The coefficient of the effective dipole describing the far flow is found readily by integrating the parabolic density of the dipole distribution, and the result is

$$\mathbf{d} = 4\pi\frac{1}{2}\left(\frac{U_x}{\delta_x}, \frac{U_y}{\delta_y}, \frac{U_z}{\delta_z} \right) \int_{-c}^{c} (c^2 - x_0^2) \, dx_0 = \frac{8}{3}\pi a^3 e^3 \left(\frac{U_x}{\delta_x}, \frac{U_y}{\delta_y}, \frac{U_z}{\delta_z} \right) \tag{7.5.3}$$

Substituting Eq. (7.5.3) into Eq. (7.2.23), we obtain the kinetic energy of the fluid

$$K = \frac{4}{3}\pi\rho a^3 e^3 \left(\frac{U_x^2}{\delta_x} + \frac{U_y^2}{\delta_y} + \frac{U_z^2}{\delta_z} \right) - \frac{2}{3}\pi\rho ab^2(U_x^2 + U_y^2 + U_z^2) \tag{7.5.4}$$

Furthermore, using Eq. (7.2.15), we find that the added-mass tensor $\boldsymbol{\alpha}$ is diagonal, and its nonvanishing components are given by

$$\alpha_{xx} = 2\frac{a^2}{b^2}\frac{e^3}{\delta_x} - 1, \qquad \alpha_{yy} = \alpha_{zz} = 2\frac{a^2}{b^2}\frac{e^3}{\delta_y} - 1, \tag{7.5.5}$$

In the limit as the eccentricity tends to vanish, we recover the results of Section 7.4 for the sphere.

Solution in prolate spheroidal coordinates

A more explicit but somewhat indirect description of the flow emerges by introducing prolate spheroidal coordinates (ζ, μ, φ) that are related to the Cartesian coordinates (x, y, z) and to the cylindrical polar coordinates (x, σ, φ) by

$$\begin{aligned} x &= c\mu\zeta, & y &= \sigma\cos\varphi \\ z &= \sigma\sin\varphi, & \sigma &= c(1 - \mu^2)^{1/2}(\zeta^2 - 1)^{1/2} \end{aligned} \tag{7.5.6}$$

where ζ ranges from unity to infinity, and μ varies between -1 and 1 (Lamb, 1932, p. 141). It is convenient to set $\mu = \cos\chi$, where the parameter χ varies between 0 and π along the contour of the generating ellipse. The surfaces of constant ζ or μ are confocal prolate spheroids or hyperboloids of two sheets, with the surface of the spheroid corresponding to $\zeta_0 = 1/e$. When $\zeta_0 = 1$, the spheroid reduces to a slender needle, and as ζ_0 tends to infinity, the spheroid tends to become a sphere.

The velocity potential and Stokes stream function of the axisymmetric flow due to a prolate spheroid translating along the x axis are given by

$$\begin{aligned} \phi &= \frac{U_x}{\delta_x}2ae\mu\left(1 - \frac{1}{2}\zeta\ln\frac{\zeta + 1}{\zeta - 1}\right) \\ \Psi &= \frac{U_x}{\delta_x}\sigma^2\left(\frac{\zeta}{\zeta^2 - 1} - \frac{1}{2}\zeta\ln\frac{\zeta + 1}{\zeta - 1}\right) \end{aligned} \tag{7.5.7}$$

where δ_x is defined in Eqs. (7.5.2).

The velocity potential of the three-dimensional flow due to a prolate spheroid translating normal to its axis is given by

$$\phi = \frac{1}{ae\delta_y}\sigma^2\left[\tfrac{1}{2}\ln\left(\frac{\zeta+1}{\zeta-1}\right) - \frac{\zeta}{\zeta^2-1}\right](U_y\cos\varphi + U_z\sin\varphi) \qquad (7.5.8)$$

where δ_y is defined in Eq. (7.5.2).

Flow Due to the Translation of an Oblate Spheroid

For an oblate spheroid, we work with oblate spheroidal coordinates (ζ, μ, φ) that are related to the Cartesian coordinates (x, y, z) and to the cylindrical polar coordinates (x, σ, φ) by

$$x = c\mu\zeta, \qquad y = \sigma\cos\varphi$$
$$z = \sigma\sin\varphi, \qquad \sigma = c(1-\mu^2)^{1/2}(\zeta^2+1)^{1/2} \qquad (7.5.9)$$

ζ varies from zero to infinity, whereas $\mu = \cos\chi$ varies between 1 and -1 as χ varies between 0 and π along the contour of the generating ellipse (Lamb, 1932, p. 144).

Consider an oblate spheroid with minor axis in the x direction equal to a and major axis in the radial direction equal to b, where $a < b$, and introduce the eccentricity $e = [1 - (a/b)^2]^{1/2}$ and focal length $c = ae$. The surface of the spheroid corresponds to $\zeta_0 = a/(be)$. The surfaces of constant ζ or μ are, respectively, planetary ellipsoids or hyperboloids of revolution of one sheet, with common focal circle at $x = 0$ and $\sigma = c$.

The velocity potential and Stokes stream function corresponding to axisymmetric flow due to axial translation are given by

$$\phi = \frac{U_x}{\delta_x}b\mu(1 - \zeta\operatorname{arccot}\zeta)$$
$$\Psi = \tfrac{1}{2}\frac{U_x}{e\delta_x}\sigma^2\left(\frac{\zeta}{\zeta^2+1} - \operatorname{arccot}\zeta\right) \qquad (7.5.10)$$

with δ_x given in Eqs. (7.5.12). The velocity potential corresponding to the three-dimensional flow due to transverse translation normal to the x axis is given by

$$\phi = \frac{1}{be\delta_y}\sigma^2\left(\frac{\zeta}{\zeta^2+1} - \operatorname{arccot}\zeta\right)(U_y\cos\varphi + U_z\sin\varphi) \qquad (7.5.11)$$

where

$$\delta_x = \sqrt{1-e^2} - \frac{1}{e}\arcsin e$$
$$\delta_y = \frac{\zeta_0^2+2}{\zeta_0(\zeta_0^2+1)} - \operatorname{arccot}\zeta_0 \qquad (7.5.12)$$

In the limit as a/b tends to zero and the eccentricity tends to unity, we obtain the flow due to the translation of a circular disk of infinitesimal thickness (Problem 7.5.3). In the opposite limit where a/b tends to unity and the eccentricity tends to vanish, we recover the results of Section 7.4 for the sphere.

Singularity Methods for Axial Translation of Axisymmetric Bodies

An efficient method of computing the generally three-dimensional flow due to the translation of an axisymmetric body proceeds by representing the flow in terms of a collection of singularities placed at selected locations along the centerline or in the interior of the body, and then computing the strength of the singularities in order to satisfy the required no-penetration boundary condition in some approximate sense.

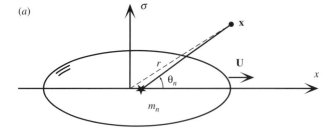

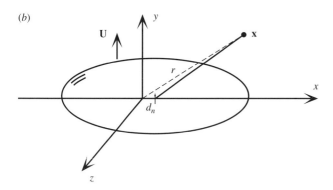

Figure 7.5.1 Representation of a flow due to the translation of an axisymmetric body in terms of singularities; (a) point sources and point sinks for axial motion; (b) point-source dipoles for transverse motion along the y axis.

In the case of axial translation, we introduce a representation in terms of a collection of N point sources or point sinks located at the centerline of the body as shown in Figure 7.5.1(a). For a body with fore-and-aft symmetry, it is appropriate to distribute the singularities symmetrically with respect to its midplane. Since the flow rate across any surface that encloses the body vanishes, the sum of the strengths of the singularities will be equal to zero.

In a frame of reference that translates with the body with velocity U, and with reference to spherical polar coordinates centered at the body axis, the Stokes stream function is given by

$$\Psi(\mathbf{x}) = -\tfrac{1}{2}Ur^2 \sin^2\theta - \frac{1}{4\pi}\sum_{n=1}^{N} m_n \cos\theta_n \qquad (7.5.13)$$

where θ_n is the meridional angle subtended by the point $\mathbf{x}$ and the location of the nth singularity as shown in Figure 7.5.1(a). The problem is reduced to computing the strengths of the singularities so that Ψ takes a constant value, which can be set equal to zero, over the contour of the body in an azimuthal plane. Once this has been accomplished, the effective coefficient of the dipole follows as

$$d_x = \sum_{n=1}^{N} x_n m_n \qquad (7.5.14)$$

Introducing, in particular, just one point source and one point sink with strengths of equal magnitude and opposite sign $\pm m$ located at $x = \pm a$, we find that Ψ vanishes over a family of axisymmetric bodies that are parametrized by the dimensionless variable $\lambda = Ua^2/m$, known as *Rankine ovoids*. As λ tends to zero, the ovoids obtain a spherical shape, and the two singularities merge to yield a point-source dipole.

Given a body of a certain shape, one way of computing the strength of the point sources or point sinks m_n is by pointwise collocation, that is, by requiring that Ψ vanish at M selected points over the contour of the body in the xy plane. Selecting $M > N$, and then solving the emerging overdetermined system of linear equations for m_n using a least-squares or a minimization method reduces the sensitivity of the solution to the location of the collocation points.

In a slight variation of this method, we represent the flow in terms of a continuous distribution of point sources deployed over a certain section of the axis inside the body between $x = -a$ and b, thereby obtaining the continuous version of Eq. (7.5.13)

$$\Psi(\mathbf{x}) = -\tfrac{1}{2}Ur^2 \sin^2 \theta - \frac{1}{4\pi} \int_{-a}^{b} \cos \theta' \, h(x') \, dx' \tag{7.5.15}$$

where $\theta' = \arctan[\sigma/(x - x')]$ and h is a distribution density function. Requiring that Ψ vanish over the contour of the body in the xy plane provides us with an integral equation for $m(x)$ that may then be solved using a standard numerical method, including the collocation method discussed previously and the method of weighted residuals (Delves and Mohamed, 1985).

It is interesting to note that the continuous distribution (7.5.15) is equivalent to a series expansion of the Stokes stream function in terms of Legendre polynomials, and the coefficients of the series are related to the moments of the density function $h(x)$ (Batchelor, 1967, p. 460).

Singularity Methods for the Transverse Translation of Axisymmetric Bodies

In the case of transverse translation, we introduce a representation in terms of a collection of N point-source dipoles with strengths d_n oriented in the direction of translation, placed at selected locations along the centerline within the body as shown in Figure 7.5.1(b). The velocity field due to translation along the y axis is

$$\mathbf{u}(\mathbf{x}) = \frac{1}{4\pi}U_y \sum_{n=1}^{N} d_n \left(-\frac{1}{|\mathbf{x} - \mathbf{x}_n|^3}\mathbf{j} + 3\frac{y}{|\mathbf{x} - \mathbf{x}_n|^5}(\mathbf{x} - \mathbf{x}_n) \right) \tag{7.5.16}$$

where the $\mathbf{x}_n = (x_n, 0, 0)$ are the locations of the singularities, and $\mathbf{j}$ is the unit vector along the y axis. Evaluating Eq. (7.5.16) at the surface of the body, requiring the no-penetration condition $\mathbf{u} \cdot \mathbf{n} = \mathbf{U} \cdot \mathbf{n} = U_y n_y$, and writing $n_y = (y/\sigma)n_\sigma$ and $n_z = (z/\sigma)n_\sigma$, where $y^2 + z^2 = \sigma^2$, we obtain

$$\sum_{n=1}^{N} d_n \left(-\frac{1}{|\mathbf{x} - \mathbf{x}_n|^3} + 3\frac{\sigma}{n_\sigma}\frac{(x - x_n)n_x + \sigma n_\sigma}{|\mathbf{x} - \mathbf{x}_n|^5} \right) = 4\pi \tag{7.5.17}$$

To compute the strengths of the dipoles d_n, we apply Eq. (7.5.17) at M selected collocation points along the contour of the body in the xy plane and then work as discussed previously for axial translation. For a sphere, we recover the exact solution with just one dipole placed at its center.

In a variation of the preceding method, we represent the flow in terms of a continuous distribution of dipoles pointing in the direction of motion, and solve for the density of the distribution using a collocation or a weighted residual method. Our previous discussion guarantees that, for a prolate spheroid, a parabolic distribution over the focal length is sufficient for obtaining the exact solution. For more general shapes, the optimal domain of distribution that yields the highest degree accuracy must be found by numerical experimentation.

Approximate Methods for Slender Axisymmetric Bodies

When the length of an axisymmetric body is much larger than the typical size of its cross section, it is appropriate to represent the flow in terms of distributions of singularities over the centerline of the body and then compute the densities of the distributions using asymptotic methods. The approximate solution obtained in this manner provides us with a reasonable, and in some cases competitive, alternative to full numerical computation.

Axial motion

Let us consider a slender axisymmetric body whose surface is described in cylindrical polar coordinates as $\sigma = f(x)$, where $-a < x < a$. For simplicity, but without loss of generality, we have placed the origin of the x axis midway between the two ends. Assuming that the body translates parallel to the x axis with velocity U_x, we introduce a representation in terms of a distribution of point sources, obtaining

$$\phi(\mathbf{x}) = -\frac{U_x}{4\pi} \int_{-a}^{a} \frac{h(x_0)}{|\mathbf{x} - \mathbf{x}_0|} \, dx_0 \tag{7.5.18}$$

where the point $\mathbf{x}_0 = (x_0, 0, 0)$ lies on the x axis. The density of the distribution h must be found so as to satisfy the no-penetration condition in an approximate sense.

Adopting first a physical rather than a mathematical approach, we write a mass balance over a control volume that is fixed in space and is confined between the surface of the body and two planes that are perpendicular to the x axis, and take the limit as the distance between the two planes vanishes to obtain

$$\frac{\partial A}{\partial t} = \frac{\partial Q}{\partial x} \tag{7.5.19}$$

where $A = \pi f^2$ is the cross-sectional area of the body and $Q(x)$ is the instantaneous volumetric flow rate through a plane that is perpendicular to the x axis over the exterior of the body. Introducing the instantaneous flow rate $q(x)$ across the *whole area of a plane* that is perpendicular to the x axis, and using Eq. (7.5.18), we find $\partial q/\partial x = U_x h$. Approximating Q in Eq. (7.5.19) with q, and using the fact that in a frame of reference moving with the body the radius of the body is constant, $\partial f/\partial t + U_x \partial f/\partial x = 0$, we find that the density of the distribution of the point sources is given by

$$h = \frac{1}{U_x} \frac{\partial A}{\partial t} = -\frac{\partial A}{\partial x} \tag{7.5.20}$$

The assumptions that underlie these equations will fail near the ends of the body, where Eq. (7.5.19) ceases to be valid, but this occurrence does not prevent Eq. (7.5.20) from providing us with a reasonable approximation of the flow over the main portion of the body.

The coefficient of the dipole $\mathbf{d}$ is oriented along the axis of the body. Using Eq. (7.5.20), and integrating by parts we find

$$d_x = U_x \int_{-a}^{a} x_0 h(x_0) \, dx_0 = -U_x \int_{-a}^{a} x_0 \frac{d(\pi f^2)}{dx_0} \, dx_0 = U_x V_B \tag{7.5.21}$$

Substituting Eq. (7.5.21) into Eqs. (7.2.15) and (7.2.23) shows that $\alpha_{xx} = 0$ and $K = 0$, which, unfortunately, are contrived results that neglect the presence of the body. A higher-order approximation is required in order to estimate the added mass and kinetic energy of the fluid.

To assess the accuracy of the slender-body theory, it is useful to compare the approximate solution given in Eq. (7.5.21) with the exact solution for a prolate spheroid given in Eq. (7.5.3). It may be shown that, as the inverse aspect ratio b/a tends to zero, the exact solution yields

$$d_x = \tfrac{4}{3}\pi U_x a^3 \left(\frac{b}{a}\right)^2 \left[1 - \left(\frac{b}{a}\right)^2 \ln\frac{b}{a}\right] + \cdots \tag{7.5.22}$$

Neglecting the second term within the square brackets yields the right-hand side of Eq. (7.5.21). This approximation introduces a relative error of order $(b/a)^2 \ln(b/a)$, which is small for bodies with moderate and high aspect ratios. Consequently, the slender-body theory provides us with a rigorous basis for approximating the coefficient of the dipole with the product of the volume of the body and velocity of translation.

Equation (7.5.21) may be derived in a more rigorous manner, working within the framework of asymptotic expansions. We begin by considering the no-penetration condition, which requires that $u_x n_x + u_\sigma n_\sigma = U_x n_x$ on the surface of the body, and introduce the approximations $n_x = -\partial f/\partial x$ and $n_\sigma = 1$. Noting that u_x and u_σ have comparable magnitudes, whereas n_x is small compared to n_σ, we derive the approximation $u_\sigma = -U_x \partial f/\partial x$. Differentiating Eq. (7.5.18) with respect to σ, we express the radial velocity as

$$u_\sigma(\mathbf{x}) = \frac{U_x}{4\pi}\sigma \int_{-a}^{a} \frac{h(x_0)}{[(x-x_0)^2 + \sigma^2]^{3/2}}\,dx_0 = \frac{U_x}{4\pi\sigma}\int_{-(a+x)/\sigma}^{(a-x)/\sigma} \frac{h(x_0)}{(\eta^2+1)^{3/2}}\,d\eta \qquad (7.5.23)$$

where $\eta = (x_0 - x)/\sigma$ is an ancillary variable of integration. Next we evaluate Eq. (7.5.23) at a point on the surface of the body where $\sigma = f(x)$, expand $h(x_0)$ in a Taylor series about x and retain only the leading constant term, and make use of the fact that, as f/a tends to zero, the limits of integration with respect to η become infinite, to obtain the approximation

$$u_\sigma(\mathbf{x}) = \frac{U_x}{4\pi f(x)}h(x)\int_{-\infty}^{\infty} \frac{d\eta}{(\eta^2+1)^{3/2}} = \frac{U_x}{2\pi f(x)}h(x) \qquad (7.5.24)$$

Substituting Eq. (7.5.24) into the approximate boundary condition $u_\sigma = -U_x \partial f/\partial x$ yields Eq. (7.5.20).

Working in a similar manner, we find that the axial component of the velocity at a point on the surface of the body is given by

$$u_x(\mathbf{x}) = \frac{U_x}{4\pi}\int_{-a}^{a} \frac{x-x_0}{[(x-x_0)^2+\sigma^2]^{3/2}}h(x_0)\,dx_0 \approx -\frac{U_x}{4\pi}\frac{dh(x)}{dx}\ln\frac{4(a^2-x^2)}{f^2(x)} \qquad (7.5.25)$$

which may be used along with Eq. (7.5.24) and Bernoulli's equation to evaluate the surface pressure.

Transverse motion

For a slender axisymmetric body translating in the y direction that is normal to its axis, we approximate the flow in the vicinity of the body with that due to a translating circular cylinder whose radius is equal to the local radius of the body $f(x)$. In Section 7.7 we shall see that the latter may be expressed in terms of a two-dimensional dipole oriented along the y axis with strength equal to $2\pi U_y f^2 = 2U_y A$, where A is the cross-sectional area of the body. Noting that the two-dimensional dipole emerges by integrating the three-dimensional dipole along the x axis suggests a representation in terms of a distribution of three-dimensional dipoles with strength per unit length equal to $2\pi U_y f^2$ in the form

$$\phi = -\frac{1}{2}U_y \int_{\text{Axis}} \frac{y}{|\mathbf{x}-\mathbf{x}_0|^3}f^2(x_0)\,dx_0 \qquad (7.5.26)$$

This representation may be derived in a more rigorous manner working as before for axial motion using asymptotic expansions [Problem 7.5.1(c)].

The coefficient of the dipole, $d_y = 2U_y V_B$, is equal to twice that for axial motion shown in Eq. (7.5.21). The added-mass coefficient and kinetic energy of the fluid are $\alpha_{yy} = 1$ and $K = \frac{1}{2}\rho V_B U_y^2$. A comparison with the exact solution for prolate spheroids confirms the consistency and accuracy of the slender-body representation (7.5.26) for bodies with moderate and large aspect ratios.

Singularity Methods for Three-Dimensional Bodies

Singularity methods for three-dimensional bodies arise as straightforward extensions of the methods for axisymmetric bodies. The types of singularities employed and their location within the body must be selected on an individual basis, exercising physical intuition. The strengths of the singularities are typically computed by pointwise collocation. When the flow is represented in terms of a collection of N point sources and K dipoles, the coefficient of the dipole is computed as

$$\mathbf{d} = \sum_{n=1}^{N} \mathbf{x}_n m_n + \sum_{k=1}^{K} \mathbf{d}_k \qquad (7.5.27)$$

In an advanced implementation of the singularity method, the locations of the singularities are not specified in the statement of the problem, but are computed as part of the solution so as to minimize an appropriate positive functional that derives from the boundary conditions (Zhou and Pozrikidis, 1995). This modification allows for a high degree of accuracy with a small number of singularities.

PROBLEMS

7.5.1 **Slender-body theory.** (a) Compute the coefficient of the dipole for a sphere according to Eq. (7.5.21) and discuss your result with reference to the exact value. (b) Compute the asymptotic limit of the coefficient of the dipole for a prolate spheroid given in Eq. (7.5.3) as b/a tends to zero, and thus verify that the leading-order term is given by $d_y = 2U_y V_B$, in agreement with slender-body theory. (c) Derive the representation (7.5.26) in the context of asymptotic expansions working as in the case of axial motion discussed in the text.

7.5.2 **Application of slender-body theory.** Derive the precise form of the singularity distribution corresponding to the axial translation of an axisymmetric body whose surface is described by $\sigma = f(x) = b(1 - x^2/a^2)$ for $-a < x < a$, where b is a constant. Then evaluate the pressure at the surface of the body.

7.5.3 **Flow due to an axially translating circular disk.** Consider the flow due to an axially translating oblate spheroid and take the limit as the eccentricity e tends to unity to derive the velocity potential due to the translation of a circular disk of infinitesimal thickness (Lamb, 1932, p. 144).

Computer Problems

7.5.4 **Rankine ovoids.** Compute and plot the contours of the Rankine ovoids in an azimuthal plane for a sequence of values of the slenderness parameter $\lambda = Ua^2/m$.

7.5.5 **Flow due to translation of an axisymmetric body.** Write a program called SPF that computes the strengths of the point sources or dipoles on the basis of Eqs. (7.5.13) and (7.5.16) using a collocation method. Run the program for a prolate spheroid of aspect ratio $a/b = 2$, and compare the computed coefficient of the dipole with the exact value given in Eq. (7.5.3). The singularities should be distributed evenly over the focal length of the spheroid. Study the convergence of the method as the number of singularities is increased.

7.6 | FLOW PAST OR DUE TO THE MOTION OF TWO-DIMENSIONAL BODIES

The domain of flow past a two-dimensional body is inevitably doubly connected, and the velocity potential is a single-valued function only when the circulation around the body vanishes. This important fact calls for special attention in studying the properties and computing the structure of two-dimensional flow.

Flow past a Stationary or Translating Rigid Body

Consider a potential flow past a two-dimensional body that translates with velocity $\mathbf{U}$ with finite circulation around it, as illustrated in Figure 7.6.1. To describe the flow, we decompose the velocity into three components, $\mathbf{u} = \mathbf{u}^\infty + \mathbf{v} + \mathbf{u}^D$; $\mathbf{u}^\infty$ corresponds to the incident flow; $\mathbf{v}$ represents the flow due to a point vortex whose strength is equal to the circulation around the body, located at some point within the body; and $\mathbf{u}^D$ represents a disturbance flow that may be described in terms of a single-valued harmonic potential ϕ^D. The no-penetration boundary condition requires that over the contour of the body C, $\mathbf{u} \cdot \mathbf{n} = \mathbf{U} \cdot \mathbf{n}$, or

$$\nabla\phi^D \cdot \mathbf{n} = \mathbf{U} \cdot \mathbf{n} - \nabla\phi^\infty \cdot \mathbf{n} - \mathbf{v} \cdot \mathbf{n} \tag{7.6.1}$$

The counterpart of the boundary integral representation (7.2.1) is

$$\phi^D(\mathbf{x}_0) = -\int_C G(\mathbf{x}_0, \mathbf{x})\,[\mathbf{U} - \nabla\phi^\infty(\mathbf{x}) - \mathbf{v}(\mathbf{x})] \cdot \mathbf{n}(\mathbf{x})\,dl(\mathbf{x})$$

$$+ \int_C \nabla G(\mathbf{x}_0, \mathbf{x}) \cdot \mathbf{n}(\mathbf{x})\,\phi^D(\mathbf{x})\,dl(\mathbf{x}) \tag{7.6.2}$$

Writing the counterpart of Eq. (7.2.2) for two-dimensional flow, combining it with Eq. (7.6.2) to simplify the single-layer potential, and then adding to both sides of the resulting equation the potential of the incident flow ϕ^∞, we obtain the counterpart of Eq. (7.2.3)

$$\Phi(\mathbf{x}_0) = \phi^\infty(\mathbf{x}_0) + \int_C G(\mathbf{x}_0, \mathbf{x})\,\mathbf{v}(\mathbf{x}) \cdot \mathbf{n}(\mathbf{x})\,dl(\mathbf{x})$$

$$- \int_C \nabla G(\mathbf{x}_0, \mathbf{x}) \cdot \mathbf{n}(\mathbf{x})\,[\mathbf{U} \cdot \mathbf{x} - \Phi(\mathbf{x})]\,dl(\mathbf{x}) \tag{7.6.3}$$

where $\Phi = \phi^\infty + \phi^D$. The total velocity potential is given by

$$\phi = \frac{\kappa}{2\pi}\theta + \Phi \tag{7.6.4}$$

where θ is the polar angle measured around the location of the point vortex. When the circulation around the body is equal to zero, $\mathbf{v}$ vanishes, and Φ reduces to ϕ. The modular cases of flow past a stationary body and flow due to the translation of a body in an otherwise quiescent fluid emerge by setting, respectively, $\mathbf{U}$ or ϕ^∞ equal to zero.

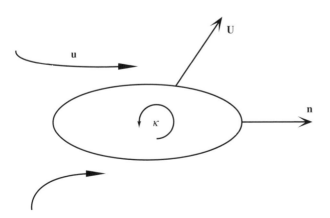

Figure 7.6.1 Illustration of flow past a translating two-dimensional rigid body with finite circulation around it.

Far-field expansion

To assess the behavior of the flow far from the body, we follow a procedure that is similar to that outlined in Section 7.2 for three-dimensional flow in a singly-connected domain. The result is

$$\phi(\mathbf{x}_0) = \phi^\infty(\mathbf{x}_0) + \frac{\kappa}{2\pi}\theta - \nabla_c G(\mathbf{x}_0, \mathbf{x}_c) \cdot \mathbf{d} + \cdots \tag{7.6.5}$$

where the coefficient of the dipole is given by

$$\mathbf{d} = A_B \mathbf{U} - \int_C [\mathbf{n}(\mathbf{x})\,\Phi(\mathbf{x}) - (\mathbf{x} - \mathbf{x}_c)\,\mathbf{v}(\mathbf{x}) \cdot \mathbf{n}(\mathbf{x})]\,dl(\mathbf{x}) \tag{7.6.6}$$

A_B is the area of the cross-section of the body in the xy plane, and $\mathbf{x}_c$ is an arbitrary point in the vicinity or in the interior of the body. It will be noted that $\mathbf{d}$ is expressed in terms of the distribution of the single-valued part of the velocity potential and the velocity due to the point vortex over the body.

Flow Due to the Motion and Deformation of a Body

Consider next the flow due to the translation, rotation, or deformation of a body in an otherwise quiescent fluid. Working as previously, we decompose the velocity into two components writing $\mathbf{u} = \mathbf{v} + \mathbf{u}^D$, where $\mathbf{u}^D$ is a disturbance flow described by the single-valued harmonic potential ϕ^D. The boundary integral formulation provides us with the integral representation

$$\phi(\mathbf{x}_0) = \frac{\kappa}{2\pi}\theta - \int_C G(\mathbf{x}_0, \mathbf{x}) \nabla\phi^D(\mathbf{x}) \cdot \mathbf{n}(\mathbf{x})\,dl(\mathbf{x})$$
$$+ \int_C \nabla G(\mathbf{x}_0, \mathbf{x}) \cdot \mathbf{n}(\mathbf{x})\,\phi^D(\mathbf{x})\,dl(\mathbf{x}) \tag{7.6.7}$$

To assess the behavior of the flow far from the body, we expand the Green's function and its dipole in a Taylor series with respect to $\mathbf{x}$ about the point $\mathbf{x}_c$ and obtain the counterpart of Eq. (7.2.6)

$$\phi(\mathbf{x}_0) = \frac{\kappa}{2\pi}\theta - G(\mathbf{x}_0, \mathbf{x}_c)\int_C \nabla\phi^D(\mathbf{x}) \cdot \mathbf{n}(\mathbf{x})\,dl(\mathbf{x}) - \nabla_c G(\mathbf{x}_0, \mathbf{x}_c) \cdot \mathbf{d} + \cdots \tag{7.6.8}$$

where the coefficient of the dipole is given by

$$\mathbf{d} = \int_C [-\phi^D(\mathbf{x})\,\mathbf{n}(\mathbf{x}) + (\mathbf{x} - \mathbf{x}_c)\,\nabla\phi^D(\mathbf{x}) \cdot \mathbf{n}(\mathbf{x})]\,dl(\mathbf{x}) \tag{7.6.9}$$

Using the divergence theorem, one may show that the value of the integral on the right-hand side of Eq. (7.6.9) remains unchanged when the contour of integration C is replaced by any other closed contour that encloses C.

Equation (7.6.8) shows that, far from the body, the flow behaves like that due to a point vortex with circulation around it equal to κ. If κ is equal to zero, the flow behaves like the flow due to a two-dimensional point source. If, in addition, the area of the body remains constant, the far flow behaves like that due to a two-dimensional potential dipole.

Motion of a rigid body

The single-valued disturbance potential due to a rigid body that translates with time-dependent linear velocity $\mathbf{U}$ and rotates with angular velocity $\mathbf{\Omega}$ around the z axis about the point $\mathbf{x}_c$ satisfies the boundary condition over C

$$\nabla\phi^D(\mathbf{x}) \cdot \mathbf{n}(\mathbf{x}) = [\mathbf{U} + \mathbf{\Omega} \times (\mathbf{x} - \mathbf{x}_c) - \mathbf{v}] \cdot \mathbf{n}(\mathbf{x}) \tag{7.6.10}$$

Substituting Eq. (7.6.10) into Eq. (7.6.9), we obtain the counterpart of Eq. (7.2.9)

$$\mathbf{d} = -\int_C \phi^D(\mathbf{x})\,\mathbf{n}(\mathbf{x})\,dl(\mathbf{x}) + A_B\mathbf{U} + \int_C \{\mathbf{\Omega} \cdot [(\mathbf{x} - \mathbf{x}_c) \times \mathbf{n}(\mathbf{x})]\}\,(\mathbf{x} - \mathbf{x}_c)\,dl(\mathbf{x})$$
$$- \int_C [\mathbf{v}(\mathbf{x}) \cdot \mathbf{n}(\mathbf{x})]\,(\mathbf{x} - \mathbf{x}_c)\,dl(\mathbf{x}) \tag{7.6.11}$$

which shows that the coefficient of the dipole may be computed from a knowledge of the distribution of the disturbance velocity potential over the body. The second integral on the right-hand side of Eq. (7.6.11) can be made to vanish by choosing the point $\mathbf{x}_c$ to be the centroid of the area of the body, which may be expressed in terms of a line integral that is analogous to the surface integral shown in Eq. (4.1.3).

Decomposition into fundamental modes of rigid-body-motion and flow due to a point vortex

The linear nature of the equations and boundary conditions governing the structure of the potential flow due to the motion of a rigid body allows us to express the harmonic potential as a linear combination of (1) the velocity of translation $\mathbf{U}$, (2) the angular velocity of rotation about a certain point $\mathbf{x}_c$, $\mathbf{\Omega}$, and (3) the strength of a point vortex that is placed at a certain location $\mathbf{x}^{PV}$ within the body and produces a desired degree of circulation. We thus write

$$\phi(\mathbf{x}) = U_x(t)\,\Phi_1[\mathbf{x}, \mathbf{x}_c, \mathbf{e}(t)] + U_y(t)\,\Phi_2[\mathbf{x}, \mathbf{x}_c, \mathbf{e}(t)] + \Omega(t)\,\Phi_3[\mathbf{x}, \mathbf{x}_c, \mathbf{e}(t)]$$
$$+ \kappa\left(\frac{1}{2\pi}\,\arg[\mathbf{x} - \mathbf{x}^{PV}(t)] + \Phi_4[\mathbf{x}, \mathbf{x}^{PV}(t), \mathbf{x}_c(t), \mathbf{e}(t)]\right) \tag{7.6.12}$$

where Φ_i are four fundamental single-valued velocity potentials, the unit vector $\mathbf{e}$ describes the instantaneous body orientation, and the square brackets contain the arguments of Φ_i.

Substituting Eq. (7.6.12) into Eq. (7.6.11) and assuming that the point $\mathbf{x}_c$ is the areal centroid of the body, thereby discarding the second integral on the right-hand side, we obtain the counterpart of Eq. (7.2.12)

$$\mathbf{d} = A_B[\mathbf{U} \cdot (\mathbf{I} + \boldsymbol{\alpha}) + \Omega\boldsymbol{\zeta} + \kappa\boldsymbol{\eta}] - \int_C [\mathbf{v}(\mathbf{x}) \cdot \mathbf{n}(\mathbf{x})]\,(\mathbf{x} - \mathbf{X}_c)\,dl(\mathbf{x}) \tag{7.6.13}$$

where we have defined

$$\alpha_{ij} = -\frac{1}{A_B}\int_C \Phi_i\,n_j\,dl(\mathbf{x}), \qquad \zeta_i = -\frac{1}{A_B}\int_C \Phi_3\,n_i\,dl(\mathbf{x})$$
$$\eta_i = -\frac{1}{A_B}\int_C \Phi_4\,n_i\,dl(\mathbf{x}) \tag{7.6.14}$$

Working as in Section 7.2, one may show that the added-mass matrix α_{ij} is symmetric and may therefore be rendered diagonal by an appropriate choice of the Cartesian axes.

Kinetic energy of the flow due to the motion of a rigid body

When the circulation around a body is finite, the velocity decays like $1/r$, and the kinetic energy of the fluid takes an infinite value. This singular behavior is a manifestation of the fact that, in practice, the onset of a finite circulation around a body is accompanied by the generation of an equal amount of vorticity with opposite sign, which keeps the kinetic energy at a finite value.

Useful information into the energetics of the flow can be obtained by considering the kinetic energy of the fluid residing within a finite area that is confined by the body and a circle of radius R. Decomposing ϕ into the single-valued component ϕ^D and a complementary component associated with the point vortex and using Eq. (2.1.17), we obtain

$$K = -\tfrac{1}{2}\rho \int_C \phi^D(\mathbf{x}) \, \nabla\phi^D(\mathbf{x}) \cdot \mathbf{n}(\mathbf{x}) \, dl(\mathbf{x}) + \cdots \tag{7.6.15}$$

When the circulation around the body is finite, as R tends to infinity, the omitted terms represented by the dots take an infinite value; otherwise they vanish. The finite part of the kinetic energy represented by the first term on the right-hand side is amenable to the analysis of Section 7.2 for three-dimensional flow with a straightforward change of notation.

Force on a Translating Rigid Body

To compute the force exerted on a translating two-dimensional body with constant circulation around it, we work with Bernoulli's equation as discussed in Section 7.3 for three-dimensional flow. In this manner, we find

$$\mathbf{F} = -\rho A_B \frac{d\mathbf{U}}{dt} \cdot \boldsymbol{\alpha} + \rho \int_C [\mathbf{u}\,(\mathbf{U} \cdot \mathbf{n}) - (\mathbf{U} \cdot \mathbf{u})\mathbf{n}]\, dl - \rho A_B \, \mathbf{g} \tag{7.6.16}$$

The three terms on the right-hand side of Eq. (7.6.16) represent, respectively, the acceleration reaction, the steady force, and the buoyancy force.

Using the divergence theorem, we find that the value of the integral on the right-hand side of Eq. (7.6.16) remains unchanged when the contour of integration is replaced by any other contour that encloses C. When the circulation around the body has a finite value, the velocity decays like $1/r$, and the integral over a circle with large radius obtains the *Kutta–Joukowski* value $\kappa\mathbf{k} \times \mathbf{U}$, where $\mathbf{k}$ is the unit vector in the direction of the z axis. Substituting this result into Eq. (7.6.16), we obtain

$$\mathbf{F} = -\rho A_B \boldsymbol{\alpha} \cdot \frac{d\mathbf{U}}{dt} + \rho\kappa\mathbf{k} \times \mathbf{U} - \rho A_B\mathbf{g} \tag{7.6.17}$$

which shows that the steady drag force exerted on a translating two-dimensional body vanishes, whereas the lift force is proportional to the circulation around the body, independently of the shape of the body! The direction of the lift force when the circulation around the body has a positive value is illustrated in Figure 7.6.2. Physically, the motion associated with a positive circulation acts to accelerate the fluid above the body and decelerate the fluid below the body, thereby causing a pressure difference that results in an upward lift force.

The lifting action of an airfoil hinges upon its ability to generate a sufficient amount of positive circulation so that the lift force counterbalances the aircraft's weight. The generation of circulation relies upon viscous effects within the boundary layers developing around the airfoil, to be discussed in Chapter 8. But once this is established, viscous stresses have a minor effect on the magnitude of the lift force.

The torque exerted on the body may be computed working as in Section 7.3 for three-dimensional flow.

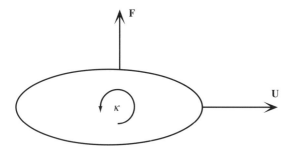

Figure 7.6.2 Direction of the lift force on a translating body with positive circulation around it.

PROBLEMS

7.6.1 **Boundary conditions for the fundamental potentials.** Write out the boundary conditions satisfied by the fundamental potentials introduced in Eq. (7.6.12).

7.6.2 **Force on a translating body.** (a) Derive Eq. (7.6.16), and (b) the Kutta–Joukowski value for the lift force from Eqs. (7.1.3) and (7.1.4).

7.7 | COMPUTATION OF TWO-DIMENSIONAL FLOW PAST OR DUE TO THE MOTION OF A BODY

Having established the general properties of two-dimensional flow past or due to the motion of a rigid body, we proceed to discuss specific solutions obtained using analytical or numerical methods.

Flow Due to the Translation and Rotation of a Circular Cylinder with Finite Circulation around It

Consider first a circular cylinder of radius a translating with time-dependent velocity $\mathbf{U}(t)$ and rotating about its center with angular velocity $\Omega(t)$, in a fluid of infinite expanse, with circulation $C(t)$ around it. Since the component of the velocity normal to the surface of the cylinder due to the rotation vanishes, the rotation does not generate fluid motion and may thus be overlooked.

The flow due to the translation and circulation may be represented, respectively, in terms of a point-source dipole with strength $\mathbf{d} = 2\pi a^2 \mathbf{U}$ and a point vortex with strength $\kappa = C$, both placed at the center of the cylinder. The harmonic potential and velocity field are given by

$$\phi = \frac{\kappa}{2\pi}\theta - \frac{a^2}{r^2}\mathbf{U}\cdot\hat{\mathbf{x}}$$

$$\mathbf{u} = \frac{\kappa}{2\pi}\frac{1}{r}\mathbf{e}_\theta + a^2\left(-\frac{1}{r^2}\mathbf{U} + \frac{2}{r^4}(\mathbf{U}\cdot\hat{\mathbf{x}})\hat{\mathbf{x}}\right)$$

(7.7.1)

where $\hat{\mathbf{x}} = \mathbf{x} - \mathbf{x}_c$, $r = |\hat{\mathbf{x}}|$, $\mathbf{x}_c$ is the instantaneous center of the cylinder, and the polar angle θ and associated unit vector $\mathbf{e}_\theta$ are defined with respect to $\mathbf{x}_c$. In plane polar coordinates with the x axis parallel to $\mathbf{U}$ and corresponding unit vector $\mathbf{i}$, the stream function is given by

$$\psi = -\frac{\kappa}{2\pi}\ln\frac{r}{a} + \mathbf{U}\cdot\mathbf{i}\frac{a^2}{r}\sin\theta$$

(7.7.2)

Comparing the first of Eqs. (7.7.1) with Eq. (7.6.12) shows that the four fundamental potentials for translation, rotation, and circulation are given by

$$\Phi_1 = -\frac{a^2}{r^2}\hat{x}, \qquad \Phi_2 = -\frac{a^2}{r^2}\hat{y}, \qquad \Phi_3 = 0, \qquad \Phi_4 = 0$$

(7.7.3)

Since the velocity due to the point vortex is parallel to the surface of the cylinder, the last term on the right-hand side of Eq. (7.6.13) vanishes. Using Eqs. (7.7.3) and the definitions in Eqs. (7.6.14), we find that the coefficients $\boldsymbol{\zeta}$ and $\boldsymbol{\eta}$ vanish, and thus obtain $\mathbf{d} = A_B\mathbf{U}\cdot(\mathbf{I} + \boldsymbol{\alpha})$. Remembering that $\mathbf{d} = 2\pi a^2\mathbf{U}$, we obtain $\boldsymbol{\alpha} = \mathbf{I}$, which can be substituted into Eq. (7.6.17) to yield the force exerted on the cylinder

$$\mathbf{F} = -\rho\pi a^2\frac{d\mathbf{U}}{dt} + \rho\kappa\mathbf{k}\times\mathbf{U} - \rho\pi a^2\mathbf{g}$$

(7.7.4)

(a)

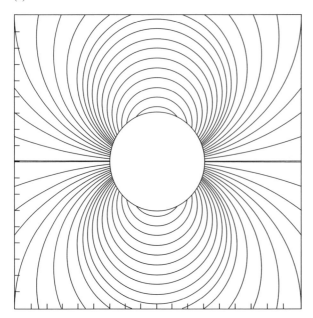

(b)

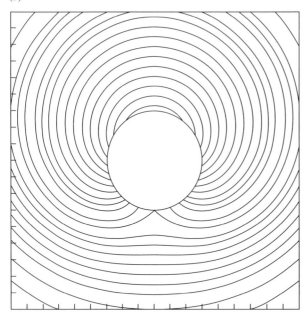

Figure 7.7.1 Streamline patterns of the flow due to the translation of a circular cylinder with circulation parameter (a) $\lambda \equiv \kappa/(4\pi aU) = 0$ and (b) 0.50. (*continued*)

(c)

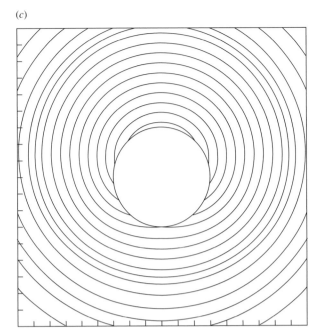

Figure 7.7.1 (*continued*)   Streamline patterns of the flow due to the
translation of a circular cylinder with circulation parameter (c) 1.0.

Equations (7.7.1) and (7.7.2) show that the structure of the flow depends upon the value
of the dimensionless circulation parameter $\lambda \equiv \kappa/(4\pi a U)$ where $U = \mathbf{U} \cdot \mathbf{i}$. Three families of
streamline patterns corresponding to positive values of λ with $\kappa > 0$ and $U > 0$ are shown in
Figure 7.7.1a–c. In all cases, the lift force points toward the positive y axis.

Uniform Flow past a Stationary Circular Cylinder with Finite Circulation around It

Unsteady uniform flow with velocity $\mathbf{V}(t)$ past a stationary or rotating circular cylinder
with circulation $C(t)$ around it, derives from the flow due to a translating cylinder, working in a
frame of reference in which the flow far from the cylinder is stationary. We thus obtain

$$\phi = \frac{\kappa}{2\pi}\theta + \left(1 + \frac{a^2}{r^2}\right)\mathbf{V} \cdot \hat{\mathbf{x}}$$

$$\mathbf{u} = \mathbf{V} + \frac{\kappa}{2\pi}\frac{1}{r}\mathbf{e}_\theta + \frac{a^2}{r^2}\left(\mathbf{V} - \frac{2}{r^2}(\mathbf{V} \cdot \hat{\mathbf{x}})\hat{\mathbf{x}}\right)$$

(7.7.5)

where $\kappa = C$, $\hat{\mathbf{x}} = \mathbf{x} - \mathbf{x}_c$, $r = |\hat{\mathbf{x}}|$, $\mathbf{x}_c$ is the instantaneous center of the cylinder, and the polar
angle θ and associated unit vector $\mathbf{e}_\theta$ are defined with respect to $\mathbf{x}_c$. In plane polar coordinates
with the x axis parallel to $\mathbf{V}$ corresponding to the unit vector $\mathbf{i}$, the stream function is given by

$$\psi = -\frac{\kappa}{2\pi}\ln\frac{r}{a} + \mathbf{V} \cdot \mathbf{i}\left(r - \frac{a^2}{r}\right)\sin\theta$$

(7.7.6)

Equation (7.7.6) shows that the structure of the flow depends upon the value of the dimen-
sionless circulation parameter $\beta \equiv -\kappa/(4\pi a V)$ where $V = \mathbf{V} \cdot \mathbf{i}$. Four families of streamline

patterns corresponding to positive values of β with $\kappa > 0$ and $V < 0$ are shown in Figure 7.7.2(a–d). The tangential component of the velocity around the surface of the cylinder is readily found by differentiating Eq. (7.7.6) with respect to r,

$$u_\theta(r = a) = \frac{\kappa}{2\pi a} - 2V \sin\theta \tag{7.7.7}$$

We observe that u_θ vanishes at the points where $\sin\theta = -\beta$, and this reveals the onset of a symmetric pair of stagnation points on the surface of the cylinder for $-1 < \beta < 1$. For values of β outside this range, the stagnation points move off the surface of the cylinder and merge within the flow to yield a free stagnation point.

It is evident from Eq. (7.7.7) that when $\beta > 0$, the magnitude of the velocity at the top of the cylinder is higher than that at the bottom. Bernoulli's equation then shows that the surface pressure at the top is lower than that at the bottom, which provides us with a physical explanation for the occurrence of a lift force toward the positive direction of the y axis, first noted by Magnus in 1853.

Representation in Terms of a Boundary Vortex Sheet

An interesting representation of a two-dimensional potential flow past a stationary body, with vanishing or finite circulation around it, emerges by assuming that the interior of the body is occupied by a stationary fluid, and then regarding the disturbance flow due to the body as though it were induced by a two-dimensional vortex sheet lining the surface of the body. The strength of the vortex sheet is equal to the tangential component of the velocity of the fluid $u_t = \mathbf{u} \cdot \mathbf{t}$, where $\mathbf{t}$ is the tangential unit vector pointing in the counterclockwise direction around the body, as shown in Figure 7.7.3. This point of view leads us to express the stream function in the form

$$\psi(\mathbf{x}_0) = \psi^\infty(\mathbf{x}_0) - \frac{1}{2\pi} \int_0^L \ln|\mathbf{x}_0 - \mathbf{x}| u_t(\mathbf{x}) \, dl(\mathbf{x}) + c \tag{7.7.8}$$

where ψ^∞ is the stream function of the incident flow, L is the total arc length of the contour of the body in the xy plane, and c is an arbitrary constant.

Applying Eq. (7.7.8) at a point on the surface of the body and requiring the no-penetration condition $\psi = 0$ yields an integral equation of the first kind for the distribution of the tangential boundary velocity

$$\frac{1}{2\pi} \int_0^L \ln|\mathbf{x}_0 - \mathbf{x}| u_t(\mathbf{x}) \, dl(\mathbf{x}) = \psi^\infty(\mathbf{x}_0) + c \tag{7.7.9}$$

which admits a continuous family of solutions parametrized by c, reflecting the fact that the circulation around the body may be set to an arbitrary value.

A simple numerical method for solving the integral equation (7.7.9) proceeds by tracing the contour of the body with a set of N marker points that are arranged in the counterclockwise sense as shown in Figure 7.7.3, and then approximating the contour of the body with the polygonal line that connects successive marker points. Assuming that the value of u_t is constant over each segment, equal to U_i, where $i = 1, \ldots, N$, we obtain the discrete version of Eq. (7.7.9)

$$A_i(\mathbf{x}_0) U_i = \psi^\infty(\mathbf{x}_0) + c \tag{7.7.10}$$

where

$$A_i(\mathbf{x}_0) = \frac{1}{2\pi} \int_{\mathbf{x}_i}^{\mathbf{x}_{i+1}} \ln|\mathbf{x}_0 - \mathbf{x}| \, dl(\mathbf{x}) \tag{7.7.11}$$

are the *influence coefficients*. One way of computing U_i is by pointwise collocation, which involves identifying $\mathbf{x}_0$ with the middle of each segment $\mathbf{y}_j = \frac{1}{2}(\mathbf{x}_j + \mathbf{x}_{j+1})$, thereby obtaining a

(a)

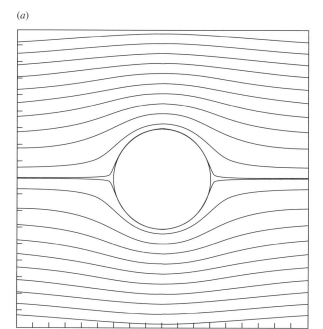

(b)

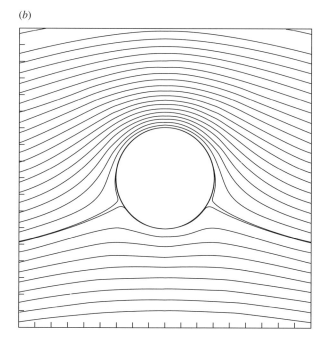

Figure 7.7.2   Streamline patterns of uniform flow past a stationary or rotating circular cylinder for values of the circulation parameter (a) $\beta \equiv -\kappa/(4\pi aV) = 0$ and (b) $\beta = 0.50$.   (*continued*)

(*c*)

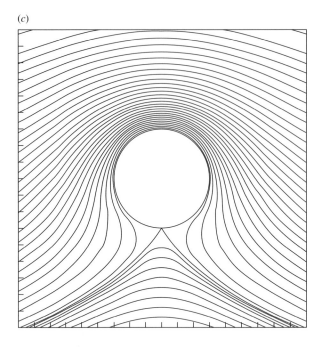

(*d*)

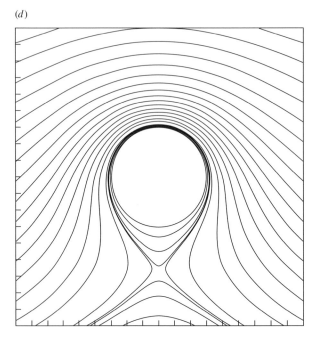

Figure 7.7.2 (*continued*)   Streamline patterns of uniform flow past a stationary or rotating circular cylinder for values of the circulation parameter (c) $\beta = 1.0$ and (d) $\beta = 1.2$.

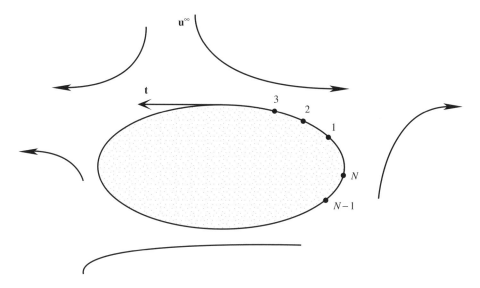

Figure 7.7.3 The flow past a stationary two-dimensional body may be represented in terms of a vortex sheet situated over the surface of the body. The strength of the vortex sheet, which is equal to the tangential component of the velocity of the fluid, may be computed by solving an integral equation of the first kind using a numerical method.

system of N linear equations for the unknowns U_i

$$A_i(\mathbf{y}_j)\, U_i = \psi^\infty(\mathbf{y}_j) + c \tag{7.7.12}$$

To ensure that the circulation around the body has a prescribed value κ, we require the additional constraint

$$\sum_{i=1}^{N} U_i \left|\mathbf{x}_{i+1} - \mathbf{x}_i\right| = \kappa \tag{7.7.13}$$

and solve the system of the $N + 1$ linear equations comprised of Eqs. (7.7.12) and (7.7.13) for the $N + 1$ unknowns U_i and c.

The influence coefficients $A_i(\mathbf{y}_j)$ for $i \neq j$ may be computed using a standard integration method, the simplest one being the trapezoidal rule (Section B.7, Appendix B). The integrals corresponding to the host collocation point $A_i(\mathbf{y}_j)$ are singular but may be computed exactly as

$$A_j(\mathbf{x}_j) = \frac{1}{\pi}\left|\mathbf{y}_j - \mathbf{x}_j\right|\left(\ln\left|\mathbf{y}_j - \mathbf{x}_j\right| - 1\right) \tag{7.7.14}$$

Plate of infinitesimal thickness

For flow past a flat plate of infinitesimal thickness with length equal to $2a$, placed on the x axis between the points $x = \pm a$, the upper and lower surface of the body coincide, and Eq. (7.7.8) reduces to

$$\psi(\mathbf{x}_0) = \psi(\mathbf{x}_0^\infty) - \frac{1}{2\pi}\int_{-a}^{a} \ln\left|\mathbf{x}_0 - x\mathbf{i}\right| \Delta u(x)\, dx + c \tag{7.7.15}$$

where $\Delta u = u_t^- - u_t^+$ is the discontinuity in the tangential velocity across the plate; the plus and minus superscripts indicate evaluations, respectively, on the upper and lower side of the plate. Correspondingly, Eq. (7.7.9) yields the integral equation

$$\frac{1}{2\pi} \int_{-a}^{a} \ln|x_0 - x| \, \Delta u(x) \, dx = \psi^{\infty}(x_0) + c \tag{7.7.16}$$

For uniform incident flow with velocity $\mathbf{V}$, $\psi^{\infty} = V_x y - V_y x$, and the solution of Eq. (7.7.16) is known to be

$$\Delta u = \left(V_y x + \frac{\kappa}{2\pi}\right) \frac{2}{\sqrt{a^2 - x^2}} \tag{7.7.17}$$

Eq. (7.10.11). Note that singularities occur at both ends of the plate. When κ assumes the *Kutta value* $2\pi a V_y$, the singularity at the trailing edge $x = -a$ disappears, which means that the fluid particles on either side of the plate leave the trailing edge in the tangential direction without making a sharp turn.

Computation of Flow past Bodies with Arbitrary Geometry

The singularity methods discussed in Section 7.5 for three-dimensional flow may be adapted in a straightforward manner to handle flow past or due to the motion of two-dimensional bodies. The efficiency of these methods, however, competes with that of a class of methods based on formulation in complex variables that uses the powerful tool of conformal mapping to be discussed in Sections 7.8–7.11, and another class of methods based on boundary-integral equations to be discussed in Chapter 10.

Slender-Body Theory

Slender-body theory can provide us with dependable results, while requiring a minimal amount of analytical or computational effort. To illustrate the application of this theory, consider the flow due to the axial translation of a slender two-dimensional body that is symmetric with respect to the x axis, subject to the assumption that the circulation around the body vanishes. The upper and lower surfaces of the body are described as $y = \pm f(x)$, where $-a < x < a$. As in the case of flow due to the motion of an axisymmetric body discussed in Section 7.5, we represent the flow in terms of two-dimensional point sources distributed over the centerline of the body, writing

$$\phi(\mathbf{x}) = \frac{U_x}{2\pi} \int_{-a}^{a} \ln|\mathbf{x} - \mathbf{x}_0| \, h(x_0) \, dx_0 \tag{7.7.18}$$

where h is the density of the distribution, and the point $\mathbf{x}_0 = (x_0, 0)$ lies on the x axis. We can proceed as in the case of axisymmetric flow to compute h based on physical arguments, but it is preferable for exposition purposes to follow a more rigorous approach based on a formal asymptotic method.

The no-penetration condition on the upper surface of the body requires that $u_x n_x + u_y n_y = U_x n_x$. Introducing the approximations $n_x = -df/dx$ and $n_y = 1$, and noting that u_x and u_y have comparable magnitudes, but n_x is small compared to n_y, we derive the approximate form

$$u_y(x, f) \cong -U_x \frac{df}{dx} \tag{7.7.19}$$

Next we differentiate Eq. (7.7.18) with respect to y and express u_y in terms of h as

$$u_y(\mathbf{x}) = \frac{U_x}{2\pi} y \int_{-a}^{a} \frac{h(x_0)}{(x - x_0)^2 + y^2} \, dx_0 = \frac{U_x}{2\pi} \int_{-(a+x)/y}^{(a-x)/y} \frac{h(x_0)}{\eta^2 + 1} \, d\eta \tag{7.7.20}$$

where $\eta = (x_0 - x)/y$ is an ancillary variable of integration. Evaluating Eq. (7.7.20) at a point $\mathbf{x} = (x, f)$ on the upper surface of the body, expanding $h(x_0)$ in a Taylor series about x and retaining only the leading constant term, and making use of the fact that as f/a tends to zero, the limits of integration with respect to η become infinite, yields the approximation

$$u_y(x, f) \cong \frac{U_x}{2\pi} h(x) \int_{-\infty}^{\infty} \frac{d\eta}{\eta^2 + 1} = \frac{1}{2} U_x h(x) \tag{7.7.21}$$

Substituting Eq. (7.7.21) into Eq. (7.7.19), we find

$$h = -2 \frac{df}{dx} \tag{7.7.22}$$

The coefficient of the dipole is given by

$$d_x = U_x \int_{-a}^{a} x_0 \, h(x_0) \, dx_0 = -2U_x \int_{-a}^{a} x_0 \frac{df}{dx_0} \, dx_0 = 2U_x A_B \tag{7.7.23}$$

which, remarkably but fortuitously, provides us with the exact answer for the circular cylinder.

PROBLEMS

7.7.1 **Lift on a circular cylinder.** Substitute Eqs. (7.7.1) into Bernoulli's equation to obtain the pressure distribution over the surface of the cylinder, and then use Eqs. (7.1.4) to compute the force exerted on the cylinder.

7.7.2 **Formulation in terms of a boundary vortex sheet.** (a) Derive the counterpart of Eqs. (7.7.8) and (7.7.9) for axisymmetric flow past a stationary axisymmetric body in terms of the Stokes stream function. (b) Discuss the implementation of the method for three-dimensional flow; that is, derive an integral equation for the two tangential components of the strength of the surface vortex sheet. *Hint:* Use the condition that the vorticity field is solenoidal.

 ## Computer Problem

7.7.3 **Boundary vortex sheet.** (a) Write a program called *BVS* that solves the system of Eqs. (7.7.12) and (7.7.13) for a specified body geometry described in terms of a collection of marker points. (b) Run the program for uniform flow along the x axis past a circular cylinder, plot the distribution of the tangential velocity along the surface of the cylinder for several values of the circulation, and compare the numerical results with the exact solution. (c) Run the program for uniform flow along the x axis past an elliptical cylinder with axes ratio $a/b = 3$ whose major axis is parallel to the x axis, and plot the distribution of the tangential velocity for several values of the circulation. (d) Repeat (c) for uniform flow along the y axis. (e) Run the program for uniform flow with $V_x = V_y$ past a flat plate, plot the distribution of the tangential velocity for several values of the circulation, and compare the numerical results with the exact solution given in Eq. (7.7.17).

7.8 | FORMULATION OF TWO-DIMENSIONAL FLOW IN COMPLEX VARIABLES

A powerful method of analyzing and computing two-dimensional irrotational flow is based on reformulating the problem in the context of complex variables. The theory of analytic functions of a complex variable allows us then to derive solutions for a variety of problems, even for domains with complicated geometries, using efficient and elegant analytical and numerical methods.

TABLE 7.8.1

The complex potential and velocity field of several two-dimensional potential flows.

Uniform flow

$$w = (U - iV)z, \qquad \phi = Ux + Vy, \qquad \psi = Uy - Vx, \qquad \frac{dw}{dz} = U - iV, \qquad u = U, v = V$$

Point source

$$w = \frac{m}{2\pi} \ln(z - z_0), \qquad \phi = \frac{m}{2\pi} \ln|z - z_0|, \qquad \psi = \frac{m}{2\pi} \arg(z - z_0), \qquad \frac{dw}{dz} = \frac{m}{2\pi} \frac{z^* - z_0^*}{|z - z_0|^2}$$

Point-source dipole

$$w = -\frac{d_x + id_y}{2\pi} \frac{1}{z - z_0}, \qquad \phi = -\frac{1}{2\pi} \frac{d_x(x - x_0) + d_y(y - y_0)}{|z - z_0|^2}$$

$$\psi = \frac{1}{2\pi} \frac{d_x(y - y_0) - d_y(x - x_0)}{|z - z_0|^2}, \qquad \frac{dw}{dz} = \frac{d_x + id_y}{2\pi} \frac{1}{(z - z_0)^2}$$

Point vortex

$$w = \frac{\kappa}{2\pi i} \ln(z - z_0), \qquad \phi = \frac{\kappa}{2\pi} \arg(z - z_0), \qquad \psi = -\frac{\kappa}{2\pi} \ln|z - z_0|, \qquad \frac{dw}{dz} = \frac{\kappa}{2\pi i} \frac{z^* - z_0^*}{|z - z_0|^2}$$

Periodic array of point sources along the x axis separated by distance a

$$w = \frac{m}{2\pi} \ln\left[\sin\left(\frac{k}{2}(z - z_0)\right)\right]$$

$$\phi = \frac{m}{4\pi} \ln\left[\cosh\left(k(y - y_0)\right) - \cos\left(k(x - x_0)\right)\right]$$

$$\psi = \frac{m}{2\pi} \arg\left[\sin\left(\frac{k}{2}(x - x_0)\right)\cosh\left(\frac{k}{2}(y - y_0)\right)\right.$$
$$\left. + i \cos\left(\frac{k}{2}(x - x_0)\right)\sinh\left(\frac{k}{2}(y - y_0)\right)\right]$$

$$\frac{dw}{dz} = \frac{m}{4\pi} k \cot\left(\frac{k}{2}(z - z_0)\right)$$

$k = 2\pi/a$ is the wave number; z_0 is the location of one point source in the array.

Periodic array of point vortices along the x axis separated by distance a

$$w = \frac{\kappa}{2\pi i} \ln\left[\sin\left(\frac{k}{2}(z - z_0)\right)\right]$$

$$\phi = \frac{\kappa}{2\pi} \arg\left[\sin\left(\frac{k}{2}(x - x_0)\right)\cosh\left(\frac{k}{2}(y - y_0)\right)\right.$$
$$\left. + i \cos\left(\frac{k}{2}(x - x_0)\right)\sinh\left(\frac{k}{2}(y - y_0)\right)\right]$$

$$\psi = -\frac{\kappa}{4\pi} \ln\left[\cosh\left(k(y - y_0)\right) - \cos\left(k(x - x_0)\right)\right]$$

$$\frac{dw}{dz} = \frac{\kappa}{4\pi i} k \cot\left(\frac{k}{2}(z - z_0)\right)$$

$k = 2\pi/a$ is the wave number; z_0 is the location of one point vortex in the array.

Flow around a corner with angle π/m

$$w = Az^m, \qquad \phi = Ar^m \cos(m\theta), \qquad \psi = Ar^m \sin(m\theta), \qquad \frac{dw}{dz} = mAz^{m-1}$$

A is a constant, origin is at the apex.

The Complex Potential

Let us consider a two-dimensional irrotational flow of an incompressible fluid and introduce the harmonic potential ϕ and associated stream function ψ. The velocity vector is tangential to the lines of constant ψ, which are the instantaneous streamlines, and perpendicular to the lines of constant ϕ. The curvilinear grid that is composed of lines of constant ϕ and ψ is sometimes called the *flow net.* Since the vorticity vanishes, Eq. (2.7.7) shows that ψ is a two-dimensional harmonic function, $\nabla^2\psi = 0$.

Equations (2.7.3) and (7.1.1) show that the partial derivatives of the harmonic functions ϕ and ψ are related by the Cauchy–Riemann equations

$$u = \frac{\partial\phi}{\partial x} = \frac{\partial\psi}{\partial y}, \qquad v = \frac{\partial\phi}{\partial y} = -\frac{\partial\psi}{\partial x} \tag{7.8.1}$$

and this suggests that ϕ and ψ constitute a pair of *conjugate harmonic functions* and may thus be considered as the real and imaginary parts of an analytic function $w(z)$ of the complex variable $z = x + iy$,

$$w(z) = \phi + i\psi \tag{7.8.2}$$

called the *complex potential.* The reader is reminded that a function of the complex variable z is called analytic at a certain point z_0, if its first derivative with respect to z is finite and independent of the direction of differentiation at every point in a certain neighborhood of z_0. An *entire function* is analytic at each point in the complex plane.

Differentiating Eq. (7.8.2) and using Eqs. (7.8.1), we obtain the two components of the velocity $u = u_x$ and $v = u_y$, corresponding to the x and y axes, in the complex form

$$\frac{dw}{dz} = u - iv \tag{7.8.3}$$

Since the velocity field is assumed to be a continuous function, the complex potential must be an analytic function of z everywhere in a flow with the possible exception of isolated singular regions, lines, or points.

Making different selections for the analytic function $w(z)$ allows us to construct various families of incompressible irrotational flows. In Table 7.8.1 we present a number of already familiar examples, including flows due to isolated or periodic arrangements of singularities; an asterisk designates the complex conjugate.

The streamline pattern due to *a point-source dipole* and a *periodic array of point vortices* are shown in Figures 2.1.3(b) and 2.10.1. The streamline pattern of the flow due to a *periodic array of point sources* and the streamline pattern of potential *flow around a corner* are shown in Figure 7.8.1. The flow due to a point source located between two parallel plane walls and the associated Green and Neumann functions are discussed in Problem 7.8.4.

Computation of the Complex Potential from the Potential Function or Stream Function

The theory of conjugate harmonic functions provides us with a method of computing the stream function from the potential function and vice versa, thereby constructing the complex potential from a knowledge of one of its harmonic components. The procedure involves integrating the Cauchy–Riemann equations along a certain path in the complex plane connecting two points, and the result is

$$\psi(x, y) = \psi(x_0, y_0) + \int_{y_0}^{y} \frac{\partial\phi}{\partial x}(x_0, y')\,dy' - \int_{x_0}^{x} \frac{\partial\phi}{\partial y}(x', y_0)\,dx' \tag{7.8.4}$$

$$\phi(x, y) = \phi(x_0, y_0) - \int_{y_0}^{y} \frac{\partial\psi}{\partial x}(x_0, y')\,dy' + \int_{x_0}^{x} \frac{\partial\psi}{\partial y}(x', y_0)\,dx' \tag{7.8.5}$$

where (x_0, y_0) is an arbitrary point where the potential function and stream function may be assigned arbitrary values (Dettman, 1965).

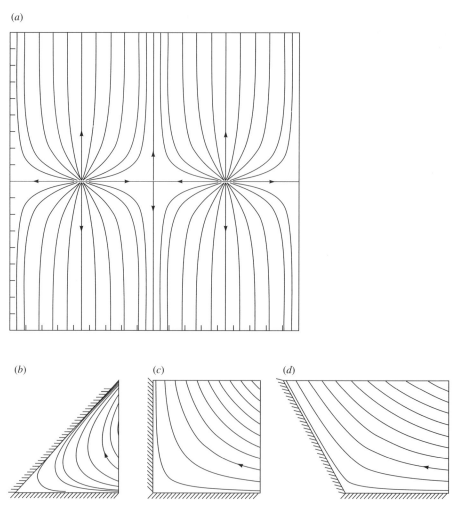

Figure 7.8.1 Streamline patterns of (a) flow due to a periodic array of point sources, and flow in a corner of aperture (b) $\pi/4$, (c) $\pi/2$, (d) $3\pi/4$. (*continued*)

Far-Field Expansion of the Complex Potential of an Infinite Flow

Insights into the structure of an infinite flow past a two-dimensional body can be obtained by studying the behavior of the complex potential at large distances from the body. First, we ask what the most general form of the complex potential is corresponding to a velocity field that either vanishes or tends to a constant value V far from the body. The answer becomes evident by noting that dw/dz is an analytic function, and it may thus be expressed in terms of a Laurent series about a certain point z_0; the series is expected to converge outside a circle of a sufficiently large radius. The series begins with the constant term and contains only negative powers of $z - z_0$. Integrating the Laurent expansion and setting, for simiplicity, the integration constant equal to zero, we find

$$w(z) = V^*(z - z_0) + \frac{m - i\kappa}{2\pi} \ln(z - z_0) + b_0 + \sum_{n=1}^{\infty} \frac{b_n}{(z - z_0)^n} \qquad (7.8.6)$$

where b_n are constant complex coefficients. Reference to Table 7.8.1 shows that second term on

(e)

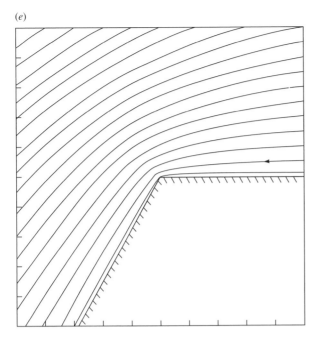

Figure 7.8.1 (*continued*)   Streamline patterns of flow in a corner of aperture (e) $5\pi/4$.

the right-hand side of Eq. (7.8.6) represents the flow due to a point source and a point vortex; m is the flow rate across a closed contour that encloses the body, and κ is the cyclic constant of the flow around the body.

As an example, for flow due to the isotropic expansion of a circular bubble centered at z_0, all constants in Eq. (7.8.6) but m are equal to zero.

Blasius Theorems

Blasius (1910) developed an elegant method of computing the force and torque exerted on a body that is held stationary in an ambient steady flow. As a first step, we introduce the complex modified force $F = F_x + iF_y$ exerted on the body, and use Bernoulli's equation to express it in the form

$$F^* = -i\oint_{\text{Body}} P\,dz^* = \tfrac{1}{2}i\rho\oint_{\text{Body}} \left(\frac{dw}{dz}\right)^2 dz \qquad (7.8.7)$$

where P is the modified pressure, the path of integration is taken in the counterclockwise direction, and an asterisk signifies the complex conjugate. In deriving the right-hand side of Eq. (7.8.7) we made use of the fact that the stream function over the contour of the body is constant. Working similarly with the magnitude of the torque T with respect to the origin, we obtain

$$T = \oint_{\text{Body}} P(x\,dx + y\,dy) = -\tfrac{1}{2}\rho\,\text{Re}\left[\oint_{\text{Body}} \left(\frac{dw}{dz}\right)^2 z\,dz\right] \qquad (7.8.8)$$

Assuming that the functions within the integrals on the right-hand sides of Eqs. (7.8.7) and (7.8.8) are analytic throughout the domain of flow, we use Cauchy's integral theorem to replace the integrals over the contour of the body with corresponding integrals over a closed contour that encloses the body. Introducing, in particular, the Laurent series of the complex potential allows us to compute the integrals in terms of the coefficients of the far-field expansion.

As an application, let us consider uniform flow $\mathbf{V}$ past a body. Substituting the Laurent series (7.8.6) with $m = 0$ into Eq. (7.8.7), we obtain

$$
\begin{aligned}
F^* &= \tfrac{1}{2} i \rho \oint_{\text{Body}} \left(V^* + \frac{\kappa}{2\pi i} \frac{1}{z - z_0} - \sum_{n=1}^{\infty} \frac{n b_n}{(z - z_0)^{n+1}} \right)^2 dz \\
&= \tfrac{1}{2} i \rho \oint_{\text{Body}} \left(V^{*2} + \frac{\kappa V^*}{\pi i} \frac{1}{z - z_0} + \cdots \right) dz
\end{aligned}
\tag{7.8.9}
$$

Evaluating the last integral by the method of residues yields

$$
F_x = \rho \kappa V_y, \qquad F_y = -\rho \kappa V_x
\tag{7.8.10}
$$

which shows that the presence of a finite circulation around a body combined with a uniform incident flow, generates a lift force that is independent of the shape of the body, in agreement with our earlier results in Section 7.6.

The force on a body that translates with velocity $\mathbf{U}$ follows from Eq. (7.8.10) by setting $\mathbf{V} = -\mathbf{U}$, and this yields

$$
F_x = -\rho \kappa U_y, \qquad F_y = \rho \kappa U_x
\tag{7.8.11}
$$

which is also in agreement with our previous results obtained in Section 7.6.

Working in a similar manner we find that the modified torque with respect to the point z_0 is given by

$$
T = -2\pi \rho \operatorname{Im}(V^* b_1)
\tag{7.8.12}
$$

The value of the complex constant b_1 depends on the shape of the body, and this makes the result for the torque less general than that for the force.

Flow Due to the Translation of a Circular Cylinder

Combining the first of Eqs. (7.7.1) with Eq. (7.7.2) and using Table 7.8.1, we find that the complex potential due to a circular cylinder of radius a that translates along the x axis with velocity U_x and along the y axis with velocity U_y, with circulation κ around it, is given by

$$
w = -U \frac{a^2}{z - z_c} + \frac{\kappa}{2\pi i} \ln \frac{z - z_c}{a}
\tag{7.8.13}
$$

where z_c is the instantaneous center of the cylinder, and $U = U_x + iU_y$. The first term on the right-hand side of Eq. (7.8.13) represents a source dipole, and the second term represents a point vortex.

Uniform Flow past a Circular Cylinder

Combining the first of Eqs. (7.7.5) with Eq. (7.7.6) and using Table 7.8.1, we find that the complex potential corresponding to uniform flow past a stationary circular cylinder is given by

$$
w = V^* z + V \frac{a^2}{z - z_c} + \frac{\kappa}{2\pi i} \ln \frac{z - z_c}{a}
\tag{7.8.14}
$$

where $V = V_x + iV_y$ is the velocity far from the cylinder. The three terms on the right-hand side of Eq. (7.8.14) represent, respectively, uniform flow, flow due to a potential dipole, and flow due to a point vortex. The solution (7.8.14) will be used in Section 7.9 to derive the complex potential of flow past bodies with a variety of cross-sectional shapes using the method of conformal mapping.

Arbitrary Flow past a Circular Cylinder

Consider next an arbitrary incident flow past a stationary circular cylinder of radius a whose center is located at the point z_c, described by the complex potential w^∞, which is expressed as a function of $z - z_c$, subject to the constraint that $w^\infty(z)$ has no singularities in the interior of the cylinder. The *circle theorem* due to Milne–Thomson (1940) provides us with an expression for the disturbance complex potential due to the cylinder in the form

$$w^D(z - z_c) = \hat{w}^\infty\left(\frac{a^2}{z - z_c}\right) \tag{7.8.15}$$

$\hat{w}^\infty$ is the complex conjugate of the function w^∞, defined as $\hat{w}^\infty(z) = w^{\infty*}(z^*)$.

Uniform flow

As an example, let us consider uniform incident flow past a circular cylinder with vanishing circulation around it, for which $w^\infty = V^* z$. The restriction of vanishing circulation results from the constraint that there are no singularities inside the cylinder. To apply the circle theorem, we write $w^\infty = V^*(z - z_c) + V^* z_c$ and use Eq. (7.8.15) to obtain the first two terms on the right-hand side of Eq. (7.8.14) added to the constant $V z_c^*$. The latter plays no role on the velocity field and may thus be disregarded with no consequences on the structure of the flow.

Flow due to a point source

As a further application, we consider an incident flow due to a point source located at the point z_s exterior to a cylinder, for which $w^\infty = (m/2\pi)\ln(z - z_s)$. To apply the circle theorem, we write $w^\infty = (m/2\pi)\ln[z - z_c - (z_s - z_c)]$, and use Eq. (7.8.15) to find that the complex potential in the presence of the cylinder is

$$w = \frac{m}{2\pi}\ln(z - z_s) + \frac{m}{2\pi}\ln\left(\frac{a^2}{z - z_c} - (z_s^* - z_c^*)\right)$$

$$= \frac{m}{2\pi}\ln\left[\frac{z - z_s}{z - z_c}\left(z_c + \frac{a^2}{z_s^* - z_c^*} - z\right)\right] + \frac{m}{2\pi}\ln(z_s^* - z_c^*) \tag{7.8.16}$$

The disturbance flow due to the cylinder is thus represented in terms of a point sink with strength equal to $-m$ located at the center of the cylinder, and another point source with strength equal to m located at the inverse point of the primary point source with respect to the cylinder, at the point $z_c + a^2/(z_s^* - z_c^*)$. The last term on the right-hand side of Eq. (7.8.16) is constant and may be discarded.

It is worth noting that the harmonic potential corresponding to Eq. (7.8.16) with $m = -1$ is the two-dimensional Green's function of the second kind or the Neumann function, $N(z, z_0)$.

Flow due to a point vortex

As a third example, we consider the flow due to a point vortex that is located at the point z_0 in the exterior of a cylinder, for which $w^\infty = (\kappa/2\pi i)\ln(z - z_0)$ or $w^\infty = (\kappa/2\pi i)\ln[z - z_c - (z_0 - z_c)]$. The complex potential in the presence of the cylinder is

$$w = \frac{\kappa}{2\pi i}\ln(z - z_0) - \frac{\kappa}{2\pi i}\ln\left(\frac{a^2}{z - z_c} - (z_0^* - z_c^*)\right)$$

$$= \frac{\kappa}{2\pi i}\ln\left(\frac{(z - z_0)(z - z_c)}{a^2 - (z - z_c)(z_0^* - z_c^*)}\right) \tag{7.8.17}$$

which reveals that the disturbance flow due to the cylinder may be represented in terms of a point vortex with strength equal to κ placed at the center of the cylinder, and another point vortex with strength equal to $-\kappa$ located at the image point of the primary point vortex with respect to the cylinder, at the point $z_c + a^2/(z_0^* - z_c^*)$. The velocity induced by the first point vortex is tangential to the cylinder and may thus be disregarded, yielding the simplified result

$$w = \frac{\kappa}{2\pi i} \ln\left(\frac{z - z_0}{z - z_c - a^2/(z_0^* - z_c^*)}\right) \tag{7.8.18}$$

Considering the exterior point vortex as the image of the interior point vortex allows us to extend the validity of Eq. (7.8.18) to the case of flow due to a point vortex located inside the cylinder.

PROBLEMS

7.8.1   **Linear irrotational flow past a circular cylinder.** Consider a linear flow past a circular cylinder $\mathbf{u}^\infty = \mathbf{A} \cdot \mathbf{x}$, where $\mathbf{A}$ is a constant symmetric matrix with vanishing trace. Derive the corresponding complex potential.

7.8.2   **Flow due to a point source.** Explain, in physical terms, why the complex potential due to a point source located inside a cylinder or within any other closed surface does not exist. What will happen if we allow for a small perforation on the contour of the cylinder?

7.8.3   **Flow due to a point-source dipole in the presence of a cylinder.** Derive the complex potential due to a point-source dipole located (a) outside and (b) inside a circular cylinder.

7.8.4   **Flow due to a point source between two parallel plates.** (a) The velocity field due to a point source placed between two parallel plane walls separated by distance h may be represented in terms of two infinite periodic arrays of point sources running perpendicular to the walls with wave length $2h$. The first array contains the point source, and the second array contains its image with respect to either the upper or lower wall. The strengths of the point sources in both arrays are equal. Show that if the walls are parallel to the x axis, the corresponding potential is given by

$$\phi = \frac{m}{4\pi} \ln(\{\cosh[k(x - x_0)] - \cos[k(y - y_0)]\}$$
$$\times \{\cosh[k(x - x_0)] - \cos[k(y - y_0^{\mathrm{Im}})]\}) \tag{7.8.19}$$

where $k = \pi/h$ and y_0^{Im} is the image of the point source with respect to either the upper or the lower wall. Derive the associated Neumann function. (b) The Green's function G in a domain bounded by two parallel plane walls may be represented in terms of the two infinite arrays discussed in (a), but now the two arrays have opposite strengths. Show that

$$G(\mathbf{x}, \mathbf{x}_0) = -\frac{1}{4\pi} \ln\left(\frac{\cosh[k(x - x_0)] - \cos[k(y - y_0)]}{\cosh[k(x - x_0)] - \cos[k(y - y_0^{\mathrm{Im}})]}\right) \tag{7.8.20}$$

7.9 | CONFORMAL MAPPING

Conformal mapping allows us to compute potential flows in two-dimensional domains with complex geometries from a knowledge of elementary flows in domains with simpler geometries.

We begin developing the method by introducing the complex variable $\zeta = \xi + i\eta$, where ξ and η are two real variables, and the complex function $F(z)$ that is analytic in a certain region of the complex z plane, and maps a point in the z plane to another point in the ζ plane, so that

$$\zeta = F(z) \qquad (7.9.1)$$

When the function $F(z)$ is multivalued, we introduce an appropriate branch cut in the z plane so as to render the mapping unique. The image of an open line or closed loop in the z plane is another open line or closed loop in the ζ plane with generally different orientation and shape.

Furthermore, we introduce the inverse mapping function that maps a point in the ζ plane back to a point in the z plane

$$z = f(\zeta) \qquad (7.9.2)$$

When the function $f(\zeta)$ is multivalued, we introduce an appropriate branch cut in the ζ plane so as to render the inverse mapping unique.

Differentiating $F(z)$ and $f(\zeta)$ with respect to their arguments, we find that the ratio of two corresponding infinitesimal differential vectors dz and $d\zeta$ in the z and ζ planes are related by

$$\frac{d\zeta}{dz} = F'(z) = |F'(z)| \exp\{i \arg[F'(z)]\}$$

$$\frac{dz}{d\zeta} = f'(\zeta) = |f'(\zeta)| \exp\{i \arg[f'(\zeta)]\} \qquad (7.9.3)$$

Since the functions $F(z)$ and $f(z)$ have been assumed to be analytic, the right-hand sides of Eqs. (7.9.3) are independent of the orientation of dz or $d\zeta$.

The first of Eqs. (7.9.3) states that the length of the differential vector $d\zeta$ is equal to the length of dz multiplied by the scalar factor $|F'(z)|$, and the direction of $d\zeta$ is rotated with respect to that of dz by an angle that is equal to the argument of $F'(z)$. A singular behavior occurs at the points where $F'(z)$ assumes an infinite value or, equivalently, $f'(\zeta)$ vanishes, and these are the *critical points* of the mapping (7.9.1). A similar interpretation of the differentials pertains to the second of Eqs. (7.9.3).

Let us select a point z_0 in the z plane and draw two infinitesimal vectors dz_1 and dz_2 that start at z_0, as shown in Figure 7.9.1. The images of these vectors are the two corresponding vectors $d\zeta_1$ and $d\zeta_2$ in the ζ plane that start at the point $\zeta_0 = F(z_0)$. Using either the first or the second of Eqs. (7.9.3), we find that the angle subtended by the segments dz_1 and dz_2 is the same as the

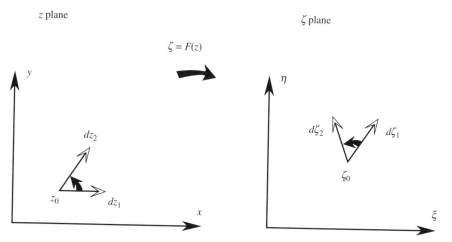

Figure 7.9.1   Conformal mapping of the z plane to the ζ plane using the mapping function $\zeta = F(z)$. The angle subtended between two infinitesimal vectors is preserved.

angle subtended by the segments $d\zeta_1$ and $d\zeta_2$, except if z_0 happens to be a critical point. The equality of these angles may be traced back to the analyticity of the complex function $F(z)$. As a consequence, the image of a tiny loop having the shape of a little horse in the z plane will look like a tiny little horse in the ζ plane, possibly rotated and amplified or shrunk, but definitely looking like a horse. This property justifies calling the mapping (7.9.1) and its inverse *conformal*.

Elementary Conformal Mappings

We shall present now certain elementary conformal mappings that find frequent usage in the theory of potential flow. Extensive discussions can be found in standard texts on complex analysis, including that by Ahlfors (1979), and handbooks of mathematical functions (Abramowitz and Stegun, 1972).

- The function $\zeta = F(z) = z + b$ shifts every point by the complex number b.
- The linear function $\zeta = F(z) = az$ multiplies the distance from the origin $|z|$ by the scalar factor $|a|$, while rotating the line that connects the origin to z by an angle that is equal to the argument of a. Circles in the z plane remain circles in the ζ plane.
- The exponential function $\zeta = F(z) = a\exp(bz)$ where a and b are two real positive constants maps the semi-infinite strip $0 < x < \infty, 0 < y < 2\pi/b$ to the exterior of a circle with radius a centered at the origin. The inverse function is $z = f(\zeta) = (1/b)\ln(\zeta/a)$.
- The logarithmic function $\zeta = F(z) = a\ln z$ where a is a real positive constant maps the whole z plane onto the infinite strip $-\infty < \xi < \infty, 0 < \eta < 2\pi a$. The inverse function is $z = \exp(J/a)$.
- The Möbius transformation and its inverse

$$\zeta = F(z) = \frac{Az + B}{Cz + D}, \qquad z = f(\zeta) = \frac{D\zeta - B}{-C\zeta + A} \tag{7.9.4}$$

where A, B, C, D are four complex constants with $AD - BC \neq 0$, map the whole complex plane onto itself. Circles and straight lines in the z plane are mapped to circles or straight lines in the ζ plane and vice versa.

Flow in the ζ Plane

The complex potential w is a function of z, but since to every value of z there is a corresponding value of ζ according to Eq. (7.9.1), we may also regard w a function of ζ. Combining Eqs. (7.8.2) and (7.9.2), we write

$$w(z) = w[f(\zeta)] = W(\zeta) = \Phi(\xi, \eta) + i\Psi(\xi, \eta) \tag{7.9.5}$$

where Φ and Ψ are two real functions. Since an analytic function of another analytic function is also analytic, the functions Φ and Ψ are harmonic with respect to their arguments; that is, they satisfy Laplace's equations

$$\frac{\partial^2 \Phi}{\partial \xi^2} + \frac{\partial^2 \Phi}{\partial \eta^2} = 0, \qquad \frac{\partial^2 \Psi}{\partial \xi^2} + \frac{\partial^2 \Psi}{\partial \eta^2} = 0 \tag{7.9.6}$$

Moreover, Φ and Ψ constitute a pair of conjugate harmonic functions. These properties allow us to identify Φ with the harmonic potential, and Ψ with the stream function of a certain flow in the ζ plane, thus setting

$$\frac{dW}{d\zeta} = U_\xi - iU_\eta \tag{7.9.7}$$

where $\mathbf{U}$ designates the velocity in the ζ plane. To find the relation between the magnitude and direction of the velocity $\mathbf{u}$ in the z plane and those of the corresponding velocity $\mathbf{U}$ in the ζ plane, we write

$$\frac{dW}{d\zeta} = \frac{dw}{dz}\frac{dz}{d\zeta} = \frac{dw}{dz}f'(\zeta) \tag{7.9.8}$$

which shows that the ratio between the two complex velocities is given by

$$\frac{U_\xi - iU_\eta}{u - iv} = f'(\zeta) = |f'(\zeta)|\exp\{i\,\arg[f'(\zeta)]\} \tag{7.9.9}$$

Equation (7.9.9) allows us to compute **u** from **U** and vice versa in terms of the mapping function f. Comparing Eq. (7.9.9) to the second of Eqs. (7.9.3) shows that differential vectors and velocities are amplified in inverse proportion, and this means that the flow rate across corresponding differential line elements remains the same in both planes.

Flow due to singularities

It is instructive to investigate the nature of the flow in the ζ plane corresponding to flow due to singularities in the z plane, including the point source and the point vortex. This can be done by expressing the harmonic potential W in a Laurent series about the point ζ_0, which is the image of the location of the singularity z_0.

Considering first the flow due to a point source in the z plane, we expand $f(\zeta)$ in a Taylor series about ζ_0, finding

$$W(\zeta) = w[f(\zeta)] = w(z) = \frac{m}{2\pi}\ln(z - z_0) = \frac{m}{2\pi}\ln[f(\zeta) - f(\zeta_0)]$$

$$\cong \frac{m}{2\pi}\ln[f'(\zeta_0)(\zeta - \zeta_0)] \cong \frac{m}{2\pi}\ln(\zeta - \zeta_0) \tag{7.9.10}$$

which shows that the flow in the ζ plane contains a point source with identical strength placed at ζ_0. An exception occurs when the point z_0 happens to be a critical point of the conformal mapping. Replacing m with $-i\kappa$, we obtain a corresponding result for the point vortex.

For the point-source dipole we find

$$W(\zeta) \cong -\frac{1}{2\pi}\frac{d_x + id_y}{f'(\zeta_0)}\frac{1}{\zeta - \zeta_0} \tag{7.9.11}$$

which shows that the flow in the ζ plane contains a point-source dipole with modified strength and orientation. Replacing **d** with $-i\boldsymbol{\lambda}$, we obtain a corresponding result for the point-vortex dipole.

Transformation of Boundary Conditions

Consider next the boundary of a flow in the z plane and the corresponding boundary of the flow in the ζ plane computed on the basis of the conformal mapping (7.9.1). It is clear from Eq. (7.9.5) that the values of $w(z)$ and $W(\zeta)$ at two corresponding points will be identical, and this shows that Dirichlet boundary conditions that specify the real or imaginary parts of $w(z)$ or $W(\zeta)$ are preserved.

The Neumann boundary conditions, however, undergo a quantitative change but maintain their character. This becomes evident by writing

$$\nabla\phi \cdot \mathbf{n} = \frac{\partial\phi}{\partial x}\frac{dy}{dl} - \frac{\partial\phi}{\partial y}\frac{dx}{dl} = \frac{\partial\phi}{\partial\xi}\left(\frac{\partial\xi}{\partial x}\frac{dy}{dl} - \frac{\partial\xi}{\partial y}\frac{dx}{dl}\right) + \frac{\partial\phi}{\partial\eta}\left(\frac{\partial\eta}{\partial x}\frac{dy}{dl} - \frac{\partial\eta}{\partial y}\frac{dx}{dl}\right)$$

$$= \frac{\partial\phi}{\partial\xi}\frac{d\eta}{dl} - \frac{\partial\phi}{\partial\eta}\frac{d\xi}{dl} = \frac{dL}{dl}\nabla\Phi \cdot \mathbf{N} = |F'(z)|\,\nabla\Phi \cdot \mathbf{N} \tag{7.9.12}$$

where **n** and **N** are the unit normal vectors in the z and ζ planes pointing into the flow, the corresponding arc lengths l and L are measured in the counterclockwise direction as shown in

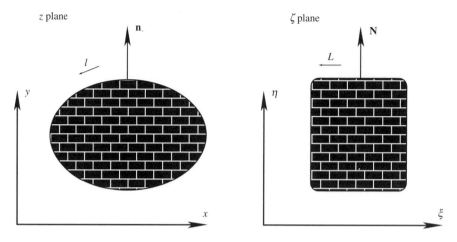

Figure 7.9.2   Mapping of a flow in the z plane to a corresponding flow in the ζ plane.

Figure 7.9.2, and we have made use of the Cauchy–Riemann equations for the function $F(z)$

$$\frac{\partial \xi}{\partial x} = \frac{\partial \eta}{\partial y}, \qquad \frac{\partial \xi}{\partial y} = -\frac{\partial \eta}{\partial x} \qquad (7.9.13)$$

Equation (7.9.13) shows that impermeable boundaries in the z plane remain so in the image plane.

Flow past a Circular Cylinder Subject to the Linear Fractional Transformation

As an application, let us consider uniform flow in the z plane past a circular cylinder of radius a centered at the origin with finite circulation around it. The corresponding complex potential is given by Eq. (7.8.14) with $z_c = 0$.

The linear fractional transformation

$$\zeta = F(z) = ib\frac{z+a}{z-a}, \qquad z = f(\zeta) = a\frac{\zeta+ib}{\zeta-ib} \qquad (7.9.14)$$

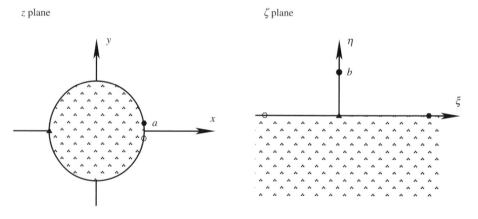

Figure 7.9.3   Mapping of the exterior of a circle to the upper half-plane using the linear fractional transformation shown in Eqs. (7.9.14). Corresponding points are shown with the same symbols. Uniform flow past a cylinder with finite circulation around it transforms to flow due to a point vortex and a point-source dipole supplemented by their images.

where a and b are real positive constants, maps the exterior or interior of a circle in the z plane to the upper or lower half-plane in the ζ plane as shown in Figure 7.9.3. Infinity is mapped to the singular point $\zeta = ib$, and the center of the circle is mapped to the point $\zeta = -ib$. Substituting the inverse transformation given by the second of Eqs. (7.9.14) into Eq. (7.8.14) with $z_c = 0$ yields the complex potential of the flow in the ζ plane

$$W(\zeta) = V^* a \frac{\zeta + ib}{\zeta - ib} + V a \frac{\zeta - ib}{\zeta + ib} + \frac{\kappa}{2\pi i} \ln \frac{\zeta + ib}{\zeta - ib} \tag{7.9.15}$$

After some straightforward manipulations and discarding a meaningless constant, we obtain the equivalent form

$$W(\zeta) = 2iab \left(\frac{V^*}{\zeta - ib} - \frac{V}{\zeta + ib} \right) + \frac{\kappa}{2\pi i} \ln \frac{\zeta + ib}{\zeta - ib} \tag{7.9.16}$$

Reference to Table 7.8.1 shows that the flow in the ζ plane is represented by a point-source dipole with strength $d = -4\pi i V^* ab$ and a point vortex with strength $-\kappa$ both placed at the singular point $\zeta = ib$, both accompanied by their images with respect to the wall. The presence of the images is necessary in order to satisfy the no-penetration condition at the wall.

Use of Conformal Mapping

To put the conformal mapping theory into practice, we consider a certain flow in the ζ plane, and then study the corresponding flow in the z plane subject to a proper mapping function $f(\zeta)$, and vice versa. In a typical application, however, we are faced with the problem of finding a mapping function that transforms the physical domain of flow in the z plane to a domain of flow with a simpler geometry in the ζ plane, such as the semi-infinite plane or the interior or exterior of a disk, so that the flow in the image plane may be computed readily, preferably in closed form. The issue of the existence of the mapping function $\zeta = F(z)$ thus arises in a natural manner.

Riemann's Mapping Theorem

Riemann's mapping theorem guarantees that any two singly connected domains, with the exception of the whole plane, may be mapped conformally onto each other. This is a consequence of the fact that, with those exceptions, any singly connected domain D in the z plane may be mapped onto a disk of unit radius in the ζ plane centered at the origin, which establishes an indirect correspondence between the two regions in terms of the individual conformal mappings.

There is a three-parameter family of functions $\zeta = F(z)$ that map a singly connected region D to another region D'. To make the transformation unique, we must specify one of the following three sets of conditions, where the definition of a point in the complex plane is generalized to include infinity:

1. Stipulate that an arbitrary point z_0 within D is mapped to $\zeta = 0$, and specify the direction of an infinitesimal vector that starts at z_0 and whose image lies on the ξ axis.
2. Stipulate that an arbitrary point z_0 within D is mapped to $\zeta = 0$ and that an arbitrary point z_1 at the boundary of D is mapped to an arbitrary point ζ_1 at the boundary of D'.
3. Stipulate that a set of three arbitrary points z_0, z_1, z_2 at the boundary of D are mapped to three arbitrary corresponding points $\zeta_0, \zeta_1, \zeta_2$ along the boundary of D' arranged in the same order.

PROBLEMS

7.9.1 **Conformal mapping.** Discuss the properties of the conformal mapping $\zeta = F(z) = az^n$, where a and n are, respectively, a complex and a positive real constant.

7.9.2   **Properties of conjugate flows.** Consider an open contour in the z plane and the corresponding contour in the ζ plane. Show that the flow rate and circulation along these contours corresponding to the complex potentials $w(z)$ and $W(\zeta)$ are identical.

7.9.3   **A point vortex inside or outside a cylindrical surface.** (a) The complex potential in the ζ plane due to a point vortex located at the point ζ_0 above a plane wall that is represented by the ξ axis, is composed of the sum of the complex potential due to the point vortex and the complex potential due to its image with respect to the wall, yielding

$$W(\zeta) = \frac{\kappa}{2\pi i} \ln \frac{\zeta - \zeta_0}{\zeta - \zeta_0^*} \tag{7.9.17}$$

Using Table 7.8.1 and the conformal mapping described in Eqs. (7.9.14), show that the complex potential for a point vortex located inside or outside a circular cylinder of radius a is given by Eq. (7.8.18). (b) Let $\zeta = F(z)$ map the interior or exterior of a simply connected domain D to the interior or exterior of a disk of radius a centered at the origin. Show that the complex potential associated with a point vortex is given by

$$w = \frac{\kappa}{2\pi i} \ln \left(\frac{F(z) - F(z_0)}{a^2 - F(z)F(z_0^*)} \right) \tag{7.9.18}$$

7.9.4   **Green's function in the presence of a cylinder with arbitrary cross-section.** Let $\zeta = F(z)$ map the interior or exterior of a simply connected domain D to the interior or exterior of a disk of radius a centered at the origin. Show that the complex potential corresponding to the Green's function of the first kind G is given by

$$w = \frac{1}{2\pi} \ln \left(\frac{a^2 - F(z)F(z_0^*)}{F(z) - F(z_0)} \right) \tag{7.9.19}$$

where $G = \mathrm{Re}(W)$. Verify that when z is on D, $G = 0$.

7.10   APPLICATIONS OF CONFORMAL MAPPING TO FLOW PAST TWO-DIMENSIONAL BODIES

To compute the velocity field corresponding to uniform flow past a two-dimensional body with an arbitrary cross-section, we map the exterior of the body in the z plane to the exterior of a disk of radius c centered at the origin in the ζ plane, and then recover the flow in the physical plane from the flow in the image plane using the exact solution (7.8.14), which in this case becomes

$$W(\zeta) = V^*\zeta + V\frac{c^2}{\zeta} + \frac{\kappa}{2\pi i} \ln \frac{\zeta}{c} \tag{7.10.1}$$

To ensure that the far flows in the two planes behave in a similar manner, so that uniform flow in the ζ plane is also uniform flow in the z plane far from the body, we require that the mapping function $\zeta = F(z)$ and its inverse $z = f(\zeta)$ behave in a linear manner far from the body as $|z|$ tends to infinity, so that their first derivatives tend to a constant.

Flow past an Elliptical Cylinder

As a first application, let us consider uniform flow past a cylinder with an elliptical cross-section. One may readily verify that the inverse mapping function

$$z = f(\zeta) = \zeta + \frac{e^2}{4} \frac{a^2}{\zeta} \tag{7.10.2}$$

where $e = [1 - (b/a)^2]^{1/2}$ is the eccentricity, maps the exterior of a disk with radius $c = \frac{1}{2}(a+b)$ centered at the origin to the exterior of an ellipse with major and minor semiaxes equal to a and *b* also centered at the origin. Furthermore, *f* exhibits the required linear behavior at infinity and is thus acceptable for the study of uniform flow.

Decomposing Eq. (7.10.2) into its real and imaginary parts, we obtain the explicit coordinate transformations

$$x = \xi\left(1 + \frac{e^2}{4}\frac{a^2}{|\zeta|^2}\right), \qquad y = \eta\left(1 - \frac{e^2}{4}\frac{a^2}{|\zeta|^2}\right) \tag{7.10.3}$$

The inverse transformations are found by solving the quadratic equation (7.10.2) for ζ. Since the root with the negative sign corresponds to a point inside the ellipse, we maintain the root with positive sign and obtain

$$\zeta = F(z) = \tfrac{1}{2}[z + (z^2 - a^2 + b^2)^{1/2}] \tag{7.10.4}$$

The value of the square root on the right-hand side of Eq. (7.10.4) becomes unique by introducing a branch cut along the *x* axis extending from $-ae$ to ae.

The complex potential of the flow in the *z* plane is found readily by substituting Eq. (7.10.4) into Eq. (7.10.1) and setting $c = \tfrac{1}{2}(a + b)$, yielding

$$w(z) = V^*\tfrac{1}{2}(z + \sqrt{z^2 - a^2 + b^2}) + V\tfrac{1}{2}\frac{(a + b)^2}{z + \sqrt{z^2 - a^2 + b^2}}$$

$$+ \frac{\kappa}{2\pi i}\ln\frac{z + \sqrt{z^2 - a^2 + b^2}}{a + b} \tag{7.10.5}$$

Setting $a = b$ produces the solution for flow past a circular cylinder.

Flow past a Flat Plate

Letting b/a tend to zero, in which case *e* tends to unity, reduces the ellipse to a flat plate of length equal to $2a$. The transformation (7.10.2) becomes

$$z = f(\zeta) = \zeta + \tfrac{1}{4}\frac{a^2}{\zeta} \tag{7.10.6}$$

which maps the exterior of a disk of radius $c = a/2$ centered at the origin to the whole complex plane; the contour of the disk is mapped to the flat plate. A different method of arriving at Eq. (7.10.6) will be discussed in Problem 7.11.1. The inverse transformation (7.10.4) becomes

$$\zeta = F(z) = \tfrac{1}{2}[z + (z^2 - a^2)^{1/2}] \tag{7.10.7}$$

The branch of the square root coincides with the length of the plate.

Substituting Eq. (7.10.7) into Eq. (7.10.1) and setting $c = a/2$, or applying Eq. (7.10.5) with $b = 0$, yields the complex potential of the flow in the *z* plane

$$w(z) = V^*\tfrac{1}{2}[z + (z^2 - a^2)^{1/2}] + V\tfrac{1}{2}\frac{a^2}{z + (z^2 - a^2)^{1/2}} + \frac{\kappa}{2\pi i}\ln\left(\frac{z + (z^2 - a^2)^{1/2}}{a}\right) \tag{7.10.8}$$

which may be simplified to

$$w(z) = V_x z - iV_y(z^2 - a^2)^{1/2} + \frac{\kappa}{2\pi i}\ln\left(\frac{z + (z^2 - a^2)^{1/2}}{a}\right) \tag{7.10.9}$$

The velocity field is given by

$$u - iv = \frac{dw}{dz} = V_x - i\left(V_y z + \frac{\kappa}{2\pi}\right)\frac{1}{(z^2 - a^2)^{1/2}} \tag{7.10.10}$$

The tangential velocities on the upper and lower surface of the plate, designated, respectively, by the plus and minus superscripts, are given by

$$u^{\pm} = V_x \mp \left(V_y x + \frac{\kappa}{2\pi}\right) \frac{1}{(a^2 - x^2)^{1/2}} \qquad (7.10.11)$$

where $-a < x < a$. Note that the velocity diverges at both ends at $x = \pm a$.

The *Kutta–Joukowski condition* requires that, in practice, the circulation established around the plate be such that the velocity becomes finite at the trailing end at $z = -a$, so that the two fluid streams above and below the airfoil merge in a smooth manner (Crighton, 1985). This implies that, in the unsteady startup process, the action of viscosity will be such that, when the final potential flow state is established, viscous effects will be significant only so far as to satisfy the Kutta–Joukowski condition. Observation has shown that the Kutta-Joukowski condition is physically relevant not only for steady, but also for unsteady flow.

Equation (7.10.11) shows that the circulation around a flat plate that is necessary in order to satisfy the Kutta–Joukowski condition has the value $\kappa = 2\pi a V_y$. The two components of the force exerted on the airfoil follow from Eqs. (7.8.10). The streamline patterns for $V_x/V_y = 1, 0$ and circulation parameter $\beta \equiv \kappa/2\pi a V_y = 0, 1$ are illustrated in Figure 7.10.1(a–d). Note that at the Kutta–Joukowski value $\beta = 1$ the streamlines merge smoothly at the trailing edge.

Joukowski's transformation

The study of potential flow past two-dimensional bodies has been historically motivated, to a large extent, by applications in aircraft design. To study the performance of an aircraft, we must have available the structure of the nearly two-dimensional potential around an airfoil with different degrees of circulation around it, including the circulation corresponding to the Kutta–Joukowski value. Knowledge of the potential flow allows us to compute the viscous drag force exerted on the airfoil on the basis of the boundary-layer theory to be discussed in Chapter 8.

The analytical computation of potential flow past an airfoil with an arbitrary shape is generally intractable. It is then not surprising that early work in aerodynamics, before the advent of high-speed computers, has concentrated on families of airfoil shapes that are produced by mapping a circle using carefully crafted transformations. One such family of shapes emerges from the Joukowski transformation

$$z = f(\zeta) = \zeta + \lambda^2/\zeta \qquad (7.10.12)$$

where λ is a real positive constant. It will be noted that Eq. (7.10.12) encompasses Eqs. (7.10.2) and (7.10.6) corresponding to an ellipse and a flat plate. The critical points of the transformation function shown in Eq. (7.10.12), at which $df/d\zeta$ vanishes, are located at $\zeta = \pm\lambda$, corresponding to $z = \pm 2\lambda$. A smooth curve in the ζ plane that passes through the first singular point and encloses the second singular point is mapped to a cusped curve in the z plane representing an airfoil; the cusp is located at the image of the first singular point, as shown in Figure 7.10.2. A circle in the ζ plane centered at the origin and passing through both singular points transforms into a flat plate in the z plane with semilength equal to $a = 2\lambda$, in which case Eq. (7.10.12) reduces to Eq. (7.10.6).

Joukowski's transformation is used to generate airfoils with a rounded leading edge and a cusped trailing edge. The airfoils are the images of circles in the ζ plane that pass through the first singular point $\zeta = -\lambda$ and enclose the second singular point $\zeta = \lambda$, as shown in Figure 7.10.2. When the center of the circle is located on the ξ axis, the airfoil is symmetric about the x axis. When the center of the circle is located in the first quadrant, the airfoil is cambered downward.

Numerical Computation of the Mapping Function

We turn next to consider the more difficult problem of computing a function $f(\zeta)$ that maps the exterior of a circle in the ζ plane to the exterior of a body with a specified shape in the z plane. To this end, we note that the mapping function corresponding to an ellipse shown

(a)

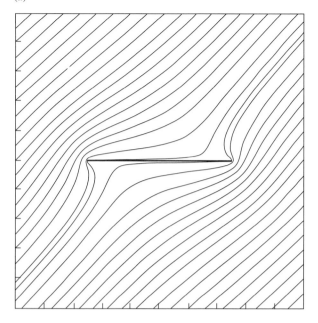

(b)

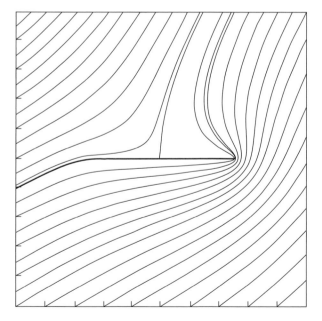

Figure 7.10.1 Uniform flow past a flat plate with (a) incident velocity ratio $V_x/V_y = 1$ and circulation parameter $\beta = \kappa/2\pi a V_y = 0$ and
(b) $V_x/V_y = 1$, $\beta = 1.0$ (Kutta–Joukowski value). (*continued*)

(*c*)

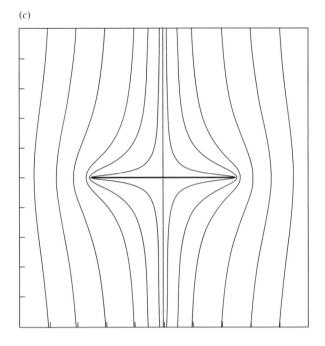

(*d*)

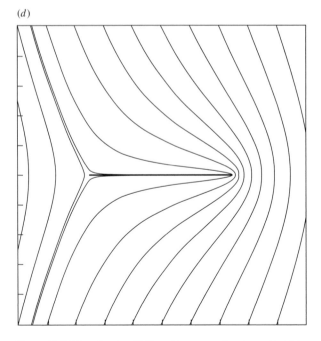

Figure 7.10.1 (*continued*)   Uniform flow past a flat plate with (c) incident velocity ratio $V_x/V_y = 0$ and circulation parameter $\beta = 0$ and
(d) $V_x/V_y = 0, \beta = 1$.

(a)

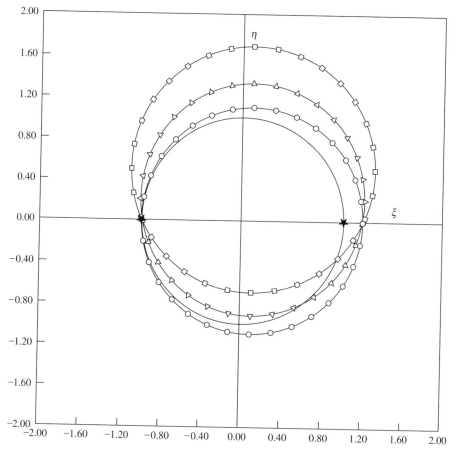

Figure 7.10.2 Mapping with Joukowski's transformation. (a) Circles in the ζ plane. (*continued*)

in Eq. (7.10.2) or to the Joukowski airfoils shown in Eq. (7.10.12) has the form of a truncated Laurent series. For a body with a more general shape, one might expect the full expansion

$$z = f(\zeta) = \zeta + a_0 + \sum_{n=1}^{\infty} \frac{a_n}{\zeta^n} \tag{7.10.13}$$

where a_n are complex coefficients that depend upon the body's shape and orientation. The linear term on the right-hand side satisfies the requirement that $df/d\zeta$ tends to unity as ζ tends to infinity in order to maintain the uniform flow. The constant a_0 determines the position of the body in the z plane and may be set equal to zero. In principle, a finite set of subsequent coefficients may be computed by stipulating that a collection of points on the contour of the body are mapped onto a corresponding collection of points on the circular contour in the ζ plane, but an arbitrary selection of pairs will result in an image body with oscillating shape.

When the body has a polygonal shape, the transformation $f(\zeta)$ that maps the exterior of a circle to the exterior of an arbitrary body is known in closed form in terms of a finite set of real coefficients. The particular form of the mapping function and the computation of the coefficients will be the subject of Section 7.11.

(*b*)

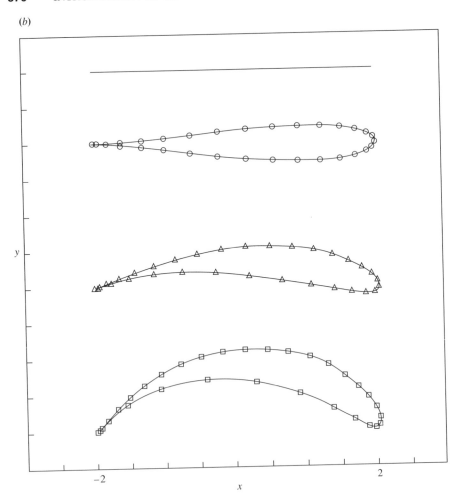

Figure 7.10.2 (*continued*) Mapping with Joukowski's transformation. (b) Corresponding airfoil shapes in the z plane.

Another way of obtaining the mapping function is to map the exterior of the body to the interior of an ancillary body using the inverse transformation $\zeta = 1/z$, where the origin has been placed within the body, and then map the interior of the ancillary body to the upper half-plane or to the interior of a disk. Note, however, that even if the original body may have a polygonal boundary, the ancillary body will have a curved boundary, which precludes the application of specialized methods for polygonal shapes. One way to circumvent this difficulty is to approximate the smooth curve with a polygonal line, and then map the interior of a polygonal domain to the upper half-plane or to a disk. Efficient iterative methods for computing the scalar coefficients involved in the second transformation are available, as will be discussed in Section 7.11.

Two additional methods of computing a function that maps the interior of a body to a disk are: (1) solving an integral equation over the contour of the body, and (2) formulating the problem in terms of a variational principle. These methods are discussed and reviewed by Carrier, Krook, and Pearson (1983, pp. 175–80) and Henrici (1986, Chapter 16).

Computer Problem

7.10.1 **Joukowski airfoils.** (a) Compute and plot a family of airfoils produced by the Joukowski transformation. The airfoils should be the images of circles in the ζ plane that pass through the first
singular point $\zeta = -\lambda$ and enclose the second singular point $\zeta = \lambda$, with the center of the circle located on the ξ axis at $\xi = \lambda$, 1.2λ, 1.5λ, 2.0λ. (b) Repeat (a) with the center of the circle
located at $\xi = \lambda$, 1.2λ, 1.5λ, and 2.0λ, and $\eta = 0.20\lambda$.

7.11 | THE SCHWARZ–CHRISTOFFEL TRANSFORMATION AND ITS APPLICATIONS

When the domain of a flow in the z plane is bounded by a set of finite straight segments, semifinite straight lines, or even infinite straight lines, the function $z = f(\zeta)$ whose inverse maps
the domain of flow onto a circular disk or half-space in the ζ plane is provided by the Schwarz–
Christoffel transformation in terms of an integral that involves a finite set of scalar coefficients.
Computing the transformation is reduced to computing the values of these coefficients, which can
be done using analytical methods for simple geometries and numerical methods for more complex
shapes.

The usefulness of the Schwarz–Christoffel transformation might appear to be limited in view
of its rather strong restrictions on the boundary geometry. We note, however, that a curved boundary may be approximated with a polygonal boundary connecting a set of marker points that are
distributed over the curved boundary; the accuracy of this approximation improves as an increasing number of marker points are used in the representation. Thus being able to map a polygonal
domain with a large number of vertices to a disk or to the half-plane lends itself to approximating
the mapping function for an arbitrary domain.

In this section we shall present the various forms of the Schwarz–Christoffel transformation
illustrating, by means of examples, their analytical or numerical computation.

Mapping the Interior of a Polygon to a Semi-Infinite Plane

First we consider mapping the interior of a closed N-sided polygon in the z plane to the
upper half-ζ-plane, as shown in Figure 7.11.1. We begin by numbering the vertices of the polygon
sequentially in the counterclockwise sense, and compute the exterior angles γ_n that are subtended
by the projections of the previous and next sides, where $-\pi < \gamma_n < \pi$. Note that γ_n is positive
when a corner is projecting into the exterior of the domain, and negative otherwise. For a convex
polygon all γ_n are positive, and for a regular polygon all γ_n are equal. In all cases, the sum of the
angles satisfies the geometrical constraint

$$\sum_{n=1}^{N} \gamma_n = 2\pi \tag{7.11.1}$$

The Schwarz–Christoffel transformation provides us with the inverse mapping function in the
form

$$z = f(\zeta) = z_0 + c \int_{\zeta_0}^{\zeta} \prod_{n=1}^{N} (\zeta' - \xi_n)^{-\gamma_n/\pi} \, d\zeta' \tag{7.11.2}$$

where c is a complex constant, z_0 and ζ_0 are a pair of corresponding points, and ξ_n are the images
of the vertices of the polygon on the ξ axis. The complex powers in Eq. (7.11.2) are computed by

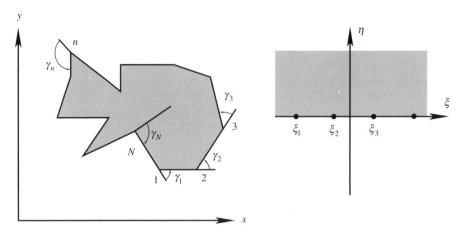

Figure 7.11.1   Mapping of the interior of a polygon in the z plane to the upper half-ζ-plane.

setting

$$(\zeta - \xi_n)^{-\gamma_n/\pi} = \exp\left[-\frac{\gamma_n}{\pi}\ln(\zeta - \xi_n)\right] \tag{7.11.3}$$

where the branch cut of the logarithmic function is the negative real axis.

Given the polygon shape, the problem is reduced to selecting or computing the images of the vertices ξ_n and the values of the three constants c, z_0, and ζ_0. Riemann's mapping theorem, discussed in Section 7.9, allows us to choose the images of three vertices ξ_n in an arbitrary manner, and then compute the rest of them so as to make the upper half-plane fit the polygon. It is permissible to set $\xi_1 = -\infty$ or $\xi_N = \infty$, in which cases the corresponding factors in the product within the integrand in Eq. (7.11.2) are simply omitted.

Mapping the interior of a triangle to the upper half-plane

In the simplest possible application, we map the interior of a triangle with vertices located at the points $z_1 = b + id$, $z_2 = 0$, $z_3 = a$ illustrated in Figure 7.11.2, to the upper half-ζ-plane, where a, b, d are three real positive constants. The Riemann mapping theorem allows us to select

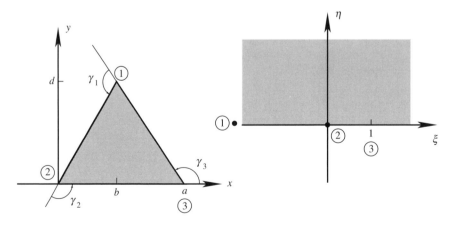

Figure 7.11.2   Mapping of the interior of a triangle in the z plane to the upper half-ζ-plane.

the images of all three vertices in an arbitrary manner, and we choose $\xi_1 = -\infty, \xi_2 = 0, \xi_3 = 1$. Using Eq. (7.11.2), we find

$$z = f(\zeta) = c \int_0^\zeta \zeta'^{-\gamma_2/\pi} (\zeta' - 1)^{-\gamma_3/\pi} \, d\zeta' \tag{7.11.4}$$

Unfortunately, the integration may not be carried out in closed form for any arbitrary triangular shape. To compute the constant c we require that $f(1) = a$, obtaining

$$a = c \int_0^1 \zeta'^{-\gamma_2/\pi} (\zeta' - 1)^{-\gamma_3/\pi} \, d\zeta' = (-1)^{-\gamma_3/\pi} c \frac{\Gamma(1 - \gamma_2/\pi) \, \Gamma(1 - \gamma_3/\pi)}{\Gamma(2 - \gamma_2/\pi - \gamma_3/\pi)} \tag{7.11.5}$$

The values of the gamma function Γ may be read off tables or approximated in terms of series expansions (Abramowitz and Stegun, 1972, p. 255).

Uniform flow along the ξ axis in the ζ plane corresponds to flow due to a point-source dipole placed at the vertex z_1 in the z plane.

Mapping the interior of a semi-infinite strip to the upper half-space

As d tends to infinity, the triangle becomes a semi-infinite strip along the y axis with width equal to a. In that case $\gamma_2 = \gamma_3 = \pi/2$, and Eq. (7.11.4) yields

$$z = f(\zeta) = c \int_0^\zeta [\zeta'(\zeta' - 1)]^{-1/2} \, d\zeta' = c \int_0^{2\zeta - 1} (\mu^2 - 1)^{-1/2} \, d\mu \tag{7.11.6}$$

$$= \pm ic \arcsin(2\zeta - 1)$$

Requiring the condition $f(1) = a$, we obtain $\pm ic = 2a/\pi$. The forward mapping function follows by inverting Eq. (7.11.6)

$$\zeta = F(z) = \frac{1}{2} \left[\sin\left(\frac{z\pi}{2a}\right) + 1 \right] \tag{7.11.7}$$

Uniform flow along the ξ axis in the ζ plane corresponds to flow coming down along the left vertical side of the strip and leaving up along the right side in the z plane (Problem 7.11.3).

Mapping the interior of a rectangle to the upper half-space

Consider next mapping the interior of a rectangle with vertices at the points $z_1 = -a + ib$, $z_2 = -a$, $z_3 = a$, and $z_4 = a + ib$ to the upper half-ζ-plane, as shown in Figure 7.11.3. Taking advantage of the symmetry of the domain, we specify $\xi_2 = -1, \xi_3 = 1$, and require that the origin of the z plane is mapped to the origin of the ζ plane, which dictates setting $z_0 = 0$ and

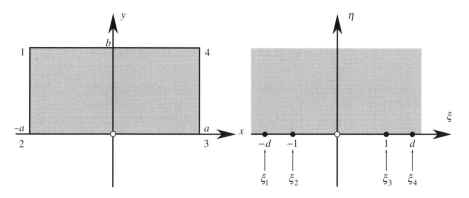

Figure 7.11.3   Mapping of the interior of a rectangle in the z plane to the upper half-ζ-plane.

$\zeta_0 = 0$. Furthermore, anticipating the symmetry of the solution, we set $\xi_1 = -d$ and $\xi_4 = d$, where d is to be computed as part of the solution. Substituting these values into Eq. (7.11.2), we obtain the transformation

$$F(\zeta) = c \int_0^\zeta [(\zeta' + d)(\zeta' + 1)(\zeta' - 1)(\zeta' - d)]^{-1/2} \, d\zeta \qquad (7.11.8)$$

Uniform flow along the ξ axis in the ζ plane corresponds to flow due to a point source and a point sink with strengths of equal magnitude placed, respectively, at the vertices z_1 and z_4 in the z plane.

To compute the values of the two constants c and d, we require that, as we move along the ξ axis from the origin up to ξ_3 and then up to ξ_4, we find ourselves at the corresponding vertices z_3 and z_4. The first condition yields the real equation

$$a = c \int_0^1 [(d^2 - \xi^2)(1 - \xi^2)]^{-1/2} \, d\xi \qquad (7.11.9)$$

which shows that the constant c is real. The second condition yields the complex equation

$$ib = c \int_1^d [(d^2 - \xi^2)(1 - \xi^2)]^{-1/2} \, d\xi \qquad (7.11.10)$$

which may be restated in the real form

$$b = c \int_1^d [(d^2 - \xi^2)(\xi^2 - 1)]^{-1/2} \, d\xi \qquad (7.11.11)$$

Equations (7.11.9) and (7.11.11) provide us with a system of two coupled nonlinear algebraic equations for the unknowns c and d, which may be solved using standard numerical methods such as the method of fixed-point iterations or Newton's method (Section B.3, Appendix B). An alternative simpler iterative procedure involves the following steps:

1. Guess a value for d that is greater than unity.
2. Compute the integral on the right-hand side of Eq. (7.11.9) and thus obtain the value of c.
3. Compute the right-hand side of Eq. (7.11.11), call it B, and then improve the guess for d by setting $d^{\text{New}} = d^{\text{Old}} + \omega(B - b)$, where ω is a relaxation parameter whose optimal value is found by numerical experimentation.

Certain aspects of the computation require special attention. First we note that the integral in Eq. (7.11.9) may be placed in the form $(1/d)F(1/d)$, where F is the complete elliptic integral of the first kind defined as

$$F(m) \equiv \int_0^1 [(1 - \xi^2)(1 - m^2\xi^2)]^{-1/2} \, d\xi = \int_0^{\pi/2} (1 - m^2 \sin^2 \theta)^{-1/2} \, d\theta \qquad (7.11.12)$$

which may be computed by numerical integration, iterative methods, or approximated with polynomial expansions as discussed in Section B.9, Appendix B.

To compute the singular integral on the right-hand side of Eq. (7.11.11), we subtract off the singularities on either end of the domain of integration by recasting the integral into the form

$$\int_1^d \left[\frac{1}{[(d^2 - \xi^2)(\xi^2 - 1)]^{1/2}} - \frac{1}{(d^2 - 1)^{1/2}} \left(\frac{1}{(\xi^2 - 1)^{1/2}} + \frac{1}{(d^2 - \xi^2)^{1/2}} \right) \right] d\xi$$

$$+ (d^2 - 1)^{-1/2} \int_1^d (\xi^2 - 1)^{-1/2} \, d\xi + (d^2 - 1)^{-1/2} \int_1^d (d^2 - \xi^2)^{-1/2} \, d\xi \qquad (7.11.13)$$

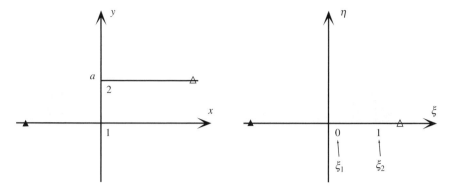

Figure 7.11.4   Mapping of a semifinite region above a wall with a step to the upper half-ζ-plane.

The first integral on the right-hand side of Eq. (7.11.13) is regular and may be computed using a standard method of numerical integration (Section B.6, Appendix B). The second and third integrals are singular, but their values may be computed analytically using elementary methods. An alternative is to use the Gauss–Chebyshev quadrature.

Mapping a semi-infinite region to the upper half-space

The Schwarz–Christoffel transformation may be extended to include domains that are bounded by generalized polygons with one or two vertices at infinity. One example is the semi-infinite strip discussed earlier in this section. Another example is the semi-infinite region above a wall with a step shown in Figure 7.11.4, whose boundary may be regarded as a generalized polygon with one vertex at infinity. Mapping the vertex at infinity in the z plane to the positive infinity of the ξ axis, and specifying that $\xi_1 = 0, \xi_2 = 1$, we obtain the transformation

$$z = f(\zeta) = c \int_0^\zeta \left(\frac{\zeta' - 1}{\zeta'} \right)^{1/2} d\zeta' \tag{7.11.14}$$

The value of the constant c is determined by requiring $f(1) = ia$, and this yields

$$ia = c \int_0^1 \left(\frac{\xi - 1}{\xi} \right)^{1/2} d\xi = ic \int_0^1 \left(\frac{1 - \xi}{\xi} \right)^{1/2} d\xi \tag{7.11.15}$$

Computing the integral on the right-hand side yields $c = 2a/\pi$.

Angles of generalized polygons

The angle γ_i corresponding to the vertex of a generalized polygon that is located at infinity is equal to $2\pi - \theta$, where θ is the external angle formed by the intersection of its sides when they are extended back from infinity. For example, for the generalized polygon illustrated in Figure 7.11.5 with two collapsed sides and one vertex at infinity, $\gamma_1 = \pi/2, \gamma_2 = 4\pi/3, \gamma_3 = 2\pi/3, \gamma_4 = \pi/2, \gamma_5 = -\pi$ (Trefethen, 1980).

Mapping the Interior of a Polygon to the Interior of the Unit Disk

In certain applications, it is preferable to map the interior of a polygon in the z plane to a unit disk centered at the origin in the complex τ plane. This is done by mapping the disk to the upper half-ζ-plane by means of the linear fractional transformation

$$\zeta = G(\tau) = i\frac{1 + \tau}{1 - \tau} \tag{7.11.16}$$

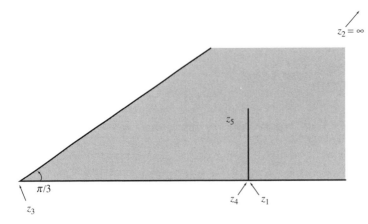

Figure 7.11.5   A generalized polygon with two sides collapsed and a vertex at infinity.

and then mapping the upper half-plane to the interior of the polygon. Substituting Eq. (7.11.16) into the right-hand side of Eq. (7.11.2), we obtain the effective transformation

$$z = f(\zeta) = f[G(\tau)] \equiv Q(\tau) = z_0 + c \int_{\tau_0}^{\tau} \prod_{n=1}^{N} \left(i \frac{1+\tau'}{1-\tau'} - \xi_n \right)^{-\gamma_n/\pi} \frac{2i}{(1-\tau')^2} \, d\tau' \tag{7.11.17}$$

Rearranging the integrand and taking into account Eq. (7.11.1) yields

$$z = Q(\tau) = z_0 + d \int_{\tau_0}^{\tau} \prod_{n=1}^{N} (\tau' - a_n)^{-\gamma_n/\pi} \, d\tau' \tag{7.11.18}$$

where the points

$$a_n = \frac{\xi_n - i}{\xi_n + i} \tag{7.11.19}$$

are located on a circle of unit radius, and d is a new constant. Placing ξ_1 or ξ_N at infinity, we obtain $a_1 = 1$ or $a_N = 1$, in which cases the corresponding multipliers do not appear in Eq. (7.11.2), but must be included within the product in Eq. (7.11.18).

Mapping the Exterior of a Polygon to the Interior of a Unit Disk

To derive the transformation that maps the exterior of a polygon to the interior of a unit disk, we regard the exterior of the polygon as a doubly connected domain that is bounded by the polygon and another polygon of large size extending to infinity. We number the vertices of the inner polygon sequentially in the *clockwise* sense, compute the exterior angles γ_n as usual, and then use the modified transformation

$$z = Q(\tau) = z_0 + d \int_{\tau_0}^{\tau} \prod_{n=1}^{N} (\tau' - a_n)^{-\gamma_n/\pi} \frac{1}{\tau'^2} \, d\tau' \tag{7.11.20}$$

where d is a complex constant, z_0 and τ_0 are two arbitrary corresponding points, and the images a_n lie on the circle bounding the unit disk (Nehari, 1952, p. 193). The infinity of the z plane is mapped to the center of the disk in the τ plane.

Mapping the Exterior of a Polygon to the Upper Half-Plane

Inverting the transformation (7.11.16) and substituting the result into Eq. (7.11.20), we obtain a transformation that maps the exterior of a polygon to the upper half-plane,

$$z = f(\zeta) = z_0 + c \int_{\zeta_0}^{\zeta} \prod_{n=1}^{N} (\zeta' - \xi_n)^{-\gamma_n/\pi} \frac{d\zeta'}{(\zeta' - b)^2 (\zeta' - b^*)^2} \qquad (7.11.21)$$

where b is an arbitrary constant with positive imaginary part representing the image of infinity in the z plane (Carrier, Krook, and Pearson, 1983, p. 153). The computation of the unknown parameters proceeds in a manner that is analogous to that for mapping the interior of a polygon discussed previously.

Numerical Computation of Schwarz–Christoffel Transformations

We have discussed several types of transformations that map the interior or exterior of a polygon to the upper half-plane or to the interior or exterior of a unit disk. For polygons with a small number of vertices, the computation of the constants involved in the mapping function may be carried out using analytical or elementary numerical methods. The difficulty of the computations increases rapidly as the number of vertices is increased, and crowding of the vertices becomes a serious problem. Fortunately, over the past decade, efficient iterative numerical methods for computing Schwarz–Christoffel and related transformations for polygonal domains with complicated geometries have been developed (Trefethen, 1980, 1989; Henrici, 1986; Floryan and Zemach, 1993; Brady and Pozrikidis, 1993). Numerical conformal mapping has emerged as a challenging topic of applied complex analysis with important applications in fluid mechanics.

PROBLEMS

7.11.1 **Mapping the exterior of a flat plate to the exterior of a disk.** Derive Eq. (7.10.6) from Eq. (7.11.20), thus obtaining the transformation that maps the exterior of a disk with radius $c = a/2$ centered at the origin, to the exterior of a flat plate subtended between the points $x = \pm a$.

7.11.2 **Mapping the interior of a regular polygon to the interior of a disk.** (a) Use Eq. (7.11.18) to show that the transformation

$$z = Q(\tau) = d \int_{\tau_0}^{\tau} \prod_{n=1}^{N} (\tau' - e^{2\pi ni/N})^{-2/N} d\tau' \qquad (7.11.22)$$

maps the interior of a unit disk centered at the origin in the τ plane to the interior of a regular N-sided polygon in the z plane. Identify the location of the vertices and the length of the sides of the polygon. (b) Show that in the case of a square, Eq. (7.11.22) reduces to

$$z = Q(\tau) = d \int_{\tau_0}^{\tau} (\tau'^4 - 1)^{-1/2} d\tau' \qquad (7.11.23)$$

Computer Problems

7.11.3 **Flow in a strip.** Compute and plot the streamlines of the flow within the semi-infinite strip corresponding to the transformation (7.11.7), subject to uniform flow along the ξ axis in the ζ plane.

7.11.4 **Semi-infinite flow above a step.** Compute and plot the streamlines of flow in the semi-infinite region above a step corresponding to the transformation (7.11.14) subject to uniform flow along the ξ axis in the ζ plane.

7.11.5 **Mapping the interior of a rectangle to the upper half-space** Write a program called $CFRC$ that computes the constants involved in the transformation that maps the rectangle shown in

Figure 7.11.3 to the upper half-space, using the iterative method discussed in the text. Then plot the streamlines of the flow in the rectangle corresponding to uniform flow along the ξ axis in the ζ plane.

References

Abramowitz, M., and Stegun, I. A., 1972, *Handbook of Mathematical Functions.* Dover.

Ahlfors, L. A., 1979, *Complex Analysis.* McGraw Hill.

Batchelor, G. K., 1967, *An Introduction to Fluid Dynamics.* Cambridge University Press.

Blake, J. R., and Gibson, D. C., 1987, Cavitation of bubbles near boundaries. *Annu. Rev. Fluid Mech.* **19**, 99–123.

Blasius, H., 1910, Funktionentheoretische Methoden in der Hydrodynamik. *Z. Math. Phys.* **58**, 90–110.

Brady, M., and Pozrikidis, C., 1993, Diffusive transport across irregular and fractal walls. *Proc. Roy. Soc. London* A, **442**, 571–83.

Butler, S. F. J., 1953, A note on Stokes's stream function for motion with a spherical boundary. *Proc. Camb. Phil. Soc.* **49**, 169–74.

Carrier, G. F., Krook, M., and Pearson, C. E., 1983, *Functions of a Complex Variable: Theory and Technique.* Hod Books.

Chwang, A. T., and Wu, T. Y., 1974, A note of potential flow involving prolate spheroids. *Schiffstech.* **21**, 19–31.

Crighton, D. G., 1985, The Kutta condition in unsteady flow. *Annu. Rev. Fluid Mech.* **17**, 411–45.

Delves, L. M., and Mohamed, J. L., 1985, *Computational Methods for Integral Equations.* Cambridge University Press.

Dettman, J. W., 1965, *Applied Complex Variables.* Dover.

Floryan, J. M., and Zemach, C., 1993, Schwarz–Christoffel methods for conformal mapping of regions with a periodic boundary. *J. Comp. Appl. Math.* **46**, 77–102.

Henrici, P., 1986, *Applied and Computational Complex Analysis,* Vol. 3. Wiley.

Joseph, D. D., Liao, T. Y., and Hu, H. H., 1993, Drag and moment in viscous potential flow. *Eur. J. Mech. B/fluids* **12**, 97–106.

Lamb, H., 1932, *Hydrodynamics.* Dover.

Levich, V. G. 1962, *Physicochemical Hydrodynamics.* Prentice Hall.

Milne-Thomson, L. M., 1940, Hydrodynamic images. *Proc. Camb. Phil. Soc.* **36**, 246–47.

Milne-Thomson, L. M., 1968, *Theoretical Hydrodynamics.* Macmillan.

Moore, D. W., 1963, The boundary layer on a spherical gas bubble. *J. Fluid Mech.* **16**, 161–76.

Nehari, Z., 1952, *Conformal Mapping.* Dover.

Plesset, M. S., and Prosperetti, A., 1977, Bubble dynamics and cavitation. *Annu. Rev. Fluid Mech.* **9**, 145–85.

Trefethen, L. N., 1980, Numerical computation of the Schwarz–Christoffel transformation. *SIAM J. Sci. Stat. Comput.* **1**, 82–102.

Trefethen, L. N., 1989, Schwarz–Christoffel mapping in the 1980's. *Numerical Analysis Report 89-1,* Dept. of Math., MIT.

Weiss, P., 1945, On hydrodynamic images, arbitrary irrotational flow disturbed by a sphere. *Proc. Camb. Phil. Soc.* **3**, 259–61.

Yih, C-S., 1979, *Fluid Mechanics.* West River Press.

Zhou, H., and Pozrikidis, C., 1995, Adaptive singularity method for Stokes flow past particles. *J. Comp. Phys.* **117**, 79–89.

C H A P T E R

Boundary Layers **8**

██████████

There is a class of flows wherein the curl of the vorticity or the vorticity itself virtually vanishes everywhere except within thin layers or columns of fluid that wrap around or trail behind solid boundaries, free surfaces, or fluid interfaces, or even within compact regions that either are attached to the boundaries or are located in the bulk of the fluid. The equation of motion shows that the viscous force exerted on a small fluid parcel is small outside these regions and may thus be neglected, but makes important contributions within these regions and should be retained.

An important consequence of dropping the viscous force in the Navier–Stokes equation, obtaining Euler's equation, is that the order of the governing system of equations with respect to the spatial partial derivatives is reduced from two to one. This makes it impossible, in general, to satisfy three scalar boundary conditions over each boundary as required for viscous fluids. The presence of boundary layers within which the motion of the fluid is governed by a second-order partial differential equation is thus necessary for the uncompromised description of the flow. Conversely, viscous forces may not be generally neglected uniformly throughout a bounded domain of a flow.

One class of boundary layers occur in high-Reynolds-number flow past a streamlined body such as an airfoil, or flow past a body that does not have too much of a bluff shape such as a cylinder or a sphere. Prandtl (1904) argued that, at sufficiently high Reynolds numbers, but not so high that the motion becomes turbulent, the bulk of the flow is nearly irrotational, and the viscous force is significant only within boundary layers that line the surface of the body, as well as within narrow wakes and possibly compact regions of recirculating flow. The computation of the flow may then be carried out in two steps: First we consider the outer irrotational flow, and then we compute the flow within the boundary layers subject to appropriate simplifications. These stem from the fact that the flow within the boundary layers is nearly unidirectional along the tangential component of the outer flow. Following Prandtl's original idea, a large body of research has demonstrated the success and physical relevance of the boundary-layer theory and opened a new era in the study of viscous flow (Tani, 1977).

There is another class of boundary layers occurring at free surfaces or interfaces between two streams of the same or different fluids. In these cases, an irrotational incident flow is not capable of satisfying both the kinematic boundary condition of continuity of velocity and the dynamic boundary condition that specifies the discontinuity in the interfacial traction, and the role of the boundary layer is to make up for this deficiency.

In the present chapter we introduce the fundamental concepts that enter the mathematical formulation of the boundary-layer theory, and discuss selected analytical and numerical methods for solving the equations that govern the motion of the fluid within boundary layers of various types. The practical importance of boundary-layer theory has spawned a large body of literature, and the interested reader will find extensive discussions in the monographs and reviews by Rosenhead (1963), Moore (1964), Schlichting (1968), Blottner (1970), Stewartson (1974), Cebeci and Bradshaw (1977, 1984), and Smith (1982).

8.1 | BOUNDARY-LAYER THEORY

The fundamental assumption and point of departure for developing a boundary-layer theory is that the flow is composed of an outer region in which the curl of the vorticity vanishes and the flow is described by the equation of motion for inviscid fluids, and a boundary layer, as illustrated in Figure 8.1.1. The presence of wakes and regions of recirculating flow is allowed but is significant only insofar as to modify the structure of the outer flow.

Prandtl Boundary Layers around Solid Boundaries

To illustrate the physical arguments that enter the formulation of the boundary-layer theory and to demonstrate the resulting mathematical simplifications, let us consider the two-dimensional boundary layer developing around a mildly curved rigid body that is held stationary in an irrotational incident flow.

Continuity equation

We begin the analysis by introducing Cartesian coordinates with the x axis tangential and the y axis perpendicular to the surface of the body at a certain point as shown in Figure 8.1.1, and apply the continuity equation at a point in the vicinity of the origin,

$$\frac{\partial u}{\partial x} + \frac{\partial v}{\partial y} = 0 \tag{8.1.1}$$

where u and v are the components of the velocity along the x and y axes. If L is the typical dimension of the body and U is the typical magnitude of the velocity of the irrotational flow, we expect that the magnitude of the derivative $\partial u/\partial x$ within the boundary layer will be comparable to the ratio U/L. Furthermore, if δ is the typical thickness of the boundary layer *defined as*

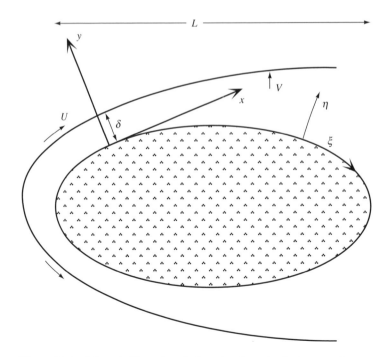

Figure 8.1.1   Schematic illustration of a boundary layer developing around a two-dimensional body.

the region around the body across which the velocity undergoes a rapid variation and within which viscous force is significant, and V is the typical magnitude of component of the velocity normal to the body at the edge of the boundary layer, we expect that the magnitude of the derivative $\partial v/\partial y$ will be comparable to the ratio V/δ. The continuity equation (8.1.1) then requires that

$$\frac{U}{L} \approx \frac{V}{\delta} \quad \text{or} \quad V \approx U\frac{\delta}{L} \tag{8.1.2}$$

Next we examine the two components of the equation of motion at a point in the vicinity of the origin.

x component of the equation of motion

Considering the x component of the Navier–Stokes equation, we scale u with U, $\partial u/\partial x$ with U/L, v with V, $\partial u/\partial y$ with U/δ, the second derivative $\partial^2 u/\partial x^2$ with U/L^2, and $\partial^2 u/\partial y^2$ with U/δ^2. Furthermore, we use the scaling shown in Eqs. (8.1.2) to eliminate V in favor of U and find the magnitudes of the various terms as shown beneath them

$$\frac{\partial u}{\partial t} + u\frac{\partial u}{\partial x} + v\frac{\partial u}{\partial y} = -\frac{1}{\rho}\frac{\partial P}{\partial x} + \nu\frac{\partial^2 u}{\partial x^2} + \nu\frac{\partial^2 u}{\partial y^2}$$

$$\frac{U^2}{L} \quad \frac{U^2}{L} \qquad\qquad \nu\frac{U}{L^2} \quad \nu\frac{U}{\delta^2} \tag{8.1.3}$$

where P is the modified pressure. At this point, there is no obvious way of scaling the modified pressure gradient on the basis of kinematics. The proper scaling of the term involving the time derivative depends upon the temporal behavior of the incident irrotational flow, which is left unspecified at this point.

The scalings shown underneath Eq. (8.1.3), combined with the assumption that $\delta < L$, lead to two important conclusions. First, the penultimate viscous term is small compared to the last viscous term and may thus be neglected, yielding the boundary-layer equation

$$\frac{\partial u}{\partial t} + u\frac{\partial u}{\partial x} + v\frac{\partial u}{\partial y} = -\frac{1}{\rho}\frac{\partial P}{\partial x} + \nu\frac{\partial^2 u}{\partial y^2} \tag{8.1.4}$$

Second, the magnitude of the last viscous term must be comparable to the magnitude of the inertial terms on the left-hand side, and this requires that

$$\frac{U^2}{L} \approx \nu\frac{U}{\delta^2} \tag{8.1.5}$$

or

$$\delta \approx \left(\frac{\nu L}{U}\right)^{1/2} = \frac{L}{Re^{1/2}} \tag{8.1.6}$$

where we have introduced the Reynolds number $Re = UL/\nu$ (see also Problem 8.1.2).

y component of the equation of motion

Considering the individual terms in the y component of the equation of motion, we scale v with V, $\partial v/\partial x$ with V/L, u with U, $\partial v/\partial y$ with V/δ, the second derivative $\partial^2 v/\partial x^2$ with V/L^2, and $\partial^2 v/\partial y^2$ with V/δ^2. Furthermore, we express the kinematic viscosity ν in terms of δ using Eq. (8.1.5), writing $\nu \approx U\delta^2/L$. We thus find that the magnitude of the various terms is as shown beneath them

$$\frac{\partial v}{\partial t} + u\frac{\partial v}{\partial x} + v\frac{\partial v}{\partial y} = -\frac{1}{\rho}\frac{\partial P}{\partial y} + \nu\frac{\partial^2 v}{\partial x^2} + \nu\frac{\partial^2 v}{\partial y^2}$$

$$\frac{U^2\delta}{L^2} \quad \frac{U^2\delta}{L^2} \qquad\qquad \frac{U^2\delta}{L^2} \quad \frac{U^2\delta}{L^2} \tag{8.1.7}$$

The magnitude of all nonlinear convective and viscous terms is of order δ, and this requires that, unless the magnitude of the temporal derivative is of order unity, the gradient of the modified pressure across the boundary layer must also be of order δ, which allows us to write the leading-order approximation

$$\frac{\partial P}{\partial y} = 0 \tag{8.1.8}$$

Thus *pressure variations across the boundary layer may be neglected,* and the pressure within the boundary layer may be regarded a function of arc length l along the contour of the body alone. In certain laboratory experiments, one measures the pressure on the boundary, and this is a good approximation of the surface pressure of the outer flow.

Boundary-layer equations

Evaluating the pressure gradient at the edge of the boundary layer using Euler's equation for the incident irrotational flow, written at a point in vicinity of the origin, and invoking the no-penetration condition, we obtain

$$-\frac{1}{\rho}\frac{\partial P}{\partial x} = \frac{\partial U}{\partial t} + U\frac{\partial U}{\partial x} \tag{8.1.9}$$

where U represents the *tangential component of the velocity of the outer flow.* The boundary-layer equation (8.1.4) then becomes

$$\frac{\partial u}{\partial t} + u\frac{\partial u}{\partial x} + v\frac{\partial u}{\partial y} = \frac{\partial U}{\partial t} + U\frac{\partial U}{\partial x} + \nu\frac{\partial^2 u}{\partial y^2} \tag{8.1.10}$$

Equations (8.1.1) and (8.1.10) provide us with a system of two second-order, *nonlinear* partial-differential equations for the two velocity components u and v, which are to be solved subject to (1) the boundary condition that u and v vanish at the surface of the body, and (2) the far-field condition that, as y/δ tends to infinity, u tends to U. Because the boundary-layer equations do not involve the second partial derivative of v with respect to y, a far-field condition for v is not required. The pressure follows from a knowledge of the structure of the outer irrotational flow and plays the role of a known forcing function.

When the flow is steady, we apply Eq. (8.1.10) at the origin and enforce the no-slip and no-penetration conditions to obtain

$$\left(\frac{\partial^2 u}{\partial y^2}\right)_{y=0} = -\frac{1}{\nu}U\frac{dU}{dx} \tag{8.1.11}$$

which shows that the sign of the curvature of the velocity profile at the wall is opposite to that of the streamwise acceleration of the outer flow dU/dx. In turn, this reveals that the flow within the boundary layer in a decelerating flow with $dU/dx < 0$ reverses direction near the boundary. In practice, this is accompanied by convection of the vorticity away from the boundary and formation of vortices within the bulk of the flow. The steady-state form of Eq. (8.1.9) shows that, when $dU/dx < 0$, $dP/dx > 0$, which means that the outer flow is subjected to an *adverse pressure gradient.* In the opposite case where $dP/dx < 0$, the boundary layer is subjected to a *favorable pressure gradient.*

Boundary-layer equations in curvilinear coordinates

The Prandtl boundary-layer equation (8.1.10) was developed with reference to the local Cartesian system of axes shown in Figure 8.1.1, and is valid at a point in the vicinity of the origin. To avoid redefining the axes at every point along the boundary, we introduce a curvilinear coordinate system with the ξ axis tangential to the boundary and the η axis perpendicular to the boundary as shown in Figure 8.1.1 and denote the corresponding components of the velocity u_ξ and u_η by u and v. Repeating the preceding arguments, we find that the boundary-layer equations (8.1.1), (8.1.8), and (8.1.10) remain valid to a leading-order approximation, provided that the Cartesian x and y coordinates are replaced by the arc lengths l_ξ and l_η corresponding to the curvilinear ξ and η coordinates. Equation (8.1.8), in particular, becomes

$$\frac{\partial P}{\partial l_\eta} = \rho \kappa U^2 \tag{8.1.12}$$

where κ is the curvature of the boundary, which shows that the pressure drop across the boundary layer is of order δ provided that κ has a moderate value, which means that the boundary is not too sharply curved. For simplicity, in the remainder of this chapter we shall denote l_ξ and l_η, respectively, by x and y.

Parabolized form of the equation of motion

The absence of a second partial derivative with respect to x renders the boundary-layer equation (8.1.10) a parabolic differential equation with respect to x; this classification has important consequences on the nature of, and methods of computing the solution. We note, in particular, that the system of equations (8.1.1) and (8.1.10) may be solved using a *marching method* with respect to x, which is typical of initial-value problems, beginning from a particular x station where the structure of the boundary layer is somehow known. In contrast, the Navier–Stokes equation is an elliptic partial differential equation with respect to x and y and must be solved simultaneously and at once at every point throughout the domain of flow.

The parabolic nature of Eq. (8.1.10) with respect to x implies that, if a perturbation is introduced at a certain point along the boundary layer, it will modify the structure of the flow downstream but will leave the structure of the flow upstream unaffected.

Dimensionless form of the equation of motion

The estimates of the magnitude of the various terms in the equation of motion discussed earlier in this section suggest a particular way of defining dimensionless variables to be used for the nondimensionalization of the equation of motion, given by

$$\hat{x} \equiv \frac{x}{L}, \qquad \hat{y} \equiv \frac{y}{\delta} = \frac{y}{L} Re^{1/2}, \qquad \hat{t} \equiv \frac{tU}{L}$$

$$\hat{u} \equiv \frac{u}{U}, \qquad \hat{v} \equiv \frac{v}{U} Re^{1/2}, \qquad \hat{P} \equiv \frac{P}{\rho U^2} \tag{8.1.13}$$

where $Re = UL/\nu$. Here we have assumed that the appropriate time scale is given by L/U. The corresponding dimensionless form of the continuity equation and two components of the Navier–Stokes equation are

$$\frac{\partial \hat{u}}{\partial \hat{x}} + \frac{\partial \hat{v}}{\partial \hat{y}} = 0$$

$$\frac{\partial \hat{u}}{\partial \hat{t}} + \hat{u} \frac{\partial \hat{u}}{\partial \hat{x}} + \hat{v} \frac{\partial \hat{u}}{\partial \hat{y}} = -\frac{\partial \hat{P}}{\partial \hat{x}} + \frac{1}{Re} \frac{\partial^2 \hat{u}}{\partial \hat{x}^2} + \frac{\partial^2 \hat{u}}{\partial \hat{y}^2} \tag{8.1.14}$$

$$\frac{1}{Re}\left(\frac{\partial \hat{v}}{\partial \hat{t}} + \hat{u} \frac{\partial \hat{v}}{\partial \hat{x}} + \hat{v} \frac{\partial \hat{v}}{\partial \hat{y}} \right) = -\frac{\partial \hat{P}}{\partial \hat{y}} + \frac{1}{Re^2} \frac{\partial^2 \hat{v}}{\partial \hat{x}^2} + \frac{1}{Re} \frac{\partial^2 \hat{v}}{\partial \hat{y}^2}$$

The magnitudes of all partial derivatives in Eq. (8.1.14) are of order unity. Taking the limit as Re tends to infinity and maintaining the leading-order terms yields the simplified system

$$\frac{\partial \hat{u}}{\partial \hat{x}} + \frac{\partial \hat{v}}{\partial \hat{y}} = 0$$

$$\frac{\partial \hat{u}}{\partial \hat{t}} + \hat{u}\frac{\partial \hat{u}}{\partial \hat{x}} + \hat{v}\frac{\partial \hat{u}}{\partial \hat{y}} = -\frac{\partial \hat{P}}{\partial \hat{x}} + \frac{\partial^2 \hat{u}}{\partial \hat{y}^2} \qquad (8.1.15)$$

$$\frac{\partial \hat{P}}{\partial \hat{y}} = 0$$

Reverting to dimensional variables and using Euler's or Bernoulli's equation, we obtain Eqs. (8.1.1), (8.1.8), and (8.1.10).

The absence of the Reynolds number in the governing system of equations (8.1.15) has the consequence that the Reynolds number is significant only insofar as to determine the physical thickness of the boundary layer and magnitudes of the components of the velocity, but does not influence the structure of the flow within the boundary layer. The assumption that the Reynolds number is sufficiently large, however, is a necessary condition for this asymptotic behavior to prevail.

Dissipation of energy

Viewed at a distance, the boundary layer resembles a vortex sheet whose strength is equal to the tangential component of the velocity of the outer flow. Since the velocity gradient at the location of a vortex sheet diverges, the associated rate of viscous dissipation takes an infinite value. Viewed from a point in its proximity, however, the boundary layer resembles a vortex layer of finite thickness within which the rate of viscous dissipation has a finite value. More specifically, the rate of viscous dissipation per unit mass of fluid within the boundary layer can be approximated with $\nu(\partial u/\partial y)^2 \approx \nu U^2/\delta^2$, which according to Eq. (8.1.5) scales like U^3/L and therefore remains finite as the viscosity tends to vanish and the Reynolds number tends to infinity. The rate of viscous dissipation within the boundary layer per unit surface area of the boundary may be defined as

$$\int_0^\delta \mu \left(\frac{\partial u}{\partial y}\right)^2 dy \approx \mu \frac{U^2}{\delta} \qquad (8.1.16)$$

which, according to Eq. (8.1.6), scales as $\rho U^3 Re^{-1/2}$. Noting that the rate of viscous dissipation per unit volume of fluid outside the boundary layer scales as $(\rho U^3/L)Re^{-1}$ reveals that the rate of viscous dissipation within the boundary layer makes a dominant contribution.

Flow separation

Boundary-layer analysis relies on several assumptions, two important ones being that the Reynolds number of the incident flow is sufficiently large, but not so large that the motion becomes turbulent, and the vorticity remains adjacent to the boundary. The validity of the second assumption depends upon the structure of the incident flow and the boundary shape. Streamlined bodies allow laminar boundary layers to develop over a large portion or even the whole of their surface, whereas bluff bodies cause the vorticity to concentrate within compact regions, forming steady or unsteady wakes. The latter are characterized by the generation of alternating compact vortex regions of opposite sign generating a *vortex street,* as illustrated in Figure 8.1.2(a) after Homann (1936) and White (1974).

The changes in the structure of a flow as a function of the Reynolds number are reflected on the values of certain global flow quantities, such as the angular frequency f of the vortices developing in the wake, usually expressed in terms of the Strouhal number $St = fL/U$, and the drag force that is exerted on an object. Figure 8.1.2(b) presents a plot of the drag coefficient $c_D = F/\rho V^2 a$, where F is the drag force per unit length exerted on a circular cylinder of radius

(a)

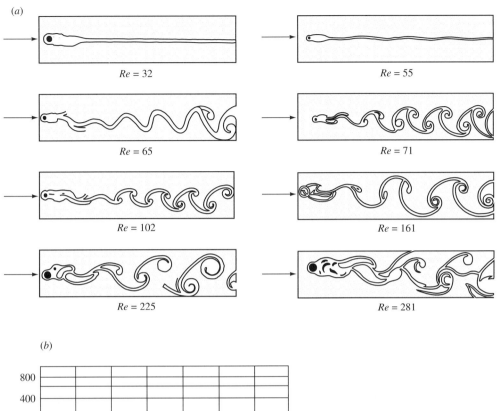

(b)

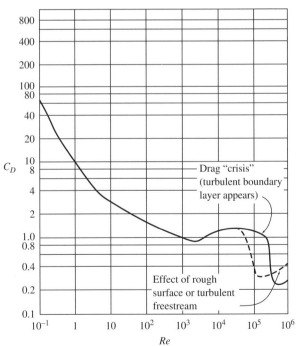

Figure 8.1.2 (a) Schematic illustration of the structure of uniform flow past a circular cylinder at different Reynolds numbers, showing boundary-layer separation and the formation of a wake. The vorticity within the enclosed clear areas has a substantially increased absolute value. (b) Drag coefficient as a function of the Reynolds number, after Homann (1936) and White (1974).

a that is held stationary in a uniform incident flow with velocity V, as a function of the Reynolds number $Re = 2Va/\nu$, after White (1974). As the Reynolds number is increased, a boundary layer is established at the front portion of the cylinder accompanied by a steady and then an unsteady wake. Finally, the flow within the wake becomes turbulent and drag "crisis" occurs when Re becomes close to 30,000.

Three-dimensional boundary layers

To analyze the structure of the flow within a boundary layer developing on a flat or curved three-dimensional surface, we introduce a system of curvilinear coordinates with two axes lying in the surface and the third axis perpendicular to the surface, and implement the boundary-layer approximations discussed earlier in this section. The analysis reveals that the pressure variation across the boundary layer may be neglected to a leading-order approximation, and the components of the Navier–Stokes equation in the tangential directions are parabolized with the elimination of the second partial derivative of the velocity with respect to distance in the normal direction. The precise form of the boundary-layer equations for three-dimensional flow will be discussed in the end of Section 8.5.

Boundary Layers along Free Surfaces and Fluid Interfaces

Consider now the behavior of the flow around a free surface over which the tangential component of the traction is required to vanish. An irrotational incident flow is not generally capable of satisfying the condition of vanishing shear stress, and a boundary layer must develop in order to rectify this deficiency.

An example of a pertinent situation is provided by the boundary layer forming along the surface of a spherical bubble that rises at high Reynolds numbers, investigated in detail by Moore (1963). The distinguishing and important difference between this boundary layer and the Prandtl boundary layer developing over a solid boundary is that, in the case of a free surface, the velocity does not have to undergo a jump across the boundary layer whose magnitude is comparable to the velocity of the incident flow, for the fluid is allowed to slip along the free surface.

To analyze the boundary layer developing on a free surface, it is useful to decompose the velocity field into the outer component, denoted by $\mathbf{u}^\infty$, and a complementary component associated with the boundary layer, denoted by $\mathbf{u}^{BL}$, setting $\mathbf{u} = \mathbf{u}^\infty + \mathbf{u}^{BL}$. Requiring that the tangential component of the traction vanish at the free surface, we find that the tangential component of the vorticity at the free surface is given by Eq. (3.6.13). If the flow $\mathbf{u}^\infty$ is irrotational, the vorticity at the edge of the boundary layer must vanish, and the right-hand side of Eq. (3.6.13) evaluated for $\mathbf{u} = \mathbf{u}^\infty$ must be equal to $[\mathbf{n} \cdot \nabla \mathbf{u}^\infty + (\nabla \mathbf{u}^\infty) \cdot \mathbf{n}] \times \mathbf{n}$.

For the vorticity within the boundary layer to be an order-one quantity, the jump in the complementary velocity $\mathbf{u}^{BL}$ across the boundary layer must be of the same order of magnitude as the boundary layer thickness δ, which scales with $Re^{-1/2}$ and will therefore be small as long as the Reynolds number is sufficiently large. The jump in the tangential component of the vorticity across the boundary layer may then be approxmiated with $-[\mathbf{n} \cdot \nabla \mathbf{u}^\infty + (\nabla \mathbf{u}^\infty) \cdot \mathbf{n}] \times \mathbf{n}$, evaluated at the location of the free surface. Furthermore, the smallness of $|\mathbf{u}^{BL}|$ allows us to linearize the equation of motion and vorticity transport equation with respect to $\mathbf{u}$ about $\mathbf{u}^\infty$.

Because $\mathbf{u}^{BL}$ is much smaller than $\mathbf{u}^\infty$, the tendency for backflow in a decelerating flow along a free surface is much weaker than along a solid wall, and the rate of viscous dissipation within the boundary layer is comparable to that within the outer flow.

To analyze the flow around a fluid interface, we introduce two boundary layers on either side of the interface, demand that the velocity remain continuous across the interface, and require that the boundary layers develop so as to satisfy the dynamic boundary condition for the jump in the tangential traction. The pertinent mathematical formulation is discussed in detail by Harper and Moore (1968) with reference to the boundary layer developing around the interface of a viscous drop rising through another fluid at high Reynolds numbers (see also Pozrikidis, 1989).

Boundary Layers in Homogeneous Fluids

Boundary-layer theory builds upon the fact that the magnitude of the derivatives of the tangential components of the velocity normal to a surface is much larger than that of the derivatives along the surface. Additional examples of pertinent situations arise during the flow of a homogeneous fluid with uniform physical properties within which the velocity exhibits sharp variations across thin layers of fluid called *free boundary* or *shear layers*. An example is the shear layer developing during the viscous spreading of a vortex sheet discussed in Section 5.2.

Another example is provided by the widening of a symmetric two-dimensional wake behind a streamlined object that is held stationary in a uniform incident flow along the x axis. Under the assumption that the rate of widening is sufficiently small, the boundary-layer approximations lead to the linear partial differential equation

$$U\frac{\partial u}{\partial x} = \nu\frac{\partial^2 u}{\partial y^2} \tag{8.1.17}$$

where U is the uniform velocity of the incident flow. Note that this is similar to the unsteady heat equation where x plays the role of time. A solution that satisfies the boundary condition that u tends to U as y tends to infinity is

$$u(x, y) = U\left[1 - \frac{c}{\sqrt{x}}\exp\left(-\frac{Uy^2}{4\nu x}\right)\right] \tag{8.1.18}$$

where c is a dimensional constant whose value is determined by the particular form of the velocity distribution at the beginning of the wake. A momentum integral balance around a rectangular control area that contains the object and has two parallel sides perpendicular to the x axis provides us with a relation between c and the drag force per unit width F exerted on the body by means of the equation

$$F = \rho\int_{-\infty}^{\infty} u\,(U - u)\,dy \tag{8.1.19}$$

Substituting the profile given in Eq. (8.1.18) into Eq. (8.1.19) yields

$$c = \frac{1}{2\sqrt{\pi}}\frac{F}{\rho U^{3/2}\nu^{1/2}} \tag{8.1.20}$$

(Problem 8.1.4). Further examples of free boundary layers will be discussed in Problems 8.2.3, 8.5.4, and 8.5.5.

PROBLEMS

8.1.1   **Boundary conditions.** Discuss the number of boundary conditions required for the computation of the velocity components within a two-dimensional Prandtl boundary layer.

8.1.2   **Lubrication flow and boundary layers.** Discuss the relation between the boundary-layer equations and the lubrication equation for nearly unidirectional viscous flow within a narrow channel discussed in Section 6.3.

8.1.3   **Prandtl's transposition theorem.** Show that if $u(x, y, t)$ and $v(x, y, t)$ satisfy the Prandtl boundary-layer equations, then $U(X, Y, t)$ and $V(X, Y, t)$ will also satisfy the boundary-layer equations, where $X = x$, $Y = y + f(x)$, $U(X, Y, t) = u(x, y, t)$, $V(X, Y, t) = v(x, y, t) + f'(x)\,u(x, y, t)$, and $f(x)$ is an arbitrary function (Rosenhead, 1963, p. 211).

8.1.4   **Widening of a two-dimensional wake.** Derive Eq. (8.1.19) and the relation between the coefficient c and F given in Eq. (8.1.20). To simplify the computations, you may approximate the integrand in Eq. (8.1.19) with $U(U - u)$.

8.1.5   **Von Mises's transformation.** Von Mises's transformation regards the flow quantities as functions of the independent variables x and ψ, instead of x and y, where ψ is the stream function whose value is set equal to zero over an impermeable boundary (Mises, 1927). Show that, in the von Mises's variables, and subject to the boundary conditions $u = 0$ and $v = 0$ at $y = 0$, and u tends to U as y/δ tends to infinity, the equation of motion (8.1.10) becomes

$$\frac{\partial u}{\partial t} + u\frac{\partial u}{\partial x} = \frac{\partial U}{\partial t} + U\frac{\partial U}{\partial x} + \nu u \frac{\partial}{\partial \psi}\left(u\frac{\partial u}{\partial \psi}\right) \qquad (8.1.21)$$

with boundary conditions $u = 0$ at $\psi = 0$, and u tends to U as ψ tends to infinity. Furthermore, show that for steady flow, Eq. (8.1.10) may be restated as a nonlinear parabolic equation,

$$\frac{\partial \chi}{\partial x} = \nu \sqrt{U^2 - \chi^2}\frac{\partial^2 \chi}{\partial \psi^2} \qquad (8.1.22)$$

where $\chi = U^2 - u^2$, with boundary conditions $\chi = U^2$ at $\psi = 0$, and χ tends to vanish as ψ tends to infinity.

8.2 | THE BOUNDARY LAYER ON A SEMI-INFINITE FLAT PLATE

Uniform flow along a stationary semi-infinite flat plate that is held parallel to the incident stream, as depicted in Figure 8.2.1(a), provides us with a classical example of a steady boundary-layer flow. Since the tangential component of the velocity of the outer flow U is constant over the plate, $dU/dx = 0$, Eq. (8.1.10) simplifies to the homogeneous equation

$$u\frac{\partial u}{\partial x} + v\frac{\partial u}{\partial y} = \nu\frac{\partial^2 u}{\partial y^2} \qquad (8.2.1)$$

which is to be solved along with the continuity equation (8.1.1) for u and v subject to appropriate boundary conditions.

Because the plate is flat, a natural length scale L cannot be found, and it is imperative to scale all variables using as length scale the streamwise position x. Equation (8.1.6) then provides us with an expression for the boundary-layer thickness

$$\delta(x) \equiv \left(\frac{\nu x}{U}\right)^{1/2} = \frac{x}{Re_x^{1/2}} \qquad (8.2.2)$$

which is expressed in terms of the *local* Reynolds number $Re_x = Ux/\nu$ whose value increases with distance from the leading edge.

Fortunately, computing the solution of the system of equations (8.2.1) and (8.1.1) may be reduced to solving a single ordinary differential equation. This is done by assuming that the flow develops in a self-similar manner, so that the streamwise velocity profile across the boundary is a function of the scaled transverse position expressed by the similarity variable $\eta = y/\delta(x)$, and writing $u(x, y) = UF(\eta)$, where

$$\eta \equiv \frac{y}{\delta(x)} = y\left(\frac{U}{\nu x}\right)^{1/2} \qquad (8.2.3)$$

is a similarity variable. Next we note that the self-similar streamwise profile derives from the stream function

$$\psi(x, y) = (\nu Ux)^{1/2} f(\eta) \qquad (8.2.4)$$

where $F = f'$. The principal advantage of introducing the stream function is that the continuity equation is satisfied automatically and does not need to be considered in the further analysis.

(*a*)

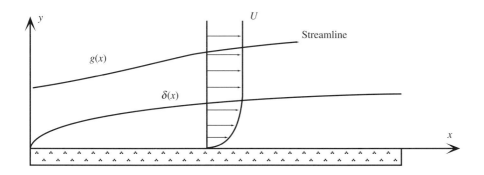

(*b*)

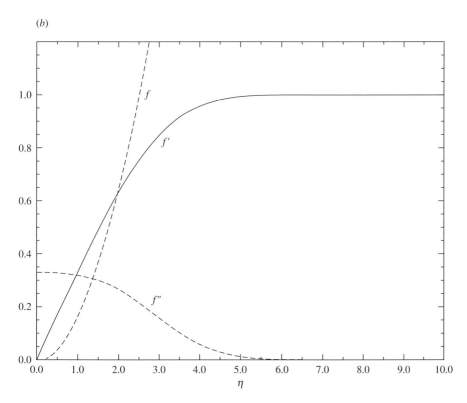

Figure 8.2.1 (a) Schematic illustration of a boundary layer developing along a stationary semi-infinite flat plate that is held parallel to a uniform incident flow. (b) Graphs of the self-similar streamwise velocity profile $u/U = f'$ and the functions f and f''.

Differentiating Eq. (8.2.4) with respect to y and x, we find

$$u(x, y) = \frac{\partial \psi}{\partial y} = U f'(\eta)$$

$$v(x, y) = -\frac{\partial \psi}{\partial x} = \frac{1}{2} \left(\frac{\nu U}{x} \right)^{1/2} [\eta f'(\eta) - f(\eta)]$$

(8.2.5)

Substituting the expressions in Eqs. (8.2.5) into Eq. (8.2.1), we derive a *third-order nonlinear ordinary differential equation* for the function f

$$f''' + \tfrac{1}{2} f f'' = 0 \tag{8.2.6}$$

which is the *Blasius equation* (Blasius, 1908). Enforcing the no-slip and no-penetration conditions, and requiring that, far from the plate, the flow in the boundary layer is consistent with the outer uniform flow, we obtain

$$f = 0 \quad \text{and} \quad f' = 0 \quad \text{at } \eta = 0 \tag{8.2.7a}$$

$$f' \to 1 \quad \text{as} \quad \eta \to \infty \tag{8.2.7b}$$

Furthermore, by applying Eq. (8.2.6) at the wall and using the boundary conditions (8.2.7a), we obtain $f'''(0) = 0$, which shows that the slope of the streamwise velocity profile vanishes at the wall, in agreement with Eq. (8.1.11).

Since boundary conditions are specified at both ends of the solution domain $(0, \infty)$, we are faced with a boundary-value problem that must be solved using a shooting method according to the following steps:

1. Guess the value of $f''(0)$.
2. Integrate Eq. (8.2.6) from $\eta = 0$ to ∞ subject to the initial conditions (8.2.7a) and the assumed value of $f''(0)$. In practice, it is found that integrating up to $\eta = 10$ yields satisfactory accuracy.
3. Check to see whether $f'(\infty) = 1$ is fulfilled, and if not, improve the guess for $f''(0)$.

Fortunately, *only one guess* is sufficient for computing the exact solution. This becomes evident by noting that, if the function $g(\eta)$ is a solution of Eq. (8.2.6), then the function $q(\eta) = \alpha g(\alpha \eta)$ will also be a solution of Eq. (8.2.6) for any value of the constant α. Requiring that the function g satisfy the boundary conditions $g(0) = 0$, $g'(0) = 0$ and setting $\alpha^2 = 1/g'(\infty)$ ensures that the function q satisfies the boundary conditions shown in Eqs. (8.2.7a,b) and therefore it represents the desired solution f.

The profile of the streamwise velocity $u/U = f'$, first computed by Blasius (1908), is shown in Figure 8.2.1(b) along with the profiles of f and f''. The numerical solution reveals that $f''(0) = 0.332$ and $u/U = 0.99$ when $\eta = 4.9$; the second value allows us to define the 99% boundary-layer thickness

$$\delta_{99}(x) = 4.9 \left(\frac{\nu x}{U} \right)^{1/2} \quad \text{or} \quad \frac{\delta_{99}(x)}{x} = \frac{4.9}{Re_x} \tag{8.2.8}$$

The 99.5% boundary-layer thickness is defined in a similar manner; the numerical solution shows that the corresponding numerical coefficient on the right-hand side of Eqs. (8.2.8) is equal to 5.3.

Shear Stress and Drag Force on the Plate

Of primary engineering importance is the wall shear stress and drag force exerted on the plate. According to the similarity solution, the former is given by

$$\sigma_{xy}(x, y = 0) = \mu \left(\frac{\partial u}{\partial y} \right)_{y=0} = \frac{f''(0)}{\sqrt{Re_x}} \rho U^2 = \frac{0.332}{\sqrt{Re_x}} \rho U^2 \tag{8.2.9}$$

which shows that the shear stress takes an infinite value at the leading edge and decreases like the inverse square root of streamwise distance or local Reynolds number. The physical significance of the singular behavior at the origin of the x axis, however, is obliterated by the fact that the assumptions that led us to the boundary-layer equations cease to be valid at the leading edge.

Even though the shear stress at the leading edge is infinite, the inverse-square-root singularity is integrable, and the drag force exerted on the plate has a finite value. Using the similarity solution, we find that the drag force exerted on *both sides* of the plate, from the leading edge up to a certain distance x, is given by

$$D(x) \equiv 2 \int_0^x \sigma_{xy}(x', y = 0) \, dx' = 0.664 \frac{\rho U^{3/2}}{\nu^{1/2}} \int_0^x \frac{1}{\sqrt{x'}} \, dx' = \frac{1.328}{\sqrt{Re_x}} \rho U^2 x \qquad (8.2.10)$$

which can be expressed in terms of the dimensionless drag coefficient

$$c_D \equiv \frac{D(x)}{\frac{1}{2}\rho U^2 x} = \frac{2.656}{\sqrt{Re_x}} \qquad (8.2.11)$$

The theoretical predictions stated in Eqs. (8.2.10) and (8.2.11) have been shown to agree well with laboratory measurements up to about $Re_x = 120{,}000$, whereupon the flow within the boundary layer develops a wavy pattern and ultimately becomes turbulent. Above the critical value of Re_x the function $c_D(Re_x)$ jumps to a different branch with a significantly higher value of c_D. The transition from the steady laminar flow considered at present to the unsteady turbulent flow will be discussed further in Section 9.7 in the context of hydrodynamic stability.

Vorticity Transport

Neglecting the velocity component along the y axis, we find that the magnitude of the vorticity within the boundary layer is given by

$$\omega(x, y) = -\frac{\partial u}{\partial y} = -\frac{f''(\eta)}{\sqrt{Re_x}} \frac{U^2}{\nu} = -f''(\eta) \frac{U}{\delta(x)} \qquad (8.2.12)$$

which shows that, at a particular location η, ω scales with the inverse of the local boundary-layer thickness due to the widening of the velocity profile.

The rate of convection of vorticity across a plane that is perpendicular to the plate is given by

$$-\int_0^\infty u \frac{\partial u}{\partial y} dy = -\frac{1}{2} U^2 \qquad (8.2.13)$$

which is independent of the downstream position x. One implication of this result is that the flux of the vorticity across the plate is equal to zero, which is consistent with the earlier observation that the gradient of the vorticity at the wall vanishes, $U \partial f'''/\partial y = \partial^2 u/\partial y^2 = -\partial \omega/\partial y = 0$. Another consequence is that all of the convected vorticity is generated at the leading edge where the boundary-layer approximation ceases to be valid. Viscous mechanisms at the leading edge generate an amount of vorticity that is precisely equal to that needed for the establishment of the self-similar flow.

Displacement Thickness

As a result of the broadening of the velocity profile in the streamwise direction, the streamlines are deflected upward away from the plate, as shown in Figure 8.2.1(a). Let us consider a streamline that lies outside the boundary layer, describe it as $y = g(x)$, and write a mass balance over a control area that is enclosed by (1) the streamline, (2) a vertical line at $x = 0$, (3) a vertical line located at a certain value of x, and (4) the plane wall. Noting that the velocity profile is flat at $x = 0$, we obtain

$$\int_0^{g(0)} U \, dy = \int_0^{g(x)} u \, dy \qquad (8.2.14)$$

Straightforward rearrangement yields

$$U[g(x) - g(0)] = \int_0^{g(x)} (U - u) \, dy \qquad (8.2.15)$$

Taking the limit as the streamline moves away from the plate, we find

$$\lim_{g(0)\to\infty} [g(x) - g(0)] = \delta^*(x) \tag{8.2.16}$$

where we have introduced the *displacement thickness*

$$\delta^*(x) \equiv \int_0^\infty \left(1 - \frac{u}{U}\right) dy \tag{8.2.17}$$

Using the numerical solution of the Blasius equation to evaluate the integral on the right-hand side of Eq. (8.2.17), we derive the value

$$\delta^*(x) = \left(\frac{\nu x}{U}\right)^{1/2} \int_0^\infty (1 - f') d\eta = 1.721 \left(\frac{\nu x}{U}\right)^{1/2} \tag{8.2.18a}$$

or

$$\frac{\delta^*(x)}{x} = \frac{1.721}{Re_x^{1/2}} \tag{8.2.18b}$$

which shows that the displacement thickness, like the 99% boundary-layer thickness, increases as the square root of streamwise distance.

The physical interpretation of the displacement thickness is evident from Eq. (8.2.16): It represents the vertical displacement of the streamlines with respect to their elevation at the leading edge far from the plate. Observations have shown that the laminar boundary layer undergoes a transition to the turbulent state when the displacement thickness reaches the value $\delta^* \approx 600\nu/U$, whereupon turbulent shear stresses due to small-scale motion become important and the present analysis for laminar flow ceases to be valid.

The displacement thickness defines the surface of a *virtual* impenetrable body that is held stationary in the incident irrotational flow. Replacing the tangential component of the velocity of the irrotational flow at the surface of the actual body, U, with the corresponding tangential component of the velocity at the surface of the virtual body improves the accuracy of the boundary-layer approximation. The irrotational flow past the virtual body must be computed *after* the displacement thickness has been found by neglecting the thickness of the boundary layer to a first approximation, as discussed in the preceding section. This iterative improvement provides us with a basis for describing the flow in the context of an asymptotic expansion (Nayfeh, 1985).

Momentum Thickness

It is both illuminating and useful to perform a momentum integral balance over the control area that was used previously to define the displacement thickness (von Kármán, 1921). Since the upper boundary of the control volume is a streamline, it does not contribute to the input of linear momentum. Assuming that the normal stresses on the vertical sides are equal in magnitude and opposite in sign, which is justified by the assumption that the pressure drop across the boundary layer is negligibly small, and neglecting the traction along the top streamline, we obtain

$$\int_0^{g(0)} U(\rho U) dy - \int_0^{g(x)} u(\rho u) dy - \tfrac{1}{2}D(x) = 0 \tag{8.2.19}$$

where $D(x)$ is the drag force exerted on *both sides* of the plate, defined in Eq. (8.2.10). Rearranging Eq. (8.2.19) and passing to the limit as $g(0)$ tends to infinity, we obtain the relation

$$D(x) = 2\rho U^2 \Theta(x) \tag{8.2.20}$$

where Θ is the *momentum thickness* defined as

$$\Theta(x) \equiv \int_0^\infty \frac{u}{U}\left(1 - \frac{u}{U}\right) dy \tag{8.2.21}$$

Using the numerical solution, we find

$$\Theta(x) = \left(\frac{\nu x}{U}\right)^{1/2} \int_0^\infty f'(1 - f')\, d\eta = 0.664 \left(\frac{\nu x}{U}\right)^{1/2} \qquad (8.2.22)$$

Shape Factor

The ratio between the displacement and momentum thickness, $H = \delta^*/\Theta$, is called the *shape factor*. Inspecting the definitions of δ^* and Θ shows that $H > 1$, as long as u/U is less than unity within a substantial portion of the boundary layer, which is the physically expected behavior. The smaller the value of H, the sharper the velocity profile across the boundary layer. Combining Eqs. (8.2.18a) and (8.2.22), we find that, for the flat plate, $H = 2.59$.

Von Kármán's Approximate Method

Equation (8.2.20) provides us with the cumulative drag force $D(x)$ in terms of the momentum thickness $\Theta(x)$. It is important to note that the latter may be computed from a knowledge of the velocity profile by performing an *integration,* which smooths out numerical inaccuracies or errors that are inherent in laboratory measurements. Differentiating Eq. (8.2.20) with respect to x and using the definition (8.2.10), we obtain a differential relationship between the wall shear stress and momentum thickness

$$\sigma_{xy}(x, y = 0) = \rho U^2 \frac{d\Theta(x)}{dx} \qquad (8.2.23)$$

To this end, we have two methods of computing the shear stress at the wall and drag force exerted on the plate from a knowledge of the velocity profile. First we may differentiate the streamwise velocity profile u with respect to y to compute the shear stress at the wall, and then integrate it with respect to x to obtain D. Second we may use the velocity profile to compute the momentum thickness, and then differentiate the momentum thickness to obtain the shear stress according to Eq. (8.2.23). The second method is more forgiving, in the sense that it does not require a precise knowledge of the velocity profile near the wall.

For the velocity profile that arises by solving the Blasius equation, the two aforementioned methods are equivalent. Indeed, substituting Eq. (8.2.22) into Eq. (8.2.23) yields Eq. (8.2.9), and this confirms that the approximations that led us to the boundary-layer equations are identical to those that led us to the simplified momentum integral balance (8.2.19).

Let us now *select* a certain self-similar velocity profile of some reasonable form involving a free parameter, which may be done by exercising physical intuition under the guidance of laboratory observations, *instead of computing* it by solving the Blasius equation. Is it possible to adjust the free parameter so that the two methods of computing the wall shear stress discussed previously produce the same answer? And if this is possible, will the boundary-layer flow obtained in this manner be a good approximation to exact solution?

To resolve the answers to these questions, let us consider the velocity profile

$$\frac{u}{U} = f'\left(\frac{y}{\Delta(x)}\right) = \begin{cases} \sin \dfrac{\pi y}{2\Delta(x)}, & \text{for } 0 < y < \Delta(x) \\ 1, & \text{for } y > \Delta(x) \end{cases} \qquad (8.2.24)$$

where $\Delta(x)$ is a free parameter that plays the role of an effective boundary-layer thickness, such as the δ_{99} thickness introduced in Eqs. (8.2.8). This velocity profile respects the required boundary conditions $f'(0) = 0$, $f'''(0) = 0$, and $f'(\infty) = 1$, but does not satisfy the Blasius equation.

Differentiating the profile (8.2.24), we obtain the wall shear stress

$$\sigma_{xy}(x, y = 0) = \frac{\pi \mu U}{2\Delta(x)} \qquad (8.2.25)$$

The corresponding displacement and momentum thicknesses are readily found to be

$$\delta^*(x) = \left(1 - \frac{2}{\pi}\right)\Delta(x) = 0.363\Delta(x)$$

$$\Theta(x) = \left(\frac{2}{\pi} - \frac{1}{2}\right)\Delta(x) = 0.137\Delta(x) \tag{8.2.26}$$

and the shape factor is $H = 2.660$. Substituting the expression for the momentum thickness and wall shear stress shown in Eqs. (8.2.25) and (8.2.26) into Eq. (8.2.23), we obtain an ordinary differential equation for $\Delta(x)$

$$\frac{\pi\mu}{2\Delta(x)} = 0.137\rho U \frac{d\Delta(x)}{dx} \tag{8.2.27}$$

Integrating subject to the initial condition $\Delta = 0$ at $x = 0$ yields

$$\Delta(x) = 4.80\left(\frac{\nu x}{U}\right)^{1/2} \tag{8.2.28}$$

Substituting this expression back into Eqs. (8.2.25) and (8.2.26), we obtain the specific expressions

$$\sigma_{xy}(x, y = 0) = 0.327\mu U\left(\frac{U}{\nu x}\right)^{1/2}$$

$$\delta^*(x) = 1.743\left(\frac{\nu x}{U}\right)^{1/2} \tag{8.2.29}$$

$$\Theta(x) = 0.655\left(\frac{\nu x}{U}\right)^{1/2}$$

which are in excellent agreement with the exact relations shown in Eqs. (8.2.9), (8.2.18) and (8.2.22). This kind of agreement, however, is fortuitous and therefore atypical of the accuracy of this approximate method (Problem 8.2.2).

PROBLEMS

8.2.1   **Prandtl boundary layer.** (a) Show that the streamwise velocity component within the boundary layer is constant along a parabola described by $x = ay^2$, where a is an arbitrary constant. (b) Show that the function $g(\eta) = f''(\eta)$ satisfies Weyl's integral equation

$$g(\eta) = \exp\left(-\frac{1}{4}\int_0^\eta (\xi - \eta)^2 g(\xi)\, d\xi\right) \tag{8.2.30}$$

(Rosenhead, 1963, p. 233).

8.2.2   **von Kármán's approximate method.** Assuming that the velocity profile is given by $u/U = \tanh[y/\Delta(x)]$ instead of that shown in Eq. (8.2.24), show that the effective boundary-layer thickness, wall shear stress, displacement thickness, and momentum thickness are given by the right-hand sides of equations (8.2.28) and (8.2.29); except that the numerical coefficients on the right-hand sides are equal, respectively, to 2.553, 0.392, 1.770, 0.664 (White, 1974, p. 247). Discuss the accuracy of these results with reference to the exact solution.

8.2.3   **Spreading of a two-dimensional jet.** Consider the spreading of a symmetric two-dimensional jet that emerges from a slit into a quiescent fluid along the x axis. Applying the momentum integral balance over an infinite control area that has two parallel sides perpendicular to the x

axis, we find that the value of the momentum integral

$$M = \rho \int_{-\infty}^{\infty} u^2 \, dy \tag{8.2.31}$$

must be independent of x and thus a constant. In this case, it is appropriate to express the stream function in the form

$$\psi(x, y) = \left(\frac{M\nu}{\rho}\right)^{1/3} x^{1/3} f(\eta), \qquad \text{where} \qquad \eta = \left(\frac{M}{\nu\mu}\right)^{1/3} \frac{y}{x^{2/3}} \tag{8.2.32}$$

f is a dimensionless function, and η is a dimensionless similarity variable. The values of these exponents emerge by inspecting the boundary-layer equation and noting the invariance of M.

Show that the steady version of the boundary-layer equation (8.1.10), with U set equal to zero, leads to the nonlinear homogeneous ordinary differential equation

$$f''' + \tfrac{1}{3} f f'' + \tfrac{1}{3} f'^2 = 0 \tag{8.2.33}$$

which is to be solved subject to the boundary conditions $f(0) = 0$, $f''(0) = 0$, $f'(\infty) = 0$ and the integral constraint expressed by Eq. (8.2.31). Derive the solution $f(\eta) = 6\alpha \tanh(\alpha\eta)$, where $\alpha = 48^{-1/3}$, and show that the axial volumetric flow rate per unit width is given by $Q = 12\alpha(M\nu x/\rho)^{1/3}$.

8.2.4   **Shear layer.** Consider the flow between two parallel streams that merge along the x axis with velocities equal to U_1 and U_2, and express the stream function in the form $\psi = (\nu x)^{1/2} f(\eta)$, where f is a dimensionless function, $\eta = y/\delta$, and $\delta = (\nu x/U_1)^{1/2}$. Show that, subject to the boundary-layer approximations, f satisfies Blasius's equation and derive the appropriate boundary conditions.

Computer Problems

8.2.5   **Prandtl boundary layer.** Compute and plot the velocity profile shown in Figure 8.2.1(b).

8.2.6   **Boundary layer on a moving flat surface.** Consider the flow of a liquid in the first quadrant above a semi-infinite belt that is moving with velocity U along the x axis, scraping a vertical stationary wall that is placed along the positive part of the y axis, so that the velocity tends to vanish far from the belt. The flow within the boundary layer is governed by the Blasius equation (8.2.6) with the modified set of boundary conditions $f = 0$ and $f' = 1$ at $\eta = 0$, and $f' = 0$ as $\eta \to \infty$. Obtain the solution using a numerical method of your choice. Plot and discuss the streamwise velocity profile, and compute the displacement thickness, momentum thickness, and shape factor (Sakiadis, 1961). [*Hint:* The shooting method converges when $f''(0) = -0.4437$.]

8.3 | BOUNDARY LAYERS IN ACCELERATING AND DECELERATING FLOW

Having examined the structure of the boundary layer on a flat plate that is aligned with a uniform incident flow, distinguished by the fact that the tangential component of the velocity of the outer flow is constant, we proceed to consider circumstances where the incident flow exhibits acceleration or deceleration with an associated favorable and adverse pressure gradient. Examples of physical situations where this might occur are illustrated in Figure 8.3.1(a–d).

Let us consider, in particular, a situation where the tangential component of the velocity of the outer flow U exhibits a power-law dependence on streamwise distance x measured along a flat boundary,

$$U = cx^m \tag{8.3.1}$$

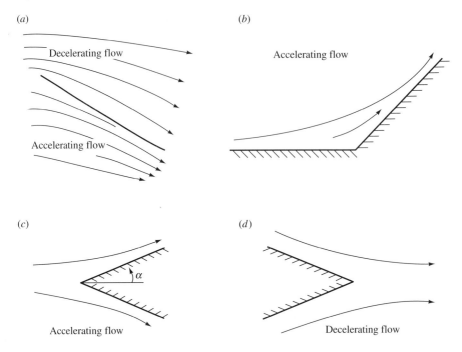

Figure 8.3.1   Examples of boundary layers in accelerating and decelerating flow. (a) Uniform flow past a flat plate at a finite angle of attack; (b) flow near a corner; (c, d) flow past a wedge-shaped body.

where c is a positive coefficient and m is a positive or negative constant, first studied by Falkner and Skan (1931). When $m = 0$, we recover the flow over a flat plate at zero angle of attack, whereas when $m = 1$ we obtain orthogonal stagnation-point flow against a flat plate. Intermediate values of m correspond to symmetric flow past a wedge of semiangle equal to $\alpha = \pi m/(m + 1)$, illustrated in Figure 8.3.1(c). Differentiating Eq. (8.3.1) yields the streamwise acceleration or deceleration of the outer flow

$$\frac{dU}{dx} = cmx^{m-1} \tag{8.3.2}$$

which shows that the outer flow accelerates when $m > 0$ and decelerates when $m < 0$. In the first case the continuity equation requires that the derivative with respect to y of the component of the velocity of the outer flow that is normal to the boundary, V, is negative, $\partial V/\partial y < 0$. Since V vanishes on the wall, it must have a negative value at the edge of the boundary layer; consequently, the motion of the fluid normal to the boundary will tend to convect the vorticity toward the boundary.

Substituting Eq. (8.3.2) into the steady-state form of Eq. (8.1.10), we obtain the particular boundary-layer equation

$$u\frac{\partial u}{\partial x} + v\frac{\partial u}{\partial y} = c^2 m\, x^{2m-1} + \nu\frac{\partial^2 u}{\partial y^2} \tag{8.3.3}$$

Proceeding as in the case of flow parallel to a flat plate discussed in Section 8.2, we identify the characteristic length L with the current streamwise position x, and use Eq. (8.1.6) to write

$$\delta(x) \approx \left(\frac{\nu x}{U}\right)^{1/2} = \left(\frac{\nu}{cx^{m-1}}\right)^{1/2} \tag{8.3.4}$$

Furthermore, we assume that the velocity profile across the boundary is self-similar; that is, u is a function of the similarity variable $\eta = y/\delta(x)$, and write $u(x, y) = UF(\eta)$, where

$$\eta \equiv \frac{y}{\delta(x)} = \sqrt{\frac{c}{\nu}} \frac{y}{x^{(1-m)/2}} \tag{8.3.5}$$

This self-similar velocity profile can be derived from the stream function shown in Eq. (8.2.4) with $F = f'$ and U given in Eq. (8.3.1). More specifically, differentiating Eq. (8.2.4) with respect to x and y and using the relation

$$\frac{\partial \eta}{\partial x} = \tfrac{1}{2}(m - 1)\frac{y}{x}\sqrt{\frac{U}{\nu x}} \tag{8.3.6}$$

we obtain the two components of the velocity

$$u(x, y) = \frac{\partial \psi}{\partial y} = U f'(\eta)$$

$$v(x, y) = -\frac{\partial \psi}{\partial x} = \tfrac{1}{2}\left(\frac{\nu U}{x}\right)^{1/2} [(1 - m)\eta f'(\eta) - (1 + m)f(\eta)] \tag{8.3.7}$$

Substituting the expressions (8.3.7) into Eq. (8.3.3) we obtain a third-order nonlinear inhomogeneous ordinary differential equation for the function f

$$f''' + \tfrac{1}{2}(m + 1)f f'' - mf'^2 + m = 0 \tag{8.3.8}$$

which is to be solved subject to the boundary conditions shown in Eq. (8.2.7a,b). When $m = 0$, Eqs. (8.3.7) and (8.3.8) reduce to the Blasius equations (8.2.5) and (8.2.6). When $m = 1$, we obtain Eq. (5.3.12), which provides us with an *exact solution* of the unsimplified form of the equation of motion describing orthogonal stagnation-point flow.

Applying Eq. (8.3.8) at the wall, located at $y = 0$, and enforcing the aforementioned boundary conditions, we find $f'''(0) = -m$, which is positive when $m < 0$, corresponding to a decelerating flow; in this case the curvature of the velocity profile is positive at the wall indicating the occurrence of back-flow. Noting that as η becomes large f''' must become negative in order for f' to tend to the constant value of unity, reveals the presence of an inflection point in the velocity profile. In Chapter 9 we shall see that this behavior renders the boundary layer susceptible to hydrodynamic instabilities mediated by the growth of small perturbations.

Since boundary conditions are specified at both ends of the solution domain $(0, \infty)$, we are faced with a boundary-value problem that must be solved using the shooting method described in Section 8.2. The solution for $m = 1$, corresponding to orthogonal stagnation-point flow, was obtained previously in Section 5.3.

Numerical solutions to the Falkner–Skan boundary-layer equation have been presented by a number of authors, following the original contributions of Hartree (1937) and Stewartson (1954). In Figure 8.3.2 we plot velocity profiles expressed by the derivative $f'(\eta)$ for several values of m. The numerical results converge when $f''(0) = 1.493$ for $m = 1.5$, $f''(0) = 1.232$ for $m = 1$, $f''(0) = 0.675$ for $m = 0.25$, $f''(0) = 0.332$ for $m = 0.25$, and $f''(0) = 0$ for $m = -0.0904$. There is a unique solution branch when $m > 0$, but multiple branches exist when $m < 0$. In the second case, we retain the solution that appears to be most physically relevant. Other solution branches are discussed by Evans (1968) and White (1974). The profiles for $m < 0$ exhibit an inflection point close to the wall, in agreement with our earlier observations.

The shear stress at the wall is given by

$$\sigma_{xy}(x, y = 0) = \mu \left(\frac{\partial u}{\partial y}\right)_{y=0} = c^{3/2} f''(0) x^{(3m-1)/2} \tag{8.3.9}$$

When $m = \tfrac{1}{3}$ the shear stress is independent of streamwise position, whereas when $m > \tfrac{1}{3}$ or $< \tfrac{1}{3}$, the shear stress increases or decreases with streamwise distance. When $m = -0.0904$, corresponding to a decelerating flow, $f''(0) = 0$, and the shear stress and thus the skin friction vanish uniformly along the wall.

The rate of convection of vorticity across a plane that is perpendicular to the wall is given by the right-hand side of Eq. (8.2.13), but because U is a function of streamwise position, vorticity must diffuse across the wall in order to satisfy the vorticity balance.

The *displacement thickness* of the boundary layer, defined in Eq. (8.2.17), is given by

$$\delta^*(x) = \int_0^\infty \left(1 - \frac{u}{U}\right) dy = \left(\frac{\nu x}{U}\right)^{1/2} \int_0^\infty (1 - f') d\eta = \alpha \left(\frac{\nu x}{U}\right)^{1/2}$$

$$= \alpha \left(\frac{\nu}{c}\right)^{1/2} x^{(1-m)/2}$$

(8.3.10)

where α is a positive dimensionless coefficient whose value depends upon the value of m. When $m < 1$, δ^* increases, whereas when $m > 1$, δ^* decreases in the streamwise direction due to the acceleration of the incident flow. When $m = 1$, corresponding to orthogonal stagnation-point flow, the displacement thickness is constant; the diffusion of vorticity due to viscous stresses is balanced by convection.

Computer Problem

8.3.1   **Solving the Falkner–Skan equation.** (a) Write a computer program called *FSBL* that solves the Falkner–Skan equation using a shooting method, reproduce Figure 8.3.2, and tabulate the

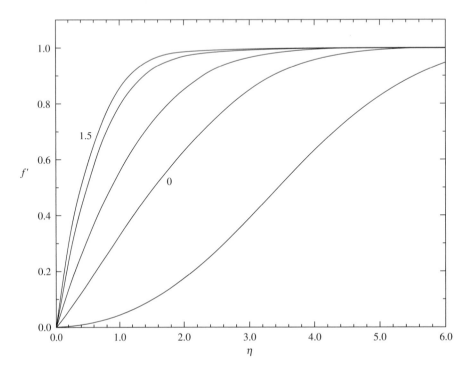

Figure 8.3.2   Velocity profile across Falkner–Skan boundary layers for several values of the acceleration parameter $m = 1.5, 1.0, 0.25, 0.0, -0.0904$.

corresponding values of the dimensionless coefficient α defined in Eq. (8.3.10). (b) Derive an expression for the momentum thickness that is similar to that shown in Eq. (8.3.10), and tabulate the counterpart of the dimensionless coefficient to α.

8.4 ## COMPUTATION OF BOUNDARY LAYERS AROUND TWO-DIMENSIONAL BODIES

A variety of approximate, asymptotic, and numerical methods are available for computing the flow within a boundary layer developing over an arbitrarily shaped two-dimensional body that either moves or is held stationary in an arbitrary irrotational flow. An authoritative review is presented in a comprehensive monograph edited by Rosenhead (1963), a concise but illuminating discussion is given by White (1974, Chapter 4), and more recent developments are discussed by Cebeci and Bradshaw (1977, 1984).

In this section we shall discuss an approximate method due to von Kármán and Pohlhausen that is easy to implement and requires a modest amount of computational effort, and shall outline the principles of a more general class of finite-difference methods.

Momentum Integral Balance

Von Kármán (1921) developed an approximate method for computing the flow within a boundary layer over a two-dimensional body that is placed in an arbitrary incident irrotational flow, based on a momentum integral balance.

We begin developing the method by considering an irrotational flow over a flat surface located at $y = 0$. As a first step, we introduce a control area that is enclosed by (1) two vertical planes located at x_1 and x_2, (2) the plane wall, and (3) a horizontal plane located at the elevation $y = h$. Maintaining our previous notation, we denote the tangential component of the velocity of the outer flow along the surface as U. Applying the momentum integral balance (3.1.17) with the Newtonian constitutive equation (3.2.17), under the additional stipulation that the physical properties of the fluid are uniform throughout the domain of flow, and neglecting the viscous stresses over the vertical and top planes, we obtain

$$\int_{x_1}^{x_2} \int_0^h \rho \frac{\partial u}{\partial t} \, dy \, dx = \int_0^h [-P + u(\rho u)]_{x=x_1} \, dy - \int_0^h [-P + u(\rho u)]_{x=x_2} \, dy$$
$$- \int_{x_1}^{x_2} [v(\rho u)]_{y=h} \, dx' - \int_{x_1}^{x_2} (\sigma_{xy})_{y=0} \, dx' \qquad (8.4.1)$$

Taking the limit as x_1 tends to x_2, assuming that P remains constant across the boundary layer, and setting $u = U$ at $y = h$, we obtain the differential relation

$$\rho \int_0^h \frac{\partial u}{\partial t} dy = h \left(\frac{dP}{dx} \right)_{y=h} - \rho \frac{d}{dx} \int_0^h u^2 dy - \rho U(v)_{y=h} - (\sigma_{xy})_{y=0} \qquad (8.4.2)$$

To reduce the number of unknowns, we eliminate $v(y = h)$ in favor of u using the continuity equation, setting

$$(v)_{x, y=h} = - \int_0^h \frac{\partial u(x, y')}{\partial x} dy' \qquad (8.4.3)$$

Equation (8.4.2) then becomes

$$\rho \int_0^h \frac{\partial u}{\partial t} dy = h \left(\frac{dP}{dx} \right)_{y=h} - \rho \frac{d}{dx} \int_0^h u^2 dy + \rho U \frac{d}{dx} \int_0^h u \, dy - (\sigma_{xy})_{y=0} \qquad (8.4.4)$$

Using Bernoulli's equation to evaluate the modified pressure and rearranging the integrals, we obtain

$$\rho \int_0^h \frac{\partial (U - u)}{\partial t} dy = -\rho \frac{d}{dx} \int_0^h u(U - u)\,dy - \rho \frac{d}{dx} \int_0^h U(U - u)\,dy$$

$$+ \rho U \frac{d}{dx} \int_0^h (U - u)\,dy + (\sigma_{xy})_{y=0} \tag{8.4.5}$$

which may be interpreted as a conservation law for the momentum deficit $\rho(U - u)$. To this end, we let h tend to infinity, and use the definitions of the displacement and momentum thicknesses to obtain the von Kármán momentum integral balance

$$\rho \frac{\partial (U\delta^*)}{\partial t} + \rho \frac{d}{dx}(U^2\Theta) + \rho \frac{d}{dx}(U^2\delta^*) - \rho U \frac{d}{dx}(U\delta^*) - (\sigma_{xy})_{y=0} = 0 \tag{8.4.6}$$

which may be rearranged to yield an expression for the wall shear stress in terms of the displacement and momentum thicknesses

$$\frac{(\sigma_{xy})_{y=0}}{\rho U^2} = \frac{\partial (U\delta^*)}{\partial t} + \frac{d\Theta}{dx} + (2\Theta + \delta^*)\frac{1}{U}\frac{dU}{dx} \tag{8.4.7}$$

(see also Problem 8.4.1). When the flow is steady, the first term on the right-hand side is absent. When U is independent of x, Eq. (8.4.7) reduces to Eq. (8.2.23), corresponding to uniform flow over a stationary flat plate that is held at zero angle of attack.

If fluid is injected into or withdrawn from the flow through a porous boundary with normal velocity equal to V, the right-hand side of Eq. (8.4.7) contains the additional term $-V/U$, where V is positive for injection (Problem 8.4.2).

The von Kármán–Pohlhausen Method

Von Kármán (1921) and Pohlhausen (1921) developed an approximate method for computing the boundary-layer thickness on the basis of Eq. (8.4.7), which proceeds according to the general principles discussed in the paragraphs following Eq. (8.2.23). The main idea is to describe the velocity profile across the boundary layer in terms of a certain reasonable functional form $u/U = F(\eta)$, where $\eta \equiv y/\Delta(x)$ and $\Delta(x)$ is an effective boundary-layer thickness that is analogous to the δ_{99} boundary-layer thickness, and then compute $\Delta(x)$ so as to satisfy the momentum integral balance (8.4.7). The implementation of this method for uniform flow over a flat plate where $F(\eta)$ within the boundary layer is described by a quarter of a period of a sinusoidal function, as shown in Eq. (8.2.24), was already discussed at the end of Section 8.2.

Pohlhausen described the function $F(\eta)$ within the boundary layer with a fourth-order polynomial, setting

$$F(\eta) = \begin{cases} \eta(a + b\eta + c\eta^2 + d\eta^3), & \text{for } 0 < \eta < 1 \\ 1, & \text{for } \eta > 1 \end{cases} \tag{8.4.8}$$

where the coefficients a, b, c, d are functions of x. It will be noted that this functional form respects the no-slip boundary condition at the wall.

To compute a, b, c, d, we require four boundary conditions. Demanding that both the first and second derivatives of the velocity remain continuous across the edge of the boundary layer yields

$$F(1) = 1, \qquad F'(1) = 0, \qquad F''(1) = 0 \tag{8.4.9a}$$

Furthermore, enforcing Eq. (8.1.11) we obtain

$$F''(0) = -\Lambda, \qquad \text{where} \qquad \Lambda = \frac{\Delta^2}{\nu}\frac{dU}{dx} \tag{8.4.9b}$$

The dimensionless parameter Λ, introduced for convenience in lieu of Δ, expresses the ratio between the magnitudes of the inertial acceleration forces of the irrotational flow and the viscous forces within the boundary layer; $\Lambda = 0$ when $dU/dx = 0$. By definition, the effective boundary-layer thickness Δ is related to Λ by

$$\Delta(x) = \left(\frac{\nu\Lambda}{U'}\right)^{1/2} \tag{8.4.10}$$

where we have set $U' = dU/dx$.

Requiring that Eq. (8.4.8) satisfy Eq. (8.4.9a,b), we obtain

$$a = 2 + \Lambda/6, \qquad b = -\Lambda/2, \qquad c = -2 + \Lambda/2, \qquad d = 1 - \Lambda/6 \tag{8.4.11}$$

Substituting these values into Eq. (8.4.8) and rearranging its terms yields the velocity profile in terms of the parameter Λ

$$F(\eta) = \begin{cases} \eta(2 - 2\eta^2 + \eta^3) + \Lambda\frac{1}{6}\eta(1 - \eta)^3, & \text{for } 0 < \eta < 1 \\ 1, & \text{for } \eta > 1 \end{cases} \tag{8.4.12}$$

A family of profiles corresponding to several values of Λ is shown in Figure 8.4.1. When $\Lambda > 12$, corresponding to a strongly accelerating flow, the profiles exhibit an overshooting that places a limit on the physical relevance of the fourth-order polynomial expansion. On the other hand, when $\Lambda = -12$, the slope of the velocity profile at the wall vanishes, which suggests that the flow is about to reverse direction; at that point, the approximations that led us to the boundary-layer equations cease to be valid, and the boundary layer is believed to separate from the wall yielding regions of recirculating flow.

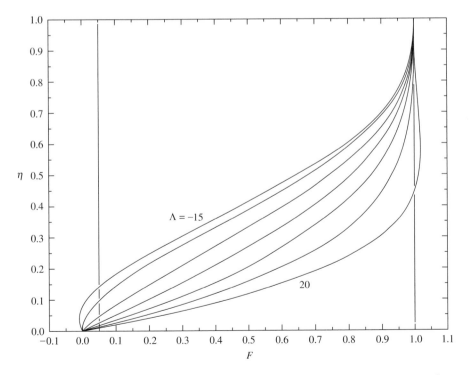

Figure 8.4.1 Profiles of the Pohlhausen polynomial for several values of the acceleration parameter $\Lambda = 20, 12, 6, 0, -6, -12, -15$.

The displacement thickness, momentum thickness, and shear stress at the wall may be computed readily in terms of Δ and Λ on the basis of Eq. (8.4.12), and are found to be

$$\delta^* = \frac{\Delta}{10}\left(3 - \frac{\Lambda}{12}\right)$$

$$\Theta = \frac{\Delta}{315}\left(37 - \frac{\Lambda}{3} - \frac{5}{144}\Lambda^2\right) \tag{8.4.13}$$

$$(\sigma_{xy})_{y=0} = \frac{\mu U}{\Delta}\left(2 + \frac{\Lambda}{6}\right)$$

Expressing Δ in terms of Λ by means of Eq. (8.4.10) yields corresponding expressions in terms of Λ alone. The central idea is to substitute the expressions (8.4.13) into Eq. (8.4.7), and thereby obtain a first-order nonlinear ordinary differential equation for Λ with respect to x. Having computed Λ, we recover Δ from Eq. (8.4.10).

Semi-infinite flat plate at zero angle of attack

In the case of uniform flow past a semi-infinite flat plate that is held at zero angle of attack studied in Section 8.2, U is a constant, Λ vanishes, and the right-hand side of Eq. (8.4.10) becomes indeterminate. This singular behavior, however, does not impose an essential difficulty; substituting the expressions in Eqs. (8.4.13) into Eq. (8.4.7) yields an ordinary differential equation for $\Delta(x)$

$$\Delta(x)\frac{d\Delta(x)}{dx} = \frac{630}{37}\frac{\nu}{U} \tag{8.4.14}$$

which can be integrated readily from the leading edge of the plate up to a point x subject to the initial condition $(\Delta)_{x=0} = 0$ to give

$$\Delta(x) = 5.836\left(\frac{\nu x}{U}\right)^{1/2} \tag{8.4.15}$$

It is instructive to compare Eq. (8.4.15) with Eq. (8.2.28), which corresponds to a sinusoidal velocity profile. Substituting Eq. (8.4.15) into the right-hand sides of Eqs. (8.4.13) yields

$$(\sigma_{xy})_{y=0} = 0.343\mu U\left(\frac{U}{\nu x}\right)^{1/2}$$

$$\delta^* = 1.751\left(\frac{\nu x}{U}\right)^{1/2} \tag{8.4.16}$$

$$\Theta = 0.685\left(\frac{\nu x}{U}\right)^{1/2}$$

with an associated shape factor $H = 2.556$, which are reasonably close to the exact values given in Eqs. (8.2.9), (8.2.18), and (8.2.22).

Numerical solutions

In the case of uniform flow past a semi-infinite flat plate held at zero angle of attack, we were able to derive an analytical solution for Δ. Under more general circumstances, it is necessary to obtain the solution using numerical methods. To this end, it is convenient to introduce the *Holstein–Bohlen dimensionless parameter*

$$\lambda \equiv \frac{\Theta^2}{\Delta^2}\Lambda = \frac{\Theta^2}{\nu}U' \tag{8.4.17}$$

whose physical interpretation is similar to that of Λ. Using the expression for the momentum thickness given in Eqs. (8.4.13) we obtain a nonlinear relationship between λ and Λ

$$\left(\frac{\lambda}{\Lambda}\right)^{1/2} = \frac{1}{315}\left(37 - \frac{\Lambda}{3} - \frac{5}{144}\Lambda^2\right) \tag{8.4.18}$$

The value $\Lambda = -12$ corresponds to $\lambda = -0.15673$, at which point the boundary layer is expected to separate.

To expedite the solution, we multiply both sides of the momentum integral balance (8.4.7) by Θ, and rearrange to obtain

$$\frac{1}{\nu}\frac{d\Theta^2}{dx} = \frac{d}{dx}\left(\frac{\lambda}{U'}\right) = \frac{2S(\lambda) - 2[2 + H(\lambda)]\lambda}{U} \tag{8.4.19}$$

where we have introduced the *shear function S* defined as

$$S \equiv \frac{(\sigma_{xy})_{y=0}\Theta}{\mu U} \tag{8.4.20}$$

Physically, the shear function expresses the ratio between the shear stress at the wall and the average value of the shear stress across the boundary layer, and is thus another measure of the sharpness of the velocity profile.

Using the expressions in Eqs. (8.4.13), we find that the shape factor and shear function are given by

$$H = \frac{315}{10}\frac{3 - \Lambda/12}{37 - \Lambda/3 - 5\Lambda^2/144}$$

$$S = \frac{1}{315}\left(2 + \frac{\Lambda}{6}\right)\left(37 - \frac{\Lambda}{3} - \frac{5}{144}\Lambda^2\right) \tag{8.4.21}$$

where Λ can be expressed in terms of λ by means of Eq. (8.4.18). The numerical procedure involves the following steps:

1. Given the value of λ at a particular position x, compute the corresponding value of Λ by solving the nonlinear algebraic equation (8.4.18).
2. Evaluate the functions S and H using Eqs. (8.4.21).
3. Compute the right-hand side of Eq. (8.4.19).
4. Advance the value of λ using, for instance, a Runge–Kutta method, and return to step 1.

The numerical integration usually begins at a stagnation point where U vanishes and the right-hand side of Eq. (8.4.19) becomes indeterminate. To avoid the occurrence of a singularity, we require that the numerator be also equal to zero, and this provides us with a nonlinear algebraic equation for Λ that has the physically acceptable solution $\Lambda = 7.052$ corresponding to $\lambda = 0.0770$. This value is used to initialize the computation. Note, however, that the assumptions that led us to the boundary-layer equations cease to be valid at the stagnation point, where the flow is not nearly unidirectional, but this occurrence does not restrict the validity of the solution far from the stagnation point.

To evaluate the right-hand side of Eq. (8.4.19) at the stagnation point, let us denote the numerator by $F(\lambda)$. Applying the l'Hôpital rule to evaluate the fraction, we obtain

$$\frac{d}{dx}\left(\frac{\lambda}{U'}\right)_{x=0} = \left(\frac{1}{U'}\frac{dF}{dx}\right)_{x=0} = \left(\frac{dF}{d\lambda}\frac{1}{U'}\frac{d\lambda}{dx}\right)_{x=0} = \left\{\frac{dF}{d\lambda}\left[\frac{d}{dx}\left(\frac{\lambda}{U'}\right) + \lambda\frac{U''}{U'^2}\right]\right\}_{x=0} \tag{8.4.22}$$

which may be rearranged to yield

$$\frac{d}{dx}\left(\frac{\lambda}{U'}\right)_{x=0} = \left\{\lambda\frac{dF}{d\lambda}\left(1 - \frac{dF}{d\lambda}\right)^{-1}\frac{U''}{U'^2}\right\}_{x=0} \tag{8.4.23}$$

Evaluating the coefficients in the right-hand side, we obtain the initial value

$$\frac{d}{dx}\left(\frac{\lambda}{U'}\right)_{x=0} = -0.0652\left(\frac{U''}{U'^2}\right)_{x=0} \tag{8.4.24}$$

Boundary layer around a curved surface

The von Kármán–Pohlhausen method was developed with reference to a flat boundary, where the x coordinate increases along the boundary in the direction of the velocity of the incident flow. To tackle the more general case of a curved boundary, we simply replace x with the arc length l measured along the boundary in the direction of the tangential velocity of the incident flow and begin the integration from a stagnation point. There might be a difficulty in computing the right-hand side of Eq. (8.4.24) at the stagnation point, due to the fact that the acceleration U' may vanish or assume an infinite value, but this can be circumvented by carrying out the first step on the basis of the Falkner–Skan similarity solution with a proper value of the exponent m (Problem 8.4.4).

Boundary layer around a circular cylinder

As an application, let us consider uniform flow past a stationary cylinder of radius a, with vanishing circulation around it. Far from the cylinder, the velocity tends to obtain the uniform value $-U_0\mathbf{i}$ where $U_0 > 0$ and $\mathbf{i}$ is the unit vector in the direction of the x axis. In this case U' and U'' designate derivatives with respect to arc length $l = a\theta$ measured around the cylinder from the front stagnation point, where θ is the polar angle.

Using the potential-flow solution given in Eq. (7.7.7) with $\kappa = 0$, we find that the tangential component of the outer velocity and its derivatives are given by

$$U = 2U_0\sin\theta, \qquad \frac{dU}{dl} = \frac{2U_0}{a}\cos\theta, \qquad \frac{d^2U}{dl^2} = -\frac{2U_0}{a^2}\sin\theta \tag{8.4.25}$$

Equation (8.4.24) yields $[d(\lambda/U')/dl]_{l=0} = 0$, which, along with the initial value $\lambda = 0.0770$, is used to initialize the computations.

Graphs of λ and Λ as functions of the polar angle θ are shown in Figure 8.4.2(a); the corresponding velocity profiles across the boundary layer at different stations around the cylinder may be inferred from Figure 8.4.1. The numerical results reveal that $\Lambda = -12$ when $\theta = 109.5°$, at which point the shear stress vanishes and the boundary layer is expected to separate. Comparing this estimate with the experimentally observed critical value of $\theta = 80.5°$, we find a serious disagreement which is attributed to the fact that, in practice, the outer flow deviates substantially from the potential flow solution given in Eq. (7.7.7) due to the presence of a wake. Using, however, laboratory measurements to describe the distribution of the outer tangential velocity U instead of the idealized distribution of Eqs. (8.4.25) yields results that are in excellent agreement with experimental observation (White, 1974, p. 323).

In Figure 8.4.2(a) we also plot with the line labeled 1 the distribution of the reduced effective boundary-layer thickness $\Delta(x)/a$, with the line labeled 2 the distribution of reduced displacement thickness $\delta^*/(\nu a/U_0)^{1/2}$, with the line labeled 3 the distribution of reduced momentum thickness $\Theta/(\nu a/U_0)^{1/2}$, and with the line labeled 4 the distribution of reduced wall shear stress $\sigma_W/(\mu U_0/a)$. In Figure 8.4.2(b) we plot the associated distributions of the shape factor H and shear function S. Note that the shear function becomes negative at the point where separation occurs, which means that the average value of the shear stress across the boundary layer is still positive. The values of all functions shown in Figure 8.4.2(a,b) at the front stagnation point are close to those predicted by the Falkner and Skan similarity solution for orthogonal stagnation-point flow discussed in Section 8.3, with $c = 2U_0/a$.

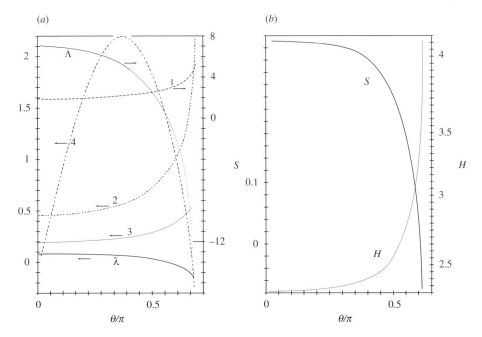

Figure 8.4.2 Development of a Prandtl boundary layer along a circular cylinder of radius a that is held stationary in an incident uniform flow with velocity U_0, with vanishing circulation around it, computed using the von Kármán–Pohlhausen method. (a) Distributions of the effective boundary-layer thickness $\Delta(x)/a$ shown with the line labeled 1, reduced displacement thickness $\delta^*/(\nu a/U_0)^{1/2}$ shown with the line labeled 2, reduced momentum thickness $\Theta/(\nu a/U_0)^{1/2}$ shown with the line labeled 3, and reduced wall shear stress $\sigma_w/(\mu U_0/a)$ shown with the line labeled 4. (b) Distribution of the shape factor and shear function.

Extensions

The von Kármán–Pohlhausen method can been modified in several ways to enhance its physical relevance and predictive capacity. In one line of extension, the fourth-order Pohlhausen polynomial is replaced with an improved function that is constructed on the basis of theoretical arguments or laboratory observations. For instance, one may use a high-degree polynomial, which is required to satisfy additional boundary conditions at the wall and at the end of the boundary layer at $y = \Delta(x)$ (Problem 8.4.3). In a related extension, the shear and shape functions $S(\lambda)$ and $H(\lambda)$ as well as the numerator on the right-hand side of Eq. (8.4.19), call it $F(\lambda)$, are described in terms of empirical correlations.

Thwaites (1949), in particular, proposed the linear form $F(\lambda) = 0.45\lambda - 6.0$, which is based on laboratory data, and allows for the analytical solution

$$\Theta^2(x) = \Theta^2(x_0) + \frac{0.45\nu}{U^6} \int_{x_0}^{x} U^5 dx \qquad (8.4.26)$$

where x_0 is an arbitrary point (Problem 8.4.4). The predictions of this equation are in excellent agreement with experimental observations.

Numerical Methods

The boundary-layer equations provide us with a system of two parabolic nonlinear partial differential equations with respect to arc length l that may be integrated numerically using standard space-marching, weighted-residual, or finite-difference methods discussed in Chapter 12. Expressing, in particular, the velocity in terms of the stream function does away with the continuity equation and allows us to concentrate on solving a single third-order parabolic partial differential equation.

Görtler, Smith, and Clutter

Smith and Clutter (1963) developed a simple yet accurate method that proceeds by regarding the stream function as a function of the independent variables x and $\eta = y(U/\nu x)^{1/2}$, and writing $\psi = (U\nu x)^{1/2} f(x, \eta)$, where f is an unknown function (Görtler 1957); x stands for the arc length along the boundary l, and y stands for the arc length in the normal direction. Substituting this functional form into the boundary-layer equation yields the third-order partial differential equation

$$f''' + \tfrac{1}{2}(m + 1)f f'' - m f'^2 + m = x\left(f'\frac{\partial f'}{\partial x} - f''\frac{\partial f}{\partial x}\right) \qquad (8.4.27)$$

where a prime designates differentiation with respect to η, and $m = (x/U)dU/dx$. The boundary conditions are $f(x, 0) = 0$, $f'(x, 0) = 0$, and $f'(x, \infty) = 1$. For the Falkner–Skan distribution of Eq. (8.3.1), m is constant, the right-hand side of (8.4.27) vanishes, and Eq. (8.4.27) reduces to the *ordinary* differential equation (8.3.8).

Approximating the partial derivatives with respect to x in Eq. (8.4.27) with first-order or second-order backward finite differences yields an ordinary differential equation for f with respect to η that can be solved subject to the aforementioned boundary conditions using a standard shooting method (Problem 8.4.6). The efficiency of this method has been demonstrated on a number of occasions and for a broad range of flows. In another implementation of the method, Eq. (8.4.27) is rewritten as a system of three first-order equations with respect to η, which is then integrated using a standard finite-difference method (Cebeci and Bradshaw, 1977, p. 214).

Finite-difference methods

The boundary-layer equations may be solved using simplified versions of more general numerical methods developed for problems of nonlinear convection–diffusion, as discussed in Chapters 12 and 13. Reviews of early and more recent implementations are presented by Blottner (1970), Cebeci and Bradshaw (1984), and Fletcher (1988, vol. 2).

PROBLEMS

8.4.1 **Momentum integral balance.** Equation (8.4.7) may be derived directly from the boundary-layer equations (8.1.1) and (8.1.10) without a reference to the integral momentum balance (Pohlhausen, 1921). Multiply the left-hand side of Eq. (8.1.1) by $U - u$, add the product to the right-hand side of Eq. (8.1.10), integrate the result from $y = 0$ to ∞, and use the required boundary conditions to obtain Eq. (8.4.7).

8.4.2 **Boundary layer with suction.** Show that if fluid is injected into or withdrawn from the flow through a porous wall with normal velocity V, the right-hand side of Eq. (8.4.7) contains the additional term $-V/U$; V is negative for suction.

8.4.3 **Boundary conditions at the wall.** Assume that the velocity profile across a steady boundary layer is described by the functional form $u/U = F(\eta)$. Show that, in addition to satisfying the no-slip condition $F(0) = 0$ and Eqs. (8.4.9a), the velocity profile should be subject to the boundary condition $F'''(0) = 0$.

8.4.4 **Thwaites's choice.** Integrate Eq. (8.4.19) when $F(\lambda)$ is expressed by the general linear form $a\lambda - b$, where a and b are two constants, and thereby derive a generalized form of Eq. (8.4.26).

8.4.5 **Similarity patching.** Develop an algorithm that allows you to integrate the boundary-layer equations by pretending that the boundary layer is composed of a sequence of finite patches of Falkner–Skan boundary layers, where the exponent m varies over the boundary (Smith, 1956).

8.4.6 **Smith and Clutter.** Derive the explicit form of an ordinary differential equation for the function f with respect to η, by approximating the partial derivatives of f with respect to x in Eq. (8.4.27) with second-order backward finite differences (Section B.5, Appendix B), and discuss the implementation of the method.

Computer Problem

8.4.7 **Boundary layer around a circular cylinder.** (a) Write a program called *KPBL* that uses the von Kármán–Pohlhausen method to compute the boundary layer around a circular cylinder with vanishing circulation around it subject to an incident uniform flow discussed in the text, and reproduce Figure 8.4.2(a, b). (b) Repeat part (a) with a finite circulation of your choice, and discuss the relative behavior of your results.

8.5 | BOUNDARY LAYERS IN AXISYMMETRIC AND THREE-DIMENSIONAL FLOWS

The derivation and solution of the boundary-layer equations for axisymmetric and three-dimensional flows are carried out according to the general principles discussed in the preceding section for two-dimensional flow, with some rather minor modifications.

Axisymmetric Flow

To carry out the boundary-layer analysis, we introduce the arc length along the trace of a boundary in an azimuthal plane l and arc length l_η in the normal direction, where the latter is associated with the orthogonal curvilinear coordinate η. The corresponding tangential and normal components of the velocity are denoted by u and v, and the azimuthal component of the velocity expressing swirling motion is denoted by w. In these curvilinear coordinates, the continuity equation takes the form

$$\frac{\partial(\sigma u)}{\partial l} + \sigma \frac{\partial v}{\partial l_\eta} = 0 \tag{8.5.1}$$

where σ is the distance from the axis of the flow, which is assumed to be much larger than the boundary-layer thickness. The normal component of the equation of motion states that, to this level of approximation, the pressure drop across the boundary layer can be neglected. Consequently, the pressure distribution across and over the boundary layer may be computed on the basis of Euler's equation for the incident flow, and is a function of arc length l alone. We thus find

$$-\frac{1}{\rho}\frac{\partial P}{\partial l} = \frac{\partial U}{\partial t} + U\frac{\partial U}{\partial l} - \frac{W^2}{\sigma}\frac{d\sigma}{dl} \tag{8.5.2}$$

$$-\frac{1}{\rho\sigma}\frac{\partial P}{\partial \varphi} \approx \frac{\partial W}{\partial t} + U\frac{\partial W}{\partial l} + \frac{UW}{\sigma}\frac{d\sigma}{dl} = 0 \tag{8.5.3}$$

where U and W are the meridional tangential and swirling components of the velocity of the outer flow. The meridional tangential and azimuthal components of the equation of motion simplify to the boundary-layer equations

$$\frac{\partial u}{\partial t} + u\frac{\partial u}{\partial l} + v\frac{\partial u}{\partial l_\eta} - \frac{w^2}{\sigma}\frac{d\sigma}{dl} = -\frac{1}{\rho}\frac{\partial P}{\partial l} + \nu\frac{\partial^2 u}{\partial l_\eta^2} \tag{8.5.4}$$

$$\frac{\partial w}{\partial t} + u\frac{\partial w}{\partial l} + v\frac{\partial w}{\partial l_\eta} + \frac{uw}{\sigma}\frac{d\sigma}{dl} = \nu\frac{\partial^2 w}{\partial l_\eta^2} \tag{8.5.5}$$

It is interesting to note that, in the absence of swirling motion, Eq. (8.5.5) is satisfied in a trivial manner, and Eq. (8.5.4) becomes identical to Eq. (8.1.10) with a straightforward change in notation; the effect of axisymmetry is manifested in the continuity equation alone.

Mangler's transformation

Mangler (1948) devised a remarkable transformation applicable to axisymmetric flow without swirling motion, which reduces the system of Eqs. (8.5.1) and (8.5.4) to the system of Eqs. (8.1.1) and (8.1.10) describing two-dimensional flow. The latter is written in terms of the judiciously chosen variables

$$\hat{x} = \frac{1}{L^2} \int_0^l \sigma^2(l')\,dl', \qquad \hat{y} = \frac{\sigma}{L} l_\eta,$$

$$\hat{u} = u, \qquad \hat{v} = \frac{L}{\sigma}\left(v + \frac{l_\eta}{\sigma} \frac{d\sigma}{dl} u \right) \tag{8.5.6}$$

Mangler's transformation reduces the problem of computing a boundary layer over an axisymmetric body to that of computing a boundary layer over a two-dimensional body with modified geometry in a fictitious flow. A practical drawback of the method is that the pressure distribution of the fictitious two-dimensional flow may be considerably more involved than that of the physical flow.

Momentum integral balance

Working as in the case of two-dimensional flow, we derive the momentum integral balance for axisymmetric flow without swirling motion,

$$\frac{1}{\rho U^2} \sigma_{xy}(l_\eta = 0) = \frac{\partial(U\delta^*)}{\partial t} + \frac{1}{\sigma}\frac{d(\sigma\Theta)}{dl} + (2\Theta + \delta^*)\frac{1}{U}\frac{dU}{dl} \tag{8.5.7}$$

which is a slightly modified version of Eq. (8.4.7) (Problem 8.5.1). The displacement and momentum thicknesses are defined as in the case of two-dimensional flow in terms of integrals with respect to arc length in the normal direction l_η.

Using the fourth-degree Pohlhausen polynomial and the Holstein–Bohlen dimensionless parameter introduced in Eq. (8.4.17), we find that the momentum integral balance for steady flow takes the form

$$\frac{1}{\sigma^2}\frac{d}{dx}\left(\frac{\sigma^2\lambda}{U'} \right) = \frac{2S(\lambda) - 2[2 + H(\lambda)]\lambda}{U} \tag{8.5.8}$$

where the functions S and H are defined in Eqs. (8.4.21) and a prime denotes a derivative with respect to l (Problem 8.5.1). At the front stagnation point on a smooth surface, σ reduces to l and U behaves like kl, where k is the local rate of extension. Making these substitutions into the left-hand side of Eq. (8.5.8) and requiring that $d\lambda/dl$ be finite yields the starting value $\lambda = 0.0571$. The numerical integration of Eq. (8.5.8) proceeds as in the case of two-dimensional flow; separation occurs when $\lambda = -0.15673$ or $\Lambda = -12$.

For uniform flow past a stationary sphere, the numerical integration of Eq. (8.5.8) shows that the boundary layer separates at a polar angle θ that is substantially higher than the experimentally observed value of 83° (Problem 8.5.6). The discrepancy is attributed to the poor representation of the outer irrotational flow due to the neglect of wakes.

Görtler, Smith, and Clutter

The counterpart of Görtler's equation (8.4.27) for axisymmetric flow is

$$f''' + \frac{1}{2}\left(m + 1 + \frac{2l}{\sigma}\frac{d\sigma}{dl} \right)ff'' - mf'^2 + m = l\left(f'\frac{\partial f'}{\partial l} - f''l \right) \tag{8.5.9}$$

The solution can be computed using a shooting or a direct finite-difference method as discussed previously for two-dimensional flow.

Three-Dimensional Flow

The computation of boundary layers in three-dimensional flow becomes complicated by the fact that the tangential component of the velocity of the outer flow may point in an arbitrary direction and must be described in terms of two scalar components. The boundary-layer equations emerge from two tangential components of the equation of motion by discarding the second partial derivatives of the velocity with respect to arc length in the tangential directions. The normal component of the equation of motion states that the pressure drop across the boundary layer is negligibly small, and this allows us to set the tangential pressure gradient equal to that of the outer irrotational flow. In global Cartesian coordinates, the boundary-layer equations for an irrotational incident flow with tangential boundary velocity $\mathbf{U}$ assume the form

$$\frac{\partial \mathbf{u}}{\partial t} \cdot \mathbf{P} + \mathbf{u} \cdot (\nabla \mathbf{u}) \cdot \mathbf{P} = \frac{\partial \mathbf{U}}{\partial t} \cdot \mathbf{P} + \tfrac{1}{2} \mathbf{P} \cdot \nabla (\mathbf{U} \cdot \mathbf{U})$$
$$+ \nu (\mathbf{nn}) : (\nabla \nabla \mathbf{u}) \cdot \mathbf{P} \qquad (8.5.10)$$

where $\mathbf{P} = \mathbf{I} - \mathbf{nn}$ is the tangential projection operator. The continuity equation retains its full form for three-dimensional flow.

In practice, Eq. (8.5.10) is decomposed into two tangential components corresponding to a pair of orthogonal or nonorthogonal surface curvilinear coordinates ξ_1 and ξ_2 that wrap around the boundary, and a third coordinate ξ_3 that varies in the normal direction and is thus orthogonal to both ξ_1 and ξ_2 (Problem 8.5.3). The surface streamlines of the incident potential flow and a family of curves that are normal to them, called the *intrinsic coordinates,* offer a convenient choice for ξ_1 and ξ_2. The two scalar component boundary-layer equations involve derivatives of the velocity with respect to ξ_1, ξ_2, and ξ_3 multiplied by the metrics of these three curvilinear coordinates. The wall shear stress, also called skin friction, and displacement thickness are two tangential vectors that may point in an arbitrary direction, while the momentum thickness is a rank-two tensor. The momentum integral balance and the von Kármán–Pohlhausen method and its variations are developed in a straightforward manner but assume more involved vectorial forms (Mager, 1964).

The increase in computational difficulty in going from a two-dimensional to a three-dimensional problem is usually substantial, and boundary layers are no exception to this rule. A discussion of the various approximate and numerical methods for computing boundary layers in three-dimensional flow can be found in dedicated monographs, including those by Rosenhead (1963), Moore (1964), and Cebeci and Bradshaw (1977).

PROBLEMS

8.5.1 **Momentum integral balance.** (a) Derive the momentum integral balance (8.5.7), and (b) the von Kármán–Pohlhausen differential equation (8.5.8).

8.5.2 **Mangler's transformation.** Compute and discuss the transformed variables according to Mangler for uniform flow past a sphere.

8.5.3 **Three-dimensional boundary layers.** Write the explicit forms of the tangential components of Eq. (8.5.10) corresponding to a system of orthogonal surface curvilinear coordinates ξ_1 and ξ_2 that wrap around a body and a third coordinate ξ_3 that is normal to the body.

8.5.4 **Spreading of an axisymmetric jet.** Consider the spatial spreading of an axisymmetric jet that emanates from a circular slot into a quiescent fluid along the x axis. Applying the momentum integral balance for a control volume that has two parallel sides perpendicular to the x axis shows that the value of the momentum integral

$$M = 2\pi \rho \int_0^\infty u^2 \sigma \, d\sigma \qquad (8.5.11)$$

is independent of x. Implementing the boundary-layer approximation in the axial component of the equation of motion written in terms of global cylindrical polar coordinates, and noting that the ambient pressure is constant, we obtain

$$u_x \frac{\partial u_x}{\partial x} + u_\sigma \frac{\partial u_x}{\partial \sigma} = \nu \frac{1}{\sigma} \frac{\partial}{\partial \sigma} \left(\sigma \frac{\partial u_x}{\partial \sigma} \right) \tag{8.5.12}$$

It is expedient to introduce the Stokes stream function and express it in the form

$$\Psi(x, \sigma) = \nu x f(\eta), \qquad \text{where } \eta = \frac{\sigma}{x} \tag{8.5.13}$$

where f is a dimensionless function and η is a dimensionless similarity variable. (a) Show that the boundary-layer equation (8.5.12) leads to the following nonlinear third-order homogeneous ordinary differential equation

$$\left(f'' - \frac{f'}{\eta} \right)' + \frac{f f''}{\eta} + \frac{f'^2}{\eta} - \frac{f f'}{\eta^2} = 0 \tag{8.5.14}$$

which is to be solved subject to the boundary conditions $f(0) = 0$, $f'(0) = 0$, $f'(\infty) = 0$, and the integral constraint (8.5.11). (b) Show that if $f(\eta)$ is a solution, then $f(c\eta)$, where c is a constant, is another solution. (c) Derive the solution $f(\eta) = 4\alpha\eta^2/(4 + \alpha\eta^2)$, where $\alpha = 3M/16\pi\mu\nu$ is a dimensionless constant. (d) Show that the axial volumetric flow rate Q is equal to $8\pi\nu x$, independently of M.

8.5.5   **Widening of an axisymmetric wake.** Consider the widening of an axisymmetric wake behind a streamlined object that is held stationary in a uniform incident stream flowing along the x axis with velocity U. Show that under the assumption that the rate of widening is sufficiently small, the boundary-layer approximations lead to the solution

$$u(x, \sigma) = U \left[1 - \frac{c}{x} \exp\left(-\frac{U\sigma^2}{4\nu x} \right) \right] \tag{8.5.15}$$

where c is a dimensional constant given by $c = F/4\pi\mu U$, and F is the drag force exerted on the object.

Computer Problem

8.5.6   **Boundary layer around a sphere.** Write a program called *KPABL* that uses the von Kármán–Pohlhausen method to compute the boundary layer around a sphere that is placed in an incident uniform flow. Plot the distribution of the wall shear stress, and estimate the polar angle at which the boundary layer is expected to separate.

8.6 | UNSTEADY BOUNDARY LAYERS

The physical processes that govern the behavior of boundary layers developing over solid surfaces in unsteady flow can be illustrated by considering the evolution of the boundary layer developing along a rigid body that is held stationary in an impulsively started incident flow or is suddenly introduced into a steady irrotational flow. At the initial instant, the velocity field is irrotational everywhere except within a thin vortex layer that lines the surface of the body; viewed at a distance, the vortex layer resembles a vortex sheet whose strength is equal to the tangential component of the velocity of the potential flow. Immediately after the motion has been initiated, the vortex sheet starts diffusing into the flow; locally around the boundary, the flow resembles that developing over an infinite flat plate subject to an impulsively started uniform flow discussed in Section 5.2. When the thickness of the vortex layer has become significant, convection of vorticity in the tangential and normal directions become important, and the leading-order similarity solution that describes the local flow over a flat plate is modified with the addition of a second-order term that is proportional to the time since startup (Stuart, 1963). Examination of the second-order term shows that the boundary shear stress vanishes and backflow occurs at the point where the irrotational flow decelerates along the boundary. The time of separation predicted by the second-order solution is in remarkably good agreement with experimental observations.

More generally, changes in the conditions of an incident flow affect both the distribution of the tangential velocity and pressure gradient that drive the flow within the boundary layer, and therefore cause vorticity to diffuse across the boundaries and enter or exit the flow. Describing the evolution of the boundary layer presents us with a challenging analytical and computational problem that has been tackled only on a limited number of occasions using asymptotic methods (Rosenhead, 1963).

Oscillatory Boundary Layers

An important class of boundary layers occurs during oscillatory flow past a stationary body or a suspended particle, pulsating flow within a channel or tube, or oscillatory flow caused by the natural or forced vibrations of a rigid or deformable body. For example, an oscillatory boundary layer may develop at the bottom of the sea due to the periodic flow generated by the propagation of free-surface or internal gravity waves.

To illustrate the physical mechanisms that govern the structure of an oscillatory boundary layer, let us consider a low-amplitude uniform pulsating flow with angular frequency Ω past a stationary rigid body. The reversal of the velocity of the incident flow over one period causes the sign of the boundary vorticity to alternate in a periodic manner. As a result, vorticity of positive and negative sign diffuses across the boundary and enters the flow in a sequential fashion, and the boundary layer consists of adjacent vortex layers that diffuse into one another and thus tend to cancel each other as they are convected by the incident flow. A simple example is provided by the Stokes boundary layer occurring during oscillatory unidirectional flow over an infinite plate discussed in Section 5.2. The Stokes boundary layer is composed of traveling bands of vorticity of alternating sign that penetrate the flow by a distance that is comparable to the Stokes boundary-layer thickness $\delta = (2\nu/\Omega)^{1/2}$. Similar boundary layers occur during oscillatory flow in channels and tubes discussed in Section 5.2.

When the frequency parameter of the incident oscillatory flow $\beta = \Omega L^2/\nu$ is large and the geometry of the boundary is sufficiently simple, the thickness of the oscillatory boundary layer is small compared to the size of the body L. Under these conditions, the main part of the flow is nearly irrotational, call it $\mathbf{u}^\infty$, and the boundary layer is driven by the tangential component of $\mathbf{u}^\infty$ which is given by $(\mathbf{n} \times \mathbf{u}^\infty) \times \mathbf{n} = (\mathbf{I} - \mathbf{nn}) \cdot \mathbf{u}^\infty \equiv U\mathbf{t}\cos(\Omega t)$, where $\mathbf{n}$ is the unit normal vector directed into the flow, $\mathbf{t}$ is a unit tangential vector, and U is the amplitude of the tangential component of the velocity. Neglecting the curvature of the boundary allows us to approximate the flow near a particular point on the boundary with the flow over an infinite flat plate due to an overpassing oscillatory streaming flow with amplitude $U\mathbf{t}$; the associated velocity profile was given in Eq. (5.2.8) with $A = -\rho\Omega U$. Introducing a local coordinate system with the y axis perpendicular to the boundary at a point and the origin at that point, we find that the velocity within the oscillatory boundary layer is given by the real part of

$$\mathbf{u} = \mathbf{t}\,U\left\{1 - \exp\left[\left(-\frac{i\Omega}{\nu}\right)^{1/2} y\right]\right\} \exp(-i\Omega t) \qquad (8.6.1)$$

where $(-i)^{1/2} = e^{3\pi i/4}$.

The force exerted on the boundary is composed of the acceleration reaction associated with the unsteady irrotational flow and a viscous drag $\mathbf{D}$ due to the Stokes boundary layer. It might appear that we can use the local solution (8.6.1) to compute $\mathbf{D}$, and this would suggest that the phase shift between $\mathbf{D}$ and $\mathbf{u}^\infty$ is equal to $-3\pi/4$ independently of the geometry of the boundary. Batchelor (1967, p. 355), however, pointed out that this is an erroneous deduction that neglects the viscous *normal* stresses due to the boundary curvature.

In the case of uniform oscillatory flow past a stationary body, where far from the body $\mathbf{u}^\infty = \mathbf{V}\cos(\Omega t)$ with $\mathbf{V}$ being a constant velocity, there is a method of computing the component of $\mathbf{D}$ in the direction of $\mathbf{V}$ that is in phase with $\mathbf{u}^\infty$, called the *damping force*, on the basis of an energy argument due to Batchelor (1967, p. 356). The computation proceeds by integrating the energy integral balance expressed by Eq. (3.3.18) over one period of the oscillation, noting that the average kinetic energy of the fluid over one period is constant, and then setting the rate of energy dissipation within the Stokes boundary layer equal to the rate of working that is necessary in order to maintain the body at a fixed position, both averaged over one cycle.

Substituting Eq. (8.6.1) into Eq. (3.3.14), we find that the rate of dissipation within a small volume of the boundary layer is given by

$$\Phi = \mu\left(\frac{\partial}{\partial y}\text{Re}(\mathbf{u}\cdot\mathbf{t})\right)^2 = \rho\Omega U^2\cos^2\left[\Omega t - \frac{3\pi}{4} - \left(\frac{\Omega}{2\nu}\right)^{1/2}y\right]\exp\left(-\left(\frac{2\Omega}{\nu}\right)^{1/2}y\right) \quad (8.6.2)$$

where $\mathbf{u}$ was approximated with the right-hand side of Eq. (8.6.1). The average rate of viscous dissipation over one period of the oscillation $T = 2\pi/\Omega$ is thus given by

$$\frac{1}{T}\int_0^T\int_0^\infty\int_{\text{Boundary}}\Phi\,dS\,dy\,dt = \tfrac{1}{2}\mu\left(\frac{\Omega}{2\nu}\right)^{1/2}\int_{\text{Body}}U^2dS \quad (8.6.3)$$

Next we express the drag force as the real part of $\mathbf{D} = \mathbf{F}\exp[-i(\Omega t - \alpha)]$, where $\mathbf{F}$ is a real vector and α is the phase shift between the drag force and the outer velocity. The average rate at which the body does work against the fluid, expressed by the second integral on the right-hand side of Eq. (3.3.18), is given by

$$\frac{1}{T}\int_0^T\mathbf{V}\cdot\mathbf{D}\cos\alpha\cos^2(\Omega t)\,dt = \tfrac{1}{2}\cos\alpha\mathbf{V}\cdot\mathbf{D} \quad (8.6.4)$$

Setting the right-hand sides of Eqs. (8.6.3) and (8.6.4) equal to one another yields the desired result

$$\mathbf{V}\cdot\mathbf{D} = \frac{1}{\cos\alpha}\mu\left(\frac{\Omega}{2\nu}\right)^{1/2}\int_{\text{Body}}U^2dS \quad (8.6.5)$$

The results of Section 6.14 and, in particular, the boundary integral equation (6.14.22) indicate that unless the boundary has sharp edges, in which case the linearized equation of motion will cease to be valid, $\mathbf{D}$ is an analytic function of $(-i\Omega/\nu)^{1/2}$, and therefore $\alpha = -\pi/4$.

As an example, using the well-known solution for uniform irrotational flow past a sphere of radius a to evaluate the integral on the right-hand side of Eq. (8.6.5), we find

$$\mathbf{V}\cdot\mathbf{D} = \left(\frac{\Omega a^2}{\nu}\right)^{1/2}6\pi\mu aV^2 \quad (8.6.6)$$

For uniform irrotational flow past a cylinder of radius a and length L, with vanishing circulation around it, we find

$$\mathbf{V}\cdot\mathbf{D} = \left(\frac{\Omega a^2}{\nu}\right)^{1/2}4\pi\mu LV^2 \quad (8.6.7)$$

PROBLEMS

8.6.1 **Damping force on a sphere and a cylinder.** Integrate the potential flow solution to compute the dissipation integral on the right-hand side of Eq. (8.6.5) for (a) a sphere and (b) a cylinder.

Computer Problem

8.6.2 **Damping force on a spheroid.** Integrate the potential flow solution to compute the dissipation integral on the right-hand side of Eq. (8.6.5) for the axial flow past a prolate spheroid, and plot your results in an appropriate dimensionless form against the spheroid aspect ratio.

References

Batchelor, G. K., 1967, *An Introduction to Fluid Dynamics.* Cambridge.

Blasius, Von H., 1908, Grenzschichten in Flüssigkeiten mit kleiner Reibung. *Z. Math. Phys.* **56,** 1–37.

Blottner, F. G., 1970, Finite difference methods of solution of the boundary layer equations. *AIAA J.* **8,** 193–205.

Cebeci, T., and Bradshaw, P., 1977, *Momentum Transfer in Boundary Layers.* Hemisphere.

Cebeci, T., and Bradshaw, P., 1984, *Physical and Computational Aspects of Convective Heat Transfer.* Springer-Verlag.

Evans, H., 1968, *Laminar Boundary Layers.* Addison–Wesley.

Falkner, V. M., and Skan, S. W., 1931, Solutions of the boundary layer equations. *Phil. Mag.* **12,** 865–96.

Fletcher, C. A. J., 1988, *Computational Techniques for Fluid Dynamics.* 2 volumes. Springer-Verlag.

Görtler, H., 1957, A new series for the calculation of steady laminar boundary layer flows. *J. Math. Mech.* **6,** 1–66.

Harper, J. F., and Moore, D. W., 1968, The motion of a spherical liquid drop at high Reynolds number. *J. Fluid Mech.* **32,** 367–91.

Hartree, D. R., 1937, On an equation occurring in Falkner and Skan's approximate treatment of the equations of the boundary layer. *Proc. Camb. Phil. Soc.* **33,** 223–39.

Homann, F., 1936, Einfluss grosser Zähigkeit bei Strömung um Zylinder. *Forschg. Ing.-Wes.* **7,** 1–10.

Mager, A., 1964, Three-dimensional laminar boundary layers. In *Theory of Laminar Flows.* F. K. Moore, Editor, Princeton University Press.

Mangler, W., 1948, Zusammenhang zwischen ebenen und rotationssymmetrischen Grenzschichten in kompressiblen Flüssigkeiten. *Z. angew. Math. Mech.* **28,** 97–103.

Mises, R. von, 1927, Bemerkungen zur Hydrodynamik. *Z. angew. Math. Mech.* **7,** 425–31.

Moore, D. W., 1963, The boundary layer on a spherical gas bubble. *J. Fluid Mech.* **16,** 161–76.

Moore, F. K., 1964, *Theory of Laminar Flows.* Princeton University Press.

Nayfeh, A. H., 1985, *Problems in Perturbation.* Wiley.

Prandtl, L., 1904, Über Flüssigkeitsbewegung bei sehr kleiner Reibung, *Verh. III. Int. math. Kongr., Heidelburg,* 484–491. (Also translated into English in NACA Techn. Mem. 452.)

Pohlhausen, E., 1921, Zur näherungsweisen Integration der Differentialgleichung der laminaren Grenzschicht. *Z. angew. Math. Mech.* **1,** 252–68.

Pozrikidis, C., 1989, Inviscid drops with internal circulation. *J. Fluid Mech.* **209,** 77–92.

Rosenhead, L., 1963, *Laminar Boundary Layers.* Oxford.

Sakiadis, B. C., 1961, Boundary-layer behavior on continuous solid surfaces: II. The boundary layer on a continuous flat surface. *AIChE J.* **7**(2), 221–25.

Schlichting, H., 1968, *Boundary Layer Theory.* McGraw–Hill.

Smith, A. M. O., 1956, Rapid laminar boundary-layer calculations by piecewise application of similar solutions. *J. Aeronaut. Sci.* **23,** 901–12.

Smith, A. M. O., and Clutter, D. W., 1963, Solution of the incompressible laminar boundary-layer equations. *AIAA J.* **1,** 2062–71.

Smith, F. T., 1982, On the high Reynolds number theory of laminar flows. *IMA J. Appl. Math.* **28,** 207–81.

Stewartson, K., 1954, Further solutions of the Falkner–Skan equation, *Proc. Camb. Phil. Soc.* **50,** 454–65.

Stewartson, K., 1974, Multistructured boundary layers on flat plates and related bodies. *Adv. Appl. Math.* **14,** 145–239.

Stuart, J. T., 1963, Unsteady boundary layers. In *Laminar Boundary Layers,* L. Rosenhead, Editor. Oxford.

Tani, I., 1977, History of boundary-layer theory. *Annu. Rev. Fluid Mech.* **9,** 87–111.

Thwaites, B., 1949, Approximate calculation of the laminar boundary layer. *Aeronaut. Quart.* **1,** 245–80.

Von Kármán, T., 1921, Über laminare und turbulente Reibung. *Z. angew. Math. Mech.* **1,** 233–52.

White, F. M., 1974, *Viscous Fluid Flow.* McGraw–Hill.

CHAPTER 9

In the preceding chapters we discussed a variety of analytical and numerical methods for computing the structure of steady and unsteady flows under a broad range of conditions, including flows at low Reynolds numbers, irrotational flows, and flows that are dominated by vortex motions. In this chapter we address the important question of whether these flows can be realized in practice.

In nature and technology, a flow is established necessarily through a transient process beginning from a certain initial condition, and subject to the imposed boundary conditions. The fact that the boundary conditions may be consistent with a particular steady or unsteady state that is describable in analytical or numerical form does not guarantee that that state will be established. For instance, imposing a constant pressure drop across the length of a circular tube does not guarantee the onset of unidirectional Poiseuille flow with a parabolic velocity profile discussed in Section 5.1. On the contrary, as early as 1883, Reynolds observed that, at sufficiently high Reynolds numbers, the flow develops wavy motions and becomes turbulent, and the assumption of steady unidirectional motion ceases to be valid.

Furthermore, flows in industrial, laboratory, or natural settings are subjected to small-amplitude disturbances due to a variety of reasons including equipment vibration and Brownian motion of microscopic suspended particles. In fact, in certain technological applications, perturbations are purposely introduced into the flow in order to initiate a desired type of action, such as to enhance fluid mixing or delay boundary-layer separation. It is then possible that natural or artificial disturbances may amplify in time or space, leading to unsteady motion or to a new state.

The behavior of a disturbance or perturbation depends upon its particular characteristics as well as upon the structure of the unperturbed flow, which, in the present context, is called the *base flow.* Moreover, disturbances are known to exhibit different types of behavior depending upon the values of the Reynolds number and other dimensionless numbers that characterize the base flow. In certain cases the perturbations grow while being convected with the flow, in which case they cause a *convective instability,* whereas in other cases they spread out and contaminate the whole domain of flow, in which case they cause an *absolute instability.* There are other more complex types of behavior discussed and classified by Huerre and Monkewitz (1990). Establishing criteria for the resilience and thus physical relevance of a particular steady or unsteady base flow is the central objective of the theory of hydrodynamic stability.

One way of assessing the stability of a flow is to subject it to a broad range of perturbations of various forms and then observe its subsequent evolution. If all perturbations decay, the flow is *stable* and thus realizable. If certain perturbations amplify, the flow is *unstable* and will not occur in practice unless some external mechanism acts to suppress the growth of the unstable perturbations. In certain cases, assessing the behavior of the perturbations can be done on the basis of simple physical arguments, but, more generally, it must be done by detailed analysis and numerical computation.

The behavior of a disturbance may be studied theoretically by solving the equation of motion and the continuity equation subject to the required boundary conditions. Since, however, the number of admissible disturbance modes is innumerable, it is futile to attempt to exhaust all possibilities, and we must proceed in an alternative fashion. One way to make progress is to assume that the magnitude of a disturbance is and remains small during a certain time period, and then linearize the equation of motion with respect to the velocity about the base state and solve it for a broad range of initial conditions using, for instance, the method of Laplace transform. This approach is the cornerstone of *linear stability analysis.* Even after linearization, we find that a general solution in analytical form can be found only for a limited class of flows by means of the *normal-mode analysis,* which examines the behavior of perturbations with exponential growth or decay in time.

If linear stability analysis indicates that certain perturbations grow in time, the flow is certainly unstable. Since, however, the neglected nonlinear effects may be responsible for unstable behavior, the converse is true only when the amplitude of the perturbations is and remains sufficiently small *at all times.* In certain cases, nonlinear effects may slow down or even suppress the growth of unstable perturbations and lead to a new steady or periodic state. Assessing the effects of nonlinearities is a much more difficult problem, but some progress can be made under the auspices of weakly-nonlinear stability theory, in which the perturbation is expressed in terms of an appropriate perturbation series, and the analysis is carried out up to the second or a higher order with respect to the perturbation parameter. A full assessment of the nonlinear effects requires the use of numerical methods such as the finite-difference methods discussed in Chapter 13.

In this chapter, we introduce the basic concepts that enter the formulation of the linear stability problem for internal, external, interfacial, and free-surface flows. Furthermore, we discuss the stability characteristics of a certain class of viscous and inviscid flows of salient fundamental significance and technological importance. Extensive discussions of linear and nonlinear stability theories can be found in the dedicated monographs and reviews of hydrodynamic stability by Lin (1955), Chandrasekhar (1961), Betchov and Criminale (1967), Drazin and Reid (1981), Maslowe (1981), Craik (1985), and Huerre and Monkewitz (1990).

9.1 | EVOLUTION EQUATIONS AND FORMULATION OF THE LINEAR STABILITY PROBLEM

To prepare the ground for computing the evolution of perturbations, we summarize and discuss the equations that govern the evolution of the velocity, vorticity, and pressure of the flow of *an incompressible fluid with uniform physical properties, in the presence of a uniform body force.* Relaxing these assumptions require only minor changes in the mathematical formulation.

Evolution of the Velocity

For the purposes of the present discussion, it is useful to regard the Navier–Stokes equation (3.4.4) as an evolution equation for the velocity, and recast it into the form

$$\frac{\partial \mathbf{u}}{\partial t} = \mathbf{F}(\mathbf{u}, p) \tag{9.1.1}$$

where $\mathbf{F}$ is a forcing function defined as

$$\mathbf{F}(\mathbf{u}, p) = -\mathbf{u} \cdot \nabla \mathbf{u} - \frac{1}{\rho} \nabla p + \nu \nabla^2 \mathbf{u} + \mathbf{g} \tag{9.1.2}$$

We note again that the viscosity has been assumed to be uniform throughout the domain of flow.

Evolution of the Pressure

An evolution equation for the pressure is not available in an explicit form. The condition of fluid incompressibility, however, requires that the instantaneous pressure field develop such that the rate of expansion remains equal to zero at all times, $\nabla \cdot \mathbf{u} = 0$. To illustrate the way in which the incompressibility condition provides us with an implicit evolution equation for the pressure, we take the divergence of Eq. (9.1.1), and under the stipulation that the density of the fluid is uniform, we find

$$\frac{\partial \nabla \cdot \mathbf{u}}{\partial t} = \nabla \cdot \mathbf{F} = -\nabla \cdot (\mathbf{u} \cdot \nabla \mathbf{u}) - \frac{1}{\rho} \nabla^2 p + \nu \nabla^2 \nabla \cdot \mathbf{u} \qquad (9.1.3)$$

which may be regarded as an evolution equation for the rate of expansion. Requiring that the left-hand side of Eq. (9.1.3) as well as $\nabla \cdot \mathbf{u}$ on the right-hand side vanish at all times, we obtain a Poisson equation for the pressure

$$\nabla^2 p = -\rho \nabla \cdot (\mathbf{u} \cdot \nabla \mathbf{u}) = -\rho \nabla \nabla : (\mathbf{uu}) \qquad (9.1.4)$$

which is to be solved subject to boundary conditions to be discussed in Section 13.2. Conversely, if the pressure field develops so that Eq. (9.1.4) is fulfilled at every instant, the gradient of the pressure on the right-hand side of Eq. (9.1.2) will be such that $\nabla \cdot \mathbf{F}$ vanishes at all times (Problem 9.1.1).

Let us now take the partial derivative of Eq. (9.1.4) with respect to time, expand the derivatives on the right-hand side, and use Eq. (9.1.1) to derive a Poisson equation for the rate of change of the pressure,

$$\nabla^2 \frac{\partial p}{\partial t} = -\rho \nabla \cdot (\mathbf{F} \cdot \nabla \mathbf{u}) - \rho \nabla \cdot (\mathbf{u} \cdot \nabla \mathbf{F}) = -2\rho \nabla \nabla : (\mathbf{u} \mathbf{F}) \qquad (9.1.5)$$

The solution may be found using the Poisson inversion formula and then written in the symbolic form of a standard evolution equation as

$$\frac{\partial p}{\partial t} = G[\mathbf{u}, \mathbf{F}(\mathbf{u}, p)] \qquad (9.1.6)$$

where the particular form of the function G depends upon the instantaneous structure of the flow and required boundary conditions. Equation (9.1.6) is the desired evolution equation for the pressure which, however, is implicit due to the inherent inversion of the Poisson equation (9.1.5).

To further illustrate the role of the pressure in maintaining the velocity field solenoidal, let us consider the right-hand side of Eq. (9.1.2) with the pressure term omitted and denote it as $\mathbf{F}^{(-p)}$. The Helmholtz decomposition theorem discussed in Section 2.6 allows us to express $\mathbf{F}^{(-p)}$ as the sum of a solenoidal and an irrotational field in the form

$$\mathbf{F}^{(-p)} = \nabla \times \mathbf{A} + \nabla f \qquad (9.1.7)$$

where $\mathbf{A}$ is a vector potential and f is a potential function. By definition we then have

$$\mathbf{F} = \nabla \times \mathbf{A} + \nabla f - \frac{1}{\rho} \nabla p \qquad (9.1.8)$$

If the fluid is incompressible, $\mathbf{F}$ is solenoidal, and we may require that

$$\nabla p = \rho \nabla f \qquad (9.1.9)$$

which shows that the gradient of the pressure projects $\mathbf{F}^{(-p)}$ into the space of solenoidal functions, thereby transforming it into the solenoidal field $\mathbf{F}$. In Section 13.4 we shall see that this interpretation provides us with a basis for a class of numerical methods used to integrate the equation of motion in time, called *projection* or *pressure-correction* methods.

Evolution of the Vorticity

An evolution equation for the vorticity emerges immediately by recasting the vorticity transport equation for an incompressible fluid with uniform physical properties, expressed by Eqs. (3.8.9) and (3.8.10), into the form

$$\frac{\partial \boldsymbol{\omega}}{\partial t} = \mathbf{H}(\boldsymbol{\omega}, \mathbf{u}) \qquad (9.1.10)$$

where

$$\begin{aligned}\mathbf{H}(\boldsymbol{\omega}, \mathbf{u}) &= -\nabla \times (\boldsymbol{\omega} \times \mathbf{u}) + \nu \nabla^2 \boldsymbol{\omega} \\ &= -\mathbf{u} \cdot \nabla \boldsymbol{\omega} + \boldsymbol{\omega} \cdot \nabla \mathbf{u} + \nu \nabla^2 \boldsymbol{\omega}\end{aligned} \qquad (9.1.11)$$

One noteworthy feature of Eq. (9.1.11) is that it does not involve the pressure. This feature is exploited for the expedient analytical or numerical computation of the structure of a steady flow or the evolution of an unsteady flow based on the vorticity transport equation to be discussed in Chapters 11 and 13.

Summary of Evolution Equations

Compiling Eqs. (9.1.1), (9.1.6), and (9.1.10), we obtain the general evolution equation

$$\frac{\partial}{\partial t} \begin{bmatrix} \mathbf{u} \\ p \\ \boldsymbol{\omega} \end{bmatrix} = \begin{bmatrix} \mathbf{F}(\mathbf{u}, p) \\ G(\mathbf{u}, p) \\ \mathbf{H}(\boldsymbol{\omega}, \mathbf{u}) \end{bmatrix} \qquad (9.1.12)$$

When the flow is steady, the left-hand side of Eq. (9.11.2) vanishes, and the structure of the velocity, pressure, and vorticity fields is governed by the equations of steady flow

$$\mathbf{F}(\mathbf{u}^S, p^S) = \mathbf{0}, \qquad G(\mathbf{u}^S, p^S) = 0, \qquad \mathbf{H}(\boldsymbol{\omega}^S, \mathbf{u}^S) = \mathbf{0} \qquad (9.1.13)$$

where the superscript S stands for *steady*.

Linearized Evolution from a Steady State

We consider next a *nearly steady flow*, that is, a flow that deviates only by a small amount from a certain steady state. The mathematical statement of this physical condition can be implemented by expressing the velocity, pressure, and vorticity fields in the forms

$$[\mathbf{u}, p, \boldsymbol{\omega}](\mathbf{x}, t) = [\mathbf{u}^S, p^S, \boldsymbol{\omega}^S](\mathbf{x}) + \varepsilon [\mathbf{u}^U, p^U, \boldsymbol{\omega}^U](\mathbf{x}, t) \qquad (9.1.14)$$

where ε is a small dimensionless number, and the superscript U signifies the unsteady component. Both the steady and unsteady components are required to satisfy the continuity equation

$$\nabla \cdot \mathbf{u}^S = 0, \qquad \nabla \cdot \mathbf{u}^U = 0 \qquad (9.1.15)$$

Substituting the expressions (9.1.14) into the right-hand sides of Eqs. (9.1.2) and (9.1.11), taking into account Eqs. (9.1.13), and discarding terms with quadratic dependence on ε, we obtain the linearized forms

$$\mathbf{F}(\mathbf{u}, p) \cong \varepsilon \left(-\mathbf{u}^S \cdot \nabla \mathbf{u}^U - \mathbf{u}^U \cdot \nabla \mathbf{u}^S - \frac{1}{\rho} \nabla p^U + \nu \nabla^2 \mathbf{u}^U \right) \qquad (9.1.16)$$

$$\begin{aligned}\mathbf{H}(\boldsymbol{\omega}, \mathbf{u}) \cong \varepsilon (&-\mathbf{u}^S \cdot \nabla \boldsymbol{\omega}^U - \mathbf{u}^U \cdot \nabla \boldsymbol{\omega}^S + \boldsymbol{\omega}^U \cdot \nabla \mathbf{u}^S \\ &+ \boldsymbol{\omega}^S \cdot \nabla \mathbf{u}^U + \nu \nabla^2 \boldsymbol{\omega}^U)\end{aligned} \qquad (9.1.17)$$

Next we substitute the expressions (9.1.14) along with Eqs. (9.1.16) and (9.1.17) into the evolution equations (9.1.1) and (9.1.10) and obtain linear evolution equations for the unsteady components of the velocity and vorticity

$$\frac{\partial \mathbf{u}^U}{\partial t} = -\mathbf{u}^S \cdot \nabla \mathbf{u}^U - \mathbf{u}^U \cdot \nabla \mathbf{u}^S - \frac{1}{\rho} \nabla p^U + \nu \nabla^2 \mathbf{u}^U \qquad (9.1.18)$$

$$\begin{aligned}\frac{\partial \boldsymbol{\omega}^U}{\partial t} = &-\mathbf{u}^S \cdot \nabla \boldsymbol{\omega}^U - \mathbf{u}^U \cdot \nabla \boldsymbol{\omega}^S + \boldsymbol{\omega}^U \cdot \nabla \mathbf{u}^S \\ &+ \boldsymbol{\omega}^S \cdot \nabla \mathbf{u}^U + \nu \nabla^2 \boldsymbol{\omega}^U\end{aligned} \qquad (9.1.19)$$

The preceding two equations may be collected into the compact form

$$\frac{\partial}{\partial t} \begin{bmatrix} \mathbf{u}^U \\ \boldsymbol{\omega}^U \end{bmatrix} = \begin{bmatrix} \mathbf{A} & \mathbf{B} & 0 \\ \mathbf{C} & 0 & \mathbf{D} \end{bmatrix} \cdot \begin{bmatrix} \mathbf{u}^U \\ p^U \\ \boldsymbol{\omega}^U \end{bmatrix} \qquad (9.1.20)$$

where $\mathbf{A}, \mathbf{B}, \mathbf{C}, \mathbf{D}$ are differential operators whose precise form depends upon the structure of the base flow, given by

$$\mathbf{A} = -\mathbf{u}^S \cdot \nabla - (\nabla \mathbf{u}^S)^T \cdot + \nu \nabla^2, \qquad \mathbf{B} = -\frac{1}{\rho} \nabla$$

$$\mathbf{C} = -(\nabla \boldsymbol{\omega}^S)^T \cdot + \boldsymbol{\omega}^S \cdot \nabla$$

$$\mathbf{D} = -\mathbf{u}^S \cdot \nabla + (\nabla \mathbf{u}^S)^T \cdot + \nu \nabla^2 \qquad (9.1.21)$$

Furthermore, we substitute the expressions (9.1.14) for the velocity and pressure into Eq. (9.1.4) and linearize the right-hand side of the resulting equation to obtain a Poisson equation for the unsteady component of the pressure,

$$\nabla^2 p^U = -\rho \nabla \cdot (\mathbf{u}^S \cdot \nabla \mathbf{u}^U + \mathbf{u}^U \cdot \nabla \mathbf{u}^S) = -2\rho \nabla \nabla : \mathbf{u}^U \mathbf{u}^S \qquad (9.1.22)$$

We can work in a similar manner with the evolution equation (9.1.6), casting it into a form that is similar to that of Eq. (9.1.20), but this will not be necessary for the purposes of the present discussion.

Linear Stability Analysis

Eqs. (9.1.20) and (9.1.22) provide us with a complete system of linear homogeneous partial differential equations for the evolution of the unsteady component of a nearly steady flow. Solving these equations allows us to study the departure of a flow from a certain steady state during an initial period of time during which the magnitude of the unsteady component is still small compared to the magnitude of the base flow.

Considering now a certain base flow, we identify the unsteady component with a disturbance. Depending on the structure of the base flow and on the form of the disturbance, the unsteady component may grow or decay in a local or global fashion. If the magnitude of a disturbance, defined in some global or local sense, grows, remains constant, or decays in time, then the disturbance is called, respectively, *unstable, marginally stable,* or *stable.* If all disturbances decay, the base flow is *linearly stable,* but if certain disturbances grow, the flow is *linearly unstable.* An unstable flow can be realized in practice only when the unstable disturbances are screened off from the physical system by some externally provided controlling mechanism.

To this end, we return to underline the limitations and advisory nature of linear stability theory. A flow that is stable according to linear stability theory will not necessarily occur in practice; nonlinear effects and small deviations from the assumed idealized geometry of the domain of flow due, for example, to wall roughness, may be responsible for unstable behaviors. An example is Poiseuille flow in a circular tube, which is stable according to linear theory but has been observed to be unstable in practice (Section 9.7).

PROBLEMS

9.1.1 **Pressure Poisson equation.** Consider an initially solenoidal velocity field, and show that if the pressure is computed on the basis of Eq. (9.1.4), it will remain solenoidal at all times. What will happen if the initial velocity field is not solenoidal?

9.1.2 **Nonlinear evolution equation in operator form.** Recast Eq. (9.1.12) into an operator form similar to that shown in Eq. (9.1.20).

9.1.3 **Linearized vorticity transport for two-dimensional flow.** Discuss the physical significance of the various terms in Eq. (9.1.19) for two-dimensional flow.

9.2 | SOLUTION OF THE INITIAL-VALUE PROBLEM AND NORMAL-MODE ANALYSIS

We proceed to discuss methods of solving the linearized equation of motion and vorticity transport equation collected in Eq. (9.1.20), subject to a certain initial condition representing a perturbation.

Couette Flow of an Inviscid Fluid

We begin by discussing a case for which the initial-value problem can be solved exactly in closed form. Consider the stability of plane Couette flow along the x axis in a channel that is confined between two plane walls located at $y = \pm a$; the velocity field of the base flow is given by $\mathbf{u}^S = (Gy, 0)$, where G is the *shear rate*. Concentrating on two-dimensional perturbations in the plane of the flow, we describe them in terms of the disturbance stream function ψ. Requiring the no-penetration condition at the two walls, and that the disturbance does not have an effect on the axial flow rate, we obtain the boundary conditions $\psi = 0$ at $y = \pm a$. The z component of the linearized vorticity transport equation (9.1.19) shows that, in the absence of viscous forces, the vorticity associated with the disturbance flow is simply convected by the base flow

$$\left(\frac{\partial}{\partial t} + Gy\frac{\partial}{\partial x}\right)\nabla^2\psi = 0 \tag{9.2.1}$$

This result is due to the vanishing of the vorticity gradient of the base flow. Inverting the hyperbolic operator on the left-hand side yields the general solution

$$\nabla^2\psi = -F(x - Gyt, y) \tag{9.2.2}$$

where $F(x, y) = \omega(x, y, t = 0)$ is the strength of the disturbance vorticity at the initial instant.

Considering, for simplicity, perturbations that are symmetric with respect to the plane $x = 0$ at the initial instant and decay sufficiently fast far from the origin, we express the stream function in terms of a sine Fourier series in the y direction and a cosine Fourier integral in the x direction; antisymmetric perturbations may be treated in a similar manner, and a general perturbation may be expressed in terms of symmetric and antisymmetric modes (Problem 9.2.1). Taking into account the boundary condition $\psi = 0$ at $y = \pm a$, we write

$$\psi(x, y, t = 0) = \sum_{n=1}^{\infty} \sin\left(\frac{n\pi}{2a}(y + a)\right)\int_{-\infty}^{\infty} b_n(k)\cos kx\, dk \tag{9.2.3}$$

where $b_n(k)$ are real Fourier coefficients with dimensions of velocity multiplied by squared length. Straightforward differentiation yields the initial vorticity field

$$F(x, y, t = 0) \equiv -\nabla^2\psi(x, y, t = 0)$$

$$= \sum_{n=1}^{\infty}\left[\left(\frac{n\pi}{2a}\right)^2 + k^2\right]\sin\left(\frac{n\pi}{2a}(y + a)\right)\int_{-\infty}^{\infty} b_n(k)\cos kx\, dk \tag{9.2.4}$$

Next we substitute Eq. (9.2.4) into Eq. (9.2.2), take the x Fourier transform of the emerging equation, solve the resulting ordinary differential equation with respect to y, and invert the Fourier transform to find the solution

$$\psi(x, y, t) = \frac{1}{2}\sum_{\substack{n=-\infty \\ n\neq 0}}^{\infty}\int_{-\infty}^{\infty}\frac{(n\pi/2a)^2 + k^2}{(n\pi/2a - Gt)^2 + k^2}b_n(k)\, f_n(x, y, t, k)\, dk \tag{9.2.5}$$

where $b_{-n}(k) = -b_n(k)$, and

$$f_n(x, y, t, k) = \sin\left(k(x - yGt) + \frac{n\pi}{2a}(y + a)\right)$$

$$- \frac{\sinh k(a - y)\sin k(x + aGt) + \sinh k(a + y)\sin[k(x - aGt) + n\pi]}{\sinh 2ak}$$

$$(9.2.6)$$

(Orr, 1907; Yih, 1979, p. 482). Evaluating the right-hand side of Eq. (9.2.6) at long times shows that the disturbance stream function decays like $1/t^2$, and this means that the base flow is stable (Eliassen, Høiland, and Riis, 1953, pp. 149-150; see Maslowe, 1981, or Drazin and Reid, 1981).

Laplace Transform Method

In the case of plane inviscid Couette flow, we were able to solve the initial-value problem governed by Eqs. (9.1.20) and (9.1.22) exactly by analytical means. To compute the solution for a more general base flow including the effects of viscosity, we resort to approximate and numerical methods pertinent to linear partial differential equations.

The Laplace transform allows us to eliminate the time derivative from the governing equations, replacing it with an algebraic dependence. Assuming that the perturbation has been introduced at the origin of time and then grows, at most, at an exponential rate, we introduce the one-sided Laplace transforms of the velocity, pressure, and vorticity fields defined as

$$\begin{bmatrix} \hat{\mathbf{u}}^U \\ \hat{p}^U \\ \hat{\boldsymbol{\omega}}^U \end{bmatrix}(\mathbf{x}, s) = \int_{0^+}^{\infty} \begin{bmatrix} \mathbf{u}^U \\ p^U \\ \boldsymbol{\omega}^U \end{bmatrix}(\mathbf{x}, t)\exp(-st)\,dt \qquad (9.2.7)$$

where s is a complex variable with a sufficiently large positive real part. Taking the Laplace transform of Eq. (9.1.20) yields a linear, inhomogeneous, system of second-order partial differential equations in the spatial variables for the Laplace-transformed functions,

$$s\begin{bmatrix} \hat{\mathbf{u}}^U \\ \hat{\boldsymbol{\omega}}^U \end{bmatrix} = \begin{bmatrix} \mathbf{A} & \mathbf{B} & 0 \\ \mathbf{C} & 0 & \mathbf{D} \end{bmatrix} \cdot \begin{bmatrix} \hat{\mathbf{u}}^U \\ \hat{p}^U \\ \hat{\boldsymbol{\omega}}^U \end{bmatrix} + \begin{bmatrix} \mathbf{u}^U \\ \boldsymbol{\omega}^U \end{bmatrix}(\mathbf{x}, t = 0^+) \qquad (9.2.8)$$

Having computed the solution subject to appropriate boundary conditions, we recover the physical variables in the time domain in terms of the Bromwich integral in the complex s plane as

$$\begin{bmatrix} \mathbf{u}^U \\ p^U \\ \boldsymbol{\omega}^U \end{bmatrix}(\mathbf{x}, t) = \frac{1}{2\pi i}\int_{\gamma - i\infty}^{\gamma + i\infty} \begin{bmatrix} \hat{\mathbf{u}}^U \\ \hat{p}^U \\ \hat{\boldsymbol{\omega}}^U \end{bmatrix}(\mathbf{x}, s)\exp(st)\,ds \qquad (9.2.9)$$

where γ is a sufficiently large real positive number so that all singularities of the Laplace-transformed functions are located on the left of the vertical path of integration in the complex plane. The integral may be evaluated by the method of residues, which involves introducing a closed contour C that encloses all singularities of the integrand. For instance, provided that there are no branch points of the Laplace-transformed variables, C may be identified with part of the vertical line $s = \gamma$ and a semicircular arc of large radius lying on the left of the vertical path.

Green's Functions

Another method of solving the initial-value problem involves using Green's functions. These are solutions of the linearized equation of motion and associated vorticity transport equation for the velocity, vorticity, and pressure, subject to the required boundary conditions, computed under the stipulation that the right-hand side of the equation of motion or of the vorticity transport equation is enhanced with a singular forcing function that is proportional to $\mathbf{f}\,\delta(x - x_0)\,\delta(y - y_0)\,\delta(z - z_0)\,\delta(t)$; $\mathbf{x}_0$ is an arbitrary point in the domain of flow, $\mathbf{f}$ is a constant vector, and δ

is the one-dimensional delta function. The Green's function tensor for the velocity $\mathbf{G}(\mathbf{x}, t; \mathbf{x}_0)$ emerges by recasting the solution for the velocity of this singularly forced problem in the form $\mathbf{G}(\mathbf{x}, t; \mathbf{x}_0) \cdot \mathbf{f}$. The solution of the initial-value problem can be expressed as a volume integral of the Green's function over the domain of flow, multiplied by an appropriate density distribution function $\mathbf{q}$ that is determined by the initial condition, and a convolution integral in time, in the form

$$\mathbf{u}(\mathbf{x}, t) = \int_0^t \int_{\text{Flow}} \mathbf{G}(\mathbf{x}, t - \tau; \mathbf{x}_0) \cdot \mathbf{q}(\mathbf{x}_0) \, dV(\mathbf{x}_0) \, d\tau \qquad (9.2.10)$$

Unfortunately, the computation of Green's functions is generally a difficult task, and the method of Green's functions finds limited usage in practice (Huerre and Monkewitz, 1990).

Normal-Mode Analysis

To study the evolution of every possible type of perturbation on a given base flow is practically impossible. One way to make progress is to express the initial disturbance as a combination of linearly independent fundamental modes and then examine the evolution of each mode alone. Of course, this approach assumes that a complete set of fundamental modes is available.

A convenient set of modes that are analogous to the eigenvectors of a matrix are *normal modes* with exponential dependence in time, for which the unsteady component of the flow assumes the form

$$\begin{bmatrix} \mathbf{u}^{\text{NM}} \\ p^{\text{NM}} \\ \boldsymbol{\omega}^{\text{NM}} \end{bmatrix} (\mathbf{x}, t) = \begin{bmatrix} \mathbf{V} \\ \Pi \\ \boldsymbol{\Omega} \end{bmatrix} (\mathbf{x}, \sigma) \exp(-i\sigma t) \qquad (9.2.11)$$

where the superscript NM stands for normal mode, σ is a complex constant called the *complex growth rate* or *complex cyclic frequency,* and $\mathbf{V}$, Π, and $\boldsymbol{\Omega}$ are complex functions of $\mathbf{x}$ and σ. All dependent variables are assumed to be complex, with the understanding that both their real and imaginary parts represent admissible modes.

Substituting the expressions (9.2.11) into Eqs. (9.1.20), we obtain a linear system of homogeneous equations governing the spatial structure of the normal modes

$$(i\sigma - \mathbf{u}^S \cdot \nabla + \nu\nabla^2) \mathbf{V} - \mathbf{V} \cdot \nabla \mathbf{u}^S - \frac{1}{\rho} \nabla\Pi = \mathbf{0} \qquad (9.2.12)$$

$$(i\sigma - \mathbf{u}^S \cdot \nabla + \nu\nabla^2) \boldsymbol{\Omega} - \mathbf{V} \cdot \nabla\boldsymbol{\omega}^S + \boldsymbol{\Omega} \cdot \nabla\mathbf{u}^S + \boldsymbol{\omega}^S \cdot \nabla\mathbf{V} = \mathbf{0} \qquad (9.2.13)$$

The continuity equation requires that

$$\nabla \cdot \mathbf{V} = 0 \qquad (9.2.14)$$

Nontrivial solutions to the system of Eqs. (9.2.12)–(9.2.14) will exist only when the complex growth rate σ takes values within a set of complex numbers called the *spectrum of eigenvalues* of the base flow.

The spectrum of eigenvalues is generally composed of a *discrete* part that contains a set of distinct complex eigenvalues, and a *continuous* part that contains a family of eigenvalues that vary in a continuous fashion along a curve with respect to some parameter. The discrete part of the spectrum may contain a finite or an infinite number of eigenvalues or no eigenvalues at all (Craik, 1985, p. 52). For instance, in the case of *viscous* unidirectional Couette flow, there are an infinite number of discrete eigenvalues forming a complete set, and no continuous spectrum. For *inviscid* unidirectional Couette flow, on the other hand, there is only a continuous spectrum composed of stable normal modes. The nature of the two components of the spectrum will be illustrated in the next section for the particular case of inviscid Couette flow.

Whether or not the normal modes provide us with a complete basis of eigenfunctions depends upon whether the domain of flow extends to infinity, and presence of singularities (Ladyshenskaya, 1975). If the normal modes provide us with a complete basis, an arbitrary disturbance may be expressed as a linear combination of (1) the discrete normal modes multiplied by appropriate complex coefficients, and (2) distributions of the continuous normal modes weighted by appropriate complex density functions. It is important to note that, even though the individual normal modes may grow or decay at an exponential rate, their superposition may exhibit a different type of temporal behavior (see, for example, Case, 1960b). In fact, a nominally "stable" perturbation may amplify by several orders of magnitude before it starts decaying.

Laplace transform and normal modes

It is instructive to apply the method of Laplace transforms to compute the evolution of the normal modes. Using Eqs. (9.2.11) and the definition (9.2.7), we find

$$\begin{bmatrix} \hat{u}^{NM} \\ \hat{p}^{NM} \\ \hat{\omega}^{NM} \end{bmatrix} (\mathbf{x}, s) = \frac{1}{s + i\sigma} \begin{bmatrix} \mathbf{V} \\ \Pi \\ \Omega \end{bmatrix} (\mathbf{x}, \sigma) \qquad (9.2.15)$$

Substituting the right-hand side of Eqs. (9.2.15) into Eq. (9.2.8) and simplifying, we recover Eqs. (9.2.12) and (9.2.13), thereby reconciling the inhomogeneous problem expressed by the former equation with the eigenvalue problem expressed by the latter equation.

Expressions (9.2.15) show that $s = -i\sigma$ are simple poles of the Laplace-transformed flow variables, and this suggests that the associated normal modes may be used in order to evaluate the Bromwich integral in Eq. (9.2.9) using the method of residues. Conversely, the poles of the Laplace-transformed variables correspond to the normal modes that fall within the discrete part of the spectrum. The continuous part of the spectrum is associated with *branch cuts of branch points* in the complex s plane.

Temporal growth rate and cyclic frequency

Decomposing σ into its real and imaginary parts, $\sigma = \sigma_R + i\sigma_I$, we express Eq. (9.2.11) in the form

$$\begin{bmatrix} \mathbf{u}^{NM} \\ p^{NM} \\ \omega^{NM} \end{bmatrix} (\mathbf{x}, t) = \begin{bmatrix} \mathbf{V} \\ \Pi \\ \Omega \end{bmatrix} (\mathbf{x}, \sigma) \exp(\sigma_I t) \exp(-i\sigma_R t) \qquad (9.2.16)$$

which shows that σ_I is the *temporal growth rate* of the normal mode, and σ_R is the *cyclic frequency* of the perturbation. When σ_I is positive, the disturbance grows exponentially in time, and the base flow is linearly unstable; when σ_I is negative, the disturbance decays, and the normal mode is linearly stable; when σ_I is equal to zero, the disturbance is neutrally stable.

If the initial perturbation is composed of a number of superimposed normal modes, the mode corresponding to the eigenvalue with the maximum growth rate σ_I, called the *most unstable* or *most dangerous* normal mode, will dominate the rest of the modes. The computation of the corresponding eigensolution is often the prime objective of linear stability analysis.

Another objective of linear stability theory is to establish the conditions under which a flow is expected to be unstable. This is done by examining the behavior of the neutrally stable perturbations with $\sigma_I = 0$, with respect to the dimensionless numbers that characterize the base flow, such as the Reynolds number or the Weber number for interfacial flow.

PROBLEMS

9.2.1   **Inviscid Couette flow.** (a) Verify that Eq. (9.2.5) reduces to Eq. (9.2.3) at the initial instant. (b) Derive the counterpart of Eq. (9.2.5) for antisymmetric perturbations where the stream function at the initial instant is given in terms of a sine Fourier integral in the x direction.

9.2.2   **Normal modes in operator form.** Recast Eqs. (9.2.12) and (9.2.13) into operator forms, similar to those shown in Eqs. (9.1.20).

9.2.3   **Eigenvalues and eigenvectors of a matrix.** Consider a square $N \times N$ matrix, and discuss (a) the number of its eigenvalues, (b) the number of its eigenvectors, (c) the conditions under which the eigenvectors form a complete basis for the N-dimensional space.

9.2.4   **Ordinary differential equations.** Consider a system of first-order linear ordinary differential equations of the form $d\mathbf{X}/dt = \mathbf{A} \cdot \mathbf{X}$, where $\mathbf{X}$ is the unknown solution vector and $\mathbf{A}$ is a square matrix. Express the general solution in terms of exponential functions in time involving the eigenvalues of $\mathbf{A}$ and discuss the occurrence of nonexponential solutions with reference to the multiplicity of the eigenvalues.

Computer Problem

9.2.5   **Inviscid Couette flow.** Plot the streamline patterns according to Eq. (9.2.3) at the initial instant and at a later time for $b_n(k) = 0$, except that $b_1(k) = Ga^3 \exp(-ka)$, and discuss the behavior of the perturbation.

9.3 | NORMAL-MODE ANALYSIS OF UNIDIRECTIONAL FLOWS

Unidirectional flows are encountered in a broad range of natural processes and engineering applications, some of which were discussed in Section 5.1. Examples include flows associated with shear layers forming between two merging streams, atmospheric boundary-layer flows, pressure-driven flows in channels and tubes, and flows of liquid films down inclined surfaces.

The velocity, pressure, and vorticity fields of a steady rectilinear unidirectional flow in the x direction are given by

$$\mathbf{u}^S = U(y)\,\mathbf{e}_x, \qquad p^S(x), \qquad \boldsymbol{\omega}^S = -\frac{\partial U(y)}{\partial y}\,\mathbf{e}_z \qquad (9.3.1)$$

where $U(y)$ represents the velocity profile, and $\mathbf{e}_x$ and $\mathbf{e}_z$ are the unit vectors along the x and z axes. In certain applications, such as in Couette flow between two planes, the pressure is constant, whereas in others, such as in Poiseuille flow through a tube, the pressure varies in a linear fashion with respect to x, and the streamwise pressure gradient is constant.

To carry out the normal-mode stability analysis, we substitute Eq. (9.3.1) into Eqs. (9.2.12) and (9.2.13) and simplify to obtain the linearized equation of motion

$$\left(i\sigma - U\frac{\partial}{\partial x} + \nu\nabla^2\right)\mathbf{V} - V_y U'\mathbf{e}_x - \frac{1}{\rho}\nabla\Pi = \mathbf{0} \qquad (9.3.2)$$

and linearized vorticity transport equation

$$\left(i\sigma - U\frac{\partial}{\partial x} + \nu\nabla^2\right)\boldsymbol{\Omega} + V_y U''\mathbf{e}_z + U'\left(\Omega_y\mathbf{e}_x - \frac{\partial\mathbf{V}}{\partial z}\right) = \mathbf{0} \qquad (9.3.3)$$

where a prime designates a derivative with respect to y. We note again that the physical properties of the fluid have been assumed to be uniform throughout the domain of flow.

A normal-mode disturbance may be expressed in terms of a double complex Fourier integral with respect to x and z. Because the governing equations (9.3.2) and (9.3.3) are linear, each Fourier mode will evolve independently and may be studied on its own. We thus set

$$[\mathbf{V}, \Pi, \mathbf{\Omega}](\mathbf{x}) = \exp(ik_x x + ik_z z)[\mathbf{F}, G, \mathbf{Q}](y) \qquad (9.3.4)$$

where $\mathbf{F}$, G, and $\mathbf{Q}$ are three complex functions of y to be determined as part of the solution, and k_x, k_z are the complex wave numbers of the perturbation in the x and z directions which may be considered as the components of the two-dimensional wave-number vector $\mathbf{k} = (k_x, 0, k_z)$.

Temporal and Spatial Instability

When $\mathbf{k}$ is real and the cyclic frequency σ is complex, we obtain a spatially periodic disturbance that evolves in time, corresponding to the *temporal stability problem*. In the opposite extreme case where $\mathbf{k}$ is complex and σ is real, we obtain a disturbance that evolves in space while its amplitude at a particular point exhibits oscillatory behavior, corresponding to the *spatial stability problem*. The fact that the perturbation may grow exponentially with distance casts some doubts on the physical relevance of this problem. In the more general case where both $\mathbf{k}$ and σ are complex, we obtain a disturbance that evolves in both time and space, corresponding to a generalized stability problem whose solution may be constructed by an appropriate superposition of temporally and spatially evolving modes.

Squire's Theorems

Before proceeding to discuss the solution of Eqs. (9.3.2) and (9.3.3) subject to the assumed form of Eq. (9.3.4), we digress to discuss Squire's theorems, which *relate the behavior of three-dimensional spatially periodic perturbations evolving in time to the behavior of corresponding two-dimensional perturbations in the xy plane* (Squire, 1933).

We begin by substituting Eq. (9.3.4) into the continuity equation (9.2.14) and carry out the differentiations to obtain

$$k_x F_x + k_z F_z - iF'_y = 0 \qquad (9.3.5)$$

Furthermore, we use the definition of the vorticity as the curl of the velocity and find that the functions $\mathbf{F}$ and $\mathbf{Q}$ are related by

$$Q_x = F'_z - ik_z F_y, \qquad Q_y = ik_z F_x - ik_x F_z, \qquad Q_z = ik_x F_y - F'_x \qquad (9.3.6)$$

Substituting Eq. (9.3.4) into Eq. (9.3.3) and using the second of Eqs. (9.3.6), we obtain the vectorial equation

$$[i\sigma - iUk_x - \nu(k_x^2 + k_z^2)]\mathbf{Q} + \nu\mathbf{Q}'' - iU'\begin{bmatrix} k_x F_z \\ k_z F_y \\ k_z F_z \end{bmatrix} + U''F_y\begin{bmatrix} 0 \\ 0 \\ 1 \end{bmatrix} = \mathbf{0} \qquad (9.3.7)$$

For a two-dimensional disturbance in the xy plane with $k_z = 0$ and $F_z = 0$, the x and y components of Eq. (9.3.7) are satisfied in a trivial manner, and the z component becomes

$$(i\sigma - iUk_x - \nu k_x^2)Q_z + \nu Q_z'' + U''F_y = 0 \qquad (9.3.8)$$

where Q_z is given by the third of Eqs. (9.3.6).

Returning to the three-dimensional problem, we introduce the unit wave-number vector and its reciprocal defined as

$$\hat{\mathbf{k}} = \frac{1}{|\mathbf{k}|}(k_x, 0, k_z), \qquad \hat{\mathbf{l}} = \frac{1}{|\mathbf{k}|}(-k_z, 0, k_x) \qquad (9.3.9)$$

Assuming that the wave number is real, and projecting Eq. (9.3.7) onto the reciprocal unit vector, we obtain

$$\left(i\frac{|\mathbf{k}|}{k_x}\sigma - iU|\mathbf{k}| - \frac{|\mathbf{k}|\nu}{k_x}|\mathbf{k}|^2\right)J + \frac{|\mathbf{k}|\nu}{k_x}J'' + U''F_y = 0 \qquad (9.3.10)$$

where we have introduced the component of the vorticity vector in the direction of the reciprocal vector

$$J \equiv \mathbf{Q} \cdot \mathbf{1} = i|\mathbf{k}|F_y - \frac{k_x F_x' + k_z F_z'}{|\mathbf{k}|} \tag{9.3.11}$$

To this end, we make the substitutions

$$\tilde{\sigma} = \frac{|\mathbf{k}|\sigma}{k_x}, \qquad \tilde{k}_x = |\mathbf{k}|, \qquad \tilde{k}_z = 0, \qquad \tilde{\nu} = \frac{|\mathbf{k}|\nu}{k_x}$$

$$\tilde{F}_x = \frac{k_x F_x + k_z F_z}{|\mathbf{k}|}, \qquad \tilde{F}_y = F_y, \qquad \tilde{F}_z = 0, \qquad \tilde{Q}_z = J \tag{9.3.12}$$

and observe that Eqs. (9.3.10) and (9.3.11) reduce to Eq. (9.3.8) and the third of Eqs. (9.3.6), written for the variables and physical parameters with tildes. This means that *the study of three-dimensional perturbations may be reduced to the study of two-dimensional perturbations with a suitable change of the wavelength of the perturbation and viscosity of the fluid, that is, the Reynolds number of the base flow.* Consequently, to assess the stability of a unidirectional flow with uniform physical properties, it is sufficient to consider only two-dimensional disturbances. Once the variables with tildes corresponding to the *equivalent two-dimensional problem* have been computed, the variables without tildes corresponding to the primary three-dimensional problem are recovered from Eqs. (9.3.12) for a specified value of k_x or k_z.

The third of Eqs. (9.3.12) shows that the kinematic viscosity of the fluid for the variables with tildes corresponding to the equivalent two-dimensional problem is higher than that for the variables without tildes corresponding to the three-dimensional problem, and this suggests that the Reynolds number of the flow in the former variables is lower than that in the latter variables. In turn, this leads us to Squire's theorem for viscous flow, stating that *to compute the maximum Reynolds number required for stability, it is sufficient to consider two-dimensional disturbances with wave-number vectors in the direction of the flow.*

Furthermore, the first of Eqs. (9.3.12) shows that the growth rate of the equivalent two-dimensional problem is higher than that of the actual three-dimensional problem. When the fluid is inviscid, the two problems occur at the same infinite Reynolds number, and we obtain Squire's theorem for inviscid flow, stating that *for every unstable three-dimensional perturbation, there is another two-dimensional perturbation with different wavelength that is more unstable.*

It should be emphasized that these results may not be valid when the boundaries of the flow deform in response to a perturbation (Hesia, Pranckh, and Preziosi, 1986), and are applicable only for spatially periodic modes corresponding to the temporal stability problem.

The Orr–Sommerfeld Equation

Motivated by Squire's theorems, we restrict our attention to two-dimensional disturbances in the xy plane and set out to compute the eigenvalues and eigenfunctions associated with the normal modes. In order to reduce the number of unknown functions, we introduce a vector potential for the velocity $\mathbf{V}$, writing

$$\mathbf{V} = \nabla \times (q\mathbf{e}_z), \qquad \Omega_z = -\nabla^2 q \tag{9.3.13}$$

where $\mathbf{e}_z$ is the unit vector along the z axis and $q(x, y)$ is a complex function that plays the role of a stream function. Conforming with the periodicity of the flow, we set

$$q(x, y) = f(y)\exp(ikx) \tag{9.3.14}$$

where $f(y)$ is a complex function with dimensions of velocity multiplied by length; to simplify the notation, we have set $k = k_x$.

Substituting Eq. (9.3.14) into Eq. (9.3.13) and then into Eq. (9.3.8), we obtain the second-order linear ordinary differential equation

$$w'' - \left(k^2 + i\frac{k}{\nu}(U - c)\right)w = i\frac{k}{\nu}U''f \tag{9.3.15}$$

The auxiliary function

$$w = -f'' + k^2 f \tag{9.3.16}$$

provides us with the strength of the disturbance vorticity. Substituting Eq. (9.3.16) into Eq. (9.3.15), we derive a fourth-order linear homogeneous ordinary differential equation, derived independently by Orr (1907) and Sommerfeld (1908)

$$f'''' - 2k^2 f'' + k^4 f = i\frac{k}{\nu}[(U - c)(f'' - k^2 f) - U'' f] \tag{9.3.17}$$

called the *Orr–Sommerfeld equation,* where

$$c \equiv \frac{\sigma}{k} \tag{9.3.18}$$

is the *complex phase velocity.*

Dimensionless form

Nondimensionalizing all variables using as characteristic length, velocity, and time L, V, and L/V, as discussed in Section 3.7, we recast the Orr–Sommerfeld equation into the dimensionless form

$$\hat{f}'''' - 2\alpha^2 \hat{f}'' + \alpha^4 \hat{f} = i\,Re\,\alpha[(\hat{U} - \hat{c})(\hat{f}'' - \alpha^2 \hat{f}) - \hat{U}'' \hat{f}] \tag{9.3.19}$$

where we have defined the following dimensionless variables and constants

$$\hat{f} \equiv \frac{f}{VL}, \qquad \hat{U} \equiv \frac{U}{V}, \qquad \hat{c} \equiv \frac{c}{V}$$
$$\hat{\mathbf{x}} \equiv \frac{\mathbf{x}}{L}, \qquad \alpha \equiv kL, \qquad Re \equiv \frac{VL}{\nu} \tag{9.3.20}$$

The primes in Eq. (9.3.19) indicate differentiations with respect to the dimensionless position vector $\hat{\mathbf{x}}$.

Temporal instability

In the temporal stability problem we specify the real wave number k and supply proper boundary conditions for f to obtain an eigenvalue problem for c. The real part of c is the *phase velocity* of the perturbation. Having obtained c, we compute the complex growth rate $\sigma = kc$ and decompose it into its real and imaginary parts $\sigma = \sigma_R + i\sigma_I$ to obtain the growth rate of the perturbation σ_I. At neutral stability $\sigma_I = 0$ and both c and σ are real.

It is instructive to note that if U in Eq. (9.3.17) is replaced by $U - U_0$, where U_0 is an arbitrary constant, the eigenvalues of the Orr–Sommerfeld equation will be shifted from c to $c - U_0$. This means that switching to a frame of reference that translates steadily in the direction of the flow with velocity U_0 changes the phase velocity of the perturbation but leaves the growth rate unaffected, in agreement with physical intuition.

In dimensionless variables, the solution of the temporal stability problem depends upon the magnitude of the Reynolds number Re and dimensionless wave number α. A typical stability phase diagram in the (Re, α) plane for plane Poiseuille flow is shown in Figure 9.3.1 after Shen (1954). We observe the occurrence of an unstable flow regime that is enclosed by the neutral-stability curve along which $\sigma_I = 0$, and the existence of a critical Reynolds number Re_c below which the flow is stable for all wave numbers.

Spatial instability

In the spatial stability problem we set $k = k_R + ik_I$, where k_R is the real wave number of the perturbation and $-k_I$ is the corresponding spatial growth rate, and $\sigma = \sigma_R$, where σ_R is the real *cyclic frequency* of the perturbation. We then specify σ_R and supply proper boundary conditions for f to obtain an eigenvalue problem for k_R and k_I. At neutral stability $k_I = 0$ and k is real. In dimensionless variables, the solution of the spatial stability problem depends upon the magnitudes of the Reynolds number Re and dimensionless real cyclic frequency $\omega = L\sigma/V$, where L and V are the characteristic length and velocity scales of the flow.

Relationship between temporal and spatial instability

The temporal and spatial stability problems are based on different scenarios for the evolution of a perturbation, growth in time versus growth in space. At neutral stability both k and σ are real, and the solutions of the two problems are identical. This means that the threshold wave numbers, cyclic frequencies, and Reynolds number that demarcate regions of stable and unstable flow coincide.

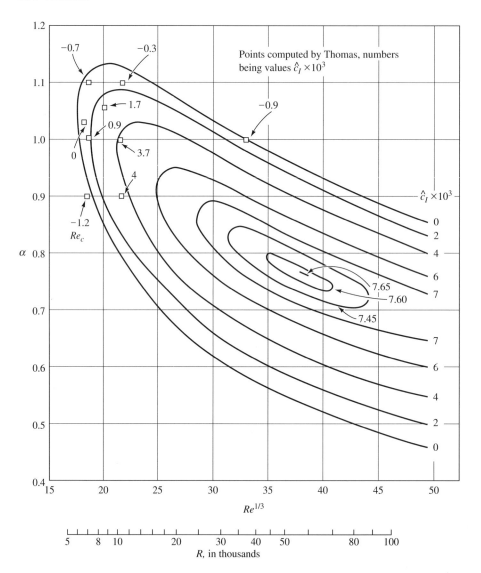

Figure 9.3.1 A phase diagram for the temporal stability of plane Poiseuille flow in a two-dimensional channel of width $2a$; $\hat{c}_i = c_i/U_{\text{Max}}$, $\alpha = ka$, and the Reynolds number is defined as $Re = aU_{\text{Max}}/\nu$ where U_{Max} is the velocity at the center line (Shen 1954).

Let us consider the general case where k and σ are both complex, introduce dimensionless variables, and write the solution of the Orr–Sommerfeld equation in the form $F(Re, \alpha, \omega) = 0$, where $\alpha = kL$ is the dimensionless complex wave number and $\omega = L\sigma/V$ is the dimensionless complex cyclic frequency. Assuming that F is an analytic function of its arguments, we write

$$dF = \frac{\partial F}{\partial \alpha} \, d\alpha + \frac{\partial F}{\partial \omega} \, d\omega + \frac{\partial F}{\partial Re} \, d\,Re = 0 \tag{9.3.21}$$

Now let us denote the real wave number of the temporal stability problem for neutral stability at a particular Reynolds number Re_0 by α_N, and the corresponding real cyclic frequency of the spatial stability problem for neutral stability at the same Reynolds number by ω_N; it is evident that $F(Re_0, \alpha_N, \omega_N) = 0$. If we change the Reynolds number by a differential amount $d\,Re$, α_N and ω_N will change by $d\alpha_N$ and $d\omega_N$ so as to satisfy the equations

$$\frac{\partial F}{\partial \alpha} \, d\alpha_N + \frac{\partial F}{\partial Re} \, d\,Re = 0, \qquad \frac{\partial F}{\partial \omega} \, d\omega_N + \frac{\partial F}{\partial Re} \, d\,Re = 0 \tag{9.3.22}$$

where the partial derivatives are evaluated at Re_0, α_N, ω_N. Combining the two equations (9.3.22) to eliminate $d\,Re$ and using Eq. (9.3.21) yields

$$d\omega_N = \frac{\partial F/\partial \alpha}{\partial F/\partial \omega} \, d\alpha_N = -\left(\frac{\partial \omega}{\partial \alpha}\right)_{Re=Re_0} d\alpha_N \equiv -c_G \, d\alpha_N \tag{9.3.23}$$

where we have introduced the *group velocity* $c_G = \partial \omega/\partial a$ of the spatially growing waves under conditions of neutral stability (Gaster, 1962). Equation (9.3.23) may be used to obtain information on the temporal stability problem using results from the spatial stability problem and vice versa.

Computation of the Disturbance Flow

Having solved the eigenvalue problem, we combine Eqs. (9.2.16), (9.3.4), (9.3.13), and (9.3.14) and thus obtain the stream function of the disturbance associated with a normal mode

$$\psi^{NM}(x, y, t) = f(y)\exp[i(kx - \sigma t)] = f(y)\exp[ik(x - ct)] \tag{9.3.24}$$

To expedite the solution of the linear stability problem, the functional form (9.3.24) is sometimes assumed at the outset. Differentiating the stream function yields the two disturbance velocity components

$$\begin{bmatrix} u_x^{NM} \\ u_y^{NM} \end{bmatrix}(x, y, t) = \begin{bmatrix} f'(y) \\ -ikf(y) \end{bmatrix}\exp[ik(x - ct)] \tag{9.3.25}$$

and the disturbance vorticity

$$\omega^{NM}(x, y, t) = [-f''(y) + k^2 f(y)]\exp[ik(x - ct)] \tag{9.3.26}$$

The Rayleigh Equation for Inviscid Flow

When the effects of viscosity are negligible, that is, the Reynolds number on the right-hand side of Eq. (9.3.19) is sufficiently large or the fluid is effectively inviscid, the Orr–Sommerfeld equation simplifies to the *Rayleigh equation*

$$(U - c)(f'' - k^2 f) - U'' f = 0 \tag{9.3.27}$$

Recasting Eq. (9.3.27) into the standard form of a second-order linear ordinary differential equation with variable coefficients yields

$$f'' - \left(k^2 + \frac{U''}{U - c}\right)f = 0 \tag{9.3.28}$$

To expedite the solution of Eq. (9.3.28), it is sometimes useful to introduce the variable $F = f'/f$ and reduce Eq. (9.3.28) to *Riccati's equation*

$$F' = -F^2 + k^2 + \frac{U''}{U - c} \qquad (9.3.29)$$

which is a well-studied first-order *nonlinear* equation. The reduction in order from two to one between Rayleigh's equation and Riccati's equation has been offset by the occurrence of a non-linearity.

For future reference, we rewrite Eq. (9.3.28) in an alternative form involving a self-adjoint operator as

$$[(U - c)^2 g']' - k^2(U - c)^2 g = 0 \qquad (9.3.30)$$

where

$$g = \frac{f}{U - c} \qquad (9.3.31)$$

One important feature of Eq. (9.3.28) is that the term in parentheses becomes singular at the *critical plane*, where $U = c$, at which point the unperturbed fluid velocity becomes equal to the phase velocity of the disturbance. This happens, in particular, when c is real, which means that the flow is neutrally stable; the associated disturbances fall within the continuous part of the spectrum (Case, 1960a). We shall see in subsequent sections that the occurrence of this singularity has important implications on the efficiency of the various numerical methods used to compute c.

Flows with Linear Velocity Profile

When the velocity profile $U(y)$ is uniform or has the linear form, $U = B + Gy$, where the shear rate G and the reference velocity B are two constants, U'' vanishes throughout the domain of flow, and the right-hand side of Eq. (9.3.15) is equal to zero. In this case, it is possible to simplify the functional form of Eq. (9.3.15) by introducing the new dimensionless variable

$$\xi \equiv \left(\frac{\nu}{kG}\right)^{2/3}\left(k^2 + i\frac{k}{\nu}(Gy + B - c)\right) \qquad (9.3.32)$$

which is a linear function of y. Substituting Eq. (9.3.32) into Eq. (9.3.15), we obtain the *Stokes equation*

$$\frac{d^2w}{d\xi^2} + \xi w = 0 \qquad (9.3.33)$$

whose solution can be obtained in terms of contour integrals in the complex plane (Feldman, 1957).

Inviscid flow

For a base flow with a uniform or linear velocity profile, Rayleigh's equation (9.3.27) assumes the simplified form

$$(U - c)(f'' - k^2 f) = 0 \qquad (9.3.34)$$

which is clearly satisfied by all solutions of the one-dimensional Helmholtz equation $f'' - k^2 f = 0$; Eq. (9.3.26) shows that the corresponding eigenfunctions represent irrotational flow. The associated eigenvalues provide us with the discrete part of the spectrum. Using Eqs. (9.3.25), we find that the corresponding velocity potential is given by

$$\phi^{NM}(x, y, t) = -\frac{i}{k}f'(y)\exp[ik(x - ct)] \qquad (9.3.35)$$

These eigensolutions will find application in Section 9.5, where we shall study the instability of a finite vortex layer with a linear velocity profile separating two uniform streams.

A second family of solutions to Eq. (9.3.34), corresponding to the continuous part of the spectrum, emerges by allowing the term in the second set of parentheses on the left-hand side to be singular when $U - c = 0$, which occurs at $y = \eta \equiv (U - B)/G$. Since $c = U$ is real, the corresponding normal modes are neutrally stable. These solutions are found by setting

$$f'' - k^2 f = GL\delta(y - \eta) \qquad (9.3.36)$$

where the elevation η varies between the lower and upper flow boundaries, L is a certain reference length, and δ is the one-dimensional delta function with dimensions of inverse length. The solutions of Eq. (9.3.36) are the Green's function of the one-dimensional Helmholtz equation $f'' - k^2 f = 0$ with poles at the points where the phase velocity becomes equal to the fluid velocity of the base linear flow. Reference to Eq. (9.3.26) shows that the corresponding disturbances represent the flow due to a flat vortex sheet with sinusoidal strength, located at $y = \eta$, traveling along the x axis.

An acceptable solution of Eq. (9.3.36) must (1) be continuous with discontinuous derivatives at $y = \eta$, (2) satisfy the homogeneous equation $f'' - k^2 f = 0$ everywhere except at $y = \eta$, (3) satisfy the required kinematic condition at the boundaries, and (4) obey the discontinuity condition $f(\eta^+) - f(\eta^-) = GL$, where the plus and minus superscripts indicate evaluations on the upper and lower sides of the level $y = \eta$ (Case, 1960a).

Inviscid Couette flow

To illustrate the application of the preceding results, we return to study the stability of inviscid plane Couette flow whose velocity profile is given by $U(y) = Gy$, discussed earlier in Section 9.2. In this case $B = 0$, y varies between $-a$ and a, and the origin has been set midway between the walls.

A solution of the Helmholtz equation $f'' - k^2 f = 0$ that is consistent with the no-penetration condition $f = 0$ at $y = \pm a$ does not exist, and this means that the discrete part of the spectrum is null. In physical terms, this is a consequence of the fact that, when the normal component of the boundary velocity is required to vanish, an irrotational velocity field in a confined domain must vanish.

The general solution of Eq. (9.3.36), corresponding to the continuous part of spectrum, can be parametrized with respect to the elevation $\eta = U/G$, which ranges between $-a$ and a. Identifying the characteristic length L with a, we obtain

$$f(y, k, \eta) = \begin{cases} -\dfrac{Ga}{k} \dfrac{\sinh k(a - \eta)}{\sinh 2ka} \sinh k(y + a), & \text{for } -a < y < \eta \\[2ex] -\dfrac{Ga}{k} \dfrac{\sinh k(a + \eta)}{\sinh 2ka} \sinh k(a - y), & \text{for } \eta < y < a \end{cases} \qquad (9.3.37)$$

It is a straightforward exercise to verify that Eq. (9.3.37) satisfies the four conditions stated after Eq. (9.3.36) with $L = a$ (Problem 9.3.1).

The stream function corresponding to an arbitrary perturbation may be expressed in terms of the eigenfunctions (9.3.37) as

$$\psi(x, y, t) = \int_{-\infty}^{\infty} \int_{-a}^{a} q(k, \eta) f(y, k, \eta) \exp[ik(x - G\eta t)] \, d\eta \, dk \qquad (9.3.38)$$

where the dimensionless function q is determined by the form of the perturbation at the initial instant. To demonstrate this dependence explicitly, we express the initial stream function in terms of a Fourier integral with respect to x in the form

$$\psi(x, y, t = 0) = \int_{-\infty}^{\infty} g(k, y) \exp(ikx) \, dk \tag{9.3.39}$$

where g is the one-dimensional Fourier transform of the initial distribution of ψ with respect to x. Comparing Eqs. (9.3.38) and (9.3.39) allows us to write

$$g(k, y) = \int_{-a}^{a} q(k, \eta) f(y, k, \eta) \, d\eta \tag{9.3.40}$$

Now, taking the second derivatives of both sides of Eq. (9.3.40) with respect to y, switching the order of the second derivative and the integral sign on the right-hand side, and using Eq. (9.3.36) with $L = a$ in conjunction with the properties of the delta function, we find

$$q(k, y) = \frac{1}{Ga} \left(\frac{\partial^2}{\partial y^2} - k^2 \right) g(k, y) \tag{9.3.41}$$

which provides us with a relation between q and g. Furthermore, remembering that $\omega = -\nabla^2 \psi$ and using Eq. (9.3.39), we find that $-Gaq(k, y)$ is the Fourier transform of the initial vorticity distribution with respect to x.

We can proceed further by expanding $g(k, y)$ in a sine Fourier series with respect to y, similar to that shown in Eq. (9.2.3), writing

$$g(k, y) = \sum_{n-1}^{\infty} c_n(k) \sin\left(\frac{n\pi}{2a}(y + a) \right) \tag{9.3.42}$$

where c_n are complex coefficients. Substituting Eq. (9.3.42) into the right-hand side of Eq. (9.4.31), we obtain a sine Fourier series with respect to y for $q(k, y)$

$$q(k, y) = -\frac{1}{Ga} \sum_{n=1}^{\infty} c_n(k) \left[\left(\frac{n\pi}{2a} \right)^2 + k^2 \right] \sin\left(\frac{n\pi}{2a}(y + a) \right) \tag{9.3.43}$$

Finally, substituting Eq. (9.3.43) into the integrand of Eq. (9.3.38) and carrying out the integration with respect to η yields the general solution in terms of c_n (Case, 1960a). In the particular case where the c_n are real, equal to b_n, the real part of the final expression yields Eq. (9.2.5), which was obtained using a different method (Problem 9.3.2). This confirms that *the set of the normal modes that fall within the continuous part of the spectrum is complete.*

PROBLEMS

9.3.1   **Green's function of the Helmholtz operator.** Verify that Eq. (9.3.37) satisfies the four conditions stated after Eq. (9.3.36) with $L = a$ (Case, 1960a).

9.3.2   **Inviscid Couette flow.** Show that in the particular case where c_n are real, equal to b_n, the real part of the expression that emerges by substituting Eq. (9.3.43) into Eq. (9.3.38) and carrying out the integration with respect to η is given in Eq. (9.2.5).

 ## Computer Problem

9.3.3   **Inviscid plane Couette flow.** Draw the streamlines corresponding to the eigenfunctions described by Eq. (9.3.37) for several values of η and discuss the structure of the normal modes.

9.4 | GENERAL THEOREMS ON THE TEMPORAL STABILITY OF INVISCID SHEAR FLOWS

There are several general theorems that allow us to assess the temporal stability of an inviscid unidirectional flow, as well as obtain estimates for the location of the phase velocity in the complex plane, from mere inspection of the velocity profile and without having to compute the spectrum of eigenvalues.

Rayleigh's Criterion on the Significance of an Inflection Point

We begin by multiplying both sides of Eq. (9.3.28) by f^*, where the asterisk designates the complex conjugate, and rearrange the terms on the left-hand side to obtain

$$(f'f^*)' - |f'|^2 = \left(k^2 + \frac{U''}{U-c}\right)|f|^2 \tag{9.4.1}$$

where a prime denotes a derivative with respect to y. Assuming that the flow is confined between two impenetrable planar boundaries located at $y = a$ and b, we integrate Eq. (9.4.1) with respect to y between a and b, and require the no-penetration condition $f(a) = f(b) = 0$ to find

$$-\int_a^b |f'|^2 \, dy = \int_a^b \left(k^2 + \frac{U''}{|U-c|^2}(U-c^*)\right)|f|^2 \, dy \tag{9.4.2}$$

Setting the real and imaginary parts of the two sides of Eq. (9.4.2) equal to one another yields

$$-\int_a^b |f'|^2 \, dy = \int_a^b \left(k^2 + U''\frac{U-c_R}{|U-c|^2}\right)|f|^2 \, dy \tag{9.4.3a}$$

$$c_I \int_a^b U'' \frac{|f|^2}{|U-c|^2} \, dy = 0 \tag{9.4.3b}$$

The last equation requires that either $c_I = 0$, in which case we obtain neutral stability, or the integral on the right-hand side vanish. But since the sign of the integrand is the same as the sign of U'', we find that U'' must change sign at least once between $y = a$ and b. Thus *for a normal mode of a unidirectional shear flow to be unstable, the velocity profile must exhibit at least one inflection point,* as pointed out by Rayleigh (1880). This is a necessary but not sufficient condition; that is, a normal-mode disturbance of a unidirectional shear flow whose velocity profile has an inflection-point profile is not necessarily unstable.

An important consequence of Rayleigh's criterion is that the normal modes of inviscid infinite simple shear flow, Couette flow, or Poiseuille flow in a channel are all stable. As curious as it may seem, viscous forces are required in order to render certain perturbations unstable. The physical explanation lies in the fact that viscous stresses spread out the disturbances, giving them a better chance to grow.

Fjørtoft's Condition for Instability

One can proceed beyond Rayleigh's criterion by combining the two equations (9.4.3) to find that, unless a perturbation is neutrally stable, it must be

$$\int_a^b U''(U-U_0)\frac{|f|^2}{|U-c|^2} \, dy = -\int_a^b (k^2|f|^2 + |f'|^2) \, dy < 0 \tag{9.4.4}$$

where U_0 is an arbitrary constant (Fjørtoft, 1950). Let us consider a flow whose velocity profile has a single inflection point, identify U_0 with the velocity at the inflection point U_I, and note that the sign of the product $U''(U-U_I)$ is constant through the domain of integration. For the integral on the left-hand side of Eq. (9.4.4) to be negative; the sign of $U''(U - U_I)$ must also be negative; otherwise the disturbance will be neutrally stable. A consequence of this observation is that *for a normal mode of a unidirectional shear flow to be unstable, the maximum of the absolute value of the vorticity of the base flow must occur at the inflection point.*

Combining Rayleigh's and Fjørtoft's theorems, we find that the normal modes of the flow shown in Figure 9.4.1(a) are stable, whereas those of the flow shown in Figure 9.4.1(b) may be either stable or unstable.

Sufficient Condition for Instability

Rayleigh's and Fjørtoft's theorems provide us with necessary but not sufficient conditions for instability. Tollmien indicated that for shear flows in channels with symmetric and monotonically varying profiles $U(y)$, similar to those observed in boundary layers, these conditions are also sufficient. The proof and further extensions of the theory are discussed in detail by Yih (1979, p. 473) and Drazin and Reid (1981, p. 132).

Rayleigh's Theorem on the Phase Velocity of Unstable Disturbances and Howard's Semicircle Theorem

Certain iterative numerical procedures for computing the eigenvalues c require an accurate initial guess. In some cases, this can be found using Gershgorin's circle theorem, which is applicable to standard algebraic eigenvalue problems (Section B.2, Appendix B). A more general but less accurate method of locating the eigenvalues is provided by Howard's semicircle theorem, which states that c corresponding to an unstable mode must fall within a half-disk in the upper half complex plane. The center of the disk lies on the real axis at the point $\frac{1}{2}(U_{\text{Max}} + U_{\text{Min}})$, and its radius is equal to $\frac{1}{2}(U_{\text{Max}} - U_{\text{Min}})$, where U_{Max} and U_{Min} are the maximum and minimum values of U (Howard, 1961).

One corollary of Howard's semicircle theorem is that there is always a point where the real part of c is equal to U, and therefore the disturbance travels with the local velocity of the fluid; the region around that point is called the *critical layer.*

To prove the semicircle theorem, we multiply Eq. (9.3.30) by g^*, where an asterisk indicates the complex conjugate, integrate the resulting equation from $y = a$ to b, and require the no-penetration condition $g(a) = 0$, $g(b) = 0$ to obtain

$$\int_a^b (U - c)^2 R \, dy = 0 \tag{9.4.5}$$

where $R = |g'|^2 + k^2|g^2|$ is a non-negative function. Decomposing the integral into its real and imaginary parts we obtain

(a)   (b)

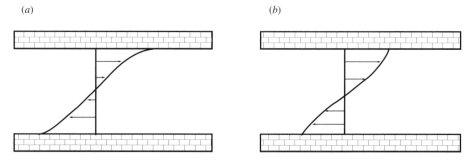

Figure 9.4.1 Applications of Rayleigh's theorem on the significance of inflection points, and Fjørtoft's theorem. Normal modes for flow (a) are stable, but those for flow (b) may be unstable.

$$\int_a^b (U^2 - 2c_R U + c_R^2 - c_I^2) R \, dy = 0$$

(9.4.6)

$$c_I \int_a^b (U - c_R) R \, dy = 0$$

The second equation is satisfied when $c_I = 0$, but this corresponds to neutrally stable perturbations. Leaving this case aside, we set the second integral equal to zero and find that the phase velocity c_R must take values between U_{Max} and U_{Min}, as first noted by Rayleigh (1880), which shows that *a disturbance that is not neutrally stable cannot move faster than the fluid.*

Proceeding, we use the second to simplify the first of Eqs. (9.4.6), and thus obtain

$$\int_a^b (U^2 - |c|^2) R \, dy = 0$$

(9.4.7)

Furthermore, we make the independent observation that, since R is non-negative,

$$\int_a^b (U - U_{Min})(U - U_{Max}) R \, dy < 0$$

(9.4.8)

and use the second of Eqs. (9.4.6) to recast Eq. (9.4.8) into the form

$$\int_a^b [U^2 - c_R(U_{Max} + U_{Min}) + U_{Max} U_{Min}] R \, dy < 0$$

(9.4.9)

Finally, we combine Eqs. (9.4.7) and (9.4.9) and obtain

$$\int_a^b [|c|^2 - c_R(U_{Max} + U_{Min}) + U_{Max} U_{Min}] R \, dy$$

$$= \int_a^b \left[\left(c_R - \frac{U_{Max} + U_{Min}}{2} \right)^2 + c_I^2 - \left(\frac{U_{Max} - U_{Min}}{2} \right)^2 \right] R \, dy < 0$$

(9.4.10)

Since R is non-negative, the term within the brackets in the integrand must be non-positive, and this shows that c must be located within a disk in the complex plane centered on the real axis at the point $\frac{1}{2}(U_{Max} + U_{Min})$ with radius equal to $\frac{1}{2}(U_{Max} - U_{Min})$. The upper half of the disk yields unstable normal modes with positive values for c_I.

PROBLEM

9.4.1 **Stability of an inviscid shear flow with sinusoidal velocity profile.** Consider a unidirectional inviscid shear flow with $U(y) = A \sin(Gy)$, where A and G are two constants, in a channel that is bounded by two plane walls located at $y = a$ and b. Show that if there are no values of $G/(n\pi)$ between a and b, where n is an integer, then the flow is stable.

9.5 | STABILITY OF A UNIFORM VORTEX LAYER SUBJECT TO SPATIALLY PERIODIC DISTURBANCES

Having established certain general theorems pertaining to the temporal stability of inviscid unidirectional shear flows, we proceed to consider in detail a prototypical flow for which the Rayleigh equation may be solved exactly in closed form (Rayleigh, 1894, Vol. II, p. 393). The results will provide us with useful insights into the nature of the normal modes and stability of more general free shear flows with monotonically varying velocity profiles.

Consider an unbounded shear flow of a fluid with uniform density, containing an infinite vortex layer of thickness $2b$ and uniform vorticity Ω oriented along the x axis, separating two streams that flow with uniform velocities $\pm U_0$, as illustrated in Figure 9.5.1, where $\Omega = -U_0/b$. In the unperturbed state, the velocity profile is uniform outside the vortex layer and varies in a linear fashion across the vortex layer. This piecewise linear velocity profile may be regarded as a rough representation of a continuous profile, which may be described, for instance, by a hyperbolic-tangent function, $U(y) = U_0 \tanh(y/b)$.

Since $U'' = 0$ everywhere within the flow except at the boundaries of the vortex layer, the normal modes are described by the simplified version of Rayleigh's equation (9.3.34). Leaving aside the neutrally stable modes with $c = U$, corresponding to the continuous part of the spectrum, we obtain the Helmholtz equation $f'' - k^2 f = 0$ expressing irrotational perturbations, whose general solution is readily found to be

$$f_n(y) = A_n e^{ky} + B_n e^{-ky} \tag{9.5.1}$$

where A_n and B_n are constants with $n = -1, 0, 1$ corresponding, respectively, to the region below, inside, and above the vortex layer. The stream function and velocity of the normal modes are given by Eqs. (9.3.24) and (9.3.25). Note that, since the disturbance flow is irrotational throughout the domain of flow, it may also be described in terms of the velocity potential given in Eq. (9.3.35), although this will not be necessary for our present purposes.

The disturbance causes the upper and lower boundaries of the vortex layer to deform in response to the flow. The location of the boundaries in the deformed state may be described by the real or imaginary parts of the right-hand sides of

$$y^{\text{Upper}} = \eta_1(x, t) = b + \varepsilon a_1 \exp[ik(x - ct)] \tag{9.5.2a}$$

$$y^{\text{Lower}} = \eta_{-1}(x, t) = -b + \varepsilon a_{-1} \exp[ik(x - ct)] \tag{9.5.2b}$$

where $a_{\pm 1}$ are two complex constants and ε is a small dimensionless number.

To compute the eight unknowns A_n, B_n, $n = -1, 0, 1$, and $a_{\pm 1}$, we require eight equations. Demanding that the disturbance vanish far from the shear layer, we obtain two equations

$$B_{-1} = A_1 = 0 \tag{9.5.3}$$

Two additional equations emerge by requiring that the y component of the velocity remain continuous across the boundaries of the vortex layer. Equation (9.3.25) suggests that this will be the case provided that $f_{-1} = f_0$ at $y = \eta_{-1}$ and $f_0 = f_1$ at $y = \eta_1$; to first order in ε, we obtain $f_{-1} = f_0$ at $y = -b$ and $f_0 = f_1$ at $y = b$. Using Eq. (9.5.1), we then find

$$A_0 + \beta B_0 = A_{-1}, \qquad \beta A_0 + B_0 = B_1 \tag{9.5.4}$$

where, for convenience, we have defined $\beta = \exp(2kb)$.

Next we require that the x component of the disturbance velocity remain continuous across the boundaries of the vortex layer. To implement this condition, we evaluate u_x^{NM} at $y = \eta_{\pm 1}$ using Eq. (9.3.25), expand the resulting expressions in a Taylor series about $y = \pm b$, and maintain only linear terms with respect to ε. For the upper boundary we find

$$u_x(y=\eta_1) = (U + \varepsilon u_x^{\text{NM}})_{y=\eta_1} \cong U_{(y=b)} + \left(\frac{dU}{dy}\right)_{y=b} (\eta_1 - b) + \varepsilon u_x^{\text{NM}}(y=b)$$

$$= -\Omega b + \varepsilon \left(a_1 \frac{dU}{dy} + f'\right)_{y=b} \exp[ik(x - ct)] \tag{9.5.5}$$

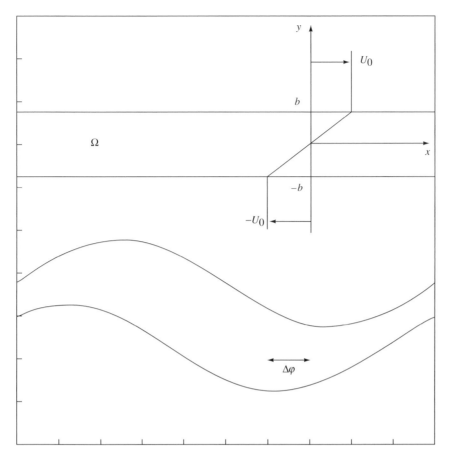

Figure 9.5.1 Illustration of a vortex layer of thickness $2b$ separating two uniform streams flowing with uniform velocity $\pm U_0$ in the unperturbed and perturbed states.

Noting that $dU/dy = -\Omega$ inside the vortex layer and $dU/dy = 0$ outside the vortex layer, and demanding continuity of velocity, we derive the equation

$$-\Omega a_1 + f_0'(y=b) = f_1'(y=b) \tag{9.5.6}$$

Working in a similar manner for the lower vortex boundary, we obtain

$$-\Omega a_{-1} + f_0'(y=-b) = f_{-1}'(y=-b) \tag{9.5.7}$$

Substituting Eq. (9.5.1) into Eqs. (9.5.6) and (9.5.7) yields the relations

$$V a_1 - \beta A_0 + B_0 = B_1, \qquad V a_{-1} - A_0 + \beta B_0 = -A_{-1} \tag{9.5.8}$$

where we have defined $V = (\Omega/k)\exp(kb)$.

Finally, we exploit the fact that the boundaries of the vortex layer are material lines and are therefore convected by the flow, introduce the kinematic constraint $D(y - \eta_{\pm 1})/Dt = 0$, expand all terms in Taylor series with respect to ε and retain only the linear contributions, and thus obtain

$$\frac{\partial \eta_{\pm 1}}{\partial t} + U(y=\pm b)\frac{\partial \eta_{\pm 1}}{\partial x} - \varepsilon u_y^{\text{NM}}(y=\pm b) = 0 \tag{9.5.9}$$

Substituting Eqs. (9.5.2) and (9.3.25) into Eq. (9.5.9) yields the two equations

$$a_{-1} = \frac{A_{-1}}{c + U_0} e^{-kb}, \qquad a_1 = \frac{B_1}{c - U_0} e^{-kb} \qquad (9.5.10)$$

Equations (9.5.3), (9.5.4), (9.5.8), and (9.5.10) provide us with the desired system of eight linear homogeneous equations for the eight unknown coefficients. For a nontrivial solution to exist, the determinant of the coefficient matrix must vanish, and this provides us with a secular equation for the computation of the complex phase velocity c. To find the solution, we substitute Eqs. (9.5.10) into Eqs. (9.5.8) and obtain

$$A_0 - \beta B_0 = A_{-1}\left(1 + \frac{\Omega}{k(c + U_0)}\right)$$

$$\beta A_0 - B_0 = B_1\left(-1 + \frac{\Omega}{k(c - U_0)}\right) \qquad (9.5.11)$$

Furthermore, we substitute Eqs. (9.5.4) into Eqs. (9.5.11) and obtain a homogeneous system of two equations

$$\begin{bmatrix} \Omega & \beta(2kc + 2kU_0 + \Omega) \\ \beta(2kc - 2kU_0 - \Omega) & -\Omega \end{bmatrix}\begin{bmatrix} A_0 \\ B_0 \end{bmatrix} = 0 \qquad (9.5.12)$$

Setting the determinant of the matrix on the left-hand side equal to zero yields a quadratic equation for the growth rate, whose solution is readily found to be

$$c = \pm U_0 \frac{1}{2kb} \sqrt{(1 - 2kb)^2 - e^{-4kb}} \qquad (9.5.13)$$

The quantity under the radical is negative when $0 < kb < 0.639$ and positive when $kb > 0.639$. In the first case c is a purely imaginary number, and the plus sign in Eq. (9.5.13) yields an unstable mode with growth rate equal to $\sigma_I = kc$ and vanishing phase velocity; the minus sign yields a stable normal mode. When $kb > 0.639$, c is real, and the normal modes translate upstream or downstream with phase velocity c while maintaining their initial amplitude; the lack of energy dissipation in an inviscid fluid prevents the decay of the kinetic energy of the perturbation.

In summary, we found that a normal-mode disturbance can be unstable only when the dimensionless wave number kb is less than 0.639 or, equivalently, the ratio between the wave length and the layer thickness $\lambda/2b$ is larger than 4.92. Normal-mode disturbances with a shorter wavelength travel along the vortex layer maintaining their initial amplitude. The significance of these results on the behavior of general spatially periodic disturbances that do not necessarily represent normal modes will be discussed at the end of this section.

In Figure 9.5.2(a) we present graphs of the dimensionless imaginary part of the phase velocity c_I/U_0 and dimensionless growth rate $4b\sigma_I/U_0 = 4kbc_I/U_0 = -4kc_I/\Omega$ for the unstable normal modes, plotted with a solid and a dashed line, respectively, against the reduced wave number kb. The dimensionless growth rate reaches a maximum at $kb = 0.398$, which corresponds to the *most unstable* or *most dangerous normal mode*.

In Figure 9.5.2(b) we present a graph of the dimensionless phase velocity $c_R/U_0 = -c_R/(\Omega b)$ of the stable traveling normal modes with $kb > 0.639$, corresponding to the plus sign in Eq. (9.5.13), plotted with a solid line against kb. As kb becomes larger, c_R/U_0 tends to unity, which means that the phase velocity tends to become equal to the velocity of the upper stream. Had we maintained the minus sign in Eq. (9.5.13), we would have found that the phase velocity shown in Figure 9.5.2(b) simply reverses its sign.

Waves on Vortex Boundaries

It is illuminating to examine the deformation of the vortex layer subject to a normal-mode perturbation. For this purpose, we compute the ratio of the complex amplitudes of the waves on

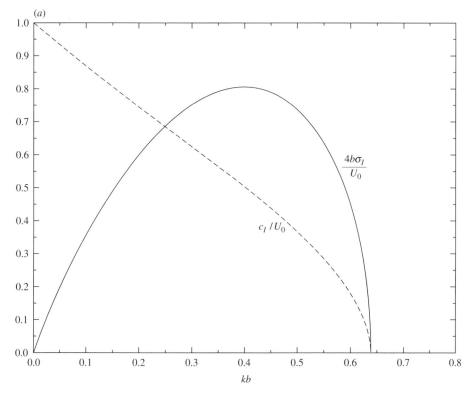

Figure 9.5.2 Graph of the (a) reduced imaginary part of the complex phase velocity c_i/U_0 (dashed line) and reduced growth rate $4b\sigma_i/U_0$ (solid line) against the reduced wave number kb in the unstable regime. (*continued*)

the upper and lower vortex boundaries, dividing the two equations (9.5.10) side by side, and then use Eqs. (9.5.4) and (9.5.12) to obtain

$$\frac{a_{-1}}{a_1} = \frac{A_{-1}}{B_1}\frac{c - U_0}{c + U_0} = \frac{(1 - 2kb)\beta^2 - 1 + 2kb\beta^2\hat{c}}{2kb\beta(1 + \hat{c})} \tag{9.5.14}$$

where $\hat{c} = c/U_0$ is computed from Eq. (9.5.13).

The ratio between the amplitudes of the waves on the two contours, $|a_{-1}/a_1|$, is equal to unity in the unstable band of kb. In Figure 9.5.2(b) we plot, with a dashed line, the magnitude $|a_{-1}/a_1|$ in the stable regime corresponding to the plus sign in Eq. (9.5.13), and observe a rapid decay with increasing kb; physically, this behavior reveals that the lower vortex contour is only slightly deformed. Had we maintained the minus sign in Eq. (9.5.13), we would have found that $|a_{-1}/a_1|$ is equal to the inverse of that plotted in Figure 9.5.2(b), which means that the upper vortex contour would be only slightly deformed; the normal modes have no spatial bias.

To every normal mode there is a corresponding phase shift $\Delta\varphi$ of the waves along the upper and lower vortex boundaries defined as $\Delta\varphi = \text{Arg}(a_{-1}/a_1)$, as illustrated in Figure 9.5.1. In the stable regime of kb, $\Delta\varphi$ is equal to π, which reveals that the vortex-boundary waves travel while being out of phase with each other. In Figure 9.5.2(c) we plot, with a solid line, $\Delta\varphi/\pi$ in the unstable regime and observe that, as kb becomes smaller, $\Delta\varphi$ tends to vanish, which means that the vortex boundary waves tend to grow in phase. Conversely, as kb is raised from small values, $\Delta\varphi$ obtains positive values and reaches the maximum value of π at the point of neutral stability.

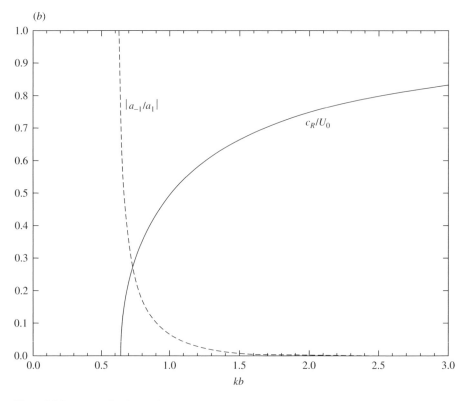

Figure 9.5.2 (*continued*)   Graph of the (b) reduced phase velocity c_R/U_0 (solid line), and ratio of amplitudes of the traveling waves along the lower and upper vortex contours $|a_{-1}/a_1|$ in the stable regime (dashed line). (*continued*)

Disturbance in circulation

The occurrence of a finite phase shift $\Delta\varphi$ between the waves on the upper and lower vortex boundaries results in a redistribution of rotational fluid within the vortex layer. The concomitant periodic accumulation of vorticity may be regarded as an engine that drives the amplification of the disturbance.

To examine the redistribution of rotational fluid within the vortex layer in a global sense, we define the *strength of the perturbed vortex layer* in terms of the thickness of the vortex layer, as

$$\gamma(x) = \Omega(\eta_1 - \eta_{-1}) = 2\Omega b - \Omega\varepsilon(a_{-1} - a_1)\exp[ik(x - ct)] \tag{9.5.15}$$

Inspecting Eq. (9.5.15), it becomes evident that the perturbation causes a disturbance in the strength of the vortex layer, whose phase shift with respect to the displacement of the upper vortex contour is given by

$$\Delta\varphi_\gamma = \text{Arg}\left(\frac{a_{-1} - a_1}{a_1}\right) \tag{9.5.16}$$

Using Eq. (9.5.14), we find

$$\frac{a_{-1} - a_1}{a_1} = -\frac{2kb\beta(1 + \beta) + 1 - \beta^2 + 2kb\beta(1 - \beta)\hat{c}}{2kb\beta(1 + \hat{c})} \tag{9.5.17}$$

(c)

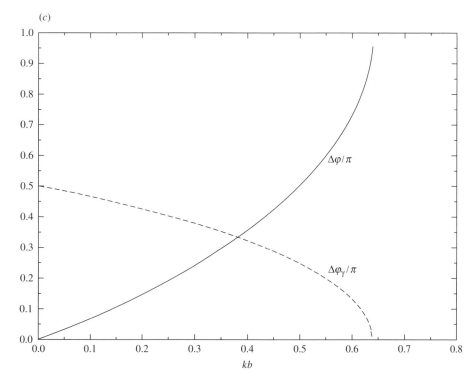

Figure 9.5.2 (*continued*)   Graph of the (c) phase shift $\Delta\varphi/\pi$ between the growing waves along the upper and lower vortex contours (solid line) and, and phase shift between the disturbance in circulation and the wave along the upper vortex boundary $\Delta\varphi_\gamma/\pi$ for unstable modes (dashed line).

which may be used to compute $\Delta\varphi_\gamma$ as a function of the reduced vortex layer thickness kb. The results show that $\Delta\varphi_\gamma$ vanishes in the stable regime of kb, because the waves on the vortex boundaries are out of phase, and $\Delta\varphi = \pi$. In Figure 9.5.2(c) we plot, with a dashed line, $\Delta\varphi_\gamma$ for unstable perturbations corresponding to the plus sign in Eq. (9.5.13), and observe that as the reduced thickness of the vortex layer kb becomes smaller, $\Delta\varphi_\gamma$ tends to the value of $\pi/2$, which means that rotational fluid tends to accumulate midway between the crests and the troughs.

Vortex sheet

Taking the limit as kb tends to zero while maintaining $U_0 = -\Omega b$ constant reduces the vortex layer to a vortex sheet of infinitesimal thickness. Expanding Eq. (9.5.13) in a Taylor series with respect to kb yields the complex growth rate

$$\sigma \equiv kc \to \pm ikU_0 \qquad (9.5.18)$$

which shows that *a vortex sheet is unstable for all wave numbers* and, furthermore, the growth rate of the normal modes is proportional to the wave number and therefore it becomes infinite as the wavelength of a disturbance becomes infinitesimally small. This singular behavior causes essential difficulties in computing the motion of vortex sheets discussed in Section 11.4 in the context of vortex methods.

Behavior of general periodic disturbances

To study the behavior of an arbitrary periodic disturbance that deforms the boundaries of the vortex layer in a sinusoidal manner with reduced wave number kb, we decompose it into two normal-mode disturbances with the same wave number [Problem 9.5.1(b)]. If $kb < 0.639$, one of the two normal-modes will be unstable, which means that the original disturbance will be unstable as long as it does not coincide with the stable normal mode. If $kb > 0.639$, the disturbance will be stable.

Consider now a periodic disturbance that deforms the vortex boundaries in a nonsinusoidal manner. The initial location of the boundaries may be expressed in terms of a Fourier series, and each term in the series may be expressed in terms of two normal modes; according to the preceding discussion, the original disturbance will be unstable to the Fourier modes with $kb < 0.639$.

Nonlinear motion

The long-term evolution of an unstable normal mode is illustrated in Figure 11.6.2. We observe an initial linear growth period followed by nonlinear evolution which is evidenced by the non-sinusoidal shapes. The motion was computed using the numerical method of *contour dynamics* discussed in Section 11.6. The amplification of the disturbance leads to the formation of a periodic sequence of compact vortices which are connected with thin filaments of rotational fluid; this is an example of an instability growing and leading to a new nearly-steady state.

PROBLEMS

9.5.1   **Vortex layer.** (a) Consider the vortex layer shown in Figure 9.5.1, but, in addition, superimpose a uniform flow in the x direction with velocity V. Compute the complex phase velocity and discuss the effect of V on the behavior of the normal modes. (b) Consider a disturbance that deforms the upper and lower vortex boundaries in a sinusoidal manner with the same wavelength but different amplitudes and a finite phase shift. Express the disturbance in terms of normal modes.

9.5.2   **Vortex layer on a wall.** Consider a vortex layer with uniform vorticity that is attached to an impermeable wall on one side and is exposed to a uniform flow on the other side, and compute the complex phase velocity of the normal modes (Pullin, 1981).

9.5.3   **Compound vortex layer.** Set up the linear stability problem for an unbounded shear flow that contains two attached vortex layers with different values of uniform vorticity and different thicknesses, separating two uniform streams flowing at different velocities (Pozrikidis and Higdon, 1987).

Computer Problems

9.5.4   **Normal modes on a vortex layer.** (a) Plot the shape of the vortex contours corresponding to stable and unstable normal modes for several values of kb in the unstable and stable regimes. Discuss the results with reference to the redistribution of rotational fluid. (b) Plot the streamlines of the perturbed flow for several values of kb in the unstable and stable regimes, and discuss the physical significance of your observations.

9.5.5   **Compound vortex layer.** With reference to Problem 9.5.3, write a program called *CVL* that computes the complex phase velocity of the normal modes as a function of kb_1, where b_1 is the thickness of the upper vortex layer, under the assumption that the vorticity of the upper layer is equal in magnitude with and opposite in sign to the vorticity of the lower layer. Prepare several plots of the growth rate of unstable disturbances as a function of kb_1 for different values of the ratio b_2/b_1, where b_2 is the thickness of the lower vortex layer, and discuss your results. Note that there may be two distinct bands of unstable wave numbers.

9.6 NUMERICAL SOLUTION OF THE ORR–SOMMERFELD AND RAYLEIGH EQUATIONS

Exact analytical solutions to the Orr–Sommerfeld and Rayleigh equations are feasible only for a limited class of creeping or ideal flows. To study the stability of arbitrary unidirectional flows at finite Reynolds numbers, we must resort to approximate and asymptotic methods. In this section we shall develop and illustrate the application of several classes of numerical methods, pointing out their relative merits and discussing the subtleties of their implementation. Extensive reviews and further analyses may be found in the articles of Gersting and Jankowski (1972) and Davey (1977), and in the monograph of Drazin and Reid (1981).

Finite-Difference Methods

We begin by developing finite-difference methods for solving Rayleigh's equation with reference to shear flow of an effectively inviscid fluid in a channel that is confined between two impermeable plates located at $y = -A$ and B. Focusing our attention on the temporal stability problem, where the real wave number k is specified and the complex phase velocity c is sought, we rewrite Rayleigh's equation (9.3.27) in the form

$$U f'' - (U k^2 + U'') f = c(f'' - k^2 f) \qquad (9.6.1)$$

Note that we have moved the unknown eigenvalue c to the right-hand side. Next we introduce a one-dimensional uniform grid of nodes located at y_i, $i = 0, \ldots, N + 1$, where $y_0 = -A$ and $y_{N+1} = B$, and approximate f'' at the ith node using central differences (Section B.5, Appendix B), thereby replacing Eq. (9.6.1) with the finite-difference equation

$$U_i f_{i-1} - [2U_i + \Delta y^2 (U_i k^2 + U_i'')] f_i + U_i f_{i+1} = c[f_{i-1} - (2 + k^2 \Delta y^2) f_i + f_{i+1}] \qquad (9.6.2)$$

where f_i is the value of the complex function f at y_i. To satisfy the no-penetration condition, we require $f_0 = 0$ and $f_{N+1} = 0$. Collecting the values of f_i, $i = 1, \ldots, N$ into the N-dimensional vector $\mathbf{X} = (f_1, f_2, \ldots, f_{N-1}, f_N)$, we form the linear system of equations

$$\mathbf{A} \cdot \mathbf{X} = c \mathbf{B} \cdot \mathbf{X} \qquad (9.6.3)$$

where $\mathbf{A}$ and $\mathbf{B}$ are two square tridiagonal matrices of size $N \times N$ whose structure may be inferred readily from the finite-difference equation (9.6.2). Two points are worth noting: The structure of the matrices $\mathbf{A}$ and $\mathbf{B}$ depends upon the finite-difference method chosen to approximate the derivatives, and the matrix $\mathbf{B}$ is independent of the particular form of the velocity profile U.

We have thus reduced the problem to computing the eigenvalues c of a generalized algebraic eigenvalue problem expressed by Eq. (9.6.3). The solution provides us with N complex eigenvalues, but we are mainly interested in the eigenvalue with the maximum growth rate σ_I, where $\sigma = kc$. As N is increased, the maximum growth rate obtained by solving the generalized eigenvalue problem will tend to the exact value corresponding to the continuous problem, which is expressed by Eq. (9.6.1).

Numerical methods for computing the spectrum of eigenvalues of the generalized problem (9.6.3) are reviewed by Kerner (1989). A simple way to proceed is to reduce the generalized eigenvalue problem to a standard algebraic eigenvalue problem expressed by the equation

$$\mathbf{D} \cdot \mathbf{X} = c \mathbf{X} \qquad (9.6.4)$$

where $\mathbf{D} = \mathbf{B}^{-1} \cdot \mathbf{A}$ and c is an eigenvalue of $\mathbf{D}$. The eigenvalue with the largest magnitude may be computed using the power method, and other eigenvalues may be obtained by successive deflations, as discussed in Section B.2, Appendix B. A good initial guess for c is provided by Gerschgorin's circle theorem, also discussed in Section B.2. A point of practical concern is that the computation of $\mathbf{B}^{-1}$, which is necessary in order to obtain $\mathbf{D}$, may be prohibitively expensive or introduce substantial numerical round-off error.

An alternative method, which is particularly effective for the present case of inviscid flow, proceeds by restating Eq. (9.6.2) in the form of the homogeneous equation $\mathbf{E} \cdot \mathbf{X} = 0$, and then identifying the eigenvalues c with the roots of the algebraic equation $\text{Det}(\mathbf{E}) = 0$. The structure of the matrix $\mathbf{E}$ emerges by recasting Eq. (9.6.2) into the form

$$f_{i-1} - \left[2 + \Delta y^2 \left(k^2 + \frac{U_i''}{U_i - c} \right) \right] f_i + f_{i+1} = 0 \tag{9.6.5}$$

The fact that the matrix $\mathbf{E}$ is tridiagonal allows us to compute its determinant using an efficient numerical method discussed in Section B.2. Having specified k, we compute the eigenvalues according to the following steps:

1. Guess a complex value for c.
2. Compute $\text{Det}(\mathbf{E})$ using a method for tridiagonal matrices.
3. Correct the value of c so as to make both the real and imaginary parts of $\text{Det}(\mathbf{E})$ vanish. The correction may be done using, for instance, Newton's method, setting

$$c^{\text{New}} = c^{\text{Old}} - \text{Det}[\mathbf{E}(c^{\text{Old}})] / \left(\frac{d\,\text{Det}[\mathbf{E}(c)]}{dc} \right)_{c=c^{\text{Old}}} \tag{9.6.6}$$

Since $\text{Det}[\mathbf{E}(c)]$ is an analytic function of c, the derivative $d\,\text{Det}[\mathbf{E}(c)]/dc$ may be approximated using a finite-difference method, setting, for instance,

$$\frac{d\,\text{Det}[\mathbf{E}(c)]}{dc} \cong \frac{\text{Det}[\mathbf{E}(c + \varepsilon)] - \text{Det}[\mathbf{E}(c)]}{\varepsilon} \tag{9.6.7}$$

where ε is a real or complex number with small magnitude.

In Figure 9.6.1 we plot, with dashed lines, the dimensionless imaginary part of the reduced phase velocity c_I/U_0 and, with solid lines, the reduced growth rate $4b\sigma_I/U_0$ of unstable disturbances corresponding to a shear flow whose velocity profile is given by $U(y) = U_0 \tanh(y/b)$, where U_0 and b are two constants, computed using this method. The depicted family of curves correspond to a sequence of channel widths with $A = B$. It is evident that the presence of walls reduces the growth rate of the perturbations. As A/b tends to infinity, we obtain unbounded shear flow for which the critical wave number for neutral stability is known to be $kb = 1.0$, and the value of c_I/U_0 at $kb = 0$ is equal to unity (Michalke, 1964). It is instructive to note the similarity of the curves shown in Figure 9.6.1 to those shown in Figure 9.5.2(a) for a vortex layer, which suggests that the detailed structure of the velocity profile—piecewise linear versus hyperbolic-tangent—does not have a critical effect on the properties of the flow.

Viscous flow

Finite-difference methods for solving the Orr–Sommerfeld equation are developed working in a similar manner. The no-penetration and no-slip boundary conditions over a stationary wall require that both f and its first derivative at the wall vanish. The finite-difference discretization of the derivatives of f yields a linear system of algebraic equations that is similar to the one shown in Eq. (9.6.3), but now, because of the presence of the fourth derivative, $\mathbf{A}$ is a complex pentadiagonal matrix, and $\mathbf{B}$ is a real tridiagonal matrix. The algebraic system may be placed in the form $\mathbf{E} \cdot \mathbf{X} = 0$, where $\mathbf{E}$ is a complex pentadiagonal matrix, but, unfortunately, the determinant of $\mathbf{E}$ may no longer be computed using an efficient specialized algorithm, and must be found using a general-purpose method such as Gaussian elimination or $\mathbf{LU}$ decomposition (Section B.1, Appendix B). The first implementation of a finite-difference method can be traced back to the pioneering work of Thomas (1953) on the stability of plane Poiseuille flow.

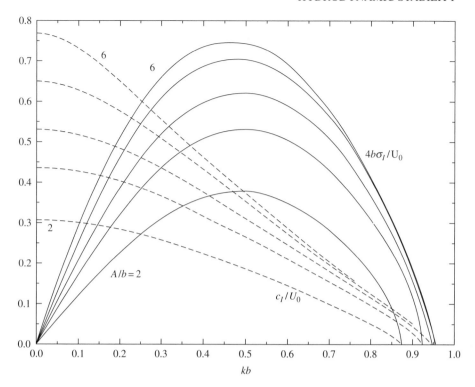

Figure 9.6.1   Graphs of the reduced imaginary part of the phase velocity c_i/U_0 (dashed lines) and reduced growth rate $4b\sigma_i/U_0$ (solid lines) with respect to kb in the unstable regime for a shear flow with velocity profile $U(y) = U_0 \tanh(y/b)$, for several values of channel width $A/b = 2.0, 2.5, 3.0, 4.0. 6.0$.

Weighted-Residual Methods

Another class of numerical methods proceeds by expanding the complex eigenfunction f in a truncated sum of a complete set of, preferably orthogonal, basis functions g_k that satisfy the prescribed boundary conditions, writing

$$f(y) = \sum_{k=1}^{N} a_k g_k(y) \tag{9.6.8}$$

where a_k are unknown constants. The method involves substituting Eq. (9.6.8) into the Rayleigh or Orr–Sommerfeld equation, multiplying the resulting equation with a sequence of weighting functions $w_k(y)$, and integrating the product with respect to y over the domain of flow, thereby obtaining a homogeneous system of algebraic equations for a_k. Identifying, in particular, w_k with g_k yields Galerkin's method. Requiring that the final system of equations have a nontrivial solution yields a generalized eigenvalue problem whose solution provides us with the values of c.

The choice of basis functions has an important effect on the accuracy and efficiency of the numerical method. For plane Poiseuille flow in a channel with width equal to $2b$, one successful choice for symmetric disturbances is $g_k = (1 - Y^2)Y^{2(k-1)}$, where $Y = y/b$ and the origin has been placed midway between the walls (Gersting and Jankowski, 1972). Orszag (1971) finds that expanding the eigenfunctions in terms of Chebyshev polynomials yields higher accuracy than expanding in other, seemingly more relevant, sets of orthogonal functions. To treat unbounded shear flows, the arguments of the basis functions are set equal to exponentially decaying functions (Murdock and Stewartson, 1977).

Solving Differential-Equation Eigenvalue Problems

The most popular class of numerical methods for computing c involves integrating the Orr–Sommerfeld equation, Rayleigh's equation, or one of their modified versions, using a standard numerical method, such as a Runge–Kutta method, while performing shooting with respect to c in order to satisfy the required boundary conditions. Use of these methods bypasses the computation of eigenvalues of large matrices and offers conceptual and analytical simplicity, but may cause certain difficulties associated with numerical instabilities.

To integrate the Orr–Sommerfeld equation, it is a standard practice to decompose it into a system of four first-order differential equations as

$$\frac{d\mathbf{F}}{dt} = \mathbf{M} \cdot \mathbf{F} \tag{9.6.9}$$

where we have introduced the four-dimensional unknown vector $\mathbf{F} = (f, f', f'' - k^2 f, f''' - k^2 f')$ and the coefficient matrix

$$\mathbf{M} = \begin{bmatrix} 0 & 1 & 0 & 0 \\ k^2 & 0 & 1 & 0 \\ 0 & 0 & 0 & 1 \\ -(ik/\nu)U'' & 0 & k^2 + (ik/\nu)(U - c) & 0 \end{bmatrix} \tag{9.6.10}$$

To illustrate the implementation of the method, consider plane Poiseuille in a channel that is confined between two walls located at $y = -b$ and b. The no-slip and no-penetration boundary conditions require that $f(\pm b)$ and $f'(\pm b)$ vanish. The method proceeds according to the following steps:

1. Guess a complex value for c.
2. Integrate Eq. (9.6.9) from $y = -b$ to b with initial condition $\mathbf{F} = (0, 0, 1, 0)$, and call the solution $\mathbf{F}_1$.
3. Repeat step 2 with initial condition $\mathbf{F} = (0, 0, 0, 1)$, and call the solution $\mathbf{F}_2$.
4. The linear combination $\mathbf{F}_3 = \mathbf{F}_1 + \beta \mathbf{F}_2$, where β is a constant, will be a solution of Eq. (9.6.9) with $f_3(b) = f_1(b) + \beta f_2(b)$, and $f_3'(b) = f_1'(b) + \beta f_2'(b)$. Require $f_3(b) = 0$ and $f_3'(b) = 0$ to obtain the compatibility condition

$$G(c) \equiv f_1(b)f_2'(b) - f_2(b)f_1'(b) = 0 \tag{9.6.11}$$

   In general, this condition will not be satisfied for the guessed value of c.
5. Improve the value of c and return to step 2. The improvement may be done using, for example, Newton's method.

In searching for symmetric eigenmodes, we require that $f'(0)$, $f'''(0)$, $f(b)$, $f'(b)$ vanish, and carry out the integrations in steps 2 and 3 from the centerline of the channel where $y = 0$ toward the upper wall with initial conditions, respectively, equal to $(1, 0, 0, 0)$ and $(0, 0, 1, 0)$.

Although seemingly innocuous, this procedure may suffer from numerical instabilities that lead to unreliable results when Eq. (9.6.9) is integrated in the vicinity of the walls, especially at large Reynolds numbers. There are several ways to rectify this difficulty, including (1) performing a forward and a backward integration and combining the solutions to eliminate the spurious oscillations, (2) filtering out the numerical instabilities, (3) orthonormalizing the solution during the numerical integration, and (4) performing parallel shooting (Gersting and Jankowski, 1972).

Similar difficulties are encountered when integrating Rayleigh's equation for inviscid flow. The numerical computations proceed smoothly when integrating from a region where the velocity profile of the base flow is nearly uniform to the main core of the shear flow, but numerical instabilities occur when the integration continues into the region of uniform flow. A physical explanation is that the perturbation decays exponentially into the region of vanishing shear rate and is thus difficult to capture in the numerical method.

The practical difficulties associated with integrating the Orr–Sommerfeld or Rayleigh equation can be avoided by working with a modified nonlinear set of equations that is constructed according to Riccati's method, or by using the compound-matrix method.

Riccati's equation for inviscid fluids

To illustrate the method, let us a consider a unidirectional shear flow of an inviscid fluid within a channel that is confined between two walls located at $y = -A$ and B. The method is implemented according to the following steps:

1. Guess a complex value for c.
2. Integrate Riccati's equation (9.3.29) from $y = -A$ to B using as initial condition the required boundary condition $F(-A) = 0$.
3. Adjust the value of c to achieve $F(B) = 0$. The correction may be done using, for example, Newton's method.

The method presents certain difficulties when the domain of the shear flow is unbounded, which means that one or both of A and B are infinite. Even if the domain of computation is truncated to a finite level, inaccuracies may cause numerical instabilities that degrade the accuracy of the results. One way of circumventing these difficulties is to map the infinite domain of flow onto a finite strip using, for example, the transformation

$$z = \tanh(y/b) \tag{9.6.12}$$

where b is a characteristic length scale that is comparable to the effective width of the shear flow and z is a dimensionless variable (Michalke, 1964). As y varies from $-\infty$ to $+\infty$, z varies from -1 to 1. Substituting Eq. (9.6.12) into Riccati's equation (9.3.29), we obtain

$$\frac{dF}{dz} = \frac{b}{1 - z^2}\left(k^2 - F^2 + \frac{U''}{U - c}\right) \tag{9.6.13}$$

where a prime indicates a derivative with respect to y. To develop appropriate boundary conditions for use with Eq. (9.6.13), we use Rayleigh's equation and find that as y tends to $\pm\infty$, $f(y)$ behaves like $\exp[-k(\pm y)]$. Using the definition $F = f'/f$, we then obtain

$$F(z = -1) = k, \qquad F(z = 1) = -k \tag{9.6.14}$$

The numerical procedure involves the following steps:

1. Guess a complex value for c.
2. Integrate Eq. (9.6.13) from $z = -1$ toward 1 with initial condition $F(z = -1) = k$.
3. Adjust the value of c to achieve $F(z = 1) = -k$ using, for instance, Newton's method.

It will be noted that the right-hand side of Eq. (9.6.13) is indeterminate at $z = -1$. A simple-minded way of avoiding this apparent difficulty is to begin the integration from $z = -1 + \varepsilon$ using as initial condition $F(z = -1 + \varepsilon) = k$, where ε is a small number. A better way is to compute the right-hand side at $z = -1$ using the l'Hôpital rule as illustrated in the following example.

Consider an unbounded shear flow with velocity profile $U(y) = U_0 \tanh(y/b) = U_0 z$, for which $U'' = -2(U_0/b^2)z(1 - z^2)$. Substituting these expressions into Eq. (9.6.13) yields the nonlinear differential equation

$$\frac{dF}{dz} = b\frac{k^2 - F^2}{1 - z^2} - \frac{2}{b}\frac{z}{z - c/U_0} \tag{9.6.15}$$

(Michalke, 1964). To compute dF/dz at $z = \pm 1$, we use the l'Hôpital rule to evaluate the first term on the right-hand side, finding

$$\left(\frac{dF}{dz}\right)_{z=\pm 1} = \mp kb\left(\frac{dF}{dz}\right)_{z=\pm 1} - \frac{2}{b}\frac{1}{1 \mp c/U_0} \tag{9.6.16}$$

which may be rearranged to yield

$$\left(\frac{dF}{dz}\right)_{z=\pm 1} = -\frac{2}{b(1 \pm kb)(1 \mp c/U_0)} \tag{9.6.17}$$

Riccati's method for viscous flow

Riccati's method reduces the Orr–Sommerfeld equation to a quadratically nonlinear system of four first-order ordinary differential equations for the four entries of a two-by-two matrix $\mathbf{R}$, which is defined by the equation $\mathbf{u} = \mathbf{R} \cdot \mathbf{v}$. The two-dimensional vectors $\mathbf{u}$ and $\mathbf{v}$ contain, respectively, the second and fourth, and the first and third entries of the vector $\mathbf{F}$ defined after Eq. (9.6.9). The implementation of this powerful method is discussed in detail by Davey (1977).

Compound-matrix method

This method involves developing differential equations for the four minors of a four-by-two solution matrix. The two columns of the solution matrix contain the values and the first three derivatives of two independent solutions of the Orr–Sommerfeld equation computed using two distinct sets of initial conditions. The implementation of the method is discussed by Ng and Reid (1979) and Drazin and Reid (1981, p. 311).

PROBLEM

9.6.1   **Neutral stability of a shear layer.** Show that $c = 0$ is an eigenvalue, and $f(y) = A \operatorname{sech}(y/b)$ where A is an arbitrary constant, is the corresponding eigenfunction of Rayleigh's equation for an infinite shear flow with velocity profile $U(y) = U_0 \tanh(y/b)$, corresponding to a neutrally stable perturbation. What are the dimensions of A?

Computer Problems

9.6.2   **Finite-difference method for Rayleigh's equation.** (a) Write a routine that computes the determinant of a complex tridiagonal matrix according to the specialized algorithm discussed in Section B.2, Appendix B. (b) Employ the routine of (a) in a program called *EGVTR* that computes the complex eigenvalue c of Rayleigh's equation for a shear flow with an arbitrary velocity profile, as a function of k. (c) Consider the growth of spatially periodic perturbations on a family of inviscid shear flows with velocity profile

$$U(y) = U_0\{\delta \tanh(y/b) + (\delta - 1)\exp[-(y/b)^2]\} \tag{9.6.18}$$

where the parameter δ ranges between 0 and 1. The limiting values $\delta = 1$ and 0 yield, respectively, a shear layer with a hyperbolic-tangent velocity profile and a symmetric wake with a Gaussian velocity profile. Assuming that the flow occurs in the bounded domain $-a < y < a$, use the program of (b) to construct a graph of the maximum growth rate versus the wave number kb, for $\delta = 0, 0.50, 1.0$ and $a/b = 2.0, 3.0, 4.0$, and discuss your results. (d) Repeat (c) for the velocity profile

$$U(y) = U_0[\delta \tanh(y/b) + (1 - \delta)\operatorname{sech}^2(y/b)] \tag{9.6.19}$$

where the parameter δ ranges between 0 and 1. The limiting values $\delta = 1$ and 0 yield, respectively, a shear layer with a hyperbolic-tangent velocity profile and the Bickley jet (Wallace and Redekopp, 1992).

9.6.3   **Riccati's method.** Repeat Problem 9.6.2 using Riccati's equation.

9.7 │ STABILITY OF CERTAIN CLASSES OF UNIDIRECTIONAL FLOWS

The notable fundamental and practical significance of unidirectional and nearly unidirectional viscous and nearly inviscid shear flows has motivated numerous studies of their stability using a host of analytical and numerical methods. Drazin and Reid (1981, p. 211) review methods and discuss results for flows in channels, free shear layers, boundary layers, and jets. Huerre and Monkewitz (1990, Table 3) illustrate in tabular form the stability properties of several classes of external shear flows and provide a comprehensive list of references. In the present section we summarize and briefly discuss the stability of a selected class of flows that are representative of their families.

Free Shear Layers

A free shear layer is the region between two fluid streams that merge in a parallel manner at different velocities. An example is provided by the infinite vortex layer studied in Section 9.5. In Figure 9.7.1 we present a contour plot of the growth rate in the *Reynolds number–wave number* plane for a shear layer with a hyperbolic-tangent velocity profile $U = U_0 \tanh(y/b)$, discussed in Section 9.6, after Betchov and Szewczyk (1963); the Reynolds number is defined as $Re = U_0 b / \nu$. The neutral stability curve corresponds to vanishing dimensionless growth rate $\sigma_I / b U_0 = 0$.

In Section 9.6 we saw that, in the limit of inviscid flow, the flow is unstable for disturbances whose wave number falls within the range $0 < kb < 1$. Figure 9.7.1 shows that the flow is, in fact, unstable at all Reynolds numbers except in the theoretical limit of Stokes flow. The range of unstable wave numbers becomes broader, and the maximum growth rate increases, as the Reynolds number is raised. These observations reveal that *viscosity has a stabilizing influence on the dynamics of free shear flows.*

It should be noted that, although the velocity profile of a free shear layer represents an exact solution of the equations of steady inviscid flow, it does not satisfy the steady version of the Navier–Stokes equation with the viscous force included, and, therefore, it does not represent an acceptable steady unidirectional viscous flow. In practice, viscosity causes the shear layer to spread out and the vorticity to diffuse away from the central region of the shear layer as discussed in Section 5.2, but these effects are usually overlooked in the formulation of the linear stability problem. In studying the stability of a nearly steady or nearly unidirectional flow, one should keep in mind that neglecting the temporal or spatial evolution of the base flow may lead to significant error (Huerre and Monkewitz, 1990).

Boundary Layers

The normal-mode analysis of the self-similar velocity profiles associated with the Falkner–Skan boundary layers (Section 8.3), conducted under the additional approximation that the flow is locally unidirectional, shows that the flow is stable as long as the Reynolds number is below a critical threshold value Re_c that depends upon the streamwise pressure gradient of the outer irrotational flow dP/dx. The velocity profile of a boundary layer with an adverse pressure gradient, $dP/dx > 0$, exhibits an inflection point, and it is thus likely to be unstable in the limit of inviscid flow; numerical solutions reveal that this is the case indeed. Boundary layers with zero or favorable pressure gradient, $dP/dx \leq 0$, are stable in the limit of inviscid flow, as required by Rayleigh's theorem on the significance of the inflection point discussed in Section 9.4.

Contour plots of the growth rate in the *Reynolds number–wave number* plane contain a loop that separates a regime of stable flow from a regime of unstable flow, schematically illustrated in Figure 9.7.2; the unstable normal modes are called *Tollmien–Schlichting waves.* For the Blasius boundary layer, in particular, corresponding to $dP/dx = 0$, the critical Reynolds number is known to be $Re_c = 520$, where $Re = U \delta * / \nu$ and $\delta *$ is the displacement thickness. When

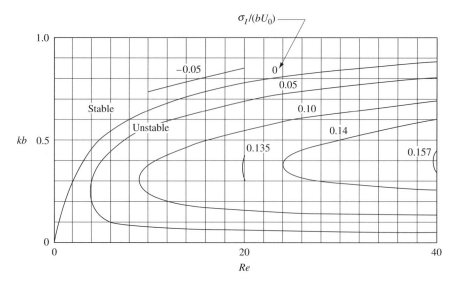

Figure 9.7.1 Contour plot of the dimensionless growth rate $\sigma_I/(bU_0)$ in the Reynolds number $Re = U_0 b/\nu$–wave number plane, for a shear layer with the hyperbolic-tangent velocity profile $U = U_0 \tanh(y/b)$ (Betchov and Szewczyk 1963).

instability first occurs at Re_c, the wave number of the marginally stable mode is given by $k\delta* = 0.35$, corresponding to a wavelength of $\lambda = 18\delta*$. In the other extreme case of a boundary layer in orthogonal stagnation-point flow with a favorable pressure gradient, $Re_c = 14{,}000$.

The Blasius boundary layer and other boundary layers with favorable pressure gradient are stable in the limit of inviscid flow but are unstable at finite and sufficiently large Reynolds numbers. This behavior demonstrates that *viscous effects may have a destabilizing influence on the dynamics of a shear flow.*

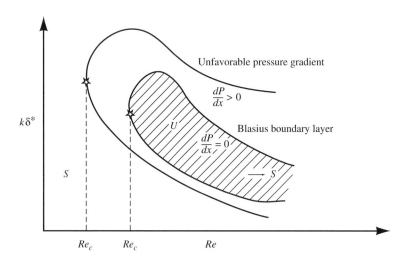

Figure 9.7.2 Regions of stable and unstable flow in the Reynolds number $Re = U\delta^*/\nu$–wave number plane, for the self-similar velocity profiles associated with boundary layers.

Jets

The properties of two-dimensional jets, concerning stability, are similar in many ways to those of free shear layers. The physical explanation lies in the fact that the edges of the jet constitute shear layers that are susceptible to the same modes of instability of the free shear flows. The interaction between the two shear layers, however, results in further modes corresponding to symmetric or sinuous, antisymmetric or varicose, displacements. For a circular jet, we obtain axisymmetric or spiral deformations.

A prototypical two-dimensional jet with a symmetric velocity profile is the *Bickley jet* corresponding to $U = U_0 \operatorname{sech}^2(y/b)$. In the limit of inviscid flow, the jet falls prey to sinuous and varicose unstable normal modes whose unstable range, growth rate, and phase velocity are shown in Figure 9.7.3(a,b) after Drazin and Reid (1981). The growth rate of the symmetric mode is higher than that of the antisymmetric mode, and this suggests that the growth of an arbitrary disturbance will lead to the formation of a staggered array of vortices in the arrangement of the *von Kármán vortex street* (Figure 11.2.4). Solving the Orr–Sommerfeld equation shows that the flow is unstable when the Reynolds number exceeds the critical value $Re_c = 4$, where $Re = Ub/\nu$. The wave number at marginal stability corresponding to Re_c is $kb = 0.20$.

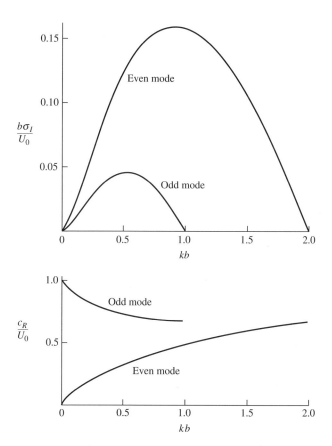

Figure 9.7.3   (a) Growth rate and (b) phase velocity of symmetric and antisymmetric normal modes for the Bickley jet with velocity profile $U = U_0 \operatorname{sech}^2(y/b)$, in the limit of inviscid flow (Drazin and Reid 1981).

Plane Couette Flow

The normal-mode analysis reveals that plane Couette flow is *stable at all Reynolds numbers and for all wave numbers.* The unstable behavior observed in practice at Reynolds numbers as low as $Re = Gb^2/\nu = 350$, where G is the shear rate or slope of the velocity profile and $2b$ is the channel width, is attributed to nonlinear effects associated with the finite amplitude of the disturbances. Unstable behavior may also be triggered by wall roughness.

Poiseuille Flow

The normal-mode analysis reveals that *plane* Poiseuille flow becomes unstable when the Reynolds number $Re = bU/\nu$ exceeds the value of 5,772, where $2b$ is the channel width and U is the velocity at the center of the channel. At lower values of Re the flow is stable. The wave number of the disturbance that first becomes unstable at the critical Reynolds number is $kb = 1.020$. In practice, the flow has been observed to become unstable when the Reynolds number exceeds the value of about 1,500. The discrepancy between theory and observation is attributed to nonlinear effects, wall roughness, and deviation from unidirectional motion near the entrance.

Poiseuille flow *in a circular tube* is stable for axisymmetric perturbations, and there is ample evidence that it is also stable for arbitrary three-dimensional perturbations. In practice, Poiseuille flow in a tube of radius a has been observed to become unstable at Reynolds numbers $Re = aU/\nu$ as low as 1,100 where U is the velocity at the axis. This behavior is attributed to the reasons stated previously for plane Poiseuille flow.

PROBLEMS

9.7.1 **Bickley jet.** Show that, in the limit of inviscid flow, the wave number, phase velocity, and eigenfunctions corresponding to symmetric and antisymmetric neutrally stable modes are given by $kb = 2$, $cU_0 = 2/3$, $f(y) = bU_0 \operatorname{sech}^2(y/b)$, and $kb = 1$, $cU_0 = 2/3$, $f(y) = bU_0 \operatorname{sech}(y/b) \tanh(y/b)$ (Drazin & Reid, 1981, p. 233).

 Computer Problems

9.7.2 **Boundary layers.** Compute and plot the temporal growth rate as a function of the wave number in the unstable regime for two self-similar boundary-layer profiles of your choice with adverse pressure gradients, in the limit of inviscid flow (Sections 8.2 and 8.3). Show that, in the limit as dP/dx tends to zero, the regime of unstable wave numbers shrinks down to zero.

9.7.3 **Bickley jet.** Compute and plot the growth rate and phase velocity curves shown in Figure 9.7.3 using a method of your choice.

9.8 | STABILITY OF A PLANAR INTERFACE IN POTENTIAL FLOW

Having investigated the stability of homogeneous unidirectional shear flows involving fluids with uniform density, we proceed to examine the behavior of inhomogeneous flows with density striations. We concentrate, in particular, on a class of flows in the presence of interfaces between two different fluids across which the density undergoes a step discontinuity and surface tension enters the interfacial force balance. In the present section we shall study the motion under the assumption that the base flow on either side of the interface is irrotational, viscous stresses are negligible, and the fluids may be considered to be effectively inviscid. In the next section we shall address the more general problem of rotational flow with non-negligible viscous stresses.

One way of setting up the mathematical formulation of the linear stability problem is to describe the evolution of the disturbance on either side of the interface individually, in the context of the linearized equation of motion, and then match the two disturbance flows by requiring appropriate kinematic and dynamic boundary conditions at the interface. As an alternative, we observe that the tangential component of the velocity at an interface between two inviscid fluids is allowed to undergo a discontinuity, and the interface may be regarded as a vortex sheet whose strength γ evolves in a manner that is consistent with the growth or decay of the disturbance. This point of view leads us to regard the perturbation as the result of the instability of the interfacial vortex sheet, and to study not only the linear, but also the nonlinear, stages of the motion using vortex methods to be discussed in Section 11.4.

Kelvin–Helmholtz Instability

Consider the behavior of the interface between two inviscid fluids with uniform densities ρ_1 and ρ_2, flowing in the horizontal direction with uniform velocities U_1 and U_2 as shown in Figure 9.8.1. For simplicity, we shall assume that the interface has a constant surface tension denoted by T; the standard symbol γ is reserved for the strength of the interfacial vortex sheet. In the unperturbed state, the interface represents a planar vortex sheet with uniform strength equal to $\gamma = U_2 - U_1$.

A two-dimensional, spatially periodic disturbance corresponding to a normal mode causes the interface to deform in a sinusoidal fashion. The location of the perturbed interface may be described by the real or imaginary part of the right-hand side of

$$y = \eta(x, t) = \varepsilon A \exp[ik(x - ct)] \tag{9.8.1}$$

where A is a complex constant, k is the real wave number, and c is the complex phase velocity.

Since $U'' = 0$ within both the upper and the lower fluids, Rayleigh's equation (9.3.27) simplifies to the Helmholtz equation $f'' - k^2 f = 0$ within both fluids. The general solution is readily found to be

$$f_n(y) = A_n e^{ky} + B_n e^{-ky} \tag{9.8.2}$$

where $n = 1, 2$, respectively, for the fluid above and below the interface, and A_n, B_n are four constants. Because the vorticity gradient of the base flow within each stream vanishes, the disturbance flow is required to be irrotational.

To compute the five constants $A_n, B_n, n = 1, 2$, and A, we require five equations. Demanding that the disturbance vanish far from the interface yields the two equations

$$A_1 = B_2 = 0 \tag{9.8.3}$$

Next we require that the velocity on either side of the interface is such that the motion of the point particles that are adjacent to the interface is consistent with the shape of the interface as expressed

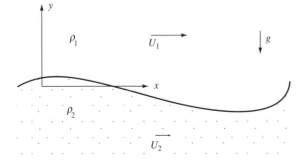

Figure 9.8.1   A vortex sheet representing the interface between two inviscid fluids in uniform motion in the horizontal direction.

by Eq. (9.8.1). Introducing the kinematic constraint $D(y - \eta)/Dt = 0$ on either side of the vortex sheet (Section 1.6), expanding all terms in Taylor series with respect to ε, and retaining only the linear contributions, we find

$$\frac{\partial \eta}{\partial t} + U_n \frac{\partial \eta}{\partial x} - \varepsilon (u_y^{\text{NM}})_{y=0} = 0 \tag{9.8.4}$$

where the subscript n of η_n indicates that the velocity is computed on the upper and lower side of the vortex sheet corresponding to $n = 1$ or 2. Substituting Eqs. (9.3.25) and (9.8.1)–(9.8.3) into Eq. (9.8.4), we derive the two equations

$$A = \frac{A_2}{c - U_2} = \frac{B_1}{c - U_1} \tag{9.8.5}$$

The fifth equation emerges by requiring that the normal stress, which is equal to the negative of the pressure, undergo a discontinuity across the interface that is balanced by surface tension, so that

$$p_2 - p_1 = -T \frac{\partial^2 \eta}{\partial x^2} \tag{9.8.6}$$

evaluated at the interface. Note that we have approximated the curvature of the interface with the negative of the second derivative of η with respect to x (Table 4.2.1). To express the pressure in terms of the velocity, we use Bernoulli's equation and linearize the quadratic terms, thereby converting Eq. (9.8.6) to

$$-\rho_2 \left(\varepsilon \frac{\partial \phi_2^{\text{NM}}}{\partial t} + \varepsilon U_2 \frac{\partial \phi_2^{\text{NM}}}{\partial x} + g\eta \right) + \rho_1 \left(\varepsilon \frac{\partial \phi_1^{\text{NM}}}{\partial t} + \varepsilon U_1 \frac{\partial \phi_1^{\text{NM}}}{\partial x} + g\eta \right)$$

$$= -T \frac{\partial^2 \eta}{\partial x^2} \tag{9.8.7}$$

where all partial derivatives on the left-hand side are evaluated at $y = 0$. Furthermore, we use Eqs. (9.3.35), (9.8.2), and (9.8.3) to write

$$\phi_1^{\text{NM}} = iB_1 e^{-ky} \exp[ik(x - ct)]$$
$$\phi_2^{\text{NM}} = -iA_2 e^{ky} \exp[ik(x - ct)] \tag{9.8.8}$$

Substituting Eqs. (9.8.8) and (9.8.1) into Eq. (9.8.7) yields

$$-\rho_2[k(U_2 - c)A_2 + gA] + \rho_1[-k(U_1 - c)B_1 + gA] = k^2 AT \tag{9.8.9}$$

Equations (9.8.3), (9.8.5), and (9.8.9) provide us with a homogeneous system of five linear algebraic equations for A_n, B_n, $n = 1, 2$, and A. Requiring the existence of a nontrivial solution, we set the determinant of the coefficient matrix equal to zero and obtain

$$c = \frac{1}{\rho_2 + \rho_1} \left[\rho_2 U_2 + \rho_1 U_1 \pm \left(\frac{g}{k}(\rho_2^2 - \rho_1^2) + kT(\rho_2 + \rho_1) \right. \right.$$

$$\left. \left. - \rho_2 \rho_1 (U_2 - U_1)^2 \right)^{1/2} \right] \tag{9.8.10}$$

If the quantity under the square root is negative, one of the two solutions will have a positive imaginary part, and certain disturbances will be unstable. The condition for such circumstances to occur may be expressed in terms of the square of the velocity difference above which the flow will be unstable

$$(U_2 - U_1)^2 > \frac{g}{k} \frac{\rho_2^2 - \rho_1^2}{\rho_2 \rho_1} \left(1 + \frac{Tk^2}{g(\rho_2 - \rho_1)}\right) \tag{9.8.11}$$

The critical velocity difference $|U_2 - U_1|_{cr}$ under which the flow is stable to all disturbances corresponds to the minimum of the right-hand side of Eq. (9.8.11) regarded as a function of k, which occurs when $k = [g(\rho_2 - \rho_1)/T]^{1/2}$. We then find

$$(U_2 - U_1)_{cr}^4 = 4(\rho_2 - \rho_1)gT \left(\frac{\rho_2 + \rho_1}{\rho_2 \rho_1}\right)^2 \tag{9.8.12}$$

Above this value, there is a band of unstable disturbances whose growth rate and phase velocity are given by

$$\sigma_I = kc_I = k|U_2 - U_1| \frac{(\rho_2 \rho_1)^{1/2}}{\rho_2 + \rho_1} \left[1 - \frac{1}{(U_2 - U_1)^2} \frac{\rho_2 + \rho_1}{\rho_2 \rho_1} \left(\frac{g}{k}(\rho_2 - \rho_1) + kT\right)\right]^{1/2} \tag{9.8.13}$$

$$c_R = \frac{\rho_2 U_2 + \rho_1 U_1}{\rho_2 + \rho_1} \tag{9.8.14}$$

The critical wave numbers for neutral stability k_c are found by setting the expression under the radical in Eq. (9.8.13) equal to zero, and this yields the quadratic equation

$$Tk_{cr}^2 - (U_2 - U_1)^2 \frac{\rho_2 \rho_1}{\rho_2 + \rho_1} k_{cr} + g(\rho_2 - \rho_1) = 0 \tag{9.8.15}$$

which has two real solutions enclosing a finite band of unstable wave numbers.

It is instructive to consider the disturbance in the strength of the vortex sheet. Recalling that the circulation along the vortex sheet is equal to the jump of the harmonic potential across the vortex sheet, and using Eqs. (9.8.8) and (9.8.5), we find that the strength of the perturbed vortex sheet is given by

$$\begin{aligned}
\gamma &= U_2 - U_1 - \varepsilon \left(\frac{\partial(\phi_1^{NM} - \phi_2^{NM})}{\partial x}\right)_{y=0} \\
&= U_2 - U_1 + \varepsilon k(B_1 + A_2) \exp[ik(x - ct)] \\
&= U_2 - U_1 + \varepsilon Ak(2c - U_1 - U_2) \exp[ik(x - ct)]
\end{aligned} \tag{9.8.16}$$

Comparing Eq. (9.8.16) with Eq. (9.8.1) shows that the perturbation in the strength of the vortex sheet has a phase shift with respect to the perturbation of the interface, which is equal to the argument of $c - \frac{1}{2}(U_1 + U_2)$.

Helmholtz Instability

When the densities of the fluids above and below the vortex sheet are matched, $\rho_1 = \rho_2 = \rho$, the right-hand side of Eq. (9.8.11) vanishes, and this means that certain perturbations will be unstable for any values of $|U_1 - U_2|$. The critical wave numbers are found from Eq. (9.8.15) to be

$$k_{cr,1} = 0, \qquad k_{cr,2} = \frac{\rho}{2T}(U_2 - U_1)^2 \tag{9.8.17}$$

Wave numbers with intermediate values yield unstable normal modes whose growth rates are given by

$$\sigma_I = \tfrac{1}{2}k|U_2 - U_1| \left(1 - \frac{2kT}{\rho(U_2 - U_1)^2}\right)^{1/2} \tag{9.8.18}$$

and phase velocity equal to the average velocity of the unperturbed streams, $c_R = \frac{1}{2}(U_1 + U_2)$. The phase shift between the disturbance in strength and shape of the vortex sheet is equal to $\pi/2$ for unstable normal modes and vanishes for neutrally stable normal modes, in agreement with the results shown in Figure 9.5.2(c). Maximum growth rate, corresponding to the most dangerous normal mode, occurs when $k = \frac{2}{3}k_{cr,2}$.

Large wave numbers with small wavelengths are stabilized by the restraining but not dampening action of surface tension. When the surface tension vanishes, all wave numbers are unstable, and Eq. (9.8.18) is in agreement with Eq. (9.5.18) with $U_1 = U_0$ and $U_2 = -U_0$, yielding a linear relationship between the growth rate and wave number.

The initial growth and long-time evolution of the Helmholtz instability in the absence of surface tension for a flow with $U_1 = -U_2 < 0$ is illustrated in Figure 11.4.3. The motion is computed using a variation of the point-vortex method discussed in Section 11.4, which is based on identifying the interface with a vortex sheet. We observe that the growth of sinusoidal waves leads to the formation of a sequence of spiral patterns. The numerical results suggest that the curvature of the vortex sheet becomes discontinuous at a certain point within each period just before the spirals begin forming, which is evidence of a singular behavior characteristic of an ill-posed problem.

Rayleigh–Taylor Instability

This type of instability occurs when the unperturbed fluids are quiescent, $U_1 = 0$ and $U_2 = 0$, and the density of the lower fluid is less than that of the upper fluid, $\rho_2 < \rho_1$, in which case the fluids are said to be *unstably* or *inversely stratified* (Rayleigh, 1883). The strength of the unperturbed vortex sheet representing the interface vanishes.

Since the right-hand side of Eq. (9.8.11) is negative, a certain class of perturbations will be unstable. Equation (9.8.15) shows that there exists a single critical wave number given by

$$k_{cr} = \left(\frac{g}{T}(\rho_1 - \rho_2) \right)^{1/2} \tag{9.8.19}$$

Smaller wave numbers are unstable, with associated growth rate given by

$$\sigma_I = kc_I = \left(k \frac{g(\rho_1 - \rho_2) - k^2 T}{\rho_1 + \rho_2} \right)^{1/2} \tag{9.8.20}$$

and vanishing phase velocity; larger wave numbers are stabilized by surface tension; and maximum growth rate occurs when $k = k_{cr}/3^{1/2}$. When the surface tension vanishes, all wave numbers become unstable, and the growth rate is given by the simplified expression

$$\sigma_I = (kg)^{1/2} \left(\frac{\rho_1 - \rho_2}{\rho_1 + \rho_2} \right)^{1/2} \tag{9.8.21}$$

The fraction enclosed by the second set of parentheses on the right-hand side of Eq. (9.8.21) is known as the *Atwood ratio*, $A = (\rho_1 - \rho_2)/(\rho_1 + \rho_2)$. In this case, the higher the wave number, the faster the growth rate of the perturbation. This behavior suggests that the interface may develop a singularity at the nonlinear stages of the motion, as it is suspected to do in the case of the Helmholtz instability discussed previously.

The initial growth and long-time evolution of the Rayleigh–Taylor instability for $A = 0.50$, in the absence of surface tension, is illustrated in Figure 9.8.2 (Tryggvason, 1988). The motion is computed using a variation of the point-vortex method discussed in Section 11.4. The occurrence of a convoluted interfacial pattern with secondary Helmholtz instabilities developing along the sides of the plunging filaments is a striking feature of the nonlinear motion.

Taylor (1950) noted that when two adjacent fluids undergo linear acceleration with velocity $\mathbf{V}(t)$ normal to the interface, the preceding analysis remains valid in a noninertial frame that

Figure 9.8.2 The evolution of an interface between a heavy fluid with density ρ_1 lying above another lighter fluid with density ρ_2, for Atwood ratio $A = (\rho_1 - \rho_2)/(\rho_1 + \rho_2) = 0.50$, in the absence of surface tension, computed using a vortex method (Tryggvason 1988).

translates with the interface, provided that the acceleration of gravity **g** is enhanced with the acceleration $-d\mathbf{V}(t)/dt$ (Problem 9.8.2). In this light, the Rayleigh–Taylor instability is regarded as the instability of the interface between two accelerating fluids.

Wall effects

The presence of container walls may have a significant stabilizing influence on the behavior of perturbations, by placing limits on the minimum wave numbers of the normal modes that are allowed to enter the physical system.

Consider, for instance, a vertical circular cylinder of radius a that is closed at the bottom, and is half-filled with a liquid labeled 2 and a second liquid labeled 1 lying above liquid 2. In the unperturbed state, the interface is flat and the pressure within the two fluids assumes the hydrostatic distribution. Introducing cylindrical polar coordinates (x, σ, φ) with the x axis coaxial with the cylinder pointing upward against the direction of gravity, we express the harmonic velocity potentials associated with a normal-mode disturbance within the two fluids as

$$\phi_n^{\text{NM}}(\sigma, \varphi, t) = B_{m,n} \exp((-1)^n kx) J_m(k\sigma) \cos m\varphi \exp(-ikct) \qquad (9.8.22)$$

where m is an integer expressing the azimuthal dependence of the perturbation, J_m is a Bessel function, $n = 1, 2$, respectively for the upper and lower fluid, and $B_{m,n}$ are complex constants determined by the form of the perturbation at the initial instant (Problem 9.8.4). To ensure that the no-penetration condition at the surface of the tube is fulfilled, we require that $\partial\phi_n/\partial\sigma = 0$ at $\sigma = a$ and find that an acceptable radial wave number k must satisfy the equation

$$\left(\frac{dJ_m(k\sigma)}{d\sigma}\right)_{\sigma = a} = 0 \qquad (9.8.23)$$

Reference to standard mathematical tables shows that: (1) for $m = 0$ corresponding to an axisymmetric perturbation, $ka = 3.83, 7.02, 10.17, \ldots$, (2) for $m = 1$, $ka = 1.84, 5.33, 8.53, \ldots$, (3) for $m = 2$, $ka = 3.05, 6.70, 9.97$. Noting that the smallest value of ka is 1.84 and using the criterion (9.8.19), whose applicability is justified by the results of Problem 9.8.3, shows that the interface will be unstable to the $m = 1$ mode, according to which half of the interface will plunge and half of it will rise, when the tube radius exceeds the critical value

$$a_{cr} = 1.84 \left(\frac{T}{g(\rho_1 - \rho_2)} \right)^{1/2} \tag{9.8.24}$$

When the radius is smaller than this critical value, the interface is stable, and the heavy fluid lies above the lighter fluid supported by interfacial tension.

Progressive and Standing Gravity–Capillary Waves

The results of the preceding analysis may be used to study the propagation of waves on the free surface of an otherwise quiescent liquid, such as water in the ocean. Setting $U_1 = 0$ and $U_2 = 0$, neglecting the density of the upper fluid by putting $\rho_1 = 0$, and setting for convenience $\rho_2 = \rho$, we find that Eq. (9.8.10) yields

$$c = \pm \left[\frac{g}{k} \left(1 + \frac{Tk^2}{g\rho} \right) \right]^{1/2} \tag{9.8.25}$$

The perturbations are neutrally stable and represent waves of constant amplitude traveling to the left or to the right with a phase velocity that is a function of the wave number k.

The initial shape of an interface that is composed of a packet of waves with different wave numbers changes due to fact that the packet disperses as each wave travels with its own phase velocity. The dispersion of the packet is determined by the functional relationship between the phase velocity and the wave number $c(k)$ shown in Eq. (9.8.25), which, in the present context, is called the *dispersion relation.*

Superposing, in particular, two waves with identical initial amplitudes traveling to opposite directions, corresponding to the two signs in Eq. (9.8.25), we obtain a standing gravity–capillary wave oscillating with angular frequency equal to kc. Whether a progressive or a standing wave will occur in practice is determined by the actual mechanism that is responsible for the disturbance, as well as on the boundary conditions at the point where the interface meets a container.

PROBLEMS

9.8.1 **Oblique waves.** Derive the counterpart of Eq. (9.8.10) when the interfacial waves are directed in an oblique manner with respect to the direction of the flow, with wave numbers in the x and z directions, respectively, equal to k and l (Craik, 1985).

9.8.2 **Instability of a flat accelerating interface.** Consider a flat horizontal interface that undergoes acceleration in the vertical direction with velocity $V(t)\mathbf{i}$, where $\mathbf{i}$ is the unit vector along the x axis pointing against the direction of gravity. Show that the stability analysis discussed in this section remains valid provided that g is replaced by $g + dV/dt$. Discuss the implications of this result on the behavior of the interface between two liquids within a freely falling capsule.

9.8.3 **Rayleigh-Taylor instability.** (a) Show that an appropriate Reynolds number of the flow due to the instability at the early stages of the motion is given by $Re = g^{1/2}\lambda^{3/2}/\nu$. The flow will be virtually irrotational as long as $Re \gg 0$. What is the condition for the flow to *remain* virtually irrotational at long times? (b) Assume that the interface deforms in a three-dimensional manner so that its position is described by the real or imaginary part of

$$y = \eta(x, z, t) = \varepsilon A S(x, z) \exp(-ikct) \tag{9.8.26}$$

where A is a complex constant. Show that the function $S(x, z)$ satisfies the Helmholtz equation

$$\left(\frac{\partial}{\partial x^2} + \frac{\partial}{\partial z^2} \right) S(x, z) = -k^2 S(x, z) \tag{9.8.27}$$

which is to be solved subject to periodicity or no-penetration conditions, and the growth rate satisfies Eq. (9.8.20).

9.8.4 **Rayleigh–Taylor instability of a flat interface within a circular cylinder.** Consider the Rayleigh–Taylor instability of two fluids within a vertical circular cylinder discussed in the text. (a) Verify that the potentials given in Eq. (9.8.22) satisfy Laplace's equation. (b) Using the kinematic boundary condition at the interface, show that the location of the interface is described by

$$x = \eta(\sigma, \varphi, t) = \varepsilon A J_m(k\sigma) \cos m\varphi \exp(-ikct) \qquad (9.8.28)$$

where A is a complex constant determined by the initial condition. (c) Compute the constants $B_{m,n}$ in terms of the constant A.

9.8.5 **Rayleigh–Taylor instability of an interface between two vertical plates.** Study the Rayleigh–Taylor instability of a flat two-dimensional interface subtended between two parallel vertical plates that are separated by distance $2a$. The contact line is free to move under the action of the flow.

9.8.6 **Equipartition of energy for progressive gravity waves.** Show that the potential energy of small-amplitude progressive gravity waves is equal to the kinetic energy of the fluid.

9.8.7 **Immiscible displacement in a porous medium.** A liquid labeled 2 is displacing another liquid labeled 1 through a homogeneous isotropic porous medium (Saffman and Taylor, 1958). Let us assume that the flow within each fluid may be described by *Darcy's law* as $\mathbf{u} = \nabla\phi$, where $\phi = (\kappa/\mu)P$, and κ is the *permeability* of each fluid through the porous medium (Bear, 1972). The continuity equation requires that the potential ϕ and thus the pressure P be harmonic functions. In the unperturbed state, the two fluids translate along the x axis against the direction of gravity with uniform velocity U, and the interface is flat. (a) Graph the pressure distribution in the xy plane in the unperturbed state. (b) Consider the growth of two-dimensional perturbations in the xy plane. Perform the normal-mode stability analysis and require that the pressure undergo a discontinuity across the interface that is balanced by surface tension T to show that the complex phase velocity is given by

$$c = -\left(\frac{\mu_1}{\kappa_1} + \frac{\mu_2}{\kappa_2}\right)^{-1} \left[g(\rho_2 - \rho_1) + U\left(\frac{\mu_2}{\kappa_2} - \frac{\mu_1}{\kappa_1}\right) + Tk^2\right] \qquad (9.8.29)$$

and thus the flow will be stable only if

$$g(\rho_2 - \rho_1) > U\left(\frac{\mu_1}{\kappa_1} - \frac{\mu_2}{\kappa_2}\right) - Tk^2 \qquad (9.8.30)$$

Based on Eq. (9.8.30) derive an expression for the *maximum displacement speed* for stable flow. (c) Discuss the physical implications of Eq. (9.8.30) in the limit of vanishing U, and show that, in the absence of surface tension, the interface of a low-viscosity fluid displacing a high-viscosity fluid of the same density is unstable. (d) Discuss the physical relevance of the dynamic boundary condition at the interface.

9.8.8 **Immiscible displacement in the Hele–Shaw cell.** (a) Reformulate Problem 9.8.7 for the case where the fluids displace each other through the Hele–Shaw cell (Section 6.3), and derive criteria for stable displacement. (b) Discuss the physical relevance of the dynamic boundary condition at the interface with reference to the curvature of the meniscus in the plane that is perpendicular to the xy plane of the flow.

9.9 | VISCOUS INTERFACIAL FLOWS

Proceeding in the direction of increasing difficulty, we consider the instability of *viscous* unidirectional flows in the presence of an interface. Apart from the shear-flow instabilities and those due to density striations encountered in the preceding sections, this class of flows exhibit additional unstable normal modes associated with differences in the viscosity between the two fluids.

Our primary goal in the present section is to illustrate the derivation of the boundary conditions that accompany the linear system of governing equations; this we shall accomplish by discussing the stability of a liquid film flowing down an inclined plane.

Stability of a Liquid Film Flowing down an Inclined Wall

Film flow down an inclined surface is encountered on many occasions in everyday life and is an integral part of a variety of engineering processes. Examples are provided by the flow of rainwater down the windshield of an automobile, the flow of a cooling film down a heated surface, and the flow of a layer of a photographic gelatin emulsion down an inclined plate awaiting subsequent deposition onto a moving substratum. The base flow whose stability we wish to investigate is described by the flat-film solution discussed earlier in Section 5.1 and illustrated in Figure 5.1.1.

To simplify the notation, we nondimensionalize all variables using as characteristic velocity $u_{x,\text{Max}}$, the velocity at the free surface, characteristic length the unperturbed film thickness H, characteristic time $H/u_{x,\text{Max}}$, and characteristic pressure and stress $\mu u_{x,\text{Max}}/H$. In terms of dimensionless variables, indicated with a caret, the base-flow steady-state velocity and pressure above the ambient level are given by

$$\hat{u}_x^S \equiv \frac{u_x}{u_{x,\text{Max}}} = \frac{2\nu u_x}{gH^2 \sin\theta_0} = \hat{y}(2 - \hat{y})$$

$$\hat{p}^S \equiv \frac{2p}{\rho g H \sin\theta_0} = 2\cot\theta_0 (1 - \hat{y}) \tag{9.9.1}$$

where $\hat{y} = y/H$ and the superscript S denotes steady flow. The dimensionless volumetric flow rate per unit width of the film and mean velocity of the fluid are given by

$$\hat{Q}^S \equiv \frac{2\nu Q}{gH^3 \sin\theta_0} = \tfrac{2}{3} \qquad \hat{u}_{\text{Mean}}^S \equiv \frac{2\nu u_{\text{Mean}}}{gH^2 \sin\theta_0} = \tfrac{2}{3} \tag{9.9.2}$$

For future reference, we introduce the associated dimensionless stress tensor in the xy plane

$$\hat{\boldsymbol{\sigma}}^S \equiv \frac{2}{\rho g H \sin\theta_0}\boldsymbol{\sigma} = 2(1 - \hat{y}) \begin{bmatrix} -\cot\theta_0 & 1 \\ 1 & -\cot\theta_0 \end{bmatrix} \tag{9.9.3}$$

In the Cartesian system of axes depicted in Figure 9.9.1, the two-dimensional Navier–Stokes equation obtains the dimensionless form

$$Re\left(\frac{\partial \hat{\mathbf{u}}}{\partial \hat{t}} + \hat{\mathbf{u}} \cdot \hat{\nabla}\hat{\mathbf{u}}\right) = -\hat{\nabla}\hat{p} + \hat{\nabla}^2\hat{\mathbf{u}} + 2\begin{bmatrix} 1 \\ -\cot\theta_0 \end{bmatrix} \tag{9.9.4}$$

where $\hat{t} = t u_{x,\text{Max}}/H$, $\hat{\nabla} = (\partial/\partial\hat{x}, \partial/\partial\hat{y})$, and $\hat{\mathbf{x}} = \mathbf{x}/H$. The Reynolds number is given by

$$Re = \frac{u_{x,\text{Max}}H}{\nu} = \frac{gH^3 \sin\theta_0}{2\nu^2} = \frac{2}{\sin\theta_o}Fr^2 \tag{9.9.5}$$

where the *Froude number* is defined as $Fr = u_{x,\text{Max}}/(gH)^{1/2}$.

Orr–Sommerfeld equation

A normal-mode disturbance changes the velocity field from $\hat{\mathbf{u}}^S$ to $\hat{\mathbf{u}} = \hat{\mathbf{u}}^S + \varepsilon\hat{\mathbf{u}}^{\text{NM}}$, where NM stands for normal mode, and ε is a small dimensionless number. Nondimensionalizing Eq. (9.3.25) using the aforementioned scales, we obtain the dimensionless disturbance velocity

$$\hat{u}_x^{\text{NM}} = \hat{f}'(\hat{y})\exp[i\alpha(\hat{x} - \hat{c}\hat{t})], \qquad \hat{u}_y^{\text{NM}} = -i\alpha\hat{f}(\hat{y})\exp[i\alpha(\hat{x} - \hat{c}\hat{t})] \tag{9.9.6}$$

where $\alpha = kH$ is the dimensionless wave number and $\hat{c} = c/u_{x,\text{Max}}$ is the dimensionless complex phase velocity. Substituting the velocity profile given in Eq. (9.9.1) into the dimensionless

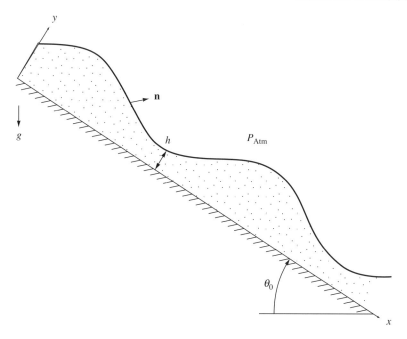

Figure 9.9.1 Growth of periodic disturbances on a liquid film flowing down an inclined plane.

version of the Orr–Sommerfeld equation (9.3.19), we obtain

$$\hat{f}'''' - 2\alpha^2 \hat{f}'' + \alpha^4 \hat{f} = i\,Re\,\alpha[(2\hat{y} - \hat{y}^2 - \hat{c})(\hat{f}'' - \alpha^2 \hat{f}) + 2\hat{f}] \tag{9.9.7}$$

where a prime denotes a derivative with respect to $\hat{y}$.

   To complete the definition of the linear stability problem, we require four boundary conditions for $\hat{f}$. Two boundary conditions arise immediately by requiring that both components of the velocity vanish at the wall, yielding

$$\hat{f}(0) = \hat{f}'(0) = 0 \tag{9.9.8}$$

Two additional boundary conditions emerge by demanding that the shear and normal stress are consistent with the jump in the interfacial traction due to surface tension at the free surface. Before we can implement these conditions, however, we must introduce an equation that describes the location of the free surface.

Evolution of the free surface

   Anticipating that the free surface will deform in a manner that is consistent with the spatial structure of the perturbation, we express the dimensionless film thickness $\hat{h}$ in terms of the real or imaginary part of the right-hand side of the equation

$$\hat{h}(\hat{x}, \hat{t}) = 1 + \varepsilon \hat{\eta}(\hat{x}, \hat{t}) = 1 + \varepsilon A \exp[i\alpha(\hat{x} - \hat{c}\hat{t})] \tag{9.9.9}$$

where A is a dimensionless complex constant. Substituting Eq. (9.9.9) into the kinematic boundary condition (1.6.5), we find

$$i\varepsilon\alpha A(-\hat{c} + \hat{u}_x)\exp[i\alpha(\hat{x} - \hat{c}\hat{t})] - \hat{u}_y = 0 \tag{9.9.10}$$

where the velocity components $\hat{u}_x$ and $\hat{u}_y$ are evaluated at the location of the free surface, at $\hat{y} = \hat{h}$.

Next we extend the domain of flow beyond the deformed free surface and up to the unperturbed free surface located at $\hat{y} = 1$, and use Taylor's expansions to relate the values of the flow variables at the free surface to those at the unperturbed position. For instance, expanding the velocity in a Taylor series around the undeformed position $\hat{y} = 1$ yields

$$\hat{\mathbf{u}}(\hat{y} = \hat{h}) = \hat{\mathbf{u}}(\hat{y} = 1) + \left(\frac{\partial \hat{\mathbf{u}}}{\partial \hat{y}}\right)_{\hat{y} = 1} (\hat{h} - 1) + \cdots$$

$$= \hat{\mathbf{u}}^S(\hat{y} = 1) + \left[\left(\frac{\partial \hat{\mathbf{u}}^S}{\partial \hat{y}}\right)_{\hat{y} = 1} (\hat{h} - 1) + \varepsilon \hat{\mathbf{u}}^{NM}(\hat{y} = 1)\right] + \cdots \qquad (9.9.11)$$

The dots represent terms with quadratic and higher-order dependence on ε. Substituting Eq. (9.9.9) into Eq. (9.9.11) and then into Eq. (9.9.10), using Eqs. (9.9.6) and maintaining terms of first order with respect to ε, we find $A = \hat{f}(1)/(\hat{c} - 1)$. Thus the deformation of the free surface is described by the real or imaginary part of

$$\hat{\eta}(\hat{x}, \hat{t}) = \frac{\hat{f}(1)}{\hat{c} - 1} \exp[i\alpha(\hat{x} - \hat{c}\hat{t})] \qquad (9.9.12)$$

Boundary conditions at the free surface

At the free surface we require that the traction undergo a jump that is balanced by surface tension or a more general type of an interfacial force field. For a clean interface with uniform surface tension γ we require

$$\hat{\boldsymbol{\sigma}} \cdot \mathbf{n} = -\frac{1}{Bo} \hat{\kappa} \mathbf{n} \qquad (9.9.13)$$

where $\hat{\boldsymbol{\sigma}}$ is the dimensionless stress tensor defined as in Eq. (9.9.3), $\mathbf{n}$ is the unit normal vector pointing into the ambient atmosphere, $\hat{\kappa}$ is the dimensionless curvature of the free surface in the xy plane, and

$$Bo = \frac{\rho g H^2 \sin \theta_0}{2\gamma} \qquad (9.9.14)$$

is the *Bond number*. It is sometimes convenient to replace Bo by the *inverse Bond number* Γ, *Weber number We*, or a *property group S* whose value depends exclusively on the physical properties of the fluid, defined as

$$\Gamma \equiv \frac{1}{Bo} = \frac{Re}{We} = \frac{S}{Re^{2/3} \sin \theta_0^{1/3}}$$

$$We = \frac{\rho H u_{x,\text{Max}}^2}{\gamma}, \qquad S = \gamma \left(\frac{2\rho}{g\mu^4}\right)^{1/3} \qquad (9.9.15)$$

For water at room temperature, S is approximately equal to 4,280.

Next we expand both sides of Eq. (9.9.13) in Taylor series with respect to ε. For this purpose, we decompose the stress tensor and normal vector into their steady and unsteady or disturbance components, writing

$$\hat{\boldsymbol{\sigma}} = \hat{\boldsymbol{\sigma}}^S + \varepsilon \hat{\boldsymbol{\sigma}}^{NM}, \qquad \mathbf{n} = \mathbf{n}^S + \varepsilon \mathbf{n}^{NM} \qquad (9.9.16)$$

where

$$\mathbf{n}^S = (0, 1), \qquad \mathbf{n}^{NM} = \left(-\frac{\partial \hat{\eta}}{\partial \hat{x}}, 0\right)$$

To compute the left-hand side of Eq. (9.9.13) in terms of flow variables at the undisturbed position of the interface, we apply once again the method of domain perturbation, writing

$$
\left(\hat{\boldsymbol{\sigma}} \cdot \mathbf{n}\right)_{\hat{y}=\hat{h}} = \left[\hat{\boldsymbol{\sigma}}(\hat{y}=1) + \varepsilon\hat{\eta}\left(\frac{\partial\hat{\boldsymbol{\sigma}}}{\partial\hat{y}}\right)_{\hat{y}=1} + \cdots\right]\cdot\mathbf{n}
$$

$$
= \left\{\hat{\boldsymbol{\sigma}}^S(\hat{y}=1) + \varepsilon\left[\hat{\boldsymbol{\sigma}}^{NM}(\hat{y}=1) + \hat{\eta}\left(\frac{\partial\hat{\boldsymbol{\sigma}}^S}{\partial\hat{y}}\right)_{\hat{y}=1}\right] + \cdots\right\}\cdot(\mathbf{n}^S + \varepsilon\,\mathbf{n}^{NM})
$$

$$
= \varepsilon\left[\hat{\boldsymbol{\sigma}}^{NM}(\hat{y}=1) + \hat{\eta}\left(\frac{\partial\hat{\boldsymbol{\sigma}}^S}{\partial\hat{y}}\right)_{\hat{y}=1}\right]\cdot\mathbf{n}^S + \cdots \tag{9.9.17}
$$

Substituting the right-hand side of Eq. (9.9.17) into the left-hand side of Eq. (9.9.13), making the approximation $\hat{\kappa} = -\varepsilon\,\partial^2\hat{\eta}/\partial\hat{x}^2$ on the right-hand side, and equating terms of first order in ε, we obtain

$$
\hat{\boldsymbol{\sigma}}^{NM}(\hat{y}=1)\cdot\begin{bmatrix}0\\1\end{bmatrix} = 2\hat{\eta}\begin{bmatrix}1\\-\cot\theta_0\end{bmatrix} + \Gamma\frac{\partial^2\hat{\eta}}{\partial\hat{x}^2}\begin{bmatrix}0\\1\end{bmatrix} \tag{9.9.18}
$$

Furthermore, we decompose Eq. (9.9.18) into its two scalar constituents expressing the tangential and normal force balance at the free surface and obtain

$$
\hat{\sigma}_{xy}^{NM}(\hat{y}=1) = 2\hat{\eta}, \qquad \hat{\sigma}_{yy}^{NM}(\hat{y}=1) = -2\hat{\eta}\cot\theta_0 + \Gamma\frac{\partial^2\hat{\eta}}{\partial\hat{x}^2} \tag{9.9.19}
$$

In addition, we write out the stress tensor in terms of its pressure and the viscous constituents, and thus obtain the two linearized boundary conditions

$$
\left(\frac{\partial\hat{u}_x^{NM}}{\partial\hat{y}} + \frac{\partial\hat{u}_y^{NM}}{\partial\hat{x}}\right)_{\hat{y}=1} = 2\hat{\eta}
$$

$$
\hat{p}^{NM}(\hat{y}=1) = 2\left(\frac{\partial\hat{u}_y^{NM}}{\partial\hat{y}}\right)_{\hat{y}=1} + 2\hat{\eta}\cot\theta_0 - \Gamma\frac{\partial^2\hat{\eta}}{\partial\hat{x}^2} \tag{9.9.20}
$$

Substituting Eqs. (9.9.6) along with Eq. (9.9.12) into the first of Eqs. (9.9.20) yields the first scalar dynamic boundary condition

$$
\hat{f}''(1) + \left(\alpha^2 - \frac{2}{\hat{c}-1}\right)\hat{f}(1) = 0 \tag{9.9.21}
$$

To derive the second scalar dynamic boundary condition, corresponding to the second of Eqs. (9.9.20), we must have available an expression for the disturbance pressure in terms of the velocity. This expression emerges by applying the dimensionless form of the linearized Navier-Stokes equation at the free surface, projecting it onto the tangent vector $\mathbf{t}$, and rearranging to find

$$
\mathbf{t}\cdot\hat{\nabla}\hat{p}^{NM} = -Re\left(\frac{\partial\hat{\mathbf{u}}^{NM}}{\partial\hat{t}} + \hat{\mathbf{u}}^S\cdot\hat{\nabla}\hat{\mathbf{u}}^{NM} + \hat{\mathbf{u}}^{NM}\cdot\hat{\nabla}\hat{\mathbf{u}}^S\right)\cdot\mathbf{t}
$$

$$
+ (\hat{\nabla}^2\hat{\mathbf{u}}^{NM})\cdot\mathbf{t} \tag{9.9.22}
$$

Differentiating the second of Eqs. (9.9.20) with respect to $\hat{x}$ yields

$$
\frac{\partial\hat{p}^{NM}}{\partial\hat{x}} = 2\frac{\partial^2\hat{u}_y^{NM}}{\partial\hat{x}\,\partial\hat{y}} + 2\frac{\partial\hat{\eta}}{\partial\hat{x}}\cot\theta_0 - \Gamma\frac{\partial^3\hat{\eta}}{\partial\hat{x}^3} \tag{9.9.23}
$$

evaluated at $\hat{y} = 1$. Maintaining only terms of first order in ε in both Eqs. (9.9.22) and (9.9.23) and setting the right-hand sides equal to each other yields

$$-Re\left(\frac{\partial \hat{u}_x^{NM}}{\partial \hat{t}} + \frac{\partial \hat{u}_x^{NM}}{\partial \hat{x}}\right) + \hat{\nabla}^2 \hat{u}_x^{NM} = 2\frac{\partial^2 \hat{u}_y^{NM}}{\partial \hat{x}\partial \hat{y}} + 2\frac{\partial \hat{\eta}}{\partial \hat{x}}\cot\theta_0 - \Gamma\frac{\partial^3 \hat{\eta}}{\partial \hat{x}^3} \tag{9.9.24}$$

evaluated at $\hat{y} = 1$. Finally, we substitute Eqs. (9.9.6) along with Eq. (9.9.12) into Eq. (9.9.24) and obtain the desired boundary condition

$$\hat{f}'''(1) + \alpha[i\,Re(\hat{c} - 1) - 3\alpha]\hat{f}'(1)$$
$$- i\frac{\alpha}{\hat{c} - 1}(2\cot\theta_0 + \Gamma\alpha^2)\hat{f}(1) = 0 \tag{9.9.25}$$

Summary of equations and boundary conditions

The problem has been reduced to solving the eigenvalue problem expressed by the Orr–Sommerfeld equation (9.9.7) subject to the four boundary conditions stated in Eqs. (9.9.8), (9.9.21), and (9.9.25). Solutions in closed form are generally not available, and we must proceed using approximate, asymptotic, and numerical methods (see Problem 9.9.1).

Long-wave solution

Useful insights into the stability of the flow can be obtained by considering disturbances with long wavelengths or small wave numbers α (Yih, 1963). As it turns out from the numerical solution of the exact eigenvalue problem, the fastest growing mode does have a long wavelength, and the approximate analysis for small wave numbers provides us with accurate estimates.

Expanding $\hat{f}$ and $\hat{c}$ in Taylor series with respect to α, we obtain

$$\hat{f} = \hat{f}_0 + \alpha\hat{f}_1 + \alpha^2\hat{f}_2 + \cdots$$
$$\hat{c} = \hat{c}_0 + \alpha\hat{c}_1 + \alpha^2\hat{c}_2 + \cdots \tag{9.9.26}$$

Substituting these expansions into the Orr–Sommerfeld equation and boundary conditions, assuming that both Re and Γ are of order one, and collecting terms of same order in α, we derive a sequence of eigenvalue problems.

The first problem is described by the differential equation

$$\hat{f}_0'''' = 0 \tag{9.9.27a}$$

with boundary conditions

$$\hat{f}_0(0) = \hat{f}_0'(0) = 0, \qquad \hat{f}_0''(1) - \frac{2}{\hat{c}_0 - 1}\hat{f}_0(1) = 0, \qquad \hat{f}_0'''(1) = 0. \tag{9.9.27b}$$

The second problem corresponds to the differential equation

$$\hat{f}_1'''' = -i\,Re[(\hat{y}^2 - 2\hat{y} + \hat{c}_0)\hat{f}_0'' - 2\hat{f}_0] \tag{9.9.28a}$$

with boundary conditions

$$\hat{f}_1(0) = \hat{f}_1'(0) = 0$$
$$\hat{f}_1''(1) - \frac{2}{\hat{c}_0 - 1}\hat{f}_1(1) = -\frac{2\hat{c}_1}{(\hat{c}_0 - 1)^2}\hat{f}_0(1) \tag{9.9.28b}$$
$$\hat{f}_1'''(1) = -i\,Re(\hat{c}_0 - 1)\hat{f}_0'(1) + 2i\frac{\cot\theta_0}{\hat{c}_0 - 1}\hat{f}_0(1)$$

Note that since α has been used as a perturbation variable, it has been scaled out from Eqs. (9.9.27) and (9.9.28).

The solution of the first problem described by Eqs. (9.9.27a,b) is readily found to be

$$\hat{f}_0 = B\hat{y}^2, \qquad \hat{c}_0 = 2 \tag{9.9.29}$$

where B is an arbitrary constant expessing the initial magnitude of the perturbation. The second of Eqs. (9.9.29) shows that *long waves translate with a phase velocity that is equal to twice the velocity of the liquid film at the free surface,* in contrast with Rayleigh's theorem for inviscid fluids discussed in Section 9.4.

Substituting Eqs. (9.9.29) into Eq. (9.9.28a), we obtain the inhomogeneous form $\hat{f}_1'''' = 4Bi\, Re(\hat{y} - 1)$. A solution that satisfies the two boundary conditions at the wall is given by the fifth-degree polynomial

$$\hat{f}_1 = Bi\, Re\frac{1}{30}\hat{y}^4(\hat{y} - 5) + C\hat{y}^3 \tag{9.9.30}$$

where C is a new constant. Substituting Eq. (9.9.30) into the third and fourth boundary conditions in Eqs. (9.9.28b), we obtain a system of two linear homogeneous equations for B and C that may be placed in the form

$$\begin{bmatrix} \hat{c}_1 - \frac{8}{15}i\, Re & 2 \\ -2i\cot\theta_0 & 6 \end{bmatrix}\begin{bmatrix} B \\ C \end{bmatrix} = \mathbf{0} \tag{9.9.31}$$

For a nontrivial solution to exist, the determinant of the coefficient matrix must vanish, and this requires that

$$\hat{c}_1 = i\frac{8}{15}(Re - \frac{5}{4}\cot\theta_0) \tag{9.9.32}$$

for which $C = Bi\frac{1}{3}\cos\theta_0$.

Collecting the solutions of the first and second problems we obtain

$$\hat{c} = 2 + i\alpha\frac{8}{15}(Re - \frac{5}{4}\cot\theta_0) + \cdots \tag{9.9.33}$$

To this end, we recall that the dimensionless growth rate of the disturbance is equal to the imaginary part of $\alpha\hat{c}$ and find that long waves will grow when

$$Re > \frac{5}{4}\cot\theta_0 \tag{9.9.34}$$

which was first deduced by Benjamin (1957) using a different method. Substituting the definition of the Reynolds number from Eq. (9.9.5), we find that long waves will grow when

$$H > \left(\frac{10}{4}\frac{\nu^2}{g}\frac{\cot\theta_0}{\sin\theta_0}\right)^{1/3} \tag{9.9.35}$$

This inequality places a limit on the maximum film thickness for stable flat-film flow. A vertical film, corresponding to $\theta_0 = \pi/2$, will be unstable at all Reynolds numbers.

Other approximate and numerical methods

Benjamin (1957) developed a different approximate method of solving the linear stability problem based upon the polynomial expansion

$$\hat{f}(\hat{y}) = \sum_{n=0}^{\infty} A_n\hat{y}^n \tag{9.9.36}$$

Substituting Eq. (9.9.36) into the Orr–Sommerfeld equation yields a recurrence relationship among the coefficients A_n, A_{n-2}, A_{n-4}, A_{n-6}. Requiring the satisfaction of the boundary conditions yields an eigenvalue problem with infinite dimensions.

In practice, the series (9.9.36) and associated linear algebraic eigenvalue problem are truncated at a finite level. Benjamin noted that truncating amounts to expressing the solution in a Taylor series with respect to α and Re, and then truncating that series at a finite level. For instance, maintaining 16 terms produces results that are accurate to third order in α and Re. Benjamin solved the eigenvalue problem analytically maintaining 16 terms and, after a considerable amount of algebra, produced an involved relationship between the critical wave number for neutral stability, Re, and We.

Anshus and Goren (1966) noted that the main difficulty in solving the Orr–Sommerfeld equation (9.9.7) is due to the presence of the nonconstant coefficient on the right-hand side involving the unperturbed velocity profile, and proposed replacing the velocity distribution with the maximum velocity at the free surface. At low and moderate Reynolds numbers, the approximate growth rates computed in this manner turn out to be close to the exact values computed using numerical methods.

Instead of implementing the long-wave approximation into the linearized system, Benney (1966) worked with an approximate form of the Navier–Stokes equation that assumes long waves of finite amplitude at the outset, developed according to our discussion in Section 6.3. By carrying out a linear stability analysis of the resulting evolution equation for the film thickness, he produced the second and third coefficients of the complex growth rate c shown in Eqs. (9.9.26)

$$\hat{c}_2 = -2 - \tfrac{32}{63} Re(Re - \tfrac{5}{4}\cot\theta_0) \tag{9.9.37}$$

$$\hat{c}_3 = i\left(-\tfrac{1}{3}\Gamma - \tfrac{157}{56}Re + \tfrac{6}{5}\cot\theta_0 - \tfrac{8}{45}Re\cot^2\theta_0 \right.$$

$$\left. + \tfrac{138,904}{155,925}Re^2\cot\theta_0 - \tfrac{1,213,952}{2,027,025}Re^3 \right) \tag{9.9.38}$$

The neutral stability curves where the imaginary part of $\hat{c}$ vanishes are given by $\alpha_{cr} = 0$ and $(-\hat{c}_1/\hat{c}_3)^{1/2}$.

The linear equations governing the temporal and spatial instability of the flat film have been solved on a number of occasions using a variety of asymptotic and numerical methods (Chang, 1994). For example, Pierson and Whitaker (1977) prepared graphs of the growth rate at low and moderate Reynolds numbers, and Chin, Abernathy, and Bertschy (1986) extended the numerical results to high Reynolds numbers.

PROBLEMS

9.9.1   **Film in creeping flow.** When the Reynolds number is small, the right-hand side of the Orr–Sommerfeld equation may be neglected, and the general solution may be found readily in closed form

$$\hat{f} = ae^{\alpha\hat{y}} + b\hat{y}e^{\alpha\hat{y}} + ce^{-\alpha\hat{y}} + d\hat{y}e^{-\alpha\hat{y}} \tag{9.9.39}$$

where a, b, c, d are four constants. (a) Show that requiring the boundary conditions leads to the following homogeneous system of four linear equations

$$c = -a \tag{9.9.40}$$

$$\begin{bmatrix} 2\alpha & 1 & 1 \\ (1-M)(\alpha^2 - \Lambda) & \alpha^2 + \alpha - \Lambda & (\alpha^2 - \alpha - \Lambda)M \\ 2\alpha^2(1+M) + i\Lambda\Pi(1-M) & 2\alpha^2 + i\Lambda\Pi & (-2\alpha^2 + i\Lambda\Pi)M \end{bmatrix} \begin{bmatrix} a \\ b \\ d \end{bmatrix} = 0$$

$$\tag{9.9.41}$$

where

$$M = e^{-2\alpha}, \qquad \Lambda = \frac{1}{\hat{c} - 1}, \qquad \Pi = 2\cot\theta_0 + \Gamma\alpha^2 \tag{9.9.42}$$

(b) To ensure that the system (9.9.41) has a nontrivial solution, we require that the determinant of the matrix of the coefficients vanish, and this provides us with a relation between the wave number and the complex phase velocity. Prepare a graph of the growth rate and phase velocity against α.

9.9.2   **Benney's method.** Show that according to Benney's analysis, the neutral stability curves along which the imaginary part of c vanishes, are given by

$$Re = \tfrac{5}{4}\cot\theta_0 + \alpha_{cr}^2(0.625\Gamma + 4.320870\cot\theta_0$$
$$- 0.000006\cot^3\theta_0) + O(\alpha_{cr}^4) \tag{9.9.43}$$

9.9.3   **Two films in a channel.** Consider the flow of two superposed layers of two fluids in plane Couette or Poiseuille flow in a channel. Formulate and discuss the equations and boundary conditions of the temporal linear stability problem (Yih, 1967; Hooper, 1989).

9.10 | CAPILLARY INSTABILITY OF A CURVED INTERFACE

A new kind of instability occurs when an unperturbed interface is not flat but has a finite mean curvature; distortions of the interface cause pressure variations due to surface tension, which may assist instead of inhibiting the amplification of a perturbation, causing a *capillary instability*. A simple manifestation of this instability occurs when a cylindrical jet of a fluid penetrates another immiscible ambient fluid, in which case the interface develops corrugations and the jet breaks up into a series of droplets as shown in Figure 9.10.2. In fact, the occurrence of this instability is exploited in ink-jet printer technology, where an ejected column of ink breaks up into an array of droplets that are subsequently guided onto printed paper with the aid of an electrical field. The capillary instability of a quiescent liquid column is responsible for the formation of a spider's web, whereas the capillary instability of an annular layer that is coated on the exterior or interior of a cylindrical surface, such as a pulmonary airway, is responsible for the formation of annular rings and may cause pulmonary closure.

Rayleigh Instability of a Uniform Jet or Quiescent Column of an Inviscid Liquid

Let us consider a cylindrical jet of an inviscid liquid with a circular cross-section of radius a, shooting with uniform velocity into an ambient gas of negligible density and constant pressure P_0, as shown in Figure 9.10.1. Assuming that the gravitational force is negligible, we find that the pressure within the jet is given by $P_0 + \gamma/a$, where γ is the surface tension; the mean curvature of the unperturbed interface is $\kappa_m = 1/(2a)$. To simplify the analysis, we introduce a frame of reference that translates with the jet, in which case the jet appears to be a stationary column of fluid. We then introduce cylindrical coordinates (x, σ, φ) that are coaxial with the jet and assume that the perturbation causes the interface to deform in an axisymmetric manner, so that its radius is described by the real part of the right-hand side of

$$\sigma = a + \eta(x, t) = a + \varepsilon A \sin kx \exp(-ikct) \tag{9.10.1}$$

where ε is a small number; the value of the complex constant A is determined by the initial amplitude of the perturbation. Confining our attention to irrotational and spatially periodic perturbations, we describe the associated flow in terms of the velocity potential

$$\phi = \varepsilon BF(\sigma)\sin kx \exp(-ikct) \tag{9.10.2}$$

where B is a complex constant and F is a complex function. Requiring that ϕ satisfy Laplace's equation in cylindrical coordinates, and demanding that the velocity be bounded, we find that $F(\sigma) = J_0(ik\sigma)$, where J_0 is a Bessel function of the first kind.

To compute constants A and B, we require two scalar constraints. Substituting the expressions on the right-hand sides of Eqs. (9.10.1) and (9.10.2) into the linearized form of the kinematic

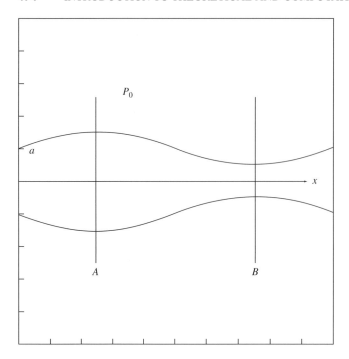

Figure 9.10.1   Capillary instability of a cylindrical jet of a liquid shooting into an ambient gas.

condition at the interface (1.6.5), $\partial\eta/\partial t - \partial\phi/\partial\sigma = 0$, we obtain

$$-ikcA = B\frac{dF}{d\sigma} = B\frac{dJ_0(ik\sigma)}{d\sigma} = -ikB\,J_1(ik\sigma) \tag{9.10.3}$$

which may be simplified to

$$cA = B\,J_1(ik\sigma) \tag{9.10.4}$$

To derive the second constraint, we use the dynamic boundary condition at the interface, requiring that the pressure jump be balanced by surface tension, $p - P_0 = 2\kappa_m\gamma$, where κ_m is the mean curvature of the interface. Linearizing the expression for the mean curvature given by the appropriate entry of Table 4.2.1, we obtain

$$p - P_0 = \frac{\gamma}{a}\left(1 - \frac{\eta}{a}\right) - \gamma\frac{\partial^2\eta}{\partial x^2} \tag{9.10.5}$$

The last term on the right-hand side expresses the effect of the interfacial curvature in an azimuthal plane, whereas the first term expresses the effect of the curvature in the conjugate plane.

Next we use Bernoulli's equation for unsteady irrotational flow (3.4.16) to evaluate p. Conforming with the linear approximation, we discard the /square of the velocity and substitute the resulting linearized form into Eq. (9.10.5) to obtain

$$\rho\frac{\partial\phi}{\partial t} = \gamma\left(\frac{\eta}{a^2} + \frac{\partial^2\eta}{\partial x^2}\right) \tag{9.10.6}$$

Substituting Eqs. (9.10.1) and (9.10.2) into Eq. (9.10.6) and evaluating the left-hand side at the location of the unperturbed interface, at $\sigma = a$, yields

$$-i\rho k\, cB\, J_0(ika) = \frac{\gamma}{a^2} A(1 - k^2 a^2) \tag{9.10.7}$$

Dividing corresponding sides of Eqs. (9.10.4) and (9.10.7) to eliminate A and B and rearranging we obtain the square of the complex phase velocity

$$c^2 = i\frac{\gamma}{\rho k a^2}\frac{J_1(ika)}{J_0(ika)}(1 - k^2 a^2) = -\frac{\gamma}{\rho k a^2}\frac{I_1(ka)}{I_0(ka)}(1 - k^2 a^2) \tag{9.10.8}$$

where we have introduced the real and positive modified Bessel function $I_1(ka) = -iJ_1(ika)$ and used the identity $I_0(ka) = J_0(ika)$ (Problem 9.10.3).

When the right-hand side of Eq. (9.10.8) is negative, c assumes two complex conjugate values; the one with the positive imaginary part yields an unstable mode with corresponding growth rate given by

$$\sigma_1^2 = k^2 c_I^2 = \frac{\gamma}{\rho a^3}\frac{I_1(ka)}{I_0(ka)}ka\,(1 - k^2 a^2) \tag{9.10.9}$$

(Rayleigh, 1879). It is then evident that, when $0 < ka < 1$, the jet falls prey to the capillary instability, which means that *small wave numbers with large wavelengths are unstable*. The fastest growth occurs when $ka = 0.679$, and the corresponding maximum growth rate is given by

$$\sigma_I = 0.34\left(\frac{\gamma}{\rho a^3}\right)^{1/2} \tag{9.10.10}$$

The growth of perturbations in the unstable regime $0 < ka < 1$ may be explained in physical terms by considering the pressure distribution within the liquid column at the initial instant, neglecting pressure variations due to the fluid acceleration. Considering station A in Figure 9.10.1, we note that the curvature of the interface in an azimuthal plane increases due to the interfacial corrugation, whereas the curvature in a plane that is perpendicular to the x axis decreases due to the increased jet radius; the inverse is true for station B. The linearized form of the expression for the mean curvature on the right-hand side of Eq. (9.10.7) then shows that the pressure at station A will be lower than the pressure at station B when $0 < ka < 1$, and this implies that the fluid will be driven toward the crest, thereby amplifying the interfacial corrugations.

To study the behavior of nonaxisymmetric disturbances, we assume that the location of the perturbed interface is described by the real part of the right-hand side of

$$\sigma = a + \eta(x, t) = a + \varepsilon A \sin(kx + m\varphi)\exp(-ikct) \tag{9.10.11}$$

where m is the azimuthal wave number. Repeating the preceding steps, we find that the complex phase velocity is given by the generalized expression

$$c^2 = -\frac{\gamma}{\rho k a^2}\frac{I_m'(ka)}{I_m(ka)}(1 - k^2 a^2 - m^2) \tag{9.10.12}$$

(Rayleigh, 1879). The ratio of the modified Bessel functions is positive for any value of m, and this shows that the right-hand side of Eq. (9.10.12) is positive for any $m > 0$, and the interface is stable to nonaxisymmetric perturbations. Thus a cylindrical column of fluid is expected to break up by developing axisymmetric corrugations, in agreement with observation.

Figure 9.10.2 illustrates the development of Rayleigh's instability subject to an axisymmetric perturbation during the early and advanced stages of the motion where nonlinear effects become important, computed using a boundary-integral method discussed in Chapter 10, after Mansour and Lundgren (1990). We observe the initial amplification of sinusoidal interfacial corrugations,

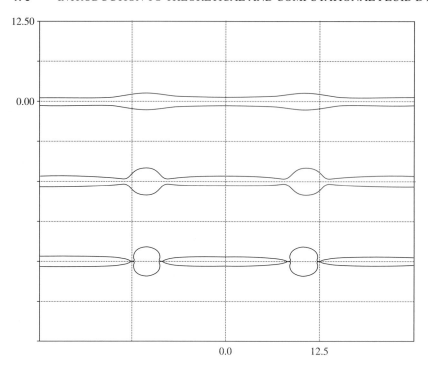

Figure 9.10.2   Development of Rayleigh's capillary instability on a cylindrical jet with an axisymmetric perturbation during the early and advanced stages of the motion, computed using a boundary-integral method (Mansour and Lundgren, 1990).

and the ultimate formation of primary drops that are connected by axisymmetric liquid bridges; the latter eventually break up to yield satellite drops.

Extensions of Rayleigh's theory to account for the density of the ambient fluid as well as for viscous stresses and nonlinear effects have been reviewed by Bogy (1979). The results show that including these effects may decelerate the growth but does not prevent the occurrence of the instability.

Instability of Viscous Jets and Annular Layers

The capillary instability of viscous jets and columns of fluid, as well as annular layers coated on the internal or external surface of a cylinder, have been studied on a number of occasions; a review was provided by Newhouse and Pozrikidis (1992). Figure 9.10.3(a,b) illustrates the evolution of an annular layer coated on the interior of a circular tube subject to an axisymmetric perturbation, computed using a boundary-integral method for Stokes flow, after Newhouse and Pozrikidis (1992). We observe the initial amplification of interfacial corrugations and the formation of primary drops that are connected by axisymmetric bridges shown in panel (a); the latter eventually break up to yield two alternating sequences of drops shown in panel (b).

PROBLEMS

9.10.1   **Coalescence of an array of drops.** Consider a one-dimensional array of touching spherical drops arranged on a straight line, representing, for, instance, a sequence of melted glass beads. Explain why one should expect (a) the drops to coalesce forming a column of liquid

(a)

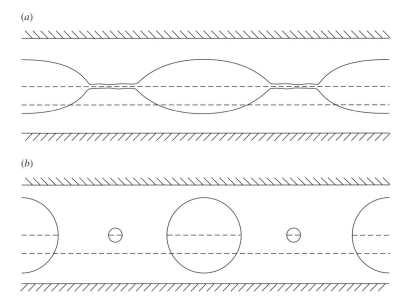

(b)

Figure 9.10.3   Evolution of an annular layer coating the interior side of a circular tube subject to an axisymmetric perturbation (Newhouse and Pozrikidis, 1992). The off-centered dotted line shows the initial position of the interface. (a) The initial amplification of sinusoidal interfacial corrugations and formation of primary drops connected with axisymmetric bridges. (b) The bridges eventually break up to yield two alternating sequences of drops.

and (b) the column to break up into a new array of drops. Compute the ratio of the radii of the initial and final drops.

9.10.2   **Viscous liquid thread.**   Formulate the equations that govern the linear stability of a cylindrical viscous liquid column suspended in an ambient fluid of negligible density and viscosity, assuming that the flow occurs under conditions of creeping motion (Newhouse and Pozrikidis, 1992).

Computer Problem

9.10.3   **Capillary instability of an inviscid jet.**   Prepare a plot of the growth rate versus reduced wave number ka based on Eq. (9.10.9). To compute the modifed Bessel functions use the polynomial approximations

$$I_0(x) = 1 + \frac{1}{2^2}x^2 + \frac{1}{2^2 \cdot 4^2}x^4 + \cdots$$

$$I_1(x) = \tfrac{1}{2}x + \frac{1}{2^2 \cdot 4}x^2 + \frac{1}{2^2 \cdot 4^2 \cdot 6}x^3 + \cdots$$

9.11 | INERTIAL INSTABILITY OF ROTATING FLUIDS

A new type of instability occurs when the base flow rotates so that the streamlines are curved. Centrifugal forces generate an effective distributed body force that causes the onset of a pressure gradient across the streamlines so as to satisfy the force balance, and this may act to destabilize the flow.

The simplest manifestation of this instability occurs in purely swirling flows. Rayleigh (1916) used a simple energy argument to show that, when viscous stresses are insignificant, a necessary and sufficient condition for a swirling flow to be stable subject to axisymmetric disturbances is that the distribution of circulation $C = 2\pi\sigma u_\varphi$ in the radial direction σ satisfy the criterion

$$\frac{dC^2}{d\sigma} > 0 \qquad (9.11.1)$$

where (x, σ, φ) are cylindrical coordinates that are concentric with the axis of the swirling motion (Problem 9.11.1).

Von Kármán (1934) provided an appealing physical interpretation of Eq. (9.11.1). Consider a fluid ring which, due to the instability, is displaced in a manner that preserves the axisymmetry of the flow from the initial radial position σ_1 to another position σ_2. Kelvin's circulation theorem requires that the circulation C around the ring be preserved; as a consequence, the old and new azimuthal components of the velocity of the ring $u_{\varphi 1}$ and $u_{\varphi 2}$ must be related by $C_1 = 2\pi\sigma_1 u_{\varphi 1} = 2\pi\sigma_2 u_{\varphi 2}$. The centrifugal force per unit mass of the fluid due to the rotation at the unperturbed and perturbed states are, respectively, $F_1 = u^2_{\varphi 1}/\sigma_1$ and $F_2 = u^2_{\varphi 2}/\sigma_2 = C_1^2/(4\pi^2\sigma_2^3)$. Assuming that the pressure field remains unchanged, we find that the radial pressure gradient at the new position has the undisturbed value $(dP/d\sigma)_2 = \rho C_2^2/4\pi^2\sigma_2^3$. Equation (9.11.1) then indicates that the flow will be stable as long as the pressure gradient is able to overcome the centrifugal acceleration and thus push the fluid ring back to its original position.

Taylor Instability

Taylor (1923) studied the stability of the circular Couette flow generated by the rotation of two coaxial cylinders discussed in Section 5.1. Observation shows, and linear stability analysis confirms, that the flow is unstable for certain combinations of the inner and cylinder angular velocities Ω_1 and Ω_2 that fall within a certain range that depends upon the radii of the inner and outer cylinders R_1 and R_2. When the inner cylinder is stationary and the outer cylinder rotates, the flow is stable, whereas in the opposite case the flow may be unstable. For fixed radii R_1 and R_2, there is a region in the (Ω_1, Ω_2) plane where unstable modes grow and then lead to a new steady state that is characterized by the presence of axisymmetric rolling patterns with coiled streamlines illustrated in Figure 9.11.1(a), known as the *Taylor vortex flow*. More complicated wavy and turbulent states occur for other combinations of Ω_1 and Ω_2. The normal-mode analysis reveals that the base flow is stable at all Reynolds numbers provided that Rayleigh's circulation criterion is fulfilled.

Görtler Instability

A more complex manifestation of the inertial instability of rotating fluids occurs in high-Reynolds-number boundary-layer flow over a concavely curved surface, observed by Görtler (1940). Under certain conditions, the boundary layer develops an alternating sequence of rolling structures illustrated in Figure 9.11.1(b). Regarding the wall as the stationary outer cylinder of a circular Couette flow device allows us to make an analogy between the instability of this flow and that of the flow between concentric cylinders. A review of the Görtler instability and its various manifestations was given by Saric (1994).

PROBLEMS

9.11.1 **Rayleigh's circulation criterion.** Derive Rayleigh's circulation criterion (9.11.1) by carrying out the following steps: (a) Consider two fluid rings of equal volume, with radii σ_1 and σ_2, compute their respective kinetic energies per unit volume, and add them to obtain the combined

(*a*)

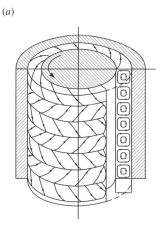

(*b*)

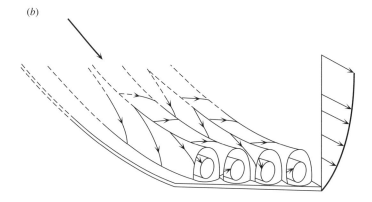

Figure 9.11.1   Instabilities of rotating flows. (a) Taylor instability of the flow between two rotating concentric cylinders and (b) Görtler instability of a high-Reynolds-number flow along a curved wall.

kinetic energy in the unperturbed state K_U in terms of the rings' radii and circulations. (b) Assume that the rings interchange radial positions, and compute the combined kinetic energy per unit volume in the perturbed state K_P subject to the restriction imposed by Kelvin's circulation theorem, in terms of the radii and circulations. (c) Requiring that $K_P > K_U$, which means that the perturbation must be supplied with a finite amount of energy, derive Rayleigh's criterion 9.11.1)

9.11.2   **Flow in a curved tube.**   Consider pressure-driven flow along a *curved* cylindrical tube, known as the *Dean flow,* and discuss whether the flow might be subject to an inertial instability due to centrifugal forces (Drazin and Reid, 1981, p. 108).

References

Abramowitz, M., and Stegun, I. A., 1972, *Handbook of Mathematical Functions.* Dover.

Anshus, B. E., and Goren, S. L., 1966, A method of getting approximate solutions to the Orr–Sommerfeld equation for flow on a vertical wall. *AIChE J.* **12**, 1004–8.

Bear, J., 1972, *Dynamics of Fluids in Porous Media.* Dover.

Benjamin, T. B., 1957, Wave formation in laminar flow down an inclined plane. *J. Fluid Mech.* **2**, 554–74.

Benney, D. J., 1966, Long waves on liquid films. *J. Math. and Phys.* **45**, 150–55.

Betchov, R., and Criminale, W. O., 1967, *Stability of Parallel Flows.* Academic Press.

Betchov, R., and Szewczyk, A., 1963, Stability of a shear layer between parallel streams. *Phys. Fluids* **6**, 1391–96.

Bogy, D. B., 1979, Drop formation in a circular liquid jet. *Annu. Rev. Fluid Mech.* **11**, 207–28.

Case, K. M., 1960a, Stability of inviscid plane Couette flow. *Phys. Fluids* **3**, 143–48.

Case, K. M., 1960b, Stability of an idealized atmosphere. I. Discussion of results. *Phys. Fluids* **3**, 149–54.

Chandrasekhar, S., 1961, *Hydrodynamic and Hydromagnetic Stability.* Dover.

Chang, H.-C., 1994, Wave evolution on a falling film. *Annu. Rev. fluid Mech.* **26**, 103–36.

Chin, R. W., Abernathy, F. H., and Bertschy, J. R., 1986, Gravity and shear wave stability of free surface flows. Part 1. Numerical calculations. *J. Fluid Mech.* **168**, 501–13.

Craik, A. D. D., 1985, *Wave Interactions and Fluid Flows.* Cambridge University Press.

Davey, A., 1977, On the numerical solution of difficult eigenvalue problems. *J. Comput. Phys.* **24**, 331–38.

Drazin, P. G., and Reid, W. H., 1981, *Hydrodynamic Stability.* Cambridge University Press.

Eliassen, A., Høiland, E., and Riis, E., 1953, Two-dimensional perturbations of a flow with constant shear of a stratified fluid. *Inst. Weather Climate Res.,* Oslo, Publ. No 1.

Feldman, S., 1957, On the hydrodynamic stability of two viscous incompressible fluids in parallel uniform shearing motion. *J. Fluid Mech.* **2**, 343–70.

Fjørtoft, R., 1950, Application of integral theorems in deriving criteria of stability for laminar flows and for the baroclinic circular vortex. *Geophys. Publ.* **17**(5).

Gaster, M., 1962, A note on the relation between temporally-increasing and spatially-increasing disturbances in hydrodynamic stability. *J. Fluid Mech.* **14**, 222–24.

Gersting, J. M., and Jankowski, D. F., 1972, Numerical methods for Orr–Sommerfeld problems. *Int. J. Num. Meth. Eng.* **4**, 195–206.

Görtler, H., 1940, On the three-dimensional instability of laminar boundary layers on concave walls. *Tech. Mem. Nat. Adv. Comm. Aero. Wash.,* No. 1375 (1975).

Hesla, T. I., Pranckh, and Preziosi, L., 1986, Squire's theorem for two stratified fluids. *Phys. Fluids* **29**, 2808–11.

Hooper, A., 1989, The stability of two superposed viscous fluids in a channel. *Phys. Fluids* A **1**, 1133–42.

Howard, L. N., 1961, Note on a paper by John W. Miles. *J. Fluid Mech.* **10**, 509–12.

Huerre, P., and Monkewitz, P. A., 1990, Local and global instabilities in spatially developing flows. *Annu. Rev. Fluid Mech.* **22**, 473–537.

Kármán, T. von, 1934, Some aspects of the turbulence problem. *Proc. 4th Int. Cong. Appl. Mech.* Cambridge, England.

Kerner, W., 1989, Large-scale complex eigenvalue problems. *J. Comput. Phys.* **85**, 1–85.

Ladyzhenskaya, O. A., 1975, Mathematical analysis of Navier–Stokes equations for incompressible liquids. *Annu. Rev. Fluid Mech.* **7**, 249–72.

Lin, C. C., 1955, *The Theory of Hydrodynamic Stability.* Cambridge University Press.

Mansour, N. N., and Lundgren, T. S., 1990, Satellite formation in capillary jet breakup. *Phys. Fluids* A **2**, 1141–44.

Maslowe, S. A., 1981, Shear flow instabilities and transition. *Hydrodynamic Instabilities and the Transition to Turbulence,* Swinney and Gollub, Editors, Springer-Verlag.

Michalke, A., 1964, On the inviscid instability of the hyperbolic-tangent velocity profile. *J. Fluid Mech.* **19**, 543–56.

Murdock, J. N., and Stewartson, K., 1977, Spectra of the Orr–Sommerfeld equation. *Phys. Fluids* **20**, 1404–11.

Newhouse, L., and Pozrikidis, C., 1992, The capillary instability of annular layers and liquid threads. *J. Fluid Mech.* **242**, 193–209.

Ng, B. S., and Reid, W. H., 1979, An initial value method for eigenvalue problems using compound matrices. *J. Comput. Phys.* **30**, 125–36.

Orr, W. M. F., 1907, The stability or instability of the steady motions of a perfect liquid and of a viscous liquid. Part I. A perfect liquid; Part II. A viscous liquid. *Proc. Irish Acad.* **27**, 9–138.

Orszag, S. A., 1971, Accurate solution of the Orr–Sommerfeld stability equation. *J. Fluid Mech.* **50**, 689–703.

Pierson, F. W., and Whitaker, S., 1977, Some theoretical and experimental observations of the wave structure of falling liquid films. *Ind. Eng. Chem. Fundam.* **16**, 401–8.

Pozrikidis, C., and Higdon, J. J. L., 1987, Instability of compound vortex layers and wakes. *Phys. Fluids* **30**, 2965–75.

Pullin, D. I., 1981, The nonlinear behavior of a constant vorticity layer at a wall. *J. Fluid Mech.* **108**, 401–21.

Rayleigh, Lord, 1879, On the instability of jets. *Proc. London Math. Soc.* **10**, 4–13.

Rayleigh, Lord, 1880, On the stability or instability of certain fluid motions. *Proc. London Math. Soc.* **11**, 57–70.

Rayleigh, Lord, 1883, Investigation of the character of the equilibrium of an incompressible heavy fluid of variable density. *Proc. London Math. Soc.* **14**, 170–77.

Rayleigh, Lord, 1894, *The Theory of Sound.* Macmillan (also Dover 1945).

Rayleigh, Lord, 1916, On the dynamics of revolving fluids. *Proc. Roy. Soc. London* A **93**, 148–54.

Saffman, P. G., and Taylor, G. I., 1958, The penetration of a fluid into a porous medium or Hele–Shaw cell containing a more viscous liquid. *Proc. Roy. Soc. London* A **245**, 312–29.

Saric, W. S., 1994, Görtler vortices, *Annu. Rev. Fluid Mech.* **26**, 379–409.

Shen, S. F., 1954, Calculated amplified oscillations in plane Poiseuille and Blasius flows. *J. Aero. Sc.* **21**, 62–64.

Sommerfeld, A., 1908, Ein Beitrag zur hydrodynamischen Erklärung der turbulenten Flüssigkeitsbewegung. *Proc. 4th Intern. Congress Math. Rome,* 116–24.

Squire, H. B., 1933, On the stability of three-dimensional disturbances of viscous flow between parallel walls. *Proc. Roy. Soc. London* A **142**, 621–28.

Taylor, G. I., 1923, Stability of a viscous liquid contained between two rotating cylinders. *Phil. Trans. Roy. Soc. London* A **223**, 289–343.

Taylor, G. I., 1950, The instability of liquid surfaces when accelerated in a direction perpendicular to their planes. *Proc. Roy. Soc. London* A **201**, 192–96.

Thomas, L. H., 1953, The stability of plane Poiseuille flow. *Phys. Rev.* **91**, 780–83.

Tryggvason, G., 1988, Numerical simulations of the Rayleigh–Taylor instability. *J. Comp. Phys.* **75**, 253–83.

Wallace, D., and Redekopp, L. G., 1992, Linear instability characteristics of wake-shear layers. *Phys. Fluids* A **4**, 189–91.

Yih, C.-S., 1963, Stability of liquid flow down an inclined plane. *Phys. Fluids* **6**, 321–34.

Yih, C.-S., 1967, Instability due to viscosity stratification. *J. Fluid Mech.* **27**, 337–52.

Yih, C.-S., 1979, *Fluid Mechanics.* West River Press.

Boundary-Integral
Methods for Potential Flow

Potential flow arises in a variety of natural contexts and engineering applications; a familiar example is high-Reynolds-number flow past a streamlined body discussed in Chapters 7 and 8. Since the vorticity is confined within thin boundary layers and narrow wakes, the main body of the flow is virtually irrotational and may thus be described in terms of a velocity potential ϕ, setting $\mathbf{u} = \nabla\phi$. The continuity equation requires ϕ to be a harmonic function, $\nabla^2\phi = 0$, and this reduces the computation of the flow to solving Laplace's equation subject to the no-penetration condition over the impermeable boundaries. A related example concerns the flow due to the propagation of gravity waves on a free surface.

Another example of potential flow from a different physical context is provided by the flow of a viscous fluid through the Hele–Shaw cell discussed in Section 6.3. Equation (6.3.2) shows that the average velocity of the fluid across the width of the channel $\mathbf{U}$ is proportional to the gradient of the modified pressure P; hence, P plays the role of a velocity potential with $\mathbf{U} = \nabla\phi$, where $\phi = -h^2 P/(12\mu)$. Conservation of mass requires that the modified pressure be a harmonic function, $\nabla^2 P = 0$, and this reduces the computation of the flow to solving Laplace's equation subject to Dirichlet, Neumann, or mixed-type boundary conditions, as discussed in the examples of Section 6.3.

A related application concerns the flow of a viscous fluid through an isotropic porous medium such as a ground rock. According to *Darcy's law,* the macroscopic velocity of the fluid $\mathbf{U}$, defined as the average velocity of the fluid over a control volume that is small compared to the global dimensions of the flow but large compared to the size of the grains or fibers, is related to the corresponding macroscopic modified pressure P by means of the equation $\mathbf{U} = -(\kappa/\mu)\nabla P$, where κ is a physical constant called the *permeability* of the fluid through the porous medium (Bear, 1972). It is evident that the macroscopic pressure plays the role of a potential function with $\phi = -(\kappa/\mu)P$. The continuity equation requires that P be a harmonic function, $\nabla^2 P = 0$, and this reduces the computation of the flow to solving Laplace's equation subject to Neumann boundary conditions over the impermeable boundaries, or Dirichlet boundary conditions over the boundaries that are exposed to the atmospheric pressure.

Motivated by the pervasiveness of Laplace's equation in the various branches of fluid mechanics, we devote this chapter to discussing a powerful class of numerical methods for computing potential flow in domains of arbitrary complexity, known under the aliases *boundary integral, boundary element, boundary-integral-equation,* and *panel methods.* The mathematical formulation and the basic numerical implementation are discussed in Sections 10.1 and 10.2, and subsequent sections are devoted to discussing extensions and advanced topics.

10.1 | THE BOUNDARY-INTEGRAL EQUATION

In Section 2.3 we developed an integral representation of a harmonic function ϕ in terms of two boundary integrals representing boundary distributions of point sources and point-source dipoles;

the densities of these distributions are proportional, respectively, to the normal derivative and to the boundary values of the harmonic function. For a point $\mathbf{x}_0$ located within a certain control volume that is bounded by a collection of surfaces denoted as D, as shown in Figure 2.2.1(a), we derived Eq. (2.3.8), repeated here for ready reference

$$\phi(\mathbf{x}_0) = -\int_D G(\mathbf{x}_0, \mathbf{x})\, \nabla\phi(\mathbf{x}) \cdot \mathbf{n}(\mathbf{x})\, dS(\mathbf{x}) + \int_D \nabla G(\mathbf{x}_0, \mathbf{x}) \cdot \mathbf{n}(\mathbf{x})\, \phi(\mathbf{x})\, dS(\mathbf{x}) \tag{10.1.1}$$

where $G(\mathbf{x}_0, \mathbf{x}) = G(\mathbf{x}, \mathbf{x}_0)$ is a Green's function of Laplace's equation, and the unit normal vector $\mathbf{n}$ points into the control volume. The first integral on the right-hand side of Eq. (10.1.1), representing a distribution of point sources, is called the *single-layer potential,* whereas the second integral, representing a distribution of point-source dipoles, is called the *double-layer potential.*

Equation (10.1.1) is also valid at a point $\mathbf{x}_0$ that is located within a partially or totally infinite domain of flow that is bounded by an interior boundary D, as shown in Figure 2.2.1(b), provided that the velocity vanishes, and therefore ϕ tends to a constant value at infinity.

The representation (10.1.1) allows us to compute the value of ϕ at any point $\mathbf{x}_0$ within a selected control volume from a knowledge of ϕ and its normal derivative $\nabla\phi \cdot \mathbf{n}$ over the boundaries of the control volume. In practice, however, physical considerations provide us with just one scalar boundary condition for the distribution of either ϕ or $\nabla\phi \cdot \mathbf{n}$ or their linear combination, but not for both; thus it might appear that Eq. (10.1.1) is of limited practical value. This pessimistic conclusion, however, is revoked by the observation that the unknown boundary distribution may be computed by taking the limit as the point $\mathbf{x}_0$ approaches D, thereby reducing Eq. (10.1.1) to a Fredholm integral equation for the unknown distribution over D.

Behavior of the Hydrodynamic Potentials

Before taking the limit of Eq. (10.1.1) as the point $\mathbf{x}_0$ approaches D, we must examine the behavior of the single-layer and double-layer potentials. In order to avoid mathematical complications, hereafter we shall assume that each surface represented by D has a continuously varying normal vector, which means that it is a *Lyapunov surface.*

Examining the singularity of the integrands in Eq. (10.1.1), we find that, as the point $\mathbf{x}_0$ approaches and then crosses a surface represented by D, the single-layer potential varies in a continuous manner, but the double-layer potential shows a discontinuous behavior. The latter becomes evident by writing

$$\lim_{\mathbf{x}_0 \to D^\pm} \int_D \nabla G(\mathbf{x}_0, \mathbf{x}) \cdot \mathbf{n}(\mathbf{x})\, \phi(\mathbf{x})\, dS(\mathbf{x})$$

$$= \lim_{\mathbf{x}_0 \to D^\pm} \int_D \nabla G(\mathbf{x}_0, \mathbf{x}) \cdot \mathbf{n}(\mathbf{x})\, [\phi(\mathbf{x}) - \phi(\mathbf{x}_0)]\, dS(\mathbf{x})$$

$$+ \phi(\mathbf{x}_0) \lim_{\mathbf{x}_0 \to D^\pm} \int_D \nabla G(\mathbf{x}_0, \mathbf{x}) \cdot \mathbf{n}(\mathbf{x})\, dS(\mathbf{x}) \tag{10.1.2}$$

where the superscripts $+$ and $-$ designate that $\mathbf{x}_0$ approaches D from within the control volume, which is indicated by the direction of the normal vector, or from the external side. Examining the singularity of the Green's function shows that the first integral on the right-hand side of Eq. (10.1.2), which we denote by $J(\mathbf{x}_0)$, varies in a continuous fashion. To assess the behavior of the second integral, we recall that D represents the collection of all boundaries of an enclosed domain of flow, as depicted in Figure 2.2.1(a), and use the identity (2.2.13) to find

$$\lim_{\mathbf{x}_0 \to D^\pm} \int_D \nabla G(\mathbf{x}_0, \mathbf{x}) \cdot \mathbf{n}(\mathbf{x})\, \phi(\mathbf{x})\, dS(\mathbf{x}) = J(\mathbf{x}_0) + \alpha^\pm \phi(\mathbf{x}_0) \tag{10.1.3}$$

where $\alpha^+ = 1$ and $\alpha^- = 0$. This demonstrates that the double-layer integral undergoes a discontinuity of magnitude $\phi(\mathbf{x}_0)$ across D.

When the point $\mathbf{x}_0$ is located precisely on D, the double-layer integral is an improper but convergent integral, called a *principal-value integral,* denoted by the initials PV. Subtracting off the singularity, as we have done in Eq. (10.1.2), and using the middle part of identity (2.2.13), we find

$$\int_D^{\text{PV}} \nabla G(\mathbf{x}_0, \mathbf{x}) \cdot \mathbf{n}(\mathbf{x}) \, \phi(\mathbf{x}) \, dS(\mathbf{x}) = J(\mathbf{x}_0) + \tfrac{1}{2}\phi(\mathbf{x}_0) \tag{10.1.4}$$

Combining Eqs. (10.1.3) and (10.1.4) yields a relationship between the limits of the double-layer potential and its principal value

$$\lim_{\mathbf{x}_0 \to D^{\pm}} \int_D \nabla G(\mathbf{x}_0, \mathbf{x}) \cdot \mathbf{n}(\mathbf{x}) \, \phi(\mathbf{x}) \, dS(\mathbf{x})$$
$$= \int_D^{\text{PV}} \nabla G(\mathbf{x}_0, \mathbf{x}) \cdot \mathbf{n}(\mathbf{x}) \, \phi(\mathbf{x}) \, dS(\mathbf{x}) \pm \tfrac{1}{2}\phi(\mathbf{x}_0) \tag{10.1.5}$$

The significance of Eq. (10.1.5) lies in the fact that the principal value of the double-layer potential is much easier to compute than the limit of the double-layer potential as the point $\mathbf{x}_0$ approaches the boundary from either side.

Boundary-integral equation

Having assessed the behavior of the single-layer and double-layer potentials, we take the limit of Eq. (10.1.1) as the point $\mathbf{x}_0$ approaches D and use Eq. (10.1.5) to derive the boundary-integral equation

$$\phi(\mathbf{x}_0) = -2 \int_D G(\mathbf{x}_0, \mathbf{x}) \nabla \phi(\mathbf{x}) \cdot \mathbf{n}(\mathbf{x}) \, dS(\mathbf{x})$$
$$+ 2 \int_D^{\text{PV}} \nabla G(\mathbf{x}_0, \mathbf{x}) \cdot \mathbf{n}(\mathbf{x}) \, \phi(\mathbf{x}) \, dS(\mathbf{x}) \tag{10.1.6}$$

It will be noted that Eq. (10.1.6) is identical to Eq. (10.1.1) except that the right-hand side is multiplied by a factor of two, and the double-layer integral assumes its principal value. Because of the symmetry property (2.2.18), the arguments of the Green's function in both the single-layer and double-layer potential may be switched.

There is an alternative method of deriving Eq. (10.1.6) that proceeds by assuming that the point $\mathbf{x}_0$ lies on a smooth surface D, and then repeating the analysis of Section 2.3, excluding from the control volume the volume of a hemisphere, instead of a sphere, centered at the point $\mathbf{x}_0$. Using this method, we find that Eq. (10.1.6) is also valid at a point $\mathbf{x}_0$ that is located at a boundary corner, provided that $\phi(\mathbf{x}_0)$ on the left-hand side is multiplied by the factor $\alpha/2\pi$, where α is the solid angle subtended by the corner on the side of the control volume (Problem 10.1.1). For a smooth boundary, $\alpha = 2\pi$.

Infinite Flow

Equations (10.1.5) and (10.1.6) are also valid for an infinite flow that is bounded by the interior closed boundary D as illustrated in Figure 2.2.1(b) provided that the potential function vanishes at infinity. In this case, we use Eq. (2.2.13), identifying V_c with the volume enclosed by D, note that the normal vector in Eq. (2.2.13) points into V_c, whereas that in Eq. (10.1.2) is directed into the flow and thus outward from D, and arrive at Eq. (10.1.3) with $\alpha^+ = 0$ and $\alpha^- = -1$. Proceeding as above, we derive Eqs. (10.1.5) and (10.1.6).

Integral Equations of the Second Kind

Specifying the distribution of the normal derivative $\nabla \phi \cdot \mathbf{n}$ over D reduces Eq. (10.1.6) to a *Fredholm integral equation of the second kind* for the boundary distribution of ϕ

$$\phi(\mathbf{x}_0) = 2 \int_D^{\text{PV}} \nabla G(\mathbf{x}_0, \mathbf{x}) \cdot \mathbf{n}(\mathbf{x}) \, \phi(\mathbf{x}) \, dS(\mathbf{x}) - 2I^S(\mathbf{x}_0) \tag{10.1.7}$$

is the Green's function $G(\mathbf{x}_0, \mathbf{x})$ and taking into consideration the singular behavior shown in Eq. (2.2.10), we find that when the boundary D is a smooth surface with a continuously varying normal vector, Eq. (10.1.12) is a weakly singular integral equation. The existence and further properties of the solution will be discussed in Section 10.4.

Axisymmetric Flow

For axisymmetric flow, the single-layer and double-layer surface integrals may be reduced to line integrals over the contour of the boundaries in an azimuthal plane. This simplification facilitates the solution of the integral equations as well as the computation of the potentials at points within the domain of flow.

To implement this simplification, we introduce cylindrical polar coordinates as illustrated in Figure 10.1.1, note that ϕ is independent of the azimuthal angle φ, and write $\mathbf{u} = (u_x, u_\sigma \cos \varphi, u_\sigma \sin \varphi)$, $\mathbf{n} = (n_x, n_\sigma \cos \varphi, n_\sigma \sin \varphi)$, $dS = \sigma \, d\varphi \, dl$, where l is the arc length along the trace of the boundary in the xy plane C, and $\mathbf{x} = (x, \sigma \cos \varphi, \sigma \sin \varphi)$ over the boundary. We then apply Eq. (10.1.6) with the free-space Green's function $G(\mathbf{x}_0, \mathbf{x}) = 1/(4\pi r)$, where $r = |\mathbf{x} - \mathbf{x}_0|$, at a point $\mathbf{x}_0$ in the xy plane, and make the aforementioned substitutions to obtain

$$\phi(\mathbf{x}_0) = -\frac{1}{2\pi} \int_C (\mathbf{u} \cdot \mathbf{n})(l)\, \sigma(l)\, K^S(\mathbf{x}_0, l)\, dl(\mathbf{x})$$

$$+ \frac{1}{2\pi} \int_C^{PV} \phi(l)\, \sigma(l)\, K^D(\mathbf{x}_0, l)\, dl \tag{10.1.13}$$

where we have introduced the *axisymmetric kernels* of the single-layer and double-layer potential

$$K^S(\mathbf{x}_0, l) \equiv \int_0^{2\pi} \frac{d\varphi(\mathbf{x})}{|\mathbf{x} - \mathbf{x}_0|} = I_{10}(\hat{x}, \sigma, \sigma_0) \tag{10.1.14}$$

$$K^D(\mathbf{x}_0, l) \equiv \int_0^{2\pi} \frac{(\mathbf{x}_0 - \mathbf{x}) \cdot \mathbf{n}(\mathbf{x})}{|\mathbf{x} - \mathbf{x}_0|^3}\, d\varphi(\mathbf{x})$$

$$= (\hat{x} n_x + \sigma_0 n_\sigma) I_{30}(\hat{x}, \sigma, \sigma_0) - \sigma n_\sigma I_{31}(\hat{x}, \sigma, \sigma_0) \tag{10.1.15}$$

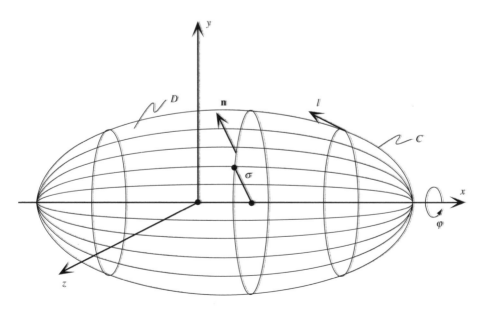

Figure 10.1.1   The boundary integrals for axisymmetric flow may be reduced to line integrals over the trace of the boundaries in an azimuthal plane.

where $\hat{x} = x_0 - x$. The functions I_{nm} are defined in Eq. (2.9.6) and are computed with the help of Eqs. (2.9.8) and (2.9.9).

Two-Dimensional Flow

The results obtained previously in this section for three-dimensional flow may be modified in a straightforward manner to correspond to two-dimensional flow. The counterpart of the boundary-integral equation (10.1.6) is

$$\phi(\mathbf{x}_0) = -2 \int_C G(\mathbf{x}_0, \mathbf{x}) \nabla\phi(\mathbf{x}) \cdot \mathbf{n}(\mathbf{x}) \, dl(\mathbf{x})$$

$$+ 2 \int_C^{PV} \nabla G(\mathbf{x}_0, \mathbf{x}) \cdot \mathbf{n}(\mathbf{x}) \, \phi(\mathbf{x}) \, dl(\mathbf{x}) \qquad (10.1.16)$$

where $G(\mathbf{x}_0, \mathbf{x})$ is a two-dimensional Green's function, C is the boundary of a selected control area of flow, l is the arc length along C, and $\mathbf{n}$ is the unit vector pointing into the selected control area. In the remainder of this subsection, we shall illustrate the application of Eq. (10.1.16) and derive similar equations applicable to several types of internal and external flow.

Flow past a rigid body with circulation around it

Considering a flow past a stationary two-dimensional body with finite circulation around it, we begin with Eq. (7.6.3), use the free-space Green's function $G(\mathbf{x}_0, \mathbf{x}) = -(1/2\pi) \ln r$, where $r = |\mathbf{x} - \mathbf{x}_0|$, and set $\mathbf{U} = \mathbf{0}$, to obtain an integral equation of the second kind for the single-valued component of the potential Φ

$$\Phi(\mathbf{x}_0) = -\frac{1}{\pi} \int_C^{PV} \frac{\mathbf{x} - \mathbf{x}_0}{r^2} \cdot \mathbf{n}(\mathbf{x}) \Phi(\mathbf{x}) \, dl(\mathbf{x}) + 2\phi^\infty(\mathbf{x}_0)$$

$$- \frac{1}{\pi} \int_C \ln r \, \mathbf{v}(\mathbf{x}) \cdot \mathbf{n}(\mathbf{x}) \, dl(\mathbf{x}) \qquad (10.1.17)$$

where $\mathbf{v}$ is the velocity due to a point vortex whose strength is equal to the circulation of the flow around the body, located inside the body. In Section 10.6 we shall show that Eq. (10.1.17) has a unique solution.

Fluid sloshing in a tank

Consider now the flow due to the sloshing of a liquid within a two-dimensional tank depicted in Figure 10.1.2(a). Applying Eq. (10.1.16) with the free-space Green's function, and requiring the no-penetration condition over the walls, we obtain the integral equation

$$\phi(\mathbf{x}_0) = -\frac{1}{\pi} \int_{W, S}^{PV} \frac{\mathbf{x} - \mathbf{x}_0}{r^2} \cdot \mathbf{n}(\mathbf{x}) \phi(\mathbf{x}) \, dl(\mathbf{x}) + \frac{1}{\pi} \int_S \ln r \, \mathbf{u}(\mathbf{x}) \cdot \mathbf{n}(\mathbf{x}) \, dl(\mathbf{x}) \qquad (10.1.18)$$

where W represents the side and bottom walls and S represents the free surface. An efficient computational procedure for describing the evolution of the free surface from a certain initial state proceeds according to the following steps (see, for example, Nakayama, 1990):

1. Trace the free surface with a collection of marker points and assign to them initial values for ϕ.
2. Solve the integral equation (10.1.18) for the normal component of the velocity over the free surface and for the potential over the walls. Note that, in the present application, Eq. (10.1.18) is an integral equation of mixed kind.
3. Differentiate the potential along the free surface to compute the tangential component of the velocity of the fluid, and move the marker points with the velocity of the fluid over the period of a small time interval. Note that, in this case, the marker points behave like point particles. To prevent clustering, it is preferable to move the marker points with the component of the

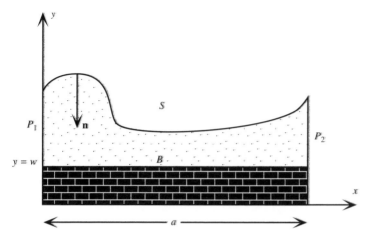

Figure 10.1.3   Flow due to the propagation of a periodic free-surface gravity wave in the presence of a flat bottom.

an ambiguity in the mathematical formulation associated with the fact that the fluid may translate in the x direction as a rigid body with an arbitrary time-dependent velocity; to remove this degree of freedom, we stipulate that the potential ϕ respects the periodicity of the free surface, that is, $\phi(x + a, y) = \phi(x, y)$.

To simplify the boundary integral formulation, it is advantageous to use a *Green's function of the second kind* that is periodic in the x direction with period equal to a, and whose normal derivative vanishes at the bottom. Working as in the case of fluid sloshing in a tank, we find that this Green's function is given by the right-hand side of Eq. (10.1.21) except that the summation is from one to two, the wave number is given by $k = 2\pi/a$, $\mathbf{x}_1 = (x_0, y_0)$, and $\mathbf{x}_2 = (x_0, -y_0 + 2w)$.

Next we select as a control area one period of the flow enclosed by the contours S, P_1, P_2, and B, as shown in Figure 10.1.3, and derive Eq. (10.1.16). The double-layer integral over B vanishes because the normal derivative of the Green's function vanishes when $\mathbf{x}$ is located on B, and the single-layer integral over B vanishes because of the no-penetration condition at the bottom. Furthermore, the integrals over the vertical segments P_1 and P_2 cancel each other because ϕ has been assumed to be a periodic function and the corresponding normal vectors point into opposite directions. These simplifications allow us to identify the contour C with one period of the free surface S. The evolution of the flow may be computed using the numerical procedure discussed previously for the problem of fluid sloshing (see also Problem 10.1.4).

PROBLEMS

10.1.1   General form of the boundary integral equation.   Show that the boundary integral representation and boundary integral equation may be cast into the general form

$$c\phi(\mathbf{x}_0) = -\int_D G(\mathbf{x}_0, \mathbf{x}) \, \nabla\phi(\mathbf{x}) \cdot \mathbf{n}(\mathbf{x}) \, dS(\mathbf{x})$$

$$+ \int_D \nabla G(\mathbf{x}_0, \mathbf{x}) \cdot \mathbf{n}(\mathbf{x}) \, \phi(\mathbf{x}) \, dS(\mathbf{x}) \tag{10.1.22}$$

where the constant c is equal to *unity* when the point $\mathbf{x}_0$ is located in the interior of a selected control volume, *zero* when it is located outside a selected control volume, *one-half* when it lies on a smooth section of D, and $\alpha/4\pi$ when it is located at a corner; α is the solid angle subtended by the corner facing the control volume. In all cases, the normal vector $\mathbf{n}$ is directed into the control volume.

10.1.2 Flow past a sphere. Consider uniform potential flow with velocity $\mathbf{V}$ past a stationary sphere of radius a. The exact solution for the velocity potential is known to be

$$\phi(\mathbf{x}) = \mathbf{V} \cdot \mathbf{x} - \tfrac{1}{2}a^3\mathbf{V} \cdot \nabla \frac{1}{|\mathbf{x}|} = \mathbf{V} \cdot \mathbf{x}\left(1 + \tfrac{1}{2}\frac{a^3}{|\mathbf{x}|^3}\right) \tag{10.1.23}$$

where the origin has been set at the center of the sphere [see Eqs. (7.4.5)]. (a) Substitute the boundary distribution $\phi = \tfrac{3}{2}\mathbf{V} \cdot \mathbf{x}$ into both sides of Eq. (10.1.10) with the free-space Green's function, and simplify to obtain the identity

$$\mathbf{x}_0 = \frac{3}{2\pi}\int_{\text{Sphere}}^{\text{PV}} \mathbf{x}\frac{1}{r^3}(\mathbf{x} - \mathbf{x}_0) \cdot \mathbf{n}(\mathbf{x})\, dS(\mathbf{x}) \tag{10.1.24}$$

(b) To confirm the validity of Eq. (10.1.24), write

$$\int_{\text{Sphere}}^{\text{PV}} \mathbf{x}\frac{1}{r^3}(\mathbf{x} - \mathbf{x}_0) \cdot \mathbf{n}(\mathbf{x})\, dS(\mathbf{x})$$

$$= \int_{\text{Sphere}} (\mathbf{x} - \mathbf{x}_0)\frac{1}{r^3}(\mathbf{x} - \mathbf{x}_0) \cdot \mathbf{n}(\mathbf{x})\, dS(\mathbf{x})$$

$$- \mathbf{x}_0\int_{\text{Sphere}}^{\text{PV}} \mathbf{n}(\mathbf{x}) \cdot \nabla\frac{1}{r}\, dS(\mathbf{x}) \tag{10.1.25}$$

Considering the first integral on the right-hand side of Eq. (10.1.25), introduce a Cartesian coordinate system with its origin at the center of the sphere and the x axis passing through the point $\mathbf{x}_0$, and observe that, due to symmetry, the y and z components of the integral vanish. Carrying out the integration in terms of the polar angle θ, show that the x component of the integral is equal to $(-4\pi/3)\mathbf{x}_0$. Use Eq. (2.2.13) to show that the second integral on the right-hand side of Eq. (10.1.25) is equal to -2π. Combining these results, demonstrate the validity of Eq. (10.1.24).

10.1.3 Fluid sloshing in a tank. Consider fluid sloshing within a two-dimensional tank that undergoes rotational oscillations around the z axis, and discuss the necessary modifications of the computational procedure discussed in the text. (*Hint:* Consider Bernoulli's equation for a flow with constant vorticity.)

10.1.4 Gravity waves in a semi-infinite fluid. Develop a boundary integral formulation for the flow due to the propagation of two-dimensional periodic gravity waves at the free surface of a fluid of infinite depth.

10.2 | BOUNDARY-ELEMENT METHODS

Analytical solutions to the boundary-integral equations discussed in the preceding section are feasible only for a limited number of boundary geometries and types of flow. To compute the solution under general circumstances, we must resort to approximate, asymptotic, and numerical methods. Reviewing the available strategies for solving the Fredholm integral equations of mathematical physics, we find a variety of approaches with varying degrees of accuracy and sophistication. Discussions of general and specialized methods are described in the monographs and review articles by Atkinson (1976, 1980, 1990), Baker (1977), Delves and Mohamed (1985), Hess (1990), and Katz and Plotkin (1991).

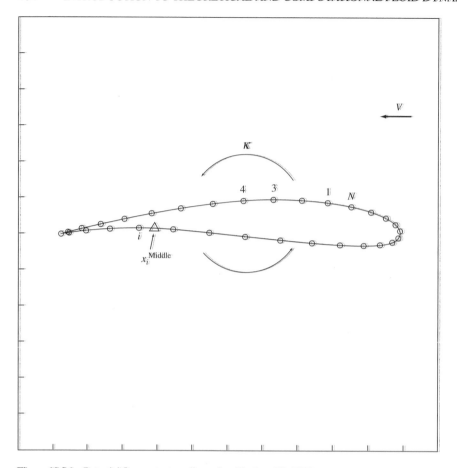

Figure 10.2.1   Potential flow past a two-dimensional Joukowski airfoil.

implied over i, we note that, when the elements are straight segments and the integration point $\mathbf{x}$ is located on the ith element, $\mathbf{x} - \mathbf{x}_i^{\mathrm{Middle}}$ is perpendicular to the normal vector, and the corresponding integrand in Eq. (10.2.4) vanishes; this yields $A_i(\mathbf{x}_i^{\mathrm{Middle}}) = 0$. When the elements are curved, we apply Taylor series expansions with respect to the midpoints and find that $A_i(\mathbf{x}_i^{\mathrm{Middle}}) = 0$ to a leading-order approximation. These simplifications justify the use of the principal value of the double-layer integral, instead of the limit of the double-layer integral as the point where the velocity is evaluated approaches the boundaries.

When $i = j$, the integral $B_i(\mathbf{x}_i^{\mathrm{Middle}})$ presents a logarithmic singularity, which, however, can be integrated analytically over a straight segment to yield

$$B_i(\mathbf{x}_i^{\mathrm{Middle}}) = -\frac{1}{\pi}|\mathbf{x}_i - \mathbf{x}_{i+1}|\left(\ln\frac{|\mathbf{x}_i - \mathbf{x}_{i+1}|}{2} - 1\right) \tag{10.2.7}$$

A similar analytical integration is possible when the boundary elements are circular arcs (Problem 10.2.1).

Having solved the linear system of Equations (10.2.5) or (10.2.6), we differentiate the boundary distribution of the velocity potential with respect to arc length using a standard numerical method to obtain the boundary distribution of the tangential velocity, and then apply Bernoulli's equation to compute the boundary distribution of pressure and pressure drag coefficient. The drag force and lift force exerted on the airfoil may be computed from the pressure by integrating along its contour, as discussed in Section 7.1. The flawless computation will show that the drag force vanishes, and the lift force has the Kutta–Joukowski value given in Eq. (7.6.17).

PROBLEM

10.2.1 **Boundary-element integrals.** (a) Derive Eq. (10.2.7), and examine whether it is possible to compute $B_j(\mathbf{x}_i^{\text{Middle}})$ analytically over straight segments when $i \neq j$. (b) Derive the counterpart of Eq. (10.2.7) when the boundary elements assume the shape of circular arcs.

Computer Problems

10.2.2 **Flow past a two-dimensional airfoil.** Compute the distribution of pressure, drag force, and lift force on a Joukowski airfoil of your choice with a certain degree of circulation of your choice around it (Section 7.10). Compare your results for the lift and the drag force with the theoretical predictions.

10.2.3 **Flow in a coating die.** Compute the distribution of velocity at the exit of the coating die shown in Figure 6.3.2 for $h = 3$ mm, $L = 20$ cm, $W = 20$ cm, sloping angle $\alpha = 20°$, and constant tube radius $R = 6$ mm. The pressure at the inlet point is 3 atm, and the working fluid is glycerine.

10.2.4 **Unidirectional flow in a pipe.** Compute the distribution of shear stress around the boundary of a cylindrical pipe with an elliptical cross section of your choice in unidirectional pressure-driven flow, and compare your results with the exact solution discussed in Section 5.1.

10.3 | GENERALIZED BOUNDARY-INTEGRAL REPRESENTATIONS

In Section 7.2 we developed a boundary-integral representation for the harmonic potential of an irrotational flow past a stationary or translating rigid body involving the double-layer potential alone, shown in Eq. (7.2.3). Two important advantages of this simplified representation are conceptual simplicity and computational ease in solving the associated integral equation. Motivated by these benefits, we set out to eliminate one of the two hydrodynamic potentials from the general boundary-integral representation and associated boundary-integral equation, thereby obtaining *generalized boundary-integral representations* and associated boundary-integral equations.

Exterior Flow

Consider first an exterior flow that vanishes at infinity and is bounded by the interior boundary D, and introduce the harmonic potential ϕ' in the interior of D so that $\phi' = \phi$ over D. Applying the reciprocal relation (2.3.2) for ϕ' at a point $\mathbf{x}_0$ that is located in the exterior of D, and subtracting Eq. (2.3.2) from Eq. (2.3.8), we obtain a generalized representation in terms of a single-layer potential alone

$$\phi(\mathbf{x}_0) = -\int_D G(\mathbf{x}_0, \mathbf{x})\, q(\mathbf{x})\, dS(\mathbf{x}) \tag{10.3.1}$$

where the density of the distribution q is given by

$$q = (\nabla\phi - \nabla\phi') \cdot \mathbf{n} \tag{10.3.2}$$

and $\mathbf{n}$ is the unit normal vector pointing into the exterior of D.

S_1 in place of D, we find that the volumetric flow rate Q_1 through S_1 is given by

$$Q_1 = \int_{S_1} \nabla_0 \phi(\mathbf{x}_0) \cdot \mathbf{N}(\mathbf{x}_0) \, dS(\mathbf{x}_0)$$
$$= \int_D \left(\int_{S_1} \nabla_0 G(\mathbf{x}_0, \mathbf{x}) \cdot \mathbf{N}(\mathbf{x}_0) \, dS(\mathbf{x}_0) \right) q(\mathbf{x}) \, dS(\mathbf{x})$$
$$= - \int_D q(\mathbf{x}) \, dS(\mathbf{x}) \tag{10.4.3}$$

where the gradient ∇_0 involves derivatives with respect to $\mathbf{x}_0$. We thus find that Q_1 vanishes only when the total strength of the distribution of the Green's function, defined as the last integral in Eq. (10.4.3), is equal to zero.

Working in a similar manner, we find that the volumetric flow rate across a closed fluid surface that does not enclose any boundaries, such as the surface S_2 illustrated in Figure 10.4.1, is equal to zero (see also Problem 10.4.1).

Derivatives of the Single-Layer Potential

The derivative of the single-layer potential at the location of the point $\mathbf{x}_0$ and in the direction of the unit vector $\mathbf{n}$ is given by

$$\nabla_0 \phi(\mathbf{x}_0) \cdot \mathbf{n}(\mathbf{x}_0) = \mathbf{n}(\mathbf{x}_0) \cdot \int_D \nabla_0 G(\mathbf{x}_0, \mathbf{x}) q(\mathbf{x}) \, dS(\mathbf{x}) \tag{10.4.4}$$

By definition, when G is a Green's function of the second kind, the normal derivative of ϕ vanishes over the boundary S_B.

To investigate the behavior of the normal derivative across the distribution domain D, we take the limit as the point $\mathbf{x}_0$ approaches D, introduce the free-space Green's function G^{FS}, note that $\nabla_0 G^{FS} = -\nabla G^{FS}$, and write

$$\nabla_0 \phi(\mathbf{x}_0) \cdot \mathbf{N}(\mathbf{x}_0) = \int_D [\mathbf{N}(\mathbf{x}_0) \cdot \nabla_0 G(\mathbf{x}_0, \mathbf{x}) q(\mathbf{x}) - \nabla_0 G^{FS}(\mathbf{x}_0, \mathbf{x}) \cdot \mathbf{N}(\mathbf{x}) q(\mathbf{x}_0)] \, dS(\mathbf{x})$$
$$- q(\mathbf{x}_0) \int_D \nabla G^{FS}(\mathbf{x}_0, \mathbf{x}) \cdot \mathbf{N}(\mathbf{x}) \, dS(\mathbf{x}) \tag{10.4.5}$$

Examining the singularity of the first integrand on the right-hand side of Eq. (10.4.5), we find that its integral, which we denote by $J(\mathbf{x}_0)$, is continuous across D. Considering the second integral and using Eq. (10.4.2), we obtain

$$\nabla_0 \phi^+(\mathbf{x}_0) \cdot \mathbf{N}(\mathbf{x}_0) = J(\mathbf{x}_0)$$
$$\nabla_0 \phi^-(\mathbf{x}_0) \cdot \mathbf{N}(\mathbf{x}_0) = J(\mathbf{x}_0) + q(\mathbf{x}_0) \tag{10.4.6}$$

where the plus sign signifies the exterior side of D, indicated by the direction of $\mathbf{N}$, and the minus sign signifies the interior side. Thus the normal derivative of the single-layer potential undergoes a discontinuity of magnitude $-q$ across the distribution domain D

$$\nabla_0 \phi^+(\mathbf{x}_0) \cdot \mathbf{N}(\mathbf{x}_0) - \nabla_0 \phi^-(\mathbf{x}_0) \cdot \mathbf{N}(\mathbf{x}_0) = q(\mathbf{x}_0) \tag{10.4.7}$$

To this end, we define the *principal value of the normal derivative of the single-layer potential* as the value of the right-hand side of Eq. (10.4.5) computed when the point $\mathbf{x}_0$ is located precisely on D. Since D has been assumed to be a smooth surface, we can use Eq. (10.4.2) once again to obtain

$$[\nabla_0 \phi(\mathbf{x}_0) \cdot \mathbf{N}(\mathbf{x}_0)]^{PV} = J(\mathbf{x}_0) + \tfrac{1}{2} q(\mathbf{x}_0) \tag{10.4.8}$$

Combining the preceding three equations yields

$$\nabla_0 \phi^\pm(\mathbf{x}_0) \cdot \mathbf{N}(\mathbf{x}_0) = [\nabla_0 \phi(\mathbf{x}_0) \cdot \mathbf{N}(\mathbf{x}_0)]^{PV} \mp \tfrac{1}{2} q(\mathbf{x}_0) \tag{10.4.9}$$

Since the single-layer potential is a continuous function, its tangential derivatives over the distribution domain D must vary in a continuous fashion. Combining this observation with Eq. (10.4.9) allows us to express the gradient of the single-layer potential on either side of D as

$$\nabla_0 \phi^{\pm}(\mathbf{x}_0) = [\nabla_0 \phi(\mathbf{x}_0)]^{\mathrm{PV}} \mp \tfrac{1}{2} q(\mathbf{x}_0) \mathbf{N}(\mathbf{x}_0) \qquad (10.4.10)$$

We have investigated the properties of the single-layer potential in sufficient detail. Next we proceed to derive integral equations from the single-layer representation (10.4.1).

Dirichlet Problem

First we consider the Dirichlet problem where the distribution of ϕ over D is specified, and the density distribution q is sought. Applying Eq. (10.4.1) at a point $\mathbf{x}_0$ on D yields a Fredholm integral equation of the first kind for q. Numerical evidence suggests that, in general, the solution of this equation is unique, and the computational problem is well posed in the sense that it is amenable to standard numerical methods, including the panel methods discussed in Section 10.2.

Neumann Problem

Next we consider the Neumann problem, in which the normal derivative of ϕ is specified over the external or internal side of D, and the density distribution q is sought. In either case we use Eq. (10.4.9) and obtain a Fredholm integral equation of the second kind for q

$$q(\mathbf{x}_0) = \pm 2 \int_D^{\mathrm{PV}} \mathbf{N}(\mathbf{x}_0) \cdot \nabla_0 G(\mathbf{x}_0, \mathbf{x}) \, q(\mathbf{x}) \, dS(\mathbf{x})$$
$$\mp 2 \nabla_0 \phi^{\pm}(\mathbf{x}_0) \cdot \mathbf{N}(\mathbf{x}_0) \qquad (10.4.11)$$

It is convenient to introduce the linear integral operator

$$O^{\pm}\langle q \rangle = q(\mathbf{x}_0) \mp 2 \int_D^{\mathrm{PV}} \mathbf{N}(\mathbf{x}_0) \cdot \nabla_0 G(\mathbf{x}_0, \mathbf{x}) \, q(\mathbf{x}) \, dS(\mathbf{x}) \qquad (10.4.12)$$

and recast Eq. (10.4.11) into the compact form

$$O^{\pm}\langle q \rangle = \mp 2 \nabla_0 \phi^{\pm}(\mathbf{x}_0) \cdot \mathbf{N}(\mathbf{x}_0) \qquad (10.4.13)$$

In Section 10.6 we shall show that, in the case of exterior flow, Eq. (10.4.11) has a unique solution, whereas in the case of interior flow, it has either no solution or an infinite number of solutions. In the second case, an infinite number of solutions will exist only when the restriction (10.6.21) is fulfilled. The satisfaction of this restriction, however, is ensured by the continuity equation for any flow that is free of point sources or point sinks with a finite net rate of discharge. In summary, *the single-layer representation is capable of representing any external flow and any nonsingular internal flow.*

Flow past a rigid body

As a specific application, let us consider an irrotational flow with velocity $\mathbf{u}^{\infty}$ past a stationary three-dimensional rigid body, and decompose the velocity potential into the incident component ϕ^{∞} and a disturbance component due to the body ϕ^D, writing $\phi = \phi^{\infty} + \phi^D$. Introducing a single-layer representation for ϕ^D, we obtain

$$\phi(\mathbf{x}_0) = \phi^{\infty}(\mathbf{x}_0) + \int_B G(\mathbf{x}_0, \mathbf{x}) \, q(\mathbf{x}) \, dS(\mathbf{x}) \qquad (10.4.14)$$

where B denotes the surface of the body. Requiring the no-penetration condition $\nabla \phi \cdot \mathbf{N} = 0$ over B yields the integral equation

$$q(\mathbf{x}_0) = 2 \int_B^{\mathrm{PV}} \mathbf{N}(\mathbf{x}_0) \cdot \nabla_0 G(\mathbf{x}_0, \mathbf{x}) \, q(\mathbf{x}) \, dS(\mathbf{x})$$
$$+ 2 \mathbf{u}^{\infty}(\mathbf{x}_0) \cdot \mathbf{N}(\mathbf{x}_0) \qquad (10.4.15)$$

According to our previous discussion, Eq. (10.4.15) has a unique solution for any incident flow.

Furthermore, we use the divergence theorem to write

$$\int_D \nabla q^-(\mathbf{x}) \cdot \mathbf{N}(\mathbf{x}) \, dS(\mathbf{x}) = \int_{V^-} \nabla^2 q^-(\mathbf{x}) \, dS(\mathbf{x}) = 0 \qquad (10.5.10)$$

where V^- is the interior of D, which shows that the total strength of the single-layer potential vanishes, thereby confirming that far from D, ϕ^{Ext} decays like a Green's function dipole.

Working in a similar manner, we find that the double-layer potential at a point $\mathbf{x}_0$ that is located in the interior of D may be expressed in the form of a single-layer integral as

$$\phi^{\text{Int}}(\mathbf{x}_0) = \int_D G(\mathbf{x}_0, \mathbf{x}) \nabla q^+(\mathbf{x}) \cdot \mathbf{N}(\mathbf{x}) \, dS(\mathbf{x}) \qquad (10.5.11)$$

Taking the limit as the point $\mathbf{x}_0$ approaches D from either side, and using Eqs. (10.5.9) and (10.5.11) in conjunction with Eq. (10.5.7), we find

$$\phi^+(\mathbf{x}_0) - \phi^-(\mathbf{x}_0) = \int_D G(\mathbf{x}_0, \mathbf{x})(\nabla q^- - \nabla q^+)(\mathbf{x}) \cdot \mathbf{N}(\mathbf{x}) \, dS(\mathbf{x}) = q(\mathbf{x}_0) \qquad (10.5.12)$$

Derivatives of the Double-Layer Potential and the Velocity Field

The gradient of the double-layer potential is continuous throughout the domain of flow but undergoes a discontinuity across the distribution domain D. To study the behavior of the normal derivative across D, we write the boundary integral equation (10.1.6) twice, first for the exterior flow and then for the interior flow, and add the two equations to obtain

$$\phi^+(\mathbf{x}_0) + \phi^-(\mathbf{x}_0) = -2 \int_D G(\mathbf{x}_0, \mathbf{x})(\nabla \phi^+ - \nabla \phi^-)(\mathbf{x}) \cdot \mathbf{N}(\mathbf{x}) \, dS(\mathbf{x})$$
$$+ 2 \int_D^{\text{PV}} \nabla G(\mathbf{x}_0, \mathbf{x}) \cdot \mathbf{N}(\mathbf{x})(\phi^+ - \phi^-)(\mathbf{x}) \, dS(\mathbf{x}) \qquad (10.5.13)$$

Using Eq. (10.5.7), we simplify the integrand of the double-layer potential on the right-hand side, and then invoke the definition (10.5.1) to identify the double-layer potential with the principal value of ϕ. Taking into account Eq. (10.5.6), we finally obtain

$$\int_D G(\mathbf{x}_0, \mathbf{x})(\nabla \phi^+ - \nabla \phi^-)(\mathbf{x}) \cdot \mathbf{N}(\mathbf{x}) \, dS(\mathbf{x}) = 0 \qquad (10.5.14)$$

Observing that this equation holds true for any arbitrary double-layer potential requires that the integrand vanish. Thus, *the normal derivative of the double-layer potential is continuous across D.*

It is evident from Eq. (10.5.7) that the tangential derivative of the double-layer potential suffers a discontinuity across D whose magnitude depends upon the particular form of q. Specifically,

$$\nabla \phi^{\pm}(\mathbf{x}) = [\nabla \phi(\mathbf{x})]^{\text{PV}} \pm \tfrac{1}{2}(\mathbf{I} - \mathbf{NN}) \cdot \nabla q$$
$$= [\nabla \phi(\mathbf{x})]^{\text{PV}} \pm \tfrac{1}{2}(\mathbf{N} \times \nabla q) \times \mathbf{N} \qquad (10.5.15)$$
$$\nabla \phi^+(\mathbf{x}) - \nabla \phi^-(\mathbf{x}) = (\mathbf{I} - \mathbf{NN}) \cdot \nabla q = (\mathbf{N} \times \nabla q) \times \mathbf{N} \qquad (10.5.16)$$

The projection operators $\mathbf{I} - \mathbf{NN} = -\mathbf{N} \times (\mathbf{N} \times)$ on the right-hand sides ensure that the derivatives of q are computed in the plane of the distribution domain D.

Equation (10.5.16) states that the tangential component of the velocity undergoes a discontinuity across D, and this allows us to regard the flow as if it were induced by a vortex sheet situated over D in the presence of the external or internal boundary S_B. Comparing Eq. (10.5.16) with Eq. (1.9.8) shows that the strength of the vortex sheet is given by

$$\boldsymbol{\zeta} = \mathbf{N} \times \nabla q \qquad (10.5.17)$$

We thus arrive at the interesting result that *the flow due to a double-layer potential may be regarded as the flow due to a vortex sheet.*

Vector Potential

Considering the free-space Green's function $G = 1/(4\pi r)$, where $r = |\mathbf{x} - \mathbf{x}_0|$, we exploit the property $\nabla G(r) = -\nabla_0 G(r)$, where the subscript 0 indicates differentiation with respect to $\mathbf{x}_0$, and find that a vector potential for the velocity is given by

$$\mathbf{A}(\mathbf{x}_0) = \int_D \mathbf{N}(\mathbf{x}) \times \nabla G(\mathbf{x}_0, \mathbf{x}) q(\mathbf{x}) \, dS(\mathbf{x}) \tag{10.5.18}$$

One may readily verify by straightforward differentiation that $\mathbf{u} = \nabla \times \mathbf{A} = \nabla \phi$.

Integral Equations for the Dirichlet Problem

Specifying the distribution of ϕ on either side of D renders Eq. (10.5.5) a Fredholm integral equation of the second kind for the density distribution q

$$q(\mathbf{x}_0) = \mp 2 \int_D^{PV} \nabla G(\mathbf{x}_0, \mathbf{x}) \cdot \mathbf{N}(\mathbf{x}) q(\mathbf{x}) \, dS(\mathbf{x}) \pm 2\phi^{\pm}(\mathbf{x}_0) \tag{10.5.19}$$

For future reference, we introduce the integral operator

$$P^{\pm}\langle q \rangle = q(\mathbf{x}_0) \pm 2 \int_D^{PV} \nabla G(\mathbf{x}, \mathbf{x}_0) \cdot \mathbf{N}(\mathbf{x}) q(\mathbf{x}) \, dS(\mathbf{x}) \tag{10.5.20}$$

and rewrite Eq. (10.5.19) in the more compact form

$$P^{\pm}\langle q \rangle = \pm 2\phi^{\pm}(\mathbf{x}_0) \tag{10.5.21}$$

Identity (10.4.2) shows that the homogeneous equation $P^{(+)}\langle q \rangle = 0$ has the nontrivial solution $q = $ constant for any shape of D, and this suggests that Eq. (10.5.19) for the exterior problem has either no solution or an infinite number of solutions that differ by an arbitrary constant.

The properties of the integral equation (10.5.19) will be discussed extensively in Section 10.6. The results will show that the exterior problem has a solution only when the flow rate of the velocity corresponding to the forcing function ϕ across D vanishes. When this restriction is met, the integral equation will have a multiplicity of solutions that differ by an arbitrary constant. In contrast, the interior flow problem will have a unique solution for any forcing function ϕ. Thus *the double-layer representation is capable of representing any internal flow and a restricted class of external flows.*

Flow past a rigid body

As an application, consider an irrotational flow $\mathbf{u}^{\infty}$ past a stationary three-dimensional rigid body denoted as B. We proceed by decomposing the potential ϕ into the incident component ϕ^{∞} and a disturbance component due to the body ϕ^D, setting $\phi = \phi^{\infty} + \phi^D$, and introduce a double-layer representation for ϕ^D, writing

$$\phi(\mathbf{x}_0) = \phi^{\infty}(\mathbf{x}_0) + \int_B \nabla G(\mathbf{x}_0, \mathbf{x}) \cdot \mathbf{N}(\mathbf{x}) q(\mathbf{x}) \, dS(\mathbf{x}) \tag{10.5.22}$$

The no-penetration condition requires that $\nabla \phi \cdot \mathbf{N} = 0$ over the exterior side of B.

To derive an integral equation, we recall that the normal derivative of the double-layer potential is continuous across the surface of B, and this suggests that the normal component of the velocity vanishes over the interior side of B. Consequently, the internal velocity computed on the basis of Eq. (10.5.22) is equal to zero or, equivalently, ϕ has the constant value c in the interior of B. Applying Eq. (10.5.22) at a point on the internal side of B and using Eq. (10.5.5), we derive an integral equation of the second kind for q

$$q(\mathbf{x}_0) = 2 \int_B^{PV} \nabla G(\mathbf{x}_0, \mathbf{x}) \cdot \mathbf{N}(\mathbf{x}) q(\mathbf{x}) \, dS(\mathbf{x}) + 2[\phi^{\infty}(\mathbf{x}_0) - c] \tag{10.5.23}$$

$$\nabla_0 \phi^{\text{Ext}}(\mathbf{x}_0) = -\nabla_0 \times \nabla_0 \times \int_{V^-} \nabla[G(r)\,q^-(\mathbf{x})]\,dV(\mathbf{x})$$

$$= -\nabla_0 \times \int_{V^-} \nabla \times [q^-(\mathbf{x})\,\nabla G(r)]\,dV(\mathbf{x})$$

$$= \nabla_0 \times \int_{V^-} \nabla \times [G(r)\,\nabla q^-(\mathbf{x})]\,dV(\mathbf{x}) \tag{10.5.34}$$

A variation of the Stokes theorem states that for any twice-differentiable vector function $\mathbf{F}$

$$\int_{V^-} \nabla \times \mathbf{F}\,dV = \int_D \mathbf{N} \times \mathbf{F}\,dS \tag{10.5.35}$$

(Section A.6, Appendix A). Applying Eq. (10.5.35) for the last volume integral in Eq. (10.5.34) yields

$$\nabla_0 \phi^{\text{Ext}}(\mathbf{x}_0) = -\nabla_0 \times \int_D G(r)\,\nabla q^-(\mathbf{x}) \times \mathbf{N}(\mathbf{x})\,dS(\mathbf{x})$$

$$= \int_D \nabla G(r) \times [\nabla q^-(\mathbf{x}) \times \mathbf{N}(\mathbf{x})]\,dS(\mathbf{x}) \tag{10.5.36}$$

Working in a similar manner, show that the gradient of the double-layer potential at a point $\mathbf{x}_0$ that is located in the interior of D is given by the corresponding expression

$$\nabla_0 \phi^{\text{Int}}(\mathbf{x}_0) = \int_D \nabla G(r) \times [\nabla q^+(\mathbf{x}) \times \mathbf{N}(\mathbf{x})]\,dS(\mathbf{x}) \tag{10.5.37}$$

(c) Equations (10.5.36) and (10.5.37) provide us with the gradient of the double-layer potential in terms of distributions of the Green's function dipoles whose distribution densities $\nabla q^{(-)} \times \mathbf{N}$ and $\nabla q^{(+)} \times \mathbf{N}$ are tangential to D, and may therefore be determined from a knowledge of the distribution of q over D independently of the structure of the field functions q^- and q^+. This observation allows us to write $\nabla q^- \times \mathbf{N} = \nabla q^+ \times \mathbf{N} = \nabla q \times \mathbf{N}$ and thereby recover Eq. (10.5.26). Evaluate Eqs. (10.5.36) and (10.5.37) at a point on either side of D and subtract off the dominant singularity of the integrands to obtain

$$\nabla_0 \phi^{\pm}(\mathbf{x}_0) = \int_D \nabla G(r) \times [\nabla q(\mathbf{x}) \times \mathbf{N}(\mathbf{x}) - \nabla q(\mathbf{x}_0) \times \mathbf{N}(\mathbf{x}_0)]\,dS(\mathbf{x})$$

$$+ \int_D \nabla G(r)\,dS(\mathbf{x}) \times [\nabla q(\mathbf{x}_0) \times \mathbf{N}(\mathbf{x}_0)] \tag{10.5.38}$$

Explain why the first integral on the right-hand side of Eq. (10.5.38) is continuous across D, but the second integral undergoes a discontinuity of magnitude $\mathbf{N}$.

Computer Problem

10.5.5 **Flow past a two-dimensional airfoil.** Solve the integral equation (10.5.24) and thereby obtain the pressure distribution, drag force, and lift force on a Joukowski airfoil discussed in Problem 10.2.2, using the boundary-element method discussed in Section 10.2.

10.6 INVESTIGATION OF INTEGRAL EQUATIONS OF THE SECOND KIND

In the previous two sections we made occasional references to the existence and uniqueness of solution of the integral equations that arise from the standard or generalized boundary-integral representations. In the present section we examine the properties of the integral equations in more detail. Since the integral equations of the first kind are amenable only to rudimentary theoretical investigations (Pogorzelski, 1966, Chapter 6), we focus our attention on integral equations of the second kind and study the existence and uniqueness of their solution, as well as the feasibility of computing the solution using iterative methods.

Integral Equations of the Second Kind

For simplicity, we assume that the flow is bounded either externally or internally by a single closed surface D; the unit normal vector pointing outward from D is denoted as $\mathbf{N}$. The integral equation (10.1.6) then becomes

$$\phi^{\pm}(\mathbf{x}_0) = \pm 2 \int_D^{PV} \nabla G(\mathbf{x}_0, \mathbf{x}) \cdot \mathbf{N}(\mathbf{x}) \phi^{\pm}(\mathbf{x}) \, dS(\mathbf{x})$$

$$\mp 2 \int_D G(\mathbf{x}_0, \mathbf{x}) \nabla \phi^{\pm}(\mathbf{x}) \cdot \mathbf{N}(\mathbf{x}) \, dS(\mathbf{x}) \tag{10.6.1}$$

where the plus and minus sign correspond, respectively, to the exterior and interior problem.

For ready reference, we also repeat Eq. (10.4.11), that arises from the single-layer representation with Neumann boundary conditions

$$q(\mathbf{x}_0) = \pm 2 \int_D^{PV} \mathbf{N}(\mathbf{x}_0) \cdot \nabla_0 G(\mathbf{x}, \mathbf{x}_0) q(\mathbf{x}) \, dS(\mathbf{x}) \mp 2 \nabla_0 \phi^{\pm}(\mathbf{x}_0) \cdot \mathbf{N}(\mathbf{x}_0) \tag{10.6.2}$$

and Eq. (10.5.19), that arises from the double-layer representation with Dirichlet boundary conditions

$$q(\mathbf{x}_0) = \mp 2 \int_D^{PV} \nabla G(\mathbf{x}_0, \mathbf{x}) \cdot \mathbf{N}(\mathbf{x}) q(\mathbf{x}) \, dS(\mathbf{x}) \pm 2 \phi^{\pm}(\mathbf{x}_0) \tag{10.6.3}$$

Generalized Homogeneous Equations

To investigate the existence and uniqueness of solution of the aforementioned integral equations, we consider the corresponding generalized homogeneous equations, which emerge by discarding the forcing functions on the right-hand sides and then multiplying the principal-value integrals with the complex constant β or its complex conjugate β^*, obtaining

$$\psi(\mathbf{x}_0) = 2\beta \int_D^{PV} \nabla G(\mathbf{x}, \mathbf{x}_0) \cdot \mathbf{N}(\mathbf{x}) \psi(\mathbf{x}) \, dS(\mathbf{x}) \tag{10.6.4}$$

$$\chi(\mathbf{x}_0) = 2\beta^* \int_D^{PV} \mathbf{N}(\mathbf{x}_0) \cdot \nabla_0 G(\mathbf{x}, \mathbf{x}_0) \chi(\mathbf{x}) \, dS(\mathbf{x}) \tag{10.6.5}$$

where ψ and χ are the eigenfunctions.

In the standard terminology of integral equation theory, β is a complex *eigenvalue of the double-layer operator*

$$L\langle q \rangle = 2 \int_D^{PV} \nabla G(\mathbf{x}, \mathbf{x}_0) \cdot \mathbf{N}(\mathbf{x}) q(\mathbf{x}) \, dS(\mathbf{x}) \tag{10.6.6}$$

and β^* is the complex eigenvalue of the adjoint of the double-layer operator

$$L^A\langle q \rangle = 2 \int_D^{PV} \mathbf{N}(\mathbf{x}_0) \cdot \nabla_0 G(\mathbf{x}, \mathbf{x}_0) q(\mathbf{x}) \, dS(\mathbf{x}) \tag{10.6.7}$$

Note that the adjoint operator L^A derives from L by interchanging the arguments $\mathbf{x}_0$ and $\mathbf{x}$ of the kernel.

Using the above definitions, we write Eqs. (10.6.4) and (10.6.5) in the concise forms

$$\psi = \beta L\langle \psi \rangle, \qquad \chi = \beta^* L^A\langle \chi \rangle \tag{10.6.8}$$

$\eta = -L^A\langle\eta\rangle + \chi$. According to Fredholm's alternative, for a solution of this equation to exist, χ must be orthogonal to ψ, but since ψ is constant, this contradicts Eq. (10.6.18). We conclude that both the algebraic and geometric multiplicity of the eigenvalue $\beta = -1$ are equal to one.

Case 3

A third exception occurs when one or both of $K^{\pm}$ are equal to zero, but this takes us back to the two cases discussed previously.

Summary and Discussion

In summary, we found that the eigenvalues of the double-layer potential are real and fall outside the closed range $(-1, 1]$; $\beta = -1$ is a marginal eigenvalue for any shape of D. The spectral radius of the double-layer operator and its adjoint is thus exactly equal to unity. Based on these results, we deduce:

- Equations (10.6.1) and (10.6.2) for the exterior problem, and Eq. (10.6.3) for the interior problem, have a unique solution for any boundary shape D.

- Equation (10.6.1) for the interior problem has either no solution or a multiplicity of solutions that differ by an arbitrary constant. According to Fredholm's alternative, an infinity of solutions will exist only if the projection of the forcing function, given by the last term on the right-hand side, onto the adjoint eigensolution χ, given in Eq. (10.6.17), vanishes, but this will be true for any flow that does not contain point sources and point sinks. To see this, we denote the single-layer potential as I^S and project it onto χ to obtain

$$
\begin{aligned}
(\chi, I^S) &= \int_D \chi(\mathbf{x}_0) \int_D G(\mathbf{x}_0, \mathbf{x}) \, \nabla\phi^-(\mathbf{x}) \cdot \mathbf{N}(\mathbf{x}) \, dS(\mathbf{x}) \, dS(\mathbf{x}_0) \\
&= \int_D \nabla\phi^-(\mathbf{x}) \cdot \mathbf{N}(\mathbf{x}) \int_D G(\mathbf{x}_0, \mathbf{x}) \, \chi(\mathbf{x}_0) \, dS(\mathbf{x}_0) \, dS(\mathbf{x}) \\
&= \int_D \nabla\phi^-(\mathbf{x}) \cdot \mathbf{N}(\mathbf{x}) \, c \, dS(\mathbf{x}) = 0
\end{aligned}
\tag{10.6.20}
$$

  where c is the constant value of the potential defined in Eq. (10.6.12).

- Equation (10.6.2) for the interior problem has either no solution or a multiplicity of solutions. According to Fredholm's alternative, an infinity of solutions will exist only if the projection of the forcing function onto the constant eigensolution of the double-layer potential corresponding to $\beta = -1$ vanishes; that is,

$$
\int_{D^-} \nabla\phi^-(\mathbf{x}) \cdot \mathbf{N}(\mathbf{x}) \, dS(\mathbf{x}) = 0
\tag{10.6.21}
$$

  This requires that the flow rate of the field corresponding to the forcing function across D vanish.

- Equation (10.6.3) for the exterior problem has either no solution or a multiplicity of solutions that differ by an arbitrary constant. According to Fredholm's alternative, an infinity of solutions will exist only if the projection of the forcing function onto the adjoint eigensolution χ corresponding to $\beta = -1$ vanishes; that is,

$$
\int_D \phi^+(\mathbf{x}) \chi(\mathbf{x}) \, dS(\mathbf{x}) = -c \int_D \nabla\phi^+(\mathbf{x}) \cdot \mathbf{N}(\mathbf{x}) \, dS(\mathbf{x}) = 0
\tag{10.6.22}
$$

  where c is the constant value of the potential defined in Eq. (10.6.12). The second expression in Eq. (10.6.22) arises from the reciprocal theorem discussed in Section 2.2.

- None of the aforementioned equations may be solved by the method of successive substitutions.

The Spectrum of the Sphere

It is instructive to compute the spectrum of the double-layer operator for the particular case where the domain D is the surface of a sphere of radius a. Noting that the single-layer potential (10.6.12) is continuous across D and using Eq. (10.6.13), we derive two matching conditions at $r = a$

$$\phi^+ = \phi^- \qquad \text{and} \qquad \frac{\partial \phi^+}{\partial r} = \kappa \frac{\partial \phi^-}{\partial r} \qquad (10.6.23)$$

where $\kappa = (1 - \beta^*)/(1 + \beta^*)$ and r is the radial distance from the center of the sphere. The problem is reduced to computing the interior and exterior harmonic potentials ϕ^+ and ϕ^- subject to Eqs. (10.6.23).

Expanding ϕ^+ and ϕ^- in series of solid spherical harmonics, we obtain

$$\phi^+(\mathbf{x}) = \sum_{n=0}^{\infty} b_n \left(\frac{a}{r}\right)^{n+1} S_n(\theta, \varphi)$$

$$\phi^-(x) = \sum_{n=0}^{\infty} c_n \left(\frac{r}{a}\right)^n S_n(\theta, \varphi) \qquad (10.6.24)$$

where S_n are spherical surface harmonics, and b_n and c_n are constants (Lamb, 1932, p. 110). Requiring the matching conditions (10.6.23), we find $b_n = c_n$ and $-b_n(n + 1) = \kappa n c_n$, which requires that $\kappa = -(n+1)/n$ and $\beta = -2n - 1$. Thus *the spectrum of the double-layer operator for the sphere is the set of all negative integers.*

PROBLEMS

10.6.1 **Adjoint of the double-layer operator.** Show that any two complex functions q_1 and q_2 satisfy $(L\langle q_1\rangle, q_2) = (q_1, L^A\langle q_2\rangle)$, where the projection $(\,,\,)$ is defined in Eq. (10.6.10).

10.6.2 **A known eigensolution.** Verify that when D is the surface of a sphere, $\chi = \text{const}$ is an eigensolution of Eq. (10.6.5) corresponding to the eigenvalue $\beta = -1$.

10.6.3 **Null space of operators.** Find the null space of the operators $O^\pm$ and $P^\pm$ introduced in Eqs. (10.4.12) and (10.5.20); that is, find all functions q that satisfy $O^\pm\langle q\rangle = 0$ and $P^\pm\langle q\rangle = 0$.

10.6.4 **Multiple boundaries.** Show that when D consists of a number of closed surfaces $D^{(k)}$, $k = 1, \ldots, K$, the most general eigensolution of Eq. (10.6.4) with $\beta = -1$ is $\psi = c_k$ over $D^{(k)}$, where the c_k are arbitrary constants. What is the most general adjoint eigensolution χ?

10.7 REGULARIZATION OF INTEGRAL EQUATIONS OF THE SECOND KIND

We saw that the integral equations (10.6.1) and (10.6.2) for interior flow and the integral equation (10.6.3) for exterior flow do not have a unique solution, and an infinity of solutions will exist only when appropriate constraints on the forcing functions are fulfilled. Furthermore, we saw that the integral equations (10.6.1) and (10.6.2) for exterior flow and the integral equation (10.6.3) for interior flow have a unique solution that, however, cannot be computed by the efficient method of successive substitutions.

The existence of multiple solutions and our inability to use the method of successive substitutions are undesirable features that, fortunately, can be eliminated by replacing the original integral equations with *regularized integral equations*. The distinguishing properties of the latter are that *they have a unique solution that may be computed using the method of successive substitutions,* and, more important, *their solutions satisfy the original integral equations.*

$$q(\mathbf{x}_0) = 2 \int_D^{\mathrm{PV}} \mathbf{N}(\mathbf{x}_0) \cdot \nabla_0 G(\mathbf{x}_0, \mathbf{x}) \, q(\mathbf{x}) \, dS(\mathbf{x}) + \frac{1}{S_D^{(k)}} \int_{D^{(k)}} q(\mathbf{x}) \, dS(\mathbf{x})$$

$$- 2\nabla\phi^+(\mathbf{x}_0) \cdot \mathbf{N}(\mathbf{x}_0) + \frac{1}{S_D^{(k)}} \int_{D^{(k)}} \nabla\phi^+(\mathbf{x}) \cdot \mathbf{N}(\mathbf{x}) \, dS(\mathbf{x}) \qquad (10.7.12)$$

Regularization of Eq. (10.6.3) for Interior Flow

Next we apply the method of eigenvalue deflation to the integral equation (10.6.3) for interior flow

$$q(\mathbf{x}_0) = 2 \int_D^{\mathrm{PV}} \nabla G(\mathbf{x}_0, \mathbf{x}) \cdot \mathbf{N}(\mathbf{x}) \, q(\mathbf{x}) \, dS(\mathbf{x}) - 2\phi^-(\mathbf{x}_0) \qquad (10.7.13)$$

which arises from the double-layer representation with Dirichlet boundary conditions. The solution is known to be unique, but, because the spectral radius of the adjoint double-layer operator is equal to unity, it may not be computed by the method of successive substitutions.

Working as before, we discretize the domain D into N boundary elements and assume that q is constant over each element to obtain the linear algebraic system (10.7.5) but with a different influence matrix $\mathbf{A}$ and constant vector $\mathbf{b}$. Instead of solving this system, we propose to solve the alternative system

$$\mathbf{x} = \mathbf{A}^{(1)} \cdot \mathbf{x} + \mathbf{c} \qquad (10.7.14)$$

where the matrix $\mathbf{A}^{(1)}$ was defined in Eq. (10.7.1). To ensure that the solution of the modified system (10.7.14) also satisfies the original system $\mathbf{x} = \mathbf{A} \cdot \mathbf{x} + \mathbf{b}$, we require

$$\mathbf{c} = \mathbf{b} + \lambda_1 \mathbf{u}(\mathbf{w} \cdot \mathbf{x}) \qquad (10.7.15)$$

Noting that $\mathbf{A}^{(1)} \cdot \mathbf{u} = \mathbf{0}$, we rewrite Eq. (10.7.14) in the form

$$\mathbf{y} = \mathbf{A}^{(1)} \cdot \mathbf{y} + \mathbf{b} \qquad (10.7.16)$$

where we have defined

$$\mathbf{y} = (\mathbf{I} - \lambda_1 \mathbf{u}\mathbf{w}) \cdot \mathbf{x} \qquad (10.7.17)$$

Projecting Eq. (10.7.17) onto $\mathbf{w}$, we obtain $\mathbf{w} \cdot \mathbf{y} = (1 - \lambda_1) \mathbf{w} \cdot \mathbf{x}$; substituting this result into Eq. (10.7.17) yields

$$\mathbf{x} = \left(\mathbf{I} + \frac{\lambda_1}{1 - \lambda_1} \mathbf{u}\mathbf{w}\right) \cdot \mathbf{y} \qquad (10.7.18)$$

Instead of solving equation $\mathbf{x} = \mathbf{A} \cdot \mathbf{x} + \mathbf{b}$ for $\mathbf{x}$, we prefer to solve Eq. (10.7.16) for $\mathbf{y}$ and then use Eq. (10.7.18) to recover $\mathbf{x}$. The crucial advantage of this seemingly cumbersome procedure hinges upon the fact that $\mathbf{A}^{(1)}$ has one non-zero eigenvalue less than $\mathbf{A}$, and if the removed eigenvalue happens to be the marginal one, then the spectral radius of $\mathbf{A}^{(1)}$ will be less than that of $\mathbf{A}$.

Returning to Eq. (10.7.13), we recall that $\beta = -1$ is an eigenvalue of the generalized homogeneous equation (10.6.4) with corresponding eigenfunction $\psi = c$, note that $\mathbf{u}$ plays the role ψ, and introduce the function η corresponding to $\mathbf{w}$. To satisfy the constraint $\mathbf{w} \cdot \mathbf{u} = 1$, we set $\eta = 1/(cS_D)$, where S_D is the surface of D. Finally, we write the counterparts of Eq. (10.7.16) and (10.7.18)

$$Q(\mathbf{x}_0) = 2 \int_D^{\mathrm{PV}} \nabla G(\mathbf{x}_0, \mathbf{x}) \cdot \mathbf{N}(\mathbf{x}) Q(\mathbf{x}) \, dS(\mathbf{x})$$

$$+ \frac{1}{S_D} \int_D Q(\mathbf{x}) \, dS(\mathbf{x}) - 2\phi^-(\mathbf{x}_0) \qquad (10.7.19)$$

$$q(\mathbf{x}_0) = Q(\mathbf{x}_0) - \frac{1}{2S_D} \int_D Q(\mathbf{x}) \, dS(\mathbf{x}) \qquad (10.7.20)$$

The procedure involves solving Eq. (10.7.19) for Q and then using Eq. (10.7.20) to recover q.

The generalized adjoint homogeneous equation corresponding to Eq. (10.7.19) is given by Eq. (10.7.11). Our previous analysis then shows that Eq. (10.7.19) has a unique solution that may be computed using the method of successive substitutions. Furthermore, straightforward substitution of Eq. (10.7.20) into Eq. (10.7.19) yields the original integral equation (10.7.13) (Problem 10.7.2).

Regularization of Eq. (10.6.1) for Exterior Flow

The deflation of Eq. (10.6.1) for exterior flow is similar to that of Eq. (10.6.3) for interior flow described previously. The counterparts of Eqs. (10.7.19) and (10.7.20) are

$$\Phi(\mathbf{x}_0) = 2 \int_D^{\text{PV}} \nabla G(\mathbf{x}_0, \mathbf{x}) \cdot \mathbf{N}(\mathbf{x}) \Phi(\mathbf{x}) \, dS(\mathbf{x})$$

$$+ \frac{1}{S_D} \int_D \Phi(\mathbf{x}) \, dS(\mathbf{x}) - 2I^S(\mathbf{x}_0) \tag{10.7.21}$$

$$\phi(\mathbf{x}_0) = \Phi(\mathbf{x}_0) - \frac{1}{2S_D} \int_D \Phi(\mathbf{x}) \, dS(\mathbf{x}) \tag{10.7.22}$$

where I^S is the single-layer integral on the right-hand side of Eq. (10.6.1).

When D is composed of a number of disconnected closed surfaces, we obtain a modified set of equations. For a point $\mathbf{x}_0$ that lies on $D^{(k)}$, we find

$$\Phi(\mathbf{x}_0) = 2 \int_D^{\text{PV}} \nabla G(\mathbf{x}, \mathbf{x}_0) \cdot \mathbf{N}(\mathbf{x}) \Phi(\mathbf{x}) \, dS(\mathbf{x})$$

$$+ \frac{1}{S_D^{(k)}} \int_{D^{(k)}} \Phi(\mathbf{x}) \, dS(\mathbf{x}) - 2I^S(\mathbf{x}_0) \tag{10.7.23}$$

$$\phi(\mathbf{x}_0) = \Phi(\mathbf{x}_0) - \frac{1}{2S_D^{(k)}} \int_{D^{(k)}} \Phi(\mathbf{x}) \, dS(\mathbf{x}) \tag{10.7.24}$$

Regularization of Eq. (10.6.3) for Exterior Flow

Considering next Eq. (10.6.3) for exterior flow, which arises from the double-layer representation with Dirichlet boundary conditions, we write the counterpart of Eq. (10.7.14)

$$\mathbf{x} = -\mathbf{A}^{(1)} \cdot \mathbf{x} + \mathbf{c} \tag{10.7.25}$$

where the influence matrix $\mathbf{A}$ is identical to the one in Eq. (10.7.14), but the vector $\mathbf{c}$ has a different value. Working as before, we replace Eq. (10.7.25) with

$$\mathbf{y} = -\mathbf{A}^{(1)} \cdot \mathbf{y} + \mathbf{b} \tag{10.7.26}$$

where

$$\mathbf{y} = (\mathbf{I} + \lambda_1 \mathbf{uw}) \cdot \mathbf{x} \tag{10.7.27}$$

In order to perform eigenvalue deflation, we must apply Eq. (10.7.27) with $\lambda_1 = -1$, but inverting this equation to recover $\mathbf{x}$ is prohibited by the fact that the matrix $\mathbf{I} - \mathbf{uw}$ is singular (Problem 10.7.3). This implies that we cannot regularize Eq. (10.6.3) for exterior flow. Similar difficulties arise when we attempt to deflate Eqs. (10.6.1) and (10.6.2) for interior flow. The reason for these difficulties will be discussed in Section 10.8.

PROBLEMS

10.7.1 Wielandt's deflation. Show that Wielandt's deflation replaces one eigenvalue by zero, leaves all other eigenvalues and adjoint eigenvectors unaffected, but changes the eigenvectors.

10.7.2 Deflated equation. Show that Eqs. (10.7.19) and (10.7.20) produce the solution of Eq. (10.7.13).

10.7.3 A singular matrix. Show that the matrix $\mathbf{I} - \mathbf{uw}$ with $\mathbf{u} \cdot \mathbf{w} = 1$ is singular. *Hint:* Show that one eigenvalue is equal to zero.

$$\chi(\mathbf{x}_0) = -2 \int_D^{PV} \mathbf{N}(\mathbf{x}_0) \cdot \nabla_0 G(\mathbf{x}_0, \mathbf{x}) \chi(\mathbf{x}) \, dS(\mathbf{x})$$

$$- 2 A \, G(\mathbf{x}_0, \mathbf{x}_s) \int_D \chi(\mathbf{x}) \, dS(\mathbf{x}) \tag{10.8.6}$$

Integrating Eq. (10.8.6) over D and using Eq. (10.4.2), we find that the last integral vanishes, and this identifies Eq. (10.8.6) with Eq. (10.6.5) with $\beta = -1$, but violates the constraint (10.6.18). As a consequence, Eq. (10.8.6) has only the trivial solution for any value of A.

PROBLEMS

10.8.1 **Completion with a single-layer potential.** One plausible choice for Φ is a single-layer potential with density ζ

$$\Phi(\mathbf{x}_0) = \int_D G(\mathbf{x}_0, \mathbf{x}) \zeta(\mathbf{x}) \, dS(\mathbf{x})$$

In order to maintain the linearity of the integral equation, we may require that ζ be a linear function of q, and set $\zeta = Aq$, where A is a constant. Show that, with this choice, Eq. (10.8.2) has a unique solution for any positive value of A.

10.8.2 **Completion with a collection of point sources.** Identify Φ with the potential due to a finite collection of point sources located in the interior of D, and discuss the uniqueness of the solution of the resulting integral equation.

10.9 | ITERATIVE SOLUTION OF INTEGRAL EQUATIONS OF THE SECOND KIND

Compared to direct methods, iterative methods for solving integral equations of the second kind have two significant advantages: They require a reduced amount of algebraic manipulations associated with grouping the unknown coefficients of the local expansions in the boundary element implementation, as discussed in Section 10.2, and they are more affordable in terms of computer memory requirements, central processing unit time, and programming effort.

To illustrate the implementation of an iterative method, let us consider the prototypical integral equation of the second kind

$$q = \alpha O\langle q \rangle + F \tag{10.9.1}$$

where O is a compact linear integral operator defined over the surface D, F is a certain forcing function, and q is an unknown function. We proceed by discretizing D into N boundary elements and approximate the operator $O\langle q \rangle$ using a quadrature over each element to obtain

$$q(\mathbf{x}) = \alpha \sum_{\substack{\text{Elements,} \\ n=1,\dots,N}} \sum_{\substack{\text{Quadrature points,} \\ k=1,\dots,K}} A_{nk}(\mathbf{x}, \mathbf{x}_{nk}) \, q(\mathbf{x}_{nk}) + F(\mathbf{x}) \tag{10.9.2}$$

where $\{\mathbf{x}_{nk}\}$ is the collection of the K quadrature base points over all elements, and $\mathbf{A}$ is an influence matrix. Adopting *Nÿstrom's method,* we apply Eq. (10.9.2) at the quadrature points over all elements and thus derive a system of NK linear algebraic equations for the unknown values $q(\mathbf{x}_{nk})$ (Atkinson 1973).

Implementing Jacobi's method, which is the discrete version of the method of successive approximation, we guess the values of $q(\mathbf{x}_{nk})$, compute the right-hand side of Eq. (10.9.2) at all base points, and then replace the assumed with the computed values. Other iterative schemes employ generalized conjugate gradient methods discussed in Section B.1, Appendix B.

To reduce the cost of the computations, it is helpful to compute the influence coefficients A_{nk} once, and then store and recall them during the subsequent iterations. When N is of order of a few hundred and the unknown function is assumed to be constant or vary in a linear manner over each element, the computer memory requirements are affordable. When the number of boundary elements is large or high-order approximations are employed, however, the size of A_{nk} may exceed the available computer memory, and this necessitates that some or all of the entries of A_{nk} be recomputed before each iteration.

In a slight variation of this procedure, in the spirit of the *spectral element method,* we express the unknown density q in terms of a polynomial or trigonometric series with respect to properly defined surface variables over each element, and identify the coefficients of the expansion with the values of q at the corresponding nodes. We assign initial values to q at the nodes, evaluate the polynomials over all elements, and apply the integral equation at the nodes. Finally, we compute the double-layer integral and add it to the forcing function F, thereby producing the new values of q at the nodes. We then repeat the procedure until the values of q at all nodes change by less than a preset minimum after one iteration. This method has the advantage that it produces the value of q at the nodes in a direct manner, and therefore circumvents the need for further interpolations, which is especially convenient when q represents a primary variable such as the velocity.

It is important to emphasize that the iterative procedures discussed here will converge only when $|\alpha|$ is less than the spectral radius of the operator O and the domain D is a Lyapunov surface. If D contains sharp edges or corners, the iterations are likely to diverge. In this case, in order to compute a solution, we must restate Eqs. (10.9.2) in the standard form of a system of linear equations and then compute the solution using a direct method.

Computer Implementations with Parallel Processing

The availability of computers with parallel-processor architecture allows us to tackle problems with increased geometric complexity. A typical computational problem requires solving a linear system of equations, such as that shown in Eq. (10.9.2), which can be done using either a direct or an iterative method. Parallelization of the computational tasks in a direct method ranges from difficult to infeasible (Bertsekas and Tsitsiklis, 1989). For instance, parallelization in Gaussian elimination is prohibited by pivoting. Iterative methods, on the other hand, are ideally suited for parallel computation.

In the context of parallel computation, it is useful to distinguish between two classes of problems according to the topology of the domain of flow. The first class includes problems in domains that are bounded by a number of distinct closed surfaces representing, for instance, the boundaries of a collection of suspended bodies. The second class includes problems in domains with continuous but complex boundaries representing, for instance, the surface of an automobile or an aircraft.

Assuming that the boundary of a flow is composed of a collection of M smooth closed surfaces, we decompose the integral operator in Eq. (10.9.1) into the sum of M operators, where each operator is supported by a distinct surface, and write

$$q(\mathbf{x}) = \alpha \sum_{m=1}^{M} O_m \langle q \rangle + F(\mathbf{x}) \qquad (10.9.3)$$

We then guess the distribution of q over each surface, call it q^0, and iterate individually on each surface using the method of successive substitutions, where each surface is assigned to a different processor. More specifically, over the ith surface, we iterate based on the equation

$$q(\mathbf{x}) = \alpha O_i \langle q \rangle + G(\mathbf{x}) \qquad (10.9.4)$$

where

$$G(\mathbf{x}) = \alpha \sum_{\substack{m=1 \\ m \neq i}}^{M} O_m \langle q \rangle + F(\mathbf{x}) \tag{10.9.5}$$

is held constant during the iterations. These local iterations are guaranteed to converge as long as $|\alpha|$ is less than the spectral radius of the corresponding operators. After a number of local iterations have been carried out, the initial guess q is updated across all processors, a second global iteration is carried out, and the procedure continues until a converged solution is achieved. The frequency and protocol of communication among the processors may play an important role in determining the overall efficiency of the method (see, for instance, David and Blyth, 1992; Fuentes and Kim, 1992).

References

Atkinson, K. E., 1973, Iterative variants of the Nÿstrom method for the numerical solution of integral equations. *Numer. Math.* **22,** 17–31.

Atkinson, K. E., 1976, *A Survey of Numerical Methods for the Solution of Fredholm Integral Equations of the Second Kind.* SIAM.

Atkinson, K. E., 1980, The numerical solution of Laplace's equation in three dimensions. II. *Numerical Treatment of Integral Methods,* Edited by J. Albrecht and L. Collatz, Birkhäuser, Basel, Switzerland.

Atkinson, K. E., 1990, A survey of boundary integral methods for the numerical solution of Laplace's equation in three dimensions. *Numerical Solution of Integral Methods,* Edited by M. A. Goldberg, Plenum Press.

Baker, C. T. H., 1977, *The Numerical Treatment of Integral Equations.* Clarendon Press.

Baker, G. R., 1983, Generalized vortex methods for free-surface flows. *Waves in Fluid Interfaces.* Academic Press.

Bear, J., 1972, *Dynamics of Fluids in Porous Media.* Dover.

Bertsekas, D. P., and Tsitsiklis, J. N., 1989, *Parallel and Distributed Computation.* Prentice Hall.

Bodewig, E., 1959, *Matrix Calculus.* North-Holland.

Colton, D., and Kress, R., 1983, *Integral Equation Methods in Scattering Theory.* John Wiley & Sons.

David, T., and Blyth, G., 1992, Parallel algorithms for panel methods. *Int. J. Num. Meth. Fluids* **14,** 95–108.

Delves, L. M., and Mohamed, J. L., 1985, *Computational Methods for Integral Equations.* Cambridge University Press.

Fuentes, Y. O., and Kim, S., 1992, Parallel computational microhydrodynamics: communication scheduling strategies. *AIChE J.* **38,** 1059–78.

Hess, J. L., 1990, Panel methods in computational fluid dynamics. *Annu. Rev. Fluid Mech.* **22,** 255–74.

Katz, J., and Plotkin, A., 1991, *Low-Speed Aerodynamics. From Wing Theory to Panel Methods.* McGraw-Hill.

Kozlov, S. V., Lifanov, I. K., and Mikhailov, A. A., 1991, A new approach to mathematical modelling of flow of ideal fluid around bodies. *Sov. J. Num. Anal. Mod.* **6,** 209–22.

Lamb, H., 1932, *Hydrodynamics.* Dover.

Nakayama, T., 1990, A computational method for simulating transient motions of an incompressible inviscid fluid with a free surface. *Int. J. Num. Meth. Fluids* **10,** 683–95.

Pogorzelski, W., 1966, *Integral Equations and their Applications.* Pergamon Press.

Pozrikidis, C., 1992, *Boundary Integral and Singularity Methods for Linearized Viscous Flow.* Cambridge University Press.

Wilkinson, J. H., 1965, *The Algebraic Eigenvalue Problem.* Oxford University Press.

In this chapter we focus our attention on a particular class of flows that are dominated by the presence or motion of compact regions of concentrated vorticity, concisely called *vortices,* including vortex filaments and vortex layers. The analysis of these flows and the development of pertinent methods of numerical computation will be conducted under the assumption that the Reynolds number is sufficiently large that viscous forces are confined within thin boundary layers, and the fluid within the main part of the flow may be considered to be truly or nearly inviscid.

In nature and technology, vortices may be generated by a variety of mechanisms, including the deposition of vorticity within compact wakes behind bluff bodies, the rollup of separated boundary layers and vortex layers that are ejected from sharp corners, and the instability of shear layers forming at the interface between two fluids that merge at different speeds. In everyday life, vortices may be seen to develop from the rollup of vortex sheets developing on either side of a blade that is made to move parallel to itself broadside on along a free surface.

To compute the evolution of a flow that is dominated by vortex motions, it is expedient to proceed in an indirect manner that involves computing the evolution of the vorticity field using the vorticity transport equation, and then obtaining the simultaneous evolution of the velocity field by inverting the definition of the vorticity $\boldsymbol{\omega} = \nabla \times \mathbf{u}$. This method of computation, concisely described as a *vortex method,* is an alternative to a direct method that proceeds by integrating in time the equation of motion and the continuity equation for the primitive variables including the velocity and the pressure. Direct formulations will be discussed in Chapter 13 in the context of finite-difference methods.

The computational advantages of the vortex methods stem, mainly, from the fact that the vorticity transport equation may undergo substantial simplifications that are not necessarily reflected in the equation of motion. For instance, the absence of vortex stretching for two-dimensional flow has the important consequence that the vorticity evolves under the action of convection and diffusion alone. As a consequence, when viscous effects are insignificant, the region where the vorticity assumes finite values remains compact at all times, and the vortices are merely convected by the flow. These features allow us to reduce the computational domain by considering only those regions of the flow where the vorticity assumes substantial values above a preset threshold.

Another important feature of vortex methods is that they bypass the computation of the pressure. The significance of this simplification will become evident in Chapter 13, where we shall discuss the subtleties involved in deriving and implementing boundary conditions for the pressure from specified boundary conditions for the velocity or surface stress.

A typical computational algorithm involved in a vortex method proceeds according to the following steps:

1. Given an initial velocity field, compute the associated initial vorticity field. If the initial vorticity field is specified in the statement of the problem, this step is skipped.

2. Use the vorticity transport equation to advance the vorticity field by one step in time.

3. Invert the definition $\boldsymbol{\omega} = \nabla \times \mathbf{u}$ to obtain the velocity field at the new time instant.

4. Return to step 2 and repeat the computation.

One way to obtain the velocity field from the vorticity field is to use the Biot–Savart integral discussed in Sections 2.5, 2.8, 2.9, and 2.10. This method is particularly effective for the inviscid or slightly viscous flows considered in the present chapter. Two other methods, to be discussed in Section 13.1 in the context of finite-difference methods, involve expressing the velocity in terms of a solenoidal vector potential $\mathbf{A}$, so that $\mathbf{u} = \nabla \times \mathbf{A}$ with $\nabla \cdot \mathbf{A} = 0$, and then solving a Poisson equation for $\mathbf{A}$, $\nabla^2 \mathbf{A} = -\boldsymbol{\omega}$, or inverting the Poisson equation for the velocity $\nabla^2 \mathbf{u} = -\nabla \times \boldsymbol{\omega}$. These methods are designated for more general viscous flows.

In the ensuing sections we shall outline and discuss the fundamental principles, governing equations, and computational algorithms associated with a general class of vortex methods. The literature on vortex dynamics and vortex methods is extensive, and the interested reader will find further information in the reviews Clements and Maull (1975), Saffman and Baker (1979), Leonard (1980, 1985), Pullin (1992), Saffman (1992), and Puckett (1993).

11.1 | INVARIANTS OF THE MOTION

When computing vortex motion in a domain of infinite expanse in the absence of interior boundaries, subject to the condition that the velocity vanishes at infinity, it is useful to check the accuracy of the numerical results by monitoring the evolution of certain known invariants of the motion. For simplicity, in this discussion we shall assume that the density of the fluid is uniform throughout the domain of flow.

Three-Dimensional Flow

In Section 2.8 we saw that, provided that the vorticity decays faster than $1/r^3$, the *integral of the vorticity over the volume of the flow* vanishes, and this suggests that

$$\int_{\text{Flow}} \boldsymbol{\omega} \, dV = \mathbf{0} \tag{11.1.1}$$

is an invariant of the motion for inviscid as well as for viscous fluids.

Two additional invariants for inviscid and viscous fluids are the *linear* and *angular impulse* required to generate the motion and thereby impart to the fluid a certain amount of linear and angular momentum. These can be expressed in terms of the vorticity as

$$\mathbf{P} = \tfrac{1}{2}\rho \int_{\text{Flow}} \mathbf{x} \times \boldsymbol{\omega} \, dV$$

$$\mathbf{A} = \tfrac{1}{3}\rho \int_{\text{Flow}} \mathbf{x} \times (\mathbf{x} \times \boldsymbol{\omega}) \, dV \tag{11.1.2}$$

(Batchelor, 1967, Section 7.2; Winckelmans and Leonard, 1993). To show that $\mathbf{P}$ and $\mathbf{A}$ remain constant, we take the time derivative of the integrals in Eqs. (11.1.2) and use the Eulerian form of the vorticity transport equation for homogeneous fluids, Eq. (3.8.10) (Problem 11.1.1).

In the absence of viscous dissipation, the *total kinetic energy* of the fluid K must be conserved, and this suggests that

$$K = \tfrac{1}{2}\rho \int_{\text{Flow}} \mathbf{B} \cdot \boldsymbol{\omega} \, dV = \rho \int_{\text{Flow}} \mathbf{u} \cdot (\mathbf{x} \times \boldsymbol{\omega}) \, dV \tag{11.1.3}$$

is another invariant of the motion for inviscid fluids (Section 2.8). Here $\mathbf{B}$ is a solenoidal vector potential for the velocity, defined by $\mathbf{u} = \nabla \times \mathbf{B}$.

Projecting Euler's equation onto the vorticity vector and then integrating the product over the volume of the flow, we find that the *helicity,* defined by the equation

$$H = \int_{\text{Flow}} \mathbf{u} \cdot \boldsymbol{\omega} \, dV \tag{11.1.4}$$

is a fifth invariant of the motion for inviscid fluids. Physically, the helicity is a measure of the net linkage of the vortex lines (Moffatt and Tsinober, 1992).

Viscous effects

When the effects of viscosity are not negligible, the total vorticity defined in Eq. (11.1.1) remains equal to zero, and the linear and angular impulses defined in Eqs. (11.1.2) remain invariant during the motion. The kinetic energy and helicity, however, do not maintain their initial values.

The kinetic energy, in particular, decreases due to viscous dissipation. Using the energy balance expressed by Eq. (3.3.18), we find

$$\frac{dK}{dt} = -2\mu \int_{\text{Flow}} \mathbf{E} : \mathbf{E} \, dV$$

$$= -\mu \int_{\text{Flow}} [\boldsymbol{\omega} \cdot \boldsymbol{\omega} + 2 \, \nabla \cdot (\mathbf{u} \cdot \mathbf{L})] \, dV$$

$$= -\mu \int_{\text{Flow}} \boldsymbol{\omega} \cdot \boldsymbol{\omega} \, dV \qquad (11.1.5)$$

where $\mathbf{L}$ and $\mathbf{E}$ are the velocity gradient and rate-of-deformation tensors (Problem 11.1.1). The integral on the right-hand side of Eq. (11.1.5) is called the *enstrophy* of the flow. Projecting the vorticity transport equation onto the vorticity vector, we derive the following evolution equation for the enstrophy

$$\frac{d}{dt} \int_{\text{Flow}} \boldsymbol{\omega} \cdot \boldsymbol{\omega} \, dV = 2 \int_{\text{Flow}} (\boldsymbol{\omega}\boldsymbol{\omega}) : \mathbf{E} \, dV - 2\nu \int_{\text{Flow}} \nabla \boldsymbol{\omega} : \nabla \boldsymbol{\omega} \, dV \qquad (11.1.6)$$

where ν is the kinematic viscosity of the fluid, which is the simplified version of a more general equation for bounded flow given in Eq. (3.8.36) (Problem 11.1.1).

Axisymmetric Flow

The vorticity of an axisymmetric flow without swirling motion points in the azimuthal direction. Writing $\boldsymbol{\omega} = \omega \mathbf{e}_\varphi$, where $\mathbf{e}_\varphi$ is the unit vector in the azimuthal direction, we derive simplified expressions for the linear momentum, angular momentum, kinetic energy, and helicity

$$\mathbf{P} = \pi\rho \int_{\text{Flow}} \omega\sigma^2 \, dx \, d\sigma \, \mathbf{i}, \qquad \mathbf{A} = \mathbf{0}$$

$$K = \pi\rho \int_{\text{Flow}} \Psi\omega \, dx \, d\sigma, \qquad H = 0 \qquad (11.1.7)$$

where $\mathbf{i}$ is the unit vector along the x axis, and Ψ is the Stokes stream function [see Eq. (2.8.13)].

Two-Dimensional Flow

For two-dimensional flow, we obtain a new set of invariants, some but not all of which are related to the invariants for three-dimensional flow. In Section 3.8 we saw that the *total circulation* of the flow is conserved but it is not necessarily equal to zero, and this guarantees that

$$C = \int_{\text{Flow}} \omega \, dA \qquad (11.1.8)$$

is an invariant of the motion for both inviscid and viscous fluids. Here $\boldsymbol{\omega} = \omega \mathbf{k}$ where $\mathbf{k}$ is the unit vector along the z axis.

Two additional invariants for inviscid fluids are the *centroid of vorticity* $\mathbf{X}$ and the square of the *dispersion length* D, defined, respectively, as

$$\mathbf{X} = \frac{1}{C} \int_{\text{Flow}} \mathbf{x}\omega \, dA, \qquad D^2 = \frac{1}{C} \int_{\text{Flow}} \mathbf{x} \cdot \mathbf{x}\omega \, dA \qquad (11.1.9)$$

These quantities are related to the linear and angular impulse that must be expended in order to generate the motion of the fluid (Batchelor, 1967, Section 7.2).

The kinetic energy of an infinite two-dimensional flow with nonvanishing total circulation C is infinite. It can be shown, however, that in the absence of viscous dissipation, the quantity

$$W = \tfrac{1}{2}\rho \int_{\text{Flow}} \psi\omega \, dA \qquad (11.1.10)$$

where ψ is the stream function, is finite, and remains constant during the motion. Physically, W expresses the part of the kinetic energy of the fluid that depends upon the particular way in which the vorticity is distributed within the flow (Batchelor, 1967, Section 7.2).

Viscous effects

When viscous effects are significant, the total circulation C and centroid $\mathbf{X}$ remain invariant, but the dispersion length increases at a constant rate according to

$$\frac{dD^2}{dt} = 4\nu \qquad (11.1.11)$$

(Batchelor, 1967, p. 536). This behavior reflects the tendency of the vorticity to diffuse away from its initial distribution and occupy the whole space. Furthermore, viscous dissipation requires that W decrease monotonically during the motion.

PROBLEM

11.1.1   **Invariants of the motion.** (a) Show that $\mathbf{P}$ and $\mathbf{A}$ for three-dimensional flow are conserved during the motion of an inviscid or viscous fluid. (b) Derive the evolution Eqs. (11.1.5) and (11.1.6). (c) Derive Eq. (11.1.11).

11.2 | POINT VORTICES

We begin the discussion of vortex motion by considering the two-dimensional flow of an unbounded inviscid fluid containing a collection of N point vortices. The vorticity transport equation guarantees that the point vortices maintain their initial strength as they move about the domain of flow. Using the Biot–Savart integral, we find that the velocity of each point vortex is equal to the sum of the individual velocities induced by all other point vortices. The four fundamental steps of vortex methods outlined in the introduction to this chapter combine to yield a system of nonlinear ordinary equations for the coordinates of the point vortices

$$\frac{dX_i}{dt} = -\frac{1}{2\pi}\sum_{\substack{j=1 \\ j\neq i}}^{N} \kappa_j \frac{Y_i - Y_j}{|\mathbf{X}_i - \mathbf{X}_j|^2}, \qquad \frac{dY_i}{dt} = \frac{1}{2\pi}\sum_{\substack{j=1 \\ j\neq i}}^{N} \kappa_j \frac{X_i - X_j}{|\mathbf{X}_i - \mathbf{X}_j|^2} \qquad (11.2.1)$$

where κ_j is the strength of the ith point vortex. An implicit supplement to this system is the condition of point-vortex strength invariance $d\kappa_j/dt = 0$. To compute the motion of the point vortices from a given initial configuration, we integrate Eqs. (11.2.1) in time using a standard numerical method such as Euler's method or a Runge–Kutta method.

Invariants of the Motion

The stream function of the flow due to a finite collection of N point vortices is computed readily by summing the individual stream functions due to each point vortex, and is found to be

$$\psi(\mathbf{x}) = -\frac{1}{2\pi}\sum_{i=1}^{N} \kappa_i \ln|\mathbf{x} - \mathbf{X}_i| \qquad (11.2.2)$$

Expressing the vorticity in terms of the two-dimensional delta function, and using Eq. (11.2.2), we find that the five scalar invariants of the motion shown in Eqs. (11.1.8)–(11.1.10) take the simplified forms

$$C = \sum_{i=1}^{N} \kappa_i, \qquad X = \frac{1}{C} \sum_{i=1}^{N} \kappa_i X_i, \qquad Y = \frac{1}{C} \sum_{i=1}^{N} \kappa_i Y_i$$

$$D^2 = \frac{1}{C} \sum_{i=1}^{N} \kappa_i [(X - X_i)^2 + (Y - Y_i)^2] \qquad (11.2.3)$$

$$W = -\frac{1}{4\pi} \rho \sum_{i=1}^{N} \sum_{\substack{j=1 \\ j \neq i}}^{N} \kappa_i \kappa_j \ln |\mathbf{X}_i - \mathbf{X}_j|$$

Hamiltonian Formulation

Kirchhoff noted that the system (11.2.1) may be written in the Hamiltonian form

$$\kappa_i \frac{dX_i}{dt} = \frac{\partial W}{\partial Y_i}, \qquad \kappa_i \frac{dY_i}{dt} = -\frac{\partial W}{\partial X_i} \qquad (11.2.4)$$

where summation is *not* implied on the left-hand sides. This formulation allows us to study the motion of point vortices in the context of classical Hamiltonian mechanics.

Elementary Motions in an Unbounded Domain

The motion of two point vortices with strengths equal to κ_1 and κ_2, separated by a distance d, can be predicted immediately from the requirement that the quantities X, Y, and D remain constant. We find that the point vortices move along concentric circles around the centroid (X, Y) with angular velocity $\Omega = (\kappa_1 + \kappa_2)/(2\pi d^2)$, while maintaining their initial separation.

If the point vortices have strengths of equal magnitude and opposite sign equal to $\pm \kappa$, the total circulation vanishes, the centroid of vorticity is shifted to infinity, and the point vortices move along parallel straight lines that are perpendicular to their initial separation with velocity equal to $\kappa/(2\pi d)$. The associated streamline pattern in a frame of reference moving with the point vortices is shown in Figure 11.2.1. The dividing streamline takes the shape of an oblate oval with major and minor axes, respectively, equal to $2.09a$ and $1.73a$, where $a = d/2$ is half the point vortex separation. It is instructive to observe the similarities and differences between the streamline patterns shown in Figures 11.2.1 and 2.9.1, the latter corresponding to axisymmetric flow due to a line vortex ring viewed from a stationary frame of reference.

The trajectories of more than three point vortices in an infinite domain are known to describe periodic orbits, but also exhibit complex features and chaotic behavior (Aref, 1983).

Point Vortices in Confined Domains

When the point vortices are placed in an internally or externally confined domain, their mutually induced velocity must be enhanced with that due to a complementary nonsingular irrotational flow that develops in response to the no-penetration boundary condition at the impermeable walls. Noting that a Green's function of the first kind of the two-dimensional Laplace equation represents the stream functions of the flow due to a point vortex of unit strength in the presence of an impermeable wall provides us with a systematic method for the computation of the complementary flow.

When the geometry of a boundary is sufficiently simple, the complementary potential flow may be expressed in terms of image point vortices with appropriate strengths located outside the domain of flow. The motion of the image point vortices is *not computed,* but is *deduced* from the instantaneous position of the primary vortices.

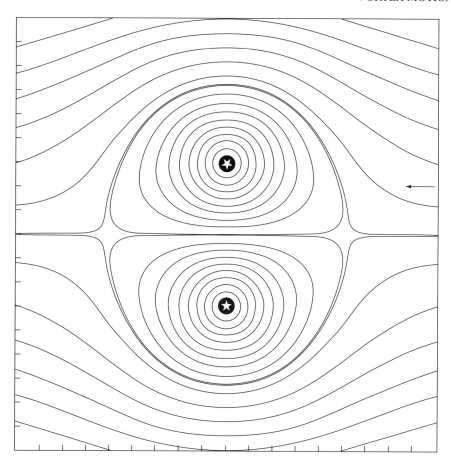

Figure 11.2.1 Streamline pattern due to a pair of point vortices with strengths of equal magnitude and opposite sign, in a frame of reference moving with the vortex pair. The plane of symmetry may be regarded as a flat wall.

Point vortex above a plane wall

The image of a point vortex with respect to a plane wall is another point vortex with strength of equal magnitude and opposite sign placed in the instantaneous mirror-image position of the original point vortex. Examining the velocity field induced by the image point vortex shows that the original point vortex moves parallel to the wall with velocity equal to $\kappa/(4\pi a)$, where a is the distance of the point vortex from the wall. The corresponding streamline pattern in a frame of reference that moves with the point vortex is identical to that shown in Figure 11.2.1 provided that we identify the plane of symmetry with the wall.

Point vortex inside a circular cylinder

The image of a point vortex located at the position $\mathbf{X}$ inside a circular cylinder of radius a is another point vortex with strength of equal magnitude and opposite sign placed at the instantaneous image point with respect to the cylinder, located at the position $\mathbf{x}_c + (\mathbf{X} - \mathbf{x}_c)a^2/|\mathbf{X} - \mathbf{x}_c|^2$, where $\mathbf{x}_c$ is the center of the cylinder. This shows that a single point vortex describes a circular path that is concentric with the cylinder, moving with azimuthal velocity

$$u_\theta = \frac{\kappa}{2\pi} \frac{|\mathbf{X} - \mathbf{x}_c|}{a^2 - |\mathbf{X} - \mathbf{x}_c|^2} \tag{11.2.5}$$

around the center of the cylinder.

Point vortex outside a circular cylinder

The image of a point vortex located at the point $\mathbf{X}$ outside a circular cylinder of radius a is another point vortex with strength of equal magnitude and opposite sign placed at the instantaneous image point. The image system may be enhanced with a third point vortex placed at the center of the cylinder whose strength κ_c is found by specifying the circulation of the flow around the cylinder. We thus find that a single point vortex describes a circular path that is concentric with the cylinder, moving with azimuthal velocity

$$u_\theta = \frac{\kappa}{2\pi} \frac{|\mathbf{X} - \mathbf{x}_c|}{a^2 - |\mathbf{X} - \mathbf{x}_c|^2} + \frac{\kappa_c}{2\pi} \frac{1}{|\mathbf{X} - \mathbf{x}_c|} \tag{11.2.6}$$

around the center of the cylinder, which is located at $\mathbf{x}_c$.

Point vortex near a corner

The image system of a point vortex in a partially infinite domain of flow that is bounded by a right-angle corner, whose boundaries are two intersecting plane walls, is the set of the three point vortices shown in Figure 11.2.2. In plane polar coordinates with the origin at the apex and the walls

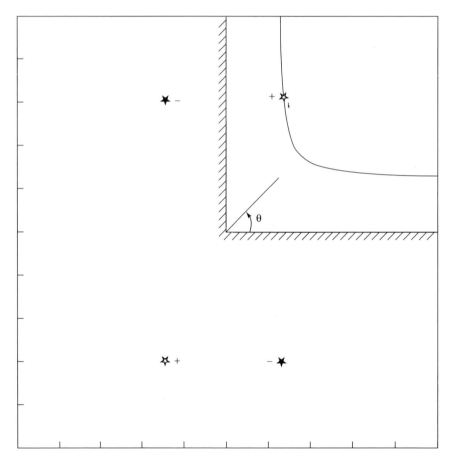

Figure 11.2.2 Trajectory and the three images of a point vortex near a right-angle corner. The strength of the point vortices shown with filled stars is equal in magnitude and opposite in sign with respect to that of the original point vortex.

located at $\theta = 0$ and $\pi/2$, the trajectory of the point vortex describes the curve $r = c/\sin(2\theta)$ where c is a constant determined by the initial position. A generalization of this configuration to corners with arbitrary aperture angle is discussed in Problem 11.2.2.

Point vortex between two parallel plane walls

Consider next the motion of a point vortex between two parallel plane walls that are separated by a distance h and are parallel to the x axis. The image system consists of two periodic arrays of point vortices arranged in the y direction with wavelength equal to $2h$. The first array contains the original point vortex, whereas the second array contains its image with respect to either the upper or the lower wall. The strength of the point vortices in the second array is equal in magnitude and opposite in sign to that of the point vortices in the first array.

Using the expression for the stream function due to a periodic array of point vortices given in the sixth entry of Table 7.8.1, we find that the stream function of the flow is given by

$$\psi(\mathbf{x}) = -\frac{\kappa}{4\pi} \ln \frac{\cosh[k(x - X)] - \cos[k(y - Y)]}{\cosh[k(x - X)] - \cos[k(y - Y^{\mathrm{Im}})]} \tag{11.2.7}$$

where $k = \pi/h$ and Y^{Im} is the y position of any one image point vortex. These results reveal that the point vortex travels along the x axis, parallel to the walls, under the influence of the velocity induced by the second array containing its image, so that

$$\frac{dX}{dt} = \frac{\kappa}{4h} \frac{\sin(2kb)}{1 - \cos(2kb)}, \qquad \frac{dY}{dt} = 0 \tag{11.2.8}$$

where b is the distance of the point vortex from the lower wall [see also Eqs. (2.10.9)]. In the limit as kb tends to zero, we recover the earlier results for the point vortex in a semi-infinite fluid above a single plane wall.

Complex boundary shapes

When the boundaries of the flow have a complex shape, the complementary potential flow may be computed using the method of conformal mapping discussed in Section 7.12, or one of the boundary integral methods discussed in Chapter 10.

Stability of a Single Row of Point Vortices

An infinite row of identical point vortices separated by a distance a is a useful model of the flow developing from the instability of a vortex layer discussed in Section 9.5. The associated streamline pattern was shown in Figure 2.10.1. When the point vortices are located precisely on the x axis, they remain stationary, and the flow is steady. To assess the stability of the periodic array, we displace the point vortices in the horizontal and vertical directions and compute their subsequent motion. As long as the displacements are small compared to a, the motion may be described in the context of the linear stability theory discussed in Chapter 9.

Following the general formalism of linear stability theory and normal-mode analysis, let us assume that the position of the mth point vortex is given by the real parts of the right-hand sides of the equations $X_m = ma + x_m$, $Y_m = y_m$, where

$$x_m = \varepsilon_1 a \exp[i(kma - \sigma t)], \qquad y_m = \varepsilon_2 a \exp[i(kma - \sigma t)] \tag{11.2.9}$$

are the complex displacements, ε_1 and ε_2 are two dimensionless complex constants with small but comparable magnitudes, k is the wave number of the perturbation, and σ is the complex growth rate. If the imaginary part of σ is positive, the vortex arrangement is unstable, as the point vortices depart from their unperturbed position at an exponential rate.

The physical requirement that the separation between the vortices be smaller than the wave length of the perturbation $L = 2\pi/k$ leads to the restriction $a/L = ka/2\pi \ll 1$. When L/a is an integer, the array is perturbed in a periodic fashion.

Linearizing the terms within the sums in Eqs. (11.2.1) with respect to ε_1 and ε_2, we obtain

$$
\begin{aligned}
\frac{Y_m - Y_n}{|\mathbf{X}_m - \mathbf{X}_n|^2} &\cong \frac{y_m - y_n}{(m - n)^2 a^2} \\
\frac{X_m - X_n}{|\mathbf{X}_m - \mathbf{X}_n|^2} &\cong \frac{(m - n)a + x_m - x_n}{(m - n)^2 a^2 + 2(m - n)a(x_m - x_n) + \cdots} \\
&\cong \frac{1}{(m - n)a} - \frac{x_m - x_n}{(m - n)^2 a^2} + \cdots
\end{aligned} \tag{11.2.10}
$$

If we attempt to apply Eqs. (11.2.1) for the infinite array, we shall run into a difficulty associated with the fact that the summation in the second equation does not converge. One way to circumvent this apparent but not essential difficulty is to retain a large but finite number of $2N + 1$ point vortices in the array. Substituting Eqs. (11.2.9) into Eqs. (11.2.10) and then into Eqs. (11.2.1), we obtain

$$
\begin{aligned}
\frac{dX_m}{dt} &= -\frac{\kappa \varepsilon_2}{2\pi a} e^{i(kma - \sigma t)} \sum_{\substack{n = -N \\ n \neq m}}^{N} \frac{1 - e^{ik(n - m)a}}{(m - n)^2} \\
\frac{dY_m}{dt} &= \frac{\kappa}{2\pi a} \sum_{\substack{n = -N \\ n \neq m}}^{N} \frac{1}{m - n} - \frac{\kappa \varepsilon_1}{2\pi a} e^{i(kma - \sigma t)} \sum_{\substack{n = -N \\ n \neq m}}^{N} \frac{1 - e^{ik(n - m)a}}{(m - n)^2}
\end{aligned} \tag{11.2.11}
$$

The first sum on the right-hand side of the second of Eqs. (11.2.11) is constant, independent of the position of the point vortices, and may thus be discarded. Taking the limit as N tends to infinity and substituting the expressions shown in Eqs. (11.2.9) into the left-hand sides, and requiring the existence of a nontrivial solution for ε_1 and ε_2, yields

$$
\sigma = \pm \frac{i\kappa}{2\pi a^2} \sum_{\substack{l = -\infty \\ l \neq 0}}^{\infty} \frac{1 - e^{ikla}}{l^2} = \pm \frac{i\kappa}{2a^2} ka \left(1 - \frac{ka}{2\pi}\right) = \pm \frac{i\pi\kappa}{a^2} \frac{a}{L} \left(1 - \frac{a}{L}\right) \tag{11.2.12}
$$

$$
\frac{\varepsilon_2}{\varepsilon_1} = \mp 1
$$

Since $a/L < 1$, the plus or minus sign may be selected according to the sign of the strength of the point vortices so that the imaginary part of σ is positive. The corresponding normal-mode displacement will cause the point vortices to move away from their original locations at an exponential rate. When the ratio L/a is an integer, call it K, we obtain

$$
\sigma = \pm i \frac{\pi\kappa}{a^2} \frac{K - 1}{K^2}, \qquad \frac{\varepsilon_2}{\varepsilon_1} = \mp 1 \tag{11.2.13}
$$

which shows that, when the strength of the point vortices is fixed, the most unstable perturbation corresponds to $K = 2$, which is associated with *pairing interactions* between adjacent point vortices.

An unstable and a stable normal-mode displacement of a vortex array with $\kappa < 0$ corresponding, respectively, to $\varepsilon_2/\varepsilon_1 = 1$ and -1 are shown in Figure 11.2.3. For the unstable mode, the point vortices are displaced in a manner that concentrates them near the regions where the interface slopes down, and this causes the local counterclockwise rotation of the array.

A different interpretation of Eq. (11.2.13) emerges by expressing the strength of the point vortices in terms of the velocity difference ΔU between the streams far above and below the vortex array and the wavelength L as $\kappa = -a\,\Delta U = -L\,\Delta U/K$. We then find

$$
\sigma = \pm i\pi \frac{\Delta U}{L} \frac{K - 1}{K} = \pm i \frac{1}{2} k\,\Delta U \frac{K - 1}{K} \tag{11.2.14}
$$

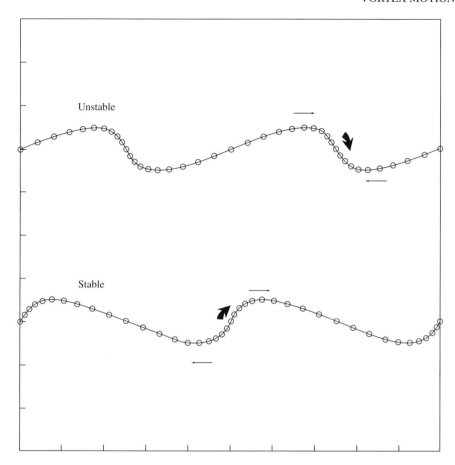

Figure 11.2.3   An unstable and a stable normal-mode displacement of an infinite array of point vortices with strength of negative sign.

which shows that keeping the velocities of the streams above and below the vortex array constant and the wavelength of the perturbation fixed, and increasing the number of point vortices per wavelength K from two to infinity, doubles the growth rate. The limit $K = \infty$ corresponds to a continuous distribution of point vortices yielding a vortex sheet. In this limit, the motion of a row of point vortices resembles that of a vortex sheet. More will be said about this connection in Section 11.4.

Periodic Arrangement of Point Vortices

Using Eqs. (2.10.9), we find that the motion of an infinite periodic collection of N point vortices that are repeated periodically in the x direction with wavelength equal to a is described by the system of equations

$$\frac{dX_i}{dt} = -\frac{1}{2a} \sum_{\substack{j=1 \\ j \neq i}}^{N} \kappa_j \frac{\sinh k(Y_i - Y_j)}{\cosh k(Y_i - Y_j) - \cos k(X_i - X_j)}$$

$$\frac{dY_i}{dt} = \frac{1}{2a} \sum_{\substack{j=1 \\ j \neq i}}^{N} \kappa_j \frac{\sin k(X_i - X_j)}{\cosh k(Y_i - Y_j) - \cos k(X_i - X_j)}$$

$$(11.2.15)$$

where $k = 2\pi/a$ is the wave number of the vortex array. Recall that we have already used these expressions previously in this section to compute the motion of a single point vortex between two parallel plane walls.

Von Kármán vortex street

As an application, let us consider the motion of the von Kármán vortex street consisting of two parallel periodic rows of point vortices arranged along the x axis and separated by a distance equal to b, as shown in Figure 11.2.4(a,b). The strength of the point vortices in the upper row, equal to κ, is equal and opposite to those in the lower row, equal to $-\kappa$.

Using Eqs. (11.2.15), we find that when the point vortices are located above each other in the arrangement of the *symmetric vortex street* shown in Figure 11.2.4(a), they translate along the x axis and their trajectories are governed by the equations

$$\frac{dX_i}{dt} = \frac{\kappa}{2a}\frac{\sinh kb}{\cosh kb - 1} = \frac{\kappa}{2a}\coth(\tfrac{1}{2}kb), \qquad \frac{dY_i}{dt} = 0 \qquad (11.2.16)$$

where $k = 2\pi/a$. Linear stability analysis shows that the symmetric vortex street is always unstable, and hence it should not be expected to occur in practice (Problem 11.2.4).

When the vortices in the two rows are shifted with respect to each other by a distance equal to half their separation, forming the *unsymmetric* or *staggered vortex street* shown in Figure 11.2.4(b), they translate along the x axis and their trajectories are governed by the equations

$$\frac{dX_i}{dt} = \frac{\kappa}{2a}\frac{\sinh kb}{\cosh kb + 1} = \frac{\kappa}{2a}\tanh(\tfrac{1}{2}kb), \qquad \frac{dY_i}{dt} = 0 \qquad (11.2.17)$$

Linear stability analysis shows that the unsymmetric vortex street is unstable, except when $\cosh(\tfrac{1}{2}kb) = \sqrt{2}$ or $b/a = 0.281$, in which case it is marginally stable (Problem 11.2.4). Non-linear effects, however, promote the instability, and, for all practical purposes, the unsymmetric vortex street is considered to be unstable.

(a)

(b)

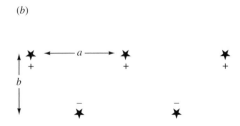

Figure 11.2.4   (a) A symmetric, and (b) an unsymmetric or staggered street of point vortices.

PROBLEMS

11.2.1 **Equation of motion of point vortices in complex variables.** Introduce the complex variable $z = x + iy$, and verify that Eqs. (11.2.1) and (11.2.15) may be expressed in the compact forms

$$\frac{dZ_i^*}{dt} = \frac{1}{2\pi i} \sum_{\substack{j=1 \\ j \neq i}}^{N} \frac{\kappa_j}{Z_i - Z_j} \tag{11.2.18}$$

$$\frac{dZ_i^*}{dt} = \frac{k}{4\pi i} \sum_{\substack{j=1 \\ j \neq i}}^{N} \kappa_j \cot\left[\frac{k}{2}(Z_i - Z_j)\right] \tag{11.2.19}$$

where an asterisk denotes the complex conjugate.

11.2.2 **Motion of a point vortex near a corner.** Consider the motion of a point vortex in the vicinity of a corner with internal angle equal to π/n. Using the theory of conformal mapping, show that, in plane polar coordinates with the origin at the apex and the walls located at $\theta = 0$ and π/n, the trajectory of the point vortex is described by $r = c/\sin(n\theta)$, where c is a constant determined by the initial position of the point vortex (Batchelor, 1967, p. 536).

11.2.3 **Point vortices behind a cylinder.** Föppl considered uniform potential flow with velocity U along the x axis past a stationary cylinder of radius a centered at the origin, with vanishing circulation around it, in the presence of a pair of point vortices with opposite strengths symmetrically located above and below the x axis behind the cylinder (Lamb, 1932, p. 223). Verify that the point vortices will remain stationary provided that they lie on a curve described by $2ry = r^2 - a^2$, and their strengths are given by $\kappa = \pm 2Uy(1 - a^4/r^4)$. This configuration, however, is unstable.

11.2.4 **Stability of the vortex street.** Carry out the linear stability analysis of (a) the symmetric and (b) the unsymmetric vortex street (Lamb, 1932, pp. 225, 228).

Computer Problems

11.2.5 **Motion of a periodic array of point vortices between two walls.** Consider the motion of a periodic array of point vortices separated by a distance a, placed between two parallel plane walls that are separated by a distance h. Compute and plot the velocity of the array as a function of the ratios b/h and a/h, where b is the distance of the point vortices from the lower wall, and discuss your results.

11.2.6 **Stability of a polygonal arrangement of point vortices.** Consider N point vortices with identical strengths equal to κ, placed at the vertices of a regular polygon that is inscribed within a circle of radius a. Havelock (1931) showed that the array rotates while maintaining its shape with angular velocity equal to $\Omega = (N - 1)\kappa/(4\pi a^2)$. The motion is stable when $N < 7$, marginally stable when $N = 7$, and unstable when $N > 7$. Write a program that computes the motion of the point vortices, and perform computations with $N = 6, 7, 8$ over an extended period of time. Discuss differences in behaviors due to the amplification of round-off error.

11.2.7 **Polygonal arrangement of point vortices in the presence of a cylinder.** Consider the polygonal arrangement of point vortices discussed in Problem 11.2.6, but this time in the presence of a circular cylinder of radius b that is concentric with the polygon. Compute and plot the angular velocity of the point vortices for $N = 2, 5, 10$ and $b/a = 0.1, 0.5, 0.9, 1.1, 1.5, 2.0$ and compare your results with the analytical solution for unbounded flow.

11.2.8 **Instability of an infinite array of point vortices.** Consider the motion of a periodic array of point vortices separated by a distance a, lying between two streams that flow with velocities

$\pm U$. The strength of each point vortex is given by $\kappa = -2U/a$. At the initial instant, the array is subjected to a sinusoidal perturbation of wavelength $L = Ka$ and amplitude b_0, where K is an integer, as described in parts (b)–(d). (a) Write a program called *PVA* that computes the motion of the point vortices, and plots the profiles of the evolving vortex array. The integration of the ordinary differential equations should be done using the modified Euler method. The program should compute the ratio between the maximum y displacement of the point vortices b to the initial amplitude b_0. (b) Consider an unstable normal-mode perturbation where the initial position of the mth point vortex is given by

$$X_m(0) = a(m-1) + b_0 \sin \varphi_m, \qquad Y_m(0) = b_0 \sin \varphi_m \qquad (11.2.20)$$

where $\varphi_m = 2\pi(m-1)/K$, $b_0 > 0$, and $\kappa < 0$. Compute the motion for $b_0/L = 0.10$ and $K = L/a = 4, 8, 16, 32$. In each case, plot the ratio b/b_0 as a function of time on a linear-log scale, and compare your results with those predicted by the linear stability analysis discussed in the text. (c) Repeat part (b) with a stable normal-mode perturbation in which the initial position of the point vortices is given by

$$X_m(0) = a(m-1) + b_0 \sin \varphi_m, \qquad Y_m(0) = -b_0 \sin \varphi_m \qquad (11.2.21)$$

(d) Repeat part (b) with a sinusoidal perturbation where the initial position of the point vortices is given by

$$X_m(0) = a(m-1), \qquad Y_m(0) = b_0 \sin \varphi_m \qquad (11.2.22)$$

Discuss the difference in behavior between cases (b) and (d).

11.3 | VORTEX BLOBS

Point vortices are mathematical idealizations arising in the limit as the cross-section of two-dimensional rectilinear vortex filaments tends to zero, while the circulation around them is held at a constant value. Unfortunately, the infinitesimal cross-section of the point vortices is known to cause irregular motion.

One way of suppressing the erratic motion is to replace the delta function that describes the vorticity distribution with a regularized distribution that is supported by a finite cross-sectional area, thereby replacing the point vortices with vortex blobs. The instantaneous vorticity associated with a vortex blob is given by

$$\omega(\mathbf{x}) = \kappa \, \zeta(\hat{\mathbf{x}}) \qquad (11.3.1)$$

where $\hat{\mathbf{x}} = |\mathbf{x} - \mathbf{X}|$, and κ and $\mathbf{X}$ are, respectively, the strength and the center of the vortex blob (Chorin, 1973). The distribution function ζ is a regularization of the two-dimensional delta function, normalized so that its integral over the entire xy plane is equal to unity.

Assuming that the vorticity of a blob is radially symmetric with respect to its center, we write

$$\zeta(\mathbf{x}) = \frac{\pi}{\sigma^2} f\left(\frac{r}{\sigma}\right) \qquad (11.3.2)$$

where $r = |\mathbf{x} - \mathbf{X}|$ is the distance of the point $\mathbf{x}$ from the center of the blob, σ is the radius of the blob, which is a measure of the extent to which the vorticity is spread out from the center of the blob, and f is a dimensionless distribution or core function that vanishes at infinity. In the limit as σ tends to zero, ζ reduces to the two-dimensional delta function. In order for the integral of the function ζ over the entire xy plane to be equal to unity, the function f must satisfy the integral constraint

$$\int_0^\infty w f(w) \, dw = \frac{1}{2\pi^2} \qquad (11.3.3)$$

A simple choice for the core function f is the piecewise constant top-hat distribution

$$f(w) = \begin{cases} 1/\pi^2 & \text{for } 0 < w < 1 \\ 0 & \text{for } w > 1 \end{cases} \tag{11.3.4}$$

which yields a *circular Rankine vortex*. Another choice is the Gaussian distribution $f(w) = (1/\pi^2)\exp(-w^2)$. Both choices satisfy the integral constraint (11.3.3).

Velocity Field Due to an Axisymmetric Vortex Blob

In plane polar coordinates centered at an axisymmetric vortex blob, the radial component of the velocity vanishes. A simple way of obtaining the angular polar component of the velocity is to use the definition of vorticity to obtain

$$\frac{1}{r}\frac{d}{dr}(ru_\theta) = \frac{\kappa\pi}{\sigma^2}f\left(\frac{r}{\sigma}\right) \tag{11.3.5}$$

Straightforward integration subject to the condition that the velocity vanishes at the center of the blob yields

$$u_\theta(x_0) = \frac{\kappa}{2\pi r}g\left(\frac{r}{\sigma}\right), \qquad \text{where} \quad g(w) = 2\pi^2\int_0^w w'f(w')\,dw' \tag{11.3.6}$$

The Cartesian components of the velocity at the point $\mathbf{x}$ due to a blob that is located at the point $\mathbf{X}$ are given by

$$\begin{bmatrix} u_x \\ u_y \end{bmatrix}(\mathbf{x}) = \frac{\kappa}{2\pi}\begin{bmatrix} -y+Y \\ x-X \end{bmatrix}\frac{g(w)}{|\mathbf{x}-\mathbf{X}|^2} \tag{11.3.7}$$

where $w = |\mathbf{x}-\mathbf{X}|/\sigma$. This velocity field differs from that due to a point vortex only by the presence of the distribution function g. Equation (11.3.3) shows that far from the center of the blob, $g(w)$ tends to unity, thereby ensuring that the velocity reduces to that due to a point vortex.

For the top-hat distribution described in Eq. (11.3.4), computing the integral in Eqs. (11.3.6) yields

$$g(w) = \begin{cases} w^2 & \text{for } 0 < w < 1 \\ 1 & \text{for } w > 1 \end{cases} \tag{11.3.8}$$

which shows that the fluid inside the blob rotates like a rigid body, whereas the flow outside the blob is identical to that due to a point vortex. For the Gaussian distribution $f(w) = (1/\pi^2)\exp(-w^2)$, we obtain $g(w) = 1 - \exp(-w^2)$.

It is instructive to rederive the velocity field due to a vortex blob using the Biot–Savart integral described by Eqs. (2.10.1). Setting the origin at the center of a blob, we find that the y component of the velocity at a point x_0 that is located on the x axis is given by

$$u_y(x_0) = \frac{\kappa}{2\pi}\int_0^{2\pi}\int_0^\infty \frac{x_0 - r\sin\theta}{x_0^2 + r^2 - 2x_0 r\cos\theta}\zeta(\mathbf{x})\,r\,dr\,d\theta \tag{11.3.9}$$

Substituting Eq. (11.3.2) into the integrand of Eq. (11.3.9), we obtain

$$\begin{aligned} u_y(x_0) &= \frac{\kappa}{2\sigma^2}\int_0^{2\pi}\int_0^\infty \frac{x_0 - r\sin\theta}{x_0^2 + r^2 - 2x_0 r\cos\theta}f\left(\frac{r}{\sigma}\right)r\,dr\,d\theta \\ &= \frac{\kappa}{2\sigma}\int_0^\infty\int_0^{2\pi} \frac{x_0/\sigma - w\sin\theta}{x_0^2/\sigma^2 + w^2 - 2(x_0/\sigma)w\cos\theta}\,d\theta\,w f(w)\,dw \end{aligned} \tag{11.3.10}$$

where $w = r/\sigma$. Computing the integral with respect to θ on the right-hand side of Eq. (11.3.10) yields the right-hand side of the first of Eqs. (11.3.6) with x_0 in place of r.

Motion of a Collection of Vortex Blobs

The motion of the center of a vortex blob that belongs to a finite collection of N radially symmetric blobs with identical structure is described by the following modified version of Eq. (11.2.1),

$$
\frac{dX_i}{dt} = -\frac{1}{2\pi} \sum_{\substack{j=1 \\ j \neq i}}^{N} \kappa_j \frac{Y_i - Y_j}{|\mathbf{X}_i - \mathbf{X}_j|^2} g\left(\frac{|\mathbf{X}_i - \mathbf{X}_j|}{\sigma_j}\right)
$$

$$
\frac{dY_i}{dt} = \frac{1}{2\pi} \sum_{\substack{j=1 \\ j \neq i}}^{N} \kappa_j \frac{X_i - X_j}{|\mathbf{X}_i - \mathbf{X}_j|^2} g\left(\frac{|\mathbf{X}_i - \mathbf{X}_j|}{\sigma_j}\right)
$$

(11.3.11)

The strength of each blob remains constant in time in order to preserve the total circulation.

Diffusing Blobs

When viscous stresses are significant, the vorticity of the blobs spreads out due to diffusion. For a single blob with Gaussian vorticity distribution $f(w) = \exp(-w^2)$, we use the vorticity transport equation, written in plane polar coordinates, and find that the vorticity distribution remains Gaussian, and its radius σ increases according to

$$
\frac{d\sigma}{dt} = 2\frac{\nu}{\sigma}
$$

(11.3.12)

where ν is the kinematic viscosity of the fluid. Thus, σ^2 increases at a linear rate.

Equations (11.3.11) and (11.3.12) may be used to describe the motion of the centers and evolution of the core sizes of a collection of Gaussian vortex blobs. In practice, the vorticity distribution of each blob loses its radial symmetry due to mutual interactions, but this effect is usually neglected in the interest of keeping the mathematical model at a simple level.

Periodic Array of Blobs

Combining Eq. (11.2.15) with Eq. (11.3.11), we find that the motion of a collection of N vortex blobs that are repeated periodically in the x direction with period a is described by the system

$$
\frac{dX_i}{dt} = -\frac{1}{2a} \sum_{\substack{j=1 \\ j \neq i}}^{N} \kappa_j \frac{\sinh k(Y_i - Y_j)}{\cosh k(Y_i - Y_j) - \cos k(X_i - X_j)}
$$

$$
- \frac{1}{2\pi} \sum_{\substack{j=1 \\ j \neq i}}^{N} \sum_{m=-\infty}^{\infty} \kappa_j \frac{Y_i - Y_j}{R_{ijm}^2} \left[g\left(\frac{R_{ijm}}{\sigma_j}\right) - 1 \right]
$$

$$
\frac{dY_i}{dt} = \frac{1}{2a} \sum_{\substack{j=1 \\ j \neq i}}^{N} \kappa_j \frac{\sin k(X_i - X_j)}{\cosh k(Y_i - Y_j) - \cos k(X_i - X_j)}
$$

$$
+ \frac{1}{2\pi} \sum_{\substack{j=1 \\ j \neq i}}^{N} \sum_{m=-\infty}^{\infty} \kappa_j \frac{X_i - X_j - ma}{R_{ijm}^2} \left[g\left(\frac{R_{ijm}}{\sigma_j}\right) - 1 \right]
$$

(11.3.13)

where

$$
R_{ijm}^2 = (X_i - X_j - ma)^2 + (Y_i - Y_j)^2
$$

The first sum on the right-hand side of each equation in (11.3.13) expresses the velocity due to a periodic array of point vortices, and the second sum expresses corrections due to the finite size of the blobs. The infinite sums over m converge rapidly and, in practice, they are truncated at a finite level.

Computer Problems

11.3.1   **Stability of a polygonal arrangement of vortex blobs.** (a) Consider Problem 11.2.6 with vortex blobs with top-hat vorticity distribution instead of point vortices. Write a program that computes the motion of the blobs, and carry out computations with $N = 6$ for a sequence of blobs with core sizes $\sigma/a = 0.01, 0.05, 0.10, 0.20, 0.50$. Plot the angular velocity of rotation of the vortex arrangement as a function of σ/a and discuss the physical significance of your results. (b) Repeat part (a) for vortex blobs with Gaussian vorticity distribution. (c) Repeat part (b) for diffusing Gaussian blobs whose radii increase according to Eq. (11.3.12). Plot the angular velocity of rotation of the vortex arrangement as a function of a proper dimensionless time and discuss the physical significance of your results.

11.3.2   **Vortex street with vortex blobs.** (a) Compute and plot the velocity of translation of vortex blobs with top-hat vorticity distribution in a symmetric and an unsymmetric vortex street as a function of the blob core size σ/a for $b/a = 0.10, 0.25, 0.50$. (b) Repeat (a) for vortex blobs with Gaussian vorticity distribution.

11.4 | TWO-DIMENSIONAL VORTEX SHEETS

A two-dimensional vortex sheet is a cylindrical surface across which the tangential component of the velocity, which is perpendicular to the generators, undergoes a discontinuity that is called the *strength* of the vortex sheet (Section 1.9). Since a finite fluid viscosity would cause the unphysical onset of infinite stresses and rate of viscous dissipation, the notion of a vortex sheet whose thickness is and remains infinitesimal at all times is acceptable only if the fluids are considered to be inviscid.

Marker Points

To describe the motion of a vortex sheet, we mark its trace in the xy plane with a continuous distribution of marker points that are identified by the label a, as depicted in Figure 11.4.1, and then follow the evolution of the vortex sheet by computing the trajectories of the marker points. Kinematic considerations require that the component of the velocity of the marker points normal to the vortex sheet be equal to the normal component of the velocity of the fluid on either

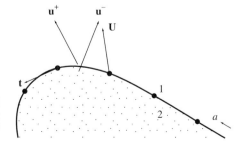

Figure 11.4.1   Schematic illustration of a two-dimensional vortex sheet described by a set of marker points. The marker points move with the velocity of the fluid normal to the vortex sheet while executing an arbitrary tangential motion.

side of the vortex sheet. The tangential component of the velocity of the marker points, however, may have an arbitrary value. Thus, if $\mathbf{X}(a)$ and $\mathbf{U}(a)$ are the position and velocity of a marker point, we write

$$\frac{d\mathbf{X}(a)}{dt} \equiv \mathbf{U}(a) = (\mathbf{u}^{\pm} \cdot \mathbf{n})\, \mathbf{n} + v(a)\, \mathbf{t} \qquad (11.4.1)$$

where $\mathbf{u}^{\pm}$ is the velocity of the fluid on either side the vortex sheet, and $v(a)$ is an arbitrary tangential velocity that may vary in time and across the line of marker points (Figure 11.4.1).

Using Eqs. (1.9.14), we recast Eq. (11.4.1) in terms of the velocity of the fluid on either side of the vortex sheet or the principal velocity of the vortex sheet, defined as the mean value of the velocity of the fluid on either side of the vortex sheet, in the form

$$\frac{d\mathbf{X}(a)}{dt} \equiv \mathbf{U}(a) = \alpha \mathbf{u}^{+} + (1 - \alpha)\, \mathbf{u}^{-} = \mathbf{u}^{PV} + (\alpha - \tfrac{1}{2})\gamma \mathbf{t} \qquad (11.4.2)$$

where γ is the strength of the vortex sheet. The scalar parameter α is related to the tangential velocity of the marker points by $v = \mathbf{u}^{PV} \cdot \mathbf{t} + (\alpha - \tfrac{1}{2})\gamma$. When $\alpha = 1, \tfrac{1}{2}, 0$, the marker points move, respectively, with the velocity of the fluid on the upper side of the vortex sheet, the principal velocity of the vortex sheet, and the velocity of the fluid on the lower side of the vortex sheet.

Substituting Eq. (2.10.12) into Eq. (11.4.2), we obtain an integro-differential equation for the motion of the marker points

$$\frac{d\mathbf{X}(a)}{dt} = \frac{1}{2\pi} \int_C^{PV} \left[\begin{array}{c} -Y(a) + y(a') \\ X(a) - x(a') \end{array} \right] \frac{\gamma(a')}{|\mathbf{X}(a) - \mathbf{x}(a')|^2} \frac{\partial l}{\partial a'}\, da'$$
$$+ (\alpha - \tfrac{1}{2})\, \gamma(a)\, \mathbf{t}(a) + \mathbf{v}(a) \qquad (11.4.3)$$

where C is the trace of the vortex sheet in the xy plane. The first two terms on the right-hand side of Eq. (11.4.3) express the *self-induced velocity* of the vortex sheet, and the last term expresses an externally imposed, generally time-dependent, rotational or irrotational motion.

Evolution of the Strength of the Vortex Sheet

To compute the motion of the vortex sheet, we require an evolution equation for the strength of the vortex sheet γ following the marker points, which is expressed by the material derivative $(\partial\gamma/\partial t)_a$. To derive this equation we write Euler's equation on either side of the vortex sheet

$$\frac{\partial \mathbf{u}^{\pm}}{\partial t} + \mathbf{u}^{\pm} \cdot \nabla \mathbf{u}^{\pm} = -\frac{1}{\rho_{\pm}} \nabla p^{\pm} + \mathbf{g} \qquad (11.4.4)$$

where the plus and minus signs indicate evaluations, respectively, on the upper and lower side of the vortex sheet. A slight rearrangement of the left-hand side yields

$$\frac{\partial \mathbf{u}^{\pm}}{\partial t} + \mathbf{U} \cdot \nabla \mathbf{u}^{\pm} + (\mathbf{u}^{\pm} - \mathbf{U}) \cdot \nabla \mathbf{u}^{\pm} = -\frac{1}{\rho_{\pm}} \nabla p^{\pm} + \mathbf{g} \qquad (11.4.5)$$

The first two terms on the left-hand side of Eq. (11.4.5) express the derivative $(\partial\mathbf{u}^{\pm}/\partial t)_a$, which represents the change in the velocity of the fluid on either side of the vortex sheet following the marker points.

Next we use the relation $\mathbf{u}^{+} - \mathbf{u}^{-} = \gamma \mathbf{t}$ in conjunction with Eq. (11.4.2), and express the velocity on either side of the vortex sheet in terms of the velocity of the marker points as

$$\mathbf{u}^{+} = \mathbf{U} + \alpha_{+}\gamma\mathbf{t}, \qquad \mathbf{u}^{-} = \mathbf{U} - \alpha_{-}\gamma\mathbf{t} \qquad (11.4.6)$$

where, for convenience of notation, we have introduced the new parameters

$$\alpha_{+} = 1 - \alpha, \qquad \alpha_{-} = \alpha \qquad (11.4.7)$$

Differentiating Eqs. (11.4.6), we obtain

$$\left(\frac{\partial \mathbf{u}^\pm}{\partial t}\right)_a = \left(\frac{\partial \mathbf{U}}{\partial t}\right)_a \pm \alpha_\pm \left(\frac{\partial \gamma \mathbf{t}}{\partial t}\right)_a \qquad (11.4.8)$$

Substituting the right-hand side of Eq. (11.4.8) in place of the first two terms in Eq. (11.4.5), replacing the difference $\mathbf{u}^\pm - \mathbf{U}$ in the third term with $\pm \alpha_\pm \gamma \mathbf{t}$ and the velocity $\mathbf{u}^\pm$ after the second gradient with $\mathbf{U} \pm \alpha_\pm \gamma \mathbf{t}$, and multiplying the resulting expression by the density, we obtain two equations involving the velocity of the marker points, the strength of the vortex sheet, and the pressure gradient on either side of the vortex sheet,

$$\rho_\pm \left(\frac{\partial \mathbf{U}}{\partial t}\right)_a \pm \rho_\pm \alpha_\pm \left(\frac{\partial \gamma \mathbf{t}}{\partial t}\right)_a \pm \rho_\pm \alpha_\pm \gamma \, \frac{\partial(\mathbf{U} \pm \alpha_\pm \gamma \mathbf{t})}{\partial l} = -\nabla p^\pm + \rho_\pm \mathbf{g} \qquad (11.4.9)$$

where l is the arc length along the vortex sheet measured in the direction of $\mathbf{t}$.

Next we take the difference of the two equations (11.4.9) evaluated on the upper and lower side of the vortex sheet, project it onto the tangential vector $\mathbf{t}$ to form the tangential derivatives of the pressure, and thus obtain

$$\left\{(\rho_+ - \rho_-)\left(\frac{\partial \mathbf{U}}{\partial t}\right)_a + (\rho_+ \alpha_+ + \rho_- \alpha_-)\left[\left(\frac{\partial \gamma \mathbf{t}}{\partial t}\right)_a + \gamma \frac{\partial \mathbf{U}}{\partial l}\right]\right.$$
$$\left. + (\rho_+ \alpha_+^2 - \rho_- \alpha_-^2)\gamma \frac{\partial \gamma \mathbf{t}}{\partial l}\right\} \cdot \mathbf{t} = -\frac{\partial(p^+ - p^-)}{\partial l} + (\rho_+ - \rho_-)\,\mathbf{g} \cdot \mathbf{t} \qquad (11.4.10)$$

Furthermore, we expand the derivatives of the terms containing $\gamma \mathbf{t}$, and note that because $\mathbf{t}$ is a unit vector $\mathbf{t} \cdot (\partial \mathbf{t}/\partial t)_a = 0$ and $\mathbf{t} \cdot \partial \mathbf{t}/\partial l = 0$. Rearranging the resulting expression, we obtain the simplified form

$$\left(\frac{\partial \gamma}{\partial t}\right)_a + \gamma \mathbf{t} \cdot \frac{\partial \mathbf{U}}{\partial l} = A_1 \mathbf{t} \cdot \left(\frac{\partial \mathbf{U}}{\partial t}\right)_a + A_2 \gamma \frac{\partial \gamma}{\partial l} + \frac{1}{\rho_+ \alpha_+ + \rho_- \alpha_-} \frac{\partial(p^- - p^+)}{\partial l} - A_1 \mathbf{g} \cdot \mathbf{t} \qquad (11.4.11)$$

where

$$A_1 = \frac{\rho_- - \rho_+}{\rho_+ \alpha_+ + \rho_- \alpha_-} = \frac{\rho_- - \rho_+}{\rho_+ + \alpha(\rho_- - \rho_+)} \qquad (11.4.12)$$

$$A_2 = \frac{\rho_- \alpha_-^2 - \rho_+ \alpha_+^2}{\rho_+ \alpha_+ + \rho_- \alpha_-} = \frac{\rho_- \alpha^2 - \rho_+ (1 - \alpha)^2}{\rho_+ + \alpha(\rho_- - \rho_+)} \qquad (11.4.13)$$

The jump in pressure across the vortex sheet on the right-hand side of Eq. (11.4.11) depends upon the properties of the interface that separates the two fluids lying on either side of the vortex sheet, as discussed in Section 3.5. Assuming, for instance, that the vortex sheet represents the interface between two fluids with constant surface tension T, we use Eq. (3.5.12) to obtain

$$p^- - p^+ = T\kappa \qquad (11.4.14)$$

where κ is the curvature of the vortex sheet in the xy plane, which is reckoned to be positive when the interface is concave when seen from the upper fluid.

Equation (11.4.11), combined with Eq. (11.4.14), provides us with the desired evolution equation for the strength of the vortex sheet following the marker points, expressed in terms of the velocity of the marker points $\mathbf{U}$ and the tangential component of the acceleration of the marker points $\mathbf{t} \cdot (\partial \mathbf{U}/\partial t)_a$. It is important to note that the tangential acceleration is an implicit function of $(\partial \gamma/\partial t)_a$. To establish this functional relationship, we differentiate the right-hand side of

Eq. (11.4.3) with respect to time, keeping a constant, project the resulting expression onto the tangential vector, and use the fact that the length of $\mathbf{t}$ remains constant in time, finding

$$\mathbf{t}(a) \cdot \left(\frac{\partial \mathbf{U}}{\partial t}\right)_a = \frac{1}{2\pi} \mathbf{t}(a) \cdot \int_C^{PV} \frac{\partial}{\partial t} \left(\begin{bmatrix} -Y(a) + y(a') \\ X(a) - x(a') \end{bmatrix} \frac{\gamma(a')}{|\mathbf{X}(a) - \mathbf{x}(a')|^2} \frac{\partial l}{\partial a'}\right)_{a,a'} da'$$
$$+ \left(\alpha - \tfrac{1}{2}\right)\left(\frac{\partial \gamma}{\partial t}\right)_a + \mathbf{t}(a) \cdot \left(\frac{\partial \mathbf{v}}{\partial t}\right)_a \qquad (11.4.15)$$

Expanding out the time derivative under the integral sign, we obtain

$$\mathbf{t}(a) \cdot \left(\frac{\partial \mathbf{U}}{\partial t}\right)_a = \frac{1}{2\pi} \mathbf{t}(a) \cdot \int_C^{PV} \begin{bmatrix} -Y(a) + y(a') \\ X(a) - x(a') \end{bmatrix} \frac{1}{|\mathbf{X}(a) - \mathbf{x}(a')|^2} \left(\frac{\partial \gamma}{\partial t}\right)_{a'} dl(a')$$
$$+ J(a) + \left(\alpha - \tfrac{1}{2}\right)\left(\frac{\partial \gamma}{\partial t}\right)_a + \mathbf{t}(a) \cdot \left(\frac{\partial \mathbf{v}}{\partial t}\right)_a \qquad (11.4.16)$$

where we have introduced the auxiliary function

$$J(a) = \frac{1}{2\pi} \mathbf{t}(a) \cdot \int_C^{PV} \frac{\partial}{\partial t}\left(\begin{bmatrix} -Y(a) + y(a') \\ X(a) - x(a') \end{bmatrix} \frac{1}{|\mathbf{X}(a) - \mathbf{x}(a')|^2} \frac{\partial l}{\partial a'}\right)_{a,a'} \gamma(a')\, da' \quad (11.4.17)$$

which expresses the tangential component of the acceleration of the point particles if they were moving while preserving their circulation.

Substituting Eq. (11.4.16) into the first term on the right-hand side of Eq. (11.4.11) and rearranging, we obtain a *linear Fredholm integral equation of the second kind* for $(\partial \gamma / \partial t)_a$

$$\left(\frac{\partial \gamma}{\partial t}\right)_a = \frac{A}{\pi} \mathbf{t}(a) \cdot \int_C^{PV} \begin{bmatrix} -Y(a) + y(a') \\ X(a) - x(a') \end{bmatrix} \frac{1}{|\mathbf{X}(a) - \mathbf{x}(a')|^2} \left(\frac{\partial \gamma}{\partial t}\right)_{a'} dl(a') + F(a) \quad (11.4.18)$$

where

$$A = \frac{\rho_- - \rho_+}{\rho_- + \rho_+} \qquad (11.4.19)$$

is the *Atwood ratio*. The forcing function $F(a)$ on the right-hand side of Eq. (11.4.18) is given by

$$F(a) = -\frac{1}{A_3} \gamma \mathbf{t} \cdot \frac{\partial \mathbf{U}}{\partial l} + 2A\, J(a) + 2A\, \mathbf{t}(a) \cdot \left(\frac{\partial \mathbf{v}}{\partial t}\right)_a$$
$$+ A_4 \gamma \frac{\partial \gamma}{\partial l} + \frac{2}{\rho_+ + \rho_-} \frac{\partial (p^- - p^+)}{\partial l} - 2A\, \mathbf{g} \cdot \mathbf{t} \qquad (11.4.20)$$

where

$$A_3 \equiv 1 + A_1(\tfrac{1}{2} - \alpha) = \frac{1}{2}\frac{\rho_- + \rho_+}{\rho_+ \alpha_+ + \rho_- \alpha_-} = \frac{1}{2}\frac{\rho_- + \rho_+}{\rho_+ + \alpha(\rho_- - \rho_+)} \qquad (11.4.21)$$

$$A_4 = \frac{A_2}{A_3} = 2\frac{\rho_- \alpha_-^2 - \rho_+ \alpha_+^2}{\rho_- + \rho_+} = 2\frac{\rho_- \alpha^2 - \rho_+ (1 - \alpha)^2}{\rho_- + \rho_+} \qquad (11.4.22)$$

Evolution of circulation along the vortex sheet

The circulation along the vortex sheet Γ is defined by the differential relationship $d\Gamma = \gamma\, dl$, where $\Gamma = 0$ at a designated marker point. Straightforward differentiation and use of Eq. (1.4.2) yields

$$\left(\frac{\partial\, d\Gamma}{\partial t}\right)_a = \left(\frac{\partial \gamma\, dl}{\partial t}\right)_a = \left(\frac{\partial \gamma}{\partial t}\right)_a dl + \gamma \mathbf{t} \cdot \frac{\partial \mathbf{U}}{\partial l} dl \qquad (11.4.23)$$

Solving Eq. (11.4.23) for $(\partial \gamma / \partial t)_a$, substituting the result into the first term on the left-hand side of Eq. (11.4.11), and then multiplying the emerging equation by $\partial l / \partial a$, we derive an integro-differential evolution equation for the circulation along the vortex sheet

$$\frac{\partial^2 \Gamma(a, t)}{\partial a \, \partial t} = A_1 \mathbf{t} \cdot \left(\frac{\partial \mathbf{U}}{\partial t}\right)_a \frac{\partial l}{\partial a} + A_2 \frac{1}{2} \frac{\partial \gamma^2}{\partial a} + \frac{1}{\rho_- \alpha_+ + \rho_- \alpha_-} \frac{\partial(p^- - p^+)}{\partial a} - A_1 \mathbf{g} \cdot \frac{\partial \mathbf{x}}{\partial a}$$

(11.4.24)

where the tangential acceleration is computed from Eq. (11.4.16).

Marker points moving with the principal velocity of the vortex sheet

When the marker points are required to move with the principal velocity of the vortex sheet, in which case $\alpha = \frac{1}{2}$, the coefficients in Eqs. (11.4.12), (11.4.13), (11.4.21), and (11.4.22) take the values

$$A_1 = 2A, \qquad A_2 = \tfrac{1}{2}A, \qquad A_3 = 1, \qquad A_4 = \tfrac{1}{2}A \qquad (11.4.25)$$

where A is the Atwood ratio defined in Eq. (11.4.19). Under these circumstances, Eq. (11.4.11) assumes the simpler form

$$\left(\frac{\partial \gamma}{\partial t}\right)_a + \gamma \mathbf{t} \cdot \frac{\partial \mathbf{U}}{\partial l} = 2A \mathbf{t} \cdot \left(\frac{\partial \mathbf{U}}{\partial t}\right)_a + \tfrac{1}{2} A \gamma \frac{\partial \gamma}{\partial l} + \frac{2}{\rho_+ + \rho_-} \frac{\partial(p^- - p^+)}{\partial l} - 2A \, \mathbf{g} \cdot \mathbf{t} \quad (11.4.26)$$

The first two terms on the right-hand side may be written in the compact form $2A \, \mathbf{t} \cdot \mathbf{a}$, where $\mathbf{a}$ is the mean value of the *acceleration of the fluid* on either side of the vortex sheet defined as

$$\mathbf{a} \equiv \tfrac{1}{2}\left(\frac{D\mathbf{u}^+}{Dt} + \frac{D\mathbf{u}^-}{Dt}\right) \qquad (11.4.27)$$

and D/Dt is the material derivative (Problem 11.4.2).

Furthermore, when the vortex sheet separates two fluids with identical densities, $A = 0$, the acceleration term disappears and Eqs. (11.4.26) and (11.4.24) provide us with explicit expressions for the rate of change of the strength of the vortex sheet and of the circulation along the vortex sheet following the marker points

$$\left(\frac{\partial \gamma}{\partial t}\right)_a = -\gamma \mathbf{t} \cdot \frac{\partial \mathbf{U}}{\partial l} + \frac{1}{\rho} \frac{\partial(p^- - p^+)}{\partial l} \qquad (11.4.28)$$

$$\frac{\partial^2 \Gamma(a, t)}{\partial a \, \partial t} = \frac{1}{\rho} \frac{\partial(p^- - p^+)}{\partial a} \qquad (11.4.29)$$

In the absence of surface tension, the pressure is continuous across the vortex sheet, and we obtain the simplified evolution equation $(\partial \Gamma / \partial t)_a = 0$, which states that the marker points preserve their initial circulation. In this case, it is convenient to identify the marker-point label a with Γ.

The Boussinesq Approximation

In certain applications, involving in particular gravity-driven flows, density variations play an important role in determining the magnitude of the body force, but make minor contributions to the magnitude of the inertial force. Under these circumstances, it is possible to simplify the computation of the flow by adopting the Boussinesq approximation, which amounts to replacing $\rho_\pm$ with the mean value $\rho = \tfrac{1}{2}(\rho_+ + \rho_-)$ in all but the gravity term in the equation of motion. Strictly speaking, the Boussinesq approximation is valid in the limit as the Atwood ratio A tends to zero while the magnitude of the body force g tends to infinity so that the product Ag remains finite.

Adopting the Boussinesq approximation, we find that the constants A_1 and A_2 in the first two terms on the right-hand side of Eq. (11.4.11) take the values $A_1 = 0$, $A_2 = 2\alpha - 1$.

Substituting these values into Eqs. (11.4.11) and (11.4.24) and rearranging, we derive the simplified expressions

$$\left(\frac{\partial\gamma}{\partial t}\right)_a + \gamma\mathbf{t}\cdot\frac{\partial\mathbf{U}}{\partial l} = (\alpha - \tfrac{1}{2})\frac{\partial\gamma^2}{\partial l} + \frac{1}{\rho}\frac{\partial(p^- - p^+)}{\partial l} - 2A\,\mathbf{g}\cdot\mathbf{t} \tag{11.4.30}$$

$$\frac{\partial^2\Gamma}{\partial a\,\partial t} = (\alpha - \tfrac{1}{2})\frac{\partial\gamma^2}{\partial a} + \frac{1}{\rho}\frac{\partial(p^- - p^+)}{\partial a} - 2A\,\mathbf{g}\cdot\frac{\partial\mathbf{x}}{\partial a} \tag{11.4.31}$$

Integrating Eq. (11.4.31) with respect to a yields the simpler form

$$\left(\frac{\partial\Gamma}{\partial t}\right)_a = (\alpha - \tfrac{1}{2})\gamma^2 + \frac{p^- - p^+}{\rho} - 2A\,\mathbf{g}\cdot\mathbf{x} + c \tag{11.4.32}$$

where c is an arbitrary constant that may be set equal to zero with no consequences on the strength of the vortex sheet. The simplifications resulting from the Boussinesq approximation are reflected in the absence of the tangential acceleration of the marker points on the right-hand sides of Eqs. (11.4.30) and (11.4.32).

In the particular case where

$$\alpha = \frac{1}{2} - \frac{p^- - p^+}{\gamma^2\rho} + \frac{2A}{\gamma^2}\,\mathbf{g}\cdot\mathbf{x} \tag{11.4.33}$$

the right-hand side of Eq. (11.4.32) vanishes, yielding the simplified evolution equation $(\partial\Gamma/\partial t)_a = 0$, which states that the marker points preserve their circulation and may thus be labeled in terms of the initial circulation, setting $a = \Gamma$. Note, however, that the computation of the right-hand side of Eq. (11.4.33) will fail at the points where the strength of the vortex sheet vanishes.

The Point-Vortex Method

Previously in this section we developed an integro-differential equation that describes the motion of points marking the location of a two-dimensional vortex sheet, as well as an integral equation for the evolution of the strength of the vortex sheet and circulation along the vortex sheet following the marker points. Numerical methods for solving this governing set of equations have been reviewed by Tryggvason (1988). The simplest such method is the point-vortex method first introduced and implemented by Rosenhead (1932).

The point-vortex method proceeds by approximating the principal-value integral along the vortex sheet in Eq. (11.4.3) using the trapezoidal rule (Section B.6, Appendix B). To implement the method, we divide the vortex sheet into N segments whose end-points are defined in terms of the marker points $\mathbf{X}_{i+1/2}$, denote the middle point of the ith segment by $\mathbf{X}_i$, its arc length by Δl_i, the difference in circulation between the end-points by $\Delta\Gamma_i$, and the unit tangential vector at the middle of the segment by $\mathbf{t}_i$, as shown in Figure 11.4.2. Using central finite differences, we obtain the simple approximation $\mathbf{t}_i = (\mathbf{X}_{i+1/2} - \mathbf{X}_{i-1/2})/\Delta l_i$. More advanced differentiation methods based, for instance, on cubic spline interpolation for the coordinates of the marker points with respect to the arc length of the polygonal line that connects adjacent marker points yield higher accuracy.

Applying Eq. (11.4.3) at the middle points, which amounts to setting $\mathbf{X} = \mathbf{X}_i$, we derive the following system of nonlinear differential equations for $\mathbf{X}_i$

$$\frac{d\mathbf{X}_i}{dt} = \frac{1}{2\pi}\sum_{\substack{j=1 \\ j\neq i}}^{N}\begin{bmatrix} -Y_i + Y_j \\ X_i - X_j \end{bmatrix}\frac{\Delta\Gamma_j}{|\mathbf{X}_i - \mathbf{X}_j|^2} + (\alpha - \tfrac{1}{2})\frac{\Delta\Gamma_i}{\Delta l_i}\,\mathbf{t}(\mathbf{X}_i) + \mathbf{v}(\mathbf{X}_i) \tag{11.4.34}$$

Apart from the last two terms, Eq. (11.4.34) is identical to Eqs. (11.2.1) describing the motion of a collection of N point vortices located at the midpoints $\mathbf{X}_i$, where the differential circulation $\Delta\Gamma_j$ plays the role of the point vortex strength κ_j. Thus, computing the evolution of the vortex sheet has been reduced to computing the motion of point vortices discussed in Section 11.2.

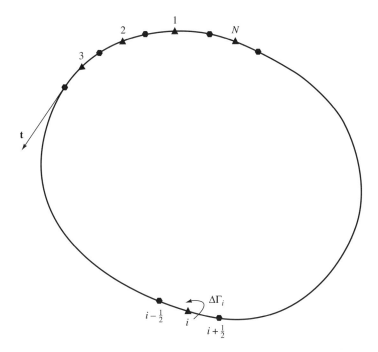

Figure 11.4.2   Discretization of a vortex sheet into a collection of elements defined by their end- and middle-points. The vortex sheet is replaced by a collection of point vortices located at the middle-points.

Having advanced the position of the middle points $\mathbf{X}_i$ on the basis of Eq. (11.4.34), we compute the new position of the end-points $\mathbf{X}_{i+1/2}$ by means of linear or higher-order interpolation. In the simplest approximation, we set $\mathbf{X}_{i+1/2}$ equal to $\frac{1}{2}(\mathbf{X}_i + \mathbf{X}_{i+1})$, but this may incur significant numerical error. A more accurate method involves advancing the position of the end-points according to the velocity field induced by the point vortices, setting

$$\frac{d\mathbf{X}_{i+1/2}}{dt} = \frac{1}{2\pi}\sum_{j=1}^{N}\begin{bmatrix} -Y_{i+1/2} + Y_j \\ X_{i+1/2} - X_j \end{bmatrix}\frac{\Delta\Gamma_j}{|\mathbf{X}_{i+1/2} - \mathbf{X}_j|^2}$$

$$+ (\alpha - \tfrac{1}{2})\,\gamma(\mathbf{X}_{i+1/2})\,\mathbf{t}(\mathbf{X}_{i+1/2}) + \mathbf{v}(\mathbf{X}_{i+1/2}) \qquad (11.4.35)$$

The unit tangential vector in the penultimate term of Eq. (11.4.35) may be computed by numerical differentiation setting, in the simplest case, $\mathbf{t}_i(\mathbf{X}_{i+1/2}) = (\mathbf{X}_{i+1} - \mathbf{X}_i)/|\mathbf{X}_{i+1} - \mathbf{X}_i|$.

Evolution of the strength of the point vortices

To derive an evolution equation for the differential circulation $\Delta\Gamma_i$, we integrate Eq. (11.4.24) with respect to a over the length of a segment, and apply the trapezoidal rule, thereby deriving the approximate form

$$\frac{d\,\Delta\Gamma_i}{dt} = A_1\left[\mathbf{t}\cdot\left(\frac{\partial\mathbf{U}}{\partial t}\right)_a\right]_{\mathbf{X}_i}\Delta l_i + A_2\frac{1}{2}\left(\frac{\partial\gamma^2}{\partial l}\right)_{\mathbf{X}_i}\Delta l_i$$

$$+ \frac{1}{\rho_-\alpha_+ + \rho_-\alpha_-}\left(\frac{\partial(p^- - p^+)}{\partial l}\right)_{\mathbf{X}_i}\Delta l_i - A_1\,\mathbf{g}\cdot\left(\frac{\partial\mathbf{x}}{\partial l}\right)_{\mathbf{X}_i}\Delta l_i$$

$$(11.4.36)$$

The derivatives with respect to arc length may be computed by numerical differentiation. The acceleration of the point vortices involved in the first term on the right-hand side is computed by differentiating the right-hand side of Eq. (11.4.34) with respect to time. To simplify the algebra, we set $\alpha = \frac{1}{2}$ and find

$$\left(\frac{\partial \mathbf{U}}{\partial t}\right)_a (\mathbf{X}_i) = \frac{1}{2\pi} \sum_{\substack{j=1 \\ j \neq i}}^{N} \frac{d\,\Delta\Gamma_j}{dt} \left[\frac{-Y_i + Y_j}{X_i - X_j}\right] \frac{1}{|\mathbf{X}_i - \mathbf{X}_j|^2}$$

$$+ \frac{1}{2\pi} \sum_{\substack{j=1 \\ j \neq i}}^{N} \Delta\Gamma_j \frac{d}{dt}\left(\left[\frac{-Y_i + Y_j}{X_i - X_j}\right]\frac{1}{|\mathbf{X}_i - \mathbf{X}_j|^2}\right) + \left(\frac{\partial \mathbf{v}}{\partial t}\right)_a (\mathbf{X}_i)$$

(11.4.37)

Carrying out the differentiation within the second sum on the right-hand side and simplifying, we obtain

$$\frac{d}{dt}\left(\left[\frac{-Y_i + Y_j}{X_i - X_j}\right]\frac{1}{|\mathbf{X}_i - \mathbf{X}_j|^2}\right) = \frac{1}{r_{ij}^4}\left[\begin{matrix} 2u_{ij}x_{ij}y_{ij} + v_{ij}(y_{ij}^2 - x_{ij}^2) \\ u_{ij}(y_{ij}^2 - x_{ij}^2) - 2v_{ij}x_{ij}y_{ij} \end{matrix}\right]$$

(11.4.38)

where

$$\begin{aligned} x_{ij} &= X_i - X_j, & y_{ij} &= Y_i - Y_j, & r_{ij}^2 &= x_{ij}^2 + y_{ij}^2 \\ u_{ij} &= U_i - U_j, & v_{ij} &= V_i - V_j \end{aligned}$$

(11.4.39)

Substituting Eq. (11.4.38) into Eq. (11.4.37) and then into Eq. (11.4.36) yields a system of linear algebraic equations for $d\,\Delta\Gamma_i/dt$, whose solution may be computed using the method of Jacobi iterations. This involves guessing the values of $d\,\Delta\Gamma_i/dt$, substituting them into the right-hand side of Eq. (11.4.37), and then computing the right-hand side of Eq. (10.4.36), thereby producing new and improved values for $d\,\Delta\Gamma_i/dt$ to be used in the subsequent iteration.

Boussinesq approximation

Adopting the Boussinesq approximation and setting $\alpha = \frac{1}{2}$ reduces Eq. (11.4.36) to an *explicit* evolution equation for the rate of the change of the strength of the point vortices

$$\frac{d\,\Delta\Gamma_i}{dt} = \frac{1}{\rho}\left(\frac{\partial(p^- - p^+)}{\partial l}\right)_{\mathbf{X}_i} \Delta l_i - 2A\,\mathbf{g}\cdot\left(\frac{\partial \mathbf{x}}{\partial l}\right)_{\mathbf{X}_i} \Delta l_i$$

(11.4.40)

The need to solve a system of equations for the rate of the change of the strength of the point vortices at each step is thus bypassed.

Regularizing the motion

In practice, unless a vortex sheet undergoes a sufficient amount of stretching, the motion of the point vortices that represent the vortex sheet is inherently unstable and leads to a disorganized and chaotic pattern. The origin of this behavior may be traced back to the amplification of the roundoff error, which is due to the unstable behavior of an array of point vortices discussed in Section 11.2.

One way to regularize the motion is to redistribute the point vortices along the vortex sheet after each or every few time steps, so that they become evenly spaced with respect to arc length (Fink and Soh, 1974). The strengths of the point vortices at the new positions are found from those at the old positions by means of interpolation with respect to either the marker labeling variable a, or arc length of the polygonal line that connects successive marker points.

Another way of regularizing the motion is to express the Cartesian coordinates of the point vortices in terms of a global Fourier series or a local polynomial expansion, using as independent variable either the Lagrangian label a or the polygonal arc length, and then recompute their position by truncating the Fourier series or discarding odd terms in the polynomial expansion (Longuett-Higgins and Cokelet, 1976; Krasny, 1986a). A simple but effective way of implementing smoothing is provided by the five-point formula

$$\mathbf{x}_i^{\text{Smoothed}} = \tfrac{1}{16}(-\mathbf{x}_{i-2} + 4\mathbf{x}_{i-1} + 10\mathbf{x}_i + 4\mathbf{x}_{i+1} - \mathbf{x}_{i+2}) \tag{11.4.41}$$

Regularization may also be done by (1) replacing the point vortices with the vortex blobs discussed in Section 11.3 (Chorin and Bernard, 1973), and (2) replacing the singular integrand in Eq. (11.4.3) with a nearly singular integrand, which results in a modified version of Eq. (11.4.34). Krasny (1986b) proposed describing the motion of the point vortices in terms of the regularized form of the Biot–Savart integral

$$\frac{d\mathbf{X}_i}{dt} = \frac{1}{2\pi} \sum_{\substack{j=1 \\ i \neq j}}^{N} \left[\frac{-Y_i + Y_j}{X_i - X_j} \right] \frac{\Delta \Gamma_j}{|\mathbf{X}_i - \mathbf{X}_j|^2 + \delta^2} + \mathbf{v}(\mathbf{X}_i) \tag{11.4.42}$$

corresponding to $\alpha = \tfrac{1}{2}$; δ is a small numerical parameter that acts to eliminate the unstable behavior by modifying the dynamical laws of vortex motion expressed by the Biot–Savart integral.

Periodic Vortex Sheets

The equations derived previously in this section for a finite or closed vortex sheet may be modified in a straightforward manner to describe a vortex sheet that is repeated periodically in the x direction with wavelength equal to a. Using Eqs. (2.10.13), we find that the motion of the

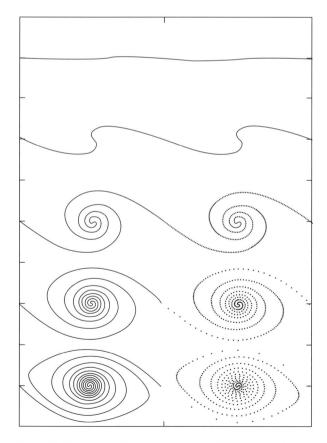

Figure 11.4.3 Stages in the evolution of a sinusoidally perturbed vortex sheet with positive strength computed using Eq. (11.4.44) with $N = 400$ and $\delta/a = 0.125$ (Krasny 1986b).

marker points is described by the modified version of Eq. (11.4.3)

$$\frac{d\mathbf{X}(a)}{dt} = \frac{1}{2a} \int_C^{PV} \begin{bmatrix} -\sinh k[Y(a) - y(a')] \\ \sin k[X(a) - X(a')] \end{bmatrix} \frac{\gamma(a')\,dl(a')}{\cosh k[Y(a) - y(a')] - \cos k[X(a) - x(a')]}$$
$$+ (\alpha - \tfrac{1}{2})\,\gamma(a)\,\mathbf{t}(a) + \mathbf{v}(a) \tag{11.4.43}$$

where $k = 2\pi/a$ is the wave number, and C is the trace of the vortex sheet in the xy plane over one period. The counterpart of the regularized system (11.4.42) is

$$\frac{d\mathbf{X}_i}{dt} = \frac{1}{2a} \sum_{\substack{j=1 \\ j \neq i}}^{N} \begin{bmatrix} -\sinh k(Y_i - Y_j) \\ \sin k(X_i - X_j) \end{bmatrix} \frac{\Delta\Gamma_j}{\cosh k(Y_i - Y_j) - \cos k(X_i - X_j) + \delta^2} + \mathbf{v}(\mathbf{X}_i)$$

$$\tag{11.4.44}$$

where N is the number of point vortices within one period. The evolution equation for the strength of the vortex sheet and strength of the point vortices remain unchanged.

Stages in the evolution of a sinusoidally perturbed vortex sheet with positive strength, separating two regions of the same fluid, computed using Eq. (11.4.44) with 400 point vortices per period and $\delta/a = 0.125$, are shown in Figure 11.4.3 after Krasny (1986b). Note that the vortex sheet develops Helmholtz's instability and rolls up to form a periodic sequence of spirals whose number of turns keeps increasing in time. The numerical results suggest that, in the limit as δ tends to zero, a cusp develops at the center of the spiral at a finite time, just before the spiral begins to form its turns.

PROBLEMS

11.4.1 **Formulation in complex variables.** Show that in complex-variable notation, Eq. (11.4.3) with $\alpha = \tfrac{1}{2}$ and $\mathbf{v} = \mathbf{0}$ assumes the form

$$\frac{dZ^*(a)}{dt} = \frac{1}{2\pi i} \int_C^{PV} \frac{d\Gamma(a')}{Z(a) - z(a')} \tag{11.4.45}$$

where i is the imaginary unit and an asterisk denotes the complex conjugate.

11.4.2 **Mean fluid acceleration on either side of a vortex sheet.** Show that the first two terms on the right-hand side of Eq. (11.4.26) may be written in the compact form $2A\,\mathbf{t} \cdot \mathbf{a}$, where $\mathbf{a}$ is given in Eq. (11.4.27).

11.4.3 **Evolution equation for the velocity potential.** Consider a vortex sheet that separates two regions of irrotational flow. In this case the circulation Γ along the vortex sheet is equal to the jump in the velocity potential across the vortex sheet. (a) Using the unsteady form of Bernoulli's equation, show that the change in the potential on either side of the vortex sheet as seen by an observer who moves with a marker point is given by

$$\left(\frac{\partial\phi^\pm}{\partial t}\right)_a + \left(\tfrac{1}{2}\mathbf{u}^\pm - \mathbf{U}\right) \cdot \mathbf{u}^\pm + \frac{p^\pm}{\rho_\pm} - \mathbf{g} \cdot \mathbf{x} = c_\pm \tag{11.4.46}$$

where $c_\pm$ are two constants. Substituting Eqs. (11.4.6) into Eq. (11.4.46), derive the equations

$$\left(\frac{\partial\phi^\pm}{\partial t}\right)_a + \tfrac{1}{2}(\alpha_\pm^2\gamma^2 - \mathbf{U} \cdot \mathbf{U}) + \frac{p^\pm}{\rho_\pm} - \mathbf{g} \cdot \mathbf{x} = c_\pm \tag{11.4.47}$$

(b) Assuming that the densities of the fluids above and below the vortex sheet are equal, combine Eqs. (11.4.47) to derive Eq. (11.4.24). (c) Discuss how Eqs. (11.4.47) can be used to derive Eq. (11.4.24) in the general case where the densities of the fluids are not equal.

11.4.4 **A fluid interface in oscillatory motion.** Consider two fluids with different densities resting upon each other within a container that executes vertical oscillatory motion normal to the interface along the x axis with velocity $V = U \sin(\Omega t)$, where U is a constant. In the absence of viscous forces, the motion of the interface may be described in terms of a vortex sheet situated over the interface. Show that, subject to the Boussinesq approximation, the evolution of the circulation along the vortex sheet with $\alpha = \frac{1}{2}$ is described by

$$\left(\frac{\partial \Gamma}{\partial t}\right)_a = \frac{T\kappa}{\rho} + 2xA[-g_x + U\Omega \cos(\Omega t)]$$ (11.4.48)

11.4.5 **Point-vortex method for an accelerating vortex sheet.** Consider the motion of a vortex sheet in a frame of reference that undergoes translational acceleration with time-dependent velocity $\mathbf{V}(t)$, and write the counterpart of the evolution equation (11.4.40).

11.4.6 **Stability of a row of regularized point vortices.** Carry out the linear stability analysis of an infinite row of point vortices with constant strength whose motion is described by the regularized evolution equation (11.4.44). Verify that, as δ tends to vanish, the results agree with those for the point vortex array discussed in Section 11.2.

Computer Problem

11.4.7 **Rayleigh–Taylor instability of a vortex sheet.** Consider the instability of an interface separating two quiescent fluids with different densities in the absence of surface tension. At the initial instant, the interface is subjected to a periodic sinusoidal perturbation in shape with wavelength L and amplitude b_0. When viscous effects are insignificant, the interface may be identified with a vortex sheet with vanishing initial strength. Write a program called *RTVS* that computes the motion of the vortex sheet using the point-vortex method subject to the Boussinesq approximation. The integration in time should be performed using the modified Euler method, and the motion should be regularized using a method of your choice. At the initial instant, assume that each point vortex is displaced along the y axis in a sinusoidal fashion while its strength remains equal to zero. Run the program for $b_0/L = 0.10$ with $K = 8, 16, 32$ point vortices within each period separated by a distance $a = L/K$, and discuss the observed differences in behavior.

11.5 | TWO-DIMENSIONAL FLOWS WITH DISTRIBUTED VORTICITY

In Section 11.4 we discussed a method of computing the motion of a two-dimensional vortex sheet that involves discretizing the vortex sheet into a collection of point vortices or vortex blobs, and then computing the motion of the point vortices or vortex blobs while updating their strengths. Extending this method to more general two-dimensional flows with continuous vorticity distribution, we express the vorticity field in terms of a collection of N point vortices setting

$$\omega(x, y) = \sum_{n=1}^{N} \kappa_n \, \delta(\mathbf{x} - \mathbf{x}_n)$$ (11.5.1)

where δ is the two-dimensional delta function (Christiansen, 1973). The motion of the point vortices is computed using the methods discussed in Section 11.2. A regularized version of the point-vortex method involves discretizing the vorticity field into vortex blobs discussed in Section 11.3 (a review is given by Puckett, 1993).

At the outset of the computation, we must decide on how to distribute the point vortices within the flow. In one method, we cover the domain of flow with a two-dimensional rectilinear or curvilinear grid forming a two-dimensional array of cells, and then place one point vortex in the middle of each cell. The strength of a point vortex is equal to the total vorticity within, or circulation of the fluid around the boundary of the corresponding cell; this ensures that the sum of the strengths of all point vortices is equal to the total vorticity of the flow or circulation around a large loop that encloses the flow. Other methods of discretizing the initial vorticity field are discussed by Chiu and Nicolaides (1988).

Vortex-in-Cell Method

When the number of point vortices is large, the point-vortex method becomes prohibitively expensive due to the fact that computing the velocity of all point vortices requires calculating N^2 mutual interactions. The vortex-in-cell method (VIC), which is a particular version of the more general cloud-in-cell method (Christiansen, 1973), helps reduce the computational cost by bypassing the direct computation of the mutual interactions. Given the instantaneous position of the point vortices, we compute the stream function by inverting the Poisson equation $\nabla^2 \psi = -\omega$, and then obtain the velocity at the position of the point vortices by differentiating the stream function using a numerical method.

The solution of the Poisson equation may be carried out using a finite-difference method or a fast Fourier transform method based on spectral expansions. Both methods require the values of the vorticity at the grid points; these are obtained from the position and strength of the point vortices according to an algorithm that preserves the local vorticity field in an approximate manner. Following Christiansen's (1973) original implementation, let us assume that the nth point vortex lies within a rectangular grid cell that is bounded by the grid lines x_i, x_{i+1}, y_j, and y_{j+1}. The contribution of that point vortex to the vorticity of the adjacent grid points is computed according to the equations

$$\omega_{i,j}^{(n)} = 4\kappa_n(x_{i+1} - x_n)(y_{j+1} - y_n)/A^2, \qquad \omega_{i+1,j}^{(n)} = 4\kappa_n(x_n - x_i)(y_{j+1} - y_n)/A^2$$

$$\omega_{i,j+1}^{(n)} = 4\kappa_n(x_{i+1} - x_n)(y_n - y_j)/A^2, \qquad \omega_{i+1,j+1}^{(n)} = 4\kappa_n(x_n - x_i)(y_n - y_j)/A^2$$

$$(11.5.2)$$

where A is the area of the cell. The contribution to all other grid points is equal to zero. Anderson (1986) discusses an extension of the vortex-in-cell method to vortex blobs.

Fast Algorithms

In recent years, a variety of computational techniques have been developed with the objective of reducing the $O(N^2)$ operations of the point vortex method to $O(N \ln N)$ or even $O(N)$ operations. Examples are *tree codes* and *multipole expansions* reviewed by Puckett (1993). One particular method proceeds by sequentially discretizing the domain of flow into smaller rectangles, which is done by subdividing a parent rectangle into four smaller rectangles, until N rectangles are formed. Near neighbors and well-separated rectangles are identified, and, for the purpose of computing the velocity, the point vortices that lie within well-separated rectangles are condensed into central point vortices whose strength is equal to the sum of the strengths of the condensed point vortices.

Viscous Effects

A physically appealing method of accounting for viscous effects is suggested by the observation that point particles in a macroscopically stationary fluid execute random motions due to thermal fluctuations. The probability distribution function of the displacement of a point particle in a particular direction over a time period Δt has a Gaussian shape with mean value equal to zero and variance equal to $2\kappa \Delta t$, where κ is the particle diffusivity (Chandrasekhar, 1943).

Moore (1969, see Milinazzo and Saffman, 1977) and Chorin (1973) argued that, since point vortices and vortex blobs in an inviscid fluid carry their original strength, viscous diffusion can be emulated by requiring that, in addition to the deterministic motion due to the mutual interactions, the point vortices or vortex blobs execute random motions. The diffusion displaces the nth point vortex over one time step by $(\Delta x_n, \Delta y_n)$, where $\{\Delta x_1, \Delta y_1, \ldots, \Delta x_N, \Delta y_N\}$ is a set of independent random numbers whose probability density function forms a Gaussian distribution with mean value equal to zero and variance equal to $2\nu \Delta t$, where ν is the kinematic viscosity of the fluid. Chorin (1973) applied the random vortex method to simulate flow at high Reynolds number past a circular cylinder. Milinazzo and Saffman (1977) and Roberts (1985) showed that casual application of the method may lead to significant numerical error due to slow convergence with respect to N; a large number of point vortices is required in order to achieve statistical equilibrium.

Another way of accounting for the effects of viscosity is to employ vortex blobs whose radius increases in time as discussed in Section 11.3. The lack of random motions justifies calling the methods based on this approach *deterministic vortex methods*. The implementation of these algorithms is the subject of current research (Puckett, 1993).

Vorticity Generation at Solid Boundaries

Let us consider a viscous flow past an impermeable solid boundary over which both the tangential and normal components of the velocity are required to vanish. In general, the velocity field induced by the point vortices that arise from the discretization of the vorticity field will satisfy neither the no-penetration nor the no-slip condition over the boundary. To annihilate the penetration velocity we introduce an appropriate complementary potential flow, but the sum of the velocity induced by the point vortices and the complementary flow will still have a finite tangential component u_t, which amounts to a boundary vortex sheet with strength equal to u_t.

Physically, the boundary vortex sheet represents a viscous boundary layer. In Chapter 8 we saw that for steady unseparated flow past a streamlined body, the thickness of the boundary layer is proportional to $Re^{-1/2}$, where Re is the Reynolds number. In a more general unsteady flow, the vorticity diffuses away from the boundaries, it is convected by the ambient flow, and enters the bulk of the fluid. Chorin (1973) proposed modeling these physical processes according to the following steps:

1. Discretize the boundary into a collection of segments with arc length equal to Δl.
2. Compute the tangential velocity u_t in the middle of each segment.
3. Introduce a vortex blob with a top-hat vorticity distribution, strength equal to $u_t \Delta l$, and radius equal to $\sigma = \Delta l/2\pi$ in the middle of each segment.
4. Move the blobs with a velocity that is equal to the sum of (i) the velocity induced by the vorticity of the flow, (ii) the velocity due to the potential flow that accounts for the no-penetration condition, and (iii) a random velocity that emulates viscous diffusion; if a vortex blob happens to cross the boundary, discard it from the flow.
5. Return to step (2) and repeat the computation for another time step.

This algorithm presents some difficulties associated with the motion of the newly created blobs. An improved version of the method, which is based on discretizing the vortex sheet into elemental vortex sheets using the Prandtl boundary-layer equations, was developed by Chorin (1978, 1980). A review of various refinements and improvements is given by Puckett (1993).

 ## Computer Problems ━━━━━━━━━━━━━━━━━━━━━━━━

11.5.1   **Kirchhoff's vortex.** Write a program called *KIRCHH* that computes the motion of an elliptical vortex with constant vorticity Ω immersed in an infinite otherwise quiescent fluid using the point-vortex method. The algorithm should employ direct summation for computing the mutually induced velocities, and the modified Euler method for integrating the differential equations. The initial position and strength of the point vortices should be computed on the basis of a rectilinear grid. Compute the motion of an elliptical vortex with axis ratio $a/b = 1.1, 1.5, 2$, and an increasing number of point vortices N, and discuss your results.

11.5.2   **Diffusing Kirchhoff's vortex.** Repeat Problem 11.5.1 including viscous effects mediated by random walks. The random displacements should be computed with the help of a random-number generator. Study the motion for axis ratios $a/b = 1.0, 1.5, 2$, and several values of the effective Reynolds number $\Omega a^2/v$, where Ω is the constant vorticity of the vortex at the initial instant. Carry out computations with an increasing number of point vortices N and discuss the behavior of your results. *Note*: Standard random-number generators provide random numbers with uniform probability density functions. To obtain a Gaussian probability, some modifications will be required (Dahlquist and Björck, 1974, chapter 11; Allen and Tildesley, 1987).

11.6 │ TWO-DIMENSIONAL VORTEX PATCHES

Two-dimensional vortex patches with constant vorticity lend themselves to analytical and numerical studies of the structure and dynamics of inviscid flows with distributed vorticity. Hence, they have been used extensively as prototypes for studying vortex dynamics in two-dimensional or even three-dimensional straining flows (Pullin, 1992).

A well-known vortex patch is a circular vortex of radius a and constant vorticity Ω immersed in an otherwise infinite quiescent fluid, known as the *Rankine vortex*. The fluid inside the vortex rotates about its center as a rigid body with azimuthal velocity equal to $\frac{1}{2}\Omega r$, where r is the distance from the center. Outside the vortex, the flow is identical to that due to a point vortex with strength equal to $\kappa = \pi \Omega a^2$ placed at the center.

The circular vortex is a special case of *Kirchhoff's elliptical vortex*. The boundary of a Kirchhoff vortex with major and minor axes equal to a and b executes rigid-body rotation with angular velocity equal to $\Omega ab/(a+b)^2$ (Lamb, 1932, p. 232). As the aspect ratio a/b is increased, the elliptical vortex reduces to a slender vortex layer resembling a vortex sheet with elliptical distribution of circulation.

The infinite vortex layer with constant vorticity, whose stability was discussed in Section 9.5, provides us with an example of an infinite vortex patch. When the boundaries of the vortex layer are parallel to each other, the velocity across the vortex layer varies in a linear manner, while the velocity above and below the vortex layer has two different constant values.

Contour Dynamics

In Section 2.10 we saw that the flow due to a vortex patch D may be computed in terms of a contour integral around the boundary of the vortex patch C as shown in Eq. (2.10.15). The vorticity transport equation for two-dimensional inviscid flow guarantees that the vorticity of the patch will be preserved. Consequently, the instantaneous velocity field may be computed from a knowledge of the location of the vortex contour alone.

In the method of contour dynamics, we follow the evolution of a vortex patch by computing the motion of the vortex contour (Zabusky, Hughes, and Roberts, 1979; Pozrikidis and Higdon, 1985; Dritschel, 1989). This is done by tracing the boundary of the patch with a set of N marker points that are arranged in the counterclockwise sense as shown in Figure 11.6.1. Assuming that the marker points are point particles moving with the velocity of the fluid, we compute their motion on the basis of the equation

$$\frac{d\mathbf{X}_i}{dt} = -\frac{\Omega}{4\pi}\int_C \ln[(X_i - x')^2 + (Y_i - y')^2]\, \mathbf{t}(\mathbf{x}')\, dl(\mathbf{x}') \qquad (11.6.1)$$

where $i = 1, \ldots, N$ and $\mathbf{t}$ is the tangential unit vector pointing into the counterclockwise direction. To compute the integral in Eq. (11.6.1), we describe the shape of the contour C in terms of the position of the marker points by means of interpolation as discussed in Section B.4, Appendix B.

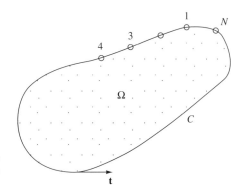

Figure 11.6.1   Illustration of a vortex patch with constant vorticity Ω. The vortex contour C is described in terms of a set of marker points.

In the simplest approach, we approximate the contour with a polygonal line that connects successive marker points. To compute the contour integral on the right-hand side of Eq. (11.6.1) over a segment that does not contain that marker point $\mathbf{X}_i$, we use a standard integration method such as the trapezoidal rule, Simpson's rule, or a Gaussian quadrature. Since, however, the integrand in Eq. (11.6.1) exhibits a logarithmic singularity, these methods are not effective for the two segments that contain $\mathbf{X}_i$ as an end-point. Fortunately, the integration in these two cases may be done analytically in closed form. Combining the trapezoidal rule for the integration over the nonsingular segments and the analytical integration for the singular segments, we obtain

$$
\frac{d\mathbf{X}_i}{dt} = -\frac{\Omega}{8\pi} \sum_{\substack{j=1 \\ j\neq i-1,i}}^{N} (\mathbf{X}_{j+1} - \mathbf{X}_j)\,(\ln|\mathbf{X}_i - \mathbf{X}_j|^2 + \ln|\mathbf{X}_i - \mathbf{X}_{j+1}|^2)
$$

$$
- \frac{\Omega}{4\pi} \sum_{j=i-1}^{i} (\mathbf{X}_{j+1} - \mathbf{X}_j)\,(\ln|\mathbf{X}_{j+1} - \mathbf{X}_j|^2 - 2)
\tag{11.6.2}
$$

Equation (11.6.2) provides us with a system of N nonlinear, coupled ordinary differential equations for the position of the marker points. The integration in time may be done using a standard numerical method such as Euler's method or a Runge–Kutta method.

When the flow contains a collection of vortex patches with different values of vorticity, the right-hand side of Eq. (11.6.1) contains the sum of the integrals over the contours of all patches multiplied by the corresponding values of the vorticity, and Eq. (11.6.2) undergoes a corresponding modification.

Periodic arrangements

Using the results of Section 2.10 [see Eq. (2.10.17)], we find that the motion of the marker points around the contour of a vortex patch that is repeated periodically in the x direction with wavelength a is governed by the modified version of Eq. (11.6.1)

$$
\frac{d\mathbf{X}_i}{dt} = -\frac{\Omega}{4\pi} \int_C \ln\{\cosh[k(Y_i - y')] - \cos[k(X_i - x')]\}\,\mathbf{t}(\mathbf{x}')\,dl(\mathbf{x}')
\tag{11.6.3}
$$

where C is the contour of *one* patch and $k = 2\pi/a$ is the wave number.

In order to compute the integral over the adjacent segments S_j of the ith marker point, corresponding to $j = i - 1$ and i, we write

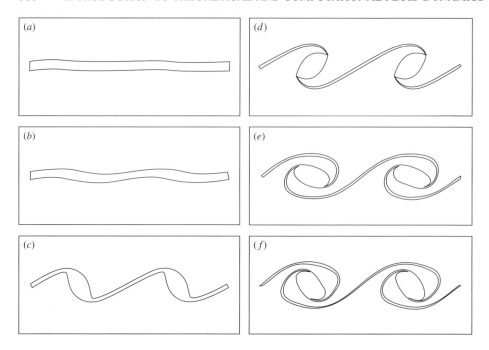

Figure 11.6.2   Successive stages in the instability of a periodically perturbed vortex layer with constant vorticity of negative sign computed using the method of contour dynamics.

$$\int_{S_j} \ln\{\cosh[k(Y_i - y')] - \cos[k(X_i - x')]\}\, \mathbf{t}(\mathbf{x'})\, dl(\mathbf{x'})$$

$$= \int_{S_j} \ln\left(\frac{\cosh[k(Y_i - y')] - \cos[k(X_i - x')]}{(X_i - x')^2 + (Y_i - y')^2}\right) \mathbf{t}(\mathbf{x'})\, dl(\mathbf{x'})$$

$$+ \int_{S_j} \ln[(X_i - x')^2 + (Y_i - y')^2]\, \mathbf{t}(\mathbf{x'})\, dl(\mathbf{x'}) \tag{11.6.4}$$

The first integral in Eq. (11.6.4) is nonsingular and may thus be computed using, for example, the trapezoidal rule. The singularity has been shifted to the second integral, which, however, is equal to the term within the second sum on the right-hand side of Eq. (11.6.2).

Figure 11.6.2 shows successive stages in the Helmholtz instability of a periodically perturbed vortex layer with $kb = 0.25$, where $k = 2\pi/a$ is the wavelength of the perturbation and $2b$ is the thickness of the unperturbed layer, computed using the method of contour dynamics (Pozrikidis and Higdon, 1985). The results show that the vortex layer rolls up into a sequence of nearly steadily rotating elliptical vortex cores that are connected by thin vortex layers. The ability of the numerical method to capture the fine scales of the motion is one of its notable features.

 Computer Problems ————————————————————

11.6.1   Kirchhoff's vortex. Write a program called *VP1* that computes the motion of an elliptical vortex patch, and perform computations for axes ratios $a/b = 1.1, 1.5, 2, 3, 4$. Verify that the patch rotates as a rigid body, plot the computed angular velocity of rotation as a function of aspect ratio, and compare the numerical results with the exact solution given in the text.

11.6.2 **Interaction of two vortex patches.** Write a program called *VP2* that computes the interaction of two identical circular vortex patches with equal vorticity. Carry out computations for three cases where the initial separation of the vortices is equal to two, three, and four times the initial vortex radius, and discuss the differences in behavior. To improve the efficiency of your program, you may wish to exploit the symmetry of the two vortex contours with respect to the midpoint of the line that connects their centers.

11.7 │ AXISYMMETRIC FLOW

The computation of axisymmetric vortex motions is similar in many ways to that of two-dimensional motions discussed in the preceding sections. The curvature of the vortex lines and the occurrence of vortex stretching, however, necessitate certain modifications. These serve to illustrate the subtleties involved in the practical implementation of algorithms for three-dimensional vortex flows.

Coaxial Line Vortex Rings

The motion of a collection of coaxial line vortex rings is analogous to that of a collection of point vortices. The velocity of each ring is equal to the sum of its self-induced velocity, which points in the axial direction, and the velocities induced by all other rings discussed in Section 2.9. Recall that, by contrast, the self-induced velocity of a rectilinear line vortex, that is, a point vortex, is equal to zero.

Denoting the self-induced velocity of the ith ring by W_i, we find that its axial position X_i and radius Σ_i evolve according to the equations

$$\frac{dX_i}{dt} = W_i(\Sigma_i) + \sum_{\substack{j=1 \\ j \neq i}}^{N} \kappa_j U_x(X_i - X_j, \Sigma_i, \Sigma_j)$$

$$\frac{d\Sigma_i}{dt} = \sum_{\substack{j=1 \\ j \neq i}}^{N} \kappa_j U_\sigma(X_i - X_j, \Sigma_i, \Sigma_j)$$

(11.7.1)

where

$$\begin{bmatrix} U_x \\ U_\sigma \end{bmatrix}(x, \sigma, \sigma') = \frac{1}{4\pi} \begin{bmatrix} -\sigma I_{31}(x, \sigma, \sigma') + \sigma' I_{30}(x, \sigma, \sigma') \\ \hat{x} \, I_{31}(x, \sigma, \sigma') \end{bmatrix}$$

(11.7.2)

and the integrals I_{nm} are defined in Eq. (2.9.8). The vorticity transport equation requires that the strength of each ring κ_i remain constant during the motion.

Self-induced velocity

The self-induced velocity W of a vortex ring depends upon the structure of the vortex core. In Section 2.8 we saw that the self-induced velocity of a line vortex ring with infinitesimal core size is infinite, and this underscores the importance of the way in which the vorticity is distributed over the core. To compute the motion of a line ring we must assume that the vorticity distribution takes a certain form, and thus transform the vortex rings into an axisymmetric vortex blob.

Heimholtz's formula for the self-induced velocity of a vortex ring whose vorticity is distributed uniformly within a circular vortex core of radius a is

$$W = \frac{\kappa}{4\pi\Sigma}\left(\ln\frac{8\Sigma}{a} - \frac{1}{4}\right) \tag{11.7.3}$$

(Lamb, 1932, p. 241). Since the volume of the fluid within the core is conserved, we require that

$$\frac{d}{dt}(\Sigma a^2) = 0 \qquad \text{or} \qquad \frac{da}{dt} = -\frac{1}{2}\frac{a}{\Sigma}\frac{d\Sigma}{dt} \tag{11.7.4}$$

The second of Eqs. (11.7.4) provides us with the rate of change of the radius of the vortex core. The motion of the vortex rings is computed by solving the system of equations (11.7.1) and the second of Eqs. (11.7.4) for the axial position, radial position, and size of the vortex core.

Hicks's (1923) formula for the self-induced velocity of a vortex ring whose vorticity is concentrated around a circular axisymmetric vortex sheet of radius a around the ring center, is identical to Eq. (11.7.3) except that the fraction $\frac{1}{4}$ is replaced by $\frac{1}{2}$. Further results and a comprehensive review of the motion of vortex rings have been given by Shariff and Leonard (1992).

Bounded domains

When the vortex rings are placed within a domain that is bounded by an axisymmetric surface, their mutually induced velocity must be enhanced with that due to a complementary potential flow that accounts for the no-penetration condition at the surface. When the geometry of the surface is sufficiently simple, the complementary flow may be expressed in terms of images of the vortex rings. The image of a vortex ring with respect to a plane wall is another ring with opposite strength placed at the instantaneous mirror-image position. The image of a ring with respect to a sphere was discussed in the end of Section 7.4.

Vortex Sheets

Computing the motion of axisymmetric vortex sheets can be done based on the general principles for two-dimensional vortex sheets discussed in Section 11.4; but the expressions for the self-induced velocity resulting from the Biot–Savart integral involve complete elliptic integrals of the first and second kind that are more difficult to evaluate. The evolution of the strength of the vortex sheet is governed by an equation that is analogous to Eq. (11.4.11), but the tangential component of the acceleration of the marker points in an azimuthal plane is given by a more complicated version than that given on the right-hand side of Eq. (11.4.16). When the marker points move with the principal velocity of the vortex sheet, and the vortex sheet separates two fluids with identical densities, the evolution of the circulation along the trace of the vortex sheet in an azimuthal plane is given by the simplified evolution equation (11.4.29).

Vortex Patches

The vorticity transport equation for inviscid axisymmetric flow takes the simple form $D(\omega/\sigma)/Dt = 0$, which shows that if the strength of the vorticity ω is equal to $\alpha\sigma$ at a particular instant, where α is a constant, it will remain equal to $\alpha\sigma$ at all times. In this case, computing the evolution of an axisymmetric vortex patch reduces to describing the motion of the enclosing vortex contour.

The numerical implementation of the contour dynamics method for axisymmetric flow is analogous to that for two-dimensional flow discussed in Section 11.6 (Pozrikidis, 1988; Pullin, 1992). Using, in particular, Eqs. (2.9.17) and (2.9.20), we find that the axial and radial position X and Σ of marker points along the vortex contour evolve according to the equation

$$\frac{d}{dt}\begin{bmatrix} X \\ \Sigma \end{bmatrix} = -\frac{\alpha}{4\pi}\int_C \begin{bmatrix} \hat{x}\, I_{10}(\hat{x}, \Sigma, \sigma')\, n_x(\mathbf{x}') + \Sigma\, I_{11}(\hat{x}, \Sigma, \sigma')\, n_\sigma(\mathbf{x}') \\ -\sigma'\, I_{11}(\hat{x}, \Sigma, \sigma')\, n_x(\mathbf{x}') \end{bmatrix} \sigma'\, dl(\mathbf{x}') \tag{11.7.5}$$

where $\mathbf{n}$ is the unit normal vector pointing outward from the vortex patch, $\hat{x} = X - x'$, and the integrals I_{nm} are defined in Eq. (2.9.8).

Close to the point **X**, the integrand in Eq. (11.7.5) behaves like the one in Eq. (11.6.1). The computation of the singular integrals over the adjacent segments of a marker point is done by subtracting off the singularity according to Eq. (11.6.4) (Problem 11.7.2).

PROBLEMS

11.7.1 **Self-induced velocity of a vortex ring with constant vorticity.** Carry out an asymptotic analysis of Eq. (2.9.5) to derive Eq. (11.7.3).

11.7.2 **Desingularization of contour dynamics integrals.** (a) Carry out an asymptotic analysis to show that, as $\hat{x} \to 0$ and σ' tends to Σ, the integrand in Eq. (11.7.5) behaves like the one shown in Eq. (11.6.1). (b) Explain how the singularity can be subtracted off from the integrand according to Eq. (11.6.4).

 ## Computer Problems

11.7.3 **Motion of coaxial line vortex rings.** Write a program called $VR2$ that computes the motion of two line vortex rings whose vorticity is distributed uniformly over the cores. Run the program to compute the interaction of two initially identical vortex rings for several ratios of the initial core-to-ring radius, and discuss their behavior.

11.7.4 **A line vortex ring approaching head-on a plane wall or a sphere.** (a) Compute the trajectory of a line vortex ring with uniform vorticity distribution approaching a plane wall with its axis perpendicular to the wall. (b) Repeat (a) for a vortex ring with initial radius $\Sigma = c$ approaching a sphere of radius a whose center is located at the ring's axis. Perform a series of computations with $c/a = 0.10, 0.50, 1.0, 2.0$, and discuss differences in behavior. The location of the image ring is discussed in the end of Section 7.4.

11.8 THREE-DIMENSIONAL FLOW

Vortex methods for three-dimensional flow arise as extensions of the methods for two-dimensional and axisymmetric flow discussed in the preceding sections. Unfortunately, these extensions are often not straightforward, and their implementation may lead to conceptual difficulties and significant numerical error.

Line Vortices

In Section 2.8 we saw that the self-induced velocity of a line vortex is infinite; this underlines the importance of the structure of the vortex core. Taking into account the finite size of the core can be done in several approximate ways with varying degrees of accuracy and sophistication.

Desingularization of the Biot–Savart integral

Rosenhead (1930) proposed desingularizing the Biot–Savart integral by replacing the second of Eqs. (2.8.19) with the modified equation

$$\mathbf{u}(\mathbf{x}) = -\frac{\kappa}{4\pi} \int_L \frac{(\mathbf{x} - \mathbf{x}') \times \mathbf{t}(\mathbf{x}')}{(|\mathbf{x} - \mathbf{x}'|^2 + \delta^2)^{3/2}} \, dl(\mathbf{x}') \tag{11.8.1}$$

where the magnitude of the numerical parameter δ is small compared to the size of the vortex core, which is assumed to be constant along the line vortex. A similar desingularization was discussed earlier in Section 11.4 in the context of the point-vortex method for vortex sheets.

Truncation of the Biot–Savart integral

Hamma (1962, 1963) maintained the exact form of the Biot–Savart integral shown in Eqs. (2.8.19), but truncated the line integral at a small distance on either side of the point where the velocity is computed. Partial justification for this approximation is provided by the fact that setting the cutoff length equal to $(b/2)\exp(\frac{1}{4})$, where b is the radius of the vortex core, reproduces the velocity of a circular vortex ring whose vorticity is distributed uniformly over the core.

Local-induction approximation

Hamma and Arms (Hamma, 1962, 1963) maintained the second term in the asymptotic form of the Biot–Savart shown in Eq. (2.8.30), but truncated the limits of integration on either side of the origin at a value that is comparable to the size of the core b, obtaining the leading-order approximation

$$\mathbf{u}(\mathbf{x}) \approx -\frac{\kappa}{4\pi} c(\mathbf{x})\, \mathbf{b}(\mathbf{x}) \ln b \tag{11.8.2}$$

where c is the curvature of the line vortex. Expression (11.8.2) is known as the local induction approximation (LIA) introduced by Da Rios in 1906 (see Ricca, 1992). According to the LIA, the velocity vector at a certain point on a line vortex is parallel to the local binormal vector $\mathbf{b}$. Hasimoto (1972) showed that the LIA may be reformulated in the context of a nonlinear Schrödinger equation, which is known to admit solutions in the form of nonlinear traveling waves called *solitons*. Da Rios (Ricca, 1992) and Betchov (1965) used the LIA to derive a coupled system of nonlinear ordinary differential equations that govern the evolution of the curvature and torsion of a line vortex.

Since, in the absence of viscous diffusion, a line vortex moves with the velocity of the fluid, the problem is reduced to solving an integro-differential equation for the position of point particles or marker points along the line vortex, $d\mathbf{X}/dt = \mathbf{u}(\mathbf{X})$. The numerical procedure is similar to that used to compute the motion of two-dimensional vortex sheets in terms of point vortices.

Three-Dimensional Vortex Sheets

To compute the motion of a three-dimensional vortex sheet, we describe it in terms of two surface curvilinear coordinates ξ and η, and then follow the motion of marker points that are located at nodes of an interfacial grid, and are thus labeled by discrete values of ξ and η. The strength of the vortex sheet $\boldsymbol{\zeta} = \mathbf{n} \times (\mathbf{u}^+ - \mathbf{u}^-)$ is tangential to its instantaneous position; the superscripts $+$ and $-$ designate evaluation at the upper or lower side of the vortex sheet, and the unit normal vector $\mathbf{n}$ points toward the upper side.

The marker points move normal to the vortex sheet with the velocity of the fluid while executing an arbitrary tangential motion. The velocity of the marker points $\mathbf{U}$ may be expressed in terms of the principal velocity of the vortex sheet, as discussed in Section 11.4 for two-dimensional flow, which is equal to the principal value of the Biot–Savart integral. To compute the evolution of the strength $\boldsymbol{\zeta}$, we work as in Section 11.4, beginning with Euler's equation on either side of the vortex sheet, written in the form of Eq. (11.4.5). The complexity of the required algebraic manipulations prevents us from presenting a detailed derivation, which is left as an exercise for the reader.

When (1) the vortex sheet separates two fluids with identical densities, (2) the flow on either side of the vortex sheet is irrotational, and (3) the marker points move with the principal velocity of the vortex sheet, we introduce the velocity potential on either side of the vortex sheet and work with the unsteady Bernoulli equation as discussed in Problem 11.4.3 to find

$$\left(\frac{\partial \phi^\pm}{\partial t}\right)_{\xi,\eta} - \tfrac{1}{2}\mathbf{u}^+ \cdot \mathbf{u}^- + \frac{p^\pm}{\rho} - \mathbf{g}\cdot\mathbf{x} = c_\pm \tag{11.8.3}$$

Setting the constants $c_\pm$ equal to one another, evaluating, and subtracting corresponding sides of Eq. (11.8.3) written for either side of the vortex sheet, we obtain

$$\left(\frac{\partial(\phi^+ - \phi^-)}{\partial t}\right)_{\xi,\eta} = \frac{p^- - p^+}{\rho} \tag{11.8.4}$$

The jump in pressure across the vortex sheet is related to surface tension T of the interface represented by the vortex sheet by $p^- - p^+ = T2\kappa_m$, where κ_m is the mean curvature of the vortex sheet. In the absence of surface tension, the difference in the potential across the vortex sheet $\phi^+ - \phi^-$ following the marker points remains constant (Kaneda, 1990).

Equation (11.8.4) is the basis for a numerical procedure according to which the position of the marker points is advanced with the principal velocity of the vortex sheet, the value of the jump $\phi^+ - \phi^-$ is updated on the basis of Eq. (11.8.4), the surface gradient $\nabla(\phi^+ - \phi^-)$ is computed over the vortex sheet, and the strength of the vortex sheet is updated by setting $\zeta = \mathbf{n} \times \nabla(\phi^+ - \phi^-)$ (Caflish, Li, and Shelley, 1993).

Particle Methods

Generalized vortex particle methods for three-dimensional flow with isolated line vortices or distributed vorticity include the *standard vortex-particle* method, which is the counterpart of the point-vortex method, the *regularized vortex-particle method,* which is the counterpart of the vortex-blob method, and the *vortex-in-cell* method.

In the standard vortex-particle method, the vorticity field is expressed in a form that is analogous to that shown in Eq. (11.5.1) as

$$\boldsymbol{\omega}(\mathbf{x},t) = \sum_{n=1}^{N} \boldsymbol{\alpha}_n(t)\, \delta(\mathbf{x} - \mathbf{x}_n(t)) \tag{11.8.5}$$

where δ is the three-dimensional delta function. Each term on the right-hand side of Eq. (11.8.5) represents a vortex particle located at the point $\mathbf{x}_n$ with strength equal to $\boldsymbol{\alpha}_n$. The velocity is obtained by taking the curl of the vector potential, which, according to Eq. (2.8.1), is given by

$$\mathbf{A}(\mathbf{x}, t) = \frac{1}{4\pi} \sum_{n=1}^{N} \frac{\boldsymbol{\alpha}_n(t)}{|\mathbf{x} - \mathbf{x}_n(t)|} \tag{11.8.6}$$

The vortex particles move with the velocity of the fluid, while their strength evolves due to vortex stretching. Using Eqs. (3.8.7) and (3.8.9), we obtain

$$\frac{d\mathbf{x}_n(t)}{dt} = \mathbf{u}[\mathbf{x}_n(t), t]$$

$$\frac{d\boldsymbol{\alpha}_n(t)}{dt} = \boldsymbol{\alpha}_n(t) \cdot [\beta\, \nabla\mathbf{u} + (1 - \beta)(\nabla\mathbf{u})^T] \tag{11.8.7}$$

where β is a free parameter. The choice $\beta = \frac{1}{2}$ leads to computational savings, whereas the choice $\beta = 0$ helps preserve the total vorticity, which is an invariant of the motion (Winckelmans and Leonard, 1993).

One problem with the representation (11.8.5) is that, in general, the discretized vorticity field and vector potential will not be solenoidal, but various improvements can be made to rectify this deficiency. The current state of the art of three-dimensional vortex methods is reviewed by Leonard (1985), Kino and Ghoniem (1990), Winckelmans and Leonard (1993), and Puckett (1993).

Reference

Allen, M. P., and Tildesley, D. J. 1987, *Computer Simulation of Liquids.* Oxford.

Anderson, C. R., 1986, A method of local corrections for computing the velocity field due to a distribution of vortex blobs. *J. Comp. Phys.* **62**, 111–23.

Aref, H., 1983, Integrable, chaotic, and turbulent vortex motion in two-dimensional flows. *Annu. Rev. Fluid Mech.* **15**, 345–89.

Batchelor, G. K., 1967, *An Introduction to Fluid Mechanics.* Cambridge University Press.

Betchov, R., 1965, On the curvature and torsion of an isolated vortex filament. *J. Fluid Mech.* **22**, 471–79.

Caflisch, R. E., Li, X., and Shelley, M. J., 1993, The collapse of an axi-symmetric, swirling vortex sheet. *Nonlinearity* **6**, 843–67.

Chandrasekhar, S., 1943, Stochastic problems in physics and astronomy. *Rev. Modern Phys.* **15**, 1–89.

Chiu, C., and Nicolaides, R. A., 1988, Convergence of a higher-order vortex method for two-dimensional Euler's equations. *Math. Comp.* **51**, 507–34.

Chorin, A. J., 1973, Numerical study of slightly viscous flow. *J. Fluid Mech.* **57**, 785–96.

Chorin, A. J., 1978, Vortex sheet approximation of boundary layers. *J. Comp. Phys.* **27**, 428–42.

Chorin, A. J., 1980, Vortex models and boundary layer instability. *SIAM J. Sci. Stat. Comput.* **1**, 1–21.

Chorin, A. J., and Bernard, P. S., 1973, Discretization of a vortex sheet, with an example of roll-up. *J. Comp. Phys.* **13**, 423–29.

Christiansen, J. P., 1973, Numerical simulation of hydrodynamics by the method of point vortices. *J. Comp. Phys.* **13**, 363–79.

Clements, R. R., and Maull, D. J., 1975, The representation of sheets of vorticity by discrete vortices. *Prog. Aerospace Sci.* **16**, 129–46.

Dahlquist, G., and Björck, Å. 1974, *Numerical Methods.* Prentice Hall.

Dritschel, D. G., 1989, Contour dynamics and contour surgery: numerical algorithms for extended, high-resolution modelling of vortex dynamics in two-dimensional, inviscid, incompressible flows. *Comp. Phys. Rep.* **10**, 77–146.

Fink, P. T., and Soh, W. K., 1974, Calculation of vortex sheets in unsteady flow and applications in ship hydrodynamics. *Proc. 10th Symp. Naval Hydrodynamics,* 463–91.

Hamma, F. R., 1962, Progressive deformation of a curved vortex filament by its own induction. *Phys. Fluids* **5**, 1156–62.

Hamma, F. R., 1963, Progressive deformation of a perturbed line vortex filament by its own induction. *Phys. Fluids* **6**, 526–34.

Hasimoto, H., 1972, A soliton on a vortex filament. *J. Fluid Mech.* **51**, 477–85.

Havelock, T. H., 1931, The stability of motion of rectilinear vortices in ring formation. *Phil. Mag. Ser. 7.* **11**(70), Suppl. 617–33.

Hicks, W. M., 1923, On the mutual threading of vortex rings. *Proc. Roy. Soc. London* A **102**, 111–31.

Kaneda, Y., 1990, On the three-dimensional motion of an infinitely thin vortex sheet in an ideal fluid. *Phys. Fluids* A **2**, 1817–26.

Kino, O. M., and Ghoniem, A. F., 1990, Numerical study of a three-dimensional vortex method. *J. Comp. Phys.* **86**, 75–106.

Krasny, R., 1986a, A study of singularity formation in a vortex sheet by the point-vortex approximation. *J. Fluid Mech.* **167**, 65–93.

Krasny, R., 1986b, Desingularization of periodic vortex sheet roll-up. *J. Comp. Phys.* **65**, 65–93.

Lamb, H., 1932, *Hydrodynamics.* Dover.

Leonard, A., 1980, Vortex methods for flow simulation. *J. Comp. Phys.* **37**, 289–335.

Leonard, A., 1985, Computing three-dimensional incompressible flows with vortex elements. *Annu. Rev. Fluid Mech.* **17**, 523–59.

Longuet-Higgins, M. S., and Cokelet, E. D., 1976, The deformation of steep surface waves on water. I. A numerical method of computation. *Proc. Roy. Soc. London* A **350**, 1–26.

Milinazzo, F., and Saffman, P. G., 1977, The calculation of large Reynolds number two-dimensional flow using discrete vortices with random walks. *J. Comp. Phys.* **23**, 380–92.

Moffatt, H. K., and Tsinober, A., 1992, Helicity in laminar and turbulent flow. *Ann. Rev. Fluid Mech.* **24**, 281–312.

Pozrikidis, C., 1988, The nonlinear instability of Hill's spherical vortex. *J. Fluid Mech.* **168**, 337–67.

Pozrikidis, C., and Higdon, J. J. L., 1985, Nonlinear Kelvin–Helmholtz instability of a finite vortex layer. *J. Fluid Mech.* **157**, 225–63.

Puckett, E. G., 1993, Vortex methods: an introduction and survey of selected research topics. *Incompressible Computational Fluids Dynamics: Trends and Advances.* Cambridge.

Pullin, D. I., 1992, Contour dynamics methods. *Annu. Rev. Fluid Mech.* **24,** 89–115.

Ricca, R. L., 1992, Physical interpretation of certain invariants for vortex filament motion under LIA. *Phys. Fluids* A **4,** 938–44.

Roberts, S. G., 1985, Accuracy of the random vortex methods for a problem with non-smooth initial conditions. *J. Comp. Phys.* **58,** 29–43.

Rosenhead, L., 1930, The spread of vorticity in the wake behind a cylinder. *Proc. Roy. Soc. London* A **127,** 590–612.

Rosenhead, L., 1932, The formation of vortices from a surface of discontinuity. *Proc. Roy. Soc. London* A **134,** 170–92.

Saffman, P. G. 1992, *Vortex Dynamics.* Cambridge University Press.

Saffman, P. G., and Baker, G. R., 1979, Vortex interactions. *Annu. Rev. Fluid Mech.* **11,** 95–122.

Shariff, K., and Leonard, A., 1992, Vortex rings. *Annu. Rev. Fluid Mech.* **24,** 235–79.

Tryggvason, G., 1988, Numerical simulations of the Rayleigh–Taylor instability. *J. Comp. Phys.* **75,** 253–83.

Winckelmans, G. S., and Leonard, A., 1993, Contributions to vortex particle methods for the computation of three-dimensional incompressible unsteady flow. *J. Comp. Phys.* **109,** 247–73.

Zabusky, N. J., Hughes, M. H., and Roberts, K. V., 1979, Contour dynamics for the Euler equations in two dimensions. *J. Comp. Phys.* **30,** 96–106.

Finite-difference methods provide us with a powerful tool for generating numerical solutions to the partial differential equations of mathematical physics including the equations of fluid flow. Before, however, we can apply these methods to solve problems in fluid dynamics we require reliable and accurate strategies for computing numerical solutions to the convection–diffusion equation shown in Eq. (12.1.1). The development of such methods and the investigation of their performance will be the theme of the present chapter.

The subject of finite-difference methods is broad and diverse, and we must necessarily confine our attention to discussing the fundamental principles and procedures, and presenting a selected class of methods that either illustrate the methodology or find extensive applications. Extended discussions can be found in specialized monographs and texts on numerical methods for partial differential equations including those by Richtmyer and Morton (1967), Mitchell (1969), Ames (1977), Mitchell and Griffiths (1980), Ferziger (1981), Sod (1985), Fletcher (1988, vol. I), Hirsch (1988), Hoffman (1992), and Hoffmann and Chiang (1993).

It is helpful to keep in mind throughout the present discussion that the particular way in which the convection–diffusion equation enters a numerical procedure for computing the structure of a steady flow or the evolution of an unsteady incompressible flow depends upon the chosen computational strategy. In certain cases, the convection–diffusion equation is integrated with reference to the equation of motion, whereas in other cases it is integrated with reference to the vorticity transport equation. Examples in each category will be discussed in Chapter 13.

12.1 | DEFINITIONS AND PROCEDURES

The most general problem addressed in the present chapter is the computation of a vector function $\mathbf{f}$ that satisfies the convection–diffusion equation

$$\frac{\partial \mathbf{f}}{\partial t} + \mathbf{U}(\mathbf{f}, \mathbf{x}, t) \cdot \nabla \mathbf{f} = \kappa \nabla^2 \mathbf{f} \tag{12.1.1}$$

within a specified one-dimensional, two-dimensional, or three-dimensional domain, subject to the initial condition $\mathbf{f}(\mathbf{x}, t = 0) = \mathbf{F}(\mathbf{x})$, where $\mathbf{F}$ is a known function. The convection velocity $\mathbf{U}$ is assumed to depend on position $\mathbf{x}$ and time t explicitly, as well as implicitly through its dependence on the solution $\mathbf{f}$.

The scalar constant κ, called the *diffusivity*, will be assumed to be uniform throughout the domain of solution. When κ has a finite value, Eq. (12.1.1) is a *parabolic* differential equation in time, whereas when κ vanishes, Eq. (12.1.1) becomes a *hyperbolic* differential equation in time.

We shall see later in this chapter that this seemingly academic classification has important consequences on the effectiveness of the various finite-difference methods.

Requiring a proper number of boundary conditions completes the statement of the computational problem. When the diffusivity κ is finite, the convection–diffusion equation is a second-order partial differential equation, and we must supply a number of boundary conditions that is equal to the dimensionality of the unknown function $\mathbf{f}$. Thus, if $\mathbf{f}$ is a three-dimensional vector, we require three scalar conditions, for example, one for each component of $\mathbf{f}$ over each boundary.

Finite-Difference Grids

The central goal of a finite-difference method is to generate the values of the unknown function $\mathbf{f}$ at the nodes of a coordinate grid that covers the domain of solution, at a sequence of discrete time levels separated by the constant or variable time step Δt.

The finite-difference grid may be defined in Cartesian coordinates (x, y, z) or other orthogonal or nonorthogonal curvilinear coordinates (ξ, η, ζ). The choice of coordinates is dictated by the geometry of the domain of solution and is selected with an objective to facilitate the implementation of the boundary conditions. For instance, when the domain of solution is the exterior or interior of a sphere, the boundary conditions are naturally described in terms of spherical polar coordinates (r, θ, φ) with the origin at the center of the sphere, and the governing equation is solved for the spherical polar coordinates of the unknown function. The use of orthogonal coordinates is desirable for the reasons of analytical simplicity and improved numerical stability.

In Cartesian coordinates, the finite-difference grid is composed of an array of straight lines that run parallel to the x, y, and z axes, with grid spacings Δx, Δy, and Δz that may vary across the domain of solution in order to allow for enhanced spatial resolution at regions where the solution is expected to exhibit sharp variations. The grid becomes finer as the grid spacings become smaller.

Once the discrete finite-difference solution has been computed, the values of $\mathbf{f}$ between grid points and time levels are obtained by applying standard methods of function interpolation, extrapolation, or approximation discussed in Sections B.4 and B.7, Appendix B.

Finite-Difference Discretizations

The distinguishing feature of a finite-difference method is the approximation of the temporal and spatial partial derivatives in the governing equation with finite differences that relate the values of the unknown functions at a set of neighboring grid points at various time levels. This approximation replaces the partial differential equation (PDE) with a finite-difference equation (FDE). Section B.5 contains a compilation of difference approximations to partial derivatives. The process of replacing the partial derivatives with algebraic differences is called the *finite-difference approximation* or *discretization* of the differential equation.

Applying the finite-difference equation sequentially at the nodes of the finite-difference grid yields a system of linear or nonlinear algebraic equations that relate the values of the unknown function at the nodes. In certain cases, the domain of solution is extended beyond the natural boundaries of the physical problem, and the finite-difference equation is applied at boundary nodes, in which case it contains the extended nodes, with an objective to increase the accuracy of the algebraic equations representing the finite-difference approximation of the boundary conditions (Problem 12.1.2(*e*)).

Consistency

The accuracy of a numerical computation based on a finite-difference method depends upon the sizes of the grid spacings and time step, which are the control parameters of the numerical method. If in the limit as both the grid spacings and time step are reduced simultaneously, but in a manner that allows them to be of the same order of magnitude, the finite-difference equation approximates the partial differential equation with increasing accuracy, then the finite-difference method is *consistent*.

The consistency of a finite-difference equation that arises by applying well-established finite-difference formulae to approximate the temporal and spatial derivatives of the partial differential equation is guaranteed. The consistency of a finite-difference equation that arises by heuristic or ad hoc modifications to well-established finite-difference approximations, however, is subject to confirmation.

The consistency of a finite-difference method may be assessed by pretending that all variables in the finite-difference equation are continuous functions of space and time, and then expanding them in a Taylor series around a selected grid point at a certain time instant. In this manner, the finite-difference equation yields a new differential equation called the *modified differential equation* (MDE) (Warming and Hyett, 1974). If in the limit as the size of the time step and grid spacings are reduced simultaneously but independently, the MDE reduces to the original PDE, then the finite-difference method is consistent. Phrased differently, if the finite-difference method is consistent, the difference between the MDE and the PDE involves terms that are proportional to powers of the grid sizes and time step. The exponents of these powers define the *order of the numerical error* or the order of the *finite-difference method.*

We shall see in subsequent sections that certain finite-difference equations emerge by applying the differential equation at a particular grid point, and then replacing it with a combination of values of the unknown function **f** at a group of neighboring grid points. In these cases, the coefficients that multiply the values of the function are computed by imposing certain restrictions, including consistency with the differential equation and a desired degree of accuracy in the approximation of the partial derivatives.

Stability

Let us assume that the exact solution of the convection–diffusion equation, or some other partial differential equation, subject to an initial condition and a proper number of boundary conditions, does not grow continuously in time, but either stays constant or decays at every point. It is not unreasonable to demand that the finite-difference solution reproduce this behavior; that is, it provide us with a bounded solution that is free of growing oscillations. If it does, the finite-difference method is stable; otherwise it is unstable. If the exact solution of a differential equation continues to grow in time, the finite-difference method is stable when it provides us with a numerical solution that grows at a rate that is equal to, or lower than that of the exact solution. A more precise definition of stability is based on the behavior of the numerical solution at a particular time instant, in the limit as the spatial and temporal steps become smaller.

The stability of relatively simple finite-difference methods for linear partial differential equations may be assessed by several methods, including the *von Neumann stability analysis,* the *projection matrix method,* and the *discrete-perturbation method.* The first method is easiest to carry out, but does not normally account for the effect of the boundary conditions. The stability of more involved finite-difference methods is more difficult to investigate, and, in practice, it is often warranted by the absence of noticeable spatial or temporal oscillations in the results of a computation.

The stability of finite-difference methods for nonlinear differential equations is typically examined by linearizing the differential equation about a particular grid point, and then studying the performance of the finite-difference method with reference to the linearized equation, as discussed in Chapter 9 in the context of hydrodynamic stability. Experience has shown that the local stability criteria obtained in this manner provide us with a reliable characterization of the overall performance of the numerical method.

Convergence

Stability imposes a modest restriction on the numerical method. Before we can claim that the numerical results bear any degree of physical relevance with respect to the physical problem described by the original partial-differential equation, we must ensure that, as the size of the grid and time step are made finer, the numerical solution converges to the exact solution.

Lax's equivalence theorem guarantees that if a numerical solution of a *linear* partial differential equation obtained using a consistent finite-difference approximation is stable, then in the limit as the grid spacings and time step tend to zero, the numerical solution will indeed converge to the exact solution (Lax and Richtmyer, 1956; Richtmyer and Morton, 1967). Thus *consistency and stability ensure convergence and vice versa.*

The convergence of finite-difference methods for nonlinear differential equations is more difficult to assess, but experience has shown that if the numerical method is consistent and locally stable, then the finite-difference solution will converge to the exact solution in the limit as the grid spacings and size of the time step are refined.

Conservative Form

When the convection velocity field is solenoidal, that is, $\nabla \cdot \mathbf{U} = 0$, Eq. (12.1.1) may be recast into the *conservative form*

$$\frac{\partial \mathbf{f}}{\partial t} + \nabla \cdot [\mathbf{U}(\mathbf{f}, \mathbf{x}, t)\,\mathbf{f}] = \kappa \nabla^2 \mathbf{f} \qquad (12.1.2)$$

In constrast, the primary equation (12.1.1) represents the *nonconservative form.* This terminology stems from the fact that, in certain finite-difference discretizations, the values of the matrix $\mathbf{Uf}$ at the grid points telescope up to the boundaries and thus conserve certain invariants of the solution. In the context of incompressible flow, both the Navier–Stokes equation and vorticity transport equation may be cast in a conservative form that is similar to that shown in Eq. (12.1.2). The conservative form is usually preferable over the nonconservative form for reasons of enhanced accuracy and numerical stability.

PROBLEMS

12.1.1 **Developing finite-difference approximations.** (a) Consider a function $f(x)$, and approximate its first and second derivatives using the difference approximations

$$f'(x) \cong af(x - \Delta x) + bf(x) + cf(x + \Delta x)$$
$$f''(x) \cong Af(x - \Delta x) + Bf(x) + Cf(x + \Delta x) \qquad (12.1.3)$$

Derive relations among the constant coefficients a, b, c and A, B, C, so that the difference equations are consistent, which means that in the limit as Δx tends to zero, they produce the exact values of the first and second derivatives. Then derive additional relations among the coefficients so that the error in the computation of the derivatives is of second order in Δx. (b) Compute the coefficients a, b, c so that the error of the following finite-difference approximations is of second order in Δx

$$f'(x) = af(x - 2\,\Delta x) + bf(x - \Delta x) + cf(x)$$
$$f'(x) = af(x) + bf(x + \Delta x) + cf(x + 2\,\Delta x) \qquad (12.1.4)$$

(c) Compute the coefficients a, b, c, d, e so that the error of the following finite-difference approximation is of fourth order in Δx

$$f'(x) = af(x - 2\,\Delta x) + bf(x - \Delta x) + cf(x) + df(x + \Delta x) + ef(x + 2\,\Delta x) \qquad (12.1.5)$$

12.1.2 **Finite-difference formulation for a linear ODE.** Consider the solution of the linear ordinary differential equation $f'' + 4f = 0$, over the domain between $x = 0$ and $x = \pi/2$, with boundary conditions $f'(0) = -2$ and $f(\pi/2) = -1$. A prime denotes a derivative with respect to x. (a) Compute the solution analytically in closed form. (b) Discretize the domain of solution into N evenly spaced intervals that are separated by the grid points $x_i = (i - 1)\Delta x$, where $\Delta x = \pi/(2N)$ and $i = 1, \ldots, N + 1$. Apply the differential equation at the ith nodal point and approximate the second derivative using central differences to derive the finite-difference equation

$$f_{i-1} - 2(1 - 2\Delta x^2)f_i + f_{i+1} = 0 \qquad (12.1.6)$$

for $i = 2, \ldots, N$, where we have set $f_i = f(x_i)$. The boundary condition at $x = \pi/2$ requires that $f_{N+1} = -1$. (c) To examine the consistency of Eq. (12.1.6), pretend that the discrete values f_{i-1}, f_i, f_{i+1} are continuous functions of x and expand them in Taylor series about the point x_i, thereby deriving the modified differential equation. Show that as Δx tends to zero, the latter reduces to the original ordinary differential equation. (d) Use a forward difference to approximate the first derivative at $x = 0$, thereby obtaining the discrete form $f_2 - f_1 = -2\Delta x$. Collect all finite-difference equations into the linear system $\mathbf{A} \cdot \mathbf{f} = \mathbf{b}$, where $\mathbf{A}$ is a tridiagonal matrix, $\mathbf{f} = (f_1, f_2, \ldots, f_N)$, and $\mathbf{b}$ is a constant vector. Provide the explicit forms of $\mathbf{A}$ and $\mathbf{b}$. (e) The finite-difference equation (12.1.6) is second-order accurate, whereas the corresponding equation for the boundary condition at $x = 0$ is first-order accurate in Δx; this limits the overall accuracy of the finite-difference method. To obtain second-order accuracy, extend the domain of solution beyond the natural boundary $x = 0$, introduce the fictitious node $x_0 = -\Delta x$, apply Eq. (12.1.6) for $i = 1$, and approximate the first boundary condition using central differences to obtain $f_2 - f_0 = -4\Delta x$. Derive the associated linear system $\mathbf{B} \cdot \mathbf{f} = \mathbf{c}$ and provide the explicit forms of the tridiagonal matrix $\mathbf{B}$ and constant vector $\mathbf{c}$.

12.1.3 **Finite-difference formulation for a quasilinear ODE.** Repeat parts (c)–(e) of Problem 12.1.2 for the equation $f'' + 4(\cos x + \sin x)f = 0$ in the same domain and subject to identical boundary conditions.

Computer Problems

12.1.4 **Finite-difference solution of a linear ODE.** Solve the tridiagonal systems of equations developed in Problem 12.1.2(c,d) using the Thomas algorithm discussed in Section B.1, Appendix B, and compare the numerical results with the exact solution developed in part (a).

12.1.5 **Finite-difference solution of a quasilinear ODE.** Solve the tridiagonal systems of equations developed in Problem 12.1.3 using the Thomas algorithm and discuss the accuracy of your computations as a function of N.

12.2 | ONE-DIMENSIONAL DIFFUSION

We begin by discussing finite-difference methods for the one-dimensional unsteady diffusion equation

$$\frac{\partial f}{\partial t} = \kappa \frac{\partial^2 f}{\partial x^2} \qquad (12.2.1)$$

which is a considerably simplified version of the more general convection–diffusion Eq. (12.1.1), subject to the initial condition $f(x, 0) = F(x)$, where $F(x)$ is a known function. The diffusivity κ is assumed to be a positive constant.

For simplicity, we assume that the domain of solution extends over the whole x axis, and require the homogeneous far-field condition $f(x = \pm\infty, t) = 0$. If the domain of solution were bounded, for instance, between $a < x < b$, we would have to require one boundary condition for either f, or $\partial f/\partial x$, or their combination, at both ends $x = a$ and b, or boundary conditions for two of them at one end.

Equation (12.2.1) is a standard *parabolic* partial differential equation with a well-known analytical solution given by

$$f(x, t) = \frac{1}{\sqrt{4\kappa\pi t}} \int_{-\infty}^{\infty} F(x + x') \exp\left(-\frac{x'^2}{4\kappa t}\right) dx' \tag{12.2.2}$$

(Carslaw and Jaeger, 1959, p. 53). Our present objective is to generate the discrete version of this solution using a finite-difference method.

As a first step towards developing finite-difference methods, we assume that, during a certain initial period of evolution, f remains infinitesimal outside a certain computational domain $a < x < b$, and introduce a two-dimensional grid that covers the semi-infinite strip $a < x < b, 0 < t < \infty$ in the time–space plane as illustrated in Figure 12.2.1. Our objective is to compute the values of the function f_i^n at the grid points $x_i, i = 1, \ldots, K + 1$, at a sequence of successive time levels t^n beginning from the initial time level $t^0 = 0$, subject to the boundary conditions $f_1^n = 0$ and $f_{K+1}^n = 0$.

Explicit FTCS Method

Applying Eq. (12.2.1) at the x_i grid point at the time instant t^n and approximating the time derivative with a forward difference and the space derivative with a central difference, we obtain the FTCS finite-difference equation

$$\frac{f_i^{n+1} - f_i^n}{\Delta t} + O(\Delta t) = \kappa \frac{f_{i+1}^n - 2f_i^n + f_{i-1}^n}{\Delta x^2} + O(\Delta x^2) \tag{12.2.3}$$

which is first-order accurate in time and second-order accurate in space. The FTCS differentiation stencil is indicated in Figure 12.2.1 with hollow circles. Solving Eq. (12.2.3) for f_i^{n+1} yields

$$f_i^{n+1} = \alpha f_{i-1}^n + (1 - 2\alpha)f_i^n + \alpha f_{i+1}^n \tag{12.2.4}$$

where

$$\alpha = \frac{\kappa \Delta t}{\Delta x^2} \tag{12.2.5}$$

is a positive dimensionless constant called the *diffusion number*. Equations (12.2.3) and (12.2.4) apply at the internal grid points $i = 2, \ldots, K$.

To confirm the consistency of Eq. (12.2.4), we regard all discrete variables as continuous functions of space and time, expand them in Taylor series about the point (x_i, t^n), and simplify to obtain the associated modified differential equation

$$f_t + \tfrac{1}{2}f_{tt}\Delta t + O(\Delta t^2) = \kappa f_{xx} + \tfrac{1}{12}\kappa f_{xxxx}\Delta x^2 + O(\Delta x^4) \tag{12.2.6}$$

where subscripts denote partial derivatives with respect to corresponding variables, and all functions are evaluated at x_i and t^n. Since in the limit as Δt and Δx tend to zero, Eq. (12.2.6) reduces to Eq. (12.2.1), the FTCS discretization is consistent indeed with the original differential equation.

Differentiating Eq. (12.2.1) once with respect to t and twice with respect to x, and combining the resulting equations, we derive the fourth-order equation $f_{tt} = \kappa^2 f_{xxxx}$. Eliminating f_{xxxx} on the right-hand side of Eq. (12.2.6) in favor of f_{tt} and combining the resulting expression with the second term on the left-hand side shows that, when $\alpha = \tfrac{1}{6}$, the accuracy of the FTCS method becomes of second order in t and fourth order in x.

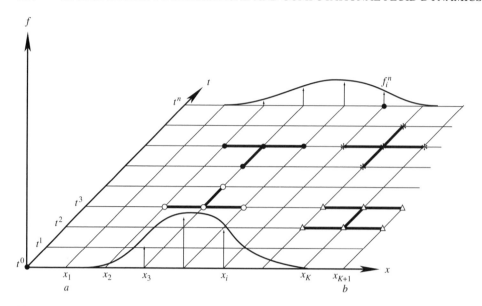

Figure 12.2.1. Discretization of the space–time domain for solving the one-dimensional diffusion equation using a finite-difference method. The initial distribution diffuses and tends to occupy the whole x axis. The hollow circles, solid circles, asterisks, and triangles indicate, respectively, the finite-difference stencils for the FTCS, BTCS, CTCS, and Crank–Nicolson methods.

Equation (12.2.4) provides us with a straightforward algorithm for computing the value of f at a grid point at the time level $n + 1$ in terms of the values of f at three grid points at the previous time level n. Since the algorithm does not require solving a system of algebraic equations, it is *explicit*.

In summary, the FTCS discretization provides us with an *explicit two-level method* whose accuracy is generally of *first order in time and second order in space*.

Successive mapping

To formalize the action of the FTCS method, we collect the values of f_i^n at the grid points $i = 2, \ldots, K$ into the vector $\mathbf{f}^n$, and use Eq. (12.2.4) in conjunction with the boundary conditions $f_1^n = 0$ and $f_{K+1}^n = 0$, to obtain $\mathbf{f}^{n+1} = \mathbf{B} \cdot \mathbf{f}^n$, where $\mathbf{B}$ is a $(K-1) \times (K-1)$ tridiagonal matrix with superdiagonal, diagonal, and subdiagonal elements, respectively, equal to α, $1 - 2\alpha$, and α. This matrix form shows that the solution vector $\mathbf{f}^{n+1}$ derives by projecting the vector $\mathbf{f}^n$ onto the matrix $\mathbf{B}$, and thus establishes a relationship between *time stepping* and *successive mapping*.

The behavior of the vector $\mathbf{f}^n$ with respect to n will depend upon the properties of the *projection matrix* $\mathbf{B}$ and, in particular, upon the spectral radius of $\mathbf{B}$ denoted by $\rho(\mathbf{B})$, defined as the maximum value of the magnitude of its eigenvalues. The theory of matrix calculus shows that if $\rho(\mathbf{B})$ is equal to unity, less than unity, or larger than unity, the length of $\mathbf{f}^n$ will stay roughly constant, decrease, or increase during the successive mappings (Wilkinson, 1965). Since, according to the exact solution, the amplitude of f diminishes due to diffusion, we tolerate the first behavior, accept the second behavior, but dismiss the third behavior as being numerically unstable.

To compute the eigenvalues of $\mathbf{B}$, we write $\mathbf{B} = \mathbf{I} + \alpha\mathbf{C}$, where $\mathbf{C}$ is a tridiagonal matrix with superdiagonal, diagonal, and subdiagonal elements, respectively, equal to 1, -2, and 1, and $\mathbf{I}$ is the identity matrix. The eigenvalues of $\mathbf{B}$ and $\mathbf{C}$, denoted, respectively, by $\lambda(\mathbf{B})$ and $\lambda(\mathbf{C})$, are related by $\lambda(\mathbf{B}) = 1 + \alpha\lambda(\mathbf{C})$. A detailed computation shows that

$$\lambda_m(\mathbf{C}) = -4 \sin^2\left(\frac{m\pi}{2K}\right) \quad \text{and} \quad \lambda_m(\mathbf{B}) = 1 - 4\alpha \sin^2\left(\frac{m\pi}{2K}\right) \quad (12.2.7)$$

where $m = 1, \ldots, K - 1$ (Problem 12.2.6). The second of Eqs. (12.2.7) shows that the spectral radius of $\mathbf{B}$ will be less than unity only when $\alpha < \frac{1}{2}$, and this shows that the FTCS method is conditionally stable.

If boundary conditions other than the homogeneous Dirichlet boundary conditions $f_1^n = 0$ and $f_{K+1}^n = 0$ were specified on one or both ends of the computational domain, the structure of the mapping matrix $\mathbf{B}$ would be altered, but the performance of the finite-difference method would still be determined by its spectral radius $\rho(\mathbf{B})$.

Consider, for instance, the Neumann boundary condition $\partial f / \partial x = g$ at the left end, at $x = a$, and maintain the homogeneous Dirichlet boundary condition at the right end, at $x = b$. To implement the first boundary condition with second-order accuracy, we extend the domain of solution beyond the physical boundary at $x = a$, introduce the fictitious node $x_0 = x_1 - \Delta x$, and use central differences to obtain $f_2^n - f_0^n = 2g \Delta x$. Having extended the domain, we apply the differential equation at the first node x_1 and write Eq. (12.2.4) with $i = 1$. To obtain the corresponding mapping matrix $\mathbf{B}$, we collect the values of f_i^n at the grid points $i = 1, \ldots, K$ into the vector $\mathbf{f}^n$, and use Eq. (12.2.4) in conjunction with the aforementioned boundary conditions to obtain $\mathbf{f}^{n+1} = \mathbf{B} \mathbf{f}^n + \mathbf{b}$, where $\mathbf{B}$ is a $K \times K$ tridiagonal matrix with superdiagonal, diagonal, and subdiagonal elements, respectively, equal to α, $1 - 2\alpha$, and α, except that the second entry in the first row is equal to 2α. All entries of the vector $\mathbf{b}$ are equal to zero except for the first one, which is equal to $-2g \Delta x$. The presence of the vector $\mathbf{b}$ does not alter the significance of the projection matrix $\mathbf{B}$ with respect to the behavior of the solution discussed previously. When $\rho(\mathbf{B})$ is equal to one, less than one, or larger than one, the length of $\mathbf{f}^n$ will stay roughly constant, decrease, or increase during the successive mappings. Unfortunately, we can no longer compute the eigenvalues of $\mathbf{B}$ in closed form.

von Neumann stability analysis

The simple structure of the projection matrix associated with the FTCS method with homogeneous Dirichlet boundary conditions at both ends allowed us to compute its eigenvalues and spectral radius exactly in closed form, and thereby assess the stability of the numerical method. Unfortunately, such closed-form solutions are not available for more advanced finite-difference discretizations and more general types of boundary conditions. An alternative is to compute the eigenvalues using a numerical method as discussed in Section B.2, Appendix B, but this is a computationally arduous task.

Another way of assessing the stability of the numerical method is to perform the von Neumann stability analysis of the finite-difference equation neglecting the boundary conditions. The basic idea is to examine the behavior of the numerical solution subject to a sinusoidal initial condition of a certain wavelength L. Motivated by the linearity of the governing equation, we separate the temporal from the spatial dependence setting

$$f_i^n = A^n \exp(I \, i \, \theta) \tag{12.2.8}$$

where $I = \sqrt{-1}$ is the imaginary unit, $\theta = 2\pi \Delta x / L$ is the *phase angle*, and A^n is a coefficient whose value depends upon the time level n. Substituting Eq. (12.2.8) into Eq. (12.2.4) and simplifying, we find

$$\frac{A^{n+1}}{A^n} \equiv G = 1 + 2\alpha(\cos \theta - 1) = 1 - 4\alpha \sin^2 \frac{\theta}{2} \tag{12.2.9}$$

where G is the *growth factor, gain,* or *amplification factor.* When $\alpha > \frac{1}{2}$, the magnitude of the right-hand side of Eq. (12.2.9) is greater than unity for a certain range of values of θ, in which case the numerical method is unstable. When $\alpha < \frac{1}{2}$, the magnitude of the right-hand side of Eq. (12.2.9) is less than unity for any value of θ, in which case the numerical method is stable. These results are consistent with our previous conclusions based on the spectral radius of the projection matrix $\mathbf{B}$.

The efficiency of the von Neumann stability analysis is now evident. One limitation of the method is that, in its simple form described above, it does not incorporate the effect of inhomogeneous boundary conditions, which may have a destabilizing effect on the finite-difference method.

Assessment of the FTCS method

Since α is proportional to the temporal step Δt and inversely proportional to the square of the spatial step Δx, the stability constraint $\alpha < \frac{1}{2}$ of the FTCS method requires the use of a time step that is excessively small and may lead to a prohibitive computational cost. The low-order accuracy combined with the conditional stability renders the method less attractive compared to its alternatives.

Explicit CTCS or Leapfrog Method

One way to achieve second-order accuracy in both time and space is to use central differences for both variables. Applying Eq. (12.2.1) at the point x_i at the time instant t^n and using central differences in both time and space, we obtain the CTCS difference equation

$$\frac{f_i^{n+1} - f_i^{n-1}}{2\Delta t} + O(\Delta t^2) = \kappa \frac{f_{i+1}^n - 2f_i^n + f_{i-1}^n}{\Delta x^2} + O(\Delta x^2) \qquad (12.2.10)$$

The corresponding finite-difference stencil is illustrated with asterisks in Figure 12.2.1. Rearranging, we derive the three-time-level explicit algorithm

$$f_i^{n+1} = f_i^{n-1} + 2\alpha f_{i-1}^n - 4\alpha f_i^n + 2\alpha f_{i+1}^n \qquad (12.2.11)$$

The solution at the first time level corresponding to $n = 1$ must be computed using a two-level method, such as the FTCS method, with a time step that is an integral fraction of Δt, of small enough size in order to prevent the onset of deleterious oscillations.

To examine the stability of the CTCS method, we substitute Eq. (12.2.8) into Eq. (12.2.11), set $A^{n+1}/A^n \equiv A^n/A^{n-1} \equiv G$, and obtain the quadratic equation for the gain, $G^2 + \beta G - 1 = 0$, where $\beta = 2\alpha(1 - \cos\theta) = 4\alpha \sin^2(\theta/2)$ is a real non-negative parameter; the solution is $G = \frac{1}{2}[-\beta \pm (\beta^2 + 4)^{1/2}]$. The magnitude of the root corresponding to the minus sign is higher than unity, and this shows that the CTCS method is *unconditionally unstable* and thus of no practical value.

The DuFort–Frankel Explicit Method

DuFort and Frankel (1953) proposed a modification of the CTCS discretization, with the objective of maintaining the second-order accuracy but improving the numerical stability. The method proceeds by replacing the middle term in the numerator on the right-hand side of Eq. (12.2.10) with an average value, yielding

$$\frac{f_i^{n+1} - f_i^{n-1}}{2\Delta t} + O(\Delta t^2) = \kappa \frac{f_{i+1}^n - 2[\frac{1}{2}(f_i^{n+1} + f_i^{n-1})] + f_{i-1}^n}{\Delta x^2} + O(\Delta x^2) \quad (12.2.12)$$

Rearranging, we obtain the three-level explicit algorithm

$$f_i^{n+1} = \frac{1 - 2\alpha}{1 + 2\alpha} f_i^{n-1} + \frac{2\alpha}{1 + 2\alpha}(f_{i-1}^n + f_{i+1}^n) \qquad (12.2.13)$$

The computation must be started using a two-level method.

Performing the von Neumann stability analysis shows that the amplification factor satisfies the quadratic equation

$$(1 + 2\alpha) G^2 - 4G\alpha \cos\theta - 1 + 2\alpha = 0 \qquad (12.2.14)$$

Examining the roots, we find that $|G| < 1$ for any value of α, and this ensures that the DuFort–Frankel method is *unconditionally stable*.

Since, however, Eq. (12.2.12) was derived on the basis of an ad hoc modification of the well-founded CTCS discretization, its consistency must be examined by comparing the associated modified differential equation with the original differential equation (12.2.1). To derive the former, we regard all discrete variables in Eq. (12.2.12) as continuous functions of space and time, expand them in Taylor series about the point (x_i, t^n), and simplify to obtain

$$f_t = \kappa f_{xx} - \kappa \left(\frac{\Delta t}{\Delta x}\right)^2 f_{tt} \tag{12.2.15}$$

Equation (12.2.15) is an accurate approximation to Eq. (12.2.1) only when the ratio $(\Delta t/\Delta x)^2$ is sufficiently small. No matter how small Δt and Δx are, if the ratio $(\Delta t/\Delta x)^2$ has a finite value, the DuFort–Frankel method solves a fictitious problem described by Eq. (12.2.15), instead of the diffusion problem described by Eq. (12.2.1). Thus, the method cannot be said to be consistent in general. The fact that Eq. (12.2.15) is classified as a hyperbolic differential equation, due to the presence of the second derivative with respect to time, whereas Eq. (12.2.1) is classified as a parabolic differential equation, suggests that *adding a term with a wavelike character has a stabilizing influence.*

Because of its advantages regarding stability, the DuFort–Frankel method has enjoyed extensive applications in practice. When using it, however, care must be taken so that the product $\kappa(\Delta t/\Delta x)^2 = \alpha \Delta t$ is sufficiently small; otherwise the results will not be physically meaningful.

Implicit BTCS or Laasonen Method

Thus far we have considered explicit methods in which the solution at a particular time level is computed directly from the solution at one or two previous time levels without solving any systems of equations. We turn now to consider *implicit discretizations that require solving systems of algebraic equations,* in hopes of achieving unconditional stability while maintaining consistency, and thus relaxing the restriction on Δt.

Applying Eq. (12.2.1) at the point x_i at the time instant t^{n+1}, and approximating the temporal derivative with a backward difference and the spatial derivative with a central difference, we obtain the BTCS difference equation

$$\frac{f_i^{n+1} - f_i^n}{\Delta t} + O(\Delta t) = \kappa \frac{f_{i+1}^{n+1} - 2f_i^{n+1} + f_{i-1}^{n+1}}{\Delta x^2} + O(\Delta x^2) \tag{12.2.16}$$

The corresponding finite-difference stencil is shown with hollow circles in Figure 12.2.1. Rearranging Eq. (12.2.16), we derive the two-level implicit algorithm

$$-\alpha f_{i-1}^{n+1} + (1 + 2\alpha)f_i^{n+1} - \alpha f_{i+1}^{n+1} = f_i^n \tag{12.2.17}$$

Recasting Eq. (12.2.17) into a matrix form and implementing the homogeneous Dirichlet boundary conditions, we obtain the system of linear equations $\mathbf{A} \cdot \mathbf{f}^{n+1} = \mathbf{f}^n$, where $\mathbf{A}$ is a tridiagonal matrix with superdiagonal, diagonal, and subdiagonal elements, respectively, equal to $-\alpha$, $1 + 2\alpha$, $-\alpha$. Solving for $\mathbf{f}^{n+1}$ yields $\mathbf{f}^{n+1} = \mathbf{A}^{-1} \cdot \mathbf{f}^n$, where $\mathbf{A}^{-1}$ is the inverse of $\mathbf{A}$, which shows that stepping in time is equivalent to mapping successively the initial vector $\mathbf{f}^0$ with the projection matrix $\mathbf{A}^{-1}$.

In practice, in order to compute the solution at the $n + 1$ time level, we solve the system of linear algebraic equations $\mathbf{A} \cdot \mathbf{f}^{n+1} = \mathbf{f}^n$, and this renders the BTCS method implicit. Since $\mathbf{A}$ is tridiagonal and diagonally dominant, the linear system may be solved uneventfully using either the Thomas algorithm or Jacobi's method discussed in Section B.1, Appendix B, both of which are economical.

To study the stability of the BTCS method, we consider the eigenvalues of the projection matrix $\mathbf{A}^{-1} = (-\alpha \mathbf{C} + \mathbf{I})^{-1}$, where the matrix $\mathbf{C}$ was defined before Eq. (12.2.7), and use the first of Eqs. (12.2.7) and the fact that the eigenvalues of the inverse of a matrix are equal to the inverse of the eigenvalues of the original matrix, to obtain

$$\lambda_m(\mathbf{A}^{-1}) = \left(1 + 4\alpha \sin^2 \frac{m\pi}{2K}\right)^{-1} \tag{12.2.18}$$

where $m = 1, \ldots, K - 1$. Since the spectral radius of the projection matrix is less than unity, the BTCS method is *unconditionally* stable. An independent way of arriving at this result is to perform the von Neumann stability analysis, obtaining the amplification factor

$$G = \frac{1}{1 + 2\alpha(1 - \cos\theta)} = \frac{1}{1 + 4\alpha \sin^2 \frac{\theta}{2}} \tag{12.2.19}$$

whose magnitude may be seen to be less than unity for any value of α or θ.

The main limitation of the BTCS method is its low-order temporal accuracy, which places a restriction on the maximum size of the time step for an accurate solution.

The Implicit Crank–Nicolson Method

Continuing our search for an efficient method, we target an algorithm that is *second-order accurate in both time and space, and unconditionally stable*. To this end, we recall that the explicit FTCS method emerged by applying Eq. (12.2.1) at the point x_i and at the time level t^n, whereas the implicit BTCS method emerged by applying Eq. (12.2.1) at the point x_i at the time level t^{n+1}. Being adventurous, the distinguishing attribute of the numerical fluid dynamicist, we apply Eq. (12.2.1) at the intermediate grid point $(x_i, t^{n+1/2})$ that is located half-way between the grid points (x_i, t^n) and (x_i, t^{n+1}), and set the spatial derivative at the $t^{n+1/2}$ level equal to the average value of the spatial derivatives at the t^n and t^{n+1} levels, to arrive at the finite-difference equation

$$\frac{f_i^{n+1} - f_i^n}{\Delta t} = \kappa \frac{1}{2} \left(\frac{f_{i+1}^{n+1} - 2f_i^{n+1} + f_{i-1}^{n+1}}{\Delta x^2} + \frac{f_{i+1}^n - 2f_i^n + f_{i-1}^n}{\Delta x^2} \right) \tag{12.2.20}$$

(Crank and Nicolson, 1947). The corresponding finite-difference stencil is shown with triangles in Figure 12.2.1. Rearranging Eq. (12.2.20), we obtain a tridiagonal system of equations

$$-\alpha f_{i-1}^{n+1} + 2(1 + \alpha)f_i^{n+1} - \alpha f_{i+1}^{n+1} = \alpha f_{i-1}^n + 2(1 - \alpha)f_i^n + \alpha f_{i+1}^n \tag{12.2.21}$$

Deriving and examining the corresponding modified differential equation shows that the Crank–Nicolson method is consistent and *second-order accurate in both time and space*.

Recasting Eq. (12.2.21) into a matrix form, we obtain the system of linear equations $\mathbf{A} \cdot \mathbf{f}^{n+1} = \mathbf{B} \cdot \mathbf{f}^n$, where $\mathbf{A}$ and $\mathbf{B}$ are tridiagonal matrices with superdiagonal, diagonal, and subdiagonal elements, respectively, equal to $-\alpha$, $2(1 + \alpha)$, $-\alpha$, and α, $2(1 - \alpha)$, α. Solving for $\mathbf{f}^{n+1}$ yields $\mathbf{f}^{n+1} = \mathbf{A}^{-1} \cdot \mathbf{B} \cdot \mathbf{f}^n$, which shows that stepping in time is equivalent to mapping with the projection matrix $\mathbf{A}^{-1} \cdot \mathbf{B}$. The spectral radius of the projection matrix may be shown to be less than unity, and this ensures that the Crank–Nicolson method is *unconditionally stable* (Problem 12.2.2).

Carrying out the von Neumann stability analysis yields the amplification factor

$$G = \frac{1 - \alpha(1 - \cos\theta)}{1 + \alpha(1 - \cos\theta)} = \frac{1 - 2\alpha \sin^2(\theta/2)}{1 + 2\alpha \sin^2(\theta/2)} \tag{12.2.22}$$

which is always less than unity, confirming that the method is unconditionally stable. Because of its notable qualities with respect to both accuracy and stability, the Crank–Nicolson method has become a standard choice in practice.

Multiple substeps

It is instructive to remark that the Crank–Nicolson method may be regarded as the result of the sequential action of a two-step method, where each substep lasts for a time interval equal to $\Delta t/2$. The first substep is carried out using the explicit FTCS method, and the second substep is carried out using the implicit BTCS method according to the finite-difference equations

$$\frac{f_i^{n+1/2} - f_i^n}{\Delta t/2} = \kappa \frac{f_{i+1}^n - 2f_i^n + f_{i-1}^n}{\Delta x^2}$$

$$\frac{f_i^{n+1} - f_i^{n+1/2}}{\Delta t/2} = \kappa \frac{f_{i+1}^{n+1} - 2f_i^{n+1} + f_{i-1}^{n+1}}{\Delta x^2} \qquad (12.2.23)$$

Adding the two equations in (12.2.23) to eliminate the intermediate variable at the time level $t^{n+1/2}$ recovers Eq. (12.2.20). The unconditional stability of the second substep prevails over the conditional stability of the first substep, and renders the overall method unconditionally stable.

Three-Level Implicit Methods

Richtmyer and Morton (1967, p. 189) present a survey of three-level implicit methods. The general form of a three-level, five-point method with a T-shaped finite-difference stencil is

$$(1 + \beta)\frac{f_i^{n+1} - f_i^n}{\Delta t} - \beta \frac{f_i^n - f_i^{n-1}}{\Delta t} + O(\Delta t)$$

$$= \kappa \frac{f_{i+1}^{n+1} - 2f_i^{n+1} + f_{i-1}^{n+1}}{\Delta x^2} + O(\Delta x^2) \qquad (12.2.24)$$

where β is an arbitrary positive constant. It can be shown that the numerical algorithm is unconditionally stable for any choice of β.

The particular choice $\beta = \frac{1}{2}$ provides us with a method that is second-order accurate in time and effective at suppressing small-scale oscillations. These features render Eq. (12.2.24) with $\beta = \frac{1}{2}$ preferable over the Crank–Nicolson method when the solution exhibits sharp spatial variations. The choice $\beta = \frac{1}{2}(1 - 1/(6\alpha))$ provides us with a method that is second-order accurate in time and fourth-order accurate in space.

PROBLEMS

12.2.1   **CTCS and DuFort–Frankel.** (a) Express Eq. (12.2.11) in vector notation in terms of the solution vectors $\mathbf{f}^{n+1}$, $\mathbf{f}^n$, $\mathbf{f}^{n-1}$. (b) Perform a consistency analysis of the DuFort–Frankel method and derive the modified differential equation (12.2.15).

12.2.2   **Spectral radius of the projection matrix of the Crank–Nicolson method.** Verify that the eigenvalues of the projection matrix corresponding to the Crank–Nicolson method are given by

$$\lambda_m(\mathbf{A}^{-1}\mathbf{B}) = \frac{1 - 2\alpha \sin^2(m\pi/2K)}{1 + 2\alpha \sin^2(m\pi/2K)} \qquad (12.2.25)$$

where $m = 1, \ldots, K - 1$, and then show that the spectral radius of the projection matrix is less than unity and therefore the method is unconditionally stable.

12.2.3   **Generalized Crank–Nicolson method.** A more general form of Eq. (12.2.20) emerges by using a weighted average to approximate the spatial derivative obtaining

$$\frac{f_i^{n+1} - f_i^n}{\Delta t} = \kappa \left(\beta \frac{f_{i+1}^{n+1} - 2f_i^{n+1} + f_{i-1}^{n+1}}{\Delta x^2} + (1 - \beta)\frac{f_{i+1}^n - 2f_i^n + f_{i-1}^n}{\Delta x^2} \right)$$

$$(12.2.26)$$

where β is a positive parameter taking values in the range $[0, 1]$. When $\beta = 0$ or $\frac{1}{2}$, we obtain, respectively, the explicit FTCS and the implicit Crank–Nicolson method. (a) Show that, in general, the method is first-order accurate in t and second-order accurate in x. (b) Show that when $\beta = \frac{1}{2}(1 - 1/(6\alpha))$ the accuracy of the method increases to second order in t and fourth order in x. (c) Provide an interpretation of Eq. (12.2.26) in terms of a sequence of two elementary substeps. (d) Perform the von Neumann stability analysis of Eq. (12.2.26), derive the amplification factor

$$G = \frac{1 - 4(1 - \beta)\alpha \sin^2(\theta/2)}{1 + 4\beta\alpha \sin^2(\theta/2)} \qquad (12.2.27)$$

and show that the method is unconditionally stable when $\frac{1}{2} < \beta < 1$ and conditionally stable when $0 < \beta < \frac{1}{2}$. Show that, in the second case, the stability restriction is $\alpha < 1/(2 - 4\beta)$ (Richtmyer and Morton, 1967, p. 189).

12.2.4 **Dispersion.** Consider the dispersion equation in one dimension

$$\frac{\partial f}{\partial t} = \beta \frac{\partial^3 f}{\partial x^3} \qquad (12.2.28)$$

where the constant β is called the *dispersion coefficient*. (a) Investigate analytically the evolution of harmonic waves. More specifically, show that

$$f(x, t) = \exp[Ik(x - k^2\beta t)] \qquad (12.2.29)$$

is an exact solution, where I is the imaginary unit, k is a real wave number, and $k^2\beta$ is the phase velocity of traveling harmonic waves. Since the phase velocity depends upon the wavelength, *the waves are dispersive,* which means that an arbitrary initial distribution composed of a finite or infinite superposition of sinusoidal waves propagates while changing its shape. (b) The FTCS discretization leads to the explicit five-point, two-level finite-difference equation

$$f_i^{n+1} = -\tfrac{1}{2}\gamma f_{i-2}^n + \gamma f_{i-1}^n + f_i^n - \gamma f_{i+1}^n + \tfrac{1}{2}\gamma f_{i+2}^n \qquad (12.2.30)$$

where $\gamma = \beta \Delta t/\Delta x^3$, which is first-order accurate in time and second-order accurate in space. Carry out the von Neumann stability analysis and derive the amplification factor

$$G = 1 - I2\gamma \sin \theta(1 - \cos \theta) \qquad (12.2.31)$$

and thus show that the method is *unconditionally unstable.* (c) Show that the implicit five-point two-level BTCS discretization leads to the finite-difference equation

$$\tfrac{1}{2}\gamma f_{i-2}^{n+1} - \gamma f_{i-1}^{n+1} + f_i^{n+1} + \gamma f_{i+1}^{n+1} - \tfrac{1}{2}\gamma f_{i+2}^{n+1} = f_i^n \qquad (12.2.32)$$

Derive the associated amplification factor

$$G = [1 + I2\gamma \sin \theta(1 - \cos \theta)]^{-1} \qquad (12.2.33)$$

and thus show that the method is *unconditionally stable.* (d) Repeat the tasks of part (b) for the counterpart of the Crank–Nicolson method.

12.2.5 **Fourth-order diffusion.** Consider the fourth-order diffusion equation in one dimension

$$\frac{\partial f}{\partial t} = -\nu \frac{\partial^4 f}{\partial x^4} \qquad (12.2.34)$$

where the positive constant ν is called the *fourth-order diffusivity*. (a) The FTCS discretization leads to the explicit five-point difference equation

$$f_i^{n+1} = \varepsilon f_{i-2}^n - 4\varepsilon f_{i-1}^n + (1 + 6\varepsilon)f_i^n - 4\varepsilon f_{i+1}^n + \varepsilon f_{i+2}^n \qquad (12.2.35)$$

where $\varepsilon = \nu \Delta t/\Delta x^4$, which is first-order accurate in time and second-order accurate in space. Carry out the von Neumann stability analysis, compute the amplification factor, and assess the stability of the method. (b) Repeat part (a) for the implicit five-point, two-level BTCS method. (c) Develop the counterpart of the Crank–Nicolson method.

12.2.6   **Eigenvalues and eigenvectors of a tridiagonal matrix with constant entries.** Consider a $(K-1) \times (K-1)$ tridiagonal matrix whose diagonal, superdiagonal, and subdiagonal elements are, respectively, equal to a, b, and c. Verify that the eigenvalues are given by

$$\lambda_m = a + 2\sqrt{bc}\cos\frac{m\pi}{K} \qquad (12.2.36)$$

and the associated eigenvectors are

$$u_l^{(m)} = \left(\frac{a}{c}\right)^{l/2}\sin\frac{lm\pi}{K} \qquad (12.2.37)$$

where $m, l = 1, \ldots, K-1$. Note that the eigenvectors are independent of the value of a.

Computer Problem

12.2.7   **Transient Couette flow.** Consider a two-dimensional channel that is confined between two parallel plates separated by a distance of 2 cm, and is filled with water. Suddenly, the upper plate starts moving parallel to itself with velocity equal to $10(1 - e^{-\varepsilon t})$ cm/sec, where ε is a constant, while the lower plate is held stationary. Compute and plot the developing velocity profiles for $\varepsilon = 0.1$ and 1 sec^{-1} using (a) the FTCS method, (b) the DuFort–Frankel method, (c) the CTCS method, and (d) the Crank–Nicolson method. In each case comment on your selection of the temporal and spatial steps.

12.3 | DIFFUSION IN TWO AND THREE DIMENSIONS

Expanding the scope of our discussion, we consider unsteady diffusion in two and three dimensions governed, respectively, by the equations

$$\frac{\partial f}{\partial t} = \kappa\left(\frac{\partial^2 f}{\partial x^2} + \frac{\partial^2 f}{\partial y^2}\right)$$

$$\frac{\partial f}{\partial t} = \kappa\left(\frac{\partial^2 f}{\partial x^2} + \frac{\partial^2 f}{\partial y^2} + \frac{\partial^2 f}{\partial z^2}\right) \qquad (12.3.1)$$

subject to the initial condition $f(\mathbf{x}, t = 0) = F(\mathbf{x})$, where F is a known function. For simplicity, we shall assume that the domain of solution extends over the whole two-dimensional plane or three-dimensional space, and the function f vanishes at infinity.

The majority of the finite-difference methods discussed in Section 12.2 can be extended in a straightforward manner to include a second or a third dimension, and several examples will be presented in the ensuing discussion. We shall see, however, that the need for computationally efficient algorithms necessitates the development of certain new procedures.

Iterative Solution of Laplace and Poisson Equations

Before proceeding to discuss specific types of discretization, we remark that the finite-difference methods for solving Eqs. (12.3.1) provide us with well-founded iterative procedures for solving the Laplace and Poisson equations in two and three dimensions. Consider, for instance, Poisson's equation in two dimensions

$$\frac{\partial^2 f}{\partial x^2} + \frac{\partial^2 f}{\partial y^2} = g(x, y) \qquad (12.3.2)$$

where g is a certain known source function, and introduce a fictitious unsteady diffusion problem in the presence of a source term, governed by the unsteady diffusion–reaction equation

$$\frac{\partial f}{\partial t} = \kappa \left(\frac{\partial^2 f}{\partial x^2} + \frac{\partial^2 f}{\partial y^2} \right) - \kappa g(x, y) \qquad (12.3.3)$$

subject to a certain initial condition. The asymptotic solution of Eq. (12.3.3) at large times is identical to the solution of Eq. (12.3.2), and this suggests that advancing the solution of the diffusion–reaction problem in time amounts to iterating the solution of the Poisson equation with a projection matrix that arises from the finite-difference method used to integrate Eq. (12.3.3).

Finite-Difference Procedures

In three dimensions, the goal of the finite-difference method is to generate the values of the function f at the nodes of a three-dimensional grid that is described in terms of the three indices i, j, k. In two dimensions, the nodes are described by two indices i, j. For simplicity, in this discussion we shall assume that the grid spacings Δx, Δy, and Δz are uniform but not necessarily identical throughout the domain of solution.

To carry out the von Neumann stability analysis, we set

$$f_{ijk}^n = A^n \exp[I(i\theta_x + j\theta_y + k\theta_z)] \qquad (12.3.4)$$

where $I = \sqrt{-1}$ is the imaginary unit, and $\theta_x = 2\pi \Delta x/L_x$, $\theta_y = 2\pi \Delta y/L_y$, and $\theta_z = 2\pi \Delta z/L_z$ are phase angles with corresponding wavelengths L_x, L_y, L_z, and then study the magnitude of the amplification factor $G \equiv A^{n+1}/A^n$. In general, the stability criteria for two-dimensional and three-dimensional diffusion are much more restrictive than those for one-dimensional diffusion, and this renders the conditionally stable methods prohibitively expensive.

Explicit FTCS Method

With the aforementioned considerations in mind, we proceed to develop the FTCS discretization for the two-dimensional problem. Approximating the twodimensional Laplacian operator with the five-point formula described in Section B.5, Appendix B, we derive the difference equation

$$\frac{f_{i,j}^{n+1} - f_{i,j}^n}{\Delta t} = \kappa \left(\frac{f_{i+1,j}^n - 2f_{i,j}^n + f_{i-1,j}^n}{\Delta x^2} + \frac{f_{i,j+1}^n - 2f_{i,j}^n + f_{i,j-1}^n}{\Delta y^2} \right) \qquad (12.3.5)$$

Rearranging we obtain the explicit algorithm

$$f_{i,j}^{n+1} = f_{i,j}^n + \alpha_x(f_{i+1,j}^n + f_{i-1,j}^n) + [1 - 2(\alpha_x + \alpha_y)]f_{i,j}^n + \alpha_y(f_{i,j+1}^n + f_{i,j-1}^n) \qquad (12.3.6)$$

where we have introduced the diffusivities for the x and y direction

$$\alpha_x = \frac{\kappa \Delta t}{\Delta x^2}, \qquad \alpha_y = \frac{\kappa \Delta t}{\Delta y^2} \qquad (12.3.7)$$

Carrying out the von Neumann stability analysis shows that the method is stable provided that $\alpha_x + \alpha_y < \frac{1}{2}$. This constraint imposes a strong restriction on the size of the time step that renders the explicit discretization inefficient in practice. Similar but more severe difficulties are encountered in three dimensions.

Implicit BTCS Method

To achieve unconditional stability, we resort to an implicit method. The BTCS discretization for the two-dimensional problem yields the two-level algorithm

$$-\alpha_x(f_{i-1,j}^{n+1} + f_{i+1,j}^{n+1}) + (1 + 2\alpha_x + 2\alpha_y)f_{i,j}^{n+1} - \alpha_y(f_{i,j-1}^{n+1} + f_{i,j+1}^{n+1}) = f_{i,j}^n \qquad (12.3.8)$$

with first-order accuracy in time and second-order accuracy in space, which may be shown to be *unconditionally stable.* Unfortunately, the numerical implementation of Eq. (12.3.8) results in a pentadiagonal system of algebraic equations whose solution can no longer be carried out using specialized methods such as the Thomas algorithm. This complication renders the BTCS method unpopular in practice.

ADI Method in Two Dimensions

To reduce the computational burden of the implicit BTCS method for two-dimensional diffusion, Peaceman and Rachford (1955) and Douglas (1955) proposed splitting each time step into two substeps of equal duration $\Delta t/2$, and approximating the spatial derivatives in a partially implicit manner, while working sequentially and alternatingly in the x and y directions. The computations proceed according to the finite-difference equations

$$\frac{f_{i,j}^{n+1/2} - f_{i,j}^n}{\Delta t/2} = \kappa \left[\frac{f_{i+1,j}^{n+1/2} - 2f_{i,j}^{n+1/2} + f_{i-1,j}^{n+1/2}}{\Delta x^2} \right] + \kappa \frac{f_{i,j+1}^n - 2f_{i,j}^n + f_{i,j-1}^n}{\Delta y^2}$$
(12.3.9a)

$$\frac{f_{i,j}^{n+1} - f_{i,j}^{n+1/2}}{\Delta t/2} = \kappa \frac{f_{i+1,j}^{n+1/2} - 2f_{i,j}^{n+1/2} + f_{i-1,j}^{n+1/2}}{\Delta x^2} + \kappa \left[\frac{f_{i,j+1}^{n+1} - 2f_{i,j}^{n+1} + f_{i,j-1}^{n+1}}{\Delta y^2} \right]$$
(12.3.9b)

where $n + \frac{1}{2}$ is an intermediate time level, and the square brackets indicate implicit discretizations. The first substep is carried out according to the implicit BTCS method for the x direction, and the second substep is carried out according to the implicit BTCS method for the y direction. To eliminate the bias associated with this particular arrangement, we alternate this order after the completion of each step. The overall accuracy of the method is of second order in both space and time.

Rearranging Eq. (12.3.9), we obtain a two-step implicit algorithm representing the x and y sweeps

$$\alpha_x f_{i-1,j}^{n+1/2} - 2(1 + \alpha_x) f_{i,j}^{n+1/2} + \alpha_x f_{i+1,j}^{n+1/2}$$
$$= -\alpha_y f_{i,j-1}^n - 2(1 - \alpha_y) f_{i,j}^n - \alpha_y f_{i,j+1}^n$$
(12.3.10a)

$$\alpha_y f_{i,j-1}^{n+1} - 2(1 + \alpha_y) f_{i,j}^{n+1} + \alpha_y f_{i,j+1}^{n+1}$$
$$= -\alpha_x f_{i-1,j}^{n+1/2} - 2(1 - \alpha_x) f_{i,j}^{n+1/2} - \alpha_x f_{i+1,j}^{n+1/2}$$
(12.3.10b)

Completing one time step requires solving two systems of tridiagonal equations, but this can be done efficiently using the Thomas algorithm described in Section B.1.

Carrying out the von Neumann stability analysis yields the amplification factor

$$G = \frac{[1 + \alpha_x(\cos\theta_x - 1)][1 + \alpha_y(\cos\theta_y - 1)]}{[1 - \alpha_x(\cos\theta_x - 1)][1 - \alpha_y(\cos\theta_y - 1)]}$$
$$= \frac{[1 - 2\alpha_x \sin^2(\theta_x/2)][1 - 2\alpha_y \sin^2(\theta_y/2)]}{[1 + 2\alpha_x \sin^2(\theta_x/2)][1 + 2\alpha_y \sin^2(\theta_y/2)]}$$
(12.3.11)

Cursory examination shows that $|G| < 1$ under any conditions, which ensures that the ADI method is unconditionally stable. The second-order accuracy combined with the unconditional stability have made the ADI method a standard choice in practice.

Crank–Nicolson Method and Approximate Factorization

The ADI method allows us to advance the solution over one time step by solving two pseudo-one-dimensional problems, and this amounts to decoupling the diffusion processes in the two spatial directions. To demonstrate this clearly, we introduce the second central-difference operators

$$\Delta_x^2 \langle f_{i,j}^k \rangle = f_{i-1,j}^k - 2f_{i,j}^k + f_{i+1,j}^k, \qquad \Delta_y^2 \langle f_{i,j}^k \rangle = f_{i,j-1}^k - 2f_{i,j}^k + f_{i,j+1}^k \quad (12.3.12)$$

and rewrite Eqs. (12.3.10) in the equivalent forms

$$(2 - \alpha_x \Delta_x^2) f_{i,j}^{n+1/2} = (2 + \alpha_y \Delta_y^2) f_{i,j}^n, \qquad (2 - \alpha_y \Delta_y^2) f_{i,j}^{n+1} = (2 + \alpha_x \Delta_x^2) f_{i,j}^{n+1/2}$$
$$(12.3.13)$$

Combining these equations to eliminate the intermediate solution at the $n + \frac{1}{2}$ level, we obtain

$$(2 - \alpha_x \Delta_x^2)(2 - \alpha_y \Delta_y^2) f_{i,j}^{n+1} = (2 + \alpha_x \Delta_x^2)(2 + \alpha_y \Delta_y^2) f_{i,j}^n \qquad (12.3.14)$$

The aforementioned decoupling is reflected in the factorial nature of the difference operators on either side of Eq. (12.3.14).

Now, the *fully implicit Crank–Nicolson discretization* of the two-dimensional diffusion equation can be expressed in the symbolic form

$$(2 - \alpha_x \Delta_x^2 - \alpha_y \Delta_y^2) f_{i,j}^{n+1} = (2 + \alpha_x \Delta_x^2 + \alpha_y \Delta_y^2) f_{i,j}^n \qquad (12.3.15)$$

which can be restated as

$$(2 - \alpha_x \Delta_x^2)(2 - \alpha_y \Delta_y^2) f_{i,j}^{n+1} = (2 + \alpha_x \Delta_x^2)(2 + \alpha_y \Delta_y^2) f_{i,j}^n$$
$$+ \alpha_x \alpha_y \Delta_x^2 \Delta_y^2 (f_{i,j}^{n+1} - f_{i,j}^n) \qquad (12.3.16)$$

This can be shown to provide us with an unconditionally stable method (Problem 12.3.1). The ADI Eq. (12.3.14) derives from Eq. (12.3.16) by discarding the last term on the right-hand side. This simplification is permissible, for the discarded term is of fourth order in the spatial steps, whereas Eq. (12.3.14) can be claimed to be accurate only up to second order. In this light, the ADI method emerges as the result of the *approximate factorization* of the difference operators on either side of Eq. (12.3.15).

Solving Poisson's Equation in Two Dimensions

The explicit and semi-implicit methods for two-dimensional diffusion discussed earlier in this section provide us with efficient iterative procedures for solving the Poisson equation (12.3.2) according to the general principles discussed at the beginning of this section. An algorithm that derives from the explicit FTCS discretization on a uniform grid, and its modified versions corresponding to the Gauss–Seidel and successive over-relaxation methods, are collected in Table 12.3.1, where $\beta = \Delta x / \Delta y$. All methods are second-order accurate in Δx and Δy.

The explicit point–Gauss–Seidel scheme given in the second entry derives from the FTCS scheme given in the first entry by setting $1/\alpha_x = 2(1 + \beta^2)$, which satisfies the stability criterion $\alpha_x + \alpha_y < \frac{1}{2}$ in a marginal way. The relaxation parameter ω for the PSOR and LSOR schemes varies between 1 and 2; when $\omega = 1$, the SOR methods reduce to the corresponding Gauss–Seidel methods. The implicit LGS and LSOR methods require solving tridiagonal systems of equations for each grid line that is parallel to the x axis, which can be done using the efficient Thomas algorithm discussed in Section B.1, Appendix B.

To develop the ADI method, we recast the ADI equations (12.3.10a,b) into the form shown in the sixth entry of Table 12.3.1, where we have introduced the new parameter $\rho = 2/\alpha_x$. Since the ADI method is unconditionally stable, it might appear that the fastest approach to the steady state will be achieved by using a large value for Δt or small value for ρ. Careful analysis, however, shows that the minimum number of iterations for a specified level of accuracy is achieved with a certain repetitive sequence of values of ρ. Unless this sequence is known, the ADI method competes in efficiency with the successive over-relaxation method (Ames, 1977; Hoffman, 1992, p. 446). A generalized form of the ADI method involving the relaxation parameter ω is given in the last entry of Table 12.3.1 (Hoffmann and Chiang, 1993, Vol. II, p. 9).

TABLE 12.3.1
Iterative methods for solving the Poisson equation $\nabla^2 f = g$ on a uniform rectangular grid with $\beta = \Delta x/\Delta y$. All methods are second-order accurate in Δx and Δy; ω is a dimensionless relaxation constant ranging between 1 and 2.

Explicit FTCS

$$f_{i,j}^{k+1} = \alpha_x\left[f_{i+1,j}^k + f_{i-1,j}^k + \left(\frac{1}{\alpha_x} - 2(1+\beta^2)\right)f_{i,j}^k + \beta^2(f_{i,j+1}^k + f_{i,j-1}^k) - \Delta x^2 g_{i,j}\right]$$

Explicit point–Gauss–Seidel (PGS)

$$f_{i,j}^{k+1} = \frac{1}{2(1+\beta^2)}[f_{i+1,j}^k + f_{i-1,j}^k + \beta^2(f_{i,j+1}^k + f_{i,j-1}^k) - \Delta x^2 g_{i,j}]$$

Explicit point-successive over-relaxation (PSOR)

$$f_{i,j}^{k+1} = (1-\omega)f_{i,j}^k + \frac{\omega}{2(1+\beta^2)}[f_{i+1,j}^k + f_{i-1,j}^k + \beta^2(f_{i,j+1}^k + f_{i,j-1}^k) - \Delta x^2 g_{i,j}]$$

Implicit line–Gauss–Seidel for the x direction (LGS)

$$f_{i-1,j}^{k+1} - 2(1+\beta^2)f_{i,j}^{k+1} + f_{i+1,j}^{k+1} = -\beta^2(f_{i,j+1}^k + f_{i,j-1}^k) + \Delta x^2 g_{i,j}$$

Implicit line-successive over-relaxation for the x direction (LSOR)

$$\omega f_{i-1,j}^{k+1} - 2(1+\beta^2)f_{i,j}^{k+1} + \omega f_{i+1,j}^{k+i} = 2(\omega-1)(1+\beta^2)f_{i,j}^k - \omega\beta^2(f_{i,j+1}^k + f_{i,j-1}^k) + \Delta x^2 g_{i,j}$$

ADI

$$f_{i-1,j}^{k+1/2} - (2+\rho)f_{i,j}^{k+1/2} + f_{i+1,j}^{k+1/2} = -\beta^2\left[f_{i,j-1}^k + \left(2-\frac{\rho}{\beta^2}\right)f_{i,j}^k + f_{i,j+1}^k\right] + \Delta x^2 g_{i,j}$$

$$f_{i,j-1}^{k+1} - \left(2+\frac{\rho}{\beta^2}\right)f_{i,j}^{k+1} + f_{i,j+1}^{k+1} = -\frac{1}{\beta^2}[f_{i-1,j}^{k+1/2} - (2-\rho)f_{i,j}^{k+1/2} + f_{i+1,j}^{k+1/2}] + \Delta y^2 g_{i,j}$$

SOR-ADI

$$\omega f_{i-1,j}^{k+1/2} - 2(1+\beta^2)f_{i,j}^{k+1/2} + \omega f_{i+1,j}^{k+1/2}$$
$$= -2(1-\omega)(1+\beta^2)f_{i,j}^k - \omega\beta^2(f_{i,j-1}^{k+1/2} + f_{i,j+1}^k) + \omega\Delta x^2 g_{i,j}$$
$$\omega\beta^2 f_{i,j-1}^{k+1} - 2(1+\beta^2)f_{i,j}^{k+1} + \omega\beta^2 f_{i,j-1}^{k+1}$$
$$= 2(1-\omega)(1+\beta^2)f_{i,j}^{k+1/2} - \omega(f_{i-1,j}^{k+1} + f_{i+1,j}^{k+1/2}) + \omega\Delta y^2 g_{i,j}$$

ADI Method in Three Dimensions

The standard implementation of the ADI method in three dimensions involves three substeps of equal duration $\Delta t/3$, where one spatial dimension is treated implicitly while the other two dimensions are treated explicitly within each substep, in the spirit of Eqs. (12.3.9). The method can be shown to be first-order accurate in time and second-order accurate in space. Unfortunately, the partial BTCS discretizations result in an algorithm that is stable only when $\alpha_x + \alpha_y + \alpha_z < \frac{3}{2}$, which places a stringent constraint on the size of the time step and thus requires an increased computational cost.

Douglas (1962) developed an ADI method that is second-order accurate in both time and space, and unconditionally stable. The method proceeds in a *predictor–corrector* sense in three substeps: the first substep produces a predicted solution using the Crank-Nicolson method for the x direction while treating the y and z directions explicitly; the second substep produces a predicted solution using the x discretization of the first substep and the Crank–Nicolson discretization for the y direction, while treating the z direction explicitly; the third substep advances the solution using the x and y discretizations of the second substep, while using the Crank–Nicolson method for the z direction. The finite-difference equations are

$$\frac{f_{i,j,k}^{n+1/3} - f_{i,j,k}^n}{\Delta t} = \kappa \left\{ \frac{1}{2}\left[\left(\frac{\partial^2 f}{\partial x^2}\right)^{n+1/3} + \left(\frac{\partial^2 f}{\partial x^2}\right)^n \right] + \left(\frac{\partial^2 f}{\partial y^2}\right)^n + \left(\frac{\partial^2 f}{\partial z^2}\right)^n \right\}$$

$$\frac{f_{i,j,k}^{n+2/3} - f_{i,j,k}^n}{\Delta t} = \kappa \left\{ \frac{1}{2}\left[\left(\frac{\partial^2 f}{\partial x^2}\right)^{n+1/3} + \left(\frac{\partial^2 f}{\partial x^2}\right)^n + \left(\frac{\partial^2 f}{\partial y^2}\right)^{n+2/3} + \left(\frac{\partial^2 f}{\partial y^2}\right)^n \right] + \left(\frac{\partial^2 f}{\partial z^2}\right)^n \right\}$$

$$\frac{f_{i,j,k}^{n+1} - f_{i,j,k}^n}{\Delta t} = \kappa \frac{1}{2}\left[\left(\frac{\partial^2 f}{\partial x^2}\right)^{n+1/3} + \left(\frac{\partial^2 f}{\partial x^2}\right)^n + \left(\frac{\partial^2 f}{\partial y^2}\right)^{n+2/3} \right.$$

$$\left. + \left(\frac{\partial^2 f}{\partial y^2}\right)^n + \left(\frac{\partial^2 f}{\partial z^2}\right)^{n+1} + \left(\frac{\partial^2 f}{\partial z^2}\right)^n \right] \tag{12.3.17}$$

where the second-order partial derivatives are discretized using central differences. To eliminate the spatial bias associated with this particular arrangement, we alternate the sequence of the three substeps in a cyclic manner.

Operator Splitting and Fractional Steps

Another way of preserving the tridiagonal nature of the one-dimensional implicit discretization is to replace the diffusion equation in three dimensions, or its two-dimensional counterpart, with a set of three or two one-dimensional evolution equations that operate successively in fractional steps (Yanenko, 1970). In two dimensions, we obtain the two component equations

$$\frac{\partial f}{\partial t} = \kappa \frac{\partial^2 f}{\partial x^2}, \qquad \frac{\partial f}{\partial t} = \kappa \frac{\partial^2 f}{\partial y^2} \tag{12.3.18}$$

each for $t^n < t < t^n + \Delta t$, which amounts to allowing diffusion to operate sequentially in the two dimensions, each time neglecting the other dimension. Each fractional step proceeds for the full time interval of Δt, and the time is reset back to the initial value at the end of the first fractional step. To preserve the spatial isotropy of the Laplacian operator, the order of Eqs. (12.3.18) is switched after the completion of a full time step.

Carrying out the fractional steps may be done using different methods for the component equations, as discussed previously in Section 12.2. The stability restrictions of the overall method are composed of the collection of the restrictions imposed on the individual steps.

The *Approximate Factorization Implicit* method (AFI) emerges by applying the implicit BTCS discretization to each fractional step in Eq. (12.3.18), and can be regarded as the result of the approximate factorization of the fully implicit BTCS discretization of Eqs. (12.3.1) (Problem 12.3.2). Each fractional step requires solving a tridiagonal system of equations, which can be done using the efficient Thomas algorithm.

Using the Crank–Nicolson method for each fractional step yields an algorithm that is second-order accurate in time and space and unconditionally stable, expressed by the component equations

$$-\alpha_x f_{i-1,j}^* + 2(1 + \alpha_x)f_{i,j}^* - \alpha_x f_{i+1,j}^* = \alpha_x f_{i-1,j}^n + 2(1 - \alpha_x)f_{i,j}^n - \alpha_x f_{i+1,j}^n \tag{12.3.19a}$$

$$-\alpha_y f_{i,j-1}^{n+1} + 2(1 + \alpha_y)f_{i,j}^{n+1} - \alpha_y f_{i,j+1}^{n+1} = \alpha_y f_{i,j-1}^* + 2(1 - \alpha_y)f_{i,j}^* + \alpha_y f_{i,j+1}^* \tag{12.3.19b}$$

The asterisk designates the solution at the end of the first fractional step. The scheme remains unconditionally stable in three dimensions.

PROBLEMS

12.3.1 **Generalized fully implicit Crank–Nicolson method in two dimensions.** A generalized version of Eq. (12.3.15) is

$$[1 - \gamma(\alpha_x\Delta_x^2 + \alpha_y\Delta_y^2)]f_{i,j}^{n+1} = [1 + (1 - \gamma)(\alpha_x\Delta_x^2 + \alpha_y\Delta_y^2)]f_{i,j}^n \qquad (12.3.20)$$

where γ is a numerical parameter; the standard Crank–Nicolson method emerges by setting $\gamma = \frac{1}{2}$. Perform the von Neumann stability analysis, derive the amplification factor

$$G = \frac{1 - 4(1 - \gamma)[\alpha_x^2 \sin^2(\theta_x/2) + \alpha_y^2 \sin^2(\theta_y/2)]}{1 + 4\gamma[\alpha_x^2 \sin^2(\theta_x/2) + \alpha_y^2 \sin^2(\theta_y/2)]}, \qquad (12.3.21)$$

and show that the method is stable when $(1 - 2\gamma)(\alpha_x^2 + \alpha_y^2) < \frac{1}{2}$. Explain why the standard Crank–Nicolson method corresponding to $\gamma = \frac{1}{2}$ is unconditionally stable.

12.3.2 **AFI.** Show that the AFI method may be regarded as the result of the approximate factorization of the fully implicit BTCS discretization of Eqs. (12.3.1).

12.3.3 **ADI method for Poisson's equation in three dimensions.** Develop an iterative method for solving Poisson's equation in three dimensions based on the ADI method of Douglas (1962).

Computer Problems

12.3.4 **ADI method for the Poisson equation in two dimensions.** Write a computer program called *ADI2* that solves the Poisson equation in two dimensions, within a rectangular domain that has been discretized into an $N \times M$ grid, defined by $N + 1$ and $M + 1$ grid lines, based on the ADI method described in Table 12.3.1, with Dirichlet boundary conditions (see Figure 13.1.1). Run the program and compute the solution within a square box subject to a forcing function and Dirichlet boundary conditions of your choice. Examine the rate of convergence of your results with respect to ρ, and test the reliability of your solution by comparing your results against a known analytical solution of your choice.

12.3.5 **ADI method for Poisson's equation in three dimensions.** Repeat Problem 12.3.4 for the Poisson equation in three dimensions within a rectangular box that has been discretized into an $N \times M \times L$ grid, based on the ADI method developed in Problem 12.3.3.

12.4 | ONE-DIMENSIONAL CONVECTION

In the present and in the next section we turn our attention to the extreme case of pure convection. We begin by considering the one-dimensional linear convection equation

$$\frac{\partial f}{\partial t} + U\frac{\partial f}{\partial x} = 0 \qquad (12.4.1)$$

which is a considerably simplified version of the general Eq. (12.1.1); U is a positive or negative constant convection velocity. The solution is to be found subject to the initial condition $f(x, 0) = F(x)$, where $F(x)$ is a known function. For simplicity, in the ensuing discussion we shall assume that the domain of solution extends over the whole x axis and shall require the far-field condition $f(\pm\infty, t) = 0$. If the domain of solution were bounded, we would have to specify one boundary condition at one end.

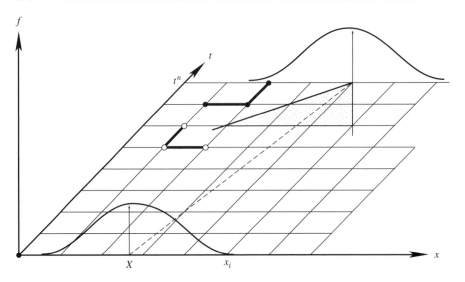

Figure 12.4.1. Discretization of the space–time domain for solving the one-dimensional convection equation using a finite-difference method. The initial distribution travels along the x axis with velocity U. The dashed line is a characteristic line along which the value of the solution remains constant. The solid and hollow circles indicate the FTBS and FTFS finite-difference stencils. The numerical cone of influence is shown as a dotted area.

One may readily verify that the exact solution to Eq. (12.4.1) is given by $f(x, t) = F(x-Ut)$, which states that the initial distribution F travels along the x axis with velocity U as illustrated in Figure 12.4.1; if U is negative, the motion is toward the negative direction of the x axis. The value of f remains constant along the characteristic line $x = X + Ut$ in the xt plane, shown with the dashed line in Figure 12.4.1, where X is an arbitrary point.

As a preliminary, we differentiate Eq. (12.4.1) twice, once with respect to t and the second time with respect to x, and combine the resulting expressions to obtain the *wave equation*

$$\frac{\partial^2 f}{\partial t^2} = U^2 \frac{\partial^2 f}{\partial x^2} \tag{12.4.2}$$

which is a prototypical second-order hyperbolic partial differential equation. The hyperbolic nature of the first-order equation (12.4.1) thus becomes apparent.

Explicit FTCS Method

Applying Eq. (12.4.1) at the point x_i at the time instant t^n, and approximating $\partial f/\partial t$ using a forward difference in time and $\partial f/\partial x$ using a central difference in space, we obtain the FTCS finite-difference equation

$$\frac{f_i^{n+1} - f_i^n}{\Delta t} + U \frac{f_{i+1}^n - f_{i-1}^n}{2\Delta x} + O(\Delta t) + O(\Delta x^2) = 0 \tag{12.4.3}$$

The associated finite-difference stencil is shown with hollow circles in Figure 12.2.1. Since Eq. (12.4.3) arose by applying standard finite-difference approximations, its consistency is guaranteed and does not need to be examined. Solving Eq. (12.4.3) for f_i^{n+1}, we obtain the explicit algorithm

$$f_i^{n+1} = \frac{c}{2} f_{i-1}^n + f_i^n - \frac{c}{2} f_{i+1}^n \tag{12.4.4}$$

where we have introduced the *convection number,* also called the *Courant number*

$$c \equiv \frac{U \, \Delta t}{\Delta x} \tag{12.4.5}$$

It is evident from Eq. (12.4.4) that the solution vector $\mathbf{f}^{n+1}$ emerges by projecting $\mathbf{f}^n$ onto a tridiagonal matrix $\mathbf{B}$ with subdiagonal, diagonal, and superdiagonal elements, respectively, equal to $-c/2$, 1, and $c/2$. As discussed in Section 12.2, when the spectral radius of $\mathbf{B}$, denoted by $\rho(\mathbf{B})$, is equal to, less than, or larger than unity, the length of $\mathbf{f}^n$ will stay roughly constant, decrease, or increase during the successive projections. Remembering that according to the exact solution the initial distribution is convected without changing its shape, we accept the first behavior, tolerate the second behavior, and dismiss the third behavior as being numerically unstable.

The eigenvalues and hence the spectral radius of $\mathbf{B}$ may be computed analytically using the results of Problem 12.2.6. An equivalent method of assessing the stability of the method is to perform the von Neumann analysis, which yields the amplification factor

$$\frac{A^{n+1}}{A^n} \equiv G = 1 - Ic \sin \theta \tag{12.4.6}$$

Since the magnitude of the right-hand side of Eq. (12.4.6) is greater than unity for any value of c and θ, the magnitude of $\mathbf{f}^n$ will amplify for any value of the wavelength L; this shows that the FTCS method is *unconditionally unstable* and must be abandoned. To this end, it is instructive to recall that, by comparison, the FTCS method for the diffusion equation was found to be conditionally stable.

FTBS Method

We proceed next to explore a different type of discretization in hopes of improving the numerical stability. Now we apply Eq. (12.4.1) at the point x_i at the time instant t^n and approximate $\partial f / \partial t$ using a forward difference and $\partial f / \partial x$ using a backward difference, and obtain the FTBS algorithm

$$\frac{f_i^{n+1} - f_i^n}{\Delta t} + U \frac{f_i^n - f_{i-1}^n}{\Delta x} + O(\Delta t) + O(\Delta x) = 0 \tag{12.4.7}$$

The corresponding finite-difference stencil is shown with solid circles in Figure 12.4.1. Rearranging Eq. (12.4.7), we derive the two-level explicit algorithm

$$f_i^{n+1} = c f_{i-1}^n + (1 - c) f_i^n \tag{12.4.8}$$

It is interesting to note that when $c = 1$ we obtain $f_i^{n+1} = f_{i-1}^n$, which reproduces the exact solution.

Carrying out the von Neumann stability analysis, we find the amplification factor

$$G = 1 - c[1 - \exp(-I\theta)] \tag{12.4.9}$$

which reveals that G is located on a circle in the complex plane with the center at the point $(1-c, 0)$ and radius equal to c. To guarantee stability, we require $|G| \leq 1$ and derive the restriction $0 \leq c \leq 1$. Clearly, if the convection velocity U happens to be negative, the method will be unstable.

FTFS Method

Returning to Eq. (12.4.1), we apply forward finite differences for both the time and space derivatives, and obtain the explicit algorithm

$$f_i^{n+1} = (1 + c) f_i^n - c f_{i+1}^n \tag{12.4.10}$$

The corresponding finite-difference stencil is shown with open circles in Figure 12.4.1. When

$c = -1$, we recover the exact analytical solution. Carrying out the von Neumann stability analysis yields the amplification factor

$$G = 1 + c[1 - \exp(-I\theta)] \tag{12.4.11}$$

which shows that $|G| \leq 1$, and thus the method is stable provided that $-1 \leq c \leq 0$. When the convection velocity U is positive, the method will be unstable.

Upwind Differencing, Numerical Diffusivity, and the CFL Condition

The complementary successes of FTBS and FTFS discretizations for positive and negative values of the convection velocity, respectively, suggest the method of *upwind differencing:* Use FTBS when $U > 0$, use FTFS when $U < 0$, and always maintain $|c| \leq 1$. The restriction $|c| \leq 1$ is known as the *Courant–Friedrichs–Lewy* or *CFL stability criterion*.

Upwind differencing is particularly effective for nonlinear problems where the convection velocity U is not constant but varies in time and space over the domain of solution. In physical terms, upwind differencing carries information regarding the structure of the solution forward from the direction of a traveling wave, and thus suppresses the growth of unwanted perturbations.

Numerical diffusivity

To explain the conditional stability of the one-sided FTBS and FTFS difference methods further, which can be contrasted with the unconditional instability of the FTCS method, we perform a consistency analysis.

Considering first Eq. (12.4.8), we pretend that the discrete variables are continuous functions of space and time, and expand them in Taylor series about the grid point (x_i, t^n) to obtain

$$f_i^n + \left(\frac{\partial f}{\partial t}\right)_i^n \Delta t + \frac{1}{2}\left(\frac{\partial^2 f}{\partial t^2}\right)_i^n \Delta t^2 + O(\Delta t^3)$$

$$= c\left[f_i^n - \left(\frac{\partial f}{\partial x}\right)_i^n \Delta x + \frac{1}{2}\left(\frac{\partial^2 f}{\partial x^2}\right)_i^n \Delta x^2\right] + (1-c)f_i^n + O(\Delta x^3) \tag{12.4.12}$$

Rearranging and using Eq. (12.4.2), we find

$$\left(\frac{\partial f}{\partial t} + U\frac{\partial f}{\partial x}\right)_i^n = \frac{1-c}{2}U\,\Delta x\left(\frac{\partial^2 f}{\partial x^2}\right)_i^n + O(\Delta t^2) + O(\Delta x^2) \tag{12.4.13}$$

As expected, in the limit as Δx tends to zero, Eq. (12.4.13) reduces to the original differential equation (12.4.1), thereby confirming the consistency of the BTCS discretization. The key observation for the present purposes is that the right-hand side of Eq. (12.4.13) involves a small diffusive term with an artificial or numerical diffusivity that is equal to $\frac{1}{2}(1 - c)U\Delta x$, which is non-negative when $0 \leq c \leq 1$.

Working similarly with the FTFS method, we derive a modified differential equation that is identical to Eq. (12.4.13), except that the numerical diffusivity is equal to $-\frac{1}{2}(1 - c)U\Delta x$, which is positive when $-1 \leq c \leq 0$. Combining these results with those of the von Neumann stability analysis suggests that *positive numerical diffusivity is instrumental in maintaining numerical stability.*

What went wrong with the FTCS method? Performing a consistency analysis, we derive an equation that is identical to Eq. (12.4.13) except that the numerical diffusivity is equal to $-\frac{1}{2}cU\Delta x = -\frac{1}{2}U^2\Delta t$, which is negative for any value of c or Δt. The negative numerical diffusivity may be regarded as the source of the numerical instability.

Numerical cone of influence

To obtain a graphical interpretation of the CFL condition, we recall that the exact solution at the grid point (x_i, t^n) may be found by traveling backward along the characteristic line, shown with the dashed line in Figure 12.4.1, until we reach the time level $t = 0$, where $x = X$; then $f_i^n = F(X)$, where $F(x)$ is the initial distribution. According to Eqs. (12.4.8) and (12.4.10), in order to compute the solution at the grid point (x_i, t^n), we use information at all grid points that are located within a planar angle with vertex at the point (x_i, t^n) shown as a dotted area in Figure 12.4.1, called the *numerical cone of influence*. The CFL condition requires that the characteristic line lie within the numerical cone of influence.

In general, the requirement that the characteristic line emanating from a certain point lie within the numerical cone of influence is necessary but not sufficient for numerical stability. Thus, if the numerical cone of influence does not contain the characteristic line that passes through the vertex, the numerical method will certainly be unstable.

Lax's Modification of the FTCS Method

Lax (1954) proposed a modification of the unconditionally unstable FTCS method that is designed to introduce a certain stabilizing diffusive action. This is done by replacing f_i^n in the temporal finite-difference approximation of Eq. (12.4.3) with the average value $\frac{1}{2}(f_{i+1}^n + f_{i-1}^n)$ obtaining

$$\frac{f_i^{n+1} - \frac{1}{2}(f_{i+1}^n + f_{i-1}^n)}{\Delta t} + U\frac{f_{i+1}^n - f_{i-1}^n}{2\,\Delta x} = 0 \tag{12.4.14}$$

Rearranging yields the explicit two-level algorithm

$$f_i^{n+1} = \frac{1}{2}(1 + c)f_{i-1}^n + \frac{1}{2}(1 - c)f_{i+1}^n \tag{12.4.15}$$

Since Eq. (12.4.14) emerged from a rather ad hoc modification of the FTCS discretization, its consistency must be examined. Expanding all variables in Taylor series about the point (x_i, t^n), we obtain the modified differential equation

$$\left(\frac{\partial f}{\partial t} + U\frac{\partial f}{\partial x}\right)_i^n = -\frac{1}{2}\,\Delta t\left(\frac{\partial^2 f}{\partial t^2}\right)_i^n + \frac{1}{2}\frac{\Delta x^2}{\Delta t}\left(\frac{\partial^2 f}{\partial x^2}\right)_i^n + O(\Delta x) + O(\Delta t) \tag{12.4.16}$$

which shows that Lax's method is consistent only when Δx and Δt are reduced simultaneously so that the ratio $\Delta x^2/\Delta t$ tends to zero. Maintaining, in particular, the ratio $\Delta x/\Delta t$ at a constant value yields a method that is consistent and first-order accurate in both time and space.

Carrying out the von Neumann stability analysis, we find the amplification factor

$$G = \cos\theta - Ic\sin\theta \tag{12.4.17}$$

which shows that Lax's method is stable provided that the CFL criterion is fulfilled, $|c| < 1$. To explain this behavior, we substitute the right-hand side of Eq. (12.4.2) for the first term on the right-hand side of Eq. (12.4.16), and group the first with the second term to obtain a diffusion term with numerical diffusivity equal to $\frac{1}{2}U\,\Delta x(1 - c^2)/c$, which is positive when $|c| < 1$ and becomes excessively large as c tends to zero. This behavior places a serious restriction on the effectiveness of Lax's method: The gain in stability was penalized by a loss of accuracy. For this reason, Lax's method finds limited usage in practice.

The Explicit Lax–Wendroff Method

Upwind differencing offers numerical stability but suffers from low-order spatial accuracy. FTFS differencing, on the other hand, offers second-order spatial accuracy but suffers from unconditional instability. Is it possible to devise a two-level explicit method that combines accuracy and stability, that is, a method that is second-order accurate in both time and space and conditionally stable?

To answer this question, we express f_i^{n+1} as a linear combination of f_{i-1}^n, f_i^n, and f_{i+1}^n, writing

$$f_i^{n+1} = a_{-1} f_{i-1}^n + a_0 f_i^n + a_1 f_{i+1}^n \tag{12.4.18}$$

where a_{-1}, a_0, and a_1 are three constant coefficients. It will be noted that Eq. (12.4.18) is a generalization of all two-level methods considered previously in this section.

To ensure the consistency of Eq. (12.4.18), we expand all variables in Taylor series about the point (x_i, t^n) and derive the modified differential equation

$$(1 - a_{-1} - a_0 - a_1) f_i^n + \Delta t \left(\frac{\partial f}{\partial t} + \frac{U}{c}(a_{-1} - a_1) \frac{\partial f}{\partial x} \right)_i^n + \frac{1}{2} \Delta t^2 \left(\frac{\partial^2 f}{\partial t^2} \right)_i^n$$

$$= \frac{1}{2} \Delta x^2 (a_{-1} + a_1) \left(\frac{\partial^2 f}{\partial x^2} \right)_i^n \tag{12.4.19}$$

Substituting the left-hand side of Eq. (12.4.2) into the last term on the left-hand side of Eq. (12.4.19), requiring that, in the limit as Δx and Δt tend to zero, Eq. (12.4.19) reduces to Eq. (12.4.1), and stipulating that the second-order temporal error cancels the second-order spatial error, we derive the following system of algebraic equations for a_{-1}, a_0, and a_1

$$\begin{bmatrix} 1 & 1 & 1 \\ 1 & 0 & -1 \\ 1 & 0 & 1 \end{bmatrix} \begin{bmatrix} a_{-1} \\ a_0 \\ a_1 \end{bmatrix} = \begin{bmatrix} 1 \\ c \\ c^2 \end{bmatrix} \tag{12.4.20}$$

whose solution is

$$a_{-1} = \tfrac{1}{2} c(c + 1), \qquad a_0 = 1 - c^2, \qquad a_1 = \tfrac{1}{2} c(c - 1) \tag{12.4.21}$$

Substituting these values into Eq. (12.4.18), we obtain the Lax–Wendroff method (Lax and Wendroff, 1960).

Carrying out the standard von Neumann stability analysis yields the amplification factor

$$G = 1 - c^2 + c^2 \cos \theta - I c \sin \theta \tag{12.4.22}$$

which shows that G traces an ellipse in the complex plane with the center at $(1 - c^2, 0)$ and semiaxes equal to c and c^2. Geometrical arguments reveal that $|G| < 1$ when $|c| < 1$, and this suggests that the method is conditionally stable. The Lax–Wendroff method provides us with an efficient algorithm: It is explicit, second-order accurate, and conditionally stable.

Explicit CTCS or Leapfrog Method

One way to guarantee second-order accuracy in time and space is to use central differences for both the temporal and spatial derivatives. Approximating both $\partial f / \partial t$ and $\partial f / \partial x$ with central differences, we obtain the CTCS difference equation

$$\frac{f_i^{n+1} - f_i^{n-1}}{2 \Delta t} + U \frac{f_{i+1}^n - f_{i-1}^n}{2 \Delta x} + O(\Delta t^2) + O(\Delta x^2) = 0 \tag{12.4.23}$$

Rearranging yields the three-level explicit algorithm

$$f_i^{n+1} = f_i^{n-1} + c f_{i-1}^n - c f_{i+1}^n \tag{12.4.24}$$

The solution at the first time level $n = 1$ must be computed using a two-level method, using a sufficiently small time step in order to avoid the occurrence of oscillations.

Performing the von Neumann stability analysis, we find that the amplification factor satisfies the quadratic equation $G^2 + 2I\delta G - 1 = 0$, where $\delta = c \sin \theta$ is a real parameter. The roots of this equation are

$$G = -I\delta \pm \sqrt{1 - \delta^2} \tag{12.4.25}$$

To assess the magnitude of the amplification factor, we distinguish between two cases: (1) If $\delta^2 > 1$, then $G = I[-\delta \pm (\delta^2 - 1)^{1/2}]$, and the magnitude of the root corresponding to the minus sign is greater than unity; (2) if $\delta^2 \leq 1$, then $|G| = 1$, which shows that the magnitude of A^n will stay constant during the successive mappings in agreement with the exact solution. Noting that $\delta^2 \leq 1$ when $|c| \leq 1$ for any value of θ shows that the CTCS method will be stable provided that the CFL condition is fulfilled. To this end, we recall that the CTCS discretization for the diffusion equation was found to be unconditionally unstable, and this suggests that the presence of diffusivity does not necessarily promote the numerical stability.

In practice, the efficiency of the CTCS method may be hindered by increased memory requirements associated with storing information at three time levels, and the occurrence of dual numerical error that grows independently at every other time step in a phenomenon that is known as *even–odd coupling.*

Implicit BTCS

We turn next to consider implicit discretizations in hopes of obtaining unconditional stability and thus relaxing the restriction on the time step Δt. The BTCS discretization yields the difference equation

$$\frac{f_i^{n+1} - f_i^n}{\Delta t} + U \frac{f_{i+1}^{n+1} - f_{i-1}^{n+1}}{2\,\Delta x} + O(\Delta t) + O(\Delta x^2) = 0 \tag{12.4.26}$$

which may be rearranged to yield

$$-c f_{i-1}^{n+1} + 2 f_i^{n+1} + c f_{i+1}^{n+1} = 2 f_i^n \tag{12.4.27}$$

In matrix notation we obtain $\mathbf{B} \cdot \mathbf{f}^{n+1} = \mathbf{f}^n$, where $\mathbf{B}$ is a tridiagonal matrix with subdiagonal, diagonal, and superdiagonal elements, respectively, equal to $-c$, 2, and c. It is interesting to note that the matrix $\mathbf{B}$ is proportional to the transpose of the matrix $\mathbf{B}$ corresponding to the FTCS discretization.

Carrying out the von Neumann stability analysis, we find the amplification factor

$$G = \frac{1}{1 + Ic \sin \theta} \tag{12.4.28}$$

Since $|G|$ is always less than unity, the BTCS method is unconditionally stable. The increase in computational effort required to solve the linear system (12.4.27) is rewarded with unconditional stability. The latter may be explained by the fact that the boundaries of the numerical cone of influence include all points at the $n + 1$ time level: The numerical cone of influence reduces to a rectangular strip that is guaranteed to contain the characteristic line passing through (x_i, t^n).

Crank–Nicolson Method

As in the case of pure diffusion, the Crank–Nicolson method proceeds by applying the differential equation midway between the time levels t^n and t^{n+1}, and approximating the spatial derivative with the average value of the two spatial derivatives at the two time levels, obtaining

$$\frac{f_i^{n+1} - f_i^n}{\Delta t} + U \frac{1}{2} \left(\frac{f_{i+1}^{n+1} - f_{i-1}^{n+1}}{2\,\Delta x} + \frac{f_{i+1}^n - f_{i-1}^n}{2\,\Delta x} \right) + O(\Delta t^2) + O(\Delta x^2) = 0 \tag{12.4.29}$$

Rearranging, we derive the implicit difference equation

$$-c f_{i-1}^{n+1} + 4 f_i^{n+1} + c f_{i+1}^{n+1} = c f_{i-1}^n + 4 f_i^n - c f_{i+1}^n \tag{12.4.30}$$

which yields a tridiagonal system of equations. The amplification factor is equal to the ratio of two complex conjugate numbers

$$G = \frac{2 - Ic \sin \theta}{2 + Ic \sin \theta} \tag{12.4.31}$$

Since $|G| = 1$, the method is unconditionally stable for any value of c.

Comparison of the Methods

The restrictions on the time step of the conditionally stable explicit methods is not usually prohibitive. Implicit methods allow the use of larger time steps, but the associated numerical error may erode the accuracy and therefore the physical relevance of the solution. Thus explicit methods are a standard choice in practice.

Modified Dynamics and Explicit Numerical Diffusion

We saw that the finite-difference discretizations introduce some type of numerical error whose leading-order term may be proportional to the second, third, or fourth spatial derivatives of the solution. The presence of this error incurs, respectively, numerical diffusion, dispersion, and fourth-order diffusion.

Numerical diffusion is indispensable for maintaining numerical stability and dampening small-scale irregularities yielding *monotone schemes,* that is, schemes that produce solutions that are free of artificial oscillations. Unfortunately, in practice, a prohibitively fine grid may be required in order to reduce the artificial smearing of sharp gradients.

A remedy would be to use a higher-order method, but the associated modified differential equation typically contains a dispersive term that causes local oscillations. The magnitude of the oscillations may be reduced by enhancing the original differential equation with an explicit diffusion-like term expressing regular diffusion or fourth-order diffusion. In one dimension, these are expressed, respectively, by the terms

$$\beta(\mathbf{x}) \frac{\partial^2 f}{\partial x^2}, \qquad -\gamma(\mathbf{x}) \frac{\partial^4 f}{\partial x^4} \qquad (12.4.32)$$

The positive diffusion coefficients β and γ are allowed to vary in space and time according to the structure of the solution. The optimal values of these coefficients depend upon the nature of the particular problem under consideration, and must be found by numerical experimentation. One justification for introducing explicit numerical diffusion is that the approximations that lead us to the convection equation have neglected certain physical diffusionlike processes anyway, and it is possible that introducing them back into the differential equation will not affect the physical relevance of the results to an alarming level.

Explicit Multistep and Predictor–Corrector Methods

One way of improving the accuracy of explicit methods is to advance the solution over each time step using a number of elementary substeps that are carried out using different numerical methods. Some of these substeps are predictive in nature, in the sense that they seek to estimate the solution at the next time level using a crude method, while others correct the predictions using more sophisticated methods. Accordingly, these multistep methods are sometimes classified as predictor–corrector methods.

The overall action of a multistep method for a problem of linear convection can be reduced to that of a single-step explicit method discussed previously in this section. Their discussion in the present context serves as a point of departure for developing methods for problems with varying convection velocity, which will be pursued later in this section.

Richtmyer's method

Richtmyer (1963) proposed carrying out a complete step in two substeps of equal duration $\Delta t/2$. The first substep is executed using Lax's method according to Eq. (12.4.15), and the second substep is executed using the CTCS method according to Eq. (12.4.24). This results in the finite-difference equations

$$f_i^{n+1/2} = \tfrac{1}{2}(1 + \tfrac{1}{2}c)f_{i-1}^n + \tfrac{1}{2}(1 - \tfrac{1}{2}c)f_{i+1}^n$$
$$f_i^{n+1} = f_i^n - \tfrac{1}{2}c(f_{i+1}^{n+1/2} - f_{i-1}^{n+1/2}) \qquad (12.4.33)$$

where $c = U \Delta t/\Delta x$. Eliminating the intermediate solution from the second equation, we find

$$f_i^{n+1} = \frac{c}{2}\left(\frac{c}{2}+1\right)f_{i-2}^n + \left(1-\frac{c^2}{4}\right)f_i^n + \frac{c}{2}\left(\frac{c}{2}-1\right)f_{i+2}^n \qquad (12.4.34)$$

which is the Lax–Wendroff formula (12.4.18) with the coefficients given in Eqs. (12.4.21) and grid spacing equal to $2\,\Delta x$. This ensures that Richtmyer's method is second-order accurate in both time and space and stable as long as $|c| \leq 2$.

Multistep Lax–Wendroff

Burstein (1967) developed a modification of Richtmyer's method according to which, in the first substep, the equation is applied at the intermediate grid point $i + \frac{1}{2}$, whereas in the second substep, the equation is applied at the regular grid point i, both times with a spatial step equal to $\Delta x/2$. The counterparts of Eqs. (12.4.33) are

$$\begin{aligned}
f_{i+1/2}^{n+1/2} &= \tfrac{1}{2}(1+c)f_i^n + \tfrac{1}{2}(1-c)f_{i+1}^n \\
f_i^{n+1} &= f_i^n - c(f_{i+1/2}^{n+1/2} - f_{i-1/2}^{n+1/2})
\end{aligned} \qquad (12.4.35)$$

where $c = U\,\Delta t/\Delta x$. Eliminating the intermediate solution from the second equation, we derive the Lax–Wendroff formula (12.4.18) with the coefficients given in Eqs. (12.4.21), and this ensures that the method is second-order accurate in both time and space, and stable as long as $|c| \leq 1$.

MacCormack's method

MacCormack (1969) developed a genuine predictor–corrector method that enjoys extensive usage in engineering practice. The predictor step provides us with an approximation to f_i^{n+1} denoted by f_i^*, which is computed using the explicit FTFS discretization according to Eq. (12.4.10),

$$f_i^* = (1+c)f_i^n - cf_{i+1}^n \qquad (12.4.36)$$

The second step uses the explicit forward time approximation, and a hybrid forward/backward space approximation that involves the predicted values, according to the difference equation

$$\frac{f_i^{n+1} - f_i^n}{\Delta t} + U\tfrac{1}{2}\left(\frac{f_{i+1}^n - f_i^n}{\Delta x} + \frac{f_i^* - f_{i-1}^*}{\Delta x}\right) = 0 \qquad (12.4.37)$$

Rearranging Eq. (12.4.37) and using Eq. (12.4.36) to eliminate f_{i+1}^n in favor of f_i^* yields the explicit formula

$$f_i^{n+1} = \tfrac{1}{2}[f_i^n + f_i^* - c(f_i^* - f_{i-1}^*)] \qquad (12.4.38)$$

Eliminating the intermediate variable f_i^* from Eqs. (12.4.36) and (12.4.38), we derive the Lax–Wendroff method, which ensures that MacCormack's method is second-order accurate in both time and step and stable as long as $|c| \leq 1$.

The bias in the solution due to the one-sided differencing involved in Eqs. (12.4.36) and (12.4.38) may be avoided by alternating the direction of the one-sided differences after completion of one time step. Thus, in the next step, we use FTBS and FTBS-FS to obtain the equations

$$f_i^* = cf_{i-1}^n + (1-c)f_i^n, \qquad f_i^{n+1} = \tfrac{1}{2}[f_i^n + f_i^* - c(f_{i+1}^* - f_i^*)] \qquad (12.4.39)$$

We shall see later in this section that MacCormack's method is particularly effective for problems with a variable convection velocity.

Flux-Corrected Transport

We saw earlier in this section that introducing explicit numerical diffusion may be necessary in order to enhance the performance of first-order methods. The idea behind the flux-corrected transport method is to use a predictor–corrector method in which the predictor step involves an artificial dampening term, while the corrector step removes the excessive dissipation by introducing anti-diffusion, that is, diffusion with a negative diffusivity. The method was developed and discussed in three parts by Boris and Book (1973); Book, Boris, and Hain (1975); and Boris and Book (1976).

Nonlinear Convection

Most of the methods discussed previously in this section for linear convection may be extended to case nonlinear case, where U is no longer a constant. The main difference is that the convection velocity must be evaluated at the grid point where the differential equation is applied to yield the finite-difference equation. When a finite-difference method is conditionally stable, the size of the time step must be kept sufficiently small in order to satisfy the stability criteria derived for linear convection with the maximum value of the velocity over the entire domain of the solution.

Implicit methods lead to a system of nonlinear algebraic equations that must be solved using iterative procedures. This complication often introduces substantial pragmatic difficulties associated with excessive computational cost, and renders these methods a mere academic alternative.

Predictor–corrector methods are efficient, easy to implement, and offer high-order accuracy. Richtmyer's and MacCormack's two-step methods are two popular choices.

Inviscid Burgers Equation

To illustrate the implementation of finite-difference methods for nonlinear problems, we consider the inviscid Burgers equation whose nonconservative and conservative forms are

$$\frac{\partial f}{\partial t} + f \frac{\partial f}{\partial x} = 0 \qquad \text{and} \qquad \frac{\partial f}{\partial t} + \frac{\partial E}{\partial x} = 0 \qquad (12.4.40)$$

where $E \equiv \frac{1}{2} f^2$. Physically, the Burgers equation describes the propagation of wave fronts with a local convection velocity that is equal to the local amplitude of the wave. The dependence of the velocity on the amplitude may cause the formation of discontinuous fronts from smooth initial distributions.

Lax's method

Lax's modification of the FTCS discretization yields the explicit two-level algorithm

$$f_i^{n+1} = \tfrac{1}{2}(f_{i-1}^n + rE_{i-1}^n) + \tfrac{1}{2}(f_{i+1}^n - rE_{i-1}^n) \qquad (12.4.41)$$

where $r = \Delta t / \Delta x$, which is first-order accurate in time and second-order accurate in space and stable as long as $|c_{\text{Max}}| = |r f_{\text{Max}}| < 1$.

Lax–Wendroff method

Proceeding as in the case of linear convection with the objective of achieving second-order accuracy, we develop the explicit Lax–Wendroff algorithm expressed by

$$\begin{aligned} f_i^{n+1} = f_i^n &- \tfrac{1}{2}r(E_{i+1}^n - E_{i-1}^n) + \tfrac{1}{4}r^2[-f_{i-1}^n(E_i^n - E_{i-1}^n) \\ &+ f_i^n(E_{i+1}^n - 2E_i^n + E_{i-1}^n) + f_{i+1}^n(E_{i+1}^n - E_i^n)] \end{aligned} \qquad (12.4.42)$$

which is second-order accurate in both space and time, and stable as long as $|c_{\text{Max}}| < 1$ (Hoffmann and Chiang, 1993, Vol. I, p. 207).

Implicit BTCS method

The implicit BTCS method results in the quadratically nonlinear system of equations

$$c(f_{i+1}^{n+1})^2 + 4f_i^{n+1} - c(f_{i-1}^{n+1})^2 = 4f_i^n \qquad (12.4.43)$$

which must be solved using an iterative method. A suitable initial guess is provided by the converged solution at the previous time step. Increased computational demands often discourage the selection of this method.

MacCormack's method

The two-step MacCormack method proceeds by predicting a solution, designated by an asterisk, and then correcting it according to the finite-difference equations

$$f_i^* = f_i^n - r(E_{i+1}^n - E_i^n), \qquad f_i^{n+1} = \tfrac{1}{2}[f_i^n + f_i^* - r(E_i^* - E_{i-1}^n)] \qquad (12.4.44)$$

where $r = \Delta t/\Delta x$. The method is stable as long as $|c_{Max}| = |r f_{Max}| < 1$. It will be noted that Eqs. (12.4.44) are straightforward extensions of Eqs. (12.4.36) and (12.4.38). MacCormack's method has become a standard choice in practice.

PROBLEMS

12.4.1   **Numerical diffusivity.** Carry out a consistency analysis of the FTCS and Lax methods and derive the corresponding numerical diffusivities.

12.4.2   **Lax–Wendroff method for the inviscid Burgers equation.** Derive Eq. (12.4.42) working by analogy with the case of linear convection.

 ## Computer Problems

12.4.3   **Burgers equation with MacCormack's method.** Use MacCormack's method to solve the inviscid Burgers equation in an infinite domain with initial conditions: (a) The Heavyside step function $F(x) = 1$ for $x < 0$ and $F(x) = 0$ for $x > 0$, (b) $F(x) = \exp(-x^2)$. Discuss the behavior of the numerical solution in each case.

12.4.4   **Burgers equation with an implicit method.** Repeat Problem 12.4.3 with the implicit BTCS method and discuss the performance of the method and required computational cost.

12.5 | CONVECTION IN TWO AND THREE DIMENSIONS

Generalizing the discussion of Section 12.4, we consider the linear convection equation in three dimensions described by

$$\frac{\partial f}{\partial t} + U\frac{\partial f}{\partial x} + V\frac{\partial f}{\partial y} + W\frac{\partial f}{\partial z} = 0 \qquad (12.5.1)$$

where the convection velocity vector $\mathbf{U} = (U, V, W)$ is assumed to be constant in time and space. Two-dimensional convection arises by setting $W = 0$. The solution of Eq. (12.5.1) is to be found subject to an appropriate initial condition $f(\mathbf{x}, t = 0) = F(\mathbf{x})$.

Assuming, for simplicity, that the domain of solution extends over the whole three-dimensional space or two-dimensional plane, we find that the exact solution is given by $F(\mathbf{x} - \mathbf{U}t)$, which states that the initial distribution F travels with constant velocity $\mathbf{U}$. The characteristic lines along which the value of the function f remains constant are described by the equation $\mathbf{x} - \mathbf{U}t = \mathbf{X}$, where $\mathbf{X}$ is an arbitrary point.

Finite-difference methods for two-dimensional and three-dimensional convection arise as direct and straightforward extensions of the methods for one-dimensional convection discussed in the preceding section.

Lax's Method

Lax's method in three dimensions emerges by replacing Eq. (12.5.1) with the finite-difference equation

$$\frac{1}{\Delta t}[f_{i,j,k}^{n+1} - \frac{1}{6}(f_{i+1,j,k}^n + f_{i-1,j,k}^n + f_{i,j+1,k}^n + f_{i,j-1,k}^n + f_{i,j,k+1}^n + f_{i,j,k-1}^n)]$$

$$+ \frac{U}{2\,\Delta x}(f_{i+1,j,k}^n - f_{i-1,j,k}^n) + \frac{V}{2\,\Delta y}(f_{i,j+1,k}^n - f_{i,j-1,k}^n)$$

$$+ \frac{W}{2\,\Delta z}(f_{i,j,k+1}^n - f_{i,j,k-1}^n) = 0 \tag{12.5.2}$$

Rearranging the various terms, we obtain the *explicit* two-level algorithm

$$f_i^{n+1} = \tfrac{1}{2}(\tfrac{1}{3} + c_x)f_{i-1,j,k}^n + \tfrac{1}{2}(\tfrac{1}{3} - c_x)f_{i+1,j,k}^n + \tfrac{1}{2}(\tfrac{1}{3} + c_y)f_{i,j-1,k}^n$$

$$+ \tfrac{1}{2}(\tfrac{1}{3} - c_y)f_{i,j+1,k}^n + \tfrac{1}{2}(\tfrac{1}{3} + c_z)f_{i,j,k-1}^n + \tfrac{1}{2}(\tfrac{1}{3} - c_z)f_{i,j,k+1}^n \tag{12.5.3}$$

where

$$c_x = \frac{U\,\Delta t}{\Delta x}, \qquad c_y = \frac{V\,\Delta t}{\Delta y}, \qquad c_z = \frac{W\,\Delta t}{\Delta z} \tag{12.5.4}$$

Performing the von Neumann stability analysis, we find that the method is stable provided that

$$c_x^2 + c_y^2 + c_z^2 < \tfrac{1}{3} \tag{12.5.5}$$

This condition imposes a strong restriction on the size of the time step that renders Lax's method inefficient in practice. For convection in two dimensions we find the analogous condition $c_x^2 + c_y^2 < \tfrac{1}{2}$, which is still a stringent constraint.

Implicit methods

To achieve unconditional stability, one may resort to a fully implicit method. Unfortunately, the finite-difference equations result in a system of algebraic equations that is sparse but not tridiagonal, and whose solution may require a prohibitive amount of computational effort. One remedy is to use an ADI method, which requires solving a one-dimensional problem within each substep. ADI methods for hyperbolic equations are developed and discussed by Lee (1962), Douglas and Gunn (1964), and Hirsch (1988, Vol. I., p. 442). The standard implementation of the ADI method leads to an algorithm that is unconditionally stable in two dimensions, but unconditionally unstable in three dimensions.

Operator splitting

Following the general idea of operator splitting and fractional steps, we replace Eq. (12.5.1) with a system of three equations that apply in a sequential manner, each for the full time interval Δt,

$$\frac{\partial f}{\partial t} + U\frac{\partial f}{\partial x} = 0, \qquad \frac{\partial f}{\partial t} + V\frac{\partial f}{\partial y} = 0, \qquad \frac{\partial f}{\partial t} + W\frac{\partial f}{\partial z} = 0 \tag{12.5.6}$$

all for $t^n < t < t^n + \Delta t$. Adopting the implicit BTCS discretization, we advance the solution over each fractional step by solving the following three tridiagonal systems of equations,

$$(1 + c_x \Delta_x)f_{i,j,k}^* = f_{i,j,k}^n, \qquad (1 + c_y \Delta_y)f_{i,j,k}^{**} = f_{i,j,k}^*$$

$$(1 + c_z \Delta_z)f_{i,j,k}^{n+1} = f_{i,j,k}^{**} \tag{12.5.7}$$

where the starred and double-starred variables designate the solution after the first and second fractional step. Δ_x is the first central-difference operator defined as

$$\Delta_x\langle f_{i,j,k}\rangle = f_{i+1,j,k} - f_{i-1,j,k} \tag{12.5.8}$$

And Δ_y and Δ_z are defined in a similar manner. Combining the three Eqs. (12.5.7), we obtain the overall finite-difference scheme

$$(1 + c_x \Delta_x)(1 + c_y \Delta_y)(1 + c_z \Delta_z)f_{i,j,k}^{n+1} = f_{i,j,k}^n \tag{12.5.9}$$

involving a factorized implicit operator in the left-hand side.

Now, the finite-difference equation corresponding to the fully implicit BTCS discretization is given by

$$(1 + c_x \Delta_x + c_y \Delta_y + c_z \Delta_z) f_{i,j,k}^{n+1} = f_{i,j,k}^n \qquad (12.5.10)$$

The left-hand sides of Eqs. (12.5.9) and (12.5.10) are identical up to first order in the spatial intervals, and this means that the fractional-step method may be regarded as the result of the approximate factorization of the explicit BTCS discretization.

PROBLEMS

12.5.1   Lax's method. (a) Perform the von Neumann stability analysis to derive the stability criteria (12.5.5), and derive the associated numerical diffusivity. (b) Analyze the stability of the method in two dimensions, and derive the stability constraint given in the text.

12.5.2   ADI. Devise an ADI method for the linear two-dimensional convection equation, and study its consistency and stability. Then indicate how the method can be extended to three dimensions.

12.6 | CONVECTION–DIFFUSION IN ONE DIMENSION

The methods developed in the preceding sections for the extreme cases of unsteady diffusion and pure convection may be combined in a straightforward fashion to tackle the mixed case of convection–diffusion. In this section we shall discuss the development of such methods for the simplest case of linear one-dimensional convection–diffusion described by the *parabolic* differential equation

$$\frac{\partial f}{\partial t} + U \frac{\partial f}{\partial x} = \kappa \frac{\partial^2 f}{\partial x^2} \qquad (12.6.1)$$

subject to the initial condition $f(x, t = 0) = F(x)$. Both the convection velocity U and the diffusivity κ are assumed to be constant.

Explicit FTCS and Lax's Method
The FTCS discretization yields the difference equation

$$\frac{f_i^{n+1} - f_i^n}{\Delta t} + U \frac{f_{i+1}^n - f_{i-1}^n}{2 \Delta x} = \kappa \frac{f_{i+1}^n - 2f_i^n + f_{i-1}^n}{\Delta x^2} + O(\Delta t) + O(\Delta x^2) \quad (12.6.2)$$

Rearranging, we obtain the explicit two-level algorithm

$$f_i^{n+1} = f_i^n - c\tfrac{1}{2}(f_{i+1}^n - f_{i-1}^n) + \alpha(f_{i+1}^n - 2f_i^n + f_{i-1}^n) \qquad (12.6.3)$$

where we recall that $c = U\Delta t/\Delta x$ is the Courant number and $\alpha = \kappa \Delta t/\Delta x^2$ is the diffusivity or diffusion number. The ratio between these two numbers

$$Re_c \equiv \frac{c}{\alpha} = \frac{U \Delta x}{\kappa} \qquad (12.6.4)$$

is called the *cell Reynolds number* or *cell Péclet number.* Physically, Re_c expresses the relative strengths of the convective and diffusive contributions to Eq. (12.6.1).

In previous sections we found that the explicit FTCS discretization is conditionally stable for pure unsteady diffusion, and unconditionally unstable for pure linear convection. Carrying out the von Neumann stability analysis for the present case of mixed convection–diffusion, we derive the amplification factor

$$G = 1 - 2\alpha + 2\alpha \cos \theta - Ic \sin \theta \qquad (12.6.5)$$

which shows that G traces an ellipse that passes through the point $(1, 0)$ in the complex plane. The center of the ellipse is located at the point $(1 - 2\alpha, 0)$, and its semiaxes are equal to 2α and c. To guarantee stability, we must ensure that the ellipse is located within the unit disk. Requiring that the lengths of the semiaxes are less than unity yields two restrictions: $\alpha < \frac{1}{2}, c < 1$. A third restriction emerges by requiring that the curvature of the ellipse at the point $(1, 0)$ be less than that of the unit circle, and this demands that $c^2 < 2\alpha$. Combining these three restrictions, we obtain

$$c^2 < 2\alpha < 1 \tag{12.6.6}$$

Recalling that the FTCS discretization for convection alone leads to an unconditionally unstable method, suggests that *the presence of the diffusion term has a stabilizing action.*

The modified differential equation corresponding to the finite-difference equation (12.6.3) is

$$\frac{\partial f}{\partial t} + U\frac{\partial f}{\partial x} = \left(\kappa - \frac{1}{2}U^2\,\Delta t\right)\frac{\partial^2 f}{\partial x^2} \tag{12.6.7}$$

involving the effective diffusivity $\kappa - U^2\Delta t/2 = \kappa(1 - c\,Re_c/2) = \kappa(1 - c^2/2\alpha)$. The stability criterion (12.6.6) thus states that a necessary condition for the method to be stable is that the positive physical diffusivity be larger than the negative numerical diffusivity in absolute value.

An equivalent form of Eq. (12.6.3) is

$$f_i^{n+1} = \frac{1}{2}\alpha(2 + Re_c)f_{i-1}^n + (1 - 2\alpha)f_i^n + \frac{1}{2}\alpha(2 - Re_c)f_{i+1}^n \tag{12.6.8}$$

Let us consider an initial condition where the value of the function f at all grid points is equal to zero except for one grid point, where it has a certain positive value. Since the initial distribution will be convected and diffuse, we expect, on physical grounds, that the value of f will be positive at all grid points at all subsequent time levels. This occurrence, however, necessitates that all coefficients on the right-hand side of Eq. (12.6.8) be positive, which suggests the *physical* restriction

$$Re_c < 2 \tag{12.6.9}$$

Violation of this inequality leads to an unphysical overshooting.

The stability restriction (12.6.6) requires that the size of the time step be excessively small, and this renders the FTCS discretization uneconomical. Lax's modification discussed in Section 12.4 leads to an unconditionally unstable method. Consequently, the FTCS method and its variations are of limited practical value.

Upwind-Differencing Methods

One might argue that the stability properties of the FTCS method will be improved by using upwind differencing for the convective derivative. Assuming that U is positive, let us use a forward difference for the time derivative, a backward difference for the first spatial derivative, and a central difference for the second spatial derivative to obtain

$$\frac{f_i^{n+1} - f_i^n}{\Delta t} + U\frac{f_i^n - f_{i-1}^n}{\Delta x} = \kappa\frac{f_{i+1}^n - 2f_i^n + f_{i-1}^n}{\Delta x^2} + O(\Delta t) + O(\Delta x) \tag{12.6.10}$$

Rearranging, we find

$$f_i^{n+1} = f_i^n - c(f_i^n - f_{i-1}^n) + \alpha(f_{i+1}^n - 2f_i^n + f_{i-1}^n) \tag{12.6.11}$$

Carrying out a consistency analysis shows that the corresponding modified differential equation is the convection–diffusion equation with an effective diffusivity that is equal to $\kappa[1+(1-c)Re_c/2]$. Since Re_c vanishes as Δx tends to zero, the method is confirmed to be consistent.

Performing the von Neumann stability analysis, we find the amplification factor

$$G = 1 - c - 2\alpha + (c + 2\alpha)\cos\theta - Ic\sin\theta \tag{12.6.12}$$

which shows that G traces an ellipse that passes through the point $(1, 0)$ in the complex plane. The center of the ellipse is located at the point $(1 - c - 2\alpha, 0)$, and its semiaxes are equal to $c + 2\alpha$ and c. To guarantee stability, we must ensure that the ellipse is located within the unit disk, and this provides us with the stability criterion

$$c^2 < c + 2\alpha < 1 \tag{12.6.13}$$

When U is negative, we use a forward difference for the convective spatial derivative, and work in a similar manner to find that the method will be stable provided that $c^2 < |c| + 2\alpha < 1$.

In practice, the numerical diffusivity associated with the upwind method may be substantial. This feature, combined with the first-order accuracy and the conditional stability, renders the upwind method inferior to its alternatives. A generalization of the method will be discussed in Problem 12.6.5.

Higher-order methods

To improve the accuracy and reduce the numerical diffusivity of the first-order upwind method, Leonard et al. (1978) proposed approximating the first spatial derivative using the third-order backward difference involving four points, while maintaining the central difference for the second spatial derivative. When U is positive, the finite-difference equation is

$$\frac{f_i^{n+1} - f_i^n}{\Delta t} + U \frac{2f_{i+1}^n + 3f_i^n - 6f_{i-1}^n + f_{i-2}^n}{6\Delta x}$$
$$= \kappa \frac{f_{i+1}^n - 2f_i^n + f_{i-1}^n}{\Delta x^2} + O(\Delta t) + O(\Delta x^2) \tag{12.6.14}$$

Rearranging, we obtain

$$f_i^{n+1} = f_i^n - \tfrac{1}{6}c(2f_{i+1}^n + 3f_i^n - 6f_{i-1}^n + f_{i-2}^n) + \alpha(f_{i+1}^n - 2f_i^n + f_{i-1}^n) \tag{12.6.15}$$

Examining the corresponding modified differential equation and carrying out the von Neumann stability analysis, we find that the method is consistent and conditionally stable, but the numerical diffusivity, which is identical to that of the FTCS method, and stability criteria are milder than those of the first-order upwind method (Hoffman, 1992, p. 679).

Explicit CTCS and the DuFort–Frankel Method

In previous sections we found that the CTCS discretization is unconditionally unstable for the unsteady diffusion equation and conditionally stable for the convection equation. Does adding convection to diffusion have a stabilizing influence? Surprisingly, we find that the answer is negative; the CTCS discretization for the convection–diffusion equation leads to an unconditionally unstable method.

The DuFort and Frankel method discussed in Section 12.2 is based on a variation of the CTCS discretization that proceeds according to the difference equation

$$\frac{f_i^{n+1} - f_i^{n-1}}{2\Delta t} + U \frac{f_{i+1}^n - f_{i-1}^n}{2\Delta x} = \kappa \frac{f_{i+1}^n - 2[\tfrac{1}{2}(f_i^{n+1} + f_i^{n-1})] + f_{i-1}^n}{\Delta x^2} \tag{12.6.16}$$

which embodies the CTCS discretization except that the middle term in the numerator on the right-hand side has been replaced by an average value. Rearranging Eq. (12.6.16), we obtain the explicit three-level algorithm

$$f_i^{n+1} = \frac{c + 2\alpha}{1 + 2\alpha} f_{i-1}^n + \frac{1 - 2\alpha}{1 + 2\alpha} f_i^{n-1} - \frac{c - 2\alpha}{1 + 2\alpha} f_{i+1}^n \tag{12.6.17}$$

A consistency analysis shows that the DuFort–Frankel method produces reliable results only when the ratio $(\Delta t/\Delta x)^2$ is sufficiently small. Carrying out the von Neumann stability analysis we find that the amplification factor satisfies the quadratic equation

$$(1 + 2\alpha)G^2 - 2(2\alpha \cos\theta - Ic\sin\theta)G - 1 + 2\alpha = 0 \qquad (12.6.18)$$

Upon detailed examination, we find that $|G| < 1$ as long as $|c| < 1$, which means that the DuFort–Frankel method is stable as long as the CFL condition is fulfilled.

It might appear that the absence of a stability restriction on α allows the use of a large time step, but, in practice, a small time step is necessary in order to obtain a solution that is sufficiently accurate and consistent with that of the original differential equation.

Implicit Methods

Implicit methods were found to be unconditionally stable for pure convection and pure diffusion equation, and remain unconditionally stable for mixed convection–diffusion.

BTCS

Implementing a backward difference for the time derivative and central differences for both the convective and diffusive spatial derivatives, we obtain the fully implicit BTCS difference equation

$$\frac{f_i^{n+1} - f_i^n}{\Delta t} + U\frac{f_{i+1}^{n+1} - f_{i-1}^{n+1}}{2\,\Delta x} = \kappa\frac{f_{i+1}^{n+1} - f_i^{n+1} + f_{i-1}^{n+1}}{\Delta x^2} + O(\Delta t) + O(\Delta x^2) \tag{12.6.19}$$

Rearranging, we derive the tridiagonal form

$$-(c + 2\alpha)f_{i-1}^{n+1} + 2(1 + 2\alpha)f_i^{n+1} + (c - 2\alpha)f_{i+1}^{n+1} = 2f_i^n \tag{12.6.20}$$

The corresponding amplification factor is found to be

$$G = \frac{1}{1 + 2\alpha(1 - \cos\theta) + Ic\sin\theta} \tag{12.6.21}$$

One may show that $|G| < 1$ for any value of α and c, which reveals that the method is unconditionally stable. The physical restriction $Re_c < 2$ must, however, be observed for the results to be physically meaningful.

Crank–Nicolson

To improve the temporal accuracy of the BTCS method, we implement the fully implicit Crank–Nicolson method according to the difference equation

$$\frac{f_i^{n+1} - f_i^n}{\Delta t} + U\frac{1}{2}\left(\frac{f_{i+1}^{n+1} - f_{i-1}^{n+1}}{2\,\Delta x} + \frac{f_{i+1}^n - f_{i-1}^n}{2\,\Delta x}\right)$$
$$= \kappa\frac{1}{2}\left(\frac{f_{i+1}^{n+1} - 2f_i^{n+1} + f_{i-1}^{n+1}}{\Delta x^2} + \frac{f_{i+1}^n - 2f_i^n + f_{i-2}^n}{\Delta x^2}\right) \tag{12.6.22}$$

The accuracy of the method is of second order in both time and space. Rearranging, we obtain the standard tridiagonal form

$$-(c + 2\alpha)f_{i-1}^{n+1} + 4(1 + \alpha)f_i^{n+1} + (c - 2\alpha)f_{i+1}^{n+1}$$
$$= (c + 2\alpha)f_{i-1}^n + 4(1 - \alpha)f_i^n - (c - 2\alpha)f_{i+1}^n \tag{12.6.23}$$

The amplification factor is given by

$$G = \frac{2 - 2\alpha(1 - \cos\theta) - Ic\sin\theta}{2 + 2\alpha(1 - \cos\theta) + Ic\sin\theta} \tag{12.6.24}$$

One may show that $|G| < 1$ for any value of α and c, and thus that the method is unconditionally stable. The physical restriction $Re_c < 2$ must, however, be fulfilled.

Three-level fully implicit method

Another way of achieving second-order accuracy in time is to use a three-level method for the approximation of the temporal derivative at the $n + 1$ time level, while maintaining the fully implicit spatial discretizations, obtaining the difference equation

$$\frac{3f_i^{n+1} - 4f_i^n + f_i^{n-1}}{2\,\Delta t} + U\frac{f_{i+1}^{n+1} - f_{i-1}^{n+1}}{2\,\Delta x}$$
$$= \kappa\frac{f_{i+1}^{n+1} - f_i^{n+1} + f_{i-1}^{n+1}}{\Delta x^2} + O(\Delta t^2) + O(\Delta x^2) \qquad (12.6.25)$$

The method is unconditionally stable and effective in dampening small-amplitude oscillations, and is thus preferable over the Crank–Nicolson method when the solution exhibits sharp variations. The physical restriction $Re_c < 2$ must, however, be fulfilled.

Multistep and Predictor–Corrector Methods

The multistep and predictor–corrector methods discussed in Section 12.4 for pure convection can be extended in a straightforward manner to handle combined convection–diffusion.

MacCormack's explicit method

This is a genuine predictor–corrector method. The predictor step is an extension of Eq. (12.4.36)

$$f_i^* = (1 + c)f_i^n - cf_{i+1}^n + \alpha(f_{i+1}^n - 2f_i^n + f_{i-1}^n) \qquad (12.6.26)$$

and the corrector step is an extension of Eq. (12.4.38)

$$f_i^{n+1} = \tfrac{1}{2}[f_i^n + f_i^* - c(f_i^* - f_{i-1}^*) + \alpha(f_{i+1}^* - 2f_i^* + f_{i-1}^*)] \qquad (12.6.27)$$

The method is second-order accurate in both time and space, and stable as long as $c < 0.90$ and $\alpha \le 0.50$ (Hoffman, 1992, p. 692).

MacCormack's implicit method

This method arises in a manner that is completely analogous to that that led us to Eqs. (12.6.26) and (12.6.27), except that the diffusion term is treated implicitly in both the predictor and corrector steps. The derivation of the difference equations is left as an exercise to the reader in Problem 12.6.2 (Hoffmann and Chiang, 1993, Vol. I, p. 263).

Operator Splitting and Fractional Steps

In previous sections we saw that certain types of discretization work well for the convection equation, while others work well for the diffusion equation. This suggests the use of a fractional-step method in which the convective and diffusive parts are treated independently by different methods according to the component equations

$$\frac{\partial f}{\partial t} + U\frac{\partial f}{\partial x} = 0, \qquad \frac{\partial f}{\partial t} = \kappa\frac{\partial^2 f}{\partial x^2} \qquad (12.6.28)$$

both for $t^n < t < t^n + \Delta t$. The time is reset to the initial value t^n at the end of the first fractional step. In general, the overall stability criteria of fractional-step methods consist of the union of the stability criteria pertaining to the individual methods used for handling the convective and diffusive steps.

Hopscotch Method

The fundamental idea behind the hopscotch method, named after a children's game, is that the solution at different grid points can be advanced using different methods, and a judicious combination of these methods leads to improved efficiency and high accuracy (Gourlay, 1970; Mitchell and Griffiths, 1980, p. 77).

The method proceeds by using the explicit FTCS discretization to advance the solution at the odd-numbered grid points x_{2j+1}, and then the implicit BTCS discretization to advance the solution at the even-numbered grid points x_{2j}, where j is an integer. The order is reversed after the completion of each time step. The crucial advantage is that, since the solution at every other grid point at the new time level is known, the implicit step does not have to be done through matrix inversion, and the method is effectively explicit.

The hopscotch method is first-order accurate in time, second-order accurate in space, and stable as long as $|c| < 1$. There is no stability restriction imposed on the diffusion number α. The efficiency of the method can be improved further by replacing the FTCS difference equation with the equivalent equation $f_i^{n+2} = 2f_i^{n+1} - f_i^n$ after the first step.

Nonlinearities

The significance and implications of nonlinearities in developing finite difference methods were discussed in Section 12.4 in the context of the pure convection equation, and the discussion carries over to the present case of combined convection–diffusion.

Burgers equation

A prototypical equation for studying the performance of finite-difference methods is the Burgers convection–diffusion equation, whose nonconservative and conservative forms are

$$\frac{\partial f}{\partial t} + f\frac{\partial f}{\partial x} = \kappa\frac{\partial^2 f}{\partial x^2} \quad \text{or} \quad \frac{\partial f}{\partial t} + \frac{\partial E}{\partial x} = \kappa\frac{\partial^2 f}{\partial x^2} \tag{12.6.29}$$

where $E = \frac{1}{2}f^2$. Remarkably, the solution in an unbounded domain subject to an arbitrary initial condition may be found analytically using the Cole–Hopf transformation $f = -(2\kappa/u)\,du/dx$ (Benton and Platzman, 1972). Note that this transformation fails when κ is equal to zero, in which case we obtain the inviscid form of the equation. It can be shown that the function u satisfies the linear unsteady diffusion equation (12.2.1), whose solution is given in closed form in Eq. (12.2.2). This fortunate occurrence allows for an unambiguous testing of the accuracy and relative merits of the various finite-difference methods.

As an example, the explicit MacCormack method arises by straightforward modifications of Eqs. (12.6.26) and (12.6.27). The predictor step is

$$f_i^* = (1 + r)E_i^n - rE_{i+1}^n + \alpha(f_{i+1}^n - 2f_i^n + f_{i-1}^n) \tag{12.6.30}$$

and the corrector step is

$$f_i^{n+1} = \tfrac{1}{2}[E_i^n + E_i^* - r(E_i^* - E_{i-1}^*) + \alpha(f_{i+1}^* - 2f_i^* + f_{i-1}^*)] \tag{12.6.31}$$

where $r = \Delta t/\Delta x$.

PROBLEMS

12.6.1 **DuFort–Frankel method.** Perform a consistency analysis of the DuFort–Frankel method and show that the corresponding partial differential equation is given by Eq. (12.2.15) enhanced with the convection term on the left-hand side.

12.6.2 **Implicit MacCormack method.** Write out the finite-difference equations for the implicit MacCormack method.

12.6.3 **The method of undetermined coefficients for a two-level method.** The general form of an implicit method for the one-dimensional convection–diffusion equation, involving three grid points and two time levels, is

$$b_{-1}f_{i-1}^{n+1} + b_0 f_i^{n+1} + b_1 f_{i+1}^{n+1} = a_{-1}f_{i-1}^n + a_0 f_i^n + a_1 f_{i+1}^n \tag{12.6.32}$$

where a_i and b_i are six constant coefficients. Requiring that Eq. (12.6.32) be consistent with Eq. (12.6.1), show that $a_{-1} + a_0 + a_1 = b_{-1} + b_0 + b_1 = 1$, where the last equality represents an arbitrary normalization. Then carry out the von Neumann stability analysis to derive the amplification factor

$$G = \frac{1 - a_1 - a_{-1} + (a_1 + a_{-1})\cos\theta + I(a_1 + a_{-1})\sin\theta}{1 - b_1 - b_{-1} + (b_1 + b_{-1})\cos\theta + I(b_1 + b_{-1})\sin\theta} \tag{12.6.33}$$

(Peyret and Taylor, 1983, p. 39). Verify that Eq. (12.6.33) is consistent with Eq. (12.6.24).

12.6.4 **Implicit BTBC-CS method.** Write out the difference equation for the BTBC-CS method which uses BS differencing for the convection term and CS differencing for the diffusion term, and discuss its stability.

12.6.5 **Generalized explicit upwind differencing.** A generalized form of the upwind differencing method is expressed by the finite-difference equation

$$\frac{f_i^{n+1} - f_i^n}{\Delta t} + \frac{1}{2}U\left((1-\beta)\frac{f_{i+1}^n - f_i^n}{\Delta x} + (1+\beta)\frac{f_i^n - f_{i-1}^n}{\Delta x}\right) = \kappa\frac{f_{i+1}^n - 2f_i^n + f_{i-1}^n}{\Delta x^2} \tag{12.6.34}$$

where β is an arbitrary constant. Setting β equal to 1 when $U > 0$ and equal to -1 when $U < 0$ yields the first-order upwind-differencing method; setting $\beta = 0$, we recover the fully explicit FTCS method. (a) Derive the modified differential equation corresponding to Eq. (12.6.34), show that the method is consistent, and compute the effective diffusivity. (b) The amplification factor corresponding to Eq. (12.6.34) may be deduced from Eq. (12.6.33). Show that the method is stable provided that

$$c^2 < 2\alpha + \beta c < 1 \tag{12.6.35}$$

(Peyret and Taylor, 1983, p. 43). Verify that these stability criteria encompass the inequalities (12.6.6) and (12.6.13).

Computer Problems

12.6.6 **Burgers equation.** An exact solution to the Burgers equation is

$$f = -2\frac{\kappa}{L}\frac{\cosh(x/L)}{\sinh(x/L) + \exp(-\kappa t/L^2)} \tag{12.6.36}$$

where L is an arbitrary length (Benton and Platzman, 1972). Compute the evolution of the solution from $t = 0$ using (a) the FTCS method, (b) the DuFort–Frankel method, (c) MacCormack's explicit method. Compare the exact with the numerical solutions.

12.6.7 **The Korteweg–de Vries equation.** A regularized form of the Korteweg–de Vries equation is

$$\frac{\partial f}{\partial t} + \varepsilon f\frac{\partial f}{\partial x} + \mu\frac{\partial^3 f}{\partial x^3} = 0 \tag{12.6.37}$$

where ε and μ are positive constants. (a) Develop a suitable explicit finite-difference method. (b) An exact solution to Eq. (12.6.37) in an unbounded domain, expressing the propagation of a solitary wave, is given by

$$f = 3c\,\text{sech}^2\left(\sqrt{\frac{\varepsilon c}{4\mu}}(x - \varepsilon ct - d)\right) \tag{12.6.38}$$

where c and d are two arbitrary constants and c is non-negative (Greig and Morris, 1976). Compute the evolution of the solution from the initial state using a finite-difference method of your choice and discuss the accuracy of your results.

12.7 | CONVECTION–DIFFUSION IN TWO AND THREE DIMENSIONS

Finite-difference methods for the linear convection–diffusion equation in three dimensions

$$\frac{\partial f}{\partial t} + U\frac{\partial f}{\partial x} + V\frac{\partial f}{\partial y} + W\frac{\partial f}{\partial z} = \kappa\left(\frac{\partial^2 f}{\partial x^2} + \frac{\partial^2 f}{\partial y^2} + \frac{\partial^2 f}{\partial z^2}\right) \tag{12.7.1}$$

and its counterpart for two dimensions emerge by straightforward extensions of the methods for the one-dimensional case discussed in the preceding section.

FTCS

The fully explicit FTCS method is consistent, first-order accurate in time, and second-order accurate in space. The stability restrictions in three dimensions are

$$\alpha_x + \alpha_y + \alpha_z < \tfrac{1}{2} \qquad \text{and} \qquad \frac{c_x^2}{\alpha_x} + \frac{c_y^2}{\alpha_y} + \frac{c_z^2}{\alpha_z} < 2 \tag{12.7.2}$$

(Hindmarsh, Gresho, and Griffiths, 1984). In two dimensions, the sums on the left-hand sides are over x and y only.

Upwind Differencing

First-order upwind differencing applied to each convective spatial derivative, combined with central differencing for the diffusive derivatives, leads to a consistent method. The stability constraint in two dimensions with $\Delta x = \Delta y$ is (Peyret and Taylor, 1983, p. 66)

$$4\alpha_x + |c_x| + |c_y| < 1 \tag{12.7.3}$$

Hopscotch Method

The implementation of the hopscotch method proceeds according to the general principles outlined in Section 12.6 for the one-dimensional problem (Gourlay, 1970). In two dimensions, we first use the explicit FTCS method to advance the solution at the grid points $x_{i,j}$, where $i + j$ is an odd integer, and then use the implicit BTCS method to advance the solution at the grid points $x_{i,j}$, where $i + j$ is an even integer. The order is reversed after the completion of a time step. After the first time step, the FTCS difference equation is replaced with the equivalent equation $f_{i,j}^{n+2} = 2f_{i,j}^{n+1} - f_{i,j}^{n}$. The method is overall explicit, first-order accurate in time, second-order accurate in space, and stable as long as $|c| < 1$.

ADI in Two Dimensions

Implicit methods are preferable because of their unconditional stability. The ADI method in two dimensions proceeds according to the finite-difference equations

$$\frac{f_{i,j}^{n+1/2} - f_{i,j}^{n}}{\Delta t/2} + U\left[\frac{f_{i+1,j}^{n+1/2} - f_{i-1,j}^{n+1/2}}{2\,\Delta x}\right] + V\frac{f_{i,j+1}^{n} - f_{i,j-1}^{n}}{2\,\Delta y}$$

$$= \kappa\left[\frac{f_{i+1,j}^{n+1/2} - 2f_{i,j}^{n+1/2} + f_{i-1,j}^{n+1/2}}{\Delta x^2}\right] + \kappa\frac{f_{i,j+1}^{n} - 2f_{i,j}^{n} + f_{i,j-1}^{n}}{\Delta y^2} \tag{12.7.4}$$

$$\frac{f_{i,j}^{n+1} - f_{i,j}^{n+1/2}}{\Delta t/2} + U\frac{f_{i+1,j}^{n+1/2} - f_{i-1,j}^{n+1/2}}{2\,\Delta x} + V\left[\frac{f_{i,j+1}^{n+1} - f_{i,j-1}^{n+1}}{2\,\Delta y}\right]$$

$$= \kappa\frac{f_{i+1,j}^{n+1/2} - 2f_{i,j}^{n+1/2} + f_{i-1,j}^{n+1/2}}{\Delta x^2} + \kappa\left[\frac{f_{i,j+1}^{n+1} - 2f_{i,j}^{n+1} + f_{i,j-1}^{n+1}}{\Delta y^2}\right] \tag{12.7.5}$$

The terms in the square brackets designate implicit discretization. The method is second-order accurate in both t and x, and unconditionally stable (Peyret and Taylor, 1983, p. 66).

ADI in Two Dimensions with Time-Dependent Velocities

When the convection velocities U and V are not constant but change in time, the ADI method becomes first-order accurate in the temporal step. To maintain second-order accuracy, we replace the constants U and V in Eq. (12.7.4) with the weighted averages

$$U = a_1 U^{n+1} + (1 - a_1 - a_2)U^n + a_2 U^{n-1}$$
$$V = b_1 V^{n+1} + (1 - b_1 - b_2)V^n + b_2 V^{n-1}$$

$$(12.7.6)$$

and the constants U and V in Eq. (12.7.5) with the weighted averages

$$U = (1 - a_1 + a_2 + a_3)U^{n+1} + (a_1 - a_2 - 2a_3)U^n + a_3 U^{n-1}$$
$$V = (1 - b_1 + b_2 + b_3)V^{n+1} + (b_1 - b_2 - 2b_3)V^n + b_3 V^{n-1}$$

$$(12.7.7)$$

where a_i, b_i are six arbitrary constants (Peyret and Taylor, 1983, p. 66).

Fractional Steps

A fractional-step method in three dimensions emerges by treating convection–diffusion in each dimension individually and separately through a sequence of three one-dimensional steps of equal duration Δt, according to the one-dimensional equations

$$\frac{\partial f}{\partial t} + U\frac{\partial f}{\partial x} = \kappa \frac{\partial^2 f}{\partial x^2}, \qquad \frac{\partial f}{\partial t} + V\frac{\partial f}{\partial y} = \kappa \frac{\partial^2 f}{\partial y^2}$$

$$\frac{\partial f}{\partial t} + W\frac{\partial f}{\partial z} = \kappa \frac{\partial^2 f}{\partial z^2},$$

$$(12.7.8)$$

all for $t^n < t < t^n + \Delta t$. The time is reset to the initial value after the completion of the first and second fractional steps. Each step is carried out using an unconditionally stable implicit method, which requires solving tridiagonal systems of equations; this can be done using the efficient Thomas algorithm (Section B.1, Appendix B).

PROBLEMS

12.7.1   **Hopscotch method.** Develop a hopscotch algorithm for the linear convection–diffusion equation in three dimensions.

12.7.2   **Fractional-step method.** Write the finite-difference equations corresponding to the Crank–Nicolson discretization of the three equations in (12.7.8).

References

Ames, W. F., 1977, *Numerical Methods for Partial Differential Equations*. Academic Press.

Benton, E. R., and Platzman, G. W., 1972, A table of solutions of the one-dimensional Burgers equation. *Quart. Appl. Math.* **30**, 195–212.

Book, D. L., Boris, J. P., and Hain, K., 1975, Flux-corrected transport. II. Generalizations of the method. *J. Comp. Phys.* **18**, 248–83.

Boris, J. P., and Book, D. L., 1973, Flux-corrected transport. I. SHASTA, a fluid transport algorithm that works. *J. Comp. Phys.* **11**, 38–69.

Boris, J. P., and Book, D. L., 1976, Flux-corrected transport. III. Minimal-error FCT algorithms. *J. Comp. Phys.* **20**, 397–431.

Burstein, S. Z., 1967, Finite-difference calculations for hydrodynamic flows containing discontinuities. *J. Comp. Phys.* **2**, 198–222.

Carslow, H. S., and Jaeger, J. C., 1959, *Conduction of Heat in Solids*. Oxford University Press.

Crank, J., and Nicolson, P., 1947, A practical method for numerical evaluation of solutions of partial differential equations of the heat-conduction type. *Proc. Camb. Phil. Soc.* **43**, 50–67.

Douglas, J., 1955, On the numerical solution of $\partial^2 u/\partial x^2 + \partial^2 u/\partial y^2 = \partial u/\partial t$ by implicit methods. *J. Soc. Indust. Appl. Math.* **3**, 42–65.

Douglas, J., 1962, Alternating direction methods for three space variables. *Numerische Mathematik,* **4,** 41–63.

Douglas, J., and Gunn, J. E., 1964, A general formulation of alternating direction methods. *Numerische Mathematik,* **6,** 428–53.

DuFort, E. C., and Frankel, S. P., 1953, Stability conditions in the numerical treatment of parabolic differential equations. *Math. Tables and Other Aids to Computation,* **7,** 135–52.

Ferziger, J. H., 1981, *Numerical Methods for Engineering Application.* Wiley.

Fletcher, C. A. J., 1988, *Computational Techniques for Fluid Dynamics.* 2 volumes. Springer-Verlag.

Gourlay, A. R., 1970, Hopscotch: a fast second-order partial differential equation solver. *J. Inst. Maths Applics.* **6,** 375–90.

Greig, I. S., and Morris, J. L., 1976, A hopscotch method for the Korteweg–de Vries equation. *J. Comp. Phys.* **20,** 64–80.

Hindmarsh, A. C., Gresho, P. M., and Griffiths, D. F., 1984, The stability of explicit Euler time-integration for certain finite difference approximations of the multi-dimensional advection–diffusion equation. *Int. J. Num. Meth. Fluids,* **4,** 853–97.

Hirsch, C., 1988, *Numerical Computation of Internal and External Flows.* 2 volumes. Wiley.

Hoffman, J. D., 1992, *Numerical Methods for Engineers and Scientists.* McGraw–Hill.

Hoffmann, K. A., and Chiang, S. T., 1993, *Computational Fluid Dynamics for Engineers, Vols. I and II,* Engineering Education System, Wichita, Kansas 67208-1078.

Lax, P. D., 1954, Weak solutions of nonlinear hyperbolic equations and their computation. *Comm. Pure Appl. Math.* **7,** 159–93.

Lax, P. D., and Richtmyer, R. D., 1956, Survey of the stability of linear finite difference equations. *Comm. Pure Appl. Math.* **9,** 267–93.

Lax, P., and Wendroff, B., 1960, Systems of conservation laws. *Comm. Pure Appl. Math.* **13,** 217–37.

Lee, M., 1962, Alternating direction methods for hyperbolic differential equations. *J. Soc. Indust. Appl. Math.* **10,** 611–16.

Leonard, B. P., Leschziner, M. A., and McGuirk, J., 1978, Third-order finite-difference method for steady two-dimensional convection. *Num. Meth. in Laminar and Turbulent Flow,* 807–19.

MacCormack, R. W., 1969, *The effect of viscosity in hypervelocity impact cratering.* AIAA paper No. 69-354.

Mitchell, A. R., 1969, *Computational Methods in Partial Differential Equations.* Wiley.

Mitchell, A. R., Griffiths, D. F., 1980, *The Finite Difference Method in Partial Differential Equations.* Wiley.

Peaceman, D. W., and Rachford, H. H., 1955, The numerical solution of parabolic and elliptic differential equations. *J. Soc. Indust. Appl. Math.* **3,** 28–41.

Peyret, R., and Taylor, T. D., 1983, *Computational Methods for Fluid Flow.* Springer-Verlag.

Richtmyer, R. D., 1963, *A survey of difference methods for non-steady fluid dynamics.* NCAR Technical Notes 63-2.

Richtmyer, R. D., and Morton, K. W., 1967, *Difference Methods for Initial-Value Problems.* Interscience.

Sod, G. A., 1985, *Numerical Methods in Fluid Dynamics. Initial and Initial Boundary-Value Problems.* Cambridge University Press.

Warming, R. F., and Hyett, B. J., 1974, The modified equation approach to the stability and accuracy analysis of finite-difference methods. *J. Comp. Phys.* **14,** 159–79.

Wilkinson, J. H., 1965, *The Algebraic Eigenvalue Problem.* Oxford University Press.

Yanenko, N. N., 1970, *The Method of Fractional Steps.* Springer-Verlag.

Finite-Difference Methods for Incompressible Newtonian Flow

Having discussed finite-difference methods for computing numerical solutions to the convection–diffusion equation in its general form, we proceed to develop corresponding methods for solving the equations of steady and unsteady incompressible Newtonian flow. The set of governing equations includes the Navier–Stokes equation and the continuity equation, and the primary unknowns are the velocity and the pressure. We recall, however, that a general rotational flow may also be described and therefore computed in terms of the secondary variables discussed in Chapter 2, including the vorticity, the stream functions, and the vector potential.

Considering the evolution of an unsteady flow, we regard the Navier–Stokes equation as an evolution equation for the velocity, providing us with the rate of change of the velocity at a particular point in the flow in terms of the instantaneous velocity and pressure. We then note that if the pressure gradient were absent, the simplified evolution equation would be identical to the nonlinear convection–diffusion equation, and could therefore be integrated in time using the finite-difference methods discussed in Chapter 12. Unfortunately, as discussed in Section 9.1, an evolution equation for the pressure is not available in an explicit form. In its place we have the restriction of incompressibility, which requires that the pressure evolve so as to ensure that the rate of expansion vanish and the velocity field remain solenoidal at all times. As we saw in Section 9.1, the restriction of incompressibility may be expressed in terms of a Poisson equation either for the pressure or for the rate of change of the pressure with a time-dependent forcing function. These equations determine the evolution of the pressure in an implicit fashion.

Computing the evolution of an incompressible Newtonian flow is thus distinguished by the necessity to solve, simultaneously, a parabolic differential equation in time, which is the equation of motion, and an elliptic differential equation in space, which is the Poisson equation for the pressure or for the rate of change of the pressure.

It is instructive to note at this point that the continuity equation for a compressible fluid has the form of an evolution equation for the density that is related to the pressure by means of an equation of state. Since the full set of governing equations is parabolic in time, it may be integrated using a standard time-marching method for initial-value problems. Shock waves aside, computing the evolution of a compressible flow is in this respect more straightforward than computing the evolution of an incompressible flow.

An additional concern that arises in computing the structure of a steady flow or the evolution of an unsteady flow in terms of the velocity and the pressure, pertains to the derivation and numerical implementation of boundary conditions for the pressure. In the vast majority of fluid-dynamics applications, these are not available in the statement of the problem, but must be derived from the equation of motion subject to the required boundary conditions for the velocity or surface stress. We shall see in this chapter that the accurate implementation of the derived pressure boundary conditions requires careful attention.

There are a number of finite-difference procedures for solving the equations of steady and unsteady incompressible Newtonian flow, and a choice must be made according to the tolerated level of programming complexity and available computational resources. In the present chapter we shall outline the fundamental principles that underlie several alternative procedures, and shall discuss the basic steps involved in their numerical implementation. Extensions and discussions of specific issues and specialized topics may be found in the cited references, as well as in general reviews and monographs on finite-difference methods in fluid dynamics including those by Orszag and Israeli (1974), Cebeci (1982), Roach (1982), Peyret and Taylor (1983), Anderson, Tannehill, and Pletcher (1984), Hirsch (1988), Fletcher (1988), Gresho (1991), and Hoffmann and Chiang (1993). Numerical methods for interfacial and free-surface flows are discussed in a comprehensive review article by Floryan and Rasmussen (1989).

13.1 | METHODS BASED ON THE VORTICITY TRANSPORT EQUATION

We begin by discussing a class of methods for computing the structure of a steady flow or the evolution of an unsteady flow on the basis of the vorticity transport equation. The numerical procedure involves computing the evolution of the vorticity field, and obtaining the simultaneous evolution of the velocity field by inverting the fundamental equation that relates the vorticity to the velocity, $\boldsymbol{\omega} = \nabla \times \mathbf{u}$, subject to the continuity equation. One advantage of this approach is that the pressure does not have to be considered, which results in computational efficiency and ease of implementation. One inevitable concern is the need to derive boundary conditions for the vorticity.

The present class of methods may be regarded as extensions of the vortex methods for inviscid or slightly viscous fluids discussed in Chapter 11. The distinguishing feature of the vortex methods is that the velocity field is obtained from the vorticity field most efficiently using the Biot–Savart integral or a related contour integral. For viscous flow, the support of the vorticity is not compact, and it is more expedient to recover the velocity from the vorticity by solving differential equations using finite-difference methods.

We begin the discussion by presenting the classical stream function—vorticity formulation for two-dimensional flow, and continue to address more general formulations for three-dimensional flow.

Stream Function–Vorticity Formulation for Two-Dimensional Flow

For two-dimensional flow in the xy plane, solving for the velocity in terms of the vorticity is done with the least amount of computational effort by introducing the stream function ψ. The two components of the velocity in the x and y direction are given by $u = \partial\psi/\partial y$ and $v = -\partial\psi/\partial x$, and the vorticity is $\boldsymbol{\omega} = \omega\mathbf{k}$, where $\mathbf{k}$ is the unit vector along the z axis, and

$$\nabla^2\psi = -\omega \tag{13.1.1}$$

The computations proceed according to the two fundamental steps of vortex methods. In the first step, we compute the evolution of the vorticity field using the simplified form of the vorticity transport equation for two-dimensional flow, written in the stream function–vorticity form as

$$\frac{\partial\omega}{\partial t} + \frac{\partial\psi}{\partial y}\frac{\partial\omega}{\partial x} - \frac{\partial\psi}{\partial x}\frac{\partial\omega}{\partial y} = \nu\,\nabla^2\omega \tag{13.1.2}$$

The sum of the second and third terms on the left-hand side of Eq. (13.1.2) is sometimes designated as the Jacobian $J(\omega, \psi)$. In the second step, we update the stream function by solving Poisson's equation (13.1.1) for ψ in terms of ω. Boundary conditions are required during both the integration of Eq. (13.1.2) and the inversion of Eq. (13.1.1).

It is instructive to note that the absence of an explicit evolution equation for the pressure in the original system of governing equations is reflected in the absence of an explicit evolution equation for the stream function.

We return to emphasize the lack of a need to compute the pressure. If the instantaneous pressure field is desired, it may be computed a posteriori by solving a Poisson equation that emerges by taking the divergence of the Navier–Stokes equation and using the continuity equation to obtain

$$\nabla^2 P = 2\rho \left[\frac{\partial^2 \psi}{\partial x^2} \frac{\partial^2 \psi}{\partial y^2} - \left(\frac{\partial^2 \psi}{\partial x \, \partial y} \right)^2 \right] \tag{13.1.3}$$

Boundary conditions for the pressure are derived by applying the equation of motion at the boundaries, projecting it onto either the normal or tangential unit vector, and then simplifying the various terms taking into account the boundary conditions for the velocity, as will be discussed in Section 13.2.

Flow in a rectangular cavity

To illustrate the practical implementation of the finite-difference method, we consider the classical problem of flow in a rectangular cavity driven by a lid that translates parallel to itself with a generally time-dependent velocity $V(t)$, as illustrated in Figure 13.1.1.

The no-penetration condition requires that the component of the velocity normal to each one of the four walls vanish. In terms of the stream function, we obtain the equivalent statement

$$\psi = c \qquad \text{over all walls} \tag{13.1.4}$$

where c is an arbitrary constant that, for simplicity, will be set equal to zero. The no-slip boundary condition requires that the tangential component of the velocity over the bottom, left, and right

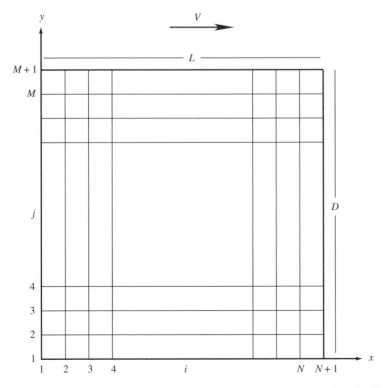

Figure 13.1.1   A non-staggered finite-difference grid for computing the two-dimensional flow in a rectangular cavity driven by a moving lid. The two components of the velocity and the pressure are defined at the grid nodes.

walls vanish, whereas the tangential component at the upper wall be equal to $V(t)$. In terms of the stream function, we obtain the statements

$$\frac{\partial \psi}{\partial y} = 0 \quad \text{at the bottom,} \qquad \frac{\partial \psi}{\partial x} = 0 \quad \text{at the sides}$$

$$\frac{\partial \psi}{\partial y} = V \quad \text{at the lid}$$

$$(13.1.5)$$

Based on these boundary conditions for the velocity, we derive simplified expressions for the boundary values of the vorticity in terms of the stream function. Beginning with Eq. (13.1.1) and noting, for example, that at the bottom wall $\partial^2 \psi / \partial x^2 = -\partial v / \partial x = 0$ because of the no-penetration condition, we find

$$\omega = -\frac{\partial^2 \psi}{\partial y^2} \quad \text{at the top and bottom walls}$$

$$\omega = -\frac{\partial^2 \psi}{\partial x^2} \quad \text{at the side walls}$$

$$(13.1.6)$$

which are restatements of the first of Eqs. (3.6.14).

To implement a finite-difference method, we introduce a two-dimensional grid of size N by M, as illustrated in Figure 13.1.1. For simplicity, we have assumed that the grid lines are evenly spaced, which means that Δx and Δy are uniform but not necessarily equal to each other. This stipulation may be relaxed with straightforward changes in the finite-difference equations. We then assign to the stream function ψ and vorticity ω discrete values at all internal and boundary grid points, and replace the differential equations (13.1.1) and (13.1.2) with difference equations as discussed in Chapter 12. The subsequent strategy of computation depends upon whether we wish to compute a steady or an unsteady flow.

Steady Flow

Two classes of distinct but somewhat related methods are available for computing a steady flow. The first class proceeds by solving the equations of steady flow using iterative methods. The second class proceeds by computing the solution of a fictitious transient flow problem that is governed by a modified set of differential equations, from a given initial condition up to the steady state. The solution of the modified problem at the steady state satisfies the original differential equations of steady two-dimensional incompressible Newtonian flow.

Direct approach

In one version of the direct approach, we regard the governing equations (13.1.1) and (13.1.2) as a coupled, nonlinear system of Poisson equations for ψ and ω, and recast them into the form

$$\nabla^2 \psi = -\omega \tag{13.1.7}$$

$$\nabla^2 \omega = \frac{1}{\nu} \left(u \frac{\partial \omega}{\partial x} + v \frac{\partial \omega}{\partial y} \right) \tag{13.1.8}$$

In the special case of Stokes flow, the right-hand side of Eq. (13.1.8) vanishes, yielding Laplace's equation for the vorticity. Equation (13.1.7) then shows that the stream function satisfies the biharmonic equation as discussed in Section 6.1.

The computational algorithm in the general case of flow at finite Reynolds numbers involves the following steps:

1. Guess the vorticity distribution.
2. Solve Poisson's equation (13.1.7) for the stream function. For boundary conditions, we have the choice between the Dirichlet boundary condition that specifies the boundary distribution

of the stream function, the Neumann boundary condition that specifies the boundary distribution of the normal derivative of the stream function, which is equal to the tangential component of the velocity, or a combination of the Dirichlet and Neumann boundary conditions over the different boundaries.

Use of the Neumann boundary condition over all boundaries is not appropriate, for a solution to the Poisson equation will exist only when the following compatibility condition is fulfilled

$$\int_{\text{Walls}} \nabla \psi \cdot \mathbf{n} \, dl = \int_{\text{Flow}} \omega \, dA \qquad (13.1.9)$$

where $\mathbf{n}$ is the unit normal vector pointing into the flow. Even though Eq. (13.1.9) may be fulfilled for a certain fortuitous guess of the vorticity distribution in step 1, the singular nature of the linear system of equations that arises from the finite-difference discretization of Eq. (13.1.7) will present additional complications. We thus prefer to enforce the Dirichlet condition over all boundaries, expressed by Eq. (13.1.4).

3. Compute the right-hand side of Eq. (13.1.8) and the boundary values of the vorticity using the required boundary conditions for the velocity, and solve Poisson's equation (13.1.8) for the vorticity.

4. Check to see whether the computed vorticity agrees with the current vorticity at all grid points, and if it does not, replace the current with the computed vorticity and return to step 2.

The details of the numerical implementation of the various steps will be now discussed with reference to the flow in a cavity illustrated in Figure 13.1.1.

1. Assign initial values for the stream function to all $(N + 1) \times (M + 1)$ internal and boundary grid points, and for the vorticity to all NM internal grid points. A simple choice is to set both the stream function and vorticity equal to zero.

2. Solve Poisson's equation (13.1.7) subject to the Dirichlet boundary condition (13.1.4) over all four walls. Since the vorticity, which is the forcing function of Poisson's equation (13.1.7), is only an approximation to the exact solution, an accurate solution for ψ at this stage is not warranted. To reduce the computational effort, we solve Poisson's equation using an iterative method and carry out only a small number of iterations.

One way of carrying out the iterations is to introduce a fictitious unsteady diffusion-reaction problem with a source term that is equal to ω and diffusivity that is equal to $\rho_1 \Delta x^2 / \Delta t$, where ρ_1 is a dimensionless constant, in which case the diffusion numbers in the x and y directions are $\alpha_x = \rho_1$ and $\alpha_y = \rho_1 \beta^2$, where $\beta = \Delta x / \Delta y$. The solution at steady state satisfies Eq. (13.1.7). Implementing the FTCS discretization, which involves using the five-point formula to approximate the Laplacian, yields

$$\psi_{i,j}^{(k+1)} = \psi_{i,j}^{(k)} + \rho_1 \left[\psi_{i+1,j}^{(k)} - 2\psi_{i,j}^{(k)} + \psi_{i-1,j}^{(k)} \right.$$
$$\left. + \beta^2 (\psi_{i,j+1}^{(k)} - 2\psi_{i,j}^{(k)} + \psi_{i,j-1}^{(k)}) + \Delta x^2 \, \omega_{i,j} \right] \qquad (13.1.10)$$

where k plays the role of an iteration number (see first entry of Table 12.3.1). The von Neumann stability analysis discussed in Section 12.3 shows that the iterations will converge provided that $\rho_1 \leq 1/[2(1 + \beta^2)]$. When the iterations are executed for the first time, the initial values $\psi_{i,j}^{(0)}$ are set equal to those guessed in step 1. After the iterations converge, the solution will be second-order accurate in Δx and Δy. Alternatives to the FTCS iterative scheme (13.1.10) are the Gauss–Siedel and successive over-relaxation schemes, as well as their SOR versions shown in Table 12.3.1.

3. Use Eqs. (13.1.6) to compute the vorticity at the boundary grid points by means of one-sided finite differences.

   To compute the vorticity at a grid point that lies on the upper wall, we expand the stream function in a Taylor series with respect to y about a grid point that lies on the lid, and evaluate the series at the Mth layer to obtain

$$\psi_{i,M} = \psi_{i,M+1} + \left(\frac{\partial \psi}{\partial y}\right)_{i,M+1}(-\Delta y) + \frac{1}{2}\left(\frac{\partial^2 \psi}{\partial y^2}\right)_{i,M+1}(-\Delta y)^2 + \cdots \qquad (13.1.11)$$

Using the no-slip boundary condition and the first of Eqs. (13.1.6), and rearranging, we obtain

$$\omega_{i,M+1} = 2\frac{\psi_{i,M+1} - \psi_{i,M}}{\Delta y^2} - 2\frac{V}{\Delta y} \qquad (13.1.12)$$

which is first-order accurate in Δy. Working in a similar manner for the bottom and side walls, we derive the analogous expressions

$$\omega_{i,1} = 2\frac{\psi_{i,1} - \psi_{i,2}}{\Delta y^2}, \qquad \omega_{1,j} = 2\frac{\psi_{1,j} - \psi_{2,j}}{\Delta x^2}$$

$$\omega_{N+1,j} = 2\frac{\psi_{N+1,j} - \psi_{N,j}}{\Delta x^2} \qquad (13.1.13)$$

The boundary condition (13.1.4) allows us to set $\psi_{i,1}, \psi_{i,M+1}, \psi_{1,j}, \psi_{N+1,j}$ equal to zero.

   To improve the accuracy to second order, we expand the stream function in a Taylor series about a grid point on the upper wall, evaluate the series at the two layers that are adjacent to the wall, and maintain terms up to third order to find

$$\psi_{i,M} = \psi_{i,M+1} + \left(\frac{\partial \psi}{\partial y}\right)_{i,M+1}(-\Delta y) + \frac{1}{2}\left(\frac{\partial^2 \psi}{\partial y^2}\right)_{i,M+1}(-\Delta y)^2$$
$$+ \frac{1}{6}\left(\frac{\partial^3 \psi}{\partial y^3}\right)_{i,M+1}(-\Delta y)^3 + \cdots \qquad (13.1.14)$$

$$\psi_{i,M-1} = \psi_{i,M+1} + \left(\frac{\partial \psi}{\partial y}\right)_{i,M+1}(-2\,\Delta y) + \frac{1}{2}\left(\frac{\partial^2 \psi}{\partial y^2}\right)_{i,M+1}(-2\,\Delta y)^2$$
$$+ \frac{1}{6}\left(\frac{\partial^3 \psi}{\partial y^3}\right)_{i,M+1}(-2\,\Delta y)^3 + \cdots \qquad (13.1.15)$$

Combining these equations to eliminate the third derivative of the stream function, solving for the second derivative, using the boundary condition in Eqs. (13.1.5), and taking into account Eq. (13.1.6), we find

$$\omega_{i,M+1} = \frac{7\psi_{i,M+1} - 8\psi_{i,M} + \psi_{i,M-1}}{2\,\Delta y^2} - 3\frac{V}{\Delta y} \qquad (13.1.16)$$

Working in a similar manner for the bottom and side walls we find

$$\omega_{i,1} = \frac{7\psi_{i,1} - 8\psi_{i,2} + \psi_{i,3}}{2\,\Delta y^2}, \qquad \omega_{1,j} = \frac{7\psi_{1,j} - 8\psi_{2,j} + \psi_{3,j}}{2\,\Delta x^2}$$

$$\omega_{N+1,j} = \frac{7\psi_{N+1,j} - 8\psi_{N,j} + \psi_{N-1,j}}{2\,\Delta x^2} \qquad (13.1.17)$$

4. Differentiate the stream function to compute the velocity at the internal grid points subject to the boundary values (13.1.4). Differentiate the vorticity to compute the right-hand side

of Eq. (13.1.8) at the internal grid points subject to the boundary values computed from Eqs. (13.1.12) and (13.1.13) or Eqs. (13.1.16) and (13.1.17). For convenience, denote the right-hand side of Eq. (13.1.8) at the i,j grid point by $N_{i,j}$, where N stands for *nonlinear.*

5. Solve Poisson's equation (13.1.8) subject to the Dirichlet boundary conditions computed in step 3. This may be done in an iterative manner according to the FTCS algorithm

$$
\omega_{i,j}^{(k+1)} = \omega_{i,j}^{(k)} + \rho_2 \left[\omega_{i+1,j}^{(k)} - 2\omega_{i,j}^{(k)} + \omega_{i-1,j}^{(k)} \right.
$$
$$
\left. + \beta^2 (\omega_{i,j+1}^{(k)} - 2\omega_{i,j}^{(k)} + \omega_{i,j-1}^{(k)}) - \Delta x^2 \, N_{i,j} \right] \tag{13.1.18}
$$

discussed in step 2, where ρ_2 is the diffusion number in the x direction, and carry out a small number of iterations. The values of the forcing function at the internal grid points $N_{i,j}$ are available from step 4. As in step 2, the algorithm (13.1.18) may be replaced with one of the algorithms shown in Table 12.3.1.

6. If the vorticity ω computed in step 5 does not agree with that previously available, return to step 2 and repeat the computations with the new grid values of ω. This outer iteration is terminated when the absolute value of the difference of the vorticity between two successive iterations at each grid point becomes less than a preestablished threshold value ε, or the sum of the absolute values of the differences in the vorticity over all internal NM grid points becomes less than $NM\varepsilon$.

One noteworthy feature of this procedure is that the corner grid points do not enter the computations, and this eliminates ambiguities stemming from the fact that the velocity undergoes a discontinuity at the upper corner points. This discontinuity may cause local oscillations and decelerate the local convergence, but does not have a deleterious effect on the global convergence of the method.

The individual steps of the numerical procedure may be modified and improved in several ways (see, for instance, Gupta, 1991). While the basic philosophy of the algorithm remains unchanged, the rate of convergence of the iterations does depend on the details of the particular implementation (Israeli, 1972). In one variation of the method, instead of iterating on Poisson's equation for the vorticity in step 5, we iterate on the full convection–diffusion equation (13.1.8), which means that we recompute the nonlinear term $N_{i,j}$ after each iteration using the updated values of the vorticity. These iterations may be done on the basis of an explicit or implicit finite-difference method for the convection–diffusion equation in two dimensions discussed in Section 12.7. At high Reynolds numbers, the flow near the center of the cavity is dominated by convection, and using upwind differencing improves the numerical stability.

Streamline patterns for flow in a square cavity are illustrated in Figure 13.1.2 for three values of the Reynolds number $Re \equiv VL/\nu$. When $Re = 1$, we obtain a nearly creeping flow, and the streamline pattern is almost symmetric with respect to the midplane of the cavity. As the Reynolds number is increased, the recirculating eddy becomes unsymmetric and shifts toward the upper right corner. Small viscous eddies are always present at the two bottom corners. At even higher Reynolds numbers, the flow is composed of a central vortex with nearly uniform vorticity, and boundary layers lining the walls.

The method of computing the boundary values of the vorticity described in step 3 works well in most cases but it has been the subject of criticism (Gresho, 1991). It has been argued that it is not proper to specify the boundary values of the vorticity in an explicit manner, but instead, the boundary distribution of the vorticity must arise in an implicit manner as part of the solution, using the natural boundary conditions for the velocity or traction. This issue will be discussed further at the end of the present section in the more general context of three-dimensional flow.

(*a*)   ***Re = 1***

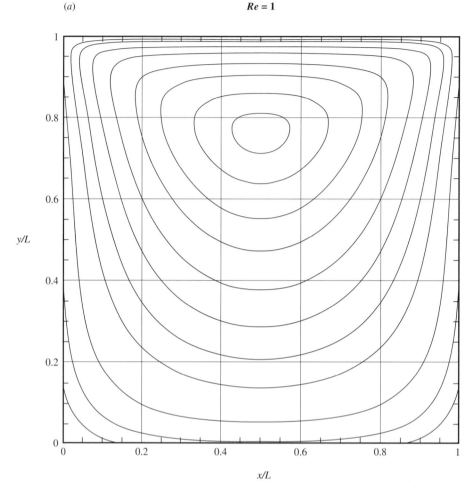

x/L

Figure 13.1.2   Streamline patterns of steady flow in a square cavity driven by a translating lid, computed using the direct method discussed in the text for three values of the Reynolds number (a) $Re = VL/\nu = 1$. (*continued*)

Method of modified dynamics or false transients

A distinguishing part of the direct approach is the solution of a Poisson equation for the stream function, which reflects the elliptic nature of the equations governing the structure of a steady flow. We saw that one way to perform the associated inner iterations is to introduce a fictitious unsteady diffusion-reaction problem and then implement the explicit FTCS discretization. This observation suggests reformulating the problem by maintaining the unsteady vorticity transport equation (13.1.2) and transforming Eq. (13.1.1) into the following evolution equation for the stream function

$$\frac{\partial \psi}{\partial t} = \alpha(\nabla^2 \psi + \omega) \tag{13.1.19}$$

where α is a positive constant that is a free parameter of the numerical method (Mallinson and de Vahl Davis, 1973). The idea is to compute the evolution of the flow from an arbitrary initial

(b)   *Re* = **100**

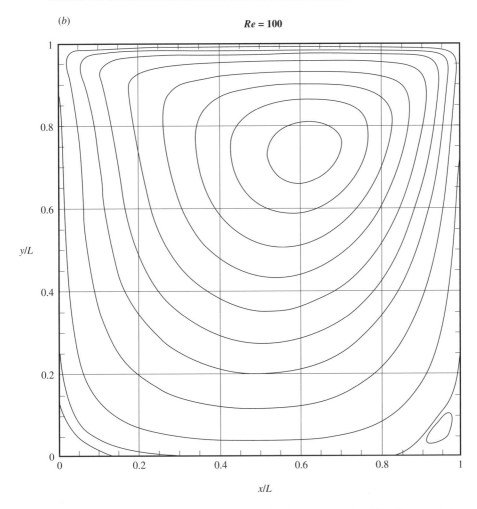

Figure 13.1.2 (*continued*)   Streamline patterns of steady flow in a square cavity driven by a translating lid, computed using the direct method discussed in the text for three values of the Reynolds number (b) *Re* = 100.   (*continued*)

condition on the basis of Eqs. (13.1.2) and (13.1.19) until a steady state is established; at that point the solution will also satisfy the original Equations (13.1.7) and (13.1.8). In the implementation of the method, the right-hand side of Eq. (13.1.2) is also multiplied with a positive factor in order to expedite the convergence. The critical advantage of this approach is that the governing equations become parabolic in time, and this allows us to use time-marching methods similar to those developed in Chapter 12 for problems of convection–diffusion.

For the problem of flow in a cavity illustrated in Figure 13.1.1, the method of modified dynamics is implemented according to the following steps:

1. Assign initial values to the stream function and vorticity at all internal and boundary grid points; a simple choice is to set them both equal to zero.
2. Differentiate the stream function to compute the two components of the velocity at all internal grid points.
3. Compute the vorticity at the boundary grid points as in step 3 of the direct approach.

(*c*)   ***Re* = 1,000**

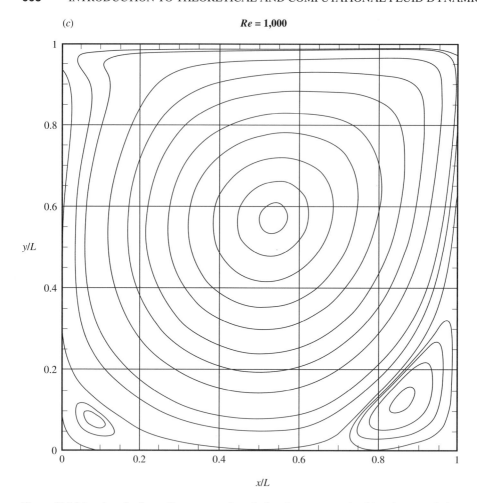

Figure 13.1.2 (*continued*)   Streamline patterns of steady flow in a square cavity driven by a translating lid, computed using the direct method discussed in the text for three values of the Reynolds number (c) *Re* = 1,000.

4. Advance the vorticity at all internal grid points on the basis of Eq. (13.1.2) using, for instance, the ADI method for the convection–diffusion equation described in Section 12.7, while maintaining the vorticity at the boundary grid points constant.

5. Advance the stream function at all internal grid points on the basis of Eq. (13.1.19) using, for instance, the ADI method for the convection–diffusion equation described in Section 12.7, subject to the boundary condition (13.1.4).

6. Return to step 2 and repeat the computations for another time step.

Unsteady Flow

To compute the evolution of an unsteady flow, we follow a procedure that combines certain features of the direct approach and certain features of the method of modified dynamics for steady flow. The algorithm involves computing the evolution of the vorticity field using Eq. (13.1.2), while obtaining the simultaneous evolution of the velocity field on the basis of the stream function using Eq. (13.1.1).

A simple strategy for computing the evolution of the flow in a cavity when the lid begins translating suddenly at a constant velocity V proceeds according to the following steps:

1. Set the stream function and velocity equal to zero at all internal and boundary grid points at the initial instant. Then set the x component of the velocity at the grid points on the lid equal to V.

2. Differentiate the velocity to compute the vorticity. For the internal grid points use central differences. For the grid points that lie on the lid use the first of Eqs. (13.1.6), take into account the boundary conditions, and apply the second-order backward difference for the first derivative of u with respect to y to obtain

$$\omega_{i,M+1} = \frac{-3V + 4u_{i,M} - u_{i,M-1}}{2\,\Delta y} \tag{13.1.20}$$

(Table B.5.1). For the grid points that lie at the bottom and side walls, use the corresponding second-order finite-difference formulae

$$\omega_{i,1} = \frac{-4u_{i,2} + u_{i,3}}{2\,\Delta y}, \qquad \omega_{1,j} = \frac{4v_{2,j} - v_{3,j}}{2\,\Delta x}$$

$$\omega_{N+1,j} = \frac{-4v_{N,j} + v_{N-1,j}}{2\,\Delta x} \tag{13.1.21}$$

3. Integrate Eq. (13.1.2) to compute the vorticity at the next time level at all internal grid points subject to the boundary conditions expressed by Eqs. (13.1.20) and (13.1.21) using, for example, an explicit method such as the FTCS method. At high Reynolds numbers, use upwind differencing.

4. Solve Poisson's equation (13.1.1) for the stream function at the next time level subject to the Dirichlet boundary condition (13.1.4).

5. Differentiate the stream function to compute the velocity at the next time level at all internal grid points.

6. Return to step 2 and repeat the computations for another time step.

To improve the temporal accuracy and enhance the stability of the method, one may update the vorticity using an implicit or a semi-implicit method such as the ADI method. In a simple implementation of the ADI method, we maintain the convection velocity constant during both substeps, equal to its value at the beginning of the first substep. In a more advanced implementation, we solve Poisson's equation for the intermediate stream function after completion of the first substep, and then set the convection velocity in the second substep equal to that computed by differentiating the intermediate stream function as described in step 5. Since, however, the convection velocity is kept constant during each step or substep, equal to its value at the beginning of the step or substep, the overall accuracy of the method will still be of first order in time.

To achieve second-order accuracy, we use the ADI method with time-dependent velocities described in Eqs. (12.7.6) and (12.7.7). Collecting the values of the vorticity at all grid points into the vector $\boldsymbol{\omega}$, we obtain the two ADI equations written in the symbolic form

$$\mathbf{A}(\mathbf{u}^{n-1}, \mathbf{u}^{n}, \mathbf{u}^{n+1}) \cdot \boldsymbol{\omega}^{n+1/2} = \mathbf{B}(\mathbf{u}^{n-1}, \mathbf{u}^{n}, \mathbf{u}^{n+1}) \cdot \boldsymbol{\omega}^{n}$$

$$\mathbf{C}(\mathbf{u}^{n-1}, \mathbf{u}^{n}, \mathbf{u}^{n+1}) \cdot \boldsymbol{\omega}^{n+1} = \mathbf{D}(\mathbf{u}^{n-1}, \mathbf{u}^{n}, \mathbf{u}^{n+1}) \cdot \boldsymbol{\omega}^{n+1/2} \tag{13.1.22}$$

where $\mathbf{A}, \mathbf{B}, \mathbf{C}$, and $\mathbf{D}$ are tridiagonal matrices that are functions of their arguments. Eqs. (13.1.22) replace the explicit FTCS equation in step 3. Steps 3, 4, and 5 described above are now combined to yield the following inner iterative loop:

i. Guess the velocities $\mathbf{u}^{n+1}$ and solve the two tridiagonal systems of Eqs. (13.1.22) with boundary conditions given in Eqs. (13.1.20) and (13.1.21) for both $\boldsymbol{\omega}^{n+1/2}$ and $\boldsymbol{\omega}^{n+1}$.

ii. Execute steps 4 and 5.

iii. Solve Eqs. (13.1.22) with the computed values of $\mathbf{u}^{n+1}$ or with a weighted average of the old and new values.

If the boundary values of the velocity change in time, we solve the first tridiagonal system of Eqs. (13.1.22) with boundary conditions $\boldsymbol{\omega}^{n+1/2} = \frac{1}{2}(\boldsymbol{\omega}^n + \boldsymbol{\omega}^{n+1})$, where $\boldsymbol{\omega}^{n+1}$ has been approximated from the previous inner iteration. To accelerate the convergence, we replace the boundary conditions for $\boldsymbol{\omega}^{n+1}$ during the inner iterations with a weighted average of its old and new values. Further details on the implementation of this method are given by Peyret and Taylor (1983, p. 197).

Methods for Three-Dimensional Flow

Algorithms based on the vorticity transport equation for three-dimensional flow involve the following two basic steps:

1. Compute the evolution of the vorticity field on the basis of the vorticity transport equation written in the conservative or Eulerian form

$$\frac{\partial \boldsymbol{\omega}}{\partial t} + \nabla \times (\boldsymbol{\omega} \times \mathbf{u}) = \nu \nabla^2 \boldsymbol{\omega} \qquad (13.1.23)$$

Taking the divergence of Eq. (13.1.23), we find that $\nabla \cdot \boldsymbol{\omega}$ satisfies the unsteady diffusion equation, and this guarantees that the computed vorticity field will be solenoidal provided that (1) it is solenoidal at the initial instant, and (2) its divergence vanishes over the boundaries of the flow at all times (see also Problem 13.1.4).

To integrate Eq. (13.1.23), we require boundary conditions for the vorticity. In the majority of numerical procedures, the boundary values of the vorticity emerge by applying the definition $\boldsymbol{\omega} = \nabla \times \mathbf{u}$ at or near the boundaries, and then simplifying them, taking into consideration the prescribed boundary conditions for the velocity. The numerical procedure is analogous to that involved in the stream function–vorticity formulation discussed earlier in this section. This approach guarantees that an initially solenoidal vorticity field will remain solenoidal at all times (Guj and Stella, 1993; Trujillo, 1994).

It has been argued, however, that it is not entirely appropriate to impose local boundary conditions for the vorticity in an explicit fashion, but instead, the boundary distribution of the vorticity must be computed as part of the solution, taking into consideration the boundary conditions for the velocity or surface stress (Gresho, 1991). Computational experiments have shown that computing, instead of imposing, boundary values for the the vorticity enhances the stability of the numerical method, but this comes at the cost of increased programming complexity and computational effort.

Quartapelle and Valz-Gris (1981), in particular, replaced the boundary conditions for the vorticity with an integral constraint. For two-dimensional flow with homogeneous boundary conditions for the velocity, this constraint requires that the vorticity be orthogonal to all nonsingular harmonic functions defined in the domain of flow, that is, the integral of the vorticity multiplied by any nonsingular harmonic function over the area of the flow vanish. The implementation of this method is discussed by Quartapelle (1981) and by Anderson (1989) in the context of Chorin's vortex sheet method (Section 11.5).

When the flow is described in a noninertial frame of reference that translates and rotates with time-dependent linear and angular velocities $\mathbf{V}$ and $\boldsymbol{\Omega}$, we work with the modified vorticity $\mathbf{W} = \boldsymbol{\omega} + 2\boldsymbol{\Omega}$, which evolves according to the standard vorticity transport equation (3.8.34) written for an inertial frame (Speziale, 1987). Since the effects of acceleration of the frame of reference enter the solution only through the boundary conditions, using the modified vorticity simplifies the numerical implementation and reduces the computational demands.

2. In the second step, we compute the evolution of the velocity field by inverting the definition $\boldsymbol{\omega} = \nabla \times \mathbf{u}$ subject to the continuity equation $\nabla \cdot \mathbf{u} = 0$. The inversion can be done in two different ways according to the *vector potential–vorticity* and *velocity–vorticity* formulation discussed in the following subsections.

Vector potential–vorticity formulation

This method proceeds by decomposing the velocity field into the sum of a solenoidal irrotational velocity field $\nabla \phi$, where ϕ is a harmonic function, and a rotational velocity field that is expressed in terms of the curl of a solenoidal vector potential $\mathbf{A}$, so that $\mathbf{u} = \nabla \phi + \nabla \times \mathbf{A}$ (Hirasaki and Hellums, 1970). The velocity potential ϕ is found by solving Laplace's equation subject to the required no-penetration condition $\nabla \phi \cdot \mathbf{n} = \mathbf{u} \cdot \mathbf{n}$, which determines ϕ uniquely up to an arbitrary but physically irrelevant constant.

To compute the vector potential, we write $\boldsymbol{\omega} = \nabla \times (\nabla \times \mathbf{A}) = \nabla(\nabla \cdot \mathbf{A}) - \nabla^2 \mathbf{A}$ and stipulate that $\mathbf{A}$ be solenoidal, thus obtaining the vectorial Poisson equation

$$\nabla^2 \mathbf{A} = -\boldsymbol{\omega} \tag{13.1.24}$$

Taking the divergence of Eq. (13.1.24) and remembering that the vorticity field is solenoidal, shows that $\nabla \cdot \mathbf{A}$ satisfies Laplace's equation, which means that the computed $\mathbf{A}$ will be solenoidal provided that the boundary conditions on $\mathbf{A}$ ensure that $\nabla \cdot \mathbf{A} = 0$ over the boundaries. For simply connected domains, one way of ensuring that this constraint is fulfilled is to require that the tangential components of $\mathbf{A}$ vanish; that is, $\mathbf{n} \times (\mathbf{A} \times \mathbf{n}) = \mathbf{0}$. This is consistent with the requirement that $(\nabla \times \mathbf{A}) \cdot \mathbf{n} = 0$. To derive a boundary condition for the normal component of $\mathbf{A}$, we introduce a local coordinate system with the x and z axes tangential to the boundary and the y axis normal to the boundary at a point, and use the condition $\nabla \cdot \mathbf{A} = 0$, to find that, at the origin,

$$
\begin{aligned}
\nabla \cdot \mathbf{A} &= \mathbf{n} \cdot (\nabla \mathbf{A}) \cdot \mathbf{n} + \frac{\partial \mathbf{A}}{\partial x} \cdot \mathbf{t}_x + \frac{\partial \mathbf{A}}{\partial z} \cdot \mathbf{t}_z \\
&= \mathbf{n} \cdot (\nabla \mathbf{A}) \cdot \mathbf{n} + \frac{\partial (\mathbf{A} \cdot \mathbf{t}_x)}{\partial x} + \frac{\partial (\mathbf{A} \cdot \mathbf{t}_z)}{\partial z} - \mathbf{A} \cdot \left(\frac{\partial \mathbf{t}_x}{\partial x} + \frac{\partial \mathbf{t}_z}{\partial z} \right) \\
&= \mathbf{n} \cdot (\nabla \mathbf{A}) \cdot \mathbf{n} + 2\kappa_m \mathbf{A} \cdot \mathbf{n} = 0
\end{aligned}
\tag{13.1.25}
$$

where $\mathbf{n} \cdot (\nabla \mathbf{A}) \cdot \mathbf{n}$ is the derivative of the normal component of $\mathbf{A}$ in a direction normal to the boundary, $\mathbf{t}_x$ and $\mathbf{t}_z$ are the unit tangential vectors in the directions of the x and z axes, and κ_m is the mean curvature of the boundary. When the boundary is flat, the mean curvature vanishes, and Eq. (13.1.25) assumes the simpler form $\mathbf{n} \cdot (\nabla \mathbf{A}) \cdot \mathbf{n} = 0$. Richardson and Cornish (1977) developed boundary conditions for multiply connected domains.

For two-dimensional or axisymmetric flow, we set all components of $\mathbf{A}$ equal to zero except for the z component or azimuthal component, which is identified, respectively, with the stream function or with the Stokes stream function divided by the radial distance σ. We then find that the vector potential–vorticity formulation reduces to the stream function–vorticity formulation discussed at the beginning of this section.

The numerical implementation of the vector potential–vorticity formulation for three-dimensional flow has been discussed by several authors, including Aziz and Hellums (1967), Aragbesola and Burley (1977), and Mallinson and de Vahl Davis (1977); the reader is referred to their works for specific details.

Velocity–vorticity formulation

In the most popular version of this formulation, the velocity is computed from the vorticity by solving the vectorial Poisson equation

$$\nabla^2 \mathbf{u} = -\nabla \times \boldsymbol{\omega} \tag{13.1.26}$$

which arises by taking the curl of the definition $\boldsymbol{\omega} = \nabla \times \mathbf{u}$ and requiring that $\mathbf{u}$ be a solenoidal function. The solution is found subject to the boundary conditions for the velocity specified in the statement of the problem.

To validate the method, we must show that the curl of the computed velocity will indeed be equal to $\boldsymbol{\omega}$, provided that $\boldsymbol{\omega}$ is solenoidal and its boundary values are computed from $\boldsymbol{\omega} = \nabla \times \mathbf{u}$ (Daube, 1992; Trujillo, 1994). For this purpose, we use the vector identity $\nabla \times (\nabla \times \mathbf{u}) = \nabla(\nabla \cdot \mathbf{u}) - \nabla^2 \mathbf{u}$ in conjunction with Eq. (13.1.26) and find $\nabla \times (\nabla \times \mathbf{u}) = \nabla(\nabla \cdot \mathbf{u}) + \nabla \times \boldsymbol{\omega}$. Taking the curl of both sides of this equation to eliminate the first term on the right-hand side yields $\nabla \times \nabla \times \mathbf{F} = \nabla(\nabla \cdot \mathbf{F}) - \nabla^2 \mathbf{F} = \mathbf{0}$, where we have defined $\mathbf{F} = \nabla \times \mathbf{u} - \boldsymbol{\omega}$. Since $\boldsymbol{\omega}$ and thus $\mathbf{F}$ is solenoidal, the components of $\mathbf{F}$ must be harmonic functions. But the boundary values of $\mathbf{F}$ are equal to zero, and this requires that $\mathbf{F}$ vanish and thus $\boldsymbol{\omega}$ be equal to $\nabla \times \mathbf{u}$ throughout the domain of flow.

One important consequence of this result is that the velocity computed from the solution of Eq. (13.1.26) will surely be solenoidal. This can be shown beginning, once again, with the identity $\nabla \times (\nabla \times \mathbf{u}) = \nabla(\nabla \cdot \mathbf{u}) - \nabla^2 \mathbf{u}$, which now shows that $\nabla(\nabla \cdot \mathbf{u}) = \mathbf{0}$. Straightforward integration in the spatial variables reveals that $\nabla \cdot \mathbf{u}$ is constant throughout the domain of flow; conservation of mass requires this constant to be equal to zero (Daube, 1992; Trujillo, 1994).

The numerical implementation of the velocity–vorticity formulation based on Eq. (13.1.26) has been discussed by several authors, including Chien (1976), Dennis, Ingham, and Cook (1979), Daube (1992), Guj and Stella (1993), Trujillo (1994). Guj and Stella (1988) developed a method of false transients for steady flow by replacing the elliptic equation (13.1.26) with the parabolic equation

$$Re \frac{\partial \mathbf{u}}{\partial t} = \nabla^2 \mathbf{u} + \nabla \times \boldsymbol{\omega} \tag{13.1.27}$$

In another version of the velocity–vorticity formulation, the velocity field is computed directly by solving the Cauchy–Riemann-type system of equations $\boldsymbol{\omega} = \nabla \times \mathbf{u}$ and $\nabla \cdot \mathbf{u} = 0$ for the velocity, subject to the no-penetration condition at the boundaries. The implementation of this method is discussed by Osswald, Ghia, and Ghia (1987) and Gatski, Grosh, and Rose (1989).

PROBLEMS

13.1.1   **Poisson's equation for the pressure.** Take the divergence of the two-dimensional Navier–Stokes equation and introduce the stream function to derive the pressure Poisson equation (13.1.3).

13.1.2   **Axisymmetric flow.** Write the counterparts of Eqs. (13.1.1)–(13.1.3) for axisymmetric flow in terms of the Stokes stream function.

13.1.3   **Boundary condition for the vorticity.** Derive the expressions given in Eqs. (13.1.20) and (13.1.21).

13.1.4   **Integration of the vorticity transport equation.** Consider the temporal integration of the vorticity transport equation written in the nonconservative form with the vortex stretching term explicit on the right-hand side, subject to the boundary condition $\boldsymbol{\omega} = \nabla \times \mathbf{u}$. Discuss whether the vorticity field will remain solenoidal during the time integration (Gatski, Grosh, and Rose, 1989).

Computer Problems

13.1.5   **Steady flow in a cavity.** (a) Write a computer program called *CV2DS1* that computes the steady flow in a square cavity with width and depth equal to L, generated by the steady translation of the lid, using the direct approach discussed in the text. The inner iterations should be conducted using the FTCS method. Carry out computations at a sequence of increasing Reynolds numbers $Re = VL/\nu = 1, 10, 100, 500, \ldots$ and discuss the changes in the structure of the flow. Study the convergence of the method as a function of the two numerical parameters ρ_1 and ρ_2 and number of iterations. Estimate the critical Reynolds number where your spatial resolution appears to be inadequate. To compute flow at the higher Reynolds numbers, it is helpful to use a continuation method in which the initial guesses for the stream function and vorticity are identified with the corresponding converged values at a lower Reynolds number. (b) Repeat part (a) but with the inner iterations carried out using the LSOR method and comment on the improvement. (c) Repeat part (a) with the method of modified dynamics.

13.1.6 **Unsteady flow in a cavity.** Write a computer program called *CV2DU1* that solves the unsteady version of Problem 13.1.5 with a lid that is set in motion impulsively at a constant velocity, using the first-order method discussed in the text.

13.2 | VELOCITY–PRESSURE FORMULATION

In this section we proceed to discuss a class of methods for computing steady and unsteady, two-dimensional and three-dimensional flows in primitive variables including the velocity and the pressure. To simplify the derivations, we shall assume that the density and the viscosity of the fluid are uniform throughout the domain of flow.

We begin developing these methods by rewriting the Navier–Stokes equation in the form of an evolution equation as

$$\frac{\partial \mathbf{u}}{\partial t} = \mathbf{N}(\mathbf{u}) - \frac{1}{\rho} \nabla P + \nu L(\mathbf{u}) \tag{13.2.1}$$

where P is the modified pressure and $\mathbf{N}$ and L are, respectively, the nonlinear-inertial and linear-viscous operators defined as

$$\mathbf{N}(\mathbf{u}) = -\mathbf{u} \cdot \nabla \mathbf{u} = -\nabla \cdot (\mathbf{u}\mathbf{u}) \tag{13.2.2}$$
$$L(\mathbf{u}) = \nabla^2 \mathbf{u} = \nabla(\nabla \cdot \mathbf{u}) - \nabla \times \boldsymbol{\omega} \tag{13.2.3}$$

Because the velocity field is solenoidal, the first term on the right-hand side of Eq. (13.2.3) vanishes and could have been discarded; maintaining it, however, will allow us to take into account and filter out the accumulation of the numerical error. The equation of motion is accompanied by the continuity equation for incompressible fluids

$$\nabla \cdot \mathbf{u} = 0 \tag{13.2.4}$$

which states that the velocity field is and must remain solenoidal at all times.

Pressure Poisson Equation

Taking the divergence of Eq. (13.2.1) and interchanging the gradient with the temporal derivative, we obtain an evolution equation for the rate of expansion

$$\frac{\partial \nabla \cdot \mathbf{u}}{\partial t} = \nabla \cdot \mathbf{N}(\mathbf{u}) - \frac{1}{\rho} \nabla^2 P + \nu \nabla \cdot L(\mathbf{u}) \tag{13.2.5}$$

The continuity equation requires that the left-hand side of Eqs. (13.2.5) vanish at all times, and this makes it necessary for the pressure to satisfy the pressure Poisson equation, (PPE),

$$\nabla^2 P = \rho \nabla \cdot \mathbf{N}(\mathbf{u}) + \mu \nabla \cdot L(\mathbf{u}) \tag{13.2.6}$$

One might argue that the since the divergence and the linear operator L commute, the second term on the right-hand side of Eq. (13.2.6) vanishes, yielding the simplified form

$$\nabla^2 P = \rho \nabla \cdot \mathbf{N}(\mathbf{u}) \tag{13.2.7}$$

We shall see, however, that issues of numerical stability require the use of the more cumbersome form (13.2.6). Following Gresho and Sani (1987), we call Eq. (13.2.6) the *consistent PPE*, and Eq. (13.2.7) the *simplified PPE*.

Alternative Systems of Governing Equations

To this end, we consider replacing the original system of governing equations (13.2.1) and (13.2.4) with either (1) the modified system of Eqs. (13.2.1) and (13.2.6), or (2) the modified system of Eqs. (13.2.1) and (13.2.7). These replacements will be acceptable as long as the modified systems guarantee that the velocity remains solenoidal at all times.

Substituting Eq. (13.2.6) into Eq. (13.2.5), we find

$$\frac{\partial \nabla \cdot \mathbf{u}}{\partial t} = 0 \tag{13.2.8}$$

which states that, if the velocity field is solenoidal at the initial time, it will remain solenoidal at all times. Thus, if the initial velocity field is solenoidal, it is permissible to replace the continuity equation with the consistent PPE (13.2.6). When the initial rate of expansion, however, is not equal to zero, the divergence of the velocity will remain finite throughout the evolution.

Substituting Eq. (13.2.7) into Eq. (13.2.5) and interchanging the divergence with the Laplacian, we obtain the unsteady diffusion equation for the rate of expansion

$$\frac{\partial \nabla \cdot \mathbf{u}}{\partial t} = \nu L(\nabla \cdot \mathbf{u}) \tag{13.2.9}$$

The general properties of the unsteady diffusion equation in a bounded domain show that the rate of expansion will vanish at all times provided that (1) the initial velocity field is solenoidal, and (2) the rate of expansion or its normal derivative over all boundaries are held equal to zero at all times. When these conditions are met, it is permissible to replace the continuity equation with the simplified PPE (13.2.6). The second condition, in particular, underlines the importance of accurately satisfying conservation of mass at the grid points near or at the boundaries. When the initial rate of expansion is not equal to zero, but its boundary distribution is kept equal to zero, the magnitude of the divergence of the velocity will keep decreasing and eventually will tend to vanish during the evolution.

Boundary Conditions for the Pressure

The consistent and modified PPEs, and their finite-difference counterparts, must be solved subject to one scalar boundary condition over each boundary of the flow. According to the preceding discussion, this condition must ensure that the boundary distribution of the divergence of the velocity vanish at all times.

The derivation of boundary conditions for the pressure is discussed in two illuminating articles by Orszag, Israeli, and Deville (1986) and Gresho and Sani (1987). Their analyses show that the Neumann boundary condition

$$\nabla P \cdot \mathbf{n} = \rho \left(-\frac{\partial \mathbf{u}}{\partial t} \cdot \mathbf{n} + \mathbf{N}(\mathbf{u}) \cdot \mathbf{n} \right) + \mu L(\mathbf{u}) \cdot \mathbf{n} \tag{13.2.10}$$

which emerges by applying Eq. (13.2.1) at the boundaries of the flow and then projecting it onto the normal vector $\mathbf{n}$, is always appropriate, for *it is another manifestation of the condition of incompressibility at the boundaries.*

Using the boundary condition (13.2.10) guarantees that replacing the continuity equation with the simplified PPE is a valid substitution. The solution for the pressure computed using Eq. (13.2.10) also satisfies the Dirichlet boundary condition, which emerges by projecting the equation of motion onto a tangential vector and then integrating it with respect to the tangential arc length. On the contrary, the pressure field that is computed by solving the PPE using the Dirichlet condition will satisfy the Neumann condition only when the initial velocity field is sufficiently regular (Gresho and Sani, 1987).

For a planar wall that translates parallel to itself with constant velocity, we use the no-slip and no-penetration conditions to find that the Neumann condition (13.2.10) simplifies to

$$\nabla P \cdot \mathbf{n} = \mu \frac{\partial^2 \mathbf{u}}{\partial l_n^2} \cdot \mathbf{n} \tag{13.2.11}$$

where $\mathbf{n}$ is the unit normal vector directed into the flow, and l_n designates the arc length normal to the wall measured toward the fluid. When the Reynolds number is sufficiently large, the right-hand side of Eq. (13.2.11) is small and is sometimes set equal to zero.

In a certain class of finite-difference methods, to be discussed in Section 13.4, the Neumann boundary condition (13.2.10) is not enforced in an explicit manner. Instead, the numerical method employs an alternative boundary condition that emerges by satisfying the continuity equation at the grid nodes that are adjacent to the boundaries or are located at the boundaries. The consistent implementation of this procedure is equivalent to requiring the Neumann condition (13.2.10).

Quartapelle and Napolitano (1986) replaced the boundary condition for the pressure with an integral constraint involving the projection of the boundary distribution of the pressure onto a nonsingular solution of the vectorial Helmholtz equation defined within the domain of flow. Assessing the efficiency of their method, however, awaits further research.

Compatibility Condition for the PPE

The emerging computational procedure involves solving a Poisson equation of the form $\nabla^2 P = g$ subject to the Neumann boundary condition $\nabla P \cdot \mathbf{n} = q$ over all boundaries. Here g can be identified with the right-hand side of either Eq. (13.2.6) or (13.2.7), and q is identified with the right-hand side of Eq. (13.2.10). Integrating the pressure Poisson equation over the domain of flow, and using the divergence theorem, we find that a solution will exist only when the following compatibility condition is fulfilled

$$\int_{\text{Flow}} \nabla^2 P \, dV = -\int_{\text{Boundaries}} \nabla P \cdot \mathbf{n} \, dS$$

or

$$\int_{\text{Flow}} g \, dV = -\int_{\text{Boundaries}} q \, dS \tag{13.2.12}$$

The counterpart of Eq. (13.2.12) in two dimensions is

$$\int_{\text{Flow}} g \, dA = -\int_{\text{Boundaries}} q \, dl \tag{13.2.13}$$

The satisfaction of Eq. (13.2.12) is guaranteed in the continuous version of the problem, but discretization errors may destroy the exact equality in the corresponding finite-difference formulation (Problem 13.2.2). In solving Poisson's equation using iterative methods, this inconsistency may result in slow convergence or even divergence of the numerical solution; in solving it using a direct method, it has the inevitable consequence that one of the linear equations expressing the Poisson equation at a particular grid point will not be satisfied, and the pressure may exhibit local oscillations.

Two ways of ensuring that the compatibility condition is fulfilled in the discrete formulation of the problem are:

1. Modify the source term g of Poisson's equation in the discrete statement of the problem by a proper amount (Ghia, Hankey, and Hodge 1977; Sheth and Pozrikidis, 1995; see also Problem 13.2.2);

2. Use a custom-made finite-difference method that coordinates the discretization of the equation of motion, the PPE, and the boundary conditions, so as to automatically satisfy the discrete version of the compatibility condition. This procedure is sometimes called the *consistent finite-difference discretization*. The method was originally developed by Abdallah (1987) for a uniform two-dimensional Cartesian grid, and was subsequently extended to three-dimensional and curvilinear grids by several authors including Mansour and Hamed (1990), Sotiropoulos and Abdallah (1991), and Babu and Korpela (1994).

When the discrete version of the compatibility condition is fulfilled, the linear system of equations associated with the discrete Poisson equation will have a multiplicity of solutions reflecting the fact that the pressure may be determined only up to an arbitrary constant. To render the solution unique, we must specify the value of the pressure at an arbitrary grid point, or set the average value of the pressure over all grid points at an arbitrary level.

An Explicit Evolution Equation for the Pressure

An evolution equation for the pressure may be obtained by differentiating the pressure Poisson equation in time and using the equation of motion to eliminate the time derivatives of the velocity in favor of the velocity and pressure, as discussed in Section 9.1. The result is the Poisson equation for $\partial P/\partial t$ shown in Eq. (9.1.5), to be solved subject to the Neumann boundary condition that arises by differentiating Eq. (13.2.10) in time and interchanging the order of the normal spatial derivative and temporal derivative on the left-hand side. Unfortunately, this method appears to be untested in practice.

In closing this section, we compare the formulation in primitive variables to the formulations based on the vorticity transport equation discussed in the preceding section and find relative weaknesses and strengths. One weakness is the need to derive boundary conditions for the pressure; one strength is ease of extension to multifluid and interfacial flows.

PROBLEMS

13.2.1   **Compatibility condition for the PPE.** Show that the compatibility condition (13.2.12) for the pressure Poisson equation Eq. (13.2.6) with boundary conditions given in Eq. (13.2.10) is fulfilled (Gresho and Sani, 1987).

13.2.2   **Solving Poisson's equation with Neumann boundary conditions.** Consider Poisson's equation $\nabla^2 \phi = g$ in a three-dimensional domain Ω with Neumann boundary conditions $\nabla \phi \cdot \mathbf{n} = q$ all around the boundaries $\partial\Omega$. For a solution to exist, the compatibility condition (13.2.12) must be fulfilled. In a discrete finite-difference formulation, the problem is expressed in terms of an $N \times N$ linear system of equations $\mathbf{A} \cdot \mathbf{x} = \mathbf{b}(g, q)$, which incorporates Poisson's equation applied at the internal grid points and the boundary conditions. This notation emphasizes that the constant vector $\mathbf{b}$ is a function of the source term g as well as of the specified boundary flux q. Unless the corresponding discrete version of Eq. (13.2.12) is fulfilled to machine accuracy, the linear system will not have a solution. One way of circumventing this difficulty is to perturb the source term g, thus modifying the constant vector $\mathbf{b}$. This can be done by replacing g with $g + \varepsilon f$, where ε is a small number to be found as part of the solution, and f is an arbitrarily specified function that is independent of g. The proper value of ε is given by

$$\varepsilon \int_\Omega f \, dV = - \int_\Omega g \, dV - \int_{\partial\Omega} q \, dS \tag{13.2.14}$$

The method can be implemented numerically according to the following steps: First, we solve the first $N - 1$ equations of the system $\mathbf{A} \cdot \mathbf{x} = \mathbf{b}(g, q)$ for the first $N - 1$ unknowns, set the last unknown equal to zero, $x_n = 0$, and call the solution $\mathbf{x}^{(1)}$. We then compute the residual of the last equation $R^{(1)} = A_{ni} x_i^{(1)} - b_n$. Next, we solve the first $N - 1$ equations of the linear system $\mathbf{A} \cdot \mathbf{x} = \mathbf{b}(f, q)$ for the first $N - 1$ unknowns, set the last unknown equal to zero, $x_n = 0$, and call the solution $\mathbf{x}^{(2)}$. Then, we compute the residual of the last equation $R^{(2)} = A_{ni} x_i^{(2)} - b_n$. Finally, we set $\varepsilon = R^{(1)}/R^{(2)}$ and compute the final solution $\mathbf{x} = \mathbf{x}^{(1)} + \varepsilon \mathbf{x}^{(2)}$. Show that this solution satisfies all N equations of the linear system $\mathbf{A} \cdot \mathbf{x} = \mathbf{b}(g + \varepsilon f, q)$.

Computer Problem

13.2.3 **Solving the Poisson equation in a rectangular domain.** Write a routine called *RPPE* that uses the method described in Problem 13.2.2 to solve Poisson's equation in a two-dimensional rectangular domain with an arbitrary source term assigned at the grid points, and arbitrary Neumann boundary conditions all around the boundaries. The Laplacian should be approximated using the five-point formula, and the normal derivatives at the boundaries should be approximated using second-order, one-side finite differences (Section B.5, Appendix, B).

13.3 | IMPLEMENTATION OF METHODS IN PRIMITIVE VARIABLES

Having discussed the basic considerations that enter the computation of an incompressible Newtonian flow in terms of the velocity and pressure, we proceed to develop specific implementations.

The Explicit Method of Harlow and Welch on a Staggered Grid

The marker and cell (MAC) method of Harlow and Welch (1965) combines a finite-difference method for solving the equations of incompressible flow and a marker-tracing method for tracking the motion of free surfaces or fluid interfaces.

The finite-difference method is based on the explicit forward-time discretization of the equation of motion (13.2.1) yielding the velocity at the next time level in terms of the velocity and the pressure at the current time level,

$$\mathbf{u}^{n+1} = \mathbf{u}^n + \Delta t \left(\mathbf{N}(\mathbf{u}^n) + \nu L(\mathbf{u}^n) - \frac{1}{\rho} \nabla P^n \right) \tag{13.3.1}$$

To compute the pressure P^n, we discretize the evolution equation for the rate of expansion, Eq. (13.2.5), using a forward difference in time and obtain

$$\frac{(\nabla \cdot \mathbf{u})^{n+1} - (\nabla \cdot \mathbf{u})^n}{\Delta t} = \nabla \cdot \mathbf{N}(\mathbf{u}^n) - \frac{1}{\rho} \nabla^2 P^n + \nu L(\nabla \cdot \mathbf{u})^n \tag{13.3.2}$$

Requiring that the divergence of the velocity at the $n + 1$ level vanish, we obtain the following Poisson equation for the pressure

$$\nabla^2 P^n = \frac{\rho}{\Delta t} (\nabla \cdot \mathbf{u})^n + \rho \nabla \cdot \mathbf{N}(\mathbf{u}^n) + \mu L(\nabla \cdot \mathbf{u})^n \tag{13.3.3}$$

which is a modified version of the consistent PPE, Eq. (13.2.6). Although small, the first and third terms on the right-hand side must be retained in order to prevent the onset of numerical instabilities. The computational algorithm involves the following steps:

1. Specify an initial solenoidal velocity field that satisfies the prescribed boundary conditions.

2. Compute the pressure field by solving Poisson's equation (13.3.3) subject to the Neumann boundary conditions (13.2.10). In practice, this is done using an alternative but equivalent set of boundary conditions to be discussed shortly.

3. Use Eq. (13.3.1) to advance the velocity field by one step in time subject to the prescribed boundary conditions. The spatial derivatives on the right-hand side of Eq. (13.3.1) are computed using central differences, as will be discussed shortly. The size of the time step must be kept sufficiently small in order to suppress the growth of numerical oscillations.

4. Return to step 2 and repeat the computations for another time step.

Staggered grid

Harlow and Welch (1965) implemented their method on the staggered grid shown in Figure 13.3.1. In two dimensions, the staggered grid is composed of two superposed grids that are displaced with respect to each other by distances equal to half the grid spacings. The *primary* grid is drawn with solid lines, and the *secondary* grid is drawn with broken lines. Note that the secondary grid conforms with the physical boundaries of the flow where the velocity is known.

Discrete values of the pressure are assigned at the primary nodes (i, j) shown with *circles,* values of the x component of the velocity are assigned at the intersections between the primary and secondary grids $(i + 1/2, j)$ shown with *squares,* and values of the y component of the velocity are assigned at the intersections between the primary and secondary grids $(i, j + 1/2)$ shown with *triangles* in Figure 13.3.1. Note that the squares and triangles are located at the faces of a primary cell.

The distinguishing feature of the staggered-grid method is that the unknown functions and governing equations are defined or enforced at different nodes, and this decoupling simplifies the numerical implementation and enhances the numerical stability of the method. We shall see, in particular, that the fact that the pressure nodes are located in the interior of the flow simplifies the implementation of the boundary conditions for the pressure. Unfortunately, the staggered-grid

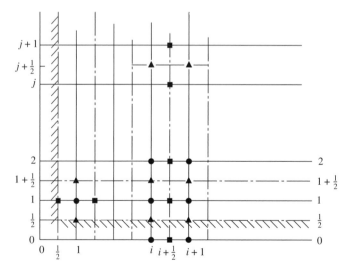

Figure 13.3.1   Staggered grid for flow in a rectangular domain near a corner. The primary grid is drawn with heavy lines, and the secondary grid is drawn with broken lines. Note that the secondary grid conforms with the physical boundaries of the flow. The pressure is defined at the circles, the x component of the velocity is defined at the squares, and the y component is defined at the triangles.

method becomes considerably more involved, and sometimes prohibitively expensive, when applied to grids that are defined in curvilinear coordinates.

Interpolation and extrapolation of the velocity

The velocity components at the vertices of the primary or secondary grids are computed by linear interpolation from the closest nodes. For instance,

$$u_{i,j} = \tfrac{1}{2}(u_{i-1/2,j} + u_{i+1/2,j}), \qquad u_{i+1/2,j+1/2} = \tfrac{1}{2}(u_{i+1/2,j} + u_{i+1/2,j+1})$$
$$v_{i,j} = \tfrac{1}{2}(v_{i,j+1/2} + v_{i,j-1/2}), \qquad v_{i+1/2,j+1/2} = \tfrac{1}{2}(v_{i,j+1/2} + v_{i+1,j+1/2}) \tag{13.3.4}$$

For a rectangular domain of flow with a grid of size $N \times M$, the velocity components at the external nodes corresponding to $i = 0$, $i = N + 1$, $j = 0$, $j = M + 1$ are computed taking into account the boundary conditions at the physical boundaries of flow that are located at $i = \frac{1}{2}$, $i = N + \frac{1}{2}$, $j = \frac{1}{2}$, $j = M + \frac{1}{2}$. For instance, requiring that $u = U$ at $j = \frac{1}{2}$, and applying linear extrapolation yields

$$u_{i+1/2,0} = 2U - u_{i+1/2,1} \tag{13.3.5}$$

which is called the *reflection formula*. Equation (13.3.5) allows us to compute the first partial derivative $\partial u/\partial y$ at the level $j = 1$ using central differences with first-order accuracy, that is, by setting $(\partial u/\partial y)_{i+1/2,1} = (u_{i+1/2,2} - u_{i+1/2,0})/(2\,\Delta y)$. To compute the corresponding second partial derivative with comparable accuracy using central differences, we must use the alternative formula

$$u_{i+1/2,0} = \tfrac{1}{3}(8U - 6u_{i+1/2,1} + u_{i+1/2,2}) \tag{13.3.6}$$

Similar equations can be derived for use with non-centered finite differences for the first or second derivatives near the boundaries (Peyret and Taylor, 1983, p. 151).

Advancing the velocity field

To advance the components of the velocity in time, we apply Eq. (13.3.1) *at the internal velocity nodes*, express the nonlinear convection term in conservative form, and approximate the spatial derivatives using central differences to obtain

$$u_{i+1/2,j}^{n+1} = u_{i+1/2,j}^{n} + \Delta t \left(\Delta u_{i+1/2,j}^{n} - \frac{1}{\rho} \frac{P_{i+1,j}^{n} - P_{i,j}^{n}}{\Delta x} \right) \tag{13.3.7}$$

and

$$v_{i,j+1/2}^{n+1} = v_{i,j+1/2}^{n} + \Delta t \left(\Delta v_{i,j+1/2}^{n} - \frac{1}{\rho} \frac{P_{i,j+1}^{n} - P_{i,j}^{n}}{\Delta y} \right) \tag{13.3.8}$$

where

$$\Delta u_{i+1/2,j} = -\frac{u_{i+1,j}^2 - u_{i,j}^2}{\Delta x} - \frac{1}{\Delta y}(u_{i+1/2,j+1/2}\,v_{i+1/2,j+1/2} - u_{i+1/2,j-1/2}\,v_{i+1/2,j-1/2})$$
$$+ \nu \left(\frac{u_{i+3/2,j} - 2u_{i+1/2,j} + u_{i-1/2,j}}{\Delta x^2} + \frac{u_{i+1/2,j+1} - 2u_{i+1/2,j} + u_{i+1/2,j-1}}{\Delta y^2} \right) \tag{13.3.9}$$

and

$$\Delta v_{i,j+1/2} = -\frac{1}{\Delta x}(u_{i+1/2,j+1/2}\,v_{i+1/2,j+1/2} - u_{i-1/2,j+1/2}\,v_{i-1/2,j+1/2}) - \frac{v_{i,j+1}^2 - v_{i,j}^2}{\Delta y}$$
$$+ \nu \left(\frac{v_{i+1,j+1/2} - 2u_{i,j+1/2} + u_{i-1,j+1/2}}{\Delta x^2} + \frac{v_{i,j+3/2} - 2v_{i,j+1/2} + v_{i,j-1/2}}{\Delta y^2} \right) \tag{13.3.10}$$

Before we can compute the right-hand sides of Eqs. (13.3.7) and (13.3.8), however, we must solve the Poisson equation for the pressure.

Pressure Poisson equation

As a first step, we approximate the rate of expansion *at the pressure nodes* using central differences and obtain

$$D_{i,j} \equiv (\nabla \cdot \mathbf{u})_{i,j} = \frac{u_{i+1/2,j} - u_{i-1/2,j}}{\Delta x} + \frac{v_{i,j+1/2} - v_{i,j-1/2}}{\Delta y} \tag{13.3.11}$$

Requiring that the right-hand side of Eq. (13.3.11) vanish at the interior pressure nodes $i = 2, \ldots, N - 1$ and $j = 2, \ldots, M - 1$, at the $n + 1$ time level, and expressing the velocities at the $n + 1$ level in terms of the velocities and pressure at the nth level using Eqs. (13.3.7) and (13.3.8), we obtain

$$D_{i,j}^n + \Delta t \left(\frac{\Delta u_{i+1/2,j}^n - \Delta u_{i-1/2,j}^n}{\Delta x} + \frac{\Delta v_{i,j+1/2}^n - \Delta v_{i,j-1/2}^n}{\Delta y} \right)$$

$$- \frac{1}{\rho} \Delta t \left(\frac{P_{i+1,j}^n - 2P_{i,j}^n + P_{i-1,j}^n}{\Delta x^2} + \frac{P_{i,j+1}^n - 2P_{i,j}^n + P_{i,j-1}^n}{\Delta y^2} \right) = 0 \tag{13.3.12}$$

Rearranging, and using the definitions (13.3.9) and (13.3.10), we derive the discrete version of the pressure Poisson equation (13.3.3), where the Laplacian is approximated using central differences over intervals equal to Δx and Δy

$$\frac{P_{i+1,j} - 2P_{i,j} + P_{i-1,j}}{\Delta x^2} + \frac{P_{i,j+1} - 2P_{i,j} + P_{i,j-1}}{\Delta y^2} = \frac{\rho}{\Delta t} D_{i,j} + \rho Q_{i,j}$$

$$+ \mu \left(\frac{D_{i+1,j} - 2D_{i,j} + D_{i-1,j}}{\Delta x^2} + \frac{D_{i,j+1} - 2D_{i,j} + D_{i,j-1}}{\Delta y^2} \right) \tag{13.3.13}$$

where we have defined

$$Q_{i,j} = - \frac{u_{i+1,j}^2 - 2u_{i,j}^2 + u_{i-1,j}^2}{\Delta x^2} - \frac{v_{i,j+1}^2 - 2v_{i,j}^2 + v_{i,j-1}^2}{\Delta y^2}$$

$$- \frac{2}{\Delta x \, \Delta y} (u_{i+1/2,j+1/2} \, v_{i+1/2,j+1/2} + u_{i-1/2,j-1/2} \, v_{i-1/2,j-1/2}$$

$$- u_{i+1/2,j-1/2} \, v_{i+1/2,j-1/2} - u_{i-1/2,j-1/2} \, v_{i-1/2,j-1/2}) \tag{13.3.14}$$

All variables in Eqs. (13.3.13) and (13.3.14) are evaluated at the nth time level.

Pressure boundary condition

For the pressure nodes that are adjacent to the boundaries corresponding to $i = 1, N$, and $j = 1, M$, we follow a slightly different approach that takes into account the boundary conditions for the velocity. Considering the first horizontal layer $j = 1$, we require that the discrete form of the rate of expansion, given by the right-hand side of Eq. (13.3.11), vanish at the $n + 1$ time level. Focusing on the pressure nodes $j = 1$ and $i = 2, \ldots, N - 1$ that are located away from the corners, we express the velocities at the nodes $(i - \frac{1}{2}, 1)(i, \frac{3}{2}), (i + \frac{1}{2}, 1)$ at the $n + 1$ level in terms of the velocities and pressure at the nth level using Eqs. (13.3.7) and (13.3.8), and obtain the counterpart of Eq. (13.3.12),

$$\frac{u_{i+1/2,1}^n - u_{i-1/2,1}^n}{\Delta x} + \frac{v_{i,3/2}^n - v_{i,1/2}^{n+1}}{\Delta y} + \Delta t \left(\frac{\Delta u_{i+1/2,1}^n - \Delta u_{i-1/2,1}^n}{\Delta x} + \frac{\Delta v_{i,3/2}^n}{\Delta y} \right)$$

$$- \frac{1}{\rho} \Delta t \left(\frac{P_{i+1,1}^n - 2P_{i,1}^n + P_{i-1,1}^n}{\Delta x^2} + \frac{P_{i,2}^n - P_{i,1}^n}{\Delta y^2} \right) = 0 \tag{13.3.15}$$

The velocity v^{n+1} at the $(i, \frac{1}{2})$ node is available from the prescribed boundary conditions. A straightforward modification of Eq. (13.3.15) is required for the corner nodes $i = 1$ and N.

Working similarly for the last horizontal layer corresponding to $j = M$ with $i = 2, \ldots, N - 1$, we find

$$\frac{u^n_{i+1/2,M} - u^n_{i-1/2,M}}{\Delta x} + \frac{v^{n+1}_{i,M+1/2} - v^n_{i,M-1/2}}{\Delta y} + \Delta t \left(\frac{\Delta u^n_{i+1/2,M} - \Delta u^n_{i-1/2,M}}{\Delta x} - \frac{\Delta v^n_{i,M-1/2}}{\Delta y} \right)$$

$$- \frac{1}{\rho} \Delta t \left(\frac{P^n_{i+1,M} - 2P^n_{i,M} + P^n_{i-1,M}}{\Delta x^2} - \frac{P^n_{i,M} - P^n_{i,M-1}}{\Delta y^2} \right) = 0 \qquad (13.3.16)$$

Straightforward modifications are necessary for the two corner nodes. Similar equations may be derived for the first and last vertical layers corresponding to $i = 1, N$ and $j = 2, \ldots, M - 1$.

Equations (13.3.15) and (13.3.16), and their counterparts for the first and last vertical layers, provide us with boundary conditions for the discrete Poisson equation (13.3.13). It appears that by using a staggered grid, we have circumvented the derivation of explicit boundary conditions for the pressure. Gresho and Sani (1987), however, demonstrated that Eq. (13.3.15) and its counterparts for the other three walls amount to the Neumann boundary condition (13.2.10).

To implement the explicit Neumann boundary condition for the pressure, let us refer to Figure 13.3.1, and introduce the external pressure nodes $P_{i,0}$. Using the continuity equation, we find that $\partial v / \partial y$ vanishes at the bottom wall, and then apply Eq. (13.2.11) and approximate the derivatives using central differences to obtain

$$(\nabla P \cdot \mathbf{n})_{i,1/2} = \left(\frac{\partial P}{\partial y} \right)_{i,1/2} \cong \frac{P_{i,1} - P_{i,0}}{\Delta y} \cong \mu \left(\frac{\partial^2 v}{\partial y^2} \right)_{i,1/2} \cong 2\mu \frac{v_{i,3/2}}{\Delta y^2} \qquad (13.3.17)$$

which may be rearranged to yield

$$P_{i,0} = P_{i,1} - 2\mu \frac{v_{i,3/2}}{\Delta y} \qquad (13.3.18)$$

Working similarly with the upper, left, and right walls, we find

$$P_{i,M+1} = P_{i,M} + 2\mu \frac{v_{i,M-1/2}}{\Delta y}, \qquad P_{0,j} = P_{1,j} - 2\mu \frac{u_{3/2,j}}{\Delta y}$$

$$P_{N+1,j} = P_{N,j} + 2\mu \frac{u_{N-1/2,j}}{\Delta y} \qquad (13.3.19)$$

Eqs. (13.3.18) and (13.3.19) provide us with an alternative set of boundary conditions for Poisson's equation (13.3.13), which is now also applied at the pressure nodes that are adjacent to the boundaries, for instance, at $j = 1$.

The Explicit Method of Harlow and Welch on a Non-staggered Grid

Consider next the implementation of the explicit method of Harlow and Welch on the non-staggered grid shown in Figure 13.1.1. The discretized equation of motion (13.3.1) and pressure Poisson equation (13.3.3) are enforced at all internal grid points, and the spatial derivatives are approximated using centered differences.

To derive boundary conditions for the pressure, we require that the discretized form of the rate of expansion D vanish at the grid points that are located on the four walls. Considering the first horizontal layer $j = 1$ with $i = 2, \ldots, N$, we use a central difference in the x direction and a second-order one-sided difference in the y direction to write

$$D_{ij} = \frac{u_{i+1,1} - u_{i-1,1}}{2\Delta x} + \frac{-v_{i,3} + 4v_{i,2} - 3v_{i,1}}{2\Delta y} \qquad (13.3.20)$$

(Chorin, 1968). Requiring that the right-hand side of Eq. (13.3.20) vanish at the $n + 1$ time level and expressing the velocities at the $(i, 2)$ and $(i, 3)$ grid points in terms of the discrete version of the right-hand side of Eq. (13.3.1) yields an equation that is similar to Eq. (13.3.15). Working in a similar manner, we derive corresponding equations for the upper, left, and right walls.

It would appear that we have again circumvented the explicit derivation of boundary conditions for the pressure, but Gresho and Sani (1987) demonstrated that Eq. (13.3.20) and its counterparts for the other three walls amount to the Neumann boundary condition (13.2.10). This is true even when first-order one-sided differences are used to approximate the second partial derivative in Eq. (13.3.20).

Solenoidality of the Discrete Velocity Field

Let us assume that the solution of the discrete version of Poisson's equation for the pressure has been computed, and the results are substituted back into Eq. (13.3.1) to advance the velocity. To this end, we must ask whether the divergence of the updated velocity $\nabla \cdot \mathbf{u}^{n+1}$ will be equal to zero to machine accuracy. The answer is negative, unless the finite-difference methods for solving the pressure Poisson equation and for computing the divergence of the velocity have been coordinated in an appropriate manner (Sotiropoulos and Abdallah, 1991).

One way to ensure that the discrete velocity field is solenoidal to machine accuracy is to solve the pressure Poisson equation using the five-point formula with intervals of size $2\,\Delta x$, $2\,\Delta y$, and $2\,\Delta z$, and then compute the divergence of the velocity using central differences with intervals of size Δx, Δy, and Δz. Unfortunately, the solution of the Poisson equation using this method may exhibit numerical instabilities due to the decoupling of the values of the pressure at neighboring grid points. In practice, a small value of the rate of expansion at the grid points is tolerated and may be reduced by grid refinement.

Higher-Order Methods

The explicit method of Harlow and Welch is first-order accurate in time and conditionally stable. To improve the accuracy and relax the stability constraints, we resort to semi-implicit, fully implicit, or predictor–corrector iterative methods.

Adopting the Crank–Nicolson method, for example, we apply the equation of motion at the intermediate $n + \frac{1}{2}$ time level and obtain

$$\mathbf{u}^{n+1} = \mathbf{u}^n + \Delta t \, \tfrac{1}{2}[\mathbf{N}(\mathbf{u}^{n+1}) + \mathbf{N}(\mathbf{u}^n) + \nu L(\mathbf{u}^{n+1}) + \nu L(\mathbf{u}^n)] - \frac{\Delta t}{\rho} \nabla P^{n+1/2}$$

$$(13.3.21)$$

which is second-order accurate in time and enjoys an enhanced numerical stability. Since the method is implicit in the nonlinear terms, carrying out each time step requires solving a system of nonlinear algebraic equations for the velocity at the $n + 1$ time level and pressure at the $n + \frac{1}{2}$ level. In practice, this is done using an iterative method, as discussed by Peyret and Taylor (1983, p. 167).

PROBLEMS

13.3.1 **Pressure boundary conditions for flow in a cavity with a staggered grid.** Derive the counterparts of Eqs. (13.3.15) and (13.3.16) for (a) the first and last vertical layers corresponding to $i = 1, N$ and $j = 2, \ldots, M - 1$, (b) the four corner pressure nodes.

13.3.2 **Neumann boundary conditions for the pressure for flow in a cavity with a nonstaggered grid.** Derive the finite-difference statement of the Neumann boundary condition for the pressure for flow in a cavity on the non-staggered grid shown in Figure 13.1.1.

Computer Problems

13.3.3 **Flow in a rectangular cavity with a staggered grid.** Write a program called *CV2DSG* that uses the explicit method of Harlow and Welch on a staggered grid, as described in the text, to compute transient flow in a square cavity due to the impulsive translation of the lid discussed in Problem 13.1.5. The pressure Poisson equation should be solved using an iterative method of your choice, including the method of the routine RPPE described in Problem 13.2.3.

13.3.4 **Flow in a rectangular cavity with a non-staggered grid.** Repeat Problem 13.3.3 with a non-staggered grid.

13.4 | OPERATOR SPLITTING, PROJECTION, AND PRESSURE-CORRECTION METHODS

The Navier–Stokes equation states that the velocity at a particular point in a flow changes because of the simultaneous action of the nonlinear convection term, the linear viscous term, and the pressure gradient. In the operator splitting method, individual terms or groups of these terms are decoupled and considered to operate sequentially for time intervals of equal duration.

Since the absence of an evolution equation for the pressure is an important consideration, it is natural to decouple the convective–diffusive term from the pressure gradient term, thereby replacing Eq. (13.2.1) with the system of two component equations

$$\frac{\partial \mathbf{u}}{\partial t} = \mathbf{N}(\mathbf{u}) + \nu L(\mathbf{u}) \tag{13.4.1}$$

$$\frac{\partial \mathbf{u}}{\partial t} = -\frac{1}{\rho} \nabla P \tag{13.4.2}$$

which are assumed to apply sequentially, each one for the full time interval Δt within each time step.

The nonlinear convection–diffusion equation (13.4.1) advances the velocity from the initial state $\mathbf{u}^n$ to an intermediate state $\mathbf{u}^*$, and the pressure-correction equation (13.4.2) advances the intermediate state $\mathbf{u}^*$ to the final state $\mathbf{u}^{n+1}$ and completes the nth time step.

Solenoidal Projection and the Role of Pressure

To analyze the nature of the fractional-step decomposition expressed by Eqs. (13.4.1) and (13.4.2), it is helpful to introduce the concept of solenoidal projections (Chorin, 1968; see also Section 9.1 of this book). First we introduce the space of all vector functions, and note that the velocity field of an unsteady incompressible flow must evolve within the subspace of solenoidal functions. Next we observe that the intermediate velocity $\mathbf{u}^*$ at the end of the convection–diffusion step will not necessarily be solenoidal. The evolution according to Eq. (13.4.1) allows for a departure from the subspace of solenoidal functions, and the pressure-correction step makes up for this departure by projecting $\mathbf{u}^*$ back into the subspace of solenoidal functions, thus producing the final velocity $\mathbf{u}^{n+1}$.

Since the pressure P has not been updated during the convection–diffusion step, it loses its physical significance and must be regarded as an auxiliary function whose main purpose is to project $\mathbf{u}^*$ onto the subspace of solenoidal functions. Thus, to be rigorous, we must replace P in Eq. (13.4.2) with a projection function ϕ, obtaining

$$\frac{\partial \mathbf{u}}{\partial t} = -\frac{1}{\rho} \nabla \phi \tag{13.4.3}$$

The relation between P and ϕ is discussed in detail by Gresho (1990).

Boundary Conditions for the Intermediate Variables

An important issue in the implementation of the fractional-step method is the choice of boundary conditions for the intermediate velocity $\mathbf{u}^*$ and for the projection function ϕ. Ideally, the boundary conditions for $\mathbf{u}^*$ should be chosen so as to minimize its divergence, subject to the constraint that the required boundary conditions will be observed at the end of a complete time step.

The derivation of boundary conditions for $\mathbf{u}^*$ and ϕ has been the subject of extensive discussions (Orszag et al., 1986; Gresho, 1990). Gresho (1990) derived several sets of optimal and simplified sets of boundary conditions with varying degrees of accuracy, sophistication, and ease of implementation. In the simplest scheme, the intermediate velocity $\mathbf{u}^*$ satisfies the boundary conditions that are required at the $n + 1$ time level, while the projection function ϕ satisfies the homogeneous Neumann condition $\nabla \phi \cdot \mathbf{n} = 0$. Far from the boundaries, ϕ reduces to P.

First-Order Projection Method

We proceed now to discuss the implementation of a particular projection method developed by Chorin (1968). We begin by splitting the convection–diffusion step expressed by Eq. (13.4.1) into four sequential convection–diffusion fractional steps expressed by the equations

$$\frac{\partial \mathbf{u}}{\partial t} + u_x \frac{\partial \mathbf{u}}{\partial x} = \nu \frac{\partial^2 \mathbf{u}}{\partial x^2}, \qquad \frac{\partial \mathbf{u}}{\partial t} + u_y \frac{\partial \mathbf{u}}{\partial y} = \nu \frac{\partial^2 \mathbf{u}}{\partial y^2}$$
$$\frac{\partial \mathbf{u}}{\partial t} + u_z \frac{\partial \mathbf{u}}{\partial z} = \nu \frac{\partial^2 \mathbf{u}}{\partial z^2} \tag{13.4.4}$$

Each component equation operates for the full time interval Δt, and the time at the end of each fractional step is reset to the initial value t^n. The three intermediate velocity fields, denoted by $\mathbf{u}^{n+1/4}$, $\mathbf{u}^{n+2/4}$, $\mathbf{u}^{n+3/4} \equiv \mathbf{u}^*$, are not generally solenoidal. For two-dimensional flow, only the first two steps in Eq. (13.4.4) are present, and the intermediate velocity fields are $\mathbf{u}^{n+1/3}$ and $\mathbf{u}^{n+2/3} \equiv \mathbf{u}^*$.

The three steps in Eqs. (13.4.4) may be carried out using either the implicit BTCS method or the Crank–Nicolson method, with first- or second-order accuracy in time, respectively, both of which are unconditionally stable and require the relatively easy task of solving tridiagonal systems of equations. Setting the convection velocity in all three steps equal to the velocity at the beginning of the step, at the time level t^n, allows us to express the components of the convection–diffusion equations in conservative form. The boundary conditions for the intermediate velocity will be discussed shortly.

The projection step expressed by Eq. (13.4.3) is carried out using the BTCS method, yielding

$$\mathbf{u}^{n+1} = \mathbf{u}^* - \frac{\Delta t}{\rho} \nabla \phi^{n+1} \tag{13.4.5}$$

Computing the right-hand side of Eq. (13.4.5) requires a knowledge of the function ϕ at the $n + 1$ time level. This is obtained by taking the divergence of Eq. (13.4.5) and requiring that $\nabla \cdot \mathbf{u}^{n+1} = 0$, thereby deriving the following Poisson equation for ϕ^{n+1}

$$\nabla^2 \phi^{n+1} = \frac{\Delta t}{\rho} \nabla \cdot \mathbf{u}^* \tag{13.4.6}$$

which is to be solved subject to the boundary condition $\nabla \phi^{n+1} \cdot \mathbf{n} = 0$. As in the case of Poisson's equation for the pressure discussed in Section 13.3, the explicit implementation of this boundary condition may be circumvented either by using a staggered grid and requiring that $\nabla \cdot \mathbf{u}^{n+1}$ vanish at the pressure nodes that are adjacent to the boundaries, or by using a non-staggered grid and requiring that $\nabla \cdot \mathbf{u}^{n+1}$ vanish at the nodes that are located at the boundaries.

Given an initial velocity field along with a prescribed boundary condition $\mathbf{u} = \mathbf{U}$, we compute the evolution of the flow according to the following steps:

1. Assign the initial values of the velocity field to the velocity nodes, and provide an estimate for ϕ^{n+1} at the pressure nodes.

2. Advance the velocity field in a sequential manner according to the three equations (13.4.4) with boundary conditions: (a) $\mathbf{u}^* \cdot \mathbf{n} = \mathbf{U} \cdot \mathbf{n}$ for the normal component of the velocity, and (b) $\mathbf{u}^* \cdot (\mathbf{I} - \mathbf{nn}) = [\mathbf{U} + (\Delta t/\rho)\nabla\phi^{n+1}] \cdot (\mathbf{I} - \mathbf{nn})$ for the tangential component of the velocity. Here $\mathbf{I}$ is the identity matrix and $\mathbf{I} - \mathbf{nn}$ is the tangential projection operator.

3. Solve Poisson's equation (13.4.6) with boundary condition $\nabla\phi^{n+1} \cdot \mathbf{n} = 0$, or one of its modified versions. Since the flow rate of $\mathbf{u}^*$ and thus of $\nabla\phi$ across the boundaries vanishes, because of the boundary conditions required for $\mathbf{u}^*$, the compatibility condition is automatically fulfilled and Poisson's equation has an infinity of solutions.

4. Compute the velocity $\mathbf{u}^{n+1}$ at all internal and boundary grid points according to Eq. (13.4.5). If the tangential boundary velocity is not equal to that required by the boundary conditions, that is, if there is a finite slip velocity, return to step 2 and repeat the computations with the new boundary distribution of ϕ. Otherwise, proceed to step 5.

5. Set the time to t^{n+1}, return to step 2, and repeat the computations for another time step.

To further illustrate the implementation of the method on a non-staggered grid, consider the familiar problem of flow in a cavity driven by a moving lid depicted in Figure 13.1.1. The two convection–diffusion equations are integrated in time using an implicit method such as the Crank–Nicolson method. When integrating in the x direction, we use the boundary conditions

$$u_{1,j} = 0, \qquad v_{1,j} = \frac{\Delta t}{\rho}\left(\frac{\partial \phi^{n+1}}{\partial y}\right)_{1,j}$$

$$u_{N+1,j} = 0, \qquad v_{N+1,j} = \frac{\Delta t}{\rho}\left(\frac{\partial \phi^{n+1}}{\partial y}\right)_{N+1,j}$$

(13.4.7)

at the side walls; boundary conditions over the upper and lower wall are not required. When integrating in the y direction, we use the boundary conditions

$$u_{i,1} = \frac{\Delta t}{\rho}\left(\frac{\partial \phi^{n+1}}{\partial x}\right)_{i,1}, \qquad v_{i,1} = 0$$

$$u_{i,M+1} = U + \frac{\Delta t}{\rho}\left(\frac{\partial \phi^{n+1}}{\partial x}\right)_{i,M+1}, \qquad v_{i,M+1} = 0$$

(13.4.8)

at the upper and lower wall; boundary conditions over the side walls are not required.

A more advanced method of carrying out the fractional steps involves integrating the convection–diffusion equation (13.4.1) by means of the ADI method. This modification reduces the temporal error due to the spatial decoupling involved in Eqs. (13.4.4), and renders the convection–diffusion step second-order accurate in Δt.

A Semi-Implicit Three-Level Method

Kim and Moin (1985) proposed advancing the velocity field according to the two fractional steps

$$\frac{\mathbf{u}^* - \mathbf{u}^n}{\Delta t} = \tfrac{1}{2}[3\mathbf{N}(\mathbf{u}^n) - \mathbf{N}(\mathbf{u}^{n-1})] + \tfrac{1}{2}\nu[L(\mathbf{u}^*) + L(\mathbf{u}^n)] \tag{13.4.9}$$

$$\frac{\mathbf{u}^{n+1} - \mathbf{u}^*}{\Delta t} = -\frac{1}{\rho}\nabla\phi^{n+1} \tag{13.4.10}$$

Equation (13.4.9) uses the explicit second-order Adams–Bashforth method for the nonlinear term and the implicit second-order Crank–Nicolson method for the viscous term. The projection function ϕ is computed by solving Poisson's equation (13.4.6), which ensures that $\mathbf{u}^{n+1}$ will be solenoidal at the end of a complete step.

It is instructive to solve Eq. (13.4.10) for $\mathbf{u}^*$ and substitute the result back into Eq. (13.4.9). We thus find

$$\frac{\mathbf{u}^{n+1} - \mathbf{u}^n}{\Delta t} = -\frac{1}{\rho}\nabla(\phi^{n+1} + \tfrac{1}{2}\nu\,\Delta t\,\nabla^2\phi^{n+1}) + \tfrac{1}{2}(3\mathbf{N}^n - \mathbf{N}^{n-1}) + \tfrac{1}{2}\nu(L(\mathbf{u}^n) + L(\mathbf{u}^{n+1})) \tag{13.4.11}$$

which shows that the function $\phi + \tfrac{1}{2}\nu\,\Delta t\,\nabla^2\phi$ plays the role of the pressure, thereby making a clear distinction between P and ϕ. Defining $\phi + (\Delta t/2)\,\nabla^2\phi$ at the $n + 1$ time level to be equal to the pressure at the $n + \tfrac{1}{2}$ level suggests that Eq. (13.4.10) is second-order accurate in Δt.

To carry out the convection–diffusion step, we rewrite Eq. (13.4.9) in the form

$$(1 - \tfrac{1}{2}\Delta t\,\nu\,\nabla^2)(\mathbf{u}^* - \mathbf{u}^n) = \tfrac{1}{2}\Delta t\,[3\mathbf{N}(\mathbf{u}^n) - \mathbf{N}(\mathbf{u}^{n-1})] + \Delta t\,\nu\,\nabla^2\mathbf{u}^n \tag{13.4.12}$$

and then factorize the operator on the left-hand side in an approximate manner to obtain

$$\left(1 - \tfrac{1}{2}\Delta t\,\nu\,\frac{\partial^2}{\partial x^2}\right)\left(1 - \tfrac{1}{2}\Delta t\,\nu\,\frac{\partial^2}{\partial y^2}\right)\left(1 - \tfrac{1}{2}\Delta t\,\nu\,\frac{\partial^2}{\partial x^2}\right)(\mathbf{u}^* - \mathbf{u}^n)$$
$$= \tfrac{1}{2}\Delta t\,[3\,\mathbf{N}(\mathbf{u}^n) - \mathbf{N}(\mathbf{u}^{n-1})] + \Delta t\,\nu\,\nabla^2\mathbf{u}^n \tag{13.4.13}$$

To compute the difference $\mathbf{u}^* - \mathbf{u}^n$, we solve three tridiagonal systems of equations, which can be done using the efficient Thomas algorithm (Problem 13.4.2). The boundary conditions for $\mathbf{u}^*$ are the same as those discussed previously for the first-order method. Poisson's equation (13.4.6) may be solved on a staggered grid in order to avoid the explicit implementation of boundary conditions for ϕ.

A Second-Order Method

Bell, Colella, and Glaz (1989) developed a projection method that is second-order accurate in Δt. The method involves introducing values for the velocity and pressure at the integral time levels n as well as at the intermediate time levels $n + \tfrac{1}{2}$, computing the auxiliary velocity $\mathbf{u}^*$ on the basis of the semi-discrete version of the modified equation of motion

$$\frac{\mathbf{u}^* - \mathbf{u}^n}{\Delta t} = \mathbf{N}(\mathbf{u}^{n+1/2}) - \frac{1}{\rho}\nabla\psi + \tfrac{1}{2}\nu[L(\mathbf{u}^*) + L(\mathbf{u}^n)] \tag{13.4.14}$$

and then obtaining the velocity at the next time level by means of the projection

$$\mathbf{u}^{n+1} = \mathbf{u}^* - \frac{\Delta t}{\rho}\nabla\phi \tag{13.4.15}$$

where ψ and ϕ are two auxiliary functions. Solving Eq. (13.4.15) for $\mathbf{u}^*$ and substituting the result into Eq. (13.4.14) yields

$$\frac{\mathbf{u}^{n+1} - \mathbf{u}^n}{\Delta t} = \mathbf{N}(\mathbf{u}^{n+1/2}) - \frac{1}{\rho}\nabla(\psi + \phi) + \tfrac{1}{2}\nu[L(\mathbf{u}^{n+1}) + L(\mathbf{u}^n)]$$
$$+ \tfrac{1}{2}\mu\,\Delta t\,\nabla L(\phi) \tag{13.4.16}$$

which represents a second-order discretization of the equation of motion provided that (1) we set $P^{n+1/2} = \psi + \phi$ and (2) the last term on the right-hand side vanishes. The computational algorithm is designed so that progressively ψ tends to $P^{n+1/2}$ and ϕ tends to vanish, and is implemented according to the following steps:

1. Given the pressure $P^{n-1/2}$ and the velocity $\mathbf{u}^n$, estimate $\mathbf{u}^{n+1/2}$ by extrapolation and set $\psi = P^{n-1/2}$.

2. Calculate $\mathbf{N}(\mathbf{u}^{n+1/2})$, compute $\mathbf{u}^*$ from Eq. (13.4.14), and call the solution $\mathbf{u}^{*,k}$ where k is an inner iteration number; the first time, set $k = 0$.

3. Introduce the discrete form of the equation of motion

$$\frac{\mathbf{u}^{n+1,k} - \mathbf{u}^n}{\Delta t} = \mathbf{N}(\mathbf{u}^{n+1/2}) - \frac{1}{\rho}\nabla P^{n+1/2,k} + \tfrac{1}{2}\nu[L(\mathbf{u}^{*,k}) + L(\mathbf{u}^n)]$$

(13.4.17)

take its divergence, and require that $\mathbf{u}^{n+1,k}$ be solenoidal to obtain a Poisson equation for $P^{n+1/2,k}$. Solve the Poisson equation subject to appropriate boundary conditions and then use Eq. (13.4.17) to compute $\mathbf{u}^{n+1,k}$.

4. Return to step 2, set $\psi = P^{n+1/2,k}$, and repeat the computations with $\mathbf{u}^{n+1} = \mathbf{u}^{n+1,k}$, where $\mathbf{u}^{n+1/2}$ is computed by interpolation or extrapolation, increasing k by one.

PROBLEMS

13.4.1 Solenoidal projection. (a) Consider a nonsolenoidal rotational velocity field that is defined over the whole three-dimensional space and vanishes at infinity, and develop a procedure for removing the nonsolenoidal component while leaving the vorticity unaffected. (b) Consider part (a) for a bounded flow, and discuss the boundary conditions for the projection function.

13.4.2 Approximate factorization. Write the three tridiagonal systems of equations corresponding to the factorized form of Eq. (13.4.13).

Computer Problem

13.4.3 Flow in a rectangular cavity. Write a program called *CV2DPR1* that uses the first-order projection method described in the text to compute the transient flow in a rectangular cavity due to the impulsive translation of a lid described in Problem 13.1.5. The Poisson equation should be solved using the routine of Problem 13.2.3.

13.5 | METHODS OF MODIFIED DYNAMICS OR FALSE TRANSIENTS

All methods discussed in the preceding two sections involve (1) advancing the velocity field using a time-marching method that is suitable for parabolic differential equations, and (2) updating the pressure field or projection function by solving the elliptic Poisson equation. The origin of this dual procedure may be traced back to the absence of an evolution equation for the pressure in the original system of governing equations. The solution of the elliptic equation consumes much of the computational effort and inhibits the development of methods with second-order temporal accuracy. If all governing equations had the form of evolution equations, a simple time-marching method would suffice.

The idea behind the methods of *modified dynamics* is to amend either the continuity equation or the equation of motion with an objective to render all governing equations parabolic in time. In certain cases, the error introduced by modifying the original equations is mild, and the transient solution obtained by solving the modified problem describes the physical evolution with acceptable accuracy. In other cases, however, the transient evolution is purely fictitious and physically irrelevant, and hence it is significant only insofar as to provide us with a vehicle that will lead us to the steady state. The modified equations are designed so that their steady solutions satisfy the equations of steady incompressible Newtonian flow with a certain degree of accuracy.

Artificial Compressibility Method for Steady Flow

One way to render the set of governing equations parabolic in time is to transform the continuity equation into an evolution equation for the pressure. In the artificial compressibility method introduced by Chorin (1967), this is achieved by replacing the continuity equation with the modified evolution equation

$$\frac{\partial P}{\partial t} + \frac{\rho}{\delta} \nabla \cdot \mathbf{u} = 0 \tag{13.5.1}$$

where δ is a small positive constant called the *artificial compressibility*. Setting $P = \rho/\delta$, where ρ is the density of the fluid, makes Eq. (13.5.1) resemble the continuity equation for a compressible fluid. At steady state, the first term on the left-hand side of Eq. (13.5.1) vanishes, and this ensures that the steady solution satisfies the equations of steady incompressible flow.

Non-staggered grid

On a non-staggered grid, the modified continuity equation (13.5.1) is discretized using the standard central-time–central-space method. At the nodes that lie on the boundaries, the divergence of the velocity is computed using central differences for the velocity component that is tangential to the boundary, and first- or second-order one-sided differences for the velocity component that is normal to the boundary. Boundary conditions for the pressure are not required.

To expedite the approach toward steady state, it is desirable to use a larger time step, but then stability issues require the use of a semi-implicit or fully implicit method. As an alternative, Chorin (1967) implemented a variant of the explicit CTCS method, modified according to the DuFort–Frankel scheme. The finite-difference equation corresponding to the x component of the equation of motion at the (i, j) grid point is

$$u_{i,j}^{n+1} = u_{i,j}^{n-1} + 2\Delta t\left(N_x(\mathbf{u}^n) + \nu \frac{u_{i+1,j}^n - 2\frac{1}{2}(u_{i,j}^{n+1} + u_{i,j}^{n-1}) + u_{i-1,j}^n}{\Delta x^2}\right.$$
$$\left. + \nu \frac{u_{i,j+1}^n - 2\frac{1}{2}(u_{i,j}^{n+1} + u_{i,j}^{n-1}) + u_{i,j-1}^n}{\Delta y^2} - \frac{1}{\rho}\frac{\partial P^n}{\partial x}\right) \tag{13.5.2}$$

The nonlinear term $\mathbf{N}$ is discretized in its conservative form using central differences. The y component of the equation of motion is discretized in an analogous fashion (Peyret and Taylor, 1983, p. 157). In Chapter 12 we saw that the modified differential equation corresponding to the DuFort–Frankel approximation is not always consistent with the original differential equation, the difference being a small term involving a second partial derivative with respect to time. The modified and original equations, however, agree at steady state.

The method may be extended in a straightforward manner to three-dimensional flow, where the DuFort–Frankel modification is applied to the three second spatial derivatives. Chorin (1967) found that, given boundary conditions for the velocity and provided that the flow remains subsonic with respect to the artificial speed of sound $1/\delta^{1/2}$, the method will be stable as long as the maximum value of the magnitude of the Courant number is less than $2(\delta/n)^{1/2}/(1 + 5^{1/2})$, where $n = 2$ or 3 for two- and three-dimensional flow.

Other explicit or implicit implementations of the artificial compressibility method can be developed in a straightforward manner. Implementing explicit upwind methods for the convection term at high Reynolds numbers and implicit methods for the viscous term, in particular, allows the use of large time steps and accelerates the approach toward steady state.

Staggered grid

Next we consider the implementation of the method on the staggered grid shown in Figure 13.3.1. Adopting an explicit formulation, we advance the velocity at the x and y velocity nodes according to Eqs. (13.3.7) and (13.3.8) subject to the boundary conditions discussed in Section 13.3. To advance the pressure, we apply Eq. (13.5.1) at the pressure nodes and introduce the implicit BTCS discretization to obtain

$$P_{i,j}^{n+1} = P_{i,j}^n - \frac{\rho \Delta t}{\delta} \left(\frac{u_{i+1/2,j}^{n+1} - u_{i-1/2,j}^{n+1}}{\Delta x} + \frac{v_{i,j+1/2}^{n+1} - v_{i,j-1/2}^{n+1}}{\Delta y} \right) \qquad (13.5.3)$$

One notable feature of the artificial compressibility method is that boundary conditions for the pressure are not required. When, however, steady state is reached, the pressure will satisfy the pressure Poisson equation with boundary conditions that result from projecting the equation of motion onto a unit vector that is normal or tangential to the boundaries.

Modified PPE

Sotiropoulos and Abdallah (1990) modified the evolution equation for the rate of expansion (13.2.5), transforming it into an evolution equation for the pressure,

$$\frac{\partial P}{\partial t} = \frac{1}{\beta} \left(\nabla^2 P - \rho \nabla \cdot \mathbf{N}(\mathbf{u}) + \rho \frac{\partial \nabla \cdot \mathbf{u}}{\partial t} \right) \qquad (13.5.4)$$

where β is a positive constant. At steady state, Eq. (13.5.4) reduces to the familiar Poisson equation for the pressure. The transient solution lacks a physical meaning, and Eq. (13.5.4) is significant only insofar as to provide us with a route toward the steady state.

Penalty-Function Formulation

The penalty-function formulation uses an artificial constitutive equation for the pressure in terms of the rate of expansion,

$$P = -\frac{1}{\varepsilon} \nabla \cdot \mathbf{u} \qquad (13.5.5)$$

where ε is a small positive constant (Temam, 1968). Since the pressure is an order-one variable, the rate of expansion is restricted to remain small at all times. Substituting Eq. (13.5.5) into the equation of motion yields a modified evolution equation for the velocity alone. In practice, in order to render the method stable, it is necessary to enhance this equation with the addition of a small term involving the rate of expansion. The governing equation of motion is

$$\frac{\partial \mathbf{u}}{\partial t} = \mathbf{N}(\mathbf{u}) + \frac{1}{\rho \varepsilon} \nabla (\nabla \cdot \mathbf{u}) - \tfrac{1}{2} \mathbf{u} \nabla \cdot \mathbf{u} + \nu L(\mathbf{u}) \qquad (13.5.6)$$

The computations proceed by integrating Eq. (13.5.6) forward in time from a given initial state subject to the specified velocity boundary conditions. The penalty-function formulation has found extensive applications predominantly in numerical procedures based on finite-element methods (Hughes, Liu, and Brooks, 1979).

PROBLEM

13.5.1 **Artificial compressibility method.** (a) Write the equivalent of Eq. (13.5.2) for the x and y components of the velocity on a staggered grid. (b) Develop an ADI method for two-dimensional flow.

References

Abdallah, S., 1987, Numerical solutions for the incompressible Navier–Stokes equations in primitive variables using a non-staggered grid II. *J. Comp. Phys.* **70**, 193–202.

Anderson, C. R., 1989, Vorticity boundary conditions and boundary vorticity generation for two-dimensional viscous incompressible flows. *J. Comp. Phys.* **80**, 72–97.

Anderson, P. A., Tannehill, J. C., and Pletcher, R. H., 1984, *Computational Fluid Dynamics and Heat Transfer.* Taylor & Francis.

Aragbesola, Y. A. S., and Burley, D. M., 1977, The vector and scalar potential method for the numerical solution of two- and three-dimensional Navier–Stokes equations. *J. Comp. Phys.* **24**, 398–415.

Aziz, K., and Hellums, J. D., 1967, Numerical solution of the three-dimensional equations of motion for laminar natural convection. *Phys. Fluids* **10**, 314–24.

Babu, V., and Korpela, S., 1994, Numerical solution of the incompressible, three-dimensional Navier–Stokes equations. *Computers & Fluids* **23**, 675–91.

Bell, J., Colella, P., and Glaz, H., 1989, A second-order projection method for the incompressible Navier–Stokes equations *J. Comp. Phys.* **85**, 257–83.

Cebeci, T., 1982, *Numerical and Physical Aspects of Aerodynamic Flows.* Springer-Verlag.

Chien, J. C., 1976, A general finite-difference formulation with applications to Navier–Stokes equations. *J. Comp. Phys.* **20**, 268–78.

Chorin, A. J., 1967, A numerical method for solving incompressible viscous flow problems *J. Comput. Phys.* **2**, 12–26.

Chorin, A., 1968, Numerical solution of the Navier–Stokes equations. *Math. Comp.* **22**, 745–62.

Daube, O., 1992, Resolution of the 2D Navier–Stokes equations in velocity–vorticity form by means of an influence matrix technique. *J. Comp. Phys.* **103**, 402–14.

Dennis, S. C. R., Ingham, D. B., and Cook, R. N., 1979, Finite-difference methods for calculating steady incompressible flows in three dimensions. *J. Comp. Phys.* **33**, 325–39.

Fletcher, C. A. J. 1988, *Computational Techniques for Fluid Dynamics.* 2 volumes. Springer-Verlag.

Floryan, J. M., and Rasmussen, H., 1989, Numerical methods for viscous flows with moving boundaries. *Appl. Mech. Rev.* **42**, 323–41.

Gatski, T. B., Grosch, C. E., and Rose, M. E., 1989, The numerical solution of the Navier–Stokes equations for 3-dimensional, unsteady, incompressible flows by compact schemes. *J. Comp. Phys.* **82**, 298–329.

Ghia, K. N., Hankey, W. L., and Hodge, J. K., 1977, Study of incompressible Navier–Stokes equations in primitive variables using implicit numerical technique. *AIAA paper 77-648,* 156–65.

Gresho, P. M., 1990, On the theory of semi-implicit projection methods for viscous incompressible flow and its implementation via a finite element method that also introduces a nearly consistent mass matrix. Part 1: Theory. *Int. J. Num. Meth. Fluids* **11**, 587–620.

Gresho, P. M., 1991, Incompressible fluid dynamics: some fundamental formulation issues. *Ann. Rev. Fluid Mech.* 1991, **23**, 413–53.

Gresho, P. M., and Sani, R. L., 1987, On pressure boundary conditions for the incompressible Navier–Stokes equations. *Int. J. Num. Meth. Fluids* **7**, 1111–45.

Guj, G., and Stella, F., 1988, Numerical solutions of high-Re recirculating flows in vorticity–velocity form. *I. J. Num. Meth. Fluids* **8**, 405–16.

Guj, G., and Stella, F., 1993, A vorticity–velocity method for the numerical solution of 3D incompressible flows. *J. Comp. Phys.* **106**, 286–98.

Gupta, M. M., 1991, High-accuracy solutions of incompressible Navier–Stokes equations. *J. Comp. Phys.* **93**, 343–59.

Harlow, H. H., and Welch, J. E., 1965, Numerical calculation of time-dependent viscous incompressible flow of fluid with free surface. *Phys. Fluids* **8**, 2182–89.

Hirasaki, G. J., and Hellums, J. D., 1970, Boundary conditions on the vector and scalar potentials in viscous three-dimensional hydrodynamics. *Quart. Appl. Math.* **28**, 293–96.

Hirsch, C., 1988, *Numerical Computation of Internal and External Flows.* 2 volumes. Wiley.

Hoffman, J. D., 1992, *Numerical Methods for Engineers and Scientists.* McGraw–Hill.

Hoffman, K. A., and Chiang, S. T., 1993, Computational Fluid Dynamics for Engineers, Vols. I and II, Engineering Education System.

Hughes, T. J. R., Liu, W. K., and Brooks, A., 1979, Finite element analysis of incompressible viscous flows by the penalty function formulation. *J. Comp. Phys.* **30**, 1–60.

Israeli, M., 1972, On the evaluation of iteration parameters for the boundary vorticity. *Stud. Appl. Math.* **51**, 67–71.

Kim, J., and Moin, P., 1985, Application of a fractional-step method to incompressible Navier–Stokes equations. *J. Comp. Phys.* **59,** 308–23.

Mallinson, G. D., and de Vahl Davis, G., 1973, The method of false transient for the solution of coupled elliptic equations. *J. Comp. Phys.* **12,** 435–61.

Mallinson, G. D., and de Vahl Davis, G., 1977, Three-dimensional natural convection in a box: a numerical study. *J. Fluid Mech.* **83,** 1–31.

Mansour, M. L., and Hamed, A., 1990, Implicit solution of the incompressible Navier–Stokes equations on a non-staggered grid. *J. Comp. Phys.* **86,** 147–67.

Orszag, S. A., and Israeli, M., 1974, Numerical simulation of viscous incompressible flows. *Annu. Rev. Fluid Mech.* **5,** 281–318.

Orszag, S. A., Israeli, M., and Deville, M. O., 1986, Boundary conditions for incompressible flows. *J. Scient. Comp.* **1,** 75–111.

Osswald, G. A., Ghia, K. N., and Ghia, U., 1987, A direct algorithm for solution of incompressible three-dimensional unsteady Navier–Stokes equations. *AIAA paper 87-1139,* 408–21.

Peyret, R., and Taylor, T. D., 1983, *Computational Methods for Fluid Flow.* Springer-Verlag.

Quartapelle, L., 1981, Vorticity conditioning in the computation of two-dimensional viscous flows. *J. Comp. Phys.* **40,** 453–77.

Quartapelle, L., and Napolitano, M., 1986, Integral conditions for the pressure in the computation of incompressible viscous flows. *J. Comp. Phys.* **62,** 340–48.

Quartapelle, L., and Valz-Gris, F., 1981, Projection conditions on the vorticity in viscous incompressible flows. *Int. J. Num. Meth. Fluids* **1,** 129–44.

Roach, P. J., 1982, *Computational Fluid Dynamics.* Hermosa Publishers, P.O. Box 8172, Albuquerque, NM 87108.

Richardson, S. M., and Cornish, A. R. H., 1977, Solution of three-dimensional incompressible flow problems. *J. Fluid Mech.* **82,** 109–319.

Sheth, K., and Pozrikidis, C., 1995, Effects of inertia on the deformation of liquid drops in simple shear flow. *Computers & Fluids* **94,** 101–10.

Sotiropoulos, F., and Abdallah, S., 1990, Coupled fully implicit solution procedure for the steady incompressible Navier–Stokes equation. *J. Comp. Phys.* **87,** 328–48.

Sotiropoulos, F., and Abdallah, S., 1991, The discrete continuity equation in primitive variable solutions of incompressible flow. *J. Comp. Phys.* **95,** 212–27.

Speziale, C. G., 1987, On the advantages of the velocity–vorticity formulation of the equations of fluid dynamics. *J. Comp. Phys.* **73,** 476–80.

Temam, R., 1968, Une méthode d'approximation de la solution des équations de Navier–Stokes. *Bull. Soc. Math. France,* **96,** 115–52.

Trujillo, J. R., 1994, *Spectral element vorticity–velocity algorithm for the incompressible Navier–Stokes equations.* Doctoral dissertation, Princeton University.

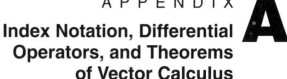

Index Notation, Differential Operators, and Theorems of Vector Calculus

A.1 | INDEX NOTATION

In index notation, a vector $\mathbf{u}$ is represented as u_i, a two-dimensional matrix $\mathbf{A}$ is represented as A_{ij}, and an N-dimensional matrix $\mathbf{B}$ is represented as $B_{ij\cdots m}$, where the number of subscripts i through m is equal to N. The same index may not appear more than twice in an index array or across index arrays in a product.

Repeated-Index Summation Convention

Einstein's summation convention reckons that, if a subscript appears twice in a variable or across a product of variables, then summation is implied over that subscript in its range. For example, under this convention

$$C_{ikl} \equiv A_{ij}B_{jkl} \equiv \sum_j A_{ij}B_{jkl}$$

$$u_i v_i \equiv u_1 v_1 + u_2 v_2 + \cdots + u_N v_N$$

$$A_{ii} \equiv A_{11} + A_{22} + \cdots A_{NN} = \text{Tr}(\mathbf{A})$$

The vector or matrix nature of an expression is determined by the number of its free indices, that is, the indices that appear only once. For example, $A_{ij}u_j$ is a vector, whereas $u_i u_j$ and $u_i A_{ijl}$ are two-dimensional matrices.

Identity Matrix

Kronecker's delta δ_{ij} represents the $N \times N$ identity or unit matrix; $\delta_{ij} = 0$ when $i \neq j$ and $\delta_{ij} = 1$ when $i = j$. On the basis of this definition we obtain

$$\delta_{ij} = \delta_{ji}, \qquad u_i\,\delta_{ij} = u_j, \qquad A_{ijk}\delta_{kl} = A_{ijl}$$

$$\delta_{kl}\,\delta_{lm}\,\delta_{mn} = \delta_{kn}, \qquad \delta_{ii} = N$$

If x_i is a set of N independent variables, then

$$\frac{\partial x_i}{\partial x_j} = \delta_{ij}, \qquad \frac{\partial x_i}{\partial x_i} = N$$

The Einstein summation convention is implied in the last case.

Alternating Matrix

The alternating matrix ε_{ijk}, where the indices take the values 1, 2, 3, is defined as follows: $\varepsilon_{ijk} = 0$ if two indices have the same values, and $\varepsilon_{ijk} = \pm 1$ otherwise. The positive value applies when the indices are arranged in a cyclic order, and the negative value otherwise. As a result of this definition

$$\varepsilon_{ijj} = 0, \qquad \varepsilon_{ijk} = \varepsilon_{kij} = \varepsilon_{jki} = -\varepsilon_{jik}$$

Two fundamental properties of ε_{ijk} are

$$\varepsilon_{ijk}\varepsilon_{ljk} = 2\delta_{il} \tag{A.1.1}$$

$$\varepsilon_{ijk}\varepsilon_{lmk} = \varepsilon_{kij}\varepsilon_{klm} = \delta_{il}\delta_{jm} - \delta_{im}\delta_{jl} \tag{A.1.2}$$

A 3×3 antisymmetric matrix A_{ij} has only three independent elements and may thus be represented in terms of a three-dimensional vector $\mathbf{v}$ as $A_{ij} = \varepsilon_{ijk}v_k$. Multiplying this definition by ε_{ijl} and using Eq.(A.1.1), we find $v_k = \frac{1}{2}\varepsilon_{kij}A_{ij}$.

Vector and Matrix Products

It is a standard practice in the theory of matrix calculus and numerical analysis to regard an N-dimensional vector $\mathbf{u}$ as a column vector, that is, an $N \times 1$ matrix. To obtain the corresponding row vector, which is a $1 \times N$ matrix, we take the transpose of the column vector to find the row vector $\mathbf{u}^T$, denoted by the superscript T. If $\mathbf{A}$ is an $N \times N$ square matrix, then the products $\mathbf{A}\mathbf{u}$ and $\mathbf{u}^T\mathbf{A}$, defined according to the usual rules of matrix multiplication, represent, respectively, an N-dimensional column and an N-dimensional row vector.

In fluid mechanics, in order to avoid using the superscript T, we designate the one-index product with a dot. For example, $\mathbf{v} = \mathbf{A}\cdot\mathbf{u}$ is equivalent to $v_i = A_{ij}u_j$, and $\mathbf{w} = \mathbf{u}\cdot\mathbf{A}$ is equivalent to $w_j = u_i A_{ij}$. The scalar $c = (\mathbf{A}\cdot\mathbf{u})\cdot\mathbf{v}$ is defined as $c = v_i A_{ij}u_j$.

If $\mathbf{u}$ and $\mathbf{v}$ are two N-dimensional vectors, then $\mathbf{A} = \mathbf{u}\mathbf{v}$ is a matrix with $A_{ij} = u_i v_j$. In contrast, $\mathbf{u}\cdot\mathbf{v} = v_i u_i$ is a scalar.

A.2 | VECTOR AND MATRIX PRODUCTS, DIFFERENTIAL OPERATORS IN CARTESIAN COORDINATES

In the following discussion we shall identify the components of a three-dimensional vector with its Cartesian coordinates in three-dimensional space. The unit vectors along the x, y, and z axes will be denoted, respectively, by $\mathbf{i}$, $\mathbf{j}$, and $\mathbf{k}$.

Scalar Product

The *scalar product* of two vectors $\mathbf{a}$ and $\mathbf{b}$ is defined as $\mathbf{a}\cdot\mathbf{b} = a_i b_i$. The scalar product of a vector with itself is a positive number that is equal to the square of the length of the vector.

Outer Product

The *outer* or *vector product* of $\mathbf{a}$ and $\mathbf{b}$ is a new vector $\mathbf{c} = \mathbf{a}\times\mathbf{b}$, where $c_i = \varepsilon_{ijk}a_j b_k$. Furthermore, $\mathbf{c}$ is given by determinant of the matrix

$$\begin{bmatrix} \mathbf{i} & \mathbf{j} & \mathbf{k} \\ a_x & a_y & a_z \\ b_x & b_y & b_z \end{bmatrix}$$

The vector $\mathbf{c}$ is perpendicular to both $\mathbf{a}$ and $\mathbf{b}$, and its length is equal to the area of the parallelepiped with two sides equal to $\mathbf{a}$ and $\mathbf{b}$. Thus, if two vectors are parallel to each other, their outer product vanishes. Furthermore, $\mathbf{a}\times\mathbf{b} = -\mathbf{b}\times\mathbf{a}$.

Triple Scalar Product

The triple scalar product of the triplet of vectors $\mathbf{a}$, $\mathbf{b}$, and $\mathbf{c}$ is a scalar defined as $[\mathbf{a},\mathbf{b},\mathbf{c}] \equiv (\mathbf{a}\times\mathbf{b})\cdot\mathbf{c}$. This is equal to the determinant of the matrix

$$\begin{bmatrix} a_x & a_y & a_z \\ b_x & b_y & b_z \\ c_x & c_y & c_z \end{bmatrix}$$

Cyclic permutation preserves the triple scalar product, that is, $[\mathbf{a},\mathbf{b},\mathbf{c}] = [\mathbf{c},\mathbf{a},\mathbf{b}] = [\mathbf{b},\mathbf{c},\mathbf{a}]$.

Triple Vector Product

The triple vector product of the triplet $\mathbf{a}$, $\mathbf{b}$, and $\mathbf{c}$ is the new vector $\mathbf{d} = \mathbf{a}\times(\mathbf{b}\times\mathbf{c})$. Writing

$$d_i = \varepsilon_{ijk}a_j(\varepsilon_{klm}b_lc_m) = \varepsilon_{ijk}\varepsilon_{klm}a_jb_lc_m = \varepsilon_{kij}\varepsilon_{klm}a_jb_lc_m$$

$$= (\delta_{il}\,\delta_{jm} - \delta_{im}\,\delta_{jl})a_jb_lc_m = a_jb_ic_j - a_jb_jc_i$$

produces the vector identity

$$\mathbf{a} \times (\mathbf{b} \times \mathbf{c}) = \mathbf{b}\,(\mathbf{a} \cdot \mathbf{c}) - \mathbf{c}\,(\mathbf{a} \cdot \mathbf{b}) \tag{A.2.1}$$

Working in a similar manner, we derive the identity

$$\mathbf{a} \times (\mathbf{b} \times \mathbf{c}) + \mathbf{c} \times (\mathbf{a} \times \mathbf{b}) + \mathbf{b} \times (\mathbf{c} \times \mathbf{a}) = \mathbf{0} \tag{A.2.2}$$

Components of a Vector

The component of a vector $\mathbf{a}$ in the direction of the unit vector $\mathbf{u}$ is given by $(\mathbf{a} \cdot \mathbf{u})\,\mathbf{u}$. The complementary component of $\mathbf{a}$ in the plane that is perpendicular to $\mathbf{u}$ is given by $\mathbf{a} \cdot (\mathbf{I} - \mathbf{uu}) = \mathbf{u} \times (\mathbf{a} \times \mathbf{u})$, where $\mathbf{I}$ is the identity matrix.

Vector-Matrix Outer Product

Given a three-dimensional vector $\mathbf{u}$ and a square 3×3 matrix $\mathbf{A}$, we define their *left outer* or *cross product* as the new matrix $\mathbf{B} = \mathbf{u} \times \mathbf{A}$, where

$$B_{ij} = \varepsilon_{ikl}u_kA_{lj}$$

Similarly, we define the *right outer product* as the new matrix $\mathbf{C} = \mathbf{A} \times \mathbf{u}$, where

$$C_{ij} = \varepsilon_{jkl}A_{ik}u_l$$

Double-Dot Product of Matrices

The *double-dot product* of a pair of two-dimensional matrices $\mathbf{A}$ and $\mathbf{B}$ is the scalar $s = \mathbf{A} : \mathbf{B} = A_{ij}B_{ij} = \mathrm{Tr}(\mathbf{B}^T \cdot \mathbf{A}) = \mathrm{Tr}(\mathbf{A}^T \cdot \mathbf{B})$. If $\mathbf{A}$ is a symmetric matrix and $\mathbf{B}$ is an antisymmetric matrix, then $\mathbf{A} : \mathbf{B} = 0$.

Del or Nabla Operator

The Cartesian components of the *del* or *nabla* operator ∇ are the partial derivatives with respect to the corresponding coordinates, $\nabla = (\partial/\partial x, \partial/\partial y, \partial/\partial z)$.

Gradient and Laplacian of a Scalar Function

If f is a scalar function of position $\mathbf{x}$, then its gradient ∇f is a vector defined as

$$\nabla f = \mathbf{i}\frac{\partial f}{\partial x} + \mathbf{j}\frac{\partial f}{\partial y} + \mathbf{k}\frac{\partial f}{\partial z} \tag{A.2.3}$$

The Laplacian of f is a scalar defined as

$$\nabla \cdot (\nabla f) \equiv \nabla^2 f = \frac{\partial^2 f}{\partial x^2} + \frac{\partial^2 f}{\partial y^2} + \frac{\partial^2 f}{\partial z^2} \tag{A.2.4}$$

Note that the Laplacian is equal to the divergence of the gradient, where the divergence is defined immediately below.

Divergence, Gradient, and Curl of a Vector Function

The divergence of a vector function of position $\mathbf{F}$ is a scalar defined as

$$\nabla \cdot \mathbf{F} = \frac{\partial F_i}{\partial x_i} = \frac{\partial F_x}{\partial x} + \frac{\partial F_y}{\partial y} + \frac{\partial F_z}{\partial z} \tag{A.2.5}$$

If $\nabla \cdot \mathbf{F}$ vanishes at every point, then $\mathbf{F}$ is called *solenoidal*.

The gradient of $\mathbf{F}$, denoted as $\mathbf{U} = \nabla\mathbf{F}$, is a two-dimensional matrix with

$$U_{ij} = \frac{\partial F_j}{\partial x_i}$$

The divergence of $\mathbf{F}$ is equal to the trace of $\mathbf{U}$.

The curl of $\mathbf{F}$, denoted as $\nabla \times \mathbf{F}$, is a vector computed according to the usual rules of the outer vector product, treating the del operator as a regular vector, yielding

$$\nabla \times \mathbf{F} = \mathbf{i}\left(\frac{\partial F_z}{\partial y} - \frac{\partial F_y}{\partial z}\right) + \mathbf{j}\left(\frac{\partial F_x}{\partial z} - \frac{\partial F_z}{\partial x}\right) + \mathbf{k}\left(\frac{\partial F_y}{\partial x} - \frac{\partial F_x}{\partial y}\right) \qquad (A.2.6)$$

If f is a scalar function and $\mathbf{F}$ and $\mathbf{G}$ are two vector functions, one can show by working in index notation that

$$\nabla \cdot (f\mathbf{F}) = f\nabla \cdot \mathbf{F} + \mathbf{F} \cdot \nabla f \qquad (A.2.7)$$

$$\nabla(\mathbf{F} \cdot \mathbf{G}) = \mathbf{F} \cdot \nabla\mathbf{G} + \mathbf{G} \cdot \nabla\mathbf{F} + \mathbf{F} \times (\nabla \times \mathbf{G}) + \mathbf{G} \times (\nabla \times \mathbf{F}) \qquad (A.2.8)$$

$$\nabla \cdot (\mathbf{F} \times \mathbf{G}) = \mathbf{G} \cdot \nabla \times \mathbf{F} - \mathbf{F} \cdot \nabla \times \mathbf{G} \qquad (A.2.9)$$

$$\nabla \times (\mathbf{F} \times \mathbf{G}) = \mathbf{F}\nabla \cdot \mathbf{G} - \mathbf{G}\nabla \cdot \mathbf{F} + \mathbf{G} \cdot \nabla\mathbf{F} - \mathbf{F} \cdot \nabla\mathbf{G} \qquad (A.2.10)$$

$$\nabla \times (\nabla f) = \mathbf{0} \qquad (A.2.11)$$

$$\nabla \cdot (\nabla \times \mathbf{F}) = 0 \qquad (A.2.12)$$

$$\nabla \times (\nabla \times \mathbf{F}) = \nabla(\nabla \cdot \mathbf{F}) - \nabla^2\mathbf{F} \qquad (A.2.13)$$

Divergence of a Matrix

The divergence $\mathbf{v} = \nabla \cdot \mathbf{Q}$ of a matrix function of position $\mathbf{Q}$ is a vector defined as

$$v_j = \frac{\partial Q_{ij}}{\partial x_i}$$

Directional Derivatives

The rate of change of the scalar and vector functions f and $\mathbf{F}$ with respect to arc length l_u measured in the direction of the *unit* vector $\mathbf{u}$ are given by

$$\frac{\partial f}{\partial l_u} = \mathbf{u} \cdot \nabla f = u_x \frac{\partial f}{\partial x} + u_y \frac{\partial f}{\partial y} + u_z \frac{\partial f}{\partial z} \qquad (A.2.14)$$

$$\frac{\partial \mathbf{F}}{\partial l_u} = \mathbf{u} \cdot \nabla\mathbf{F} = \frac{\partial \mathbf{F}}{\partial l_u} = u_x \frac{\partial \mathbf{F}}{\partial x} + u_y \frac{\partial \mathbf{F}}{\partial y} + u_z \frac{\partial \mathbf{F}}{\partial z} \qquad (A.2.15)$$

A.3 | ORTHOGONAL CURVILINEAR COORDINATES

Consider three distinct families of surfaces that fill the entire three-dimensional space, where two surfaces in the same family do not intersect, and introduce three continuous scalar variables a_1, a_2, or a_3 that are functions of position $\mathbf{x}$, so that only two of these variables vary over each surface and the third one remains constant. It is evident that the position vector $\mathbf{x}$ is a function of a_1, a_2, and a_3 and vice versa.

Placing the origin at an arbitrary point in space, we identify three surfaces passing through the origin, one surface from each family, and consider their intersections; two variables are constant along each intersection, and the third one varies in a continuous fashion. The vectors $\partial\mathbf{x}/\partial a_1$, $\partial\mathbf{x}/\partial a_2$, and $\partial\mathbf{x}/\partial a_3$ are tangential to the intersections at the origin and point in the directions of increasing a_1, a_2, or a_3; the corresponding unit vectors will be denoted by $\mathbf{e}_1$, $\mathbf{e}_2$, and $\mathbf{e}_3$. If these vectors form a right-handed system of Cartesian axes, then a_1, a_2, and a_3 are orthogonal curvilinear coordinates. In that case, the scalar product of any two of $\mathbf{e}_i$ with one another, evaluated at any point, will vanish.

A differential vector in space can be described as

$$dx = \frac{\partial x}{\partial a_i} \, da_i \tag{A.3.1}$$

and the square of its length is given by

$$dx \cdot dx = g_{ij} \, da_i \, da_j \tag{A.3.2}$$

where g is a *diagonal* metric tensor with

$$g_{ii} = \frac{\partial x}{\partial a_i} \cdot \frac{\partial x}{\partial a_i} \equiv h_i^2 \tag{A.3.3}$$

and summation is not implied over the repeated index i; the h_i are the *metric coefficients*.

The differential arc length along a line over which a_2 and a_3 are constant but a_1 varies is given by $dl_1 = h_1 \, da_1$; analogous relations pertain to the other two axes.

A differential volume in space is given by

$$dV = h_1 h_2 h_3 \, da_1 \, da_2 \, da_3 \tag{A.3.4}$$

The size of a differential surface element that lies in the surface over which either a_3 or a_2 or a_1 is constant is given, respectively, by

$$dS_1 = h_2 h_3 \, da_2 \, da_3, \qquad dS_2 = h_3 h_1 \, da_3 \, da_1, \qquad dS_3 = h_1 h_2 \, da_1 \, da_2 \tag{A.3.5}$$

Gradient and Laplacian of a Scalar Function

The gradient and Laplacian of a scalar function f are given by

$$\begin{aligned}
\nabla f &= e_1 \frac{\partial f}{\partial l_1} + e_2 \frac{\partial f}{\partial l_2} + e_3 \frac{\partial f}{\partial l_3} \\
&= e_1 \frac{1}{h_1} \frac{\partial f}{\partial a_1} + e_2 \frac{1}{h_2} \frac{\partial f}{\partial a_2} + e_3 \frac{1}{h_3} \frac{\partial f}{\partial a_3}
\end{aligned} \tag{A.3.6}$$

$$\begin{aligned}
\nabla^2 f &= \frac{1}{h_1 h_2 h_3} \left[\frac{\partial}{\partial a_1} \left(\frac{h_2 h_3}{h_1} \frac{\partial f}{\partial a_1} \right) + \frac{\partial}{\partial a_2} \left(\frac{h_3 h_1}{h_2} \frac{\partial f}{\partial a_3} \right) \right. \\
&\left. + \frac{\partial}{\partial a_3} \left(\frac{h_1 h_2}{h_3} \frac{\partial f}{\partial a_3} \right) \right]
\end{aligned} \tag{A.3.7}$$

Note that the Laplacian of f is equal to the divergence of its gradient, where the divergence is defined in Eq. (A.3.8).

Divergence and Curl of a Vector Function

The component of a vector function F in the direction of the unit vector e_i is given by the scalar product $F_i = F \cdot e_i$, so that $F = F_i e_i$.

The divergence of F is given by

$$\nabla \cdot F = \frac{1}{h_1 h_2 h_3} \left(\frac{\partial}{\partial a_1} (h_2 h_3 f_1) + \frac{\partial}{\partial a_2} (h_3 h_1 f_2) + \frac{\partial}{\partial a_3} (h_1 h_2 f_3) \right) \tag{A.3.8}$$

The curl of F is given by the determinant of the matrix

$$\frac{1}{h_1 h_2 h_3} \begin{bmatrix} h_1 e_1 & h_2 e_2 & h_3 e_3 \\ \dfrac{\partial}{\partial a_1} & \dfrac{\partial}{\partial a_2} & \dfrac{\partial}{\partial a_3} \\ h_1 F_1 & h_2 F_2 & h_3 F_3 \end{bmatrix} \tag{A.3.9}$$

where the gradient in the second row is treated as a regular vector.

A.4	**DIFFERENTIAL OPERATORS IN CYLINDRICAL AND PLANE POLAR COORDINATES**

The cylindrical polar coordinates (x, σ, φ), corresponding to the orthogonal curvilinear coordinates (a_1, a_2, a_3) discussed in Section A.3, are defined in Figure A.4.1(a). Note that the azimuthal angle φ revolves around the x axis. The base vectors and associated metric coefficients are $\mathbf{e}_x = \mathbf{i}$, $\mathbf{e}_\sigma$, and $\mathbf{e}_\varphi$, with $h_1 = 1$, $h_2 = 1$, and $h_3 = \sigma$.

Gradient and Laplacian of a Scalar Function

The gradient and Laplacian of a scalar function f are given by

$$\nabla f = \mathbf{e}_x \frac{\partial f}{\partial x} + \mathbf{e}_\sigma \frac{\partial f}{\partial \sigma} + \mathbf{e}_\varphi \frac{1}{\sigma} \frac{\partial f}{\partial \varphi} \tag{A.4.1}$$

$$\nabla^2 f = \frac{\partial^2 f}{\partial x^2} + \frac{1}{\sigma} \frac{\partial}{\partial \sigma} \left(\sigma \frac{\partial f}{\partial \sigma} \right) + \frac{1}{\sigma^2} \frac{\partial^2 f}{\partial \varphi^2} \tag{A.4.2}$$

Divergence, Curl, Laplacian, and Directional Derivative of a Vector Function

The divergence, curl, and Laplacian of a vector function $\mathbf{F}$ are given by

$$\nabla \cdot \mathbf{F} = \frac{\partial F_x}{\partial x} + \frac{1}{\sigma} \frac{\partial (\sigma F_\sigma)}{\partial \sigma} + \frac{1}{\sigma} \frac{\partial F_\varphi}{\partial \varphi} \tag{A.4.3}$$

$$\nabla \times \mathbf{F} = \mathbf{e}_x \frac{1}{\sigma} \left(\frac{\partial (\sigma F_\varphi)}{\partial \sigma} - \frac{\partial F_\sigma}{\partial \varphi} \right) + \mathbf{e}_\sigma \left(\frac{1}{\sigma} \frac{\partial F_x}{\partial \varphi} - \frac{\partial F_\varphi}{\partial x} \right)$$

$$+ \mathbf{e}_\varphi \left(\frac{\partial F_\sigma}{\partial x} - \frac{\partial F_x}{\partial \sigma} \right) \tag{A.4.4}$$

$$\nabla^2 \mathbf{F} = \mathbf{e}_x \nabla^2 F_x + \mathbf{e}_\sigma \left(\nabla^2 F_\sigma - \frac{F_\sigma}{\sigma^2} - \frac{2}{\sigma^2} \frac{\partial F_\varphi}{\partial \varphi} \right)$$

$$+ \mathbf{e}_\varphi \left(\nabla^2 F_\varphi + \frac{2}{\sigma^2} \frac{\partial F_\sigma}{\partial \varphi} - \frac{F_\varphi}{\sigma^2} \right) \tag{A.4.5}$$

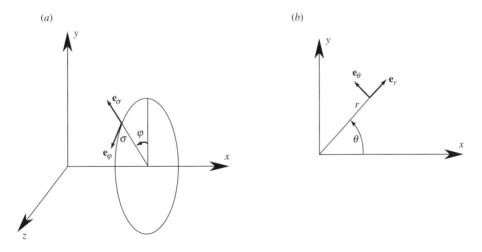

Figure A.4.1   (a) Cylindrical and (b) plane polar coordinates.

The derivative of the function $\mathbf{F}$ in the direction of the unit vector $\mathbf{u}$ is given by

$$\mathbf{u} \cdot \nabla \mathbf{F} = \frac{\partial \mathbf{F}}{\partial l_u} = \mathbf{e}_x(\mathbf{u} \cdot \nabla F_x) + \mathbf{e}_\sigma \left(\mathbf{u} \cdot \nabla F_\sigma - \frac{u_\varphi F_\varphi}{\sigma} \right)$$

$$+ \mathbf{e}_\varphi \left(\mathbf{u} \cdot \nabla F_\varphi + \frac{u_\varphi F_\sigma}{\sigma} \right) \tag{A.4.6}$$

where the cylindrical polar coordinates of the gradient of the components of $\mathbf{F}$ are computed using Eq. (A.4.1).

Plane Polar Coordinates

The differential operators in plane polar coordinates (r, θ), shown in Figure A.4.1(b), derive from those in cylindrical polar coordinates by discarding the dependence on x and renaming σ as r and φ as θ. The gradient and Laplacian of a function f are thus given by

$$\nabla f = \mathbf{e}_r \frac{\partial f}{\partial r} + \mathbf{e}_\theta \frac{1}{r} \frac{\partial f}{\partial \theta} \tag{A.4.7}$$

$$\nabla^2 f = \frac{1}{r} \frac{\partial}{\partial r} \left(r \frac{\partial f}{\partial r} \right) + \frac{1}{r^2} \frac{\partial^2 f}{\partial \theta^2} \tag{A.4.8}$$

where $\mathbf{e}_r$ and $\mathbf{e}_\theta$ are the unit vectors in the radial and angular directions.

The divergence, curl, and Laplacian of a vector function $\mathbf{F}$ that lies in the xy plane are given by

$$\nabla \cdot \mathbf{F} = \frac{1}{r} \frac{\partial(rF_r)}{\partial r} + \frac{1}{r} \frac{\partial F_\theta}{\partial \theta} \tag{A.4.9}$$

$$\nabla \times \mathbf{F} = \mathbf{e}_r \times \mathbf{e}_\theta \frac{1}{r} \left(\frac{\partial(rF_\theta)}{\partial r} - \frac{\partial F_r}{\partial \theta} \right) \tag{A.4.10}$$

$$\nabla^2 \mathbf{F} = \mathbf{e}_r \left(\nabla^2 F_r - \frac{F_r}{r^2} - \frac{2}{r^2} \frac{\partial F_\theta}{\partial \theta} \right) + \mathbf{e}_\theta \left(\nabla^2 F_\theta + \frac{2}{r^2} \frac{\partial F_r}{\partial \theta} - \frac{F_\theta}{r^2} \right) \tag{A.4.11}$$

The derivative of the vector $\mathbf{F}$ in the direction of the unit vector $\mathbf{u}$ is given by

$$\mathbf{u} \cdot \nabla \mathbf{F} = \frac{\partial \mathbf{F}}{\partial l_u} = \mathbf{e}_r \left(\mathbf{u} \cdot \nabla F_r - \frac{u_\theta F_\theta}{r} \right) + \mathbf{e}_\theta \left(\mathbf{u} \cdot \nabla F_\theta + \frac{u_\theta F_r}{r} \right) \tag{A.4.12}$$

where the plane polar coordinates of the gradient of the components of $\mathbf{F}$ are computed using Eq. (A.4.7).

A.5 | DIFFERENTIAL OPERATORS IN SPHERICAL POLAR COORDINATES

The spherical polar coordinates (r, θ, φ), corresponding to the orthogonal curvilinear coordinates (a_1, a_2, a_3) discussed in Section A.3, are defined in Figure A.5.1. The radial, meridional, and azimuthal base vectors and associated metric coefficients are $\mathbf{e}_r, \mathbf{e}_\theta,$ and $\mathbf{e}_\varphi$, with $h_1 = 1, h_2 = r,$ and $h_3 = r \sin \theta$.

The gradient and Laplacian of a scalar function f are given by

$$\nabla f = \mathbf{e}_r \frac{\partial f}{\partial r} + \mathbf{e}_\theta \frac{1}{r} \frac{\partial f}{\partial \theta} + \mathbf{e}_\varphi \frac{1}{r \sin \theta} \frac{\partial f}{\partial \varphi} \tag{A.5.1}$$

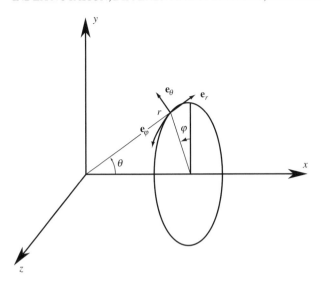

Figure A.5.1   Spherical polar coordinates.

$$\nabla^2 f = \frac{1}{r^2}\frac{\partial}{\partial r}\left(r^2\frac{\partial f}{\partial r}\right) + \frac{1}{r^2\sin\theta}\frac{\partial}{\partial\theta}\left(\sin\theta\frac{\partial f}{\partial\theta}\right) + \frac{1}{r^2\sin^2\theta}\frac{\partial^2 f}{\partial\varphi^2} \tag{A.5.2}$$

The *divergence, curl,* and *Laplacian* of a vector function **F** are given by

$$\nabla\cdot\mathbf{F} = \frac{1}{r^2}\frac{\partial(r^2 F_r)}{\partial r} + \frac{1}{r\sin\theta}\frac{\partial(\sin\theta\, F_\theta)}{\partial\theta} + \frac{1}{r\sin\theta}\frac{\partial F_\varphi}{\partial\varphi} \tag{A.5.3}$$

$$\nabla\times\mathbf{F} = \mathbf{e}_r\frac{1}{r\sin\theta}\left(\frac{\partial(\sin\theta\, F_\varphi)}{\partial\theta} - \frac{\partial F_\theta}{\partial\varphi}\right) + \mathbf{e}_\theta\frac{1}{r}\left(\frac{1}{\sin\theta}\frac{\partial F_r}{\partial\varphi} - \frac{\partial(rF_\varphi)}{\partial r}\right)$$

$$+ \mathbf{e}_\varphi\frac{1}{r}\left(\frac{\partial(rF_\theta)}{\partial r} - \frac{\partial F_r}{\partial\theta}\right) \tag{A.5.4}$$

$$\nabla^2\mathbf{F} = \mathbf{e}_r\left[\nabla^2 F_r - \frac{2}{r^2}F_r - \frac{2}{r^2\sin\theta}\left(\frac{\partial(\sin\theta\, F_\theta)}{\partial\theta} + \frac{\partial F_\varphi}{\partial\varphi}\right)\right]$$

$$+ \mathbf{e}_\theta\left[\nabla^2 F_\theta + \frac{2}{r^2}\frac{\partial F_r}{\partial\theta} - \frac{1}{r^2\sin\theta}\left(F_\theta + 2\cos\theta\frac{\partial F_\varphi}{\partial\varphi}\right)\right]$$

$$+ \mathbf{e}_\varphi\left[\nabla^2 F_\varphi + \frac{1}{r^2\sin\theta}\left(2\frac{\partial F_r}{\partial\theta} + 2\cos\theta\frac{\partial F_\theta}{\partial\varphi} - F_\varphi\right)\right] \tag{A.5.5}$$

The derivative of the vector **F** in the direction of the unit vector **u** is given by

$$\mathbf{u}\cdot\nabla\mathbf{F} = \frac{\partial\mathbf{F}}{\partial l_u} = \mathbf{e}_r\left(\mathbf{u}\cdot\nabla F_r - \frac{1}{r}(u_\theta F_\theta + u_\varphi F_\varphi)\right)$$

$$+ \mathbf{e}_\theta\left(\mathbf{u}\cdot\nabla F_\theta + \frac{1}{r}(u_\theta F_r - u_\varphi F_\varphi\cot\theta)\right)$$

$$+ \mathbf{e}_\varphi\left(\mathbf{u}\cdot\nabla F_\varphi + \frac{1}{r}u_\varphi(F_r + F_\theta\cot\theta)\right) \tag{A.5.6}$$

where the spherical polar coordinates of the gradient of the components of **F** are computed using Eq. (A.5.1).

A.6 | INTEGRAL THEOREMS OF VECTOR CALCULUS

Divergence or Gauss Theorem

Let V_c be an arbitrary closed control volume that is bounded by the surface D, and $\mathbf{n}$ be the unit vector that is normal to D pointing outward from V_c. The Gauss divergence theorem states that the volume integral of the divergence of any differentiable vector function $\mathbf{F}$ over V_c is equal to the flow rate of $\mathbf{F}$ across D

$$\int_{V_c} \nabla \cdot \mathbf{F} \, dV = \int_{D} \mathbf{F} \cdot \mathbf{n} \, dS \tag{A.6.1}$$

Making the three sequential choices $\mathbf{F} = (f, 0, 0)$, $\mathbf{F} = (0, f, 0)$, $\mathbf{F} = (0, 0, f)$, where f is a differentiable scalar function, we obtain the vector form of the divergence theorem

$$\int_{V_c} \nabla f \, dV = \int_{D} f \mathbf{n} \, dS \tag{A.6.2}$$

The particular choices $f = x$, $f = y$, or $f = z$ yield the volume of V_c in terms of a surface integral of the x, y, or z component of the normal vector.

Setting $\mathbf{F} = \mathbf{a} \times \mathbf{G}$, where $\mathbf{a}$ is a constant vector and $\mathbf{G}$ is a differentiable function, and then discarding the arbitrary constant $\mathbf{a}$, we obtain the new identity

$$\int_{V_c} \nabla \times \mathbf{G} \, dV = \int_{D} \mathbf{n} \times \mathbf{G} \, dS \tag{A.6.3}$$

Stokes's Theorem

Let C be an arbitrary closed loop with unit tangential vector $\mathbf{t}$, D be an arbitrary surface that is bounded by C, and $\mathbf{n}$ be the unit vector that is normal to D and is oriented according to the right-handed convention with respect to $\mathbf{t}$ and with reference to a designated side of D. Stokes's theorem states that the circulation of any differentiable vector function $\mathbf{F}$ along C is equal to the flow rate of the curl of $\mathbf{F}$ across D

$$\int_{C} \mathbf{F} \cdot \mathbf{t} \, dl = \int_{D} (\nabla \times \mathbf{F}) \cdot \mathbf{n} \, dS \tag{A.6.4}$$

where l is the arc length around C.

Setting $\mathbf{F} = \mathbf{a} \times \mathbf{G}$, where $\mathbf{a}$ is a constant vector and $\mathbf{G}$ is a differentiable function, expanding the right-hand side of Eq. (A.6.4), and then discarding the arbitrary constant $\mathbf{a}$, we obtain the new identity

$$\int_{C} \mathbf{G} \times \mathbf{t} \, dl = \int_{D} [\mathbf{n} \nabla \cdot \mathbf{G} - (\nabla \mathbf{G}) \cdot \mathbf{n}] \, dS \tag{A.6.5}$$

APPENDIX B

Primer of Numerical Methods

B.1 | LINEAR ALGEBRAIC EQUATIONS

We seek to compute a vector $\mathbf{x}$ which, when multiplied by the square matrix $\mathbf{A}$, yields the known vector $\mathbf{b}$; that is, $\mathbf{A} \cdot \mathbf{x} = \mathbf{b}$. In the ensuing discussion summation will *not* be implied over a repeated index but will be stated explicitly when required.

Diagonal and Triangular Systems

When the matrix $\mathbf{A}$ is *diagonal,* the unknown vector $\mathbf{x}$ is computed by the simple algorithm

$$x_i = \frac{b_i}{A_{ii}} \tag{B.1.1}$$

When $\mathbf{A}$ is *lower triangular,* we use the *forward substitution* algorithm

$$x_1 = \frac{b_1}{A_{11}}, \qquad x_i = \frac{1}{A_{ii}}\left(b_i - \sum_{j=1}^{i-1} A_{ij}x_j \right), \qquad \text{for } i = 2, \ldots, N \tag{B.1.2}$$

When $\mathbf{A}$ is *upper triangular,* we use the *backward substitution* algorithm

$$x_N = \frac{b_N}{A_{NN}}, \qquad x_i = \frac{1}{A_{ii}}\left(b_i - \sum_{j=i+1}^{N} A_{ij}x_j \right), \qquad \text{for } i = N-1, \ldots, 1 \tag{B.1.3}$$

Note that, in this case, the last unknown is computed first.

Gauss Elimination and LU Decomposition

When $\mathbf{A}$ does not have a particular structure, the solution can be found using either a direct or an iterative method; the second class of methods are designated for systems of large size involving sparse matrices.

Gauss elimination is the simplest and most popular direct method. The associated algorithm with *row pivoting* proceeds according to the following steps:

Setting up

0. Form the $N \times (N+1)$ partitioned *augmented matrix* $\mathbf{C}^{(1)} \equiv [\mathbf{A} \mid \mathbf{b}]$ and introduce the matrix $\mathbf{L}$ whose elements are initialized to zero.

First pass

1. Assume that the maximum absolute value of the elements in the first column $|C_{i1}^{(1)}|$, $i = 1, \ldots, N$ is $|C_{k1}^{(1)}|$, corresponding to the kth row.

2. Interchange the first row with the kth row of $\mathbf{C}^{(1)}$; repeat for $\mathbf{L}$. If $k = 1$, skip this step.

3. Re-evaluate the first column of $\mathbf{L}$, setting $L_{i1} = C_{i1}^{(1)}/C_{11}^{(1)}$ for $i = 2, \ldots, N$.

4. Subtract from the ith row of $\mathbf{C}^{(1)}$ the first row multiplied by L_{i1} for $i = 2, \ldots, N$. This operation yields the new augmented matrix $\mathbf{C}^{(2)} \equiv [\mathbf{A}^{(2)} \mid \mathbf{b}^{(2)}]$.

Second pass

5. Assume that the maximum absolute value of the elements in the truncated second column $|C_{i2}^{(2)}|$, over $i = 2, \ldots, N$ is $|C_{k2}^{(2)}|$, corresponding to the kth row.

6. Interchange the second row with the kth row of $\mathbf{C}^{(2)}$; repeat for $\mathbf{L}$. If $k = 2$, skip this step.

7. Re-evaluate the second column of $\mathbf{L}$, setting $L_{i2} = C_{i2}^{(1)}/C_{22}^{(1)}$ for $i = 3, \ldots, N$.

8. Subtract from the ith row of $\mathbf{C}^{(2)}$ the first row multiplied by L_{i2} for $i = 3, \ldots, N$. This operation yields the new augmented matrix $\mathbf{C}^{(3)} \equiv [\mathbf{C}^{(3)} \mid \mathbf{b}^{(3)}]$.

$\vdots$

mth pass

4m − 3. Assume that the maximum absolute value of the elements in the truncated mth column $|C_{im}^{(m)}|$, over $i = m, \ldots, N$ is $|C_{km}^{(m)}|$, corresponding to the kth row.

4m − 2. Interchange the mth row with the kth row of $\mathbf{C}^{(m)}$; repeat for $\mathbf{L}$. If $k = m$, skip this step.

4m − 1. Re-evaluate the mth column of $\mathbf{L}$, setting $L_{im} = C_{im}^{(m)}/C_{mm}^{(1)}$ for $i = m + 1, \ldots, N$.

4m. Subtract from the ith row of $\mathbf{C}^{(m)}$ the mth row multiplied by L_{im} for $i = m + 1, \ldots, N$. This operation yields the new augmented matrix $\mathbf{C}^{(m+1)} \equiv [\mathbf{A}^{(m+1)} \mid \mathbf{b}^{(m+1)}]$.

In the end of the $N - 1$ pass, corresponding to $m = N - 1$, the augmented matrix $\mathbf{C}^{(N)}$ will have the form $\mathbf{B}^{(N)} \equiv [\mathbf{A}^{(N)} \mid \mathbf{b}^{(N)}]$, where $\mathbf{A}^{(N)} \equiv \mathbf{U}$ is an upper triangular matrix. Solve the upper triangular system $\mathbf{U} \cdot \mathbf{x} = \mathbf{b}^{(N)}$ using the backward substitution algorithm to obtain the solution of the original system of equations. Furthermore, replace the zeros at the diagonal line of $\mathbf{L}$ with ones.

LU decomposition

The matrices $\mathbf{L}$ and $\mathbf{U}$ provide us with the **LU** decomposition of $\mathbf{A}$, that is, $\mathbf{L} \cdot \mathbf{U} = \mathbf{A}^{\text{Mod}}$, where $\mathbf{A}^{\text{Mod}}$ is identical to $\mathbf{A}$, except that the rows might have been switched due to pivoting. If pivoting is not done, $\mathbf{A}^{\text{Mod}} = \mathbf{A}$.

The determinant of $\mathbf{A}$ follows from $\pm \text{Det}(\mathbf{A}) = \text{Det}(\mathbf{A}^{\text{Mod}}) = \text{Det}(\mathbf{L}) \text{Det}(\mathbf{U}) = U_{11} U_{22} \cdots U_{NN}$, where the plus sign applies when an even number of row interchanges have been done due to pivoting, and the minus sign otherwise.

There are other algorithms that perform the **LU** decomposition of a matrix directly without a reference to a particular linear system of equations; examples are Crout's and Doolittle's algorithms (Dahlquist and Björck, 1974, p. 157). Once the $\mathbf{L}$ and $\mathbf{U}$ matrices have been obtained, the solution of the linear system $\mathbf{A} \cdot \mathbf{x} = \mathbf{b}$ is found in two stages: Solve $\mathbf{L} \cdot \mathbf{y} = \mathbf{b}$ for $\mathbf{y}$ using the forward-substitution algorithm, and then solve $\mathbf{U} \cdot \mathbf{x} = \mathbf{y}$ for $\mathbf{x}$ using the backward-substitution algorithm.

Thomas Algorithm for Tridiagonal Systems

The *Thomas algorithm* converts the tridiagonal system $\mathbf{A} \cdot \mathbf{x} = \mathbf{b}$ to the upper bidiagonal system $\mathbf{C} \cdot \mathbf{x} = \mathbf{d}$, where the diagonal elements of the matrix $\mathbf{C}$ are equal to unity. The upper diagonal elements of $\mathbf{C}$ and constant vector $\mathbf{d}$ are computed according to the algorithm

$$\delta_1 = \frac{\beta_1}{\alpha_1}, \qquad d_1 = \frac{b_1}{\alpha_1},$$

$$\delta_{i+1} = \frac{\beta_{i+1}}{\alpha_{i+1} - \gamma_{i+1}\delta_i} \qquad d_{i+1} = \frac{b_{i+1} - \gamma_{i+1}d_i}{\alpha_{i+1} - \gamma_{i+1}\delta_i}, \qquad i = 1, \ldots, N-1 \tag{B.1.4}$$

where we have defined

$$\alpha_i = A_{i,i}, \qquad \beta_i = A_{i,i+1}, \qquad \gamma_i = A_{i,i-1}, \qquad \delta_i = C_{i,i+1}$$

Once C and d have been obtained, the solution is found by applying the backward-substitution algorithm according to which

$$x_N = d_N, \qquad \text{and} \qquad x_i = d_i - \delta_i x_{i+1}, \qquad \text{for } i = N-1, \ldots, 1 \tag{B.1.5}$$

The Thomas algorithm can be generalized to pentadiagonal, multi-diagonal, and block tridiagonal systems (Fletcher 1988, vol. I, pp. 187–9).

Fixed-Point Iterations and Successive Substitutions

Fixed-point iteration or successive-substitution algorithms arise by recasting the original linear system $A \cdot x = b$ into the form

$$M \cdot x = N \cdot x + b \qquad \text{or} \qquad x = P \cdot x + c \tag{B.1.6}$$

where $A = M - N$ is a certain splitting of the matrix A, $P = M^{-1} \cdot N$, and $c = M^{-1} \cdot b$. The algorithm involves computing the sequence of vectors $x^{(k)}$ on the basis of the equations

$$M \cdot x^{(k+1)} = N \cdot x^{(k)} + b \qquad \text{or} \qquad x^{(k+1)} = P \cdot x^{(k)} + c \tag{B.1.7}$$

beginning with a certain initial vector $x^{(0)}$. If the spectral radius of P is less than unity, that is, the magnitude of each eigenvalue of P is less than unity, the sequence of vectors will converge to the fixed point of the mapping, denoted by X, that satisfies the equations $M \cdot X = N \cdot X + b$, $X = P \cdot X + c$ and therefore $A \cdot X = b$. In that case, the sequence $x^{(k)}$ contains successive approximations to the solution.

Jacobi's method

In Jacobi's method the iteration matrix P and constant vector c are constructed by solving the individual scalar equations of $A \cdot x = b$ for the diagonal unknowns, yielding

$$x_i^{(k+1)} = \frac{1}{A_{ii}} \left(b_i - \sum_{\substack{j=1 \\ j \neq i}}^{N} A_{ij} x_j^{(k)} \right) \tag{B.1.8}$$

The matrix M defined in Eqs. (B.1.6) is the diagonal part of A, and the matrix N, also defined in Eq. (B.1.6), is the negative of A with the diagonal elements set equal to zero. Furthermore, $P_{ij} = -A_{ij}/A_{ii}$ for $i \neq j$, $P_{ij} = 0$ for $i = j$, and $c_i = b_i/A_{ii}$; summation is *not* implied over the repeated index i.

A *sufficient* but not necessary condition for the successive substitutions to converge is that the matrix A be diagonally dominant, that is,

$$|A_{ii}| > \sum_{\substack{j=1 \\ j \neq i}}^{N} |A_{ij}| \tag{B.1.9}$$

for any i; summation over i is *not* implied on the left-hand side.

Gauss–Seidel method

The Gauss–Seidel method is a variation of Jacobi's method. Its distinguishing feature is that the components of $\mathbf{x}^{(k+1)}$ replace the corresponding components of $\mathbf{x}^{(k)}$ as soon as they are computed. The effective algorithm is

$$x_i^{(k+1)} = \frac{1}{A_{ii}} \left(b_i - \sum_{j=1}^{i-1} A_{ij} x_j^{(k+1)} - \sum_{j=i+1}^{N} A_{ij} x_j^{(k)} \right) \tag{B.1.10}$$

The associated mapping takes the form

$$(\mathbf{D} + \mathbf{L}) \cdot \mathbf{x}^{(k+1)} = -\mathbf{U} \cdot \mathbf{x}^{(k)} + \mathbf{b} \qquad \text{or} \qquad \mathbf{x}^{(k+1)} = \mathbf{P} \cdot \mathbf{x}^{(k)} + \mathbf{c} \tag{B.1.11}$$

where

$$\mathbf{P} = -(\mathbf{D} + \mathbf{L})^{-1} \cdot \mathbf{U}, \qquad \mathbf{c} = (\mathbf{D} + \mathbf{L})^{-1} \cdot \mathbf{b} \tag{B.1.12}$$

$\mathbf{D}$ is a diagonal matrix, $\mathbf{L}$ is a lower triangular matrix, and $\mathbf{U}$ is an upper triangular matrix, containing, respectively, the diagonal, lower triangular, and upper triangular parts of $\mathbf{A}$.

A *sufficient* but not necessary condition for the successive substitutions to converge is that the matrix $\mathbf{A}$ be symmetric and positive definite (Golub and Van Loan, 1989, p. 509).

Successive over-relaxation

We begin developing this method by recasting the Jacobi algorithm into the residual-correction form

$$\mathbf{x}^{(k+1)} = \mathbf{x}^{(k)} + \mathbf{r}^{(k)}, \qquad \text{where} \quad \mathbf{r}^{(k)} = (\mathbf{P} - \mathbf{I}) \cdot \mathbf{x}^{(k)} + \mathbf{c} \tag{B.1.13}$$

and then control the correction by introducing the relaxation parameter ω, setting

$$\mathbf{x}^{(k+1)} = \mathbf{x}^{(k)} + \omega \mathbf{r}^{(k)} \tag{B.1.14}$$

The associated algorithm is expressed by

$$x_i^{(k+1)} = (1 - \omega) x_i^{(k)} + \frac{\omega}{A_{ii}} \left(b_i - \sum_{\substack{j=1 \\ j \neq i}}^{N} A_{ij} x_j^{(k)} \right) \tag{B.1.15}$$

When $\omega = 1$ we obtain Eq. (B.1.8).

Updating the components of $\mathbf{x}^{(k)}$ as soon as their new values are available in the spirit of the Gauss–Seidel method yields the Successive Over-Relaxation (SOR) algorithm

$$x_i^{(k+1)} = (1 - \omega) x_i^{(k)} + \frac{\omega}{A_{ii}} \left(b_i - \sum_{j=1}^{i-1} A_{ij} x_j^{(k+1)} - \sum_{j=i+1}^{N} A_{ij} x_j^{(k)} \right) \tag{B.1.16}$$

which can be expressed as

$$(\mathbf{D} + \omega \mathbf{L}) \cdot \mathbf{x}^{(k+1)} = [(1 - \omega)\mathbf{D} - \omega \mathbf{U}] \cdot \mathbf{x}^{(k)} + \omega \mathbf{b} \tag{B.1.17}$$

When $\omega = 1$ we obtain Eqs. (B.1.11). Ideally, ω should take the value that minimizes the spectral radius of the iteration matrix $\mathbf{P} = \mathbf{M}^{-1} \cdot \mathbf{N}$, but unfortunately this is generally unknown, although it can be constructed during the iterations (Hageman and Young, 1981, Chapter 9). A necessary condition for the iterations to converge is that $0 < \omega < 2$.

Minimization and Search Methods

When $\mathbf{A}$ is real, symmetric, and positive definite, computing the solution of the linear system $\mathbf{A} \cdot \mathbf{x} = \mathbf{b}$ is equivalent to finding the vector $\mathbf{x}$ that minimizes the quadratic functional

$$f(\mathbf{x}) = \tfrac{1}{2} \mathbf{x} \cdot \mathbf{A} \cdot \mathbf{x} - \mathbf{b} \cdot \mathbf{x} \tag{B.1.18}$$

(Atkinson, 1989, p. 563). If $\mathbf{A}$ is not symmetric, we multiply the equation $\mathbf{A} \cdot \mathbf{x} = \mathbf{b}$ by the transpose of $\mathbf{A}$ and obtain the preconditioned system $\mathbf{B} \cdot \mathbf{x} = \mathbf{c}$, where $\mathbf{B} = \mathbf{A}^T \cdot \mathbf{A}$ is symmetric and positive definite, and where $\mathbf{c} = \mathbf{A}^T \cdot \mathbf{b}$.

Steepest descent search

In the method of steepest descent, we compute the solution of the minimization problem according to the algorithm

$$\mathbf{x}^{(k+1)} = \mathbf{x}^{(k)} + \alpha^{(k+1)} \mathbf{r}^{(k)} \tag{B.1.19}$$

where

$$\mathbf{r}^{(k)} = \mathbf{b} - \mathbf{A} \cdot \mathbf{x}^{(k)}, \qquad \alpha^{(k+1)} = \frac{\mathbf{r}^{(k)} \cdot \mathbf{r}^{(k)}}{\mathbf{r}^{(k)} \cdot \mathbf{A} \cdot \mathbf{r}^{(k)}} \tag{B.1.20}$$

Unfortunately, the rate of convergence is usually slow.

Directional search

In this method the minimization problem is solved by selecting a set of search directions expressed by the vectors $\{\mathbf{p}^{(1)}, \mathbf{p}^{(2)}, \mathbf{p}^{(3)}, \ldots\}$, and then advancing the solution according to the algorithm

$$\mathbf{x}^{(k)} = \mathbf{x}^{(k-1)} + \alpha^{(k)}\mathbf{p}^{(k)} \tag{B.1.21}$$

where

$$\alpha^{(k)} = \frac{\mathbf{p}^{(k)} \cdot \mathbf{r}^{(k-1)}}{\mathbf{p}^{(k)} \cdot \mathbf{A} \cdot \mathbf{p}^{(k)}}, \qquad \mathbf{r}^{(k)} = \mathbf{b} - \mathbf{A} \cdot \mathbf{x}^{(k)} \tag{B.1.22}$$

Note that the steepest-descent search emerges by setting $\mathbf{p}^{(k)} = \mathbf{r}^{(k-1)}$.

Conjugate gradients

In this method the search vectors are computed so that they are $\mathbf{A}$-conjugate with each other, that is, $\mathbf{p}^{(i)} \cdot \mathbf{A} \cdot \mathbf{p}^{(j)} = 0$ for $i \neq j$. Furthermore, the direction $\mathbf{p}^{(k)}$ is aligned as much as possible with that of $\mathbf{r}^{(k-1)}$ subject to the $\mathbf{A}$-conjugation constraint. The algorithm is

$$\mathbf{x}^{(0)} = \mathbf{0}, \qquad \mathbf{r}^{(0)} = \mathbf{b}$$

$$\mathbf{p}^{(1)} = \mathbf{r}^{(0)} \tag{B.1.23}$$

$$\mathbf{p}^{(k)} = \mathbf{r}^{(k-1)} + \beta^{(k)}\mathbf{p}^{(k-1)} \qquad \text{for } k > 1$$

where

$$\beta^{(k)} = \frac{|\mathbf{r}^{(k-1)}|^2}{|\mathbf{r}^{(k-2)}|^2}, \qquad \mathbf{r}^{(k)} \equiv \mathbf{b} - \mathbf{A} \cdot \mathbf{x}^{(k)} \tag{B.1.24}$$

and

$$\mathbf{x}^{(k)} = \mathbf{x}^{(k-1)} + \alpha^{(k)}\mathbf{p}^{(k)} \tag{B.1.25}$$

where

$$\alpha^{(k)} = \frac{|\mathbf{r}^{(k-1)}|^2}{\mathbf{p}^{(k)} \cdot \mathbf{A} \cdot \mathbf{p}^{(k)}} \tag{B.1.26}$$

(Golub and Van Loan, 1989, p. 523).

An advanced version of the conjugate-gradients method that includes preconditioning is described by Golub and Van Loan (1989, p. 527). Other related methods are discussed by Eisenstat, Elman, and Schultz (1983) and Saad and Schultz (1986).

B.2 │ COMPUTATION OF EIGENVALUES OF A MATRIX

The eigenvalues of a diagonal or triangular matrix are equal to the diagonal elements. The eigenvalues of a tridiagonal matrix with constant sub-diagonal, diagonal, and super-diagonal elements are given in Eq. (12.2.36). *Gerschgorin's circle theorem* locates the eigenvalues of a general $N \times N$ matrix within the union of N disks in the complex plane. The ith disk is centered at the diagonal element A_{ii}, and the corresponding radius is equal to the minimum of

$$\sum_{\substack{j=1 \\ i \neq j}}^{N} |A_{ij}| \quad \text{and} \quad \sum_{\substack{i=1 \\ i \neq j}}^{N} |A_{ij}| \tag{B.2.1}$$

Other theorems for locating eigenvalues are discussed by Wilkinson (1965). Three general classes of methods are available for computing eigenvalues.

Compute the Roots of the Characteristic Polynomial

The first class of methods produces the eigenvalues by computing the roots of the characteristic polynomial $P(\lambda) \equiv \text{Det}(\mathbf{A} - \lambda \mathbf{I}) = 0$, which can be done using a general-purpose numerical method for solving nonlinear algebraic equations to be discussed in Section B.3. One way to compute $P(\lambda)$ is to perform the **LU** decomposition of the matrix $\mathbf{B} = \mathbf{A} - \lambda \mathbf{I}$ for a trial value of λ, as discussed in Section B.1, using, for instance, the method of Gauss elimination.

Determinant of a tridiagonal matrix

An efficient algorithm for computing the determinant of the tridiagonal matrix $\mathbf{B}$ proceeds by denoting the nonzero elements $B_{i,i-1} = \gamma_i$, $B_{ii} = A_{ii} - \lambda = \alpha_i$, where summation is not implied over i, and $B_{i,i+1} = \beta_i$, and then computing the sequence of numbers P_i as

$$P_0 = 1, \qquad P_1 = \alpha_1, \qquad P_i = \alpha_i P_{i-1} - \beta_{i-1} \gamma_i P_{i-2}, \qquad i = 2, \ldots, N \tag{B.2.2}$$

It can be shown by straightforward substitution that $P_N = \text{Det}(\mathbf{B})$.

Power Method

This method proceeds by successively projecting an arbitrary vector onto the matrix $\mathbf{A}$, until it becomes an eigenvector corresponding to the *eigenvalue with the maximum norm*. The computational algorithm involves selecting the initial vector $\mathbf{x}^{(0)}$, and then computing the sequence

$$\mathbf{x}^{(k)} = \mathbf{A} \cdot \mathbf{x}^{(k-1)}, \qquad \lambda^{(k)} = \frac{\mathbf{x}^{(k)} \cdot \mathbf{x}^{(k)}}{\mathbf{x}^{(k-1)} \cdot \mathbf{x}^{(k)}}, \qquad \mathbf{x}^{(k)} \leftarrow \frac{\mathbf{x}^{(k)}}{\mathbf{x}^{(k)} \cdot \mathbf{x}^{(k)}} \tag{B.2.3}$$

for $k = 1, 2, \ldots$ until $\lambda^{(k)}$ and $\mathbf{x}^{(k)}$ become an eigenvalue–eigenvector pair.

Suppose that we have computed the eigenvalue of $\mathbf{A}$ with the maximum norm λ_{A1}. To obtain another eigenvalue, we apply the power method to the modified matrix $\mathbf{B} = \mathbf{A} - \lambda_{A1}\mathbf{I}$. If λ_{B1} is the eigenvalue of $\mathbf{B}$ with the maximum norm, then $\lambda_{A2} = \lambda_{B1} + \lambda_{A1}$ will be another eigenvalue of $\mathbf{A}$.

To compute the eigenvalue of $\mathbf{A}$ with the minimum norm, we apply the power method to the inverse matrix $\mathbf{B} = \mathbf{A}^{-1}$. If λ_{B1} is the eigenvalue of $\mathbf{B}$ with the maximum norm, then $\lambda_{A1} = 1/\lambda_{B1}$ will be the eigenvalue of $\mathbf{A}$ with the minimum norm.

To compute the eigenvalue of $\mathbf{A}$ that is closest to the complex number c, we apply the power method to the modified matrix $\mathbf{B} = (\mathbf{A} - c\mathbf{I})^{-1}$. If λ_{B1} is the eigenvalue of $\mathbf{B}$ with the maximum norm, then $\lambda_{A1} = c + 1/\lambda_{B1}$ is the eigenvalue of $\mathbf{A}$ closest to c.

Deflation

Suppose that we have computed one eigenvalue of $\mathbf{A}$, call it λ_1, and the corresponding eigenvector with unit length, call it $\mathbf{u}$. Let us then introduce the orthogonal *Housholder transformation matrix* defined as $P_{ij} = \delta_{ij} - 2w_i w_j$, with

$$w_1 = \sqrt{\frac{1 \pm u_1}{2}}, \qquad w_i = \pm\frac{u_i}{2w_1} \qquad \text{for } i = 2, \dots, N \qquad \text{(B.2.4)}$$

It can be shown that the matrix $\mathbf{B} = \mathbf{P} \cdot \mathbf{A} \cdot \mathbf{P}$ has zeros in the first column except that the first entry is $B_{11} = \lambda_1$. The eigenvalues of the bottom $(N - 1) \times (N - 1)$ diagonal block of $\mathbf{B}$ are also eigenvalues of $\mathbf{A}$. Applying the power method to the reduced matrix $\mathbf{B}$ yields one additional eigenvalue. The deflation continues until the sequentially deflated matrix is reduced to a scalar.

Methods Based on Similarity Transformations

If $\mathbf{P}$ is a nonsingular matrix, then the transformed matrix $\mathbf{B}$ computed by the similarity transformation $\mathbf{B} = \mathbf{P}^{-1} \cdot \mathbf{A} \cdot \mathbf{P}$ has the same eigenvalues as $\mathbf{A}$. This observation suggests searching for a transformation matrix $\mathbf{P}$ that makes $\mathbf{B}$ as simple as possible, ideally diagonal or triangular.

Jacobi's method

Jacobi's method seeks to reduce the norm of the off-diagonal elements of a symmetric matrix $\mathbf{A}$ by performing consecutive similarity transformations. Considering the A_{ij} off-diagonal element, we set the matrix $\mathbf{P}$ equal to the identity matrix $\mathbf{I}$ except that $P_{ij} = \sin\theta$. $P_{ji} = -\sin\theta$, and $P_{ii} = P_{jj} = \cos\theta$, where θ is a solution of the equation $\tan 2\theta = 2A_{ij}/(A_{ii} - A_{jj})$. It will be noted that $\mathbf{P}$ is an orthogonal matrix, and therefore $\mathbf{P}^{-1} = \mathbf{P}^T$. With these definitions, $\mathbf{B}$ turns out to be a symmetric matrix that is identical to $\mathbf{A}$, except that the entries in the ith and jth columns and rows have been altered. More important, B_{ij} and B_{ji} are equal to zero. The algorithm proceeds by sweeping the off-diagonal elements according to a certain protocol. As the similarity transformations continue, $\mathbf{A}$ tends to become diagonal.

QR decomposition

The $\mathbf{QR}$ decomposition method applies to an arbitrary, not necessarily symmetric, matrix. The method involves decomposing the matrix $\mathbf{A}$ as $\mathbf{A} = \mathbf{Q} \cdot \mathbf{R}$, where $\mathbf{Q}$ is an orthogonal matrix and $\mathbf{R}$ is an upper triangular matrix. Then $\mathbf{P} = \mathbf{Q}$ and $\mathbf{B} = \mathbf{Q}^T \cdot \mathbf{A} \cdot \mathbf{Q} = \mathbf{R} \cdot \mathbf{Q}$. As the transformations continue, the evolving form of $\mathbf{A}$ tends to become precisely or nearly upper triangular (Atkinson, 1989, p. 623).

B.3 | NONLINEAR ALGEBRAIC EQUATIONS

Consider a system of N nonlinear algebraic equations $f_i(x_1, x_2, \dots, x_N) = 0, i = 1, \dots, N$, containing N unknowns, and introduce the vector of independent variables $\mathbf{x}$ and the vector function $\mathbf{f}(\mathbf{x})$.

To compute the solution of the vector equation $\mathbf{f}(\mathbf{x}) = \mathbf{0}$, we recast it into the form $\mathbf{x} = \mathbf{g}(\mathbf{x})$ and perform *fixed-point iterations*, setting $\mathbf{x}^{(k+1)} = \mathbf{g}(\mathbf{x}^{(k)})$, beginning with a certain initial guess $\mathbf{x}^{(0)}$. It is possible that the sequence $\mathbf{x}^{(k)}$ will converge to the fixed point $\mathbf{X}$ of the mapping function $\mathbf{g}$, which, by definition, satisfies the equation $\mathbf{X} = \mathbf{g}(\mathbf{X})$ or $\mathbf{f}(\mathbf{X}) = \mathbf{0}$. A *sufficient* condition for the fixed-point iterations to converge is that

$$\sum_{j=1}^{N} |J_{ij}(\mathbf{X})| < 1, \qquad \text{where } J_{ij} \equiv \frac{\partial f_i}{\partial x_j} \qquad \text{(B.3.1)}$$

for all values of i. The *Jacobian matrix* $\mathbf{J}$ is equal to the transpose of the gradient of $\mathbf{f}$, $\mathbf{J} = (\nabla\mathbf{f})^T$. A necessary condition for the iterations to converge is that the spectral radius of $\mathbf{J}$ evaluated at the fixed point is less than unity.

Newton's and Related Methods

In Newton's method, the iteration function $\mathbf{g}$ is defined as $\mathbf{g} = \mathbf{x} - \mathbf{J}^{-1} \cdot \mathbf{f}$ where $\mathbf{J}^{-1}$ is the inverse of $\mathbf{J}$. The algorithm involves solving the linear system $\mathbf{J}(\mathbf{x}^{(k)}) \cdot \mathbf{e}^{(k)} = -\mathbf{f}(\mathbf{x}^{(k)})$ for the correction vector $\mathbf{e}^{(k)}$, and then setting $\mathbf{x}^{(k+1)} = \mathbf{x}^{(k)} + \mathbf{e}^{(k)}$.

For a single equation, $N = 1$, we obtain the algorithm $x^{(k+1)} = x^{(k)} - f(x^{(k)})/f'(x^{(k)})$, where $f' = df/dx$. In practice, the derivative f' is often computed by numerical differentiation setting, for instance, $f' = [f(x + \varepsilon) - f(x)]/\varepsilon$, where ε is a small number, as discussed in Section B.5.

Newton's method converges as long as the initial guess is sufficiently close to a root. Furthermore, in general, the rate of convergence is quadratic, which means that each time we carry out an iteration, the error is raised to the second power. If the root happens to be multiple, the rate of convergence becomes linear; each time we carry out an iteration, the error is multiplied by a constant factor that is less than unity; for a double root, the factor is equal to one-half. The second-order convergence, however, can be rectified with a simple modification of the basic algorithm (Dahlquist and Björck, 1974, p. 242).

The *secant method* is a modification of Newton's method that circumvents the computation of the partial derivatives. For a single equation, $N = 1$, the algorithm is

$$x^{(k+1)} = x^{(k)} - f^{(k)} \frac{x^{(k)} - x^{(k-1)}}{f^{(k)} - f^{(k-1)}} \tag{B.3.2}$$

Note that two guesses are required in order to initialize the computations. Each time an iteration is carried out, the error is raised roughly to the power of 1.6.

Quasi-Newton methods are modifications of Newton's methods distinguished by the fact that the Jacobian matrix is either kept constant, or constructed during the iterations (Broyden, 1965; Dahlquist and Björck, 1974, p. 251).

Bairstow's Method for Polynomials

This powerful method allows us to compute simultaneously a pair of roots of the Nth-degree polynomial $P(x) = a_1 x^N + a_2 x^{N-3} + a_3 x^{N-4} + \cdots + a_N x + a_{N+1}$. The method proceeds by factorizing the polynomial as $P(x) = (x^2 - rx - s)(b_1 x^{N-2} + b_2 x^{N-3} + b_3 x^{N-4} + \cdots + b_{N-2} x + b_{N-1}) + b_N(x-r) + b_{N+1}$, and then computing the constants r and s so that the coefficients b_N and b_{N+1} vanish. When this has been achieved, the roots of the binomial $x^2 - rx - s$ are also roots of the polynomial $P(x)$.

The problem is reduced to solving the nonlinear system of two equations $b_N(r, s) = 0$ and $b_{N+1}(r, s) = 0$, where the coefficients b_N and b_{N+1} are polynomial functions of r and s, which may be done using Newton's method. To compute the values of b_N and b_{N+1} as well as their partial derivatives with respect to r and s, we first compute the algebraic sequence

$$\begin{aligned}
b_1 &= a_1 \\
b_2 &= a_1 + rb_1 \\
b_3 &= a_3 + rb_2 + sb_1, \\
&\vdots \\
b_{N+1} &= a_{N+1} + rb_N + sb_{N-1}
\end{aligned} \tag{B.3.3}$$

and then the sequence

$$\begin{aligned}
c_1 &= 0 \\
c_2 &= b_1 \\
c_3 &= b_2 + rc_1 \\
c_4 &= b_3 + rc_3 + sc_2, \\
&\vdots \\
c_{N+1} &= b_N + rc_N + sc_{N-1}
\end{aligned} \tag{B.3.4}$$

It can be shown by straightforward substitution that $\partial b_k/\partial r = c_k$, and $\partial b_k/\partial s = c_{k-1}$, for $k = 1, \ldots, N$.

Deflation

Once a root of the equation $\mathbf{f}(\mathbf{x}) = \mathbf{0}$ has been computed, call it $\mathbf{X}$, it can be put aside by considering the modified equation $\mathbf{F}(\mathbf{x}) = \mathbf{0}$, where $\mathbf{F}(\mathbf{x}) = \mathbf{f}(\mathbf{x})/|\mathbf{x} - \mathbf{X}|^m$, m being the multiplicity of $\mathbf{X}$.

Initial Guess

A successful estimate of the location of a root can be made by examining whether any terms of the scalar components of the vector function $\mathbf{f}$ take small or large values for small or large values of the components of $\mathbf{x}$. If they do, we either ignore them or discard the rest of the terms, compute the solution of the simplified system, and check whether the approximations that lead us to the approximate solution are justified a posteriori.

For stiff problems, a successful initial guess can be made using *continuation* or *imbedding* methods. This involves perturbing the equation $\mathbf{f}(\mathbf{x}) = \mathbf{0}$, thus forming the modified equation $\mathbf{F}(\mathbf{x}, \varepsilon) = \mathbf{0}$, where $\mathbf{F}(\mathbf{x}, \varepsilon = 1) = \mathbf{f}(\mathbf{x})$, and the equation $\mathbf{F}(\mathbf{x}, \varepsilon = 0) = \mathbf{0}$ is easy to solve. The procedure involves solving a sequence of equations, for instance, $\mathbf{F}(\mathbf{x}, \varepsilon = 0) = \mathbf{0}$, $\mathbf{F}(\mathbf{x}, \varepsilon = 0.10) = \mathbf{0}, \ldots, \mathbf{F}(\mathbf{x}, \varepsilon = 1.0) = \mathbf{0}$; the initial guess for each equation is identified with the converged solution of the previous equation. In fluid mechanics, ε may be identified, for instance, with the Reynolds number.

B.4 FUNCTION INTERPOLATION

Given the values of a function $f(x)$ at the $N + 1$ data points $x_i, i = 1, \ldots, N + 1$, we want to compute its values at intermediate points.

The Interpolating Polynomial

A popular method of function interpolation is based upon replacing the function $f(x)$ with the Nth-degree polynomial $P_N(x) = a_1 x^N + a_2 x^{N-1} + a_3 x^{N-2} + \cdots + a_N x + a_{N+1}$ that passes through the data points; that is, $f(x_i) = P_N(x_i)$. It can be shown that the error incurred in polynomial interpolation is

$$P_N(x) - f(x) = -\frac{f^{(N+1)}(\xi)}{(N + 1)!}(x - x_1)(x - x_2)\cdots(x - x_N)(x - x_{N+1}) \qquad \text{(B.4.1)}$$

where $f^{(N+1)}(\xi)$ is the $N + 1$ derivative of the function f evaluated at a certain point ξ that lies somewhere between x_1 and x_{N+1}; its actual location depends upon the value of x. The error does not necessarily vanish as N tends to infinity, especially when the data points are evenly spaced. If the data points coincide with the zeros of an orthogonal polynomial, however, then the error generally diminishes uniformly as N is increased (Davis, 1975, Chapter 4).

Vandermonde matrix

To compute the interpolating polynomial, we enforce the constraint $f(x_i) = P_N(x_i)$ at the $N + 1$ data points, and thereby derive a system of $N + 1$ equations for the $N + 1$ unknown coefficients a_i. The matrix of the linear system that multiplies the unknown vector is called the *Vandermonde matrix*. The determinant of this matrix can be shown to be equal to the product of $x_i - x_j$ with $i > j$, and this shows that the solution of the linear system is unique as long as the data points are distinct. Unfortunately, the Vandermonde matrix becomes nearly singular even at moderate values of N, placing limits on the accuracy and practicality of the numerical method.

Lagrange interpolation

Lagrange's method constructs the interpolating polynomial in a manner that circumvents the explicit computation of the coefficients, as

$$P_N(x) = \sum_{i=1}^{N+1} f(x_i)\, L_i(x) \tag{B.4.2}$$

where

$$L_i(x) \equiv \frac{(x - x_1)(x - x_2)\cdots(x - x_{i-1})(x - x_{i+1})\cdots(x - x_N)(x - x_{N+1})}{(x_i - x_1)(x_i - x_2)\cdots(x_i - x_{i-1})(x_i - x_{i+1})\cdots(x_i - x_N)(x_i - x_{N+1})} \tag{B.4.3}$$

are the Lagrange polynomials. Note that the denominators of L_i are constant, whereas the numerators are Nth-degree polynomials in x.

Local Polynomial Interpolation

When the number of data points is large, or the data have an appreciable margin of error, polynomial interpolation may lead to substantial error in certain regions, especially near the ends of the domain of interpolation. To avoid this occurrence, we replace the global interpolation with a collection of local interpolations, each involving a small group of data points.

Linear interpolation employs two consecutive data points and yields the first-degree polynomial

$$P(x) = f(x_i) + (x - x_i)\frac{f(x_{i+1}) - f(x_i)}{x_{i+1} - x_i} \tag{B.4.4}$$

which is valid for $x_i < x < x_{i+1}$.

Quadratic interpolation employs three consecutive data points and yields the second-degree polynomial

$$P(x) = f(x_i) + (x - x_i)[b + a(x - x_i)] \tag{B.4.5}$$

which is valid for $x_{i-1} < x < x_{i+1}$ The coefficients a and b are computed sequentially as

$$a = \frac{1}{x_{i+1} - x_{i-1}}\left(\frac{f(x_{i+1}) - f(x_i)}{x_{i+1} - x_i} - \frac{f(x_{i-1}) - f(x_i)}{x_{i-1} - x_i}\right)$$

$$b = \frac{f(x_{i+1}) - f(x_i)}{x_{i+1} - x_i} - a(x_{i+1} - x_i) \tag{B.4.6}$$

Cubic-spline interpolation

Cubic-spline interpolation fits a third-degree polynomial over each interval between two consecutive data points, and matches the first and second derivatives of adjacent polynomials at the data points. Denoting the ith cubic by

$$P_i(x) = f(x_i) + (x - x_i)\Big[c_i + (x - x_i)\big(b_i + a_i(x - x_i)\big)\Big], \qquad \text{for } x_i < x < x_{i+1} \tag{B.4.7}$$

we find

$$a_i = \tfrac{1}{3}\frac{b_{i+1} - b_i}{h_i}, \qquad c_i = \frac{f(x_{i+1}) - f(x_i)}{h_i} - \frac{h_i}{3}(b_{i+1} + 2b_i), \qquad h_i = x_{i+1} - x_i \tag{B.4.8}$$

The collection of the b_i satisfy the $N - 2$ equations

$$\frac{h_i}{3}b_i + \tfrac{2}{3}(h_i + h_{i+1})\,b_{i+1} + \frac{h_{i+1}}{3}b_{i+2} = \frac{f(x_{i+2}) - f(x_{i+1})}{h_{i+1}} - \frac{f(x_{i+1}) - f(x_i)}{h_i} \tag{B.4.9}$$

for $i = 1, \ldots, N - 2$. Making two additional stipulations regarding the shape of the cubic splines provides us with a system of N equations for the N values of b_i. For instance, we may specify the slope of the first and last cubics at the first and last data points. If the interpolated function is periodic, we require that the first and second derivatives of the first cubic at the first point are equal to those of the last cubic at the last point.

Polynomial Interpolation in Two Variables

Polynomial interpolation in two variables is similar in spirit and practice to that in one variable; the counterpart of local linear interpolation is *bilinear interpolation*. Suppose a point $\mathbf{x}$ lies within the rectangle that is confined by the grid lines $x_i, x_{i+1}, y_j, y_{j+1}$. Then the value of a function f at $\mathbf{x}$ is computed as

$$f(\mathbf{x}) = [(x - x_{i+1})(y - y_{j+1})f_{i,j} - (x - x_i)(y - y_{j+1})f_{i+1,j}$$
$$- (x - x_{i+1})(y - y_j)f_{i,j+1} + (x - x_i)(y - y_j)f_{i+1,j+1}]/(x_{i+1} - x_i)(y_{j+1} - y_j)$$
$$\text{(B.4.10)}$$

Bilinear interpolation yields a continuous function with discontinuous first derivatives across the grid lines. Other bivariate interpolation formulae are tabulated by Abramowitz and Stegun (1972, p. 882).

Trigonometric Interpolation

Consider the function $f(x)$ over the finite interval $a < x < b$ of length $L = b - a$. Trigonometric interpolation approximates the function over that interval with the truncated Fourier series as

$$F_N(x) = \tfrac{1}{2}a_0 + \sum_{l=1}^{M} a_l \cos(lkx) + \sum_{l=1}^{M} b_l \sin(lkx) = \sum_{l=-M}^{M} c_l \exp(ilkx) \qquad \text{(B.4.11)}$$

where $k = 2\pi/L$ is the wave number, i is the imaginary unit, M is a certain truncation level, a_l and b_l are real Fourier coefficients, and c_l are complex Fourier coefficients. Outside the interval (a, b), the Fourier series yields the periodic repetition of the section of f between (a, b).

The real and complex Fourier coefficients are related by

$$c_l = \tfrac{1}{2}(a_l - ib_l), \qquad c_l = c_{-l}^*, \qquad a_l = 2\operatorname{Re}(c_l), \qquad b_l = -2\operatorname{Im}(c_l) \qquad \text{(B.4.12)}$$

where $b_0 = 0$, and an asterisk denotes the complex conjugate. Using the identity

$$\int_a^b \exp[i(l - m)kx]\, dx = \delta_{lm}L \qquad \text{(B.4.13)}$$

where l and m are two integers, we find that the Fourier coefficients are given by

$$c_l = \frac{1}{L}\int_a^b f(x)\exp(-ilkx)\, dx, \qquad a_l = \frac{2}{L}\int_a^b f(x)\cos(lkx)\, dx$$
$$b_l = \frac{2}{L}\int_a^b f(x)\sin(lkx)\, dx \qquad \text{(B.4.14)}$$

The integrals on the right-hand sides of these equations may be computed by numerical integration as discussed in Section B.6. In the limit as M tends to infinity, the representation (B.4.11) becomes exact (Tolstov, 1962).

Let us divide the interval L into N subintervals that are separated by the $N + 1$ data points $x_n = a + (n - 1)h, n = 1, \ldots, N + 1, h = L/N$, where $x_1 = a$ and $x_{N+1} = b$. Using the trapezoidal rule to evaluate the first complex Fourier integral in Eqs. (B.4.14), we obtain

$$c_l = \frac{\exp(-ilka)}{N}[\tfrac{1}{2}f(x_1) + \omega^l f(x_2) + \cdots + \omega^{(N-1)l}f(x_N) + \tfrac{1}{2}f(x_{N+1})] \quad \text{(B.4.15)}$$

where $\omega = \exp(-ikh)$. When N is odd, $M = (N-1)/2$, and when N is even, $M = N/2$. With the Fourier coefficients computed in this manner, and with the help of the identity

$$\sum_{l=1}^{N} \exp[-ilk(x_m - x_n)] = \delta_{nm}N \quad \text{(B.4.16)}$$

it can be shown that $f(x_n) = F_N(x_n)$ for $n = 2, \ldots, N+1$; that is, *the truncated Fourier series is an interpolating function.* If the function f happens to be periodic with period L, Eq. (B.4.15) assumes the simpler form

$$c_l = \frac{\exp(-ilka)}{N} \sum_{n=1}^{N} \omega^{l(n-1)} f(x_n) \quad \text{(B.4.17)}$$

In practice, when N is large, the summation is computed most efficiently using the fast Fourier transform method, which requires $N \log_2 N$ computations (Henrici, 1986).

B.5 | COMPUTATION OF DERIVATIVES

Given the values of a function $f(x)$ at the $N+1$ data points x_i, $i = 1, \ldots, N+1$, we want to compute the derivatives of the function at the data points or at intermediate points. This can be done by approximating the function with a local interpolating polynomial, as discussed in Section B.4, and then differentiating the interpolating polynomial, thus obtaining approximations to the derivatives.

Consider the computation of the derivatives of the function at the ith data point; if we use an equal number of data points on either side of x_i to construct the interpolating polynomial, we obtain *central differences;* otherwise we obtain spatially biased *forward* or *backward differences.*

Table B.5.1 summarizes finite-difference formulae for computing the derivatives of a function of one variable x at the data point x_i, using the values of the function at a collection of evenly spaced data points that are separated by a distance h. A more comprehensive table can be found in the handbook of Abramowitz and Stegun (1972, p. 914).

Table B.5.2 summarizes finite-difference formulae for computing the Laplacian of a function of two variables x and y at the data point (x_i, y_i), using the values of the function at a number of neighboring grid points. Further formulae are collected by Abramowitz and Stegun (1970, p. 883).

B.6 | FUNCTION INTEGRATION

Given the values of a function $f(x)$ at the $N+1$ *data* or *base* points x_i, $i = 1, \ldots, N+1$, that lie within the closed interval $[a, b]$, we want to compute the value of the integral

$$I = \int_a^b f(x)\,dx \quad \text{(B.6.1)}$$

A typical method of numerical integration proceeds by approximating $f(x)$ over $[a, b]$ with a global interpolating polynomial or with the union of a collection of local interpolating polynomials, as discussed in Section B.4, and then integrating the interpolating polynomials over their domain of definition.

TABLE B.5.1

Finite-difference formulae for computing derivatives of a function of a variable, $f(x)$, at the data point x_i, from the values of the function at a set of evenly spaced data points separated by distance $\Delta x = h$.

Backward differences with accuracy $O(h)$

$$
\begin{bmatrix} hf_i' \\ h^2 f_i'' \\ h^3 f_i''' \\ h^4 f_i'''' \end{bmatrix} =
\begin{bmatrix} 0 & 0 & 0 & -1 & 1 \\ 0 & 0 & 1 & -2 & 1 \\ 0 & -1 & 3 & -3 & 1 \\ 1 & -4 & 6 & -4 & 1 \end{bmatrix} \cdot
\begin{bmatrix} f_{i-4} \\ f_{i-3} \\ f_{i-2} \\ f_{i-1} \\ f_i \end{bmatrix}
$$

Backward differences with accuracy $O(h^2)$

$$
\begin{bmatrix} 2hf_i' \\ h^2 f_i'' \\ 2h^3 f_i''' \\ h^4 f_i'''' \end{bmatrix} =
\begin{bmatrix} 0 & 0 & 0 & 1 & -4 & 3 \\ 0 & 0 & -1 & 4 & -5 & 2 \\ 0 & 3 & -14 & 24 & -18 & 5 \\ -2 & 11 & -24 & 26 & -14 & 3 \end{bmatrix} \cdot
\begin{bmatrix} f_{i-5} \\ f_{i-4} \\ f_{i-3} \\ f_{i-2} \\ f_{i-1} \\ f_i \end{bmatrix}
$$

Central differences with accuracy $O(h^2)$

$$
\begin{bmatrix} 2hf_i' \\ h^2 f_i'' \\ 2h^3 f_i''' \\ h^4 f_i'''' \end{bmatrix} =
\begin{bmatrix} 0 & -1 & 0 & 1 & 0 \\ 0 & 1 & -2 & 1 & 0 \\ -1 & 2 & 0 & -2 & 1 \\ 1 & -4 & 6 & -4 & 1 \end{bmatrix} \cdot
\begin{bmatrix} f_{i-2} \\ f_{i-1} \\ f_i \\ f_{i+1} \\ f_{i+2} \end{bmatrix}
$$

Central differences with accuracy $O(h^4)$

$$
\begin{bmatrix} 12hf_i' \\ 12h^2 f_i'' \\ 8h^3 f_i''' \\ 6h^4 f_i'''' \end{bmatrix} =
\begin{bmatrix} 0 & 1 & -8 & 0 & 8 & -1 & 0 \\ 0 & -1 & 16 & -30 & 16 & -1 & 0 \\ 1 & -8 & 13 & 0 & -13 & 8 & -1 \\ -1 & 12 & -39 & 56 & -39 & 12 & -1 \end{bmatrix} \cdot
\begin{bmatrix} f_{i-3} \\ f_{i-2} \\ f_{i-1} \\ f_i \\ f_{i+1} \\ f_{i+2} \\ f_{i+3} \end{bmatrix}
$$

Forward differences with accuracy $O(h)$

$$
\begin{bmatrix} hf_i' \\ h^2 f_i'' \\ h^3 f_i''' \\ h^4 f_i'''' \end{bmatrix} =
\begin{bmatrix} -1 & 1 & 0 & 0 & 0 \\ 1 & -2 & 1 & 0 & 0 \\ -1 & 3 & -3 & 1 & 0 \\ 1 & -4 & 6 & -4 & 1 \end{bmatrix} \cdot
\begin{bmatrix} f_i \\ f_{i+1} \\ f_{i+2} \\ f_{i+3} \\ f_{i+4} \end{bmatrix}
$$

Forward differences with accuracy $O(h^2)$

$$
\begin{bmatrix} 2hf_i' \\ h^2 f_i'' \\ 2h^3 f_i''' \\ h^4 f_i'''' \end{bmatrix} =
\begin{bmatrix} -3 & 4 & -1 & 0 & 0 & 0 \\ 2 & -5 & 4 & -1 & 0 & 0 \\ -5 & 18 & -24 & 14 & -3 & 0 \\ 3 & -14 & 26 & -24 & 11 & 11 \end{bmatrix} \cdot
\begin{bmatrix} f_i \\ f_{i+1} \\ f_{i+2} \\ f_{i+3} \\ f_{i+4} \\ f_{i+5} \end{bmatrix}
$$

TABLE B.5.2
Finite-difference formulae for computing the Laplacian of a function of two variables, $f(x, y)$, at the data point (x_i, y_i), from the values of the function at a number of neighboring data points with uniform spacing Δx and Δy; $\beta = \Delta x / \Delta y$.

Five-point formula with accuracy $O(\Delta x^2)$ and $O(\Delta y^2)$

$$\nabla^2 f_{i,j} = \frac{1}{\Delta x^2}[f_{i+1,j} - 2(1 + \beta^2)f_{i,j} + f_{i-1,j} + \beta^2 f_{i,j+1} + \beta^2 f_{i,j-1}]$$

Nine-point formula with accuracy $O(\Delta x^2)$ and $O(\Delta y^2)$

$$\nabla^2 f_{i,j} = \frac{1}{6\Delta x^2}[4f_{i+1,j} - 10(1 + \beta^2)f_{i,j} + 4f_{i-1,j} + 4\beta^2 f_{i,j+1} + 4\beta^2 f_{i,j-1}$$
$$+ \beta(f_{i-1,j-1} + f_{i-1,j+1} + f_{i+1,j-1} + f_{i+1,j+1})]$$

Trapezoidal Rule

Approximating $f(x)$ with a straight line between two consecutive data points, corresponding to a first-degree local interpolating polynomial, yields the trapezoidal rule. When the base points are distributed evenly with constant separation $h = (b - a)/N$, where $x_1 = a$ and $x_{N+1} = b$, we obtain

$$I^{Tr}(h) = \frac{1}{2}h[f(x_1) + 2f(x_2) + 2f(x_3) + \cdots + 2f(x_N) + f(x_{N+1})]$$
$$- \frac{1}{12}(b - a)h^2 f''(\xi) \tag{B.6.2}$$

where the point ξ lies somewhere within the domain of integration.

Romberg integration uses the results of two computations with different interval sizes h to improve the accuracy to fourth order in h. For example,

$$I^{Rom}(h) = \frac{1}{3}[4I^{Tr}(\frac{1}{2}h) - I^{Tr}(h)] \tag{B.6.3}$$

Simpson's Rule

Approximating $f(x)$ with a parabola subtended across triplets of consecutive data points, corresponding to a second-degree local interpolating polynomial, yields Simpson's one-third rule. Assuming that the number of intervals N is even, we obtain

$$I^{Simp}(h) = \frac{1}{3}h[f(x_1) + 4f(x_2) + 2f(x_3) + 4f(x_4) + \cdots + 4f(x_N)$$
$$+ f(x_{N+1})] - \frac{1}{180}(b - a)h^4 f^{(4)}(\xi) \tag{B.6.4}$$

where the point ξ lies somewhere within the domain of integration. *Romberg integration* improves the accuracy to sixth order in h. For example

$$I^{Rom}(h) = \frac{1}{15}[16 I^{Simp}(\frac{1}{2}h) - I^{Simp}(h)] \tag{B.6.5}$$

Gauss Quadratures

Gauss quadratures require that the data points be distributed in a particular manner over the domain of integration. Global polynomial interpolation, followed by analytical integration, leads to the numerical approximation

$$I^{Gauss} = \frac{1}{2}(b - a) \sum_{i=1}^{N+1} f(x_i) w_i \tag{B.6.6}$$

The position of the base points x_i is determined by the zeros z_i of a properly selected class of $(N + 1)$-degree orthogonal polynomial to be discussed in Section B.7; w_i are the corresponding weights.

The *Gauss–Legendre* quadrature is designed for functions that do not have any singularities over the domain of integration. The base points are located at $x_i = \frac{1}{2}[a + b + z_i(b - a)]$, where z_i are the zeros of the $(N + 1)$-degree Legendre polynomial, which is defined over the interval $[-1, 1]$. Table B.6.1 summarizes the zeros z_i and weights w_i for several values of N. Note that, for each value of N, the w_i add up to two, which is necessary for Eq. (B.6.6) to be valid when f is a constant function. Polynomials of degree $2N + 1$ or less are integrated exactly.

Other Gaussian quadratures applicable to singular integrals and infinite integration domains are discussed by Carnahan, Luther, and Wilkes (1969) and Abramowitz and Stegun (1972, p. 887).

TABLE B.6.1
The zeros z_i and corresponding weights w_i of the Gauss–Legendre quadrature

Zeros z_i	Weights w_i
Two-point formula $N = 1$	
±0.57735 02691 89626	1.00000 00000 00000
Three-point formula $N = 2$	
0.00000 00000 00000	0.88888 88888 88889
±0.77459 66692 41483	0.55555 55555 55556
Four-point formula $N = 3$	
±0.33998 10435 84856	0.65214 51548 62546
±0.86113 63115 94053	0.34785 48451 37454
Five-point formula $N = 4$	
0.00000 00000 00000	0.56888 88888 88889
±0.53846 93101 05683	0.47862 86704 99366
±0.90617 98459 38664	0.23692 68850 56189
Six-point formula $N = 5$	
±0.23861 91860 83197	0.46791 39345 72691
±0.66120 93864 66265	0.36076 15730 48139
±0.93246 95142 03152	0.17132 44923 79170
Ten-point formula $N = 9$	
±0.14887 43389 81631	0.29552 42247 14753
±0.43339 53941 29247	0.26926 67193 09996
±0.67940 95682 99024	0.21908 63625 15982
±0.86506 33666 88985	0.14945 13491 50581
±0.97390 65285 17172	0.06667 13443 08688
Fifteen-point formula $N = 14$	
0.00000 00000 00000	0.20257 82419 25561
±0.20119 40937 97435	0.19843 14853 27111
±0.39415 13470 77563	0.18616 10001 15562
±0.57097 21726 08539	0.16626 92058 16994
±0.72441 77313 60170	0.13957 06779 26154
±0.84820 65834 10427	0.10715 92204 67172
±0.93727 33924 00706	0.07036 60474 88108
±0.98799 25180 20485	0.03075 32419 96117

TABLE B.6.2
The location of the base points and corresponding weights w_i of the quadrature shown in Eq. (B.6.8), for integration over a triangular domain; L is the typical side length of the triangle.

Base-point coefficients $\alpha_i, \beta_i, \gamma_i$ defined in Eq. (B.6.9)			Weights w_i
One-point formula with accuracy $O(L^2)$			
$\dfrac{1}{3}$	$\dfrac{1}{3}$	$\dfrac{1}{3}$	1.0
Three-point formula with accuracy $O(L^3)$			
$\dfrac{1}{2}$	$\dfrac{1}{2}$	0	$\dfrac{1}{3}$
0	$\dfrac{1}{2}$	$\dfrac{1}{2}$	$\dfrac{1}{3}$
$\dfrac{1}{2}$	0	$\dfrac{1}{2}$	$\dfrac{1}{3}$
Four-point formula with accuracy $O(L^4)$			
$\dfrac{1}{3}$	$\dfrac{1}{3}$	$\dfrac{1}{3}$	$-\dfrac{27}{48}$
$\dfrac{3}{5}$	$\dfrac{1}{5}$	$\dfrac{1}{5}$	$\dfrac{25}{48}$
$\dfrac{1}{5}$	$\dfrac{3}{5}$	$\dfrac{1}{5}$	$\dfrac{25}{48}$
$\dfrac{1}{5}$	$\dfrac{1}{5}$	$\dfrac{3}{5}$	$\dfrac{25}{48}$
Seven-point formula with accuracy $O(L^4)$			
$\dfrac{1}{3}$	$\dfrac{1}{3}$	$\dfrac{1}{3}$	$\dfrac{27}{60}$
$\dfrac{1}{2}$	$\dfrac{1}{2}$	0	$\dfrac{8}{60}$
0	$\dfrac{1}{2}$	$\dfrac{1}{2}$	$\dfrac{8}{60}$
$\dfrac{1}{2}$	0	$\dfrac{1}{2}$	$\dfrac{8}{60}$
1	0	0	$\dfrac{3}{60}$
0	1	0	$\dfrac{3}{60}$
0	0	1	$\dfrac{3}{60}$
Seven-point formula with accuracy $O(L^6)$			
$\dfrac{1}{3}$	$\dfrac{1}{3}$	$\dfrac{1}{3}$	0.225
a	b	b	e
b	a	b	e
b	b	a	e
c	d	d	f
d	c	d	f
d	d	c	f

where
$a = 0.05971587,$ $b = 0.47014206,$
$c = 0.79742699,$ $d = 0.10128651,$
$e = 0.13239415,$ $f = 0.12593918$

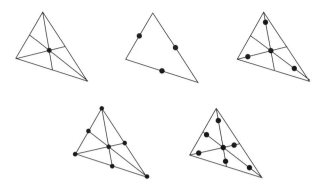

Figure B.6.1 Location of base points for integration over a triangular domain. The precise location of the base points and corresponding weights are given in Table B.6.2.

Two-Dimensional Integrals

The choice of a successful numerical method for computing the integral of a function of two variables $f(x, y)$ over a two-dimensional domain depends upon the smoothness of $f(x, y)$ and the geometry of the domain of integration (Stroud, 1971; Abramowitz and Stegun, 1972, p. 891). For example, when the function $f(x, y)$ is smooth and the domain is a rectangular area confined within $a < x < b$ and $c < y < d$, the successive application of two Gauss–Legendre quadratures for integration in each direction yields the compound quadrature

$$\int_c^d \int_a^b f(x, y)\, dx\, dy \approx \tfrac{1}{4}(b - a)(d - c) \sum_{i=1}^{N+1} \sum_{j=1}^{M+1} f(x_i, y_j)\, w_i w_j \tag{B.6.7}$$

where N and M are two independent integers and x_i, y_j correspond to the Gauss–Legendre base points.

To compute the integral of a smooth function over a flat triangle defined by the vertices $\mathbf{X}_1$, $\mathbf{X}_2$, and $\mathbf{X}_3$, we use the quadrature

$$\int_{\text{Triangle}} f(x, y)\, dx\, dy \approx A \sum_{i=1}^{N+1} f(\mathbf{x}_i)\, w_i \tag{B.6.8}$$

where A is the area of the triangle and

$$\mathbf{x}_i = \alpha_i \mathbf{X}_1 + \beta_i \mathbf{X}_2 + \gamma_i \mathbf{X}_3 \tag{B.6.9}$$

is the position of the base points. The coefficients α_i, β_i, γ_i and corresponding weights w_i are given in Table B.6.2. The location of the base points is illustrated in Figure B.6.1 (Zienkiewicz, 1971, p. 151; Cowper, 1973).

B.7 | FUNCTION APPROXIMATION

Given a certain amount of information about a function $f(x)$ within the interval $[a, b]$, we want to approximate the function with an Nth-degree *approximating polynomial* $P_N(x)$. *Weierstrass's theorem* guarantees that if $f(x)$ is a continuous function, then there is a polynomial of sufficiently high degree N so that $|f(x) - P_N(x)| < \varepsilon$ for any value of x between a and b, where ε is an arbitrarily small number. The issue at hand is how to compute this optimal polynomial.

TABLE B.7.1

The Legendre family of orthogonal polynomials. The zeros are given in Table B.5.1.

$$p_0(x) = 1, \qquad p_1(x) = x, \qquad p_2(x) = \tfrac{3}{2}x^2 - \tfrac{1}{2}x, \dots,$$

$$p_{i+1}(x) = \frac{2i+1}{i+1}\,x p_i(x) - \frac{i}{i+1}\,p_{i-1}(x)$$

Domain of definition: $c = -1, d = 1$
Weighting function: $w(x) = 1$

Orthogonality condition:

$$\int_{-1}^{1} p_i(x)\,p_j(x)\,w(x)\,dx = \delta_{ij}\,\frac{2}{2i+1}$$

In the *least-squares method*, we compute the coefficients of the polynomial by minimizing the integral of $|f(x) - P_N(x)|^2$ multiplied by a certain weighting function $w(x)$ over $[a, b]$. Unfortunately, this method results in a nearly-singular system of algebraic equations when N has a small or moderate value.

Orthogonal Polynomials

In an alternative method, we introduce a triangular family of *orthogonal polynomials* $p_i(x)$, where $i = 0, 1, 2, \dots$ and $p_i(x)$ is an ith-degree polynomial defined over a certain interval $[c, d]$ with associated weighting function $w(x)$. By construction,

$$(p_i, p_j) \equiv \int_{c}^{d} p_i(x)\,p_j(x)\,w(x)\,dx = D_{ij} \tag{B.7.1}$$

where **D** is a diagonal matrix. Two popular families of orthogonal polynomials are the Legendre polynomials and the Chebyshev polynomials both defined over the interval $[-1, 1]$, as shown in Tables B.7.1 and B.7.2. An important feature of the latter is that their magnitude is less than unity for any value of x.

Next we introduce the scaled variable $x' = c + (d - c)(x - a)/(b - a)$, which varies between c and d as x varies between a and b, and express the approximating polynomial $P_N(x)$ in terms of a weighted sum of orthogonal polynomials of a certain class as

TABLE B.7.2

The Chebyshev family of orthogonal polynomials.

$$T_0(x) = 1, \qquad T_1(x) = x, \qquad T_2(x) = 2x^2 - 1, \qquad T_3(x) = 4x^3 - 3x, \dots,$$

$$T_n(x) = 2x\,T_{n-1}(x) - T_{n-2}(x) = \cos(n\,\mathrm{arccos}\,x)$$

Domain of definition: $c = -1, d = 1$
Weighting function:

$$w(x) = \frac{1}{\sqrt{1 - x^2}}$$

Orthogonality condition:

$$\int_{-1}^{1} T_i(x)\,T_j(x)\,w(x)\,dx = \delta_{ij}\,\alpha\,\pi,$$

$$\text{where } \alpha = 1 \quad \text{if } i = 0, \text{ and } \alpha = \tfrac{1}{2} \text{ if } i \neq 0$$

$$P_N(x') = \sum_{i=0}^{N} a_i p_i(x') \tag{B.7.2}$$

where $x = a + (x' - c)(b - a)/(d - c)$. Finally, we compute the coefficients a_i using the orthogonality condition for $p_i(x')$ stated in Eq. (B.7.1), obtaining

$$a_i \equiv \frac{1}{D_{ii}} \int_c^d f(x') p_i(x') w(x') \, dx' \tag{B.7.3}$$

where summation over i is *not* implied in the denominator. The integral in Eq. (B.7.3) may be computed from the values of f at a set of data points using methods of numerical integration as discussed in Section B.6.

B.8   INTEGRATION OF ORDINARY DIFFERENTIAL EQUATIONS

Consider a system of N ordinary differential equations $dx_i/dt = f_i(x_1, x_2, \ldots, x_N, t)$, where $i = 1, \ldots, N$, subject to the initial condition $x_i(t=0)$. In terms of the vector of unknowns functions $\mathbf{x}$ and phase-space velocity $\mathbf{f}$, we obtain the compact form $d\mathbf{x}/dt = \mathbf{f}(\mathbf{x}, t)$. If the function $\mathbf{f}$ does not depend explicitly on t, the system is called *autonomous;* otherwise it is called *nonautonomous.* The space of the variables $\mathbf{x}$ comprises the *phase space* of the solution.

Euler's Method

Euler's method advances the solution over a small interval Δt by moving in the phase space by a small distance with the initial phase-space velocity,

$$\mathbf{x}^{n+1} = \mathbf{x}^n + \Delta t \, \mathbf{f}(\mathbf{x}^n, t^n) \tag{B.8.1}$$

Each step incurs a numerical error of order Δt^2.

Second-Order Runge-Kutta Method

This is a predictor–corrector method requiring two velocity evaluations for each time step. The solution is advanced according to the following five substeps:

1. Compute $\mathbf{f}(\mathbf{x}^n, t^n)$.
2. Set $\mathbf{x}^{\text{Temp}} = \mathbf{x}^n + \kappa \, \Delta t \, \mathbf{f}(\mathbf{x}^n, t^n)$.
3. Compute $\mathbf{f}^{\text{Temp}} = \mathbf{f}(\mathbf{x}^{\text{Temp}}, t^n + \kappa \, \Delta t)$.
4. Set $\mathbf{f}^{\text{Final}} = \alpha \mathbf{f}(\mathbf{x}^n, t^n) + (1 - \alpha) \mathbf{f}^{\text{Temp}}$.
5. Compute $\mathbf{x}^{n+1} = \mathbf{x}^n + \Delta t \, \mathbf{f}^{\text{Final}}$

where α is an arbitrary constant, and $\kappa = 1/[2(1 - \alpha)]$. Setting $\alpha = \frac{1}{2}$ and $\kappa = 1$, we obtain the *modified Euler method.* Each complete step in time incurs a numerical error of order Δt^3.

Fourth-Order Runge-Kutta Method

This advanced predictor–corrector method requires four velocity evaluations at each step. The solution is updated according to the following nine substeps:

1. Compute $\mathbf{f}(\mathbf{x}^n, t^n)$.
2. Set $\mathbf{x}^{\text{Temp1}} = \mathbf{x}^n + 0.5 \, \Delta t \, \mathbf{f}(\mathbf{x}^n, t^n)$.
3. Compute $\mathbf{f}^{\text{Temp1}} = \mathbf{f}(\mathbf{x}^{\text{Temp1}}, t^n + 0.5 \, \Delta t)$.
4. Set $\mathbf{x}^{\text{Temp2}} = \mathbf{x}^n + 0.5 \, \Delta t \, \mathbf{f}^{\text{Temp1}}$.
5. Compute $\mathbf{f}^{\text{Temp2}} = \mathbf{f}(\mathbf{x}^{\text{Temp2}}, t^n + 0.5 \, \Delta t)$.

6. Set $x^{Temp3} = x^n + \Delta t \, f^{Temp2}$.

7. Compute $f^{Temp3} = f(x^{Temp3}, t^n + \Delta t)$.

8. Compute $f^{Final} = \frac{1}{6}[f(x^n, t^n) + 2f^{Temp1} + 2f^{Temp2} + f^{Temp3}]$.

9. Set $x^{n+1} = x^n + \Delta t \, f^{Final}$.

Each complete step in time incurs a numerical error of order Δt^5.

B.9   COMPUTATION OF SPECIAL FUNCTIONS

The computation of special function is discussed in a comprehensive handbook of mathematical functions by Abramowitz and Stegun (1972); computer routines are provided by Press et al. (1986). Following, we discuss the computation of two such functions that find frequent usage in fluid mechanics.

Complete Elliptic Integrals of the First and Second Kind

The complete elliptic integrals of the first and second kind, denoted, respectively, by F and E, are defined as

$$F(k) = \int_0^{\pi/2} \frac{d\varphi}{\sqrt{1 - k^2 \sin^2 \varphi}}$$

$$E(k) = \int_0^{\pi/2} \sqrt{1 - k^2 \sin^2 \varphi} \, d\varphi \tag{B.9.1}$$

An efficient method of computing F and E is provided by the recursive formulae

$$F = \frac{\pi}{2}(1 + K_1)(1 + K_2)(1 + K_3) \cdots$$

$$E = F\left(1 - \frac{k^2}{2}P\right) \tag{B.9.2}$$

where

$$K_0 = k, \qquad K_p = \frac{1 - (1 - K_{p-1}^2)^{1/2}}{1 + (1 - K_{p-1}^2)^{1/2}}$$

$$P = 1 + \frac{K_1}{2}\left[1 + \frac{K_2}{2}\left(1 + \frac{K_3}{2}(\cdots)\cdots\right)\right] \tag{B.9.3}$$

Abramowitz and Stegun (1972, Chapter 17) provide alternative polynomial approximations.

Error Function

The error function is defined in terms of an integral as

$$\text{erf}(x) \equiv \frac{2}{\sqrt{\pi}} \int_0^x e^{-t^2} \, dt \tag{B.9.4}$$

where $\text{erf}(0) = 0$, $\text{erf}(\infty) = 1$, and $\text{erf}(x) = -\text{erf}(-x)$. The complementary error function is defined as $\text{erfc}(x) = 1 - \text{erf}(x)$, where $\text{erfc}(0) = 1$, $\text{erf}(\infty) = 0$.

In practice, the complementary error function is computed accurately and efficiently using polynomial approximations (Abramowitz and Stegun, 1972, p. 299). The formula

$$\text{erfc}(x) = t(0.254829592 + t\{-0.284496736 + t[1.421413741 \\ + t(-1.453152027 + 1.061405429t)]\})e^{-x^2} \tag{B.9.5}$$

where

$$t = \frac{1}{1 + 0.3275911x} \tag{B.9.6}$$

and $x > 0$, yields results that are accurate at least up to the seventh decimal place. The error function follows the form $\mathrm{erf}(x) = 1 - \mathrm{erfc}(x)$.

References

Abramowitz, M., and Stegun, I. E., 1972, *Handbook of Mathematical Functions.* Dover.

Atkinson, K. E., 1989, *An Introduction to Numerical Analysis.* Wiley.

Broyden, C. G., 1965, A class of methods for solving nonlinear simultaneous equations. *Math. Comp.* **19,** 577–93.

Carnahan, B., Luther, H. A., and Wilkes, J. O., 1969, *Applied Numerical Methods.* Wiley.

Cowper, G. R., 1973, Gaussian quadrature formulas for triangles. *Int. J. Num. Meth. Eng.* **7,** 405–408.

Dahlquist, G., and Björck, Å. 1974, *Numerical Methods.* Prentice Hall.

Davis, P. J., 1975, *Interpolation and Approximation.* Dover.

Eisenstat, S. C., Elman, H. C., and Schultz, M. H., 1983, Variational iterative methods for nonsymmetric systems of linear equations. *SIAM J. Numer. Anal.* **20,** 345–57.

Fletcher, C. A. J. 1988, *Computational Techniques for Fluid Dynamics.* 2 volumes. Springer-Verlag.

Golub, G. H., and Van Loan, C. F., 1989, *Matrix Computations.* Johns Hopkins University Press.

Hageman, L. A., and Young, D. M. 1981, *Applied Iterative Methods.* Academic.

Henrici, P., 1986, *Applied and Computational Complex Analysis,* Vol. 3. Wiley.

Press, W. J., Flannery, B. P., Teukolsky, S. A., and Vetterling, W. T., 1986, *Numerical Recipes, The Art of Scientific Computing.* Cambridge.

Saad, Y., and Schultz, M. H., 1986, GMRES: A generalized minimal residual algorithm for solving non-symmetric linear systems. *SIAM J. Sci. Stat. Comput.* **7,** 856–69.

Stroud, A. H., 1971, *Approximate Calculation of Multiple Integrals.* Prentice Hall.

Tolstov, G. P., 1962, *Fourier Series.* Dover.

Wilkinson, J. H., 1965, *The Algebraic Eigenvalue Problem.* Oxford.

Zienkiewicz, O. C. 1971, *The Finite Element Method for Engineers,* McGraw-Hill.

Index

A

Acceleration
  of a body in irrotational flow, 330
    of a sphere, 333
  of a point particle, 8
  reaction of a moving body
    in irrotational flow, 327
    in unsteady Stokes flow, 310–11
Acoustic streaming motion, 307
Added mass tensor, 322, 327
  grand, 323, 328
ADI method,
  for the convection equation, 588
  for the convection-diffusion equation, 596
  for the diffusion equation, 573, 575
AFI method, 576
Airfoil,
  flow around computed by a panel method,
    493
  Joukowski, 372
  lift force exerted on, 348
  shape by conformal mapping, 372
  slender-body theory, 356
Airy stress function, 296
Alternating matrix, 632
Amplification factor, 565
Analytic function, 359
Angular momentum
  of a fluid parcel, 13
  balance, 99,140
Annular
  effect, 199
  layer, instability of, 476
Annulus, flow through, 187, 202
Approximate factorization, 573
Approximation of a function, 657
Artificial compressibility method, 628
Atwood ratio, 462, 538
Axisymmetric flow, 1
  induced by vorticity, 83, 522, 551
  vorticity transport equation, 144

B

Bairstow's method, 648
Baroclinic generation of vorticity, 140
Barotropic fluid, 115, 141
Basis functions, 492
Basset-Boussinesq force, 310–11
Beltrami field or flow, 26, 30, 63
  extended, 151
  generalized, 141, 148–9
Bernoulli
  equations, 117, 315
  function and constant, 116
Bickley jet, 454, 457–8
Biharmonic equation and function,
  definitions, 60
  mean-value theorems for, 60–61
Bilinear interpolation, 651
Binormal vector, 27
Biot-Savart integral, 64
Blasius
  boundary-layer equation, 394
  theorems for the force and torque, 361
Body force, 95
Bond number, and its inverse, 138, 169, 173
  for film flow, 250, 468
Boundary condition,
  Dirichlet, 367, 602
  for a flow, 120
    at a contact line, 122
    derivative, 126
    in-flow and out-flow, 126
    at infinity, 126
    at an interface, 123
  no-penetration, 120
  for the pressure, 614
  no-slip, 121
  slip, 121
  Neumann, 367, 603
Boundary-element method,
  for potential flow, 482, 491
  for Stokes flow, 290

Boundary-integral representation and
    equations,
  see also integral equations
  for potential flow, 56, 59, 482
    axisymmetric, 486
    completed for exterior flow, 516
    generalized, 495
    three-dimensional,
      due to an oscillating drop, 504
      past a stationary or translating body,
      319, 485, 499, 503–5
    two-dimensional, 487
      past a stationary body, 487, 500
  for Stokes flow, 280, 287, 297
    for the pressure, 285
    past a solid body or particle, 284
    two-dimensional, 285
Boundary layer, 385
  at a free surface, 392
  in a homogeneous fluid, 393
  at an interface, 393
  Prandtl, on a solid surface, 385–6
    in accelerating or decelerating flow,
      401
    axisymmetric, 413
    computation of, 405
    around a cylinder, 410
    equations, 388
    of Falkner and Skan type, 401
    along a stationary flat plate, 394
    along a moving flat plate, 401
    99% thickness, 396
    around a sphere, 414
    stability of, 397, 455, 458
    oscillatory, 417
    separation, 314, 407
    three-dimensional, 392, 415
    unsteady, 416
Boussinesq approximation, 539
Boussinesq-Basset force, 310–11
Bromwich integral, 427
Brush paint model, 190
BTCS method
  for the convection equation, 583,
    586
  for the convection-diffusion equation,
    592
  for the diffusion equation, 567, 572
Bubble
  rising in potential flow, 334
  moving in Stokes flow, 269, 287
    Faxen laws for, 294
  expanding or contracting, 316
Butler's theorem, 335
Buoyancy force, 156
Burgers
  equation, 586, 595
  vortex, 142

C

Capillary
  cell, 161–2
  instability, 473
  length, 161
  tube, 162
  waves, 464
Catenoid and catenary, 165
Cat's eye pattern, 90
Cauchy's equation of motion,
  see equation of motion
Cauchy-Green strain tensor, 9
Cauchy-Riemann equations, 296, 359
Cavity
  Stokes flow within, 289
  flow inside, computed with a finite-difference
    method, 601, 609
Centrifugal force, 102
CFL stability criterion, 580
Channel, *see also* tube with parallel-sided
    walls
  oscillatory flow within, 197
  steady flow within, 181
  unsteady flow within, 194
  transient flow within, 198
  square, free surface flow, 191
Characteristic polynomial, 646
Chebyshev polynomials, 658
Circle,
  expanding and stretching, 19
  Milne-Thomson theorem, 363
  Gershgorin's theorem, 646
Circular cylinder, *see* cylinder
Circulation
  around a loop, 31, 139
  along a two-dimensional vortex sheet, 37
Coating die, 245, 495
Collocation method, 341, 352, 492
Compatibility condition for the PPE, 615
Complex
  lamellar field, 30, 63
  potential for irrotational flow, 358
Complex variable formulation,
  of two-dimensional potential flow, 357
  of two-dimensional Stokes flow, 296
Compound matrix method, 454
Compressibility, artificial, 628
Concave shape, 161
Cone of influence, 581
Conformal mapping, 364
Conjugate-gradients method, 645
Conservation of mass, 11
Conservative form of a transport equation, 561
Consistency, of a finite-difference method, 559
Constitutive equation
  for the stress tensor, 105
  for the jump in the interfacial traction, 124

Contact angle, 122, 159
  hysteresis, 123
Contact line, 122, 159
Continuity equation, 11
Contour dynamics method,
  for axisymmetric flow, 552
  for two-dimensional flow, 548
Convection
  equation, 577, 587
  number, 578
Convection-diffusion equation, 558, 589, 596
Convergence, of a finite-difference method, 560
Convex shape, 161
Coordinate invariance principle, 105
Coriolis force, 102
Corner,
  flow due to a point source, 214
  potential flow around, 359
  Stokes flow around or within, 227, 234
Couette flow,
  circular, 188, 478
  in a two-dimensional channel, 181, 197
    stability of, 426, 428, 437, 458
Couple, *see* torque
Couplet, *see* rotlet
Courant number, 579
Crank-Nicolson method
  for the diffusion equation, 568, 573, 576-7,
    583
  for the convection-diffusion equation, 592
Creeping flow, *see* Stokes flow
Critical plane and layer, 436, 440
CTCS method,
  for the convection equation, 582
  for the convection-diffusion equation, 591
  for the diffusion equation, 566
Curvature
  of a line, 15
  of a surface, 21, 160, 162
    mean, 21, 125, 162
Curvilinear coordinates
  in space, 7, 635
  in a surface, 15, 163, 177
Cyclic
  constant of the flow around
    a vortex tube, 32
    a boundary, 42
  frequency of a disturbance, 428
Cylinder,
  circular,
    boundary layer around, 410
    drag coefficient, 390
    irrotational flow past, 351, 363
    translating or rotating in a viscous fluid,
      201, 272, 274, 300
    translating in potential flow, 349, 362
    uniform potential flow past, 351, 362
  elliptical, potential flow past, 370

D

D'Alembert's paradox, 329
Darcy's law, 465, 482
Dean flow, 479
Deflation
  of the eigenvalue spectrum,
    of a matrix, 512, 646
    of an integral equation, 512–4
  of nonlinear algebraic equations, 649
Deformation gradient, 9
Delta method,
  for line vortices, 553
  for vortex sheets, 543
Derivative
convective, substantial, substantive, material, 7
  numerical computation of, 652
Determinant,
  of an arbitrary matrix, 642
  of a tridiagonal matrix, 646
Diagonally dominant matrix, 643
Differential equations
  ordinary,
    eigenvalues of, 452
    numerical solution of, 659
  partial,
    finite-difference methods for, 558
Diffusion
  equation,
    in one dimension, 562
    in two and three dimensions, 571
  fourth-order, 570
  number, 563
  numerical, 584
Diffusivity, 558
Direction cosines, 5
Dirichlet boundary condition, 367
Discretization of a differential equation, 559
Disk,
  rotating in a viscous fluid, 211
  translating in potential flow, 339
Dispersion
  equation, 570
  length of point vortices, 522
  relation, 464
Displacement thickness, 397
Dissipation, *see* energy dissipation
Disturbance,
  unstable, marginally stable, and stable, 425
  most unstable or most dangerous, 429
Divergence theorem, 640
Double-layer potential or integral
  of potential flow, 56, 483
    completed representation, 516
    eigenvalues and eigenfunctions of, 508
    generalized representation, 495
    properties of, 501–3
  of Stokes flow, 282, 284, 287

Drag
  crisis, 390
  coefficient of
    the boundary-layer on a flat plate, 397
    a cylinder, 390
    a sphere, 273, 300
    a spherical bubble, 334
Drop
  oscillations, 504
  resting on a flat plate, 173, 177
  in Stokes flow, 269, 286–7
    Faxen laws for, 294
Dufort-Frankel method
  for the convection-diffusion equation, 591
  for the diffusion equation, 566

E

Eddies at a corner, 237
Eigenvalues
  of an integral operator, 507
  of a matrix, 430, 646
  deflation, 512
  of the linearized equation of motion, 428
Ekman flow, 191
Elephant, 138
Elliptic integrals, complete, 380, 660
Energy
  dissipation, 99
    in a boundary layer, 390, 392
    in irrotational flow, 316
    in a Newtonian fluid, 108, 111
    minimum principle for Stokes flow,
      226
  integral balance, 100
    for a Newtonian fluid, 112
    in Stokes flow, 225
  internal, 100
  kinetic, 99
  potential, 99
Enstrophy, 148, 522
Entire function, 359
Equation of motion,
  Cauchy's for a fluid, 97
  for a Newtonian fluid, 114
    in cylindrical polar coordinates, 102
    in the presence of interfaces, 127
    in a noninertial frame, 101
    exact solutions of, 179
    in spherical polar coordinates, 103
  for a translating sphere, 311, 333
Equivalence theorem, Lax's, 561
Error function, 660
Ertel's theorem, 148
Euler's
  equation, 115, 137
  method and modified method, 659
  theorem in differential geometry, 21

Eulerian description,
  of a flow, 6
  of a surface or interface, 23
Even-odd coupling, 583
Evolution equations of a flow, 422
Extensional flow, 30
  uniaxial, 142
Extensive variable, 12

F

Fading memory principle, 106
False transients methods, 606, 627
Fast Fourier Transform, FFT, 652
Faxen relations, 278, 292, 309
Fictitious forces, 102
Field of a function
  Beltrami, 26, 30, 63, 149
  complex lamellar, 30, 63
  convected by the flow, 8
  irrotational or potential, 3
  solenoidal, 12
Film
  flowing down an inclined plane, 181, 249
    stability of, 466
  leveling on a horizontal surface, 250
  two-layered, 191
    stability of, 473
Finite-difference
  equation, FDE, 560
  methods for
    the boundary-layer equations, 411
    the convection-diffusion equation, 558
    developing flow in a tube, 200
    incompressible Newtonian flow, 599
    the Orr-Sommerfeld equation, 450
    Rayleigh's equation, 449
Fixed-point iterations
  for linear equations, 643
  for nonlinear equations, 647
Fjørtoft's condition for instability, 439
Flow
  axisymmetric, 1
  at high Reynolds numbers, 137
  at low Reynolds numbers, 136
  nearly steady, 424
  net, 359
  rate through a material surface, 19
  stable or unstable, 425
  steady, 1, 424
  two-dimensional, 1
Fluid parcel
  angular velocity of, 3
  mass of, 11
  translation of, 3
  rotation of, 3
  expansion of, 4
  properties of, 10

Flux-corrected transport, 585
Force exerted on a
  body
    under hydrostatics, 156
    oscillating at high frequencies, 418
    in potential flow
      three-dimensional, 325, 330
      two-dimensional, 348
    in Stokes flow, 277
  boundary, 97
    in a Newtonian fluid, 111, 128
  fluid parcel, 95
  of a Newtonian fluid, 110
Form drag, 111
Fourier interpolation, 651
Fractional step method, 576, 597
Free
  surface
    definition of, 110
    vorticity and stress at, 131–2
  shear layer, 393, 401, 455
  see also shear layer
Fredholm
  alternative, 508
  theorems for integral equations, 289
  see also Riesz-Fredholm theory
Frenet or Frenet-Serret relations, 15, 28
Froude number, 136, 466
FTBS method
  for the convection equation, 579
FTCS method,
  for the convection equation, 578
  for the convection-diffusion equation,
    589
  for the diffusion equation, 563, 572
FTFS method
  for the convection equation, 579
Fundamental
  form of a surface, 20
  potentials of irrotational flow due to a
    sphere, 332
    three-dimensional body, 322
    two-dimensional body, 347
  solutions; *see* singularities
  theorem of vector analysis, 70

G

Gain, 565
Galerkin's method, 451, 492
Gauss
  elimination, 641
  quadratures, 654
  -Legendre quadrature, 655
  -Seidel method, 644
Generalized Beltrami flow, 141,
    148–9
Gershgorin's circle theorem, 449, 646
Gibbs surface elasticity, 124

Görtler
  instability, 478
  transformation, 412
Gravity
  -driven flow, 181
  waves, 464, 489
Green's functions
  of Laplace's equation,
  see also point source
    definition of, 49
    in free space, 50, 54
    with multiple poles, 53
    multipoles of, 53
    periodic, 53
    properties of, 51
    two-dimensional, 53, 364, 370
  of Stokes flow, 253
    properties of, 257
    symmetry of, 278
  of unsteady Stokes flow, 302, 309
  of the vorticity transport equation, 427
Green's identities,
  first and second, 49
  third, 57, 61
Grid, finite-difference, 559
Group velocity, 435
Growth
  factor of a finite-difference method, 565
  rate of a disturbance, 428–9

H

Hadamard-Rybczynski solution, 270
Hagen-Poiseuille flow, *see* Poiseuille flow
Harlow and Welch method, 617
Harmonic function or potential, 44, 315
  behavior at infinity, 58
  extrema of, 57
  mean-value theorems for, 57
Hele-Shaw cell, 244, 482
  instability of an interface within, 465
Helical line vortex, 83
Helicity, 521
Helmholtz
  decomposition theorem, 70, 423
  equation, 150, 164
  instability, 461
    computed using a point-vortex method, 544
    of a finite vortex layer, 550
  theorems of vortex motion, 139
  velocity of a vortex ring, 551
Hill's vortex, 86, 118
Hinged plates, flow between, 243
Hodge decomposition theorem, 70
Holstein-Bohlen parameter, 408
Hopscotch method, 593, 596
Howard's semi-circle theorem, 440
Hydrostatics, 155
  pressure distribution, 119, 155

I

Ideal fluid, 108
Identity matrix, 632
Impulse of fluid motion, 82
  in terms of the vorticity, 521
Incompressible fluid, 12
  equation of motion for, 120
Index notation, 632
Inflection-point instability, 439
Influence coefficients, 290, 492
Initial-value problem
  in hydrodynamic stability, 426
Instability, *see also* stability
  absolute and convective, 421
  spatial and temporal, 431
Integral equations
  see also boundary integral equations
  of potential flow, 484
    regularization of, 511
    theoretical investigation of, 506
  of Stokes flow, 287
Integral representation of the velocity
  see also boundary integral representation
  in terms of the vorticity, 64
    and rate of expansion, 70
Integration, numerical, 652
Intensive variable, 12
Interfaces
  instability of, 458–77
  jump of traction across, 123
  generation of vorticity at, 145
  hydrostatic, 159
    axisymmetric, 171
    three-dimensional, 177
    two-dimensional, 166
Interfacial flow, computation of, 600
Interpolation of a function, 649
Invariants
  of tensors, 5
  of vortex motion, 521
Inviscid fluids, 108
Irreducible loop, 40
Irrotational flow, *see also* potential flow
  definition and applications of, 3, 314, 482
  description of, 40
  distinction with inviscid flow, 316
  equations of, 315
  three-dimensional
    past or due to a body, 319, 325, 337, 499
    due to the motion and deformation of a
      body, 321
  two-dimensional,
    past or due to a body, 245, 344, 348, 357,
      370
Isoparametric representation, 492
Iterations, fixed-point,
  for algebraic equations, 643
  for integral equations, 508, 517

J

Jacobi's method
  for eigenvalues, 647
  for linear systems, 643
Jacobian
  of the Lagrangian mapping function, 8, 10
  of a vector function, 648
Jeffery-Hamel flow, 214, 243
Jet
  axisymmetric, 415
  Bickley, 454, 457–8
  capillary instability of, 473
  stability of, 457
  two-dimensional, 400
  viscous, 476
Joukowski airfoil and transformation, 372, 493

K

Kelvin
  circulation theorem, 139
  functions, 199
  minimum kinetic energy theorem, 45
Kelvin-Helmholtz instability, 459
Kinematics of a flow, 1, 39
Kinetic energy
  of a potential flow, 44
    due to a body, 322, 324, 347
  in terms of the vorticity, 78, 521
Kirchhoff's vortex, 547
Korteweg-de Vries equation, 595
Kovasznay flow, 152
Kutta-Joukowski condition, 372
Kutta value of the circulation, 356

L

Label of a point particle,
  in a parcel, 6
  along a material line, 14
Lagrange interpolation, 650
Lagrangian
  labels, 6
  description of a flow, 6
  mapping, 8
Laplace's
  equation, 44
  transform, 427
Laplace-Young equation, 159
Laplacian, numerical computation of, 654
Lapse rate, 159
Laurent expansion of the complex potential, 360
Lax's method,
  for the convection equation, 581, 587
  for the convection-diffusion equation, 589
Lax-Wendroff method, 581
  multi-step, 585
Least-squares method, 658

Legendre polynomials, 658
Leonard method, 591
Line
    vortex, 33
        flow due to, 79
        motion of, 553
        helical, 83
    vortex ring, 37
        velocity due to, 84
Linear
    algebraic equations, 641
    momentum
        integral balance, 98
        of a fluid parcel, 13
Local-induction approximation, LIA, 82, 554
Lorentz reciprocal theorem or identity
    generalized for Navier-Stokes flow, 112
    for Stokes flow, 276
LU decomposition, 641
Lubrication flow
    assumptions of, 244
    in a rocker bearing, 246
Lyapunov surface, 289, 483

M

MAC method, 617
MacCormack method, 585, 593–4
Magnus force, 352
Mangler's transformation, 414
Mapping
    conformal, 364
    Lagrangian, 8
    function, isochoric, 12
Mass conservation, 11
Material
    derivative, 7
    line, 14
    objectivity, 106
    surface, 15
        Eulerian description of, 23
    vector, 13, 20
Mean curvature, see curvature
Mean-value theorems for
    biharmonic functions, 60–1
    harmonic functions, 57, 60
    Stokes flow, 223, 226, 294
Meniscus of a liquid
    in a circular tube, 162
    in a container, 170
    between two vertical plates, 160, 169
    pinned to an inclined plate, 164, 166
Metric coefficient
    for the arc length, 14
    for a surface, 17
    tensor for a surface, 20
    tensor for curvilinear coordinates, 636
Meusnier's theorem, 21

Mises, von, transformation, 394
Möbius transformation, 366
Modified
    dynamics, 606, 627
    pressure, 111
    differential equation, 560
Momentum, see linear or angular momentum;
    thickness, 398
Monotone scheme, 584
Motion, equation of, 97
Multiply-connected domain, 42
Multipole expansion,
    for Stokes flow past a body, particle, or drop,
        284, 286

N

Navier-Stokes equation, 114
    in cylindrical polar coordinates, 114
    dimensionless form of, 136
    exact solutions of, 179
    linearized, 137
    scaling of, 135
    in spherical polar coordinates, 115
Neumann
    boundary condition, 367
    function, 50
    series, 508
    stability analysis, 565
Newton's
    method for solving nonlinear equations, 648
    second law of motion, 94
    third law of action and reaction, 97
Newton-Raphson method, see Newton's
    method
Newtonian fluid, 94, 107–9
    generalized, 107
No-penetration condition, 120
No-slip condition, 121
Nonlinear algebraic equations, 647
Normal
    mode,
        analysis, 422, 426, 428
        stable and unstable, 429
    plane to a line, 27
    vector
        to a line, 15
        to a surface, 16
Numerical methods, 641
Nÿstrom method, 517

O

Operator splitting, 576, 588, 593, 623
Ordinary differential equations, 659
Orr-Sommerfeld equation, 432
    numerical solution of, 449
Orthogonal polynomials, 658

Oscillatory flow
  in a circular tube, 197
  in a channel with parallel-sided walls,
    197
  due to the rotation of a cylinder, 202
  above a plate, 193
  unsteady Stokes flow, 301
  due to a vibrating plate, 192
Osculating plane, 27
Oseen
  flow, 138, 222, 299
  vortex, 145
Oseen-Burgers tensor
  for three-dimensional flow, 254
  for two-dimensional flow, 257
  unsteady, 302

P

Pairing interaction of point vortices,
    528
Panel method, 482
Parallel computing, 518
Particle,
  fluid, flow past, 286–7
  methods, for vortex motion, 555
  point, *see* point particle
  solid,
    oscillatory flow past, 304
    Stokes flow past, 277, 283
    in unsteady motion, 311
  vibrating, 304
Path lines, 29
Péclet number, 589
Penalty-function formulation, 629
Periphractic domain, 58
Permanence of irrotational flow, 139
Permanent translation, 331
Permeability of a porous medium, 465,
    482
Phase
  angle, 565
  space, 659
  velocity, 433
Pipe, *see* tube or channel
Plate
  impulsively moving in a viscous fluid,
    194
  oscillating in a viscous fluid, 192
  potential flow past, 356, 371
  scraping a wall, 230
  stagnation-point flow on, 205
Pohlhausen method and polynomial, 406–7
Point force at finite Reynolds numbers, 216
Point force in Stokes flow,
  three-dimensional, 253, 260
    dipole, 260
    in free space, 254

    above a plane wall, 255
    quadrupole, 263
  unsteady, 302
  two-dimensional, 257
  in free space, 257, 297
Point particle, 1
  acceleration, 8
  angular velocity, 3
  velocity, 1
Point source, 46
  complex potential of, 358
  periodic array of, 358
  in potential flow
    in the presence of a body, 326
    in the presence of a cylinder, 363
    in the presence of a sphere, 334, 336
    between two parallel walls, 364
      with a bottom, 489
    in Stokes flow, 259, 284
  oscillatory, 305
  above a plane wall, 264, 285
  stream-function of, 73–4
Point-source dipole, 46, 260
  complex potential of, 358
  in the presence of a body, 326
Point vortex,
  in a confined domain, 524
  near a corner, 526
  in the presence of a cylinder, 363, 370, 526
    a pair of, 531
  complex potential of, 358
  definition and strength, 36
  diffusing, 144
  dipole, 90
  flow due to, 89, 523
  invariants of the motion, 523
  method
    for vortex sheets, 540
    for flows with distributed vorticity, 545
  motion of, 523
  pair of, 524
  polygonal array of, 531
  between two parallel walls, 527, 531
  infinite periodic array of, 89, 358
    complex potential of, 358
    stability of, 527
  above a wall, 525
Poise, 107
Poiseuille flow through
  an annular tube, 187
  a two-dimensional channel, 181
    stability of, 458
  a circular tube,
    oscillatory, 198
    stability of, 425, 458
    steady, 183
    transient, 199
  an elliptical tube, 185

a rectangular tube, 186
a triangular tube, 185
with partitions, 190
Poiseuille law, 184
Poisson
equation, 571
finite-difference solution of, 575
integral, 59, 62, 185, 187
Porous
medium, instability of immiscible fluids, 465
plate, flow over, 153
Potential
dipole, *see also* point-source dipole
coefficient of, flow past or due to the motion of a body, 321
relation to added mass, 324
double-layer
see also double-layer potential
for irrotational flow, 56, 483, 500
for Stokes flow, 282, 284
flow, 3, 314, 482
see also irrotational flow
integral representation of, 55
maximum of the velocity, 58
function, 12, 40
vector, 39, 62
harmonic, 12, 39, 43
single-layer
see also single-layer potential
for irrotational flow, 56, 483
for Stokes flow, 282
Power
-law fluid, 107
method for eigenvalues, 646
Prandtl
-Batchelor theorem, 119
boundary-layer theory, 314, 385
transposition theorem, 393
Predictor-corrector method for the convection equation, 584
Pressure
boundary conditions, 614
correction method, 423, 623
-driven flow in a channel or tube, 181
hydrodynamic or reaction, 106
jump across an interface, 125
across a layer, 156
minimum in irrotational flow, 318
modified, 111
Poisson equation, PPE, 423, 613
projection methods, 423
within a rotating fluid, 157
thermodynamic, 105
within a vibrating container, 157
Primitive variables, 520, 617

Principal value
of the velocity of a vortex sheet, 36
of the double-layer potential
for potential flow, 52, 484
for Stokes flow, 288
Projection methods, 623

Q

QR decomposition, 647
Quadrupole
of the point source, 260
of the point force, 263
Quasi-Newton methods, 648

R

Random walk, 546
Rankine vortex, 548
Rate of
change of circulation around a loop, 33
deformation tensor, 2
expansion or dilation
of a parcel, 4, 43
of a material surface, 17
elongation or extension of a fluid, 110
extension of a material line, 14
relative displacement, 2
strain tensor, 2
Rayleigh
capillary instability of a cylindrical column, 473
equation,
for an expanding bubble, 317
of hydrodynamic stability, 435, 449
stability criteria, 439
for swirling flow, 477
-Taylor instability, 462
computed using a point-vortex method, 545
within a tube, 463–5
Reciprocal identity or theorem
for harmonic functions, 49
for Navier-Stokes flow, 112
for solenoidal fields, 12
for Stokes flow, 276, 279
Rectifying plane, 27
Rectilinear
line vortex, *see* point vortex
flow in channels and tubes,
steady, 180
unsteady, 192
Reducible loop, 40
Reiner-Rivlin fluid, 106
Resistance tensor of Stokes flow,
for the force and torque, 278
for the traction, 277
Resistive force theory, 275
Reversibility of Stokes flow, 224

Reynolds
 lubrication equation, 251
 number, 136
  cell, 589
 transport theorem, 10
Riccati equation and method, 436, 453
Richtmyer's method, 584
Riemann mapping theorem, 369
Riesz-Fredholm theory, 508
Rigid-body motion, 3
 potential flow due to, 321, 346
  Stokes flow due to, 264, 278, 283, 292
Rod,
 coated, 191
 slender, moving in a viscous fluid, 274
Romberg integration method, 654
Rotating
 disk in a viscous fluid, 211
 fluid, instability of, 477
Rotation matrix, 4
Rotational flow, 3
Rotlet, 64, 260
 unsteady, 307
Roton, 263
Runge-Kutta method, 659

S

Scaling of the Navier-Stokes equation,
  135
Schwarz-Christoffel transformation, 377
Scraper, flow due to, 230
Secondary field, 39
Semi-circle theorem, 440
Separated flow, 314, 390
Shear flow or layer, 393
 stability of, 439, 441, 450, 453–5
Shear-driven flow, 181
Sheet vortex, *see* vortex sheet
Similarity
 solution for
  a diffusing point vortex, 144
  a diffusing vortex sheet, 142
  an impulsively moving plate, 194
 transformation, 647
Similitude of flows, 137
Simple
 fluid, 106
 shear flow, 118, 181
Simpson's rule, 654
Single-layer potential or integral
 of potential flow, 56, 483, 496
  generalized representation of, 495
  properties of, 497–8
 of Stokes flow, 282
Singly-connected domain, 40
Singularities
 of potential flow, 46

of oscillatory Stokes flow, 305
 of Stokes flow, 259
Singularity representations
 of potential flow,
  three-dimensional, 337
  two-dimensional, 356
 of Stokes flow, 264, 273
 of unsteady Stokes flow, 309
Skin friction, 111
 vector and lines, 129
Slab, settling in a channel, 189
Slender-body theory
 for potential flow
  three-dimensional, 341
  two-dimensional, 356
 for Stokes flow, 274
Slip boundary condition and coefficient, 122
Sloshing of a liquid in a tank, 487
Smoothing, numerical, 543
Solenoidal
 projection, 623
 vector field, 12
SOR, 644
Spatial instability, 433
Spectral element method, 518
Spectral radius
 of a matrix, 564
 of the double-layer operator, 508
Spectrum of eigenvalues,
 of normal modes, 428
  continuous part of, 437
 of the double-layer operator, 508
Sphere
 in axisymmetric potential flow, 335
 expanding and stretching, 19
 in linear Stokes flow, 267
 in paraboloidal Stokes flow, 279
 potential flow past, 333
 translating in
  Oseen flow, 299
  potential flow, 332
  Stokes flow, 264
 translational oscillations of, 310
 rotating in Stokes flow, 268
 rotational oscillations of, 310
 settling or rising in
  potential flow, 333
  unsteady Stokes flow, 311
 unsteady motion in potential flow, 333
Spherical drop
 translating in potential flow, 392
 translating and rotating
  in Stokes flow, 269, 286–7
   Faxen laws, 295
 oscillations of, 310
Spheroidal coordinates, 338–9
Spline interpolation, cubic, 165, 650
Squire's theorems, 431

Stability,
hydrodynamic, 421
see also instability
linear analysis and theory, 422, 425
flow with a linear velocity profile, 436
temporal and spatial, 431
numerical, of a finite-difference method, 560
Staggered grid, 618
Stagnation point
definition, 24
at a wall in two-dimensional flow, 131
Stagnation-point flow, 203
in Stokes flow, 228
at a free surface, 134
at an interface, 134
Steepest descent method, 645
Stokes
boundary layer, 193
differential equation, 436
equation, 136, 222
law, 266
number, 136
theorem of vector calculus, 640
Stokes flow,
steady and quasi-steady, 136, 222
equations and properties of, 222
singularities of, 259
near a corner, 234
due to the in-plane motion of a wall, 232
due or past bodies and drops, 264
due to a point source in a plane, 232
due to a plunging tape, 232
unsteady, 137, 301
equations and properties of, 301
Stokeslet
three-dimensional, 254
doublet, 260
unsteady, 302
symmetric quadrupole, 306
two-dimensional, 258
Stokeson, 263
oscillating, 307
Streak line, 29
Stream function
Helmholtz for two-dimensional flow, 72
in terms of the vorticity, 73, 89
in Stokes flow, 224
Stokes for axisymmetric flow, 74
in terms of the vorticity, 84
in Stokes flow, 224
for three-dimensional flow, 75
-vorticity formulation, 600
Stream surface and tube, 25
Streaming motion in oscillatory flow, 307
Streamline
definition and computation of, 24

dividing, in two-dimensional flow, 132
coordinates, 26
velocity variation around, 28
in irrotational flow, 317
Stress tensor
constitutive equations, 105
in cylindrical polar coordinates, 109
deviatoric part of, 106
modified, 111
for a Newtonian fluid, 108–9
in spherical polar coordinates, 108–9
symmetry of the stress tensor, 99
is a tensor, 97
in terms of tractions, 96
Stress-momentum tensor, 98
Stresslet, 262, 285
Stresson, 263
Strouhal number, 136, 390
Symmetry of Green's functions,
of potential flow, 52
of Stokes flow, 278
Substitution
forward and backward for linear systems, 641
successive, *see* successive substitutions
Successive
approximations or substitutions
for algebraic equations, 643
for integral equations, 508, 517
over-relaxation, SOR, 644
Superparametric representation, 492
Surface
differential geometry of, 15, 20
divergence, 17
elasticity, 124
material, *see* material surface
stress, 95
Swirling flow
instability of, 477
due to the rotation of a cylinder, 188, 202
in Stokes flow, 224, 226

T

Taylor
cellular flow, 149
instability of circular Couette flow, 478
Temporal instability, 433
Tension, surface or interfacial, 124
Tensors, 4
Terminal velocity of a
settling sphere in Stokes flow, 266
settling bubble or drop in Stokes flow, 270
rising drop in potential flow, 334
Thomas algorithm, 642
Thread, instability of, 476
Thwaites correlation, 411
Tollmien-Schlichting waves, 455

Torque exerted
  on a fluid parcel, 98
    of a Newtonian fluid, 110
  on a body or particle
    in potential flow, 324
    in Stokes flow, 278
  on a boundary, 98
    in a Newtonian fluid, 111
  on a rigid boundary, 128
Torsion of a line, 27
Traction, 95
  discontinuity across an interface, 123
  in a Newtonian fluid, 110
  rate of working of, 112
  on a rigid boundary, 128
  normal to a stream line, 135
Transformation,
  Möbius, 366
  similarity, 5
Trapezoidal rule, 654
Tridiagonal
  matrix, eigenvalues of, 571
  systems of equations, 642
Trigonometric interpolation, 651
Tube, flow through,
  see also channel
  annular, 187, 202
  with arbitrary cross-section, 187
  circular, 183, 185, 198, 458
  elliptical, 185, 192, 495
  two-layered, 191
  rectangular, 186
  tapered, 191
  triangular, 185
  with partitions, 190
Two-dimensional flow, 1, 131
  induced by vorticity, 88, 522
    surface vorticity, skin friction, and dividing
      streamlines, 131–4
  vorticity at an interface, 147
Two-way flow, 141

U

Unidirectional flow,
  nearly, 189, 243
  stability of, 430, 439
  steady, 180
  unsteady, 192
Uniqueness of solution
  of irrotational incompressible flow, 45
  of Stokes flow, 225
Unit matrix, 632
Unsteadiness parameter, 136
Unsteady
  Stokes equation and flow, 137, 301
    oscillatory, 301
  unidirectional flow, 192

Upwind differencing,
  for the convection equation, 580
  for the convection-diffusion equation, 590

V

Vandermonde matrix, 649
Vector potential, 62
  -vorticity formulation, 611
Vector products, 633
Velocity
  of the fluid, 1
  -pressure formulation, 613
  -vorticity formulation, 611
Velocity gradient tensor, 2
Vibrating container, 157
VIC, 546, 555
Virtual inertia or mass, 327
Viscosity
  dilatational, 108
  expansion, second coefficient of, 108
  dynamic, 107
  kinematic, 114
Viscous fluid, 106
Volume potential, 59
von Kármán
  momentum integral balance, 399, 405, 413
von Kármán-Pohlhausen method, 406
Vortex, 76, 314, 520
  axisymmetric, 83
    with linear vorticity distribution, 86, 117,
      552
  blob, 145, 532
  flow due to, 76
  force, 116
  line, 31
  layer of constant vorticity, 118, 550
    compound, 448
    at a wall, 448
    stability of, 441
  methods, 520
    deterministic, 547
    for three-dimensional flow, 553
    for viscous fluids, 600
  motion, 520
  patch
    axisymmetric, 86, 117, 552
    two-dimensional, 91, 118, 441, 448, 548
  point, see: *point vortex*
  ring, 88
    line, 37
      computation of its motion, 551
      in the presence of a sphere, 336
  street, 457
    of point vortices, 530
  stretching, 140
  tube, 31
Vortex-in-cell method, VIC, 546, 555

Vortex sheet
  axisymmetric, 38, 552
  definition of, 33
  diffusing in a straining flow, 142
  relation to the double-layer potential,
    502
  flow due to, 78
  jump in the potential across, 42
  motion of 535, 554
  self-induced velocity, 536
  two-dimensional, 36
    diffusing, 196
    flow due to, 91, 535
    instability of, 458
    represention of flow past a body, 352
  arising from a vortex layer, 447
Vorticity
  of an axisymmetric flow, 31, 37
    in terms of the Stokes stream function, 75
  baroclinic generation of, 140
  definition of, 2
  diffusion, 140
  dynamics of, 139
  intrinsic, 148
  field, 30, 64
  flow induced by, 76
    axisymmetric, 83
    two-dimensional, 88
  at a free surface, 132–3
  modified, 148
  at a rigid boundary, 128
  source or generation of
    at an interface, 145
    in a viscous fluid, 145

  tensor, 2
  transport equation, 94, 140
    as a basis for numerical computation, 600
    for axisymmetric flow, 144
    in Eulerian form, 141
    in a noninertial frame, 148
    for two-dimensional flow, 144
  of three-dimensional flow,
    in terms of the stream functions, 75
  of two-dimensional flow, 31
    in terms of the stream function, 73
  vector, 2
  relation to viscous forces, 115
Vorton, 64, 555, *see also* rotlet or couplet

W

Wake behind an object,
  axisymmetric, 416
  stability of, 448, 454
  two-dimensional, 390
Wave equation, 578
Wavy wall, shear flow over, 227
Weakly singular integral equation, 485
Weber number, 138
Weierstrass theorem, 657
Weighted-residual method, 451, 492
Weilandt's deflation theorem, 512
Weiss's theorem, 334
Weyl's integral equation, 400
Whitehead's paradox, 299
Womersley number, 198–9